SOLUTIONS MANUAL

MARY BETH KRAMER • *University of Delaware*
KATHLEEN THRUSH SHAGINAW • *Particular Solutions, Inc.*

Principles of Chemistry
A MOLECULAR APPROACH

NIVALDO J. TRO

Prentice Hall
New York Boston San Francisco
London Toronto Sydney Tokyo Singapore Madrid
Mexico City Munich Paris Cape Town Hong Kong Montreal

Project Editor: Jennifer Hart
Acquisitions Editor: Dan Kaveney
Editor in Chief, Chemistry and Geosciences: Nicole Folchetti
Marketing Manager: Erin Gardner
Managing Editor, Chemistry and Geosciences: Gina M. Cheselka
Project Manager, Science: Ed Thomas
Operations Specialist: Amanda A. Smith
Supplement Cover Manager: Paul Gourhan
Supplement Cover Designer: Tina Krivoshein
Cover Art: Quade Paul

Pearson Prentice Hall
Pearson Education, Inc.
Upper Saddle River, NJ 07458

Printed in the United States of America

10 9 8 7 6 5 4 3 2 1

ISBN-13: 978-0-321-58639-1
ISBN-10: 0-321-58639-5

Prentice Hall
is an imprint of

PEARSON

www.pearsonhighered.com

Table of Contents

Chapter 1
Matter, Measurement, and Problem Solving

1. a) This statement is a theory because it attempts to explain why. It is not possible to observe individual atoms.

   b) This statement is an observation.

   c) This statement is a law, because it summarizes many observations and can explain future behavior.

   d) This statement is an observation.

2. a) This statement is a law, because it summarizes many observations and can explain future behavior.

   b) This statement is a law, because it summarizes many observations and can explain future behavior.

   c) This statement is a law, because it summarizes many observations and can explain future behavior.

   d) This statement is a theory because it attempts to explain why.

3. a) If we divide the mass of the oxygen by the mass of the carbon the result is always 4/3.

   b) If we divide the mass of the oxygen by the mass of the hydrogen the result is always 16.

   c) These observations suggest that the masses of elements in molecules are ratios of whole numbers (4 and 3; and 16 and 1, respectively).

   d) Atoms combine in small whole number ratios and not as random weight ratios.

4. There are many hypotheses that may be developed. One hypothesis is that a large explosion generated galaxies with fragments that are still moving away from each other.

5. a) Sweat is a homogeneous mixture of water, sodium chloride, and other components.

   b) Carbon dioxide is a pure substance that is a compound (two or more elements bonded together).

   c) Aluminum is a pure substance that is an element (element #13 in the periodic table).

   d) Vegetable soup is a heterogeneous mixture of broth, chunks of vegetables, and extracts from the vegetables.

6. a) Wine is a generally homogeneous mixture of water, ethyl alcohol, and other components from the grapes. In some cases, there may be sediment present and so it would be a heterogeneous mixture.

   b) Beef stew is a heterogeneous mixture of thick broth and chunks of vegetables.

   c) Iron is a pure substance that is an element (element #26 in the periodic table).

   d) Carbon monoxide is a pure substance that is a compound (two or more elements bonded together).

7.

substance	pure or mixture	Type (element or compound)
aluminum	pure	element
apple juice	mixture	neither – homogeneous mixture
hydrogen peroxide	pure	compound
chicken soup	mixture	neither – heterogeneous mixture

8.

substance	pure or mixture	Type (element or compound)
water	pure	compound
coffee	mixture	neither – heterogeneous mixture

ice	pure	compound
carbon	pure	element

9. a) pure substance that is a compound (one type of molecule that contains two different elements)

   b) heterogeneous mixture (two different molecules that are segregated into regions)

   c) homogeneous mixture (two different molecules that are randomly mixed)

   d) pure substance is an element (individual atoms of one type)

10. a) pure substance that is an element (individual atoms of one type)

    b) homogeneous mixture (two different molecules that are randomly mixed)

    c) pure substance is a compound (one type of molecule that contains two different elements)

    d) pure substance is a compound (one type of molecule that contains two different elements)

11. a) physical property (color can be observed without making or breaking chemical bonds)

    b) chemical property (must observe by making or breaking chemical bonds)

    c) physical property (the phase can be observed without making or breaking chemical bonds)

    d) physical property (density can be observed without making or breaking chemical bonds)

    e) physical property (mixing does not involve making or breaking chemical bonds, so this can be observed without making or breaking chemical bonds)

12. a) physical property (color can be observed without making or breaking chemical bonds)

    b) physical property (odor can be observed without making or breaking chemical bonds)

    c) chemical property (must observe by making or breaking chemical bonds)

    d) chemical property (decomposition involves breaking bonds, so bonds must be broken to observe this property)

    e) physical property (the phase of a substance can be observed without making or breaking chemical bonds)

13. a) chemical property (burning involves breaking and making bonds, so bonds must be broken and made to observe this property)

    b) physical property (sublimation is a phase change and so can be observed without making or breaking chemical bonds)

    c) physical property (odor can be observed without making or breaking chemical bonds)

    d) chemical property (burning involves breaking and making bonds, so bonds must be broken and made to observe this property)

14. a) physical property (vaporization is a phase change and so can be observed without making or breaking chemical bonds)

    b) physical property (sublimation is a phase change and so can be observed without making or breaking chemical bonds)

    c) chemical property (rusting involves the reaction of iron with oxygen to form iron oxide; observing this process involves making and breaking chemical bonds)

d) physical property (color can be observed without making or breaking chemical bonds)

15. a) chemical change (new compounds are formed as methane and oxygen react to form carbon dioxide and water)

b) physical change (vaporization is a phase change and does not involve the making or breaking of chemical bonds)

c) chemical change (new compounds are formed as propane and oxygen react to form carbon dioxide and water)

d) chemical change (new compounds are formed as the metal in the frame is converted to oxides)

16. a) chemical change (new compounds are formed as the sugar burns)

b) physical change (dissolution is a phase change and does not involve the making or breaking of chemical bonds)

c) physical change (this is simply the rearrangement of the atoms)

d) chemical change (new compounds are formed as the silver converts to an oxide)

17. a) physical change (vaporization is a phase change and does not involve the making or breaking of chemical bonds)

b) chemical change (new compounds are formed)

c) physical change (vaporization is a phase change and does not involve the making or breaking of chemical bonds)

18. a) physical change (vaporization of butane is a phase change and does not involve the making or breaking of chemical bonds)

b) chemical change (new compounds are formed as the butane combusts)

c) physical change (vaporization of water is a phase change and does not involve the making or breaking of chemical bonds)

19. a) To convert from °F to °C, first find the equation that relates these two quantities. $°C = \frac{°F - 32}{1.8}$ Now substitute °F into the equation and compute the answer. Note: The number of digits reported in this answer follow significant figure conventions, covered in section 1.6. $°C = \frac{°F - 32}{1.8} = \frac{0.}{1.8} = 0.\ °C$

b) To convert from K to °F, first find the equations that relate these two quantities.

$K = °C + 273.15$ and $°C = \frac{°F - 32}{1.8}$ Since these equations do not directly express K in terms of °F, you must combine the equations and then solve the equation for °F. Substituting for °C:

$K = \frac{°F - 32}{1.8} + 273.15$ rearrange $K - 273.15 = \frac{°F - 32}{1.8}$ rearrange $1.8\ (K - 273.15) = (°F - 32)$

finally $°F = 1.8\ (K - 273.15) + 32$ Now substitute K into the equation and compute the answer.

$°F = 1.8\ (77\ K - 273.15) + 32 = 1.8\ (-196\ K) + 32 = -353 + 32 = -321\ °F$

c) To convert from °F to °C, first find the equation that relates these two quantities. $°C = \frac{°F - 32}{1.8}$ Now substitute °F into the equation and compute the answer. $°C = \frac{-109 °F - 32 °F}{1.8} = \frac{-141}{1.8} = -78.3\ °C$

d) To convert from °F to K, first find the equations that relate these two quantities.

$K = °C + 273.15$ and $°C = \frac{°F - 32}{1.8}$ Since these equations do not directly express K in terms of °F, you must combine the equations and then solve the equation for K. Substituting for °C

$$K = \frac{°F - 32}{1.8} + 273.15$$

Now substitute °F into the equation and compute the answer.

$$K = \frac{(98.6 - 32)}{1.8} + 273.15 = \frac{66.6}{1.8} + 273.15 = 37.0 + 273.15 = 310.2\ K$$

20. a) To convert from °F to °C, first find the equation that relates these two quantities. $°C = \frac{°F - 32}{1.8}$

Now substitute °F into the equation and compute the answer. Note: The number of digits reported in this answer follow significant figure conventions, covered in section 1.6.

$$°C = \frac{212°F - 32°F}{1.8} = \frac{180.}{1.8} = 100.\ °C$$

b) Begin by finding the equation that relates the quantity that is given (°C) and the quantity you are trying to find (K). $K = °C + 273.15$ Since this equation gives the temperature in K directly, simply substitute in the correct value for the temperature in °C and compute the answer.

$K = 22\ °C + 273.15 = 295\ K$

c) To convert from K to °F, first find the equations that relate these two quantities. $K = °C + 273.15$ and $°C = \frac{°F - 32}{1.8}$ Since these equations do not directly express K in terms of °F, you must combine the equations and then solve the equation for °F. Substituting for °C: $K = \frac{°F - 32}{1.8} + 273.15$, rearrange $K - 273.15 = \frac{°F - 32}{1.8}$, rearrange $1.8\ (K - 273.15) = (°F - 32)$, rearrange $°F = 1.8\ (K - 273.15) + 32$.

Now substitute K into the equation and compute the answer.

$°F = 1.8\ (0.00\ K - 273.15) + 32 = 1.8\ (-273.15\ K) + 32 = -491.67 + 32 = -459.67\ °F$

d) Begin by finding the equation that relates the quantity that is given (°C) and the quantity you are trying to find (K). $K = °C + 273.15$ Since this equation does not directly express °C in terms of K, you must solve the equation for °C. $°C = K - 273.15$ Now substitute K into the equation and compute the answer. $°C = 2.735 - 273.15 = -270.42\ °C$

21. To convert from °F to °C, first find the equation that relates these two quantities. $°C = \frac{°F - 32}{1.8}$ Now substitute °F into the equation and compute the answer. Note: The number of digits reported in this answer follow significant figure conventions, covered in section 1.6. $°C = \frac{-80.°F - 32°F}{1.8} = \frac{-112}{1.8} = -62.2\ °C$

Begin by finding the equation that relates the quantity that is given (°C) and the quantity you are trying to find (K). $K = °C + 273.15$ Since this equation gives the temperature in K directly, simply substitute in the correct value for the temperature in °C and compute the answer. $K = -62.2\ °C + 273.15 = 210.9\ K$.

22. To convert from °F to °C, first find the equation that relates these two quantities. $°C = \frac{°F - 32}{1.8}$ Now substitute °F into the equation and compute the answer. Note: The number of digits reported in this answer follow significant figure conventions, covered in section 1.6.

$$°C = \frac{134°F - 32°F}{1.8} = \frac{102}{1.8} = 56.\underline{6}667\ °C = 56.7\ °C$$

Begin by finding the equation that relates the quantity that is given (°C) and the quantity you are trying to find

(K). K = °C + 273.15 Since this equation gives the temperature in K directly, simply substitute in the correct value for the temperature in °C and compute the answer. K = 56.6667 °C + 273.15 = 329.8 K

23. Use Table 1.2 to determine the appropriate prefix multiplier and substitute the meaning into the expressions.
    a) 10^{-9} implies "nano" so 1.2 $\times 10^{-9}$ m = 1.2 nanometers = 1.2 nm

    b) 10^{-15} implies "femto" so 22 $\times 10^{-15}$ s = 22 femtoseconds = 22 fs

    c) 10^{9} implies "giga" so 1.5 $\times 10^{9}$ g = 1.5 gigagrams = 1.5 Gg

    d) 10^{6} implies "mega" so 3.5 $\times 10^{6}$ L = 3.5 megaliters = 3.5 ML

24. Use Table 1.2 to determine the appropriate prefix multiplier and substitute the meaning into the expressions.
    a) 10^{-9} implies "nano" so 4.5 ns = 4.5 nanoseconds = 4.5 $\times 10^{-9}$ s

    b) 10^{-15} implies "femto" so 18 fs = 18 femtoseconds = 18 $\times 10^{-15}$ s = 1.8 $\times 10^{-14}$ s
    Remember that in scientific notation the first number should be smaller than 10.

    c) 10^{-12} implies "pico" so 128 pm = 128 $\times 10^{-12}$ m = 1.28 x 10^{-10} m
    Remember that in scientific notation the first number should be smaller than 10.

    d) 10^{-6} implies "micro" so 35 μm = 35 micrograms = 35 $\times 10^{-6}$ g = 3.5 x 10^{-5} m
    Remember that in scientific notation the first number should be smaller than 10.

25. b) **Given:** 515 km **Find:** dm

Conceptual Plan: **km → m → dm**

$$\frac{1000\ \text{m}}{1\text{km}} \qquad \frac{10\ \text{dm}}{1\text{m}}$$

Solution: $$515\ \text{km} \times \frac{1000\ \text{m}}{1\ \text{km}} \times \frac{10\ \text{dm}}{1\ \text{m}} = 5.15 \times 10^{6}\ \text{dm}$$

Check: The units (dm) are correct. The magnitude of the answer (10^6) makes physical sense because a decimeter is a much smaller unit than a kilometer.

Given: 515 km **Find:** cm

Conceptual Plan: **km → m → cm**

$$\frac{1000\ \text{m}}{1\text{km}} \qquad \frac{100\ \text{cm}}{1\text{m}}$$

Solution: $$515\ \text{km} \times \frac{1000\ \text{m}}{1\ \text{km}} \times \frac{100\ \text{cm}}{1\ \text{m}} = 5.15 \times 10^{7}\ \text{cm}$$

Check: The units (cm) are correct. The magnitude of the answer (10^7) makes physical sense because a centimeter is a much smaller unit than a kilometer and a decimeter.

c) **Given:** 122.355 s **Find:** ms

Conceptual Plan: **s → ms**

$$\frac{1000\ \text{ms}}{1\text{s}}$$

Solution: $$122.355\ \text{s} \times \frac{1000\ \text{ms}}{1\ \text{s}} = 1.22355 \times 10^{5}\ \text{ms}$$

Check: The units (ms) are correct. The magnitude of the answer (10^5) makes physical sense because a millisecond is a much smaller unit than a second.

Given: 122.355 s **Find:** ks

Conceptual Plan: **s → ks**

$$\frac{1\ \text{ks}}{1000\ \text{s}}$$

Solution: $$122.355\ \text{s} \times \frac{1\ \text{ks}}{1000\ \text{s}} = 1.22355 \times 10^{-1}\ \text{ks} = 0.122355\ \text{ks}$$

Check: The units (ks) are correct. The magnitude of the answer (10^{-1}) makes physical sense because a kilosecond is a much larger unit than a second.

d) **Given:** 3.345 kJ **Find:** J
Conceptual Plan: **kJ → J**

$$\frac{1000\text{ J}}{1\text{kJ}}$$

Solution: $3.345\ \cancel{\text{kJ}} \times \frac{1000\text{ J}}{1\cancel{\text{kJ}}} = 3.345 \times 10^3\text{ J}$

Check: The units (J) are correct. The magnitude of the answer (10^3) makes physical sense because a joule is a much smaller unit than a kilojoule.
Given: 3.345×10^3 J (from above) **Find:** mJ
Conceptual Plan: **J → mJ**

$$\frac{1000\text{ mJ}}{1\text{ J}}$$

Solution: $3.345 \times 10^3\ \cancel{\text{J}} \times \frac{1000\text{ mJ}}{1\cancel{\text{J}}} = 3.345 \times 10^6\text{ mJ}$

Check: The units (mJ) are correct. The magnitude of the answer (10^6) makes physical sense because a millijoule is a much smaller unit than a joule.

26. a) **Given:** 254,998 m **Find:** km
Conceptual Plan: **m → km**

$$\frac{1\text{ km}}{1000\text{ m}}$$

Solution: $254{,}998\ \cancel{\text{m}} \times \frac{1\text{ km}}{1000\ \cancel{\text{m}}} = 2.54998 \times 10^2\text{ km} = 254.998\text{ km}$

Check: The units (km) are correct. The magnitude of the answer (10^2) makes physical sense because a kilometer is a much larger unit than a meter.

b) **Given:** 254,998 m **Find:** Mm
Conceptual Plan: **m → Mm**

$$\frac{1\text{ Mm}}{10^6\text{ m}}$$

Solution: $254{,}998\ \cancel{\text{m}} \times \frac{1\text{ Mm}}{10^6\ \cancel{\text{m}}} = = 2.54998 \times 10^{-1}\text{ Mm} = 0.254998\text{ Mm}$

Check: The units (Mm) are correct. The magnitude of the answer (10^{-1}) makes physical sense because a megameter is a much larger unit than a meter or kilometer.

c) **Given:** 254,998 m **Find:** mm
Conceptual Plan: **m → mm**

$$\frac{1000\text{ mm}}{1\text{ m}}$$

Solution: $254{,}998\ \cancel{\text{m}} \times \frac{1000\text{ mm}}{1\ \cancel{\text{m}}} = = 2.54998 \times 10^8\text{ mm}$

Check: The units (mm) are correct. The magnitude of the answer (10^8) makes physical sense because a millimeter is a much smaller unit than a meter.

d) **Given:** 254,998 m **Find:** cm

Conceptual Plan: **m → cm**

$$\frac{100\text{ cm}}{1\text{ m}}$$

Solution: $254{,}998\ \cancel{\text{m}} \times \frac{100\text{ cm}}{1\ \cancel{\text{m}}} = = 2.54998 \times 10^7\text{ cm}$

Check: The units (cm) are correct. The magnitude of the answer (10^7) makes physical sense because a centimeter is a much smaller unit than a meter, but larger than a millimeter.

27. **Given:** 1 m square 1 m^2 **Find:** cm^2
Conceptual Plan: $\mathbf{1\ m^2 \rightarrow cm^2}$

$$\frac{100\text{ cm}}{1\text{ m}}$$

Notice that for squared units, the conversion factors must be squared.

Solution: $1\ \cancel{m}^2 \times \dfrac{(100\text{ cm})^2}{(1\ \cancel{m})^2} = 1 \times 10^4\ \text{cm}^2$

Check: The units of the answer are correct and the magnitude makes sense. The unit centimeter is smaller than a meter, so the value in square centimeters should be larger than in square meters.

28. **Given:** 4 cm on each edge cube **Find:** cm^3
Conceptual Plan: Read the information given carefully. The cube is 4 cm on each side.

$$\boldsymbol{l, w, h} \rightarrow \boldsymbol{V}$$

$$V = l\,w\,h$$

$$\textit{in a cube } l = w = h$$

Solution: 4 cm x 4 cm x 4 cm = $(4\text{ cm})^3 = \underline{6}4\text{ cm}^3 = 60\text{ cm}^3$
Check: The units of the answer are correct and the magnitude makes sense. The unit 4 centimeter is larger than 1 centimeter, so the value in cubic centimeters should be larger.

29. **Given:** $m = 2.49$ g, $V = 0.349\text{ cm}^3$ **Find:** d in g/cm^3 and compare to pure copper.
Conceptual Plan: $\boldsymbol{m, V \rightarrow d}$

$$d = m/V$$

Compare to the published value. d (pure copper) = 8.96 g/cm^3 (This value is in Table 1.4.)

Solution: $d = \dfrac{2.49\text{ g}}{0.349\text{ cm}^3} = 7.13\ \dfrac{\text{g}}{\text{cm}^3}$

The density of the penny is much smaller than the density of pure copper (7.13 g/cm^3 < 8.96 g/cm^3) so the penny is not pure copper.
Check: The units (g/cm^3) are correct. The magnitude of the answer seems correct. Many coins are layers of metals, so it is not surprising that the penny is not pure copper.

30. **Given:** $m = 1.41$ kg, $V = 0.314$ L **Find:** d in g/cm^3 and compare to pure titanium.
Conceptual Plan: $\boldsymbol{m, V \rightarrow d}$ **then kg → g then L → $\mathbf{cm^3}$**

$$d = m/V \qquad \frac{1000\text{ g}}{1\text{ kg}} \qquad \frac{1000\text{ cm}^3}{1\text{ L}}$$

Compared to the published value. d (pure titanium) = 4.51 g/cm^3 (This value is in Table 1.4.)

Solution: $d = \dfrac{1.41\ \cancel{kg}}{0.314\text{ L}} \times \dfrac{\cancel{1000}\ \cancel{g}}{1\ \cancel{kg}} \times \dfrac{1\text{ L}}{\cancel{1000}\ \cancel{cm^3}} = 4.49\ \dfrac{\text{g}}{\text{cm}^3}$

Check: The units (g/cm^3) are correct. The magnitude of the answer seems correct. The density of the frame is almost exactly the density of pure titanium (4.49 g/cm^3 versus 4.51 g/cm^3) so the frame could be titanium.

31. **Given:** $m = 4.10 \times 10^3$ g, $V = 3.25$ L **Find:** d in g/cm^3
Conceptual Plan: $\boldsymbol{m, V \rightarrow d}$ **then L → $\mathbf{cm^3}$**

$$d = m/V \qquad \frac{1000\text{ cm}^3}{1\text{ L}}$$

Solution: $d = \dfrac{4.10 \times 10^3\text{ g}}{3.25\ \cancel{L}} \times \dfrac{1\ \cancel{L}}{1000\text{ cm}^3} = 1.26\ \dfrac{\text{g}}{\text{cm}^3}$

Check: The units (g/cm^3) are correct. The magnitude of the answer seems correct.

32. **Given:** $m = 371$ grams, $V = 19.3$ mL **Find:** d in g/cm^3 and compare to pure gold.
Conceptual Plan: $\boldsymbol{m, V \rightarrow d}$ $\quad d = m/V$
Compare to the published value. d (pure gold) = 19.3 g/mL (This value is in Table 1.4.)

Solution: $d = \dfrac{371\text{ g}}{19.3\text{ mL}} = 19.2\ \dfrac{\text{g}}{\text{mL}}$

The density of the nugget is essentially the same as the density of pure gold (19.2 g/mL versus 19.3 g/mL) so the nugget could be gold.

Check: The units (g/cm^3) are correct. The magnitude of the answer seems correct and is essentially the same as the density of pure gold.

33. a) **Given:** $d = 1.11\text{ g/cm}^3$, $V = 417$ mL **Find:** m

Conceptual Plan: $d, V \rightarrow m$ then $cm^3 \rightarrow mL$

$$d = m/V \qquad \frac{1\text{ mL}}{1\text{ cm}^3}$$

Solution: $d = m/V$ Rearrange by multiplying both sides of equation by V. $m = d \times V$

$$m = 1.11\ \frac{\text{g}}{\text{cm}^3} \times \frac{1\text{ cm}^3}{1\text{ mL}} \times 417\text{ mL} = 4.63 \times 10^2\text{ g}$$

Check: The units (g) are correct. The magnitude of the answer seems correct considering the value of the density is about 1 g/cm^3.

b) **Given:** $d = 1.11\text{ g/cm}^3$, $m = 4.1$ kg **Find:** V in L

Conceptual Plan: $d, V \rightarrow m$ then $kg \rightarrow g$ and $cm^3 \rightarrow L$

$$d = m/V \qquad \frac{1000\text{ g}}{1\text{ kg}} \qquad \frac{1\text{ L}}{1000\text{ cm}^3}$$

Solution: $d = m/V$ Rearrange by multiplying both sides of equation by V and dividing both sides of the equation by d.

$$V = \frac{m}{d} = \frac{4.1\text{ kg}}{1.11\ \frac{\text{g}}{\text{cm}^3}} \times \frac{1000\text{ g}}{1\text{ kg}} = 3.7 \times 10^3\text{ cm}^3 \times \frac{1\text{ L}}{1000\text{ cm}^3} = 3.7\text{ L}$$

Check: The units (L) are correct. The magnitude of the answer seems correct considering the value of the density is about 1 g/cm^3.

34. a) **Given:** $d = 0.7857\text{ g/cm}^3$, $V = 28.56$ mL **Find:** m

Conceptual Plan: $d, V \rightarrow m$

$$d = m/V$$

Solution: $d = m/V$ Rearrange by multiplying both sides of equation by V. $m = d \times V$

$$m = \left(0.7857\ \frac{\text{g}}{\text{cm}^3}\right) \times \frac{1\text{ cm}^3}{1\text{ mL}} \times (28.56\text{ mL}) = 22.44\text{ g}$$

Check: The units (g) are correct. The magnitude of the answer seems correct considering the value of the density is less than 1 g/cm^3.

b) **Given:** $d = 0.7857\text{ g/cm}^3$, $m = 6.54$ g **Find:** V

Conceptual Plan: $d, m \rightarrow V$ then $cm^3 \rightarrow mL$

$$d = m/V \qquad \frac{1\text{ mL}}{1\text{ cm}^3}$$

Solution: $d = m/V$ Rearrange by multiplying both sides of equation by V and dividing both sides of the equation by d.

$$V = \frac{m}{d} = \frac{6.54\text{ g}}{0.7857\ \frac{\text{g}}{\text{cm}^3}} = 8.32\text{ cm}^3 \times \frac{1\text{ mL}}{1\text{ cm}^3} = 8.32\text{ mL}$$

Check: The units (mL) are correct. The magnitude of the answer seems correct considering the value of the density is less than 1 g/cm^3.

35. In order to obtain the readings, look to see where the bottom of the meniscus lies. Estimate the distance between two markings on the device.

a) 73.5 mL – the meniscus appears to be about half way between the 73 mL and the 74 mL marks.

b) 88.2 °C – the mercury is between the 84 °C mark and the 85 °C mark, but it is closer to the lower number.

c) 645 mL – the meniscus appears to be just above the 640 mL mark.

36. In order to obtain the readings, look to see where the bottom of the meniscus lies. Estimate the distance between two markings on the device. Use all digits on a digital device.

a) 4.50 mL – the meniscus appears to be on the 4.5 mL mark.

b) 27.43 °C – the mercury is just above the 27.4 °C mark. Note that the 10's digit is only labeled every 10 °C.

c) 0.873 g – read all the places on the digital display.

37. Remember that

1. Interior zeroes (zeroes between two numbers) are significant.
2. Leading zeroes (zeroes to the left of the first non-zero number) are not significant. They only serve to locate the decimal point.
3. Trailing zeroes (zeroes at the end of a number) are categorized as follows:
   - trailing zeroes after a decimal point are always significant.
   - trailing zeroes before an implied decimal point are ambiguous and should be avoided by using scientific notation or by inserting a decimal point at the end of the number.

a) 1,050,501 km

b) 0.0020 m

c) 0.000000000000002 s

d) 0.001090 cm

38. Remember that

1. Interior zeroes (zeroes between two numbers) are significant.
2. Leading zeroes (zeroes to the left of the first non-zero number) are not significant. They only serve to locate the decimal point.
3. Trailing zeroes (zeroes at the end of a number) are categorized as follows:
   - trailing zeroes after a decimal point are always significant.
   - trailing zeroes before an implied decimal point are ambiguous and should be avoided by using scientific notation or by inserting a decimal point at the end of the number.

a) 180,701 mi

b) 0.001040 m

c) 0.005710 km

d) 90,201 m

39. Remember all of the rules from section 1.7.

a) Three significant figures. The 3, 1, and the 2 are significant (rule 1). The leading zeroes only mark the decimal place and are therefore not significant (rule 3).

b) Ambiguous. The 3, 1, and the 2 are significant (rule 1). The trailing zeroes occur before an implied decimal point and are therefore ambiguous (rule 4). Without more information, we would assume 3 significant figures. It is better to write this as 3.12×10^5 to indicate three significant figures or as 3.12000×10^5 to indicate six (rule 4).

c) Three significant figures. The 3, 1, and the 2 are significant (rule 1).

d) Five significant figures. The 1's, 3, 2, and 7 are significant (rule 1).

e) Ambiguous. The 2 is significant (rule 1). The trailing zeroes occur before an implied decimal point and are therefore ambiguous (rule 4). Without more information, we would assume one significant figure. It is better to write this as 2×10^3 to indicate one significant figure or as 2.000×10^3 to indicate four (rule 4).

40. Remember all of the rules from section 1.7.
   a) Four significant figures. The 1's are significant (rule 1). The leading zeroes only mark the decimal place and are therefore not significant (rule 3).

   b) One significant figure. The 7 is significant (rule 1). The leading zeroes only mark the decimal place and are therefore not significant (rule 3).

   c) Ambiguous. The 1, 8, and the 7 are significant (rule 1). The first 0 is significant, since it is an interior 0 (rule 2). The trailing zeroes occur before an implied decimal point and are therefore ambiguous (rule 4). Without more information, we would assume 4 significant figures. It is better to write this as 1.087×10^5 to indicate 4 significant figures or as 1.08700×10^5 to indicate six (rule 4).

   d) Seven significant figures. The 1, 5, 6, and 3's are significant (rule 1). The trailing zeros are significant because they are to the right of the decimal point and non-zero numbers (rule 4)

   e) Ambiguous. The 3 and 8 are significant (rule 1). The first 0 is significant because the first one is an interior zero. The trailing zeroes occur before an implied decimal point and are therefore ambiguous (rule 4). Without more information, we would assume three significant figures. It is better to write this as 3.08×10^4 to indicate three significant figures or as 3.0800×10^4 to indicate five (rule 4).

41. a) This is not exact because π is an irrational number. 3.14 only shows three of the infinite number of significant figures that π has.

   b) This is an exact conversion, because it comes from a definition of the units, and so has an unlimited number of significant figures.

   c) This is a measured number and so it is not an exact number. There are 2 significant figures.

   d) This is an exact conversion, because it comes from a definition of the units, and so has an unlimited number of significant figures.

42. a) This is a measured number and so it is not an exact number. There are 9 significant figures.

   b) This is a an exact conversion, so has an unlimited number of significant figures

   c) This is a measured number and so it is not an exact number. There are 3 significant figures.

   d) This is an exact conversion, because it comes from a definition of the units and so has an unlimited number of significant figures.

43. a) 156.9 – the 8 is rounded up since the next digit is a 5.

   b) 156.8 – the last two digits are dropped since 4 is less than 5.

   c) 156.8 – the last two digits are dropped since 4 is less than 5.

   d) 156.9 – the 8 is rounded up since the next digit is a 9 which is greater than 5.

44. a) 7.98×10^4 – the last digits are dropped since 4 is less than 5.

   b) 1.55×10^7 – the 8 is rounded up since the next digit is a 9 which is greater than 5.

   c) 2.35 – the 4 is rounded up since the next digit is a 9 which is greater than 5.

   d) 0.0000454 – the 3 is rounded up since the next digit is an 8 which is greater than 5.

45. a) $9.15 \div 4.970 = 1.84$ – Three significant figures are allowed to reflect the three significant figures in the least precisely known quantity (9.15).

b) $1.54 \times 0.03060 \times 0.69 = 0.033$ – Two significant figures are allowed to reflect the two significant figures in the least precisely known quantity (0.69). The intermediate answer (0.03251556) is rounded up since the first non-significant digit is a 5.

c) $27.5 \times 1.82 \div 100.04 = 0.500$ – Three significant figures are allowed to reflect the three significant figures in the least precisely known quantity (27.5 and 1.82). The intermediate answer (0.50029988) is truncated since the first non-significant digit is a 2, which is less than 5.

d) $(2.290 \times 10^6) \div (6.7 \times 10^4) = 34$ – Two significant figures are allowed to reflect the two significant figures in the least precisely known quantity (6.7×10^4). The intermediate answer (34.17910448) is truncated since the first non-significant digit is a 1, which is less than 5.

46. a) $89.3 \times 77.0 \times 0.08 = 6 \times 10^2$ – One significant figure is allowed to reflect the one significant figure in the least precisely known quantity (0.08). The intermediate answer (5.50088×10^2) is rounded up since the first non-significant digit is a 5.

b) $(5.01 \times 10^5) \div (7.8 \times 10^2) = 6.4 \times 10^2$ – Two significant figures are allowed to reflect the two significant figures in the least precisely known quantity (7.8×10^2). The intermediate answer (6.423076923×10^2) is truncated since the first non-significant digit is a 2, which is less than 5.

c) $4.005 \times 74 \times 0.007 = 2$ – One significant figure is allowed to reflect the one significant figure in the least precisely known quantity (0.007). The intermediate answer (2.07459) is truncated since the first non-significant digit is a 0, which is less than 5.

d) $453 \div 2.031 = 223$ – Three significant figures are allowed to reflect the three significant figures in the least precisely known quantity (453). The intermediate answer (223.042836) is truncated since the first non-significant digit is a 0, which is less than 5.

47. a)

$$\begin{array}{r} 43.7 \\ -\ 2.341 \\ \hline 41.359 \end{array} = 41.4$$

Round the intermediate answer to one decimal place to reflect the quantity with the fewest decimal places (43.7). Round the last digit up since the first non-significant digit is 5.

b)

$$\begin{array}{r} 17.6 \\ +\ 2.838 \\ +\ 2.3 \\ +\ 110.77 \\ \hline 133.508 \end{array} = 133.5$$

Round the intermediate answer to one decimal place to reflect the quantity with the fewest decimal places (2.3). Truncate non-significant digits since the first non-significant digit is 0.

c)

$$\begin{array}{r} 19.6 \\ +\ 58.33 \\ -\ 4.974 \\ \hline 72.956 \end{array} = 73.0$$

Round the intermediate answer to one decimal place to reflect the quantity with the fewest decimal places (19.6). Round the last digit up since the first non-significant digit is 5.

d)

$$\begin{array}{r} 5.99 \\ -\ 5.572 \\ \hline 0.418 \end{array} = 0.42$$

Round the intermediate answer to two decimal places to reflect the quantity with the fewest decimal places (5.99). Round the last digit up since the first non-significant digit is 8.

48. a)

$$\begin{array}{r} 0.004 \\ +\ 0.09879 \\ \hline 0.10279 \end{array} = 0.103$$

Round the intermediate answer to three decimal places to reflect the quantity with the fewest decimal places (0.004). Round the last digit up since the first non-significant digit is 9.

b)
$$\begin{array}{r} 1239.3 \\ +\ \ 9.73 \\ +\ \ \underline{3.42} \\ 1252.45 \end{array} = 1252.5$$

Round the intermediate answer to one decimal place to reflect the quantity with the fewest decimal places (1239.3). Round the last digit up since the first non-significant digit is 5.

c)
$$\begin{array}{r} 2.4 \\ -\ \underline{1.777} \\ 0.623 \end{array} = 0.6$$

Round the intermediate answer to one decimal place to reflect the quantity with the fewest decimal places (2.4). Truncate non-significant digits since the first non-significant digit is 2.

d)
$$\begin{array}{r} 532 \\ +\ \ 7.3 \\ -\ \underline{48.523} \\ 490.777 \end{array} = 491$$

Round the intermediate answer to zero decimal places to reflect the quantity with the fewest decimal places (532). Round the last digit up since the first non-significant digit is 7.

49. Perform operations in parentheses first. Keep track of significant figures in each step, by noting which is the last significant digit in an intermediate result.

a) $(24.6681 \times 2.38) + 332.58 = \begin{array}{r} 58.\underline{7}10078 \\ +\ \underline{332.58} \\ 391.290078 \end{array} = 391.3$

The first intermediate answer has one significant digit to the right of the decimal, because it is allowed three significant figures (reflecting the quantity with the fewest significant figures (2.38)). Underline the most significant digit in this answer. Round the next intermediate answer to one decimal place to reflect the quantity with the fewest decimal places (58.7). Round the last digit up since the first non-significant digit is 9.

b) $\dfrac{(85.3 - 21.489)}{0.0059} = \dfrac{63.\underline{8}11}{0.0059} = 1.\underline{0}81542 \times 10^4 = 1.1 \times 10^4$

The first intermediate answer has one significant digit to the right of the decimal, to reflect the quantity with the fewest decimal places (85.3). Underline the most significant digit in this answer. Round the next intermediate answer to two significant figures to reflect the quantity with the fewest significant figures (0.0059). Round the last digit up since the first non-significant digit is 8.

c) $(512 \div 986.7) + 5.44 = \begin{array}{r} 0.51\underline{8}9014 \\ +\ \ \ \underline{5.44} \\ 5.9589014 \end{array} = 5.96$

The first intermediate answer has three significant figures and three significant digits to the right of the decimal, reflecting the quantity with the fewest significant figures (512). Underline the most significant digit in this answer. Round the next intermediate answer to two decimal places to reflect the quantity with the fewest decimal places (5.44). Round the last digit up since the first non-significant digit is 8.

d) $[(28.7 \times 10^5) \div 48.533] + 144.99 = \begin{array}{r} 59\underline{1}35.01 \\ +\ \ \ \underline{144.99} \\ 59280.01 \end{array} = 59300 = 5.93 \times 10^4$

The first intermediate answer has three significant figures, reflecting the quantity with the fewest significant figures (28.7×10^5). Underline the most significant digit in this answer. Since the number is so large this means that when the addition is performed, the most significant digit is the 100's place. Round the next intermediate answer to the 100's places and put in scientific notation to remove any ambiguity. Note that the last digit is rounded up since the first non-significant digit is 8.

50. Perform operations in parentheses first. Keep track of significant figures in each step, by noting which is the last significant digit in an intermediate result.

a) $[(1.7 \times 10^6) \div [(2.63 \times 10^5)] + 7.33 =$

$$\begin{array}{r} 6.\underline{4}63878 \\ + \quad 7.33 \\ \hline 13.793878 \end{array} = 13.8$$

The first intermediate answer has one significant digit to the right of the decimal, because it is allowed two significant figures (reflecting the quantity with the fewest significant figures (1.7×10^6)). Underline the most significant digit in this answer. Round the next intermediate answer to one decimal place to reflect the quantity with the fewest decimal places (6.5). Round the last digit up since the first non-significant digit is 9.

b) $(568.99 - 232.1) \div 5.3 = 336.\underline{8}9 \div 5.3 = 63.564151 = 64$

The first intermediate answer has one significant digit to the right of the decimal, to reflect the quantity with the fewest decimal places (232.1). Underline the most significant digit in this answer. Round the next intermediate answer to two significant figures to reflect the quantity with the fewest significant figures (5.3). Round the last digit up since the first non-significant digit is 5.

c) $(9443 + 45 - 9.9) \times 8.1 \times 10^6 = = 947\underline{8}.1 \times 8.1 \times 10^6 = 7.67726 \times 10^{10} = 7.7 \times 10^{10}$

The first intermediate answer only has significant digits to the left of the decimal, reflecting the quantity with the fewest significant figures (9443 and 45). Underline the most significant digit in this answer. Round the next intermediate answer to two significant figures to reflect the quantity with the fewest significant figures (8.1×10^6). Round the last digit up since the first non-significant digit is 7.

d) $(3.14 \times 2.4367) - 2.34 =$

$$\begin{array}{r} 7.6\underline{5}1238 \\ -2.34 \\ \hline 5.311238 \end{array} = 5.31$$

The first intermediate answer has three significant figures, reflecting the quantity with the fewest significant figures (3.14). Underline the most significant digit in this answer. This number has two significant digits to the right of the decimal point. Round the next intermediate answer to two significant digits to the right of the decimal point, since both numbers have two significant digits to the right of the decimal point. Note that the last digit is truncated since the first non-significant digit is 1.

51. a) **Given:** 154 cm **Find:** in

Conceptual Plan: **cm → in**

$$\frac{1\ \text{in}}{2.54\ \text{cm}}$$

Solution: $154\ \cancel{\text{cm}} \times \dfrac{1\ \text{in}}{2.54\ \cancel{\text{cm}}} = 60.\underline{6}2992\ \text{in} = 60.6\ \text{in}$

Check: The units (in) are correct. The magnitude of the answer (60.6) makes physical sense because an inch is a larger unit than a cm. Three significant figures are allowed because 154 cm has three significant figures.

b) **Given:** 3.14 kg **Find:** g

Conceptual Plan: **kg → g**

$$\frac{1000\ \text{g}}{1\ \text{kg}}$$

Solution: $3.14\ \cancel{\text{kg}} \times \dfrac{1000\ \text{g}}{1\ \cancel{\text{kg}}} = 3.14 \times 10^3\ \text{g}$

Check: The units (g) are correct. The magnitude of the answer (10^3) makes physical sense because a kg is a much larger unit than a gram. Three significant figures are allowed because 3.14 kg has three significant figures.

c) **Given:** 3.5 L **Find:** qt

Conceptual Plan: **L → qt**

$$\frac{1.057\ \text{qt}}{1\ \text{L}}$$

Solution: $3.5\ \cancel{L} \times \frac{1.057\ qt}{1\ \cancel{L}} = 3.6995\ qt = 3.7\ qt$

Check: The units (qt) are correct. The magnitude of the answer (3.7) makes physical sense because a L is a smaller unit than a qt. Two significant figures are allowed because 3.5 L has two significant figures. Round the last digit up because the first non-significant digit is a 9.

d) **Given:** 109 mm **Find:** in
Conceptual Plan: **mm → m → in**

$\frac{1\ m}{1000\ mm} \quad \frac{39.37\ in}{1\ m}$

Solution: $109\ \cancel{mm} \times \frac{1\ \cancel{m}}{1000\ \cancel{mm}} \times \frac{39.37\ in}{1\ \cancel{m}} = 4.2\underline{9}133\ in = 4.29\ in$

Check: The units (in) are correct. The magnitude of the answer (4) makes physical sense because a mm is a much smaller unit than an in. Three significant figures are allowed because 109 mm has three significant figures.

52. a) **Given:** 1.4 in **Find:** mm
Conceptual Plan: **in → cm → m → mm**

$\frac{2.54\ cm}{1\ in} \quad \frac{1\ m}{100\ cm} \quad \frac{1000\ mm}{1\ m}$

Solution: $1.4\ \cancel{in} \times \frac{2.54\ \cancel{cm}}{1\ \cancel{in}} \times \frac{1\ \cancel{m}}{100\ \cancel{cm}} \times \frac{1000\ mm}{1\ \cancel{m}} = 3\underline{5}.56\ mm = 36\ mm$

Check: The units (mm) are correct. The magnitude of the answer (36) makes physical sense because a mm is smaller than an in. Two significant figures are allowed because 1.4 in has two significant figures. Round the last digit up because the first non-significant digit is a 5.

b) **Given:** 116 ft **Find:** cm
Conceptual Plan: **ft → in → cm**

$\frac{12\ in}{1\ ft} \quad \frac{2.54\ cm}{1\ in}$

Solution: $116\ \cancel{ft} \times \frac{12\ \cancel{in}}{1\ \cancel{ft}} \times \frac{2.54\ cm}{1\ \cancel{in}} = 3.5\underline{3}57 \times 10^3\ cm = 3.54 \times 10^3\ cm$

Check: The units (cm) are correct. The magnitude of the answer (10^3) makes physical sense because a ft is a much larger unit than a cm. Three significant figures are allowed because 116 ft has three significant figures. Round the last digit up because the first non-significant digit is a 5.

c) **Given:** 1845 kg **Find:** lb
Conceptual Plan: **kg → g → lb**

$\frac{1000\ g}{1\ kg} \quad \frac{1\ lb}{453.6\ g}$

Solution: $1845\ \cancel{kg} \times \frac{1000\ \cancel{g}}{1\ \cancel{kg}} \times \frac{1\ lb}{453.6\ \cancel{g}} = 4.06\underline{7}5 \times 10^3\ lb = 4.068 \times 10^3\ lb$

Check: The units (lb) are correct. The magnitude of the answer (10^3) makes physical sense because a lb is a smaller unit than a kg. Four significant figures are allowed because 1845 kg and 453.6 g/lb each have four significant figures. Round the last digit up because the first non-significant digit is a 5.

d) **Given:** 815 yd **Find:** km
Conceptual Plan: **yd → m → km**

$\frac{1\ m}{1.094\ yd} \quad \frac{1\ km}{1000\ m}$

Solution: $815\ \cancel{yd} \times \frac{1\ \cancel{m}}{1.094\ \cancel{yd}} \times \frac{1\ km}{1000\ \cancel{m}} = 0.74\underline{4}9726\ km = 0.745\ km$

Check: The units (km) are correct. The magnitude of the answer (0.7) makes physical sense because a yd is a much smaller unit than a km. Three significant figures are allowed because 815 yd has three significant figures. Round the last digit up because the first non-significant digit is a 9.

53. **Given:** 10.0 km **Find:** minutes **Other:** running pace = 7.5 miles per hour
Conceptual Plan: **km → mi → hr → min**

$$\frac{0.6214 \text{ mi}}{1 \text{ km}} \quad \frac{1 \text{hr}}{7.5 \text{ mi}} \quad \frac{60 \text{ min}}{1 \text{ hr}}$$

Solution: $10.0 \text{ km} \times \frac{0.6214 \text{ mi}}{1 \text{ km}} \times \frac{1 \text{ hr}}{7.5 \text{ mi}} \times \frac{60 \text{ min}}{1 \text{ hr}} = 49.712 \text{ min} = 50. \text{ min} = 5.0 \times 10^1 \text{ min}$

Check: The units (min) are correct. The magnitude of the answer (50) makes physical sense because she is running almost 7.5 miles (which would take her 60 min = 1 hr). Two significant figures are allowed because of the limitation of 7.5 mi/hr (two significant figures). Round the last digit up because the first non-significant digit is a 7.

54. **Given:** 195 km **Find:** hours **Other:** riding pace = 24 miles per hour
Conceptual Plan: km → mi → hr

$$\frac{0.6214 \text{ mi}}{1 \text{ km}} \quad \frac{1 \text{hr}}{24 \text{ mi}}$$

Solution: $195 \text{ km} \times \frac{0.6214 \text{ mi}}{1 \text{ km}} \times \frac{1 \text{hr}}{24 \text{ mi}} = 5.048875 \text{ hr} = 5.0 \text{ hr}$

Check: The units (hr) are correct. The magnitude of the answer (5) makes physical sense because she is riding over 100 miles (which would take her over 4 hr). Two significant figures are allowed because of the limitation of 24 mi/hr (two significant figures). Truncate after the last digit up because the first non-significant digit is a 4.

55. **Given:** 14 km/L **Find:** miles per gallon
Conceptual Plan: $\frac{\textbf{km}}{\textbf{L}} \rightarrow \frac{\textbf{mi}}{\textbf{L}} \rightarrow \frac{\textbf{mi}}{\textbf{gal}}$

$$\frac{0.6214 \text{ mi}}{1 \text{ km}} \quad \frac{3.785 \text{ L}}{1 \text{ gallon}}$$

Solution: $\frac{14 \text{ km}}{1 \text{ L}} \times \frac{0.6214 \text{ mi}}{1 \text{ km}} \times \frac{3.785 \text{ L}}{1 \text{ gallon}} = 32.927986 \frac{\text{miles}}{\text{gallon}} = 33 \frac{\text{miles}}{\text{gallon}}$

Check: The units (mi/gal) are correct. The magnitude of the answer (33) makes physical sense because the dominating factor is that a L is much smaller than a gallon, so the answer should go up. Two significant figures are allowed because of the limitation of 14 km/L (two significant figures). Round the last digit up because the first non-significant digit is a 9.

56. **Given:** 5.0 gallons **Find:** cm^3
Conceptual Plan: **gal → L → cm^3**

$$\frac{3.785 \text{ L}}{1 \text{ gallon}} \quad \frac{1000 \text{ cm}^3}{1 \text{ L}}$$

Solution: $5.0 \text{ gallons} \times \frac{3.785 \text{ L}}{1 \text{ gallon}} \times \frac{1000 \text{ cm}^3}{1 \text{ L}} = 1.8925 \times 10^4 \text{ cm}^3 = 1.9 \times 10^4 \text{ cm}^3$

Check: The units (cm^3) are correct. The magnitude of the answer (10^4) makes physical sense because cm^3 is much smaller than a gallon, so the answer should go up several orders of magnitude. Two significant figures are allowed because of the limitation of 15.0 gallons (two significant figures). Round the last digit up because the first non-significant digit is a 9.

57. a) **Given:** 195 m^2 **Find:** km^2
Conceptual Plan: **m^2 → km^2**

$$\frac{(1 \text{ km})^2}{(1000 \text{ m})^2}$$

Notice that for squared units, the conversion factors must be squared.

Solution: $195\ \text{m}^2 \times \dfrac{(1\ \text{km})^2}{(1000\ \text{m})^2} = 1.95 \times 10^{-4}\ \text{km}^2$

Check: The units (km^2) are correct. The magnitude of the answer (10^{-4}) makes physical sense because a kilometer is a much larger unit than a meter.

b) **Given:** 195 m^2 **Find:** dm^2
Conceptual Plan: $\mathbf{m^2 \rightarrow dm^2}$

$$\frac{(10\ \text{dm})^2}{(1\ \text{m})^2}$$

Notice that for squared units, the conversion factors must be squared.

Solution: $195\ \text{m}^2 \times \dfrac{(10\ \text{dm})^2}{(1\ \text{m})^2} = 1.95 \times 10^{4}\ \text{dm}^2$

Check: The units (dm^2) are correct. The magnitude of the answer (10^4) makes physical sense because a decimeter is a much smaller unit than a meter.

c) **Given:** 195 m^2 **Find:** cm^2
Conceptual Plan: $\mathbf{m^2 \rightarrow cm^2}$

$$\frac{(100\ \text{cm})^2}{(1\ \text{m})^2}$$

Notice that for squared units, the conversion factors must be squared.

Solution: $195\ \text{m}^2 \times \dfrac{(100\ \text{cm})^2}{(1\ \text{m})^2} = = 1.95 \times 10^{6}\ \text{cm}^2$

Check: The units (cm^2) are correct. The magnitude of the answer (10^6) makes physical sense because a centimeter is a much smaller unit than a meter.

58. a) **Given:** 115 m^3 **Find:** km^3
Conceptual Plan: $\mathbf{m^3 \rightarrow km^3}$

$$\frac{(1\ \text{km})^3}{(1000\ \text{m})^3}$$

Notice that for cubed units, the conversion factors must be cubed.

Solution: $115\ \text{m}^3 \times \dfrac{(1\ \text{km})^3}{(1000\ \text{m})^3} = 1.15 \times 10^{-7}\ \text{km}^3$

Check: The units (km^3) are correct. The magnitude of the answer (10^{-7}) makes physical sense because a kilometer is a much larger unit than a meter.

b) **Given:** 115 m^3 **Find:** dm^3
Conceptual Plan: $\mathbf{m^3 \rightarrow mm^3}$

$$\frac{(10\ \text{dm})^3}{(1\ \text{m})^3}$$

Notice that for cubed units, the conversion factors must be cubed.

Solution: $115\ \text{m}^3 \times \dfrac{(10\ \text{dm})^3}{(1\ \text{m})^3} = 1.15 \times 10^{5}\ \text{dm}^3$

Check: The units (dm^3) are correct. The magnitude of the answer (10^5) makes physical sense because a decimeter is a much smaller unit than a meter.

c) **Given:** 115 m^3 **Find:** cm^3
Conceptual Plan: $\mathbf{m^3 \rightarrow cm^3}$

$$\frac{(100\ \text{cm})^3}{(1\ \text{m})^3}$$

Notice that for cubed units, the conversion factors must be cubed.

Solution: $115\ \text{m}^3 \times \dfrac{(100\ \text{cm})^3}{(1\ \text{m})^3} = 1.15 \times 10^{8}\ \text{cm}^3$

Check: The units (cm^3) are correct. The magnitude of the answer (10^8) makes physical sense because a centimeter is a much smaller unit than a meter.

59. **Given:** 435 acres **Find:** square miles **Other:** 1 acre = 43,560 ft^2, 1 mile = 5280 ft
Conceptual Plan: **acres → ft^2 → mi^2**

$$\frac{43560\ ft^2}{1\ acre} \quad \frac{(1\ mi)^2}{(5280\ ft)^2}$$

Notice that for squared units, the conversion factors must be squared.

Solution: $435\ \cancel{acres} \times \frac{43560\ \cancel{ft^2}}{1\ \cancel{acre}} \times \frac{(1\ mi)^2}{(5280\ \cancel{ft})^2} = 0.6\underline{7}96875\ mi^2 = 0.680\ mi^2$

Check: The units (mi^2) are correct. The magnitude of the answer (0.7) makes physical sense because an acre is much smaller than a mi^2, so the answer should go down several orders of magnitude. Three significant figures are allowed because of the limitation of 435 acres (three significant figures). Round the last digit up because the first non-significant digit is a 7.

60. a) **Given:** 954 million acres **Find:** square miles **Other:** 1 acre = 43,560 ft^2, 1 mile = 5280 ft
Conceptual Plan: **Substitute 10^6 for million then acres → ft^2 → mi^2**

$$\frac{43560\ ft^2}{1\ acre} \quad \frac{(1\ mi)^2}{(5280\ ft)^2}$$

Notice that for squared units, the conversion must be squared.
Solution: 954 million acres = 954×10^6 acres

$954 \times 10^6\ \cancel{acres} \times \frac{43560\ \cancel{ft^2}}{1\ \cancel{acre}} \times \frac{(1\ mi)^2}{(5280\ \cancel{ft})^2} = 1.4\underline{9}0625 \times 10^6\ mi^2 = 1.49 \times 10^6\ mi^2$

Check: The units (mi^2) are correct. The magnitude of the answer (10^6) makes physical sense because an acre is much smaller than a mi^2, so the answer should go down several orders of magnitude. Three significant figures are allowed because of the limitation of 435 acres (three significant figures). Truncate the last digit up because the first non-significant digit is a 0.

b) **Given:** 3.537 million square miles **Find:** percentage of U.S. land is farmland
Conceptual Plan: **Substitute 10^6 for million then $\%\ farm = \frac{farmland}{total\ land} \times 100\%$**

Note - units of farmland and total land must be the same.

Solution: 3.537 million mi^2 = $3.537 \times 10^6\ mi^2$

$\%\ farmland = \frac{1.49 \times \cancel{10^6}\ \cancel{mi^2}}{3.537 \times \cancel{10^6}\ \cancel{mi^2}} \times 100\ \% = 42.\underline{1}437659\ \%\ farmland = 42.1\ \%\ farmland$

Check: The units (%) are correct. The magnitude of the answer (4 %) makes physical sense because less and less of our land is devoted to farmland. Three significant figures are allowed because of the limitation of 435 acres (three significant figures). Truncate the last digit up because the first non-significant digit is a 4.

61. **Given:** 14 lbs **Find:** mL **Other:** 80 mg/0.80 mL and 15 mg/kg body
Conceptual Plan: **lb → kg body → mg → mL**

$$\frac{1\ kg\ body}{2.205\ lb} \quad \frac{15\ mg}{1\ kg\ body} \quad \frac{0.80\ mL}{80\ mg}$$

Solution: $14\ \cancel{lb} \times \frac{1\ \cancel{kg\ body}}{2.205\ \cancel{lb}} \times \frac{15\ \cancel{mg}}{1\ \cancel{kg\ body}} \times \frac{0.80\ mL}{80\ \cancel{mg}} = 0.9\underline{5}23809524\ mL = 0.95\ mL$

Check: The units (cm^3) are correct. The magnitude of the answer (1 mL) makes physical sense because it is reasonable amount of liquid to give to a baby. Two significant figures are allowed because of the statement in the problem. Truncate the last digit because the first non-significant digit is a 2.

62. **Given:** 18 lbs **Find:** mL **Other:** 100 mg/5.0 mL and 10 mg/kg body
Conceptual Plan: **lb → kg body → mg → mL**

$$\frac{1\ kg\ body}{2.205\ lb} \quad \frac{10\ mg}{1\ kg\ body} \quad \frac{5.0\ mL}{100\ mg}$$

Solution: $18\ \cancel{lb} \times \frac{1\ \cancel{kg\ body}}{2.205\ \cancel{lb}} \times \frac{10\ \cancel{mg}}{1\ \cancel{kg\ body}} \times \frac{5.0\ mL}{100\ \cancel{mg}} = 4.0\underline{8}1632653\ mL = 4.1\ mL$

Check: The units (cm^3) are correct. The magnitude of the answer (4 mL) makes physical sense because it is reasonable amount of liquid to give to a baby.
Two significant figures are allowed because of the statement in the problem. Round up the last digit because the first non-significant digit is an 8.

63. **Given:** solar year **Find:** seconds **Other:** 60 seconds/minute; 60 minutes/ hour; 24 hours/solar day; and 365.24 solar days/solar year

Conceptual Plan: **yr → day → hr → min → sec**

$$\frac{365.24\ day}{1\ solar\ yr} \quad \frac{24\ hr}{1 day} \quad \frac{60\ min}{1\ hr} \quad \frac{60\ sec}{1\ min}$$

Solution:

$$1\ \cancel{solar\ yr} \times \frac{365.24\ \cancel{day}}{1\ \cancel{solar\ yr}} \times \frac{24\ \cancel{hr}}{1\ \cancel{day}} \times \frac{60\ \cancel{min}}{1\ \cancel{hr}} \times \frac{60\ sec}{1\ \cancel{min}} = 3.155\underline{6}736 \times 10^7\ sec = 3.1557 \times 10^7\ sec$$

Check: The units (seconds) are correct. The magnitude of the answer (10^7) makes physical sense because each conversion factor increases the value of the answer – a second is many orders of magnitude smaller than a year. Five significant figures are allowed because all conversion factors are assumed to be exact, except for the 365.24 days/ solar year (five significant figures). Round up the last digit because the first non-significant digit is a 7.

64. a) "Million" translates to 10^6 in Table 1.2. Substitute this quantity into the expression and move the decimal point to be in proper scientific notation. Fifty million Frenchmen = 50×10^6 Frenchmen = 5×10^7 Frenchmen (assuming one significant figure in fifty.)

b) This can be expressed as two different ratios: 10 jokes / 100 enemies = 1.0×10^{-1} jokes per enemy or 100 enemies / 10 jokes = 1×10^1 enemies per joke

c) "Hundred" translates as 10^{-2} (since this modifies millionth, something less than 1) and "millionth" translates to 10^{-6} in Table 1.2. Substitute this quantity into the expression and move the decimal point to be in proper scientific notation. $1.8 \times 10^{-2} \times 10^{-6}$ cm = 1.8×10^{-8} cm

d) "Thousand" translates to 10^3 in Table 1.2. Substitute this quantity into the expression and move the decimal point to be in proper scientific notation. Sixty thousand dollars = 60×10^3 dollars = 6×10^4 dollars (assuming one significant figure in sixty.)

e) The density of platinum (Table 1.4) = 21.4 g/mL = 2.14×10^1 g/mL moving the decimal point to be in proper scientific notation.

65. a) Extensive – the volume of a material depends on how much there is present.

b) Intensive – the boiling point of a material is independent of how much material you have, so these values can be published in reference tables.

c) Intensive – the temperature of a material depends on how much there is present.

d) Intensive – the electrical; conductivity of a material is independent of how much material you have, so these values can be published in reference tables.

e) Extensive - the energy contained in material depends on how much there is present. Many times energy is expressed in terms of Joules/mole, which then turns this quantity into an intensive property.

66. **Given:** $°C = \frac{°F - 32}{1.8}$ **Find:** temperature where °F = °C

Conceptual Plan: $°C = \frac{°F - 32}{1.8}$ **set °C = °F = x and solve for x**

Solution: $x = \dfrac{x - 32°F}{1.8}$ → $1.8\,x = x - 32$ → $1.8\,x - x = -32$ → $0.8\,x = -32$ →

$x = -32/0.8 = -40.$ → $-40.\ °F = -40.\ °C$

Check: The units (°F and °C) are correct. Plugging the result back into the equation confirms that the calculations were done correctly. The magnitude of the answer seems correct, since it is known that the result is not between 0°C and 100 °C. The numbers are getting closer together as the temperature is dropped.

67. **Given:** 130 °X = 212 °F and 10 °X = 32 °F **Find:** temperature where °X = °F.

Conceptual Plan: Use data to derive an equation relating °X and °F. Then set °F = °X = z and solve for z.

Solution: Assume a linear relationship between the two temperatures ($y = mx + b$).

Let y = °F and let x = °X.

The slope of the line (m) is the relative change in the two temperature scales:

$$m = \frac{\Delta\ °F}{\Delta\ °X} = \frac{212\ °F - 32\ °F}{130\ °X - 10\ °X} = \frac{180\ °F}{120\ °X} = 1.5$$

Solve for intercept (b) by plugging one set of temperatures into the equation:

$y = 1.5\,x + b$ → $32 = (1.5)(10) + b$ → $32 = 15 + b$ → $b = 17$ → $°F = (1.5)\ °X + 17$

Set °F = °X = z and solve for z.

$z = 1.5\,z + 17$ → $-17 = 1.5\,z - z$ → $-17 = 0.5\,z$ → $z = -34$ → $-34°F = -34\ °X$

Check: The units (°F and °X) are correct. Plugging the result back into the equation confirms that the calculations were done correctly. The magnitude of the answer seems correct, since it is known that the result is not between 32°F and 212 °F. The numbers are getting closer together as the temperature is dropped.

68. **Given:** 17 °J = 0 °H and 97 °J = 120 °H **Find:** temperature where methyl alcohol boils in °J

Other: methyl alcohol boils at 84° H

Conceptual Plan: Use data to derive an equation relating °J and °H. Then set °H = 84 °H and solve for °J.

Solution: Assume a linear relationship between the two temperatures ($y = mx + b$).

Let y = °J and let x = °H.

The slope of the line (m) is the relative change in the two temperature scales:

$$m = \frac{\Delta\ °J}{\Delta °H} = \frac{97\ °J - 17\ °J}{120\ °H - 0\ °H} = \frac{80\ °J}{120\ °H} = 0.667$$

Solve for intercept (b) by plugging one set of temperatures into the equation:

$y = 0.6\underline{6}7\,x + b$ → $17 = (0.6\underline{6}7)(0) + b$ → $b = 17$ → $°J = (0.6\underline{6}7)\ °H + 17$

Set °H = 84 °H and solve for °J.

$°J = (0.6\underline{6}7)(84) + 17$ → $°J = 56 + 17 = 73\ °J$

Check: The units (°J) are correct. Plugging the original data points back into the equation confirms that the calculations were done correctly. The magnitude of the answer seems correct, since the result should be between 17°J and 97 °J, and closer to 97°J than 17°J.

69. a) $1.76 \times 10^{-3}/8.0 \times 10^{2} = 2.2 \times 10^{-6}$ Two significant figures are allowed to reflect the quantity with the fewest significant figures (8.0×10^{2}).

b) Write all figures so that the decimal points can be aligned:

$$\begin{array}{r} 0.0187 \\ +\ 0.0002 \\ -\ 0.0030 \\ \hline 0.0159 \end{array}$$

All quantities are known to four places to the right of the decimal place, so the answer should be reported to four places to the right of the decimal place or three significant figures.

c) $[(136000)(0.000322)/0.082](129.2) = 6.899910244 \times 10^{4} = 6.9 \times 10^{4}$ Round the intermediate answer to two significant figures to reflect the quantity with the fewest significant figures (0.082). Round up the last digit since the first non-significant digit is 9.

70. **Given:** one gallon of gasoline **Find:** US dollars **Other:** 1 Euro = $1.57 US and 1 liter of gasoline in France = 1.35 Euro

Conceptual Plan: **gal → L → Euro → $ US**

$$\frac{3.785\ L}{1\ gallon} \quad \frac{1.35\ Euro}{1\ L} \quad \frac{\$\,1.57\ US}{1\ Euro}$$

Solution: $1 \text{ gallon} \times \dfrac{3.785 \text{ L}}{1 \text{ gallon}} \times \dfrac{1.35 \text{ Euro}}{1 \text{ L}} \times \dfrac{\$\,1.57 \text{ US}}{1 \text{ Euro}} = \$\,8.0223075 \text{ US} = \$\,8.02 \text{ US}$

Check: The units (\$ US) are correct. The magnitude of the answer (\$8 US) makes physical sense because the dominating conversion factor is ~ 4. Three significant figures are allowed because of the limitation of 1.35 Euro and \$1.35 (three significant figures each). Truncate the non-significant digits because the first non-significant digit is a 2.

71. a) **Given:** cylinder dimensions: length = 22 cm, radius = 3.8 cm, *d*(gold) = 19.3 g/cm^3 and *d*(sand) = 3.00 g/cm^3
Find: *m*(gold) and *m*(sand)
Conceptual Plan: ***l, r*** **→ *V*** **then** ***d, V*** **→ *m***

$V = l\,\pi\,r^2 \qquad\qquad d = m/V$

Solution: $V(\text{gold}) = V(\text{sand}) = (22 \text{ cm})(\pi)(3.8 \text{ cm})^2 = 998.0212 \text{ cm}^3$ $d = m/V$ Rearrange by multiplying both sides of equation by *V*. → $m = d \times V$

$$m(\text{gold}) = \left(19.3 \dfrac{\text{g}}{\text{cm}^3}\right) \times (998.0212 \text{ cm}^3) = 1.926181 \times 10^4 \text{ g} = 1.9 \times 10^4 \text{ g}$$

Check: The units (g) are correct. The magnitude of the answer seems correct considering the value of the density is ~20 g/cm^3. Two significant figures are allowed to reflect the significant figures in 22 cm and 3.8 cm. Truncate the non-significant digits because the first non-significant digit is a 2.

$$\text{m(sand)} = \left(3.00 \dfrac{\text{g}}{\text{cm}^3}\right) \times (998.0212 \text{ cm}^3) = 2.99206 \times 10^3 \text{ g} = 3.0 \times 10^3 \text{ g}$$

Check: The units (g) are correct. The magnitude of the answer seems correct considering the value of the density is 3 g/cm^3. This number is much lower than the gold mass. Two significant figures are allowed to reflect the significant figures in 22 cm and 3.8 cm. Round the last digit up because the first non-significant digit is a 9.

b) Comparing the two values 1.9×10^4 g versus 3.0×10^3 g shows a difference in weight of almost a factor of 10. This difference should be enough to trip the alarm and alert the authorities to the presence of the thief.

72. **Given:** $r = 1.0 \times 10^{-13}$ cm, $m = 1.7 \times 10^{-24}$ g **Find:** density **Other:** $V = (4/3)\,\pi\,r^3$
Conceptual Plan: ***r*** **→ *V*** **then** ***m, V*** **→ *d***

$V = (4/3)\,\pi r^3 \qquad\qquad d = m/V$

Solution: $V = (4/3)\,\pi\,r^3 = (4/3)(\pi)(1.0 \times 10^{-13} \text{ cm})^3 = 4.188790205 \times 10^{-39} \text{ cm}^3$

$$d = \frac{m}{V} = \frac{1.7 \times 10^{-24} \text{ g}}{4.188790205 \times 10^{-39} \text{ cm}^3} = 4.058451049 \times 10^{14} \frac{\text{g}}{\text{cm}^3} = 4.1 \times 10^{14} \frac{\text{g}}{\text{cm}^3}$$

Check: The units (g/cm^3) are correct. The magnitude of the answer seems correct considering how small a nucleus is compared to an atom. Two significant figures are allowed to reflect the significant figures in 1.0×10^{-13} cm. Round the last digit up because the first non-significant digit is a 5.

73. **Given:** 3.5 lb of titanium **Find:** volume in in^3 **Other:** density of titanium is 4.5 g/cm^3
Conceptual Plan: **lb → g** **then** ***m, d*** **→ *V*** **then cm^3 → in^3**

$\dfrac{453.6 \text{ g}}{1 \text{ lb}} \qquad d = m/V \qquad \dfrac{(1 \text{ in})^3}{(2.54 \text{ cm})^3}$

Solution: $3.5 \text{ lb} \times \dfrac{453.6 \text{ g}}{1 \text{ lb}} = 1.5876 \times 10^3 \text{ g}$

$d = m/V$ Rearrange by multiplying both sides of the equation by *V* and dividing both sides of the equation by *d*. → $V = \dfrac{m}{d} = \dfrac{1.5876 \times 10^3 \text{ g}}{4.5 \dfrac{\text{g}}{\text{cm}^3}} = 3.528 \times 10^2 \text{ cm}^3 = 3.5 \times 10^2 \text{ cm}^3 \times \dfrac{(1 \text{ in})^3}{(2.54 \text{ cm})^3} = 22 \text{ in}^3$

Check: The units (in^3) are correct. The magnitude of the answer seems correct considering many grams we have. Two significant figures are allowed to reflect the significant figures in 3.5 lb. Truncate the non-significant digits because the first non-significant digit is a 2.

74. **Given:** density (g/cm^3) **Find:** density (lb/in^3) **Other:** density of iron is 7.86 g/cm^3
Conceptual Plan: $\frac{\text{g}}{\text{cm}^3} \rightarrow \frac{\text{lb}}{\text{cm}^3} \rightarrow \frac{\text{lb}}{\text{in}^3}$

$$\frac{1\text{ lb}}{453.6\text{ g}} \qquad \frac{(2.54\text{ cm})^3}{(1\text{ in})^3}$$

Solution: $$\frac{7.86\text{ g}}{\text{cm}^3} \times \frac{1\text{ lb}}{453.6\text{ g}} \times \frac{(2.54\text{ cm})^3}{(1\text{ in})^3} = 0.2839557386\frac{\text{lb}}{\text{in}^3} = 0.284\frac{\text{lb}}{\text{in}^3}$$

Check: The units (lb/in^3) are correct. The magnitude of the answer seems correct considering that the dominating factor is that a gram is smaller than a pound, so the answer should go down. Three significant figures are allowed to reflect the significant figures in 7.86 lb. Round the last digit up because the first non-significant digit is a 9.

75. **Given:** cylinder dimensions: length = 2.16 in, radius = 0.22 in, m= 41 g **Find:** density (g /cm^3)
Conceptual Plan: in → cm then $l, r \rightarrow V$ then $m, V \rightarrow d$

$$\frac{2.54\text{ cm}}{1\text{ in}} \qquad V = l\pi r^2 \qquad d = m/V$$

Solution: $2.16\text{ in} \times \frac{2.54\text{ cm}}{1\text{ in}} = 5.4864\text{ cm} = l$ $\qquad 0.22\text{ in} \times \frac{2.54\text{ cm}}{1\text{ in}} = 0.5588\text{ cm} = r$

$$V = l\pi r^2 = (5.4864\text{ cm})(\pi)(0.5588\text{ cm})^2 = 5.3820798\text{ cm}^3$$

$$d = \frac{m}{V} = \frac{41\text{ g}}{5.3820798\text{ cm}^3} = 7.6178729\frac{\text{g}}{\text{cm}^3} = 7.6\frac{\text{g}}{\text{cm}^3}$$

Check: The units (g/cm^3) are correct. The magnitude of the answer seems correct considering the value of the density of iron (a major component in steel) is 7.86 g/cm^3. Two significant figures are allowed to reflect the significant figures in 0.22 in and 41 g. Truncate the non-significant digits because the first non-significant digit is a 2.

76. **Given:** m = 85 g **Find:** radius of the sphere (inches) **Other:** density (aluminum) = 2.7 g /cm^3
Conceptual Plan: $m, d \rightarrow V$ then $V \rightarrow r$ then cm → in

$$d = m/V \qquad V = (4/3)\pi r^3 \qquad \frac{1\text{ in}}{2.54\text{ cm}}$$

Solution: $d = m/V$ Rearrange by multiplying both sides of the equation by V and dividing both sides of the equation by d.

$$V = \frac{m}{d} = \frac{85\text{ g}}{2.7\frac{\text{g}}{\text{cm}^3}} = 31.48148148\text{ cm}^3$$

$V = (4/3)\pi r^3$ Rearrange by dividing both sides of the equation by (4/3) π. $r^3 = \frac{3V}{4\pi}$

Take the cube root of both sides of the equation.

$$r = \left(\frac{3V}{4\pi}\right)^{1/3} = \left(\frac{(3)(31.48148148\text{ cm}^3)}{4\pi}\right)^{1/3} = (7.51565009\text{ cm}^3)^{1/3} = 1.958794386\text{ cm}$$

$$1.958794386\text{ cm} \times \frac{1\text{ in}}{2.54\text{ cm}} = 0.7711788923\text{ in} = 0.77\text{ in}$$

Check: The units (in) are correct. The magnitude of the answer seems correct. The magnitude of the volume is about a third of the mass (density is about 3 g/cm^3). The radius in cm seems right considering the geometry involved. The magnitude goes down when we convert from cm to inches, because an inch is bigger than a cm. Two significant figures are allowed to reflect the significant figures in 2.7 g/cm^3 and 85 g. Truncate the non-significant digits because the first non-significant digit is a 1.

77. **Given:** 185 cubic yards (yd^3) of H_2O **Find:** mass of the H_2O (pounds) **Other:** $d(H_2O) = 1.00\ g/cm^3$ at 0°C
Conceptual Plan: $yd^3 \rightarrow m^3 \rightarrow cm^3 \rightarrow g \rightarrow lb$

$$\frac{(1\ m)^3}{(1.094\ yd)^3} \quad \frac{(100\ cm)^3}{(1m)^3} \quad \frac{1.00\ g}{1.00\ cm^3} \quad \frac{1\ lb}{453.59\ g}$$

Solution:

$$185\ yd^3 \times \frac{(1\ m)^3}{(1.094\ yd)^3} \times \frac{(100\ cm)^3}{(1\ m)^3} \times \frac{1.00\ g}{1.00\ cm^3} \times \frac{1\ lb}{453.59\ g} = 3.114987377 \times 10^5\ lbs = 3.11 \times 10^5\ lbs$$

Check: The units (lb) are correct. The magnitude of the answer (10^5) makes physical sense because a pool is not a small object. Three significant figures are allowed because the conversion factor with the least precision is the density (1.00 g/cm^3 - 3 significant figures) and the initial size has three significant figures. Truncate after the last digit because the first non-significant digit is a 4.

78. **Given:** 7655 cubic feet (ft^3) of ice **Find:** mass of the ice (kg) **Other:** 1.00 cm^3 ice = 0.917 g ice at 0°C
Conceptual Plan: $ft^3 \rightarrow cm^3 \rightarrow g \rightarrow kg$

$$\frac{(30.48\ cm)^3}{(1\ ft)^3} \quad \frac{0.917\ g}{1.00\ cm^3} \quad \frac{1\ kg}{1000\ g}$$

Solution: $7655\ ft^3 \times \frac{(30.48\ cm)^3}{(1\ ft)^3} \times \frac{0.917\ g}{1.00\ cm^3} \times \frac{1\ kg}{1000\ g} = 1.987739274 \times 10^5\ kg = 1.99 \times 10^5\ kg$

Check: The units (kg) are correct. The magnitude of the answer (10^5) makes physical sense because an iceberg is a large object. Three significant figures are allowed because the conversion factor with the least precision is the density (0.917 g/cm^3 - 3 significant figures). Round up the last digit because the first non-significant digit is a 7.

79. **Given:** 15 liters of gasoline **Find:** kilometers **Other:** 52 mi/gal in the city
Conceptual Plan: $L \rightarrow gal \rightarrow mi \rightarrow km$

$$\frac{1\ gallon}{3.785\ L} \quad \frac{52\ mi}{1.0\ gallon} \quad \frac{1\ km}{0.6214\ mi}$$

Solution: $15\ L \times \frac{1\ gallon}{3.785\ L} \times \frac{52\ mi}{1.0\ gallon} \times \frac{1\ km}{0.6214\ mi} = 3.316327941 \times 10^2\ km = 3.3 \times 10^2\ km$

Check: The units (km) are correct. The magnitude of the answer (10^2) makes physical sense because the dominating conversion factor is the mileage, which increases the answer. Two significant figures are allowed because the conversion factor with the least precision is 52 mi/gallon (2 significant figures) and the initial volume (15 L) has 2 significant figures. Truncate the last digit because the first non-significant digit is a 1. It is best to put the answer in scientific notation so that it is unambiguous how many significant figures are expressed.

80. **Given:** 355 mL of gasoline **Find:** kilometers **Other:** 57 mi/gal in the city .
Conceptual Plan: $mL \rightarrow L \rightarrow gal \rightarrow mi \rightarrow km$

$$\frac{1\ L}{1000\ mL} \quad \frac{1\ gallon}{3.785\ L} \quad \frac{57\ mi}{1.0\ gallon} \quad \frac{1\ km}{0.6214\ mi}$$

Solution: $355\ mL \times \frac{1\ L}{1000\ mL} \times \frac{1\ gallon}{3.785\ L} \times \frac{57\ mi}{1.0\ gallon} \times \frac{1\ km}{0.6214\ mi} = 8.603319984\ km = 8.6\ km$

Check: The units (km) are correct. The magnitude of the answer (8.6) makes physical sense because the dominating conversion factor is the conversion from mL to L, which decreases the answer. Two significant figures are allowed because the conversion factor with the least precision is 57 mi/gallon (2 significant figures). Truncate the last digit because the first non-significant digit is a 0.

81. **Given:** radius of nucleus of the hydrogen atom = 1.0×10^{-13} cm; radius of the hydrogen atom = 52.9 pm.
Find: percent of volume occupied by nucleus (%)
Conceptual Plan: $cm \rightarrow m$ then $pm \rightarrow m$ then $r \rightarrow V$ then $V_{atom}, V_{nucleus} \rightarrow \%\ V_{nucleus}$

$$\frac{1\ m}{100\ cm} \quad \frac{1\ m}{10^{12}\ pm} \quad V = (4/3)\pi r^3 \quad \%\ V_{nucleus} = \frac{V_{nucleus}}{V_{atom}} \times 100\%$$

Solution: 1.0×10^{-13} ~~cm~~ x $\frac{1 \text{ m}}{100 \text{ cm}}$ = 1.0×10^{-15} m and 52.9 ~~pm~~ x $\frac{1 \text{ m}}{10^{12} \text{ pm}}$ = 5.29×10^{-11} m

$V = (4/3)\pi r^3$ Substitute into $\%V$ equation.

$$\% V_{nucleus} = \frac{V_{nucleus}}{V_{atom}} \times 100\% \rightarrow \quad \% V_{nucleus} = \frac{(4/3)\ \pi r^3_{nucleus}}{(4/3)\ \pi r^3_{atom}} \times 100\%$$ Simplify equation.

$$\% V_{nucleus} = \frac{r^3_{nucleus}}{r^3_{atom}} \times 100\%$$ Substitute numbers and calculate result.

$$\% V_{nucleus} = \frac{(1.0 \times 10^{-15}\ \text{m})^3}{(5.29 \times 10^{-11}\ \text{m})^3} \times 100\% = (1.\underline{8}90359168 \times 10^{-5})^3 \times 100\% = 6.\underline{7}55118685 \times 10^{-13}\ \%$$

$= 6.8 \times 10^{-13}$ %

Check: The units (%) are correct. The magnitude of the answer seems correct (10^{-13} %), since a proton is so small. Two significant figures are allowed to reflect the significant figures in 1.0×10^{-13} cm. Round up the last digits because the first non-significant digit is a 5.

82. **Given:** radius of neon = 69 pm; 2.69×10^{22} atoms per liter **Find:** percent of volume occupied by neon (%)
Conceptual Plan: Assume 1L total volume.
pm → m → cm then $r \rightarrow V$ then cm^3 → L then L/atom → L then V_{Ne}, V_{Total} → $\%V_{Ne}$

$\frac{1 \text{ m}}{10^{12} \text{ pm}}$ $\frac{100 \text{ cm}}{1 \text{ m}}$ $V = (4/3)\pi r^3 = 1 \text{ atom}$ $\frac{1 \text{ L}}{1000 \text{ cm}^3}$ 2.69×10^{22} atoms $\% V_{Ne} = \frac{V_{Ne}}{V_{Total}} \times 100\%$

Solution: 69 ~~pm~~ x $\frac{1 \text{ m}}{10^{12} \text{ pm}}$ x $\frac{100 \text{ cm}}{1 \text{ m}}$ = 6.9×10^{-9} cm

$V = (4/3)\ \pi r^3 = (4/3)\ \pi\ (6.9 \times 10^{-9}\ \text{cm})^3 = 1.\underline{3}7605528 \times 10^{-24}\ \text{cm}^3$

$1.\underline{3}7605528 \times 10^{-24}$ ~~cm³~~ x $\frac{1 \text{ L}}{1000 \text{ cm}^3}$ = $1.\underline{3}7605528 \times 10^{-27}$ L

$\frac{1.\underline{3}7605528 \times 10^{-27}\ \text{L}}{\text{atom}} \times 2.69 \times 10^{22}$ atoms = $3.\underline{7}01588707 \times 10^{-5}$ L Substitute into $\%V$ equation.

$$\% V_{Ne} = \frac{V_{Ne}}{V_{Total}} \times 100\% = \frac{3.\underline{7}01588707 \times 10^{-5}\ \text{L}}{1\ \text{L}} \times 100\% = 3.\underline{7}01588707 \times 10^{-3}\ \% = 3.7 \times 10^{-3}\ \%$$

Check: This says that the separation between atoms is very large in the gas phase.
The units (%) are correct. The magnitude of the answer seems correct (10^{-3} %), it is known that gases are primarily empty space. Two significant figures are allowed to reflect the significant figures in 69 pm. Truncate the non-significant digits because the first non-significant digit is a 0.

83. **Given:** mass of black hole (BH) = 1×10^3 suns; radius of black hole = one-half the radius of our moon.
Find: density (g/cm^3) **Other:** radius of our sun = 7.0×10^5 km; average density of our sun = 1.4×10^3 kg/m^3; diameter of the moon = 2.16×10^3 miles
Conceptual Plan: $d_{BH} = m_{BH}/V_{BH}$
Calculate m_{BH} : $r_{sun} \rightarrow V_{sun}$ $km^3_{sun} \rightarrow m^3_{sun}$ $V_{sun}, d_{sun} \rightarrow m_{sun}$ $m_{sun} \rightarrow m_{BH}$ kg → g

$V = (4/3)\pi r^3$ $\frac{(1000 \text{ m})^3}{(1 \text{ km})^3}$ $d_{sun} = \frac{m_{sun}}{V_{sun}}$ $m_{BH} = (1 \times 10^3) \times m_{sun}$ $\frac{1000 \text{ g}}{1 \text{ kg}}$

Calculate V_{BH}: $d_{moon} \rightarrow r_{moon} \rightarrow r_{BH}$ mi → km → m → cm $r \rightarrow V$

$r_{moon} = ½\ d_{moon}$ $r_{BH} = ½\ r_{moon}$ $\frac{1 \text{ km}}{0.6214 \text{ mi}}$ $\frac{1000 \text{ m}}{1 \text{ km}}$ $\frac{100 \text{ cm}}{1 \text{ m}}$ $V = (4/3)\pi r^3$

Substitute into $d_{BH} = m_{BH}/V_{BH}$
Solution: Calculate m_{BH} $V_{sun} = (4/3)\ \pi\ r_{sun}^3 = (4/3)\ \pi\ (7.0 \times 10^5\ \text{km})^3 = 1.\underline{4}3675504 \times 10^{18}\ \text{km}^3$

$1.\underline{4}3675504 \times 10^{18}$ ~~km³~~ x $\frac{(1000 \text{ m})^3}{(1 \text{ km})^3}$ = $1.\underline{4}3675504 \times 10^{27}$ m^3 $d_{sun} = m_{sun}/V_{sun}$ Solve for m by multiplying both sides of the equation by V_{sun}

$m_{sun} = V_{sun} \times d_{sun}$
$m_{sun} = (1.\underline{4}3675504 \times 10^{27}\ \text{m}^3)(1.4 \times 10^3\ \text{kg/m}^3) = 2.\underline{0}11457056 \times 10^{30}\ \text{kg}$
$m_{BH} = (1 \times 10^3) \times m_{sun} = (1 \times 10^3) \times (2.\underline{0}11457056 \times 10^{30}\ \text{kg}) = 2.\underline{0}11457056 \times 10^{33}\ \text{kg}$

$$2.\underline{0}11457056 \times 10^{33}\ \text{kg} \times \frac{1000\ \text{g}}{1\ \text{kg}} = 2.\underline{0}11457056 \times 10^{36}\ \text{g}$$

Calculate V_{BH} $r_{moon} = \frac{1}{2} d_{moon} = \frac{1}{2}(2.16 \times 10^3 \text{ miles}) = 1.08 \times 10^3$ miles

$r_{BH} = \frac{1}{2} r_{moon} = \frac{1}{2}(1.08 \times 10^3 \text{ miles}) = 540.$ miles

$$540.\ \text{miles} \times \frac{1\ \text{km}}{0.6214\ \text{mi}} \times \frac{1000\ \text{m}}{1\ \text{km}} \times \frac{100\ \text{cm}}{1\ \text{m}} = 8.6\underline{9}00547 \times 10^7\ \text{cm}$$

$$V = (4/3)\,\pi\, r^3 = (4/3)\,\pi\,(8.6\underline{9}00547 \times 10^7\ \text{cm})^3 = 2.7\underline{4}888228 \times 10^{24}\ \text{cm}^3$$

Substitute into $$d_{BH} = \frac{m_{BH}}{V_{BH}} = \frac{2.\underline{0}11457056 \times 10^{36}\ \text{g}}{2.7\underline{4}888228 \times 10^{24}\ \text{cm}^3} = 7.\underline{3}1737339 \times 10^{11}\ \frac{\text{g}}{\text{cm}^3} = 7.3 \times 10^{11}\ \frac{\text{g}}{\text{cm}^3}$$

Check: The units (g/cm^3) are correct. The magnitude of the answer seems correct (10^{12}), since we expect extremely high numbers for black holes. Two significant figures are allowed to reflect the significant figures in the radius of our sun (7.0×10^5 km) and the average density of the sun (1.4×10^3 kg/m^3). Truncate the non-significant digits because the first non-significant digit is a 4.

85. **Given:** cubic nanocontainers with an edge length = 25 nanometers
Find: a) volume of one nanocontainer; b) grams of oxygen could be contained by each nanocontainer; c) grams of oxygen inhaled per hour; d) minimum number of nanocontainers per hour; and e) minimum volume of nanocontainers.
Other: (pressurized oxygen) = 85 g/L; 0.28 g of oxygen per liter; average human inhales about 0.50 L of air per breath and takes about 20 breaths per minute; and adult total blood volume = ~ 5 L.
Conceptual Plan:

a) **nm → m → cm then $l \rightarrow V$ then cm^3 → L**

$$\frac{1\ \text{m}}{10^9\ \text{nm}} \quad \frac{100\ \text{cm}}{1\ \text{m}} \qquad V = l^3 \qquad \frac{1\ \text{L}}{1000\ \text{cm}^3}$$

b) **L → g pressurized oxygen**

$$\frac{85\ \text{g oxygen}}{1\ \text{L nanocontainers}}$$

c) **hr → min → breaths → L_{air} → g_{O2}**

$$\frac{60\ \text{min}}{1\ \text{hr}} \quad \frac{20\ \text{breath}}{1\ \text{min}} \quad \frac{0.50\ L_{air}}{1\ \text{breath}} \quad \frac{0.28\ g_{CO}}{1\ L_{air}}$$

d) **grams oxygen → number nanocontainers**

$$\frac{1\ \text{nanocontainer}}{\text{part (b) grams of oxygen}}$$

e) **number nanocontainers → volume nanocontainers**

$$\frac{\text{part (a) volume}}{\text{of 1 nanocontainer}}$$

Solution:

a) $$25\ \text{nm} \times \frac{1\ \text{m}}{10^9\ \text{nm}} \times \frac{100\ \text{cm}}{1\ \text{m}} = 2.5 \times 10^{-6}\ \text{cm}$$

$$V = l^3 = (2.5 \times 10^{-6}\ \text{cm})^3 = 1.\underline{5}625 \times 10^{-17}\ \text{cm}^3 \times \frac{1\ \text{L}}{1000\ \text{cm}^3} = 1.\underline{5}625 \times 10^{-20}\ \text{L} = 1.6 \times 10^{-20}\ \text{L}$$

b) $$1.\underline{5}625 \times 10^{-20}\ \text{L} \times \frac{85\ \text{g oxygen}}{1\ \text{L nanocontainers}} = 1.\underline{3}28125 \times 10^{-18}\ \frac{\text{g pressurized } O_2}{\text{nanocontainer}}$$

$$= 1.3 \times 10^{-18}\ \frac{\text{g pressurized } O_2}{\text{nanocontainer}}$$

c) $$1\ \text{hr} \times \frac{60\ \text{min}}{1\ \text{hr}} \times \frac{20\ \text{breath}}{1\ \text{min}} \times \frac{0.50\ L_{air}}{1\ \text{breath}} \times \frac{0.28\ gO_2}{1\ L_{air}} = 1.\underline{6}8 \times 10^2\ \text{g oxygen} = 1.7 \times 10^2\ \text{g oxygen}$$

d) $1.68 \times 10^2 \text{ g oxygen} \times \dfrac{1 \text{ nanocontainer}}{1.3 \times 10^{-18} \text{ g of oxygen}} = 1.292307692 \times 10^{20}$ nanocontainers

$= 1.3 \times 10^{20}$ nanocontainers

e) $1.292307692 \times 10^{20} \text{ nanocontainers} \times \dfrac{1.5625 \times 10^{-20} \text{ L}}{\text{nanocontainer}} = 2.019230769 \text{ L} = 2.0 \text{ L}$

This volume is much too large to be feasible, since the volume of blood in the average human is 5 L.

Check:

a) The units (L) are correct. The magnitude of the answer (10^{-20}) makes physical sense because these are very, very tiny containers. Two significant figures are allowed, reflecting the significant figures in the starting dimension (25 nm – 2 significant figures). Round up the last digit because the first non-significant digit is a 6.

b) The units (g) are correct. The magnitude of the answer (10^{-18}) makes physical sense because these are very, very tiny containers and very few molecules can fit inside. Two significant figures are allowed, reflecting the significant figures in the starting dimension (25 nm) and the given concentration (85 g/L) – 2 significant figures in each. Truncate the non-significant digits because the first non-significant digit is a 2.

c) The units (g oxygen) are correct. The magnitude of the answer (10^2) makes physical sense because of the conversion factors involved and the fact that air is not very dense. Two significant figures are allowed, because it is stated in the problem. Round up the last digit because the first non-significant digit is an 8.

d) The units (nanocontainers) are correct. The magnitude of the answer (10^{20}) makes physical sense because these are very, very tiny containers and we need a macroscopic quantity of oxygen in these containers. Two significant figures are allowed, reflecting the significant figures in both of the quantities in the calculation – 2 significant figures. Round up the last digit because the first non- significant digit is a 9.

e) The units (L) are correct. The magnitude of the answer (2) makes physical sense because, the magnitudes of the numbers in this step. Two significant figures are allowed reflecting the significant figures in both of the quantities in the calculation – 2 significant figures. Truncate the non-significant digits because the first non-significant digit is a 1.

87. c) is the best representation. When solid carbon dioxide (dry ice) sublimes, it changes phase from a solid to a gas. Phase changes are physical changes, so no molecular bonds are broken. This diagram shows molecules with one carbon atom and two oxygen atoms bonded together in every molecule. The other diagrams have no carbon dioxide molecules.

88. This problem is similar to problem #60, only the dimension is changes to 7 cm on each edge.
Given: 7 cm on each edge cube **Find:** cm^3
Conceptual plan: Read the information given carefully. The cube is 7 cm on each side.

$$l, w, h \rightarrow V$$
$$V = l\,w\,h$$
$$\text{in a cube } l = w = h$$

Solution: $7 \text{ cm} \times 7 \text{ cm} \times 7 \text{ cm} = (7 \text{ cm})^3 = \; = 343 \text{ cm}^3$ or 343 cubes

89. In order to determine which number is large, the units need to be compared. There is a factor of 1000 between grams and kg, in the numerator. There is a factor of $(100)^3$ or 1,000,000 between cm^3 and m^3. This second factor more than compensates for the first factor. Thus Substance A with a density of 1.7 g/cm^3 is denser than Substance B with a density of 1.7 kg/m^3.

90. Remember that density = mass/volume.
a) The darker colored box has a heavier mass, but a smaller volume, so it is denser than the lighter-colored box.

b) The lighter colored box is heavier than the darker colored box and both boxes have the same volume, so the lighter colored box is denser.

c) The larger box is the heavier box, so it can not be determined with this information which box is denser.

91. Remember that: An observation is the information collected when studying phenomena. A law is a concise statement that summarizes observed behaviors and observations and predicts future observations. A theory attempts to explain why the observed behavior is happening.

a) This statement is most like a law, because it summarizes many observations and can explain future behavior – many places and many days.

b) This statement is a theory because it attempts to explain why – gravitational forces.

c) This statement is most like an observation, because it is information collected in order to understand tidal behavior.

d) This statement is most like a law, because it summarizes many observations and can explain future behavior – many places and many days and months.

Chapter 2
Atoms and Elements

1. **Given:** 1.50 g hydrogen; 12.0 g oxygen **Find:** grams water vapor
Conceptual Plan: total mass reactants = total mass products
Solution: Mass of reactants = 1.50 g hydrogen + 12.0 g oxygen = 13.5 grams
Mass of products = mass of reactants = 13.5 grams water vapor.
Check: According to the Law of Conservation of Mass, matter is not created or destroyed in a chemical reaction, so, since water vapor is the only product the masses of hydrogen and oxygen must combine to form the mass of water vapor.

2. **Given:** 21 kg gasoline; 84 kg oxygen **Find:** mass of carbon dioxide and water
Conceptual Plan: total mass reactants = total mass products
Solution: Mass of reactants = 21 kg gasoline + 84 kg oxygen = 105 kg mass
Mass of products = mass of reactants = 105 kg of mass of carbon dioxide and water.
Check: According to the Law of Conservation of Mass, matter is not created or destroyed in a chemical reaction, so, since carbon dioxide and water are the only products the masses of gasoline and oxygen must combine to form the mass of carbon dioxide and water.

3. **Given:** sample 1: 38.9 g carbon, 448 g chlorine; sample 2: 14.8 g carbon, 134 g chlorine
Find: consistent with definite proportions.
Conceptual Plan: determine mass ratio of sample 1 and 2 and compare.
$$\frac{\text{mass of chlorine}}{\text{mass of carbon}}$$
Solution: Sample 1: $\frac{448\,\text{g chorine}}{38.9\,\text{g carbon}} = 11.5$ Sample 2: $\frac{134\,\text{g chlorine}}{14.8\,\text{g carbon}} = 9.05$
Results are not consistent with the law of definite proportions because the ratio of chlorine to carbon is not the same.
Check: According to the Law of Definite Proportions, the mass ratio of one element to another is the same for all samples of the compound.

4. **Given:** sample 1: 6.98 grams sodium, 10.7 grams chlorine; sample 2: 11.2 g sodium, 17.3 grams chlorine
Find: consistent with definite proportions.
Conceptual Plan: determine mass ratio of sample 1 and 2 and compare.
$$\frac{\text{mass of chlorine}}{\text{mass of sodium}}$$
Solution: Sample 1: $\frac{10.7\,\text{g chorine}}{6.98\,\text{g sodium}} = 1.53$ Sample 2: $\frac{17.3\,\text{g chlorine}}{11.2\,\text{g sodium}} = 1.54$
Results are consistent with the law of definite proportions.
Check: According to the Law of Definite Proportions, the mass ratio of one element to another is the same for all samples of the compound.

5. **Given:** mass ratio sodium to fluorine = 1.21:1; sample = 28.8 g sodium **Find:** g fluorine
Conceptual Plan: g sodium → g fluorine
$$\frac{\text{mass of fluorine}}{\text{mass of sodium}}$$
Solution: $28.8\,\text{g}\,\cancel{\text{sodium}} \times \frac{1\,\text{g fluorine}}{1.21\,\text{g}\,\cancel{\text{sodium}}} = 23.8\,\text{g fluorine}$
Check: The units of the answer, g fluorine are correct. The magnitude of the answer is reasonable since it is less than the grams of sodium.

6. **Given:** sample 1: 1.65 kg magnesium, 2.57 kg fluorine; sample 2: 1.32 kg magnesium
Find:g fluorine in sample 2.
Conceptual Plan: mass magnesium and mass fluorine → mass ratio → mass fluorine(kg) → mass fluorine(g)
$$\frac{\text{mass of fluorine}}{\text{mass of magnesium}} \qquad \frac{1000\,\text{g}}{\text{kg}}$$

Solution: $\text{mass ratio} = \dfrac{2.57\text{ kg fluorine}}{1.65\text{ kg magnesium}} = \dfrac{1.56\text{ kg fluorine}}{1.00\text{ kg magnesium}}$

$$1.32\ \text{kg }\cancel{\text{magnesium}} \times \frac{1.56\text{ kg fluorine}}{1.00\ \text{kg }\cancel{\text{magnesium}}} \times \frac{1000\text{ g}}{\text{kg}} = 2.06 \times 10^{3}\text{ g fluorine}$$

Check: The units of the answer, g fluorine, are correct. The magnitude of the answer is reasonable since it is greater than the mass of magnesium and the ratio is greater than 1.

7. **Given:** 1 gram osmium: sample 1 = 0.168 g oxygen; sample 2 = 0.3369 g oxygen
Find: consistent with multiple proportions.
Conceptual Plan: determine mass ratio of oxygen

$$\frac{\text{mass of oxygen sample 2}}{\text{mass of oxygen sample 1}}$$

Solution: $\dfrac{0.3369\text{ g oxygen}}{0.168\text{ g oxygen}} = 2.00$ Ratio is a small whole number. Results are consistent with multiple proportions

Check: According to the Law of Multiple Proportions, when two elements form two different compounds, the masses of element B that combine with 1 g of element A can be expressed as a ratio of small whole numbers.

8. **Given:** 1 g palladium: compound A: 0.603 g S, compound B: 0.301 g S; compound C: 0.151 g S
Find: consistent with multiple proportions
Conceptual Plan: determine mass ratio of sulfur in the 3 compounds

$$\frac{\text{mass of sulfur sample A}}{\text{mass of sulfur sample B}} \quad \frac{\text{mass of sulfur sample A}}{\text{mass of sulfur sample C}} \quad \frac{\text{mass of sulfur sample B}}{\text{mass of sulfur sample C}}$$

Solution: $\dfrac{0.603\text{ g S in compound A}}{0.301\text{ g S in compound B}} = 2.00$ $\quad \dfrac{0.603\text{ g S in compound A}}{0.151\text{ g S in compound C}} = 3.99 \sim 4$

$$\frac{0.301\text{ g S in compound B}}{0.151\text{ g S in compound C}} = 1.99 \sim 2$$

Ratio of each is a small whole number. Results are consistent with multiple proportions.

Check: According to the Law of Multiple Proportions, when two elements form two different compounds, the masses of element B that combine with 1 g of element A can be expressed as a ratio of small whole numbers.

9. **Given:** sulfur dioxide = 3.49 g oxygen and 3.50 g sulfur, sulfur trioxide = 6.75 g oxygen and 4.50 g sulfur.
Find: mass oxygen per g S for each compound and then determine the mass ratio of oxygen

$$\frac{\text{mass of oxygen in sulfur dioxide}}{\text{mass of sulfur in sulfur dioxide}} \quad \frac{\text{mass of oxygen in sulfur trioxide}}{\text{mass of sulfur in sulfur trioxide}} \quad \frac{\text{mass of oxyen in sulfur trioxide}}{\text{mass of oxyen in sulfur dioxide}}$$

Solution: $\text{sulfur dioxide} = \dfrac{3.49\text{ g oxygen}}{3.50\text{ g sulfur}} = \dfrac{0.997\text{ g oxygen}}{1\text{ g sulfur}}$ $\quad \text{sulfur trioxide} = \dfrac{6.75\text{ g oxygen}}{4.50\text{ g sulfur}} = \dfrac{1.50\text{ g oxygen}}{1\text{ g sulfur}}$

$$\frac{1.50\text{ g oxygen in sulfur trioxide}}{0.997\text{ g oxygen in sulfur dioxide}} = \frac{1.50}{1} = \frac{3}{2}$$

Ratio is in small whole numbers and is consistent with multiple proportions.

Check: According to the Law of Multiple Proportions, when two elements form two different compounds, the masses of element B that combine with 1 g of element A can be expressed as a ratio of small whole numbers.

10. **Given:** sulfur hexafluoride = 4.45 g fluorine and 1.25 g sulfur, sulfur tetrafluoride = 4.43 g fluorine and 1.87 g sulfur.
Find: mass fluorine per g S for each compound and then determine the mass ratio of fluorine

$$\frac{\text{mass of fluorine in sulfur hexafluoride}}{\text{mass of sulfur in sulfur hexafluoride}} \quad \frac{\text{mass of fluorine in sulfur tetrafluoride}}{\text{mass of sulfur in sulfur tetrafluoride}} \quad \frac{\text{mass of oxyen in sulfur hexafluoride}}{\text{mass of oxyen in sulfur tetrafluoride}}$$

Solution: $\text{sulfur hexafluoride} = \dfrac{4.45\text{ g fluorine}}{1.25\text{ g sulfur}} = \dfrac{3.56\text{ g fluorine}}{1\text{ g sulfur}}$

$$\text{sulfur tetrafluoride} = \frac{4.43\text{ g fluorine}}{1.87\text{ g sulfur}} = \frac{2.3\underline{6}9\text{ g fluorine}}{1\text{ g sulfur}}$$

$$\frac{3.56\text{ g fluorine in sulfur hexafluoride}}{2.3\underline{6}9\text{ g fluorine in sulfur tetrafluoride}} = \frac{1.50}{1} = \frac{3}{2}$$

Ratio is in small whole numbers and is consistent with multiple proportions.

Check: According to the Law of Multiple Proportions, when two elements form two different compounds, the masses of element B that combine with 1 g of element A can be expressed as a ratio of small whole numbers.

11. a) Sulfur and oxygen atoms have the same mass. INCONSISTENT with Dalton's atomic theory because only atoms of the same element have the same mass.

    b) All cobalt atoms are identical. CONSISTENT with Dalton's atomic theory because all atoms of a given element have the same mass and other properties that distinguish them from atoms of other elements.

    c) Potassium and chlorine atoms combine in a 1:1 ratio to form potassium chloride. CONSISTENT with Dalton's atomic theory because atoms combine in simple whole number ratios to form compounds.

    d) Lead atoms can be converted into gold. INCONSISTENT with Dalton's atomic theory because atoms of one element cannot change into atoms of another element.

12. a) All carbon atoms are identical. CONSISTENT with Dalton's atomic theory because all atoms of a given element have the same mass and other properties that distinguish them from atoms of other elements.

    b) An oxygen atom combines with 1.5 hydrogen atoms to form a water molecule. INCONSISTENT with Dalton's atomic theory because atoms combine in simple whole number ratios to form compounds. An oxygen atom actually combines with 2 hydrogen atoms to form a water molecule.

    c) Two oxygen atoms combine with a carbon atom to form a carbon dioxide molecule. CONSISTENT with Dalton's atomic theory because atoms combine in simple whole number ratios to form compounds.

    d) The formation of a compound often involves the destruction of one or more atoms. INCONSISTENT with Dalton's atomic theory because atoms change the way that they are bound together with other atoms when they form a new substance.

13. a) The volume of an atom is mostly empty space. CONSISTENT with Rutherford's nuclear theory because most of the volume of the atom is empty space, throughout which tiny, negatively charged electrons are dispersed.

    b) The nucleus of an atom is small compared to the size of the atom. CONSISTENT with Rutherford's nuclear theory because most of the atom's mass and all of its positive charge are contained in a small core called the nucleus.

    c) Neutral lithium atoms contain more neutrons than protons. INCONSISTENT with Rutherford's nuclear theory because it did not distinguish where the mass of the nucleus came from other than from the protons.

    d) Neutral lithium atoms contain more protons than electrons. INCONSISTENT with Rutherford's nuclear theory because there are as many negatively charged particles outside the nucleus as there are positively charged particles within the nucleus.

14. a) Since electrons are smaller than protons, and since a hydrogen atom contains only one proton and one electron, it must follow that the volume of a hydrogen atom is mostly due to the proton. INCONSISTENT with Rutherford's nuclear theory because most of the volume of the atom is empty space, throughout which tiny, negatively charged electrons are dispersed.

    b) A nitrogen atom has seven protons in its nucleus and seven electrons outside of its nucleus. CONSISTENT with Rutherford's nuclear theory because there are as many negatively charged particles outside the nucleus as there are positively charged particles within the nucleus.

    c) A phosphorus atom has 15 protons in its nucleus and 150 electrons outside of its nucleus. INCONSISTENT with Rutherford's nuclear theory because there are as many negatively charged particles outside the nucleus as there are positively charged particles within the nucleus.

d) The majority of the mass of a fluorine atom is due to its nine electrons. INCONSISTENT with Rutherford's nuclear theory because most of the atom's mass and all of its positive charge are contained in a small core called the nucleus.

15. **Given:** drop A = -6.9×10^{-19} C; drop B = -9.2×10^{-19} C; drop C = -11.5×10^{-19} C; drop D = -4.6×10^{-19} C
Find: The charge on a single electron
Conceptual Plan: determine the ratio of charge for each set of drops.

$$\frac{\text{charge on drop 1}}{\text{charge on drop 2}}$$

Solution: $\frac{-6.9 \times 10^{-19}\text{C drop A}}{-4.6 \times 10^{-19}\text{ C drop D}} = 1.5$ $\quad\frac{-9.2 \times 10^{-19}\text{C drop B}}{-4.6 \times 10^{-19}\text{ C drop D}} = 2$ $\quad\frac{-11.5 \times 10^{-19}\text{C drop C}}{-4.6 \times 10^{-19}\text{ C drop D}} = 2.5$

The ratios obtained are not whole numbers, but can be converted to whole numbers by multiplying by 2. Therefore, the charge on the electron has to be 1/2 the smallest value experimentally obtained. The charge on the electron = -2.3×10^{-19} C.

Check: The units of the answer, Coulombs, are correct. The magnitude of the answer is reasonable since all the values experimentally obtained are integer multiples of -2.3×10^{-19}.

16. **Given:** drop A = -4.8×10^{-9} z; drop B = -9.6×10^{-9} z; drop C = -6.4×10^{-9} z; drop D = -12.8×10^{-9} z
Find: The charge on a single electron
Conceptual Plan:

determine the ratio of charge for each set of drops then determine the charge on an electron

$$\frac{\text{charge on drop 1}}{\text{charge on drop 2}}$$

then determine the number of electrons in each drop

$$\frac{\text{charge on drop}}{\text{charge on one electron}}$$

Solution: $\frac{-9.6 \times 10^{-9}\text{z drop B}}{-4.8 \times 10^{-9}\text{ z drop A}} = 2$ $\quad\frac{-6.4 \times 10^{-9}\text{z drop C}}{-4.8 \times 10^{-9}\text{ z drop A}} = 1.33$ $\quad\frac{-12.8 \times 10^{-9}\text{z drop B}}{-4.8 \times 10^{-9}\text{ z drop A}} = 2.66$

The ratios obtained are not whole numbers, but can be converted to whole numbers by multiplying by 3. Therefore, the charge on the electron has to be 1/3 the smallest value experimentally obtained. The charge on the electron $= \frac{1}{3} \times -4.8 \times 10^{-9}\text{ z} = -1.6 \times 10^{-9}\text{ z}$.

Number of electrons in:

Drop A $\frac{-4.8 \times 10^{-9}\text{ z}}{-1.6 \times 10^{-9}\text{ z}} = 3$ electrons : Drop B: $\frac{-9.6 \times 10^{-9}\text{ z}}{-1.6 \times 10^{-9}\text{ z}} = 6$ electrons

Drop C: $\frac{-6.4 \times 10^{-9}\text{ z}}{-1.6 \times 10^{-9}\text{ z}} = 4$ electrons Drop D: $\frac{-12.8 \times 10^{-9}\text{ z}}{-1.6 \times 10^{-9}\text{ z}} = 8$ electrons

Check: The units of the answer, zorg, are correct. The magnitude of the answer is reasonable since all the values experimentally obtained are integer multiples of -1.6×10^{-9}.

17. **Given:** charge on body = $-15\ \mu$C. **Find:** number of electrons, mass of the electrons
Conceptual Plan: μC → C → number of electrons → mass of electrons

$$\frac{1\text{ C}}{10^6\ \mu\text{C}} \quad \frac{1\text{ electron}}{-1.60 \times 10^{-19}\text{C}} \quad \frac{9.10 \times 10^{-28}\text{ g}}{1\text{ electron}}$$

Solution: $-15\ \mu\text{C} \times \frac{1\text{ C}}{10^6\ \mu\text{C}} \times \frac{1\text{ electron}}{-1.60 \times 10^{-19}\text{ C}} = 9.\underline{3}75 \times 10^{13}\text{ electrons} = 9.4 \times 10^{13}\text{ electrons}$

$$9.\underline{3}75 \times 10^{13}\text{ electrons} \times \frac{9.10 \times 10^{-28}\text{ g}}{1\text{ electron}} = 8.5 \times 10^{-14}\text{ g}$$

Check: The units of the answers, number of electrons and grams, are correct. The magnitude of the answers are reasonable since the charge on an electron and the mass of an electron are very small.

18. **Given:** charge = - 1.0 C. **Find:** number of electrons, mass of the electrons
Conceptual Plan: C → number of electrons → mass of electrons

$$\frac{1 \text{ electron}}{-1.60 \times 10^{-19}\text{C}} \qquad \frac{9.10 \times 10^{-28} \text{ g}}{1 \text{ electron}}$$

Solution: $-1.0 \text{ C} \times \dfrac{1 \text{ electron}}{-1.60 \times 10^{-19} \text{ C}} = 6.\underline{2}5 \times 10^{18} \text{ electrons} = 6.3 \times 10^{18} \text{ electrons}$

$$6.\underline{2}5 \times 10^{18} \text{ electrons} \times \frac{9.10 \times 10^{-28} \text{ g}}{1 \text{ electron}} = 5.7 \times 10^{-9} \text{ g}$$

Check: The units of the answers, number of electrons and grams, are correct. The magnitude of the answers are reasonable since the charge on an electron and the mass of an electron are very small.

19. a) True: protons and electrons have equal and opposite charges.

b) True: protons and electrons have opposite charge so they will attract each other.

c) True: the mass of the electron is much less than the mass of the neutron.

d) False: the mass of the proton and the mass of the neutron are about the same.

20. a) True: protons and electrons have equal and opposite charges.

b) True: the mass of the proton and the mass of the neutron are about the same.

c) False: all atoms contain protons. The lightest element, hydrogen, contains 1 proton.

d) False: protons have a positive charge, neutrons are neutral.

21. **Given:** mass of proton **Find:** number of electron in equal mass
Conceptual Plan: mass of protons → number of electrons

$$\frac{1 \text{ electron}}{0.00091 \times 10^{-27} \text{ kg}}$$

Solution: $1.67262 \times 10^{-27} \text{ kg} \times \dfrac{1 \text{ electron}}{0.00091 \times 10^{-27} \text{ kg}} = 1.8 \times 10^{3} \text{ electrons}$

Check: The units of the answer, electrons, are correct. The magnitude of the answer is reasonable since the mass of the electron is much less than the mass of the proton.

22. **Given:** helium nucleus **Find:** number of electrons in equal mass
Conceptual Plan:
 # protons → mass of protons and # neutrons → mass of neutrons → total mass → number of electrons

$$\frac{1.67262 \times 10^{-27} \text{ kg}}{1 \text{ proton}} \qquad \frac{1.67493 \times 10^{-27} \text{ kg}}{1 \text{ neutron}} \qquad \text{mass protons + mass neutrons} \qquad \frac{1 \text{ electron}}{0.00091 \times 10^{-27} \text{ kg}}$$

Solution: $2 \text{ protons} \times \dfrac{1.67262 \times 10^{-27} \text{ kg}}{\text{proton}} = 3.34524 \times 10^{-27} \text{ kg}$

$$2 \text{ neutrons} \times \frac{1.67493 \times 10^{-27} \text{ kg}}{\text{neutron}} = 3.34986 \times 10^{-27} \text{ kg}$$

$$(3.34524 \times 10^{-27} \text{ kg} + 3.34986 \times 10^{-27} \text{ kg}) \times \frac{1 \text{ electron}}{0.00091 \times 10^{-27} \text{ kg}} = 7.4 \times 10^{3} \text{ electrons}$$

Check: The units of the answer, electrons, are correct. The magnitude of the answer is reasonable since the mass of the electrons is much less than the mass of the proton and neutron.

23. For each of the isotopes: determine Z (the number of protons) from the periodic table then determine A (protons + neutrons). Then, write the symbol in the form ${}^{A}_{Z}X$.

a) The sodium isotope with 12 neutrons: Z = 11; A = 11 + 12 = 23. ${}^{23}_{11}Na$

b) The oxygen isotope with 8 neutrons: Z = 8; A = 8 + 8 = 16. $^{16}_{8}O$

c) The aluminum isotope with 14 neutrons: Z = 13; A = 13 + 14 = 27 $^{27}_{13}Al$

d) The iodine isotope with 74 neutrons: Z = 53; A = 53 + 74 = 127 $^{127}_{53}I$

24. For each of the isotopes: determine Z (the number of protons) from the periodic table then determine A (protons + neutrons). Then, write the symbol in the form X – A.

a) The argon isotope with 22 neutrons: Z = 18; A = 18 + 22 =40. Ar – 40

b) The plutonium isotope with 145 neutrons: Z = 94: A = 94 + 145 = 239. Pu – 239

c) The phosphorus isotope with 16 neutrons: Z = 15; A = 15 + 16 = 31. P – 31

d) The fluorine isotope with 10 neutrons: Z = 9: A = 9 + 10 = 19 F – 19

25. a) $^{14}_{7}N$: Z = 7 ; A = 14: protons = Z = 7; neutrons = A – Z = 14 – 7 = 7

b) $^{23}_{11}Na$: Z = 11: A = 23: protons = Z = 11; neutrons = A – Z = 23 – 11 = 12

c) $^{222}_{86}Rn$: Z = 86: A = 222: protons = Z = 86; neutrons = A – Z = 222 – 86 = 136

d) $^{208}_{82}Pb$: Z = 82: A = 208: protons = Z = 82; neutrons = A – Z = 208 – 82 = 126

26. a) $^{40}_{19}K$: Z = 19 ; A = 40: protons = Z = 19; neutrons = A – Z = 40 – 19 = 21

b) $^{226}_{88}Ra$: Z = 88 ; A = 226: protons = Z = 88; neutrons = A – Z = 226 – 88 = 138

c) $^{99}_{43}Tc$: Z = 43 ; A = 99: protons = Z = 43; neutrons = A – Z = 99 – 43 = 56

d) $^{33}_{15}P$: Z = 15 ; A = 33: protons = Z = 15; neutrons = A – Z = 33 – 15 = 18

27. Carbon – 14: A = 14, Z = 6: $^{14}_{6}C$ # protons = Z = 6 # neutrons = A – Z = 14 – 6 = 8

28. Uranium – 235: A = 235, Z = 92: $^{235}_{92}U$ # protons = Z = 92 # neutrons = A – Z = 235 – 92 = 143

29. In a neutral atom the number of protons = the number of electrons = Z. For an ion, electrons are lost (cations) or gained (anions)

a) Ni^{2+}: Z = 28 = protons; Z – 2 = 26 = electrons

b) S^{2-}: Z = 16 = protons; Z + 2 = 18 = electrons

c) Br^{-}: Z = 35 = protons; Z + 1 = 36 = electrons

d) Cr^{3+}: Z = 24 = protons; Z – 3 = 21 = electrons

30. In a neutral atom the number of protons = the number of electrons = Z. For an ion, electrons are lost (cations) or gained (anions)

a) Al^{3+}: Z = 13 = protons; Z – 3 = 10 = electrons

b) Se^{2-}: Z = 34 = protons; Z + 2 = 36 = electrons

c) Ga^{3+}: Z = 31 = protons; Z – 3 = 28 = electrons

d) Sr^{2+}: Z = 38 = protons; Z – 2 = 36 = electrons

31. Main group metal atoms will lose electrons to form a cation with the same number of electrons as the nearest, previous noble gas.
Nonmetal atoms will gain electrons to form an anion with the same number of electrons as the nearest noble gas.

a) O^{2-} O is a nonmetal and has 8 electrons. It will gain electrons to form an anion. The nearest noble gas is neon with 10 electrons, so O will gain 2 electrons.

b) K^{+} K is a main group metal and has 19 electrons. It will lose electrons to form a cation. The nearest noble gas is argon with 18 electrons, so K will lose 1 electron.

c) Al^{3+} Al is a main group metal and has 13 electrons. It will lose electrons to form a cation. The nearest noble gas is neon with 10 electrons, so Al will lose 3 electrons.

d) Rb^{+} Rb is a main group metal and has 37 electrons. It will lose electrons to form a cation. The nearest noble gas is krypton with 36 electrons, so Rb will lose 1 electron.

32. Main group metal atoms will lose electrons to form a cation with the same number of electrons as the nearest, previous noble gas.
Nonmetal atoms will gain electrons to form an anion with the same number of electrons as the nearest noble gas.

a) Mg^{2+} Mg is a main group metal and has 12 electrons. It will lose electrons to form a cation. The nearest noble gas is neon with 10 electrons, so Mg will lose 2 electrons

b) N^{3-} N is a nonmetal and has 7 electrons. It will gain electrons to form an anion. The nearest noble gas is neon with 10 electrons, so N will gain 3 electrons.

c) F^{-} F is a nonmetal and has 9 electrons. It will gain electrons to form an anion. The nearest noble gas is neon with 10 electrons, so F will gain 1 electron.

d) Na^{+} Na is a main group metal and has 11 electrons. It will lose electrons to form a cation. The nearest noble gas is neon with 10 electrons, so Na will lose 1 electron.

33. Main group metal atoms will lose electrons to form a cation with the same number of electrons as the nearest, previous noble gas.
Nonmetal atoms will gain electrons to form an anion with the same number of electrons as the nearest noble gas.

Symbol	Ion Formed	Number of Electrons in Ion	Number of Protons in Ion
Ca	Ca^{2+}	**18**	**20**
Be	Be^{2+}	2	**4**
Se	**Se^{2-}**	**36**	34
In	**In^{3+}**	**46**	49

34. Main group metal atoms will lose electrons to form a cation with the same number of electrons as the nearest, previous noble gas.
Nonmetal atoms will gain electrons to form an anion with the same number of electrons as the nearest noble gas.

Symbol	Ion Formed	Number of Electrons in Ion	Number of Protons in Ion
Cl	**Cl^{-}**	**18**	17
Te	**Te^{2-}**	54	**52**
Br	Br^{-}	**36**	**35**
Sr	Sr^{2+}	**36**	38

35. a) Na Sodium is a metal

b) Mg Magnesium is a metal

c) Br Bromine is a nonmetal

d) N Nitrogen is a nonmetal

e) As Arsenic is a metalloid

36. a) lead Pb is a metal

b) iodine I is a nonmetal

c) potassium K is a metal

d) silver Ag is a metal

e) xenon Xe is a nonmetal

37. a) tellurium Te is in group 6A and is a main group element

b) potassium K is in group 1A and is a main group element

c) vanadium V is in group 5B and is a transition element

d) manganese Mn is in group 7B and is a transition element

38. a) Cr chromium is in group 6B and is a transition element

b) Br bromine is in group 7A and is a main group element

c) Mo molybdenum is in group 6B and is a transition element

d) Cs cesium is in group 1A and is a main group element

39. a) sodium Na is in group 1A and is an alkali metal

b) iodine I is in group 7A and is a halogen

c) calcium Ca is in group 2A and is an alkaline earth metal

d) barium Ba is in group 2A and is an alkaline earth metal

e) krypton Kr is in group 8A and is a noble gas

40. a) F fluorine is in group 7A and is a halogen

b) Sr strontium is in group 2A and is an alkaline earth metal

c) K potassium is in group 1A and is an alkali metal

d) Ne neon is in group 8A and is a noble gas

e) At astatine is in group 7A and is a halogen

41. a) N and Ni would not be similar. Nitrogen is a nonmetal, nickel is a metal.

b) Mo and Sn would not be most similar. Although both are metals, molybdenum is a transition metal and tin is a main group metal.

c) Na and Mg would not be similar. Although both are main group metals, sodium is in group 1A and magnesium is in group 2A.

d) Cl and F would be most similar. Chlorine and fluorine are both in group 7A. Elements in the same group have similar chemical properties.

e) Si and P would not be most similar. Silicon is a metalloid and phosphorus is a nonmetal.

42. a) Nitrogen and oxygen would not be most similar. Although both are nonmetals, N is in group 5A and O is in group 6A.

b) Titanium and gallium would not be most similar. Although both are metals, Ti is a transition metal and Ga is a main group metal.

c) Lithium and sodium would be most similar. Li and Na are both in group 1A. Elements in the same group have similar chemical properties.

d) Germanium and arsenic would not be the most similar. Ge and As are both metalloids and would share some properties, but Ge is in group 4A and As is in group 5A.

e) Argon and bromine would not be most similar. Although both are nonmetals, Ar is in group 8A and Br is in group 7A.

43. **Given:** Rb – 85; mass = 84.9118 amu; 72.15%: Rb – 87; mass = 86.9092 amu; 27.85 % **Find:** atomic mass Rb
Conceptual Plan: % abundance → fraction and then find atomic mass

$$\frac{\%\,\text{abundance}}{100} \qquad \text{Atomic mass} = \sum_n (\text{fraction of isotope n}) \times (\text{mass of isotope n})$$

Solution: $\text{Fraction Rb - 85} = \frac{72.15}{100} = 0.7215 \qquad \text{Fraction Rb - 87} = \frac{27.85}{100} = 0.2785$

$$\text{Atomic mass} = \sum_n (\text{fraction of isotope n}) \times (\text{mass of isotope n})$$

$$= 0.7215(84.9116\ \text{amu}) + 0.2785(86.9092\ \text{amu}) = 85.47\ \text{amu}$$

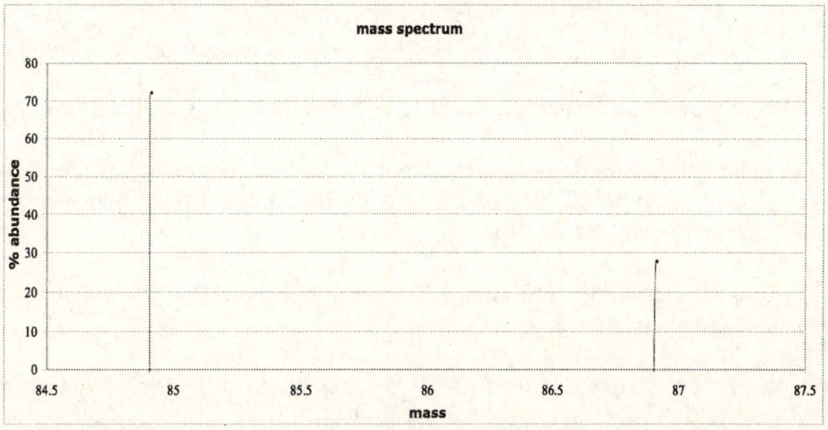

Check: Units of the answer, amu, are correct. The magnitude of the answer is reasonable because it lies between 84.9116 amu and 86.9092 amu and is closer to 84.9118, which has the higher % abundance. The mass spectrum is reasonable because it has two mass lines corresponding to the two isotopes and the line at 84.9116 is about 2.5 times larger than the line at 86.9092.

44. **Given:** Si – 28: mass = 27.9769 amu; 92.2%: Si – 29: mass = 28.9765 amu; 4.67 %: Si – 30: mass = 29.9737 amu; 3.10%
Find: atomic mass Si
Conceptual Plan: % abundance → fraction and then find atomic mass

$$\frac{\%\,\text{abundance}}{100} \qquad \text{Atomic mass} = \sum_n (\text{fraction of isotope n}) \times (\text{mass of isotope n})$$

Solution:

Fraction Si - 28 = $\frac{92.2}{100}$ = 0.922 Fraction Si - 29 = $\frac{4.67}{100}$ = 0.0467 Fraction Si - 30 = $\frac{3.10}{100}$ = 0.0310

Atomic mass = $\sum_n$ (fraction of isotope n) x (mass of isotope n)

= 0.922(27.9769 amu) + 0.0467(28.9765 amu) + 0.0310(29.9737 amu) = 28.1 amu

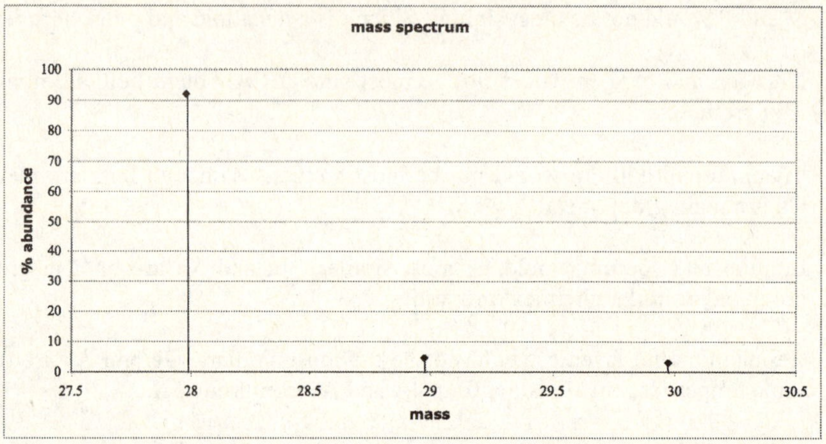

Check: Units of the answer, amu, are correct. The magnitude of the answer is reasonable because it lies between 27.9769 amu and 29.9737 amu and is closer to 27.9769 which has the highest % abundance. The mass spectrum is reasonable because it has three mass lines corresponding to the three isotopes and the line at 27.9769 is about 20 times larger than the other two lines.

45. **Given:** Isotope – 1; mass = 120.9038 amu; 57.4%: Isotope – 2; mass = 122.9042 amu;
Find: atomic mass of the atom and identify the atom
Conceptual Plan:
% abundance isotope 2 → and then % abundance → fraction and then find atomic mass

100% – % abundance Isotope 1 $\frac{\%\text{ abundance}}{100}$ Atomic mass = $\sum_n$ (fraction of isotope n) x (mass of isotope n)

Solution: 100.0% - 57.4 % Isotope 1 = 42.6 % Isotope 2

Fraction Isotope 1 = $\frac{57.4}{100}$ = 0.574 Fraction Isotope 2 = $\frac{42.6}{100}$ = 0.426

Atomic mass = $\sum_n$ (fraction of isotope n) x (mass of isotope n) = 0.574(120.9038 amu) + 0.426(122.9042 amu) = 121.8 amu

From the periodic table Sb has a mass of 121.757 amu, so it is the closest mass and the element is antimony.

Check: The units of the answer, amu, are correct. The magnitude of the answer is reasonable because it lies between 120.9038 and 122.9042 and is slightly less than halfway between the two values because the lower value has a slightly greater abundance.

46. **Given:** Br – 81; mass = 80.9163 amu, 49.31 %, atomic mass Br = 79.904 amu
Find: mass and abundance Br – 79

Conceptual Plan: % abundance Br – 79 → then % abundance → fraction → mass Br – 79

100% – % Br – 81 $\frac{\%\text{ abundance}}{100}$ Atomic mass = $\sum_n$ (fraction of isotope n) x (mass of isotope n)

Solution: 100.00% - 49.31 % = 50.69% Br – 79

Fraction Br - 79 = $\frac{50.69}{100}$ = 0.5069 Fraction Br - 81 = $\frac{49.31}{100}$ = 0.4931

Let X be the mass of Br – 79

Atomic mass = $\sum_n$ (fraction of isotope n) x (mass of isotope n)

79.904 amu = 0.5069(X amu) + 0.4931(80.9163 amu)

X = 78.92 amu = mass Br - 79

Check: The units of the answer, amu, are correct. The magnitude of the answer is reasonable because it is less than the mass of the atom and the second isotope (Br – 81) has a mass greater than the mass of the atom.

47. **Given:** 3.8 mol sulfur **Find:** atoms of sulfur
Conceptual Plan: mol S → atoms S

$$\frac{6.022 \times 10^{23}\ \text{atoms}}{\text{mol}}$$

Solution: $3.8\ \cancel{\text{mol S}} \times \dfrac{6.022 \times 10^{23}\ \text{atoms S}}{\cancel{\text{mol S}}} = 2.3 \times 10^{24}\ \text{atoms S}$

Check: The units of the answer, atoms S, are correct. The magnitude of the answer is reasonable since there is more than 1 mole of material present.

48. **Given:** 5.8×10^{24} aluminum atoms **Find:** mol Al
Conceptual Plan: atoms Al → mol Al

$$\frac{1\ \text{mol}}{6.022 \times 10^{23}\ \text{atoms}}$$

Solution: $5.8 \times 10^{24}\ \cancel{\text{atoms Al}} \times \dfrac{1\ \text{mol Al}}{6.022 \times 10^{23}\ \cancel{\text{atoms Al}}} = 9.6\ \text{mol Al}$

Check: The units of the answer, mol Al, are correct. The magnitude of the answer is reasonable since there is greater than Avogadro's number of atoms present.

49. a) **Given:** 11.8 g Ar **Find:** mol Ar
Conceptual Plan: g Ar → mol Ar

$$\frac{1\ \text{mol Ar}}{39.948\ \text{g Ar}}$$

Solution: $11.8\ \cancel{\text{g Ar}} \times \dfrac{1\ \text{mol Ar}}{39.948\ \cancel{\text{g Ar}}} = 0.295\ \text{mol Ar}$

Check: The units of the answer, mol Ar, are correct. The magnitude of the answer is reasonable since there is less than the mass of 1 mol present.

b) **Given:** 3.55 g Zn **Find:** mol Zn
Conceptual Plan: g Zn → mol Zn

$$\frac{1\ \text{mol Zn}}{65.39\ \text{g Zn}}$$

Solution: $3.55\ \cancel{\text{g Zn}} \times \dfrac{1\ \text{mol Zn}}{65.39\ \cancel{\text{g Zn}}} = 0.0543\ \text{mol Zn}$

Check: The units of the answer, mol Zn, are correct. The magnitude of the answer is reasonable since there is less than the mass of 1 mol present.

c) **Given:** 26.1 g Ta **Find:** mol Ta
Conceptual Plan: g Ta → mol Ta

$$\frac{1\ \text{mol Ta}}{180.948\ \text{g Ta}}$$

Solution: $26.1\ \cancel{\text{g Ta}} \times \dfrac{1\ \text{mol Ta}}{180.948\ \cancel{\text{g Ta}}} = 0.144\ \text{mol Ta}$

Check: The units of the answer, mol Ta, are correct. The magnitude of the answer is reasonable since there is less than the mass of 1 mol present.

d) **Given:** 0.211 g Li **Find:** mol Li
Conceptual Plan: g Li → mol Li

$$\frac{1\ \text{mol Li}}{6.941\ \text{g Li}}$$

Solution: $0.211\ \cancel{\text{g Li}} \times \dfrac{1\ \text{mol Li}}{6.941\ \cancel{\text{g Li}}} = 0.0304\ \text{mol Li}$

Check: The units of the answer, mol Li, are correct. The magnitude of the answer is reasonable since there is less than the mass of 1 mol present.

50. a) **Given:** 2.3×10^{-3} mol Sb **Find:** grams Sb

Conceptual Plan: mol Sb → g Sb

$$\frac{121.757 \text{ g Sb}}{1 \text{ mol Sb}}$$

Solution: $2.3 \times 10^{-3} \text{ mol Sb} \times \frac{121.757 \text{ g Sb}}{1 \text{ mol Sb}} = 0.28 \text{ grams Sb}$

Check: The units of the answer, grams Sb, are correct. The magnitude of the answer is reasonable since there is less than 1 mol of Sb present.

b) **Given:** 0.0355 mol Ba **Find:** grams Ba

Conceptual Plan: mol Ba → g BA

$$\frac{137.327 \text{ g Ba}}{1 \text{ mol Ba}}$$

Solution: $0.0355 \text{ mol Ba} \times \frac{137.327 \text{ g Ba}}{1 \text{ mol Ba}} = 4.88 \text{ grams Ba}$

Check: The units of the answer, grams Ba, are correct. The magnitude of the answer is reasonable since there is less than 1 mol of Ba present.

c) **Given:** 43.9 mol Xe **Find:** grams Xe

Conceptual Plan: mol Xe → g Xe

$$\frac{131.29 \text{ g Xe}}{1 \text{ mol Xe}}$$

Solution: $43.9 \text{ mol Xe} \times \frac{131.29 \text{ g Xe}}{1 \text{ mol Xe}} = 5.76 \times 10^{3} \text{ grams Xe}$

Check: The units of the answer, grams Xe, are correct. The magnitude of the answer is reasonable since there is much more than 1 mol of Xe present.

d) **Given:** 1.3 mol W **Find:** grams W

Conceptual Plan: mol W → g W

$$\frac{183.85 \text{ g W}}{1 \text{ mol W}}$$

Solution: $1.3 \text{ mol W} \times \frac{183.85 \text{ g W}}{1 \text{ mol W}} = 2.4 \times 10^{2} \text{ grams W}$

Check: The units of the answer, grams W, are correct. The magnitude of the answer is reasonable since there is slightly over 1 mol of W present.

51. **Given:** 3.78 g silver **Find:** atoms Ag

Conceptual Plan: g Ag → mol Ag → atoms Ag

$$\frac{1 \text{ mol Ag}}{107.868 \text{ g Ag}} \qquad \frac{6.022 \times 10^{23} \text{ atoms}}{\text{mol}}$$

Solution: $3.78 \text{ g Ag} \times \frac{1 \text{ mol Ag}}{107.868 \text{ g Ag}} \times \frac{6.022 \times 10^{23} \text{ atoms Ag}}{1 \text{ mol Ag}} = 2.11 \times 10^{22} \text{ atoms Ag}$

Check: The units of the answer, atoms Ag, are correct. The magnitude of the answer is reasonable since there is less than the mass of 1 mol of Ag present.

52. **Given:** 4.91×10^{21} Pt atoms **Find:** g Pt

Conceptual Plan: atoms Pt → mol Pt → g Pt

$$\frac{1 \text{ mol}}{6.022 \times 10^{23} \text{ atoms}} \qquad \frac{195.08 \text{ g Pt}}{1 \text{ mol Pt}}$$

Solution: $4.91 \times 10\ \cancel{\text{atoms Pt}} \times \dfrac{1\ \cancel{\text{mol Pt}}}{6.022 \times 10^{23}\ \cancel{\text{atoms Pt}}} \times \dfrac{195.08\text{ g Pt}}{1\ \cancel{\text{mol Pt}}} = 1.59\text{ g Pt}$

Check: The units of the answer, g Pt, are correct. The magnitude of the answer is reasonable since there is less than 1 mol of Pt atoms present.

53. a) **Given:** 5.18 g P **Find:** atoms P

Conceptual Plan: g P → mol P → atoms P

$\dfrac{1\text{ mol P}}{30.9738\text{ g P}} \quad \dfrac{6.022 \times 10^{23}\text{ atoms}}{\text{mol}}$

Solution: $5.18\ \cancel{\text{g P}} \times \dfrac{1\ \cancel{\text{mol P}}}{30.9738\ \cancel{\text{g P}}} \times \dfrac{6.022 \times 10^{23}\text{ atoms P}}{1\ \cancel{\text{mol P}}} = 1.01 \times 10^{23}\text{ atoms P}$

Check: The units of the answer, atoms P, are correct. The magnitude of the answer is reasonable since there is slightly less than the mass of 1 mol of P present.

b) **Given:** 2.26 g Hg **Find:** atoms Hg

Conceptual Plan: g Hg → mol Hg → atoms Hg

$\dfrac{1\text{ mol Hg}}{200.59\text{ g Hg}} \quad \dfrac{6.022 \times 10^{23}\text{ atoms}}{\text{mol}}$

Solution: $2.26\ \cancel{\text{g Hg}} \times \dfrac{1\ \cancel{\text{mol Hg}}}{200.59\ \cancel{\text{g Hg}}} \times \dfrac{6.022 \times 10^{23}\text{ atoms Hg}}{1\ \cancel{\text{mol Hg}}} = 6.78 \times 10^{21}\text{ atoms Hg}$

Check: The units of the answer, atoms Hg, are correct. The magnitude of the answer is reasonable since there is less than the mass of 1 mol of Hg present.

c) **Given:** 1.87 g Bi **Find:** atoms Bi

Conceptual Plan: g Bi → mol Bi → atoms Bi

$\dfrac{1\text{ mol Bi}}{208.98\text{ g Bi}} \quad \dfrac{6.022 \times 10^{23}\text{ atoms}}{\text{mol}}$

Solution: $1.87\ \cancel{\text{g Bi}} \times \dfrac{1\ \cancel{\text{mol Bi}}}{208.98\ \cancel{\text{g Bi}}} \times \dfrac{6.022 \times 10^{23}\text{ atoms Bi}}{1\ \cancel{\text{mol Bi}}} = 5.39 \times 10^{21}\text{ atoms Bi}$

Check: The units of the answer, atoms Bi, are correct. The magnitude of the answer is reasonable since there is less than the mass of 1 mol of Bi present.

d) **Given:** 0.082 g Sr **Find:** atoms Sr

Conceptual Plan: g Sr → mol Sr → atoms Sr

$\dfrac{1\text{ mol Sr}}{87.62\text{ g Sr}} \quad \dfrac{6.022 \times 10^{23}\text{ atoms}}{\text{mol}}$

Solution: $0.082\ \cancel{\text{g Sr}} \times \dfrac{1\ \cancel{\text{mol Sr}}}{87.62\ \cancel{\text{g Sr}}} \times \dfrac{6.022 \times 10^{23}\text{ atoms Sr}}{1\ \cancel{\text{mol Sr}}} = 5.6 \times 10^{20}\text{ atoms Sr}$

Check: The units of the answer, atoms Sr, are correct. The magnitude of the answer is reasonable since there is less than the mass of 1 mol of Sr present.

54. a) **Given:** 1.1×10^{23} gold atoms **Find:** grams Au

Conceptual Plan: atoms Au → mol Au → g Au

$\dfrac{1\text{mol}}{6.022 \times 10^{23}\text{ atoms}} \quad \dfrac{196.97\text{ g Au}}{1\text{ mol Au}}$

Solution: $1.1 \times 10^{23}\ \cancel{\text{atoms Au}} \times \dfrac{1\ \cancel{\text{mol Au}}}{6.022 \times 10^{23}\ \cancel{\text{atoms Au}}} \times \dfrac{196.97\text{ g Au}}{1\ \cancel{\text{mol Au}}} = 36\text{ g Au}$

Check: The units of the answer, g Au, are correct. The magnitude of the answer is reasonable since there is less than Avogadro's number of atoms in the sample.

b) **Given:** 2.82×10^{22} helium atoms **Find:** grams He
Conceptual Plan: atoms He → mol He → g He

$$\frac{1\,\text{mol}}{6.022 \times 10^{23}\,\text{atoms}} \quad \frac{4.002\,\text{g He}}{1\,\text{mol He}}$$

Solution: $2.82 \times 10^{22}\ \cancel{\text{atoms He}} \times \frac{1\ \cancel{\text{mol He}}}{6.022 \times 10^{23}\ \cancel{\text{atoms He}}} \times \frac{4.002\ \text{g He}}{1\ \cancel{\text{mol He}}} = 0.187\ \text{g He}$

Check: The units of the answer, g He, are correct. The magnitude of the answer is reasonable since there is less than Avogadro's number of atoms in the sample.

c) **Given:** 1.8×10^{23} lead atoms **Find:** grams Pb
Conceptual Plan: atoms Pb → mol Pb → g Pb

$$\frac{1\,\text{mol}}{6.022 \times 10^{23}\,\text{atoms}} \quad \frac{207.2\,\text{g Pb}}{1\,\text{mol Pb}}$$

Solution: $1.8 \times 10^{23}\ \cancel{\text{atoms Pb}} \times \frac{1\ \cancel{\text{mol Pb}}}{6.022 \times 10^{23}\ \cancel{\text{atoms Pb}}} \times \frac{207.2\ \text{g Pb}}{1\ \cancel{\text{mol Pb}}} = 62\ \text{g Pb}$

Check: The units of the answer, g Pb, are correct. The magnitude of the answer is reasonable since there is less than Avogadro's number of atoms in the sample.

d) **Given:** 7.9×10^{21} uranium atoms **Find:** grams U
Conceptual Plan: atoms U → mol U → g U

$$\frac{1\,\text{mol}}{6.022 \times 10^{23}\,\text{atoms}} \quad \frac{238.029\,\text{g U}}{1\,\text{mol U}}$$

Solution: $7.9 \times 10^{21}\ \cancel{\text{atoms U}} \times \frac{1\ \cancel{\text{mol U}}}{6.022 \times 10^{23}\ \cancel{\text{atoms U}}} \times \frac{238.029\ \text{g U}}{1\ \cancel{\text{mol U}}} = 3.1\ \text{g U}$

Check: The units of the answer, g U, are correct. The magnitude of the answer is reasonable since there is less than Avogadro's number of atoms in the sample.

55. **Given:** 52 mg diamond (carbon) **Find:** atoms C
Conceptual Plan: mg C → g C → mol C → atoms C

$$\frac{1\,\text{g C}}{1000\,\text{mg C}} \quad \frac{1\,\text{mol C}}{12.011\,\text{g C}} \quad \frac{6.022 \times 10^{23}\,\text{atoms}}{\text{mol}}$$

Solution: $52\ \cancel{\text{mg C}} \times \frac{1\ \cancel{\text{g C}}}{1000\ \cancel{\text{mg C}}} \times \frac{1\ \cancel{\text{mol C}}}{12.011\ \cancel{\text{g C}}} \times \frac{6.022 \times 10^{23}\ \text{atoms C}}{1\ \cancel{\text{mol C}}} = 2.6 \times 10^{21}\ \text{atoms C}$

Check: The units of the answer, atoms C, are correct. The magnitude of the answer is reasonable since there is less than the mass of 1 mol of C present.

56. **Given:** 536 kg helium **Find:** atoms He
Conceptual Plan: kg He → g He → mol He → atoms He

$$\frac{1000\,\text{g He}}{1\,\text{kg He}} \quad \frac{1\,\text{mol He}}{4.0026\,\text{g He}} \quad \frac{6.022 \times 10^{23}\,\text{atoms}}{\text{mol}}$$

Solution: $536\ \cancel{\text{kg He}} \times \frac{1000\ \cancel{\text{g He}}}{1\ \cancel{\text{kg He}}} \times \frac{1\ \cancel{\text{mol He}}}{4.0026\ \cancel{\text{g He}}} \times \frac{6.022 \times 10^{23}\ \text{atoms He}}{1\ \cancel{\text{mol He}}} = 8.06 \times 10^{28}\ \text{atoms He}$

Check: The units of the answer, atoms He, are correct. The magnitude of the answer is reasonable since there is much more than the mass of 1 mol of He present.

57. **Given:** 1 atom platinum **Find:** g Pt
Conceptual Plan: atoms Pt → mol Pt → g Pt

$$\frac{1\,\text{mol}}{6.022 \times 10^{23}\,\text{atoms}} \quad \frac{195.08\,\text{g Pt}}{1\,\text{mol Pt}}$$

Solution: $1\ \cancel{\text{atom Pt}} \times \frac{1\ \cancel{\text{mol Pt}}}{6.022 \times 10^{23}\ \cancel{\text{atoms Pt}}} \times \frac{195.08\ \text{g Pt}}{1\ \cancel{\text{mol Pt}}} = 3.239 \times 10^{-22}\ \text{g Pt}$

Check: The units of the answer, g Pt, are correct. The magnitude of the answer is reasonable since there is only 1 atom in the sample.

58. **Given:** 35 atoms xenon **Find:** g Xe
Conceptual Plan: atoms Xe → mol Xe → g Xe

$$\frac{1\,\text{mol}}{6.022 \times 10^{23}\,\text{atoms}} \qquad \frac{131.29\,\text{g Xe}}{1\,\text{mol Xe}}$$

Solution: $35\,\cancel{\text{atom Xe}} \times \frac{1\,\cancel{\text{mol Xe}}}{6.022 \times 10^{23}\,\cancel{\text{atoms Xe}}} \times \frac{131.29\,\text{g Xe}}{1\,\cancel{\text{mol Xe}}} = 7.631 \times 10^{-21}\,\text{g Xe}$

Check: The units of the answer, g Xe, are correct. The magnitude of the answer is reasonable since there are only 35 atoms in the sample.

59. **Given:** 7.83 g HCN sample 1: 0.290 g H; 4.06 g N. 3.37 g HCN sample 2 **Find:** g C in sample 2
Conceptual Plan:
g HCN sample 1 → g C in HCN sample 1 → ratio g C to g HCN → g C in HCN sample 2

$$\text{g HCN - g H - g N} \qquad \frac{\text{g C}}{\text{g HCN}} \qquad \text{g HCN} \times \frac{\text{g C}}{\text{g HCN}}$$

Solution: 7.83 g HCN - 0.290 g H - 4.06 g N = 3.48 g C

$$3.37\,\cancel{\text{g HCN}} \times \frac{3.48\,\text{g C}}{7.83\,\cancel{\text{g HCN}}} = 1.50\,\text{g C}$$

Check: The units of the answer, g C, are correct. The magnitude of the answer is reasonable since the sample size is about half the original sample size, the g C are about half the original g C.

60. a) **Given:** mass ratio S:O = 1.0:1.0 in SO_2 **Find:** mass ratio S:O in SO_3
Conceptual Plan: determine the ratio of O:O in SO_3 and SO_2 then determine g O per g S in SO_3

Solution: for a fixed amount of S, the ratio of O is $\frac{3\,\text{O}}{2\,\text{O}} = 1.5$. So, for 1 gram S, SO_3 would have 1.5 g O. The mass ratio of S:O = 1.0:1.5 in SO_3.

Check: The answer is reasonable since the ratio is smaller than the ratio for SO_2 and SO_3 has to contain more O per gram of S.

b) **Given:** mass ratio S:O = 1.0:1.0 in SO_2 **Find:** mass ratio S:O in S_2O
Conceptual Plan: determine the ratio of S:S in S_2O and SO_2 then determine g O per g S in S_2O

Solution: for a fixed amount of O, the ratio of S is $\frac{2\,\text{S}}{0.5\,\text{S}} = 4.0$. So, for 1 gram O, S_2O would have 4 gram S. The mass ratio of S:O = 4.0:1.0 in S_2O.

Check: The answer is reasonable since the ratio is larger than the ratio for SO_2 and S_2O has to contain more S per gram of O.

61. **Given:** In CO mass ratio O:C = 1.33:1; In compound X, mass ratio O:C = 2:1. **Find:** formula of X
Conceptual Plan: determine the mass ratio of O:O in the 2 compounds.

Solution: For 1 gram of C: $\frac{2\,\text{g O in compound X}}{1.33\,\text{g O in CO}} = 1.5$

So, the ratio of O to C in compound X has to be 1.5:1 and the formula is C_2O_3.

Check: The answer is reasonable since it fulfills the criteria of multiple proportions and the mass ratio of O:C is 2:1.

62. **Given:** mass ratio 1 atom N:1 atom^{12}C = 7:6; mass ratio 2 mol N:1 mol O in N_2O = 7:4
Find: mass of 1 mol O
Conceptual Plan: determine the mass ratio of O to ^{12}C from the mass ratio of N to ^{12}C and the mass ratio of N to O and then determine the mol ratio of ^{12}C to O then use the mass of 1 mol ^{12}C to determine mass 1 mol O

$$\frac{12.00\,\text{g}\,^{12}\text{C}}{1\,\text{mol}\,^{12}\text{C}}$$

Solution: From the mass ratios: for every 7 grams N there are 6 grams ^{12}C and for every 7 grams N there are 4 grams O. So, the mass ratio of O:^{12}C is 4:6.

$$\frac{1 \text{ atom } ^{12}C}{1 \text{ atom N}} \times \frac{6.022 \times 10^{23} \text{ atom N}}{1 \text{ mol N}} \times \frac{1 \text{ mol } ^{12}C}{6.022 \times 10^{23} \text{ atom } ^{12}C} \times \frac{2 \text{ mol N}}{1 \text{ mol O}} = \frac{2 \text{ mol } ^{12}C}{1 \text{ mol O}}$$

$$\frac{2 \text{ mol } ^{12}C}{1 \text{ mol O}} \times \frac{12.00 \text{ g } ^{12}C}{1 \text{ mol } ^{12}C} \times \frac{4 \text{ g O}}{6 \text{ g } ^{12}C} = 16.00 \text{ g O/ mol O}$$

Check: The units of the answer, g O/mol O, are correct. The magnitude of the answer is reasonable since it is close to the value on the periodic table.

63. **Given:** $^4He^{2+} = 4.00151$ amu; **Find:** charge to mass ratio C/kg
Conceptual Plan: determine total charge on $^4He^{2+}$ and then amu $^4He^{2+} \rightarrow$ g $^4He^{2+} \rightarrow$ kg $^4He^{2+}$

$$\frac{+1.60218 \times 10^{-19} \text{ C}}{\text{proton}} \qquad \frac{1 \text{ g}}{1.66054 \times 10^{-24} \text{ amu}} \quad \frac{1 \text{ kg}}{1000 \text{ g}}$$

Solution:

$$\frac{2 \text{ protons}}{1 \text{ atom } ^4He^{2+}} \times \frac{+1.60218 \times 10^{-19} \text{ C}}{\text{proton}} = \frac{3.20436 \times 10^{-19} \text{ C}}{\text{atom } ^4He^{2+}}$$

$$\frac{4.00151 \text{ amu}}{1 \text{ atom } ^4He^{2+}} \times \frac{1.66054 \times 10^{-24} \text{ g}}{1 \text{ amu}} \times \frac{1 \text{ kg}}{1000 \text{ g}} = \frac{6.64466\underline{7}42 \times 10^{-27} \text{ kg}}{1 \text{ atom } ^4He^{2+}}$$

$$\frac{3.20436 \times 10^{-19} \text{ C}}{\text{atom } ^4He^{2+}} \times \frac{1 \text{ atom } ^4He^{2+}}{6.64466\underline{7}42 \times 10^{-27} \text{ kg}} = 4.82245 \times 10^{7} \text{ C/kg}$$

Check: the units of the answer, C/kg, are correct. The magnitude of the answer is reasonable when compared to the charge to mass ratio of the electron.

64. **Given:** 12.3849 g sample I; atomic mass I = 126.9045 amu; 1.00070g ^{129}I; mass ^{129}I = 128.9050 amu
Find: mass of contaminated sample
Conceptual Plan: total mass of sample $\rightarrow$ fraction I and ^{129}I in the sample $\rightarrow$ apparent "atomic mass"

$$\text{mass I} + \text{mass } ^{129}I \qquad \frac{\text{g I}}{\text{g sample}}; \frac{\text{g } ^{129}I}{\text{g sample}} \qquad \text{Atomic mass} = \sum_n (\text{fraction of isotope n}) \times (\text{mass of isotope n})$$

Solution: 12.3849 g I + 1.00070 g ^{129}I = 13.3856 g sample

$$\frac{12.3849 \text{ g}}{13.3856 \text{ g}} = 0.925240\underline{5}57 \text{ fraction I} \qquad \frac{1.00070 \text{ g}}{13.3856 \text{ g}} = 0.074759\underline{4}4 \text{ fraction } ^{129}I$$

$$\text{Atomic mass} = \sum_n (\text{fraction of isotope n}) \times (\text{mass of isotope n})$$

$$= (0.925240\underline{5}57)(126.9045 \text{ amu}) + (0.074759\underline{4}4)(128.9050 \text{ amu})$$

$$= 127.055 \text{ amu}$$

Check: The units of the answer, amu, are correct. The magnitude of the answer is reasonable because it is between 126.9045 and 128.9050 and only slightly higher than the naturally occurring value.

65. $^{236}_{90}Th$ A – Z = number of neutrons. 236 – 90 = 146 neutrons. So, any nucleus with 146 neutrons is an isotone of $^{236}_{90}Th$.

Some would be: $^{238}_{92}U$; $^{239}_{93}Np$; $^{241}_{95}Am$; $^{237}_{91}Pa$; $^{235}_{89}Ac$; $^{244}_{98}Cf$ etc.

66.

Symbol	Z	A	Number protons	Number electrons	Number neutrons	Charge
Si	14	**28**	**14**	14	14	**0**
S^{2-}	**16**	32	**16**	**18**	**16**	2–
Cu^{2+}	**29**	**63**	**29**	**27**	34	2+
P	15	**28**	**15**	15	13	**0**

67.

Symbol	Z	A	Number protons	Number electrons	Number neutrons	Charge
O^{2-}	8	**16**	**8**	**10**	8	2 –
Ca^{2+}	20	**40**	**20**	**18**	20	**2+**
Mg^{2+}	**12**	25	**12**	**10**	13	2+
N^{3-}	**7**	14	7	10	**7**	**3 –**

68. **Given:** r (neutron) = 1.0×10^{-13} cm; r(star piece) = 0.10 mm **Find:** density of neutron, mass (kg) of star piece

Conceptual Plan:
r (neutron) → vol (neutron) → density (neutron) and then r (star piece) → vol (star piece)

$V = \frac{4}{3}\pi r^3$ $\quad d = \frac{m}{v}$ $\quad V = \frac{4}{3}\pi r^3$

→ mass (star piece)

$m = dv$

Solution:
For the neutron:

$$\text{Vol(neutron)} = \frac{4}{3}\pi(1.0 \times 10^{-13}\,\text{cm})^3 = 4.19 \times 10^{-39}\,\text{cm}^3 \qquad d = \frac{1.00727\ \text{amu}}{4.19 \times 10^{-39}\,\text{cm}^3} \times \frac{1.661 \times 10^{-24}\,\text{g}}{\text{amu}} = 3.99 \times 10^{14}\,\text{g/cm}^3$$

For the star piece:

$$\text{Vol(star piece)} = \frac{4}{3}\pi(0.10\ \text{mm})^3\,\frac{(1\,\text{cm})^3}{(10\ \text{mm})^3} = 4.19 \times 10^{-4}\,\text{cm}^3 \qquad m = 4.19 \times 10^{-4}\,\text{cm}^3 \times \frac{3.99 \times 10^{14}\,\text{g}}{\text{cm}^3} \times \frac{1\,\text{kg}}{1000\,\text{g}} = 1.7 \times 10^{8}\,\text{k}$$

Check: The units of the answer, kg, are correct. The magnitude of the answer shows the great mass of the neutron star.

69. **Given:** r(nucleus) = 2.7 fm; r(atom) = 70 pm (assume 2 significant figures)
Find: vol(nucleus); vol(atom), % vol(nucleus)

Conceptual Plan:
r(nucleus)(fm) → r(nucleus)(pm) → vol(nucleus) and then r(atom) → vol(atom) and then % vol

$\frac{10^{-15}\,\text{m}}{1\,\text{fm}}$ $\quad \frac{1\,\text{pm}}{10^{-12}\,\text{m}}$ $\quad V = \frac{4}{3}\pi r^3$ $\quad V = \frac{4}{3}\pi r^3$ $\quad \frac{\text{vol(nucleus)}}{\text{vol(atom)}} \times 100$

$$\textbf{Solution: } 2.7\ \text{fm} \times \frac{10^{-15}\,\text{m}}{\text{fm}} \times \frac{1\ \text{pm}}{10^{-12}\,\text{m}} = 2.7 \times 10^{-3}\text{pm} \qquad V_{\text{nucleus}} = \frac{4}{3}\pi\,(2.7 \times 10^{-3}\text{pm})^3 = 8.2 \times 10^{-8}\ \text{pm}^3$$

$$V_{\text{atom}} = \frac{4}{3}\pi\,(70\ \text{pm})^3 = 1.4 \times 10^{6}\ \text{pm}^3 \qquad \frac{8.2 \times 10^{-8}\ \text{pm}^3}{1.4 \times 10^{6}\ \text{pm}^3} \times 100\% = 5.9 \times 10^{-12}\%$$

Check: The units of the answer, % vol, are correct. The magnitude of the answer is reasonable because the nucleus only occupies a very small % of the vol of the atom.

70. **Given:** 1 penny = 1.0 mm **Find:** height in km of Avogadro's number of pennies
Conceptual Plan: height of 1 penny → height of Avogadro's number of pennies

6.022×10^{23}

$$\textbf{Solution: } \frac{1.0\ \text{mm}}{\text{penny}} \times 6.022 \times 10^{23}\ \text{pennies} \times \frac{1\ \text{m}}{1000\ \text{mm}} \times \frac{1\,\text{km}}{1000\ \text{m}} = 6.022 \times 10^{17}\,\text{km}$$

Check: The units of the answer, km, are correct. The magnitude of the answer shows just how large Avogadro's number is.

71. **Given:** 6.022×10^{23} pennies **Find:** the amount in dollars; the dollars/person
Conceptual Plan: pennies → dollars → dollars/person

$\frac{1\,\text{dollar}}{100\,\text{pennies}}$ $\quad$ 6.5 billion people

Solution:

$$6.022 \times 10^{23} \text{ pennies} \times \frac{1 \text{ dollar}}{100 \text{ pennies}} = 6.022 \times 10^{21} \text{ dollars} \qquad \frac{6.022 \times 10^{21} \text{ dollars}}{6.5 \times 10^{9} \text{ people}} = 9.3 \times 10^{11} \text{ dollars / person},$$

They are billionaires

72. **Given:** 1 mol blueberries, m = 0.75 g; m(automobile) = 2.0 x 10^3 kg **Find:** number of autos for 1 mol blueberries
Conceptual Plan:

mol blueberries → mass blueberries (g) → mass blueberries (kg) → number of automobiles

$$\frac{0.75 \text{ g}}{\text{blueberry}} \qquad \frac{\text{kg}}{1000 \text{ g}} \qquad \frac{1 \text{ automobile}}{2.0 \times 10^3 \text{ kg}}$$

Solution:

$$1 \text{ mol blueberries} \times \frac{6.022 \times 10^{23} \text{ blueberries}}{\text{mol blueberries}} \times \frac{0.75 \text{ g}}{\text{blueberry}} \times \frac{1 \text{ kg}}{1000 \text{ g}} \times \frac{1 \text{ automobile}}{2.0 \times 10^3 \text{ kg}} = 2.3 \times 10^{17} \text{ automobiles}$$

Check: The units of the answer, automobiles, are correct. The magnitude of the answer is reasonable because Avogadro's number is so large.

73. **Given:** O = 15.9994 amu when C = 12.011 amu **Find:** mass O when C = 12.00 amu
Conceptual Plan: determine ratio O:C for ^{12}C system then use the same ratio when C = 12.00

$$\frac{\text{mass O}}{\text{mass C}}$$

Solution: Based on ^{12}C = 12.00; O = 15.9994 and C = 12.011 so, $\frac{\text{mass O}}{\text{mass C}} = \frac{15.9994 \text{ amu}}{12.011 \text{ amu}} = \frac{1.33206 \text{ amu O}}{1 \text{ amu C}}$

Based on C = 12.00, the ratio has to be the same,

$$12.000 \text{ amu C} \times \frac{1.33206 \text{ amu O}}{1 \text{ amu C}} = 15.985 \text{ amu O}$$

Check: The units of the answer, amu O, are correct. The magnitude of the answer is reasonable because the value for the new mass basis is smaller then the original mass basis, therefore, the mass of O should be less.

74. **Given:** Ti cube: d = 4.50 g/cm^3; e = 2.78 in. **Find:** number Ti atoms
Conceptual Plan: e in inch → e in cm → vol cube → g Ti → mol Ti → atoms Ti

$$\frac{2.54 \text{ cm}}{1 \text{ inch}} \qquad V = e^3 \qquad \frac{4.50 \text{ g}}{\text{cm}^3} \qquad \frac{1 \text{ mol Ti}}{47.867 \text{ g}} \qquad \frac{6.022 \times 10^{23} \text{ atoms}}{\text{mol}}$$

Solution: $2.78 \text{ in} \times \frac{2.54 \text{ cm}}{\text{in}} = 7.0\underline{6}1 \text{ cm}$

$$(7.0\underline{6}1 \text{ cm})^3 \times \frac{4.50 \text{ g}}{\text{cm}^3} \times \frac{1 \text{ mol Ti}}{47.867 \text{ g}} \times \frac{6.022 \times 10^{23} \text{ atoms Ti}}{1 \text{ mol Ti}} = 4.00 \times 10^{23} \text{ atoms Ti}$$

Check: The units of the answer, atoms Ti, are correct. The magnitude of the answer is reasonable because there is about 30 mol of Ti in the cube.

75. **Given:** Cu sphere: r = 0.935 in; d = 8.96 g/cm^3 **Find:** number of Cu atoms
Conceptual Plan: r in inch → r in cm → vol sphere → g Cu → mol Cu → atoms Cu

$$\frac{2.54 \text{ cm}}{1 \text{ inch}} \qquad V = \frac{4}{3}\pi r^3 \qquad \frac{8.96 \text{ g}}{\text{cm}^3} \qquad \frac{1 \text{ mol Cu}}{63.546 \text{ g}} \qquad \frac{6.022 \times 10^{23} \text{ atoms}}{\text{mol}}$$

Solution: $0.935 \text{ in} \times \frac{2.54 \text{ cm}}{\text{in}} = 2.3\underline{7}49 \text{ cm}$

$$\frac{4}{3}\pi(2.3\underline{7}49 \text{ cm})^3 \times \frac{8.96 \text{ g}}{\text{cm}^3} \times \frac{1 \text{ mol Cu}}{63.546 \text{ g}} \times \frac{6.022 \times 10^{23} \text{ atoms Cu}}{1 \text{ mol Cu}} = 4.76 \times 10^{24} \text{ atoms Cu}$$

Check: The units of the answer, atoms Cu, are correct. The magnitude of the answer is reasonable because there are about 8 mol Cu present.

76. **Given:** B – 10 = 10.01294 amu; B – 11 = 11.00931 amu; B = 10.811 amu
Find: % abundance B – 10 and B – 11
Conceptual Plan: Let x = fraction B – 10 then 1 – x = fraction B – 11 → abundances

$$\text{Atomic mass} = \sum_n (\text{fraction of isotope n}) \times (\text{mass of isotope n})$$

Solution:

$$\text{Atomic mass} = \sum_n (\text{fraction of isotope n}) \times (\text{mass of isotope n})$$

$$10.811 = (x)(10.01294 \text{ amu}) + (1 - x)(11.00931 \text{ amu})$$

$$0.19\underline{8}31 = 0.99637\ x$$

$$x = 0.19\underline{9}0 \qquad 1 - x = 0.80\underline{0}96$$

$$B - 10 = 0.199 \times 100 = 19.9\ \%\ \text{and}\ B - 11 = 0.801 \times 100 = 80.1\ \%$$

Check: The units of the answer, %, which gives the relative abundance of each isotope, are correct. The relative abundances are reasonable because B has an atomic mass closer to the mass of B - 11 than to B - 10.

77. **Given:** Li – 6 = 6.01512 amu; Li – 7 = 7.01601 amu; B = 6.941 amu
Find: % abundance Li – 6 and Li – 7
Conceptual Plan: Let x = fraction Li – 6 then 1 – x = fraction Li – 7 → abundances

$$\text{Atomic mass} = \sum_n (\text{fraction of isotope n}) \times (\text{mass of isotope n})$$

Solution:

$$\text{Atomic mass} = \sum_n (\text{fraction of isotope n}) \times (\text{mass of isotope n})$$

$$6.941 = (x)(6.01512 \text{ amu}) + (1 - x)(7.01601 \text{ amu})$$

$$0.07501 = 1.00089\ x$$

$$x = 0.07494 \qquad 1 - x = 0.92506$$

$$Li - 6 = 0.07494 \times 100 = 7.494\ \%\ \text{and}\ Li - 7 = 0.92506 \times 100 = 92.506\ \%$$

Check: The units of the answer, %, which gives the relative abundance of each isotope, are correct. The relative abundances are reasonable because Li has an atomic mass closer to the mass of Li – 7 than to Li – 6.

78. **Given:** 1 mol sand grains; e(cube) = 0.10 mm; area Texas = 268,601 sq mi. **Find:** height of sand ft
Conceptual Plan:
mol sand → grains sand → vol sand mm^3 → vol sand ft^3 and then area Texas mi^2 → area ft^2

$$\frac{6.022 \times 10^{23}\ \text{grains}}{\text{mol}} \qquad V = e^3 \qquad \text{mm}^3 \times \left(\frac{\text{cm}}{10\,\text{mm}}\right)^3 \left(\frac{1\,\text{in}}{2.54\,\text{cm}}\right)^3 \left(\frac{1\,\text{ft}}{12\,\text{in}}\right)^3 \qquad \left(\frac{5280\,\text{ft}}{1\,\text{mi}}\right)^2$$

and then → height ft.

$$h = \frac{\text{Volume}}{\text{Area}}$$

Solution:

$$1\ \text{mol sand grains} \times \frac{6.022 \times 10^{23}\ \text{grains}}{\text{mol}} \times \frac{(0.10\ \text{mm})^3}{\text{grain}} \times \left(\frac{\text{cm}}{10\ \text{mm}}\right)^3 \times \left(\frac{1\ \text{in}}{2.54\ \text{cm}}\right)^3 \times \left(\frac{1\ \text{ft}}{12\ \text{in}}\right)^3 = 2.12\underline{6}6 \times 10^{13}\ \text{ft}^3\ \text{sand}$$

$$\frac{2.12\underline{6}6 \times 10^{13}\ \text{ft}^3\ \text{sand}}{268{,}601\ \text{mi}^2} \times \left(\frac{1\ \text{mi}}{5280\ \text{ft}}\right)^2 = 2.84\ \text{ft of sand}$$

Check: The units of the answer, ft sand, are correct. The magnitude of the answer seems reasonable.

79. **Given:** sun: d = 1.4 g/cm^3; r = 7 x 10^8 m; 100 billion stars/galaxy; 10 billion galaxies/universe
Find: number of atoms in the universe
Conceptual Plan: r (star) in m → r (star) in cm → vol (star) → g H/star → mol H star → atoms H/star

$$\frac{100\,\text{cm}}{\text{m}} \qquad V = \frac{4}{3}\pi r^3 \qquad \frac{1.4\,\text{g H}}{\text{cm}^3} \qquad \frac{1\,\text{mol H}}{1.008\,\text{g}} \qquad \frac{6.022 \times 10^{23}\ \text{atoms}}{\text{mol}}$$

→ atoms H/galaxy → atoms H/universe

$$\frac{100 \times 10^9\ \text{stars}}{\text{galaxy}} \qquad \frac{10 \times 10^9\ \text{galaxies}}{\text{universe}}$$

Solution: $7 \times 10^8 \cancel{m} \times \frac{100 \text{ cm}}{\cancel{m}} = 7 \times 10^{10}$ cm

$$\frac{4}{3}\pi \frac{(7 \times 10^{10} \cancel{cm})^3}{\cancel{star}} \times \frac{1.4 \cancel{g H}}{\cancel{cm^3}} \times \frac{1 \cancel{mol H}}{1.008 \cancel{g H}} \times \frac{6.022 \times 10^{23} \cancel{atoms H}}{\cancel{mol H}} \times \frac{100 \times 10^9 \cancel{stars}}{\cancel{galaxy}} \times \frac{10 \times 10^9 \cancel{galaxies}}{\text{universe}}$$

$= 1 \times 10^{78}$ atoms/universe

Check: The units of the answer, atoms/universe, are correct.

80. a) **Given:** 37 Wt – 296; 2 Wt – 297; 12 Wt – 298 **Find:** % abundance of each

Conceptual Plan: total atoms → fraction of each isotope → % abundance

Sum of atoms $\quad \frac{\text{number of each isotope}}{\text{total atoms}} \quad$ fraction x 100

Solution: Total atoms = 37 + 2 + 12 = 51

$$\frac{37}{51} \times 100 = 72.55\% \text{ Wt} - 296 \qquad \frac{2}{51} \times 100 = 3.922\% \text{ Wt} - 297 \qquad \frac{12}{51} \times 100 = 23.53\% \text{ Wt} - 298$$

Check: The units of the answers, % abundance, are correct. The values of the answers are reasonable since they add up to 100 %

b)

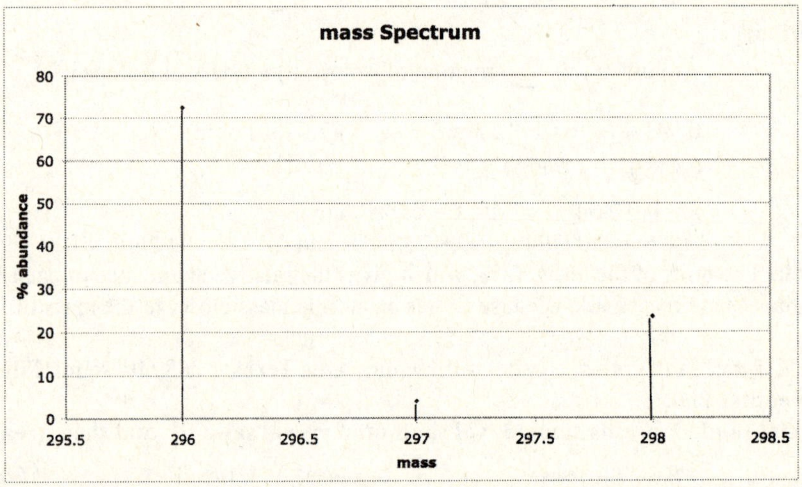

c) **Given:** Wt – 296, m =24.6630 x mass ^{12}C; 72.55%: Wt – 297, m = 24.7490 x mass ^{12}C; 3.922%: Wt – 298, m = 24.8312 x mass ^{12}C; 23.53 % **Find:** atomic mass Wt

Conceptual Plan: mass of isotope relative to ^{12}C → mass of isotope and then % abundance →

(Mass relative to ^{12}C)(12.00 amu) $\quad \frac{\%\text{ abundance}}{100}$

fraction abundance then determine atomic mass

Atomic mass = $\sum_n$ (fraction of isotope n) x (mass of isotope n)

Solution:

Wt – 296 = 24.6630 x 12.00 amu = 295.956 amu; Wt – 297 = 24.7490 x 12.00 amu = 296.988 amu;
Wt – 298 = 24.8312 x 12.00 amu = 297.974 amu

$$\text{fraction Wt} - 296 = \frac{72.55}{100} = 0.7255 \qquad \text{fraction Wt} - 297 = \frac{3.922}{100} = 0.03922 \qquad \text{fraction Wt} - 298 = \frac{23.53}{100} = 0.2353$$

Atomic mass = $\sum_n$ (fraction of isotope n) x (mass of isotope n)

= (0.7255)(295.956 amu) + (0.03922)(296.988 amu) + (0.2353)(297.974 amu)

= 296.477 amu

Check: The units of the answer, amu, are correct. The magnitude of the answer is reasonable because it lies between 295.956 and 297.974 and is closer to 296 which has the highest abundance.

81. a) This is the Law of Definite Proportions: All samples of a given compound, regardless of their source or how they were prepared, have the same proportions of their constituent elements.

b) This is the Law of Conservation of Mass: In a chemical reaction, matter is neither created nor destroyed.

c) This is the Law of Multiple Proportions: When two elements form two different compounds, the masses of element B that combine with 1 g of element A can be expressed as a ratio of small whole numbers. In this example the ratio of O from hydrogen peroxide to O from water = 16:8 → 2:1, a small whole number ratio.

82. If the amu and mole were not based on the same isotope, the numerical values obtained for an atom of material and a mole of material would not be the same. If, for example, the mole was based on the number of particles in C – 12 but the amu was changed to a fraction of the mass of an atom of Ne – 20 the number of particles and the number of amu that make up one mole of material would no longer be the same. We would no longer have the relationship where the mass of an atom in amu is numerically equal to the mass of a mole of those atoms in grams.

83. **Given:** a. Cr: 55.0 g; atomic mass = 52 g/mol b. Ti: 45.0 g; atomic mass = 48 g/mol and
c. Zn: 60.0 g; atomic mass = 65 g/mol
Find: which has the greatest mol, and which has the greatest mass
Conceptual Plan: without calculation, compare grams of material to g/mol for each.
Solution: Cr would have the greatest mole amount of the elements. It is the only one whose mass is greater than the molar mass. Zn would be the greatest mass amount because it is the largest mass value.

84. The different isotopes of the same element have the same number of protons and electrons, so the attractive forces between the nucleus and the electrons is constant and there is no difference in the radii of the isotopes. Ions, on the other hand, have a different number of electrons than the parent atom from which they are derived. Cations have fewer electrons than the parent atom. The attractive forces are greater because there is a larger positive charge in the nucleus than the negative charge in the electron cloud. So, cations are smaller than the parent atom from which they are derived. Anions have more electrons than the parent. The electron cloud has a greater negative charge than the nucleus, so the anions have larger radii than the parent.

Chapter 3
Molecules, Compounds, and Chemical Equations

1. The chemical formula gives you the kind of atom and the number of each atom in the compound.

a) $Ca_3(PO_4)_2$ contains: 3 calcium atoms, 2 phosphorus atoms, and 8 oxygen atoms

b) $SrCl_2$ contains: 1 strontium atom and 2 chlorine atoms

c) KNO_3 contains: 1 potassium atom, 1 nitrogen atom, and 3 oxygen atoms

d) $Mg(NO_2)_2$ contains: 1 magnesium atom, 2 nitrogen atoms, and 4 oxygen atoms

2. The chemical formula gives you the kind of atom and the number of each atom in the compound.

a) $Ba(OH)_2$ contains: 1 barium atom, 2 oxygen atoms, and 2 hydrogen atoms

b) NH_4Cl contains: 1 nitrogen atom, 4 hydrogen atoms, and 1 chlorine atom

c) NaCN contains: 1 sodium atom, 1 carbon atom and, 1 nitrogen atom

d) $Ba(HCO_3)_2$ contains: 1 barium atom, 2 hydrogen atoms, 2 carbon atoms, and 6 oxygen atoms

3. a) 1 blue = nitrogen, 3 white = hydrogen: NH_3

b) 2 black = carbon, 6 white = hydrogen: C_2H_6

c) 1 yellow – green = sulfur, 3 red = oxygen: SO_3

4. a) 1 blue = nitrogen, 2 red = oxygen: NO_2

b) 1 yellow – green = sulfur, 2 white = hydrogen: SH_2

c) 1 black = carbon, 4 white = hydrogen: CH_4

5. a) Neon is an element and it is not one of the elements that exist as diatomic molecules, therefore it is an atomic element.

b) Fluorine is one of the elements that exist as diatomic molecules, therefore it is a molecular element.

c) Potassium is not one of the elements that exist as diatomic molecules, therefore it is an atomic element.

d) Nitrogen is one of the elements that exist as diatomic molecules, therefore it is a molecular element.

6. a) Hydrogen is one of the elements that exist as diatomic molecules, therefore it has a molecule as its basic unit.

b) Iodine is one of the elements that exist as diatomic molecules, therefore it has a molecule as its basic unit.

c) Lead is not one of the elements that exist as a diatomic molecule, therefore it does not have a molecule as its basic unit.

d) Oxygen is one of the elements that exist as diatomic molecules, therefore it has a molecule as its basic unit.

7. a) CO_2 is a compound composed of a nonmetal and a nonmetal, therefore it is a molecular compound.

b) $NiCl_2$ is a compound composed of a metal and a nonmetal, therefore it is an ionic compound.

c) NaI is a compound composed of a metal and a nonmetal, therefore it is an ionic compound.

d) PCl_3 is a compound composed of a nonmetal and a nonmetal, therefore it is a molecular compound.

8. a) CF_2Cl_2 is a compound composed of a nonmetal and 2 other nonmetals, therefore it is a molecular compound.

   b) CCl_4 is a compound composed of a nonmetal and a nonmetal, therefore it is a molecular compound.

   c) PtO_2 is a compound composed of a metal and a nonmetal, therefore it is an ionic compound.

   d) SO_3 is a compound composed of a nonmetal and a nonmetal, therefore it is a molecular compound.

9. a) white – hydrogen: a molecule composed of two of the same element, therefore it is a molecular element.

   b) blue – nitrogen, white – hydrogen: a molecule composed of a nonmetal and a nonmetal, therefore it is a molecular compound.

   c) purple – sodium: a substance composed of all the same atoms, therefore it is an atomic element.

10. a) green – chlorine, purple - sodium: a compound composed of metal and nonmetal, therefore it is an ionic compound.

    b) green – chlorine: a molecule composed of two of the same element, therefore it is a molecular element.

    c) red – oxygen, black – carbon, white – hydrogen: a molecule composed of nonmetals, therefore it is a molecular compound.

11. To write the formula for an ionic compound: 1) Write the symbol for the metal cation and its charge and the symbol for the nonmetal anion and its charge. 2) Adjust the subscript on each cation and anion to balance the overall charge. 3) Check that the sum of the charges of the cations equals the sum of the charges of the anions.

a) magnesium and sulfur:	Mg^{2+}	S^{2-}	MgS	cations 2+; anions 2-
b) barium and oxygen:	Ba^{2+}	O^{2-}	BaO	cations 2+; anions 2-
c) strontium and bromine:	Sr^{2+}	Br^-	$SrBr_2$	cation 2+; anions 2(1-) = 2-
d) beryllium and chlorine:	Be^{2+}	Cl^-	$BeCl_2$	cations 2+; anions 2(1-) = 2-

12. To write the formula for an ionic compound: 1) Write the symbol for the metal cation and its charge and the symbol for the nonmetal anion and its charge. 2) Adjust the subscript on each cation and anion to balance the overall charge. 3) Check that the sum of the charges of the cations equals the sum of the charges of the anions.

a) aluminum and sulfur:	Al^{3+}	S^{2-}	Al_2S_3	cation 2(3+) = 6+; anions 3(2-) = 6-
b) aluminum and oxygen:	Al^{3+}	O^{2-}	Al_2O_3	cation 2(3+) = 6+; anion 3(2-) = 6-
c) sodium and oxygen:	Na^+	O^{2-}	Na_2O	cation 2(1+) = 2+: anion 2-
d) strontium and iodine:	Sr^{2+}	I^-	SrI_2	cation 2+; anion 2(1-) = 2-

13. To write the formula for an ionic compound: 1) Write the symbol for the metal cation and its charge and the symbol for the polyatomic anion and its charge. 2) Adjust the subscript on each cation and anion to balance the overall charge. 3) Check that the sum of the charges of the cations equals the sum of the charges of the anions.
Cation = barium: Ba^{2+}

a) hydroxide:	OH^-	$Ba(OH)_2$	cation 2+, anion 2(1-) = 2-
b) chromate:	CrO_4^{2-}	$BaCrO_4$	cation 2+; anion 2-
c) phosphate:	PO_4^{3-}	$Ba_3(PO_4)_2$	cation 3(2+)=6+; anion 2(3-) = 6-
d) cyanide:	CN^-	$Ba(CN)_2$	cation 2+; anion 2(1-) = 2-

14. To write the formula for an ionic compound: 1) Write the symbol for the metal cation and its charge and the symbol for the nonmetal anion and its charge. 2) Adjust the subscript on each cation and anion to balance the overall charge. 3) Check that the sum of the charges of the cations equals the sum of the charges of the anions.
Cation = sodium: Na^+

a) carbonate:	CO_3^{2-}	Na_2CO_3	cation 2(1+) = 2+; anion 2-
b) phosphate:	PO_4^{3-}	Na_3PO_4	cation 3(1+) = 3+; anion 3-
c) hydrogen phosphate:	HPO_4^{2-}	Na_2HPO_4	cation 2(1+) = 2+; anion 2-
d) acetate:	$C_2H_3O_2^-$	$NaC_2H_3O_2$	cation 1+; anion 1-

15. To name a binary ionic compound: name the metal cation followed by the base name of the anion + ide.
a) Mg_3N_2: the cation is magnesium, the anion is from nitrogen which becomes nitride: magnesium nitride.

b) KF: the cation is potassium, the anion is from fluorine which becomes fluoride: potassium fluoride.

c) Na_2O: the cation is sodium, the anion is from oxygen which becomes oxide: sodium oxide.

d) Li_2S: the cation is lithium, the anion is from sulfur which becomes sulfide: lithium sulfide.

16. To name a binary ionic compound: name the metal cation followed by the base name of the anion + ide.
a) CsF: the cation is cesium, the anion is from fluorine which becomes fluoride: cesium fluoride.

b) KI the cation is potassium, the anion is from iodine which becomes iodide: potassium iodide.

c) $SrCl_2$: the cation is strontium, the anion is from chlorine which becomes chloride: strontium chloride.

d) $BaCl_2$: the cation is barium, the anion is from chlorine which becomes chloride: barium chloride.

17. To name an ionic compound with a metal cation that can have more than one charge: name the metal cation followed by parentheses with the charge in roman numerals followed by the base name of the anion + ide.
a) $SnCl_4$: the charge on Sn must be 4+ for the compound to be charge neutral: the cation is tin(IV), the anion is from chlorine which becomes chloride; tin(IV) chloride.

b) PbI_2: the charge on Pb must be 2+ for the compound to be charge neutral: the cation is lead(II), the anion is from iodine which becomes iodide; lead(II) iodide.

c) Fe_2O_3: the charge on Fe must be 3+ for the compound to be charge neutral; the cation is iron(III), the anion is from oxygen which becomes oxide: iron(III) oxide.

d) CuI_2: the charge on Cu must be 2+ for the compound to be charge neutral; the cation is copper(II), the anion is from iodine which becomes iodide: copper(II) iodide.

18. To name an ionic compound with a metal cation that can have more than one charge: name the metal cation followed by parentheses with the charge in roman numerals followed by the base name of the anion + ide.
a) SnO_2: the charge on Sn must be 4+ for the compound to be charge neutral: the cation is tin(IV), the anion is from oxygen which becomes oxide: tin(IV) oxide.

b) $HgBr_2$: the charge of Hg must be 2+ for the compound to charge neutral: the cation is mercury(II), the anion is from bromine which becomes bromide: mercury(II) bromide

c) $CrCl_2$: the charge on Cr must be 2+ for the compound to be charge neutral: the cation is chromium(II), the anion is from chlorine which becomes chloride: chromium(II) chloride.

d) $CrCl_3$: the charge on Cr must be 3+ for the compound to be charge neutral; the cation is chromium(III), the anion is from chlorine which becomes chloride: chromium(III) chloride.

19. To name these compounds you must first decide if the metal cation is invariant or can have more than one charge. Then, name the metal cation followed by the base name of the anion + ide.

a) SnO: Sn can have more than one charge. The charge on Sn must be 2+ for the compound to be charge neutral: the cation is tin(II), the anion is from oxygen which becomes oxide: tin(II) oxide.

b) Cr_2S_3: Cr can have more than one charge. The charge on Cr must be 3+ for the compound to be charge neutral: the cation is chromium(III), the anion is from sulfur which becomes sulfide: chromium(III) sulfide.

c) RbI: Rb is invariant. The cation is rubidium, the anion is from iodine which becomes iodide: rubidium iodide.

d) $BaBr_2$: Ba is invariant. The cation is barium, the anion is from bromine which becomes bromide: barium bromide.

20. To name these compounds you must first decide if the metal cation is invariant or can have more than one charge. Then, name the metal cation followed by the base name of the anion + ide.

a) BaS: Ba is invariant. The cation is barium, the anion is from sulfur which becomes sulfide: barium sulfide.

b) $FeCl_3$: Fe can have more than one charge. The charge on Fe must be 3+ for the compound to be charge neutral. The cation is iron(III), the anion is from chlorine which becomes chloride: iron(III) chloride.

c) PbI_4: Pb can have more than one charge. The charge on Pb must be 4+ for the compound to be charge neutral. The cation is lead(IV), the anion is from iodine which becomes iodide: lead(IV) iodide.

d) $SrBr_2$: Sr is invariant. The cation is strontium, the anion is from bromine which becomes bromide: strontium bromide.

21. To name these compounds you must first decide if the metal cation is invariant or can have more than one charge. Then, name the metal cation followed by the name of the polyatomic anion.

a) $CuNO_2$: Cu can have more than one charge. The charge on Cu must be 1+ for the compound to be charge neutral. The cation is copper(I), the anion is nitrite: copper(I) nitrite

b) $Mg(C_2H_3O_2)_2$: Mg is invariant. The cation is magnesium, the anion is acetate: magnesium acetate.

c) $Ba(NO_3)_2$: Ba is invariant. The cation is barium, the anion is nitrate: barium nitrate.

d) $Pb(C_2H_3O_2)_2$: Pb can have more than one charge. The charge on Pb must be 2+ for the compound to be charge neutral. The cation is lead(II), the anion is acetate: lead(II) acetate.

e) $KClO_3$: K is invariant. The cation is potassium, the anion is chlorate: potassium chlorate.

f) $PbSO_4$: Pb can have more than one charge. The charge on Pb must be 2+ for the compound to be charge neutral. The cation is lead(II), the anion is sulfate: lead(II) sulfate.

22. To name these compounds you must first decide if the metal cation is invariant or can have more than one charge. Then, name the metal cation followed by the name of the polyatomic anion.

a) $Ba(OH)_2$: Ba is invariant. The cation is barium, the anion is hydroxide; barium hydroxide.

b) NH_4I: The cation is ammonium, the anion is from iodine which becomes iodide; ammonium iodide.

c) $NaBrO_4$: Na is invariant. The cation is sodium, the anion is perbromate; sodium perbromate.

d) $Fe(OH)_3$: Fe can have more than one charge. The charge on Fe must be 3+ for the compound to be charge neutral. The cation is iron(III), the anion is hydroxide; iron(III) hydroxide.

e) $CoSO_4$: Co can have more than one charge. The charge on Co must be 2+ for the compound to be charge neutral. The cation is cobalt(II), the anion is sulfate: cobalt(II) sulfate.

f) KClO: K is invariant. The cation is potassium, the anion is hypochlorite; potassium hypochlorite.

23. To write the formula for an ionic compound: 1) Write the symbol for the metal cation and its charge and the symbol for the nonmetal anion or polyatomic anion and its charge. 2) Adjust the subscript on each cation and anion to balance the overall charge. 3) Check that the sum of the charges of the cations equals the sum of the charges of the anions.

a) sodium hydrogen sulfite:	Na^+	HSO_3^-	$NaHSO_3$	cation 1+; anion 1-
b) lithium permanganate:	Li^+	MnO_4^-	$LiMnO_4$	cation 1+; anion 1-
c) silver nitrate:	Ag^+	NO_3^-	$AgNO_3$	cation 1+; anion 1-
d) potassium sulfate:	K^+	SO_4^{2-}	K_2SO_4	cation 2(1+) = 2+; anion 2-
e) rubidium hydrogen sulfate:	Rb^+	HSO_4^-	$RbHSO_4$	cation 1+; anion 1-
f) potassium hydrogen carbonate:	K^+	HCO_3^-	$KHCO_3$	cation 1+; anion 1-

24. To write the formula for an ionic compound: 1) Write the symbol for the metal cation and its charge and the symbol for the nonmetal anion or polyatomic anion and its charge. 2) Adjust the subscript on each cation and anion to balance the overall charge. 3) Check that the sum of the charges of the cations equals the sum of the charges of the anions.

a) copper(II) chloride:	Cu^{2+}	Cl^-	$CuCl_2$	cation 2+; anion 2(1-) = 2-
b) copper(I) iodate:	Cu^+	IO_3^-	CuCl	cation 1+; anion 1-
c) lead(II) chromate:	Pb^{2+}	CrO_4^{2-}	$PbCrO_4$	cation 2+; anion 2-
d) calcium fluoride:	Ca^{2+}	F^-	CaF_2	cation 2+; anion 2(1-) = 2-
e) potassium hydroxide:	K^+	OH^-	KOH	cation 1+; anion 1-
f) iron(II) phosphate:	Fe^{2+}	PO_4^{3-}	$Fe_3(PO_4)_2$	cation 3(2+)=6+; anion 2(3-) =6-

25. Hydrates are named the same way as other ionic compounds with the addition of the term *prefix*hydrate, where the prefix is the number of water molecules associated with each formula unit.

a) $CoSO_4•7H_2O$ — cobalt(II) sulfate heptahydrate

b) iridium(III) bromide tetrahydrate — $IrBr_3•4H_2O$

c) $Mg(BrO_3)_2•6H_2O$ — magnesium bromate hexahydrate

d) potassium carbonate dihydrate — $K_2CO_3•2H_2O$

26. Hydrates are named the same way as other ionic compounds with the addition of the term *prefix*hydrate, where the prefix is the number of water molecules associated with each formula unit.

a) cobalt(II) phosphate octahydrate — $Co_3(PO_4)_2•8H_2O$

b) $BeCl_2•2H_2O$ — beryllium chloride dihydrate

c) chromium(III) phosphate trihydrate — $CrPO_4•3H_2O$

d) $LiNO_2•H_2O$ — lithium nitrite monohydrate

27. a) CO The name of the compound is the name of the first element, *carbon*, following by the base name of the second element, *ox*, prefixed by *mono-* to indicate one and given the suffix – *ide*; carbon monoxide.

b) NI_3 The name of the compound is the name of the first element, *nitrogen*, followed by the base name of the second element, *iod*, prefixed by *tri-* to indicate three and given the suffix – *ide*: nitrogen triiodide.

c) $SiCl_4$ The name of the compound is the name of the first element, *silicon*, followed by the base name of the second element, *chlor*, prefixed by *tetra-* to indicate four and given the suffix – *ide*: silicon tetrachloride.

d) N_4Se_4 The name of the compound is the name of the first element, *nitrogen*, prefixed by *tetra-* to indicate four followed by the base name of the second element, *selen*, prefixed by *tetra-* to indicate four and given the suffix – *ide*: tetranitrogen tetraselenide.

e) I_2O_5 The name of the compound is the name of the first element, *iodine*, prefixed by di to indicate two followed by the base name of the second element, *ox*, prefixed by *penta-* to indicate five and given the suffix – *ide*: diiodine pentaoxide.

28. a) SO_3 The name of the compound is the name of the first element, *sulfur*, followed by the base name of the second element, *ox*, prefixed by *tri-* to indicate three and given the suffix – *ide*: sulfur trioxide

b) SO_2 The name of the compound is the name of the first element, *sulfur*, followed by the base name of the second element, *ox*, prefixed by *di-* to indicate two and given the suffix – *ide*: sulfur dioxide.

c) BrF_5 The name of the compound is the name of the first element, *bromine*, followed by the base name of the second element, *fluor*, prefixed by *penta-* to indicate five and given the suffix – *ide*: bromine pentafluoride.

d) NO The name of the compound is the name of the first element, *nitrogen*, followed by the base name of the second element, *ox*, prefixed by *mono-* to indicate one and given the suffix – *ide*: nitrogen monoxide.

e) XeO_3 The name of the compound is the name of the first element, *xenon*, followed by the base name of the second element, *ox*, prefixed by *tri-* to indicate three and given the suffix – *ide*: xenon trioxide.

29. a) phosphorus trichloride: PCl_3

b) chlorine monoxide: ClO

c) disulfur tetrafluoride: S_2F_4

d) phosphorus pentafluoride: PF_5

e) diphosphorus pentasulfide: P_2S_5

30. a) boron tribromide: BBr_3

b) dichlorine monoxide: Cl_2O

c) xenon tetrafluroide: XeF_4

d) carbon tetrabromide: CBr_4

e) diboron tetrachloride: B_2Cl_4

31. a) HI: the base name of I is *iod* so the name is hydroiodic acid

b) HNO_3: the oxyanion is *nitrate,* which ends in *–ate;* therefore, the name of the acid is nitric acid.

c) H_2CO_3: the oxyanion is *carbonate,* which ends in *–ate;* therefore, the name of the acid is carbonic acid.

d) $HC_2H_3O_2$: the oxyanion is *acetate,* which ends in *–ate;* therefore, the name of the acid is acetic acid.

32. a) HCl: the base name of Cl is *chlor* so the name is hydrochloric acid.

b) $HClO_2$: the oxyanion is *chlorite,* which ends in *–ite;* therefore, the name of the acid is chlorous acid.

c) H_2SO_4: the oxyanion is *sulfate,* which ends in *–ate;* therefore, the name of the acid is sulfuric acid.

d) HNO_2: the oxyanion is *nitrite,* which ends in *–ite;* therefore, the name of the acid is nitrous acid.

33. a) hydrofluoric acid: HF

b) hydrobromic acid: HBr

c) sulfurous acid: H_2SO_3

34. a) phosphoric acid: H_3PO_4

b) hydrocyanic acid: HCN

c) chlorous acid: $HClO_2$

35. To find the formula mass, we sum the atomic masses of each atom in the chemical formula.

a) NO_2 formula mass
= 1 x (atomic mass N) + 2 x (atomic mass O)
= 1 x (14.01 amu) + 2 x (16.00 amu)
= 46.01 amu

b) C_4H_{10} formula mass
= 4 x (atomic mass C) + 10 x (atomic mass H)
= 4 x (12.01 amu) + 10 x (1.008 amu)
= 58.12 amu

c) $C_6H_{12}O_6$ formula mass
= 6 x (atomic mass C) + 12 x (atomic mass H) + 6 x (atomic mass O)
= 6 x (12.01 amu) + 12 x (1.008 amu) + 6 x (16.00 amu)
= 180.16 amu

d) $Cr(NO_3)_3$ formula mass
= 1 x (atomic mass Cr) + 3 x (atomic mass N) + 9 x (atomic mass O)
= 1 x (52.00 amu) + 3 x (14.01 amu) + 9 x (16.00 amu)
= 238.0 amu

36. To find the formula mass, we sum the atomic masses of each atom in the chemical formula.

a) $MgBr_2$ formula mass
= 1 x (atomic mass Mg) + 2 x (atomic mass Br)
= 1 x (24.30 amu) + 2 x (79.90 amu)
= 184.1 amu

b) HNO_2 formula mass
= 1 x (atomic mass H) + 1 x (atomic mass N) + 2 x (atomic mass O)
= 1 x (1.008 amu) + 1 x (14.01 amu) + 2 x (16.00 amu)
= 47.02 amu

c) CBr_4 formula mass = 1 x (atomic mass C) + 4 x (atomic mass Br)
= 1 x (12.01 amu) + 4 x (79.90 amu)
= 331.6 amu

d) $Ca(NO_3)_2$ formula mass = 1 x (atomic mass Ca) + 2 x (atomic mass N) + 6 x (atomic mass O)
= 1 x (40.08 amu) + 2 x (14.01 amu) + 6 x (16.00 amu)
= 164.10 amu

37. a) **Given:** 6.5 g H_2O **Find:** number of molecules
Conceptual Plan: g H_2O → mole H_2O → number H_2O molecules

$$\frac{1 \text{ mol}}{18.02 \text{ g } H_2O} \qquad \frac{6.022 \times 10^{23} \text{ } H_2O \text{ molecules}}{\text{mol } H_2O}$$

Solution: $6.5 \text{ g } H_2O \times \frac{1 \text{ mol } H_2O}{18.02 \text{ g } H_2O} \times \frac{6.022 \times 10^{23} \text{ } H_2O \text{ molecules}}{\text{mol } H_2O} = 2.2 \times 10^{23} \text{ } H_2O \text{ molecules}$

Check: Units of the answer, H_2O molecules, are correct. The magnitude is appropriate because it is smaller than Avogadro's number as expected since we have less than 1 mole of H_2O.

b) **Given:** 389 g CBr_4 **Find:** number of molecules
Conceptual Plan: g CBr_4 → mole CBr_4 → number CBr_4 molecules

$$\frac{1 \text{ mol}}{331.6 \text{ g } CBr_4} \qquad \frac{6.022 \times 10^{23} \text{ } CBr_4 \text{ molecules}}{\text{mol } CBr_4}$$

Solution: $389 \text{ g } CBr_4 \times \frac{1 \text{ mol } CBr_4}{331.6 \text{ g } CBr_4} \times \frac{6.022 \times 10^{23} \text{ } CBr_4 \text{ molecules}}{\text{mol } CBr_4} = 7.06 \times 10^{23} \text{ } CBr_4 \text{ molecules}$

Check: Units of the answer, CBr_4 molecules, are correct. The magnitude is appropriate because it is large than Avogadro's number as expected since we have more than 1 mole of CBr_4.

c) **Given:** 22.1 g O_2**Find:** number of molecules
Conceptual Plan: g O_2 → mole O_2 → number O_2 molecules

$$\frac{1 \text{ mol}}{32.00 \text{ g } O_2} \qquad \frac{6.022 \times 10^{23} \text{ } O_2 \text{ molecules}}{\text{mol } O_2}$$

Solution: $22.1 \text{ g } O_2 \times \frac{1 \text{ mol } O_2}{32.00 \text{ g } O_2} \times \frac{6.022 \times 10^{23} \text{ } O_2 \text{ molecules}}{\text{mol } O_2} = 4.16 \times 10^{23} \text{ } O_2 \text{ molecules}$

Check: Units of the answer, O_2 molecules, are correct. The magnitude is appropriate because it is smaller than Avogadro's number as expected since we have less than 1 mole of O_2.

d) **Given:** 19.3 g C_8H_{10} **Find:** number of molecules
Conceptual Plan: g C_8H_{10} → mole C_8H_{10} → number C_8H_{10} molecules

$$\frac{1 \text{ mol}}{106.16 \text{ g } C_8H_{10}} \qquad \frac{6.022 \times 10^{23} \text{ } C_8H_{10} \text{ molecules}}{\text{mol } C_8H_{10}}$$

Solution:

$$19.3 \text{ g } C_8H_{10} \times \frac{1 \text{ mol } C_8H_{10}}{106.16 \text{ g } C_8H_{10}} \times \frac{6.022 \times 10^{23} \text{ } C_8H_{10} \text{ molecules}}{\text{mol } C_8H_{10}} = 1.09 \times 10^{23} \text{ } C_8H_{10} \text{ molecules}$$

Check: Units of the answer, C_8H_{10} molecules, are correct. The magnitude is appropriate because it is smaller than Avogadro's number as expected since we have less than 1 mole of C_8H_{10}.

38. a) **Given:** 5.94×10^{20} SO_3 molecules **Find:** mass in g
Conceptual Plan: number SO_3 molecules → mole SO_3 → g SO_3

$$\frac{1 \text{ mol } SO_3}{6.022 \times 10^{23} \text{ } SO_3 \text{ molecules}} \qquad \frac{80.07 \text{ g } SO_3}{1 \text{ mol } SO_3}$$

Solution: $5.94 \times 10^{20} \text{ } SO_3 \text{ molecules} \times \frac{1 \text{ mol } SO_3}{6.022 \times 10^{23} \text{ } SO_3 \text{ molecules}} \times \frac{80.07 \text{ g } SO_3}{1 \text{ mol } SO_3} = 0.0790 \text{ g } SO_3$

Check: Units of the answer, grams SO_3 are correct. The magnitude is appropriate because there is less than Avogadro's number of molecules so we have less than 1 mole of SO_3.

b) **Given:** 2.8×10^{22} H_2O molecules **Find:** mass in g
Conceptual Plan: number H_2O molecules → mole H_2O → g H_2O

$$\frac{1 \text{ mol } H_2O}{6.022 \times 10^{23} \text{ } H_2O \text{ molecules}} \qquad \frac{18.02 \text{ g } H_2O}{1 \text{ mol } H_2O}$$

Solution: $2.8 \times 10^{22} \text{ } \cancel{H_2O \text{ molecules}} \times \frac{1 \text{ } \cancel{\text{mol } H_2O}}{6.022 \times 10^{23} \text{ } \cancel{H_2O \text{ molecules}}} \times \frac{18.02 \text{ g } H_2O}{1 \text{ } \cancel{\text{mol } H_2O}} = 0.84 \text{ g } H_2O$

Check: Units of the answer, grams H_2O, are correct. The magnitude is appropriate because there is less than Avogadro's number of molecules so we have less than 1 mole of H_2O.

c) **Given:** 4.5×10^{25} O_3 molecules **Find:** mass in g
Conceptual Plan: number O_3 molecules → mole O_3 → g O_3

$$\frac{1 \text{ mol } O_3}{6.022 \times 10^{23} \text{ } O_3 \text{ molecules}} \qquad \frac{48.00 \text{ g } O_3}{1 \text{ mol } O_3}$$

Solution: $4.5 \times 10^{25} \text{ } \cancel{O_3 \text{ molecules}} \times \frac{1 \text{ } \cancel{\text{mol } O_3}}{6.022 \times 10^{23} \text{ } \cancel{O_3 \text{ molecules}}} \times \frac{48.00 \text{ g } O_3}{1 \text{ } \cancel{\text{mol } O_3}} = 3.6 \times 10^{3} \text{ g } O_3$

Check: Units of the answer, grams O_3, are correct. The magnitude is appropriate because there is more than Avogadro's number of molecules so we have more than 1 mole of O_3.

d) **Given:** 9.85×10^{19} CCl_2F_2 molecules **Find:** mass in g
Conceptual Plan: number CCl_2F_2 molecules → mole CCl_2F_2 → g CCl_2F_2

$$\frac{1 \text{ mol } O_3}{6.022 \times 10^{23} \text{ } O_3 \text{ molecules}} \qquad \frac{48.00 \text{ g } O_3}{1 \text{ mol } O_3}$$

Solution:

$$9.85 \times 10^{19} \text{ } \cancel{CCl_2F_2 \text{ molecules}} \times \frac{1 \text{ } \cancel{\text{mol } CCl_2F_2}}{6.022 \times 10^{23} \text{ } \cancel{CCl_2F_2 \text{ molecules}}} \times \frac{120.91 \text{ g } CCl_2F_2}{1 \text{ } \cancel{\text{mol } CCl_2F_2}} = 1.98 \times 10^{-2} \text{ g } CCl_2F_2$$

Check: Units of the answer, grams CCl_2F_2, are correct. The magnitude is appropriate because there is less than Avogadro's number of molecules so we have less than 1 mole of CCl_2F_2.

39. **Given:** 1 H_2O molecule **Find:** mass in g
Conceptual Plan: number H_2O molecules → mole H_2O → g H_2O

$$\frac{1 \text{ mol } H_2O}{6.022 \times 10^{23} \text{ } H_2O \text{ molecules}} \qquad \frac{18.02 \text{ g } H_2O}{1 \text{ mol } H_2O}$$

Solution: $1 \text{ } \cancel{H_2O \text{ molecule}} \times \frac{1 \text{ } \cancel{\text{mol } H_2O}}{6.022 \times 10^{23} \text{ } \cancel{H_2O \text{ molecules}}} \times \frac{18.02 \text{ g } H_2O}{1 \text{ } \cancel{\text{mol } H_2O}} = 2.992 \times 10^{-23} \text{ g } H_2O$

Check: Units of the answer, grams H_2O, are correct. The magnitude is appropriate because there is much less than Avogadro's number of molecules so we have much less than 1 mole of H_2O.

40. **Given:** 1 $C_6H_{12}O_6$ molecule **Find:** mass in g
Conceptual Plan: number $C_6H_{12}O_6$ molecules→ mole $C_6H_{12}O_6$→ g $C_6H_{12}O_6$

$$\frac{1 \text{ mol } C_6H_{12}O_6}{6.022 \times 10^{23} \text{ } C_6H_{12}O_6 \text{ molecules}} \qquad \frac{180.16 \text{ g } C_6H_{12}O_6}{1 \text{ mol } C_6H_{12}O_6}$$

Solution:

$$1 \text{ } \cancel{C_6H_{12}O_6 \text{ molecule}} \times \frac{1 \text{ } \cancel{\text{mol } C_6H_{12}O_6}}{6.022 \times 10^{23} \text{ } \cancel{C_6H_{12}O_6 \text{ molecules}}} \times \frac{180.16 \text{ g } C_6H_{12}O_6}{1 \text{ } \cancel{\text{mol } C_6H_{12}O_6}} = 2.992 \times 10^{-22} \text{ g } C_6H_{12}O_6$$

Check: Units of the answer, grams $C_6H_{12}O_6$, are correct. The magnitude is appropriate because there is much less than Avogadro's number of molecules so we have much less than 1 mole of $C_6H_{12}O_6$.

41. **Given:** 1.8×10^{17} $C_{12}H_{22}O_{11}$ molecule **Find:** mass in mg

Conceptual Plan: **number $C_{12}H_{22}O_{11}$ molecules → mole $C_{12}H_{22}O_{11}$ → g $C_{12}H_{22}O_{11}$ → mg $C_{12}H_{22}O_{11}$**

$$\frac{1 \text{ mol } C_{12}H_{22}O_{11}}{6.022 \times 10^{23}\ C_{12}H_{22}O_{11} \text{ molecules}} \quad \frac{342.3 \text{ g } C_{12}H_{22}O_{11}}{1 \text{ mol } C_{12}H_{22}O_{11}} \quad \frac{1 \times 10^{3} \text{ mg } C_{12}H_{22}O_{11}}{1 \text{ g } C_{12}H_{22}O_{11}}$$

Solution:

$$1.8 \times 10^{17}\ \cancel{C_{12}H_{22}O_{11} \text{ molecules}} \times \frac{1 \ \cancel{\text{mol } C_{12}H_{22}O_{11}}}{6.022 \times 10^{23}\ \cancel{C_{12}H_{22}O_{11} \text{ molecules}}} \times \frac{342.3 \ \cancel{\text{g } C_{12}H_{22}O_{11}}}{1 \ \cancel{\text{mol } C_{12}H_{22}O_{11}}} \times \frac{1 \times 10^{3} \text{ mg } \cancel{C_{12}H_{22}O_{11}}}{1 \ \cancel{\text{g } C_{12}H_{22}O_{11}}}$$

$= 0.10 \text{ mg } C_{12}H_{22}O_{11}$

Check: Units of the answer, milligrams $C_{12}H_{22}O_{11}$, are correct. The magnitude is appropriate because there is much less than Avogadro's number of molecules so we have much less than 1 mole of $C_{12}H_{22}O_{11}$.

42. **Given:** 0.12 mg NaCl **Find:** number of formula units

Conceptual Plan: **mg NaCl → g NaCl → mole NaCl → number of formula units NaCl**

$$\frac{1 \text{ g NaCl}}{1 \times 10^{3} \text{ mg NaCl}} \quad \frac{1 \text{ mol NaCl}}{58.44 \text{ g NaCl}} \quad \frac{6.022 \times 10^{23} \text{ NaCl formula units}}{1 \text{ mol NaCl}}$$

Solution:

$$0.12 \ \cancel{\text{mg NaCl}} \times \frac{1 \ \cancel{\text{g NaCl}}}{1 \times 10^{3} \ \cancel{\text{mg NaCl}}} \times \frac{1 \ \cancel{\text{mol NaCl}}}{58.44 \ \cancel{\text{g NaCl}}} \times \frac{6.022 \times 10^{23} \text{ formula units NaCl}}{1 \ \cancel{\text{mol NaCl}}} = 1.2 \times 10^{18} \text{ formula units NaCl}$$

Check: Units of the answer, formula units NaCl, are correct. The magnitude is appropriate because there is less than 1 mole of NaCl so we have less than Avogadro's number of formula units.

43. a) **Given:** CH_4 **Find:** mass percent C

Conceptual Plan: $\text{mass } \%C = \frac{1 \text{ x molar mass C}}{\text{molar mass } CH_4} \text{ x } 100$

Solution:

$1 \text{ x molar mass C} = 1(12.01 \text{g/mol}) = 12.01 \text{ g C}$

$\text{molar mass } CH_4 = 1(12.01 \text{ g/mol}) + 4(1.008 \text{ g/mol}) = 16.04 \text{ g/mol}$

$$\text{mass } \% \text{ C} = \frac{1 \text{ x molar mass C}}{\text{molar mass } CH_4} \text{ x } 100\%$$

$$= \frac{12.01 \ \cancel{\text{g/mol}}}{16.04 \ \cancel{\text{g/mol}}} \text{ x } 100\%$$

$= 74.87 \%$

Check: Units of the answer, %, are correct. The magnitude is reasonable because it is between 0 and 100% and carbon is the heaviest element.

b) **Given:** C_2H_6 **Find:** mass percent C

Conceptual Plan: $\text{mass } \%C = \frac{2 \text{ x molar mass C}}{\text{molar mass } C_2H_6} \text{ x } 100$

Solution:

$2 \text{ x molar mass C} = 2(12.01 \text{g/mol}) = 24.02 \text{ g C}$

$\text{molar mass } C_2H_6 = 2(12.01 \text{ g/mol}) + 6(1.008 \text{ g/mol}) = 30.07 \text{ g/mol}$

$$\text{mass } \% \text{ C} = \frac{2 \text{ x molar mass C}}{\text{molar mass } C_2H_6} \text{ x } 100\%$$

$$= \frac{24.02 \ \cancel{\text{g/mol}}}{30.07 \ \cancel{\text{g/mol}}} \text{ x } 100\%$$

$= 79.89 \%$

Check: Units of the answer, %, are correct. The magnitude is reasonable because it is between 0 and 100% and carbon is the heaviest element.

c) **Given:** C_2H_2 **Find:** mass percent C

Conceptual Plan: $\text{mass } \%C = \frac{2 \text{ x molar mass C}}{\text{molar mass } C_2H_2} \text{ x } 100$

Solution:

$$2 \text{ x molar mass C} = 2(12.01\text{g/mol}) = 24.02 \text{ g C}$$

$$\text{molar mass } C_2H_2 = 2(12.01 \text{ g/mol}) + 2(1.008 \text{ g/mol}) = 26.04 \text{ g/mol}$$

$$\text{mass \% C} = \frac{2 \text{ x molar mass C}}{\text{molar mass } C_2H_2} \text{ x } 100\%$$

$$= \frac{24.02 \text{ g/mol}}{26.04 \text{ g/mol}} \text{ x } 100\%$$

$$= 92.26 \text{ \%}$$

Check: Units of the answer, %, are correct. The magnitude is reasonable because it is between 0 and 100% and carbon is the heaviest element.

d) **Given:** C_2H_5Cl **Find:** mass percent C

Conceptual Plan: $\text{mass \%C} = \frac{2 \text{ x molar mass C}}{\text{molar mass } C_2H_5Cl} \text{ x } 100$

Solution:

$$2 \text{ x molar mass C} = 2(12.01\text{g/mol}) = 24.02 \text{ g C}$$

$$\text{molar mass } C_2H_5Cl = 2(12.01 \text{ g/mol}) + 5(1.008 \text{ g/mol}) + 1(35.45 \text{ g/mol}) = 64.51 \text{ g/mol}$$

$$\text{mass \% C} = \frac{2 \text{ x molar mass C}}{\text{molar mass } C_2H_5Cl} \text{ x } 100\%$$

$$= \frac{24.02 \text{ g/mol}}{64.51 \text{ g/mol}} \text{ x } 100\%$$

$$= 37.23 \text{ \%}$$

Check: Units of the answer, %, are correct. The magnitude is reasonable because it is between 0 and 100% and chlorine is heavier than carbon.

44. a) **Given:** N_2O **Find:** mass percent N

Conceptual Plan: $\text{mass \%N} = \frac{2 \text{ x molar mass N}}{\text{molar mass } N_2O} \text{ x } 100$

Solution:

$$2 \text{ x molar mass N} = 2(14.01\text{g/mol}) = 28.02 \text{ g N}$$

$$\text{molar mass } N_2O = 2(14.01 \text{ g/mol}) + (16.00 \text{ g/mol}) = 44.02 \text{ g/mol}$$

$$\text{mass \% N} = \frac{2 \text{ x molar mass N}}{\text{molar mass } N_2O} \text{ x } 100\%$$

$$= \frac{28.02 \text{ g/mol}}{44.02 \text{ g/mol}} \text{ x } 100\%$$

$$= 63.65 \text{ \%}$$

Check: Units of the answer, %, are correct. The magnitude is reasonable because it is between 0 and 100% and there are 2 nitrogen.

b) **Given:** NO **Find:** mass percent N

Conceptual Plan: $\text{mass \%N} = \frac{1 \text{ x molar mass N}}{\text{molar mass NO}} \text{ x } 100$

Solution:

$$1 \text{ x molar mass N} = 1(14.01\text{g/mol}) = 14.01 \text{ g N}$$

$$\text{molar mass NO} = (14.01 \text{ g/mol}) + (16.00 \text{ g/mol}) = 30.01 \text{ g/mol}$$

$$\text{mass \% N} = \frac{1 \text{ x molar mass N}}{\text{molar mass NO}} \text{ x } 100\%$$

$$= \frac{14.01 \text{ g/mol}}{30.01 \text{ g/mol}} \text{ x } 100\%$$

$$= 46.68 \text{ \%}$$

Check: Units of the answer, %, are correct. The magnitude is reasonable because it is between 0 and 100% and the mass of nitrogen is less than the mass of oxygen.

c) **Given:** NO_2 **Find:** mass percent N

Conceptual Plan: $\text{mass \% N} = \frac{1 \text{ x molar mass N}}{\text{molar mass } NO_2} \text{ x } 100$

Solution:

$$1 \text{ x molar mass N} = 1(14.01 \text{g/mol}) = 14.01 \text{ g N}$$

$$\text{molar mass } NO_2 = (14.01 \text{ g/mol}) + 2(16.00 \text{ g/mol}) = 46.01 \text{ g/mol}$$

$$\text{mass \% N} = \frac{1 \text{ x molar mass N}}{\text{molar mass } NO_2} \text{ x } 100\%$$

$$= \frac{14.01 \cancel{\text{g/mol}}}{46.01 \cancel{\text{g/mol}}} \text{ x } 100\%$$

$$= 30.45 \%$$

Check: Units of the answer, %, are correct. The magnitude is reasonable because it is between 0 and 100% and the mass of nitrogen is less than the mass of oxygen and there are 2 oxygen.

d) **Given:** HNO_3 **Find:** mass percent N

Conceptual Plan: $\text{mass \% N} = \frac{1 \text{ x molar mass N}}{\text{molar mass } HNO_3} \text{ x } 100$

Solution:

$$1 \text{ x molar mass N} = 1(14.01 \text{g/mol}) = 14.01 \text{ g N}$$

$$\text{molar mass } HNO_3 = (1.008 \text{ g/mol}) + (14.01 \text{ g/mol}) + 3(16.00 \text{ g/mol}) = 63.02 \text{ g/mol}$$

$$\text{mass \% N} = \frac{1 \text{ x molar mass N}}{\text{molar mass } HNO_3} \text{ x } 100\%$$

$$= \frac{14.01 \cancel{\text{g/mol}}}{63.02 \cancel{\text{g/mol}}} \text{ x } 100\%$$

$$= 22.23 \%$$

Check: Units of the answer, %, are correct. The magnitude is reasonable because it is between 0 and 100% and the mass of nitrogen is less than the mass of oxygen and there are 3 oxygen.

45. **Given:** NH_3 **Find:** mass percent N

Conceptual Plan: $\text{mass \% N} = \frac{1 \text{ x molar mass N}}{\text{molar mass } NH_3} \text{ x } 100$

Solution:

$$1 \text{ x molar mass N} = 1(14.01 \text{g/mol}) = 14.01 \text{ g N}$$

$$\text{molar mass } NH_3 = 3(1.008 \text{ g/mol}) + (14.01 \text{ g/mol}) = 17.03 \text{ g/mol}$$

$$\text{mass \% N} = \frac{1 \text{ x molar mass N}}{\text{molar mass } NH_3} \text{ x } 100\%$$

$$= \frac{14.01 \cancel{\text{g/mol}}}{17.03 \cancel{\text{g/mol}}} \text{ x } 100\%$$

$$= 82.27 \%$$

Check: Units of the answer, %, are correct. The magnitude is reasonable because it is between 0 and 100% and nitrogen is the heaviest.

Given: $CO(NH_2)_2$ **Find:** mass percent N

Conceptual Plan: $\text{mass \% N} = \frac{2 \text{ x molar mass N}}{\text{molar mass } CO(NH_2)_2} \text{ x } 100$

Solution:

$2 \text{ x molar mass N} = 1(14.01\text{g/mol}) = 28.02 \text{ g N}$

$\text{molar mass CO(NH}_2)_2 = (12.01 \text{ g/mol}) + (16.00 \text{ g/mol}) + 2(14.01 \text{ g/mol}) + 4(1.008 \text{ g/mol}) = 60.06 \text{ g/mol}$

$$\text{mass \% N} = \frac{2 \text{ x molar mass N}}{\text{molar mass CO(NH}_2)_2} \text{ x } 100\%$$

$$= \frac{28.02 \cancel{\text{g/mol}}}{60.06 \cancel{\text{g/mol}}} \text{ x } 100\%$$

$$= 46.65 \%$$

Check: Units of the answer, %, are correct. The magnitude is reasonable because it is between 0 and 100% and there are two nitrogen and only one carbon and one oxygen.

Given: NH_4NO_3 **Find:** mass percent N

Conceptual Plan: $\text{mass \% N} = \frac{2 \text{ x molar mass N}}{\text{molar mass NH}_4\text{NO}_3} \text{ x } 100$

Solution:

$2 \text{ x molar mass N} = 2(14.01\text{g/mol}) = 28.02 \text{ g N}$

$\text{molar mass NH}_4\text{NO}_3 = 2 (14.01 \text{ g/mol}) + 4(1.008 \text{ g/mol}) + 3(16.00 \text{ g/mol}) = 80.05 \text{ g/mol}$

$$\text{mass \% N} = \frac{2 \text{ x molar mass N}}{\text{molar mass NH}_4\text{NO}_3} \text{ x } 100\%$$

$$= \frac{28.02 \cancel{\text{g/mol}}}{80.05 \cancel{\text{g/mol}}} \text{ x } 100\%$$

$$= 35.00 \%$$

Check: Units of the answer, %, are correct. The magnitude is reasonable because it is between 0 and 100% and the mass of nitrogen is less than the mass of oxygen and there are two nitrogen and three oxygen.

Given: $(NH_4)_2SO_4$ **Find:** mass percent N

Conceptual Plan: $\text{mass \% N} = \frac{2 \text{ x molar mass N}}{\text{molar mass (NH}_4)_2\text{SO}_4} \text{ x } 100$

Solution:

$2 \text{ x molar mass N} = 2(14.01\text{g/mol}) = 28.02 \text{ g N}$

$\text{molar mass (NH}_4)_2\text{SO}_4 = 2(14.01 \text{ g/mol}) + 8(1.008 \text{ g/mol}) + (32.07 \text{ g/mol}) + 4(16.00 \text{ g/mol}) = 132.15 \text{ g/mol}$

$$\text{mass \% N} = \frac{2 \text{ x molar mass N}}{\text{molar mass (NH}_4)_2\text{SO}_4} \text{ x } 100\%$$

$$= \frac{28.02 \cancel{\text{g/mol}}}{132.15 \cancel{\text{g/mol}}} \text{ x } 100\%$$

$$= 21.20 \%$$

Check: Units of the answer, %, are correct. The magnitude is reasonable because it is between 0 and 100% and the mass of nitrogen is less than the mass of oxygen and sulfur.

The fertilizer with the highest nitrogen content is NH_3 because it has the highest %Nat 82.27% N.

46. **Given:** Fe_2O_3 **Find:** mass percent Fe

Conceptual Plan: $\text{mass \% Fe} = \frac{2 \text{ x molar mass Fe}}{\text{molar mass Fe}_2\text{O}_3} \text{ x } 100$

Solution:

$$2 \text{ x molar mass Fe} = 2(55.85\text{g/mol}) = 111.7 \text{ g Fe}$$

$$\text{molar mass } Fe_2O_3 = 2(55.85 \text{ g/mol}) + 3(16.00 \text{ g/mol}) = 159.7 \text{ g/mol}$$

$$\text{mass \% Fe} = \frac{2 \text{ x molar mass Fe}}{\text{molar mass } Fe_2O_3} \text{ x } 100\%$$

$$= \frac{111.7 \cancel{\text{g/mol}}}{159.7 \cancel{\text{g/mol}}} \text{ x } 100\%$$

$$= 69.94 \%$$

Check: Units of the answer, %, are correct. The magnitude is reasonable because it is between 0 and 100% and iron is the heaviest.

Given: Fe_3O_4 **Find:** mass percent Fe

Conceptual Plan: $\text{mass \% Fe} = \frac{3 \text{ x molar mass Fe}}{\text{molar mass } Fe_3O_4} \text{ x } 100$

Solution:

$$3 \text{ x molar mass Fe} = 3(55.85\text{g/mol}) = 167.6 \text{ g Fe}$$

$$\text{molar mass } Fe_3O_4 = 3(55.85 \text{ g/mol}) + 4(16.00 \text{ g/mol}) = 231.6 \text{ g/mol}$$

$$\text{mass \% Fe} = \frac{3 \text{ x molar mass Fe}}{\text{molar mass } Fe_3O_4} \text{ x } 100\%$$

$$= \frac{167.6 \cancel{\text{g/mol}}}{231.6 \cancel{\text{g/mol}}} \text{ x } 100\%$$

$$= 72.37 \%$$

Check: Units of the answer, %, are correct. The magnitude is reasonable because it is between 0 and 100% and iron is the heaviest.

Given: $FeCO_3$ **Find:** mass percent Fe

Conceptual Plan: $\text{mass \% Fe} = \frac{1 \text{ x molar mass Fe}}{\text{molar mass } FeCO_3} \text{ x } 100$

Solution:

$$1 \text{ x molar mass Fe} = (55.85\text{g/mol}) = 55.85 \text{ g Fe}$$

$$\text{molar mass } FeCO_3 = 1(55.85 \text{ g/mol}) + 1(12.01 \text{ g/mol}) + 3(16.00 \text{ g/mol}) = 115.86 \text{ g/mol}$$

$$\text{mass \% Fe} = \frac{1 \text{ x molar mass Fe}}{\text{molar mass } FeCO_3} \text{ x } 100\%$$

$$= \frac{55.85 \cancel{\text{g/mol}}}{115.86 \cancel{\text{g/mol}}} \text{ x } 100\%$$

$$= 48.20 \%$$

Check: Units of the answer, %, are correct. The magnitude is reasonable because it is between 0 and 100% and iron is slightly less than the sum of carbon and oxygen.

The ore with the highest iron content is Fe_3O_4 because it has the highest % Fe, which is 72.37% Fe.

47. **Given:** 55.5 g CuF_2: 37.42 % F **Find:** g F in CuF_2

Conceptual Plan: **g CuF_2 → g F**

$$\frac{37.42 \text{ g F}}{100.0 \text{ g } CuF_2}$$

Solution: $55.5 \text{ g} \cancel{CuF_2} \text{ x } \frac{37.42 \text{ g F}}{100.0 \text{ g} \cancel{CuF_2}} = 20.77 = 20.8 \text{ g F}$

Check: Units of the answer, g F, are correct. The magnitude is reasonable because it is less than the original mass.

48. **Given:** 155 mg Ag; 75.27 % Ag in AgCl **Find:** mg AgCl
Conceptual Plan: **mg Ag → g Ag → g AgCl → mg AgCl**

$$\frac{1\ g\ Ag}{1000\ mg\ Ag} \quad \frac{100.0\ g\ AgCl}{75.27\ g\ Ag} \quad \frac{1000\ mg\ AgCl}{1\ g\ AgCl}$$

Solution: $155\ \cancel{mg\ Ag} \times \frac{1\ \cancel{g\ Ag}}{1000\ \cancel{mg\ Ag}} \times \frac{100.0\ \cancel{g\ AgCl}}{75.27\ \cancel{g\ Ag}} \times \frac{1000\ mg\ AgCl}{1\ \cancel{g\ AgCl}} = 206\ mg\ AgCl$

Check: Units of the answer, g AgCl, are correct. The magnitude is reasonable because it is greater than the original mass.

49. **Given:** 150 μg I; 76.45% I in KI **Find:** μg KI
Conceptual Plan: **μg I → g I → g KI → μg KI**

$$\frac{1\ g\ I}{1 \times 10^6\ \mu g\ I} \quad \frac{100.0\ g\ KI}{76.45\ g\ I} \quad \frac{1 \times 10^6\ \mu g\ KI}{1\ g\ KI}$$

Solution: $150\ \cancel{\mu g\ I} \times \frac{1\ \cancel{g\ I}}{1x10^6\ \cancel{\mu g\ I}} \times \frac{100.0\ \cancel{g\ KI}}{76.45\ \cancel{g\ I}} \times \frac{1 \times 10^6\ \mu g\ KI}{1\ \cancel{g\ KI}} = 196\ \mu g\ KI$

Check: Units of the answer, μg KI, are correct. The magnitude is reasonable because it is greater than the original mass.

50. **Given:** 3.0 mg F; 45.24 % F in NaF **Find:** mg NaF
Conceptual Plan: **mg F → g F → g NaF → mg NaF**

$$\frac{1\ g\ F}{1000\ mg\ F} \quad \frac{100.0\ g\ NaF}{45.24\ g\ F} \quad \frac{1000\ mg\ NaF}{1\ g\ NaF}$$

Solution: $3.0\ \cancel{mg\ F} \times \frac{1\ \cancel{g\ F}}{1000\ \cancel{mg\ F}} \times \frac{100.0\ \cancel{g\ NaF}}{45.24\ \cancel{g\ NaF}} \times \frac{1000\ mg\ NaF}{1\ \cancel{g\ NaF}} = 6.6\ mg\ NaF$

Check: Units of the answer, mg NaF, are correct. The magnitude is reasonable because it is greater than the original mass.

51. a) red – oxygen, white – hydrogen: 2H:O H_2O
b) black – carbon, white – hydrogen: 4H:C CH_4
c) black – carbon, white – hydrogen, red – oxygen: 2C:O:6H CH_3CH_2OH or C_2H_6O

52. a) black – carbon, red – oxygen: 2O:C CO_2
b) red – oxygen, white – hydrogen : 2H:2O H_2O_2
c) red – oxygen, white – hydrogen: 2H:O H_2O

53. a) **Given:** 0.0885 mol C_4H_{10} **Find:** mol H atoms
Conceptual Plan: **mol C_4H_{10} → mole H atom**

$$\frac{10\ mol\ H}{1\ mol\ C_4H_{10}}$$

Solution: $0.0885\ \cancel{mol\ C_4H_{10}} \times \frac{10\ mol\ H}{1\ \cancel{mol\ C_4H_{10}}} = 0.885\ mol\ H\ atoms$

Check: Units of the answer, mol H atoms, are correct. The magnitude is reasonable because it is greater than the original mol C_4H_{10}.

b) **Given:** 1.3 mol CH_4 **Find:** mol H atoms
Conceptual Plan: **mol CH_4 → mole H atom**

$$\frac{4\ mol\ H}{1\ mol\ CH_4}$$

Solution: $1.3\ \cancel{mol\ CH_4} \times \frac{4\ mol\ H}{1\ \cancel{mol\ CH_4}} = 5.2\ mol\ H\ atoms$

Check: Units of the answer, mol H atoms, are correct. The magnitude is reasonable because it is greater than the original mol CH_4.

c) **Given:** 2.4 mol C_6H_{12} **Find:** mol H atoms
Conceptual Plan: **mol C_6H_{12} → mole H atom**

$$\frac{12 \text{ mol H}}{1 \text{ mol } C_6H_{12}}$$

Solution: $2.4 \cancel{\text{mol } C_6H_{12}} \times \frac{12 \text{ mol H}}{1 \cancel{\text{mol } C_6H_{12}}} = 29 \text{ mol H atoms}$

Check: Units of the answer, mol H atoms, are correct. The magnitude is reasonable because it is greater than the original mol C_6H_{12}.

d) **Given:** 1.87 mol C_8H_{18} **Find:** mol H atoms
Conceptual Plan: **mol C_8H_{18} → mole H atom**

$$\frac{18 \text{ mol H}}{1 \text{ mol } C_8H_{18}}$$

Solution: $1.87 \cancel{\text{mol } C_8H_{18}} \times \frac{18 \text{ mol H}}{1 \cancel{\text{mol } C_8H_{18}}} = 33.7 \text{ mol H atoms}$

Check: Units of the answer, mol H atoms, are correct. The magnitude is reasonable because it is greater than the original mol C_8H_{18}.

54. a) **Given:** 4.88 mol H_2O_2 **Find:** mol O atoms
Conceptual Plan: **mol H_2O_2 → mole O atom**

$$\frac{2 \text{ mol O}}{1 \text{ mol } H_2O_2}$$

Solution: $4.88 \cancel{\text{mol } H_2O_2} \times \frac{2 \text{ mol O}}{1 \cancel{\text{mol } H_2O_2}} = 9.8 \text{ mol O atoms}$

Check: Units of the answer, mol O atoms, are correct. The magnitude is reasonable because it is greater than the original mol H_2O_2.

b) **Given:** 2.15 mol N_2O **Find:** mol O atoms
Conceptual Plan: **mol N_2O → mole O atom**

$$\frac{1 \text{ mol O}}{1 \text{ mol } N_2O}$$

Solution: $2.15 \cancel{\text{mol } N_2O} \times \frac{1 \text{ mol O}}{1 \cancel{\text{mol } N_2O}} = 2.15 \text{ mol O atoms}$

Check: Units of the answer, mol O atoms, are correct. The magnitude is reasonable because it is the same as the original mol N_2O.

c) **Given:** 0.0237 mol H_2CO_3 **Find:** mol O atoms
Conceptual Plan: **mol H_2CO_3 → mole O atom**

$$\frac{3 \text{ mol O}}{1 \text{ mol } H_2CO_3}$$

Solution: $0.0237 \cancel{\text{mol } H_2CO_3} \times \frac{3 \text{ mol O}}{1 \cancel{\text{mol } H_2CO_3}} = 0.711 \text{ mol O atoms}$

Check: Units of the answer, mol O atoms, are correct. The magnitude is reasonable because it is greater than the original mol H_2CO_3.

d) **Given:** 24.1 mol CO_2 **Find:** mol O atoms
Conceptual Plan: **mol CO_2 → mole O atom**

$$\frac{2 \text{ mol O}}{1 \text{ mol } CO_2}$$

Solution: $24.1 \cancel{\text{mol } CO_2} \times \frac{2 \text{ mol O}}{1 \cancel{\text{mol } CO_2}} = 48.2 \text{ mol O atoms}$

Check: Units of the answer, mol O atoms, are correct. The magnitude is reasonable because it is greater than the original mol CO_2.

55. a) **Given:** 8.5 g NaCl **Find:** g Na
Conceptual Plan: **g NaCl → mole NaCl → mol Na → g Na**

$$\frac{1 \text{ mol NaCl}}{58.44 \text{ g NaCl}} \quad \frac{1 \text{ mol Na}}{1 \text{ mol NaCl}} \quad \frac{22.99 \text{ g Na}}{1 \text{ mol Na}}$$

Solution: $8.5 \text{ g NaCl} \times \frac{1 \text{ mol NaCl}}{58.44 \text{ g NaCl}} \times \frac{1 \text{ mol Na}}{1 \text{ mol NaCl}} \times \frac{22.99 \text{ g Na}}{1 \text{ mol Na}} = 3.3 \text{ g Na}$

Check: Units of the answer, g Na, are correct. The magnitude is reasonable because it is less than the original g NaCl.

b) **Given:** 8.5 g Na_3PO_4 **Find:** g Na
Conceptual Plan: **g Na_3PO_4 → mole Na_3PO_4 → mol Na → g Na**

$$\frac{1 \text{ mol } Na_3PO_4}{163.94 \text{ g } Na_3PO_4} \quad \frac{3 \text{ mol Na}}{1 \text{ mol } Na_3PO_4} \quad \frac{22.99 \text{ g Na}}{1 \text{ mol Na}}$$

Solution: $8.5 \text{ g } Na_3PO_4 \times \frac{1 \text{ mol } Na_3PO_4}{163.94 \text{ g } Na_3PO_4} \times \frac{3 \text{ mol Na}}{1 \text{ mol } Na_3PO_4} \times \frac{22.99 \text{ g Na}}{1 \text{ mol Na}} = 3.6 \text{ g Na}$

Check: Units of the answer, g Na, are correct. The magnitude is reasonable because it is less than the original g Na_3PO_4.

c) **Given:** 8.5 g $NaC_7H_5O_2$ **Find:** g Na
Conceptual Plan: **g $NaC_7H_5O_2$ → mole $NaC_7H_5O_2$ → mol Na → g Na**

$$\frac{1 \text{ mol } NaC_7H_5O_2}{144.10 \text{ g } NaC_7H_5O_2} \quad \frac{1 \text{ mol Na}}{1 \text{ mol } NaC_7H_5O_2} \quad \frac{22.99 \text{ g Na}}{1 \text{ mol Na}}$$

Solution: $8.5 \text{ g } NaC_7H_5O_2 \times \frac{1 \text{ mol } NaC_7H_5O_2}{144.10 \text{ g } NaC_7H_5O_2} \times \frac{1 \text{ mol Na}}{1 \text{ mol } NaC_7H_5O_2} \times \frac{22.99 \text{ g Na}}{1 \text{ mol Na}} = 1.4 \text{ g Na}$

Check: Units of the answer, g Na, are correct. The magnitude is reasonable because it is less than the original g $NaC_7H_5O_2$.

d) **Given:** 8.5 g $Na_2C_6H_6O_7$ **Find:** g Na
Conceptual Plan: **g $Na_2C_6H_6O_7$ → mole $Na_2C_6H_6O_7$ → mol Na → g Na**

$$\frac{1 \text{ mol } Na_2C_6H_6O_7}{236.1 \text{ g } Na_2C_6H_6O_7} \quad \frac{2 \text{ mol Na}}{1 \text{ mol } Na_2C_6H_6O_7} \quad \frac{22.99 \text{ g Na}}{1 \text{ mol Na}}$$

Solution:

$$8.5 \text{ g } Na_2C_6H_6O_7 \times \frac{1 \text{ mol } Na_2C_6H_6O_7}{236.1 \text{ g } Na_2C_6H_6O_7} \times \frac{2 \text{ mol Na}}{1 \text{ mol } Na_2C_6H_6O_7} \times \frac{22.99 \text{ g Na}}{1 \text{ mol Na}} = 1.7 \text{ g } Na_2C_6H_6O_7$$

Check: Units of the answer, g Na, are correct. The magnitude is reasonable because it is less than the original g $Na_2C_6H_6O_7$.

56. a) **Given:** 25 kg CF_2Cl_2 **Find:** kg Cl
Conceptual Plan: **kg CF_2Cl_2 → g CF_2Cl_2 → mole CF_2Cl_2 → mol Cl → g Cl → kg Cl**

$$\frac{1000 \text{ g } CF_2Cl_2}{1 \text{ kg } CF_2Cl_2} \quad \frac{1 \text{ mol } CF_2Cl_2}{120.91 \text{ g } CF_2Cl_2} \quad \frac{2 \text{ mol Cl}}{1 \text{ mol } CF_2Cl_2} \quad \frac{35.45 \text{ g Cl}}{1 \text{ mol Cl}} \quad \frac{1 \text{ kg Cl}}{1000 \text{ g Cl}}$$

Solution:

$$25 \text{ kg } CF_2Cl_2 \times \frac{1000 \text{ g } CF_2Cl_2}{1 \text{ kg } CF_2Cl_2} \times \frac{1 \text{ mol } CF_2Cl_2}{120.91 \text{ g } CF_2Cl_2} \times \frac{2 \text{ mol Cl}}{1 \text{ mol } CF_2Cl_2} \times \frac{35.45 \text{ g Cl}}{1 \text{ mol Cl}} \times \frac{1 \text{ kg Cl}}{1000 \text{ g Cl}}$$

$= 15 \text{ kg Cl}$

Check: Units of the answer, kg Cl, are correct. The magnitude is reasonable because it is less than the original kg CF_2Cl_2.

b) **Given:** 25 kg $CFCl_3$ **Find:** kg Cl
Conceptual Plan: **kg $CFCl_3$ → g $CFCl_3$ → mole $CFCl_3$ → mol Cl → g Cl → kg Cl**

$$\frac{1000 \text{ g } CFCl_3}{1 \text{ kg } CFCl_3} \quad \frac{1 \text{ mol } CFCl_3}{137.4 \text{ g } CFCl_3} \quad \frac{3 \text{ mol Cl}}{1 \text{ mol } CFCl_3} \quad \frac{35.45 \text{ g Cl}}{1 \text{ mol Cl}} \quad \frac{1 \text{ kg Cl}}{1000 \text{ g Cl}}$$

Solution:

$$25 \text{ kg } CFCl_3 \times \frac{1000 \text{ g } CFCl_3}{1 \text{ kg } CFCl_3} \times \frac{1 \text{ mol } CFCl_3}{137.4 \text{ g } CFCl_3} \times \frac{3 \text{ mol Cl}}{1 \text{ mol } CFCl_3} \times \frac{35.45 \text{ g Cl}}{1 \text{ mol Cl}} \times \frac{1 \text{ kg Cl}}{1000 \text{ g Cl}} = 19 \text{ kg Cl}$$

Check: Units of the answer, kg Cl, are correct. The magnitude is reasonable because it is less than the original kg CF_2Cl_2.

c) **Given:** 25 kg $C_2F_3Cl_3$ **Find:** kg Cl
Conceptual Plan: **kg $C_2F_3Cl_3$ → g $C_2F_3Cl_3$ → mole $C_2F_3Cl_3$ → mol Cl → g Cl → kg Cl**

$$\frac{1000 \text{ g } C_2F_3Cl_3}{1 \text{ kg } C_2F_3Cl_3} \quad \frac{1 \text{ mol } C_2F_3Cl_3}{187.4 \text{ g } C_2F_3Cl_3} \quad \frac{3 \text{ mol Cl}}{1 \text{ mol } C_2F_3Cl_3} \quad \frac{35.45 \text{ g Cl}}{1 \text{ mol Cl}} \quad \frac{1 \text{ kg Cl}}{1000 \text{ g Cl}}$$

Solution:

$$25 \text{ kg } C_2F_3Cl_3 \times \frac{1000 \text{ g } C_2F_3Cl_3}{1 \text{ kg } C_2F_3Cl_3} \times \frac{1 \text{ mol } C_2F_3Cl_3}{187.4 \text{ g } C_2F_3Cl_3} \times \frac{3 \text{ mol Cl}}{1 \text{ mol } C_2F_3Cl_3} \times \frac{35.45 \text{ g Cl}}{1 \text{ mol Cl}} \times \frac{1 \text{ kg Cl}}{1000 \text{ g Cl}}$$

= 14 kg Cl

Check: Units of the answer, kg Cl, are correct. The magnitude is reasonable because it is less than the original kg $C_2F_3Cl_3$.

d) **Given:** 25 kg CF_3Cl **Find:** kg Cl
Conceptual Plan: **kg CF_3Cl → g CF_3Cl → mole CF_3Cl → mol Cl → g Cl → kg Cl**

$$\frac{1000 \text{ g } CF_3Cl}{1 \text{ kg } CF_3Cl} \quad \frac{1 \text{ mol } CF_3Cl}{104.46 \text{ g } CF_3Cl} \quad \frac{1 \text{ mol Cl}}{1 \text{ mol } CF_3Cl} \quad \frac{35.45 \text{ g Cl}}{1 \text{ mol Cl}} \quad \frac{1 \text{ kg Cl}}{1000 \text{ g Cl}}$$

Solution:

$$25 \text{ kg } CF_3Cl \times \frac{1000 \text{ g } CF_3Cl}{1 \text{ kg } CF_3Cl} \times \frac{1 \text{ mol } CF_3Cl}{104.46 \text{ g } CF_3Cl} \times \frac{1 \text{ mol Cl}}{1 \text{ mol } CF_3Cl} \times \frac{35.45 \text{ g Cl}}{1 \text{ mol Cl}} \times \frac{1 \text{ kg Cl}}{1000 \text{ g Cl}} = 8.5 \text{ kg Cl}$$

Check: Units of the answer, kg Cl, are correct. The magnitude is reasonable because it is less than the original kg CF_3Cl.

57. a) **Given:** 1.651 g Ag; 0.1224 g O **Find:** empirical formula
Conceptual Plan:
convert mass to mol of each element → write pseudoformula → write empirical formula

$$\frac{1 \text{ mol Ag}}{107.9 \text{ g Ag}} \quad \frac{1 \text{ mol O}}{16.00 \text{ g O}}$$

divide by smallest number

Solution: $1.651 \text{ g Ag} \times \frac{1 \text{ mol Ag}}{107.9 \text{ g Ag}} = 0.01530 \text{ mol Ag}$

$$0.1224 \text{ g O} \times \frac{1 \text{ mol O}}{16.00 \text{ g O}} = 0.007650 \text{ mol O}$$

$$Ag_{0.01530}O_{0.007650}$$

$$Ag_{\frac{0.01530}{0.007650}}O_{\frac{0.007650}{0.007650}} \rightarrow Ag_2O$$

The correct empirical formula is Ag_2O

b) **Given:** 0.672 g Co; 0.569 g As; 0.486 g O **Find:** empirical formula
Conceptual Plan:
convert mass to mol of each element → write pseudoformula → write empirical formula

$$\frac{1 \text{ mol Co}}{58.93 \text{ g Co}} \quad \frac{1 \text{ mol As}}{74.92 \text{ g As}} \quad \frac{1 \text{ mol O}}{16.00 \text{ g O}}$$

divide by smallest number

Solution: $0.672 \text{ g Co} \times \frac{1 \text{ mol Co}}{58.93 \text{ g Co}} = 0.0114 \text{ mol Co}$

$$0.569 \text{ g As} \times \frac{1 \text{ mol As}}{74.92 \text{ g As}} = 0.00759 \text{ mol O}$$

$$0.486 \text{ g O} \times \frac{1 \text{ mol O}}{16.00 \text{ g O}} = 0.0304 \text{ mol O}$$

$Co_{0.0114}As_{0.00759}\ O_{0.0304}$

$Co_{\frac{0.0114}{0.00759}}As_{\frac{0.00759}{0.00759}}O_{\frac{0.0304}{0.00759}} \rightarrow Co_{1.5}As_1O_4$

$Co_{1.5}As_1O_4 \times 2 \rightarrow Co_3As_2O_8$

The correct empirical formula is $Co_3As_2O_8$

c) **Given:** 1.443 g Se; 5.841 g Br **Find:** empirical formula

Conceptual Plan:

convert mass to mol of each element → write pseudoformula → write empirical formula

$\frac{1\ \text{mol Se}}{78.96\ \text{g Se}}\quad \frac{1\ \text{mol Br}}{79.90\ \text{g Br}}$ divide by smallest number

Solution: $1.443\ \cancel{\text{g Se}} \times \frac{1\ \text{mol Se}}{78.96\ \cancel{\text{g Se}}} = 0.01828\ \text{mol Se}$

$5.841\ \cancel{\text{g Br}} \times \frac{1\ \text{mol Br}}{79.90\ \cancel{\text{g Br}}} = 0.07310\ \text{mol Br}$

$Se_{0.01828}\ Br_{0.07310}$

$Se_{\frac{0.01828}{0.01828}}Br_{\frac{0.07310}{0.01828}} \rightarrow SeBr_4$

The correct empirical formula is $SeBr_4$

58. a) **Given:** 1.245 g Ni; 5.381 g I **Find:** empirical formula

Conceptual Plan:

convert mass to mol of each element → write pseudoformula → write empirical formula

$\frac{1\ \text{mol Ni}}{58.69\ \text{g Ni}}\quad \frac{1\ \text{mol I}}{126.9\ \text{g I}}$ divide by smallest number

Solution: $1.245\ \cancel{\text{g Ni}} \times \frac{1\ \text{mol Ni}}{58.69\ \cancel{\text{g Ni}}} = 0.02121\ \text{mol Ni}$

$5.381\ \cancel{\text{g I}} \times \frac{1\ \text{mol I}}{126.9\ \cancel{\text{g I}}} = 0.04240\ \text{mol I}$

$Ni_{0.02121}\ I_{0.04240}$

$Ni_{\frac{0.02121}{0.02121}}I_{\frac{0.04240}{0.02121}} \rightarrow NiI_2$

The correct empirical formula is NiI_2

b) **Given:** 2.677 g Ba; 3.115 g Br **Find:** empirical formula

Conceptual Plan:

convert mass to mol of each element → write pseudoformula → write empirical formula

$\frac{1\ \text{mol Ba}}{137.3\ \text{g Ba}}\quad \frac{1\ \text{mol Br}}{79.90\ \text{g Br}}$ divide by smallest number

Solution: $2.677\ \cancel{\text{g Ba}} \times \frac{1\ \text{mol Ba}}{137.3\ \cancel{\text{g Ba}}} = 0.01950\ \text{mol Ba}$

$3.115\ \cancel{\text{g Br}} \times \frac{1\ \text{mol Br}}{79.90\ \cancel{\text{g Br}}} = 0.03899\ \text{mol Br}$

$Ba_{0.01950}\ Br_{0.03899}$

$Ba_{\frac{0.01942}{0.01950}}Br_{\frac{0.03899}{0.01950}} \rightarrow BaBr_2$

The correct empirical formula is $BaBr_2$

c) **Given:** 2.128 g Be; 7.557 g S; 15.107 g O **Find:** empirical formula

Conceptual Plan:

convert mass to mol of each element → write pseudoformula → write empirical formula

$\frac{1\ \text{mol Be}}{9.012\ \text{g Be}}\quad \frac{1\ \text{mol S}}{32.07\ \text{g S}}\quad \frac{1\ \text{mol O}}{16.00\ \text{g O}}$ divide by smallest number

Solution: $2.128\ \cancel{g\,Be} \times \frac{1\ mol\ Be}{9.012\ \cancel{g\,Be}} = 0.2361\ mol\ Be$

$7.557\ \cancel{g\,S} \times \frac{1\ mol\ S}{32.07\ \cancel{g\,S}} = 0.2356\ mol\ S$

$15.107\ \cancel{g\,O} \times \frac{1\ mol\ O}{16.00\ \cancel{g\,O}} = 0.9442\ mol\ O$

$Be_{0.2361}\ S_{0.2356}O_{0.9442}$

$Be_{\frac{0.2361}{0.2356}}S_{\frac{0.2356}{0.2356}}O_{\frac{0.9442}{0.2356}} \rightarrow BeSO_4$

The correct empirical formula is $BeSO_4$

59. a) **Given:** In a 100 g sample: 74.03 g C, 8.70 g H, 17.27 g N **Find:** empirical formula

Conceptual Plan:

convert mass to mol of each element → write pseudoformula → write empirical formula

$\frac{1\ mol\ C}{12.01\ g\ C}$ $\frac{1\ mol\ H}{1.008\ g\ H}$ $\frac{1\ mol\ N}{14.01\ g\ N}$ divide by smallest number

Solution: $74.03\ \cancel{g\,C} \times \frac{1\ mol\ C}{12.01\ \cancel{g\,C}} = 6.164\ mol\ C$

$8.70\ \cancel{g\,H} \times \frac{1\ mol\ H}{1.008\ \cancel{g\,H}} = 8.63\ mol\ H$

$17.27\ \cancel{g\,N} \times \frac{1\ mol\ N}{14.01\ \cancel{g\,N}} = 1.233\ mol\ N$

$C_{6.164}H_{8.63}N_{1.233}$

$C_{\frac{6.164}{1.233}}H_{\frac{8.63}{1.233}}N_{\frac{1.233}{1.233}} \rightarrow C_5H_7N$

The correct empirical formula is C_5H_7N

b) **Given:** In a 100 g sample: 49.48 g C, 5.19 g H, 28.85 g N, 16.48 g O **Find:** empirical formula

Conceptual Plan:

convert mass to mol of each element → write pseudoformula → write empirical formula

$\frac{1\ mol\ C}{12.01\ g\ C}$ $\frac{1\ mol\ H}{1.008\ g\ H}$ $\frac{1\ mol\ N}{14.01\ g\ N}$ $\frac{1\ mol\ O}{16.00\ g\ O}$ divide by smallest number

Solution: $49.48\ \cancel{g\,C} \times \frac{1\ mol\ C}{12.01\ \cancel{g\,C}} = 4.120\ mol\ C$

$5.19\ \cancel{g\,H} \times \frac{1\ mol\ H}{1.008\ \cancel{g\,H}} = 5.15\ mol\ H$

$28.85\ \cancel{g\,N} \times \frac{1\ mol\ N}{14.01\ \cancel{g\,N}} = 2.059\ mol\ N$

$16.48\ \cancel{g\,O} \times \frac{1\ mol\ O}{16.00\ \cancel{g\,O}} = 1.030\ mol\ O$

$C_{4.120}H_{5.15}N_{2.059}O_{1.030}$

$C_{\frac{4.120}{1.030}}H_{\frac{5.15}{1.030}}N_{\frac{2.059}{1.030}}O_{\frac{1.030}{1.030}} \rightarrow C_4H_5N_2O$

The correct empirical formula is $C_4H_5N_2O$

60. a) **Given:** In a 100 g sample: 58.80g C, 9.87 g H, 31.33 g O **Find:** empirical formula

Conceptual Plan:

convert mass to mol of each element → write pseudoformula → write empirical formula

$\frac{1\ mol\ C}{12.01\ g\ C}$ $\frac{1\ mol\ H}{1.008\ g\ H}$ $\frac{1\ mol\ O}{16.00\ g\ O}$ divide by smallest number

Solution: $58.80\ \cancel{g\,C} \times \frac{1\ mol\ C}{12.01\ \cancel{g\,C}} = 4.896\ mol\ C$

$$9.87 \text{ g H} \times \frac{1 \text{ mol H}}{1.008 \text{ g H}} = 9.79 \text{ mol H}$$

$$31.33 \text{ g O} \times \frac{1 \text{ mol O}}{16.00 \text{ g O}} = 1.958 \text{ mol O}$$

$$C_{4.896}H_{9.79}O_{1.958}$$

$$C_{\frac{4.896}{1.958}}H_{\frac{9.79}{1.958}}O_{\frac{1.958}{1.958}} \rightarrow C_{2.5}H_5O$$

$$C_{2.5}H_5O \times 2 = C_5H_{10}O_2$$

The correct empirical formula is $C_5H_{10}O_2$

b) **Given:** In a 100 g sample: 63.15 g C, 5.30 g H, 31.55 g O **Find:** empirical formula
Conceptual Plan:
convert mass to mol of each element → write pseudoformula → write empirical formula

$\frac{1 \text{ mol C}}{12.01 \text{ g C}}$ $\frac{1 \text{ mol H}}{1.008 \text{ g H}}$ $\frac{1 \text{ mol O}}{16.00 \text{ g O}}$ divide by smallest number

Solution: $$63.15 \text{ g C} \times \frac{1 \text{ mol C}}{12.01 \text{ g C}} = 5.258 \text{ mol C}$$

$$5.30 \text{ g H} \times \frac{1 \text{ mol H}}{1.008 \text{ g H}} = 5.26 \text{ mol H}$$

$$31.35 \text{ g O} \times \frac{1 \text{ mol O}}{16.00 \text{ g O}} = 1.972 \text{ mol O}$$

$$C_{5.258}H_{5.26}O_{1.972}$$

$$C_{\frac{5.258}{1.972}}H_{\frac{5.26}{1.972}}O_{\frac{1.959}{1.972}} \rightarrow C_{2.67}H_{2.67}O$$

$$C_{2.67}H_{2.67}O \times 3 = C_8H_8O_3$$

The correct empirical formula is $C_8H_8O_3$

61. **Given:** 0.77 mg N, 6.61 mg N_xCl_y **Find:** empirical formula
Conceptual Plan:
Find mg Cl → convert mg to g for each element → convert mass to mol of each element →

mg N_xCl_y - mg N $\frac{1 \text{ g}}{1000 \text{ mg}}$ $\frac{1 \text{ mol N}}{14.01 \text{ g N}}$ $\frac{1 \text{ mol Cl}}{35.45 \text{ g Cl}}$

write pseudoformula → write empirical formula
divide by smallest number

Solution: 6.61 mg N_xCl_y – 0.77 mg N = 5.84 mg Cl

$$0.77 \text{ mg N} \times \frac{1 \text{ g N}}{1000 \text{ mg N}} \times \frac{1 \text{ mol N}}{14.01 \text{ g N}} = 5.5 \times 10^{-5} \text{ mol N}$$

$$5.84 \text{ mg Cl} \times \frac{1 \text{ g Cl}}{1000 \text{ mg Cl}} \times \frac{1 \text{ mol Cl}}{35.45 \text{ g Cl}} = 1.6 \times 10^{-4} \text{ mol Cl}$$

$$N_{5.5 \times 10^{-5}}Cl_{1.6 \times 10^{-4}}$$

$$N_{\frac{5.5x10^{-5}}{5.5x10^{-5}}}Cl_{\frac{1.6 \times 10^{-4}}{5.5x10^{-5}}} \rightarrow NCl_3$$

The correct empirical formula is NCl_3

62. **Given:** 45.2 mg P, 131.6 mg P_xSe_y **Find:** empirical formula
Conceptual Plan: Find mg Se → convert mg to g for each element → convert mass to mol of each element

mg P_xSe_y - mg P $\frac{1 \text{ g}}{1000 \text{ mg}}$ $\frac{1 \text{ mol P}}{30.97 \text{ g P}}$ $\frac{1 \text{ mol Se}}{78.96 \text{ g Se}}$

→write pseudoformula → write empirical formula
divide by smallest number

Solution: 131.6 mg P_xSe_y – 45.2 mg P = 86.4 mg Se

$$45.2 \cancel{\text{mg P}} \times \frac{1 \cancel{\text{g P}}}{1000 \cancel{\text{mg P}}} \times \frac{1 \text{ mol P}}{30.97 \cancel{\text{g P}}} = 0.00146 \text{ mol P}$$

$$86.4 \cancel{\text{g Se}} \times \frac{1 \cancel{\text{g Se}}}{1000 \cancel{\text{mg Se}}} \times \frac{1 \text{ mol Se}}{78.96 \cancel{\text{g Se}}} = 0.00109 \text{ mol Se}$$

$P_{0.00146}\ Se_{0.00109}$

$P_{\frac{0.00146}{0.00109}}Se_{\frac{0.00109}{0.00109}} \rightarrow P_{1.33}Se$

$P_{1.33}Se \times 3 = P_4Se_3$

The correct empirical formula is P_4Se_3

63. a) **Given:** empirical formula = C_6H_7N, molar mass = 186.24 g/mol **Find:** molecular formula

Conceptual Plan: molecular formula = empirical formula x n $n = \frac{\text{molar mass}}{\text{empirical formula mass}}$

Solution: empirical formula mass = 6(12.01 g/mol) + 7(1.008 g/mol) + 1(14.01 g/mol) = 93.13 g/mol

$$n = \frac{\text{molar mass}}{\text{formula molar mass}} = \frac{186.24 \text{ g/mol}}{93.13 \text{ g/mol}} = 1.998 = 2$$

molecular formula $= C_6H_7N \times 2$
$= C_{12}H_{14}N_2$

b) **Given:** empirical formula = C_2HCl, molar mass = 181.44 g/mol **Find:** molecular formula

Conceptual Plan: molecular formula = empirical formula x n $n = \frac{\text{molar mass}}{\text{empirical formula mass}}$

Solution: empirical formula mass = 2(12.01 g/mol) + 1(1.008 g/mol) + 1(35.45 g/mol) = 60.48 g/mol

$$n = \frac{\text{molar mass}}{\text{formula molar mass}} = \frac{181.44 \text{ g/mol}}{60.48 \text{ g/mol}} = 3$$

molecular formula $= C_2HCl \times 3$
$= C_6H_3Cl_3$

c) **Given:** empirical formula = $C_5H_{10}NS_2$, molar mass = 296.54 g/mol **Find:** molecular formula

Conceptual Plan: molecular formula = empirical formula x n $n = \frac{\text{molar mass}}{\text{empirical formula mass}}$

Solution: empirical formula mass = 5(12.01 g/mol) + 10(1.008 g/mol) + 1(14.01 g/mol) + 2(32.07) = 148.28 g/mol

$$n = \frac{\text{molar mass}}{\text{formula molar mass}} = \frac{296.54 \text{ g/mol}}{148.28 \text{ g/mol}} = 2$$

molecular formula $= C_5H_{10}NS_2 \times 2$
$= C_{10}H_{20}N_2S_4$

64. a) **Given:** empirical formula = C_4H_9, molar mass = 114.22 g/mol**Find:** molecular formula

Conceptual Plan: molecular formula = empirical formula x n $n = \frac{\text{molar mass}}{\text{empirical formula mass}}$

Solution: empirical formula mass = 4(12.01 g/mol) + 9(1.008 g/mol) = 57.11 g/mol

$$n = \frac{\text{molar mass}}{\text{formula molar mass}} = \frac{114.22 \text{ g/mol}}{57.11 \text{ g/mol}} = 2$$

molecular formula $= C_4H_9 \times 2$
$= C_8H_{18}$

b) **Given:** empirical formula = CCl, molar mass = 284.77 g/mol **Find:** molecular formula

Conceptual Plan: molecular formula = empirical formula x n $n = \frac{\text{molar mass}}{\text{empirical formula mass}}$

Solution: empirical formula mass = 1(12.01 g/mol) + 1(35.45 g/mol) = 47.46 g/mol

$$n = \frac{\text{molar mass}}{\text{formula molar mass}} = \frac{284.77 \text{ g/mol}}{47.46 \text{ g/mol}} = 6$$

molecular formula $= CCl \times 6$
$= C_6Cl_6$

c) **Given:** empirical formula = C_3H_2N, molar mass = 312.29 g/mol **Find:** molecular formula

Conceptual Plan: molecular formula = empirical formula x n $n = \frac{\text{molar mass}}{\text{empirical formula mass}}$

Solution: empirical formula mass = 3(12.01 g/mol) + 2(1.008 g/mol) + 1(14.01 g/mol) = 52.06 g/mol

$$n = \frac{\text{molar mass}}{\text{formula molar mass}} = \frac{312.29 \text{ g/mol}}{52.06 \text{ g/mol}} = 6$$

molecular formula $= C_3H_2N \times 6$
$= C_{18}H_{12}N_6$

65. **Given:** 33.01 g CO_2, 13.51 g H_2O **Find:** empirical formula

Conceptual Plan:

mass CO_2, H_2O → mol CO_2, H_2O → mol C, mol H → pseudoformula → empirical formula

$\frac{1 \text{ mol } CO_2}{44.01 \text{ g } CO_2}$ $\frac{1 \text{ mol } H_2O}{18.02 \text{ g } H_2O}$ $\frac{1 \text{ mol C}}{1 \text{ mol } CO_2}$ $\frac{2 \text{ mol H}}{1 \text{ mol } H_2O}$ divide by smallest number

Solution:

$$33.01 \text{ g } CO_2 \times \frac{1 \text{ mol } CO_2}{44.01 \text{ g } CO_2} = 0.7500 \text{ mol } CO_2$$

$$13.51 \text{ g } H_2O \times \frac{1 \text{ mol } H_2O}{18.02 \text{ g } H_2O} = 0.7497 \text{ mol } H_2O$$

$$0.7500 \text{ mol } CO_2 \times \frac{1 \text{ mol C}}{1 \text{ mol } CO_2} = 0.7500 \text{ mol C}$$

$$0.7497 \text{ mol } H_2O \times \frac{2 \text{ mol H}}{1 \text{ mol } H_2O} = 1.499 \text{ mol H}$$

$C_{0.7500}H_{1.499}$

$C_{\frac{0.7500}{0.7500}}H_{\frac{1.499}{0.7500}} \rightarrow CH_2$

The correct empirical formula is CH_2

66. **Given:** 8.80 g CO_2, 1.44 g H_2O **Find:** empirical formula

Conceptual Plan:

mass CO_2, H_2O → mol CO_2, H_2O → mol C, mol H → pseudoformula → empirical formula

$\frac{1 \text{ mol } CO_2}{44.01 \text{ g } CO_2}$ $\frac{1 \text{ mol } H_2O}{18.02 \text{ g } H_2O}$ $\frac{1 \text{ mol C}}{1 \text{ mol } CO_2}$ $\frac{2 \text{ mol H}}{1 \text{ mol } H_2O}$ divide by smallest number

Solution:

$$8.80 \text{ g } CO_2 \times \frac{1 \text{ mol } CO_2}{44.01 \text{ g } CO_2} = 0.200 \text{ mol } CO_2$$

$$1.44 \text{ g } H_2O \times \frac{1 \text{ mol } H_2O}{18.02 \text{ g } H_2O} = 0.0799 \text{ mol } H_2O$$

$$0.200 \text{ mol } CO_2 \times \frac{1 \text{ mol C}}{1 \text{ mol } CO_2} = 0.200 \text{ mol C}$$

$$0.0799 \text{ mol } H_2O \times \frac{2 \text{ mol H}}{1 \text{ mol } H_2O} = 0.160 \text{ mol H}$$

$C_{0.200}H_{0.160}$

$C_{\frac{0.200}{0.160}}H_{\frac{0.160}{0.160}} \rightarrow C_{1.25}H_1$

$C_{1.25}H_1 \times 4 = C_5H_4$

The correct empirical formula is C_5H_4

67. **Given:** 4.30 g sample, 8.59 g CO_2, 3.52 g H_2O **Find:** empirical formula
Conceptual Plan:
mass CO_2, H_2O → mol CO_2, H_2O → mol C, mol H → mass C, mass H, mass O → mol O →

$$\frac{1 \text{ mol } CO_2}{44.01 \text{ g } CO_2} \quad \frac{1 \text{ mol } H_2O}{18.02 \text{ g } H_2O} \quad \frac{1 \text{ mol C}}{1 \text{ mol } CO_2} \quad \frac{2 \text{ mol H}}{1 \text{ mol } H_2O} \quad \frac{12.01 \text{ g C}}{1 \text{ mol C}} \quad \frac{1.008 \text{ g H}}{1 \text{ mol H}} \quad \text{g sample - gC - g H} \quad \frac{1 \text{ mol O}}{16.00 \text{ g O}}$$

pseudoformula → empirical formula
divide by smallest number

Solution:

$$8.59 \text{ g } CO_2 \times \frac{1 \text{ mol } CO_2}{44.01 \text{ g } CO_2} = 0.195 \text{ mol } CO_2$$

$$3.52 \text{ g } H_2O \times \frac{1 \text{ mol } H_2O}{18.02 \text{ g } H_2O} = 0.195 \text{ mol } H_2O$$

$$0.195 \text{ mol } CO_2 \times \frac{1 \text{ mol C}}{1 \text{ mol } CO_2} = 0.195 \text{ mol C}$$

$$0.195 \text{ mol } H_2O \times \frac{2 \text{ mol H}}{1 \text{ mol } H_2O} = 0.390 \text{ mol H}$$

$$0.195 \text{ mol C} \times \frac{12.01 \text{ g C}}{1 \text{ mol C}} = 2.34 \text{ g C}$$

$$0.390 \text{ mol } H_2O \times \frac{1.008 \text{ g H}}{1 \text{ mol H}} = 0.393 \text{ g H}$$

4.30 g - 2.34 g - 0.393 g = 1.57 g O

$$1.57 \text{ g O} \times \frac{1 \text{ mol O}}{16.00 \text{ g O}} = 0.0979 \text{ mol O}$$

$$C_{0.195}H_{0.390}O_{0.0979}$$

$$C_{\frac{0.195}{0.0979}}H_{\frac{0.390}{0.0979}}O_{\frac{0.0979}{0.0979}} \rightarrow C_2H_4O$$

The correct empirical formula is C_2H_4O

68. **Given:** 12.01 g sample, 14.01 g CO_2, 4.32 g H_2O **Find:** empirical formula
Conceptual Plan:
mass CO_2, H_2O → mol CO_2, H_2O → mol C, mol H → mass C, mass H, mass O → mol O →

$$\frac{1 \text{ mol } CO_2}{44.01 \text{ g } CO_2} \quad \frac{1 \text{ mol } H_2O}{18.02 \text{ g } H_2O} \quad \frac{1 \text{ mol C}}{1 \text{ mol } CO_2} \quad \frac{2 \text{ mol H}}{1 \text{ mol } H_2O} \quad \frac{12.01 \text{ g C}}{1 \text{ mol C}} \quad \frac{1.008 \text{ g H}}{1 \text{ mol H}} \quad \text{g sample - gC - g H} \quad \frac{1 \text{ mol O}}{16.00 \text{ g O}}$$

pseudoformula → empirical formula
divide by smallest number

Solution:

$$14.01 \text{ g } CO_2 \times \frac{1 \text{ mol } CO_2}{44.01 \text{ g } CO_2} = 0.3183 \text{ mol } CO_2$$

$$4.32 \text{ g } H_2O \times \frac{1 \text{ mol } H_2O}{18.02 \text{ g } H_2O} = 0.2397 \text{ mol } H_2O$$

$$0.3183 \text{ mol } CO_2 \times \frac{1 \text{ mol C}}{1 \text{ mol } CO_2} = 0.3183 \text{ mol C}$$

$$0.2397 \text{ mol } H_2O \times \frac{2 \text{ mol H}}{1 \text{ mol } H_2O} = 0.4795 \text{ mol H}$$

$$0.3183 \text{ mol C} \times \frac{12.01 \text{ g C}}{1 \text{ mol C}} = 3.823 \text{ g C}$$

$$0.4795 \text{ mol H} \times \frac{1.008 \text{ g H}}{1 \text{ mol H}} = 0.4833 \text{ g H}$$

12.01 g - 3.823 g - 0.4833 g = 7.70 g O

$$7.70 \text{ g O} \times \frac{1 \text{ mol O}}{16.00 \text{ g O}} = 0.481 \text{ mol O}$$

$$C_{0.3183}H_{0.4795}O_{0.481}$$

$$C_{\frac{0.3183}{0.3183}}H_{\frac{0.4795}{0.3183}}O_{\frac{0.481}{0.3183}} \rightarrow C\,H_{1.5}O_{1.5}$$

$$C\,H_{1.5}O_{1.5} \times 2 = C_2H_3O_3$$

The correct empirical formula is $C_2H_3O_3$

69. **Conceptual Plan: write a skeletal reaction → balance atoms in more complex compounds → balance elements that occur as free elements → clear fractions**

Solution:	Skeletal reaction:	$SO_2(g) + O_2(g) + H_2O(l) \rightarrow H_2SO_4(aq)$	
	Balance O:	$SO_2(g) + 1/2O_2(g) + H_2O(l) \rightarrow H_2SO_4(aq)$	
	Clear fraction:	$2SO_2(g) + O_2(g) + 2H_2O(l) \rightarrow 2H_2SO_4(aq)$	
Check:		left side	right side
		2 S atoms	2 S atoms
		8 O atoms	8 O atoms
		4 H atoms	4 H atoms

70. **Conceptual Plan: write a skeletal reaction → balance atoms in more complex compounds → balance elements that occur as free elements → clear fractions**

Solution:	Skeletal reaction:	$NO_2(g) + O_2(g) + H_2O(l) \rightarrow HNO_3(aq)$	
	Balance H:	$NO_2(g) + O_2(g) + H_2O(l) \rightarrow 2HNO_3(aq)$	
	Balance N:	$2NO_2(g) + O_2(g) + H_2O(l) \rightarrow 2HNO_3(aq)$	
	Balance O:	$2NO_2(g) + 1/2O_2(g) + H_2O(l) \rightarrow 2HNO_3(aq)$	
	Clear fraction:	$4NO_2(g) + O_2(g) + 2H_2O(l) \rightarrow 4HNO_3(aq)$	
Check:		left side	right side
		4 N atoms	4 N atoms
		12 O atoms	12 O atoms
		4 H atoms	4 H atoms

71. **Conceptual Plan: write a skeletal reaction → balance atoms in more complex compounds → balance elements that occur as free elements → clear fractions**

Solution:	Skeletal reaction:	$Na(s) + H_2O(l) \rightarrow H_2(g) + NaOH(aq)$	
	Balance H:	$Na(s) + H_2O(l) \rightarrow 1/2H_2(g) + NaOH(aq)$	
	Clear fraction:	$2Na(s) + 2H_2O(l) \rightarrow H_2(g) + 2NaOH(aq)$	
Check:		left side	right side
		2 Na atoms	2 Na atoms
		4 H atoms	4 H atoms
		2 O atoms	2 O atoms

72. **Conceptual Plan: write a skeletal reaction → balance atoms in more complex compounds → balance elements that occur as free elements → clear fractions**

Solution:	Skeletal reaction:	$Fe(s) + O_2(g) \rightarrow Fe_2O_3(s)$	
	Balance O:	$Fe(s) + 3O_2(g) \rightarrow 2Fe_2O_3(s)$	
	Balance Fe:	$4Fe(s) + 3O_2(g) \rightarrow 2Fe_2O_3(s)$	
Check:		left side	right side
		4 Fe atoms	4 Fe atoms
		6 O atoms	6 O atoms

73. **Conceptual Plan: write a skeletal reaction → balance atoms in more complex compounds → balance elements that occur as free elements → clear fractions**

Solution:	Skeletal reaction:	$C_{12}H_{22}O_{11}(aq) + H_2O(l) \rightarrow C_2H_5OH(aq) + CO_2(g)$	
	Balance H:	$C_{12}H_{22}O_{11}(aq) + H_2O(l) \rightarrow 4C_2H_5OH(aq) + CO_2(g)$	
	Balance C:	$C_{12}H_{22}O_{11}(aq) + H_2O(l) \rightarrow 4C_2H_5OH(aq) + 4CO_2(g)$	
Check:		left side	right side
		12 C atoms	12 C atoms
		24 H atoms	24 H atoms
		12 O atoms	12 O atoms

74. **Conceptual Plan: write a skeletal reaction → balance atoms in more complex compounds → balance elements that occur as free elements → clear fractions**

Solution: Skeletal reaction: $CO_2(g) + H_2O(l) \rightarrow C_6H_{12}O_6(aq) + O_2(g)$

Balance C: $6CO_2(g) + H_2O(l) \rightarrow C_6H_{12}O_6(aq) + O_2(g)$

Balance H: $6CO_2(g) + 6H_2O(l) \rightarrow C_6H_{12}O_6(aq) + O_2(g)$

Balance O: $6CO_2(g) + 6H_2O(l) \rightarrow C_6H_{12}O_6(aq) + 6O_2(g)$

Check:

left side	right side
6 C atoms	6 C atoms
18 O atoms	18 O atoms
12 H atoms	12 H atoms

75. a) **Conceptual Plan: write a skeletal reaction → balance atoms in more complex compounds → balance elements that occur as free elements → clear fractions**

Solution: Skeletal reaction: $PbS(s) + HBr(aq) \rightarrow PbBr_2(s) + H_2S(g)$

Balance Br: $PbS(s) + 2HBr(aq) \rightarrow PbBr_2(s) + H_2S(g)$

Check:

left side	right side
1 Pb atom	1 Pb atom
1 S atom	1 S atom
2 H atoms	2 H atoms
2 Br atoms	2 Br atoms

b) **Conceptual Plan: write a skeletal reaction → balance atoms in more complex compounds → balance elements that occur as free elements → clear fractions**

Solution: Skeletal reaction: $CO(g) + H_2(g) \rightarrow CH_4(g) + H_2O(l)$

Balance H: $CO(g) + 3H_2(g) \rightarrow CH_4(g) + H_2O(l)$

Check:

left side	right side
1 C atom	1 C atom
1 O atom	1 O atom
6 H atoms	6 H atoms

c) **Conceptual Plan: write a skeletal reaction → balance atoms in more complex compounds → balance elements that occur as free elements → clear fractions**

Solution: Skeletal reaction: $HCl(aq) + MnO_2(s) \rightarrow MnCl_2(aq) + H_2O(l) + Cl_2(g)$

Balance Cl: $4HCl(aq) + MnO_2(s) \rightarrow MnCl_2(aq) + H_2O(l) + Cl_2(g)$

Balance O: $4HCl(aq) + MnO_2(s) \rightarrow MnCl_2(aq) + 2H_2O(l) + Cl_2(g)$

Check:

left side	right side
4 H atoms	4 H atoms
4 Cl atoms	4 Cl atoms
1 Mn atom	1 Mn atom
2 O atoms	2 O atoms

d) **Conceptual Plan: write a skeletal reaction → balance atoms in more complex compounds → balance elements that occur as free elements → clear fractions**

Solution: Skeletal reaction: $C_5H_{12}(l) + O_2(g) \rightarrow CO_2(g) + H_2O(l)$

Balance C: $C_5H_{12}(l) + O_2(g) \rightarrow 5CO_2(g) + H_2O(l)$

Balance H: $C_5H_{12}(l) + O_2(g) \rightarrow 5CO_2(g) + 6H_2O(l)$

Balance O: $C_5H_{12}(l) + 8O_2(g) \rightarrow 5CO_2(g) + 6H_2O(l)$

Check:

left side	right side
5 C atoms	5 C atoms
12 H atoms	12 H atoms
16 O atoms	16 O atoms

76. a) **Conceptual Plan: write a skeletal reaction → balance atoms in more complex compounds → balance elements that occur as free elements → clear fractions**

Solution: Skeletal reaction: $Cu(s) + S(s) \rightarrow Cu_2S(s)$

Balance Cu: $2Cu(s) + S(s) \rightarrow Cu_2S(s)$

Check:

left side	right side
2 Cu atoms	2 Cu atoms
1 S atom	1 S atom

b) **Conceptual Plan: write a skeletal reaction → balance atoms in more complex compounds → balance elements that occur as free elements → clear fractions**

Solution: Skeletal reaction: $Fe_2O_3(s) + H_2(g) \rightarrow Fe(s) + H_2O(l)$

Balance O: $Fe_2O_3(s) + H_2(g) \rightarrow Fe(s) + 3H_2O(l)$

Balance Fe: $Fe_2O_3(s) + H_2(g) \rightarrow 2Fe(s) + 3H_2O(l)$

Balance H: $Fe_2O_3(s) + 3H_2(g) \rightarrow 2Fe(s) + 3H_2O(l)$

Check:

left side	right side
2 Fe atoms	2 Fe atoms
3 O atoms	3 O atoms
6 H atoms	6 H atoms

c) **Conceptual Plan: write a skeletal reaction → balance atoms in more complex compounds → balance elements that occur as free elements → clear fractions**

Solution: Skeletal reaction: $SO_2(g) + O_2(g) \rightarrow SO_3(g)$

Balance O: $SO_2(g) + 1/2O_2(g) \rightarrow SO_3(g)$

Clear fraction: $2SO_2(g) + O_2(g) \rightarrow 2SO_3(g)$

Check:

left side	right side
2 S atoms	2 S atoms
6 O atoms	6 O atoms

d) **Conceptual Plan: write a skeletal reaction → balance atoms in more complex compounds → balance elements that occur as free elements → clear fractions**

Solution: Skeletal reaction: $NH_3(g) + O_2(g) \rightarrow NO(g) + H_2O(g)$

Balance H: $2NH_3(g) + O_2(g) \rightarrow NO(g) + 3H_2O(g)$

Balance O: $2NH_3(g) + 5/2O_2(g) \rightarrow 2NO(g) + 3H_2O(g)$

Clear fraction: $4NH_3(g) + 5O_2(g) \rightarrow 4NO(g) + 6H_2O(g)$

Check:

left side	right side
4 N atoms	4 N atoms
6 H atoms	6 H atoms
10 O atoms	10 O atoms

77. a) **Conceptual Plan: balance atoms in more complex compounds → balance elements that occur as free elements → clear fractions**

Solution: Skeletal reaction: $CO_2(g) + CaSiO_3(s) + H_2O(l) \rightarrow SiO_2(s) + Ca(HCO_3)_2(aq)$

Balance C: $2CO_2(g) + CaSiO_3(s) + H_2O(l) \rightarrow SiO_2(s) + Ca(HCO_3)_2(aq)$

Check:

left side	right side
2 C atoms	2 C atoms
8 O atoms	8 O atoms
1 Ca atom	1 Ca atom
1 Si atom	1 Si atom
2 H atoms	2 H atoms

b) **Conceptual Plan: balance atoms in more complex compounds → balance elements that occur as free elements → clear fractions**

Solution: Skeletal reaction: $Co(NO_3)_3(aq) + (NH_4)_2S(aq) \rightarrow Co_2S_3(s) + NH_4NO_3(aq)$

Balance S: $Co(NO_3)_3(aq) + 3(NH_4)_2S(aq) \rightarrow Co_2S_3(s) + NH_4NO_3(aq)$

Balance Co: $2Co(NO_3)_3(aq) + 3(NH_4)_2S(aq) \rightarrow Co_2S_3(s) + NH_4NO_3(aq)$

Balance N: $2Co(NO_3)_3(aq) + 3(NH_4)_2S(aq) \rightarrow Co_2S_3(s) + 6NH_4NO_3(aq)$

Check:

left side	right side
2 Co atoms	2 Co atoms
12 N atoms	12 N atoms
18 O atoms	18 O atoms
24 H atoms	24 H atoms
3 S atoms	3 S atoms

c) **Conceptual Plan: balance atoms in more complex compounds → balance elements that occur as free elements → clear fractions**

Solution: Skeletal reaction: $Cu_2O(s) + C(s) \rightarrow Cu(s) + CO(g)$

Balance Cu: $Cu_2O(s) + C(s) \rightarrow 2Cu(s) + CO(g)$

Check:

left side	right side
2 Cu atoms	2 Cu atoms
1 O atom	1 O atom
1 C atom	1 C atom

d) **Conceptual Plan: balance atoms in more complex compounds → balance elements that occur as free elements → clear fractions**

Solution: Skeletal reaction: $H_2(g) + Cl_2(g) \rightarrow HCl(g)$

Balance Cl: $H_2(g) + Cl_2(g) \rightarrow 2HCl(g)$

Check:

left side	right side
2 H atom	2 H atom
2 Cl atom	2 Cl atom

78. a) **Conceptual Plan: balance atoms in more complex compounds → balance elements that occur as free elements → clear fractions**

Solution: Skeletal reaction: $Na_2S(aq) + Cu(NO_3)_2(aq) \rightarrow NaNO_3(aq) + CuS(s)$

Balance Na: $Na_2S(aq) + Cu(NO_3)_2(aq) \rightarrow 2NaNO_3(aq) + CuS(s)$

Check:

left side	right side
2 Na atom	2 Na atom
1 S atom	1 S atom
1 Cu atom	1 Cu atom
2 N atom	2 N atom
6 O atom	6 O atom

b) **Conceptual Plan: balance atoms in more complex compounds → balance elements that occur as free elements → clear fractions**

Solution: Skeletal reaction: $N_2H_4(l) \rightarrow NH_3(g) + N_2(g)$

Balance H: $3N_2H_4(l) \rightarrow 4NH_3(g) + N_2(g)$

Check:

left side	right side
6 N atoms	6 N atoms
12 H atoms	12 H atoms

c) **Conceptual Plan: balance atoms in more complex compounds → balance elements that occur as free elements → clear fractions**

Solution: Skeletal reaction: $HCl(aq) + O_2(g) \rightarrow H_2O(l) + Cl_2(g)$

Balance Cl: $2HCl(aq) + O_2(g) \rightarrow H_2O(l) + Cl_2(g)$

Balance O: $2HCl(aq) + 1/2O_2(g) \rightarrow H_2O(l) + Cl_2(g)$

Clear fraction: $4HCl(aq) + O_2(g) \rightarrow 2H_2O(l) + 2Cl_2(g)$

Check:

left side	right side
4 H atoms	4 H atoms
4 Cl atoms	4 Cl atoms
2 O atoms	2 O atoms

d) **Conceptual Plan: balance atoms in more complex compounds → balance elements that occur as free elements → clear fractions**

Solution: Skeletal reaction: $FeS(s) + HCl(aq) \rightarrow FeCl_2(aq) + H_2S(g)$

Balance Cl: $FeS(s) + 2HCl(aq) \rightarrow FeCl_2(aq) + H_2S(g)$

Check:

left side	right side
1 Fe atom	1 Fe atom
1 S atom	1 S atom
2 H atoms	2 H atoms
2 Cl atoms	2 Cl atoms

79. a) composed of metal cation and polyatomic anion – inorganic compound
b) composed of carbon and hydrogen – organic compound
c) composed of carbon, hydrogen, and oxygen – organic compound
d) composed of metal cation and nonmetal anion – inorganic compound

80. a) composed of carbon and hydrogen – organic compound
b) composed of carbon, hydrogen, and nitrogen – organic compound

c) composed of metal cation and nonmetal anion – inorganic compound
d) composed of metal cation and polyatomic anion – inorganic compound

81. **Given:** 145 mL C_2H_5OH, d = 0.789g/cm^3 **Find:** number of molecules
Conceptual Plan: cm^3 → mL: mL C_2H_5OH → g C_2H_5OH → mol C_2H_5OH → molecules C_2H_5OH

$$\frac{1\ \text{cm}^3}{1\ \text{mL}} \quad \frac{1\ \text{mL}\ C_2H_5OH}{0.789\ \text{g}\ C_2H_5OH} \quad \frac{1\ \text{mol}\ C_2H_5OH}{46.07\ \text{g}\ C_2H_5OH} \quad \frac{6.022 \times 10^{23}\ \text{molecules}\ C_2H_5OH}{1\ \text{mol}\ C_2H_5OH}$$

Solution:

$$145\ \text{mL}\ C_2H_5OH \times \frac{0.789\ \text{g}\ C_2H_5OH}{\text{cm}^3} \times \frac{1\ \text{cm}^3}{1\ \text{mL}} \times \frac{1\ \text{mol}\ C_2H_5OH}{46.07\ \text{g}\ C_2H_5OH} \times \frac{6.022 \times 10^{23}\ \text{molecules}\ C_2H_5OH}{1\ \text{mol}\ C_2H_5OH}$$

$= 1.50 \times 10^{24}$ molecules C_2H_5OH

Check: Units of answer, molecules C_2H_5OH, are correct. The magnitude is reasonable because we had more than 2 moles of C_2H_5OH and we have more than 2 times Avogadro's number of molecules.

82. **Given:** 0.05 mL H_2O, d = 1.0 g/cm^3 **Find:** number of molecules
Conceptual Plan: cm^3 → mL: mL H_2O → g H_2O → mol H_2O → molecules H_2O

$$\frac{1\ \text{cm}^3}{1\ \text{mL}} \quad \frac{1\ \text{mL}\ H_2O}{1.0\ \text{g}\ H_2O} \quad \frac{1\ \text{mol}\ H_2O}{18.02\ \text{g}\ H_2O} \quad \frac{6.022 \times 10^{23}\ \text{molecules}\ H_2O}{1\ \text{mol}\ H_2O}$$

Solution:

$$0.05\ \text{mL}\ H_2O \times \frac{1\ \text{cm}^3}{1\ \text{mL}} \times \frac{1.0\ \text{g}}{\text{cm}^3} \times \frac{1\ \text{mol}\ H_2O}{18.02\ \text{g}\ H_2O} \times \frac{6.022 \times 10^{23}\ \text{molecules}\ H_2O}{1\ \text{mol}\ H_2O} = 2 \times 10^{21}\ \text{molecules}\ H_2O$$

Check: Units of answer, molecules H_2O, are correct. The magnitude is reasonable because we have less then 1 mole H_2O and we have less than Avogadro's number of molecules.

83. a) To write the formula for an ionic compound: 1) Write the symbol for the metal cation and its charge and the symbol for the nonmetal anion or polyatomic anion and its charge. 2) Adjust the subscript on each cation and anion to balance the overall charge. 3) Check that the sum of the charges of the cations equals the sum of the charges of the anions.
potassium chromate: K^+ CrO_4^{2-}; K_2CrO_4 cation 2(1+) = 2+; anion 2-
Given: K_2CrO_4 **Find:** mass percent of each element
Conceptual Plan: %K, then %Cr, then %O

$$\text{mass \% K} = \frac{2 \times \text{molar mass K}}{\text{molar mass } K_2CrO_4} \times 100 \quad \text{mass \% Cr} = \frac{1 \times \text{molar mass Cr}}{\text{molar mass } K_2CrO_4} \times 100 \quad \text{mass \% O} = \frac{4 \times \text{molar mass O}}{\text{molar mass } K_2CrO_4} \times 100$$

molar mass of K = 39.10 g/mol, molar mass Cr = 52.00 g/mol, molar mass O = 16.00 g/mol
Solution: molar mass K_2CrO_4 = 2(39.10 g/mol) + 1(52.00 g/mol) + 4(16.00 g/mol) = 194.20 g/mol

2 x molar mass K = 2(39.10 g/mol) = 78.20 g K

$$\text{mass \% K} = \frac{2 \times \text{molar mass K}}{\text{molar mass } K_2CrO_4} \times 100\% = \frac{78.20\ \text{g/mol}}{194.20\ \text{g/mol}} \times 100\% = 40.27\ \%$$

1 x molar mass Cr = 1(52.00 g/mol) = 52.00 g Cr

$$\text{mass \% Cr} = \frac{1 \times \text{molar mass Cr}}{\text{molar mass } K_2CrO_4} \times 100\% = \frac{52.00\ \text{g/mol}}{194.20\ \text{g/mol}} \times 100\% = 26.78\ \%$$

4 x molar mass O = 4(16.00 g/mol) = 64.00 g O

$$\text{mass \% O} = \frac{4 \times \text{molar mass O}}{\text{molar mass } K_2CrO_4} \times 100\% = \frac{64.00\ \text{g/mol}}{194.20\ \text{g/mol}} \times 100\% = 32.96\ \%$$

Check: Units of the answer, %, are correct. The magnitude is reasonable because each is between 0 and 100% and the total is 100%.

b) To write the formula for an ionic compound: 1) Write the symbol for the metal cation and its charge and the symbol for the nonmetal anion or polyatomic anion and its charge. 2) Adjust the subscript on each cation and anion to balance the overall charge. 3) Check that the sum of the charges of the cations equals the sum of the charges of the anions.

Lead(II)phosphate: Pb^{2+} PO_4^{3-}; $Pb_3(PO_4)_2$ cation 3(2+) = 6+; anion 2(3-) = 6-

Given: $Pb_3(PO_4)_2$ **Find:** mass percent of each element

Conceptual Plan: %Pb, then % P, then %O

$$\text{mass \% Pb} = \frac{3 \text{ x molar mass Pb}}{\text{molar mass } Pb_3(PO_4)_2} \text{ x } 100 \quad \text{mass \% P} = \frac{2 \text{ x molar mass P}}{\text{molar mass } Pb_3(PO_4)_2} \text{ x } 100 \quad \text{mass \% O} = \frac{8 \text{ x molar mass O}}{\text{molar mass } Pb_3(PO_4)_2} \text{ x } 100$$

Solution: molar mass $Pb_3(PO_4)_2$ = 3(207.2 g/mol) + 2(30.97 g/mol) + 8(16.00 g/mol) = 811.5 g/mol

3 x molar mass Pb = 3(207.2 g/mol) = 621.6 g Pb 2 x molar mass P = 2(30.97 g/mol) = 61.94 g P

$$\text{mass \% Pb} = \frac{3 \text{ x molar mass Pb}}{\text{molar mass } Pb_3(PO_4)_2} \text{ x } 100\% = \frac{621.6 \cancel{\text{g/mol}}}{811.5 \cancel{\text{g/mol}}} \text{ x } 100\% = 76.60\ \%$$

$$\text{mass \% P} = \frac{2 \text{ x molar mass P}}{\text{molar mass } Pb_3(PO_4)_2} \text{ x } 100\% = \frac{61.94 \cancel{\text{g/mol}}}{811.5 \cancel{\text{g/mol}}} \text{ x } 100\% = 7.632\ \%$$

4 x molar mass O = 8(16.00 g/mol) = 128.0 g O

$$\text{mass \% O} = \frac{8 \text{ x molar mass O}}{\text{molar mass } Pb_3(PO_4)_2} \text{ x } 100\% = \frac{128.0 \cancel{\text{g/mol}}}{811.5 \cancel{\text{g/mol}}} \text{ x } 100\% = 15.77\ \%$$

Check: Units of the answer, %, are correct. The magnitude is reasonable because each is between 0 and 100% and the total is 100%.

c) sulfurous acid: H_2SO_3

Given: H_2SO_3 **Find:** mass percent of each element

Conceptual Plan: %H, then %S, then %O

$$\text{mass \% H} = \frac{2 \text{ x molar mass H}}{\text{molar mass } HSO_3} \text{ x } 100 \quad \text{mass \% S} = \frac{1 \text{ x molar mass S}}{\text{molar mass } HSO_3} \text{ x } 100 \quad \text{mass \% O} = \frac{3 \text{ x molar mass O}}{\text{molar mass } HSO_3} \text{ x } 100$$

Solution: molar mass H_2SO_3 = 2(1.008 g/mol) + 1(32.07 g/mol) + 3(16.00 g/mol) = 82.0$\underline{8}$6 g/mol

2 x molar mass H = 1(1.008 g/mol) = 2.016 g H 1 x molar mass S = 1(32.07 g/mol) = 32.06 g S

$$\text{mass \% H} = \frac{2 \text{ x molar mass H}}{\text{molar mass } H_2SO_3} \text{ x } 100\% = \frac{2.016 \cancel{\text{g/mol}}}{82.0\underline{8}6 \cancel{\text{g/mol}}} \text{ x } 100\% = 2.456\ \%$$

$$\text{mass \% S} = \frac{1 \text{ x molar mass S}}{\text{molar mass } H_2SO_3} \text{ x } 100\% = \frac{32.07 \cancel{\text{g/mol}}}{82.0\underline{8}6 \cancel{\text{g/mol}}} \text{ x } 100\% = 39.07\ \%$$

3 x molar mass O = 3(16.00 g/mol) = 48.00 g O

$$\text{mass \% O} = \frac{3 \text{ x molar mass O}}{\text{molar mass } H_2SO_3} \text{ x } 100\% = \frac{48.00 \cancel{\text{g/mol}}}{82.0\underline{8}6 \cancel{\text{g/mol}}} \text{ x } 100\% = 58.48\ \%$$

Check: Units of the answer, %, are correct. The magnitude is reasonable because each is between 0 and 100% and the total is 100%.

d) To write the formula for an ionic compound: 1) Write the symbol for the metal cation and its charge and the symbol for the nonmetal anion or polyatomic anion and its charge. 2) Adjust the subscript on each cation and anion to balance the overall charge. 3) Check that the sum of the charges of the cations equals the sum of the charges of the anions.

cobalt(II)bromide: Co^{2+} Br^-; $CoBr_2$ cation 2+ = 2+; anion 2(1-) = 2-

Given: $CoBr_2$ **Find:** mass percent of each element

Conceptual Plan: %Co, then %Br

$$\text{mass \% Co} = \frac{1 \text{ x molar mass Co}}{\text{molar mass CoBr}_2} \text{ x } 100 \qquad \text{mass \% Br} = \frac{2 \text{ x molar mass Br}}{\text{molar mass CoBr}_2} \text{ x } 100$$

Solution: molar mass $CoBr_2$ = (58.93 g/mol) + 2(79.90 g/mol) = 218.7 g/mol

2 x molar mass Co = 1(58.93 g/mol) = 58.93 g Co 1 x molar mass Br = 2(79.90 g/mol) = 159.8 g Br

$$\text{mass \% Co} = \frac{1 \text{ x molar mass Co}}{\text{molar mass CoBr}_2} \text{ x } 100\% \qquad \text{mass \% Br} = \frac{2 \text{ x molar mass Br}}{\text{molar mass CoBr}_2} \text{ x } 100\%$$

$$= \frac{58.93 \cancel{\text{g/mol}}}{218.7 \cancel{\text{g/mol}}} \text{ x } 100\% \qquad = \frac{159.8 \cancel{\text{g/mol}}}{218.7 \cancel{\text{g/mol}}} \text{ x } 100\%$$

= 26.94 % = 73.07 %

Check: Units of the answer, %, are correct. The magnitude is reasonable because each is between 0 and 100% and the total is 100%.

84. a) perchloric acid: $HClO_4$

Given: $HClO_4$

Find: mass percent of each element

Conceptual Plan: %H, then %Cl, then %O

$$\text{mass \% H} = \frac{1 \text{ x molar mass H}}{\text{molar mass HClO}_4} \text{ x } 100 \qquad \text{mass \% Cl} = \frac{1 \text{ x molar mass Cl}}{\text{molar mass HClO}_4} \text{ x } 100 \qquad \text{mass \% O} = \frac{4 \text{ x molar mass O}}{\text{molar mass HClO}_4} \text{ x } 100$$

Solution: molar mass $HClO_4$ = 1(1.008 g/mol) + 1(35.45 g/mol) + 4(16.00 g/mol) = 100.46 g/mol

1 x molar mass H = 1(1.008 g/mol) = 1.008 g H 1 x molar mass Cl = 1(35.45 g/mol) = 35.45 g Cr

$$\text{mass \% H} = \frac{1 \text{ x molar mass H}}{\text{molar mass HClO}_4} \text{ x } 100\% \qquad \text{mass \% Cl} = \frac{1 \text{ x molar mass Cl}}{\text{molar mass HClO}_4} \text{ x } 100\%$$

$$= \frac{1.008 \cancel{\text{g/mol}}}{100.46 \cancel{\text{g/mol}}} \text{ x } 100\% \qquad = \frac{35.45 \cancel{\text{g/mol}}}{100.46 \cancel{\text{g/mol}}} \text{ x } 100\%$$

= 1.003 % = 35.29 %

4 x molar mass O = 4(16.00 g/mol) = 64.00 g K

$$\text{mass \% O} = \frac{4 \text{ x molar mass O}}{\text{molar mass HSO}_3} \text{ x } 100\%$$

$$= \frac{64.00 \cancel{\text{g/mol}}}{100.46 \cancel{\text{g/mol}}} \text{ x } 100\%$$

= 63.71 %

Check: Units of the answer, %, are correct. The magnitude is reasonable because each is between 0 and 100% and the total is 100%.

b) phosphorus pentachloride: PCl_5

Given: PCl_5 **Find:** mass percent of each element

Conceptual Plan: %P, then %Cl

$$\text{mass \% P} = \frac{1 \text{ x molar mass P}}{\text{molar mass PCl}_5} \text{ x } 100 \qquad \text{mass \% Cl} = \frac{5 \text{ x molar mass Cl}}{\text{molar mass PCl}_5} \text{ x } 100$$

Solution: molar mass PCl_5 = 1(30.97 g/mol) + 5(35.45 g/mol) = 208.2 g/mol

1 x molar mass P = 1(30.97 g/mol) = 30.97 g P 1 x molar mass Cl = 5(35.45 g/mol) = 177.25 g Cr

$$\text{mass \% P} = \frac{1 \text{ x molar mass P}}{\text{molar mass PCl}_5} \text{ x } 100\% \qquad \text{mass \% Cl} = \frac{5 \text{ x molar mass Cl}}{\text{molar mass PCl}_5} \text{ x } 100\%$$

$$= \frac{30.97 \cancel{\text{g/mol}}}{208.2 \cancel{\text{g/mol}}} \text{ x } 100\% \qquad = \frac{177.25 \cancel{\text{g/mol}}}{208.2 \cancel{\text{g/mol}}} \text{ x } 100\%$$

= 14.87 % = 85.13 %

Check: Units of the answer, %, are correct. The magnitude is reasonable because each is between 0 and 100% and the total is 100%.

c) nitrogen triiodide: NI_3

Given: NI_3 **Find:** mass percent of each element

Conceptual Plan: %N, then %I

$$\text{mass }\%N=\frac{1 \text{ x molar mass N}}{\text{molar mass NI}_3}\text{ x }100 \qquad \text{mass }\%I=\frac{3 \text{ x molar mass I}}{\text{molar mass NI}_3}\text{ x }100$$

Solution: molar mass NI_3 = 1(14.01 g/mol) + 3(126.9 g/mol) = 394.7 g/mol
1 x molar mass N = 1(14.01 g/mol) = 14.01 g N 1 x molar mass I = 3(126.9 g/mol) = 380.7 g I

$$\text{mass }\%N=\frac{1 \text{ x molar mass N}}{\text{molar mass NI}_3}\text{ x }100\% = \frac{14.01 \text{ g/mol}}{394.7 \text{ g/mol}}\text{ x }100\% = 3.549\ \%$$

$$\text{mass }\%\ I=\frac{3 \text{ x molar mass I}}{\text{molar mass NI}_3}\text{ x }100\% = \frac{380.7 \text{ g/mol}}{394.7 \text{ g/mol}}\text{ x }100\% = 96.45\ \%$$

Check: Units of the answer, %, are correct. The magnitude is reasonable because each is between 0 and 100% and the total is 100%.

d) carbon dioxide: CO_2
Given: CO_2 **Find:** mass percent of each element
Conceptual Plan: %C, then %O

$$\text{mass }\%C=\frac{1 \text{ x molar mass C}}{\text{molar mass CO}_2}\text{ x }100 \qquad \text{mass }\%O=\frac{2 \text{ x molar mass O}}{\text{molar mass CO}_2}\text{ x }100$$

Solution: molar mass CO_2 = 1(12.01 g/mol) + 2(16.00 g/mol) = 44.01 g/mol
1 x molar mass C = 1(12.01 g/mol) = 12.01 g C 2 x molar mass O = 2(16.00 g/mol) = 32.00 g O

$$\text{mass }\%C=\frac{1 \text{ x molar mass C}}{\text{molar mass CO}_2}\text{ x }100\% = \frac{12.01 \text{ g/mol}}{44.01 \text{ g/mol}}\text{ x }100\% = 27.29\ \%$$

$$\text{mass }\%\ O=\frac{2 \text{ x molar mass O}}{\text{molar mass CO}_2}\text{ x }100\% = \frac{32.00 \text{ g/mol}}{44.01 \text{ g/mol}}\text{ x }100\% = 72.71\ \%$$

Check: Units of the answer, %, are correct. The magnitude is reasonable because each is between 0 and 100% and the total is 100%.

85. **Given:** 25 g CF_2Cl_2/mo. **Find:** g Cl /yr.
Conceptual Plan:g CF_2Cl_2/mo → g Cl/mo → g Cl/yr

$$\frac{70.09 \text{ g Cl}}{120.91 \text{ g CF}_2\text{Cl}_2} \qquad \frac{12 \text{ mo.}}{1 \text{ yr}}$$

Solution: $\frac{25 \text{ g CF}_2\text{Cl}_2}{\text{mo}}\text{ x }\frac{70.90 \text{ g Cl}}{120.91 \text{ g CF}_2\text{Cl}_2}\text{ x }\frac{12 \text{ mo}}{1 \text{ yr}} = 1.8 \text{ x } 10^2 \text{ g Cl/yr}$

Check: Units of answer, g Cl, is correct. Magnitude is reasonable because it is less than the total CF_2Cl_2 /yr.

86. **Given:** 12 kg CHF_2Cl/mo. **Find:** kg Cl /yr.
Conceptual Plan:kg CF_2Cl_2/mo → kg Cl/mo → kg Cl/yr

$$\frac{35.45 \text{ g Cl}}{86.47 \text{ g CHF}_2\text{Cl}} = \frac{35.45 \text{ kg Cl}}{86.47 \text{ kg CHF}_2\text{Cl}} \qquad \frac{12 \text{ mo.}}{1 \text{ yr}}$$

Solution: $\frac{12 \text{ kg CHF}_2\text{Cl}}{\text{mo}}\text{ x }\frac{35.45 \text{ kg Cl}}{86.47 \text{ kg CHF}_2\text{Cl}}\text{ x }\frac{12 \text{ mo}}{1 \text{ yr}} = 59 \text{ kg Cl/yr}$

Check: Units of answer, kg Cl, is correct. Magnitude is reasonable because it is less than the total CHF_2Cl /yr.

87. **Given:** MCl_3, 65.57% Cl **Find:** identify M
Conceptual Plan: g Cl → mol Cl → mol M → atomic mass M

$$\frac{1 \text{ mol Cl}}{35.45 \text{ g Cl}} \qquad \frac{1 \text{ mol M}}{3 \text{ mol Cl}} \qquad \frac{\text{g M}}{\text{mol M}}$$

Solution: in 100 g sample: 65.57 g Cl, 34.43 g M

$$65.57 \text{ g Cl x }\frac{1 \text{ mol Cl}}{35.45 \text{ g Cl}}\text{ x }\frac{1 \text{ mol M}}{3 \text{ mol Cl}} = 0.6165 \text{ mol M} \qquad \frac{34.43 \text{ g M}}{0.6165 \text{ mol M}} = 55.84 \text{ g/mol M}$$

molar mass of 55.84 = Fe
The identity of M = Fe

88. **Given:** M_2O, 16.99% O **Find:** identify M
Conceptual Plan: g O → mol O → mol M → atomic mass M

$$\frac{1 \text{ mol O}}{16.00 \text{ g O}} \quad \frac{2 \text{ mol M}}{1 \text{ mol O}} \quad \frac{\text{g M}}{\text{mol M}}$$

Solution: in 100 g sample: 16.99 g O, 83.01 g M

$$16.99 \text{ g O} \times \frac{1 \text{ mol O}}{16.00 \text{ g O}} \times \frac{2 \text{ mol M}}{1 \text{ mol O}} = 2.124 \text{ mol M} \qquad \frac{83.01 \text{ g M}}{2.124 \text{ mol M}} = 39.08 \text{ g/mol M}$$

molar mass of 39.08 = K
The identity of M = K

89. **Given:** In a 100 g sample: 79.37 g C, 8.88 g H, 11.75 g O, molar mass = 272.37g/mol
Find: molecular formula
Conceptual Plan:
convert mass to mol of each element → pseudoformula → empirical formula → molecular formula

$$\frac{1 \text{ mol C}}{12.01 \text{ g C}} \quad \frac{1 \text{ mol H}}{1.008 \text{ g H}} \quad \frac{1 \text{ mol O}}{16.00 \text{ g O}}$$ divide by smallest number — empirical formula x n

Solution:

$$79.37 \text{ g C} \times \frac{1 \text{ mol C}}{12.01 \text{ g C}} = 6.609 \text{ mol C}$$

$$8.88 \text{ g H} \times \frac{1 \text{ mol H}}{1.008 \text{ g H}} = 8.81 \text{ mol H}$$

$$11.75 \text{ g O} \times \frac{1 \text{ mol O}}{16.00 \text{ g O}} = 0.7344 \text{ mol O}$$

$$C_{6.609}H_{8.81}O_{0.7344}$$

$$C_{\frac{6.609}{0.7344}}H_{\frac{8.81}{0.7344}}O_{\frac{0.7344}{0.7344}} \rightarrow C_9H_{12}O$$

The correct empirical formula is $C_9H_{12}O$

empirical formula mass = 9(12.01 g/mol) + 12(1.008 g/mol) + 1(16.00 g/mol) = 136.19 g/mol

$$n = \frac{\text{molar mass}}{\text{formula molar mass}} = \frac{272.37 \text{ g/mol}}{136.19 \text{ g/mol}} = 2$$

molecular formula $= C_9H_{12}O \times 2 = C_{18}H_{24}O_2$

90. **Given:** In a 100 g sample: 40.00 g C, 6.72 g H, 53.29 g O, molar mass = 180.16g/mol
Find: molecular formula
Conceptual Plan:
convert mass to mol of each element → pseudoformula → empirical formula → molecular formula

$$\frac{1 \text{ mol C}}{12.01 \text{ g C}} \quad \frac{1 \text{ mol H}}{1.008 \text{ g H}} \quad \frac{1 \text{ mol O}}{16.00 \text{ g O}}$$ divide by smallest number — empirical formula x n

Solution:

$$40.00 \text{ g C} \times \frac{1 \text{ mol C}}{12.01 \text{ g C}} = 3.331 \text{ mol C}$$

$$6.72 \text{ g H} \times \frac{1 \text{ mol H}}{1.008 \text{ g H}} = 6.67 \text{ mol H}$$

$$53.29 \text{ g O} \times \frac{1 \text{ mol O}}{16.00 \text{ g O}} = 3.331 \text{ mol O}$$

$$C_{3.331}H_{6.67}O_{3.331}$$

$$C_{\frac{3.331}{3.331}}H_{\frac{6.67}{3.331}}O_{\frac{3.331}{3.331}} \rightarrow CH_2O$$

The correct empirical formula is CH_2O
empirical formula mass = 1(12.01 g/mol) + 2(1.008 g/mol) + 1(16.00 g/mol) = 30.03 g/mol

$$n = \frac{\text{molar mass}}{\text{formula molar mass}} = \frac{180.16 \text{ g/mol}}{30.03 \text{ g/mol}} = 6$$

molecular formula $= CH_2O \times 6 = C_6H_{12}O_6$

91. **Given:** 13.42 g sample, 39.61 g CO_2, 9.01 g H_2O, molar mass = 268.34 g/mol
Find: molecular formula
Conceptual Plan:
mass CO_2, H_2O → mol CO_2, H_2O → mol C, mol H → mass C, mass H, mass O → mol O →

$$\frac{1\text{ mol }CO_2}{44.01\text{ g }CO_2}\quad\frac{1\text{ mol }H_2O}{18.02\text{ g }H_2O}\qquad\frac{1\text{ mol C}}{1\text{ mol }CO_2}\quad\frac{2\text{ mol H}}{1\text{ mol }H_2O}\quad\frac{12.01\text{ g C}}{1\text{ mol C}}\quad\frac{1.008\text{ g H}}{1\text{ mol H}}\quad\text{g sample - gC - g H}\quad\frac{1\text{ mol O}}{16.00\text{ g O}}$$

pseudoformula → empirical formula → molecular formula
divide by smallest number   empirical formula x n

$$39.61\text{ g }CO_2 \times \frac{1\text{ mol }CO_2}{44.01\text{ g }CO_2} = 0.9000\text{ mol }CO_2$$

$$9.01\text{ g }H_2O \times \frac{1\text{ mol }H_2O}{18.02\text{ g }H_2O} = 0.5000\text{ mol }H_2O$$

$$0.9000\text{ mol }CO_2 \times \frac{1\text{ mol C}}{1\text{ mol }CO_2} = 0.9000\text{ mol C}$$

$$0.5000\text{ mol }H_2O \times \frac{2\text{ mol H}}{1\text{ mol }H_2O} = 1.000\text{ mol H}$$

$$0.9000\text{ mol C} \times \frac{12.01\text{ g C}}{1\text{ mol C}} = 10.81\text{ g C}$$

$$1.000\text{ mol }H_2O \times \frac{1.008\text{ g H}}{1\text{ mol H}} = 1.008\text{ g H}$$

13.42 g - 10.81 g - 1.008 g = 1.60 g O

$$1.60\text{ g O} \times \frac{1\text{ mol O}}{16.00\text{ g O}} = 0.100\text{ mol O}$$

$$C_{0.9000}H_{1.000}O_{0.100}$$

$$C_{\frac{0.9000}{0.100}}H_{\frac{1.000}{0.100}}O_{\frac{0.100}{0.100}} \rightarrow C_9H_{10}O$$

The correct empirical formula is $C_9H_{10}O$
empirical formula mass = 9(12.01 g/mol) + 10(1.008 g/mol) + 1(16.00 g/mol) = 134.2 g/mol

$$n = \frac{\text{molar mass}}{\text{formula molar mass}} = \frac{268.34\text{ g/mol}}{134.2\text{ g/mol}} = 2$$

molecular formula $= C_9H_{10}O \times 2 = C_{18}H_{20}O_2$

92. **Given:** 1.893 g sample, 5.545 g CO_2, 1.388 g H_2O, molar mass = 270.36 g/mol
Find: molecular formula
Conceptual Plan:
mass CO_2, H_2O → mol CO_2, H_2O → mol C, mol H → mass C, mass H, mass O → mol O →

$$\frac{1\text{ mol }CO_2}{44.01\text{ g }CO_2}\quad\frac{1\text{ mol }H_2O}{18.02\text{ g }H_2O}\qquad\frac{1\text{ mol C}}{1\text{ mol }CO_2}\quad\frac{2\text{ mol H}}{1\text{ mol }H_2O}\quad\frac{12.01\text{ g C}}{1\text{ mol C}}\quad\frac{1.008\text{ g H}}{1\text{ mol H}}\quad\text{g sample - gC - g H}\quad\frac{1\text{ mol O}}{16.00\text{ g O}}$$

pseudoformula → empirical formula → molecular formula
divide by smallest number   empirical formula x n

Solution:

$$5.545\text{ g }CO_2 \times \frac{1\text{ mol }CO_2}{44.01\text{ g }CO_2} = 0.1260\text{ mol }CO_2$$

$$1.388\text{ g }H_2O \times \frac{1\text{ mol }H_2O}{18.02\text{ g }H_2O} = 0.07703\text{ mol }H_2O$$

$$0.1260\text{ mol }CO_2 \times \frac{1\text{ mol C}}{1\text{ mol }CO_2} = 0.1260\text{ mol C}$$

$$0.07703\text{ mol }H_2O \times \frac{2\text{ mol H}}{1\text{ mol }H_2O} = 0.1541\text{ mol H}$$

$$0.1260\text{ mol C} \times \frac{12.01\text{ g C}}{1\text{ mol C}} = 1.513\text{ g C}$$

$$0.1541\ \cancel{\text{mol } H_2O} \times \frac{1.008 \text{ g H}}{1\ \cancel{\text{mol H}}} = 0.1553 \text{ g H}$$

1.893 g - 1.513 g - 0.1553 g = 0.2247 g O

$$0.2247\ \cancel{\text{g O}} \times \frac{1 \text{ mol O}}{16.00\ \cancel{\text{g O}}} = 0.01404 \text{ mol O}$$

$$C_{0.1260}H_{0.1541}O_{0.01404}$$

$$C_{\frac{0.1260}{0.01404}}H_{\frac{0.1541}{0.01404}}O_{\frac{0.01404}{0.01404}} \rightarrow C_9H_{11}O$$

The correct empirical formula is $C_9H_{11}O$

empirical formula mass = 9(12.01 g/mol) + 1(1.008 g/mol) + 1(16.00 g/mol) = 135.2 g/mol

$$n = \frac{\text{molar mass}}{\text{formula molar mass}} = \frac{270.36 \text{ g/mol}}{135.2 \text{ g/mol}} = 2$$

molecular formula $= C_9H_{11}O \times 2$
$= C_{18}H_{22}O_2$

93. **Given:** 4.93 g $MgSO_4{\bullet}xH_2O$, 2.41 g $MgSO_4$ **Find:** value of x
Conceptual Plan: g $MgSO_4$ → mol $MgSO_4$ $\frac{1 \text{ mol } MgSO_4}{120.38 \text{ g } MgSO_4}$ **g H_2O → mol H_2O** $\frac{1 \text{ mol } H_2O}{18.02 \text{ g } H_2O}$ **Determine mole ratio** $\frac{\text{mol } HO_2}{\text{mol } MgSO_4}$

Solution:

$$2.41\ \cancel{\text{g } MgSO_4} \times \frac{1 \text{ mol } MgSO_4}{120.38\ \cancel{\text{g } MgSO_4}} = 0.0200 \text{ mol } MgSO_4$$

Determine g H_2O: 4.93 g $MgSO_4{\bullet}xH_2O$ - 2.41 g $MgSO_4$ = 2.52 g H_2O

$$2.52\ \cancel{\text{g } H_2O} \times \frac{1 \text{ mol } H_2O}{18.02\ \cancel{\text{g } H_2O}} = 0.140 \text{ mol } H_2O$$

$$\frac{0.140 \text{ mol } H_2O}{0.0200 \text{ mol } MgSO_4} = 7$$

x = 7

94. **Given:** 3.41 g $CuCl_2{\bullet}xH_2O$, 2.69 g $CuCl_2$ **Find:** value of x
Conceptual Plan: g $CuCl_2$ → mol $CuCl_2$ $\frac{1 \text{ mol } CuCl_2}{134.45 \text{ g } CuCl_2}$ **g H_2O → mol H_2O** $\frac{1 \text{ mol } H_2O}{18.02 \text{ g } H_2O}$ **Determine mole ratio** $\frac{\text{mol } HO_2}{\text{mol } CuCl_2}$

Solution:

$$2.69\ \cancel{\text{g } CuCl_2} \times \frac{1 \text{ mol } CuCl_2}{134.45\ \cancel{\text{g } CuCl_2}} = 0.0200 \text{ mol } CuCl_2$$

Determine g H_2O: 3.41 g $CuCl_2{\bullet}xH_2O$ - 2.69 g $CuCl_2$ = 0.72 g H_2O

$$0.72\ \cancel{\text{g } H_2O} \times \frac{1 \text{ mol } H_2O}{18.02\ \cancel{\text{g } H_2O}} = 0.040 \text{ mol } H_2O$$

$$\frac{0.040 \text{ mol } H_2O}{0.0200 \text{ mol } CuCl_2} = 2$$

x = 2

95. **Given:** molar mass = 177 g/mol, g C = 8(g H) **Find:** molecular formula
Conceptual Plan: C_xH_yBrO
Solution: in 1 mol compound, let x = mol C and y = mol H, assume mol Br = 1, mol O = 1
177 g/mol = x(12.01 g/mol) + y(1.008 g/mol) + 1(79.90 g/mol) + 1(16.00 g/mol)
x(12.01 g/mol) = 8 {y(1.008 g/mol)}
177 g/mol = 8y(1.008 g/mol) + y(1.008 g/mol) + 79.90 g/mol + 16.00 g/mol
81 = 9y(1.008)
y = 9 = mol H
x(12.01) = 9(1.008)
x = 6 = mol C

molecular formula = C_6H_9BrO

Check: molar mass = 6(12.01 g/mol) + 9(1.008 g/mol) + 1(79.90 g/mol) + 1(16.00 g/mol) = 177.0 g/mol

96. **Given:** 3.54 g sample yields 8.49 g CO_2 and 2.14 g H_2O; 2.35 g sample yields 0.199 g N; molar mass = 165
Find: molecular formula
Conceptual Plan:

mass N → mol N; then mass CO_2, H_2O → mol CO_2, H_2O → mol C, mol H → mass C, mass H;

$$\frac{1\text{ mol N}}{14.01\text{ g N}} \qquad \frac{1\text{ mol CO}_2}{44.01\text{ g CO}_2}\ \frac{1\text{ mol H}_2\text{O}}{18.02\text{ g H}_2\text{O}} \qquad \frac{1\text{ mol C}}{1\text{ mol CO}_2}\ \frac{2\text{ mol H}}{1\text{ mol H}_2\text{O}} \qquad \frac{12.01\text{ g C}}{1\text{ mol C}}\ \frac{1.008\text{ g H}}{1\text{ mol H}}$$

mass O → mol O → pseudoformula → empirical formula → molecular formula

g sample - gC - g H $\quad \frac{1\text{ mol O}}{16.00\text{ g O}}$ $\quad$ divide by smallest number $\quad$ empirical formula x n

Solution:

$$\frac{0.199\text{ g N}}{2.35\text{ g sample}} = \frac{x\text{ g N}}{3.54\text{ g sample}};\ x = 0.300\text{ g N}$$

$$0.300\ \cancel{\text{g N}} \times \frac{1\text{ mol N}}{14.01\ \cancel{\text{g N}}} = 0.0214\text{ mol N}$$

$$8.49\ \cancel{\text{g CO}_2} \times \frac{1\text{ mol CO}_2}{44.01\ \cancel{\text{g CO}_2}} = 0.193\text{ mol CO}_2$$

$$2.14\ \cancel{\text{g H}_2\text{O}} \times \frac{1\text{ mol H}_2\text{O}}{18.02\ \cancel{\text{g H}_2\text{O}}} = 0.119\text{ mol H}_2\text{O}$$

$$0.193\ \cancel{\text{mol CO}_2} \times \frac{1\text{ mol C}}{1\ \cancel{\text{mol CO}_2}} = 0.193\text{ mol C}$$

$$0.119\ \cancel{\text{mol H}_2\text{O}} \times \frac{2\text{ mol H}}{1\ \cancel{\text{mol H}_2\text{O}}} = 0.238\text{ mol H}$$

$$0.193\ \cancel{\text{mol C}} \times \frac{12.01\text{ g C}}{1\ \cancel{\text{mol C}}} = 2.32\text{ g C}$$

$$0.238\ \cancel{\text{mol H}_2\text{O}} \times \frac{1.008\text{ g H}}{1\ \cancel{\text{mol H}}} = 0.240\text{ g H}$$

3.54 g - 2.32 gC - 0.240 gH - 0.300 gN = 0.680 g O

$$0.680\ \cancel{\text{g O}} \times \frac{1\text{ mol O}}{16.00\ \cancel{\text{g O}}} = 0.0425\text{ mol O}$$

$$C_{0.193}H_{0.240}N_{0.0214}O_{0.0425}$$

$$C_{\frac{0.193}{0.0214}}H_{\frac{0.240}{0.0214}}N_{\frac{0.0214}{0.0214}}O_{\frac{0.0425}{0.0214}} \rightarrow C_9H_{11}NO_2$$

The correct empirical formula is $C_9H_{11}NO_2$

empirical formula mass =

9(12.01 g/mol) + 11(1.008 g/mol) + 1(14.01 g/mol) + 2(16.00 g/mol) = 165.19 g/mol

$$n = \frac{\text{molar mass}}{\text{formula molar mass}} = \frac{165\text{ g/mol}}{165.19\text{ g/mol}} = 1$$

molecular formula $= C_9H_{11}NO_2 \times 1$
$= C_9H_{11}NO_2$

97. **Given:** 23.5 mg $C_{17}H_{22}ClNO_4$ **Find:** total number of atoms
Conceptual Plan: mg compound → g compound → mol compound → mol atoms → number of atoms

$$\frac{1\text{ g}}{1000\text{ mg}} \qquad \frac{1\text{ mol}}{339.8\text{ g}} \qquad \frac{45\text{ mol atoms}}{1\text{ mol compound}} \qquad \frac{6.022\times10^{23}\text{ atoms}}{1\text{ mol atoms}}$$

Solution: $23.5\ \cancel{\text{mg}} \times \frac{1\ \cancel{\text{g}}}{1000\ \cancel{\text{mg}}} \times \frac{1\ \cancel{\text{mol cpd}}}{339.8\ \cancel{\text{g}}} \times \frac{45\ \cancel{\text{mol atoms}}}{1\ \cancel{\text{mol cpd}}} \times \frac{6.022\times10^{23}\text{ atoms}}{\cancel{\text{mol}}} = 1.87\times10^{21}\text{ atoms}$

Check: The units of the answer, number of atoms, is correct. The magnitude of the answer is reasonable since the molecule is so complex.

98. **Given:** In a 100 g sample: 76 g V, 24 g O **Find:** formula and name
Conceptual Plan:
convert mass to mol of each element → write pseudoformula → write empirical formula

$\frac{1 \text{ mol V}}{50.94 \text{ g V}}$ $\frac{1 \text{ mol O}}{16.00 \text{ g O}}$ divide by smallest number

Solution:

$$76 \text{ g V} \times \frac{1 \text{ mol V}}{50.94 \text{ g V}} = 1.5 \text{ mol V}$$

$$24 \text{ g O} \times \frac{1 \text{ mol O}}{16.00 \text{ g O}} = 1.5 \text{ mol O}$$

$V_{1.5}O_{1.5}$

$V_{\frac{1.5}{1.5}}O_{\frac{1.5}{1.5}} \rightarrow VO$

The correct formula is VO; vanadium(II) oxide

Given: In a 100 g sample: 68 g V, 32 g O **Find:** formula and name
Conceptual Plan: convert mass to mol of each element → write pseudoformula → write empirical formula

$\frac{1 \text{ mol V}}{50.94 \text{ g V}}$ $\frac{1 \text{ mol O}}{16.00 \text{ g O}}$ divide by smallest number

Solution:

$$68 \text{ g V} \times \frac{1 \text{ mol V}}{50.94 \text{ g V}} = 1.33 \text{ mol V}$$

$$32 \text{ g O} \times \frac{1 \text{ mol O}}{16.00 \text{ g O}} = 2 \text{ mol O}$$

$V_{1.33}O_2$

$V_{\frac{1.33}{1.33}}O_{\frac{2}{1.33}} \rightarrow VO_{1.5} \rightarrow V_2O_3$

The correct formula is V_2O_3; vanadium(III) oxide

Given: In a 100 g sample: 61 g V, 39 g O **Find:** formula and name
Conceptual Plan:
convert mass to mol of each element → write pseudoformula → write empirical formula

$\frac{1 \text{ mol V}}{50.94 \text{ g V}}$ $\frac{1 \text{ mol O}}{16.00 \text{ g O}}$ divide by smallest number

Solution:

$$76 \text{ g V} \times \frac{1 \text{ mol V}}{50.94 \text{ g V}} = 1.2 \text{ mol V}$$

$$39 \text{ g O} \times \frac{1 \text{ mol O}}{16.00 \text{ g O}} = 2.4 \text{ mol O}$$

$V_{1.2}O_{2.4}$

$V_{\frac{1.2}{1.2}}O_{\frac{2.4}{1.2}} \rightarrow VO_2$

The correct formula is VO_2; vanadium(IV) oxide

Given: In a 100 g sample: 56 g V, 44 g O **Find:** formula and name
Conceptual Plan:
convert mass to mol of each element → write pseudoformula → write empirical formula

$\frac{1 \text{ mol V}}{50.94 \text{ g V}}$ $\frac{1 \text{ mol O}}{16.00 \text{ g O}}$ divide by smallest number

Solution:

$$56 \text{ g V} \times \frac{1 \text{ mol V}}{50.94 \text{ g V}} = 1.1 \text{ mol V}$$

$$44 \text{ g O} \times \frac{1 \text{ mol O}}{16.00 \text{ g O}} = 2.75 \text{ mol O}$$

$V_{1.1}O_{2.75}$

$\frac{V_{1.1}}{1.1}\frac{O_{2.75}}{1.1} \rightarrow VO_{2.5} \rightarrow V_2O_5$

The correct formula is V_2O_5; vanadium(V) oxide

99. **Given:** MCl_3, 2.395 g sample, 3.606×10^{-2} mol Cl **Find:** atomic mass M

Conceptual Plan: mol Cl → g Cl → g X

$\frac{35.45 \text{ g Cl}}{1 \text{ mol Cl}}$ g sample - g Cl

mol Cl → mol M → atomic mass M

$\frac{1 \text{ mol M}}{3 \text{ mol Cl}}$ $\frac{\text{g M}}{\text{mol M}}$

Solution:

$$3.606 \times 10^{-2} \cancel{\text{mol Cl}} \times \frac{35.45 \text{ g}}{1 \cancel{\text{mol Cl}}} = 1.278 \text{ g Cl}$$

$$2.395 \text{ g} - 1.278 \text{ g} = 1.117 \text{ g M}$$

$$3.606 \times 10^{-2} \cancel{\text{mol Cl}} \times \frac{1 \text{ mol M}}{3 \cancel{\text{mol Cl}}} = 1.202 \times 10^{-2} \text{ mol M}$$

$$\frac{1.117 \text{ g M}}{0.01202 \text{ mol M}} = 92.93 \text{ g/mol M}$$

molar mass of M = 92.93 g/mol

100. **Given:** g NaCl + g NaBr = 2.00 g, g Na = 0.75 g **Find:** g NaBr

Conceptual Plan: Let x = mol NaCl, y = mol NaBr, then x(molar mass NaCl) = g NaCl, y(molar mass NaBr) = g NaBr

Solution: x(58.4) + y(102.9) = 2.00

x(23.0) + y(23.0) = 0.75 y = 0.0326 – x

58.4x + 102.9(0.0326-x) = 2.00

58.4x + 3.354 – 102.9x = 2.00

44.5x = 1.354

x = 0.03043 mol NaCl

y = 0.0326 – 0.03043 = 0 00217 mol NaBr

g NaBr = (0.00217)(102.9g/mol) = 0.224 g NaBr

Check: The units of the answer, g NaBr, are correct. The magnitude is reasonable since it is less than the total mass.

101. **Given:** Sample 1:1.00 g X, 0.472 g Z, X_2Z_3; Sample 2: 1.00 g X, 0.630 g Z; Sample 3: 1.00 g X, 0.789 g Z

Find: empirical formula for samples 2 and 3

Conceptual Plan: moles X remains constant, determine relative moles of Z for the 3 samples.

Solution: Let X = atomic mass X, Z = atomic mass Z

$$n_X = \frac{1.00 \text{ g X}}{X} \qquad n_Z = \frac{0.472 \text{ g Z}}{Z}$$

for sample 1: $\frac{n_X}{n_Z} = \frac{2}{3}$

for sample 2: $\frac{0.630 \text{ g}}{0.472 \text{ g}} = 1.33$, so, mol = $1.33n_Z$

mol ratio: $\frac{n_X}{1.33n_Z} = \frac{2}{(1.33)3} = \frac{2}{4} = \frac{1}{2}$

Empirical formula sample 2: XZ_2

for sample 3: $\frac{0.789 \text{ g}}{0.472 \text{ g}} = 1.67$, so, mol = $1.67n_Z$

mol ratio: $\frac{n_X}{1.67n_Z} = \frac{2}{(1.67)3} = \frac{2}{5}$

Empirical formula sample 3: X_2Z_5

102. **Given:** Sample of $CaCO_3$ and $(NH_4)_2CO_3$ is 61.9% $CO_3^{\,2-}$ **Find:** % $CaCO_3$
Conceptual Plan: Let x = $CaCO_3$, y = $(NH_4)_2CO_3$, then x(molar mass $CaCO_3$) = g $CaCO_3$, y(molar mass $(NH_4)_2CO_3$) = g $(NH_4)_2CO_3$
then, a 100.0 g sample contains: x(100.0) g $CaCO_3$; y(96.1) g $(NH_4)_2CO_3$; and 61.9 g $CO_3^{\,2-}$
Solution: x(100.0) + y(96.1) = 100.0
x(60.0) + y(60.0) = 61.9 y = 1.032 – x

100.0x + 96.1(1.032-x) = 100
100.0x + 99.14 – 96.1x = 100
3.9x = 0.96
x = 0.22 mol $CaCO_3$
y = 1.032 – 0.22 = 0.81 mol $(NH_4)_2CO_3$
g $CaCO_3$ = (0.22 mol)(100.0g/mol) = 22.0 g $CaCO_3$ in a 100 g sample:
mass % $CaCO_3$ = 22.0%
Check: The units of the answer, mass % $CaCO_3$, are correct. The magnitude is reasonable since it is between 0 and 100%

103. **Given:** 50.0 g S, 100 g Cl_2, 150 g mixture S_2Cl_2 and SCl_2 **Find:** % S_2Cl_2
Conceptual Plan: total mol S = 2(mol S_2Cl_2) + mol SCl_2; mol S_2Cl_2 → g S_2Cl_2 → %S_2Cl_2

$$\frac{134.9\text{ g}}{1\text{ mol }S_2Cl_2} \qquad \frac{\text{g }S_2Cl_2}{150\text{ g sample}}\times 100$$

then, S_2Cl_2 = 134.9 g/mol, SCl_2 = 103.0 g/mol, let x = mol S_2Cl_2, y = mol SCl_2
x(134.9) = g S in S_2Cl_2, y(103.0) = g S in SCl_2
Solution:

$$\text{mol S} = 50.0\text{ g S} \times \frac{1\text{ mol S}}{32.1\text{ g S}} = 1.56\text{ mol S}$$

2x = mol S in S_2Cl_2, y = mol S in SCl_2
2x + y = 1.56
x(134.9) + y(103.0) = 150.0
134.9x+ 103.0(1.56 – 2x) = 150.0
71.1x = 10.44
x = 0.147
y = 1.27

$$0.147\text{ mol }S_2Cl_2 \times \frac{134.9\text{ g }S_2Cl_2}{1\text{ mol }S_2Cl_2} = 19.8\text{ g }S_2Cl_2$$

$$\frac{19.8\text{ g }S_2Cl_2}{150\text{ g sample}} \times 100 = 13.2\ \%S_2Cl_2$$

Check: Units of answer, % S_2Cl_2, are correct. Magnitude is reasonable since it is between 0 and 100% and mol of S_2Cl_2 are less than mol SCl_2.

104. **Given:** 1.1 kg CF_2Cl_2/automobile, 25% leak/year, 100 x 10^6 automobiles **Find:** kg Cl/yr
Conceptual Plan: **kg CF_2Cl_2 /auto → kg CF_2Cl_2 leaked/yr → kg Cl/yr/auto → kg Cl**

$$\frac{25\text{ kg }CF_2Cl_2}{100\text{ kg }CF_2Cl_2} \qquad \frac{70.9\text{ g Cl}}{120.91\text{ g }CF_2Cl_2} \qquad 100 \times 10^6\text{ auto}$$

Solution:

$$\frac{1.1\text{ kg }CF_2Cl_2}{\text{auto}} \times \frac{25\text{ kg }CF_2Cl_2}{100\text{ kg }CF_2Cl_2} \times \frac{70.9\text{ kg Cl}}{120.91\text{ kg }CF_2Cl_2} \times 100 \times 10^6\text{ auto} = 1.6 \times 10^7\text{ kg Cl/yr}$$

Check: Units of the answer, kg Cl, are correct. The magnitude is reasonable because it is less than the kg CF_2Cl_2 leaked per year.

105. **Given:** coal = 2.55%S, H_2SO_4,1.0 metric ton coal **Find:** metric ton H_2SO_4 produced
Conceptual Plan: metric ton coal → kg coal → kg S → kg H_2SO_4 → metric ton H_2SO_4

$$\frac{1000\text{ kg}}{\text{metric ton}} \qquad \frac{2.55\text{ kg S}}{100\text{ kg coal}} \qquad \frac{98.09\text{ kg }H_2SO_4}{32.07\text{ kg S}} \qquad \frac{\text{metric ton}}{1000\text{ kg}}$$

Solution:

$$1.0 \text{ metric ton coal} \times \frac{1000 \text{ kg coal}}{1 \text{ metric ton coal}} \times \frac{2.55 \text{ kg S}}{100 \text{ kg coal}} \times \frac{98.09 \text{ kg } H_2SO_4}{32.07 \text{ kg S}} \times \frac{1 \text{ metric ton } H_2SO_4}{1000 \text{ kg } H_2SO_4}$$

$= 0.078$ metric ton H_2SO_4

Check: Units of the answer, metric ton H_2SO_4, are correct. Magnitude is reasonable since it is more than 2.55% of a metric ton and the mass of H_2SO_4 is greater than the mass of S.

106. **Given:** rock contains: 38.0% PbS, 25.0% $PbCO_3$, 17.4% $PbSO_4$, **Find:** kg rock needed for 5.0 metric ton Pb
Conceptual Plan: determine kg Pb/ 100 kg rock then ton Pb→ kg Pb → kg rock

$$\frac{1000 \text{ kg}}{\text{metric ton}} \qquad \frac{100 \text{ kg rock}}{64.2 \text{ kg rock}}$$

Solution: in 100 kg rock:

$$\left(38.0 \text{ kg PbS} \times \frac{207.2 \text{ kg Pb}}{239.3 \text{ kg PbS}}\right) + \left(25.0 \text{ kg } PbCO_3 \times \frac{207.2 \text{ kg Pb}}{267.2 \text{ kg } PbCO_3}\right) + \left(17.4 \text{ kg } PbSO_4 \times \frac{207.2 \text{ kg Pb}}{303.2 \text{ kg } PbSO_4}\right)$$

$= 64.2$ kg Pb

$$5.0 \text{ metric ton Pb} \times \frac{1000 \text{ kg Pb}}{\text{metric ton Pb}} \times \frac{100 \text{ kg rock}}{64.2 \text{ kg Pb}} = 7.8 \times 10^3 \text{ kg rock}$$

Check: Units of answer, kg rock, are correct. Magnitude is reasonable since it is greater than the amount of Pb needed.

107. The sphere in the molecular models represents the electron cloud of the atom. On this scale, the nucleus would be too small to see.

108. a) Atomic mass O > atomic mass C, % O would be higher.
b) Atomic mass N and O close, molecule contains 2N to 1 O, % N would be higher.
c) Atomic mass O > atomic mass C, same number of atoms, % O would be higher.
d) Atomic mass N much greater than atomic mass H, % N would be higher.

109. The statement is incorrect because a chemical formula is based on the ratio of atoms combined, not the ratio of grams combined. The statement should read: "The chemical formula for ammonia (NH_3) indicates that ammonia contains three hydrogen atoms to each nitrogen atom."

110. The statement is incorrect because equations are balanced based on the number and kind of atoms, not molecules. The statement should read: "When a chemical equation is balanced, the number of atoms of each type on both sides of the equation will be equal."

111. H_2SO_4: Atomic mass S is approximately twice atomic mass O, both are much greater than atomic mass H. The order of % mass: % O > % S > % H

Chapter 4
Chemical Quantities and Aqueous Reactions

1. **Given:** 4.9 moles C_6H_{14} **Find:** balanced reaction, moles O_2 required
Conceptual Plan: balance the reaction then mol $C_6H_{14} \rightarrow$ mol O_2

$2\ C_6H_{14}(g) + 19\ O_2(g) \rightarrow 12\ CO_2(g) + 14\ H_2O(g)$ $\qquad \frac{19 \text{ mol } O_2}{2 \text{ mol } C_6H_{14}}$

Solution: $4.9 \text{ mol } C_6H_{14} \times \frac{19 \text{ mol } O_2}{2 \text{ mol } C_6H_{14}} = 47 \text{ mol } O_2$

Check: The units, mol O_2, are correct. The magnitude is reasonable because much more O_2 is needed than C_6H_{14}.

2. **Given:** 0.107 moles $HC_2H_3O_2$ **Find:** balanced reaction, moles $Ba(OH)_2$ required
Conceptual Plan: balance the reaction then mol $HC_2H_3O_2 \rightarrow$ mol $Ba(OH)_2$

$2\ HC_2H_3O_2\ (aq) + Ba(OH)_2(aq) \rightarrow 2\ H_2O(l) + Ba(C_2H_3O_2)_2(aq)$ $\qquad \frac{1 \text{ mol } Ba(OH)_2}{2 \text{ mol } HC_2H_3O_2}$

Solution: $0.107 \text{ mol } HC_2H_3O_2 \times \frac{1 \text{ mol } Ba(OH)_2}{2 \text{ mol } HC_2H_3O_2} = 0.0535 \text{ mol } Ba(OH)_2$

Check: The units, mol $Ba(OH)_2$, are correct. The magnitude is reasonable because much less $Ba(OH)_2$ is needed than $HC_2H_3O_2$.

3. a) **Given:** 1.3 mol N_2O_5 **Find:** mol NO_2
Conceptual Plan: mol $N_2O_5 \rightarrow$ mol NO_2

$\frac{4\ NO_2}{2\ N_2O_5}$

Solution: $1.3 \text{ mol } N_2O_5 \times \frac{4 \text{ mol } NO_2}{2 \text{ mol } N_2O_5} = 2.6 \text{ mol } NO_2$

Check: The units of the answer, mol NO_2, are correct. The magnitude is reasonable since it is greater than mol N_2O_5.

b) **Given:** 5.8 mol N_2O_5 **Find:** mol NO_2
Conceptual Plan: mol $N_2O_5 \rightarrow$ mol NO_2

$\frac{4\ NO_2}{2\ N_2O_5}$

Solution: $5.8 \text{ mol } N_2O_5 \times \frac{4 \text{ mol } NO_2}{2 \text{ mol } N_2O_5} = 11.6 \text{ mol } NO_2 = 12 \text{ mol } NO_2$

Check: The units of the answer, mol NO_2, are correct. The magnitude is reasonable since it is greater than mol N_2O_5.

c) **Given:** 10.5 g N_2O_5 **Find:** mol NO_2
Conceptual Plan: g $N_2O_5 \rightarrow$ mol $N_2O_5 \rightarrow$ molNO$_2$

$\frac{1 \text{ mol } N_2O_5}{108.02 \text{ g } N_2O_5} \qquad \frac{4\ NO_2}{2\ N_2O_5}$

Solution: $10.5 \text{ g } N_2O_5 \times \frac{1 \text{ mol } N_2O_5}{108.02 \text{ g } N_2O_5} \times \frac{4 \text{ mol } NO_2}{2 \text{ mol } N_2O_5} = 0.194 \text{ mol } NO_2$

Check: The units of the answer, mol NO_2, are correct. The magnitude is reasonable since 10 g is about 0.1 mol N_2O_5 and the answer is greater than mol N_2O_5.

d) **Given:** 1.55 kg N_2O_5 **Find:** mol NO_2

Conceptual Plan: kg N_2O_5 → g N_2O_5 → mol N_2O_5 → molNO_2

$$\frac{1000\text{ g } N_2O_5}{\text{kg } N_2O_5} \qquad \frac{1\text{ mol } N_2O_5}{108.02\text{ g } N_2O_5} \qquad \frac{4\ NO_2}{2\ N_2O_5}$$

Solution: $1.55\ \cancel{\text{kg } N_2O_5} \times \frac{1000\ \cancel{\text{g } N_2O_5}}{\cancel{\text{kg } N_2O_5}} \times \frac{1\ \cancel{\text{mol } N_2O_5}}{108.02\ \cancel{\text{g } N_2O_5}} \times \frac{4\text{ mol } NO_2}{2\ \cancel{\text{mol } N_2O_5}} = 28.7\text{ mol } NO_2$

Check: The units of the answer, mol NO_2, are correct. The magnitude is reasonable since 1.5 kg is about 14 mol N_2O_5 and the answer is greater than mol N_2O_5.

4. a) **Given:** 5.3 mol N_2H_4 **Find:** mol NH_3

Conceptual Plan: mol N_2H_4 → mol NH_3

$$\frac{4\ NH_3}{3\ N_2H_4}$$

Solution: $5.3\ \cancel{\text{mol } N_2H_4} \times \frac{4\text{ mol } NH_2}{3\ \cancel{\text{mol } N_2H_4}} = 7.1\text{ mol } NH_3$

Check: The units of the answer, mol NH_3, are correct. The magnitude is reasonable since it is greater than mol N_2H_4.

b) **Given:** 2.28 mol N_2H_4 **Find:** mol NH_3

Conceptual Plan: mol N_2H_4 → mol NH_3

$$\frac{4\ NH_3}{3\ N_2H_4}$$

Solution: $2.28\ \cancel{\text{mol } N_2H_4} \times \frac{4\text{ mol } NH_2}{3\ \cancel{\text{mol } N_2H_4}} = 3.04\text{ mol } NH_3$

Check: The units of the answer, mol NH_3, are correct. The magnitude is reasonable since it is greater than mol N_2H_4.

c) **Given:** 32.5 g N_2H_4 **Find:** mol NH_3

Conceptual Plan: g N_2H_4 → mol N_2H_4 → mol NH_3

$$\frac{1\text{ mol } N_2H_4}{32.05\text{ g } N_2H_4} \qquad \frac{4\ NH_3}{3\ N_2H_4}$$

Solution: $32.5\ \cancel{\text{g } N_2H_4} \times \frac{1\ \cancel{\text{mol } N_2H_4}}{32.05\ \cancel{\text{g } N_2H_4}} \times \frac{4\text{ mol } NH_2}{3\ \cancel{\text{mol } N_2H_4}} = 1.35\text{ mol } NH_3$

Check: The units of the answer, mol NH_3, are correct. The magnitude is reasonable since there is about 1 mol N_2H_4 and the answer is greater than mol N_2H_4.

d) **Given:** 14.7 kg N_2H_4 **Find:** mol NH_3

Conceptual Plan: kg N_2H_4 → g N_2H_4 → mol N_2H_4 → mol NH_3

$$\frac{1000\text{ g } N_2H_4}{\text{kg } N_2H_4} \qquad \frac{1\text{ mol } N_2H_4}{32.05\text{ g } N_2H_4} \qquad \frac{4\ NH_3}{3\ N_2H_4}$$

Solution: $14.7\ \cancel{\text{kg } N_2H_4} \times \frac{1000\ \cancel{\text{g } N_2H_4}}{\cancel{\text{kg } N_2H_4}} \times \frac{1\ \cancel{\text{mol } N_2H_4}}{32.05\ \cancel{\text{g } N_2H_4}} \times \frac{4\text{ mol } NH_2}{3\ \cancel{\text{mol } N_2H_4}} = 612\text{ mol } NH_3$

Check: The units of the answer, mol NH_3, are correct. The magnitude is reasonable since 15 kg is about 500 mol N_2H_4 and the answer is greater than mol N_2H_4.

5. **Given:** 3 mol SiO_2 **Find:** mol C, mol SiC, mol CO

Conceptual Plan: mol SiO_2 → mol C → mol SiC → mol CO

$$\frac{3\ C}{SiO_2} \qquad \frac{SiC}{SiO_2} \qquad \frac{2\ CO}{SiO_2}$$

Solution: $3\ \cancel{\text{mol } SiO_2} \times \frac{3\text{ mol C}}{\cancel{\text{mol } SiO_2}} = 9\text{ mol C}$ $\qquad 3\ \cancel{\text{mol } SiO_2} \times \frac{\text{mol SiC}}{\cancel{\text{mol } SiO_2}} = 3\text{ mol SiC}$

$$3\ \cancel{mol\ SiO_2} \times \frac{2\ mol\ CO}{\cancel{mol\ SiO_2}} = 6\ mol\ CO$$

Given: 6 mol C **Find:** mol SiO_2, mol SiC, mol CO

Conceptual Plan: **mol C → mol SiO_2 → mol SiC → mol CO**

$$\frac{SiO_2}{3\ C} \qquad \frac{SiC}{3\ C} \qquad \frac{2\ CO}{3\ C}$$

Solution:

$$6\ \cancel{mol\ C} \times \frac{mol\ SiO_2}{3\ \cancel{mol\ C}} = 2\ mol\ SiO_2 \qquad 6\ \cancel{mol\ C} \times \frac{mol\ SiC}{3\ \cancel{mol\ C}} = 2\ mol\ SiC$$

$$6\ \cancel{mol\ C} \times \frac{2\ mol\ CO}{3\ \cancel{mol\ C}} = 4\ mol\ CO$$

Given: 10 mol CO **Find:** mol SiO_2, mol C, mol SiC

Conceptual Plan: **mol CO → mol SiO_2 → mol C → mol SiC**

$$\frac{SiO_2}{2\ CO} \qquad \frac{3\ C}{2\ CO} \qquad \frac{SiC}{2\ CO}$$

Solution:

$$10\ \cancel{mol\ CO} \times \frac{mol\ SiO_2}{2\ \cancel{mol\ CO}} = 5.0\ mol\ SiO_2 \qquad 10\ \cancel{mol\ C} \times \frac{3\ mol\ C}{2\ \cancel{mol\ CO}} = 15\ mol\ C$$

$$10\ \cancel{mol\ CO} \times \frac{mol\ SiC}{2\ \cancel{mol\ CO}} = 5.0\ mol\ SiC$$

Given: 2.8 mol SiO_2 **Find:** mol C, mol SiC, mol CO

Conceptual Plan: **mol SiO_2 → mol C → mol SiC → mol CO**

$$\frac{3\ C}{SiO_2} \qquad \frac{SiC}{SiO_2} \qquad \frac{2\ CO}{SiO_2}$$

Solution:

$$2.8\ \cancel{mol\ SiO_2} \times \frac{3\ mol\ C}{\cancel{mol\ SiO_2}} = 8.4\ mol\ C \qquad 2.8\ \cancel{mol\ SiO_2} \times \frac{mol\ SiC}{\cancel{mol\ SiO_2}} = 2.8\ mol\ SiC$$

$$2.8\ \cancel{mol\ SiO_2} \times \frac{2\ mol\ CO}{\cancel{mol\ SiO_2}} = 5.6\ mol\ CO$$

Given: 1.55 mol C **Find:** mol SiO_2, mol SiC, mol CO

Conceptual Plan: **mol C → mol SiO_2 → mol SiC → mol CO**

$$\frac{SiO_2}{3\ C} \qquad \frac{SiC}{3\ C} \qquad \frac{2\ CO}{3\ C}$$

Solution:

$$1.55\ \cancel{mol\ C} \times \frac{3\ mol\ SiO_2}{3\ \cancel{mol\ C}} = 0.517\ mol\ SiO_2 \qquad 1.55\ \cancel{mol\ C} \times \frac{mol\ SiC}{3\ \cancel{mol\ C}} = 0.517\ mol\ SiC$$

$$1.55\ \cancel{mol\ C} \times \frac{2\ mol\ CO}{3\ \cancel{mol\ C}} = 1.03\ mol\ CO$$

SiO_2	**C**	**SiC**	**CO**
3	9	3	6
2	**6**	2	4
5.0	15	5.0	**10**
2.8	8.4	2.8	5.6
0.517	**1.55**	0.517	1.03

6. **Given:** 2 mol N_2H_4 **Find:** mol N_2O_4, mol N_2, mol H_2O

Conceptual Plan: mol N_2H_4 → mol N_2O_4 → mol N_2 → mol H_2O

$$\frac{N_2O_4}{2\,N_2H_4} \qquad \frac{3\,N_2}{2\,N_2H_4} \qquad \frac{4\,H_2O}{2\,N_2H_4}$$

Solution:

$$2\ \cancel{\text{mol } N_2H_4} \times \frac{1 \text{ mol } N_2O_4}{2\ \cancel{\text{mol } N_2H_4}} = 1 \text{ mol } N_2O_4 \qquad 2\ \cancel{\text{mol } N_2H_4} \times \frac{3 \text{ mol } N_2}{2\ \cancel{\text{mol } N_2H_4}} = 3 \text{ mol } N_2$$

$$2\ \cancel{\text{mol } N_2H_4} \times \frac{4 \text{ mol } H_2O}{2\ \cancel{\text{mol } N_2H_4}} = 4 \text{ mol } H_2O$$

Given: 5 mol N_2O_4 **Find:** mol N_2H_4, mol N_2, mol H_2O

Conceptual Plan: mol N_2O_4 → mol N_2H_4 → mol N_2 → mol H_2O

$$\frac{2\,N_2H_4}{N_2O_4} \qquad \frac{3\,N_2}{N_2O_4} \qquad \frac{4\,H_2O}{N_2O_4}$$

Solution:

$$5\ \cancel{\text{mol } N_2O_4} \times \frac{2 \text{ mol } N_2H_4}{\cancel{\text{mol } N_2O_4}} = 10 \text{ mol } N_2H_4 \qquad 5\ \cancel{\text{mol } N_2O_4} \times \frac{3 \text{ mol } N_2}{\cancel{\text{mol } N_2O_4}} = 15 \text{ mol } N_2$$

$$5\ \cancel{\text{mol } N_2O_4} \times \frac{4 \text{ mol } H_2O}{\cancel{\text{mol } N_2O_4}} = 20 \text{ mol } H_2O$$

Given: 10 mol H_2O **Find:** mol N_2O_4, mol N_2O_4, mol N_2

Conceptual Plan: mol H_2O → mol N_2H_4 → mol N_2 → mol N_2O_4

$$\frac{2\,N_2H_4}{4\,H_2O} \qquad \frac{3\,N_2}{4\,H_2O} \qquad \frac{1\,N_2O_4}{4\,H_2O}$$

Solution:

$$10\ \cancel{\text{mol } H_2O} \times \frac{2 \text{ mol } N_2H_4}{4\ \cancel{\text{mol } H_2O}} = 5.0 \text{ mol } N_2H_4 \qquad 10\ \cancel{\text{mol } H_2O} \times \frac{3 \text{ mol } N_2}{4\ \cancel{\text{mol } H_2O}} = 7.5 \text{ mol } N_2$$

$$10\ \cancel{\text{mol } H_2O} \times \frac{1 \text{ mol } N_2O_4}{4\ \cancel{\text{mol } H_2O}} = 2.5 \text{ mol } H_2O$$

Given: 2.5 mol N_2H_4 **Find:** mol N_2O_4, mol N_2, mol H_2O

Conceptual Plan: mol N_2H_4 → mol N_2O_4 → mol N_2 → mol H_2O

$$\frac{N_2O_4}{2\,N_2H_4} \qquad \frac{3\,N_2}{2\,N_2H_4} \qquad \frac{4\,H_2O}{2\,N_2H_4}$$

Solution:

$$2.5\ \cancel{\text{mol } N_2H_4} \times \frac{1 \text{ mol } N_2O_4}{2\ \cancel{\text{mol } N_2H_4}} = 1.3 \text{ mol } N_2O_4 \qquad 2.5\ \cancel{\text{mol } N_2H_4} \times \frac{3 \text{ mol } N_2}{2\ \cancel{\text{mol } N_2H_4}} = 3.8 \text{ mol } N_2$$

$$2.5\ \cancel{\text{mol } N_2H_4} \times \frac{4 \text{ mol } H_2O}{2\ \cancel{\text{mol } N_2H_4}} = 5.0 \text{ mol } H_2O$$

Given: 4.2 mol N_2O_4 **Find:** mol N_2H_4, mol N_2, mol H_2O

Conceptual Plan: mol N_2O_4 → mol N_2H_4 → mol N_2 → mol H_2O

$$\frac{2\,N_2H_4}{N_2O_4} \qquad \frac{3\,N_2}{N_2O_4} \qquad \frac{4\,H_2O}{N_2O_4}$$

Solution:

$$4.2\ \cancel{\text{mol } N_2O_4} \times \frac{2 \text{ mol } N_2H_4}{\cancel{\text{mol } N_2O_4}} = 8.4 \text{ mol } N_2H_4 \qquad 4.2\ \cancel{\text{mol } N_2O_4} \times \frac{3 \text{ mol } N_2}{\cancel{\text{mol } N_2O_4}} = 12.6 \text{ mol } N_2 = 13 \text{ mol } N_2$$

$$4.2\ \cancel{\text{mol } N_2O_4} \times \frac{4 \text{ mol } H_2O}{\cancel{\text{mol } N_2O_4}} = 16.8 \text{ mol } H_2O = 17 \text{ mol } H_2O$$

Given: 11.8 mol N_2 **Find:** mol N_2H_4, mol N_2O_4, mol H_2O

Conceptual Plan: mol N_2 → mol N_2H_4 → mol N_2O_4 → mol H_2O

$$\frac{2\, N_2H_4}{3\, N_2} \qquad \frac{2\, N_2O_4}{3\, N_2} \qquad \frac{4\, H_2O}{3\, N_2}$$

Solution:

$$11.8\ \cancel{\text{mol } N_2} \times \frac{2 \text{ mol } N_2H_4}{3\ \cancel{\text{mol } N_2}} = 7.87 \text{ mol } N_2H_4 \qquad 11.8\ \cancel{\text{mol } N_2} \times \frac{\text{mol } N_2O_4}{3\ \cancel{\text{mol } N_2}} = 3.93 \text{ mol } N_2$$

$$11.8\ \cancel{\text{mol } N_2} \times \frac{4 \text{ mol } H_2O}{3\ \cancel{\text{mol } N_2}} = 15.7 \text{ mol } H_2O$$

N_2H_2	**N_2O_4**	**N_2**	**H_2O**
2	1	3	4
10	**5**	15	20
5	2.5	7.5	**10**
2.5	1.2	3.8	5.0
8.4	**4.2**	13	17
7.87	3.93	**11.8**	15.7

7. **Given:** 3.2 g Fe **Find:** g HBr; g H_2

Conceptual Plan: **g Fe → mol Fe → mol HBr → g HBr**

$$\frac{\text{mol Fe}}{55.8 \text{ g Fe}} \qquad \frac{2 \text{ mol HBr}}{\text{mol Fe}} \qquad \frac{80.9 \text{ g HBr}}{\text{mol HBr}}$$

g Fe → mol Fe → mol H_2 → g H_2

$$\frac{\text{mol Fe}}{55.8 \text{ g Fe}} \qquad \frac{1 \text{ mol } H_2}{\text{mol Fe}} \qquad \frac{2.02 \text{ g } H_2}{\text{mol } H_2}$$

Solution:

$$3.2\ \cancel{\text{g Fe}} \times \frac{1\ \cancel{\text{mol Fe}}}{55.8\ \cancel{\text{g Fe}}} \times \frac{2\ \cancel{\text{mol HBr}}}{1\ \cancel{\text{mol Fe}}} \times \frac{80.9 \text{ g HBr}}{1\ \cancel{\text{mol HBr}}} = 9.3 \text{ g HBr}$$

$$3.2\ \cancel{\text{g Fe}} \times \frac{1\ \cancel{\text{mol Fe}}}{55.8\ \cancel{\text{g Fe}}} \times \frac{1\ \cancel{\text{mol } H_2}}{1\ \cancel{\text{mol Fe}}} \times \frac{2.02 \text{ g } H_2}{1\ \cancel{\text{mol } H_2}} = 0.12 \text{ g } H_2$$

Check: Units of answers, g HBr, g H_2, are correct. The magnitude of the answers is reasonable because molar mass HBr is greater than Fe and molar mass H_2 is much less than Fe.

8. **Given:** 15.2 g Al **Find:** g H_2SO_4; g H_2

Conceptual Plan: **g Al → mol Al → mol H_2SO_4 → g H_2SO_4**

$$\frac{\text{mol Al}}{26.98 \text{ g Al}} \qquad \frac{3 \text{ mol } H_2SO_4}{2 \text{ mol Al}} \qquad \frac{98.09 \text{ g } H_2SO_4}{\text{mol } H_2SO_4}$$

g Al → mol Al → mol H_2 → g H_2

$$\frac{\text{mol Al}}{26.98 \text{ g Al}} \qquad \frac{3 \text{ mol } H_2}{2 \text{ mol Al}} \qquad \frac{2.016 \text{ g } H_2}{\text{mol } H_2}$$

Solution:

$$15.2\ \cancel{\text{g Al}} \times \frac{1\ \cancel{\text{mol Al}}}{26.98\ \cancel{\text{g Al}}} \times \frac{3\ \cancel{\text{mol } H_2SO_4}}{2\ \cancel{\text{mol Al}}} \times \frac{98.09 \text{ g } H_2SO_4}{1\ \cancel{\text{mol } H_2SO_4}} = 82.9 \text{ g } H_2SO_4$$

$$15.2\ \cancel{\text{g Al}} \times \frac{1\ \cancel{\text{mol Al}}}{26.98\ \cancel{\text{g Al}}} \times \frac{3\ \cancel{\text{mol } H_2}}{2\ \cancel{\text{mol Al}}} \times \frac{2.016 \text{ g } H_2}{1\ \cancel{\text{mol } H_2}} = 1.70 \text{ g } H_2$$

Check: Units of answers, g H_2SO_4, g H_2, are correct. The magnitude of the answers is reasonable because molar mass H_2SO_4 is greater than Al and molar mass H_2 is much less than Al.

9. a) **Given:** 2.5 g Ba **Find:** g $BaCl_2$

Conceptual Plan: **g Ba → mol Ba → mol $BaCl_2$ → g $BaCl_2$**

$$\frac{\text{mol Ba}}{137.33\text{ g Ba}} \quad \frac{1\text{ mol BaCl}_2}{1\text{ mol Ba}} \quad \frac{208.23\text{ g BaCl}_2}{1\text{ mol BaCl}_2}$$

Solution: $2.5\text{ g Ba} \times \frac{1\text{ mol Ba}}{137.33\text{ g Ba}} \times \frac{1\text{ mol BaCl}_2}{1\text{ mol Ba}} \times \frac{208.23\text{ g BaCl}_2}{1\text{ mol BaCl}_2} = 3.8\text{ g BaCl}_2$

Check: Units of answer, g $BaCl_2$, are correct. The magnitude of the answer is reasonable because it is larger than grams Ba.

b) **Given:** 2.5 g CaO **Find:** g $CaCO_3$

Conceptual Plan: **g CaO → mol CaO → mol $CaCO_3$ → g $CaCO_3$**

$$\frac{\text{mol CaO}}{56.08\text{ g CaO}} \quad \frac{\text{mol CaCO}_3}{1\text{ mol CaO}} \quad \frac{100.09\text{ g CaCO}_3}{\text{mol CaCO}_3}$$

Solution: $2.5\text{ g CaO} \times \frac{1\text{ mol CaO}}{56.08\text{ g CaO}} \times \frac{1\text{ mol CaCO}_3}{1\text{ mol CaO}} \times \frac{100.09\text{ g CaCO}_3}{1\text{ mol CaCO}_3} = 4.5\text{ g CaCO}_3$

Check: Units of answer, g $CaCO_3$, are correct. The magnitude of the answer is reasonable because it is larger than grams CaO.

c) **Given:** 2.5 g Mg **Find:** g MgO

Conceptual Plan: **g Mg → mol Mg → mol MgO → g MgO**

$$\frac{\text{mol Mg}}{24.30\text{ g Mg}} \quad \frac{\text{mol MgO}}{\text{mol Mg}} \quad \frac{40.30\text{ g MgO}}{\text{mol MgO}}$$

Solution: $2.5\text{ g Mg} \times \frac{1\text{ mol Mg}}{24.30\text{ g Mg}} \times \frac{1\text{ mol MgO}}{1\text{ mol Mg}} \times \frac{40.30\text{ g MgO}}{1\text{ mol MgO}} = 4.1\text{ g MgO}$

Check: Units of answer, g MgO, are correct. The magnitude of the answer is reasonable because it is larger than grams Mg.

d) **Given:** 2.5 g Al **Find:** g Al_2O_3

Conceptual Plan: **g Al → mol Al → mol Al_2O_3 → g Al_2O_3**

$$\frac{\text{mol Al}}{26.98\text{ g Al}} \quad \frac{2\text{ mol Al}_2\text{O}_3}{4\text{ mol Al}} \quad \frac{101.96\text{ g Al}_2\text{O}_3}{\text{mol Al}_2\text{O}_3}$$

Solution: $2.5\text{ g Al} \times \frac{1\text{ mol Al}}{26.98\text{ g Al}} \times \frac{1\text{ mol Al}_2\text{O}_3}{1\text{ mol Al}} \times \frac{101.96\text{ g Al}_2\text{O}_3}{1\text{ mol Al}_2\text{O}_3} = 4.7\text{ g Al}_2\text{O}_3$

Check: Units of answer, g Al_2O_3, are correct. The magnitude of the answer is reasonable because it is larger than grams Al.

10. a) **Given:** 10.4 g K **Find:** g KCl

Conceptual Plan: **g K → mol K → mol KCl → g KCl**

$$\frac{\text{mol K}}{39.10\text{ g K}} \quad \frac{2\text{ mol KCl}}{2\text{ mol K}} \quad \frac{74.55\text{ g KCl}}{\text{mol KCl}}$$

Solution: $10.4\text{ g K} \times \frac{1\text{ mol K}}{39.10\text{ g K}} \times \frac{2\text{ mol KCl}}{2\text{ mol K}} \times \frac{74.55\text{ g KCl}}{1\text{ mol KCl}} = 19.8\text{ g KCl}$

Check: Units of answer, g KCl, are correct. The magnitude of the answer is reasonable because it is larger than grams K.

b) **Given:** 10.4 g K **Find:** g KBr

Conceptual Plan: **g K → mol K → mol KBr → g KBr**

$$\frac{\text{mol K}}{39.10\text{ g K}} \quad \frac{2\text{ mol KBr}}{2\text{ mol K}} \quad \frac{119.00\text{ g KBr}}{\text{mol KBr}}$$

Solution: $$10.4\ \text{g K} \times \frac{1\ \text{mol K}}{39.10\ \text{g K}} \times \frac{2\ \text{mol KBr}}{2\ \text{mol K}} \times \frac{119.00\ \text{g KBr}}{1\ \text{mol KBr}} = 31.7\ \text{g KBr}$$

Check: Units of answer, g KBr, are correct. The magnitude of the answer is reasonable because it is larger than grams K.

c) **Given:** 10.4 g Cr **Find:** g Cr_2O_3
Conceptual Plan: **g Cr → mol Cr → mol Cr_2O_3 → g Cr_2O_3**

$$\frac{\text{mol Cr}}{52.00\ \text{g Cr}} \quad \frac{2\ \text{mol } Cr_2O_3}{4\ \text{mol Cr}} \quad \frac{152.00\ \text{g } Cr_2O_3}{\text{mol } Cr_2O_3}$$

Solution: $$10.4\ \text{g Cr} \times \frac{1\ \text{mol Cr}}{52.00\ \text{g Cr}} \times \frac{2\ \text{mol } Cr_2O_3}{4\ \text{mol Cr}} \times \frac{152.00\ \text{g } Cr_2O_3}{1\ \text{mol } Cr_2O_3} = 15.2\ \text{g } Cr_2O_3$$

Check: Units of answer, g Cr_2O_3, are correct. The magnitude of the answer is reasonable because it is larger than g Cr.

d) **Given:** 10.4 g Sr **Find:** g SrO
Conceptual Plan: **g Sr → mol Sr → mol SrO → g SrO**

$$\frac{\text{mol Sr}}{87.62\ \text{g Sr}} \quad \frac{2\ \text{mol SrO}}{2\ \text{mol Sr}} \quad \frac{103.62\ \text{g SrO}}{\text{mol SrO}}$$

Solution: $$10.4\ \text{g Sr} \times \frac{1\ \text{mol Sr}}{87.62\ \text{g Sr}} \times \frac{2\ \text{mol SrO}}{2\ \text{mol Sr}} \times \frac{103.62\ \text{g SrO}}{1\ \text{mol SrO}} = 12.3\ \text{g SrO}$$

Check: Units of answer, g SrO, are correct. The magnitude of the answer is reasonable because it is larger than g Sr.

11. a) **Given:** 4.85 g NaOH **Find:** g HCl
Conceptual Plan: g NaOH → mol NaOH → mol HCl → g HCl

$$\frac{\text{mol NaOH}}{40.01\ \text{g NaOH}} \quad \frac{1\ \text{mol HCl}}{1\ \text{mol NaOH}} \quad \frac{36.46\ \text{g HCl}}{1\ \text{mol HCl}}$$

Solution: $$4.85\ \text{g NaOH} \times \frac{1\ \text{mol NaOH}}{40.01\ \text{g NaOH}} \times \frac{1\ \text{mol HCl}}{1\ \text{mol NaOH}} \times \frac{36.46\ \text{g HCl}}{1\ \text{mol HCl}} = 4\ 42\ \text{g HCl}$$

Check: Units of answer, g HCl, are correct. The magnitude of the answer is reasonable since it is less than g NaOH.

b) **Given:** 4.85 g $Ca(OH)_2$ **Find:** g HNO_3
Conceptual Plan: g $Ca(OH)_2$ → mol $Ca(OH)_2$ → mol HNO_3 → g HNO_3

$$\frac{\text{mol } Ca(OH)_2}{74.10\ \text{g } Ca(OH)_2} \quad \frac{2\ \text{mol } HNO_3}{1\ \text{mol } Ca(OH)_2} \quad \frac{63.02\ \text{g } HNO_3}{1\ \text{mol } HNO_3}$$

Solution: $$4.85\ \text{g } Ca(OH)_2 \times \frac{1\ \text{mol } Ca(OH)_2}{74.10\ \text{g } Ca(OH)_2} \times \frac{2\ \text{mol } HNO_3}{1\ \text{mol } Ca(OH)_2} \times \frac{63.02\ \text{g } HNO_3}{1\ \text{mol } HNO_3} = 8.25\ \text{g } HNO_3$$

Check: Units of answer, g HNO_3, are correct. The magnitude of the answer is reasonable since it is more than g $Ca(OH)_2$.

c) **Given:** 4.85 g KOH **Find:** g H_2SO_4
Conceptual Plan: g KOH → mol KOH → mol H_2SO_4 → g H_2SO_4

$$\frac{\text{mol KOH}}{56.11\ \text{g KOH}} \quad \frac{1\ \text{mol } H_2SO_4}{2\ \text{mol KOH}} \quad \frac{98.09\ \text{g } H_2SO_4}{1\ \text{mol } H_2SO_4}$$

Solution: $$4.85\ \text{g NaOH} \times \frac{1\ \text{mol KOH}}{56.11\ \text{g KOH}} \times \frac{1\ \text{mol } H_2SO_4}{2\ \text{mol KOH}} \times \frac{98.09\ \text{g } H_2SO_4}{1\ \text{mol } H_2SO_4} = 4\ 24\ \text{g } H_2SO_4$$

Check: Units of answer, g H_2SO_4, are correct. The magnitude of the answer is reasonable since it is less than g KOH.

12. a) **Given:** 55.8 g $Pb(NO_3)_2$ **Find:** g KI
Conceptual Plan: **g $Pb(NO_3)_2$ → mol $Pb(NO_3)_2$ → mol KI → g KI**

$$\frac{\text{mol Pb(NO}_3)_2}{331.2\text{ g Pb(NO}_3)_2} \quad \frac{2\text{ mol KI}}{1\text{ mol Pb(NO}_3)_2} \quad \frac{166.00\text{ g KI}}{1\text{ mol KI}}$$

Solution: $55.8 \text{ g Pb(NO}_3)_2 \times \frac{1\text{ mol Pb(NO}_3)_2}{331.2\text{ g Pb(NO}_3)_2} \times \frac{2\text{ mol KI}}{1\text{ mol Pb(NO}_3)_2} \times \frac{166.00\text{ g KI}}{1\text{ mol KI}} = 55.9\text{ g KI}$

Check: Units of answer, g KI, are correct. The magnitude of the answer is reasonable since there are 2 mol KI for each $Pb(NO_3)_2$.

b) **Given:** 55.8 g $CuCl_2$ **Find:** g Na_2CO_3
Conceptual Plan: **g $CuCl_2$ → mol $CuCl_2$ → mol Na_2CO_3 → g Na_2CO_3**

$$\frac{\text{mol CuCl}_2}{134.45\text{ g CuCl}_2} \quad \frac{1\text{ mol Na}_2\text{CO}_3}{1\text{ mol CuCl}_2} \quad \frac{106.01\text{ g Na}_2\text{CO}_3}{1\text{ mol Na}_2\text{CO}_3}$$

Solution: $55.8 \text{ g CuCl}_2 \times \frac{1\text{ mol CuCl}_2}{134.45\text{ g CuCl}_2} \times \frac{1\text{ mol Na}_2\text{CO}_3}{1\text{ mol CuCl}_2} \times \frac{106.01\text{ g Na}_2\text{CO}_3}{1\text{ mol Na}_2\text{CO}_3} = 44.0\text{ g Na}_2\text{CO}_3$

Check: Units of answer, g Na_2CO_3, are correct. The magnitude of the answer is reasonable since it is less than g $CuCl_2$.

c) **Given:** 55.8 g $Sr(NO_3)_2$ **Find:** g K_2SO_4
Conceptual Plan: **g $Sr(NO_3)_2$ → mol $Sr(NO_3)_2$ → mol K_2SO_4 → g K_2SO_4**

$$\frac{\text{mol Sr(NO}_3)_2}{211.64\text{ g Sr(NO}_3)_2} \quad \frac{1\text{ mol K}_2\text{SO}_4}{1\text{ mol Sr(NO}_3)_2} \quad \frac{174.27\text{ g K}_2\text{SO}_4}{1\text{ mol K}_2\text{SO}_4}$$

Solution: $55.8 \text{ g Sr(NO}_3)_2 \times \frac{1\text{ mol Sr(NO}_3)_2}{211.64\text{ g Sr(NO}_3)_2} \times \frac{1\text{ mol K}_2\text{SO}_4}{1\text{ mol Sr(NO}_3)_2} \times \frac{174.27\text{ g K}_2\text{SO}_4}{1\text{ mol K}_2\text{SO}_4} = 45.9\text{ g K}_2\text{SO}_4$

Check: Units of answer, g K_2SO_4, are correct. The magnitude of the answer is reasonable since it is less than g $Sr(NO_3)_2$.

13. a) **Given:** 2 mol Na; 2 mol Br_2 **Find:** Limiting reactant
Conceptual Plan: **mol Na → mol NaBr**

$$\frac{2\text{ mol NaBr}}{2\text{ mol Na}}$$

→ smallest mol amount determines limiting reactant

mol Br_2 → mol NaBr

$$\frac{2\text{ mol NaBr}}{1\text{ mol Br}_2}$$

Solution:

$$2\text{ mol Na} \times \frac{2\text{ mol NaBr}}{2\text{ mol Na}} = 2\text{ mol NaBr}$$

$$2\text{ mol Br}_2 \times \frac{2\text{ mol NaBr}}{1\text{ mol Br}_2} = 4\text{ mol NaBr}$$

Na is limiting reactant

Check: Answer is reasonable since Na produced smallest amount of product.

b) **Given:** 1.8 mol Na; 1.4 mol Br_2 **Find:** Limiting reactant
Conceptual Plan: **mol Na → mol NaBr**

$$\frac{2\text{ mol NaBr}}{2\text{ mol Na}}$$

→ smallest mol amount determines limiting reactant

mol Br_2 → mol NaBr

$$\frac{2\text{ mol NaBr}}{1\text{ mol Br}_2}$$

Solution: $1.8\text{ mol Na} \times \frac{2\text{ mol NaBr}}{2\text{ mol Na}} = 1.8\text{ mol NaBr}$

$$1.4\ \cancel{mol\ Br_2} \times \frac{2\ mol\ NaBr}{1\ \cancel{mol\ Br_2}} = 2.8\ mol\ NaBr$$

Na is limiting reactant

Check: Answer is reasonable since Na produced smallest amount of product.

c) **Given:** 2.5 mol Na; 1 mol Br_2 **Find:** Limiting reactant

Conceptual Plan: mol Na → mol NaBr

$$\frac{2\ mol\ NaBr}{2\ mol\ Na}$$

→ smallest mol amount determines limiting reactant

mol Br_2 → mol NaBr

$$\frac{2\ mol\ NaBr}{1\ mol\ Br_2}$$

Solution:

$$2.5\ \cancel{mol\ Na} \times \frac{2\ mol\ NaBr}{2\ \cancel{mol\ Na}} = 2.5\ mol\ NaBr$$

$$1\ \cancel{mol\ Br_2} \times \frac{2\ mol\ NaBr}{1\ \cancel{mol\ Br_2}} = 2\ mol\ NaBr$$

Br_2 is limiting reactant

Check: Answer is reasonable since Br_2 produced smallest amount of product.

d) **Given:** 12.6 mol Na; 6.9 mol Br_2 **Find:** Limiting reactant

Conceptual Plan: mol Na → mol NaBr

$$\frac{2\ mol\ NaBr}{2\ mol\ Na}$$

→ smallest mol amount determines limiting reactant

mol Br_2 → mol NaBr

$$\frac{2\ mol\ NaBr}{1\ mol\ Br_2}$$

Solution:

$$12.6\ \cancel{mol\ Na} \times \frac{2\ mol\ NaBr}{2\ \cancel{mol\ Na}} = 12.6\ mol\ NaBr$$

$$6.9\ \cancel{mol\ Br_2} \times \frac{2\ mol\ NaBr}{1\ \cancel{mol\ Br_2}} = 13.8\ mol\ NaBr$$

Na is limiting reactant

Check: Answer is reasonable since Na produced smallest amount of product.

14\. a) **Given:** 1 mol Al; 1 mol O_2 **Find:** Limiting reactant

Conceptual Plan: mol Al → mol Al_2O_3

$$\frac{2\ mol\ Al_2O_3}{4\ mol\ Al}$$

→ smallest mol amount determines limiting reactant

mol O_2 → mol Al_2O_3

$$\frac{2\ mol\ Al_2O_3}{3\ mol\ O_2}$$

Solution:

$$1\ \cancel{mol\ Al} \times \frac{2\ mol\ Al_2O_3}{4\ \cancel{mol\ Al}} = 0.5\ mol\ Al_2O_3$$

$$1\ \cancel{mol\ O_2} \times \frac{2\ mol\ Al_2O_3}{3\ \cancel{mol\ O_2}} = 0.67\ mol\ Al_2O_3$$

Al is limiting reactant

Check: Answer is reasonable since Al produced smallest amount of product.

b) **Given:** 4 mol Al; 2.6 mol O_2 **Find:** Limiting reactant

Conceptual Plan: **mol Al → mol Al_2O_3**

$$\frac{2 \text{ mol } Al_2O_3}{4 \text{ mol Al}}$$ **→ smallest mol amount determines limiting reactant**

mol O_2 → mol Al_2O_3

$$\frac{2 \text{ mol } Al_2O_3}{3 \text{ mol } O_2}$$

Solution:

$$4 \cancel{\text{mol Al}} \times \frac{2 \text{ mol } Al_2O_3}{4 \cancel{\text{mol Al}}} = 2 \text{ mol } Al_2O_3$$

$$2.6 \cancel{\text{mol } O_2} \times \frac{2 \text{ mol } Al_2O_3}{3 \cancel{\text{mol } O_2}} = 1.7 \text{ mol } Al_2O_3$$

O_2 is limiting reactant

Check: Answer is reasonable since O_2 produced smallest amount of product.

c) **Given:** 16 mol Al; 13 mol O_2 **Find:** Limiting reactant

Conceptual Plan: **mol Al → mol Al_2O_3**

$$\frac{2 \text{ mol } Al_2O_3}{4 \text{ mol Al}}$$ **→ smallest mol amount determines limiting reactant**

mol O_2 → mol Al_2O_3

$$\frac{2 \text{ mol } Al_2O_3}{3 \text{ mol } O_2}$$

Solution:

$$16 \cancel{\text{mol Al}} \times \frac{2 \text{ mol } Al_2O_3}{4 \cancel{\text{mol Al}}} = 8.0 \text{ mol } Al_2O_3$$

$$13 \cancel{\text{mol } O_2} \times \frac{2 \text{ mol } Al_2O_3}{3 \cancel{\text{mol } O_2}} = 8.67 \text{ mol } Al_2O_3$$

Al is limiting reactant

Check: Answer is reasonable since Al produced smallest amount of product.

d) **Given:** 7.4 mol Al; 6.5 mol O_2 **Find:** Limiting reactant

Conceptual Plan: **mol Al → mol Al_2O_3**

$$\frac{2 \text{ mol } Al_2O_3}{4 \text{ mol Al}}$$ **→ smallest mol amount determines limiting reactant**

mol O_2 → mol Al_2O_3

$$\frac{2 \text{ mol } Al_2O_3}{3 \text{ mol } O_2}$$

Solution:

$$7.4 \cancel{\text{mol Al}} \times \frac{2 \text{ mol } Al_2O_3}{4 \cancel{\text{mol Al}}} = 3.7 \text{ mol } Al_2O_3$$

$$6.5 \cancel{\text{mol } O_2} \times \frac{2 \text{ mol } Al_2O_3}{3 \cancel{\text{mol } O_2}} = 4.3 \text{ mol } Al_2O_3$$

Al is limiting reactant

Check: Answer is reasonable since Al produced smallest amount of product.

15. The greatest number of Cl_2 molecules will be formed from reaction mixture b and would be 3 molecules Cl_2.

a) **Given:** 7 molecules HCl, 1 molecule O_2 **Find:** Theoretical yield Cl_2

Conceptual Plan: **molecule HCl → molecules Cl_2**

$$\frac{2 \text{ molecule } Cl_2}{4 \text{ molecule HCl}}$$ **→ smallest molecule amount determines limiting reactant**

molecules O_2 → molecules Cl_2

$$\frac{2 \text{ molecules } Cl_2}{1 \text{ molecules } O_2}$$

Solution:

$$7 \text{ molecules HCl} \times \frac{2 \text{ molecules } Cl_2}{4 \text{ molecules HCl}} = 3 \text{ molecules } Cl_2$$

$$1 \text{ molecules } O_2 \times \frac{2 \text{ molecules } Cl_2}{1 \text{ molecules } O_2} = 2 \text{ molecules } Cl_2$$

Theoretical Yield = 2 molecules Cl_2

b) **Given:** 6 molecules HCl, molecule O_2 **Find:** Theoretical yield Cl_2

Conceptual Plan: molecule HCl → molecules Cl_2

$$\frac{2 \text{ molecule } Cl_2}{4 \text{ molecule HCl}}$$

→ smallest molecule amount determines limiting reactant

molecules O_2 → molecules Cl_2

$$\frac{2 \text{ molecules } Cl_2}{1 \text{ molecules } O_2}$$

Solution:

$$6 \text{ molecules HCl} \times \frac{2 \text{ molecules } Cl_2}{4 \text{ molecules HCl}} = 3 \text{ molecules } Cl_2$$

$$3 \text{ molecules } O_2 \times \frac{2 \text{ molecules } Cl_2}{1 \text{ molecules } O_2} = 6 \text{ molecules } Cl_2$$

Theoretical Yield = 3 molecules Cl_2

c) **Given:** 4 molecules HCl, 5 molecule O_2 **Find:** Theoretical yield Cl_2

Conceptual Plan: molecule HCl → molecules Cl_2

$$\frac{2 \text{ molecule } Cl_2}{4 \text{ molecule HCl}}$$

→ smallest molecule amount determines limiting reactant

molecules O_2 → molecules Cl_2

$$\frac{2 \text{ molecules } Cl_2}{1 \text{ molecules } O_2}$$

Solution:

$$4 \text{ molecules HCl} \times \frac{2 \text{ molecules } Cl_2}{4 \text{ molecules HCl}} = 2 \text{ molecules } Cl_2$$

$$5 \text{ molecules } O_2 \times \frac{2 \text{ molecules } Cl_2}{1 \text{ molecules } O_2} = 10 \text{ molecules } Cl_2$$

Theoretical Yield = 2 molecules Cl_2

Check: The units of the answer, molecules Cl_2, is correct. The answer is reasonable based on the limiting reactant in each mixture.

16. The greatest number of CO_2 molecules will be formed from reaction mixture a, and would be 2 molecules CO_2.

a) **Given:** 3 molecules CH_3OH, 3 molecule O_2 **Find:** Theoretical yield CO_2

Conceptual Plan: molecule CH_3OH → molecules CO_2

$$\frac{2 \text{ molecule } CO_2}{2 \text{ molecule } CH_3OH}$$

→ smallest molecule amount determines limiting reactant

molecules O_2 → molecules CO_2

$$\frac{2 \text{ molecules } CO_2}{3 \text{ molecules } O_2}$$

Solution:

$$3 \cancel{\text{molecules } CH_3OH} \times \frac{2 \text{ molecules } CO_2}{2 \cancel{\text{molecules } CH_3OH}} = 3 \text{ molecules } CO_2$$

$$3 \cancel{\text{molecules } O_2} \times \frac{2 \text{ molecules } CO_2}{3 \cancel{\text{molecules } O_2}} = 2 \text{ molecules } CO_2$$

Theoretical Yield = 2 molecules CO_2

b) **Given:** 1 molecules CH_3OH, 6 molecule O_2 **Find:** Theoretical yield CO_2
Conceptual Plan: molecule CH_3OH → molecules CO_2

$$\frac{2 \text{ molecule } CO_2}{2 \text{ molecule } CH_3OH}$$

→ smallest molecule amount determines limiting reactant

molecules O_2 → molecules CO_2

$$\frac{2 \text{ molecules } CO_2}{3 \text{ molecules } O_2}$$

Solution:

$$1 \cancel{\text{molecules } CH_3OH} \times \frac{2 \text{ molecules } CO_2}{2 \cancel{\text{molecules } CH_3OH}} = 1 \text{ molecules } CO_2$$

$$6 \cancel{\text{molecules } O_2} \times \frac{2 \text{ molecules } CO_2}{3 \cancel{\text{molecules } O_2}} = 4 \text{ molecules } CO_2$$

Theoretical Yield = 1 molecules CO_2

c) **Given:** 4 molecules CH_3OH, 2 molecule O_2 **Find:** Theoretical yield CO_2
Conceptual Plan: molecule CH_3OH → molecules CO_2

$$\frac{2 \text{ molecule } CO_2}{2 \text{ molecule } CH_3OH}$$

→ smallest molecule amount determines limiting reactant

molecules O_2 → molecules CO_2

$$\frac{2 \text{ molecules } CO_2}{3 \text{ molecules } O_2}$$

Solution:

$$4 \cancel{\text{molecules } CH_3OH} \times \frac{2 \text{ molecules } CO_2}{2 \cancel{\text{molecules } CH_3OH}} = 4 \text{ molecules } CO_2$$

$$2 \cancel{\text{molecules } O_2} \times \frac{2 \text{ molecules } CO_2}{3 \cancel{\text{molecules } O_2}} = 1 \text{ molecules } CO_2$$

Theoretical Yield = 1 molecules CO_2

Check: The units of the answer, molecules CO_2, is correct. The answer is reasonable based on the limiting reactant in each mixture.

17. a) **Given:** 4 mol Ti, 4 mol Cl_2 **Find:** Theoretical yield $TiCl_4$
Conceptual Plan: mol Ti → mol $TiCl_4$

$$\frac{1 \text{ mol } TiCl_4}{1 \text{ mol Ti}}$$

→ smallest mol amount determines limiting reactant

mol Cl_2 → mol $TiCl_4$

$$\frac{1 \text{ mol } TiCl_4}{2 \text{ mol } Cl_2}$$

Solution:

$$4\ \cancel{mol\ Ti} \times \frac{1\ mol\ TiCl_4}{1\ \cancel{mol\ Ti}} = 4\ mol\ TiCl_4$$

$$4\ \cancel{mol\ Cl_2} \times \frac{1\ mol\ TiCl_4}{2\ \cancel{mol\ Cl_2}} = 2\ mol\ TiCl_4$$

Theoretical Yield = 2 mol $TiCl_4$

Check: Units of the answer, mol $TiCl_4$, are correct. Answer is reasonable since Cl_2 produced smallest amount of product and is the limiting reactant.

b) **Given:** 7 mol Ti, 17 mol Cl_2 **Find:** Theoretical yield $TiCl_4$

Conceptual Plan: mol Ti → mol $TiCl_4$

$$\frac{1\ mol\ TiCl_4}{1\ mol\ Ti}$$

→ smallest mol amount determines limiting reactant

mol Cl_2 → mol $TiCl_4$

$$\frac{1\ mol\ TiCl_4}{2\ mol\ Cl_2}$$

Solution:

$$7\ \cancel{mol\ Ti} \times \frac{1\ mol\ TiCl_4}{1\ \cancel{mol\ Ti}} = 7\ mol\ TiCl_4$$

$$17\ \cancel{mol\ Cl_2} \times \frac{1\ mol\ TiCl_4}{2\ \cancel{mol\ Cl_2}} = 8.5\ mol\ TiCl_4$$

Theoretical Yield = 7 mol $TiCl_4$

Check: Units of the answer, mol $TiCl_4$, are correct. Answer is reasonable since Ti produced smallest amount of product and is the limiting reactant.

c) **Given:** 12.4 mol Ti, 18.8 mol Cl_2 **Find:** Theoretical yield $TiCl_4$

Conceptual Plan: mol Ti → mol $TiCl_4$

$$\frac{1\ mol\ TiCl_4}{1\ mol\ Ti}$$

→ smallest mol amount determines limiting reactant

mol Cl_2 → mol $TiCl_4$

$$\frac{1\ mol\ TiCl_4}{2\ mol\ Cl_2}$$

Solution:

$$12.4\ \cancel{mol\ Ti} \times \frac{1\ mol\ TiCl_4}{1\ \cancel{mol\ Ti}} = 12.4\ mol\ TiCl_4$$

$$18.8\ \cancel{mol\ Cl_2} \times \frac{1\ mol\ TiCl_4}{2\ \cancel{mol\ Cl_2}} = 9.4\ mol\ TiCl_4$$

Theoretical Yield = 9.4 mol $TiCl_4$

Check: Units of the answer, mol $TiCl_4$, are correct. Answer is reasonable since Cl_2 produced smallest amount of product and is the limiting reactant.

18. a) **Given:** 3 mol Mn, 3 mol O_2 **Find:** Theoretical yield MnO_2

Conceptual Plan: mol Mn → mol MnO_2

$$\frac{2\ mol\ MnO_2}{2\ mol\ Mn}$$

→ smallest mol amount determines limiting reactant

mol O_2 → mol MnO_2

$$\frac{2\ mol\ MnO_2}{2\ mol\ O_2}$$

Solution:

$$3\ \cancel{mol\ Mn} \times \frac{1\ mol\ MnO_2}{1\ \cancel{mol\ Mn}} = 3\ mol\ MnO_2$$

$$3 \text{ mol } O_2 \times \frac{1 \text{ mol } MnO_2}{1 \text{ mol } O_2} = 3 \text{ mol } MnO_2$$

Theoretical Yield = 3 mol MnO_2

Check: Units of the answer, mol MnO_2, are correct. Answer is reasonable since equal mol are produced for both reactants

b) **Given:** 4 mol Mn, 7 mol O_2 **Find:** Theoretical yield MnO_2

Conceptual Plan: mol Mn → mol MnO_2

$$\frac{2 \text{ mol } MnO_2}{2 \text{ mol Mn}}$$

→ smallest mol amount determines limiting reactant

mol O_2 → mol MnO_2

$$\frac{2 \text{ mol } MnO_2}{2 \text{ mol } O_2}$$

Solution:

$$4 \text{ mol Mn} \times \frac{2 \text{ mol } MnO_2}{2 \text{ mol Mn}} = 4 \text{ mol } MnO_2$$

$$7 \text{ mol } O_2 \times \frac{2 \text{ mol } MnO_2}{2 \text{ mol } O_2} = 7 \text{ mol } MnO_2$$

Theoretical Yield = 4 mol MnO_2

Check: Units of the answer, mol MnO_2, are correct. Answer is reasonable since Mn produced smallest amount of product and is the limiting reactant.

c) **Given:** 27.5 mol Mn, 43.8 mol O_2 **Find:** Theoretical yield MnO_2

Conceptual Plan: mol Mn → mol MnO_2

$$\frac{2 \text{ mol } MnO_2}{2 \text{ mol Mn}}$$

→ smallest mol amount determines limiting reactant

mol O_2 → mol MnO_2

$$\frac{2 \text{ mol } MnO_2}{2 \text{ mol } O_2}$$

Solution:

$$27.5 \text{ mol Mn} \times \frac{2 \text{ mol } MnO_2}{2 \text{ mol Mn}} = 27.5 \text{ mol } MnO_2$$

$$43.8 \text{ mol } O_2 \times \frac{2 \text{ mol } MnO_2}{2 \text{ mol } O_2} = 43.8 \text{ mol } MnO_2$$

Theoretical Yield = 27.5 mol MnO_2

Check: Units of the answer, mol MnO_2, are correct. Answer is reasonable since Mn produced smallest amount of product and is the limiting reactant.

19. a) **Given:** 2.0 g Al, 2.0 g Cl_2 **Find:** Theoretical yield in g $AlCl_3$

Conceptual Plan: g Al → mol Al → mol $AlCl_3$

$$\frac{1 \text{ mol Al}}{26.98 \text{ g Al}} \quad \frac{2 \text{ mol } AlCl_3}{2 \text{ mol Al}}$$

→ smallest mol amount determines limiting reactant

g Cl_2 → mol Cl_2 → mol $AlCl_3$

$$\frac{1 \text{ mol } Cl_2}{70.91 \text{ g } Cl_2} \quad \frac{2 \text{ mol } AlCl_3}{3 \text{ mol } Cl_2}$$

then: mol $AlCl_3$ → g $AlCl_3$

$$\frac{133.34 \text{ g } AlCl_3}{\text{mol } AlCl_3}$$

Solution:

$$2.0 \text{ g Al} \times \frac{1 \text{ mol Al}}{26.98 \text{ g Al}} \times \frac{2 \text{ mol } AlCl_3}{2 \text{ mol Al}} = 0.074 \text{ mol } AlCl_3$$

$$2.0\ \cancel{g\ Cl_2} \times \frac{1\ \cancel{mol\ Cl_2}}{70.90\ \cancel{g\ Cl_2}} \times \frac{2\ mol\ AlCl_3}{3\ \cancel{mol\ Cl_2}} = 0.01\underline{8}8\ mol\ AlCl_3$$

$$0.01\underline{8}8\ mol\ AlCl_3 \times \frac{133.34\ g\ AlCl_3}{mol\ AlCl_3} = 2.5\ g\ AlCl_3$$

Check: Units of the answer, g $AlCl_3$, are correct. Answer is reasonable since Cl_2 produced smallest amount of product and is the limiting reactant.

b) **Given:** 7.5 g Al, 24.8 g Cl_2 **Find:** Theoretical yield in g $AlCl_3$

Conceptual Plan: g Al → mol Al → mol $AlCl_3$

$$\frac{1\ mol\ Al}{26.98\ g\ Al} \quad \frac{2\ mol\ AlCl_3}{2\ mol\ Al}$$ **→ smallest mol amount determines limiting reactant**

g Cl_2 → mol Cl_2 → mol $AlCl_3$

$$\frac{1\ mol\ Cl_2}{70.91\ g\ Cl_2} \quad \frac{2\ mol\ AlCl_3}{3\ mol\ Cl_2}$$

then: mol $AlCl_3$ → g $AlCl_3$

$$\frac{133.34\ g\ AlCl_3}{mol\ AlCl_3}$$

Solution:

$$7.5\ \cancel{g\ Al} \times \frac{1\ \cancel{mol\ Al}}{26.98\ \cancel{g\ Al}} \times \frac{2\ mol\ AlCl_3}{2\ \cancel{mol\ Al}} = 0.2780\ mol\ AlCl_3$$

$$24.8\ \cancel{g\ Cl_2} \times \frac{1\ \cancel{mol\ Cl_2}}{70.90\ \cancel{g\ Cl_2}} \times \frac{2\ mol\ AlCl_3}{3\ \cancel{mol\ Cl_2}} = 0.23\underline{3}2\ mol\ AlCl_3$$

$$0.23\underline{3}2\ mol\ AlCl_3 \times \frac{133.34\ g\ AlCl_3}{mol\ AlCl_3} = 31.1\ g\ AlCl_3$$

Check: Units of the answer, g $AlCl_3$, are correct. Answer is reasonable since Cl_2 produced smallest amount of product and is the limiting reactant.

c) **Given:** 0.235 g Al, 1.15 g Cl_2 **Find:** Theoretical yield in g $AlCl_3$

Conceptual Plan: g Al → mol Al → mol $AlCl_3$

$$\frac{1\ mol\ Al}{26.98\ g\ Al} \quad \frac{2\ mol\ AlCl_3}{2\ mol\ Al}$$ **→ smallest mol amount determines limiting reactant**

g Cl_2 → mol Cl_2 → mol $AlCl_3$

$$\frac{1\ mol\ Cl_2}{70.91\ g\ Cl_2} \quad \frac{2\ mol\ AlCl_3}{3\ mol\ Cl_2}$$

then: mol $AlCl_3$ → g $AlCl_3$

$$\frac{133.34\ g\ AlCl_3}{mol\ AlCl_3}$$

Solution:

$$0.235\ \cancel{g\ Al} \times \frac{1\ \cancel{mol\ Al}}{26.98\ \cancel{g\ Al}} \times \frac{2\ mol\ AlCl_3}{2\ \cancel{mol\ Al}} = 0.0087\underline{1}0\ mol\ AlCl_3$$

$$1.15\ \cancel{g\ Cl_2} \times \frac{1\ \cancel{mol\ Cl_2}}{70.90\ \cancel{g\ Cl_2}} \times \frac{2\ mol\ AlCl_3}{3\ \cancel{mol\ Cl_2}} = 0.01081\ mol\ AlCl_3$$

$$0.0087\underline{1}0\ mol\ AlCl_3 \times \frac{133.34\ g\ AlCl_3}{mol\ AlCl_3} = 1.16\ g\ AlCl_3$$

Check: Units of the answer, g $AlCl_3$, are correct. Answer is reasonable since Al produced smallest amount of product and is the limiting reactant.

20. a) **Given:** 5.0 g Ti, 5.0 g F_2 **Find:** Theoretical yield in g TiF_4

Conceptual Plan: g Ti → mol Ti → mol TiF_4

$$\frac{1 \text{ mol Ti}}{47.87 \text{ g Ti}} \quad \frac{1 \text{ mol TiF}_4}{1 \text{ mol Ti}}$$ **→ smallest mol amount determines limiting reactant**

g F_2 → mol F_2 → mol TiF_4

$$\frac{1 \text{ mol F}_2}{38.00 \text{ g F}_2} \quad \frac{1 \text{ mol TiF}_4}{2 \text{ mol F}_2}$$

then: mol TiF_4 → g TiF_4

$$\frac{123.87 \text{ g TiF}_4}{\text{mol TiF}_4}$$

Solution:

$$5.0 \text{ g Ti} \times \frac{1 \text{ mol Ti}}{47.87 \text{ g Ti}} \times \frac{1 \text{ mol TiF}_4}{1 \text{ mol Ti}} = 0.1\underline{0}4 \text{ mol TiF}_4$$

$$5.0 \text{ g F}_2 \times \frac{1 \text{ mol F}_2}{38.00 \text{ g F}_2} \times \frac{1 \text{ mol TiF}_4}{2 \text{ mol F}_2} = 0.06\underline{5}8 \text{ mol TiF}_4$$

$$0.06\underline{5}8 \text{ mol TiF}_4 \times \frac{123.87 \text{ g TiF}_4}{\text{mol TiF}_4} = 8.1 \text{ g TiF}$$

Check: Units of the answer, g TiF_4, are correct. Answer is reasonable since F_2 produced smallest amount of product and is the limiting reactant.

b) **Given:** 2.4 g Ti, 1.6 g F_2 **Find:** Theoretical yield in g TiF_4

Conceptual Plan: g Ti → mol Ti → mol TiF_4

$$\frac{1 \text{ mol Ti}}{47.87 \text{ g Ti}} \quad \frac{1 \text{ mol TiF}_4}{1 \text{ mol Ti}}$$ **→ smallest mol amount determines limiting reactant**

g F_2 → mol F_2 → mol TiF_4

$$\frac{1 \text{ mol F}_2}{38.00 \text{ g F}_2} \quad \frac{1 \text{ mol TiF}_4}{2 \text{ mol F}_2}$$

then: mol TiF_4 → g TiF_4

$$\frac{123.87 \text{ g TiF}_4}{\text{mol TiF}_4}$$

Solution:

$$2.4 \text{ g Ti} \times \frac{1 \text{ mol Ti}}{47.87 \text{ g Ti}} \times \frac{1 \text{ mol TiF}_4}{1 \text{ mol Ti}} = 0.0501 \text{ mol TiF}_4$$

$$1.6 \text{ g F}_2 \times \frac{1 \text{ mol F}_2}{38.00 \text{ g F}_2} \times \frac{1 \text{ mol TiF}_4}{2 \text{ mol F}_2} = 0.02\underline{1}0 \text{ mol TiF}_4$$

$$0.02\underline{1}0 \text{ mol TiF}_4 \times \frac{123.87 \text{ g TiF}_4}{\text{mol TiF}_4} = 2.6 \text{ g TiF}_4$$

Check: Units of the answer, g TiF_4, are correct. Answer is reasonable since F_2 produced smallest amount of product and is the limiting reactant.

c) **Given:** 0.233 g Ti, 0.288 g F_2 **Find:** Theoretical yield in g TiF_4

Conceptual Plan: g Ti → mol Ti → mol TiF_4

$$\frac{1 \text{ mol Ti}}{47.87 \text{ g Ti}} \quad \frac{1 \text{ mol TiF}_4}{1 \text{ mol Ti}}$$ **→ smallest mol amount determines limiting reactant**

g F_2 → mol F_2 → mol TiF_4

$$\frac{1 \text{ mol F}_2}{38.00 \text{ g F}_2} \quad \frac{1 \text{ mol TiF}_4}{2 \text{ mol F}_2}$$

then: mol TiF_4 → g TiF_4

$$\frac{123.87 \text{ g TiF}_4}{\text{mol TiF}_4}$$

Solution:

$$0.233 \text{ g Ti} \times \frac{1 \text{ mol Ti}}{47.87 \text{ g Ti}} \times \frac{1 \text{ mol TiF}_4}{1 \text{ mol Ti}} = 0.004867 \text{ mol TiF}_4$$

$$0.288 \text{ g F}_2 \times \frac{1 \text{ mol F}_2}{38.00 \text{ g F}_2} \times \frac{1 \text{ mol TiF}_4}{2 \text{ mol F}_2} = 0.003\underline{7}89 \text{ mol TiF}_4$$

$$0.003\underline{7}89 \text{ mol TiF}_4 \times \frac{123.87 \text{ g TiF}_4}{\text{mol TiF}_4} = 0.469 \text{ g TiF}_4$$

Check: Units of the answer, g TiF_4, are correct. Answer is reasonable since F_2 produced smallest amount of product and is the limiting reactant.

21. **Given:** 28.5 g KCl; 25.7 g Pb^{2+}; 29.4 g $PbCl_2$ **Find:** limiting reactant, theoretical yield $PbCl_2$, % yield

Conceptual Plan: **g KCl → mol KCl → mol $PbCl_2$**

$$\frac{1 \text{ mol KCl}}{74.55 \text{ g KCl}} \quad \frac{1 \text{ mol PbCl}_2}{2 \text{ mol KCl}}$$

→ smallest mol amount determines limiting reactant

g Pb^{2+} → mol Pb^{2+} → mol $PbCl_2$

$$\frac{1 \text{ mol Pb}^{2+}}{207.2 \text{ g Pb}^{2+}} \quad \frac{1 \text{ mol PbCl}_2}{1 \text{ mol Pb}^{2+}}$$

then: mol $PbCl_2$ → g $PbCl_2$

$$\frac{278.1 \text{ g PbCl}_2}{\text{mol PbCl}_2}$$

then: determine % yield

$$\frac{\text{actual yield g PbCl}_2}{\text{theoretical yield g PbCl}_2} \times 100$$

Solution:

$$28.5 \text{ g KCl} \times \frac{1 \text{ mol KCl}}{74.55 \text{ g KCl}} \times \frac{1 \text{ mol PbCl}_2}{2 \text{ mol KCl}} = 0.19\underline{1}1 \text{ mol PbCl}_2$$

$$25.7 \text{ g Pb}^{2+} \times \frac{1 \text{ mol Pb}^{2+}}{207.2 \text{ g Pb}^{2+}} \times \frac{1 \text{ mol PbCl}_2}{1 \text{ mol Pb}^{2+}} = 0.1240 \text{ mol PbCl}_2$$

$$0.12\underline{4}0 \text{ mol PbCl}_2 \times \frac{278.1 \text{ g PbCl}_2}{1 \text{ mol PbCl}_2} = 34.\underline{5} \text{ g PbCl}_2$$

$$\frac{29.4 \text{ g PbCl}_2}{34.\underline{5} \text{ g PbCl}_2} \times 100\% = 85.2\%$$

Check: The theoretical yield has the correct units, g $PbCl_2$, and has a reasonable magnitude compared to the mass of Pb^{2+}, the limiting reactant. The % yield is reasonable, under 100%.

22. **Given:** 10.1 g Mg; 10.5 g O_2; 11.9 g MgO **Find:** limiting reactant, theoretical yield $PbCl_2$, % yield

Conceptual Plan: **g Mg → mol Mg → mol MgO**

$$\frac{1 \text{ mol Mg}}{24.30 \text{ g Mg}} \quad \frac{2 \text{ mol MgO}}{2 \text{ mol Mg}}$$

→ smallest mol amount determines limiting reactant

g O_2 → mol O_2 → mol MgO

$$\frac{1 \text{ mol O}_2}{32.00 \text{ g O}_2} \quad \frac{2 \text{ mol MgO}}{1 \text{ mol O}_2}$$

then: mol MgO → g MgO

$$\frac{40.30 \text{ g MgO}}{1 \text{ mol MgO}}$$

then: determine % yield

$$\frac{\text{actual yield g MgO}}{\text{theoretical yield g MgO}} \times 100$$

Solution:

$$10.1 \text{ g Mg} \times \frac{1 \text{ mol Mg}}{24.30 \text{ g Mg}} \times \frac{2 \text{ mol MgO}}{2 \text{ mol Mg}} = 0.41\underline{5}6 \text{ mol MgO}$$

$$10.5\ g\ O_2 \times \frac{1\ mol\ O_2}{32.00\ g\ O_2} \times \frac{2\ mol\ MgO}{1\ mol\ O_2} = 0.65\underline{6}2\ mol\ MgO$$

$$0.41\underline{5}6\ mol\ MgO \times \frac{40.30\ g\ MgO}{1\ mol\ MgO} = 16.\underline{7}5\ g\ MgO$$

$$\frac{11.9\ g\ MgO}{16.\underline{7}5\ g\ MgO} \times 100\% = 71.0\%$$

Check: The theoretical yield has the correct units, g MgO, and has a reasonable magnitude compared to the mass of Mg, the limiting reactant. The % yield is reasonable, under 100%.

23. **Given:** 136.4 kg NH_3; 211.4 kg CO_2; 168.4 kg CH_4N_2O **Find:** limiting reactant, theoretical yield CH_4N_2O, % yield

Conceptual Plan: kg NH_3 → g NH_3 → mol NH_3 → mol CH_4N_2O

$$\frac{1000\ g}{1\ kg} \quad \frac{1\ mol\ NH_3}{17.03\ g\ NH_3} \quad \frac{1\ mol\ CH_4N_2O}{2\ mol\ NH_3}$$

→ smallest amount determines limiting reactant

kg CO_2 → g CO_2 → mol CO_2 → mol CH_4N_2O

$$\frac{1000\ g}{1\ kg} \quad \frac{1\ mol\ CO_2}{44.01\ g\ CO_2} \quad \frac{1\ mol\ CH_4N_2O}{1\ mol\ CO_2}$$

then: mol CH_4N_2O → g CH_4N_2O → kg CH_4N_2O

$$\frac{60.06\ g\ CH_4N_2O}{1\ mol\ CH_4N_2O} \quad \frac{1\ kg}{1000\ g}$$

then: determine % yield

$$\frac{actual\ yield\ kg\ CH_4N_2O}{theoretical\ yield\ kg\ CH_4N_2O} \times 100$$

Solution:

$$136.4\ kg\ NH_3 \times \frac{1000\ g}{kg} \times \frac{1\ mol\ NH_3}{17.03\ g\ NH_3} \times \frac{1\ mol\ CH_4N_2O}{2\ mol\ NH_3} = 400\underline{4}.7\ mol\ CH_4N_2O$$

$$211.4\ kg\ CO_2 \times \frac{1000\ g}{kg} \times \frac{1\ mol\ CO_2}{44.01\ g\ CO_2} \times \frac{1\ mol\ CH_4N_2O}{1\ mol\ CO_2} = 480\underline{3}.2\ mol\ CH_4N_2O$$

$$400\underline{4}.7\ mol\ CH_4N_2O \times \frac{60.06\ g\ CH_4N_2O}{1\ mol\ CH_4N_2O} \times \frac{kg}{1000\ g} = 240.\underline{5}2\ kg\ CH_4N_2O$$

$$\frac{168.4\ kg\ CH_4N_2O}{240.\underline{5}2\ kg\ CH_4N_2O} \times 100\% = 70.01\%$$

Check: The theoretical yield has the correct units, kg CH_4N_2O, and has a reasonable magnitude compared to the mass of NH_3, the limiting reactant. The % yield is reasonable, under 100%.

24. **Given:** 155.8 kg SiO_2; 78.3 kg C; 66.1 kg Si **Find:** limiting reactant, theoretical yield Si, % yield

Conceptual Plan: write and balance the reaction, then

kg SiO_2 → g SiO_2 → mol SiO_2 → mol Si

$$\frac{1000\ g}{1\ kg} \quad \frac{1\ mol\ SiO_2}{60.085\ g\ SiO_2} \quad \frac{1\ mol\ Si}{2\ mol\ SiO_2}$$

→ smallest amount determines limiting reactant

kg C → g C → mol C → mol Si

$$\frac{1000\ g}{1\ kg} \quad \frac{1\ mol\ CO_2}{44.01\ g\ CO_2} \quad \frac{1\ mol\ Si}{2\ mol\ C}$$

then: mol Si → g Si → kg CH_4N_2O

$$\frac{26.981\ g\ Si}{1\ mol\ Si} \quad \frac{1\ kg}{1000\ g}$$

then: determine % yield

$$\frac{actual\ yield\ kg\ Si}{theoretical\ yield\ kg\ Si} \times 100$$

Solution: $SiO_2(s) + 2C(s) \rightarrow Si(s) + 2CO(g)$

$$155.8\ kg\ SiO_2 \times \frac{1000\ g}{kg} \times \frac{1\ mol\ SiO_2}{60.085\ g\ SiO_2} \times \frac{1\ mol\ Si}{1\ mol\ SiO_2} = 259\underline{2}.9\ mol\ Si$$

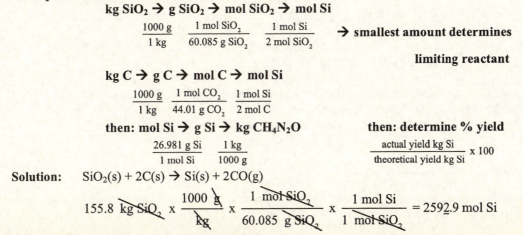

$$78.3 \text{ kg C} \times \frac{1000 \text{ g}}{\text{kg}} \times \frac{1 \text{ mol C}}{12.01 \text{ g C}} \times \frac{1 \text{ mol Si}}{2 \text{ mol C}} = 32\underline{5}9.8 \text{ mol Si}$$

$$259\underline{2}.9 \text{ mol Si} \times \frac{28.085 \text{ g Si}}{1 \text{ mol Si}} \times \frac{\text{kg}}{1000 \text{ g}} = 72.8\underline{2}2 \text{ kg Si}$$

$$\frac{66.1 \text{ kg Si}}{72.8\underline{2} \text{ kg Si}} \times 100\% = 90.8\%$$

Check: The theoretical yield has the correct units, kg Si, and has a reasonable magnitude compared to the mass of SiO_2, the limiting reactant. The % yield is reasonable, under 100%.

25. a) **Given:** 4.3 mol LiCl; 2.8 L solution **Find:** Molarity LiCl

Conceptual Plan: mol LiCl, L solution → Molarity

$$\text{molarity (M)} = \frac{\text{amount of solute (in moles)}}{\text{volume of solution (in L)}}$$

Solution: $\frac{4.3 \text{ mol LiCl}}{2.8 \text{ L solution}} = 1.5 \text{ M}$

Check: The units of the answer, M, are correct. The magnitude of the answer is reasonable. Concentrations are usually between 0 M and 18 M.

b) **Given:** 22.6 g $C_6H_{12}O_6$; 1.08 L solution **Find:** Molarity $C_6H_{12}O_6$

Conceptual Plan: g $C_6H_{12}O_6$ → mol $C_6H_{12}O_6$, L solution → Molarity

$$\frac{\text{mol } C_6H_{12}O_6}{180.16 \text{ g } C_6H_{12}O_6} \qquad \text{molarity (M)} = \frac{\text{amount of solute (in moles)}}{\text{volume of solution (in L)}}$$

Solution: $22.6 \text{ g } C_6H_{12}O_6 \times \frac{1 \text{ mol } C_6H_{12}O_6}{180.16 \text{ g } C_6H_{12}O_6} = 0.12\underline{5}4 \text{ mol } C_6H_{12}O_6$

$$\frac{0.12\underline{5}4 \text{ mol } C_6H_{12}O_6}{1.08 \text{ L solution}} = 0.116 \text{ M}$$

Check: The units of the answer, M, are correct. The magnitude of the answer is reasonable. Concentrations are usually between 0 M and 18 M.

c) **Given:** 45.5 mg NaCl; 154.4 mL solution **Find:** Molarity NaCl

Conceptual Plan: mg NaCl → g NaCl → mol NaCl, and mL solution → L solution then Molarity

$$\frac{\text{g NaCl}}{1000 \text{ mg NaCl}} \qquad \frac{\text{mol NaCl}}{58.45 \text{ g NaCl}} \qquad \frac{\text{L solution}}{1000 \text{ mL solution}} \qquad \text{molarity (M)} = \frac{\text{amount of solute (in moles)}}{\text{volume of solution (in L)}}$$

Solution: $45.5 \text{ mg NaCl} \times \frac{1 \text{ g}}{1000 \text{ mg}} \times \frac{1 \text{ mol NaCl}}{58.45 \text{ g NaCl}} = 7.7\underline{8}4 \times 10^{-4} \text{ mol NaCl}$

$$154.4 \text{ mL solution} \times \frac{1 \text{ L}}{1000 \text{ mL}} = 0.1544 \text{ L}$$

$$\frac{7.7\underline{8}4 \times 10^{-4} \text{ mol NaCl}}{0.1544 \text{ L}} = 0.00504 \text{ M NaCl}$$

Check: The units of the answer, M, are correct. The magnitude of the answer is reasonable. Concentrations are usually between 0 M and 18 M.

26. a) **Given:** 0.11 mol $LiNO_3$; 5.2 L solution **Find:** Molarity $LiNO_3$

Conceptual Plan: mol $LiNO_3$, L solution → Molarity

$$\text{molarity (M)} = \frac{\text{amount of solute (in moles)}}{\text{volume of solution (in L)}}$$

Solution: $\frac{0.11 \text{ mol } LiNO_3}{5.2 \text{ L solution}} = 0.021 \text{ M}$

Check: The units of the answer, M, are correct. The magnitude of the answer is reasonable. Concentrations are usually between 0 M and 18 M.

b) **Given:** 61.3 g C_2H_6O; 2.44 L solution **Find:** Molarity $C_6H_{12}O_6$

Conceptual Plan: g C_2H_6O →mol C_2H_6O, L solution → Molarity

$$\frac{\text{mol } C_2H_6O}{46.068 \text{ g } C_2H_6O} \qquad \text{molarity (M)} = \frac{\text{amount of solute (in moles)}}{\text{volume of solution (in L)}}$$

Solution: $61.3 \text{ g } C_2H_6O \times \frac{1 \text{ mol } C_2H_6O}{46.068 \text{ g } C_2H_6O} = 0.13\underline{3}1 \text{ mol } C_2H_6O$

$$\frac{0.13\underline{3}1 \text{ mol } C_2H_6O}{2.44 \text{ L solution}} = 0.545 \text{ M}$$

Check: The units of the answer, M, are correct. The magnitude of the answer is reasonable. Concentrations are usually between 0 M and 18 M.

c) **Given:** 15.2 mg KI; 102 mL solution **Find:** Molarity KI

Conceptual Plan: mg KI → g KI → mol KI, and mL solution →L solution then Molarity

$$\frac{\text{g KI}}{1000 \text{ mg KI}} \quad \frac{\text{mol KI}}{166.00 \text{ g KI}} \quad \frac{\text{L solution}}{1000 \text{ mL solution}} \quad \text{molarity (M)} = \frac{\text{amount of solute (in moles)}}{\text{volume of solution (in L)}}$$

Solution: $15.2 \text{ mg KI} \times \frac{1 \text{ g}}{1000 \text{ mg}} \times \frac{1 \text{ mol KI}}{166.00 \text{ g KI}} = 9.1\underline{5}7 \times 10^{-5} \text{ mol KI}$

$$102 \text{ mL solution} \times \frac{1 \text{ L}}{1000 \text{ mL}} = 0.102 \text{ L}$$

$$\frac{9.1\underline{5}7 \times 10^{-5} \text{ mol KI}}{0.102 \text{ L}} = 8.98 \times 10^{-4} \text{ M KI}$$

Check: The units of the answer, M, are correct. The magnitude of the answer is reasonable. Concentrations are usually between 0 M and 18 M.

27\. a) **Given:** 0.556 L; 2.3 M KCl **Find:** mol KCl

Conceptual Plan: volume solution x M = mol

volume solution (L) x M = mol

Solution: $0.556 \text{ L solution} \times \frac{2.3 \text{ mol KCl}}{\text{L solution}} = 1.3 \text{ mol KCl}$

Check: Units of answer, mol KCl, are correct. The magnitude is reasonable since it is less than 1 L solution.

b) **Given:** 1.8 L; 0.85 M KCl **Find:** mol KCl

Conceptual Plan: volume solution x M = mol

volume solution (L) x M = mol

Solution: $1.8 \text{ L solution} \times \frac{0.85 \text{ mol KCl}}{\text{L solution}} = 1.5 \text{ mol KCl}$

Check: Units of answer, mol KCl, are correct. The magnitude is reasonable since it is less than 2 L solution.

c) **Given:** 114 mL; 1.85 M KCl **Find:** mol KCl

Conceptual Plan: mL solution → L solution, then volume solution x M = mol

$$\frac{1 \text{ L}}{1000 \text{ mL}} \qquad \text{volume solution (L) x M = mol}$$

Solution: $114 \text{ mL solution} \times \frac{1 \text{ L}}{1000 \text{ mL}} \times \frac{1.85 \text{ mol KCl}}{\text{L solution}} = 0.211 \text{ mol KCl}$

Check: Units of answer, mol KCl, are correct. The magnitude is reasonable since it is less than 1 L solution.

28. a) **Given:** 0.45 mol C_2H_5OH, 0.200 M C_2H_5OH **Find:** volume solution
Conceptual Plan: mol C_2H_5OH → volume solution

$$\frac{\text{mol } C_2H_5OH}{\text{M } C_2H_5OH}$$

Solution:

$$\frac{0.45 \text{ mol } C_2H_5OH}{0.200 \frac{\text{mol } C_2H_5OH}{\text{L solution}}} = 2.3 \text{ L } C_2H_5OH$$

Check: Units of answer, L C_2H_5OH, are correct. The magnitude is reasonable for the amount and volume of solution.

b) **Given:** 1.22 mol C_2H_5OH, 0.200 M C_2H_5OH **Find:** volume solution
Conceptual Plan: mol C_2H_5OH → volume solution

$$\frac{\text{mol } C_2H_5OH}{\text{M } C_2H_5OH}$$

Solution:

$$\frac{1.22 \text{ mol } C_2H_5OH}{0.200 \frac{\text{mol } C_2H_5OH}{\text{L solution}}} = 6.10 \text{ L } C_2H_5OH$$

Check: Units of answer, L C_2H_5OH, are correct. The magnitude is reasonable for the amount and volume of solution.

c) **Given:** 1.2×10^{-2} mol C_2H_5OH, 0.200 M C_2H_5OH **Find:** volume solution
Conceptual Plan: mol C_2H_5OH → volume solution

$$\frac{\text{mol } C_2H_5OH}{\text{M } C_2H_5OH}$$

Solution:

$$\frac{1.2 \times 10^{-2} \text{ mol } C_2H_5OH}{0.200 \frac{\text{mol } C_2H_5OH}{\text{L solution}}} = 0.060 \text{ L } C_2H_5OH$$

Check: Units of answer, L C_2H_5OH, are correct. The magnitude is reasonable for the amount and volume of solution.

29. **Given:** 400.0 mL; 1.1 M $NaNO_3$ **Find:** g $NaNO_3$
Conceptual Plan: mL solution → L solution, then volume solution x M = mol $NaNO_3$

$$\frac{\text{L solution}}{1000 \text{ mL solution}} \qquad \text{volume solution (L) x M = mol}$$

then mol $NaNO_3$ → g $NaNO_3$

$$\frac{85.01 \text{ g } NaNO_3}{\text{mol } NaNO_3}$$

Solution: $400.0 \text{ mL solution} \times \frac{1 \text{ L}}{1000 \text{ mL}} \times \frac{1.1 \text{ mol } NaNO_3}{\text{L solution}} \times \frac{85.01 \text{ g}}{\text{mol } NaNO_3} = 37 \text{ g } NaNO_3$

Check: Units of answer, g $NaNO_3$, are correct. The magnitude is reasonable for the concentration and volume of solution.

30. **Given:** 5.5 L; 0.300 M $CaCl_2$ **Find:** g $CaCl_2$
Conceptual Plan: volume solution x M = mol $CaCl_2$ then mol $CaCl_2$ → g $CaCl_2$

$$\text{volume solution (L) x M = mol} \qquad \frac{110.98 \text{ g } CaCl_2}{\text{mol } CaCl_2}$$

Solution: $5.5 \text{ L solution} \times \frac{0.300 \text{ mol } CaCl_2}{\text{L solution}} \times \frac{110.98 \text{ g}}{\text{mol } CaCl_2} = 1.8 \times 10^2 \text{ g } CaCl_2$

Check: Units of answer, g $CaCl_2$, are correct. The magnitude is reasonable for the concentration and volume of solution.

31. **Given:** $V_1 = 123$ mL; $M_1 = 1.1$ M; $V_2 = 500.0$ mL **Find:** M_2
Conceptual Plan: mL → L then $V_1, M_1, V_2 \rightarrow M_2$

$$\frac{1\text{ L}}{1000\text{ mL}} \qquad V_1M_1 = V_2M_2$$

Solution: $123\text{ mL} \times \frac{1\text{ L}}{1000\text{ mL}} = 0.123\text{ L} \qquad 500.0\text{ mL} \times \frac{1\text{ L}}{1000\text{ mL}} = 0.5000\text{ L}$

$$M_2 = \frac{V_1M_1}{V_2} = \frac{(0.123\text{ L})(1.1\text{ M})}{(0.5000\text{ L})} = 0.27\text{ M}$$

Check: Units of the answer, M, are correct. The magnitude of the answer is reasonable since it is less than the original concentration.

32. **Given:** $V_1 = 3.5$ L; $M_1 = 4.8$ M; $V_2 = 45$ L **Find:** M_2
Conceptual Plan: $V_1, M_1, V_2 \rightarrow M_2$

$$V_1M_1 = V_2M_2$$

Solution: $M_2 = \frac{V_1M_1}{V_2} = \frac{(3.5\text{ L})(4.8\text{ M})}{(45\text{ L})} = 0.37\text{ M}$

Check: Units of the answer, M, are correct. The magnitude of the answer is reasonable since it is less than the original concentration.

33. **Given:** $V_1 = 50$ mL; $M_1 = 12$ M; $M_2 = 0.100$ M **Find:** V_2
Conceptual Plan: mL → L then $V_1, M_1, M_2 \rightarrow V_2$

$$\frac{1\text{ L}}{1000\text{ mL}} \qquad V_1M_1 = V_2M_2$$

Solution: $50\text{ mL} \times \frac{1\text{ L}}{1000\text{ mL}} = 0.050\text{ L}$

$$V_2 = \frac{V_1M_1}{M_2} = \frac{(0.050\text{ L})(12\text{ M})}{(0.100\text{ M})} = 6.0\text{ L}$$

Check: Units of the answer, L, are correct. The magnitude of the answer is reasonable since the new concentration is much less than the original, the volume must be larger.

34. **Given:** $V_1 = 25$ mL; $M_1 = 10.0$ M; $M_2 = 0.150$ M **Find:** V_2
Conceptual Plan: mL → L then $V_1, M_1, M_2 \rightarrow V_2$

$$\frac{1\text{ L}}{1000\text{ mL}} \qquad V_1M_1 = V_2M_2$$

Solution: $25\text{ mL} \times \frac{1\text{ L}}{1000\text{ mL}} = 0.025\text{ L}$

$$V_2 = \frac{V_1M_1}{M_2} = \frac{(0.025\text{ L})(10.0\text{ M})}{(0.150\text{ M})} = 1.7\text{ L}$$

Check: Units of the answer, L, are correct. The magnitude of the answer is reasonable since the new concentration is much less than the original, the volume must be larger.

35. **Given:** 95.4 mL, 0.102 M $CuCl_2$; 0.175 M Na_3PO_4 **Find:** volume Na_3PO_4
Conceptual Plan: mL $CuCl_2$ → L $CuCl_2$ → mol $CuCl_2$ → mol Na_3PO_4 → L Na_3PO_4 → mL Na_3PO_4

$$\frac{1\text{ L}}{1000\text{ mL}} \quad \frac{0.102\text{ mol }CuCl_2}{\text{L}} \quad \frac{2\text{ mol }Na_3PO_4}{3\text{ mol }CuCl_2} \quad \frac{1\text{ L}}{0.175\text{ mol }Na_3PO_4} \quad \frac{1000\text{ mL}}{\text{L}}$$

Solution:

$$95.4 \text{ mL CuCl}_2 \times \frac{1 \text{ L}}{1000 \text{ mL}} \times \frac{0.102 \text{ mol CuCl}_2}{1 \text{ L}} \times \frac{2 \text{ mol Na}_3\text{PO}_4}{3 \text{ mol CuCl}_2} \times \frac{1 \text{ L}}{0.175 \text{ mol Na}_3\text{PO}_4} \times \frac{1000 \text{ mL}}{1 \text{ L}}$$

$= 37.1 \text{ mL Na}_3\text{PO}_4$

Check: Units of answer, mL Na_3PO_4, are correct. The magnitude of the answer is reasonable since the concentration of Na_3PO_4 is greater.

36. **Given:** 125 mL, 0.150 M $Co(NO_3)_2$; 0.150 M Li_2S **Find:** volume Li_2S
Conceptual Plan: mL $Co(NO_3)_2$ → L $Co(NO_3)_2$ → mol $Co(NO_3)_2$ → mol Li_2S → L Li_2S → mL Li_2S

$$\frac{1 \text{ L}}{1000 \text{ mL}} \quad \frac{0.150 \text{ mol Co(NO}_3)_2}{\text{L}} \quad \frac{1 \text{ mol Li}_2\text{S}}{1 \text{ mol Co(NO}_3)_2} \quad \frac{1 \text{ L}}{0.155 \text{ mol Li}_2\text{S}} \quad \frac{1000 \text{ mL}}{\text{L}}$$

Solution:

$$125 \text{ mL Co(NO}_3)_2 \times \frac{1 \text{ L}}{1000 \text{ mL}} \times \frac{0.150 \text{ mol Co(NO}_3)_2}{1 \text{ L}} \times \frac{1 \text{ mol Li}_2\text{S}}{1 \text{ mol Co(NO}_3)_2} \times \frac{1 \text{ L}}{0.150 \text{ mol Li}_2\text{S}} \times \frac{1000 \text{ mL}}{1 \text{ L}}$$

$= 125 \text{ mL Li}_2\text{S}$

Check: Units of answer, mL Li_2S, are correct. The magnitude of the answer is reasonable since the concentrations are the same and the mole ratio is 1:1.

37. **Given:** 25.0 g H_2; 6.0 M H_2SO_4 **Find:** volume H_2SO_4
Conceptual Plan: g H_2 → mol H_2 → mol H_2SO_4 → L H_2SO_4

$$\frac{2.016 \text{ g H}_2}{1 \text{ mol H}_2} \quad \frac{3 \text{ mol H}_2\text{SO}_4}{3 \text{ mol H}_2} \quad \frac{1 \text{ L}}{6.0 \text{ mol H}_2\text{SO}_4}$$

Solution: $25.0 \text{ g H}_2 \times \frac{1 \text{ mol H}_2}{2.016 \text{ g H}_2} \times \frac{3 \text{ mol H}_2\text{SO}_4}{3 \text{ mol H}_2} \times \frac{1 \text{ L}}{6.0 \text{ mol H}_2\text{SO}_4} = 2.1 \text{ L H}_2\text{SO}_4$

Check: The units, L H_2SO_4, are correct. The magnitude is reasonable since there are approximately 12 mol H_2 and the mole ratio is 1:1.

38. **Given:** 25.0 g Zn, 275 mL solution **Find:** M $ZnCl_2$
Conceptual Plan: g Zn → mol Zn → mol $ZnCl_2$ → M $ZnCl_2$

$$\frac{65.38 \text{ g Zn}}{1 \text{ mol Zn}} \quad \frac{1 \text{ mol ZnCl}_2}{1 \text{ mol Zn}} \quad \frac{\text{mol ZnCl}_2}{\text{volume solution}}$$

Solution:

$$25.0 \text{ g Zn} \times \frac{1 \text{ mol Zn}}{65.38 \text{ g Zn}} \times \frac{1 \text{ mol ZnCl}_2}{1 \text{ mol Zn}} = 0.38\underline{2}4 \text{ mol ZnCl}_2$$

$$\frac{0.38\underline{2}4 \text{ mol ZnCl}_2}{275 \text{ mL}} \times \frac{1000 \text{ mL}}{\text{L}} = 1.39 \text{ M ZnCl}_2$$

Check: Units of the answer, M $ZnCl_2$, are correct. The magnitude is reasonable because the stoichiometry is 1:1 and the mol Zn is less than 0.5.

39. a) CsCl is an ionic compound. An aqueous solution is an electrolyte solution, so it conducts electricity.

b) CH_3OH is a molecular compound that does not dissociate. An aqueous solution is a nonelectrolyte solution, so it does not conduct electricity.

c) $Ca(NO_3)_2$ is an ionic compound. An aqueous solution is an electrolyte solution, so it conducts electricity.

d) $C_6H_{12}O_6$ is a molecular compound that does not dissociate. An aqueous solution is a nonelectrolyte solution, so it does not conduct electricity.

40. a) $MgBr_2$ is an ionic compound. An aqueous solution is a strong electrolyte.

b) $C_{12}H_{22}O_{11}$ is a molecular compound that does not dissociate. An aqueous solution is a nonelectrolyte.

c) Na_2CO_3 is an ionic compound. An aqueous solution is a strong electrolyte.

d) KOH is a strong base. An aqueous solution is a strong electrolyte.

41. a) $AgNO_3$ is soluble. Compounds containing NO_3^- are always soluble with no exceptions. The ions in solution are $Ag^+(aq)$ and $NO_3^-(aq)$.

b) $Pb(C_2H_3O_2)_2$ is soluble. Compounds containing $C_2H_3O_2^-$ are always soluble with no exceptions. The ions in solution are $Pb^{2+}(aq)$ and $C_2H_3O_2^-(aq)$.

c) KNO_3 is soluble. Compounds containing K^+ are always soluble with no exceptions. The ions in solution are $K^+(aq)$ and $NO_3^-(aq)$.

d) $(NH_4)_2S$ is soluble. Compounds containing NH_4^+ are always soluble with no exceptions. The ions in solution are $NH_4^+(aq)$ and $S^{2-}(aq)$.

42. a) AgI is insoluble. Compounds containing I^- are normally soluble but Ag^+ is an exception.

b) $Cu_3(PO_4)_2$ is insoluble. Compounds containing PO_4^{3-} are normally insoluble and Cu^{2+} is not an exception.

c) $CoCO_3$ is insoluble. Compounds containing CO_3^{2-} are normally insoluble and Co^{2+} is not an exception.

d) K_3PO_4 is soluble. Compounds containing PO_4^{3-} are normally insoluble, but K^+ is an exception. The ions in solution are $K^+(aq)$ and $PO_4^{3-}(aq)$.

43. a) $LiI(aq) + BaS(aq) \rightarrow$ Possible products: Li_2S and BaI_2. Li_2S is soluble. Compounds containing S^{2-} are normally insoluble but Li^+ is an exception. BaI_2 is soluble. Compounds containing I^- are normally soluble and Ba^{2+} is not an exception. $LiI(aq) + BaS(aq) \rightarrow$ No Reaction

b) $KCl(aq) + CaS(aq) \rightarrow$ Possible products: K_2S and $CaCl_2$. K_2S is soluble. Compounds containing S^{2-} are normally insoluble but K^+ is an exception. $CaCl_2$ is soluble. Compounds containing Cl^- are normally soluble and Ca^{2+} is not an exception. $KCl(aq) + CaS(aq) \rightarrow$ No Reaction

c) $CrBr_2(aq) + Na_2CO_3(aq) \rightarrow$ Possible products: $CrCO_3$ and NaBr. $CrCO_3$ is insoluble. Compounds containing CO_3^{2-} are normally insoluble and Cr^{2+} is not an exception. NaBr is soluble. Compounds containing Br^- are normally soluble and Na^+ is not an exception.
$CrBr_2(aq) + Na_2CO_3(aq) \rightarrow CrCO_3(s) + 2\ NaBr(aq)$

d) $NaOH(aq) + FeCl_3(aq) \rightarrow$ Possible products NaCl and $Fe(OH)_3$. NaCl is soluble. Compounds containing Na^+ are normally soluble, no exceptions. $Fe(OH)_3$ is insoluble. Compounds containing OH^- are normally insoluble and Fe^{3+} is not an exception.
$3\ NaOH(aq) + FeCl_3(aq) \rightarrow 3\ NaCl(aq) + Fe(OH)_3(s)$

44. a) $NaNO_3(aq) + KCl(aq) \rightarrow$ Possible products: NaCl and KNO_3. NaCl is soluble. Compounds containing Na^+ are always soluble, no exceptions. KNO_3 is soluble. Compounds containing K^+ are always soluble, no exceptions. $NaNO_3(aq) + KCl(aq) \rightarrow$ No Reaction

b) $NaCl(aq) + Hg_2(C_2H_3O_2)_2(aq) \rightarrow$ Possible products: $NaC_2H_3O_2$ and Hg_2Cl_2. $NaC_2H_3O_2$ is soluble. Compounds containing Na^+ are always soluble, no exceptions. Hg_2Cl_2 is insoluble. Compounds containing Cl^- are normally soluble but Hg_2^{2+} is an exception.
$2\ NaCl(aq) + Hg_2(C_2H_3O_2)_2(aq) \rightarrow 2\ NaC_2H_3O_2(aq) + Hg_2Cl_2(s)$

c) $(NH_4)_2SO_4(aq) + SrCl_2(aq) \rightarrow$ Possible products: NH_4Cl and $SrSO_4$. NH_4Cl is soluble. Compounds containing NH_4^+ are always soluble, no exceptions. $SrSO_4$ is insoluble. Compounds containing SO_4^{2-} are normally soluble but Sr^{2+} is an exception. $(NH_4)_2SO_4(aq) + SrCl_2(aq) \rightarrow 2\ NH_4Cl(aq) + SrSO_4(s)$

d) $NH_4Cl(aq) + AgNO_3(aq) \rightarrow$ Possible products: NH_4NO_3 and AgCl. NH_4NO_3 is soluble. Compounds containing NH_4^+ are always soluble, no exceptions. AgCl is insoluble. Compounds containing Cl^- are normally soluble, but Ag^+ is an exception. $NH_4Cl(aq) + AgNO_3(aq) \rightarrow NH_4NO_3(aq) + AgCl(s)$

45. a) $K_2CO_3(aq) + Pb(NO_3)_2(aq) \rightarrow$ Possible products: KNO_3 and $PbCO_3$. KNO_3 is soluble. Compounds containing K^+ are always soluble, no exceptions. $PbCO_3$ is insoluble. Compounds containing CO_3^{2-} are normally insoluble and Pb^{2+} is not an exception.
$K_2CO_3(aq) + Pb(NO_3)_2(aq) \rightarrow 2\ KNO_3(aq) + PbCO_3(s)$

b) $Li_2SO_4(aq) + Pb(C_2H_3O_2)_2(aq) \rightarrow$ Possible products: $LiC_2H_3O_2$ and $PbSO_4$. $LiC_2H_3O_2$ is soluble. Compounds containing Li^+ are always soluble, no exceptions. $PbSO_4$ is insoluble. Compounds containing SO_4^{2-} are normally soluble but, Pb^{2+} is an exception.
$Li_2SO_4(aq) + Pb(C_2H_3O_2)_2(aq) \rightarrow 2\ LiC_2H_3O_2(aq) + PbSO_4(s)$

c) $Cu(NO_3)_2(aq) + MgS(s) \rightarrow$ Possible products: CuS and $Mg(NO_3)_2$. CuS is insoluble. Compounds containing S^{2-} are normally insoluble and Cu^{2+} is not an exception. $Mg(NO_3)_2$ is soluble. Compounds containing NO_3^- are always soluble, no exceptions. $Cu(NO_3)_2(aq) + MgS(s) \rightarrow CuS(s) + Mg(NO_3)_2(aq)$

d) $Sr(NO_3)_2(aq) + KI(aq) \rightarrow$ Possible products: SrI_2 and KNO_3. SrI_2 is soluble. Compounds containing I^- are normally soluble and Sr^{2+} is not an exception. KNO_3 is soluble. Compounds containing K^+ are always soluble, no exceptions. $Sr(NO_3)_2(aq) + KI(aq) \rightarrow$ No Reaction

46. a) $NaCl(aq) + Pb(C_2H_3O_2)_2(aq) \rightarrow$ Possible products $NaC_2H_3O_2$ and $PbCl_2$. $NaC_2H_3O_2$ is soluble. Compounds containing Na^+ are always soluble, no exceptions. $PbCl_2$ is insoluble. Compounds containing Cl^- are normally soluble but Pb^{2+} is an exception.
$2\ NaCl(aq) + Pb(C_2H_3O_2)_2(aq) \rightarrow 2\ NaC_2H_3O_2(aq) + PbCl_2(s)$

b) $K_2SO_4(aq) + SrI_2(aq) \rightarrow$ Possible products: KI and $SrSO_4$. KI is soluble. Compounds containing K^+ are always soluble, no exceptions. $SrSO_4$ is insoluble. Compounds containing SO_4^{2-} are normally soluble, but Sr^{2+} is an exception. $K_2SO_4(aq) + SrI_2(aq) \rightarrow 2\ KI(aq)$ and $SrSO_4(s)$

c) $CsCl(aq) + CaS(aq) \rightarrow$ Possible products: Cs_2S and $CaCl_2$. Cs_2S is soluble. Compounds containing S^{2-} are normally insoluble but Cs^+ is an exception. $CaCl_2$ is soluble. Compounds containing Cl^- are normally soluble and Ca^{2+} is not an exception. $CsCl(aq) + CaS(aq) \rightarrow$ No Reaction

d) $Cr(NO_3)_3 + Na_3PO_4 \rightarrow$ Possible products: $CrPO_4$ and $NaNO_3$. $CrPO_4$ is insoluble. Compounds containing PO_4^{3-} are normally insoluble and Cr^{3+} is not an exception. $NaNO_3$ is soluble. Compounds containing Na^+ are always soluble, no exceptions.
$Cr(NO_3)_3(aq) + Na_3PO_4(aq) \rightarrow CrPO_4(s) + 3\ NaNO_3(aq)$

47. a) $H^+(aq)$ + ~~Cl^-~~(aq) + ~~Li^+~~$(aq) + OH^-(aq) \rightarrow H_2O(l)$ + ~~Li^+~~(aq) + ~~Cl^-~~(aq)
$H^+(aq) + OH^-(aq) \rightarrow H_2O(l)$

b) ~~Mg^{2+}~~$(aq) + S^{2-}(aq) + Cu^{2+}(aq) + 2$ ~~Cl^-~~$(aq) \rightarrow CuS(s)$ + ~~Mg^{2+}~~$(aq) + 2$ ~~Cl^-~~(aq)
$Cu^{2+}(aq) + S^{2-}(aq) \rightarrow CuS(s)$

c) ~~Na^+~~$(aq) + OH^-(aq) + H^+(aq)$ + ~~NO_3^-~~$(aq) \rightarrow H_2O(l)$ + ~~Na^+~~(aq) + ~~NO_3^-~~(aq)
$H^+(aq) + OH^-(aq) + \rightarrow H_2O(l)$

d) 6 ~~Na^+~~$(aq) + 2\ PO_4^{3-}(aq) + 3\ Ni^{2+}(aq) + 6$ ~~Cl^-~~$(aq) \rightarrow Ni_3(PO_4)_2(s) + 6$ ~~Na^+~~$(aq) + 6$ ~~Cl^-~~(aq)
$3\ Ni^{2+}(aq) + 2\ PO_4^{3-}(aq) \rightarrow Ni_3(PO_4)_2(s)$

48. a) 2 ~~K^+~~$(aq) + SO_4^{2-}(aq) + Ca^{2+}(aq) + 2$ ~~I^-~~$(aq) \rightarrow CaSO_4(s) + 2$ ~~K^+~~$(aq) + 2$ ~~I^-~~(aq)
$Ca^{2+}(aq) + SO_4^{2-}(aq) \rightarrow CaSO_4(s)$

b) $NH_4^+(aq)$ + ~~Cl^-~~(aq) + ~~Na^+~~$(aq) + OH^-(aq) \rightarrow H_2O(l) + NH_3(g)$ + ~~Na^+~~(aq) + ~~Cl^-~~(aq)
$NH_4^+(aq) + OH^-(aq) \rightarrow H_2O(l) + NH_3(g)$

c) $Ag^+(aq)$ + ~~NO_3^-~~(aq) + ~~Na^+~~$(aq) + Cl^-(aq) \rightarrow AgCl(s)$ + ~~Na^+~~(aq) + ~~NO_3^-~~(aq)
$Ag^+(aq) + Cl^-(aq) \rightarrow AgCl(s)$

d) $2H^+(aq) + \cancel{C_2H_3O_2}^-(aq) + \cancel{2K}^+(aq) + CO_3^{2-}(aq) \rightarrow H_2O(l) + CO_2(g) + \cancel{2K}^+(aq) + \cancel{C_2H_3O_2}^-(aq)$
$2H^+(aq) + CO_3^{2-}(aq) \rightarrow H_2O(l) + CO_2(g)$

49. $Hg_2^{2+}(aq) + \cancel{2NO_3}^-(aq) + \cancel{2Na}^+(aq) + 2\ Cl^-(aq) \rightarrow Hg_2Cl_2(s) + \cancel{2Na}^+(aq) + \cancel{2NO_3}^-(aq)$
$Hg_2^{2+}(aq) + 2\ Cl^-(aq) \rightarrow Hg_2Cl_2(s)$

50. $Pb^{2+}(aq) + \cancel{2NO_3}^-(aq) + \cancel{2K}^+(aq) + SO_4^{2-}(aq) \rightarrow PbSO_4(s) + \cancel{2K}^+(aq) + \cancel{2NO_3}^-(aq)$
$Pb^{2+}(aq) + SO_4^{2-}(aq) \rightarrow PbSO_4(s)$

51. Skeletal reaction: $HBr(aq) + KOH(aq) \rightarrow H_2O(l) + KBr(aq)$
acid base water salt
Net ionic equation: $H^+(aq) + OH^-(aq) \rightarrow H_2O(l)$

52. Skeletal reaction: $HNO_3(aq) + Ca(OH)_2(aq) \rightarrow H_2O(l) + Ca(NO_3)_2(aq)$
acid base water salt
Balanced reaction: $2\ HNO_3(aq) + Ca(OH)_2(aq) \rightarrow 2\ H_2O(l) + Ca(NO_3)_2(aq)$
Net ionic equation: $H^+(aq) + OH^-(aq) \rightarrow H_2O(l)$

53. a) Skeletal reaction: $H_2SO_4(aq) + Ca(OH)_2(aq) \rightarrow H_2O(l) + CaSO_4(s)$
acid base water salt
Balanced reaction: $H_2SO_4(aq) + Ca(OH)_2(aq) \rightarrow 2\ H_2O(l) + CaSO_4(s)$

b) Skeletal reaction: $HClO_4(aq) + KOH(aq) \rightarrow H_2O(l) + KClO_4(aq)$
acid base water salt
Balanced reaction: $HClO_4(aq) + KOH(aq) \rightarrow H_2O(l) + KClO_4(aq)$

c) Skeletal reaction: $H_2SO_4(aq) + NaOH(aq) \rightarrow H_2O(l) + Na_2SO_4(aq)$
acid base water salt
Balanced reaction: $H_2SO_4(aq) + 2\ NaOH(aq) \rightarrow 2\ H_2O(l) + Na_2SO_4(aq)$

54. a) Skeletal reaction: $HI(aq) + LiOH(aq) \rightarrow H_2O(l) + LiI(aq)$
acid base water salt
Balanced reaction: $HI(aq) + LiOH(aq) \rightarrow H_2O(l) + LiI(aq)$

b) Skeletal reaction: $HC_2H_3O_2(aq) + Ca(OH)_2(aq) \rightarrow H_2O(l) + Ca(C_2H_3O_2)_2(aq)$
acid base water salt
Balanced reaction: $2\ HC_2H_3O_2(aq) + Ca(OH)_2(aq) \rightarrow 2\ H_2O(l) + Ca(C_2H_3O_2)_2(aq)$

c) Skeletal reaction: $HCl(aq) + Ba(OH)_2(aq) \rightarrow H_2O(l) + BaCl_2(aq)$
acid base water salt
Balanced reaction: $2\ HCl(aq) + Ba(OH)_2(aq) \rightarrow 2\ H_2O(l) + BaCl_2(aq)$

55. a) Skeletal reaction: $HBr(aq) + NiS(S) \rightarrow NiBr_2(aq) + H_2S(g)$
gas
Balanced reaction: $2\ HBr(aq) + NiS(s) \rightarrow NiBr_2(aq) + H_2S(g)$

b) Skeletal reaction: $NH_4I(aq) + NaOH(aq) \rightarrow NH_4OH(aq) + NaI(aq) \rightarrow H_2O(l) + NH_3(g) + NaI(aq)$
decomposes gas
Balanced reaction: $NH_4I(aq) + NaOH(aq) \rightarrow H_2O(l) + NH_3(g) + NaI(aq)$

c) Skeletal reaction: $HBr(aq) + Na_2S(aq) \rightarrow NaBr(aq) + H_2S(g)$
gas
Balanced reaction: $2\ HBr(aq) + Na_2S(aq) \rightarrow 2\ NaBr(aq) + H_2S(g)$

d) Skeletal reaction:
$HClO_4(aq) + Li_2CO_3(aq) \rightarrow H_2CO_3(aq) + LiClO_4(aq) \rightarrow H_2O(l) + CO_2(g) + LiClO_4(aq)$
decomposes gas
Balanced reaction: $2\ HClO_4(aq) + Li_2CO_3(aq) \rightarrow H_2O(l) + CO_2(g) + 2\ LiClO_4(aq)$

56. a) Skeletal reaction:
$HNO_3(aq) + Na_2SO_3(aq) \rightarrow H_2SO_3(aq) + NaNO_3(aq) \rightarrow H_2O(l) + SO_2(g) + NaNO_3(aq)$
decomposes gas
Balanced reaction: $2\ HNO_3(aq) + Na_2SO_3(aq) \rightarrow H_2O(l) + SO_2(g) + 2\ NaNO_3(aq)$

b) Skeletal reaction: $HCl(aq) + KHCO_3(aq) \rightarrow H_2CO_3(aq) + KCl(aq) \rightarrow H_2O(l) + CO_2(g) + KCl(aq)$
decomposes gas
Balanced reaction: $HCl(aq) + KHCO_3(aq) \rightarrow H_2O(l) + CO_2(g) + KCl(aq)$

c) Skeletal reaction:
$HC_2H_3O_2(aq) + NaHSO_3(aq) \rightarrow NaC_2H_3O_2(aq) + H_2SO_3(aq) \rightarrow H_2O(l) + SO_2(g) + NaC_2H_3O_2(aq)$
decomposes gas
Balanced reaction: $HC_2H_3O_2(aq) + NaHSO_3(aq) \rightarrow H_2O(l) + SO_2(g) + NaC_2H_3O_2(aq)$

d) Skeletal reaction:
$(NH_4)_2SO_4(aq) + Ca(OH)_2(aq) \rightarrow NH_4OH(aq) + CaSO_4(s) \rightarrow H_2O(l) + NH_3(g) + CaSO_4(s)$
decomposes gas
Balanced reaction: $(NH_4)_2SO_4(aq) + Ca(OH)_2(aq) \rightarrow 2H_2O(l) + 2NH_3(g) + CaSO_4(s)$

57. a) Ag. The oxidation state of Ag = 0. The oxidation state of an atom in a free element is 0.

b) Ag^+. The oxidation state of Ag^+ = +1. The oxidation state of a monatomic ion is equal to its charge.

c) CaF_2. The oxidation state of Ca = +2, the oxidation state of F = – 1. The oxidation state of a Group 2A metal always has an oxidation state of +2, the oxidation of F is – 1 since the sum of the oxidation states in a neutral formula unit = 0.

d) H_2S. The oxidation state of H = + 1, the oxidation state of S = – 2. The oxidation state of H when listed first is +1, the oxidation state of S is – 2 since S is in group 6A and the sum of the oxidation states in a neutral molecular unit = 0.

e) $CO_3^{\,2-}$. The oxidation state of C = +4, the oxidation state of O = – 2. The oxidation state of O is normally – 2, the oxidation state of C is deduced from the formula since the sum of the oxidation states must equal the charge on the ion. (C ox state) + 4(O ox state) = – 2: (C ox state) + 2(– 2) = – 2, so C ox state = + 4.

f) $CrO_4^{\,2-}$. The oxidation state of Cr = +6, the oxidation state of O = – 2. The oxidation state of O is normally – 2, the oxidation state of Cr is deduced from the formula since the sum of the oxidation states must equal the charge on the ion. (Cr ox state) + 4(O ox state) = – 2; (Cr ox state) + 2(– 2) = – 2, so Cr ox state = + 6.

58. a) Cl_2. The oxidation of both Cl atoms = 0. Since Cl_2 is a free element, the oxidation state of Cl = 0.

b) Fe^{3+}. The oxidation of Fe = +3. The oxidation state of a monatomic ion is equal to its charge.

c) $CuCl_2$. The oxidation state of Cu = +2, the oxidation state of each Cl = – 1. The oxidation state of group 7A atoms is normally – 1, the oxidation state of Cu is deduced from the formula since the sum of the oxidation states in a neutral formula unit = 0.

d) CH_4. The oxidation state of C = – 4, the oxidation state of H = +1. The oxidation state of H is normally +1, the oxidation state of C is deduced for the formula since the sum of the oxidation states in a neutral molecular unit = 0. (C ox state) + 4(H ox state) = 0; (C ox state) + 4(+ 1) = 0, so C ox state = – 4.

e) $Cr_2O_7^{\,2-}$. The oxidation state of Cr = +6, the oxidation state of O = – 2. The oxidation state of O is normally – 2, the oxidation state of Cr is deduced from the formula since the sum of the oxidation states must equal the charge of the ion. 2(Cr ox state) + 4(O ox state) = – 2; 2(Cr ox state) + 4(– 2) = – 2, so Cr ox state = +6.

f) $HSO_4^{\,-}$. The oxidation state of H = +1, the oxidation state of S = +6, the oxidation state of O = – 2. The oxidation state of H is normally +1, the oxidation state of O is normally – 2, the oxidation state of S

is deduced from the formula since the sum of the oxidation states must equal the charge of the ion. (H ox state) + (S ox state) + 4(O ox state) = – 1;
(+1) + (S ox state) + 4(– 2) = – 1, so S ox state = +6.

59. a) CrO. The oxidation state of Cr = +2, the oxidation state of O = – 2. The oxidation state of O is normally – 2, the oxidation state of Cr is deduced from the formula since the sum of the oxidation states must = 0.
(Cr ox state) + (O ox state) = 0; (Cr ox state) + (– 2) = 0, so Cr = +2.

b) CrO_3. The oxidation state of Cr = +6, the oxidation state of O = – 2. The oxidation state of O is normally – 2, the oxidation state of Cr is deduced from the formula since the sum of the oxidation states must = 0.
(Cr ox state) + 3(O ox state) = 0; (Cr ox state) +3 (– 2) = 0, so Cr = +6.

c) Cr_2O_3. The oxidation state of Cr = +3, the oxidation state of O = – 2. The oxidation state of O is normally – 2, the oxidation state of Cr is deduced from the formula since the sum of the oxidation states must = 0.
2(Cr ox state) +3 (O ox state) = 0; 2(Cr ox state) + 3(– 2) = 0, so Cr = +3.

60. a) ClO^-. The oxidation state of Cl = +1, the oxidation state of O = – 2. The oxidation state of O is normally – 2, the oxidation state of Cl is deduced from the formula since the sum of the oxidation states must equal the charge of the ion. (Cl ox state) + (O ox state) = – 1; (Cl ox state) + (– 2) = – 1, so Cl = +1.

b) ClO_2^-. The oxidation state of Cl = +3, the oxidation state of O = – 2. The oxidation state of O is normally – 2, the oxidation state of Cl is deduced from the formula since the sum of the oxidation states must equal the charge of the ion. (Cl ox state) + 2(O ox state) = – 1; (Cl ox state) +2 (– 2) = – 1, so Cl = +3.

c) ClO_3^-. The oxidation state of Cl = +5, the oxidation state of O = – 2. The oxidation state of O is normally – 2, the oxidation state of Cl is deduced from the formula since the sum of the oxidation states must equal the charge of the ion. (Cl ox state) + 3(O ox state) = – 1; (Cl ox state) + 3(– 2) = – 1, so Cl = +5.

d) ClO_4^-. The oxidation state of Cl = +7, the oxidation state of O = – 2. The oxidation state of O is normally – 2, the oxidation state of Cl is deduced from the formula since the sum of the oxidation states must equal the charge of the ion. (Cl ox state) + 4(O ox state) = – 1; (Cl ox state) + 4(– 2) = – 1, so Cl = +7.

61. a) $4\ Li(s) + O_2(g) \rightarrow 2\ Li_2O(s)$
oxidation states; 0 0 +1 – 2
This is a redox reaction since Li increases in oxidation number (oxidation) and O decreases in number (reduction). O_2 is the oxidizing agent, Li is the reducing agent.

b) $Mg(s) + Fe^{2+}(aq) \rightarrow Mg^{2+}(aq) + Fe(s)$
oxidation states; 0 +2 +2 0
This is a redox reaction since Mg increases in oxidation number (oxidation) and Fe decreases in number (reduction). Fe^{2+} is the oxidizing agent, Mg is the reducing agent.

c) $Pb(NO_3)_2(aq) + Na_2SO_4(aq) \rightarrow PbSO_4(s) + 2\ NaNO_3(aq)$
oxidation states; +2 +5 - 2 +1 +6 - 2 +2 +6 -2 +1 +5 -2
This is a not a redox reaction since none of the atoms undergoes a change in oxidation number.

d) $HBr(aq) + KOH(aq) \rightarrow H_2O(l) + KBr(aq)$
oxidation states; +1 - 1 +1 – 2 +1 +1 -2 +1 -2
This is a not a redox reaction since none of the atoms undergoes a change in oxidation number.

62. a) $Al(s) + 3\ Ag^+(aq) \rightarrow Al^{3+}(aq) + 3\ Ag(s)$
oxidation states; 0 +1 +3 0

This is a redox reaction since Al increases in oxidation number (oxidation) and Ag decreases in number (reduction). Ag^+ is the oxidizing agent, Al is the reducing agent.

b) $SO_3(g) + H_2O(l) \rightarrow H_2SO_4(aq)$

oxidation states; +6 -2 +1 – 2 +1 +6 -2

This is a not a redox reaction since none of the atoms undergoes a change in oxidation number.

c) $Ba(s) + Cl_2(g) \rightarrow BaCl_2(s)$

oxidation states; 0 0 +2 -1

This is a redox reaction since Ba increases in oxidation number (oxidation) and Cl decreases in number (reduction). Cl_2 is the oxidizing agent, Ba is the reducing agent.

d) $Mg(s) + Br_2(l) \rightarrow MgBr_2(s)$

oxidation states; 0 0 +2 -1

This is a redox reaction since Mg increases in oxidation number (oxidation) and Br decreases in number (reduction). Mg is the oxidizing agent, Br_2 is the reducing agent.

63. a) Skeletal reaction: $S(s) + O_2(g) \rightarrow SO_2(g)$
Balanced reaction: $S(s) + O_2(g) \rightarrow SO_2(g)$

b) Skeletal reaction: $C_3H_6(g) + O_2(g) \rightarrow CO_2(g) + H_2O(g)$
Balance C: $C_3H_6(g) + O_2(g) \rightarrow 3CO_2(g) + H_2O(g)$
Balance H: $C_3H_6(g) + O_2(g) \rightarrow 3CO_2(g) + 3H_2O(g)$
Balance O: $C_3H_6(g) + 9/2\ O_2(g) \rightarrow 3CO_2(g) + 3H_2O(g)$
Clear fraction: $2C_3H_6(g) + 9O_2(g) \rightarrow 6CO_2(g) + 6H_2O(g)$

c) Skeletal reaction: $Ca(s) + O_2(g) \rightarrow CaO$
Balance O: $Ca(s) + O_2(g) \rightarrow 2CaO$
Balance Ca: $2Ca(s) + O_2(g) \rightarrow 2CaO$

d) Skeletal reaction: $C_5H_{12}S(l) + O_2(g) \rightarrow CO_2(g) + H_2O(g) + SO_2(g)$
Balance C: $C_5H_{12}S(l) + O_2(g) \rightarrow 5CO_2(g) + H_2O(g) + SO_2(g)$
Balance H: $C_5H_{12}S(l) + O_2(g) \rightarrow 5CO_2(g) + 6H_2O(g) + SO_2(g)$
Balance S: $C_5H_{12}S(l) + O_2(g) \rightarrow 5CO_2(g) + 6H_2O(g) + SO_2(g)$
Balance O: $C_5H_{12}S(l) + 9O_2(g) \rightarrow 5CO_2(g) + 6H_2O(g) + SO_2(g)$

64. a) Skeletal reaction: $C_4H_6(g) + O_2(g) \rightarrow CO_2(g) + H_2O(g)$
Balance C: $C_4H_6(g) + O_2(g) \rightarrow 4CO_2(g) + H_2O(g)$
Balance H: $C_4H_6(g) + O_2(g) \rightarrow 4CO_2(g) + 3H_2O(g)$
Balance O: $C_4H_6(g) + 11/2\ O_2(g) \rightarrow 4CO_2(g) + 3H_2O(g)$
Clear fraction: $2C_4H_6(g) + 11\ O_2(g) \rightarrow 8CO_2(g) + 6H_2O(g)$

b) Skeletal reaction: $C(s) + O_2(g) \rightarrow CO_2(g)$
Balanced reaction: $C(s) + O_2(g) \rightarrow CO_2(g)$

c) Skeletal reaction: $CS_2(s) + O_2(g) \rightarrow CO_2(g) + SO_2(g)$
Balance C: $CS_2(s) + O_2(g) \rightarrow CO_2(g) + SO_2(g)$
Balance S: $CS_2(s) + O_2(g) \rightarrow CO_2(g) + 2SO_2(g)$
Balance O: $CS_2(s) + 3O_2(g) \rightarrow CO_2(g) + 2SO_2(g)$

d) Skeletal reaction: $C_3H_8O(l) + O_2(g) \rightarrow CO_2(g) + H_2O(g)$
Balance C: $C_3H_8O(l) + O_2(g) \rightarrow 3CO_2(g) + H_2O(g)$
Balance H: $C_3H_8O(l) + O_2(g) \rightarrow 3CO_2(g) + 4H_2O(g)$
Balance O: $C_3H_8O(l) + 9/2\ O_2(g) \rightarrow 3CO_2(g) + 4H_2O(g)$
Clear fraction: $2C_3H_8O(l) + 9O_2(g) \rightarrow 6CO_2(g) + 8H_2O(g)$

65. **Given:** In 100 g solution, 20.0 g $C_2H_6O_2$; density of solution = 1.03 g/mL **Find:** M of solution

Conceptual Plan: **g $C_2H_6O_2 \rightarrow$ mol $C_2H_6O_2$ and g solution $\rightarrow$ mL solution $\rightarrow$ L solution**

$$\frac{1\text{ mol } C_2H_6O_2}{62.06\text{ g } C_2H_6O_2} \qquad \frac{1.00\text{ mL}}{1.03\text{ g}} \qquad \frac{1\text{ L}}{1000\text{ mL}}$$

then M $C_2H_6O_2$

$$M = \frac{mol\ C_2H_6O_2}{L\ solution}$$

Solution:

$$20.0\ g\ C_2H_6O_2 \times \frac{1\ mol\ C_2H_6O_2}{62.06\ g\ C_2H_6O_2} = 0.32\underline{2}2\ mol\ C_2H_6O_2$$

$$100.0\ g\ solution \times \frac{1.00\ mL\ solution}{1.03\ g\ solution} \times \frac{1\ L}{1000\ mL} = 0.097\underline{0}8\ L$$

$$M = \frac{0.32\underline{2}2\ mol\ C_2H_6O_2}{0.097\underline{0}8\ L} = 3.32\ M$$

Check: The units of the answer, M $C_2H_6O_2$, are correct. The magnitude of the answer is reasonable since the concentration of solutions is usually between 0 and 18 M.

66. **Given:** 1.35 M NaCl; density of solution = 1.05 g /mL **Find:** % NaCl by mass
Conceptual Plan: **mol NaCl → g NaCl and L solution → mL solution → g solution then % NaCl**

$$\frac{58.45\ g\ NaCl}{1\ mol\ NaCl} \qquad \frac{1000\ mL}{1\ L} \qquad \frac{1.05\ g}{1.00\ mL} \qquad \frac{g\ NaCl}{g\ solution} \times 100$$

Solution: $1.35\ M = \frac{1.35\ mol\ NaCl}{1\ L\ solution}$

$$1.35\ mol\ NaCl \times \frac{58.45\ g\ NaCl}{1\ mol\ NaCl} = 78.\underline{9}1\ g\ NaCl \qquad 1\ L\ solution \times \frac{1000\ mL}{L} \times \frac{1.05\ g\ solution}{mL\ solution} = 10\underline{5}0\ g\ solution$$

$$\frac{78.91\ g\ NaCl}{1050\ g\ solution} \times 100 = 7.52\ \%\ NaCl$$

Check: The units of the answer, % NaCl, are correct. The magnitude of the answer is reasonable for the concentration of the solution.

67. **Given:** 2.5 g $NaHCO_3$ **Find:** g HCl
Conceptual Plan: **g $NaHCO_3$ → mol $NaHCO_3$ → mol HCl → g HCl**

$$\frac{1\ mol\ NaHCO_3}{84.02\ g\ NaHCO_3} \qquad \frac{1\ mol\ HCl}{1\ mol\ NaHCO_3} \qquad \frac{36.46\ g\ HCl}{1\ mol\ HCl}$$

Solution: $HCl(aq) + NaHCO_3(aq) \rightarrow H_2O(l) + CO_2(g) + NaCl(aq)$

$$2.5\ g\ NaHCO_3 \times \frac{1\ mol\ NaHCO_3}{84.02\ g\ NaHCO_3} \times \frac{1\ mol\ HCl}{1\ mol\ NaHCO_3} \times \frac{36.46\ g\ HCl}{1\ mol\ HCl} = 1.1\ g\ HCl$$

Check: The units of the answer, g HCl, are correct. The magnitude of the answer is reasonable since the molar mass of HCl is less than the molar mass of $NaHCO_3$.

68. **Given:** 3.8 g HCl **Find:** g $CaCO_3$
Conceptual Plan: **g HCl → mol HCl → mol $CaCO_3$ → g $CaCO_3$**

$$\frac{1\ mol\ HCl}{36.46\ g\ HCl} \qquad \frac{1\ mol\ CaCO_3}{2\ mol\ HCl} \qquad \frac{100.09\ g\ CaCO_3}{1\ mol\ CaCO_3}$$

Solution: $2\ HCl(aq) + CaCO_3(s) \rightarrow H_2O(l) + CO_2(g) + CaCl_2(aq)$

$$3.8\ g\ HCl \times \frac{mol\ HCl}{36.46\ g\ HCl} \times \frac{1\ mol\ CaCO_3}{2\ mol\ HCl} \times \frac{100.09\ g\ CaCO_3}{mol\ CaCO_3} = 5.2\ g\ CaCO_3$$

Check: Units of the answer, g $CaCO_3$, are correct. The magnitude of the answer is reasonable since the molar mass of $CaCO_3$ is greater than the molar mass of HCl.

69. **Given:** 1.0 kg C_8H_{18} **Find:** kg CO_2
Conceptual Plan: **kg C_8H_{18} → g C_8H_{18} → mol C_8H_{18} → mol CO_2 → g CO_2 → kg CO_2**

$$\frac{1000\text{ g}}{\text{kg}} \quad \frac{1\text{ mol }C_8H_{18}}{114.22\text{ g }C_8H_{18}} \quad \frac{16\text{ mol }CO_2}{2\text{ mol }C_8H_{18}} \quad \frac{44.01\text{ g }CO_2}{1\text{ mol }CO_2} \quad \frac{\text{kg}}{1000\text{ g}}$$

Solution: $2\ C_8H_{18}(g) + 25\ O_2(g) \rightarrow 16\ CO_2(g) + 18\ H_2O(g)$

$$1.0\text{ kg }C_8H_{18} \times \frac{1000\text{ g}}{\text{kg}} \times \frac{1\text{ mol }C_8H_{18}}{114.22\text{ g }C_8H_{18}} \times \frac{16\text{ mol }CO_2}{2\text{ mol }C_8H_{18}} \times \frac{44.01\text{ g }CO_2}{1\text{ mol }CO_2} \times \frac{\text{kg}}{1000\text{ g}} = 3.1\text{ kg }CO_2$$

Check: The units of the answer, kg CO_2, are correct. The magnitude of the answer is reasonable since the ratio of CO_2 to C_8H_{18} is 8:1.

70. **Given:** 18.9 L C_3H_8, d = 0.621 g/ mL **Find:** kg CO_2
Conceptual Plan: **L C_3H_8 → mL C_3H_8 → g C_3H_8 → mol C_3H_8 → mol CO_2 → g CO_2 → kg CO_2**

$$\frac{1000\text{ mL}}{\text{L}} \quad \frac{0.621\text{ g}}{\text{mL}} \quad \frac{1\text{ mol }C_3H_8}{44.09\text{ g }C_3H_8} \quad \frac{3\text{ mol }CO_2}{1\text{ mol }C_3H_8} \quad \frac{44.01\text{ g }CO_2}{1\text{ mol }CO_2} \quad \frac{\text{kg}}{1000\text{ g}}$$

Solution: $C_3H_8(g) + 5\ O_2(g) \rightarrow 3\ CO_2(g) + 4\ H_2O(g)$

$$18.9\text{ L }C_3H_8 \times \frac{1000\text{ mL}}{\text{L}} \times \frac{0.621\text{ g}}{\text{mL}} \times \frac{1\text{ mol }C_3H_8}{44.09\text{ g }C_3H_8} \times \frac{3\text{ mol }CO_2}{1\text{ mol }C_3H_8} \times \frac{44.01\text{ g }CO_2}{\text{mol }CO_2} \times \frac{\text{kg}}{1000\text{ g}}$$

$= 35.1\text{ kg }CO_2$

Check: The units of the answer, kg CO_2, are correct. The magnitude of the answer is reasonable since the molar mass of CO_2 and C_3H_8 are close and there is a mole ratio of 1:3.

71. **Given:** 3.00 mL $C_4H_6O_3$, d = 1.08 g/mL; 1.25 g $C_7H_6O_3$; 1.22 g $C_9H_8O_4$ **Find:** limiting reactant, theoretical yield $C_9H_8O_4$ and % yield $C_9H_8O_4$
Conceptual Plan: **mL $C_4H_6O_3$ → g $C_4H_6O_3$ → mol $C_4H_6O_3$ → mol $C_9H_8O_4$**

$$\frac{1.08\text{ g }C_4H_6O_3}{1.00\text{ mL }C_4H_6O_3} \quad \frac{1\text{ mol }C_4H_6O_3}{102.09\text{ g }C_4H_6O_3} \quad \frac{1\text{ mol }C_9H_8O_4}{1\text{ mol }C_4H_6O_3}$$

→ smallest amount determines limiting reactant

g $C_7H_6O_3$ → mol $C_7H_6O_3$ → mol $C_9H_8O_4$

$$\frac{1\text{ mol }C_7H_6O_3}{138.12\text{ g }C_7H_6O_3} \quad \frac{1\text{ mol }C_9H_8O_4}{1\text{ mol }C_7H_6O_3}$$

then: mol $C_9H_8O_4$ → g $C_9H_8O_4$

$$\frac{180.1\text{ g }C_9H_8O_4}{\text{mol }C_9H_8O_4}$$

then: determine % yield

$$\frac{\text{actual yield g }C_9H_8O_4}{\text{theoretical yield g }C_9H_8O_4} \times 100$$

Solution:

$$3.00\text{ mL }C_4H_6O_3 \times \frac{1.08\text{ g }C_4H_6O_3}{\text{mL }C_4H_6O_3} \times \frac{1\text{ mol }C_4H_6O_3}{102.09\text{ g }C_4H_6O_3} \times \frac{1\text{ mol }C_9H_8O_4}{1\text{ mol }C_4H_6O_3} = 0.031\underline{7}4\text{ mol }C_9H_8O_4$$

$$1.25\text{ g }C_7H_6O_3 \times \frac{1\text{ mol }C_7H_6O_3}{138.12\text{ g }C_7H_6O_3} \times \frac{1\text{ mol }C_9H_8O_4}{1\text{ mol }C_7H_6O_3} = 0.0090\underline{5}0\text{ mol }C_9H_8O_4$$

$$0.0090\underline{5}0\text{ mol }C_9H_8O_4 \times \frac{180.1\text{ g }C_9H_8O_4}{1\text{ mol }C_9H_8O_4} = 1.6\underline{3}0\text{ g }C_9H_8O_4$$

$$\frac{1.22\text{ g }C_9H_8O_4}{1.6\underline{3}0\text{ g }C_9H_8O_4} \times 100 = 74.8\%$$

Check: The theoretical yield has the correct units, g $C_9H_8O_4$, and has a reasonable magnitude compared to the mass of $C_7H_6O_3$, the limiting reactant. The % yield is reasonable, under 100%.

72. **Given:** 4.62 mL C_2H_5OH, d = 0.789 g/mL; 15.55 g O_2; 3.72 mL H_2O, d = 1.00g/mL
Find: limiting reactant, theoretical yield H_2O and % yield H_2O
Conceptual Plan: **mL C_2H_5OH → g C_2H_5OH → mol C_2H_5OH → mol H_2O**

$$\frac{0.789\text{ g }C_2H_5OH}{1.00\text{ mL }C_2H_5OH} \quad \frac{1\text{ mol }C_2H_5OH}{46.07\text{ g }C_2H_5OH} \quad \frac{3\text{ mol }H_2O}{1\text{ mol }C_2H_5OH}$$

→ **smallest amount determines limiting reactant**

$g\ O_2 \rightarrow mol\ O_2 \rightarrow mol\ H_2O$

$\frac{1\ mol\ O_2}{32.00\ g\ O_2}$ $\frac{3\ mol\ H_2O}{3\ mol\ O_2}$

then: $mol\ H_2O \rightarrow g\ H_2O$ **then: determine % yield**

$\frac{18.02\ g\ H_2O}{mol\ H_2O}$ $\frac{\text{actual yield } g\ C_9H_8O_4}{\text{theoretical yield } g\ C_9H_8O_4} \times 100$

Solution:

$C_2H_5OH(l) + 3\ O_2(g) \rightarrow 2\ CO_2(g) + 3\ H_2O(l)$

$$4.62\ mL\ C_2H_5OH \times \frac{0.789\ g\ C_2H_5OH}{mL\ C_2H_5OH} \times \frac{1\ mol\ C_2H_5OH}{46.07\ g\ C_2H_5OH} \times \frac{3\ mol\ H_2O}{1\ mol\ C_2H_5OH} = 0.23\underline{7}4\ mol\ H_2O$$

$$15.55\ g\ O_2 \times \frac{1\ mol\ O_2}{31.998\ g\ O_2} \times \frac{1\ mol\ H_2O}{1\ mol\ O_2} = 0.485\underline{9}7\ mol\ H_2O$$

$$0.23\underline{7}4\ mol\ O_2 \times \frac{18.02\ g\ H_2O}{1\ mol\ O_2} = 4.2\underline{7}8\ g\ H_2O$$

$$\frac{3.72\ g\ H_2O}{4.2\underline{7}8\ g\ H_2O} \times 100 = 87.0\%$$

Check: The theoretical yield has the correct units, g H_2O, and has a reasonable magnitude compared to the mass of C_2H_5OH, the limiting reactant. The % yield is reasonable, under 100%.

73. **Given:** (a) 11 molecules H_2, 2 molecules O_2; (b) 8 molecules H_2, 4 molecules O_2; (c) 4 molecules H_2, 5 molecules O_2; (d) 3 molecules H_2, 6 molecules O_2 **Find:** loudest explosion based on equation
Conceptual Plan: loudest explosion will occur in the balloon with the mol ratio closest to the balanced equation and that contains the most H_2
Solution: $2H_2(g) + O_2(g) \rightarrow H_2O(l)$
Balloon (a) has enough O_2 to react with 4 molecules H_2; balloon (b) has enough O_2 to react with 8 molecules H_2; balloon (c) has enough O_2 to react with 10 molecules H_2; and balloon (d) has enough O_2 for 3 molecules of H_2 to react. Therefore, balloon (b) will have the loudest explosion because it has the most H_2 that will react.
Check: Answer seems correct since it has the most H_2 with enough O_2 in the balloon to completely react.

74. **Given:** Beaker contain 4 ions H^+, 4 ions Cl^-. **Find:** which NaOH beaker will just neutralize HCl beaker
Conceptual Plan: molecules $H^+ \rightarrow$ molecules OH^- then compare to 4 beakers.

$\frac{1\ ion\ OH^-}{1\ ion\ H^+}$

Solution: $HCl(aq) + NaOH(aq) \rightarrow NaCl(aq) + H_2O(l)$
Net Ionic: $H^+(aq) + OH^-(aq) \rightarrow H_2O(l)$

$$4\ ions\ H^+ \times \frac{1\ ion\ OH^-}{1\ ion\ H^+} = 4\ ions\ OH^-$$

Beaker (a) contains 2 ions OH^-; Beaker (b) contains 4 ions OH^-;
Beaker (c) contains 5 ions OH^-; Beaker (d) contains 8 ions OH^-.
Beaker (b) will completely neutralize the HCl beaker with no excess.

Check: The answer is correct because it will completely neutralize the H^+ with no excess OH^-.

75. a) Skeletal reaction: $HCl(aq) + Hg_2(NO_3)_2(aq) \rightarrow Hg_2Cl_2(s) + HNO_3(aq)$
Balance Cl: $2HCl(aq) + Hg_2(NO_3)_2(aq) \rightarrow Hg_2Cl_2(s) + 2HNO_3(aq)$

b) Skeletal reaction: $KHSO_3(aq) + HNO_3(aq) \rightarrow H_2O(l) + SO_2(g) + KNO_3(aq)$
Balanced reaction: $KHSO_3(aq) + HNO_3(aq) \rightarrow H_2O(l) + SO_2(g) + KNO_3(aq)$

c) Skeletal reaction: $NH_4Cl(aq) + Pb(NO_3)_2(aq) \rightarrow PbCl_2(s) + NH_4NO_3(aq)$
Balance Cl: $2NH_4Cl(aq) + Pb(NO_3)_2(aq) \rightarrow PbCl_2(s) + NH_4NO_3(aq)$
Balance N: $2NH_4Cl(aq) + Pb(NO_3)_2(aq) \rightarrow PbCl_2(s) + 2NH_4NO_3(aq)$

d) Skeletal reaction: $NH_4Cl(aq) + Ca(OH)_2(aq) \rightarrow NH_3(g) + H_2O(l) + CaCl_2(aq)$
Balance Cl: $2NH_4Cl(aq) + Ca(OH)_2(aq) \rightarrow NH_3(g) + H_2O(l) + CaCl_2(aq)$
Balance N: $2NH_4Cl(aq) + Ca(OH)_2(aq) \rightarrow 2NH_3(g) + H_2O(l) + CaCl_2(aq)$
Balance H: $2NH_4Cl(aq) + Ca(OH)_2(aq) \rightarrow 2NH_3(g) + 2H_2O(l) + CaCl_2(aq)$

76. a) Skeletal reaction: $H_2SO_4(aq) + HNO_3(aq) \rightarrow$ No Reaction

b) Skeletal reaction: $Cr(NO_3)_3(aq) + LiOH(aq) \rightarrow Cr(OH)_3(s) + LiNO_3(aq)$
Balance OH: $Cr(NO_3)_3(aq) + 3LiOH(aq) \rightarrow Cr(OH)_3(s) + LiNO_3(aq)$
Balance Li: $Cr(NO_3)_3(aq) + 3LiOH(aq) \rightarrow Cr(OH)_3(s) + 3LiNO_3(aq)$

c) Skeletal reaction: $C_5H_{12}O(l) + O_2(g) \rightarrow CO_2(g) + H_2O(g)$
Balance C: $C_5H_{12}O(l) + O_2(g) \rightarrow 5CO_2(g) + H_2O(g)$
Balance H: $C_5H_{12}O(l) + O_2(g) \rightarrow 5CO_2(g) + 6H_2O(g)$
Balance O: $C_5H_{12}O(l) + 15/2\ O_2(g) \rightarrow 5CO_2(g) + 6H_2O(g)$
Clear fraction: $2C_5H_{12}O(l) + 15O_2(g) \rightarrow 10CO_2(g) + 12H_2O(g)$

d) Skeletal reaction: $SrS(aq) + CuSO_4(aq) \rightarrow SrSO_4(aq) + CuS(s)$
Balanced reaction: $SrS(aq) + CuSO_4(aq) \rightarrow SrSO_4(aq) + CuS(s)$

77. **Given:** 1.5 L solution, 0.050 M $CaCl_2$, 0.085 M $Mg(NO_3)_2$ **Find:** g Na_3PO_4
Conceptual Plan: V,M $CaCl_2 \rightarrow$ mol $CaCl_2$ and V,M $Mg(NO_3)_2 \rightarrow$ mol $Mg(NO_3)_2$
$V \times M = mol$ $\quad$ $V \times M = mol$
then (mol $CaCl_2$ + mol $Mg(NO_3)_2$) $\rightarrow Na_3PO_4 \rightarrow$ g Na_3PO_4

$$\frac{2\ \text{mol}\ Na_3PO_4}{3\ \text{mol}\ (CaCl_2 + Mg(NO_3)_2)} \qquad \frac{163.97\ \text{g}\ Na_3PO_4}{1\ \text{mol}\ Na_3PO_4}$$

Solution: $3CaCl_2(aq) + 2Na_3PO_4(aq) \rightarrow Ca_3(PO_4)_2(s) + 6\ NaCl(aq)$
$3\ Mg(NO_3)_2(aq) + 2Na_3PO_4(aq) \rightarrow Mg_3(PO_4)_2(s) + 6\ NaCl(aq)$

$$1.5\ \text{L} \times 0.050\ \text{M}\ CaCl_2 = 0.07\underline{5}\ \text{mol}\ CaCl_2$$

$$1.5\ \text{L} \times 0.085\ \text{M}\ Mg(NO_3)_2 = 0.1\underline{2}75\ \text{mol}\ Mg(NO_3)_2$$

$$0.2\underline{0}25\ \cancel{\text{mol}\ CaCl_2\ \text{and}\ Mg(NO_3)_2} \times \frac{2\ \cancel{\text{mol}\ Na_2PO_4}}{3\ \text{mol}\ \cancel{CaCl_2\ \text{and}\ Mg(NO_3)_2}} \times \frac{163.97\ \text{g mol}\ Na_2PO_4}{\cancel{\text{mol}\ Na_2PO_4}} = 22\ \text{g mol}\ Na_2PO_4$$

Check: The units of the answer, g Na_3PO_4, are correct. The magnitude of the answer is reasonable since it is needed to remove both the Ca and Mg ions.

78. **Given:** 500.0 mL 0.100 M HCl and 0.200 M H_2SO_4; 0.150 M KOH **Find:** volume KOH to neutralize the acid

Conceptual Plan:
mL $\rightarrow$ L, then VM(HCl) $\rightarrow$ mol HCl $\rightarrow$ mol H^+ and VM(H_2SO_4) $\rightarrow$ mol $H_2SO_4 \rightarrow$ mol H^+

$$\frac{1\ \text{L}}{1000\ \text{mL}} \qquad V \times M = mol \qquad \frac{\text{mol}\ H^+}{\text{mol HCl}} \qquad V \times M = mol \qquad \frac{2\ \text{mol}\ H^+}{\text{mol}\ H_2SO_4}$$

Total mol $H^+ \rightarrow$ mol $OH^- \rightarrow$ mol KOH $\rightarrow$ volume KOH

$$\frac{\text{mol}\ OH^-}{\text{mol}\ H^+} \qquad \frac{\text{mol KOH}}{\text{mol}\ OH^-} \qquad V = \frac{\text{mol KOH}}{\text{M KOH}}$$

Solution:

$$500.0\ \cancel{\text{mL}} \times \frac{1\ \cancel{\text{L}}}{1000\ \cancel{\text{mL}}} \times \frac{0.100\ \cancel{\text{mol HCl}}}{1\ \cancel{\text{L}}} \times \frac{1\ \text{mol}\ H^+}{1\ \cancel{\text{mol HCl}}} = 0.050\underline{0}0\ \text{mol}\ H^+$$

$$500.0\ \cancel{\text{mL}} \times \frac{1\ \cancel{\text{L}}}{1000\ \cancel{\text{mL}}} \times \frac{0.200\ \cancel{\text{mol}\ H_2SO_4}}{1\ \cancel{\text{L}}} \times \frac{2\ \text{mol}\ H^+}{1\ \cancel{\text{mol}\ H_2SO_4}} = 0.20\underline{0}0\ \text{mol}\ H^+$$

$$(0.20\underline{0}0 + 0.050\underline{0}0)\ \text{mol}\ H^+ \times \frac{1\ \cancel{\text{mol}\ OH^-}}{1\ \cancel{\text{mol}\ H^+}} \times \frac{1\ \cancel{\text{mol KOH}}}{1\ \cancel{\text{mol}\ OH^-}} \times \frac{1\ \text{L solution}}{0.150\ \cancel{\text{mol KOH}}} = 1.67\ \text{L KOH solution}$$

Check: The units of the answer, L KOH, are correct. The magnitude of the answer is reasonable because the average concentration of acid is greater than the concentration of base.

79. **Given:** 1.0 L, 0.10 M OH^- **Find:** g Ba
Conceptual Plan: **VM → mol OH^- → mol $Ba(OH)_2$ → mol BaO → mol Ba → g Ba**

$$V \times M = mol \quad \frac{1\ mol\ Ba(OH)_2}{2\ mol\ OH^-} \quad \frac{1\ mol\ BaO}{1\ mol\ Ba(OH)_2} \quad \frac{1\ mol\ Ba}{1\ mol\ BaO} \quad \frac{137.3\ g\ Ba}{1\ mol\ Ba}$$

Solution: $BaO(s) + H_2O(l) \rightarrow Ba(OH)_2(aq)$

$$1.0\ \cancel{L} \times \frac{0.10\ \cancel{mol\ OH^-}}{\cancel{L}} \times \frac{1\ \cancel{mol\ Ba(OH)_2}}{2\ \cancel{mol\ OH^-}} \times \frac{1\ \cancel{mol\ BaO}}{1\ \cancel{mol\ Ba(OH)_2}} \times \frac{1\ \cancel{mol\ Ba}}{1\ \cancel{mol\ BaO}} \times \frac{137.3\ g\ Ba}{1\ \cancel{mol\ Ba}} = 6.9\ g\ Ba$$

Check: The units of the answer, g Ba, are correct. The magnitude is reasonable since the molar mass of Ba is large and there are 2 moles hydroxide per mole Ba.

80. **Given:** 1.00 L, 1.51 M NaF; 49.6 g sample; mixture Cr^{3+} and Mg^{2+} **Find:** g Cr^{3+}
Conceptual Plan: **V,M NaF → mol NaF → mol F^- and let x = mol CrF_3 and y = mol MgF_2 → mol F**

$$V \times M = mol \quad \frac{1\ mol\ F^-}{1\ mol\ NaF} \qquad \frac{3\ mol\ F}{mol\ CrF_3} \quad \frac{2\ mol\ F}{mol\ MgF_2}$$

and → g sample then solve for x = mol CrF_3 → mol Cr^{3+} → g Cr^{3+}

$$x\ (\text{molar mass } CrF_3) = g\ CrF_3 \quad y\ (\text{molar mass } MgF_2) = g\ MgF_2 \quad \frac{1\ mol\ Cr^{3+}}{1\ mol\ CrF_3} \quad \frac{52.00\ g\ Cr^{3+}}{1\ mol\ Cr^{3+}}$$

Solution:

$$1.00\ \cancel{L} \times \frac{1.51\ \cancel{mol\ NaF}}{\cancel{L}} \times \frac{1\ mol\ F^-}{1\ \cancel{mol\ NaF}} = 1.51\ mol\ F^-$$

Let x = mol CrF_3 and y = mol MgF_2;

3x = mol F^- from CrF_3 ; 2y = mol F^- from MgF_2

$3x + 2y = 1.51\ mol\ F^-$

$$x\ mol\ CrF_3 \times \frac{109.00\ g\ CrF_3}{mol\ CrF_3} = g\ CrF_3$$

$$y\ mol\ MgF_2 \times \frac{62.30\ g\ MgF_2}{mol\ MgF_2} = g\ MgF_2$$

$x(109.00) + y(62.30) = 49.6$ g sample

Solve simultaneous equations: $x = 0.16\underline{4}8$ = mol CrF_3

$$0.16\underline{4}8\ mol\ CrF_3 \times \frac{1\ mol\ Cr^{3+}}{1\ mol\ CrF_3} \times \frac{52.00\ g\ Cr^{3+}}{1\ mol\ Cr^{3+}} = 8.57\ g\ Cr^{3+}$$

Check: Units of answer, g Cr^{3+}, are correct. The magnitude of the answer is reasonable since it is less than the mass of the sample.

81. **Given:** 30.0% $NaNO_3$, \$9.00/ 100 lb; 20.0 % $(NH_4)_2SO_4$, \$8.10/ 100 lb **Find:** cost / lb N
Conceptual Plan: **mass fertilizer → mass $NaNO_3$ → mass N → cost/lb N**

$$\frac{30.0\ lb\ NaNO_3}{100\ lb\ fertilizer} \quad \frac{16.48\ lb\ N}{100\ lb\ NaNO_3} \quad \frac{\$9.00}{100\ lb\ fertilizer}$$

and: mass fertilizer → mass $(NH_4)_2SO_4$ → mass N → cost/lb N

$$\frac{20.0\ lb\ (NH_4)_2SO_4}{100\ lb\ fertilizer} \quad \frac{21.2\ lb\ N}{100\ lb\ (NH_4)_2SO_4} \quad \frac{\$8.10}{100\ lb\ fertilizer}$$

Solution:

$$100\ \cancel{lb\ fertilizer} \times \frac{30.0\ \cancel{lb\ NaNO_3}}{100\ \cancel{lb\ fertilizer}} \times \frac{16.48\ lb\ N}{100\ \cancel{lb\ NaNO_3}} = 4.9\underline{4}4\ lb\ N$$

$$\frac{\$9.00}{100\ \cancel{lb\ fertilizer}} \times \frac{100\ \cancel{lb\ fertilizer}}{4.9\underline{4}4\ lb\ N} = \$1.82/\ lb\ N$$

$$100\ \cancel{lb\ fertilizer} \times \frac{20.0\ \cancel{lb\ (NH_4)_2SO_4}}{100\ \cancel{lb\ fertilizer}} \times \frac{21.2\ lb\ N}{100\ \cancel{lb\ (NH_4)_2SO_4}} = 4.2\underline{4}0\ lb\ N$$

$$\frac{\$8.10}{100\ \text{lb fertilizer}} \times \frac{100\ \text{lb fertilizer}}{4.2\underline{4}\ \text{lb N}} = \$1.91/\ \text{lb N}$$

The more economical fertilizer is the $NaNO_3$ because it costs less/ lb N.

Check: The units of the cost, \$/lb N, are correct. The answer is reasonable because you compare the cost/lb N directly.

82. **Given:** 0.110 M HCl; 1.52 g $Al(OH)_3$ **Find:** volume HCl needed to neutralize

Conceptual Plan: **g $Al(OH)_3$ → mol $Al(OH)_3$ → mol HCl → vol HCL**

$$\frac{\text{mol Al(OH)}_3}{78.00\ \text{g Al(OH)}_3} \qquad \frac{3\ \text{mol HCl}}{1\ \text{mol Al(OH)}_3} \qquad \frac{\text{mol HCl}}{\text{M HCl}}$$

Solution: $3\ HCl(aq) + Al(OH)_3(aq) \rightarrow 3\ H_2O(l) + AlCl_3(aq)$

$$1.52\ \text{g Al(OH)}_3 \times \frac{1\ \text{mol Al(OH)}_3}{78.00\ \text{g Al(OH)}_3} \times \frac{3\ \text{mol HCl}}{1\ \text{mol Al(OH)}_3} \times \frac{1\ \text{L}}{0.110\ \text{mol HCl}} = 0.531\ \text{L HCl}$$

Check: Units of the answer, L HCl, are correct. The magnitude of the answer is reasonable since the mole ratio of HCl to $Al(OH)_3$ is 3:1.

83. **Given:** 24.5 g Au, 24.5 g BrF_3, 24.5 g KF **Find:** g $KAuF_4$

Conceptual Plan: **g Au → mol Au → mol $KAuF_4$**

$$\frac{1\ \text{mol Au}}{196.97\ \text{g Au}} \qquad \frac{2\ \text{mol KAuF}_4}{2\ \text{mol Au}}$$

g BrF_3 → mol BrF_3 → mol $KAuF_4$ **→ smallest mol amount determines limiting reactant**

$$\frac{1\ \text{mol BrF}_3}{136.9\ \text{g BrF}_3} \qquad \frac{2\ \text{mol KAuF}_4}{2\ \text{mol BrF}_3}$$

g KF → mol KF → mol $KAuF_4$

$$\frac{1\ \text{mol KF}}{58.10\ \text{g KF}} \qquad \frac{2\ \text{mol KAuF}_4}{2\ \text{mol KF}}$$

then: mol $KAuF_4$ → g $KAuF_4$

$$\frac{312.07\ \text{g KAuF}_4}{\text{mol KAuF}_4}$$

$2\ Au(s) + 2BrF_3(l) + 2KF(s) \rightarrow Br_2(l) + 2KAuF_4(s)$

oxidation states; 0 +3 – 1 +1 – 1 0 +1 +3 – 1

This is a redox reaction since Au increases in oxidation number (oxidation) and Br decreases in number (reduction). BrF_3 is the oxidizing agent, Au is the reducing agent.

Solution:

$$24.5\ \text{g Au} \times \frac{1\ \text{mol Au}}{196.97\ \text{g Au}} \times \frac{2\ \text{mol KAuF}_4}{2\ \text{mol Au}} = 0.12\underline{4}4\ \text{mol KAuF}_4$$

$$24.5\ \text{g BrF}_3 \times \frac{1\ \text{mol BrF}_3}{136.90\ \text{g BrF}_3} \times \frac{2\ \text{mol KAuF}_4}{2\ \text{mol BrF}_3} = 0.17\underline{9}0\ \text{mol KAuF}$$

$$24.5\ \text{g KF} \times \frac{1\ \text{mol KF}}{58.10\ \text{g KF}} \times \frac{2\ \text{mol KAuF}_4}{2\ \text{mol KF}} = 0.42\underline{1}7\ \text{mol KAuF}_4$$

$$0.12\underline{4}4\ \text{mol KAuF}_4 \times \frac{312.07\ \text{g KAuF}_4}{1\ \text{mol KAuF}_4} = 38.8\ \text{g KAuF}_4$$

Check: Units of the answer, g $KAuF_4$, are correct. The magnitude of the answer is reasonable compared to the mass of the limiting reactant Au.

84. **Given:** 0.10 L, 0.12 M NaCl; 0.23 L, 0.18 M $MgCl_2$; 0.20 M $AgNO_3$ **Find:** volume $AgNO_3$ to precipitate all the Cl^-.

Conceptual Plan: VM(NaCl) → mol NaCl → mol Cl^- and VM($MgCl_2$) → mol $MgCl_2$ → mol Cl^-

$$\text{vol} \times \text{M} = \text{mol} \qquad \frac{1\ \text{mol Cl}^-}{1\ \text{mol NaCl}} \qquad \text{vol} \times \text{M} = \text{mol} \qquad \frac{2\ \text{mol Cl}^-}{1\ \text{mol MgCl}_2}$$

Then: total mol Cl^- → mol Ag^+ → mol $AgNO_3$ → vol $AgNO_3$

$$\frac{1\ \text{mol Ag}^+}{1\ \text{mol Cl}^-} \qquad \frac{1\ \text{mol AgNO}_3}{1\ \text{mol Ag}^+} \qquad \frac{1\ \text{L AgNO}_3}{0.20\ \text{mol AgNO}_3}$$

Solution:

$$0.10\ \text{L NaCl} \times \frac{0.12\ \text{mol NaCl}}{\text{L NaCl}} \times \frac{1\ \text{mol Cl}^-}{1\ \text{mol NaCl}} = 0.01\underline{2}\ \text{mol Cl}^-$$

$$0.23\ \text{L NaCl} \times \frac{0.18\ \text{mol MgCl}_2}{\text{L MgCl}_2} \times \frac{2\ \text{mol Cl}^-}{1\ \text{mol MgCl}_2} = 0.08\underline{2}8\ \text{mol Cl}^-$$

$$0.09\underline{4}8\ \text{mol Cl}^- \times \frac{1\ \text{mol Ag}^+}{1\ \text{mol Cl}^-} \times \frac{1\ \text{mol AgNO}_3}{1\ \text{mol Ag}^+} \times \frac{\text{L AgNO}_3}{0.20\ \text{mol AgNO}_3} = 0.47\ \text{L AgNO}_3$$

Check; Units of the answer, L $AgNO_3$, are correct. The magnitude of the answer is reasonable since the Cl^- comes from two sources.

85. **Given:** solution may contain Ag^+, Ca^{2+}, and Cu^{2+} **Find:** determine which ions are present.
Conceptual Plan: test the solution sequentially with NaCl, Na_2SO_4, and Na_2CO_3 and see if precipitates form.
Solution: original solution + NaCl yields no reaction: Ag^+ not present: since chlorides are normally soluble but Ag^+ is an exception.
Original solution with Na_2SO_4 yields a precipitate and solution 2. The precipitate is $CaSO_4$, so Ca^{2+} is present. Sulfates are normally soluble but Ca^{2+} is an exception.
Solution 2 with Na_2CO_3 yields a precipitate. The precipitate is $CuCO_3$, so Cu^{2+} is present. All carbonates are insoluble.
NET IONIC EQUATIONS:

$$Ca^{2+}(aq) + SO_4^{2-}(aq) \rightarrow CaSO_4(s)$$
$$Cu^{2+}(aq) + CO_3^{2-}(aq) \rightarrow CuCO_3(s)$$

Check: the answer is reasonable since two different precipitates formed and all the Ca^{2+} was removed before the carbonate was added.

86. **Given:** solution may contain Hg_2^{2+}, Ba^{2+}, and Fe^{2+} **Find:** determine which ions are present.
Conceptual Plan: test the solution sequentially with KCl, K_2SO_4, and K_2CO_3 and see if precipitates form.
Solution: original solution + KCl yields a precipitate and solution 2: The precipitate is Hg_2Cl_2 so Hg_2^{2+} is present. Chlorides are normally soluble but Hg_2^{2+} is an exception.
Solution 2 with K_2SO_4 yields no precipitate. So Ba^{2+} is not present. Sulfates are normally soluble but Ba^{2+} is an exception.
Solution 2 with K_2CO_3 yields a precipitate. The precipitate is $FeCO_3$, so Fe^{2+} is present. All carbonates are insoluble.
NET IONIC EQUATIONS:

$$Hg_2^{2+}(aq) + 2Cl^-(aq) \rightarrow Hg_2Cl_2(s)$$
$$Fe^{2+}(aq) + CO_3^{2-}(aq) \rightarrow FeCO_3(s)$$

Check: the answer is reasonable since two different precipitates formed and all the Hg_2^{2+} was removed before the carbonate was added.

87 **Given:** 15.2 billion L lake water, 1.8×10^{-5} M H_2SO_4, 8.7×10^{-6} M HNO_3 **Find:** kg $CaCO_3$ needed to neutralize
Conceptual Plan: Vol lake → mol H_2SO_4 → mol H^+ and vol lake → mol HNO_3 → mol H^+

$$\text{vol} \times \text{M} = \text{mol} \qquad \frac{2\ \text{mol H}^+}{1\ \text{mol H}_2\text{SO}_4} \qquad \text{vol} \times \text{M} = \text{mol} \qquad \frac{1\ \text{mol H}^+}{1\ \text{mol HNO}_3}$$

Then: total mol H^+ → mol CO_3^{2-} → mol $CaCO_3$ → g $CaCO_3$ → kg $CaCO_3$

$$\frac{1\ \text{mol CO}_3^{2-}}{2\ \text{mol H}^+} \qquad \frac{1\ \text{mol CaCO}_3}{1\ \text{mol CO}_3^{2-}} \qquad \frac{100.09\ \text{g CaCO}_3}{1\ \text{mol CaCO}_3} \qquad \frac{\text{kg}}{1000\ \text{g}}$$

Solution: $2H^+(aq) + CO_3^{2-}(aq) \rightarrow H_2O(l) + CO_2(g)$

$$15.2 \times 10^9\ \text{L} \times \frac{1.8 \times 10^{-5}\ \text{mol H}_2\text{SO}_4}{\text{L soln}} \times \frac{2\ \text{mol H}^+}{\text{mol H}_2\text{SO}_4} = 5\underline{4}7200\ \text{mol H}^+$$

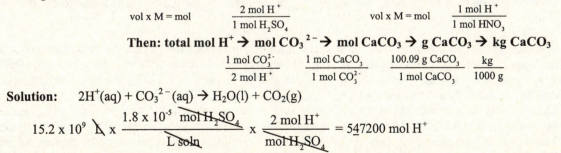

$$15.2 \times 10^9\ \text{L} \times \frac{8.7 \times 10^{-6}\ \text{mol HNO}_3}{\text{L soln}} \times \frac{1\ \text{mol H}^+}{\text{mol HNO}_3} = 13\underline{2}240\ \text{mol H}^+$$

$$6\underline{7}9440\ \text{mol H}^+ \times \frac{1\ \text{mol CO}_3^{2-}}{2\ \text{mol H}^+} \times \frac{1\ \text{mol CaCO}_3}{\text{mol CO}_3^{2-}} \times \frac{100.09\ \text{g CaCO}_3}{1\ \text{mol CaCO}_3} \times \frac{\text{kg}}{1000\ \text{g}} = 3.4 \times 10^4\ \text{kg CaCO}_3$$

Check: Units of the answer, kg $CaCO_3$, are correct. The magnitude of the answer is reasonable based on the size of the lake.

88. **Given:** 3.5×10^{-3} M Ca^{2+}. 1.1×10^{-3} M Mg^{2+}, 19.5 gal H_2O; 0.65 kg detergent/load **Find:** % by mass Na_2CO_3

Conceptual Plan: gal H_2O → L H_2O then VM → mol Ca^{2+} and VM → mol Mg^{2+}

$\frac{3.785\ \text{L}}{1\ \text{gal}}$ vol x M = mol vol x M = mol

Then total moles ions → mol CO_3^{2-} → mol $CaCO_3$ → g $CaCO_3$ → kg $CaCO_3$ → % $CaCO_3$

$\frac{1\ \text{mol CO}_3^{2-}}{1\ \text{mol ion}}$ $\frac{1\ \text{mol CaCO}_3}{1\ \text{mol CO}_3^{2-}}$ $\frac{100.09\ \text{g CaCO}_3}{1\ \text{mol CaCO}_3}$ $\frac{\text{kg}}{1000\ \text{g}}$ $\frac{\text{kg CaCO}_3}{\text{kg detergent x 100}}$

Solution: $19.5\ \text{gal} \times \frac{3.785\ \text{L}}{1\ \text{gal}} \times \frac{3.5 \times 10^{-3}\ \text{mol Ca}^{2+}}{\text{L}} = 0.2\underline{5}8\ \text{mol Ca}^{2+}$

$$19.5\ \text{gal} \times \frac{3.785\ \text{L}}{1\ \text{gal}} \times \frac{1.1 \times 10^{-3}\ \text{mol Mg}^{2+}}{\text{L}} = 0.08\underline{1}19\ \text{mol Mg}^{2+}$$

$$0.3\underline{3}92\ \text{mol ions} \times \frac{1\ \text{mol CO}_3^{2-}}{\text{mol ions}} \times \frac{1\ \text{mol Na}_2\text{CO}_3}{\text{mol CO}_3^{2-}} \times \frac{106.01\ \text{g CaCO}_3}{1\ \text{mol CaCO}_3} \times \frac{\text{kg CaCO}_3}{1000\ \text{g CaCO}_3} = 0.03\underline{5}96\ \text{kg Na}_2\text{CO}_3$$

$$\frac{0.03\underline{5}96\ \text{kg Na}_2\text{CO}_3}{0.65\ \text{kg detergent}} \times 100 = 5.5\ \%\ \text{Na}_2\text{CO}_3$$

Check: Units of the answer, % Na_2CO_3, are correct. The magnitude of the answer is reasonable. % is less than 100 %.

89. **Given:** 45 μg Pb/dL blood, Vol = 5.0 L, 1 mol succimer($C_4H_6O_4S_2$) = 1 mol Pb **Find:** mass $C_4H_6O_4S_2$ in mg

Conceptual Plan: Volume blood L → Volume blood dL → μg Pb → g Pb → mol Pb →

$\frac{10\ \text{dL}}{\text{L}}$ $\frac{45\ \mu\text{g}}{\text{dL}}$ $\frac{10^6\ \mu\text{g}}{\text{g}}$ $\frac{\text{mol Pb}}{207.2\ \text{g Pb}}$ $\frac{1\ \text{mol succimer}}{1\ \text{mol Pb}}$

mol succimer → g succimer → mg succimer

$\frac{182.23\ \text{g succimer}}{1\ \text{mol succimer}}$ $\frac{1000\ \text{mg succimer}}{1\ \text{g succimer}}$

Solution:

$$5.0\ \text{L blood} \times \frac{10\ \text{dL}}{\text{L}} \times \frac{45\ \mu\text{g}}{\text{dL}} \times \frac{1\ \text{g}}{10^6\ \mu\text{g}} \times \frac{1\ \text{mol Pb}}{207.2\ \text{g}} \times \frac{1\ \text{mol succimer}}{1\ \text{mol Pb}} \times \frac{182.23\ \text{g succimer}}{1\ \text{mol succimer}} \times \frac{1000\ \text{mg}}{\text{g}}$$

= 2.0 mg succimer

Check: The units of the answer, mg succimer, are correct. The magnitude is reasonable for the volume of blood and the concentration.

90. In designing the unit you would need to consider the theoretical yield and % yield of the reaction, how changing the limiting reactant would affect the reaction, the stoichiometry between KO_2 and O_2 to determine the mass of KO_2 required to produce enough O_2 for 10 minutes. You might also consider the speed of the reaction and whether or not the reaction produced heat. Additionally, because your body does not use 100% of the oxygen taken in with each breath, the apparatus would only need to replenish the oxygen used. The percentage of oxygen in air is about 20% and the percentage in exhaled air is about 16%, so we will assume that 4% of the air would need to be replenished with oxygen. (NOTE: The problem can also be solved by finding the amount of KO_2 that would be required to react with all of the exhaled CO_2.)
Given: air = 4% O_2, volume = 5 – 8 L/ min, 1 mol gas = 22.4 L gas **Find:** O_2 for 10 min breathing time

Conceptual Plan: 10 min → vol air → vol O_2 → mol O_2 → mol KO_2 → g KO_2

$$\frac{8\text{ L air}}{1\text{ min}} \quad \frac{4\text{ L }O_2}{100\text{ L air}} \quad \frac{1\text{ mol }O_2}{22.4\text{ L }O_2} \quad \frac{4\text{ mol }KO_2}{3\text{ mol }O_2} \quad \frac{71.10\text{ g }KO_2}{1\text{ mol }KO_2}$$

Solution: $10\ \cancel{\text{min}}\ \text{x}\ \frac{8\ \cancel{\text{L air}}}{\cancel{\text{min}}}\ \text{x}\ \frac{4\ \cancel{\text{L }O_2}}{100\ \cancel{\text{L air}}}\ \text{x}\ \frac{1\ \cancel{\text{mol }O_2}}{22.4\ \cancel{\text{L }O_2}}\ \text{x}\ \frac{4\ \cancel{\text{mol }KO_2}}{3\ \cancel{\text{mol }O_2}}\ \text{x}\ \frac{71.10\text{ g }KO_2}{1\ \cancel{\text{mol }KO_2}} = 14\text{ g }KO_2$

Check: The units of the answer, g KO_2, are correct. The magnitude of the answer is reasonable since it is an amount that could be carried in a portable device.

91. **Given:** 250 g sample, 67.2 mol % Al **Find:** Theoretical yield in g of Mn

Conceptual Plan: mol % Al → g Al and mol % MnO_2 → g MnO_2, then mass % Al

$$\frac{26.98\text{ g Al}}{\text{mol Al}} \qquad \frac{86.94\text{ g }MnO_2}{\text{mol }MnO_2} \qquad \frac{\text{g Al}}{\text{total g}}\text{ x }100$$

then: sample → g Al → mol Al → mol Mn

$$\frac{38.86\text{ g Al}}{100\text{ g sample}} \quad \frac{\text{mol Al}}{26.98\text{ g Al}} \quad \frac{3\text{ mol Mn}}{4\text{ mol Al}}$$

→ smallest mol amount determines limiting reactant

sample → g MnO_2 → mol MnO_2 → mol Mn

$$\frac{61.14\text{ g }MnO_2}{100\text{ g sample}} \quad \frac{\text{mol }MnO_2}{86.94\text{ g }MnO_2} \quad \frac{1\text{ mol Mn}}{\text{mol }MnO_2}$$

then mol Mn → g Mn

$$\frac{54.94\text{ g Mn}}{\text{mol Mn}}$$

Solution: $4Al(s) + 3\ MnO_2(s) \rightarrow 3Mn + 2Al_2O_3(s)$

Assume 1 mole: $0.672\text{ mol Al x }\frac{26.98\text{ g Al}}{\text{mol Al}} = 18.13\text{ g Al}$

$$0.328\text{ mol }MnO_2\text{ x }\frac{86.94\text{ g }MnO_2}{\text{mol }MnO_2} = 28.52\text{ g }MnO_2$$

$$\frac{18.13\text{ g Al}}{(18.13\text{ g Al} + 28\ 52\text{ g }MnO_2)}\text{ x }100 = 38.86\text{ \% Al}$$

So: 61.14 % MnO_2

$$250\ \cancel{\text{g sample}}\text{ x }\frac{38.86\ \cancel{\text{g Al}}}{100\ \cancel{\text{g sample}}}\text{ x }\frac{\cancel{\text{mol Al}}}{26.98\ \cancel{\text{g Al}}}\text{ x }\frac{3\text{ mol Mn}}{4\ \cancel{\text{mol Al}}} = 2.7\underline{0}1\text{ mol Mn}$$

$$250\ \cancel{\text{g sample}}\text{ x }\frac{61.14\ \cancel{\text{g }MnO_2}}{100\ \cancel{\text{g sample}}}\text{ x }\frac{\cancel{\text{mol }MnO_2}}{86.94\ \cancel{\text{g }MnO_2}}\text{ x }\frac{1\text{ mol Mn}}{1\ \cancel{\text{mol }MnO_2}} = 1.7\underline{5}8\text{ mol Mn}$$

$$1.7\underline{5}8\ \cancel{\text{mol Mn}}\text{ x }\frac{54.94\text{ g Mn}}{1\ \cancel{\text{mol Mn}}} = 96.6\text{ g Mn}$$

Check: the units of the answer, g Mn, are correct. The magnitude of the answer is reasonable based on the amount of the limiting reactant, MnO_2.

92. **Given:** 151 g $Na_2B_4O_7$ **Find:** g B_5H_9

Conceptual Plan: g $Na_2B_4O_7$ → mol $Na_2B_4O_7$ → mol B_5H_9 → g B_5H_9

$$\frac{\text{mol }Na_2B_4O_7}{201.22\text{ g }Na_2B_4O_7} \quad \frac{4\text{ mol }B_5H_9}{5\text{ mol }Na_2B_4O_7} \quad \frac{63.13\text{ g }B_5H_9}{\text{mol }B_5H_9}$$

Solution: All the B in B_5H_9 goes to the $Na_2B_4O_7$ so the mole ratio between the two can be used.

$$151\ \cancel{\text{g }Na_2B_4O_7}\text{ x }\frac{1\ \cancel{\text{mol }Na_2B_4O_7}}{201.22\ \cancel{\text{g }Na_2B_4O_7}}\text{ x }\frac{4\ \cancel{\text{mol }B_5H_9}}{5\ \cancel{\text{mol }Na_2B_4O_7}}\text{ x }\frac{63.13\text{ g }B_5H_9}{1\ \cancel{\text{mol }B_5H_9}} = 37.9\text{ g }B_5H_9$$

Check: The units of the answer, g B_5H_9, are correct. The magnitude of the answer is reasonable since the molar mass of B_5H_9 is less than the molar mass of $Na_2B_4O_7$.

93. The correct answer is d. The molar mass of K and O_2 are comparable. Since the stoichiometry has a ratio of 4 mol K to 1 mol O_2, K will be the limiting reactant when mass of K is less than 4 times the mass of O_2.

94. **Given:** 5 mol NO, 10 mol H_2 **Find: conditions of product mixture**
Conceptual Plan: mol H_2 → mol NO and mol H_2 → mol NH_3 and mol H_2 → mol H_2O
Solution: The correct answer is a. Since the mol ratio of H_2 to NO is 5:2, the 10 mol of H_2 will require 4 mol NO and H_2 is the limiting reactant. This eliminates answers b and c. Since there is excess NO, this eliminates d, leaving answer a.

95. **Given:** 1 M solution contains 8 particles **Find:** amount of solute or solvent needed to obtain new concentration
Conceptual Plan: determine amount of solute particles in each new solution, then determine if solute (if the number is greater) or solvent (if the number is less) needs to be added to obtain the new concentration.
Solution: Solution (a) contains 12 particles solute. Concentration is greater than the original, so solute needs to be added. $12\ \text{particles} \times \frac{1\ \text{mol}}{8\ \text{particles}} = 1.5\ \text{mol}$ $(1.5\ \text{mol} - 1.0\ \text{mol}) = 0.5\ \text{mol}$ solute added.

$$0.5\ \text{mol solute} \times \frac{8\ \text{particles}}{1\ \text{mol solute}} = 4 \text{ solute particles added}$$

Solution (a) is obtained by adding 4 particles solute to 1 L of original solution

Solution (b) contains 4 particles. Concentration is less than the original so solvent needs to be added.

$$4\ \text{particles} \times \frac{1\ \text{mol}}{8\ \text{particles}} = 0.5\ \text{mol solute}$$ So, 1 L solution contains 0.5 mol = 0.5M

$(1\ \text{M})(1\ \text{L}) = (0.5\ \text{M})(x)$ $x = 2\ \text{L}$
Solution (b) is obtained by diluting 1 L of the original solution to 2 L

Solution (c) contains 6 particles. Concentration is less than the original so solvent needs to be added.

$$6\ \text{particles} \times \frac{1\ \text{mol}}{8\ \text{particles}} = 0.75\ \text{mol solute}$$ So, 1 L solution contains 0.75 mol = 0.75M

$(1\ \text{M})(1\ \text{L}) = (0.75\ \text{M})(x)$ $x = 2\ \text{L}$
Solution (c) is obtained by diluting 1 L of the original solution to 1.3 L

96. **Given:** 6 molecules N_2H_4; 4 molecules N_2O_4 ; (a) contains: 9 molecules N_2, 12 molecules H_2O, and 1 molecule N_2O_4: solution (b) contains: 12 molecules N_2, 16 molecules H_2O , and 2 molecules N_2O_4; solution (c) contains: 9 molecules N_2, 12 molecules H_2O **Find:** theoretical yield N_2, H_2O
Conceptual Plan: **molecules N_2H_4 → molecules N_2**

$\frac{3\ \text{molecules}\ N_2}{2\ \text{molecules}\ N_2H_4}$ **→ smallest molecules amount determines limiting reactant**

molecules N_2O_4 → molecules N_2

$\frac{3\ \text{molecules}\ N_2}{1\ \text{molecules}\ N_2O_4}$

molecules N_2H_4 → molecules H_2O

$\frac{4\ \text{molecules}\ H_2O}{2\ \text{molecules}\ N_2H_4}$

molecules N_2H_4 → molecules N_2O_4

$\frac{1\ \text{molecules}\ N_2O_4}{2\ \text{molecules}\ N_2H_4}$

Solution:

$$6\ \text{molecules}\ N_2H_4 \times \frac{3\ \text{molecules}\ N_2}{2\ \text{molecules}\ N_2H_4} = 9\ \text{molecules}\ N_2$$

$$6\ \text{molecules}\ N_2O_4 \times \frac{3\ \text{molecules}\ N_2}{1\ \text{molecules}\ N_2O_4} = 18\ \text{molecules}\ N_2$$

Limiting reactant = N_2H_4 because it produced the least molecules of N_2

$$6\ \cancel{\text{molecules } N_2H_4} \times \frac{4 \text{ molecules } H_2O}{2\ \cancel{\text{molecules } N_2H_4}} = 12 \text{ molecules } H_2O$$

$$6\ \cancel{\text{molecules } N_2H_4} \times \frac{1 \text{ molecules } N_2O_4}{2\ \cancel{\text{molecules } N_2H_4}} = 3 \text{ molecules } N_2O_4 \text{ used}$$

Reaction mixture should contain: 9 molecules N_2, 12 molecules H_2O, and 1 molecule N_2O_4, this is best represented by (a).

Chapter 5
Gases

1. a) **Given:** 24.9 in Hg **Find:** atm
Conceptual Plan: **in Hg → atm**

$$\frac{1\text{ atm}}{29.92\text{ in Hg}}$$

Solution: $24.9\ \cancel{\text{in Hg}} \times \frac{1\text{ atm}}{29.92\ \cancel{\text{in Hg}}} = 0.832\text{ atm}$

Check: The units (atm) are correct. The magnitude of the answer (<1) makes physical sense because we started with less than 29.92 in Hg.

b) **Given:** 24.9 in Hg **Find:** mmHg
Conceptual Plan: **Use answer from part a) then convert atm → mmHg**

$$\frac{760\text{ mm Hg}}{1\text{ atm}}$$

Solution: $0.832\ \cancel{\text{atm}} \times \frac{760\text{ mmHg}}{1\ \cancel{\text{atm}}} = 632\text{ mmHg}$

Check: The units (mmHg) are correct. The magnitude of the answer (< 760 mmHg) makes physical sense because we started with less than 1 atm.

c) **Given:** 24.9 in Hg **Find:** psi
Conceptual Plan: **Use answer from part a) then convert atm → psi**

$$\frac{14.7\text{ psi}}{1\text{ atm}}$$

Solution: $0.832\ \cancel{\text{atm}} \times \frac{14.7\text{ psi}}{1\ \cancel{\text{atm}}} = 12.2\text{ psi}$

Check: The units (psi) are correct. The magnitude of the answer (< 14.7 psi) makes physical sense because we started with less than 1 atm.

d) **Given:** 24.9 in Hg **Find:** Pa
Conceptual Plan: **Use answer from part a) then convert atm → Pa**

$$\frac{101{,}325\text{ Pa}}{1\text{ atm}}$$

Solution: $0.832\ \cancel{\text{atm}} \times \frac{101{,}325\text{ Pa}}{1\ \cancel{\text{atm}}} = 8.43\text{x}10^4\text{ Pa}$

Check: The units (mmHg) are correct. The magnitude of the answer (< 760 mmHg) makes physical sense because we started with less than 1 atm.

2. a) **Given:** 235 mmHg **Find:** torr
Conceptual Plan: **mmHg → torr**

$$\frac{1\text{ torr}}{1\text{ mm Hg}}$$

Solution: $235\ \cancel{\text{mmHg}} \times \frac{1\text{ torr}}{1\ \cancel{\text{mmHg}}} = 235\text{ torr}$

Check: The units (torr) are correct. The magnitude of the answer (235) makes physical sense because both units are of the same size.

b) **Given:** 235 mmHg **Find:** psi
Conceptual Plan: **mmHg → atm → psi**

$$\frac{760\text{ mmHg}}{1\text{ atm}} \qquad \frac{14.7\text{ psi}}{1\text{ atm}}$$

Solution: $235\ \cancel{\text{mmHg}} \times \frac{1\ \cancel{\text{atm}}}{760\ \cancel{\text{mmHg}}} \times \frac{14.7\text{ psi}}{1\ \cancel{\text{atm}}} = 4.55\text{ psi}$

Check: The units (psi) are correct. The magnitude of the answer (< 14.7 psi) makes physical sense because we started with less than 760 mmHg = 1 atm.

c) **Given:** 235 mmHg **Find:** in Hg
Conceptual Plan: **mmHg → in Hg**

$$\frac{1\text{ in Hg}}{25.4\text{ mmHg}}$$

Solution: $235\,\cancel{\text{mmHg}} \times \dfrac{1\text{ in Hg}}{25.4\,\cancel{\text{mmHg}}} = 9.25\text{ in Hg}$

Check: The units (in Hg) are correct. The magnitude of the answer (9) makes physical sense because inches are larger than mm.

d) **Given:** 235 mmHg **Find:** atm
Conceptual Plan: **mmHg → atm**

$$\frac{760\text{ mmHg}}{1\text{atm}}$$

Solution: $235\,\cancel{\text{mmHg}} \times \dfrac{1\text{ atm}}{760\,\cancel{\text{mmHg}}} = 0.309\text{ atm}$

Check: The units (atm) are correct. The magnitude of the answer (< 1) makes physical sense because we started with less than 760 mmHg.

3. a) **Given:** 31.85 in Hg **Find:** mmHg
Conceptual Plan: **in Hg → mmHg**

$$\frac{25.4\text{ mmHg}}{1\text{ in Hg}}$$

Solution: $31.85\,\cancel{\text{inHg}} \times \dfrac{25.4\text{ mmHg}}{1\,\cancel{\text{inHg}}} = 809.0\text{ mmHg}$

Check: The units (mmHg) are correct. The magnitude of the answer (809) makes physical sense because inches are larger than mm.

b) **Given:** 31.85 in Hg **Find:** atm
Conceptual Plan: **Use answer from part a) then convert** **mmHg → atm**

$$\frac{1\text{atm}}{760\text{mmHg}}$$

Solution: $809.0\,\cancel{\text{mmHg}} \times \dfrac{1\text{atm}}{760\,\cancel{\text{mmHg}}} = 1.064\text{ atm}$

Check: The units (atm) are correct. The magnitude of the answer (>1) makes physical sense because we started with more than 760 mmHg.

c) **Given:** 31.85 in Hg **Find:** torr
Conceptual Plan: **Use answer from part a) then convert** **mmHg → torr**

$$\frac{1\text{torr}}{1\text{mmHg}}$$

Solution: $809.0\,\cancel{\text{mmHg}} \times \dfrac{1\text{torr}}{1\,\cancel{\text{mmHg}}} = 809.0\text{ torr}$

Check: The units (torr) are correct. The magnitude of the answer (809) makes physical sense because both units are of the same size.

d) **Given:** 31.85 in Hg **Find:** kPa
Conceptual Plan: **Use answer from part b) then convert** **atm → Pa → kPa**

$$\frac{101{,}325\text{ Pa}}{1\text{atm}} \quad \frac{1\text{ kPa}}{1000\text{ Pa}}$$

Solution: $1.064\,\cancel{\text{atm}} \times \dfrac{101{,}325\,\cancel{\text{Pa}}}{1\,\cancel{\text{atm}}} \times \dfrac{1\text{ kPa}}{1000\,\cancel{\text{Pa}}} = 107.8\text{ kPa}$

Check: The units (kPa) are correct. The magnitude of the answer (108) makes physical sense because we started with more than 1 atm and there are ~101 kPa in an atm.

4. a) **Given:** 652.5 mmHg **Find:** torr
Conceptual Plan: **mmHg → torr**

$$\frac{1\text{ torr}}{1\text{ mmHg}}$$

Solution: $652.5\ \cancel{\text{mmHg}} \times \frac{1\ \text{torr}}{1\ \cancel{\text{mmHg}}} = 652.5\ \text{torr}$

Check: The units (torr) are correct. The magnitude of the answer (653) makes physical sense because both units are of the same size.

b) **Given:** 652.5 mmHg **Find:** atm
Conceptual Plan: **mmHg → atm**

$$\frac{760\ \text{mmHg}}{1\ \text{atm}}$$

Solution: $652.5\ \cancel{\text{mmHg}} \times \frac{1\ \text{atm}}{760\ \cancel{\text{mmHg}}} = 0.8586\ \text{atm}$

Check: The units (psi) are correct. The magnitude of the answer (< 14.7 psi) makes physical sense because we started with less than 760 mmHg = 1 atm.

c) **Given:** 652.5 mmHg **Find:** in Hg
Conceptual Plan: **mmHg → in Hg**

$$\frac{1\ \text{in Hg}}{25.4\ \text{mmHg}}$$

Solution: $652.5\ \cancel{\text{mmHg}} \times \frac{1\ \text{in Hg}}{25.4\ \cancel{\text{mmHg}}} = 25.69\ \text{in Hg}$

Check: The units (in Hg) are correct. The magnitude of the answer (26) makes physical sense because inches are larger than mm.

d) **Given:** 652.5 mmHg **Find:** psi
Conceptual Plan: **Use answer from part b) then convert atm → psi**

$$\frac{14.70\ \text{psi}}{1\ \text{atm}}$$

Solution: $0.8586\ \cancel{\text{atm}} \times \frac{14.70\ \text{psi}}{1\ \cancel{\text{atm}}} = 12.62\ \text{psi}$

Check: The units (psi) are correct. The magnitude of the answer (< 14.7 psi) makes physical sense because we started with less than 1 atm.

5. **Given:** $V_1 = 2.8$ L, $P_1 = 755$ mmHg, and $V_2 = 3.7$ L **Find:** P_2
Conceptual Plan: $\mathbf{V_1, P_1, V_2 \rightarrow P_2}$

$$P_1 V_1 = P_2 V_2$$

Solution:

$P_1 V_1 = P_2 V_2$ Rearrange to solve for P_2. $P_2 = P_1 \frac{V_1}{V_2} = 755\ \text{mmHg} \times \frac{2.8\ \cancel{\text{L}}}{3.7\ \cancel{\text{L}}} = 570\ \text{mmHg} = 5.7 \times 10^2\ \text{mmHg}$

Check: The units (mmHg) are correct. The magnitude of the answer (570 mmHg) makes physical sense because Boyle's Law indicates that as the volume increases, the pressure decreases.

6. **Given:** $V_1 = 32.6$ L, $P_1 = 1.3$ atm, and $V_2 = 13.8$ L **Find:** P_2
Conceptual Plan: $\mathbf{V_1, P_1, V_2 \rightarrow P_2}$

$$P_1 V_1 = P_2 V_2$$

Solution: $P_1 V_1 = P_2 V_2$ Rearrange to solve for P_2. $P_2 = P_1 \frac{V_1}{V_2} = 1.3\ \text{atm} \times \frac{32.6\ \cancel{\text{L}}}{13.8\ \cancel{\text{L}}} = 3.1\ \text{atm}$

Check: The units (atm) are correct. The magnitude of the answer (3 atm) makes physical sense because Boyle's Law indicates that as the volume decreases, the pressure increases.

7. **Given:** $V_1 = 48.3$ mL, $T_1 = 22$ °C, and $T_2 = 87$ °C **Find:** V_2
Conceptual Plan: **°C → K then $\mathbf{V_1, T_1, T_2 \rightarrow V_2}$**

$$\text{K} = {}^\circ\text{C} + 273.15 \qquad \frac{V_1}{T_1} = \frac{V_2}{T_2}$$

Solution: $T_1 = 22\ ^\circ\text{C} + 273.15 = 295\ \text{K}$ and $T_2 = 87\ ^\circ\text{C} + 273.15 = 360.\ \text{K}$

$\frac{V_1}{T_1} = \frac{V_2}{T_2}$ Rearrange to solve for V_2. $V_2 = V_1 \frac{T_2}{T_1} = 48.3\ \text{mL} \times \frac{360\ \cancel{\text{K}}}{295\ \cancel{\text{K}}} = 58.9\ \text{mL}$

Check: The units (mL) are correct. The magnitude of the answer (59 mL) makes physical sense because Charles' Law indicates that as the volume increases, the temperature increases.

8. **Given:** $V_1 = 1.55$ mL, $T_1 = 95.3$ °C, and $T_2 = 0.0$ °C **Find:** V_2
Conceptual Plan: **°C → K then V_1, T_1, T_2 → V_2**

$$\text{K} = °\text{C} + 273.15 \qquad \frac{V_1}{T_1} = \frac{V_2}{T_2}$$

Solution: $T_1 = 95.3\ °\text{C} + 273.15 = 368.5$ K and $T_2 = 0.0\ °\text{C} + 273.15 = 273.2$ K

$\frac{V_1}{T_1} = \frac{V_2}{T_2}$ Rearrange to solve for V_2. $V_2 = V_1 \frac{T_2}{T_1} = 1.55 \text{ mL} \times \frac{273.2 \text{ K}}{368.5 \text{ K}} = 1.15 \text{ mL}$

Check: The units (mL) are correct. The magnitude of the answer (1.15 mL) makes physical sense because Charles' Law indicates that as the volume decreases, the temperature decreases.

9. **Given:** $V_1 = 2.76$ L, $n_1 = 0.128$ mol, and $\Delta n = 0.073$ mol **Find:** V_2
Conceptual Plan: **n_1 → n_2 then V_1, n_1, n_2 → V_2**

$$n_1 + \Delta n = n_2 \qquad \frac{V_1}{n_1} = \frac{V_2}{n_2}$$

Solution: $n_2 = 0.128 \text{ mol} + 0.073 \text{ mol} = 0.201 \text{ mol}$

$\frac{V_1}{n_1} = \frac{V_2}{n_2}$ Rearrange to solve for V_2. $V_2 = V_1 \frac{n_2}{n_1} = 2.76 \text{ L} \times \frac{0.201 \text{ mol}}{0.128 \text{ mol}} = 4.33 \text{ L}$

Check: The units (L) are correct. The magnitude of the answer (4.33 L) makes physical sense because Avogadro's Law indicates that as the number of moles increases, the volume increases.

10. **Given:** $V_1 = 334$ mL, $n_1 = 0.87$ mol, and $\Delta n = 0.22$ mol **Find:** V_2
Conceptual Plan: **n_1 → n_2 then V_1, n_1, n_2 → V_2**

$$n_1 + \Delta n = n_2 \qquad \frac{V_1}{n_1} = \frac{V_2}{n_2}$$

Solution: $n_2 = 0.87 \text{ mol} + 0.22 \text{ mol} = 1.09 \text{ mol}$

$\frac{V_1}{n_1} = \frac{V_2}{n_2}$ Rearrange to solve for V_2. $V_2 = V_1 \frac{n_2}{n_1} = 334 \text{ mL} \times \frac{1.09 \text{ mol}}{0.87 \text{ mol}} = 420 \text{ mL} = 4.2 \times 10^2 \text{ mL}$

Check: The units (mL) are correct. The magnitude of the answer (420 L) makes physical sense because Avogadro's Law indicates that as the number of moles increases, the volume increases.

11. **Given:** $n = 0.118$ mol, $P = 0.97$ atm, and $T = 305$ K **Find:** V
Conceptual Plan: **n, P, T → V**

$$PV = nRT$$

Solution: $PV = nRT$ Rearrange to solve for V. $V = \frac{nRT}{P} = \frac{0.118 \text{ mol} \times 0.08206 \frac{\text{L} \cdot \text{atm}}{\text{mol} \cdot \text{K}} \times 305 \text{ K}}{0.97 \text{ atm}} = 3.0 \text{ L}$

Check: The units (L) are correct. The magnitude of the answer (3 L) makes sense because, as you will see in the next section, one mole of an ideal gas under standard conditions (273 K and 1 atm) occupies 22.4 L. Although these are not standard conditions, they are close enough for a ballpark check of the answer. Since this gas sample contains 0.118 moles, a volume of 3 L is reasonable.

12. **Given:** $V = 10.0$ L, $n = 0.448$ mol, and $T = 315$ K **Find:** P
Conceptual Plan: **n, V, T → P**

$$PV = nRT$$

Solution:

$PV = nRT$ Rearrange to solve for P. $P = \frac{nRT}{V} = \frac{0.448 \text{ mol} \times 0.08206 \frac{\text{L} \cdot \text{atm}}{\text{mol} \cdot \text{K}} \times 315 \text{ K}}{10.0 \text{ L}} = 1.16 \text{ atm}$

Check: The units (atm) are correct. The magnitude of the answer (~1 L) makes sense because, as you will see in the next section, one mole of an ideal gas under standard conditions (273 K and 1 atm) occupies 22.4 L. Although these are not standard conditions, they are close enough for a ballpark check of the answer. Since this gas sample contains 0.448 moles in a volume of 10 L, a pressure of 1 atm is reasonable.

13. **Given:** $V = 28.5$ L, $P = 1.8$ atm, and $T = 298$ K **Find:** n
Conceptual Plan: **V, P, T → n**

$$PV = nRT$$

Solution: $PV = nRT$ Rearrange to solve for n. $n = \frac{PV}{RT} = \frac{1.8 \text{ atm} \times 28.5 \text{ L}}{0.08206 \frac{\text{L} \cdot \text{atm}}{\text{mol} \cdot \text{K}} \times 298 \text{ K}} = 2.1 \text{ mol}$

Check: The units (mol) are correct. The magnitude of the answer (2 mol) makes sense because, as you will see in the next section, one mole of an ideal gas under standard conditions (273 K and 1 atm) occupies 22.4 L. Although these are not standard conditions, they are close enough for a ballpark check of the answer. Since this gas sample has a volume of 28.5 L, and a pressure of 1.8 atm, ~ 2 mol is reasonable.

14. **Given:** $V = 11.1$ L, $P = 1.3$ atm, and $n = 0.52$ mol **Find:** T
Conceptual Plan: $V, P, n \rightarrow T$

$PV = nRT$

Solution: $PV = nRT$ Rearrange to solve for T. $T = \frac{PV}{nT} = \frac{1.3 \text{ atm} \times 11.8 \text{ L}}{0.52 \text{ mol} \times 0.08206 \frac{\text{L} \cdot \text{atm}}{\text{mol} \cdot \text{K}}} = 360 \text{ K}$

Check: The units (T) are correct. The magnitude of the answer (360 K) makes sense because, as you will see in the next section, one mole of an ideal gas under standard conditions (273 K and 1 atm) occupies 22.4 L. Although these are not standard conditions, they are close enough for a ballpark check of the answer. Since this gas sample has 0.52 mol, a volume of 11.8 L, and a pressure of 1.3 atm, 360 K is reasonable.

15. **Given:** $P_1 = 36.0$ psi (gauge P), $V_1 = 11.8$ L, $T_1 = 12.0$ °C, $V_2 = 12.2$ L, and $T_2 = 65.0$ °C
Find: P_2 and compare to $P_{max} = 38.0$ psi (gauge P)
Conceptual Plan: **°C → K and gauge P → psi → atm then $P_1, V_1, T_1, V_2, T_2 \rightarrow P_2$**

$K = °C + 273.15$ psi = gauge P + 14.7 $\frac{1 \text{ atm}}{14.7 \text{ psi}}$ $\frac{P_1V_1}{T_1} = \frac{P_2V_2}{T_2}$

Solution: $T_1 = 12.0 \text{ °C} + 273.15 = 285.2 \text{ K}$ and $T_2 = 65.0 \text{ °C} + 273.15 = 338.2 \text{ K}$

$P_1 = 36.0 \text{ psi (gauge P)} + 14.7 = 50.7 \text{ psi} \times \frac{1 \text{ atm}}{14.7 \text{ psi}} = 3.4\underline{4}898 \text{ atm}$

$P_{max} = 38.0 \text{ psi (gauge P)} + 14.7 = 52.7 \text{ psi} \times \frac{1 \text{ atm}}{14.7 \text{ psi}} = 3.59 \text{ atm}$

$\frac{P_1V_1}{T_1} = \frac{P_2V_2}{T_2}$ Rearrange to solve for P_2. $P_2 = P_1 \frac{V_1}{V_2} \frac{T_2}{T_1} = 3.4\underline{4}898 \text{ atm} \times \frac{11.8 \text{ L}}{12.2 \text{ L}} \times \frac{338.2 \text{ K}}{285.2 \text{ K}} = 3.96 \text{ atm}$

This exceeds the maximum tire rating of 3.59 atm or 38.0 psi (gauge P).
Check: The units (atm) are correct. The magnitude of the answer (3.95 atm) makes physical sense because the relative increase in T is greater than the relative increase in V, so P should increase.

16. **Given:** $P_1 = 748$ mmHg, $V_1 = 28.5$ L, $T_1 = 28.0$ °C, $P_2 = 385$ mmHg, and $T_2 = -15.0$ °C **Find:** V_2
Conceptual Plan: **°C → K and mmHg → atm then $P_1, V_1, T_1, V_2, T_2 \rightarrow P_2$**

$K = °C + 273.15$ $\frac{1 \text{ atm}}{760 \text{ mm Hg}}$ $\frac{P_1V_1}{T_1} = \frac{P_2V_2}{T_2}$

Solution: $T_1 = 28.0 \text{ °C} + 273.15 = 301.2 \text{ K}$ and $T_2 = -15.0 \text{ °C} + 273.15 = 258.2 \text{ K}$

$P_1 = 748 \text{ mmHg} \times \frac{1 \text{ atm}}{760 \text{ mmHg}} = 0.98\underline{4}211 \text{ atm}$ $P_2 = 385 \text{ mmHg} \times \frac{1 \text{ atm}}{760 \text{ mmHg}} = 0.50\underline{6}579 \text{ atm}$

$\frac{P_1V_1}{T_1} = \frac{P_2V_2}{T_2}$ Rearrange to solve for V_2. $V_2 = V_1 \frac{P_1}{P_2} \frac{T_2}{T_1} = 28.5 \text{ L} \times \frac{0.98\underline{4}211 \text{ atm}}{0.50\underline{6}579 \text{ atm}} \times \frac{258.2 \text{ K}}{301.2 \text{ K}} = 47.5 \text{ L}$

Check: The units (L) are correct. The magnitude of the answer (47 L) makes physical sense because the relative decrease in P is greater than the relative decrease in T, so V should increase.

17. **Given:** m (CO_2) = 28.8 g, $P = 742$ mmHg, and $T = 22$ °C **Find:** V
Conceptual Plan: **°C → K and mmHg → atm and g → mol then $n, P, T \rightarrow V$**

$K = °C + 273.15$ $\frac{1 \text{ atm}}{760 \text{ mm Hg}}$ $\frac{1 \text{ mol}}{44.01 \text{ g}}$ $PV = nRT$

Solution: $T_1 = 22 \text{ °C} + 273.15 = 295 \text{ K}$, $P = 742 \text{ mmHg} \times \frac{1 \text{ atm}}{760 \text{ mmHg}} = 0.97\underline{6}316 \text{ atm}$,

$n = 28.8 \text{ g} \times \frac{1 \text{ mol}}{44.01 \text{ g}} = 0.65\underline{4}397 \text{ mol}$ $PV = nRT$ Rearrange to solve for V.

$$V = \frac{nRT}{P} = \frac{0.65\underline{4}397 \text{ mol} \times 0.08206 \frac{\text{L} \cdot \text{atm}}{\text{mol} \cdot \text{K}} \times 295 \text{ K}}{0.97\underline{6}316 \text{ atm}} = 16.2 \text{ L}$$

Check: The units (L) are correct. The magnitude of the answer (16 L) makes sense because one mole of an ideal gas under standard conditions (273 K and 1 atm) occupies 22.4 L. Although these are not standard conditions, they are close enough for a ballpark check of the answer. Since this gas sample contains 0.65 moles, a volume of 16 L is reasonable.

18. **Given:** 1.0 L of liquid N_2 w/ $d = 0.807$ g/mL, $T = 25.0$ °C, $P = 1.0$ atm, and closet is 1.0 m x 1.0 m x 2.0 m
Find: V% of closet displaced by evaporated liquid
Conceptual Plan: °C → K and L → mL → g → mol then *n, P, T* → V_{evap}

$K = °C + 273.15$; $\frac{1000 \text{ mL}}{1 \text{ L}}$; $d = m/V$; $\frac{1 \text{ mol}}{28.02 \text{ g}}$; $PV = nRT$

then l, w, h → V_{closet} m³ → cm³ → L finally V_{evap}, V_{closet} → % V displaced

$V = l\,w\,h$; $\frac{(100 \text{ cm})^3}{(1 \text{ m})^3}$; $\frac{1 \text{ L}}{1000 \text{ mL}}$; $\%V\ displaced = \frac{V_{evap}}{V_{closet}} \times 100\%$

Solution: $T_1 = 25.0 \text{ °C} + 273.15 = 298.2 \text{ K}$, $1.0 \text{ L} \times \frac{1000 \text{ mL}}{1 \text{ L}} = 1.0 \times 10^3 \text{ mL}$, $d = m/V$ Rearrange to solve for m. $m = d \times V = 0.807 \frac{\text{g}}{\text{mL}} \times 1.0 \times 10^3 \text{ mL} = 8.1 \times 10^2 \text{ g} \times \frac{1 \text{ mol}}{28.02 \text{ g}} = 2\underline{8}.908 \text{ mol}$ $PV = nRT$

Rearrange to solve for V.

$$V_{evap} = \frac{nRT}{P} = \frac{2\underline{8}.908 \text{ mol} \times 0.08206 \frac{\text{L} \cdot \text{atm}}{\text{mol} \cdot \text{K}} \times 298.2 \text{ K}}{1.0 \text{ atm}} = 7.\underline{0}739 \times 10^2 \text{ L}$$

$$V_{closet} = l\,w\,h = 1.0 \text{ m} \times 1.0 \text{ m} \times 2.0 \text{ m} = 2.0 \text{ m}^3 \quad V_{closet} = 2.0 \text{ m}^3 \times \frac{(100 \text{ cm})^3}{(1 \text{ m})^3} \times \frac{1 \text{ L}}{1000 \text{ cm}^3} = 2.0 \times 10^3 \text{ L}$$

$$\%V\ displaced = \frac{V_{evap}}{V_{closet}} \times 100\% = \frac{7.\underline{0}739 \times 10^2 \text{ L}}{2.0 \times 10^3 \text{ L}} \times 100\% = 35\ \%$$

Check: The units (%) are correct. The magnitude of the answer (35 %) makes sense because, should be between 0 and 100 %. Looking at the two volumes, when a liquid evaporates, the volume increases by several orders of magnitude; when converting from cubic meters to L, there is an increase of 3 orders of magnitude.

19. **Given:** sample a = 5 gas particles, sample b = 10 gas particles, and sample c = 8 gas particles, with all temperatures and volumes the same **Find:** sample with largest P
Conceptual Plan: *n, V, T* → *P*

$PV = nRT$

Solution: $PV = nRT$ Since V and T are constant, this means that $P \propto n$. The sample with the largest number of gas particles will have the highest P. $P_b > P_c > P_a$.

20. **Given:** $P_1 = 1$ atm, $V_1 = 1$ L, $T_1 = 25$ °C, $V_2 = 0.5$ L, and $T_2 = 250.$ °C **Find:** Draw picture and P_2
Conceptual Plan: °C → K then P_1, V_1, T_1, V_2, T_2 → P_2

$K = °C + 273.15$; $\frac{P_1V_1}{T_1} = \frac{P_2V_2}{T_2}$

Solution: $T_1 = 25 \text{ °C} + 273.15 = 298 \text{ K}$ and $T_2 = 250. \text{ °C} + 273.15 = 520. \text{ K}$

$\frac{P_1V_1}{T_1} = \frac{P_2V_2}{T_2}$ Rearrange to solve for P_2. $P_2 = P_1 \frac{V_1}{V_2} \frac{T_2}{T_1} = 1 \text{ atm} \times \frac{1 \text{ L}}{0.5 \text{ L}} \times \frac{520. \text{ K}}{298 \text{ K}} = 3 \text{ atm}$

Insert a drawing that looks like the one in the original problem, with the following changes: (1) decrease the volume by a factor of 2, and (2) lengthen the tails on the gas particles to indicate that they are moving faster. Keep the same number of particles.

Check: The units (atm) are correct. The magnitude of the answer (3 atm) makes physical sense because there is an increase in T and a decrease in V, both of which increase P.

21. **Given:** $P_1 = 755$ mmHg, $T_1 = 25$ °C, and $T_2 = 1155$ °C **Find:** P_2
Conceptual Plan: °C → K and mmHg → atm then P_1, T_1, T_2 → P_2

$K = °C + 273.15$; $\frac{1 \text{ atm}}{760 \text{ mmHg}}$; $\frac{P_1}{T_1} = \frac{P_2}{T_2}$

Solution: $T_1 = 25 \text{ °C} + 273.15 = 298 \text{ K}$ and $T_2 = 1155 \text{ °C} + 273.15 = 1428 \text{ K}$

$$P = 755\ \text{mmHg} \times \frac{1\ \text{atm}}{760\ \text{mmHg}} = 0.99\underline{3}421\ \text{atm} \qquad \frac{P_1}{T_1} = \frac{P_2}{T_2}$$ Rearrange to solve for P_2.

$$P_2 = P_1 \frac{T_2}{T_1} = 0.99\underline{3}421\ \text{atm} \times \frac{1428\ \text{K}}{298\ \text{K}} = 4.76\ \text{atm}$$

Check: The units (atm) are correct. The magnitude of the answer (5 atm) makes physical sense because there is a significant increase in *T*, which will increase *P* significantly.

22. **Given:** $V_1 = 1.75$ L, $P_1 = 1.35$ atm, $T_1 = 25$ °C, $V_2 = 1.75$ L, and $T_2 = 355$ °C **Find:** P_2
Conceptual Plan: **°C → K then P_1, T_1, T_2 → P_2**

$K = °C + 273.15$ $\qquad \frac{P_1}{T_1} = \frac{P_2}{T_2}$

Solution: $T_1 = 25\ °C + 273.15 = 298$ K and $T_2 = 355\ °C + 273.15 = 628$ K

$\frac{P_1}{T_1} = \frac{P_2}{T_2}$ Rearrange to solve for P_2. $P_2 = P_1 \frac{T_2}{T_1} = 1.35\ \text{atm} \times \frac{628\ \text{K}}{298\ \text{K}} = 2.84\ \text{atm}$

Check: The units (atm) are correct. The magnitude of the answer (3 atm) makes physical sense because there is a significant increase in *T*, which will increase *P* significantly.

23. **Given:** STP and *m* (Ne) = 10.0 g **Find:** *V*
Conceptual Plan: **g → mol → *V***

$\frac{1\ \text{mol}}{20.18\ \text{g}}$ $\qquad \frac{22.414\ \text{L}}{1\ \text{mol}}$

Solution: $10.0\ \text{g} \times \frac{1\ \text{mol}}{20.18\ \text{g}} \times \frac{22.414\ \text{L}}{1\ \text{mol}} = 11.1\ \text{L}$

Check: The units (L) are correct. The magnitude of the answer (11 L) makes sense because one mole of an ideal gas under standard conditions (273 K and 1 atm) occupies 22.4 L and we have about 0.5 mol.

24. **Given:** STP and CO_2 **Find:** *d*
Conceptual Plan: **mol → g then *m*,*V* → *d***

$\frac{44.01\ \text{g}}{1\ \text{mol}}$ $\qquad d = \frac{m}{V}$

Solution: $1\ \text{mol} \times \frac{44.01\ \text{g}}{1\ \text{mol}} = 44.01\ \text{g} = m$ at STP $V = 22.414$ L $\qquad d = \frac{m}{V} = \frac{44.01\ \text{g}}{22.414\ \text{L}} = 1.964\ \text{g/L}$

Check: The units (g/L) are correct. The magnitude of the answer (2 g/L) is reasonable for a gas density.

25. **Given:** H_2, $P = 1655$ psi, and $T = 20.0$ °C **Find:** *d*
Conceptual Plan: **°C → K and psi → atm then *P*, *T*, $\mathfrak{M}$ → *d***

$K = °C + 273.15$ $\qquad \frac{1\ \text{atm}}{14.7\ \text{psi}}$ $\qquad d = \frac{P\mathfrak{M}}{RT}$

Solution: $T = 20.0\ °C + 273.15 = 293.2$ K $\quad P = 1655\ \text{psi} \times \frac{1\ \text{atm}}{14.7\ \text{psi}} = 11\underline{2}.585\ \text{atm}$

$$d = \frac{P\mathfrak{M}}{RT} = \frac{11\underline{2}.585\ \text{atm} \times 2.016\ \frac{\text{g}}{\text{mol}}}{0.08206\ \frac{\text{L atm}}{\text{K mol}} \times 293.2\ \text{K}} = 9.43\ \frac{\text{g}}{\text{L}}$$

Check: The units (g/L) are correct. The magnitude of the answer (9 g/L) makes physical sense because this is a high pressure, so the gas density will be on the high side.

26. **Given:** N_2O, $d = 2.85$ g/L, and $T = 298$ K **Find:** *P* (mmHg)

Conceptual Plan: ***d*, *T*, $\mathfrak{M}$ → *d* then atm → mmHg**

$d = \frac{P\mathfrak{M}}{RT}$ $\qquad \frac{760\ \text{mmHg}}{1\ \text{atm}}$

Solution: $d = \frac{P\mathfrak{M}}{RT}$ Rearrange to solve for *P*. $P = \frac{dRT}{\mathfrak{M}} = \frac{2.85\ \frac{\text{g}}{\text{L}} \times 0.08206\ \frac{\text{L atm}}{\text{K mol}} \times 298\ \text{K}}{44.02\ \frac{\text{g}}{\text{mol}}} = 1.58\underline{3}22\ \text{atm}$

$$P = 1.58\underline{3}22\ \text{atm} \times \frac{760\ \text{mmHg}}{1\ \text{atm}} = 1.20 \times 10^3\ \text{mmHg}$$

Check: The units (mmHg) are correct. The magnitude of the answer (1200 mmHg) makes physical sense because the gas density is reasonable and so we expect a $P \sim 1$ atm.

27. **Given:** $V = 248$ mL, $m = 0.433$ g, $P = 745$ mmHg, and $T = 28$ °C **Find:** $\mathfrak{M}$
Conceptual Plan: **°C → K mmHg → atm mL → L then $V, m \rightarrow d$ then $d, P, T \rightarrow \mathfrak{M}$**

$$K = °C + 273.15 \qquad \frac{1 \text{ atm}}{760 \text{ mmHg}} \qquad \frac{1 \text{ L}}{1000 \text{ mL}} \qquad d = \frac{m}{V} \qquad d = \frac{P\mathfrak{M}}{RT}$$

Solution: $T = 28\ °C + 273.15 = 301$ K $\qquad P = 745 \text{ mmHg} \times \frac{1 \text{ atm}}{760 \text{ mmHg}} = 0.98\underline{0}263$ atm

$V = 248 \text{ mL} \times \frac{1 \text{ L}}{1000 \text{ mL}} = 0.248$ L $\qquad d = \frac{m}{V} = \frac{0.433 \text{ g}}{0.248 \text{ L}} = 1.7\underline{4}597$ g/L $\qquad d = \frac{P\mathfrak{M}}{RT}$ Rearrange to solve for $\mathfrak{M}$.

$$\mathfrak{M} = \frac{dRT}{P} = \frac{1.7\underline{4}597 \frac{\text{g}}{\text{L}} \times 0.08206 \frac{\text{L} \cdot \text{atm}}{\text{K} \cdot \text{mol}} \times 301 \text{ K}}{0.98\underline{0}263 \text{ atm}} = 44.0 \text{ g/mol}$$

Check: The units (g/mol) are correct. The magnitude of the answer (44 g/mol) makes physical sense because this is a reasonable number for a molecular weight of a gas.

28. **Given:** $V = 113$ mL, $m = 0.171$ g, $P = 721$ mmHg, and $T = 32$ °C **Find:** $\mathfrak{M}$
Conceptual Plan: **°C → K mmHg → atm mL → L then $V, m \rightarrow d$ then $d, P, T \rightarrow \mathfrak{M}$**

$$K = °C + 273.15 \qquad \frac{1 \text{ atm}}{760 \text{ mmHg}} \qquad \frac{1 \text{ L}}{1000 \text{ mL}} \qquad d = \frac{m}{V} \qquad d = \frac{P\mathfrak{M}}{RT}$$

Solution: $T = 32\ °C + 273.15 = 305$ K $\qquad P = 721 \text{ mmHg} \times \frac{1 \text{ atm}}{760 \text{ mmHg}} = 0.94\underline{8}684$ atm

$V = 113 \text{ mL} \times \frac{1 \text{ L}}{1000 \text{ mL}} = 0.113$ L $\qquad d = \frac{m}{V} = \frac{0.171 \text{ g}}{0.113 \text{ L}} = 1.5\underline{1}327$ g/L $\qquad d = \frac{P\mathfrak{M}}{RT}$ Rearrange to solve for $\mathfrak{M}$.

$$\mathfrak{M} = \frac{dRT}{P} = \frac{1.5\underline{1}327 \frac{\text{g}}{\text{L}} \times 0.08206 \frac{\text{L} \ \text{atm}}{\text{K} \ \text{mol}} \times 305 \text{ K}}{0.94\underline{8}684 \text{ atm}} = 39.9 \text{ g/mol}$$

Check: The units (g/mol) are correct. The magnitude of the answer (40 g/mol) makes physical sense because this is a reasonable number for a molecular weight of a gas.

29. **Given:** $m = 38.8$ mg, $V = 224$ mL, $T = 55$ °C, and $P = 886$ torr **Find:** $\mathfrak{M}$
Conceptual Plan: **mg → g mL → L °C → K torr → atm then $V, m \rightarrow d$ then $d, P, T \rightarrow \mathfrak{M}$**

$$\frac{1 \text{ g}}{1000 \text{ mg}} \qquad \frac{1 \text{ L}}{1000 \text{ mL}} \qquad K = °C + 273.15 \qquad \frac{1 \text{ atm}}{760 \text{ torr}} \qquad d = \frac{m}{V} \qquad d = \frac{P\mathfrak{M}}{RT}$$

Solution: $m = 38.8 \text{ mg} \times \frac{1 \text{ g}}{1000 \text{ mg}} = 0.0388$ g $\quad V = 224 \text{ mL} \times \frac{1 \text{ L}}{1000 \text{ mL}} = 0.224$ L $\quad T = 55\ °C + 273.15 = 328$ K

$P = 886 \text{ torr} \times \frac{1 \text{ atm}}{760 \text{ torr}} = 1.1\underline{6}5789$ atm $\qquad d = \frac{m}{V} = \frac{0.0388 \text{ g}}{0.224 \text{ L}} = 0.17\underline{3}214$ g/L $\qquad d = \frac{P\mathfrak{M}}{RT}$

Rearrange to solve for $\mathfrak{M}$. $\qquad \mathfrak{M} = \frac{dRT}{P} = \frac{0.17\underline{3}214 \frac{\text{g}}{\text{L}} \times 0.08206 \frac{\text{L} \ \text{atm}}{\text{K} \ \text{mol}} \times 328 \text{ K}}{1.1\underline{6}5789 \text{ atm}} = 4.00 \text{ g/mol}$

Check: The units (g/mol) are correct. The magnitude of the answer (4 g/mol) makes physical sense because this is a reasonable number for a molecular weight of a gas, especially since the density is on the low side.

30. **Given:** $m = 0.555$ g, $V = 117$ mL, $T = 85$ °C, and $P = 753$ mmHg **Find:** $\mathfrak{M}$
Conceptual Plan: **mL → L °C → K mmHg → atm then $V, m \rightarrow d$ then $d, P, T \rightarrow \mathfrak{M}$**

$$\frac{1 \text{ L}}{1000 \text{ mL}} \qquad K = °C + 273.15 \qquad \frac{1 \text{ atm}}{760 \text{ mmHg}} \qquad d = \frac{m}{V} \qquad d = \frac{P\mathfrak{M}}{RT}$$

Solution: $V = 117 \text{ mL} \times \frac{1 \text{ L}}{1000 \text{ mL}} = 0.117$ L $\qquad T = 85\ °C + 273.15 = 358$ K

$P = 753 \text{ mmHg} \times \frac{1 \text{ atm}}{760 \text{ mmHg}} = 0.99\underline{0}7895$ atm $\qquad d = \frac{m}{V} = \frac{0.555 \text{ g}}{0.117 \text{ L}} = 4.7\underline{4}359$ g/L $\qquad d = \frac{P\mathfrak{M}}{RT}$

Rearrange to solve for $\mathfrak{M}$. $\mathfrak{M} = \dfrac{dRT}{P} = \dfrac{4.74359\ \frac{\text{g}}{\text{L}} \times 0.08206 \frac{\text{L} \cdot \text{atm}}{\text{K} \cdot \text{mol}} \times 358\ \text{K}}{0.9907895\ \text{atm}} = 141\ \text{g/mol}$

Check: The units (g/mol) are correct. The magnitude of the answer (141 g/mol) makes physical sense because this is a reasonable number for a molecular weight of a gas, especially since the density is on the high side.

31. **Given:** $P_{N2} = 325$ torr, $P_{O2} = 124$ torr, $P_{He} = 209$ torr, $V = 1.05$ L, and $T = 25.0$ °C
Find: m_{N2}, m_{O2}, m_{He}
Conceptual Plan: °C → K and torr → atm and P, V, T → n then mol → g

$K = °C + 273.15$ $\quad \dfrac{1\ \text{atm}}{760\ \text{torr}}$ $\quad PV = nRT$ $\quad \mathfrak{M}$

and P_{N2}, P_{O2}, P_{He} → P_{Total}

$P_{Total} = P_{N2} + P_{O2} + P_{He}$

Solution: $T_1 = 25.0\ °C + 273.15 = 298.2$ K, $\quad PV = nRT \quad$ Rearrange to solve for n.

$n = \dfrac{PV}{RT}$ $\quad P_{N2} = 325\ \text{torr} \times \dfrac{1\ \text{atm}}{760\ \text{torr}} = 0.427632\ \text{atm}$ $\quad n_{N2} = \dfrac{0.427632\ \text{atm} \times 1.05\ \text{L}}{0.08206 \frac{\text{L} \cdot \text{atm}}{\text{mol} \cdot \text{K}} \times 298.2\ \text{K}} = 0.0183493\ \text{mol}$

$0.0183493\ \text{mol} \times \dfrac{28.02\ \text{mol}}{1\ \text{mol}} = 0.514\ \text{g}\ N_2$

$P_{O2} = 124\ \text{torr} \times \dfrac{1\ \text{atm}}{760\ \text{torr}} = 0.163158\ \text{atm}$ $\quad n_{O2} = \dfrac{0.163158\ \text{atm} \times 1.05\ \text{L}}{0.08206 \frac{\text{L} \cdot \text{atm}}{\text{mol} \cdot \text{K}} \times 298.2\ \text{K}} = 0.00700097\ \text{mol}$

$0.00700097\ \text{mol} \times \dfrac{32.00\ \text{mol}}{1\ \text{mol}} = 0.224\ \text{g}\ O_2$

$P_{He} = 209\ \text{torr} \times \dfrac{1\ \text{atm}}{760\ \text{torr}} = 0.275\ \text{atm}$ $\quad n_{He} = \dfrac{0.275\ \text{atm} \times 1.05\ \text{L}}{0.08206 \frac{\text{L} \cdot \text{atm}}{\text{mol} \cdot \text{K}} \times 298.2\ \text{K}} = 0.0118000\ \text{mol}$

$0.0118000\ \text{mol} \times \dfrac{4.003\ \text{mol}}{1\ \text{mol}} = 0.0472\ \text{g He}$ and

$P_{Total} = P_{N2} + P_{O2} + P_{He} = 0.428\ \text{atm} + 0.163\ \text{atm} + 0.275\ \text{atm} = 0.866\ \text{atm}$

Check: The units (g and atm) are correct. The magnitude of the answer (1 g) makes sense because gases are not very dense and these pressures are < 1 atm. Since all of the pressures are small, the total is < 1 atm.

32. **Given:** $P_{Total} = 755$ mmHg, $P_{CO2} = 255$ mmHg, $P_{Ar} = 124$ mmHg, $P_{O2} = 167$ mmHg, $V = 10.0$ L, and $T =$ 273 K **Find:** P_{He} and m_{He}
Conceptual Plan: $P_{Total}, P_{CO2}, P_{Ar}, P_{O2}$ → P_{He} then mmHg → atm then P, V, T → n then

$P_{Total} = P_{CO2} + P_{Ar} + P_{O2} + P_{He}$ $\quad \dfrac{1\ \text{atm}}{760\ \text{mmHg}}$ $\quad PV = nRT$

mol → g

$\dfrac{4.003\ \text{g}}{1\ \text{mol}}$

Solution: $P_{Total} = P_{CO2} + P_{Ar} + P_{O2} + P_{He}$ $\quad$ Rearrange to solve for P_{He}.

$P_{He} = P_{Total} - P_{CO2} + P_{Ar} + P_{O2} = 755\ \text{mmHg} - 255\ \text{mmHg} - 124\ \text{mmHg} - 167\ \text{mmHg} = 209\ \text{mmHg}$

$P_{He} = 209\ \text{mmHg} \times \dfrac{1\ \text{atm}}{760\ \text{mmHg}} = 0.275\ \text{atm}$ $\quad PV = nRT \quad$ Rearrange to solve for n. $\quad n = \dfrac{PV}{RT}$

$n_{He} = \dfrac{0.275\ \text{atm} \times 10.0\ \text{L}}{0.08206 \frac{\text{L} \cdot \text{atm}}{\text{mol} \cdot \text{K}} \times 273\ \text{K}} = 0.1227548\ \text{mol}$ $\quad 0.1227548\ \text{mol} \times \dfrac{4.003\ \text{g}}{1\ \text{mol}} = 0.491\ \text{g He}$

Check: The units (g) are correct. The magnitude of the answer (1 g) makes sense because gases are not very dense and these pressures are < 1 atm.

33. **Given:** $m\ (CO_2) = 1.20$ g, $V = 755$ mL, $P_{N2} = 725$ mmHg, and $T = 25.0$ °C **Find:** P_{Total}
Conceptual Plan: mL → L and °C → K and g → mol and n, P, T → V then atm → mmHg

$\dfrac{1\ \text{L}}{1000\ \text{mL}}$ $\quad K = °C + 273.15$ $\quad \dfrac{1\ \text{mol}}{44.01\ \text{g}}$ $\quad PV = nRT$ $\quad \dfrac{760\ \text{mmHg}}{1\ \text{atm}}$

finally P_{CO2}, P_{N2} → P_{Total}

$P_{Total} = P_{CO2} + P_{N2}$

Solution: $V = 755\ \text{mL} \times \dfrac{1\ \text{L}}{1000\ \text{mL}} = 0.755\ \text{L}$ $\quad T = 25.0\ °C + 273.15 = 298.2\ \text{K}$,

$n = 1.20\ \text{g} \times \dfrac{1\ \text{mol}}{44.01\ \text{g}} = 0.02\underline{7}2665\ \text{mol}$, $\quad PV = nRT$ Rearrange to solve for P.

$$P = \frac{nRT}{V} = \frac{0.02\underline{7}2665\ \text{mol} \times 0.08206\ \frac{\text{L}\cdot\text{atm}}{\text{mol}\cdot\text{K}} \times 298.2\ \text{K}}{0.755\ \text{L}} = 0.88\underline{3}735\ \text{atm}$$

$$P_{CO2} = 0.88\underline{3}735\ \text{atm} \times \frac{760\ \text{mmHg}}{1\ \text{atm}} = 672\ \text{mmHg}$$

$P_{Total} = P_{CO2} + P_{N2} = 672\ \text{mmHg} + 725\ \text{mmHg} = 1397\ \text{mmHg}$

Check: The units (mmHg) are correct. The magnitude of the answer (1400 mmHg) makes sense because it must be greater than 725 mmHg.

34. **Given:** $V_{1He} = 275$ mL, $P_{1He} = 752$ torr, $V_{1Ar} = 475$ mL, and $P_{1Ar} = 722$ torr **Find:** P_{2He}, P_{2Ar}, and P_{Total}

Conceptual Plan: $V_{1He}, V_{1Ar} \rightarrow V_2$ $\quad$ $V_1, P_1, V_2 \rightarrow P_2$ $\quad$ **then** $\quad$ $P_{2He}, P_{2Ar} \rightarrow P_{Total}$

$V_{1He} + V_{1Ar} = V_2$ $\quad$ $P_1 V_1 = P_2 V_2$ $\quad$ $P_{Total} = P_{2He} + P_{2Ar}$

Solution: $V_{1He} + V_{1Ar} = V_2 = 275\ \text{mL} + 475\ \text{mL} = 750.\ \text{mL}$, $P_1 V_1 = P_2 V_2$ Rearrange to solve for P_2.

$$P_2 = P_1 \frac{V_1}{V_2} \qquad P_{2He} = P_{1He} \frac{V_{1He}}{V_2} = 752\ \text{torr} \times \frac{275\ \text{mL}}{750.\ \text{mL}} = 27\underline{5}.733\ \text{torr}$$

$$P_{2Ar} = P_{1Ar} \frac{V_{1Ar}}{V_2} = 722\ \text{torr} \times \frac{475\ \text{mL}}{750.\ \text{mL}} = 45\underline{7}.267\ \text{torr}$$

$P_{Total} = P_{2He} + P_{2Ar} = 27\underline{5}.733\ \text{torr} + 45\underline{7}.267\ \text{torr} = 733\ \text{torr}$

Check: The units (torr) are correct. The magnitude of the answers makes physical sense because Boyle's Law indicates that as the volume increases, the pressure decreases. Since both initial pressures are ~700 torr, the final total pressure should be about the same pressure.

35. **Given:** $m\ (N_2) = 1.25$ g, $m\ (O_2) = 0.85$ g, $V = 1.55$ L, and $T = 18$ °C **Find:** χ_{N2}, χ_{O2}, P_{N2}, P_{O2}

Conceptual Plan: **g → mol then** $n_{N2}, n_{O2} \rightarrow \chi_{N2}$ **and** $n_{N2}, n_{O2} \rightarrow \chi_{O2}$ **°C → K**

$\mathfrak{M}$ $\quad$ $\chi_{N2} = \dfrac{n_{N2}}{n_{N2} + n_{O2}}$ $\quad$ $\chi_{O2} = \dfrac{n_{O2}}{n_{N2} + n_{O2}}$ $\quad$ K = °C + 273.15

then $n, V, T \rightarrow P$

$PV = nRT$

Solution: $n_{N2} = 1.25\ \text{g} \times \dfrac{1\ \text{mol}}{28.02\ \text{g}} = 0.044\underline{6}110\ \text{mol}$, $n_{O2} = 0.85\ \text{g} \times \dfrac{1\ \text{mol}}{32.00\ \text{g}} = 0.02\underline{6}563\ \text{mol}$,

$T = 18\ °\text{C} + 273.15 = 291\ \text{K}$, $\chi_{N2} = \dfrac{n_{N2}}{n_{N2} + n_{O2}} = \dfrac{0.044\underline{6}110\ \text{mol}}{0.044\underline{6}110\ \text{mol} + 0.02\underline{6}563\ \text{mol}} = 0.6\underline{2}6792 = 0.627$,

$\chi_{O2} = \dfrac{n_{O2}}{n_{N2} + n_{O2}} = \dfrac{0.02\underline{6}563\ \text{mol}}{0.044\underline{6}110\ \text{mol} + 0.02\underline{6}563\ \text{mol}} = 0.37$ We can also calculate this as

$\chi_{O2} = 1 - \chi_{N2} = 1 - 0.6\underline{2}6792 = 0.3\underline{7}3208 = 0.373$ $\quad PV = nRT$ $\quad$ Rearrange to solve for P. $\quad P = \dfrac{nRT}{V}$

$$P_{N2} = \frac{0.044\underline{6}11\ \text{mol} \times 0.08206\ \frac{\text{L}\cdot\text{atm}}{\text{mol}\cdot\text{K}} \times 291\ \text{K}}{1.55\ \text{L}} = 0.687\ \text{atm}$$

$$P_{O2} = \frac{0.02\underline{6}563\ \text{mol} \times 0.08206\ \frac{\text{L}\cdot\text{atm}}{\text{mol}\cdot\text{K}} \times 291\ \text{K}}{1.55\ \text{L}} = 0.409\ \text{atm}$$

Check: The units (none and atm) are correct. The magnitude of the answers makes sense because the mole fractions should total 1 and since the weight of N_2 is greater than O_2, its mole fraction is larger. The number of moles is <<1, so we expect the pressures to be <1 atm, given the V (1.55 L).

36. **Given:** Table 5.2, $m\ (O_2) = 10.0$ g, $T = 273$ K, and $P = 1.00$ atm **Find:** χ_{O2} and V_{air}

Conceptual Plan: **%V →** χ_{O2} **and** **g → mol →** V_{O2} **→** V_{air}

$\dfrac{1}{100\ \%}$ $\quad$ $\dfrac{1\ \text{mol}}{32.00\ \text{g}}$ $\quad$ $\dfrac{22.414\ \text{L}}{1\ \text{mol}}$ $\quad$ $\dfrac{100\ \text{L air}}{21\ \text{L}\ O_2}$

Solution: from Table 5.3 $\%V_{O2} = 21\ \%$ $\quad 21\ \% \times \dfrac{1}{100\ \%} = 0.21 = \chi_{O2}$

$$10.0\ \text{g} \times \frac{1\ \text{mol}}{32.00\ \text{g}} \times \frac{22.414\ \text{L}}{1\ \text{mol}} = 7.00\ \text{L} = V_{O2} \quad \text{finally} \quad 7.00\ \text{L}\ O_2 \times \frac{100\ \text{L air}}{21\ \text{L}\ O_2} = 33.3\ \text{L air}$$

Check: The units (none and L) are correct. The magnitude of the answer (0.21) makes sense because most of air is nitrogen. The magnitude of the answer (33 L) makes sense because one mole of an ideal gas under standard conditions (273 K and 1 atm) occupies 22.4 L and we have about 1/3 mol of O_2 and so over a mole of air.

37. **Given:** $T = 30.0\ °C$, $P_{Total} = 732$ mmHg, and $V = 722$ mL **Find:** P_{H2} and m_{H2}
Conceptual Plan: $T \rightarrow P_{H2O}$ (*Table 5.3*) **then** $P_{Total}, P_{H2O} \rightarrow P_{H2}$ ($P_{Total} = P_{H2O} + P_{H2}$) **then mmHg → atm** ($\frac{1\ \text{atm}}{760\ \text{mmHg}}$) **and mL → L** ($\frac{1\ \text{L}}{1000\ \text{mL}}$)
and °C → K ($K = °C + 273.15$) $P, V, T \rightarrow n$ ($PV = nRT$) **then mol → g** ($\frac{2.016\ \text{g}}{1\ \text{mol}}$)

Solution: Table 5.3 states that $P_{H2O} = 31.86$ mmHg $P_{Total} = P_{H2O} + P_{H2}$
Rearrange to solve for P_{H2}. $P_{H2} = P_{Total} - P_{H2O} = 732\ \text{mmHg} - 31.86\ \text{mmHg} = 700.\ \text{mmHg}$

$$P_{H2} = 700.\ \text{mmHg} \times \frac{1\ \text{atm}}{760\ \text{mmHg}} = 0.92\underline{1}052\ \text{atm} \quad V = 722\ \text{mL} \times \frac{1\ \text{L}}{1000\ \text{mL}} = 0.722\ \text{L},$$

$T = 30.0\ °C + 273.15 = 303.2$ K, $PV = nRT$ Rearrange to solve for n. $n = \frac{PV}{RT}$

$$n_{H2} = \frac{0.92\underline{1}052\ \text{atm} \times 0.722\ \text{L}}{0.08206\ \frac{\text{L} \cdot \text{atm}}{\text{mol} \cdot \text{K}} \times 303.2\ \text{K}} = 0.026\underline{7}277\ \text{mol} \quad \text{then} \quad 0.026\underline{7}277\ \text{mol} \times \frac{2.016\ \text{g}}{1\ \text{mol}} = 0.0539\ \text{g}\ H_2$$

Check: The units (g) are correct. The magnitude of the answer (<< 1 g) makes sense because gases are not very dense, hydrogen is light, the volume is small, and the pressure is ~1 atm.

38. **Given:** $T = 25\ °C$, $V = 5.45$ L, and $P_{Total} = 745$ mmHg **Find:** n
Conceptual Plan: $T \rightarrow P_{H2O}$ (*Table 5.3*) **then** $P_{Total}, P_{H2O} \rightarrow P_{air}$ ($P_{Total} = P_{H2O} + P_{air}$) **then mmHg → atm** ($\frac{1\ \text{atm}}{760\ \text{mmHg}}$) **and °C → K** ($K = °C + 273.15$)
$P, V, T \rightarrow n$ ($PV = nRT$)

Solution: Table 5.3 states that $P_{H2O} = 23.78$ mmHg $P_{Total} = P_{H2O} + P_{air}$ Rearrange to solve for P_{air}. $P_{air} = P_{Total} - P_{H2O} = 745\ \text{mmHg} - 23.78\ \text{mmHg} = 721\ \text{mmHg}$

$$P_{air} = 721\ \text{mmHgr} \times \frac{1\ \text{atm}}{760\ \text{mmHg}} = 0.94\underline{8}684\ \text{atm} \quad T = 25\ °C + 273.15 = 298\ \text{K}, \quad PV = nRT$$

Rearrange to solve for n. $$n = \frac{PV}{RT} = \frac{0.94\underline{8}684\ \text{atm} \times 5.45\ \text{L}}{0.08206\ \frac{\text{L} \cdot \text{atm}}{\text{mol} \cdot \text{K}} \times 298\ \text{K}} = 0.211\ \text{mol}$$

Check: The units (mol) are correct. The magnitude of the answer (0.2 mol) makes sense because 22.4 L of a gas at STP contains 1 mol. We have only 5.45 L, so the answer makes sense.

39. **Given:** $T = 25\ °C$, $P_{Total} = 748$ mmHg, and $V = 0.951$ L **Find:** P_{H2} and m_{H2}
Conceptual Plan: $T \rightarrow P_{H2O}$ (*Table 5.3*) **then** $P_{Total}, P_{H2O} \rightarrow P_{H2}$ ($P_{Total} = P_{H2O} + P_{H2}$) **then mmHg → atm** ($\frac{1\ \text{atm}}{760\ \text{mmHg}}$) **and mL → L** ($\frac{1\ \text{L}}{1000\ \text{mL}}$)
and °C → K ($K = °C + 273.15$) $P, V, T \rightarrow n$ ($PV = nRT$) **then mol → g** ($\frac{2.016\ \text{g}}{1\ \text{mol}}$)

Solution: Table 5.3 states that $P_{H2O} = 23.78$ mmHg $P_{Total} = P_{H2O} + P_{H2}$
Rearrange to solve for P_{H2}. $P_{H2} = P_{Total} - P_{H2O} = 748\ \text{mmHg} - 23.78\ \text{mmHg} = 724\ \text{mmHg}$

$$P_{H2} = 724\ \text{mmHg} \times \frac{1\ \text{atm}}{760\ \text{mmHg}} = 0.95\underline{2}632\ \text{atm} \quad T = 25\ °C + 273.15 = 298\ \text{K}, \quad PV = nRT$$

Rearrange to solve for n. $$n_{H2} = \frac{PV}{RT} = \frac{0.95\underline{2}632\ \text{atm} \times 0.951\ \text{L}}{0.08206\ \frac{\text{L} \cdot \text{atm}}{\text{mol} \cdot \text{K}} \times 298\ \text{K}} = 0.037\underline{0}474\ \text{mol}$$

$$0.037\underline{0}474\ \text{mol} \times \frac{2.016\ \text{g}}{1\ \text{mol}} = 0.0747\ \text{g}\ H_2$$

Check: The units (g) are correct. The magnitude of the answer (<< 1 g) makes sense because gases are not very dense, hydrogen is light, the volume is small, and the pressure is ~1 atm.

40. **Given:** $m\,(O_2) = 2.0$ g, $m\,(\text{He}) = 98.0$ g, $P_{\text{Total}} = 8.5$ atm **Find:** P_{O2}
Conceptual Plan: **g → mol then n_{O2}, n_{He} → χ_{O2} then $\chi_{O2}, P_{\text{Total}}$ → P_{O2}**

$\mathfrak{M}$ $\quad \chi_{O2} = \dfrac{n_{O2}}{n_{O2} + n_{He}} \quad P_{O2} = \chi_{O2} P_{Total}$

Solution: $n_{O2} = 2.0\text{ g} \times \dfrac{1\text{ mol}}{32.00\text{ g}} = 0.0625\text{ mol}$, $n_{He} = 98.0\text{ g} \times \dfrac{1\text{ mol}}{4.003\text{ g}} = 24.\underline{4}816\text{ mol}$,

$$\chi_{O2} = \frac{n_{O2}}{n_{O2}+n_{He}} = \frac{0.0625\text{ mol}}{0.0625\text{ mol} + 24.\underline{4}816\text{ mol}} = 0.002\underline{5}464,\ P_{O2} = \chi_{O2} P_{Total} = 0.002\underline{5}464 \times 8.5\text{ atm} = 0.022\text{ atm}$$

Check: The units (atm) are correct. The magnitude of the answer (0.22 atm) makes sense because, at these depths, high oxygen pressures can cause toxicity.

41. **Given:** $m\,(\text{C}) = 15.7$ g, $P = 1.0$ atm, and $T = 355$ K **Find:** V
Conceptual Plan: **g C → mol C → mol H_2 then n (mol H_2), P, T → V**

$\dfrac{1\text{ mol}}{12.01\text{ g C}} \quad \dfrac{1\text{ mol H}_2}{1\text{ mol C}} \quad PV = nRT$

Solution: $15.7\text{ g C} \times \dfrac{1\text{ mol C}}{12.01\text{ g C}} \times \dfrac{1\text{ mol H}_2}{1\text{ mol C}} = 1.3\underline{0}724\text{ mol H}_2,\ PV = nRT$ Rearrange to solve for V.

$$V = \frac{nRT}{P} = \frac{1.3\underline{0}724\text{ mol} \times 0.08206\ \frac{\text{L}\cdot\text{atm}}{\text{mol}\cdot\text{K}} \times 355\text{ K}}{1.0\text{ atm}} = 38\text{ L}$$

Check: The units (L) are correct. The magnitude of the answer (38 L) makes sense because we have more than one mole of gas, and so we expect more than 22 L.

42. **Given:** $V_{O2} = 1.4$ L, $T = 315$ K, $P_{O2} = 0.957$ atm **Find:** g H_2O
Conceptual Plan: ***P*(mol O_2), *V*(mol O_2), *T* → *n*(mol O_2) then mol O_2 → mol H_2O → g H_2O**

$PV = nRT \quad \dfrac{2\text{ mol H}_2\text{O}}{1\text{ mol O}_2} \quad \dfrac{18.02\text{ g H}_2\text{O}}{1\text{ mol H}_2\text{O}}$

Solution: $PV = nRT$ Rearrange to solve for n. $n = \dfrac{PV}{RT} = \dfrac{0.957\text{ atm} \times 1.4\text{ L}}{0.08206\ \frac{\text{L}\cdot\text{atm}}{\text{mol}\cdot\text{K}} \times 315\text{ K}} = 0.05\underline{1}832\text{ mol O}_2$

$$0.05\underline{1}832\text{ mol O}_2 \times \frac{2\text{ mol H}_2\text{O}}{1\text{ mol O}_2} \times \frac{18.02\text{ g H}_2\text{O}}{1\text{ mol H}_2\text{O}} = 1.9\text{ g H}_2\text{O}$$

Check: The units (g) are correct. The magnitude of the answer (2 g) makes sense because we have much less than a mole of oxygen.

43. **Given:** $P = 748$ mmHg, $T = 86$ °C, and $m\,(CH_3OH) = 25.8$ g, and **Find:** V_{H2} and V_{CO}
Conceptual Plan: **g CH_3OH → mol CH_3OH → mol H_2 and mmHg → atm and °C → K**

$\dfrac{1\text{ mol CH}_3\text{OH}}{32.04\text{ g CH}_3\text{OH}} \quad \dfrac{2\text{ mol H}_2}{1\text{ mol CH}_3\text{OH}} \quad \dfrac{1\text{ atm}}{760\text{ mmHg}} \quad \text{K} = {}^\circ\text{C} + 273.15$

then n (mol H_2), P, T → V and mol H_2 → mol CO then n (mol CO), P, T → V

$PV = nRT \quad \dfrac{1\text{ mol CO}}{2\text{ mol H}_2} \quad PV = nRT$

Solution: $25.8\text{ g CH}_3\text{OH} \times \dfrac{1\text{ mol CH}_3\text{OH}}{32.04\text{ g CH}_3\text{OH}} \times \dfrac{2\text{ mol H}_2}{1\text{ mol CH}_3\text{OH}} = 1.6\underline{1}049\text{ mol H}_2$,

$P_{H2} = 748\text{ mmHg} \times \dfrac{1\text{ atm}}{760\text{ mmHg}} = 0.98\underline{4}211\text{ atm}$, $T = 86\text{ °C} + 273.15 = 359\text{ K}$, $PV = nRT$

Rearrange to solve for V. $V = \dfrac{nRT}{P}$ $\quad V_{H2} = \dfrac{1.6\underline{1}049\text{ mol} \times 0.08206\ \frac{\text{L}\cdot\text{atm}}{\text{mol}\cdot\text{K}} \times 359\text{ K}}{0.98\underline{4}211\text{ atm}} = 48.2\text{ L H}_2$

$1.6\underline{1}049\text{ mol H}_2 \times \dfrac{1\text{ mol CO}}{2\text{ mol H}_2} = 0.80\underline{5}25\text{ mol CO}$, $V_{CO} = \dfrac{0.80\underline{5}25\text{ mol} \times 0.08206\ \frac{\text{L}\cdot\text{atm}}{\text{mol}\cdot\text{K}} \times 359\text{ K}}{0.98\underline{4}211\text{ atm}} = 24.1\text{ L CO}$

Check: The units (L) are correct. The magnitude of the answer (48 L and 24 L) makes sense because we have more than one mole of hydrogen gas and half that of CO and so we expect significantly more than 22 L for hydrogen and half that for CO.

44. **Given:** $P = 782$ mmHg, $T = 25$ °C, and m (Al) = 53.2 g, and **Find:** V_{O2}
Conceptual Plan: **g Al → mol Al → mol O_2 and mmHg → atm and °C → K then**

$\frac{1 \text{ mol Al}}{26.98 \text{ g Al}}$ $\frac{3 \text{ mol } O_2}{4 \text{ mol Al}}$ $\frac{1 \text{ atm}}{760 \text{ mmHg}}$ $K = °C + 273.15$

n (mol O_2), P, T → V
$PV = nRT$

Solution: $53.2 \text{ g Al} \times \frac{1 \text{ mol Al}}{26.98 \text{ g } O_2} \times \frac{3 \text{ mol } O_2}{4 \text{ mol Al}} = 1.4\underline{7}8873 \text{ mol } O_2$,

$P_{O2} = 782 \text{ mmHg} \times \frac{1 \text{ atm}}{760 \text{ mmHg}} = 1.0\underline{2}8947 \text{ atm}$, $T = 25 \text{ °C} + 273.15 = 298 \text{ K}$, $PV = nRT$

Rearrange to solve for V. $V_{O2} = \frac{nRT}{P} = \frac{1.4\underline{7}8873 \text{ mol} \times 0.08206 \frac{\text{L} \cdot \text{atm}}{\text{mol} \cdot \text{K}} \times 298 \text{ K}}{1.0\underline{2}8947 \text{ atm}} = 35.1 \text{ L } O_2$

Check: The units (L) are correct. The magnitude of the answer (35 L) makes sense because we have more than one mole of oxygen gas and more than 1 atm, so we expect significantly more than 22 L.

45. **Given:** $V = 11.8$ L, and STP **Find:** m (NaN_3)
Conceptual Plan: **V_{N2} → mol N_2 → mol NaN_3 → g NaN_3**

$\frac{1 \text{ mol } N_2}{22.414 \text{ L } N_2}$ $\frac{2 \text{ mol } NaN_3}{3 \text{ mol } N_2}$ $\frac{65.01 \text{ g } NaN_3}{1 \text{ mol } NaN_3}$

Solution: $11.8 \text{ L } N_2 \times \frac{1 \text{ mol } N_2}{22.414 \text{ L } N_2} \times \frac{2 \text{ mol } NaN_3}{3 \text{ mol } N_2} \times \frac{65.01 \text{ g } NaN_3}{1 \text{ mol } NaN_3} = 22.8 \text{ g } NaN_3$

Check: The units (g) are correct. The magnitude of the answer (23 g) makes sense because, we have about a half a mole of nitrogen gas, which translates to even fewer moles of NaN_3 and so we expect significantly less than 65 g.

46. **Given:** $V = 58.5$ mL, and STP **Find:** m (Li)
Conceptual Plan: **mL$_{N2}$ → L$_{N2}$ → mol N_2 → mol Li → g Li**

$\frac{1 \text{ L}}{1000 \text{ mL}}$ $\frac{1 \text{ mol } N_2}{22.414 \text{ L } N_2}$ $\frac{6 \text{ mol Li}}{1 \text{ mol } N_2}$ $\frac{6.941 \text{ g Li}}{1 \text{ mol Li}}$

Solution: $58.5 \text{ mL } N_2 \times \frac{1 \text{ L } N_2}{1000 \text{ mL } N_2} \times \frac{1 \text{ mol } N_2}{22.414 \text{ L } N_2} \times \frac{6 \text{ mol Li}}{1 \text{ mol } N_2} \times \frac{6.941 \text{ g Li}}{1 \text{ mol Li}} = 0.109 \text{ g Li}$

Check: The units (g) are correct. The magnitude of the answer (0.1 g) makes sense because we have such a small volume, which translates to a small fraction of a mole of Li, and so we expect significantly less than 6.9 g.

47. **Given:** $V_{CH4} = 25.5$ L, $P_{CH4} = 732$ torr, and $T = 25$ °C; mixed with $V_{H2O} = 22.8$ L, $P_{H2O} = 702$ torr, and $T =$ 125 °C; forms $P_{H2} = 26.2$ L at STP **Find:** % Yield
Conceptual Plan: **CH_4: torr → atm and °C → K and P, V, T → n_{CH4} → n_{H2}**

$\frac{1 \text{ atm}}{760 \text{ torr}}$ $K = °C + 273.15$ $PV = nRT$ $\frac{3 \text{ mol } H_2}{1 \text{ mol } CH_4}$

H_2O: torr → atm and °C → K and P, V, T → n_{H2O} → n_{H2}

$\frac{1 \text{ atm}}{760 \text{ torr}}$ $K = °C + 273.15$ $PV = nRT$ $\frac{3 \text{ mol } H_2}{1 \text{ mol } H_2O}$

Select smaller n_{H2} as theoretical yield.
then L_{H2} → mol H_2 (actual yield) finally actual yield, theoretical yield → % Yield

$\frac{1 \text{ mol } H_2}{22.414 \text{ L } H_2}$ $\% \text{ Yield} = \frac{\text{actual yield}}{\text{theoretical yield}} \times 100 \%$

Solution: CH_4: $P_{CH4} = 732 \text{ torr} \times \frac{1 \text{ atm}}{760 \text{ torr}} = 0.96\underline{3}158 \text{ atm}$, $T = 25 \text{ °C} + 273.15 = 298 \text{ K}$, $PV = nRT$

Rearrange to solve for n. $n = \frac{PV}{RT}$ $n_{CH4} = \frac{0.96\underline{3}158 \text{ atm} \times 25.5 \text{ L}}{0.08206 \frac{\text{L} \cdot \text{atm}}{\text{mol} \cdot \text{K}} \times 298 \text{ K}} = 1.0\underline{0}436 \text{ mol } CH_4$

$$1.00436 \text{ mol CH}_4 \times \frac{3 \text{ mol H}_2}{1 \text{ mol CH}_4} = 3.01308 \text{ mol H}_2$$

H_2O: $P_{H2O} = 702 \text{ torr} \times \frac{1 \text{ atm}}{760 \text{ torr}} = 0.923684 \text{ atm}$, $T = 125\ °C + 273.15 = 398$ K, $n = \frac{PV}{RT}$

$$n_{H2O} = \frac{0.923684 \text{ atm} \times 22.8 \text{ L}}{0.08206 \frac{\text{L} \cdot \text{atm}}{\text{mol} \cdot \text{K}} \times 398 \text{ K}} = 0.644828 \text{ mol H}_2\text{O} \qquad 0.644828 \text{ mol H}_2\text{O} \times \frac{3 \text{ mol H}_2}{1 \text{ mol H}_2\text{O}} = 1.93448 \text{ mol H}_2$$

Water is the limiting reagent since the moles of hydrogen generated is lower.

Theoretical yield = 1.93448 mol H_2. $26.2 \text{ L H}_2 \times \frac{1 \text{ mol H}_2}{22.414 \text{ L H}_2} = 1.16891 \text{ mol H}_2$ = actual yield

$$\%\ Yield = \frac{actual\ yield}{theoretical\ yield} \times 100\ \% = \frac{1.16891 \text{ mol H}_2}{1.93448 \text{ mol H}_2} \times 100\ \% = 60.4\ \%$$

Check: The units (%) are correct. The magnitude of the answer (60 %) makes sense because it is between 0 and 100 %.

48. **Given:** $P = 25.0$ mmHg, $T = 225$ K, and $m\ (CF_3Cl) = 15.0$ g, and 10 cycles **Find:** V_{O3}
Conceptual Plan: **g CF_3Cl → mol CF_3Cl → mol O_3 → mol O_3 and mmHg → atm then**
$\frac{1 \text{ mol CF}_3\text{Cl}}{104.46 \text{ g CF}_3\text{Cl}}$ $\quad \frac{2 \text{ mol O}_3\text{/ cycle}}{1 \text{ mol CF}_3\text{Cl}}$ $\quad$ **10 cycles** $\quad \frac{1 \text{ atm}}{760 \text{ mmHg}}$

n, P, T* → *V
$PV = nRT$

Solution: $15.0 \text{ g CF}_3\text{Cl} \times \frac{1 \text{ mol CF}_3\text{Cl}}{104.46 \text{ g CF}_3\text{Cl}} \times \frac{2 \text{ mol O}_3 \text{ / cycle}}{1 \text{ mol CF}_3\text{Cl}} \times 10 \text{ cycles} = 2.871913 \text{ mol O}_3$,

$P_{O3} = 25.0 \text{ torr} \times \frac{1 \text{ atm}}{760 \text{ torr}} = 0.0328947 \text{ atm}$, $PV = nRT$ Rearrange to solve for V. $V = \frac{nRT}{P}$

$$V_{O3} = \frac{2.871913 \text{ mol} \times 0.08206 \frac{\text{L} \cdot \text{atm}}{\text{mol} \cdot \text{K}} \times 225 \text{ K}}{0.0328947 \text{ atm}} = 1.61 \times 10^3 \text{ L O}_3$$

Check: The units (L) are correct. The magnitude of the answer (1600 L) makes sense because we have ~ 3 moles of ozone gas and the pressure is so low (0.03 atm), so we expect a large volume.

49. a) Yes, since the average kinetic energy of a particle is proportional to the temperature in kelvins and the two gases are at the same temperature, they have the same average kinetic energy.

b) No, since the helium atoms are lighter, they must move faster to have the same kinetic energy as argon atoms.

c) No, since the Ar atoms are moving slower to compensate for their larger mass, they will exert the same pressure on the walls of the container.

d) Since He is lighter, it will have the faster rate of effusion.

50. a) Since both gases have a mole fraction of 0.5, they will have the same partial pressure.

b) The nitrogen molecules will have a greater velocity since they are lighter than Xe atoms.

c) Since the average kinetic energy of a particle is proportional to the temperature in kelvins and the two gases are at the same temperature, they have the same average kinetic energy.

d) Since nitrogen is lighter, it will have the faster rate of effusion.

51. **Given:** F_2, Cl_2, Br_2, and $T = 298$ K **Find:** u_{rms} KE_{avg} for each gas and relative rates of effusion
Conceptual Plan: $\mathfrak{M}$, ***T*** → u_{rms} → KE_{avg}

$$u_{rms} = \sqrt{\frac{3RT}{\mathfrak{M}}} \qquad KE_{avg} = \frac{1}{2} N_A m u_{rms}^2 = \frac{3}{2} RT$$

Solution:

F_2: $\mathfrak{M} = \frac{38.00\ \text{g}}{1\ \text{mol}} \times \frac{1\ \text{kg}}{1000\ \text{g}} = 0.03800\ \text{kg/mol}$, $u_{rms} = \sqrt{\frac{3RT}{\mathfrak{M}}} = \sqrt{\frac{3 \times 8.314 \frac{\text{J}}{\text{K} \times \text{mol}} \times 298\ \text{K}}{0.03800 \frac{\text{kg}}{\text{mol}}}} = 442\ \text{m/s}$

Cl_2: $\mathfrak{M} = \frac{70.90\ \text{g}}{1\ \text{mol}} \times \frac{1\ \text{kg}}{1000\ \text{g}} = 0.07090\ \text{kg/mol}$, $u_{rms} = \sqrt{\frac{3RT}{\mathfrak{M}}} = \sqrt{\frac{3 \times 8.314 \frac{\text{J}}{\text{K} \times \text{mol}} \times 298\ \text{K}}{0.07090 \frac{\text{kg}}{\text{mol}}}} = 324\ \text{m/s}$

Br_2: $\mathfrak{M} = \frac{159.80\ \text{g}}{1\ \text{mol}} \times \frac{1\ \text{kg}}{1000\ \text{g}} = 0.15980\ \text{kg/mol}$, $u_{rms} = \sqrt{\frac{3RT}{\mathfrak{M}}} = \sqrt{\frac{3 \times 8.314 \frac{\text{J}}{\text{K} \times \text{mol}} \times 298\ \text{K}}{0.15980 \frac{\text{kg}}{\text{mol}}}} = 216\ \text{m/s}$

All molecules have the same kinetic energy:

$KE_{avg} = \frac{3}{2}RT = \frac{3}{2} \times 8.314 \frac{\text{J}}{\text{K} \cdot \text{mol}} \times 298\ \text{K} = 3.72 \times 10^3\ \text{J/mol}$

Since rate of effusion is proportional to $\sqrt{\frac{1}{\mathfrak{M}}}$, F_2 will have the fastest rate and Br_2 will have the slowest rate.

Check: The units (m/s) are correct. The magnitude of the answer (200 – 450 m/s) makes sense because it is consistent with what was seen in the text, and the heavier the molecule, the slower the molecule.

52. **Given:** CO, CO_2, SO_3, and $T = 298$ K **Find:** u_{rms} KE_{avg} for each gas, and rate greatest u_{rms} KE_{avg} and rates of effusion

Conceptual Plan: $\mathfrak{M}$, T → u_{rms} → KE_{avg}

$u_{rms} = \sqrt{\frac{3RT}{\mathfrak{M}}}$ $KE_{avg} = \frac{1}{2}N_A m u_{rms}^2 = \frac{3}{2}RT$

Solution:

CO: $\mathfrak{M} = \frac{28.01\ \text{g}}{1\ \text{mol}} \times \frac{1\ \text{kg}}{1000\ \text{g}} = 0.02801\ \text{kg/mol}$, $u_{rms} = \sqrt{\frac{3RT}{\mathfrak{M}}} = \sqrt{\frac{3 \times 8.314 \frac{\text{J}}{\text{K} \times \text{mol}} \times 298\ \text{K}}{0.02801 \frac{\text{kg}}{\text{mol}}}} = 515.1\ \text{m/s}$

CO_2: $\mathfrak{M} = \frac{44.01\ \text{g}}{1\ \text{mol}} \times \frac{1\ \text{kg}}{1000\ \text{g}} = 0.04401\ \text{kg/mol}$, $u_{rms} = \sqrt{\frac{3RT}{\mathfrak{M}}} = \sqrt{\frac{3 \times 8.314 \frac{\text{J}}{\text{K} \times \text{mol}} \times 298\ \text{K}}{0.04401 \frac{\text{kg}}{\text{mol}}}} = 411.0\ \text{m/s}$

SO_3: $\mathfrak{M} = \frac{80.07\ \text{g}}{1\ \text{mol}} \times \frac{1\ \text{kg}}{1000\ \text{g}} = 0.08007\ \text{kg/mol}$, $u_{rms} = \sqrt{\frac{3RT}{\mathfrak{M}}} = \sqrt{\frac{3 \times 8.314 \frac{\text{J}}{\text{K} \times \text{mol}} \times 298\ \text{K}}{0.08007 \frac{\text{kg}}{\text{mol}}}} = 304.7\ \text{m/s}$

All molecules have the same kinetic energy:

$KE_{avg} = \frac{3}{2}RT = \frac{3}{2} \times 8.314 \frac{\text{J}}{\text{K} \cdot \text{mol}} \times 298\ \text{K} = 3.72 \times 10^3\ \text{J/mol}$ CO has the fastest speed; all molecules have the same kinetic energy; and since rate of effusion is proportional to $1/\sqrt{\mathfrak{M}}$, CO will have the fastest rate.

Check: The units (m/s) are correct. The magnitude of the answer (300 – 520 m/s) makes sense because it is consistent with what was seen in the text, and the heavier the molecule, the slower the molecule.

53. **Given:** $^{238}UF_6$ and $^{235}UF_6$ U-235 = 235.054 amu, U-238 = 238.051 amu

Find: ratio of effusion rates $^{238}UF_6 / ^{235}UF_6$

Conceptual Plan: $\mathfrak{M}(^{238}UF_6)$, $\mathfrak{M}(^{235}UF_6)$ → **Rate ($^{238}UF_6$) / Rate ($^{235}UF_6$)**

$$\frac{Rate(^{238}UF_6)}{Rate(^{235}UF_6)} = \sqrt{\frac{\mathfrak{M}(^{235}UF_6)}{\mathfrak{M}(^{238}UF_6)}}$$

Solution: $^{238}UF_6$: $\mathfrak{M} = \frac{352.05\ \text{g}}{1\ \text{mol}} \times \frac{1\ \text{kg}}{1000\ \text{g}} = 0.35205\ \text{kg/mol}$,

$^{235}UF_6$: $\mathfrak{M} = \frac{349.05\ \text{g}}{1\ \text{mol}} \times \frac{1\ \text{kg}}{1000\ \text{g}} = 0.34905\ \text{kg/mol}$,

$$\frac{Rate(^{238}UF_6)}{Rate(^{235}UF_6)} = \sqrt{\frac{\mathfrak{M}(^{235}UF_6)}{\mathfrak{M}(^{238}UF_6)}} = \sqrt{\frac{0.34905\ \text{kg/mol}}{0.35205\ \text{kg/mol}}} = 0.99574$$

Check: The units (none) are correct. The magnitude of the answer (<1) makes sense because, the heavier molecule has the lower effusion rate since it moves slower.

54. **Given:** Ar and Kr **Find:** ratio of effusion rates Ar / Kr
Conceptual Plan: $\mathfrak{M}$ **(Ar) ,** $\mathfrak{M}$ **(Kr) → Rate (Ar) / Rate (Kr)**

$$\frac{Rate(Ar)}{Rate(Kr)} = \sqrt{\frac{\mathfrak{M}(Kr)}{\mathfrak{M}(Ar)}}$$

Solution: Ar: $\mathfrak{M} = \frac{39.95\ \text{g}}{1\ \text{mol}} \times \frac{1\ \text{kg}}{1000\ \text{g}} = 0.03995\ \text{kg/mol}$, Kr: $\mathfrak{M} = \frac{83.80\ \text{g}}{1\ \text{mol}} \times \frac{1\ \text{kg}}{1000\ \text{g}} = 0.08380\ \text{kg/mol}$,

$$\frac{Rate(Ar)}{Rate(Kr)} = \sqrt{\frac{\mathfrak{M}(Kr)}{\mathfrak{M}(Ar)}} = \sqrt{\frac{0.08380\ \text{kg/mol}}{0.03995\ \text{kg/mol}}} = 1.448$$

Check: The units (none) are correct. The magnitude of the answer (>1) makes sense because the lighter molecule has the higher effusion rate since it moves faster.

55. **Given:** Ne and unknown gas; and Ne effusion in 76 s and unknown in 155 s
Find: identify unknown gas
Conceptual Plan: $\mathfrak{M}$ **(Ne), Rate (Ne), Rate (Unk) →** $\mathfrak{M}$ **(Kr)**

$$\frac{Rate(Ne)}{Rate(Unk)} = \sqrt{\frac{\mathfrak{M}(Unk)}{\mathfrak{M}(Ne)}}$$

Solution: Ne: $\mathfrak{M} = \frac{20.18\ \text{g}}{1\ \text{mol}} \times \frac{1\ \text{kg}}{1000\ \text{g}} = 0.02018\ \text{kg/mol}$, $\frac{Rate(Ne)}{Rate(Unk)} = \sqrt{\frac{\mathfrak{M}(Unk)}{\mathfrak{M}(Ne)}}$ Rearrange to solve for $\mathfrak{M}(Unk)$. $\mathfrak{M}(Unk) = \mathfrak{M}(Ne)\left(\frac{Rate(Ne)}{Rate(Unk)}\right)^2$ Since Rate α 1/(effusion time),

$$\mathfrak{M}(Unk) = \mathfrak{M}(Ne)\left(\frac{Time(Unk)}{Time(Ne)}\right)^2 = 0.02018\frac{\text{kg}}{\text{mol}}\ x\left(\frac{155\ \text{s}}{76\ \text{s}}\right)^2 = 0.084\frac{\text{kg}}{\text{mol}} \times \frac{1000\ \text{g}}{1\ \text{kg}} = 84\ \text{g/mol}\ \text{or Kr.}$$

Check: The units (g/mol) are correct. The magnitude of the answer (>Ne) makes sense because, Ne effused faster and so must be lighter.

56. **Given:** NO_2 and I_2 gas; and NO_2 effusion in 42 s **Find:** effusion time for I_2 gas
Conceptual Plan: $\mathfrak{M}$ **(NO_2),** $\mathfrak{M}$ **(Kr), Rate (NO_2), → Rate (I_2)**

$$\frac{Rate(N_2O)}{Rate(I_2)} = \sqrt{\frac{\mathfrak{M}(I_2)}{\mathfrak{M}(N_2O)}}$$

Solution: N_2O: $\mathfrak{M} = \frac{44.02\ \text{g}}{1\ \text{mol}} \times \frac{1\ \text{kg}}{1000\ \text{g}} = 0.04402\ \text{kg/mol}$, I_2: $\mathfrak{M} = \frac{253.8\ \text{g}}{1\ \text{mol}} \times \frac{1\ \text{kg}}{1000\ \text{g}} = 0.2538\ \text{kg/mol}$,

$\frac{Rate(N_2O)}{Rate(I_2)} = \sqrt{\frac{\mathfrak{M}(I_2)}{\mathfrak{M}(N_2O)}}$ Since Rate α 1/(effusion time) $\frac{Time(I_2)}{Time(N_2O)} = \sqrt{\frac{\mathfrak{M}(I_2)}{\mathfrak{M}(N_2O)}}$ Rearrange to solve for *Time (I_2)*. $Time(I_2) = Time(N_2O)\ x\ \sqrt{\frac{\mathfrak{M}(I_2)}{\mathfrak{M}(N_2O)}} = 42\ \text{s} \times \sqrt{\frac{0.2538\ \text{kg/mol}}{0.04402\ \text{kg/mol}}} = 1.0 \times 10^2\ \text{s}$

Check: The units (s) are correct. The magnitude of the answer (100 s) makes sense because the mass ratio of I_2 to N_2O is ~6, so the time should be over twice as long.

57. Gas A has the higher molar mass, since it has the slower average velocity. Gas B will have the higher effusion rate, since it has the higher velocity.

58. T_2 is the higher temperature, since it has the higher average velocity.

59. The postulate that the volume of the gas particles is small compared to the space between them breaks down at high pressure. At high pressures the number of molecules increases, so the volume of the gas particles becomes larger and since the spacing between the particles is smaller, the molecules themselves occupy a significant portion of the volume.

60. The postulate that the forces between the gas particles are not significant breaks down at low temperatures. At low temperatures, the molecules are not moving as fast as at higher temperatures, so when they collide they have a greater opportunity to interact.

61. **Given:** Ne, $n = 1.000$ mol, $P = 500.0$ atm, and $T = 355.0$ K **Find:** V(ideal) and V(van der Waals)
Conceptual Plan: ***n, P, T* → *V*** **and** ***n, P, T* → *V***

$$PV = nRT \qquad \left(P + \frac{an^2}{V^2}\right)(V - nb) = nRT$$

Solution: $PV = nRT$ Rearrange to solve for V.

$$V = \frac{nRT}{P} = \frac{1.000\ \text{mol} \times 0.08206\ \frac{\text{L} \cdot \text{atm}}{\text{mol} \cdot \text{K}} \times 355.0\ \text{K}}{500.0\ \text{atm}} = 0.05826\ \text{L}$$

$\left(P + \frac{an^2}{V^2}\right)(V - nb) = nRT$ Rearrange to solve to: $V = \dfrac{nRT}{\left(P + \frac{an^2}{V^2}\right)} + nb$

Using a = 0.211 L^2 atm/mol^2 and b = 0.0171 L/mol from Table 5.5, and the V from the ideal gas law above, solve for V by successive approximations.

$$V = \frac{1.000\ \text{mol} \times 0.08206\ \frac{\text{L} \cdot \text{atm}}{\text{mol} \cdot \text{K}} \times 355.0\ \text{K}}{500.0\ \text{atm} + \frac{0.211 \frac{\text{L}^2 \cdot \text{atm}}{\text{mol}^2} \times (1.000\ \text{mol})^2}{(0.05826\ \text{L})^2}} + \left(1.000\ \text{mol} \times 0.0171 \frac{\text{L}}{\text{mol}}\right) = 0.06\underline{8}915\ \text{L}$$

Plug in this new value.

$$V = \frac{1.000\ \text{mol} \times 0.08206\ \frac{\text{L} \cdot \text{atm}}{\text{mol} \cdot \text{K}} \times 355.0\ \text{K}}{500.0\ \text{atm} + \frac{0.211 \frac{\text{L}^2 \cdot \text{atm}}{\text{mol}^2} \times (1.000\ \text{mol})^2}{(0.068915\ \text{L})^2}} + \left(1.000\ \text{mol} \times 0.0171 \frac{\text{L}}{\text{mol}}\right) = 0.070\underline{6}09\ \text{L}$$

Plug in this new value.

$$V = \frac{1.000\ \text{mol} \times 0.08206\ \frac{\text{L} \cdot \text{atm}}{\text{mol} \cdot \text{K}} \times 355.0\ \text{K}}{500.0\ \text{atm} + \frac{0.211 \frac{\text{L}^2 \cdot \text{atm}}{\text{mol}^2} \times (1.000\ \text{mol})^2}{(0.070609\ \text{L})^2}} + \left(1.000\ \text{mol} \times 0.0171 \frac{\text{L}}{\text{mol}}\right) = 0.070\underline{8}17\ \text{L}$$

Plug in this new value.

$$V = \frac{1.000\ \text{mol} \times 0.08206\ \frac{\text{L} \cdot \text{atm}}{\text{mol} \cdot \text{K}} \times 355.0\ \text{K}}{500.0\ \text{atm} + \frac{0.211 \frac{\text{L}^2 \cdot \text{atm}}{\text{mol}^2} \times (1.000\ \text{mol})^2}{(0.070817\ \text{L})^2}} + \left(1.000\ \text{mol} \times 0.0171 \frac{\text{L}}{\text{mol}}\right) = 0.070\underline{8}42\ \text{L} = 0.0708\ \text{L}$$

The two values are different because we are at very high pressures. The pressure is corrected from 500.0 atm to 542.1 atm and the final volume correction is 0.0171 L.
Check: The units (L) are correct. The magnitude of the answer (~0.06 L) makes sense because we are at such a high pressure and have one mole of gas.

62. **Given:** Cl_2, $n = 1.000$ mol, L = 5.000 L, and $T = 273.0$ K **Find:** P(ideal) and P(van der Waals)
Conceptual Plan: ***n, V, T* → *P*** **and** ***n, V, T* → *P***

$$PV = nRT \qquad \left(P + \frac{an^2}{V^2}\right)(V - nb) = nRT$$

Solution: $PV = nRT$ Rearrange to solve for P.

$$P = \frac{nRT}{V} = \frac{1.000\ \text{mol} \times 0.08206\ \frac{\text{L} \cdot \text{atm}}{\text{mol} \cdot \text{K}} \times 273.0\ \text{K}}{5.000\ \text{L}} = 4.480\ \text{atm}$$

$\left(P + \frac{an^2}{V^2}\right)(V - nb) = nRT$ Rearrange to solve for P. $P = \frac{nRT}{(V - nb)} - \frac{an^2}{V^2}$

Using a = 6.49 L^2 atm/mol^2 and b = 0.1383 L/mol from Table 5.5:

$$P = \frac{1.000 \text{ mol} \times 0.08206 \frac{\text{L} \cdot \text{atm}}{\text{mol} \cdot \text{K}} \times 273.0 \text{ K}}{5.000 \text{ L} - \left(1.000 \text{ mol} \times 0.1383 \frac{\text{L}}{\text{mol}}\right)} - \frac{6.49 \frac{\text{L}^2 \cdot \text{atm}}{\text{mol}^2} \times (1.000 \text{ mol})^2}{(5.000 \text{ L})^2} = 2.66 \text{ atm}$$

The pressure values differ because the pressure is high, the temperature is moderate, and chlorine is not the smallest gas molecule, all of these contribute to non-ideal behavior.

Check: The units (atm) are correct. The magnitude of the answer (4 atm and 3 atm) makes sense because we are at such a low temperature and small volume with one mole of gas – we expect a P >1 atm.

63. **Given:** m (penny) = 2.482 g, T = 25 °C, V = 0.899 L, and P_{Total} = 791 mmHg
Find: % Zn in penny
Conceptual Plan: $T \rightarrow P_{H2O}$ **then** $P_{\text{Total}}, P_{H2O} \rightarrow P_{H2}$ **then** mmHg → atm **and** °C → K

Table 5.3 $\quad P_{Total} = P_{H2O} + P_{H2} \quad \frac{1 \text{ atm}}{760 \text{ mmHg}} \quad$ K = °C + 273.15

and $P, V, T \rightarrow n_{H2} \rightarrow n_{Zn} \rightarrow g_{Zn} \rightarrow$ **% Zn**

$PV = nRT \quad \frac{1 \text{ mol Zn}}{1 \text{ mol } H_2} \quad \frac{65.39 \text{ g Zn}}{1 \text{ mol Zn}} \quad \% Zn = \frac{g_{Zn}}{g_{penny}} \times 100\ \%$

Solution: Table 5.3 states that P_{H2O} = 23.78 mmHg at 25 °C $P_{Total} = P_{H2O} + P_{H2}$ Rearrange to solve for P_{H2}.
$P_{H2} = P_{Total} - P_{H2O}$ = 791 mmHg – 23.78 mmHg = 767 mmHg

$P_{H2} = 767 \text{ mmHg} \times \frac{1 \text{ atm}}{760 \text{ mmHg}} = 1.0\underline{0}95$ atm then T = 25 °C + 273.15 = 298 K, $PV = nRT$

Rearrange to solve for n. $n_{H2} = \frac{PV}{RT} = \frac{1.0\underline{0}95 \text{ atm} \times 0.899 \text{ L}}{0.08206 \frac{\text{L} \cdot \text{atm}}{\text{mol} \cdot \text{K}} \times 298 \text{ K}} = 0.037\underline{1}123$ mol

$0.037\underline{1}123 \text{ mol } H_2 \times \frac{1 \text{ mol Zn}}{1 \text{ mol } H_2} \times \frac{65.39 \text{ g Zn}}{1 \text{ mol Zn}} = 2.4\underline{2}677$ g Zn

$\% Zn = \frac{g_{Zn}}{g_{penny}} \times 100\ \% = \frac{2.4\underline{2}677 \text{ g}}{2.482 \text{ g}} \times 100\ \% = 97.8\ \%$ Zn

Check: The units (% Zn) are correct. The magnitude of the answer (98 %) makes sense because it should be between 0 and 100 %. We expect about 1/22 a mole of gas, since our conditions are close to STP and we have ~ 1 L of gas.

64. **Given:** m (CFC) = 2.85 g, V = 564 mL, P = 752 mmHg, and T = 298 K **Find:** % Cl in CFC
Conceptual Plan: **mmHg → atm and mL → L and** $P, V, T \rightarrow n_{Cl2} \rightarrow g_{Cl}$

$\frac{1 \text{ atm}}{760 \text{ mmHg}} \quad \frac{1 \text{ L}}{1000 \text{ mL}} \quad PV = nRT \quad \frac{70.90 \text{ g Cl}}{1 \text{ mol } Cl_2}$

then $g_{Cl}, g_{CFC} \rightarrow$ **% Cl**

$\% Cl = \frac{g_{Cl}}{g_{CFC}} \times 100\ \%$

Solution: $P_{Cl2} = 752 \text{ mmHg} \times \frac{1 \text{ atm}}{760 \text{ mmHg}} = 0.98\underline{9}47$ atm, $V_{Cl2} = 564 \text{ mL} \times \frac{1 \text{ L}}{1000 \text{ mL}} = 0.564$ L,

$PV = nRT$ Rearrange to solve for n. $n_{Cl2} = \frac{PV}{RT} = \frac{0.98\underline{9}47 \text{ atm} \times 0.564 \text{ L}}{0.08206 \frac{\text{L} \cdot \text{atm}}{\text{mol} \cdot \text{K}} \times 298 \text{ K}} = 0.022\underline{8}21$ mol Cl_2

$0.022\underline{8}21 \text{ mol } Cl_2 \times \frac{70.90 \text{ g Cl}}{1 \text{ mol } Cl_2} = 1.6\underline{1}80$ g Cl, $\% Cl = \frac{g_{Cl}}{g_{CFC}} \times 100\ \% = \frac{1.6\underline{1}80 \text{ g}}{2.85 \text{ g}} \times 100\ \% = 56.8\ \%$ Cl

Check: The units (% Cl) are correct. The magnitude of the answer (57 %) makes sense because it should be between 0 and 100 %. Since there will be carbon in the compound, we do not expect it to be extremely close to 100 %.

65. **Given:** V = 255 mL, m (flask) = 143.187 g, m (flask + gas) = 143.289 g, P = 267 torr, and T = 25 °C
Find: $\mathfrak{M}$

Conceptual Plan: °C → K torr → atm mL → L *m* (flask), *m* (flask + gas) → *m* (gas)

$K = °C + 273.15$ $\frac{1\text{ atm}}{760\text{ torr}}$ $\frac{1\text{ L}}{1000\text{ mL}}$ $m\,(gas) = m\,(flask + gas) - m\,(flask)$

then *V*, *m* → *d* **then** *d*, *P*, *T*, → $\mathfrak{M}$

$d = \frac{m}{V}$ $d = \frac{P\mathfrak{M}}{RT}$

Solution: $T = 25\ °C + 273.15 = 298\ K$, $P = 267\ \text{torr} \times \frac{1\text{ atm}}{760\text{ torr}} = 0.351316\text{ atm}$,

$V = 255\ \text{mL} \times \frac{1\text{ L}}{1000\text{ mL}} = 0.255\text{ L}$,

$m\,(gas) = m\,(flask + gas) - m\,(flask) = 143.289\text{ g} - 143.187\text{ g} = 0.102\text{ g}$, $d = \frac{m}{V} = \frac{0.102\text{ g}}{0.255\text{ L}} = 0.400\text{ g/L}$,

$d = \frac{P\mathfrak{M}}{RT}$ Rearrange to solve for $\mathfrak{M}$.

$$\mathfrak{M} = \frac{dRT}{P} = \frac{0.400\frac{\text{g}}{\text{L}} \times 0.08206\frac{\text{L atm}}{\text{K mol}} \times 298\text{ K}}{0.351316\text{ atm}} = 27.8\text{ g/mol}$$

Check: The units (g/mol) are correct. The magnitude of the answer (28 g/mol) makes physical sense because this is a reasonable number for a molecular weight of a gas.

66. **Given:** $V = 118$ mL, *m* (flask) = 97.129 g, *m* (flask + gas) = 97.171 g, $P = 768$ torr, and $T = 35$ °C
Find: Is gas pure?
Conceptual Plan: °C → K torr → atm mL → L *m* (flask), *m* (flask + gas) → *m* (gas)

$K = °C + 273.15$ $\frac{1\text{ atm}}{760\text{ torr}}$ $\frac{1\text{ L}}{1000\text{ mL}}$ $m\,(gas) = m\,(flask + gas) - m\,(flask)$

then *V*, *m* → *d* **then** *d*, *P*, *T*, → $\mathfrak{M}$

$d = \frac{m}{V}$ $d = \frac{P\mathfrak{M}}{RT}$

Solution: $T = 35\ °C + 273.15 = 308\ K$, $P = 768\ \text{torr} \times \frac{1\text{ atm}}{760\text{ torr}} = 1.01053\text{ atm}$,

$V = 118\ \text{mL} \times \frac{1\text{ L}}{1000\text{ mL}} = 0.118\text{ L}$

$m\,(gas) = m\,(flask + gas) - m\,(flask) = 97.171\text{ g} - 97.129\text{ g} = 0.042\text{ g}$,

$d = \frac{m}{V} = \frac{0.042\text{ g}}{0.118\text{ L}} = 0.35593\text{ g/L}$, $d = \frac{P\mathfrak{M}}{RT}$ Rearrange to solve for $\mathfrak{M}$.

$$\mathfrak{M} = \frac{dRT}{P} = \frac{0.35593\frac{\text{g}}{\text{L}} \times 0.08206\frac{\text{L atm}}{\text{K mol}} \times 308\text{ K}}{1.01053\text{ atm}} = 8.9\text{ g/mol}$$

The gas is not pure He, since the molar mass is not 4.003 g/ml.
Check: The units (g/mol) are correct. The magnitude of the answer (9 g/mol) makes physical sense because this is a reasonable number for a molecular weight of a gas.

67. **Given:** $V = 158$ mL, *m* (gas) = 0.275 g, $P = 556$ mmHg, $T = 25$ °C, gas = 82.66 % C and 17.34 % H.
Find: Molecular formula
Conceptual Plan: °C → K mmHg → atm mL → L **then** *V*, *m* → *d*

$K = °C + 273.15$ $\frac{1\text{ atm}}{760\text{ mmHg}}$ $\frac{1\text{ L}}{1000\text{ mL}}$ $d = \frac{m}{V}$

then *d*, *P*, *T*, → $\mathfrak{M}$ **then** % C, % H, $\mathfrak{M}$ → **formula**

$d = \frac{P\mathfrak{M}}{RT}$ $\#C = \frac{\mathfrak{M}\ 0.8266\ g\,C}{12.01\frac{g\,C}{mol\,C}}$ $\#H = \frac{\mathfrak{M}\ 0.1734\ g\,H}{1.008\frac{g\,H}{mol\,H}}$

Solution: $T = 25\ °C + 273.15 = 298\ K$, $P = 556\ \text{mmHg} \times \frac{1\text{ atm}}{760\text{ mmHg}} = 0.731579\text{ atm}$,

$V = 158\ \text{mL} \times \frac{1\text{ L}}{1000\text{ mL}} = 0.158\text{ L}$, $d = \frac{m}{V} = \frac{0.275\text{ g}}{0.158\text{ L}} = 1.74051\text{ g/L}$, $d = \frac{P\mathfrak{M}}{RT}$ Rearrange to solve for $\mathfrak{M}$.

$$\mathfrak{M} = \frac{dRT}{P} = \frac{1.74051\frac{\text{g}}{\text{L}} \times 0.08206\frac{\text{L atm}}{\text{K mol}} \times 298\text{ K}}{0.731579\text{ atm}} = 58.2\text{ g/mol},$$

$$\#C = \frac{\mathfrak{M} \text{ x } 0.8266\ g\ C}{12.01 \frac{g\ C}{mol\ C}} = \frac{58.2 \frac{\text{g HC}}{\text{mol HC}} \text{ x } \frac{0.8266 \text{ g C}}{1 \text{ g HC}}}{12.01 \frac{\text{g C}}{\text{mol C}}} = 4.00 \frac{\text{mol C}}{\text{mol HC}}$$

$$\#H = \frac{\mathfrak{M} \text{ x } 0.1734\ g\ H}{1.008 \frac{g\ H}{mol\ H}} = \frac{58.2 \frac{\text{g HC}}{\text{mol HC}} \text{ x } \frac{0.1734 \text{ g H}}{1 \text{ g HC}}}{1.008 \frac{\text{g H}}{\text{mol H}}} = 10.0 \frac{\text{mol H}}{\text{mol HC}}$$ Formula is C_4H_{10} or butane.

Check: The answer came up with integer number of C and H atoms in the formula and a molecular weight (58 g/mol) that is reasonable for a gas.

68. **Given: STP,** $V = 258$ **mL,** m **(gas) = 0.646 g, gas = 85.63 % C and 14.37 % H.** **Find:** $\mathfrak{M}$
Conceptual Plan: mL → L then $V, m \rightarrow d$ **then** $d, P, T, \rightarrow \mathfrak{M}$

$\frac{1\text{ L}}{1000\text{ mL}}$ $\quad d = \frac{m}{V} \quad d = \frac{P\mathfrak{M}}{RT}$

then % C, % H, $\mathfrak{M}$ **→ formula**

$\#C = \frac{\mathfrak{M}\ 0.8563\ g\ C}{12.01 \frac{g\ C}{mol\ C}} \quad \#H = \frac{\mathfrak{M}\ 0.1437\ g\ H}{1.008 \frac{g\ H}{mol\ H}}$

Solution: $V = 258 \text{ mL x} \frac{1\text{ L}}{1000\text{ mL}} = 0.258\text{ L}$, $d = \frac{m}{V} = \frac{0.646\text{ g}}{0.258\text{ L}} = 2.50388\text{ g/L}$, $d = \frac{P\mathfrak{M}}{RT}$ Rearrange to solve for $\mathfrak{M}$. $\mathfrak{M} = \frac{dRT}{P} = \frac{2.50388 \frac{\text{g}}{\text{L}} \text{ x } 0.08206 \frac{\text{L atm}}{\text{K mol}} \text{ x } 273.15\text{ K}}{1\text{ atm}} = 56.12\text{ g/mol}$,

$$\#C = \frac{\mathfrak{M} \text{ x } 0.8563\ g\ C}{12.01 \frac{g\ C}{mol\ C}} = \frac{56.12 \frac{\text{g HC}}{\text{mol HC}} \text{ x } \frac{0.8563 \text{ g C}}{1 \text{ g HC}}}{12.01 \frac{\text{g C}}{\text{mol C}}} = 4.00 \frac{\text{mol C}}{\text{mol HC}}$$

$$\#H = \frac{\mathfrak{M} \text{ x } 0.1437\ g\ H}{1.008 \frac{g\ H}{mol\ H}} = \frac{56.12 \frac{\text{g HC}}{\text{mol HC}} \text{ x } \frac{0.1437 \text{ g H}}{1 \text{ g HC}}}{1.008 \frac{\text{g H}}{\text{mol H}}} = 8.00 \frac{\text{mol H}}{\text{mol HC}}$$ Formula is C_4H_8 or butene.

Check: The answer came up with integer number of C and H atoms in the formula and a molecular weight (56 g/mol) that is reasonable for a gas.

69. **Given:** m **(NiO) = 24.78 g,** $T = 40.0$ **°C, and** $P_{Total} = 745$ **mmHg** **Find:** V_{O2}
Conceptual Plan: $T \rightarrow P_{H2O}$ **then** $P_{Total}, P_{H2O} \rightarrow P_{O2}$ **then mmHg → atm and °C → K**

Table 5.3 $\quad P_{Total} = P_{H2O} + P_{O2} \quad \frac{1\text{ atm}}{760\text{ mmHg}} \quad \text{K} = °\text{C} + 273.15$

and $g_{NiO} \rightarrow n_{NiO} \rightarrow n_{O2}$ **then** $P, V, T \rightarrow n_{O2}$

$\frac{1\text{ mol NiO}}{74.69\text{ g NiO}} \quad \frac{1\text{ mol } O_2}{2\text{ mol NiO}} \quad PV = nRT$

Solution: Table 5.3 states that $P_{H2O} = 55.40$ mmHg at 40°C $P_{Total} = P_{H2O} + P_{O2}$ Rearrange to solve for P_{O2}.
$P_{O2} = P_{Total} - P_{H2O} = 745\text{ mmHg} - 55.40\text{ mmHg} = 689.6\text{ mmHg}$

$P_{O_2} = 689.6 \text{ mmHg x} \frac{1\text{ atm}}{760\text{ mmHg}} = 0.907368\text{ atm}$ $\quad T = 40.0\text{ °C} + 273.15 = 313.2\text{ K}$,

$24.78 \text{ g NiO x} \frac{1\text{ mol NiO}}{74.69\text{ mol NiO}} \text{ x} \frac{1\text{ mol } O_2}{2\text{ mol NiO}} = 0.1658857\text{ mol } O_2$ $\quad PV = nRT$

Rearrange to solve for V. $V_{O_2} = \frac{nRT}{P} = \frac{0.1658857\text{ mol x } 0.08206 \frac{\text{L} \cdot \text{atm}}{\text{mol} \cdot \text{K}} \text{ x } 313.2\text{ K}}{0.907368\text{ atm}} = 4.70\text{ L}$

Check: The units (L) are correct. The magnitude of the answer (5 L) makes sense because we have much less than 0.5 mole of NiO, so we get less than a mole of oxygen. Thus we expect a volume much less than 22 L.

70. **Given:** m (Ag) = 15.8 g, T = 25 °C, and P_{Total} = 752 mmHg **Find:** V_{O2}
Conceptual Plan: $T \rightarrow P_{H2O}$ then $P_{Total}, P_{H2O} \rightarrow P_{O2}$ then mmHg → atm and °C → K

Table 5.3 $\quad P_{Total} = P_{H2O} + P_{O2} \quad \frac{1 \text{ atm}}{760 \text{ mmHg}} \quad K = °C + 273.15$

and $g_{Ag} \rightarrow n_{Ag} \rightarrow n_{O2}$ then $P, V, T \rightarrow n_{O2}$

$\frac{1 \text{ mol Ag}}{107.9 \text{ g Ag}} \quad \frac{1 \text{ mol } O_2}{4 \text{ mol Ag}} \quad PV = nRT$

Solution: Table 5.3 states that P_{H2O} = 23.78 mmHg at 25 °C $P_{Total} = P_{H2O} + P_{O2}$ Rearrange to solve for P_{O2}.
$P_{O2} = P_{Total} - P_{H2O}$ = 752 mmHg − 23.78 mmHg = 728.22 mmHg

$$P_{O2} = 728.22 \text{ mmHg} \times \frac{1 \text{ atm}}{760 \text{ mmHg}} = 0.958184 \text{ atm} \quad T = 25 \text{ °C} + 273.15 = 298 \text{ K},$$

$$15.8 \text{ g Ag} \times \frac{1 \text{ mol Ag}}{107.9 \text{ mol Ag}} \times \frac{1 \text{ mol } O_2}{4 \text{ mol Ag}} = 0.03660780 \text{ mol } O_2 \qquad PV = nRT \quad \text{Rearrange to solve for } V.$$

$$V_{O2} = \frac{nRT}{P} = \frac{0.03660780 \text{ mol} \times 0.08206 \frac{\text{L} \cdot \text{atm}}{\text{mol} \cdot \text{K}} \times 298 \text{ K}}{0.958184 \text{ atm}} = 0.934 \text{ L}$$

Check: The units (L) are correct. The magnitude of the answer (1 L) makes sense because, we have ~ 0.1 mole of Ag, so we get less than 0.5 mol of oxygen. Thus we expect a volume much, much less than 22 L.

71. **Given:** HCl, K_2S to H_2S, V_{H2S} = 42.9 mL, P_{H2S} = 752 mmHg, and T = 25.8 °C **Find:** $m(K_2S)$
Conceptual Plan: read description of reaction and convert words to equation then °C → K

$K = °C + 273.15$

and mmHg → atm and mL → L then $P, V, T \rightarrow n_{H2S} \rightarrow n_{K2S} \rightarrow g_{K2S}$

$\frac{1 \text{ atm}}{760 \text{ mmHg}} \quad \frac{1 \text{ L}}{1000 \text{ mL}} \quad PV = nRT \quad \frac{1 \text{ mol } K_2S}{1 \text{ mol } H_2S} \quad \frac{1 \text{ mol } K_2S}{110.27 \text{ g } K_2S}$

Solution: 2 HCl (aq) + K_2S (s) → H_2S (g) + 2 KCl (aq)

$$T = 25.8 \text{ °C} + 273.15 = 299.0 \text{ K}, P_{H2S} = 752 \text{ mmHg} \times \frac{1 \text{ atm}}{760 \text{ mmHg}} = 0.989474 \text{ atm},$$

$$V_{H2S} = 42.9 \text{ mL} \times \frac{1 \text{ L}}{1000 \text{ mL}} = 0.0429 \text{ L} \quad PV = nRT \quad \text{Rearrange to solve for } n_{H2S}.$$

$$n_{H2S} = \frac{PV}{RT} = \frac{0.989474 \text{ atm} \times 0.0429 \text{ L}}{0.08206 \frac{\text{L} \cdot \text{atm}}{\text{mol} \cdot \text{K}} \times 299.0 \text{ K}} = 0.00173005 \text{ mol}$$

$$0.00173005 \text{ mol } H_2S \times \frac{1 \text{ mol } K_2S}{1 \text{ mol } H_2S} \times \frac{110.27 \text{ g } K_2S}{1 \text{ mol } K_2S} = 0.191 \text{ g } K_2S$$

Check: The units (g) are correct. The magnitude of the answer (0.1 g) makes sense because we have such a small volume of gas generated.

72. a) **Given:** T = 315 K, P = 50.0 mmHg, V_{SO2} = 285.5 mL, V_{O2} = 158.9 mL
Find: limiting reagent and theoretical yield
Conceptual Plan: mmHg → atm and $\text{mL}_{SO2} \rightarrow \text{L}_{SO2}$ **then** $P, V_{SO2}, T \rightarrow n_{SO2} \rightarrow n_{SO3}$

$\frac{1 \text{ atm}}{760 \text{ mmHg}} \quad \frac{1 \text{ L}}{1000 \text{ mL}} \quad PV = nRT \quad \frac{2 \text{ mol } SO_3}{2 \text{ mol } SO_2}$

and $\text{mL}_{O2} \rightarrow \text{L}_{O2}$ **then** $P, V_{SO2}, T \rightarrow n_{O2} \rightarrow n_{SO3}$

$\frac{1 \text{ L}}{1000 \text{ mL}} \quad PV = nRT \quad \frac{2 \text{ mol } SO_3}{1 \text{ mol } O_2}$

Solution: $50.0 \text{ mmHg} \times \frac{1 \text{ atm}}{760 \text{ mmHg}} = 0.0657895 \text{ atm}$ and $285.5 \text{ mL } SO_2 \times \frac{1 \text{L } SO_2}{1000 \text{ mL } SO_2} = 0.2855 \text{ L}$

then $PV = nRT$ Rearrange to solve for n.

$$n_{SO2} = \frac{PV}{RT} = \frac{0.0657895 \text{ atm} \times 0.2855 \text{ L}}{0.08206 \frac{\text{L} \cdot \text{atm}}{\text{mol} \cdot \text{K}} \times 315 \text{ K}} = 7.26642 \times 10^{-4} \text{ mol } SO_2$$

$$7.26642 \times 10^{-4} \text{ mol } SO_2 \times \frac{2 \text{ mol } SO_3}{2 \text{ mol } SO_2} = 7.26642 \times 10^{-4} \text{ mol } SO_3$$

then $158.9\ \cancel{mL\ O_2} \times \frac{1\ L\ O_2}{1000\ \cancel{mL\ O_2}} = 0.1589\ L\ O_2$ then $PV = nRT$ Rearrange to solve for n.

$$n_{O2} = \frac{PV}{RT} = \frac{0.0657895\ \cancel{atm} \times 0.1589\ \cancel{L}}{0.08206\ \frac{\cancel{L} \cdot \cancel{atm}}{mol \cdot \cancel{K}} \times 315\ \cancel{K}} = 4.04425 \times 10^{-4}\ mol\ O_2$$ then

$4.04425 \times 10^{-4}\ \cancel{mol\ O_2} \times \frac{2\ mol\ SO_3}{1\ \cancel{mol\ O_2}} = 8.08851 \times 10^{-4}\ mol\ SO_3$ Since the amount generated from the SO_2 is less, it is the limiting reagent and the theoretical yield is 7.27×10^{-4} mol SO_3.

Check: The units (mol) are correct. The magnitude of the answer (0.0007 mol) makes sense because we have small volumes of gas involved (compared to 22 L).

b) **Given:** above info and $V_{SO3} = 187.2$ mL, $T = 315$ K, and $P = 50.0$ mmHg **Find:** % Yield

Conceptual Plan: mmHg → atm and mL$_{SO3}$ → L$_{SO3}$ then P, V_{SO2}, T → n_{SO2} → n_{SO3}

$\frac{1\ atm}{760\ mmHg}$ $\frac{1\ L}{1000\ mL}$ $PV = nRT$

then actual yield, theoretical yield → % yield

$$\%\ Yield = \frac{actual\ yield}{theoretical\ yield} \times 100\ \%$$

Solution: $50.0\ \cancel{mmHg} \times \frac{1\ atm}{760\ \cancel{mmHg}} = 0.0657895\ atm$ and

$187.2\ \cancel{mL\ SO_3} \times \frac{1\ L\ SO_3}{1000\ \cancel{mL\ SO_3}} = 0.1872\ L\ SO_3$ then $PV = nRT$ Rearrange to solve for n.

$$n_{SO3} = \frac{PV}{RT} = \frac{0.0657895\ \cancel{atm} \times 0.1872\ \cancel{L}}{0.08206\ \frac{\cancel{L} \cdot \cancel{atm}}{mol \cdot \cancel{K}} \times 315\ \cancel{K}} = 4.76453 \times 10^{-4}\ mol\ SO_3$$

then $\%\ Yield = \frac{actual\ yield}{theoretical\ yield} \times 100\ \% = \frac{4.76453 \times \cancel{10^{-4}\ mol\ SO_3}}{7.26642 \times \cancel{10^{-4}\ mol\ SO_3}} \times 100\ \% = 65.57\ \%$

Check: The units (%) are correct. The magnitude of the answer (66 %) makes sense because it should be between 0 and 100 %. The volume of product is a bit over half of the volume of the limiting reagent and since they and a 2:2 mole ratio, we expect a number a bit over 50 %.

73. **Given:** $T = 22$ °C, $P = 1.02$ atm, and $m = 11.83$ g **Find:** V_{Total}

Conceptual Plan: °C → K and g$_{(NH4)2CO3}$ → $n_{(NH4)2CO3}$ → n_{Gas} then P, n, T → V

$K = °C + 273.15$ $\frac{1\ mol\ (NH_4)_2CO_3}{96.09\ g\ (NH_4)_2CO_3}$ $\frac{(2+1+1=4)\ mol\ gas}{1\ mol\ NH_4CO_3}$ $PV = nRT$

Solution: $T = 22\ °C + 273.15 = 295$ K,

$11.83\ \cancel{g\ NH_4CO_3} \times \frac{1\ \cancel{mol\ (NH_4)_2CO_3}}{96.09\ \cancel{g\ (NH_4)_2CO_3}} \times \frac{4\ mol\ gas}{1\ \cancel{mol\ (NH_4)_2CO_3}} = 0.492455\ mol\ gas$

$PV = nRT$ Rearrange to solve for V_{gas}. $V_{gas} = \frac{nRT}{P} = \frac{0.492455\ \cancel{mol\ gas} \times 0.08206\ \frac{L \cdot \cancel{atm}}{\cancel{mol} \cdot \cancel{K}} \times 295\ \cancel{K}}{1.02\ \cancel{atm}} = 11.7\ L$

Check: The units (L) are correct. The magnitude of the answer (12 L) makes sense because we have about a half a mole of gas generated.

74. **Given:** $T = 125$ °C, $P = 748$ mmHg, and $m = 1.55$ kg **Find:** V_{Total}

Conceptual Plan: °C → K and mmHg → atm and kg$_{NH4NO3}$ → g$_{NH4NO3}$ → n_{NH4NO3} → n_{Gas}

$K = °C + 273.15$ $\frac{1\ atm}{760\ mmHg}$ $\frac{1000\ g}{1\ kg}$ $\frac{1\ mol\ NH_4NO_3}{80.05\ g\ NH_4NO_3}$ $\frac{(2+1+4=7)\ mol\ gas}{2\ mol\ NH_4NO_3}$

then P, n, T → V

$PV = nRT$

Solution: $T = 125\ °C + 273.15 = 398$ K, $748\ \cancel{mmHg} \times \frac{1\ atm}{760\ \cancel{mmHg}} = 0.984211\ atm$

$$1.55\ \cancel{\text{kg NH}_4\text{NO}_3} \times \frac{1000\ \cancel{\text{g NH}_4\text{NO}_3}}{1\ \cancel{\text{kg NH}_4\text{NO}_3}} \times \frac{1\ \cancel{\text{mol NH}_4\text{NO}_3}}{80.05\ \cancel{\text{g NH}_4\text{NO}_3}} \times \frac{7\ \text{mol gas}}{2\ \cancel{\text{mol NH}_4\text{NO}_3}} = 67.\underline{7}701\ \text{mol gas}$$

$PV = nRT$ Rearrange to solve for V_{gas}. $V_{gas} = \frac{nRT}{P} = \frac{67.\underline{7}701\ \cancel{\text{mol gas}} \times 0.08206\ \frac{\text{L}\cdot\cancel{\text{atm}}}{\cancel{\text{mol}}\cdot\cancel{\text{K}}} \times 398\ \cancel{\text{K}}}{0.98\underline{4}211\ \cancel{\text{atm}}} = 2250\ \text{L}$

Check: The units (L) are correct. The magnitude of the answer (2250 L) makes sense because we have about 67 moles of gas generated, so we expect a volume a bit above 67 x 22 L.

75. **Given:** He and air; $V = 855$ mL, $P = 125$ psi, $T = 25$ °C, $\mathfrak{M}$ (air) = 28.8 g/mol **Find:** $\Delta = m(\text{air}) - m(\text{He})$

Conceptual Plan: mL → L and psi → atm and °C → K then *P, T,* $\mathfrak{M}$ → *d*

$\frac{1\ \text{L}}{1000\ \text{mL}}$ $\quad \frac{1\ \text{atm}}{14.7\ \text{psi}}$ $\quad K = °C + 273.15$ $\quad d = \frac{P\mathfrak{M}}{RT}$

then *d, V* → *m* *m*(air), *m*(He) → Δ

$d = \frac{m}{V}$ $\quad \Delta = m(air) - m(He)$

Solution: $V = 855\ \cancel{\text{mL}} \times \frac{1\ \text{L}}{1000\ \cancel{\text{mL}}} = 0.855\ \text{L}$, $P = 125\ \cancel{\text{psi}} \times \frac{1\ \text{atm}}{14.7\ \cancel{\text{psi}}} = 8.5\underline{0}340\ \text{atm}$, $T = 25\ °\text{C} + 273.15 = 298\text{K}$,

$d_{air} = \frac{P\mathfrak{M}}{RT} = \frac{8.5\underline{0}340\ \cancel{\text{atm}} \times 28.8\ \frac{\text{g air}}{\cancel{\text{mol air}}}}{0.08206\ \frac{\text{L}\cdot\cancel{\text{atm}}}{\cancel{\text{K}}\cdot\cancel{\text{mol}}} \times 298\ \cancel{\text{K}}} = 10.\underline{0}147\ \frac{\text{g air}}{\text{L}}$, $d = \frac{m}{V}$ Rearrange to solve for m. $m = dV$

$m_{air} = 10.\underline{0}147\ \frac{\text{g air}}{\cancel{\text{L}}} \times 0.8554\ \cancel{\text{L}} = 8.5\underline{6}657\ \text{g air}$, $d_{He} = \frac{P\mathfrak{M}}{RT} = \frac{8.5\underline{0}340\ \cancel{\text{atm}} \times 4.03\ \frac{\text{g He}}{\cancel{\text{mol He}}}}{0.08206\ \frac{\text{L}\cdot\cancel{\text{atm}}}{\cancel{\text{K}}\cdot\cancel{\text{mol}}} \times 298\ \cancel{\text{K}}} = 1.4\underline{0}136\ \frac{\text{g He}}{\text{L}}$,

$m_{air} = 1.4\underline{0}136\ \frac{\text{g He}}{\cancel{\text{L}}} \times 0.855\ \cancel{\text{L}} = 1.1\underline{9}816\ \text{g He}$, $\Delta = m(air) - m(He) = 8.5\underline{6}657\ \text{g air} - 1.1\underline{9}816\ \text{g He} = 7.37\ \text{g}$

Check: The units (g) are correct. We expect the difference to be less than the difference in the molecular weights since we have less than a mole of gas.

76. **Given:** $V_1 = 2.95$ L, $P = 0.998$ atm, $T_1 = 25.0$ °C, and $T_2 = -196$ °C **Find:** V_2 and compare to 0.61 L

Conceptual Plan: °C → K then V_1, T_1, T_2 → V_2 and then compare to 0.61 L

$K = °C + 273.15$ $\quad \frac{V_1}{T_1} = \frac{V_2}{T_2}$

Solution: $T_1 = 25\ °\text{C} + 273.15 = 298\text{K}$, $T_2 = -196\ °\text{C} + 273.15 = 77\ \text{K}$, $\frac{V_1}{T_1} = \frac{V_2}{T_2}$ Rearrange to solve for V_2.

$V_2 = V_1 \times \frac{T_2}{T_1} = 2.95\ \text{L} \times \frac{77\ \cancel{\text{K}}}{298\ \cancel{\text{K}}} = 0.76\ \text{L}$ This is 25 % larger than the measured volume. We expect gases to behave non-ideally as the temperature drops. We are at the boiling point of the material, so the velocity dramatically decreases and some nitrogen will be condensing.

Check: The units (L) are correct. We expect the volume to dramatically decrease since the temperature has dropped significantly.

77. **Given:** $\text{Flow}_{NO} = 335$ L/s, $P_{NO} = 22.4$ torr, $T_{NO} = 955$ K, $P_{NH3} = 755$ torr, and $T_{NO} = 298$ K, and NH_3 purity = 65.2 % **Find:** Flow_{NH3}

Conceptual Plan: torr → atm then P_{NO}, V_{NO}/s, T_{NO} → n_{NO}/s → n_{NH3}/s (pure)

$\frac{1\ \text{atm}}{760\ \text{torr}}$ $\quad PV = nRT$ $\quad \frac{4\ \text{mol NH}_3}{4\ \text{mol NO}}$

then n_{NH3}/s (pure) → n_{NH3}/s (impure) then n_{NH3}/s (impure), P_{NH3}, T_{NH3} → V_{NH3}/s

$\frac{100\ \text{mol NH}_3\ \text{impure}}{65.2\ \text{mol NH}_3\ \text{pure}}$ $\quad PV = nRT$

Solution: $P_{NO} = 22.4\ \cancel{\text{torr}} \times \frac{1\ \text{atm}}{760\ \cancel{\text{torr}}} = 0.029\underline{4}737\ \text{atm}$, $P_{NH3} = 755\ \cancel{\text{torr}} \times \frac{1\ \text{atm}}{760\ \cancel{\text{torr}}} = 0.99\underline{3}421\ \text{atm}$

$PV = nRT$ Rearrange to solve for n_{NO}. Note that we can substitute V/s for V and get n/s as a result.

$$\frac{n_{NO}}{s}=\frac{PV}{RT}=\frac{0.0294737\ \cancel{\text{atm}} \times 335\ \cancel{\text{L}}/\text{s}}{0.08206\ \frac{\cancel{\text{L}}\cdot\cancel{\text{atm}}}{\text{mol}\cdot\cancel{\text{K}}} \times 955\ \cancel{\text{K}}}=0.125992\ \frac{\text{mol NO}}{\text{s}}$$

$$0.125992\ \frac{\cancel{\text{mol NO}}}{\text{s}} \times \frac{\cancel{4}\ \cancel{\text{mol NH}_3}}{\cancel{4}\ \cancel{\text{mol NO}}} \times \frac{100\ \text{mol NH}_3\ \text{impure}}{65.2\ \cancel{\text{mol NH}_3\ \text{pure}}}=0.193240\ \frac{\text{mol NH}_3\ \text{impure}}{\text{s}} \qquad PV=nRT$$

Rearrange to solve for V_{NH3}. Note that we can substitute n/s for n and get V/s as a result.

$$\frac{V_{NH3}}{s}=\frac{nRT}{P}=\frac{0.193240\ \frac{\cancel{\text{mol NH}_3\ \text{impure}}}{\text{s}} \times 0.08206\ \frac{\text{L}\cdot\cancel{\text{atm}}}{\cancel{\text{mol}}\cdot\cancel{\text{K}}} \times 298\ \cancel{\text{K}}}{0.993421\ \cancel{\text{atm}}}=4.76\ \frac{\text{L}}{\text{s}}\ \text{impure NH}_3$$

Check: The units (L) are correct. The magnitude of the answer (5 L/s) makes sense because we expect it to be less than for the NO. The NO is at a very low concentration and a high temperature, when this converts to a much higher pressure and lower temperature this will go down significantly, even though the ammonia is impure. From a practical standpoint, you would like a low flow rate to make it economical.

78. **Given:** Flow_{NO} = 2.55 L/s, P_{NO} = 12.4 torr, T_{NO} = 655 K, and 8.0 hours **Find:** m (urea)

Conceptual Plan: **torr → atm and hr → min → s then P_{NO}, V_{NO}/s, T_{NO} → n_{NO}/s**

$\frac{1\ \text{atm}}{760\ \text{torr}}$; $\frac{60\ \text{s}}{1\ \text{min}}$; $\frac{60\ \text{min}}{1\ \text{hr}}$; $PV=nRT$

n_{NO}/s → n_{urea}/s then s → n_{urea} → g_{urea}

$\frac{2\ \text{mol urea}}{4\ \text{mol NO}}$; $n_{urea}=(n_{urea}/s)(s)$; $\frac{60.06\ \text{g urea}}{1\ \text{mol urea}}$

Solution: $P_{NO}=12.4\ \cancel{\text{torr}} \times \frac{1\ \text{atm}}{760\ \cancel{\text{torr}}}=0.01631579\ \text{atm}$, $8.0\ \cancel{\text{hr}} \times \frac{60\ \cancel{\text{min}}}{1\ \cancel{\text{hr}}} \times \frac{60\text{s}}{1\ \cancel{\text{min}}}=28800\ \text{s}$

$PV=nRT$ Rearrange to solve for n_{NO}. Note that we can substitute V/s for V and get n/s as a result.

$$\frac{n_{NO}}{s}=\frac{PV}{RT}=\frac{0.01631579\ \text{atm}\ \cancel{\text{atm}} \times 2.55\ \cancel{\text{L}}/\text{s}}{0.08206\ \frac{\cancel{\text{L}}\cdot\cancel{\text{atm}}}{\text{mol}\cdot\cancel{\text{K}}} \times 655\ \cancel{\text{K}}}=0.000774061\ \frac{\text{mol NO}}{\text{s}},$$

$$0.000774061\ \frac{\cancel{\text{mol NO}}}{\text{s}} \times \frac{2\ \cancel{\text{mol urea}}}{4\ \cancel{\text{mol NO}}}=0.000387031\ \frac{\text{mol urea}}{\text{s}},$$

$$28800\ \cancel{\text{s}} \times 0.000387031\ \frac{\cancel{\text{mol urea}}}{\cancel{\text{s}}} \times \frac{60.06\ \text{g urea}}{1\ \cancel{\text{mol urea}}}=670\ \text{g urea}$$

Check: The units (g) are correct. The magnitude of the answer (670 g) is not unreasonable mass to add to a car because many more grams of gasoline are burned in 8 hours of driving.

79. **Given:** l = 30.0 cm, w = 20.0 cm, h = 15.0 cm, 14.7 psi **Find:** Force (lbs)

Conceptual Plan: **l, w, h → Surface Area, SA (cm^2) → Surface Area(in^2) → Force**

$SA=2(lh)+2(wh)+2(lw)$; $\frac{(1\ \text{in})^2}{(2.54\ \text{cm})^2}$; $\frac{14.7\ \text{lbs}}{1\ \text{in}^2}$

Solution:

$SA=2(lh)+2(wh)+2(lw)=2(30.0\ \text{cm} \times 15.0\ \text{cm})+2(20.0\ \text{cm} \times 15.0\ \text{cm})+2(30.0\ \text{cm} \times 20.0\ \text{cm})=2700\ \text{cm}^3$

$2700\ \cancel{\text{cm}^2} \times \frac{(1\ \text{in})^2}{(2.54\ \cancel{\text{cm}})^2}=418.50\ \text{in}^2$, $418.50\ \cancel{\text{in}^2} \times \frac{14.7\ \text{lbs}}{1\ \cancel{\text{in}^2}}=6150\ \text{lbs}$ The can would be crushed.

Check: The units (lbs) are correct. The magnitude of the answer (6150 lbs) is not unreasonable since there is a large surface area.

80. **Given:** l = 20.0 cm, r = 10.0 cm, 25 mL with d = 0.807 g/ml, P_1 = 760.0 mmHg = 1.000 atm

Find: Force (lbs)

Conceptual Plan: **mL → g → mol then l, r → V(cm^3) → V(L) then V, n, T → P_{N2} then**

$d=\frac{m}{V}$; $\frac{1\ \text{mol}}{28.02\ \text{g}}$; $V=\pi r^2 l$; $\frac{1\ \text{L}}{1000\ \text{cm}^3}$; $PV=nRT$

P_{N2}, P_{atm} → P_{Total} then atm → psi l, r → Surface Area(cm^2) → Surface Area(in^2) → Force

$P_{Total}=P_{atm}+P_{N2}$; $\frac{14.7\ \text{psi}}{1\ \text{atm}}$; $SA=2\pi rl+2\pi r^2$; $\frac{(1\ \text{in})^2}{(2.54\ \text{cm})^2}$; $P=\frac{F}{A}$

Solution: $d=\frac{m}{V}$ Rearrange to solve for m. $m=dV=0.807\ \frac{\text{g}}{\cancel{\text{mL}}} \times 25\ \cancel{\text{mL}}=20.175\ \text{g}$,

$20.175\ \text{g} \times \frac{1\ \text{mol}}{28.02\ \text{g}} = 0.72002\ \text{mol}$, $V = \pi r^2 l = \pi \times (10.0\ \text{cm})^2 \times 20.0\ \text{cm} = 6283.19\ \text{cm}^3$,

$6283.19\ \text{cm}^3 \times \frac{1\ \text{L}}{1000\ \text{cm}^3} = 6.28319\ \text{L}$, $PV = nRT$ Rearrange to solve for P.

$$P_{N2} = \frac{nRT}{V} = \frac{0.72002\ \text{mol} \times 0.08206\ \frac{\text{L}\cdot\text{atm}}{\text{mol}\cdot\text{K}} \times 298\ \text{K}}{6.28319\ \text{L}} = 2.80229\ \text{atm},$$

$P_{\text{Total}} = P_{\text{atm}} + P_{\text{N2}} = 1.000\ \text{atm} + 2.80229\ \text{atm} = 3.80229\ \text{atm}$, $3.80229\ \text{atm} \times \frac{14.7\ \text{psi}}{1\ \text{atm}} = 55.894\ \text{psi}$

$SA = 2\pi rl + 2\pi r^2 = (2 \times \pi \times 10.0\ \text{cm} \times 20.0\ \text{cm}) + (2 \times \pi \times (10.0\ \text{cm})^2) = 1884.956\ \text{cm}^2$,

$1884.956\ \text{cm}^2 \times \frac{(1\ \text{in})^2}{(2.54\ \text{cm})^2} = 292.169\ \text{in}^2$, $P = \frac{F}{A}$ Rearrange to solve for F.

$292.169\ \text{in}^2 \times \frac{55.894\ \text{lbs}}{1\ \text{in}^2} = 1.6 \times 10^4\ \text{lbs}$

Check: The units (lbs) are correct. The magnitude of the answer (16,000 lbs) is not unreasonable since there is a large surface area and this is a high pressure.

81. **Given:** $V_1 = 160.0$ L, $P_1 = 1855$ psi, 3.5 L/balloon, $P_2 = 1.0$ atm = 14.7 psi and $T = 298$ K
Find: # balloons
Conceptual Plan: $V_1, P_1, P_2 \rightarrow V_2$ then L → # balloons

$P_1V_1 = P_2V_2$; $\frac{1\ \text{balloon}}{3.5\ \text{L}}$

Solution: $P_1V_1 = P_2V_2$ Rearrange to solve for V_2. $V_2 = \frac{P_1}{P_2}V_1 = \frac{1855\ \text{psi}}{14.7\ \text{psi}} \times 160.0\ \text{L} = 20190.5\ \text{L}$,

$20190.5\ \text{L} \times \frac{1\ \text{balloon}}{3.5\ \text{L}} = 5800\ \text{balloons}$

Check: The units (balloons) are correct. The magnitude of the answer (5800) is reasonable since a store does not want to buy a new helium tank very often.

82. **Given:** 11.5 mL with $d = 0.573$ g/ml, $T = 28.5$ °C, $P = 892$ torr **Find:** V
Conceptual Plan: mL → g → mol and °C, → K and torr → atm then $P, n, T \rightarrow V$

$d = \frac{m}{V}$; $\frac{1\ \text{mol}}{58.12\ \text{g}}$; K = °C + 273.15 ; $\frac{1\ \text{atm}}{760\ \text{torr}}$; $PV = nRT$

Solution: $d = \frac{m}{V}$ Rearrange to solve for m. $m = dV = 0.573\ \frac{\text{g}}{\text{mL}} \times 11.5\ \text{mL} = 6.5895\ \text{g}$,

$6.5895\ \text{g} \times \frac{1\ \text{mol}}{58.12\ \text{g}} = 0.113377\ \text{mol}$, $T = 28.5\text{C} + 273.15 = 301.7\ \text{K}$, $892\ \text{torr} \times \frac{1\ \text{atm}}{760\ \text{torr}} = 1.17368\ \text{atm}$

$$PV = nRT \quad \text{Rearrange to solve for } V.\quad V = \frac{nRT}{P} = \frac{0.113377\ \text{mol} \times 0.08206\ \frac{\text{L}\cdot\text{atm}}{\text{mol}\cdot\text{K}} \times 301.7\ \text{K}}{1.17368\ \text{atm}} = 2.39\ \text{L}$$

Check: The units (L) are correct. The magnitude of the answer (1 L) is reasonable since there is a lot less than one mole of butane.

83. **Given:** $r_1 = 2.5$ cm, $P_1 = 4.00$ atm, $T = 298$ K, and $P_2 = 1.00$ atm **Find:** r_2
Conceptual Plan: $r_1 \rightarrow V_1$ $V_1, P_1, P_2 \rightarrow V_2$ then $V_2 \rightarrow r_2$

$V = \frac{4}{3}\pi r^3$; $P_1V_1 = P_2V_2$; $V = \frac{4}{3}\pi r^3$

Solution: $V = \frac{4}{3}\pi r^3 = \frac{4}{3} \times \pi \times (2.5\ \text{cm})^3 = 65.450\ \text{cm}^3$ $P_1V_1 = P_2V_2$ Rearrange to solve for V_2.

$V_2 = \frac{P_1}{P_2}V_1 = \frac{4.00\ \text{atm}}{1.00\ \text{atm}} \times 65.450\ \text{cm}^3 = 261.80\ \text{cm}^3$, $V = \frac{4}{3}\pi r^3$

Rearrange to solve for r. $r = \sqrt[3]{\frac{3V}{4\pi}} = \sqrt[3]{\frac{3 \times 261.80\ \text{cm}^3}{4 \times \pi}} = 4.0\ \text{cm}$

Check: The units (cm) are correct. The magnitude of the answer (4 cm) is reasonable since the bubble will expand as the pressure is decreased.

84. **Given:** max SA = 1257 cm^2, V_1 = 3.0 L, P_1 = 755 torr, T_1 = 298 K, T_2 = 273 K
Find: P_2 to burst balloon
Conceptual Plan: **torr → atm and A → r → V(cm^3) → V(L) then**

$$\frac{1 \text{ atm}}{760 \text{ torr}} \qquad SA = 4\pi r^2 \qquad V = \frac{4}{3}\pi r^3 \qquad \frac{1 \text{ L}}{1000 \text{ cm}^3}$$

$P_1, V_1, P_2, T_1, V_2, T_2$ → P_2

$$\frac{P_1V_1}{T_1} = \frac{P_2V_2}{T_2}$$

Solution: $P_1 = 755 \text{ torr} \times \frac{1 \text{ atm}}{760 \text{ torr}} = 0.993421 \text{ atm}$, $SA = 4\pi r^2$ Rearrange to solve for r.

$r = \sqrt{\frac{SA}{4\pi}} = \sqrt{\frac{1257 \text{cm}^2}{4\pi}} = 10.00144 \text{ cm}$, $V = \frac{4}{3}\pi r^3 = \frac{4}{3} \times \pi \times (10.00144 \text{ cm})^3 = 4190.600 \text{ cm}^3$

$4190.600 \text{ cm}^3 \times \frac{1 \text{ L}}{1000 \text{ cm}^3} = 4.190600 \text{ L}$ $\quad \frac{P_1V_1}{T_1} = \frac{P_2V_2}{T_2}$ Rearrange to solve for P_2.

$$P_2 = P_1 \frac{V_1}{V_2}\frac{T_2}{T_1} = 0.993421 \text{ atm} \times \frac{3.00 \text{ L}}{4.190600 \text{ L}} \times \frac{273 \text{ K}}{298 \text{ K}} = 0.652 \text{ atm}$$

Check: The units (atm) are correct. The magnitude of the answer (0.65 atm) is reasonable since the pressure must decrease in order for the balloon to expand.

85. **Given:** 2.0 mol CO : 1.0 mol O_2, V = 2.45 L, P_1 = 745 torr, P_2 = 552 torr, and T = 5552 °C
Find: % reacted
Conceptual Plan: from $PV = nRT$ we know that P α n, looking at the chemical reaction we see that 2 + 1 = 3 moles of gas gets converted to 2 moles of gas. If all the gas reacts, P_2 = 2/3 P_1.
Calculate –ΔP for 100 % reacted and for actual case. Then calculate % reacted.

$$-\Delta P \text{ 100 \% reacted} = P_1 - \frac{2}{3}P_1 \qquad -\Delta P \cdot \text{actual} = P_1 - P_2 \qquad \text{\% reacted} = \frac{\Delta P \text{ actual}}{\Delta P \text{ 100\% reacted}} \times 100\%$$

Solution: $-\Delta P \text{ 100 \% reacted} = P_1 - \frac{2}{3}P_1 = 745 \text{ torr} - \frac{2}{3} 745 \text{ torr} = 248.333 \text{ torr}$,

$-\Delta P \text{ actual} = P_1 - P_2 = 745 \text{ torr} - 552 \text{ torr} = 193 \text{ torr}$,

$$\text{\% reacted} = \frac{\Delta P \text{ actual}}{\Delta P \text{ 100\% reacted}} \times 100\% = \frac{193 \text{ torr}}{248.333 \text{ torr}} \times 100\% = 77.7\%$$

Check: The units (%) are correct. The magnitude of the answer (78 %) makes sense because the pressure dropped most of the way to the pressure if all of the reactants had reacted. Note: There are many ways to solve this problem, including calculating the moles of reactants and products using $PV = nRT$.

86. **Given:** N_2, V_1 = 1.0 L, P_1 = 1.0 atm, T_1 = 300. K, and V_2 = 3.0 L, **Find:** d_2
Conceptual Plan: **$\mathfrak{M}$, V_1, P_1, T_1 → d_1 → d_2**

$$d = \frac{P\mathfrak{M}}{RT} \qquad d = \frac{m}{V}$$

Solution: $d_1 = \frac{P\mathfrak{M}}{RT} = \frac{1.0 \text{ atm} \times 28.02 \frac{\text{g}}{\text{mol}}}{0.08206 \frac{\text{L atm}}{\text{K mol}} \times 300. \text{ K}} = 1.13819 \frac{\text{g}}{\text{L}}$, $d = \frac{m}{V}$

Since we have a sealed container, $m_1 = m_2$. Rearrange to solve for m. $m = dV$ or $m = d_1V_1 = d_2V_2$

Rearrange to solve for d_2. $d_2 = d_1\frac{V_1}{V_2} = 1.13819 \frac{\text{g}}{\text{L}} \times \frac{1.0 \text{ L}}{3.0 \text{ L}} = 0.379 \frac{\text{g}}{\text{L}}$

Check: The units (g/L) are correct. The magnitude of the answer (0.4 g/L) is a typical gas density. The density dropped as the volume went up.

87. **Given:** P(Total)$_1$ = 2.2 atm = CO + O_2, P(Total)$_2$ = 1.9 atm = CO + O_2 + CO_2, V = 1.0 L, T = 1.0 x 10^3 K
Find: CO_2 made
Conceptual Plan: P(Total)$_1$ = 2.2 atm = P(CO)$_1$ + P(O_2)$_1$, P(Total)$_2$ = 1.9 atm = P(CO)$_2$ + P(O_2)$_2$ + P(CO_2)$_2$, Let x = amount of P(O_2) reacted. From stoichiometry: P(CO)$_2$ = P(CO)$_1$ – 2x, P(O_2)$_2$ = P(O_2)$_1$ – x, P(CO_2)$_2$ = 2x. Thus P(Total)$_2$ = 1.9 atm = P(CO)$_1$ – 2x + P(O_2)$_1$ – x + 2x = P(Total)$_1$ – x. Using the initial conditions: 1.9 atm = 2.2 atm – x. So x = 0.3 atm and since 2x = P(CO_2)$_2$ = 0.6 atm then

P, V, T* → *n

$PV = nRT$

Solution: $PV = nRT$ Rearrange to solve for *n*. $n = \frac{PV}{RT} = \frac{0.6\ \cancel{\text{atm}} \times 1.0\ \cancel{\text{L}}}{0.08206\ \frac{\cancel{\text{L}} \cdot \cancel{\text{atm}}}{\text{mol} \cdot \cancel{\text{K}}} \times 1000\ \cancel{\text{K}}} = 0.007\ \text{mol}$

Check: The units (mol) are correct. The magnitude of the answer (0.007 mol) makes sense because we have such a small volume, at a very high temperature and such a small pressure. All of these lead us to expect a very small number of moles.

88. **Given:** $r = 1.3 \times 10^{-8}$ cm, $V = 100.$ mL, $P = 1.0$ atm, $T_1 = 273$ K
Find: *V* fraction occupied by Xe atoms
Conceptual Plan: mL → L then *V* → *n* → atoms then *r* → *V* (cm³)/atom

$\frac{1\ \text{L}}{1000\ \text{mL}}$ $\frac{1\ \text{mol}}{22.414\ \text{L}}$ at STP 6.022×10^{23} atoms/mol $V = \frac{4}{3}\pi r^3$

then atoms, *V* (cm³)/atom → *V*(Xe) then *V*(Xe), *V*(container) → Fraction Xe

$V(Xe) = (V / atom)(atoms)$ $\%\ V(Xe) = \frac{V(Xe)}{V(container)} \times 100\ \%$

Solution: $100\ \cancel{\text{mL}} \times \frac{1\ \text{L}}{1000\ \cancel{\text{mL}}} = 0.100\ \text{L}$,

$0.100\ \cancel{\text{L}} \times \frac{1\ \text{mol}}{22.414\ \cancel{\text{L}}} = 0.0044\underline{6}149\ \cancel{\text{mol}} \times 6.022 \times 10^{23}\ \frac{\text{atoms}}{\cancel{\text{mol}}} = 2.6\underline{8}671 \times 10^{21}$ atoms

$V = \frac{4}{3}\pi r^3 = \frac{4}{3} \times \pi \times (1.3 \times 10^{-8}\ \text{cm})^3 = 9.\underline{2}028 \times 10^{-24}\ \text{cm}^3 / \text{atom}$,

$V(Xe) = (V / atom)(atoms) = \frac{9.\underline{2}028 \times 10^{-24}\text{cm}^3}{\cancel{\text{atom}}} \times 2.6\underline{8}671 \times 10^{21}\ \cancel{\text{atoms}} = 0.02\underline{4}725\ \text{cm}^3$

$\%\ V(Xe) = \frac{V(Xe)}{V(container)} \times 100\ \% = \frac{0.02\underline{4}725\ \cancel{\text{cm}^3}}{100\ \cancel{\text{cm}^3}} \times 100\ \% = 0.025\ \%\text{V}$

Check: The units (%*V*) are correct. The magnitude of the answer (0.025 %*V*) is reasonable since we expect the molecules to take up very little of the volume of a container of a gas.

89. **Given:** $h_1 = 22.6$ m, $T_1 = 22$ °C, and $h_2 = 23.8$ m **Find:** T_2
Conceptual Plan: °C → K since $V_{cylinder}\ \alpha\ h$ we do not need to know r to use $V_1, T_1, T_2 \rightarrow V_2$

$K = °C + 273.15$ $V = \pi r^2 h$ $\frac{V_1}{T_1} = \frac{V_2}{T_2}$

Solution: $T_1 = 22\ °C + 273.15 = 295K$, $\frac{V_1}{T_1} = \frac{V_2}{T_2}$ Rearrange to solve for T_2.

$T_2 = T_1 \times \frac{V_2}{V_1} = T_1 \times \frac{\cancel{\pi r^2} l_2}{\cancel{\pi r^2} l_1} = 295\ \text{K} \times \frac{23.8\ \cancel{\text{m}}}{22.6\ \cancel{\text{m}}} = 311\ \text{K}$

Check: The units (K) are correct. We expect the temperature to increase since the volume increased.

90. **Given:** $m\ (CH_4) = 8.0$ g, m (Xe) = 8.0 g, $P_{Total} = 0.44$ atm **Find:** P_{CH4}
Conceptual Plan: g → mol then $n_{CH4}, n_{Xe} \rightarrow \chi_{CH4}$ then $\chi_{CH4}, P_{Total} \rightarrow P_{CH4}$

$\mathfrak{M}$ $\chi_{CH4} = \frac{n_{CH4}}{n_{CH4} + n_{Xe}}$ $P_{CH4} = \chi_{CH4} P_{Total}$

Solution: $n_{CH4} = 8.0\ \cancel{\text{g}} \times \frac{1\ \text{mol}}{16.04\ \cancel{\text{g}}} = 0.4\underline{9}869\ \text{mol}$, $n_{Xe} = 8.0\ \cancel{\text{g}} \times \frac{1\ \text{mol}}{131.3\ \cancel{\text{g}}} = 0.06\underline{0}929\ \text{mol}$,

$\chi_{CH4} = \frac{n_{CH4}}{n_{CH4} + n_{Xe}} = \frac{0.4\underline{9}869\ \text{mol}}{0.4\underline{9}869\ \text{mol} + 0.06\underline{0}929\ \text{mol}} = 0.8\underline{9}112$,

$P_{CH4} = \chi_{CH4} P_{Total} = 0.8\underline{9}112 \times 0.44\ \text{atm} = 0.39\ \text{atm}$

Check: The units (atm) are correct. The magnitude of the answer (0.22 atm) makes sense because the molecular weight of methane is so much lighter than xenon, so we have so many more moles of methane. The partial pressure of methane is almost as large as the total pressure.

91. **Given:** He, $V = 0.35$ L, $P_{max} = 88$ atm, and $T = 299$ K **Find:** m_{He}

Conceptual Plan: $P, V, T \rightarrow n$ then mol $\rightarrow$ g

$PV = nRT$ $\mathfrak{M}$

Solution: $PV = nRT$ Rearrange to solve for n.

$$n_{He} = \frac{PV}{RT} = \frac{88\ \text{atm} \times 0.35\ \text{L}}{0.08206\ \frac{\text{L}\cdot\text{atm}}{\text{mol}\cdot\text{K}} \times 299\ \text{K}} = 1.2553\ \text{mol}, \quad 1.2553\ \text{mol} \times \frac{4.003\ \text{g}}{1\ \text{mol}} = 5.0\ \text{g He}$$

Check: The units (g) are correct. The magnitude of the answer (5 g) makes sense because the high pressure and the low volume cancel out (remember 22 L / mol at STP) and so we expect ~ 1 mol and so ~ 4 g.

92. **Given:** NaH + water, $V = 0.490$ L, $P_{Total} = 758$ mmHg, and $T = 35$ °C **Find:** m_{H2} and m_{NaH}

Other: $P_{H2O} = 42.23$ mmHg at 35 °C

Conceptual Plan: °C $\rightarrow$ K and P_{Total}(mmHg) $\rightarrow$ P_{H2}(mmHg) $\rightarrow$ P_{H2}(atm) then

K = °C + 273.15 $P_{Total} = P_{H2O} + P_{H2}$ $\frac{1\ \text{atm}}{760\ \text{mm Hg}}$

Write balanced reaction

NaH *(s)* + H_2O *(l)* $\rightarrow$ NaOH *(aq)* + H_2 *(g)*

$P, V, T \rightarrow n_{H2}$ then $\text{mol}_{H2} \rightarrow g_{H2}$ then $\text{mol}_{H2} \rightarrow \text{mol}_{NaH} \rightarrow g_{NaH}$

$PV = nRT$ $\frac{2.016\ \text{g}}{1\ \text{mol}}$ $\frac{1\ \text{mol NaH}}{1\ \text{mol H}_2}$ $\frac{24.0\ \text{g}}{1\ \text{mol}}$

Solution: $T = 35°\text{C} + 273.15 = 308$ K, $P_{Total} = P_{H2O} + P_{H2}$ Rearrange to solve for P_{H2}.

$$P_{H2} = 758\ \text{mmHg} - 42.23\ \text{mmHg} = 715.77\ \text{mmHg}, \quad 715.77\ \text{mmHg} \times \frac{1\ \text{atm}}{760\ \text{mmHg}} = 0.941803\ \text{atm}$$

$PV = nRT$ Rearrange to solve for n.
$$n_{H2} = \frac{PV}{RT} = \frac{0.941803\ \text{atm} \times 0.490\ \text{L}}{0.08206\ \frac{\text{L}\cdot\text{atm}}{\text{mol}\cdot\text{K}} \times 308\ \text{K}} = 0.0182589\ \text{mol},$$

$$0.0182589\ \text{mol} \times \frac{2.016\ \text{g}}{1\ \text{mol}} = 0.0368\ \text{g H}_2 \text{ and } 0.0182589\ \text{mol H}_2 \times \frac{1\ \text{mol NaH}}{1\ \text{mol H}_2} = 0.0182589\ \text{mol NaH},$$

$$0.0182589\ \text{mol NaH} \times \frac{24.0\ \text{g}}{1\ \text{mol}} = 0.438\ \text{g NaH}$$

Check: The units (g) are correct. The magnitude of the answer (0.04 g and 0.4 g) makes sense because we have much less than a mole of each material (remember 22 L / mol at STP) and so we expect < 2 g gas and < 24 g solid.

93. **Given:** 15.0 mL HBr in 1.0 min; and 20.3 mL unknown hydrocarbon gas in 1.0 min

Find: formula of unknown gas

Conceptual Plan: Since these are gases under the same conditions $V \alpha n$, V, time $\rightarrow$ Rate then

$Rate = \frac{V}{time}$

$\mathfrak{M}$ (HBr), Rate (HBr), Rate (Unk) $\rightarrow$ $\mathfrak{M}$ (Unk)

$\frac{Rate(HBr)}{Rate(U)} = \sqrt{\frac{\mathfrak{M}(U)}{\mathfrak{M}(HBr)}}$

Solution: $Rate(HBr) = \frac{V}{time} = \frac{15.0\ \text{mL}}{1.0\ \text{min}} = 15.0\ \frac{\text{mL}}{\text{min}}$, $Rate(Unk) = \frac{V}{time} = \frac{20.3\ \text{mL}}{1.0\ \text{min}} = 20.3\ \frac{\text{mL}}{\text{min}}$,

$\frac{Rate(HBr)}{Rate(Unk)} = \sqrt{\frac{\mathfrak{M}(Unk)}{\mathfrak{M}(HBr)}}$ Rearrange to solve for $\mathfrak{M}(Unk)$.

$$\mathfrak{M}(Unk) = \mathfrak{M}(HBr)\left(\frac{Rate(HBr)}{Rate(Unk)}\right)^2 = 80.91\ \frac{\text{g}}{\text{mol}} \times \left(\frac{15.0\ \frac{\text{mL}}{\text{min}}}{20.3\ \frac{\text{mL}}{\text{min}}}\right)^2 = 44.2\ \frac{\text{g}}{\text{mol}}$$ The formula is C_3H_8, propane.

Check: The units (g/mol) are correct. The magnitude of the answer (< HBr) makes sense because the unknown diffused faster and so must be lighter.

94. **Given:** 9.0×10^{12} kg/yr octane; atm = 387 ppm CO_2 by volume; atm thickness = 15 km; $r_{Earth} = 6371$ km, $P_{atm} = 381$ torr; $T_{atm} = 275$ K **Find:** $m(CO_2)$ and % increase in CO_2

Conceptual Plan: Write a balanced chemical reaction $\text{kg}_{C3H8} \rightarrow \text{g}_{C3H8} \rightarrow \text{mol}_{C3H8} \rightarrow \text{mol}_{CO2}$

$2\ C_3H_8\ (g) + 25\ O_2\ (g) \rightarrow 16\ CO_2\ (g) + 18\ H_2O\ (l)$ $\frac{1000\ \text{g}}{1\ \text{kg}}$ $\frac{1\ \text{mol}}{114.22\ \text{g}}$ $\frac{16\ \text{mol CO}_2}{2\ \text{mol C}_3\text{H}_8}$

then mol_{CO2} → g_{CO2} and ppm_{CO2} → χ_{CO2} → P_{CO2} torr → atm then r_{Earth} → V_{Earth} and

$\frac{44.01\ g}{1\ mol}$ $\frac{1\ part}{10^6\ parts}$ $P_{CO2} = \chi_{CO2} P_{atm}$ $\frac{1\ atm}{760\ torr}$ $V = \frac{4}{3}\pi r^3$

r_{Earth}, atm thickness → $r_{Earth+atm}$ then $r_{Earth+atm}$ → $V_{Earth+atm}$ $V_{Earth+atm}$, V_{Earth} → V_{atm} then

$r_{Earth+atm} = r_{Earth} + r_{atm}$ $V = \frac{4}{3}\pi r^3$ $V_{atm} = V_{Earth+atm} - V_{Earth}$

km^3 → m^3 → cm^3 → L then V_{atm}, P_{CO2}, T_{atm} → n_{CO2} → g_{CO2} and

$\left(\frac{1000\ m}{1\ km}\right)^3$ $\left(\frac{100\ cm}{1\ m}\right)^3$ $\frac{1\ L}{1000\ cm^3}$ $PV = nRT$ $\frac{44.01\ g}{1\ mol}$

$g_{CO2added}$, $g_{CO2initially}$ → % $increase_{CO2}$

$\%\ increase = \frac{added}{initial} \times 100\ \%$

Solution: $9.0 \times 10^{12}\ kg \times \frac{1000\ g}{1\ kg} \times \frac{1\ mol}{114.22\ g} \times \frac{16\ mol\ CO_2}{2\ mol\ C_8H_{18}} \times \frac{44.01\ g}{1\ mol} = 2.7742 \times 10^{16}\ g\ CO_2$ added,

$387\ parts\ CO_2 \frac{1\ part}{10^6\ parts} = 3.87 \times 10^{-4} = \chi_{CO2}$, $P_{CO2} = \chi_{CO2} P_{atm} = 3.87 \times 10^{-4} \times 381\ torr = 0.147447\ torr$,

$0.147447\ torr \times \frac{1\ atm}{760\ torr} = 0.00019401\ atm$, $V_{Earth} = \frac{4}{3}\pi r^3 = \frac{4}{3} \times \pi \times (6371\ km)^3 = 1.08321 \times 10^{12}\ km^3$,

$r_{Earth+atm} = r_{Earth} + r_{atm} = 6371\ km + 15\ km = 6386\ km$,

$V_{Earth+atm} = \frac{4}{3}\pi r^3 = \frac{4}{3} \times \pi \times (6386\ km)^3 = 1.09086 \times 10^{12}\ km^3$,

$V_{atm} = V_{Earth+atm} - V_{Earth} = 1.09086 \times 10^{12}\ km^3 - 1.08321 \times 10^{12}\ km^3 = 7.666 \times 10^9\ km^3$,

$7.666 \times 10^9\ km^3 \times \left(\frac{1000\ m}{1\ km}\right)^3 \times \left(\frac{100\ cm}{1\ m}\right)^3 \times \frac{1\ L}{1000\ cm^3} = 7.666 \times 10^{21}\ L$, $PV = nRT$

Rearrange to solve for n. $n_{CO2\ initial} = \frac{PV}{RT} = \frac{0.00019401\ atm \times 7.666 \times 10^{21}\ L}{0.08206 \frac{L \cdot atm}{mol \cdot K} \times 275\ K} = 6.591 \times 10^{16}\ mol$,

$6.591 \times 10^{16}\ mol \times \frac{44.01\ g}{1\ mol} = 2.9005 \times 10^{18}\ g\ CO_2$

$\%\ increase = \frac{added}{initial} \times 100\ \% = \frac{2.7742 \times 10^{16}\ g\ CO_2}{2.9005 \times 10^{18}\ g\ CO_2} \times 100\ \% = 1\%$ increase

Check: The units (g and %) are correct. The magnitude of the answer (10^{16} g) is reasonable since we started with so much octane and the mass of CO_2 will be larger than the original octane weight since there is so much added oxygen. The % increase is reasonable since the volume of the atmosphere is so large.

95. **Given:** CH_4: V = 155 mL at STP; O_2: V = 885 mL at STP; NO: V = 55.5 mL at STP; mixed in a flask: V = 2.0 L, T = 275 K, and 90.0 % of limiting reagent used. **Find:** P's of all components and P_{Total}.
Conceptual Plan: CH_4: mL → L → mol_{CH4} → mol_{CO2} and

$\frac{1\ L}{1000\ mL}$ $\frac{1\ mol}{22.414\ L}$ $\frac{1\ mol\ CO_2}{5\ mol\ NO}$

O_2: mL → L → mol_{O2} → mol_{CO2} and NO: mL → L → mol_{NO} → mol_{CO2}

$\frac{1\ L}{1000\ mL}$ $\frac{1\ mol}{22.414\ L}$ $\frac{1\ mol\ CO_2}{5\ mol\ O_2}$ $\frac{1\ L}{1000\ mL}$ $\frac{1\ mol}{22.414\ L}$ $\frac{1\ mol\ CO_2}{5\ mol\ NO}$

the smallest yield determines the limiting reagent then initial mol_{NO} → reacted mol_{NO} → final mol_{NO}

NO is the limiting reagent 90.0 % $0.100 \times$ initial mol_{no}

reacted mol_{NO} → reacted mol_{CH4} then initial mol_{CH4}, reacted mol_{CH4} → final mol_{CH4} then

$\frac{1\ mol\ CH_4}{5\ mol\ NO}$ initial mol_{CH4} – reacted mol_{CH4} = final mol_{CH4}

final mol_{CH4}, V, T → final P_{CH4} and reacted mol_{NO} → reacted mol_{O2} then

$PV = nRT$ $\frac{5\ mol\ O_2}{5\ mol\ NO}$

initial mol_{O2}, reacted mol_{O2} → final mol_{O2} then final mol_{O2}, V, T → final P_{O2} and

initial mol_{O2} – reacted mol_{O2} = final mol_{O2} $PV = nRT$

final mol$_{NO}$, V, T → final P_{NO} and theoretical mol$_{CO2}$ from NO → final mol$_{CO2}$

$PV = nRT$ 90.0 %

final mol$_{CO2}$, V, T → P_{CO2} then final mol$_{CO2}$ → mol$_{H2O}$ then mol$_{H2O}$, V, T → P_{H2O} and

$PV = nRT$ $\frac{1 \text{ mol } H_2O}{1 \text{ mol } CO_2}$ $PV = nRT$

final mol$_{CO2}$ → mol$_{NO2}$ then mol$_{NO2}$, V, T → P_{NO2} and final mol$_{CO2}$ → mol$_{OH}$ then

$\frac{1 \text{ mol } NO_2}{1 \text{ mol } CO_2}$ $PV = nRT$ $\frac{2 \text{ mol OH}}{1 \text{ mol } CO_2}$

mol$_{OH}$, V, T → P_{OH} finally P_{CH4}, P_{O2}, P_{NO}, P_{CO2}, P_{H2O}, P_{NO2}, P_{OH} → P_{Ttoal}

$PV = nRT$ $P_{Total} = \sum P$

Solution: CH_4: $155 \cancel{mL} \times \frac{1 \cancel{L}}{1000 \cancel{mL}} \times \frac{1 \cancel{mol\,CH_4}}{22.414 \cancel{L}} \times \frac{1 \text{ mol } CO_2}{1 \cancel{mol\,CH_4}} = 0.0069\underline{1}532 \text{ mol } CO_2$,

O_2: $885 \cancel{mL} \times \frac{1 \cancel{L}}{1000 \cancel{mL}} \times \frac{1 \cancel{mol\,O_2}}{22.414 \cancel{L}} = 0.039\underline{4}842 \text{ mol } O_2 \times \frac{1 \cancel{mol\,CO_2}}{5 \cancel{mol\,O_2}} = 0.0078\underline{9}685 \text{ mol } CO_2$

NO: $55.5 \cancel{mL} \times \frac{1 \cancel{L}}{1000 \cancel{mL}} \times \frac{1 \cancel{mol\,NO}}{22.414 \cancel{L}} \times \frac{1 \text{ mol } CO_2}{5 \cancel{mol\,NO}} = 0.00049\underline{5}226 \text{ mol } CO_2$.

$0.00049\underline{5}226 \text{ mol } CO_2$ is the smallest yield, so NO is the limiting reagent.

$55.5 \cancel{mL} \times \frac{1 \cancel{L}}{1000 \cancel{mL}} \times \frac{1 \text{ mol NO}}{22.414 \cancel{L}} = 0.0024\underline{7}613 \text{ mol NO}$

reacted mol NO = 0.900 x *mol NO* = 0.900 x 0.0024$\underline{7}$613 mol NO = 0.0022$\underline{2}$852 mol NO ,

unreacted mol NO = 0.100 x *mol NO* = 0.100 x 0.0024$\underline{7}$613 mol NO = 0.00024$\underline{7}$613 mol NO ,

$0.0022\underline{2}852 \cancel{mol\,NO} \times \frac{1 \text{ mol } CH_4}{5 \cancel{mol\,NO}} = 0.00044\underline{5}704 \text{ mol } CH_4$ reacted ,

$0.0069\underline{1}532 \text{ mol } CH_4 - 0.00044\underline{5}704 \text{ mol } CH_4 \text{ reacted} = 0.0064\underline{6}962 \text{ mol } CH_4$ then $PV = nRT$

Rearrange to solve for P. $P = \frac{nRT}{V} = \frac{0.0064\underline{6}962 \cancel{mol} \times 0.08206 \frac{\cancel{L} \cdot atm}{\cancel{mol} \cdot \cancel{K}} \times 275 \cancel{K}}{2.0 \cancel{L}} = 0.0730 \text{ atm } CH_4$

$0.0022\underline{2}852 \cancel{mol\,NO} \times \frac{5 \text{ mol } O_2}{5 \cancel{mol\,NO}} = 0.0022\underline{2}852 \text{ mol } O_2$ reacted ,

$0.039\underline{4}842 \text{ mol } O_2 - 0.0022\underline{2}852 \text{ mol } O_2 \text{ reacted} = 0.037\underline{2}557 \text{ mol } O_2$

$$P = \frac{nRT}{V} = \frac{0.037\underline{2}557 \cancel{mol} \times 0.08206 \frac{\cancel{L} \cdot atm}{\cancel{mol} \cdot \cancel{K}} \times 275 \cancel{K}}{2.0 \cancel{L}} = 0.420 \text{ atm } O_2$$

$$P = \frac{nRT}{V} = \frac{0.00024\underline{7}613 \cancel{mol} \times 0.08206 \frac{\cancel{L} \cdot atm}{\cancel{mol} \cdot \cancel{K}} \times 275 \cancel{K}}{2.0 \cancel{L}} = 0.00279 \text{ atm NO} ,$$

$0.0022\underline{2}852 \cancel{mol\,NO} \times \frac{1 \text{ mol } CO_2}{5 \cancel{mol\,NO}} = 0.00044\underline{5}704 \text{ mol } CO_2$

$$P = \frac{nRT}{V} = \frac{0.00044\underline{5}704 \cancel{mol} \times 0.08206 \frac{\cancel{L} \cdot atm}{\cancel{mol} \cdot \cancel{K}} \times 275 \cancel{K}}{2.0 \cancel{L}} = 0.00503 \text{ atm } CO_2$$

$0.0022\underline{2}852 \cancel{mol\,NO} \times \frac{1 \text{ mol } H_2O}{5 \cancel{mol\,NO}} = 0.00044\underline{5}704 \text{ mol } H_2O$

$$P = \frac{nRT}{V} = \frac{0.00044\underline{5}704 \cancel{mol} \times 0.08206 \frac{\cancel{L} \cdot atm}{\cancel{mol} \cdot \cancel{K}} \times 275 \cancel{K}}{2.0 \cancel{L}} = 0.00503 \text{ atm } H_2O$$

$0.0022\underline{2}852 \cancel{mol\,NO} \times \frac{5 \text{ mol } NO_2}{5 \cancel{mol\,NO}} = 0.0022\underline{2}852 \text{ mol } NO_2$

$$P = \frac{nRT}{V} = \frac{0.00222852 \text{ mol} \times 0.08206 \frac{\text{L} \cdot \text{atm}}{\text{mol} \cdot \text{K}} \times 275 \text{ K}}{2.0 \text{ L}} = 0.0251 \text{ atm } NO_2$$

$$0.00222852 \text{ mol NO} \times \frac{2 \text{ mol OH}}{5 \text{ mol NO}} = 0.000891408 \text{ mol OH}$$

$$P = \frac{nRT}{V} = \frac{0.000891408 \text{ mol} \times 0.08206 \frac{\text{L} \cdot \text{atm}}{\text{mol} \cdot \text{K}} \times 275 \text{ K}}{2.0 \text{ L}} = 0.0101 \text{ atm OH}$$

$$P_{Total} = \sum P$$

$= 0.0730 \text{ atm} + 0.420 \text{ atm} + 0.00279 \text{ atm} + 0.00503 \text{ atm} + 0.00503 \text{ atm} + 0.0251 \text{ atm} + 0.0101 \text{ atm}$

$= 0.541 \text{ atm}$

Check: The units (atm) are correct. The magnitude of the answers is reasonable. The limiting reagent has the lowest pressure. The product pressures are in line with the ratios of the stoichiometric coefficients.

96. **Given:** He and air **Find:** % He diffused through balloon wall
Conceptual Plan: $\mathfrak{M}$ **(N_2),** $\mathfrak{M}$ **(O_2) → $\mathfrak{M}$ (air) then $\mathfrak{M}$ (air), $\mathfrak{M}$ (He), % air diffused → % He diffused**

$$\mathfrak{M}(air) = \chi(N_2)\mathfrak{M}(N_2) + \chi(O_2)\mathfrak{M}(O_2) \qquad \frac{Rate(He)}{Rate(air)} = \sqrt{\frac{\mathfrak{M}(air)}{\mathfrak{M}(He)}}$$

Solution: $\mathfrak{M}(air) = \chi(N_2)\mathfrak{M}(N_2) + \chi(O_2)\mathfrak{M}(O_2) = \left(\frac{4}{5} \times 28.02 \text{ g/mol}\right) + \left(\frac{1}{5} \times 32.00 \text{ g/mol}\right) = 28.82 \text{ g/mol}$

$\frac{Rate(He)}{Rate(air)} = \sqrt{\frac{\mathfrak{M}(air)}{\mathfrak{M}(He)}}$ Since Rate α % diffused, substitute % diffused for rate and rearrange to solve for % He diffused. $\% He\ diffused = \% air\ diffused \sqrt{\frac{\mathfrak{M}(air)}{\mathfrak{M}(He)}} = 5.0\% \sqrt{\frac{28.82 \text{ g/mol}}{4.003 \text{ g/mol}}} = 13\%$

Check: The units (%) are correct. The magnitude of the answer (>5%) makes sense because He is lighter and so has the higher diffusion rate.

97. **Given:** $P_{CH4} + P_{C2H6} = 0.53$ atm, $P_{CO2} + P_{H2O} = 2.2$ atm **Find:** χ_{CH4}
Conceptual Plan: Write balanced reactions to determine change in moles of gas for CH_4 and C_2H_6.

$2\ CH_4\ (g) + 4\ O_2\ (g) \rightarrow 4\ H_2O\ (g) + 2\ CO_2\ (g)$ and $2\ C_2H_6\ (g) + 7\ O_2\ (g) \rightarrow 6\ H_2O\ (g) + 4\ CO_2\ (g)$ thus $\frac{6 \text{ mol gases}}{2 \text{ mol } CH_4}$ $\frac{10 \text{ mol gases}}{2 \text{ mol } C_2H_6}$

write expression for final pressure, substituting in data given → χ_{CH4}

$$\chi_{CH4} = \frac{n_{CH4}}{n_{CH4} + n_{C2H6}} \text{ and } \chi_{C2H6} = 1 - \chi_{CH4}$$

$$P_{CH4} = \chi_{CH4} P_{Total} \quad P_{C2H6} = \chi_{C2H6} P_{Total} \quad P_{Final} = \left(\chi_{CH4} P_{Total} \times \frac{6 \text{ mol gases}}{2 \text{ mol } CH_4}\right) + \left((1 - \chi_{CH4}) P_{Total} \times \frac{10 \text{ mol gases}}{2 \text{ mol } C_2H_6}\right)$$

Solution: $P_{Final} = \left(\chi_{CH4} \times 0.53 \text{ atm} \times \frac{6 \text{ mol gases}}{2 \text{ mol } CH_4}\right) + \left((1 - \chi_{CH4}) \times 0.53 \text{ atm} \times \frac{10 \text{ mol gases}}{2 \text{ mol } C_2H_6}\right) = 2.2 \text{ atm}$

Rearrange to solve for $\chi_{CH4} = 0.42$.
Check: The units (none) are correct. The magnitude of the answer (0.42) makes sense because if it were all methane the final pressure would have been 1.59 atm and if it were all ethane the final pressure would have been 2.65 atm. Since we are closer to the latter pressure, we expect the mole fraction of methane to be less than 0.5.

98. **Given:** $P_{C2H2} = 7.8$ kPa initially, $P_{C2H2} + P_{C6H6} = 3.9$ kPa **Find:** fraction of C_2H_2 reacted
Conceptual Plan: Write balanced reaction to determine change in moles of gas.

$3C_2H_2\ (g) \rightarrow C_6H_6\ (g)$ thus $\frac{1 \text{ mol } C_6H_6}{3 \text{ mol } C_2H_2 \textit{ reacted}}$ Since P_{C2H2} α n_{C2H2}, the pressure will drop 2 kPa for every 3 kPa of ethylene that reacts.

$P_{initial}$, P_{final} → P_{drop} **write expression for reacted P_{C2H2}, then**

$$P_{drop} = P_{initial} - P_{final} \qquad reacted\ P_{C2H2} = \Delta P \frac{3 \text{ kPa } C_2H_2 \textit{ reacted}}{2 \text{ kPa } \textit{pressure drop}}$$

reacted P_{C2H2}, initial P_{C2H2} → % C_2H_2 reacted

$$\% \ C_2H_2 \ reacted = \frac{reacted \ P_{C_2H_2}}{initial \ P_{C_2H_2}} \text{ x } 100 \ \%$$

Solution: $P_{drop} = P_{initial} - P_{final} = 7.8 \text{ kPa} - 3.9 \text{ kPa} = 3.9 \text{ kPa}\frac{1}{2}$,

$$reacted \ P_{C2H2} = \Delta P \frac{3 \text{ kPa } C_2H_2 \ reacted}{2 \text{ kPa } pressure \ drop} = 3.9 \text{ kPa x } \frac{3 \text{ kPa } C_2H_2 \ reacted}{2 \text{ kPa } pressure \ drop} = 5.\underline{8}5 \text{ kPa}$$

$$\% \ C_2H_2 \ reacted = \frac{reacted \ P_{C_2H_2}}{initial \ P_{C_2H_2}} \text{ x } 100 \ \% = \frac{5.\underline{8}5 \text{ kPa}}{7.8 \text{ kPa}} \text{ x } 100 \ \% = 75 \ \%$$

Check: The units (%) are correct. The magnitude of the answer (75 %) makes sense because if all the ethylene reacted, the final pressure would have been 2.6 kPa. Since we are most of the way to that, we expect the amount reacted to be higher than 50 %.

99. Since the passengers have more mass than the balloon, they have more momentum than the balloon. The passengers will continue to travel in their original direction longer. The car is slowing so the relative position of the passengers is to move forward and the balloons to move backwards. The opposite happens upon acceleration.

100. B is the limiting reactant (2.0 L of B requires 1.0 L A to completely react). The final container will have 0.5 L A and 2.0 L C, so the final volume will be 2.5 L. The change will be ((2.5 L/3.5 L) x 100 %) – 100 % = – 29 %.

101. Since each gas will occupy 22.414 L / mole at STP and we have 2 moles of gas, we will have a volume of 44.828 L.

102. a) False – all gases have the same average kinetic energy at the same temperature.

b) False – the gases will have the same partial pressures since we have the same number of moles of each.

c) False – the average velocity of the B molecules will be less than that of the A molecules since the B's are heavier.

d) True – since B molecules are heavier they will contribute more to the density ($d = m/V$).

103. Br_2 would deviate the most from ideal behavior since it is the largest of the three.

Chapter 6
Thermochemistry

1. a) **Given:** 3.55×10^4 J **Find:** cal
Conceptual Plan: **J → cal**

$$\frac{1\text{ cal}}{4.184\text{ J}}$$

Solution: $3.55 \times 10^4 \text{ J} \times \frac{1\text{ cal}}{4.184\text{ J}} = 8.48 \times 10^3\text{ cal}$

Check: The units (cal) are correct. The magnitude of the answer (8000) makes physical sense because a calorie is larger than a Joule, so the answer decreases.

b) **Given:** 1025 Cal **Find:** J
Conceptual Plan: **Cal → J**

$$\frac{4184\text{ J}}{1\text{ Cal}}$$

Solution: $1025\text{ Cal} \times \frac{4184\text{ J}}{1\text{ Cal}} = 4.289 \times 10^6\text{ J}$

Check: The units (J) are correct. The magnitude of the answer (10^6) makes physical sense because a Calorie is much larger than a Joule, so the answer increases.

c) **Given:** 355 kJ **Find:** cal
Conceptual Plan: **kJ → J → cal**

$$\frac{1000\text{ J}}{1\text{ kJ}} \quad \frac{1\text{ cal}}{4.184\text{ J}}$$

Solution: $355\text{ kJ} \times \frac{1000\text{ J}}{1\text{ kJ}} \times \frac{1\text{ cal}}{4.184\text{ J}} = 8.48 \times 10^4\text{ cal}$

Check: The units (cal) are correct. The magnitude of the answer (10^4) makes physical sense because a calorie is much smaller than a kJ, so the answer increases.

d) **Given:** 125 kWh **Find:** J
Conceptual Plan: **kWh → J**

$$\frac{3.60 \times 10^6\text{ J}}{1\text{ kWh}}$$

Solution: $125\text{ kWh} \times \frac{3.60 \times 10^6\text{ J}}{1\text{ kWh}} = 4.50 \times 10^8\text{ J}$

Check: The units (J) are correct. The magnitude of the answer (10^8) makes physical sense because a kWh is much larger than a Joule, so the answer increases.

2. a) **Given:** 1.58×10^3 kJ **Find:** kcal
Conceptual Plan: **kJ → J → cal → kcal**

$$\frac{1000\text{ J}}{1\text{ kJ}} \quad \frac{1\text{ cal}}{4.184\text{ J}} \quad \frac{1\text{ kcal}}{1000\text{ cal}}$$

Solution: $1.58 \times 10^3\text{ kJ} \times \frac{1000\text{ J}}{1\text{ kJ}} \times \frac{1\text{ cal}}{4.184\text{ J}} \times \frac{1\text{ kcal}}{1000\text{ cal}} = 378\text{ kcal}$

Check: The units (kcal) are correct. The magnitude of the answer (400) makes physical sense because a kcal is larger than a kJ, so the answer decreases.

b) **Given:** 865 cal **Find:** kJ
Conceptual Plan: **cal → J → kJ**

$$\frac{4.184\text{ J}}{1\text{ cal}} \quad \frac{1\text{ kJ}}{1000\text{ J}}$$

Solution: $865\text{ cal} \times \frac{4.184\text{ J}}{1\text{ cal}} \times \frac{1\text{ kJ}}{1000\text{ J}} = 3.62\text{ kJ}$

Check: The units (kJ) are correct. The magnitude of the answer (4) makes physical sense because a kJ is much larger than a cal, so the answer decreases.

c) **Given:** 1.93×10^4 J **Find:** Cal
Conceptual Plan: **J → Cal**

$$\frac{1\,\text{Cal}}{4184\,\text{J}}$$

Solution: $$1.93 \times 10^4\,\text{J} \times \frac{1\,\text{Cal}}{4184\,\text{J}} = 4.61\,\text{Cal}$$

Check: The units (Cal) are correct. The magnitude of the answer (5) makes physical sense because a J is much smaller than a Cal, so the answer increases.

d) **Given:** 1.8×10^4 kJ **Find:** kWh
Conceptual Plan: **kJ → J → kWh**

$$\frac{1000\,\text{J}}{1\,\text{kJ}} \quad \frac{1\,\text{kWh}}{3.60 \times 10^6\,\text{J}}$$

Solution: $$1.8 \times 10^8\,\text{kJ} \times \frac{1000\,\text{J}}{1\,\text{kJ}} \times \frac{1\,\text{kWh}}{3.60 \times 10^6\,\text{J}} = 5.0 \times 10^4\,\text{kWh}$$

Check: The units (kWh) are correct. The magnitude of the answer (5) makes physical sense because a kWh is much larger than a Joule, so the answer decreases.

3. a) **Given:** 2155 Cal **Find:** J
Conceptual Plan: **Cal → J**

$$\frac{4184\,\text{J}}{1\,\text{Cal}}$$

Solution: $$2155\,\text{Cal} \times \frac{4184\,\text{J}}{1\,\text{Cal}} = 9.017 \times 10^6\,\text{J}$$

Check: The units (J) are correct. The magnitude of the answer (10^6) makes physical sense because a Calorie is much larger than a Joule, so the answer increases.

b) **Given:** 2155 Cal **Find:** kJ
Conceptual Plan: **Cal → J → kWh**

$$\frac{4184\,\text{J}}{1\,\text{Cal}} \quad \frac{1\,\text{kJ}}{1000\,\text{J}}$$

Solution: $$2155\,\text{Cal} \times \frac{4184\,\text{J}}{1\,\text{Cal}} \times \frac{1\,\text{kJ}}{1000\,\text{J}} = 9.017 \times 10^3\,\text{kJ}$$

Check: The units (kJ) are correct. The magnitude of the answer (10^3) makes physical sense because a Calorie is larger than a kJ, so the answer increases.

c) **Given:** 2155 Cal **Find:** kWh
Conceptual Plan: **Cal → J → kWh**

$$\frac{4184\,\text{J}}{1\,\text{Cal}} \quad \frac{1\,\text{kWh}}{3.60 \times 10^6\,\text{J}}$$

Solution: $$2155\,\text{Cal} \times \frac{4184\,\text{J}}{1\,\text{Cal}} \times \frac{1\,\text{kWh}}{3.60 \times 10^6\,\text{J}} = 2.50\,\text{kWh}$$

Check: The units (kWh) are correct. The magnitude of the answer (3) makes physical sense because a Calorie is much smaller than a kWh, so the answer decreases.

4. a) **Given:** 655 kWh **Find:** J
Conceptual Plan: **kWh → J**

$$\frac{3.60 \times 10^6\,\text{J}}{1\,\text{kWh}}$$

Solution: $$655\,\text{kWh} \times \frac{3.60 \times 10^6\,\text{J}}{1\,\text{kWh}} = 2.36 \times 10^9\,\text{J}$$

Check: The units (J) are correct. The magnitude of the answer (10^8) makes physical sense because a kWh is much larger than a Joule, so the answer increases.

b) **Given:** 655 kWh **Find:** kJ
Conceptual Plan: **kWh → J → kJ**

$$\frac{3.60 \times 10^6 \text{ J}}{1 \text{kWh}} \qquad \frac{1 \text{kJ}}{1000 \text{ J}}$$

Solution: $$655 \cancel{\text{kWh}} \times \frac{3.60 \times 10^6 \cancel{\text{J}}}{1 \cancel{\text{kWh}}} \times \frac{1 \text{ kJ}}{1000 \cancel{\text{J}}} = 2.36 \times 10^6 \text{ J}$$

Check: The units (J) are correct. The magnitude of the answer (10^8) makes physical sense because a kWh is much larger than a Joule, so the answer increases.

c) **Given:** 655 kWh **Find:** Cal
Conceptual Plan: **kWh → J → Cal**

$$\frac{3.60 \times 10^6 \text{ J}}{1 \text{kWh}} \qquad \frac{1 \text{Cal}}{4184 \text{ J}}$$

Solution: $$655 \cancel{\text{kWh}} \times \frac{3.60 \times 10^6 \cancel{\text{J}}}{1 \cancel{\text{kWh}}} \times \frac{1 \text{Cal}}{4184 \cancel{\text{J}}} = 5.64 \times 10^5 \text{ Cal}$$

Check: The units (Cal) are correct. The magnitude of the answer (10^5) makes physical sense because a kWh is much larger than a Cal, so the answer increases.

5. d) $\Delta E_{sys} = - \Delta E_{surr}$

6. The sign is positive since the energy is being taken in or deposited into the system.

7. a) The energy exchange is primarily heat since the skin (part of the surroundings) is cooled. There is a small expansion (work) since water is being converted from a liquid to a gas. The sign of ΔE_{sys} is positive since the surroundings cool.

   b) The energy exchange is primarily work. The sign of ΔE_{sys} is negative since the system is expanding (doing work on the surroundings).

   c) The energy exchange is primarily heat. The sign of ΔE_{sys} is positive since the system is being heated by the flame.

8. a) The energy exchange is primarily work since there is a lot of motion. There is a small amount of heat transferred since there is some friction as the balls roll. The sign of ΔE_{sys} is negative since the kinetic energy of the first ball is transferred to the second ball.

   b) The energy exchange is primarily work. The sign of ΔE_{sys} is negative since the potential energy of the book is decreased as it falls.

   c) The energy exchange is primarily work. The sign of ΔE_{sys} is positive since the father is doing work to move the girl and the swing.

9. **Given:** 415 kJ heat released; 125 kJ work done on surroundings **Find:** ΔE_{sys}
Conceptual Plan: interpret language to determine the sign of the two terms then $q, w \rightarrow \Delta E_{sys}$
$\Delta E = q + w$

Solution: since heat is released from the system to the surroundings, $q = -$ 415 kJ; since the system is doing work on the surroundings, $w = -$ 125 kJ. $\Delta E = q + w = -$ 415 kJ $-$ 125 kJ $= -$ 540. kJ $= -5.40 \times 10^2$ kJ

Check: The units (kJ) are correct. The magnitude of the answer (–540) makes physical sense because both terms are negative.

10. **Given:** 214 kJ heat absorbed; surroundings do 110 kJ work **Find:** ΔE_{sys}
Conceptual Plan: interpret language to determine the sign of the two terms then $q, w \rightarrow \Delta E_{sys}$
$\Delta E = q + w$

Solution: since heat is absorbed by the system, $q = +$ 214 kJ; since the surroundings are doing work on the system, $w = +$ 110 kJ. $\Delta E = q + w =$ 214 kJ + 110 J = 3$\underline{2}$4 kJ = 320 kJ

Check: The units (kJ) are correct. The magnitude of the answer (+300) makes physical sense because both terms are positive.

11. **Given:** 655 J heat absorbed; 344 J work done on surroundings **Find:** ΔE_{sys}
Conceptual Plan: interpret language to determine the sign of the two terms then $q, w \rightarrow \Delta E_{sys}$

$$\Delta E = q + w$$

Solution: since heat is absorbed by the system, $q = +655$ J; since the system is doing work on the surroundings, $w = -344$ J. $\Delta E = q + w = 655\ \text{J} - 344\ \text{J} = 311\ \text{J}$

Check: The units (J) are correct. The magnitude of the answer (+300) makes physical sense because heat term dominates over the work term.

12. **Given:** 155 J heat absorbed; 77 kJ work done on surroundings **Find:** ΔE_{sys}
Conceptual Plan: interpret language to determine the sign of the two terms J → kJ then $q, w \rightarrow \Delta E_{sys}$

$$\frac{1\text{kJ}}{1000\ \text{J}} \qquad \Delta E = q + w$$

Solution: since heat is absorbed by the system, $q = +155$ J; since the system is doing work on the surroundings, $w = -77$ kJ. Thus $-77\ \text{kJ} \times \frac{1000\ \text{J}}{1\ \text{kJ}} = -7\underline{7}000\ \text{J}$ and

$\Delta E = q + w = 155\ \text{J} - 7\underline{7}000\ \text{J} = -7\underline{7}000\ \text{J} = -77\ \text{kJ}$

Check: The units (kJ) are correct. The magnitude of the answer (– 77 kJ) makes physical sense because work term dominates over the heat term. In fact the heat term is negligible compared to the work term.

13. Cooler A had more ice after 3 hours because most of the ice in cooler B was melted in order to cool the soft drinks that started at room temperature. In cooler A the drinks were already cold and so the ice only needed to maintain this cool temperature.

14. Since the specific heat capacity of water is much larger than the specific heat capacity of aluminum, much more heat needs to be released by the water than the aluminum, for each 1 °C of temperature drop. This means that more heat is stored in each kg of water than aluminum.

15. **Given:** 1.50 L water, $T_i = 25.0$ °C, $T_f = 100.0$ °C, d = 1.0 g/mL **Find:** q
Conceptual Plan:
L → mL → g and pull C_s from Table 6.4 and T_i, T_f → ΔT then m, C_s, ΔT → q

$$\frac{1000\text{mL}}{1\ \text{L}} \quad \frac{1.0\text{g}}{1.0\ \text{mL}} \quad 4.18\frac{\text{J}}{\text{g}\cdot{}^\circ\text{C}} \qquad \Delta T = T_f - T_i \qquad q = m\,C_s\,\Delta T$$

Solution: $1.50\ \text{L} \times \frac{1000\ \text{mL}}{1\ \text{L}} \times \frac{1.0\ \text{g}}{1.0\ \text{mL}} = 1\underline{5}00\ \text{g}$ and $\Delta T = T_f - T_i = 100.0\ {}^\circ\text{C} - 25.0\ {}^\circ\text{C} = 75.0\ {}^\circ\text{C}$

then $q = m\,C_s\,\Delta T = 1\underline{5}00\ \text{g} \times 4.18\frac{\text{J}}{\text{g}\cdot{}^\circ\text{C}} \times 75.0\ {}^\circ\text{C} = 4.7 \times 10^5\ \text{J}$

Check: The units (J) are correct. The magnitude of the answer (10^6) makes physical sense because there is such a large mass, a significant temperature change, and a high specific heat capacity material.

16. **Given:** 1.50 kg sand, $T_i = 25.0$ °C, $T_f = 100.0$ °C **Find:** q
Conceptual Plan: kg → g and pull C_s from Table 6.4 and T_i, T_f → ΔT then m, C_s, ΔT → q

$$\frac{1000\text{g}}{1\ \text{g}} \quad 0.84\frac{\text{J}}{\text{g}\cdot{}^\circ\text{C}} \qquad \Delta T = T_f - T_i \qquad q = m\,C_s\,\Delta T$$

Solution: $1.50\ \text{L} \times \frac{1000\ \text{mL}}{1\ \text{L}} \times \frac{1.0\ \text{g}}{1.0\ \text{mL}} = 1\underline{5}00\ \text{g}$ and $\Delta T = T_f - T_i = 100.0\ {}^\circ\text{C} - 25.0\ {}^\circ\text{C} = 75.0\ {}^\circ\text{C}$

then $q = m\,C_s\,\Delta T = 1\underline{5}00\ \text{g} \times 0.84\frac{\text{J}}{\text{g}\cdot{}^\circ\text{C}} \times 75.0\ {}^\circ\text{C} = 9.45 \times 10^4\ \text{J}$

Check: The units (J) are correct. The magnitude of the answer (~10^5) makes physical sense because there is such a large mass and a significant temperature change.

17. a) **Given:** 25 g gold, $T_i = 27.0$ °C, $q = 2.35$ kJ **Find:** T_f

Conceptual Plan:

kJ → J and pull C_s from Table 6.4 then m, C_s, q → ΔT then $T_i, \Delta T$ → T_f

$\frac{1000\text{g}}{1\text{g}}$ $\quad 0.128\frac{\text{J}}{\text{g}\cdot{}^\circ\text{C}}$ $\quad q = m\,C_s\,\Delta T$ $\quad \Delta T = T_f - T_i$

Solution: $2.35\,\text{kJ} \times \frac{1000\text{ J}}{1\text{ kJ}} = 2350\text{ J}$ then $q = m\,C_s\,\Delta T$ Rearrange to solve for ΔT.

$$\Delta T = \frac{q}{m\,C_s} = \frac{2350\text{ J}}{25\text{ g} \times 0.128\frac{\text{J}}{\text{g}\cdot{}^\circ\text{C}}} = 734.375\ {}^\circ\text{C}$$ finally $\Delta T = T_f - T_i$ Rearrange to solve for T_f.

$T_f = \Delta T + T_i = 734.375\ {}^\circ\text{C} + 27.0\ {}^\circ\text{C} = 760\ {}^\circ\text{C}$

Check: The units (°C) are correct. The magnitude of the answer (760) makes physical sense because there is such a large heat absorbed, such a small mass, and specific heat capacity. The temperature change should be very large.

b) **Given:** 25 g silver, $T_i = 27.0$ °C, $q = 2.35$ kJ **Find:** T_f

Conceptual Plan:

kJ → J and pull C_s from Table 6.4 then m, C_s, q → ΔT then $T_i, \Delta T$ → T_f

$\frac{1000\text{g}}{1\text{g}}$ $\quad 0.235\frac{\text{J}}{\text{g}\cdot{}^\circ\text{C}}$ $\quad q = m\,C_s\,\Delta T$ $\quad \Delta T = T_f - T_i$

Solution: $2.35\,\text{kJ} \times \frac{1000\text{ J}}{1\text{ kJ}} = 2350\text{ J}$ then $q = m\,C_s\,\Delta T$ Rearrange to solve for ΔT.

$$\Delta T = \frac{q}{m\,C_s} = \frac{2350\text{ J}}{25\text{ g} \times 0.235\frac{\text{J}}{\text{g}\cdot{}^\circ\text{C}}} = 400\ {}^\circ\text{C}$$ finally $\Delta T = T_f - T_i$ Rearrange to solve for T_f.

$T_f = \Delta T + T_i = 400\ {}^\circ\text{C} + 27.0\ {}^\circ\text{C} = 430\ {}^\circ\text{C}$

Check: The units (°C) are correct. The magnitude of the answer (430) makes physical sense because there is such a large amount of heat absorbed, such a small mass, and specific heat capacity. The temperature change should be very large. The temperature change should be less than that of the gold because the specific heat capacity is greater.

c) **Given:** 25 g aluminum, $T_i = 27.0$ °C, $q = 2.35$ kJ **Find:** T_f

Conceptual Plan:

kJ → J and pull C_s from Table 6.3 then m, C_s, q → ΔT then $T_i, \Delta T$ → T_f

$\frac{1000\text{g}}{1\text{g}}$ $\quad 0.903\frac{\text{J}}{\text{g}\cdot{}^\circ\text{C}}$ $\quad q = m\,C_s\,\Delta T$ $\quad \Delta T = T_f - T_i$

Solution: $2.35\,\text{kJ} \times \frac{1000\text{ J}}{1\text{ kJ}} = 2350\text{ J}$ then $q = m\,C_s\,\Delta T$ Rearrange to solve for ΔT.

$$\Delta T = \frac{q}{m\,C_s} = \frac{2350\text{ J}}{25\text{ g} \times 0.903\frac{\text{J}}{\text{g}\cdot{}^\circ\text{C}}} = 104.10\ {}^\circ\text{C}$$ finally $\Delta T = T_f - T_i$ Rearrange to solve for T_f.

$T_f = \Delta T + T_i = 104.10\ {}^\circ\text{C} + 27.0\ {}^\circ\text{C} = 130\ {}^\circ\text{C}$

Check: The units (°C) are correct. The magnitude of the answer (130) makes physical sense because there is such a large heat absorbed, and such a small mass. The temperature change should be less than that of the silver because the specific heat capacity is greater.

d) **Given:** 25 g water, $T_i = 27.0$ °C, $q = 2.35$ kJ **Find:** T_f

Conceptual Plan:

kJ → J and pull C_s from Table 6.4 then m, C_s, q → ΔT then $T_i, \Delta T$ → T_f

$\frac{1000\text{g}}{1\text{g}}$ $\quad 4.18\frac{\text{J}}{\text{g}\cdot{}^\circ\text{C}}$ $\quad q = m\,C_s\,\Delta T$ $\quad \Delta T = T_f - T_i$

Solution: $2.35\,\text{kJ} \times \dfrac{1000\ \text{J}}{1\ \text{kJ}} = 23\underline{5}0\ \text{J}$ then $q = m\,C_s\,\Delta T$ Rearrange to solve for ΔT.

$$\Delta T = \frac{q}{m\,C_s} = \frac{23\underline{5}0\ \text{J}}{25\ \text{g} \times 4.18\dfrac{\text{J}}{\text{g}\cdot{}^\circ\text{C}}} = 2\underline{2}.491\ {}^\circ\text{C}$$ finally $\Delta T = T_f - T_i$ Rearrange to solve for T_f.

$$T_f = \Delta T + T_i = 2\underline{2}.491\ {}^\circ\text{C} + 27.0\ {}^\circ\text{C} = 49\ {}^\circ\text{C}$$

Check: The units (°C) are correct. The magnitude of the answer (130) makes physical sense because there is such a large heat absorbed, and such a small mass. The temperature change should be less than that of the aluminum because the specific heat capacity is greater.

18. a) **Given:** Pyrex glass, $q = 1.95 \times 10^3$ J, $T_i = 23.0$ °C, $T_f = 55.4$ °C **Find:** m

Conceptual Plan: pull C_s from Table 6.4 and $T_i, T_f \rightarrow \Delta T$ then $\Delta T, C_s, q \rightarrow m$

$0.75\ \dfrac{\text{J}}{\text{g}\cdot{}^\circ\text{C}}$ $\qquad \Delta T = T_f - T_i \qquad q = m\,C_s\,\Delta T$

Solution: $\Delta T = T_f - T_i = 55.4\ {}^\circ\text{C} - 23.0\ {}^\circ\text{C} = 32.4\ {}^\circ\text{C}$ then $q = m\,C_s\,\Delta T$ Rearrange to solve for m.

$$m = \frac{q}{C_s\,\Delta T} = \frac{1.95 \times 10^3\ \text{J}}{0.75\dfrac{\text{J}}{\text{g}\cdot{}^\circ\text{C}} \times 32.4\ {}^\circ\text{C}} = 80.\ \text{g or } 8.0 \times 10^1\ \text{g}$$

Check: The units (g) are correct. The magnitude of the answer (80) makes physical sense because there is such a large heat absorbed and a moderate temperature rise and specific heat capacity.

b) **Given:** sand, $q = 1.95 \times 10^3$ J, $T_i = 23.0$ °C, $T_f = 62.1$ °C **Find:** m

Conceptual Plan: pull C_s from Table 6.4 and $T_i, T_f \rightarrow \Delta T$ then $\Delta T, C_s, q \rightarrow m$

$0.84\ \dfrac{\text{J}}{\text{g}\cdot{}^\circ\text{C}}$ $\qquad \Delta T = T_f - T_i \qquad q = m\,C_s\,\Delta T$

Solution: $\Delta T = T_f - T_i = 62.1\ {}^\circ\text{C} - 23.0\ {}^\circ\text{C} = 39.1\ {}^\circ\text{C}$ then $q = m\,C_s\,\Delta T$ Rearrange to solve for m.

$$m = \frac{q}{C_s\,\Delta T} = \frac{1.95 \times 10^3\ \text{J}}{0.84\dfrac{\text{J}}{\text{g}\cdot{}^\circ\text{C}} \times 39.1\ {}^\circ\text{C}} = 59\ \text{g}$$

Check: The units (g) are correct. The magnitude of the answer (60) makes physical sense because there is such a large heat absorbed and a moderate temperature rise and specific heat capacity.

c) **Given:** ethanol, $q = 1.95 \times 10^3$ J, $T_i = 23.0$ °C, $T_f = 44.2$ °C **Find:** m

Conceptual Plan: pull C_s from Table 6.4 and $T_i, T_f \rightarrow \Delta T$ then $\Delta T, C_s, q \rightarrow m$

$2.42\ \dfrac{\text{J}}{\text{g}\cdot{}^\circ\text{C}}$ $\qquad \Delta T = T_f - T_i \qquad q = m\,C_s\,\Delta T$

Solution: $\Delta T = T_f - T_i = 44.2\ {}^\circ\text{C} - 23.0\ {}^\circ\text{C} = 21.2\ {}^\circ\text{C}$ then $q = m\,C_s\,\Delta T$ Rearrange to solve for m.

$$m = \frac{q}{C_s\,\Delta T} = \frac{1.95 \times 10^3\ \text{J}}{2.42\dfrac{\text{J}}{\text{g}\cdot{}^\circ\text{C}} \times 21.2\ {}^\circ\text{C}} = 38.0\ \text{g}$$

Check: The units (g) are correct. The magnitude of the answer (40) makes physical sense because there is such a large heat absorbed and small temperature rise and specific heat capacity.

d) **Given:** water, $q = 1.95 \times 10^3$ J, $T_i = 23.0$ °C, $T_f = 32.4$ °C **Find:** m

Conceptual Plan: pull C_s from Table 6.4 and $T_i, T_f \rightarrow \Delta T$ then $\Delta T, C_s, q \rightarrow m$

$4.18\ \dfrac{\text{J}}{\text{g}\cdot{}^\circ\text{C}}$ $\qquad \Delta T = T_f - T_i \qquad q = m\,C_s\,\Delta T$

Solution: $\Delta T = T_f - T_i = 32.4\ {}^\circ\text{C} - 23.0\ {}^\circ\text{C} = 9.4\ {}^\circ\text{C}$ then $q = m\,C_s\,\Delta T$ Rearrange to solve for m.

$$m = \frac{q}{C_S \, \Delta T} = \frac{1.95 \times 10^3 \text{ J}}{4.18 \frac{\text{J}}{\text{g} \cdot {}^\circ\text{C}} \times 9.4 \, {}^\circ\text{C}} = 50.\text{ g or } 5.0 \times 10^1 \text{ g}$$

Check: The units (g) are correct. The magnitude of the answer (50) makes physical sense because there is such a large heat absorbed and small temperature rise and very specific heat capacity.

19. **Given:** $V_i = 0.0$ L, $V_f = 2.5$ L, $P = 1.1$ atm **Find:** w (J)
Conceptual Plan: $V_i, V_f \rightarrow \Delta V$ **then** **then** $P, \Delta V \rightarrow w$ **(L atm)** $\rightarrow w$ **(J)**

$\Delta V = V_f - V_i$ $w = -P\,\Delta V$ $\frac{101.3 \text{ J}}{1 \text{ L} \cdot \text{atm}}$

Solution: $\Delta V = V_f - V_i = 2.5 \text{ L} - 0.0 \text{ L} = 2.5 \text{ L}$ then

$$w = -P\,\Delta V = -1.1 \text{ atm} \times 2.5 \text{ L} \times \frac{101.3 \text{ J}}{1 \text{ L} \cdot \text{atm}} = -280 \text{ J}$$

Check: The units (J) are correct. The magnitude of the answer (–280) makes physical sense because this is an expansion (negative work) and we have ~ atmospheric pressure and a small volume of expansion.

20. **Given:** $\Delta V = 0.50$ L, $P = 1.0$ atm **Find:** w (J)
Conceptual Plan: $P, \Delta V \rightarrow w$ **(L atm)** $\rightarrow w$ **(J)**

$w = -P\,\Delta V$ $\frac{101.3 \text{ J}}{1 \text{ L} \cdot \text{atm}}$

Solution: $w = -P\,\Delta V = -1.0 \text{ atm} \times 0.50 \text{ L} \times \frac{101.3 \text{ J}}{1 \text{ L} \cdot \text{atm}} = -51 \text{ J}$

Check: The units (J) are correct. The magnitude of the answer (–51) makes physical sense because this is a small expansion (negative work) and we do not expect breathing to take much energy.

21. **Given:** $q = 565$ J absorbed, $V_i = 0.10$ L, $V_f = 0.85$ L, $P = 1.0$ atm **Find:** ΔE_{sys}
Conceptual Plan: $V_i, V_f \rightarrow \Delta V$ **and interpret language to determine the sign of the heat then**

$\Delta V = V_f - V_i$ $q = +565 \text{ J}$

then $P, \Delta V \rightarrow w$ **(L atm)** $\rightarrow w$ **(J)** **finally** $q, w \rightarrow \Delta E_{sys}$

$w = -P\,\Delta V$ $\frac{101.3 \text{ J}}{1 \text{ L atm}}$ $\Delta E = q + w$

Solution: $\Delta V = V_f - V_i = 0.85 \text{ L} - 0.10 \text{ L} = 0.75 \text{ L}$ then

$$w = -P\,\Delta V = -1.0 \text{ atm} \times 0.75 \text{ L} \times \frac{101.3 \text{ J}}{1 \text{ L} \cdot \text{atm}} = -75.975 \text{ J} \quad \Delta E = q + w = +565 \text{ J} - 75.975 \text{ J} = 489 \text{ J}$$

Check: The units (J) are correct. The magnitude of the answer (500) makes physical sense because the heat absorbed dominated the small expansion work (negative work).

22. **Given:** $q = 124$ J released, $V_i = 5.55$ L, $V_f = 1.22$ L, $P = 1.00$ atm **Find:** ΔE_{sys}
Conceptual Plan: $V_i, V_f \rightarrow \Delta V$ **and interpret language to determine the sign of the heat then**

$\Delta V = V_f - V_i$ $q = -124 \text{ J}$

then $P, \Delta V \rightarrow w$ **(L atm)** $\rightarrow w$ **(J)** **finally** $q, w \rightarrow \Delta E_{sys}$

$w = -P\,\Delta V$ $\frac{101.3 \text{ J}}{1 \text{ L} \cdot \text{atm}}$ $\Delta E = q + w$

Solution: $\Delta V = V_f - V_i = 1.22 \text{ L} - 5.55 \text{ L} = -4.33 \text{ L}$ then

$$w = -P\,\Delta V = -1.00 \text{ atm} \times (-4.33 \text{ L}) \times \frac{101.3 \text{ J}}{1 \text{ L} \cdot \text{atm}} = +438.629 \text{ J} \text{ then}$$

$\Delta E = q + w = -124 \text{ J} + 438.629 \text{ J} = 315 \text{ J}$

Check: The units (J) are correct. The magnitude of the answer (300) makes physical sense because the compression work dominated the small heat released.

23. **Given:** 1 mol fuel, 3452 kJ heat produced; 11 kJ work done on surroundings **Find:** ΔE_{sys}, ΔH
Conceptual Plan:
interpret language to determine the sign of the two terms then $q \rightarrow \Delta H$ and $q, w \rightarrow \Delta E_{sys}$

$\Delta H = q_P$ $\quad$ $\Delta E = q + w$

Solution: since heat is produced by the system to the surroundings, $q = -$ 3452 kJ; since the system is doing work on the surroundings, $w = -$ 11 kJ. $\Delta H = q_P = -3452$ kJ and

$\Delta E = q + w = -3452 \text{ kJ} - 11 \text{kJ} = -3463 \text{ kJ}$

Check: The units (kJ) are correct. The magnitude of the answer (–3500) makes physical sense because both terms are negative. We expect significant amounts of energy from fuels.

24. **Given:** 1 mol octane, $P = 1.0$ atm, $\Delta E_{sys} = 5084.3$ kJ; $\Delta H = 5074.1$ kJ **Find:** w
Conceptual Plan:
interpret language to determine the sign of the two terms then $q \rightarrow \Delta H$ and $q, \Delta E_{sys} \rightarrow w$

$\Delta E_{sys} = -5084.3$ kJ; $\Delta H = -5074.1$ kJ $\quad$ $\Delta H = q_P$ $\quad$ $\Delta E = q + w$

Solution: since heat is produced by the system to the surroundings, $\Delta H = q_P = -5074.1$ kJ ; pistons are used to produce energy so $\Delta E_{sys} = -5084.3$ kJ $\Delta E = q + w$ Rearrange to solve for w.

$w = \Delta E - q = -5084.3 \text{ kJ} - -5074.1 \text{ kJ} = -10.2 \text{ kJ}$

Check: The units (kJ) are correct. The magnitude of the answer (–10) makes physical sense because the work should be negative in an expansion. We expect more heat than work in an engine.

25. a) combustion is an exothermic process, ΔH is negative

    b) evaporation requires an input of energy and so it is endothermic, ΔH is positive

    c) condensation is the reverse of evaporation, so it is exothermic, ΔH is negative

26. a) sublimation requires an input of energy and so it is endothermic, ΔH is positive

    b) combustion is an exothermic process, ΔH is negative

    c) since the temperature drops this is an endothermic process, ΔH is positive

27. **Given:** 177 mL acetone (C_3H_6O), $\Delta H^\circ_{rxn} = -1790$ kJ; d = 0.788 g/mL **Find:** q
Conceptual Plan: **mL acetone → g acetone → mol acetone → q**

$\frac{0.788 \text{ g}}{1 \text{ mL}}$ $\quad$ $\frac{1 \text{ mol}}{58.08 \text{ g}}$ $\quad$ $\frac{-1790 \text{ kJ}}{1 \text{ mol}}$

Solution: $177 \text{ mL} \times \frac{0.788 \text{ g}}{1 \text{ mL}} \times \frac{1 \text{ mol}}{58.08 \text{ g}} \times \frac{-1790 \text{ kJ}}{1 \text{ mol}} = -4.30 \times 10^3$ kJ or 4.30×10^3 kJ released

Check: The units (kJ) are correct. The magnitude of the answer (-10^3) makes physical sense because the enthalpy change is negative and we have more than a mole of acetone. We expect more than 1790 kJ to be released.

28. **Given:** natural gas (CH_4), $\Delta H^\circ_{rxn} = -802.3$ kJ; $q = 267$ kJ **Find:** m
Conceptual Plan: **q → mol natural gas → g natural gas**

$\frac{1 \text{ mol}}{-802.3 \text{ kJ}}$ $\quad$ $\frac{16.04 \text{ g}}{1 \text{ mol}}$

Solution: $-267 \text{ kJ} \times \frac{1 \text{ mol}}{-802.3 \text{ kJ}} \times \frac{16.04 \text{ g}}{1 \text{ mol}} = 5.34 \text{ g}$

Check: The units (g) are correct. The magnitude of the answer (5) makes physical sense because the enthalpy change per mole is so large and we need to burn less than a mole.

29. **Given:** pork roast, $\Delta H^\circ_{rxn} = -2217$ kJ; q used = 1.6 x 10^3 kJ, 10 % efficiency **Find:** $m(CO_2)$
Conceptual Plan: ***q* used → *q* generated → mol CO_2 → g CO_2**

$$\frac{100\text{ kJ generated}}{10\text{ kJ used}} \quad \frac{3\text{ mol}}{2217\text{ kJ}} \quad \frac{44.01\text{ g}}{1\text{ mol}}$$

Solution: $$1.6 \times 10^3\ \text{kJ} \times \frac{100\ \text{kJ generated}}{10\ \text{kJ used}} \times \frac{3\ \text{mol } CO_2}{2217\ \text{kJ}} \times \frac{44.01\ \text{g } CO_2}{1\ \text{mol } CO_2} = 950\ \text{g } CO_2$$

Check: The units (g) are correct. The magnitude of the answer (~1000) makes physical sense because the process is not very efficient and a lot of energy is needed.

30. **Given:** carbon, $\Delta H^\circ_{rxn} = -393.5$ kJ; q needed = 5.00 x 10^2 kJ **Find:** $m(CO_2)$
Conceptual Plan: ***q* needed → mol CO_2 → g CO_2**

$$\frac{1\text{ mol}}{393.5\text{ kJ}} \quad \frac{44.01\text{ g}}{1\text{ mol}}$$

Solution: $$5.00 \times 10^2\ \text{kJ} \times \frac{1\ \text{mol } CO_2}{393.5\ \text{kJ}} \times \frac{44.01\ \text{g } CO_2}{1\ \text{mol } CO_2} = 55.9\ \text{g } CO_2$$

Check: The units (g) are correct. The magnitude of the answer (~1000) makes physical sense because the process is not very efficient and a lot of energy is needed.

31. $\Delta H_{rxn} = q_P$ and $\Delta E_{rxn} = q_V = \Delta H - P\,\Delta V$. Since combustions always involve expansions, expansions do work and so have a negative value. Combustions are always exothermic and so have a negative value. This means that ΔE_{rxn} is more negative than ΔH°_{rxn} and so A (–25.9 kJ) is the constant volume process and B (–23.3 kJ) is the constant pressure process.

32. Constant volume conditions should be used. Since $\Delta E = q + w$ and $w = -P\,\Delta V$, this means that at constant V w = 0 and all of the energy is released at heat ($\Delta E_{rxn} = q_V$). At constant P $\Delta H_{rxn} = q_P$ and $\Delta E_{rxn} = q_P = \Delta H - P\,\Delta V$. Since combustions always involve expansions; expansions do work and so have a negative value. Combustions are always exothermic and so have a negative value. This means that ΔE_{rxn} is more negative than ΔH°_{rxn} and so more heat will be generated in a constant V process.

33. **Given:** 0.514 g biphenyl ($C_{12}H_{10}$), bomb calorimeter, T_i = 25.8 °C, T_f = 29.4 °C, C_{cal} = 5.86 kJ/°C
Find: ΔE_{rxn}
Conceptual Plan:
T_i, T_f → ΔT then ΔT, C_{cal} → q_{cal} → q_{rxn} then g $C_{12}H_{10}$ → mol $C_{12}H_{10}$

$$\Delta T = T_f - T_i \qquad q_{cal} = C_{cal}\,\Delta T \qquad q_{cal} = -q_{rxn} \qquad \frac{1\text{ mol}}{154.20\text{ g}}$$

then q_{rxn}, mol $C_{12}H_{10}$ → ΔE_{rxn}

$$\Delta E_{rxn} = \frac{q_V}{mol\ C_{12}H_{10}}$$

Solution: $\Delta T = T_f - T_i = 29.4\ °C - 25.8\ °C = 3.6\ °C$ then $q_{cal} = C_{cal}\,\Delta T = 5.86\ \frac{\text{kJ}}{°C} \times 3.6\ °C = 21.096$ kJ

then $q_{cal} = -q_{rxn} = -21.096$ kJ and $0.514\ \text{g } C_{12}H_{10} \times \frac{1\ \text{mol } C_{12}H_{10}}{154.20\ \text{g } C_{12}H_{10}} = 0.00333333\ \text{mol } C_{12}H_{10}$ then

$$\Delta E_{rxn} = \frac{q_V}{mol\ C_{12}H_{10}} = \frac{-21.096\ \text{kJ}}{0.00333333\ \text{mol } C_{12}H_{10}} = -6.3 \times 10^3\ \text{kJ/mol}$$

Check: The units (kJ/mol) are correct. The magnitude of the answer (–6000) makes physical sense because there is such a large heat generated from a very small amount of biphenyl.

34. **Given:** 1.025 g naphthalene ($C_{10}H_8$), bomb calorimeter, T_i = 24.25°C, T_f = 32.33 °C, C_{cal} = 5.11 kJ/°C
Find: ΔE_{rxn}
Conceptual Plan:
T_i, T_f → ΔT then ΔT, C_{cal} → q_{cal} → q_{rxn} then g $C_{10}H_8$ → mol $C_{10}H_8$

$$\Delta T = T_f - T_i \qquad q_{cal} = C_{cal}\,\Delta T \qquad q_{cal} = -q_{rxn} \qquad \frac{1\text{ mol}}{128.16\text{ g}}$$

then q_{rxn}, mol $C_{10}H_8$ → ΔE_{rxn}

$$\Delta E_{rxn} = \frac{q_V}{mol\ C_{10}H_8}$$

Solution: $\Delta T = T_f - T_i = 32.33\ °C - 24.25\ °C = 8.08\ °C$ then

$$q_{cal} = C_{cal}\,\Delta T = 5.11\ \frac{kJ}{\cancel{°C}} \times 8.08\ \cancel{°C} = 41.2\underline{8}88\ kJ \text{ then } q_{cal} = -q_{rxn} = -41.2\underline{8}88\ kJ \text{ and}$$

$$1.025\ \cancel{g\ C_{10}H_8} \times \frac{1\ mol\ C_{10}H_8}{128.16\ \cancel{g\ C_{10}H_8}} = 0.00799\underline{7}815\ mol\ C_{10}H_8 \text{ then}$$

$$\Delta E_{rxn} = \frac{q_V}{mol\ C_{10}H_8} = \frac{-41.2\underline{8}88\ kJ}{0.00799\underline{7}815\ mol\ C_{10}H_8} = -5.16 \times 10^3\ kJ/mol$$

Check: The units (kJ/mol) are correct. The magnitude of the answer (–5000) makes physical sense because there is such a large heat generated from a very small amount of naphthalene.

35. **Given:** 0.103 g zinc, coffee-cup calorimeter, $T_i = 22.5°C$, $T_f = 23.7\ °C$, 50.0 mL solution, d (solution) = 1.0 g/mL, $C_{sol'n} = 4.18$ kJ/g °C **Find:** ΔH_{rxn}

Conceptual Plan:

T_i, T_f → ΔT and mL sol'n → g sol'n then ΔT, C_{cal} → q_{cal} → q_{rxn} then

$\Delta T = T_f - T_i$; $\frac{1.0\ g}{1.0\ mL}$; $q_{cal} = m\,C_{sol'n}\,\Delta T$; $q_{sol'n} = -q_{rxn}$

g Zn → mol Zn then q_{rxn}, mol Zn → ΔH_{rxn}

$\frac{1\ mol}{65.37\ g}$; $\Delta H_{rxn} = \frac{q_P}{mol\ Zn}$

Solution: $\Delta T = T_f - T_i = 23.7\ °C - 22.5\ °C = 1.2\ °C$ and $50.0\ \cancel{mL} \times \frac{1.0\ g}{1.0\ \cancel{mL}} = 50.0\ g$ then

$$q_{sol'n} = m\,C_{sol'n}\,\Delta T = 50.0\ \cancel{g} \times 4.18\ \frac{J}{\cancel{g} \cdot \cancel{°C}} \times 1.2\ \cancel{°C} = 2\underline{5}0.8\ J \text{ then } q_{sol'n} = -q_{rxn} = -2\underline{5}0.8\ J \text{ and}$$

$$0.103\ \cancel{g\ Zn} \times \frac{1\ mol\ Zn}{65.37\ \cancel{g\ Zn}} = 0.0015\underline{7}565\ mol\ Zn \text{ then}$$

$$\Delta H_{rxn} = \frac{q_P}{mol\ Zn} = \frac{-2\underline{5}0.8\ J}{0.0015\underline{7}565\ mol\ Zn} = -1.6 \times 10^5\ J/mol = -1.6 \times 10^2\ kJ/mol$$

Check: The units (kJ/mol) are correct. The magnitude of the answer (–160) makes physical sense because there is such a large heat generated from a very small amount of zinc.

36. **Given:** 1.25 g NH_4NO_3, coffee-cup calorimeter, $T_i = 25.8°C$, $T_f = 21.9\ °C$, 25.0 mL solution, d (solution) = 1.0 g/mL, $C_{sol'n} = 4.18$ kJ/g °C **Find:** ΔH_{rxn}

Conceptual Plan:

T_i, T_f → ΔT and mL sol'n → g sol'n then ΔT, C_{cal} → q_{cal} → q_{rxn} then

$\Delta T = T_f - T_i$; $\frac{1.0\ g}{1.0\ mL}$; $q_{cal} = m\,C_{sol'n}\,\Delta T$; $q_{sol'n} = -q_{rxn}$

g NH_4NO_3 → mol NH_4NO_3 then q_{rxn}, mol NH_4NO_3 → ΔH_{rxn}

$\frac{1\ mol}{80.05\ g}$; $\Delta H_{rxn} = \frac{q_P}{mol\ NH_4NO_3}$

Solution: $\Delta T = T_f - T_i = 21.9\ °C - 25.8\ °C = -3.9\ °C$ and $25.0\ \cancel{mL} \times \frac{1.0\ g}{1.0\ \cancel{mL}} = 25.0\ g$ then

$$q_{sol'n} = m\,C_{sol'n}\,\Delta T = 25.0\ \cancel{g} \times 4.18\ \frac{J}{\cancel{g} \cdot \cancel{°C}} \times (-3.9\ \cancel{°C}) = -4\underline{0}7.55\ J \text{ then } q_{sol'n} = -q_{rxn} = 4\underline{0}7.55\ J \text{ and}$$

$$1.25\ \cancel{g\ NH_4NO_3} \times \frac{1\ mol\ NH_4NO_3}{80.05\ \cancel{g\ NH_4NO_3}} = 0.015\underline{6}152\ mol\ NH_4NO_3 \text{ then}$$

$$\Delta H_{rxn} = \frac{q_P}{mol\ NH_4NO_3} = \frac{4\underline{0}7.55\ J}{0.015\underline{6}152\ mol\ NH_4NO_3} = 2.6 \times 10^4\ J/mol = 26\ kJ/mol$$

Check: The units (kJ/mol) are correct. The magnitude of the answer (26) makes physical sense because there is such a small amount of heat absorbed.

37. a) Since A + B → 2 C has ΔH_1 then 2 C → A + B will have a $\Delta H_2 = -\Delta H_1$. When the reaction direction is reversed, it changes from exothermic to endothermic (or vice versa), so the sign of ΔH changes.

    b) Since A + ½ B → C has ΔH_1 then 2 A + B → 2 C will have a $\Delta H_2 = 2\ \Delta H_1$. When the reaction amount doubles, the amount of heat (or ΔH) doubles.

    c) Since A → B + 2 C has ΔH_1 then ½ A → ½ B + C will have a $\Delta H_1' = ½\ \Delta H_1$. When the reaction amount is cut in half, the amount of heat (or ΔH) is cut in half. Then ½ B + C → ½ A will have a $\Delta H_2 = -\Delta H_1' = -½\ \Delta H_1$. When the reaction direction is reversed, it changes from exothermic to endothermic (or vice versa), so the sign of ΔH changes.

38. a) Since A + 2 B → C + 3 D has ΔH = 155 kJ then 3A + 6 B → 3 C + 9 D will have a $\Delta H' = 3\ \Delta H$ = 3 (155 kJ) = 465 kJ . When the reaction amount triples, the amount of heat (or ΔH) triples.

    b) Since A + 2 B → C + 3 D has ΔH = 155 kJ then C + 3 D → 3A + 6 B will have a $\Delta H' = -\Delta H$ = – 155 kJ . When the reaction direction is reversed, it changes from endothermic to exothermic, so the sign of ΔH changes.

    c) Since A + 2 B → C + 3 D has ΔH = 155 kJ then ½ A + B → ½ C + 3/2 D will have a $\Delta H'$ = ½ ΔH = ½ (155 kJ) = 77.5 kJ. When the reaction amount is cut in half, the amount of heat (or ΔH) is cut in half. Then ½ C + 3/2 D → ½ A + B will have a $\Delta H'' = -\Delta H'$ = – 77.5 kJ. When the reaction direction is reversed, the sign of it changes from endothermic to exothermic, so the sign of ΔH changes.

39. Since the first reaction has Fe_2O_3 as a product and the reaction of interest has it as a product, we need to reverse the first reaction. When the reaction direction is reversed, ΔH changes.

$Fe_2O_3\ (s) \rightarrow 2\ Fe\ (s) + 3/2\ O_2\ (g)$ $\qquad \Delta H$ = + 824.2 kJ

Since the second reaction has 1 mole CO as a reactant and the reaction of interest has 3 moles of CO as a reactant, we need to multiply the second reaction and the ΔH by 3.

$3\,[CO\ (g) + 1/2\ O_2\ (g) \rightarrow CO_2\ (g)]$ $\qquad \Delta H = 3(-282.7$ kJ) = – 848.1 kJ

Hess' Law states the ΔH of the net reaction is the sum of the ΔH of the steps.
The rewritten reactions are:

$Fe_2O_3\ (s) \rightarrow 2\ Fe\ (s) + \cancel{3/2\ O_2\ (g)}$ $\qquad \Delta H$ = + 824.2 kJ

$3\ CO\ (g) + \cancel{3/2\ O_2\ (g)} \rightarrow 3\ CO_2\ (g)$ $\qquad \Delta H$ = – 848.1 kJ

$Fe_2O_3\ (s) + 3\ CO\ (g) \rightarrow 2\ Fe\ (s) + 3\ CO_2\ (g)$ $\qquad \Delta H_{rxn}$ = – 23.9 kJ

40. Since the first reaction has $CaCO_3$ as a product and the reaction of interest has it as a product, we simply write the first reaction and the ΔH unchanged.

$Ca\ (s) + CO_2\ (g) + 1/2\ O_2\ (g) \rightarrow CaCO_3\ (s)$ $\qquad \Delta H$ = – 812.8 kJ

Since the second reaction has 2 moles CaO as a product and the reaction of interest has 1 mole of CaO as a reactant, we need to reverse the direction of the reaction of the second reaction and multiply it by ½. The sign of the ΔH in the second reaction is changed and is multiplied by ½.

$1/2[2\ CaO\ (s) \rightarrow 2\ Ca\ (s) + O_2\ (g)]$ $\qquad \Delta H = -1/2(-1269.8$ kJ) = + 634.9 k

Hess' Law states the ΔH of the net reaction is the sum of the ΔH of the steps.
The rewritten reactions are:

$\cancel{Ca\ (s)} + CO_2\ (g) + \cancel{1/2\ O_2\ (g)} \rightarrow CaCO_3\ (s)$ $\qquad \Delta H$ = – 812.8 kJ

$CaO\ (s) \rightarrow \cancel{Ca\ (s)} + \cancel{1/2\ O_2\ (g)}$ $\qquad \Delta H$ = + 634.9 kJ

$CaO\ (s) + CO_2\ (g) \rightarrow CaCO_3\ (s)$ $\qquad \Delta H_{rxn}$ = – 177.9 kJ

41. Since the first reaction has C_5H_{12} as a reactant and the reaction of interest has it as a product, we need to reverse the first reaction. When the reaction direction is reversed, ΔH changes.

$5\ CO_2\ (g) + 6\ H_2O\ (g) \rightarrow C_5H_{12}\ (l) + 8\ O_2\ (g)$ $\qquad \Delta H$ = + 3505.8 kJ

Since the second reaction has 1 mole C as a reactant and the reaction of interest has 5 moles of C as a reactant, we need to multiply the second reaction and the ΔH by 5.

$5\,[C\,(s) + O_2\,(g) \rightarrow CO_2\,(g)]$ $\quad\Delta H = 5(-393.5\text{ kJ}) = -1967.5\text{ kJ}$

Since the third reaction has 2 moles H_2 as a reactant and the reaction of interest has 6 moles of H_2 as a reactant, we need to multiply the third reaction and the ΔH by 3.

$3[2\,H_2\,(g) + O_2\,(g) \rightarrow 2\,H_2O\,(g)]$ $\quad\Delta H = 3(-483.5\text{ kJ}) = -1450.5\text{ kJ}$

Hess' Law states the ΔH of the net reaction is the sum of the ΔH of the steps.
The rewritten reactions are:

$5\,CO_2\,(g) + 6\,H_2O\,(g) \rightarrow C_5H_{12}\,(l) + 8\,O_2\,(g)$ $\quad\Delta H = +3505.8\text{ kJ}$

$5\,C\,(s) + 5\,O_2\,(g) \rightarrow 5\,CO_2\,(g)$ $\quad\Delta H = -1967.5\text{ kJ}$

$6\,H_2\,(g) + 3\,O_2\,(g) \rightarrow 6\,H_2O\,(g)$ $\quad\Delta H = -1450.5\text{ kJ}$

$5\,C\,(s) + 6\,H_2\,(g) \rightarrow C_5H_{12}\,(l)$ $\quad\Delta H_{rxn} = +87.8\text{ kJ}$

42. Since the first reaction has CH_4 as a product and the reaction of interest has it as a reactant, we need to reverse the first reaction. When the reaction direction is reversed, ΔH changes.

$CH_4\,(g) \rightarrow C\,(s) + 2\,H_2\,(g)$ $\quad\Delta H = +74.6\text{ kJ}$

Since the first reaction has CCl_4 as a product and the reaction of interest has it as a product, we simply write the first reaction and the ΔH unchanged.

$C\,(s) + 2\,Cl_2\,(g) \rightarrow CCl_4\,(g)$ $\quad\Delta H = -95.7\text{ kJ}$

Since the third reaction has 2 moles HCl as a product and the reaction of interest has 4 moles of HCl as a product, we need to multiply the third reaction and the ΔH by 2.

$2[H_2\,(g) + Cl_2\,(g) \rightarrow 2\,HCl\,(g)]$ $\quad\Delta H = 2(-92.3\text{ kJ}) = -184.6\text{ k}$

Hess' Law states the ΔH of the net reaction is the sum of the ΔH of the steps.
The rewritten reactions are:

$CH_4\,(g) \rightarrow C\,(s) + 2\,H_2\,(g)$ $\quad\Delta H = +74.6\text{ kJ}$

$C\,(s) + 2\,Cl_2\,(g) \rightarrow CCl_4\,(g)$ $\quad\Delta H = -95.7\text{ kJ}$

$2\,H_2\,(g) + 2\,Cl_2\,(g) \rightarrow 4\,HCl\,(g)$ $\quad\Delta H = -184.6\text{ kJ}$

$CH_4\,(g) + 4\,Cl_2\,(g) \rightarrow CCl_4\,(g) + 4\,HCl\,(g)$ $\quad\Delta H_{rxn} = -205.7\text{ kJ}$

43. a) $\frac{1}{2}N_2\,(g) + 3/2\,H_2\,(g) \rightarrow NH_3\,(g)$ $\quad\Delta H^\circ_f = -45.9\text{ kJ/mol}$

b) $C\,(s) + O_2\,(g) \rightarrow CO_2\,(g)$ $\quad\Delta H^\circ_f = -393.5\text{ kJ/mol}$

c) $Fe\,(s) + 3/2\,O_2\,(g) \rightarrow Fe_2O_3\,(s)$ $\quad\Delta H^\circ_f = -824.2\text{ kJ/mol}$

d) $C\,(s) + 2\,H_2\,(g) \rightarrow CH_4\,(g)$ $\quad\Delta H^\circ_f = -74.6\text{ kJ/mol}$

44. a) $\frac{1}{2}N_2\,(g) + O_2\,(g) \rightarrow NO_2\,(g)$ $\quad\Delta H^\circ_f = 33.2\text{ kJ/mol}$

b) $Mg\,(s) + C\,(s) + 3/2\,O_2\,(g) \rightarrow MgCO_3\,(s)$ $\quad\Delta H^\circ_f = -1095.8\text{ kJ/mol}$

c) $2\,C\,(s) + 2\,H_2\,(g) \rightarrow C_2H_4\,(g)$ $\quad\Delta H^\circ_f = 52.4\text{ kJ/mol}$

d) $C\,(s) + 2\,H_2\,(g) + \frac{1}{2}O_2\,(g) \rightarrow CH_3OH\,(l)$ $\quad\Delta H^\circ_f = -238.6\text{ kJ/mol}$

45. **Given:** $N_2H_4\,(l) + N_2O_4\,(g) \rightarrow 2\,N_2O\,(g) + 2\,H_2O\,(g)$ **Find:** ΔH°_{rxn}

Conceptual Plan: $\Delta H^0_{rxn} = \sum n_P \Delta H^0_f(products) - \sum n_R \Delta H^0_f(reactants)$

Solution:

Reactant/Product	ΔH_f^0 (kJ/mol from Appendix IIB)
N_2H_4 (*l*)	50.6
N_2O_4 (*g*)	11.1
N_2O (*g*)	81.6
H_2O (*g*)	– 241.8

Be sure to pull data for the correct formula and phase.

$$\Delta H^0_{rxn} = \sum n_P \Delta H_f^0(products) - \sum n_R \Delta H_f^0(reactants)$$
$$= [2(\Delta H_f^0(N_2O\ (g))) + 2(\Delta H_f^0(H_2O\ (g)))] - [1(\Delta H_f^0(N_2H_4\ (l))) + 1(\Delta H_f^0(N_2O_4\ (g)))]$$
$$= [2(81.6\text{ kJ}) + 2(-241.8\text{ kJ})] - [1(50.6\text{ kJ}) + 1(11.1\text{ kJ})]$$
$$= [-320.4\text{ kJ}] - [61.7\text{ kJ}]$$
$$= -382.1\text{ kJ}$$

Check: The units (kJ) are correct. The answer is negative, which means that the reaction is exothermic. The answer is dominated by the negative heat of formation of water.

46. **Given:** C_5H_{12} (*l*) + 8 O_2 (*g*) → 5 CO_2 (*g*) + 6 H_2O (*g*) **Find:** ΔH°_{rxn}

Conceptual Plan: $\Delta H^0_{rxn} = \sum n_P \Delta H_f^0(products) - \sum n_R \Delta H_f^0(reactants)$

Solution:

Reactant/Product	ΔH_f^0 (kJ/mol from Appendix IIB)
C_5H_{12} (*l*)	–146.8
O_2 (*g*)	0.0
CO_2 (*g*)	– 393.5
H_2O (*g*)	– 241.8

Be sure to pull data for the correct formula and phase.

$$\Delta H^0_{rxn} = \sum n_P \Delta H_f^0(products) - \sum n_R \Delta H_f^0(reactants)$$
$$= [5(\Delta H_f^0(CO_2\ (g))) + 6(\Delta H_f^0(H_2O\ (g)))] - [1(\Delta H_f^0(C_5H_{12}\ (l))) + 8(\Delta H_f^0(O_2\ (g)))]$$
$$= [5(-393.5\text{ kJ}) + 6(-241.8\text{ kJ})] - [1(-146.8\text{ kJ}) + 8(0.0\text{ kJ})]$$
$$= [-3418.3\text{ kJ}] - [-146.8\text{ kJ}]$$
$$= -3271.5\text{ kJ}$$

Check: The units (kJ) are correct. The answer is negative, which means that the reaction is exothermic, which is typical for combustion reactions.

47. a) **Given:** C_2H_4 (*g*) + H_2 (*g*) → C_2H_6 (*g*) **Find:** ΔH°_{rxn}

Conceptual Plan: $\Delta H^0_{rxn} = \sum n_P \Delta H_f^0(products) - \sum n_R \Delta H_f^0(reactants)$

Solution:

Reactant/Product	ΔH_f^0 (kJ/mol from Appendix IIB)
C_2H_4 (*g*)	52.4
H_2 (*g*)	0.0
C_2H_6 (*g*)	– 84.68

Be sure to pull data for the correct formula and phase.

$$\Delta H^0_{rxn} = \sum n_P \Delta H_f^0(products) - \sum n_R \Delta H_f^0(reactants)$$
$$= [1(\Delta H_f^0(C_2H_6\ (g)))] - [1(\Delta H_f^0(C_2H_4\ (g))) + 1(\Delta H_f^0(H_2\ (g)))]$$
$$= [1(-84.68\text{ kJ})] - [1(52.4\text{ kJ}) + 1(0.0\text{ kJ})]$$
$$= [-84.68\text{ kJ}] - [52.4\text{ kJ}]$$
$$= -137.1\text{ kJ}$$

Check: The units (kJ) are correct. The answer is negative, which means that the reaction is exothermic. Both hydrocarbon terms are negative, so the final answer is negative.

b) **Given:** $CO\ (g) + H_2O\ (g) \rightarrow H_2\ (g) + CO_2\ (g)$ **Find:** ΔH°_{rxn}

Conceptual Plan: $\Delta H^0_{rxn} = \sum n_P \Delta H^0_f(products) - \sum n_R \Delta H^0_f(reactants)$

Solution:

Reactant/Product	ΔH^0_f (kJ/mol from Appendix IIB)
$CO\ (g)$	–110.5
$H_2O\ (g)$	– 241.8
$H_2\ (g)$	0.0
$CO_2\ (g)$	– 393.5

Be sure to pull data for the correct formula and phase.

$$\Delta H^0_{rxn} = \sum n_P \Delta H^0_f(products) - \sum n_R \Delta H^0_f(reactants)$$
$$= [1(\Delta H^0_f(H_2\ (g))) + 1(\Delta H^0_f(CO_2\ (g)))] - [1(\Delta H^0_f(CO\ (g))) + 1(\Delta H^0_f(H_2O\ (g)))]$$
$$= [1(0.0\text{ kJ}) + 1(-393.5\text{ kJ})] - [1(-110.5\text{ kJ}) + 1(-241.8\text{ kJ})]$$
$$= [-393.5\text{ kJ}] - [-352.3\text{ kJ}]$$
$$= -41.2\text{ kJ}$$

Check: The units (kJ) are correct. The answer is negative, which means that the reaction is exothermic.

c) **Given:** $3\ NO_2\ (g) + H_2O\ (l) \rightarrow 2\ HNO_3\ (aq) + NO\ (g)$ **Find:** ΔH°_{rxn}

Conceptual Plan: $\Delta H^0_{rxn} = \sum n_P \Delta H^0_f(products) - \sum n_R \Delta H^0_f(reactants)$

Solution:

Reactant/Product	ΔH^0_f (kJ/mol from Appendix IIB)
$NO_2\ (g)$	33.2
$H_2O\ (l)$	– 285.8
$HNO_3\ (aq)$	–207
$NO\ (g)$	91.3

Be sure to pull data for the correct formula and phase.

$$\Delta H^0_{rxn} = \sum n_P \Delta H^0_f(products) - \sum n_R \Delta H^0_f(reactants)$$
$$= [2(\Delta H^0_f(HNO_3\ (aq))) + 1(\Delta H^0_f(NO\ (g)))] - [3(\Delta H^0_f(NO_2\ (g))) + 1(\Delta H^0_f(H_2O\ (l)))]$$
$$= [2(-207\text{ kJ}) + 1(91.3\text{ kJ})] - [3(33.2\text{ kJ}) + 1(-285.8\text{ kJ})]$$
$$= [-32\underline{2}.7\text{ kJ}] - [-186.2\text{ kJ}]$$
$$= -137\text{ kJ}$$

Check: The units (kJ) are correct. The answer is negative, which means that the reaction is exothermic.

d) **Given:** $Cr_2O_3\ (s) + 3\ CO\ (g) \rightarrow 2\ Cr\ (s) + 3\ CO_2\ (g)$ **Find:** ΔH°_{rxn}

Conceptual Plan: $\Delta H^0_{rxn} = \sum n_P \Delta H^0_f(products) - \sum n_R \Delta H^0_f(reactants)$

Solution:

Reactant/Product	ΔH^0_f (kJ/mol from Appendix IIB)
$Cr_2O_3\ (s)$	– 1139.7
$CO\ (g)$	–110.5
$Cr\ (s)$	0.0
$CO_2\ (g)$	– 393.5

Be sure to pull data for the correct formula and phase.

$$\Delta H^0_{rxn} = \sum n_P \Delta H^0_f(products) - \sum n_R \Delta H^0_f(reactants)$$
$$= [2(\Delta H^0_f(\text{Cr }(s))) + 3(\Delta H^0_f(\text{CO}_2\ (g)))] - [1(\Delta H^0_f(\text{Cr}_2\text{O}_3\ (s))) + 3(\Delta H^0_f(\text{CO }(g)))]$$
$$= [2(0.0 \text{ kJ}) + 3(-393.5 \text{ kJ})] - [1(-1139.7 \text{ kJ}) + 3(-110.5 \text{ kJ})]$$
$$= [-1180.5 \text{ kJ}] - [-1471.2 \text{ kJ}]$$
$$= 290.7 \text{ kJ}$$

Check: The units (kJ) are correct. The answer is positive, which means that the reaction is endothermic.

48. a) **Given:** $2\ H_2S\ (g) + 3\ O_2\ (g) \rightarrow 2\ H_2O\ (l) + 2\ SO_2\ (g)$ **Find:** ΔH°_{rxn}

Conceptual Plan: $\Delta H^0_{rxn} = \sum n_P \Delta H^0_f(products) - \sum n_R \Delta H^0_f(reactants)$

Solution:

Reactant/Product	ΔH^0_f (kJ/mol from Appendix IIB)
$H_2S\ (g)$	– 20.6
$O_2\ (g)$	0.0
$H_2O\ (l)$	– 285.8
$SO_2\ (g)$	–296.8

Be sure to pull data for the correct formula and phase.

$$\Delta H^0_{rxn} = \sum n_P \Delta H^0_f(products) - \sum n_R \Delta H^0_f(reactants)$$
$$= [2(\Delta H^0_f(\text{H}_2\text{O }(l))) + 2(\Delta H^0_f(\text{SO}_2\ (g)))] - [2(\Delta H^0_f(\text{H}_2\text{S }(g))) + 3(\Delta H^0_f(\text{O}_2\ (g)))]$$
$$= [2(-285.8 \text{ kJ}) + 2(-296.8 \text{ kJ})] - [2(-20.6 \text{ kJ}) + 3(0.0 \text{ kJ})]$$
$$= [-1165.2 \text{ kJ}] - [-41.2 \text{ kJ}]$$
$$= -1124.0 \text{ kJ}$$

Check: The units (kJ) are correct. The answer is negative, which means that the reaction is exothermic.

b) **Given:** $SO_2\ (g) + 1/2\ O_2\ (g) \rightarrow SO_3\ (g)$ **Find:** ΔH°_{rxn}

Conceptual Plan: $\Delta H^0_{rxn} = \sum n_P \Delta H^0_f(products) - \sum n_R \Delta H^0_f(reactants)$

Solution:

Reactant/Product	ΔH^0_f (kJ/mol from Appendix IIB)
$SO_2\ (g)$	–296.8
$O_2\ (g)$	0.0
$SO_3\ (g)$	–395.7

Be sure to pull data for the correct formula and phase.

$$\Delta H^0_{rxn} = \sum n_P \Delta H^0_f(products) - \sum n_R \Delta H^0_f(reactants)$$
$$= [1(\Delta H^0_f(\text{SO}_3\ (g)))] - [1(\Delta H^0_f(\text{SO}_2\ (g))) + 1/2(\Delta H^0_f(\text{O}_2\ (g)))]$$
$$= [1(-395.7 \text{ kJ})] - [1(-296.8 \text{ kJ}) + 1/2(0.0 \text{ kJ})]$$
$$= [-395.7 \text{ kJ}] - [-296.8 \text{ kJ}]$$
$$= -98.9 \text{ kJ}$$

Check: The units (kJ) are correct. The answer is negative, which means that the reaction is exothermic. The SO_3 has a lower heat of formation than SO_2, so we expect an exothermic reaction.

c) **Given:** $C\ (s) + H_2O\ (g) \rightarrow CO\ (g) + H_2\ (g)$ **Find:** ΔH°_{rxn}

Conceptual Plan: $\Delta H^0_{rxn} = \sum n_P \Delta H^0_f(products) - \sum n_R \Delta H^0_f(reactants)$

Solution:

Reactant/Product	ΔH_f^0 (kJ/mol from Appendix IIB)
C (*s*)	0.0
H_2O (*g*)	– 241.8
CO (*g*)	– 110.5
H_2 (*g*)	0.0

Be sure to pull data for the correct formula and phase.

$$\Delta H^0_{rxn} = \sum n_P \Delta H_f^0(products) - \sum n_R \Delta H_f^0(reactants)$$
$$= [1(\Delta H_f^0(\text{CO }(g))) + 1(\Delta H_f^0(\text{H}_2\,(g)))] - [1(\Delta H_f^0(\text{C }(s))) + 1(\Delta H_f^0(\text{H}_2\text{O }(g)))]$$
$$= [1(-110.5\text{ kJ}) + 1(0.0\text{ kJ})] - [1(0.0\text{ kJ}) + 1(-241.8\text{ kJ})]$$
$$= [-110.5\text{ kJ}] - [-241.8\text{ kJ}]$$
$$= 131.2\text{ kJ}$$

Check: The units (kJ) are correct. The answer is positive, which means that the reaction is endothermic. The CO has a smaller (less negative) heat of formation than H_2O, so we expect an endothermic reaction.

d) **Given:** N_2O_4 (*g*) + 4 H_2 (*g*) → N_2 (*g*) + 4 H_2O (*g*) **Find:** $\Delta H°_{rxn}$

Conceptual Plan: $\Delta H^0_{rxn} = \sum n_P \Delta H_f^0(products) - \sum n_R \Delta H_f^0(reactants)$

Solution:

Reactant/Product	ΔH_f^0 (kJ/mol from Appendix IIB)
N_2O_4 (*g*)	11.1
H_2 (*g*)	0.0
N_2 (*g*)	0.0
H_2O (*g*)	– 241.8

Be sure to pull data for the correct formula and phase.

$$\Delta H^0_{rxn} = \sum n_P \Delta H_f^0(products) - \sum n_R \Delta H_f^0(reactants)$$
$$= [1(\Delta H_f^0(\text{N}_2\,(g))) + 4(\Delta H_f^0(\text{H}_2\text{O }(g)))] - [1(\Delta H_f^0(\text{N}_2\text{O}_4\,(g))) + 4(\Delta H_f^0(\text{H}_2\,(g)))]$$
$$= [1(0.0\text{kJ}) + 4(-241.8\text{ kJ})] - [1(11.1\text{ kJ}) + 4(0.0\text{ kJ})]$$
$$= [-967.2\text{ kJ}] - [11.1\text{ kJ}]$$
$$= -978.3\text{ kJ}$$

Check: The units (kJ) are correct. The answer is negative, which means that the reaction is exothermic. The H_2O has a lower heat of formation than N_2O_4, so we expect an exothermic reaction.

49. **Given:** form glucose ($C_6H_{12}O_6$) and oxygen from sunlight, carbon dioxide and water **Find:** $\Delta H°_{rxn}$

Conceptual Plan: write balanced reaction then $\Delta H^0_{rxn} = \sum n_P \Delta H_f^0(products) - \sum n_R \Delta H_f^0(reactants)$

Solution: 6 CO_2 (*g*) + 6 H_2O (*l*) → $C_6H_{12}O_6$ (*s*) + 6 O_2 (*g*)

Reactant/Product	ΔH_f^0 (kJ/mol from Appendix IIB)
CO_2 (*g*)	– 393.5
H_2O (*l*)	– 285.8
$C_6H_{12}O$ (*s*)	– 1273.3
O_2 (*g*)	0.0

Be sure to pull data for the correct formula and phase.

$$\Delta H^0_{rxn} = \sum n_P \Delta H_f^0(products) - \sum n_R \Delta H_f^0(reactants)$$
$$= [1(\Delta H_f^0(\text{C}_6\text{H}_{12}\text{O }(s))) + 6(\Delta H_f^0(\text{O}_2\,(g)))] - [6(\Delta H_f^0(\text{CO}_2\,(g))) + 6(\Delta H_f^0(\text{H}_2\text{O }(l)))]$$
$$= [1(-1273.3\text{ kJ}) + 6(0.0\text{ kJ})] - [6(-393.5\text{ kJ}) + 6(-285.8\text{ kJ})]$$
$$= [-1273.3\text{ kJ}] - [-4076\text{ kJ}]$$
$$= +2803\text{ kJ}$$

Check: The units (kJ) are correct. The answer is positive, which means that the reaction is endothermic. The reaction requires the input of light energy, so we expect that this will be an endothermic reaction.

50. **Given:** ethanol (C_2H_5OH) combustion **Find:** ΔH°_{rxn}

Conceptual Plan: write balanced reaction then $\Delta H^0_{rxn} = \sum n_P \Delta H^0_f(products) - \sum n_R \Delta H^0_f(reactants)$

Solution: combustion is combination with oxygen to form carbon dioxide and water

$C_2H_5OH\ (l)\ +\ 3\ O_2\ (g) \rightarrow\ 2\ CO_2\ (g)\ +\ 3\ H_2O\ (g)$

Reactant/Product	ΔH^0_f (kJ/mol from Appendix IIB)
$C_2H_5OH\ (l)$	– 277.6
$O_2\ (g)$	0.0
$CO_2\ (g)$	– 393.5
$H_2O\ (g)$	– 241.8

Be sure to pull data for the correct formula and phase.

$$\begin{aligned}\Delta H^0_{rxn} &= \sum n_P \Delta H^0_f(products) - \sum n_R \Delta H^0_f(reactants)\\ &= [2(\Delta H^0_f(CO_2\ (g))) + 3(\Delta H^0_f(H_2O\ (g)))] - [1(\Delta H^0_f(C_2H_5OH\ (l))) + 3(\Delta H^0_f(O_2\ (g)))]\\ &= [2(-393.5\ kJ) + 3(-241.8\ kJ)] - [1(-277.6\ kJ) + 3(0.0\ kJ)]\\ &= [-1512.4\ kJ] - [-277.6\ kJ]\\ &= -1234.8\ kJ\end{aligned}$$

Check: The units (kJ) are correct. The answer is negative, which means that the reaction is exothermic, which is typical for combustion reactions.

51. **Given:** $2\ CH_3NO_2\ (l)\ +\ 3/2\ O_2\ (g) \rightarrow\ 2\ CO_2\ (g)\ +\ 3\ H_2O\ (l)\ +\ N_2\ (g)$ and $\Delta H^\circ_{rxn} = -1418.4$ kJ/mol

Find: $\Delta H^\circ_f\ (CH_3NO_2\ (l))$

Conceptual Plan: fill known values into $\Delta H^0_{rxn} = \sum n_P \Delta H^0_f(products) - \sum n_R \Delta H^0_f(reactants)$ **and rearrange to solve for** $\Delta H^\circ_f\ (CH_3NO_2\ (l))$

Solution:

Reactant/Product	ΔH^0_f (kJ/mol from Appendix IIB)
$O_2\ (g)$	0.0
$CO_2\ (g)$	– 393.5
$H_2O\ (l)$	– 285.8
$N_2\ (g)$	0.0

Be sure to pull data for the correct formula and phase.

$$\begin{aligned}H^0_{rxn} &= \sum n_P \Delta H^0_f(products) - \sum n_R \Delta H^0_f(reactants)\\ &= [2(\Delta H^0_f(CO_2\ (g))) + 3(\Delta H^0_f(H_2O(l))) + 1(\Delta H^0_f(N_2\ (g)))] - [2(\Delta H^0_f(CH_3NO_2\ (l))) + 3/2(\Delta H^0_f(O_2\ (g)))]\end{aligned}$$

$$-1418.4\ kJ = [2(-393.5\ kJ) + 3(-285.8\ kJ) + 1(0.0\ kJ)] - [2(\Delta H^0_f(CH_3NO_2\ (l)) + 3/2(0.0\ kJ)]$$

$$-1418.4\ kJ = [-1644.4\ kJ] - [2(\Delta H^0_f(CH_3NO_2\ (l))]$$

$$\Delta H^0_f(CH_3NO_2\ (l)) = -113.0\ kJ/mol$$

Check: The units (kJ/mol) are correct. The answer is negative, but not as negative as water and carbon dioxide, which is consistent with an exothermic combustion reaction. Also, the answer agrees with the published value for the heat of formation (– 113 kJ/mol).

52. **Given:** $4\ C_3H_5N_3O_9\ (l) \rightarrow\ 12\ CO_2\ (g)\ +\ 10\ H_2O\ (g)\ +\ 6\ N_2\ (g) +\ O_2\ (g)$ and $\Delta H^\circ_{rxn} = -5678$ kJ/mol

Find: $\Delta H^\circ_f\ (C_3H_5N_3O_9\ (l))$

Conceptual Plan: fill known values into $\Delta H^0_{rxn} = \sum n_P \Delta H^0_f(products) - \sum n_R \Delta H^0_f(reactants)$ **and rearrange to solve for** $\Delta H^\circ_f\ (C_3H_5N_3O_9\ (l))$

Solution:

Reactant/Product	ΔH_f^0(kJ/mol from Appendix IIB)
CO_2 (g)	– 393.5
H_2O (g)	– 241.8
N_2 (g)	0.0
O_2 (g)	0.0

Be sure to pull data for the correct formula and phase.

$$H^0_{rxn} = \sum n_P \Delta H^0_f(products) - \sum n_R \Delta H^0_f(reactants)$$
$$= [12(\Delta H^0_f(CO_2(g))) + 10(\Delta H^0_f(H_2O(g))) + 6(\Delta H^0_f(N_2(g))) + 1(\Delta H^0_f(O_2(g)))] - [4(\Delta H^0_f(C_3H_5N_3O_9(l)))]$$
$$-5678 \text{ kJ} = [12(-393.5 \text{ kJ}) + 10(-241.8 \text{ kJ}) + 6(0.0 \text{ kJ}) + 1(0.0 \text{ kJ})] - [4(\Delta H^0_f(C_3H_5N_3O_9(l)))]$$
$$-5678 \text{ kJ} = [-7140. \text{ kJ}] - [4(\Delta H^0_f(C_3H_5N_3O_9(l))]$$
$$\Delta H^0_f(CH_3NO_2(l)) = -365.5 \text{ kJ/mol}$$

Check: The units (kJ/mol) are correct. The answer is negative, which is not surprising since this is a combustion reaction.

53. **Given:** billiard ball$_A$ = system: $m_A = 0.17$ kg, $v_{A1} = 4.5$ m/s slows to $v_{A2} = 3.8$ m/s and $v_{A3} = 0$; ball$_B$: $m_B = 0.17$ kg, $v_{B1} = 0$ and $v_{B2} = 3.8$ m/s and $KE = ½\, mv^2$ **Find:** w, q, ΔE_{sys}

Conceptual Plan:

m, v* → *KE then ***KE$_{A3}$, KE$_{A1}$* → Δ*E$_{sys}$*** and ***KE$_{A2}$, KE$_{A1}$* → *q*** and ***KE$_{B2}$, KE$_{B1}$* → *w*$_B$**

$KE = ½\, mv^2$ $\quad \Delta E_{sys} = KE_{A3} - KE_{A1}$ $\quad q = KE_{A2} - KE_{A1}$ $\quad w_B = KE_{B2} - KE_{B1}$

Δ*E$_{sys}$*, *q* → *w*$_A$ verify that *w*$_A$ = −*w*$_B$ so that no heat is transferred to ball$_B$

$\Delta E = q + w$

Solution: $KE = ½\, mv^2$ since m is in kg and v is in m/s, KE will be in kg·m^2/s^2 which is joule.

$$KE_{A1} = ½\,(0.17 \text{ kg})\left(4.5\frac{\text{m}}{\text{s}}\right)^2 = 1.\underline{7}213\frac{\text{kg}\cdot\text{m}^2}{\text{s}^2} = 1.\underline{7}213 \text{ J},$$

$$KE_{A2} = ½\,(0.17 \text{ kg})\left(3.8\frac{\text{m}}{\text{s}}\right)^2 = 1.\underline{2}274\frac{\text{kg}\cdot\text{m}^2}{\text{s}^2} = 1.\underline{2}274 \text{ J},$$

$$KE_{A3} = ½\,(0.17 \text{ kg})\left(0\frac{\text{m}}{\text{s}}\right)^2 = 0\frac{\text{kg}\cdot\text{m}^2}{\text{s}^2} = 0 \text{ J}, \quad KE_{B1} = ½\,(0.17 \text{ kg})\left(0\frac{\text{m}}{\text{s}}\right)^2 = 0\frac{\text{kg}\cdot\text{m}^2}{\text{s}^2} = 0 \text{ J and}$$

$$KE_{B2} = ½\,(0.17 \text{ kg})\left(3.8\frac{\text{m}}{\text{s}}\right)^2 = 1.\underline{2}274\frac{\text{kg}\cdot\text{m}^2}{\text{s}^2} = 1.\underline{2}274 \text{ J}.$$

$$\Delta E_{sys} = KE_{A3} - KE_{A1} = 0 \text{ J} - 1.\underline{7}213 \text{ J} = -1.\underline{7}213 \text{ J} = -1.7 \text{ J},$$

$$q = KE_{A2} - KE_{A1} = 1.\underline{2}274 \text{ J} - 1.\underline{7}213 \text{ J} = -0.\underline{4}939 \text{ J} = -0.5 \text{ J},$$

$$w_B = KE_{B2} - KE_{B1} = 1.\underline{2}274 \text{ J} - 0 \text{ J} = 1.\underline{2}274 \text{ J} \quad \text{and } w = \Delta E - q = -1.\underline{7}213 \text{ J} - -0.\underline{4}939 \text{ J} = -1.\underline{2}274 \text{ J} = -1.2 \text{ J}.$$

Since $w_A = -w_B$ so that no heat is transferred to ball$_B$.

Check: The units (J) are correct. Since the ball is initially moving and is stopped at the end, it has lost energy (negative ΔE_{sys}). As the ball slows due to friction, it is releasing heat (negative q). The kinetic energy is transferred to a second ball, so it does work (w negative).

54. **Given:** 100-W lightbulb in a piston; bulb on for 0.015 hr, $V_i = 0.85$ L, $V_f = 5.88$ L, $P = 1.0$ atm
Find: w, q, ΔE_{sys}

Conceptual Plan:

bulb wattage, time → ΔE_{sys}(Wh) → ΔE_{sys}(Wh) → ΔE_{sys}(J) and **V_i, V_f → ΔV then**

$$\Delta E = (wattage)(time) \qquad \frac{1\ \text{kW}}{1000\ \text{W}} \qquad \frac{3.60 \times 10^6\ \text{J}}{1\text{kWh}} \qquad \Delta V = V_f - V_i$$

P, ΔV → w (L atm) → w (J) **finally ΔE_{sys}, w → q**

$$w = -P\Delta V \qquad \frac{101.3\ \text{J}}{1\ \text{L atm}} \qquad \Delta E = q + w$$

Solution: $\Delta E = (wattage)(time) = (100\ \text{W})(0.015\ \text{hr}) = 1.5\ \text{Wh} \times \frac{1\ \text{kW}}{1000\ \text{W}} \times \frac{3.60 \times 10^6\ \text{J}}{1\ \text{kWh}} = 5400\ \text{J}$ and

$\Delta V = V_f - V_i = 5.88\ \text{L} - 0.85\ \text{L} = 5.03\ \text{L}$ then

$w = -P\Delta V = -1.0\ \text{atm} \times 5.03\ \text{L} \times \frac{101.3\ \text{J}}{1\ \text{L atm}} = -50\underline{9}.539\ \text{J} = -5.0 \times 10^2\ \text{J} \quad \Delta E = q + w$

Rearrange to solve for q. $q = \Delta E_{sys} - w = +5\underline{4}00\ \text{J} - -50\underline{9}.539\ \text{J} = 5900\ \text{J}$

Check: The units (J) are correct. Electricity is added so energy is added (positive ΔE_{sys}). The piston expands and so does work (negative work). In order for the lightbulb to generate light, it must be heated or absorb energy (positive q).

55. **Given:** $H_2O\ (l) \rightarrow H_2O\ (g)$ $\Delta H°_{rxn} = +44.01$ kJ/mol; $\Delta T_{body} = -0.50$ °C, $m_{body} = 95$ kg, $C_{body} = 4.0$ J/g °C
Find: m_{H2O}
Conceptual Plan:
kg → g then m_{body}, ΔT, C_{body} → q_{body} → q_{rxn} (J) → q_{rxn} (kJ) → mol H_2O → g H_2O

$$\frac{1000\ \text{g}}{1\ \text{kg}} \qquad q_{body} = m_{body}\, C_{body}\, \Delta T_{body} \quad q_{rxn} = -q_{body} \qquad \frac{1\ \text{kJ}}{1000\ \text{J}} \qquad \frac{1\ \text{mol}}{44.01\ \text{kJ}} \qquad \frac{18.01\ \text{g}}{1\ \text{mol}}$$

Solution: $95\ \text{kg} \times \frac{1000\ \text{g}}{1\ \text{kg}} = 95000\ \text{g}$ then

$q_{body} = m_{body}\, C_{body}\, \Delta T_{body} = 95000\ \text{g} \times 4.0 \frac{\text{J}}{\text{g} \cdot °\text{C}} \times (-0.50\ °\text{C}) = -1\underline{9}0000\ \text{J}$ then

$q_{rxn} = -q_{body} = 1\underline{9}0000\ \text{J} \times \frac{1\ \text{kJ}}{1000\ \text{J}} \times \frac{1\ \text{mol}}{44.01\ \text{kJ}} \times \frac{18.01\ \text{g}}{1\ \text{mol}} = 78\ \text{g}\,H_2O$

Check: The units (g) are correct. The magnitude of the answer (78) makes physical sense because a person can sweat this much on a hot day.

56. **Given:** LP gas combustion, $\Delta H°_{rxn} = -2044$ kJ; 1.5 L water, $T_{H2Oi} = 25.0$ °C, $T_{H2Of} = 100.0$ °C, 15 % efficiency
Find: $m_{LP\ gas}$
Conceptual Plan:
L → mL → g and T_i, T_f → ΔT then m_{H2O}, ΔT_{H2O}, C_{H2O} → q_{H2O} → q_{rxn}

$$\frac{1000\ \text{mL}}{1\ \text{L}} \qquad \frac{1.0\ \text{g}}{1.0\ \text{mL}} \qquad \Delta T = T_f - T_i \qquad q_{H2O} = m_{H2O}\, C_{H2O}\, \Delta T_{H2O} \quad q_{rxn} = -q_{H2O}$$

then q_{rxn} needed → q_{rxn} generated (J) → q_{rxn} (kJ) → mol LP gas → g LP gas

$$\frac{100\ \text{J generated}}{15\ \text{J needed}} \qquad \frac{1\ \text{kJ}}{1000\ \text{J}} \qquad \frac{1\ \text{mol}}{-2044\ \text{kJ}} \qquad \frac{44.09\ \text{g}}{1\ \text{mol}}$$

Solution: $1.5\ \text{L} \times \frac{1000\ \text{mL}}{1\ \text{L}} \times \frac{1.0\ \text{g}}{1.0\ \text{mL}} = 1500\ \text{g}$ and $\Delta T = T_f - T_i = 100.0\ °\text{C} - 25.0\ °\text{C} = 75.0\ °\text{C}$ then

$q_{H2O} = m_{H2O}\, C_{H2O}\, \Delta T_{H2O} = 1500\ \text{g} \times 4.184 \frac{\text{J}}{\text{g} \cdot °\text{C}} \times (75.0\ °\text{C}) = 4\underline{7}0700\ \text{J}$ then

$q_{rxn} = -q_{H2O} = -4\underline{7}0700\ \text{J needed} \times \frac{100\ \text{J generated}}{15\ \text{J needed}} \times \frac{1\ \text{kJ}}{1000\ \text{J}} \times \frac{1\ \text{mol}}{-2044\ \text{kJ}} \times \frac{44.09\ \text{g}}{1\ \text{mol}} = 68\ \text{g LP gas}$

Check: The units (g) are correct. The magnitude of the answer (68) makes physical sense because a tank of LP gas contains many orders of magnitude more than this amount.

57. **Given:** $H_2O\ (s) \rightarrow H_2O\ (l)$ $\Delta H°_f\ (H_2O\ (s)) = -291.8$ kJ/mol; 355 mL beverage $T_{Bevi} = 25.0$ °C, $T_{Bevf} = 0.0$ °C, $C_{Bev} = 4.184$ J/g °C, $d_{Bev} = 1.0$ g/mL **Find:** $\Delta H°_{rxn}$ (ice melting) and m_{ice}

Conceptual Plan: $\Delta H^0_{rxn} = \sum n_P \Delta H^0_f(products) - \sum n_R \Delta H^0_f(reactants)$ **mL → g and T_i, T_f → ΔT then**

$\frac{1.0\text{ g}}{1.0\text{ mL}}$ $\Delta T = T_f - T_i$

m_{H2O}, ΔT_{H2O}, C_{H2O} → q_{H2O} → q_{rxn} (J) → q_{rxn} (kJ) → mol ice → g ice

$q_{Bev} = m_{Bev}\, C_{Bev}\, \Delta T_{Bev}$ $q_{rxn} = -\, q_{Bev}$ $\frac{1\text{ kJ}}{1000\text{ J}}$ $\frac{1\text{ mol}}{\Delta H^0_{rxn}}$ $\frac{18.01\text{ g}}{1\text{ mol}}$

Solution:

Reactant/Product	ΔH^0_f (kJ/mol from Appendix IIB)
H_2O (*s*)	– 291.8
H_2O (*l*)	– 285.8

Be sure to pull data for the correct formula and phase.

$$\begin{aligned}\Delta H^0_{rxn} &= \sum n_P \Delta H^0_f(products) - \sum n_R \Delta H^0_f(reactants)\\ &= [1(\Delta H^0_f(\text{H}_2\text{O }(l)))] - [1(\Delta H^0_f(\text{H}_2\text{O }(s)))]\\ &= [1(-285.8\text{ kJ})] - [1(-291.8\text{ kJ})]\\ &= +6.0\text{ kJ}\end{aligned}$$

$355\text{ mL} \times \frac{1.0\text{ g}}{1.0\text{ mL}} = 355\text{ g}$ and $\Delta T = T_f - T_i = 0.0\text{ °C} - 25.0\text{ °C} = -25.0\text{ °C}$ then

$q_{Bev} = m_{Bev}\, C_{Bev}\, \Delta T_{Bev} = 355\text{ g} \times 4.184 \frac{\text{J}}{\text{g}\cdot\text{°C}} \times (-25.0\text{ °C}) = -37\underline{1}33\text{ J}$ then

$q_{rxn} = -q_{Bev} = -37\underline{1}33\text{ J} \times \frac{1\text{ kJ}}{1000\text{ J}} \times \frac{1\text{ mol}}{-6.0\text{ kJ}} \times \frac{18.01\text{ g}}{1\text{ mol}} = 110\text{ g ice}$

Check: The units (kJ and g) are correct. The answer is positive, which means that the reaction is endothermic. We expect an endothermic reaction because we know that heat must be added to melt ice. The magnitude of the answer (110 g) makes physical sense because it is much smaller than the weight of the beverage and it would fit in a glass with the beverage.

58. **Given:** CO_2 (*s*) → CO_2 (*g*) $\Delta H°_f$ (CO_2 (*s*)) = – 427.4 kJ/mol; 15.0 L water T_{H2Oi} = 85 °C, T_{H2Of} = 25 °C
Find: $\Delta H°_{rxn}$ (dry ice sublimation) and m_{dryice}
Conceptual Plan:

$\Delta H^0_{rxn} = \sum n_P \Delta H^0_f(products) - \sum n_R \Delta H^0_f(reactants)$ **L → mL → g and T_i, T_f → ΔT then**

$\frac{1000\text{ mL}}{1\text{ L}}$ $\frac{1.00\text{ g}}{1.00\text{ mL}}$ $\Delta T = T_f - T_i$

m_{H2O}, ΔT_{H2O}, C_{H2O} → q_{H2O} → q_{rxn} (J) → q_{rxn} (kJ) → mol ice → g ice

$q_{H2O} = m_{H2O}\, C_{H2O}\, \Delta T_{H2O}$ $q_{rxn} = -\, q_{H2O}$ $\frac{1\text{ kJ}}{1000\text{ J}}$ $\frac{1\text{ mol}}{\Delta H^0_{rxn}}$ $\frac{44.01\text{ g}}{1\text{ mol}}$

Solution:

Reactant/Product	ΔH^0_f (kJ/mol from Appendix IIB)
CO_2 (*s*)	– 427.4
CO_2 (*g*)	– 393.5

Be sure to pull data for the correct formula and phase.

$$\begin{aligned}\Delta H^0_{rxn} &= \sum n_P \Delta H^0_f(products) - \sum n_R \Delta H^0_f(reactants)\\ &= [1(\Delta H^0_f(\text{CO}_2\ (g)))] - [1(\Delta H^0_f(\text{CO}_2\ (s)))]\\ &= [1(-393.5\text{ kJ})] - [1(-427.4\text{ kJ})]\\ &= +33.9\text{ kJ}\end{aligned}$$

$15.0\text{ L} \times \frac{1000\text{ mL}}{1\text{ L}} \times \frac{1.00\text{ g}}{1.00\text{ mL}} = 15\underline{0}00\text{ g}$ and $\Delta T = T_f - T_i = 25\text{ °C} - 85\text{ °C} = -60.\text{ °C}$ then

$q_{H2O} = m_{H2O}\, C_{H2O}\, \Delta T_{H2O} = 15\underline{0}00\text{ g} \times 4.184 \frac{\text{J}}{\text{g}\cdot\text{°C}} \times (-60.\text{ °C}) = -3\underline{7}65600\text{ J}$ then

$$q_{rxn} = -q_{H2O} = 3\underline{7}65600 \text{ J} \times \frac{1 \text{ kJ}}{1000 \text{ J}} \times \frac{1 \text{ mol}}{33.9 \text{ kJ}} \times \frac{44.01 \text{ g}}{1 \text{ mol}} = 4900 \text{ g dry ice}$$

Check: The units (kJ and g) are correct. The answer is positive, which means that the reaction is endothermic. We expect an endothermic reaction because we know that heat must be added to sublime dry ice. The magnitude of the answer (4900 g) makes physical sense because the temperature change of the water is fairly large, and the volume of water is large. It is a reasonable amount to put in a cooler.

59. **Given:** 25.5 g aluminum, $T_{Ali} = 65.4$ °C, 55.2 g water, $T_{H2Oi} = 22.2$ °C, **Find:** T_f
Conceptual Plan: **pull C_s values from table** **then** $m, C_s, T_i \rightarrow T_f$

Al: $0.903 \frac{\text{J}}{\text{g}\cdot°\text{C}}$ H_2O: $4.18 \frac{\text{J}}{\text{g}\cdot°\text{C}}$ $q = m\, C_s(T_f - T_i)$ then set $q_{Al} = -q_{H2O}$

Solution: $q = m\, C_s(T_f - T_i)$ substitute in values and set $q_{Al} = -q_{H2O}$.

$$q_{Al} = m_{Al}\, C_{Al}(T_f - T_{Ali}) = 25.5 \text{ g} \times 0.903 \frac{\text{J}}{\text{g}\cdot°\text{C}} \times (T_f - 65.4\ °\text{C}) =$$

$$-q_{H2O} = -m_{H2O}\, C_{H2O}(T_f - T_{H2Oi}) = -55.2 \text{ g} \times 4.18 \frac{\text{J}}{\text{g}\cdot°\text{C}} \times (T_f - 22.2°\text{C})$$

Rearrange to solve for T_f.

$$23.\underline{0}265 \frac{\text{J}}{°\text{C}} \times (T_f - 65.4\ °\text{C}) = -23\underline{0}.736 \frac{\text{J}}{°\text{C}} \times (T_f - 22.2°\text{C}) \rightarrow$$

$$23.\underline{0}265 \frac{\text{J}}{°\text{C}} T_f - 15\underline{0}5.93 \text{ J} = -23\underline{0}.736 \frac{\text{J}}{°\text{C}} T_f + 51\underline{2}2.34 \text{ J} \rightarrow$$

$$-51\underline{2}2.34 \text{ J} - 15\underline{0}5.93 \text{ J} = -23\underline{0}.736 \frac{\text{J}}{°\text{C}} T_f - 23.\underline{0}265 \frac{\text{J}}{°\text{C}} T_f \rightarrow 66\underline{2}8.27 \text{ J} = 25\underline{3}.7625 \frac{\text{J}}{°\text{C}} T_f \rightarrow$$

$$T_f = \frac{66\underline{2}8.27 \text{ J}}{25\underline{3}.7625 \frac{\text{J}}{°\text{C}}} = 26.1\ °\text{C}$$

Check: The units (°C) are correct. The magnitude of the answer (26) makes physical sense because the heat transfer is dominated by the water (larger mass and larger specific heat capacity). The final temperature should be closer to the initial temperature of water than of aluminum.

60. **Given:** ethanol: 50.0 mL; $d = 0.789$ g/mL, $T_{EtOHi} = 7.0$ °C, water: 50.0 mL; $d = 1.0$ g/mL $T_{H2Oi} = 28.4$ °C,
Find: T_f
Conceptual Plan: **pull C_s values from Table 6.4** **mL → g** **then** $m, C_s, T_i \rightarrow T_f$

EtOH: $2.42 \frac{\text{J}}{\text{g}\cdot°\text{C}}$ H_2O: $4.18 \frac{\text{J}}{\text{g}\cdot°\text{C}}$ EtOH: $\frac{0.789 \text{ g}}{1.0 \text{ mL}}$ H_2O: $\frac{1.0 \text{ g}}{1.0 \text{ mL}}$ $q = m\, C_s(T_f - T_i)$ then set $q_{Al} = -q_{H2O}$

Solution: $50.0 \text{ mL} \times \frac{0.789 \text{ g}}{1.0 \text{ mL}} = 39.\underline{4}5 \text{ g EtOH}$ and $50.0 \text{ mL} \times \frac{1.0 \text{ g}}{1.0 \text{ mL}} = 50.0 \text{ g } H_2O$ then

$q = m\, C_s(T_f - T_i)$ substitute in values and set $q_{EtOH} = -q_{H2O}$.

$$q_{EtOH} = m_{EtOH}\, C_{EtOH}(T_f - T_{EtOHi}) = 39.\underline{4}5 \text{ g} \times 2.42 \frac{\text{J}}{\text{g}\cdot°\text{C}} \times (T_f - 7.0\ °\text{C}) =$$

Rearrange to solve for T_f.

$$-q_{H2O} = -m_{H2O}\, C_{H2O}(T_f - T_{H2Oi}) = -50.0 \text{ g} \times 4.18 \frac{\text{J}}{\text{g}\cdot°\text{C}} \times (T_f - 28.4°\text{C})$$

$$95.\underline{4}69 \frac{\text{J}}{°\text{C}} \times (T_f - 7.0\ °\text{C}) = -20\underline{9}.0 \frac{\text{J}}{°\text{C}} \times (T_f - 28.4°\text{C}) \rightarrow$$

$$95.\underline{4}69 \frac{\text{J}}{°\text{C}} T_f - 6\underline{6}8.283 \text{ J} = -20\underline{9}.0 \frac{\text{J}}{°\text{C}} T_f + 59\underline{3}5.6 \text{ J} \rightarrow$$

$$-6\underline{6}8.283 \text{ J} - 59\underline{3}5.6 \text{ J} = -20\underline{9}.2 \frac{\text{J}}{°\text{C}} T_f - 95.\underline{4}69 \frac{\text{J}}{°\text{C}} T_f \rightarrow 66\underline{0}9.563 \text{ J} = 30\underline{4}.669 \frac{\text{J}}{°\text{C}} T_f \rightarrow$$

$$T_f = \frac{6603.783 \text{ J}}{304.669 \frac{\text{J}}{°\text{C}}} = 21.7\ °\text{C}$$

Check: The units (°C) are correct. The magnitude of the answer (22) makes physical sense because the heat transfer is dominated by the water (larger mass and larger specific heat capacity). The final temperature should be closer to the initial temperature of water than of ethanol.

61. **Given:** palmitic acid ($C_{16}H_{32}O_2$) combustion $\Delta H°_f$ ($C_{16}H_{32}O_2$ (*s*)) = – 208 kJ/mol; sucrose ($C_{12}H_{22}O_{11}$) combustion $\Delta H°_f$ ($C_{12}H_{22}O_{11}$ (*s*)) = – 2226.1 kJ/mol **Find:** $\Delta H°_{rxn}$ in kJ/mol and Cal/g

Conceptual Plan: write balanced reaction then $\Delta H^0_{rxn} = \sum n_P \Delta H^0_f(products) - \sum n_R \Delta H^0_f(reactants)$ **then**

kJ/mol → J/mol → Cal/mol → Cal/g

$\frac{1000 \text{ J}}{1 \text{ kJ}}$ $\frac{1 \text{ Cal}}{4184 \text{ J}}$ PA: $\frac{1 \text{ mol}}{256.42 \text{ g}}$ S: $\frac{1 \text{ mol}}{342.30 \text{ g}}$

Solution: combustion is combination with oxygen to form carbon dioxide and water (*l*)

$C_{16}H_{32}O_2$ (*s*) + 23 O_2 (*g*) → 16 CO_2 (*g*) + 16 H_2O (*l*)

Reactant/Product	ΔH^0_f(kJ/mol from Appendix IIB)
$C_{16}H_{32}O_2$ (*s*)	– 208
O_2 (*g*)	0.0
CO_2 (*g*)	– 393.5
H_2O (*g*)	– 285.8

Be sure to pull data for the correct formula and phase.

$$\Delta H^0_{rxn} = \sum n_P \Delta H^0_f(products) - \sum n_R \Delta H^0_f(reactants)$$
$$= [16(\Delta H^0_f(CO_2\ (g))) + 16(\Delta H^0_f(H_2O\ (l)))] - [1(\Delta H^0_f(C_{16}H_{32}O_2\ (s))) + 23(\Delta H^0_f(O_2\ (g)))]$$
$$= [16(-393.5 \text{ kJ}) + 16(-285.8 \text{ kJ})] - [1(-208 \text{ kJ}) + 23(0.0 \text{ kJ})]$$
$$= [-10868.8 \text{ kJ}] - [-208 \text{ kJ}]$$
$$= -10,66\underline{0}.8 \text{ kJ/mol} = -10,661 \text{ kJ/mol}$$

$$-10,66\underline{0}.8 \frac{\text{kJ}}{\text{mol}} \times \frac{1000 \text{ J}}{1 \text{ kJ}} \times \frac{1 \text{ Cal}}{4184 \text{ J}} \times \frac{1 \text{ mol}}{256.42 \text{ g}} = -9.9378 \text{ Cal/g}$$

$C_{12}H_{22}O_{11}$ (*s*) + 12 O_2 (*g*) → 12 CO_2 (*g*) + 11 H_2O (*l*)

Reactant/Product	ΔH^0_f(kJ/mol from Appendix IIB)
$C_{12}H_{22}O_{11}$ (*s*)	– 2226.1
O_2 (*g*)	0.0
CO_2 (*g*)	– 393.5
H_2O (*g*)	– 285.8

Be sure to pull data for the correct formula and phase.

$$\Delta H^0_{rxn} = \sum n_P \Delta H^0_f(products) - \sum n_R \Delta H^0_f(reactants)$$
$$= [12(\Delta H^0_f(CO_2\ (g))) + 11(\Delta H^0_f(H_2O\ (l)))] - [1(\Delta H^0_f(C_{12}H_{22}O_{11}\ (s))) + 12(\Delta H^0_f(O_2\ (g)))]$$
$$= [12(-393.5 \text{ kJ}) + 11(-285.8 \text{ kJ})] - [1(-2226.1 \text{ kJ}) + 12(0.0 \text{ kJ})]$$
$$= [-7865.8 \text{ kJ}] - [-2226.1 \text{ kJ}]$$
$$= -5639.7 \text{ kJ/mol}$$

$$-5639.7 \frac{\text{kJ}}{\text{mol}} \times \frac{1000 \text{ J}}{1 \text{ kJ}} \times \frac{1 \text{ Cal}}{4184 \text{ J}} \times \frac{1 \text{ mol}}{342.30 \text{ g}} = -3.938 \text{ Cal/g}$$

Check: The units (kJ/mol and Cal/g) are correct. The magnitudes of the answers are consistent with the food labels we see every day.

62. **Given:** hydrogen, methanol (CH_3OH), and octane combustion **Find:** q released in kJ/kg
Conceptual Plan: write balanced reaction then $\Delta H^0_{rxn} = \sum n_P \Delta H^0_f(products) - \sum n_R \Delta H^0_f(reactants)$
then kJ/mol → kJ/g → kJ/kg

$$H_2: \frac{1\,\text{mol}}{2.016\ \text{g}} \quad MeOH: \frac{1\,\text{mol}}{32.04\ \text{g}} \quad O: \frac{1\ \text{mol}}{114.22\,\text{g}} \qquad \frac{1000\ \text{g}}{1\ \text{kg}}$$

Solution: combustion is combination with oxygen to form carbon dioxide and water
$H_2\,(g) + \frac{1}{2}\,O_2\,(g) \rightarrow H_2O\,(g)$ This reaction is the heat of formation of gaseous water so $\Delta H^0_{rxn} = -\,241.8$ kJ/mol.

$$-241.8\,\frac{\text{kJ}}{\text{mol}} \times \frac{1\,\text{mol}}{2.016\ \text{g}} \times \frac{1000\ \text{g}}{1\,\text{kg}} = -\,1.199\times10^5\,\text{kJ/g}\,H_2 \quad \text{and}$$

$CH_3OH\,(l) + 3/2\,O_2\,(g) \rightarrow CO_2\,(g) + 2\,H_2O\,(g)$

Reactant/Product	ΔH^0_f (kJ/mol from Appendix IIB)
$CH_3OH\,(l)$	− 238.6
$O_2\,(g)$	0.0
$CO_2\,(g)$	− 393.5
$H_2O\,(g)$	− 241.8

Be sure to pull data for the correct formula and phase.

$$\Delta H^0_{rxn} = \sum n_P \Delta H^0_f(products) - \sum n_R \Delta H^0_f(reactants)$$
$$= [1(\Delta H^0_f(CO_2\,(g))) + 2(\Delta H^0_f(H_2O\,(g)))] - [1(\Delta H^0_f(CH_3OH\,(l))) + 3/2(\Delta H^0_f(O_2\,(g)))]$$
$$= [1(-\,393.5\ \text{kJ}) + 2(-\,241.8\ \text{kJ})] - [1(-238.6\ \text{kJ}) + 3/2(0.0\ \text{kJ})]$$
$$= [-\,877.1\ \text{kJ}] - [-238.6\ \text{kJ}]$$
$$= -\,638.5\ \text{kJ}$$

$$-638.5\,\frac{\text{kJ}}{\text{mol}} \times \frac{1\,\text{mol}}{32.04\ \text{g}} \times \frac{1000\ \text{g}}{1\,\text{kg}} = -\,1.993\times10^4\,\text{kJ/kg}\,CH_3OH \quad \text{and}$$

$$-5074.1\,\frac{\text{kJ}}{\text{mol}} \times \frac{1\,\text{mol}}{114.22\ \text{g}} \times \frac{1000\ \text{g}}{1\,\text{kg}} = -\,4.4424\times10^4\,\text{kJ/kg}\,C_8H_{18}$$

Hydrogen delivers the most energy per weight of fuel. This is not surprising since hydrogen is so light.
Check: The units (kJ/kg fuel) are correct. The magnitude of the answers ($10^4 - 10^5$) makes physical sense because there are many moles of fuel in a kg and the heat of reactions are high.

63. At constant P $\Delta H_{rxn} = q_P$ and at constant V $\Delta E_{rxn} = q_V = \Delta H_{rxn} - P\,\Delta V$. $PV = nRT$ at constant P, and a constant number of moles of gas, as we change the T the only variable that can change is V, so $P\Delta V = nR\Delta T$. Substituting into the equation for ΔE_{rxn} we get $\Delta E_{rxn} = \Delta H_{rxn} - nR\Delta T$ or $\Delta H_{rxn} = \Delta E_{rxn} + nR\Delta T$.

64. **Given:** $SO_2\,(g) + 1/2\,O_2\,(g) \rightarrow SO_3\,(g)$, $\Delta H_{rxn} = +89.5$ kJ, and $\Delta H_f(SO_3\,(g)) = -\,204.2$ kJ
Find: $\Delta H_{rxn}(SO_2\,(g))$
Conceptual Plan: fill known values into $\Delta H^0_{rxn} = \sum n_P \Delta H^0_f(products) - \sum n_R \Delta H^0_f(reactants)$ **and rearrange to solve for $\Delta H^\circ_f\,(SO_2\,(g))$**
Solution:

Reactant/Product	ΔH^0_f (kJ/mol)
$O_2\,(g)$	0.0
$SO_3\,(g)$	−204.2

$$\Delta H^0_{rxn} = \sum n_P \Delta H^0_f(products) - \sum n_R \Delta H^0_f(reactants)$$
$$= [1(\Delta H^0_f(SO_3(g)))] - [1(\Delta H^0_f(SO_2(g))) + 1/2(\Delta H^0_f(O_2(g)))]$$
$$= [1(-204.2 \text{ kJ})] - [1(\Delta H^0_f(SO_2(g))) + 1/2(0.0 \text{ kJ})]$$
$$+89.5 \text{ kJ} = [-204.2 \text{ kJ}] - [\Delta H^0_f(SO_2(g)]$$
$$\Delta H^0_f(SO_2(g)) = -293.7 \text{ kJ}$$

Check: The units (kJ) are correct. The answer is more negative than $\Delta H_f(SO_3(g))$ which makes sense since the reaction is endothermic.

65. **Given:** 16 g peanut butter, bomb calorimeter, $T_i = 22.2\ °C$, $T_f = 25.4\ °C$, $C_{cal} = 120.0$ kJ/°C
Find: calories in peanut butter
Conceptual Plan:
$T_i, T_f \rightarrow \Delta T$ then $\Delta T, C_{cal} \rightarrow q_{cal} \rightarrow q_{rxn}$ (kJ) $\rightarrow q_{rxn}$ (kJ) $\rightarrow q_{rxn}$ (Cal)
$\Delta T = T_f - T_i$ $\quad q_{cal} = C_{cal}\Delta T$ $\quad q_{rxn} = -q_{cal}$ $\quad \frac{1000 \text{ J}}{1 \text{ kJ}}$ $\quad \frac{1 \text{ Cal}}{4184 \text{ J}}$

then q_{rxn} (Cal) $\rightarrow$ Cal/g
÷16 g peanut butter

Solution: $\Delta T = T_f - T_i = 25.4\ °C - 22.2\ °C = 3.2\ °C$ then $q_{cal} = C_{cal}\Delta T = 120.0 \frac{\text{kJ}}{°C} \times 3.2\ °C = 38\underline{4}$ kJ

then $q_{rxn} = -q_{cal} = -38\underline{4} \text{ kJ} \times \frac{1000 \text{ J}}{1 \text{ kJ}} \times \frac{1 \text{ Cal}}{4184 \text{ J}} = 9\underline{1}.778$ Cal then $\frac{9\underline{1}.778 \text{ Cal}}{16 \text{ g}} = 5.7$ Cal/g

Check: The units (Cal/g) are correct. The magnitude of the answer (6) makes physical sense because there is a significant percentage of fat and sugar in peanut butter. The answer is in line with the answers in #93.

66. **Given:** 2.0 mol H_2 (*g*) + 1.0 mol O_2 (*g*) at 25°C **Find:** temperature of water
Conceptual Plan: write balanced reaction then $\Delta H^0_{rxn} = \sum n_P \Delta H^0_f(products) - \sum n_R \Delta H^0_f(reactants)$
then kJ/mol $\rightarrow q$(kJ) $\rightarrow q$(J) then 2.0 mol H_2 + 1.0 mol O_2 $\rightarrow$ mol H_2O $\rightarrow$ g H_2O then
x mol of limiting reagent $\quad \frac{1000 \text{ J}}{1 \text{ kJ}}$ $\quad \frac{1 \text{ mol } H_2O}{1 \text{ mol } H_2}$ $\quad \frac{18.01 \text{ g}}{1 \text{ mol}}$

$q_{rxn} \rightarrow q_{H2O}$ then pull C_s for H_2O (*l*) then $q, m, C_s, T_i \rightarrow T_f$
$q_{rxn} = -q_{H2O}$ $\quad 4.18 \frac{\text{J}}{\text{g} \cdot °C}$ $\quad q = m\,C_s(T_f - T_i)$

Solution: combustion is combination with oxygen to form water, choose liquid water because $T = 25°C$.
H_2 (*g*) + ½ O_2 (*g*) $\rightarrow$ H_2O (*l*) This reaction is the heat of formation of gaseous water so
$\Delta H^0_{rxn} = -285.8$ kJ/mol The two reactants are in the stoichiometric ratio, so either amount can be used.

$$-285.8 \frac{\text{kJ}}{1 \text{ mol } H_2} \times 2.0 \text{ mol } H_2 \times \frac{1000 \text{ J}}{1 \text{ kJ}} = -5\underline{7}1600 \text{ J} \quad \text{and} \quad q_{H2O} = -q_{rxn} = 5\underline{7}1600 \text{ J}$$

$$2.0 \text{ mol } H_2 \times \frac{1 \text{ mol } H_2O}{1 \text{ mol } H_2} \times \frac{18.01 \text{ g}}{1 \text{ mol } H_2O} = 36.02 \text{ g } H_2O$$ then $q = m\,C_s(T_f - T_i)$ Rearrange to solve for T_f.

$$T_f = \frac{m\,C_s\,T_i + q}{m\,C_s} = \frac{\left(36.02 \text{ g} \times 4.18 \frac{\text{J}}{\text{g} \cdot °C} \times 25\ °C\right) + 5\underline{7}1600 \text{ J}}{36.02 \text{ g} \times 4.18 \frac{\text{J}}{\text{g} \cdot °C}} = 3821°C$$ This is much higher than the boiling point of water. The heat needed to raise the water to 100 °C is

$$q = m\,C_s(T_f - T_i) = 36.02 \text{ g} \times 4.18 \frac{\text{J}}{\text{g} \times °C} \times (100\ °C - 25\ °C) = 1\underline{1}292 \text{ J}$$ so $5\underline{7}1600 \text{ J} - 1\underline{1}292 \text{ J} = 5\underline{6}0308 \text{ J}$

is still available. 2.0 moles H_2O utilizes 88,000 J (= 44 kJ/mol) so $4\underline{7}2308$ J (= $5\underline{6}0308$ J – 88000 J) is available to heat steam. Note: Cs (steam) = 2.04 J/g · °C.

Using equation from above $T_f = \frac{m\,C_s\,T_i + q}{m\,C_s} = \frac{\left(36.02\ \text{g} \times 2.04\ \frac{\text{J}}{\text{g}\cdot{}^\circ\text{C}} \times 100.\ {}^\circ\text{C}\right) + 47\underline{2}308\ \text{J}}{36.02\ \text{g} \times 2.04\ \frac{\text{J}}{\text{g}\cdot{}^\circ\text{C}}} = 6500\,{}^\circ\text{C}$

Check: The units (°C) are correct. The temperature is extremely high. A large amount of heat is liberated and only a relatively small amount of mass absorbs it.

67. **Given:** $V_1 = 20.0$ L at $P_1 = 3.0$ atm; $P_2 = 1.5$ atm let expand at constant T. **Find:** w, q, ΔE_{sys}

Conceptual Plan: $V_1, P_1, P_2 \rightarrow V_2$ **then** $V_1, V_2 \rightarrow \Delta V$ **then** $P, \Delta V \rightarrow w$ **(L atm)** $\rightarrow w$ **(J)**

$P_1 V_1 = P_2 V_2 \qquad \Delta V = V_2 - V_1 \qquad w = -P\,\Delta V \qquad \frac{101.3\ \text{J}}{1\ \text{L atm}}$

for an ideal gas ΔE_{sys} α T, so since this is a constant temperature process $\Delta E_{sys} = 0$ finally $\Delta E_{sys}, w \rightarrow q$

$\Delta E = q + w$

Solution: $P_1 V_1 = P_2 V_2$ Rearrange to solve for V_2. $V_2 = V_1 \frac{P_1}{P_2} = (20.0\ \text{L}) \times \frac{3.0\ \text{atm}}{1.5\ \text{atm}} = 40.\ \text{L}$ and

$\Delta V = V_2 - V_1 = 40.\ \text{L} - 20.0\ \text{L} = 20.\ \text{L}$ then

$w = -P\,\Delta V = -1.5\ \text{atm} \times 20.\ \text{L} \times \frac{101.3\ \text{J}}{1\ \text{L}\cdot\text{atm}} = -3\underline{0}39\ \text{J} = -3.0 \times 10^3\ \text{J} \quad \Delta E = q + w$

Rearrange to solve for q. $q = \Delta E_{sys} - w = +0\ \text{J} - (-3\underline{0}39\ \text{J}) = 3.0 \times 10^3\ \text{J}$

Check: The units (J) are correct. Since there is no temperature change, we expect no energy change ($\Delta E_{sys} = 0$). The piston expands and so does work (negative work) and so heat is absorbed (positive q).

68. **Given:** 10.00 g P_4 (s) + O_2 (g) to form P_4O_{10}(s); q released heats 2950 g water from $T_i = 18.0$ °C to $T_f = 38.0$ °C
Find: $\Delta H^\circ_f(P_4O_{10}(s))$

Conceptual Plan: write balanced reaction then $\Delta H^0_{rxn} = \sum n_P \Delta H^0_f(products) - \sum n_R \Delta H^0_f(reactants)$ **then**
m, C_s, T_i, T_f → q_{H2O} → q_{rxn} (J) → q(kJ) then g (P_4) → mol (P_4) finally

$q = m\,C_s\left(T_f - T_i\right) \qquad q_{rxn} = -q_{H2O} \qquad \frac{1\ \text{kJ}}{1000\ \text{J}} \qquad \frac{123.90\ \text{g}}{1\ \text{mol}}$

q(kJ), mol (P_4) → ΔH°_f (P_4O_{10} (s))

$\Delta H^0_f\ (P_4O_{10}\ (s)) = \frac{q}{\text{mol } P_4}$

Solution: P_4 (s) + 5 O_2 (g) → P_4O_{10} (s) This reaction is the heat of formation of P_4O_{10} (s) so

$\Delta H^0_{rxn} = \Delta H^0_f\ (P_4O_{10}\ (s))$ then $q = m\,C_s\left(T_f - T_i\right) = 2950\ \text{g} \times 4.18\ \frac{\text{J}}{\text{g}\cdot{}^\circ\text{C}} \times (38.0\ {}^\circ\text{C} - 18.0\ {}^\circ\text{C}) = 24\underline{6}620\ \text{J}$

$q_{rxn} = -q_{H2O} = -24\underline{6}620\ \text{J} \times \frac{1\ \text{kJ}}{1000\ \text{J}} = -24\underline{6}.620\ \text{kJ}$ then $10.00\ \text{g}\ P_4 \times \frac{1\ \text{mol}\ P_4}{123.90\ \text{g}\ P_4} = 0.08071\underline{0}\ \text{mol}\ P_4$

then $\Delta H^0_f\ (P_4O_{10}\ (s)) = \frac{q}{\text{mol } P_4} = \frac{-24\underline{6}.620\ \text{kJ}}{0.08071\underline{0}\ \text{mol}} = -3060\ \text{kJ/mol}$

Check: The units (kJ/mol) are correct. The negative sign is consistent with he fact that there was heat released to heat a large amount of water. The magnitude (3000) is not surprising since a small amount of phosphorous heated a lot of water (a high heat capacity material).

69. **Given:** 655 kWh/yr, coal is 3.2 % S, remainder is C, S emitted as SO_2 (g) and gets converted to H_2SO_4 when reacting with water **Find:** m (H_2SO_4)/yr

Conceptual Plan: write balanced reaction then $\Delta H^0_{rxn} = \sum n_P \Delta H^0_f(products) - \sum n_R \Delta H^0_f(reactants)$
(since the form of sulfur is not given, assume all heat is from combustion of only carbon) then
kWh → J → kJ → mol (C) → g (C) → g (S) → mol (H_2SO_4) → mol (H_2SO_4) → g (H_2SO_4)

$\frac{3.60 \times 10^6\ \text{J}}{1\text{kWh}} \quad \frac{1\ \text{kJ}}{1000\ \text{J}} \quad \frac{\text{mol C}}{\Delta H^0_f\ (CO_2\ (g))} \quad \frac{12.01\ \text{g}}{1\ \text{mol}} \quad \frac{3.2\ \text{g S}}{(100.0 - 3.2)\ \text{g C}} \quad \frac{1\ \text{mol}}{32.06\ \text{g}} \quad \frac{1\ \text{mol}\ H_2SO_4}{1\ \text{mol S}} \quad \frac{98.09\ \text{g}}{1\ \text{mol}}$

Solution: C (s) + O_2 (g) → CO_2 (g) This reaction is the heat of formation of CO_2 (g) so

$\Delta H^0_{rxn} = \Delta H^0_f\,(CO_2\,(g)) = -\,393.5$ kJ/mol then

$$655\ \cancel{kWh} \times \frac{3.60 \times 10^6\ \cancel{J}}{1\ \cancel{kWh}} \times \frac{1\ \cancel{kJ}}{1000\ \cancel{J}} \times \frac{\cancel{mol\ C}}{393.5\ \cancel{kJ}} \times \frac{12.01\ \cancel{g\ C}}{1\ \cancel{mol\ C}} \times \frac{3.2\ \cancel{g\ S}}{(100.0 - 3.2)\ \cancel{g\ C}} \times \frac{1\ \cancel{mol\ S}}{32.06\ \cancel{g\ S}} \times$$

$$\times \frac{1\ \cancel{mol\ H_2SO_4}}{1\ \cancel{mol\ S}} \times \frac{98.09\ g\ H_2SO_4}{1\ \cancel{mol\ H_2SO_4}} = 7.3 \times 10^3\ g\ H_2SO_4$$

Check: The units (g) are correct. The magnitude (7300) is reasonable, considering this is just 1 home.

70. **Given:** 2.5×10^3 kg SUV, $v_1 = 0.0$ mph, $v_2 = 65.0$ mph, octane combustion, 30 % efficiency
Find: m(CO_2)
Conceptual Plan: **mi/hr → m/hr → m/min → m/s** **then** ***m, v* → *KE*** **then**

$$\frac{1000\ m}{0.6214\ mi} \quad \frac{1\ hr}{60\ min} \quad \frac{1\ min}{60\ sec} \qquad KE = \tfrac{1}{2} mv^2$$

KE₁, KE₂ **→ *ΔE* used → *ΔE* generated**

$$\Delta E_{sys} = KE_2 - KE_1 \qquad \frac{100\ J\ generated}{30\ J\ used}$$

use reaction from #56 $C_8H_{18}\,(l) + 25/2\,O_2\,(g) \rightarrow 8\,CO_2\,(g) + 9\,H_2O\,(g)$
with $\Delta H^\circ_{rxn} = -\,5074.1$ kJ

ΔE generated (J) → kJ → mol (C_8H_{18}) → mol CO_2 → g CO_2

$$\frac{1\ kJ}{1000\ J} \quad \frac{1\ mol\ C_8H_{18}}{5074.1\ kJ} \quad \frac{8\ mol\ CO_2}{1\ mol\ C_8H_{18}} \quad \frac{44.01\ g}{1\ mol}$$

Solution: $v_1 = 0.0$ m/s, $65.0\ \frac{\cancel{mi}}{\cancel{hr}} \times \frac{1000\ m}{0.6214\ \cancel{mi}} \times \frac{1\ \cancel{hr}}{60\ \cancel{min}} \times \frac{1\ \cancel{min}}{60\ sec} = 29.0563\ \frac{m}{s}$ then $KE = \tfrac{1}{2} mv^2$

$$KE_1 = \tfrac{1}{2}(2.5 \times 10^3\ kg)(0)^2 = 0$$

$$KE_2 = \tfrac{1}{2}(2.5 \times 10^3\ kg)\left(29.0563\ \frac{m}{s}\right)^2 = 1.05533 \times 10^6\ \frac{kg\ m^2}{s^2} = 1.05533 \times 10^6\ J$$

$$\Delta E_{sys} = KE_2 - KE_1 = 1.05533 \times 10^6\ J - 0\ J = 1.05533 \times 10^6\ \cancel{J\ used} \times \frac{100\ J\ generated}{30\ \cancel{J\ used}} =$$

$= 3.51777 \times 10^6$ J generated

$$3.51777 \times 10^6\ \cancel{J\ generated} \times \frac{1\ \cancel{kJ}}{1000\ \cancel{J}} \times \frac{1\ \cancel{mol\ C_8H_{18}}}{5074.1\ \cancel{kJ}} \times \frac{8\ \cancel{mol\ CO_2}}{1\ \cancel{mol\ C_8H_{18}}} \times \frac{44.01\ g\ CO_2}{1\ \cancel{mol\ CO_2}} = 240\ g\ CO_2$$

Check: The units (g) are correct. The magnitude (240) is reasonable, considering the vehicle is so heavy and we generate 8 moles of CO_2 for each mole of octane.

71. **Given:** methane combustion, 100 % efficiency, $\Delta T = 10.0$ °C, house = 30.0 m x 30.0 m x 3.0 m, Cs (air) = 30 J/K·mol, 1.00 mol air = 22.4 L **Find:** m(CH_4)
Conceptual Plan:
l, *w*, h → V(m³) → V(cm³) → V(L) → mol (air) **then m, C_s, Δ*T* → q_{air} (J)**

$$V = l\,w\,h \quad \frac{(100\ cm)^3}{(1\ m)^3} \quad \frac{1\ L}{1000\ cm^3} \quad \frac{1\ mol\ air}{22.4\ L} \qquad q = m\,C_s\,\Delta T$$

then q_{air} (J) → q_{rxn} (J) → *q*(kJ) then write balanced reaction for methane combustion

$$q_{rxn} = -\,q_{air} \qquad \frac{1\ kJ}{1000\ J}$$

then $\Delta H^0_{rxn} = \sum n_P \Delta H^0_f(products) - \sum n_R \Delta H^0_f(reactants)$ **then *q*(kJ) → mol (CH_4) → g (CH_4)**

$$\Delta H^0_{rxn} \qquad \frac{16.04\ g}{1\ mol}$$

Solution: $V = l\,w\,h = 30.0\ m \times 30.0\ m \times 3.0\ m = 2700\ m^3$ then

$2\underline{7}00 \text{ m}^3 \times \frac{(100 \text{ cm})^3}{(1 \text{ m})^3} \times \frac{1 \text{ L}}{1000 \text{ cm}^3} \times \frac{1 \text{mol air}}{22.4 \text{ L}} = 1.\underline{2}0536 \times 10^5 \text{ mol air}$ then

$q = m\, C_s\, \Delta T = 1.\underline{2}0536 \times 10^5 \text{ mol} \times 30 \frac{\text{J}}{\text{mol} \cdot {}^\circ\text{C}} \times 10.0\,{}^\circ\text{C} = \underline{3}.6161 \times 10^7 \text{ J} \times \frac{1 \text{ kJ}}{1000 \text{ J}} = \underline{3}.6161 \times 10^4 \text{ J}$

$CH_4\ (g) + 2\ O_2\ (g) \rightarrow CO_2\ (g) + 2\ H_2O\ (g)$

Reactant/Product	ΔH_f^0(kJ/mol from Appendix IIB)
$CH_4\ (g)$	− 74.6
$O_2\ (g)$	0.0
$CO_2\ (g)$	− 393.5
$H_2O\ (g)$	− 241.8

Be sure to pull data for the correct formula and phase.

$$\Delta H^0_{rxn} = \sum n_P \Delta H_f^0(products) - \sum n_R \Delta H_f^0(reactants)$$
$$= [1(\Delta H_f^0(CO_2\ (g))) + 2(\Delta H_f^0(H_2O\ (g)))] - [1(\Delta H_f^0(CH_4\ (g))) + 2(\Delta H_f^0(O_2\ (g)))]$$
$$= [(-393.5 \text{ kJ}) + 2(-241.8 \text{ kJ})] - [1(-74.6 \text{ kJ}) + 2(0.0 \text{ kJ})]$$
$$= [-877.1 \text{ kJ}] - [-74.6 \text{ kJ}]$$
$$= -802.5 \text{ kJ}$$

$$q_{rxn} = -q_{air} = -\underline{3}.6161 \times 10^4 \text{ kJ} \times \frac{1 \text{ mol } CH_4}{-802.5 \text{ kJ}} \times \frac{16.04 \text{ g } CH_4}{1 \text{ mol } CH_4} = \underline{7}22.8 \text{ g } CH_4 = 700 \text{ g } CH_4$$

Check: The units (g) are correct. The magnitude (700) is not surprising since the volume of a house is large.

72. **Given:** water: $V = 35$ L, $T_i = 25.0$ °C, $T_f = 100.0$ °C; fuel = C_7H_{16}, 15 % efficiency, $d = 0.78$ g/ml
Find: V (fuel)
Conceptual Plan: write balanced reaction then $\Delta H^0_{rxn} = \sum n_P \Delta H_f^0(products) - \sum n_R \Delta H_f^0(reactants)$ **then**
L → mL → g then T_i, T_f → ΔT then m, C_s, ΔT → q_{H2O} (J) → q_{H2O} (kJ) → q_{rxn} (kJ)

$\frac{1000 \text{ mL}}{1 \text{ L}}$ $\frac{1.0 \text{ g}}{1.0 \text{ mL}}$ $\Delta T = T_f - T_i$ $q_{H2O} = m_{H2O} C_{H2O} \Delta T_{H2O}$ $\frac{1 \text{kJ}}{1000 \text{ J}}$ $q_{rxn} = -q_{H2O}$

then q_{rxn} generated (J) → q_{rxn} used (kJ) → mol C_7H_{16} → g C_7H_{16} → mL C_7H_{16}

$\frac{100 \text{ J generated}}{15 \text{ J needed}}$ $\frac{1 \text{mol } C_7H_{16}}{\Delta H^0_{rxn}}$ $\frac{100.21 \text{ g}}{1 \text{mol}}$ $\frac{1.0 \text{ mL}}{0.78 \text{ g}}$

Solution: combustion is combination with oxygen to form carbon dioxide and water
$C_7H_{16}\ (l) + 11\ O_2\ (g) \rightarrow 7\ CO_2\ (g) + 8\ H_2O\ (g)$

Reactant/Product	ΔH_f^0(kJ/mol from Appendix IIB)
$C_7H_{16}\ (l)$	− 224.4
$O_2\ (g)$	0.0
$CO_2\ (g)$	− 393.5
$H_2O\ (g)$	− 241.8

Be sure to pull data for the correct formula and phase.

$$\Delta H^0_{rxn} = \sum n_P \Delta H_f^0(products) - \sum n_R \Delta H_f^0(reactants)$$
$$= [7(\Delta H_f^0(CO_2\ (g))) + 8(\Delta H_f^0(H_2O\ (g)))] - [1(\Delta H_f^0(C_7H_{16}\ (l))) + 11(\Delta H_f^0(O_2\ (g)))]$$
$$= [7(-393.5 \text{ kJ}) + 8(-241.8 \text{ kJ})] - [1(-224.4 \text{ kJ}) + 11(0.0 \text{ kJ})]$$
$$= [-4688.9 \text{ kJ}] - [-224.4 \text{ kJ}]$$
$$= -4464.5 \text{ kJ}$$

$35 \text{ L} \times \frac{1000 \text{ mL}}{1 \text{ L}} \times \frac{1.0 \text{ g}}{1.0 \text{ mL}} = 3\underline{5}000 \text{ g}$ then $\Delta T = T_f - T_i = 100.0\text{°C} - 25.0\text{ °C} = 75.0\text{ °C}$ then

$q_{H2O} = m_{H2O} C_{H2O} \Delta T_{H2O} = 3\underline{5}000 \text{ g} \times 4.18 \frac{\text{J}}{\text{g} \cdot {}^\circ\text{C}} \times 75.0\,{}^\circ\text{C} = 1.\underline{0}9725 \times 10^7 \text{ J} \times \frac{1 \text{kJ}}{1000 \text{ J}} = 1.\underline{0}9725 \times 10^4 \text{ kJ}$

$q_{rxn} = -q_{H2O} = -1.0\underline{9}725 \times 10^4 \cancel{kJ} \times \frac{100 \text{ kJ generated}}{15 \text{ kJ used}} \times \frac{1 \text{ mol } C_7H_{16}}{-4467.3 \text{ kJ}} \times \frac{100.21 \text{ g } C_7H_{16}}{1 \text{ mol } C_7H_{16}} \times \frac{1.0 \text{ mL } C_7H_{16}}{0.78 \text{ g } C_7H_{16}} =$

$= 2100 \text{ mL } C_7H_{16} = 2.1 \text{ L } C_7H_{16}$

Check: The units (mL) are correct. The magnitude (2 L) is a reasonable volume to have to take on a backpacking trip.

73. **Given:** m (ice) = 9.0 g; coffee: T_1 = 90.0 °C, m = 120.0 g, $C_s = C_{H2O}$, $\Delta H°_{fus}$ = 6.0 kJ/mol
Find: T_f of coffee
Conceptual Plan: $q_{ice} = -q_{coffee}$ **so g (ice) → mol (ice) →** q_{fus}**(kJ) →** q_{fus} **(J) →** q_{coffee} **(J) then**

$\frac{1 \text{ mol}}{18.01 \text{g}}$ $\quad \frac{6.0 \text{ kJ}}{1 \text{ mol}}$ $\quad \frac{1000 \text{ J}}{1 \text{ kJ}}$ $\quad q_{coffee} = -q_{ice}$

q, m, C_s **→** ΔT **then** $T_i, \Delta T$ **→** T_2 **now we have slightly cooled coffee in contact with 0.0 °C water**

$q = mC_S\Delta T \qquad \Delta T = T_2 - T_i$

so $q_{ice} = -q_{coffee}$ **with** m, C_s, T_i **→** T_f

$q = m\,C_S(T_f - T_i)$ then set $q_{Al} = -q_{H2O}$

Solution: $9.0 \cancel{g} \times \frac{1 \cancel{mol}}{18.01 \cancel{g}} \times \frac{6.0 \cancel{kJ}}{1 \text{ mol}} \times \frac{1000 \text{ J}}{1 \cancel{kJ}} = 2.\underline{9}983 \times 10^3 \text{ J}, \quad q_{coffee} = -q_{ice} = -2.\underline{9}983 \times 10^3 \text{ J}$

$q = mC_S\Delta T$ Rearrange to solve for ΔT. $\Delta T = \frac{q}{mC_S} = \frac{-2.\underline{9}983 \times 10^3 \cancel{J}}{120.0 \cancel{g} \times 4.18 \frac{\cancel{J}}{\cancel{g} \cdot °C}} = -5.\underline{9}775 \text{ °C}$ then

$\Delta T = T_2 - T_i$. Rearrange to solve for T_2. $T_2 = \Delta T + T_i = -5.\underline{9}775 \text{ °C} + 90.0 \text{ °C} = 84.\underline{0}225 \text{ °C}$

$q = m\,C_s(T_f - T_i)$ substitute in values and set $q_{H2O} = -q_{coffee}$.

$q_{H2O} = m_{H2O}\,C_{H2O}(T_f - T_{H2Oi}) = 9.0 \text{ g} \times 4.18 \frac{\text{J}}{\text{g} \cdot °\text{C}} \times (T_f - 0.0 \text{ °C}) =$

$-q_{coffee} = -m_{coffee}\,C_{coffee}(T_f - T_{coffee2}) = -120.0 \text{ g} \times 4.18 \frac{\text{J}}{\text{g} \cdot °\text{C}} \times (T_f - 84.\underline{0}225 \text{°C})$

Rearrange to solve for T_f.

$9.0 \text{ g } T_f = -120.0 \text{ g}(T_f - 84.\underline{0}225 \text{°C}) \rightarrow 9.0 \text{ g } T_f = -120.0 \text{ g } T_f + 10\underline{0}82.7 \text{ g} \rightarrow$

$-10\underline{0}82.7 \text{ g} = -129.0 \frac{\text{g}}{°\text{C}} T_f \rightarrow T_f = \frac{-10\underline{0}82.7 \cancel{g}}{-129.0 \frac{\cancel{g}}{°\text{C}}} = 78.2 \text{ °C}$

Check: The units (°C) are correct. The temperature is closer to the original coffee temperature since the mass of coffee is so much larger than the ice mass.

74. **Given:** liquid water at – 10.0 °C, Cs (ice) = 2.04 J/g·°C; $\Delta H°_{fus}$ = – 322 J/g (@ 0.0 °C)
Find: ΔH, ΔE, q, and w for freezing at –10.0 °C
Conceptual Plan: Assume exactly 1 g H_2O **for all calculations (report answers as J/g) and constant** P **= 1 atm. Construct the following path: According to Hess' Law** $\Delta H_1 + \Delta H_2 + \Delta H_3 = \Delta H_4$
= $\Delta H°_{fus}$ **@ –10.0 °C**

	#2		
Liquid @ 0.0 °C	→	**solid @ 0.0 °C**	
↑ #1		**↓ #3**	**at constant** P, $\Delta H = q$
Liquid @ –10.0 °C	→	**solid @ –10.0 °C**	**for steps #1 and #3** $q = mC_s\Delta T$
	#4		

Look up the density of liquid and solid water at 0.0 °C (Assume the density of each phase does not change significantly at – 10.0 °C.)
d_L = 0.9998 g/mL and d_S = 0.917 g/mL

g → mL → L then V_L, V_S → ΔV then P, ΔV → w (L atm) → w (J) then q, w → ΔE

L: $\frac{1\text{ mL}}{0.9998\text{ g}}$ S: $\frac{1\text{ mL}}{0.917\text{ g}}$ $\frac{1\text{ L}}{1000\text{ mL}}$ $\Delta V = V_S - V_L$ $w = -P\,\Delta V$ $\frac{101.3\text{ J}}{1\text{ L atm}}$ $\Delta E = q + w$

Solution: $\Delta H_1 = q_1 = mC_S\Delta T = 1\text{ g} \times 4.18\frac{\text{J}}{\text{g}\cdot{}^\circ\text{C}} \times (0.00\,{}^\circ\text{C} - (-10.0\,{}^\circ\text{C})) = +41.8\text{ J}$,

$\Delta H_2 = q_2 = m\Delta H = 1\text{ g} \times -332\frac{\text{J}}{\text{g}} = -332\text{ J}$,

$\Delta H_3 = q_3 = mC_S\Delta T = 1\text{ g} \times 2.04\frac{\text{J}}{\text{g}\cdot{}^\circ\text{C}} \times (-10.00\,{}^\circ\text{C} - 0.0\,{}^\circ\text{C}) = -20.4\text{ J}$ so

$\Delta H_4 = q_4 = \Delta H_1 + \Delta H_2 + \Delta H_3 = +41.8\text{ J} - 332\text{ J} - 20.4\text{ J} = -31\underline{0}.6\text{ J} = -311\text{ J/g}$.

$V_L = 1\text{ g} \times \frac{1\text{ mL}}{0.9998\text{ g}} \times \frac{1\text{ L}}{1000\text{ mL}} = 0.00100\underline{0}20004\text{ L}$ and $V_S = 1\text{ g} \times \frac{1\text{ mL}}{0.917\text{ g}} \times \frac{1\text{ L}}{1000\text{ mL}} = 0.00109\underline{0}5\text{ L}$

then $\Delta V = V_S - V_L = 0.00109\underline{0}5\text{ L} - 0.00100\underline{0}20004\text{ L} = \underline{9}.02999 \times 10^{-5}\text{ L}$

then, $w = -P\,\Delta V = -1\text{ atm} \times \underline{9}.02999 \times 10^{-5}\text{ L} \times \frac{101.3\text{ J}}{1\text{ L atm}} = -0.00\underline{9}147\text{ J} = -0.009\text{ J/g}$

and $\Delta E = q + w = -31\underline{0}.6\text{ J} - 0.00\underline{9}147\text{ J} = -311\text{ J/g}$

Check: The units (J/g) are correct. We expect freezing to release less energy at –10 °C because we are below the normal freezing point. The work is negligible since the volume change is so small.

75. $KE = \frac{1}{2}mv^2$, for an ideal gas $v = u_{rms} = \sqrt{\frac{3RT}{\mathfrak{M}}}$ and so $KE_{avg} = \frac{1}{2}N_A m u_{rms}^2 = \frac{3}{2}RT$ then

$\Delta E_{sys} = KE_2 - KE_1 = \frac{3}{2}RT_2 - \frac{3}{2}RT_1 = \frac{3}{2}R\Delta T$. At constant V $\Delta E_{sys} = C_V\Delta T$ so $C_V = \frac{3}{2}R$.

At constant P, $\Delta E_{sys} = q + w = q_P - P\Delta V = \Delta H - P\Delta V$ but since $PV = nRT$, for one mole of an ideal gas at constant P $P\Delta V = R\Delta T$ so $\Delta E_{sys} = q + w = q_P - P\Delta V = \Delta H - P\Delta V = \Delta H - R\Delta T$ then

$\frac{3}{2}R\Delta T = \Delta H - R\Delta T$ or $\Delta H = \frac{5}{2}R\Delta T = C_P\Delta T$ so $C_P = \frac{5}{2}R$.

76. **Given:** fixed amount of an ideal gas; Step #1: $V_1 = 12.0$ L to $V_2 = 24.0$ L at constant $P = 1.0$ atm; Step #2: gas cooled at constant $V = 24.0$ L to original T; Step #3: $V_1 = 24.0$ L to $V_2 = 12.0$ L **Find:** q for entire process
Solution: For the expansion: $w_1 = -(24.0\text{ L} - 12.0\text{ L})(1.0\text{ atm}) = -12.0$ L atm; For the constant V step: $w_2 = 0$ since there is no PV work at constant volume; For the contraction: $w_3 = -nRT\ln(V_2/V_1) = -PV\ln(V_2/V_1) = -(1.0\text{ atm})(12.0\text{ L})\ln(12.0\text{ L}/24.0\text{ L}) = 8.3\text{ L}\cdot\text{atm}$, then $w_{\text{total}} = w_1 + w_2 + w_3 = -12.0$ L atm $+ 0 + 8.3$ L atm

$w_{\text{total}} = -3.7\text{ L}\cdot\text{atm} \times \frac{101\text{ J}}{\text{L}\cdot\text{atm}} = -370\text{ J}$. Since the system ends where it started $\Delta E = 0$ and, therefore,

$q = -w = 370$ J

Check: The units (J) are correct. The total energy change over the entire cycle is 0 since we end where we started, but this does not mean that q has to be 0.

77. d) only one answer is possible. $\Delta E_{sys} = -\Delta E_{surr}$

78. a) False. An isothermal process has $\Delta E_{sys} = 0$.

b) False. $w < 0$ for expansions

c) True. If $\Delta E_{sys} = 0$ and $\Delta E_{sys} = q + w$ and $w < 0$ then $q > 0$.

d) False. An isothermal process has $\Delta E_{sys} = 0$.

79. a) At constant P, $\Delta E_{sys} = q + w = q_P + w = \Delta H + w$ so $\Delta E_{sys} - w = \Delta H$

80. Since the mass of B is greater, it will work harder on day 1 to cool it down. On day 2 the water will stabilize the temperature with the loss of heat from the water (a high heat capacity material).

81. The aluminum cylinder will be cooler after one hour because it has a lower heat capacity than water (less heat needs to be pulled out for every °C temperature change).

82. **Given:** 2418 J heat produced; 5 kJ work done on surroundings at constant P **Find:** ΔE, ΔH, q, and w
Conceptual Plan: interpret language to determine the sign of the two terms then $q, w \rightarrow \Delta E_{sys}$
$\Delta E = q + w$

Solution: since heat is released from the system to the surroundings, $q = -2418\ \text{J} = -2.418\ \text{kJ}$; since the system is doing work on the surroundings, $w = -5$ kJ. At constant P, $\Delta H = q = -2.418$ kJ;
$\Delta E = q + w = -2.418\ \text{kJ} - 5\ \text{kJ} = -7\ \text{kJ}$

Check: The units (kJ) are correct. The magnitude of the answer (–7) makes physical sense because both terms are negative and small.

83. The internal energy of a chemical system is the sum of its kinetic energy and its potential energy. It is this potential energy that is the energy source in an exothermic chemical reaction. Under normal circumstances, chemical potential energy (or simply chemical energy) arises primarily from the electrostatic forces between the protons and electrons that compose the atoms and molecules within the system. In an exothermic reaction, some bonds break and new ones form, and the protons and electrons go from an arrangement of higher potential energy to one of lower potential energy. As they rearrange, their potential energy is converted into kinetic energy. Heat is emitted in the reaction and so it feels hot to the touch.

84. b) If ΔV is positive the $w = -P\Delta V < 0$. Since $\Delta E_{sys} = q + w = q_P + w = \Delta H + w$ if w is negative then $\Delta H > \Delta E_{sys}$.

Chapter 7
The Quantum-Mechanical Model of the Atom

1. **Given:** distance to sun = 1.496×10^8 km **Find:** time for light to travel from sun to Earth
Conceptual Plan: distance km → distance m → time

$$\frac{1000\ \text{m}}{\text{km}} \qquad \text{time} = \frac{\text{distance}}{3.00 \times 10^8\ \text{m/s}}$$

Solution: $1.496 \times 10^8\ \cancel{\text{km}} \times \frac{1000\ \cancel{\text{m}}}{\cancel{\text{km}}} \times \frac{\text{s}}{3.00 \times 10^8\ \cancel{\text{m}}} = 499\ \text{s}$

Check: The units of the answer, seconds, is correct. The magnitude of the answer is reasonable, since it corresponds to about 8 min.

2. **Given:** 4.3 light years to star **Find:** distance in km
Conceptual Plan: light years → days → hours → seconds → m → km

$$\frac{365\ \text{days}}{\text{yr}} \quad \frac{24\ \text{hr}}{\text{day}} \quad \frac{3600\ \text{s}}{\text{hr}} \quad \frac{3.00 \times 10^8\ \text{m}}{\text{s}} \quad \frac{\text{km}}{1000\ \text{m}}$$

Solution: $4.3\ \text{light}\ \cancel{\text{yr}} \times \frac{365\ \cancel{\text{days}}}{\cancel{\text{yr}}} \times \frac{24\ \cancel{\text{hrs}}}{\cancel{\text{day}}} \times \frac{3600\ \cancel{\text{s}}}{\cancel{\text{hr}}} \times \frac{3.00 \times 10^8\ \cancel{\text{m}}}{\cancel{\text{s}}} \times \frac{\text{km}}{1000\ \cancel{\text{m}}} = 4.1 \times 10^{13}\ \text{km}$

Check: The units of the answer, km, are correct. The magnitude of the answer is reasonable since it takes much longer for the light to reach Earth from Proxima Centauri than from the sun, so the distance should be much greater.

3. i) by increasing wavelength the order is: d) ultraviolet < c) infrared < b) microwave < a) radio waves

   ii) by increasing energy the order is: a) radio waves < b) microwaves < c) infrared < d) ultraviolet

4. i) by increasing frequency the order is: b) radio waves < c) microwaves < d) visible light < a) gamma rays

   ii) by decreasing energy the order is: a) gamma rays > d) visible light > c) microwaves > b) radio waves

5. a) **Given:** $\lambda = 632.8$ nm **Find:** frequency (ν)
Conceptual Plan: nm → m → ν

$$\frac{\text{m}}{10^9\ \text{nm}} \qquad \nu = \frac{c}{\lambda}$$

Solution: $632.8\ \cancel{\text{nm}} \times \frac{\text{m}}{10^9\ \cancel{\text{nm}}} = 6.328 \times 10^{-7}\ \text{m}$ $\qquad \nu = \frac{3.00 \times 10^8\ \cancel{\text{m}}}{\text{s}} \times \frac{1}{6.328 \times 10^{-7}\ \cancel{\text{m}}} = 4.74 \times 10^{14}\ \text{s}^{-1}$

Check: The units of the answer, s^{-1}, are correct. The magnitude of the answer seems reasonable since wavelength and frequency are inversely proportional.

b) **Given:** $\lambda = 503$ nm **Find:** frequency (ν)
Conceptual Plan: nm → m → ν

$$\frac{\text{m}}{10^9\ \text{nm}} \qquad \nu = \frac{c}{\lambda}$$

Solution: $503\ \cancel{\text{nm}} \times \frac{\text{m}}{10^9\ \cancel{\text{nm}}} = 5.03 \times 10^{-7}\ \text{m}$ $\qquad \nu = \frac{3.00 \times 10^8\ \cancel{\text{m}}}{\text{s}} \times \frac{1}{5.03 \times 10^{-7}\ \cancel{\text{m}}} = 5.96 \times 10^{14}\ \text{s}^{-1}$

Check: The units of the answer, s^{-1}, are correct. The magnitude of the answer seems reasonable since wavelength and frequency are inversely proportional.

c) **Given:** $\lambda = 0.052$ nm **Find:** frequency (ν)
Conceptual Plan: nm → m → ν

$$\frac{\text{m}}{10^9\ \text{nm}} \qquad \nu = \frac{c}{\lambda}$$

Solution: $0.052\ \text{nm} \times \frac{\text{m}}{10^9\ \text{nm}} = 5.2 \times 10^{-9}\ \text{m}$ $\nu = \frac{3.00 \times 10^8\ \text{m}}{\text{s}} \times \frac{1}{5.2 \times 10^{-9}\ \text{m}} = 5.8 \times 10^{18}\ \text{s}^{-1}$

Check: The units of the answer, s^{-1}, are correct. The magnitude of the answer seems reasonable since wavelength and frequency are inversely proportional.

6. a) **Given:** ν = 100.2 MHz **Find:** wavelength (λ)

Conceptual Plan: MHz → Hz → s^{-1} → λ

$\frac{10^6\ \text{Hz}}{\text{MHz}}$ $1\ \text{Hz} = 1\ \text{s}^{-1}$ $\lambda = \frac{c}{\nu}$

Solution: $100.2\ \text{MHz} \times \frac{10^6\ \text{Hz}}{\text{MHz}} \times \frac{\text{s}^{-1}}{\text{Hz}} = 1.002 \times 10^8\ \text{s}^{-1}$ $\lambda = \frac{3.00 \times 10^8\ \text{m}}{\text{s}} \times \frac{\text{s}}{1.002 \times 10^8} = 2.99\ \text{m}$

Check: The units of the answer, m, are correct. The magnitude of the answer is reasonable because FM wavelengths are generally in the 3 – 8 m range.

b) **Given:** ν = 1070 kHz **Find:** wavelength (λ)

Conceptual Plan: kHz → Hz → s^{-1} → λ

$\frac{10^3\ \text{Hz}}{\text{kHz}}$ $1\ \text{Hz} = 1\ \text{s}^{-1}$ $\lambda = \frac{c}{\nu}$

Solution: $1070\ \text{kHz} \times \frac{10^3\ \text{Hz}}{\text{kHz}} \times \frac{\text{s}^{-1}}{\text{Hz}} = 1.070 \times 10^6\ \text{s}^{-1}$ $\lambda = \frac{3.00 \times 10^8\ \text{m}}{\text{s}} \times \frac{\text{s}}{1.070 \times 10^6} = 280.\ \text{m}$

Check: The units of the answer, m, are correct. The magnitude of the answer is reasonable because AM wavelengths are generally in the 100 – 1000 m range.

c) **Given:** ν = 835.6 MHz **Find:** wavelength (λ)

Conceptual Plan: MHz → Hz → s^{-1} → λ

$\frac{10^6\ \text{Hz}}{\text{MHz}}$ $1\ \text{Hz} = 1\ \text{s}^{-1}$ $\lambda = \frac{c}{\nu}$

Solution:

$835.6\ \text{MHz} \times \frac{10^6\ \text{Hz}}{\text{MHz}} \times \frac{\text{s}^{-1}}{\text{Hz}} = 8.356 \times 10^8\ \text{s}^{-1}$ $\lambda = \frac{3.00 \times 10^8\ \text{m}}{\text{s}} \times \frac{\text{s}}{8.356 \times 10^8} = 3.59 \times 10^{-1}\ \text{m}$

Check: The units of the answer, m, are correct. The magnitude of the answer is reasonable because cell phone wavelengths are generally in the 0.1 – 0.5 m range.

7. a) **Given:** frequency (ν) from 5 a. = $4.74 \times 10^{14}\ \text{s}^{-1}$ **Find:** Energy

Conceptual Plan: ν → E

$E = h\nu$ $h = 6.626 \times 10^{-34}\ \text{J s}$

Solution: $6.626 \times 10^{-34}\ \text{J s} \times \frac{4.74 \times 10^{14}}{\text{s}} = 3.14 \times 10^{-19}\ \text{J}$

Check: The units of the answer, J, are correct. The magnitude of the answer is reasonable since we are talking about the energy of one photon.

b) **Given:** frequency (ν) from 5 b. = $5.96 \times 10^{14}\ \text{s}^{-1}$ **Find:** Energy

Conceptual Plan: ν → E

$E = h\nu$ $h = 6.626 \times 10^{-34}\ \text{J s}$

Solution: $6.626 \times 10^{-34}\ \text{J s} \times \frac{5.96 \times 10^{14}}{\text{s}} = 3.95 \times 10^{-19}\ \text{J}$

Check: The units of the answer, J, are correct. The magnitude of the answer is reasonable since we are talking about the energy of one photon.

c) **Given:** frequency (ν) from 5 c. = $5.8 \times 10^{18}\ \text{s}^{-1}$ **Find:** Energy

Conceptual Plan: ν → E

$E = h\nu$ $h = 6.626 \times 10^{-34}\ \text{J s}$

Solution: $6.626 \times 10^{-34}\ \text{J}\ \cancel{\text{s}} \times \frac{5.8 \times 10^{18}}{\cancel{\text{s}}} = 3.8 \times 10^{-15}\ \text{J}$

Check: The units of the answer, J, are correct. The magnitude of the answer is reasonable since we are talking about the energy of one photon.

8. a) **Given:** frequency (ν) from 6 a. = 100.2 MHz **Find:** Energy

Conceptual Plan: MHz → Hz → s^{-1} → E

$\frac{10^6\ \text{Hz}}{\text{MHz}}$ $1\ \text{Hz} = 1\ \text{s}^{-1}$ $E = h\nu$ $h = 6.626 \times 10^{-34}\ \text{J s}$

Solution: $100.2\ \cancel{\text{MHz}} \times \frac{10^6\ \cancel{\text{Hz}}}{\cancel{\text{MHz}}} \times \frac{\text{s}^{-1}}{\cancel{\text{Hz}}} = 1.002 \times 10^8\ \text{s}^{-1}$ $6.626 \times 10^{-34}\ \text{J}\ \cancel{\text{s}} \times \frac{1.002 \times 10^8}{\cancel{\text{s}}} = 6.639 \times 10^{-26}\ \text{J}$

Check: The units of the answer, J, are correct. The magnitude of the answer is reasonable since we are talking about the energy of one photon and have a relatively long wavelength.

b) **Given:** ν = 1070 kHz **Find:** Energy

Conceptual Plan: kHz → Hz → s^{-1} → E

$\frac{10^3\ \text{Hz}}{\text{kHz}}$ $1\ \text{Hz} = 1\ \text{s}^{-1}$ $E = h\nu$ $h = 6.626 \times 10^{-34}\ \text{J s}$

Solution: $1070\ \cancel{\text{kHz}} \times \frac{10^3\ \cancel{\text{Hz}}}{\cancel{\text{kHz}}} \times \frac{\text{s}^{-1}}{\cancel{\text{Hz}}} = 1.070 \times 10^6\ \text{s}^{-1}$ $6.626 \times 10^{-34}\ \text{J}\ \cancel{\text{s}} \times \frac{1.070 \times 10^6}{\cancel{\text{s}}} = 7.090 \times 10^{-28}\ \text{J}$

Check: The units of the answer, J, are correct. The magnitude of the answer is reasonable since we are talking about the energy of one photon and have a relatively long wavelength.

c) **Given:** ν = 835.6 MHz **Find:** Energy

Conceptual Plan: MHz → Hz → s^{-1} → E

$\frac{10^6\ \text{Hz}}{\text{MHz}}$ $1\ \text{Hz} = 1\ \text{s}^{-1}$ $E = h\nu$ $h = 6.626 \times 10^{-34}\ \text{J s}$

Solution: $835.6\ \cancel{\text{MHz}} \times \frac{10^6\ \cancel{\text{Hz}}}{\cancel{\text{MHz}}} \times \frac{\text{s}^{-1}}{\cancel{\text{Hz}}} = 8.356 \times 10^8\ \text{s}^{-1}$ $6.626 \times 10^{-34}\ \text{J}\ \cancel{\text{s}} \times \frac{8.356 \times 10^8}{\cancel{\text{s}}} = 5.537 \times 10^{-25}\ \text{J}$

Check: The units of the answer, J, are correct. The magnitude of the answer is reasonable since we are talking about the energy of one photon and have a relatively long wavelength.

9. **Given:** λ = 532 nm and $E_{pulse} = 4.67$ mJ **Find:** number of photons

Conceptual Plan: nm → m → E_{photon} → number of photons

$\frac{\text{m}}{10^9\ \text{nm}}$ $E = \frac{hc}{\lambda}$; $h = 6.626 \times 10^{-34}\ \text{J s}$ $\frac{E_{pulse}}{E_{photon}}$

Solution:

$532\ \cancel{\text{nm}} \times \frac{\text{m}}{10^9\ \cancel{\text{nm}}} = 5.32 \times 10^{-7}\ \text{m}$ $E = \frac{6.626 \times 10^{-34}\ \text{J}\ \cancel{\text{s}} \times \frac{3.00 \times 10^8\ \cancel{\text{m}}}{\cancel{\text{s}}}}{5.32 \times 10^{-7}\ \cancel{\text{m}}} = 3.7\underline{3}64 \times 10^{-19}\ \text{J / photon}$

$4.67\ \cancel{\text{mJ}} \times \frac{\cancel{\text{J}}}{1000\ \cancel{\text{mJ}}} \times \frac{1\ \text{photon}}{3.7\underline{3}64 \times 10^{-19}\ \cancel{\text{J}}} = 1.25 \times 10^{16}\ \text{photons}$

Check: The units of the answer, number of photons, are correct. The magnitude of the answer is reasonable for the amount of energy involved.

10. **Given:** λ = 6.5 μm; power = 25.5 watts **Find:** photons/second

Conceptual Plan: μm → m → E_{photon} and then watts → J/s → number of photons

$\frac{\text{m}}{10^6\ \mu\text{m}}$ $E = \frac{hc}{\lambda}$; $h = 6.626 \times 10^{-34}\ \text{J s}$ $\frac{\text{J / sec}}{\text{watt}}$ $\frac{\text{J / sec}}{\text{J / photon}}$

Solution:

$$6.5\ \mu m \times \frac{m}{10^6\ \mu m} = 6.5 \times 10^{-6}\ m \qquad E = \frac{6.626 \times 10^{-34}\ J\ s \times \frac{3.00 \times 10^8\ m}{s}}{6.5 \times 10^{-6}\ m} = 3.\underline{0}58 \times 10^{-20}\ J / photon$$

$$25.5\ watts \times \frac{J/s}{1\ watt} \times \frac{1\,photon}{3.\underline{0}58 \times 10^{-20}\ J} = 8.3 \times 10^{20}\ photons / s$$

Check: The units of the answer, photons/s, are correct. The magnitude of the answer is reasonable for the amount of energy involved.

11. a) **Given:** $\lambda = 1500$ nm **Find:** E for 1 mol photons

Conceptual Plan: nm → m → E_{photon} → $E(J)_{mol}$ → $E(kJ)_{mol}$

$$\frac{m}{10^9\ nm} \quad E = \frac{hc}{\lambda};\ h = 6.626 \times 10^{-34}\ J\,s \quad \frac{mol}{6.022 \times 10^{23}\ photons} \quad \frac{kJ}{1000\ J}$$

Solution:

$$1500\ nm \times \frac{m}{10^9\ nm} = 1.500 \times 10^{-6}\ m \qquad E = \frac{6.626 \times 10^{-34}\ J\ s \times \frac{3.00 \times 10^8\ m}{s}}{1.500 \times 10^{-6}\ m} = 1.3\underline{2}52 \times 10^{-19}\ J / photon$$

$$\frac{1.3\underline{2}52 \times 10^{-19}\ J}{photon} \times \frac{6.022 \times 10^{23}\ photons}{mol} \times \frac{kJ}{1000\ J} = 79.8\ kJ / mol$$

Check: The units of the answer, kJ/mol, are correct. The magnitude of the answer is reasonable for a wavelength in the infrared region.

b) **Given:** $\lambda = 500$ nm **Find:** E for 1 mol photons

Conceptual Plan: nm → m → E_{photon} → E_{mol} → $E(kJ)_{mol}$

$$\frac{m}{10^9\ nm} \quad E = \frac{hc}{\lambda};\ h = 6.626 \times 10^{-34}\ J\,s \quad \frac{mol}{6.022 \times 10^{23}\ photons} \quad \frac{kJ}{1000\ J}$$

Solution:

$$500\ nm \times \frac{m}{10^9\ nm} = 5.00 \times 10^{-7}\ m \qquad E = \frac{6.626 \times 10^{-34}\ J\ s \times \frac{3.00 \times 10^8\ m}{s}}{5.00 \times 10^{-7}\ m} = 3.9\underline{7}56 \times 10^{-19}\ J / photon$$

$$\frac{3.9\underline{7}56 \times 10^{-19}\ J}{photon} \times \frac{6.022 \times 10^{23}\ photons}{mol} \times \frac{kJ}{1000\ J} = 239\ kJ / mol$$

Check: The units of the answer, kJ/mol, are correct. The magnitude of the answer is reasonable for a wavelength in the visible region.

c) **Given:** $\lambda = 150$ nm **Find:** E for 1 mol photons

Conceptual Plan: nm → m → E_{photon} → E_{mol} → $E(kJ)_{mol}$

$$\frac{m}{10^9\ nm} \quad E = \frac{hc}{\lambda};\ h = 6.626 \times 10^{-34}\ J\,s \quad \frac{mol}{6.022 \times 10^{23}\ photons} \quad \frac{kJ}{1000\ J}$$

Solution:

$$1.50\ nm \times \frac{m}{10^9\ nm} = 1.50 \times 10^{-7}\ m \qquad E = \frac{6.626 \times 10^{-34}\ J\ s \times \frac{3.00 \times 10^8\ m}{s}}{1.50 \times 10^{-7}\ m} = 1.3\underline{2}52 \times 10^{-18}\ J / photon$$

$$\frac{1.3\underline{2}52 \times 10^{-18}\ J}{photon} \times \frac{6.022 \times 10^{23}\ photons}{mol} \times \frac{kJ}{1000\ J} = 798\ kJ / mol$$

Check: The units of the answer, kJ/mol, are correct. The magnitude of the answer is reasonable for a wavelength in the ultraviolet region. Note: the energy increases from the IR to the Vis to the UV as expected.

12. a) **Given:** $\lambda = 0.155$ nm **Find:** E for 1 mol photons

Conceptual Plan: **nm → m → E_{photon} → E_{mol} → $E(kJ)_{mol}$**

$$\frac{m}{10^9\ nm} \quad E=\frac{hc}{\lambda};\ h=6.626\times10^{-34}\ J\ s \quad \frac{mol}{6.022\times10^{23}\ photons} \quad \frac{kJ}{1000\ J}$$

Solution:

$$0.155\ nm \times \frac{m}{10^9\ nm} = 1.55\times10^{-10}\ m \qquad E=\frac{6.626\times10^{-34}\ J\ s \times \dfrac{3.00\times10^8\ m}{s}}{1.55\times10^{-10}\ m}=1.2\underline{8}24\times10^{-15}\ J/photon$$

$$\frac{1.2\underline{8}24\times10^{-15}\ J}{photon}\times\frac{6.022\times10^{23}\ photons}{mol}\times\frac{kJ}{1000\ J}=7.72\times10^5\ kJ/mol$$

Check: The units of the answer, kJ/mol, are correct. The magnitude of the answer is reasonable for a wavelength in the x-ray region.

b) **Given:** $\lambda = 2.55\times10^{-5}$ nm **Find:** E for 1 mol photons

Conceptual Plan: **nm → m → E_{photon} → E_{mol} → $E(kJ)_{mol}$**

$$\frac{m}{10^9\ nm} \quad E=\frac{hc}{\lambda};\ h=6.626\times10^{-34}\ J\ s \quad \frac{mol}{6.022\times10^{23}\ photons} \quad \frac{kJ}{1000\ J}$$

Solution:

$$2.55\times10^{-5}\ nm \times \frac{m}{10^9\ nm} = 2.55\times10^{-14}\ m \qquad E=\frac{6.626\times10^{-34}\ J\ s \times \dfrac{3.00\times10^8\ m}{s}}{2.55\times10^{-14}\ m}=7.7\underline{9}53\times10^{-12}\ J/photon$$

$$\frac{7.7\underline{9}53\times10^{-12}\ J}{photon}\times\frac{6.022\times10^{23}\ photons}{mol}\times\frac{kJ}{1000\ J}=4.69\times10^9\ kJ/mol$$

Check: The units of the answer, kJ/mol, are correct. The magnitude of the answer is reasonable for a wavelength in the gamma ray region.

13. The interference pattern would be a series of light and dark lines.

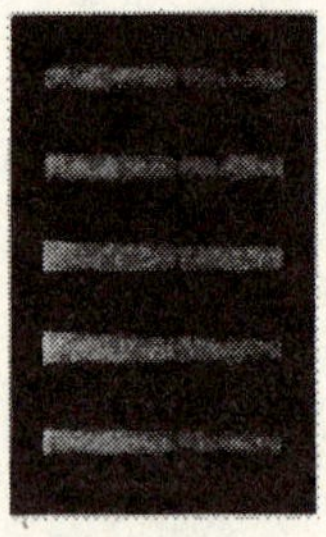

14. Since the interference pattern is caused by single electrons interfering with themselves, the pattern remains the same even when the rate of the electrons passing through the slits is one electron per hour. It will simply take longer for the full pattern to develop. When a laser is placed behind the slits to determine which hole the electron passes through, the laser flashes to indicate which hole the electron passed through, but the interference pattern is now absent. With the laser on the electrons hit positions directly behind each slit, as if they were ordinary particles.

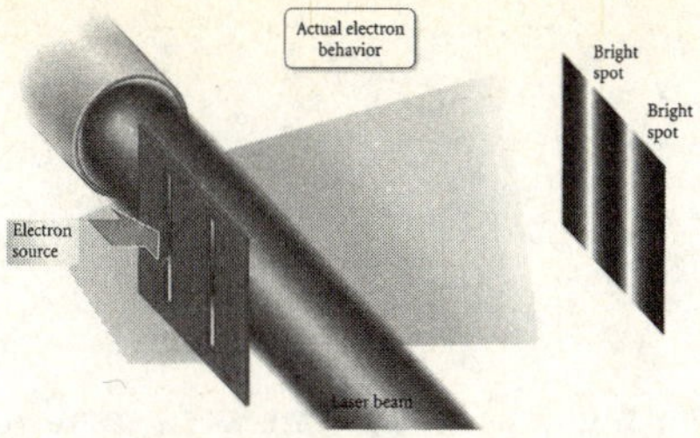

15. **Given:** $v = 1.15 \times 10^5$ m/s **Find:** λ

Conceptual Plan: $v \rightarrow \lambda$

$$\lambda = \frac{h}{mv}$$

Solution: $$\frac{6.626 \times 10^{-34} \frac{\cancel{kg} \cdot \cancel{m^2}}{\cancel{s^2}} \cdot \cancel{s}}{(9.11 \times 10^{-31}\ \cancel{kg})(1.15 \times 10^5\ \frac{\cancel{m}}{\cancel{s}})} = 6.32 \times 10^{-9}\ m$$

Check: The units of the answer, m, are correct. The magnitude of the answer is very small, as expected for the wavelength of an electron.

16. **Given:** $\lambda = 225$ nm **Find:** v

Conceptual Plan: $\lambda \rightarrow v$

$$v = \frac{h}{m\lambda}$$

Solution: $$\frac{6.626 \times 10^{-34} \frac{\cancel{kg} \cdot m^{\cancel{2}}}{s^{\cancel{2}}} \cdot \cancel{s}}{(9.11 \times 10^{-31}\ \cancel{kg})(225\ \cancel{nm})\left(\frac{\cancel{m}}{10^9\ \cancel{nm}}\right)} = 3.23 \times 10^3\ m/s$$

Check: The units of the answer, m/s, are correct. The magnitude of the answer is large, as expected for the speed of an electron.

17. **Given:** m = 143 g; v = 95 mph **Find:** λ

Conceptual Plan: m,v $\rightarrow \lambda$

$$\lambda = \frac{h}{mv}$$

Solution: $$\frac{6.626 \times 10^{-34} \frac{\cancel{kg} \cdot \cancel{m^2}}{\cancel{s^2}} \cdot \cancel{s}}{(143\ \cancel{g})\left(\frac{\cancel{kg}}{1000\ \cancel{g}}\right)\left(\frac{95\ \cancel{mi}}{\cancel{hr}}\right)\left(\frac{1.609\ \cancel{km}}{\cancel{mi}}\right)\left(\frac{1000\ \cancel{m}}{\cancel{km}}\right)\left(\frac{\cancel{hr}}{3600\ \cancel{s}}\right)} = 1.1 \times 10^{-34}\ m$$

The value of the wavelength, 1.1×10^{-34} m, is so small it will not have an effect on the trajectory of the baseball.

Check: The units of the answer, m, are correct. The magnitude of the answer is very small as would be expected for the de Broglie wavelength of a baseball.

18. **Given:** m = 27 g; v = 765 m/s **Find:** λ

Conceptual Plan: m,v $\rightarrow \lambda$

$$\lambda = \frac{h}{mv}$$

Solution:

$$\frac{6.626\times10^{-34}\,\frac{\text{kg}\cdot\text{m}^2}{\text{s}^2}\cdot\text{s}}{(27\,\text{g})\left(\frac{\text{kg}}{1000\,\text{g}}\right)\left(\frac{765\,\text{m}}{\text{s}}\right)} = 3.2\times10^{-35}\ \text{m}$$

The value of the wavelength, 3.2 x 10^{-35} m, is so small it will not have an effect on the trajectory of the bullet.

Check: The units of the answer, m, are correct. The magnitude of the answer is very small as would be expected for the de Broglie wavelength of a bullet.

19. Since the size of the orbital is determined by the *n* quantum, with the size increasing with increasing *n*, an electron in a 2*s* orbital is closer, on average, to the nucleus than an electron in a 3*s* orbital.

20. Since the size of the orbital is determined by the *n* quantum, with the size increasing with increasing *n*, an electron in a 4*p* orbital is further away, on average, from the nucleus than an electron in a 3*p* orbital.

21. The value of *l* is an integer that lies between 0 and $n - 1$.
    a) When $n = 1$, *l* can only be: $l = 0$.

    b) When $n = 2$, *l* can be: $l = 0$ or $l = 1$.

    c) When $n = 3$, *l* can be: $l = 0$, $l = 1$, or $l = 2$.

    d) When $n = 4$, *l* can be: $l = 0$, $l = 1$, $l = 2$, or $l = 3$.

22. The value of m_l is an integer that lies between $-l$ and $+l$.
    a) When $l = 0$; m_l can only be: $m_l = 0$.

    b) When $l = 1$, m_l can be: $m_l = -1$, $m_l = 0$, $m_l = +1$.

    c) When $l = 2$, m_l can be: $m_l = -2$, $m_l = -1$; $m_l = 0$, $m_l = +1$, $m_l = +2$.

    d) When $l = 3$, m_l can be: $m_l = -3$, $m_l = -2$, $m_l = -1$, $m_l = 0$, $m_l = +1$, $m_l = +2$, $m_l = +3$.

23. Set c cannot occur together as a set of quantum numbers to specify an orbital. *l* must lie between 0 and $n - 1$, so for $n = 3$, l can only be as high as 2.

24. a) 1*s* is a real orbital, $n = 1$, $l = 0$.

    b) 2*p* is a real orbital, $n = 2$, $l = 1$.

    c) 4*s* is a real orbital, $n = 4$, $l = 0$.

    d) 2*d* is an impossible representation. $n = 2$, $l = 2$ is not allowed. *l* must lie between 0 and $n - 1$, so for $n = 2$ *l* can only be as high as 1(*p*).

25. The 2*s* orbital would be the same shape as the 1*s* orbital but would be larger in size and the 3*p* orbitals would have the same shape as the 2*p* orbitals but would be larger in size. Also, the 2*s* and 3*p* orbitals would have more nodes.

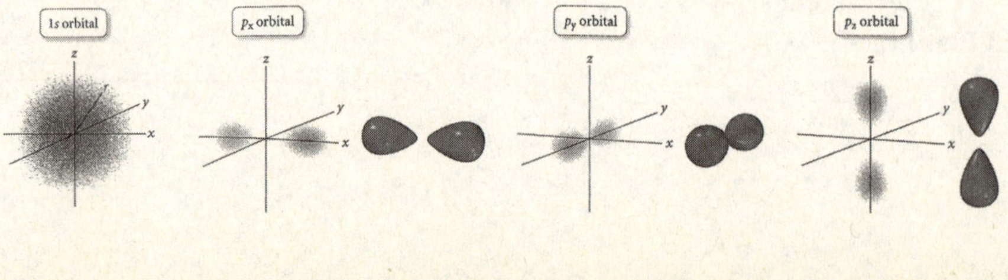

26. The 4*d* orbitals would be the same shape as the 3*d* orbitals but would be larger in size and the 4*d* orbital would have more nodes.

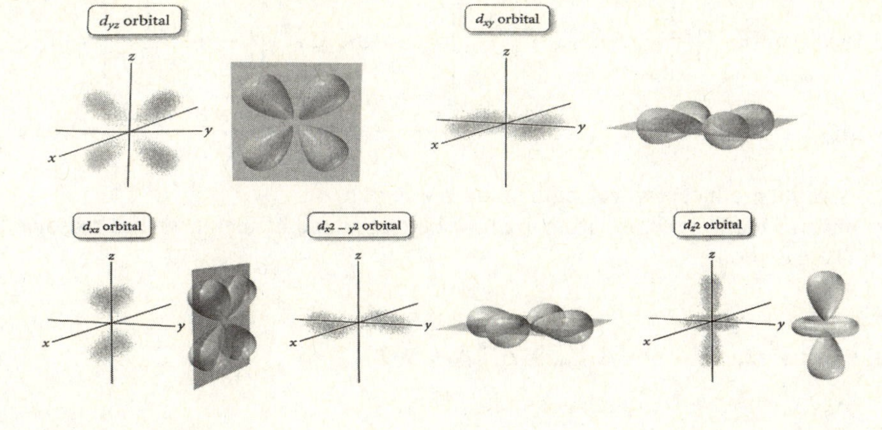

▲ FIGURE 7.27 The 3*d* Orbitals

27. When the atom emits the photon of energy that was needed to raise the electron to the $n = 2$ level, the photon has the same energy as the energy absorbed to move the electron to the excited state. Therefore, the electron has to be in $n = 1$ (the ground state) following the emission of the photon.

28. a) from $n = 3 \rightarrow n = 1$, the electron is moving to a lower energy, therefore, there is an emission of energy.

b) from $n = 2 \rightarrow n = 4$, the electron is moving to a higher energy, therefore, there is an absorption of energy.

c) from $n = 4 \rightarrow n = 3$, the electron is moving to a lower energy, therefore, there is an emission of energy.

29. According to the quantum-mechanical model, the higher the *n* level the higher the energy. So, the transition from $3p \rightarrow 1s$ would be a greater energy difference than a transition from $2p \rightarrow 1s$. The lower energy transition would have the longer wavelength. Therefore, the $2p \rightarrow 1s$ transition would produce a longer wavelength.

30. According to the quantum-mechanical model, the higher the n level the higher the energy and the higher in energy, the closer the levels are to each other. So, the transition from $3p \rightarrow 2s$ would be a greater energy difference than the transition from $4p \rightarrow 3p$. The lower energy transition would have the longer wavelength. Therefore, the $4p \rightarrow 3p$ transition would produce a longer wavelength.

31. a) **Given:** $n = 2 \rightarrow n = 1$ **Find:** λ

Conceptual Plan: $n = 1, n = 2 \rightarrow \Delta E_{atom} \rightarrow \Delta E_{photon} \rightarrow \lambda$

$\Delta E_{atom} = E_1 - E_2$ $\quad \Delta E_{atom} \rightarrow -\Delta E_{photon}$ $\quad E = \frac{hc}{\lambda}$

Solution:

$$\Delta E = E_1 - E_2$$

$$= -2.18 \times 10^{-18}\,\text{J}\left(\frac{1}{1^2}\right) - \left[-2.18 \times 10^{-18}\left(\frac{1}{2^2}\right)\right] = -2.18 \times 10^{-18}\,\text{J}\left[\left(\frac{1}{1^2}\right) - \left(\frac{1}{2^2}\right)\right] = -1.6\underline{3}5 \times 10^{-18}\,\text{J}$$

$$\Delta E_{photon} = -\Delta E_{atom} = 1.6\underline{3}5 \times 10^{-18}\,\text{J} \quad \lambda = \frac{hc}{E} = \frac{(6.626 \times 10^{-34}\,\cancel{\text{J}} \cdot \cancel{\text{s}})(3.00 \times 10^8\,\text{m}/\cancel{\text{s}})}{1.6\underline{3}5 \times 10^{-18}\,\cancel{\text{J}}} = 1.22 \times 10^{-7}\,\text{m}$$

This transition would produce a wavelength in the UV region.

Check: The units of the answer, m, are correct. The magnitude of the answer is reasonable since it is in the region of UV radiation.

b) **Given:** $n = 3 \rightarrow n = 1$ **Find:** λ

Conceptual Plan: $n = 1, n = 3 \rightarrow \Delta E_{atom} \rightarrow \Delta E_{photon} \rightarrow \lambda$

$\Delta E_{atom} = E_1 - E_3$ $\quad \Delta E_{atom} \rightarrow -\Delta E_{photon}$ $\quad E = \frac{hc}{\lambda}$

Solution:

$\Delta E = E_1 - E_3$

$$= -2.18\times10^{-18}\,J\left(\frac{1}{1^2}\right) - \left[-2.18\times10^{-18}\left(\frac{1}{3^2}\right)\right] = -2.18\times10^{-18}\,J\left[\left(\frac{1}{1^2}\right)-\left(\frac{1}{3^2}\right)\right] = -1.9\underline{3}8\times10^{-18}\,J$$

$$\Delta E_{photon} = -\Delta E_{atom} = 1.9\underline{3}8\times10^{-18}\,J \quad \lambda = \frac{hc}{E} = \frac{(6.626\times10^{-34}\,\cancel{J}\cdot\cancel{s})(3.00\times10^8\,m/\cancel{s})}{1.9\underline{3}8\times10^{-18}\,\cancel{J}} = 1.03\times10^{-7}\,m$$

This transition would produce a wavelength in the UV region.

Check: The units of the answer, m, are correct. The magnitude of the answer is reasonable since it is in the region of UV radiation.

c) **Given:** $n = 4 \rightarrow n = 2$ **Find:** λ

Conceptual Plan: $n = 2$, $n = 4 \rightarrow \Delta E_{atom} \rightarrow \Delta E_{photon} \rightarrow \lambda$

$\Delta E_{atom} = E_2 - E_4$ $\quad \Delta E_{atom} \rightarrow -\Delta E_{photon}$ $\quad E = \frac{hc}{\lambda}$

Solution:

$\Delta E = E_2 - E_4$

$$= -2.18\times10^{-18}\,J\left(\frac{1}{2^2}\right) - \left[-2.18\times10^{-18}\left(\frac{1}{4^2}\right)\right] = -2.18\times10^{-18}\,J\left[\left(\frac{1}{2^2}\right)-\left(\frac{1}{4^2}\right)\right] = -4.0\underline{8}7\times10^{-19}\,J$$

$$\Delta E_{photon} = -\Delta E_{atom} = 4.0\underline{8}7\times10^{-19}\,J \quad \lambda = \frac{hc}{E} = \frac{(6.626\times10^{-34}\,\cancel{J}\cdot\cancel{s})(3.00\times10^8\,m/\cancel{s})}{4.0\underline{8}7\times10^{-19}\,\cancel{J}} = 4.86\times10^{-7}\,m$$

This transition would produce a wavelength in the visible region.

Check: The units of the answer, m, are correct. The magnitude of the answer is reasonable since it is in the region of visible light.

d) **Given:** $n = 5 \rightarrow n = 2$ **Find:** λ

Conceptual Plan: $n = 2$, $n = 5 \rightarrow \Delta E_{atom} \rightarrow \Delta E_{photon} \rightarrow \lambda$

$\Delta E_{atom} = E_2 - E_5$ $\quad \Delta E_{atom} \rightarrow -\Delta E_{photon}$ $\quad E = \frac{hc}{\lambda}$

Solution:

$\Delta E = E_2 - E_5$

$$= -2.18\times10^{-18}\,J\left(\frac{1}{2^2}\right) - \left[-2.18\times10^{-18}\left(\frac{1}{5^2}\right)\right] = -2.18\times10^{-18}\,J\left[\left(\frac{1}{2^2}\right)-\left(\frac{1}{5^2}\right)\right] = -4.5\underline{7}8\times10^{-19}\,J$$

$$\Delta E_{photon} = -\Delta E_{atom} = 4.5\underline{7}8\times10^{-19}\,J \quad \lambda = \frac{hc}{E} = \frac{(6.626\times10^{-34}\,\cancel{J}\cdot\cancel{s})(3.00\times10^8\,m/\cancel{s})}{4.5\underline{7}8\times10^{-19}\,\cancel{J}} = 4.34\times10^{-7}\,m$$

This transition would produce a wavelength in the visible region.

Check: The units of the answer, m, are correct. The magnitude of the answer is reasonable since it is in the region of visible light.

32. a) **Given:** $n = 4 \rightarrow n = 3$ **Find:** ν

Conceptual Plan: $n = 3$, $n = 4 \rightarrow \Delta E_{atom} \rightarrow \Delta E_{photon} \rightarrow \nu$

$\Delta E_{atom} = E_3 - E_4$ $\quad \Delta E_{atom} \rightarrow -\Delta E_{photon}$ $\quad E = h\nu$

Solution:

$\Delta E = E_3 - E_4$

$$= -2.18\times10^{-18}\,J\left(\frac{1}{3^2}\right) - \left[-2.18\times10^{-18}\left(\frac{1}{4^2}\right)\right] = -2.18\times10^{-18}\,J\left[\left(\frac{1}{3^2}\right)-\left(\frac{1}{4^2}\right)\right] = -1.0\underline{6}0\times10^{-19}\,J$$

$$\Delta E_{photon} = -\Delta E_{atom} = 1.0\underline{6}0\times10^{-19}\,J \quad \nu = \frac{E}{h} = \frac{(1.0\underline{6}0\times10^{-19}\,\cancel{J})}{6.626\times10^{-34}\,\cancel{J}\cdot s)} = 1.60\times10^{14}\,s^{-1}$$

Check: The units of the answer, s^{-1}, are correct. The magnitude of the answer is reasonable since it is a transition between two close levels and the levels become closer as the n value increases. Therefore, the energy difference is smaller and the frequency is smaller.

b) **Given:** $n = 5 \rightarrow n = 1$ **Find:** ν

Conceptual Plan: $n = 1, n = 5 \rightarrow \Delta E_{atom} \rightarrow \Delta E_{photon} \rightarrow \nu$

$\Delta E_{atom} = E_1 - E_5$ $\quad$ $\Delta E_{atom} \rightarrow -\Delta E_{photon}$ $\quad$ $E = h\nu$

Solution:

$$\Delta E = E_1 - E_5$$

$$= -2.18\times10^{-18}\,J\left(\frac{1}{1^2}\right) - \left[-2.18\times10^{-18}\left(\frac{1}{5^2}\right)\right] = -2.18\times10^{-18}\,J\left[\left(\frac{1}{1^2}\right)-\left(\frac{1}{5^2}\right)\right] = -2.093\times10^{-18}\,J$$

$$\Delta E_{photon} = -\Delta E_{atom} = 2.093\times10^{-18}\,J \quad \nu = \frac{E}{h} = \frac{2.093\times10^{-18}\,J}{6.626\times10^{-34}\,J\cdot s} = 3.16\times10^{15}\,s^{-1}$$

Check: The units of the answer, s^{-1}, are correct. The magnitude of the answer is reasonable since it is a transition that will produce a wavelength in the UV region and the frequency is correct for the UV region.

c) **Given:** $n = 5 \rightarrow n = 4$ **Find:** ν

Conceptual Plan: $n = 4, n = 5 \rightarrow \Delta E_{atom} \rightarrow \Delta E_{photon} \rightarrow \nu$

$\Delta E_{atom} = E_4 - E_5$ $\Delta E_{atom} \rightarrow -\Delta E_{photon}$ $\quad$ $E = h\nu$

Solution:

$$\Delta E = E_4 - E_5$$

$$= -2.18\times10^{-18}\,J\left(\frac{1}{4^2}\right) - \left[-2.18\times10^{-18}\left(\frac{1}{5^2}\right)\right] = -2.18\times10^{-18}\,J\left[\left(\frac{1}{4^2}\right)-\left(\frac{1}{5^2}\right)\right] = -4.905\times10^{-20}\,J$$

$$\Delta E_{photon} = -\Delta E_{atom} = 4.905\times10^{-20}\,J \quad \nu = \frac{E}{h} = \frac{4.905\times10^{-20}\,J}{6.626\times10^{-34}\,J\cdot s} = 7.40\times10^{13}\,s^{-1}$$

Check: The units of the answer, s^{-1}, are correct. The magnitude of the answer is reasonable since it is a transition between two close levels and the levels become closer as the *n* value increases. Therefore, the energy difference is smaller and the frequency is smaller.

d) **Given:** $n = 6 \rightarrow n = 5$ **Find:** ν

Conceptual Plan: $n = 5, n = 6 \rightarrow \Delta E_{atom} \rightarrow \Delta E_{photon} \rightarrow \nu$

$\Delta E_{atom} = E_5 - E_6$ $\quad$ $\Delta E_{atom} \rightarrow -\Delta E_{photon}$ $\quad$ $E = h\nu$

Solution:

$$\Delta E = E_5 - E_6$$

$$= -2.18\times10^{-18}\,J\left(\frac{1}{5^2}\right) - \left[-2.18\times10^{-18}\left(\frac{1}{6^2}\right)\right] = -2.18\times10^{-18}\,J\left[\left(\frac{1}{5^2}\right)-\left(\frac{1}{6^2}\right)\right] = -2.664\times10^{-20}\,J$$

$$\Delta E_{photon} = -\Delta E_{atom} = 2.664\times10^{-20}\,J \quad \nu = \frac{E}{h} = \frac{2.664\times10^{-20}\,J}{6.626\times10^{-34}\,J\cdot s} = 4.02\times10^{13}\,s^{-1}$$

Check: The units of the answer, s^{-1}, are correct. The magnitude of the answer is reasonable since it is a transition between two close levels and the levels become closer as the *n* value increases. Therefore, the energy difference is smaller and the frequency is smaller.

33. **Given:** *n*(initial) = 7 $\lambda = 397$ nm **Find:** *n*(final)

Conceptual Plan: $\lambda \rightarrow \Delta E_{photon} \rightarrow \Delta E_{atom} \rightarrow n = x, n = 7$

$E = \frac{hc}{\lambda}$ $\quad$ $\Delta E_{photon} \rightarrow -\Delta E_{atom}$ $\quad$ $\Delta E_{atom} = E_x - E_7$

Solution:

$$E = \frac{hc}{\lambda} = \frac{(6.626\times10^{-34}\,J\cdot s)(3.00\times10^{8}\,m/s)}{(397\,nm)\left(\frac{m}{10^9\,nm}\right)} = 5.007\times10^{-19}\,J$$

$$\Delta E_{atom} = -\Delta E_{photon} = -5.007\times10^{-19}\,J$$

$$\Delta E = E_x - E_7 \quad = -5.007 \times 10^{-19} = -2.18 \times 10^{-18}\, J\left(\frac{1}{x^2}\right) - \left[-2.18 \times 10^{-18}\left(\frac{1}{7^2}\right)\right] = -2.18 \times 10^{-18}\, J\left[\left(\frac{1}{x^2}\right) - \left(\frac{1}{7^2}\right)\right]$$

$$0.2297 = \left(\frac{1}{x^2}\right) - \left(\frac{1}{7^2}\right) \qquad 0.25229 = \left(\frac{1}{x^2}\right) \qquad x^2 = 3.998 \qquad x = 2$$

Check: The answer is reasonable since it is an integer less than the initial value of 7.

34. **Given:** n(final) = 4, ν = 11.4 THz **Find:** n(initial)
Conceptual Plan: **ν → ΔE_{photon} → ΔE_{atom} → n = 4, n = x**

$E = h\nu \qquad \Delta E_{photon} \rightarrow -\Delta E_{atom} \qquad \Delta E_{atom} = E_4 - E_x$

Solution: $E = h\nu = (6.626 \times 10^{-34}\, J \cdot s)(114\, THz)\left(\frac{10^{12}\, Hz}{T}\right)\left(\frac{s^{-1}}{Hz}\right) = 7.553 \times 10^{-20}\, J$

$\Delta E_{atom} = -\Delta E_{photon} = -7.553 \times 10^{-20}\, J$

$$\Delta E = E_4 - E_x \quad = -7.553 \times 10^{-20} J = -2.18 \times 10^{-18}\, J\left(\frac{1}{4^2}\right) - \left[-2.18 \times 10^{-18}\left(\frac{1}{x^2}\right)\right] = -2.18 \times 10^{-18}\, J\left[\left(\frac{1}{4^2}\right) - \left(\frac{1}{x^2}\right)\right]$$

$$0.03465 = \left(\frac{1}{4^2}\right) - \left(\frac{1}{x^2}\right) \qquad 0.02785 = \left(\frac{1}{x^2}\right) \qquad x^2 = 35.9 = 36$$

Check: The answer is reasonable since it is an integer greater than the final value of 4.

35. **Given:** 348 kJ/mol **Find:** λ
Conceptual Plan: **kJ/mol → kJ/molec → J/molec → λ**

$\frac{6.022 \times 10^{23}\, C-C\ bonds}{mol\ C-C\ bonds} \quad \frac{1000\, J}{kJ} \qquad E = \frac{hc}{\lambda}$

Solution: $\frac{348\, kJ}{mol\ C-C\ bonds} \times \frac{mol\ C-C\ bonds}{6.022 \times 10^{23}\, C-C\ bonds} \times \frac{1000\, J}{kJ} = 5.779 \times 10^{-19}\, J$

$$\lambda = \frac{(6.626 \times 10^{-34}\, J \cdot s)(3.00 \times 10^8\, m/s)}{5.779 \times 10^{-19}\, J} = 3.44 \times 10^{-7}\, m = 344\, nm$$

Check: The units of the answer, m or nm, is correct. The magnitude of the answer is reasonable since this wavelength is in the UV region.

36. **Given:** 164 kJ/mol **Find:** λ
Conceptual Plan: **kJ/mol → kJ/molec → J/molec → λ**

$\frac{6.022 \times 10^{23}\, molec}{mol} \quad \frac{1000\, J}{kJ} \qquad E = \frac{hc}{\lambda}$

Solution: $\frac{164\, kJ}{mol} \times \frac{mol}{6.022 \times 10^{23}\, molecules} \times \frac{1000\, J}{kJ} = 2.723 \times 10^{-19}\, J$

$$\lambda = \frac{(6.626 \times 10^{-34}\, J \cdot s)(3.00 \times 10^8\, m/s)}{2.723 \times 10^{-19}\, J} = 7.30 \times 10^{-7}\, m = 730\, nm$$

Check: The units of the answer, m or nm, is correct. The magnitude of the answer is reasonable since this wavelength is in the red region of visible light.

37. **Given:** E_{pulse} = 5.0 watts; d = 5.5 mm; hole = 1.2 mm; λ = 532 nm **Find:** photons/s
Conceptual Plan: fraction of beam through hole → fraction of power and then E_{photon} → number photons/s

$\frac{area\ hole}{area\ beam} \qquad fraction \times power \qquad E = \frac{hc}{\lambda} \quad \frac{power/s}{E/photon}$

Solution: $A = \pi r^2 \qquad \frac{\pi(0.60 mm)^2}{\pi(2.75 mm)^2} = 0.0476 \qquad 0.0476 \times 5.0\, watts \times \frac{J/s}{watt} = 0.238\, J/s$

$$E_{photon} = \frac{(6.626 \times 10^{-34}\ J \cdot \cancel{s})(3.00 \times 10^{8}\ \cancel{m}/\cancel{s})}{(532\ \cancel{nm})\left(\frac{\cancel{m}}{10^{9}\ \cancel{nm}}\right)} = 3.7\underline{3}6 \times 10^{-19}\ J/photon$$

$$\frac{0.2\underline{3}8\ \cancel{J}/s}{3.7\underline{3}6 \times 10^{-19}\ \cancel{J}/photon} = 6.4 \times 10^{17}\ photons/s$$

Check: The units of the answer, number of photons/s, is correct. The magnitude of the answer is reasonable.

38. **Given:** $A_{leaf} = 2.50\ cm^2$; $E_{rad} = 1000\ W/m^2$; $\lambda = 504$ nm **Find:** photons/s

Conceptual Plan: $E_{rad}/s \rightarrow E_{leaf}/s$ **and then** $E_{photon} \rightarrow$ **number photons/s**

$$E_{rad} \times A_{leaf} \qquad E = \frac{hc}{\lambda} \qquad \frac{E_{leaf}/s}{E/photon}$$

Solution: $$E_{rad} = 2.50\ \cancel{cm^2} \times \frac{1000\ \cancel{W}}{\cancel{m^2}} \times \frac{\cancel{m^2}}{(100\ \cancel{cm})^2} \times \frac{J/s}{\cancel{W}} = 0.250\ J/s$$

$$E_{photon} = \frac{hc}{\lambda} = \frac{(6.626 \times 10^{-34}\ J \cdot \cancel{s})(3.00 \times 10^{8}\ \cancel{m}/\cancel{s})}{(504\ \cancel{nm})\left(\frac{\cancel{m}}{10^{9}\ \cancel{nm}}\right)} = 3.9\underline{4}4 \times 10^{-19}\ J/photon$$

$$\frac{E_{rad}}{E_{photon}} = \frac{0.250\ \cancel{J}/s}{3.9\underline{4}4 \times 10^{-19}\ \cancel{J}/photon} = 6.34 \times 10^{17}\ photons/s$$

Check: The units of the answer, photons/s, are correct. The magnitude of the answer is reasonable compared to the radiation from the sun.

39. **Given:** KE = 506 eV **Find:** λ

Conceptual Plan: $KE_{ev} \rightarrow KE_J \rightarrow v \rightarrow \lambda$

$$\frac{1.602 \times 10^{-19}\ J}{eV} \qquad KE = 1/2\,mv^2 \qquad \lambda = \frac{h}{mv}$$

Solution:

$$506\,eV\left(\frac{1.602 \times 10^{-19}\ J}{eV}\right)\left(\frac{\frac{kg \cdot m^2}{s^2}}{J}\right) = \frac{1}{2}\left(9.11 \times 10^{-31}\ kg\right) v^2$$

$$v^2 = \frac{506\ \cancel{eV}\left(\frac{1.602 \times 10^{-19}\ \cancel{J}}{\cancel{eV}}\right)\left(\frac{\frac{\cancel{kg} \cdot m^2}{s^2}}{\cancel{J}}\right)}{\frac{1}{2}\left(9.11 \times 10^{-31}\ \cancel{kg}\right)} = 1.7796 \times 10^{14}\ \frac{m^2}{s^2}$$

$$v = 1.33 \times 10^{7}\ m/s \qquad \lambda = \frac{h}{mv} = \frac{6.626 \times 10^{-34}\ \frac{\cancel{kg} \cdot m^{\cancel{2}}}{\cancel{s}^{\cancel{2}}} \cdot \cancel{s}}{(9.11 \times 10^{-31}\ \cancel{kg})(1.33 \times 10^{7}\ \cancel{m}/\cancel{s})} = 5.47 \times 10^{-11}\ m = 0.0547\ nm$$

Check: The units of the answer, m or nm, are correct. The magnitude of the answer is reasonable because a deBroglie wavelength is usually a very small number.

40. **Given:** $\lambda = 0.989$ nm; KE = 969 eV **Find:** BE/ mol

Conceptual Plan: $\lambda \rightarrow E_{photon} \rightarrow BE_{photon} \rightarrow BE_{mol}$

$$E = \frac{hc}{\lambda} \qquad BE_{photon} = E_{photon} - KE \qquad \frac{mol}{6.022 \times 10^{23}\ photons}$$

Solution: $E_{photon} = \frac{hc}{\lambda} = \frac{(6.626 \times 10^{-34}\ J \cdot \cancel{s})(3.00 \times 10^{8}\ \cancel{m} / \cancel{s})}{(0.989\ \cancel{nm})\left(\frac{\cancel{m}}{10^{9}\ \cancel{nm}}\right)} = 2.01\underline{0} \times 10^{-16}\ J / photon$

$$BE_{photon} = 2.01\underline{0} \times 10^{-16}\ J / photon - \left[(969\ \cancel{eV})\left(\frac{1.602 \times 10^{-19}\ J}{\cancel{eV}}\right)\right] = 4.5\underline{7}6 \times 10^{-17}\ J / photon$$

$$\frac{4.5\underline{7}6 \times 10^{-17}\ \cancel{J}}{\cancel{photon}} \times \frac{6.022 \times 10^{23}\ \cancel{photons}}{mol} \times \frac{kJ}{1000\ \cancel{J}} = 2.76 \times 10^{4}\ kJ / mol$$

Check: The units of the answer, kJ/mol, are correct. The magnitude of the answer is reasonable since it should require a large amount of energy to remove an electron from a metal surface.

41. **Given:** $n = 1 \rightarrow n = \infty$ **Find:** E; λ
Conceptual Plan: $n = \infty, n = 1 \rightarrow \Delta E_{atom} \rightarrow \Delta E_{photon} \rightarrow \lambda$

$\Delta E_{atom} = E_{\infty} - E_1$ $\quad \Delta E_{atom} \rightarrow \Delta E_{photon}$ $\quad E = \frac{hc}{\lambda}$

Solution: $\Delta E = E_{\infty} - E_1 = 0 - \left[-2.18 \times 10^{-18}\left(\frac{1}{1^2}\right)\right] = +2.18 \times 10^{-18}\ J$

$$\Delta E_{photon} = -\Delta E_{atom} = +2.18 \times 10^{-18}\ J$$

$$\lambda = \frac{hc}{E} = \frac{(6.626 \times 10^{-34}\ \cancel{J} \cdot \cancel{s})(3.00 \times 10^{8}\ m / \cancel{s})}{2.18 \times 10^{-18}\ \cancel{J}} = 9.12 \times 10^{-8}\ m = 91.2\ nm$$

Check: The units of the answers, J for E and m or nm for part 1, are correct. The magnitude of the answer is reasonable because it would require more energy to completely remove the electron than just moving it to a higher n level. This results in a shorter wavelength.

42. **Given:** E = 496 kJ/mol **Find:** ν
Conceptual Plan: **kJ/mol → kJ/molecule → J/molecule → ν**

$\frac{6.022 \times 10^{23}\ molecules}{mol}$ $\quad \frac{1000\ J}{kJ}$ $\quad E = h\nu$

Solution: $\nu = \frac{E}{h} = \frac{\left(\frac{496\ \cancel{kJ}}{\cancel{mol}}\right)\left(\frac{\cancel{mol}}{6.022 \times 10^{23}\ atom}\right)\left(\frac{1000\ \cancel{J}}{\cancel{kJ}}\right)}{6.626 \times 10^{-34}\ \cancel{J} \cdot s} = 1.24 \times 10^{15}\ s^{-1}$

Check: The units of the answer, s^{-1}, are correct. The magnitude of the answer is reasonable because the frequency is slightly higher than the visible region of the spectrum and this is expected because the excitation of sodium produces a line in the visible region.

43. a) **Given:** $n = 1$ **Find:** number of orbitals if $l = 0 \rightarrow n$
Conceptual Plan: **value n → values l → values m_l → number of orbitals**
$l = 0 \rightarrow n$ $\quad m_l = -1 \rightarrow +1$ total m_l

Solution:

$n =$	1		
$l =$	0	1	
$m_l =$	0	- 1, 0, +1	

total 4 orbitals

Check: The total orbitals will be equal to the number of l sublevels^2

b) **Given:** $n = 2$ **Find:** number of orbitals if $l = 0 \rightarrow n$
Conceptual Plan: **value n → values l → values m_l → number of orbitals**
$l = 0 \rightarrow n$ $\quad m_l = -1 \rightarrow +1$ total m_l

Solution:

$n =$	2		
$l =$	0	1	2
$m_l =$	0	- 1, 0, +1	- 2,- 1,0,1,2

total 9 orbitals

Check: The total orbitals will be equal to the number of l sublevels^2

c) **Given:** $n = 3$ **Find:** number of orbitals if $l = 0 \rightarrow n$

Conceptual Plan: value n → values l → values m_l → number of orbitals

$l = 0 \rightarrow n$ $m_l = -1 \rightarrow +1$ total m_l

Solution:

n =	3			
l =	0	1	2	3
m_l =	0	- 1, 0, +1	- 2,- 1,0,1,2	- 3,- 2,- 1,0,1,2,3

total 16 orbitals

Check: The total orbitals will be equal to the number of l sublevels^2

44. a) **Given:** s sublevel **Find:** number of orbital if $m_l = -l\text{-}1 \rightarrow l+1$

Conceptual Plan: values l → values m_l → number of orbitals

$m_l = -l-1 \rightarrow +l+1$ total m_l

Solution: sublevel $s \rightarrow l = 0$

$m_l = -1, 0, +1$

total 3 orbitals

b) **Given:** p sublevel **Find:** number of orbital if $m_l = -l\text{-}1 \rightarrow l+1$

Conceptual Plan: values l → values m_l → number of orbitals

$m_l = -l-1 \rightarrow +l+1$ total m_l

Solution: sublevel $p \rightarrow l = 1$

$m_l = -2, -1, 0, +1, +2$

total 5 orbitals

c) **Given:** d sublevel **Find:** number of orbital if $m_l = -l\text{-}1 \rightarrow l+1$

Conceptual Plan: values l → values m_l → number of orbitals

$m_l = -l-1 \rightarrow +l+1$ total m_l

Solution: sublevel $d \rightarrow l = 2$

$m_l = -3, -2, -1, 0, +1, +2, +3$

total 7 orbitals

45. **Given:** $\lambda = 1875$ nm; 1282 nm; 1093 nm **Find:** equivalent transitions

Conceptual Plan: $\lambda \rightarrow E_{photon} \rightarrow E_{atom} \rightarrow n$

$$E = \frac{hc}{\lambda} \qquad E_{photon} = -E_{atom} \qquad E = -2.18 \times 10^{-18}\,J\left(\frac{1}{n_f^2} - \frac{1}{n_i^2}\right)$$

Solution: Since the wavelength of the transitions are longer wavelengths than those obtained in the visual region, the electron must relax to a higher n level. Therefore, we can assume that the electron returns to the n = 3 level.

For $\lambda = 1875$ nm:

$$E = \frac{(6.626 \times 10^{-34}\,J \cdot s)(3.00 \times 10^8\,m/s)}{1875\,nm\left(\dfrac{m}{10^9\,nm}\right)} = 1.060 \times 10^{-19}\,J \qquad 1.060 \times 10^{-19}\,J = -1.060 \times 10^{-19}\,J$$

$$-1.060 \times 10^{-19}\,J = -2.18 \times 10^{-18}\left(\frac{1}{3^2} - \frac{1}{n^2}\right);\ n = 4$$

For $\lambda = 1282$ nm:

$$E = \frac{(6.626 \times 10^{-34}\,J \cdot s)(3.00 \times 10^8\,m/s)}{1282\,nm\left(\dfrac{m}{10^9\,nm}\right)} = 1.551 \times 10^{-19}\,J \qquad 1.551 \times 10^{-19}\,J = -1.551 \times 10^{-19}\,J$$

$$-1.551 \times 10^{-19} = -2.18 \times 10^{-18}\left(\frac{1}{3^2} - \frac{1}{n^2}\right);\ n = 5$$

For $\lambda = 1093$ nm:

$$E = \frac{(6.626 \times 10^{-34}\,J \cdot s)(3.00 \times 10^8\,m/s)}{1093\,nm\left(\dfrac{m}{10^9\,nm}\right)} = 1.819 \times 10^{-19}\,J \qquad 1.819 \times 10^{-19}\,J = -1.819 \times 10^{-19}\,J$$

$$-1.819 \times 10^{-19} J = -2.18 \times 10^{-18}\left(\frac{1}{3^2} - \frac{1}{n^2}\right);\ n = 6$$

Check: The values obtained are all integers, which is correct. The values of *n*; 4,5,6, are reasonable. The values of n increase as the wavelength decreases because the two n levels involved are further apart and more energy is released as the electron relaxes to the *n* = 3 level.

46. **Given:** λ = 121.5 nm;102.6 nm; 97.23 nm **Find:** equivalent transitions
Conceptual Plan: **λ → E_{photon} → E_{atom} → *n***

$$E = \frac{hc}{\lambda} \qquad E_{photon} = -E_{atom} \qquad E = -2.18 \times 10^{-18} J\left(\frac{1}{n_f^2} - \frac{1}{n_i^2}\right)$$

Solution: Since the wavelengths of the transitions are shorter wavelengths than those obtained in the visual region, the electron must relax to a lower n level. Therefore, we can assume that the electron returns to the *n* = 1 level.
For λ = 121.5 nm:

$$E = \frac{(6.626 \times 10^{-34} J \cdot \cancel{s})(3.00 \times 10^8 \cancel{m}/\cancel{s})}{121.5\ \cancel{nm}\left(\frac{\cancel{m}}{10^9\ \cancel{nm}}\right)} = 1.636 \times 10^{-18} J \qquad 1.636 \times 10^{-18} J = -1.636 \times 10^{-18} J$$

$$-1.636 \times 10^{-18} J = -2.18 \times 10^{-18}\left(\frac{1}{1^2} - \frac{1}{n^2}\right);\ n = 2$$

For λ = 102.6 nm:

$$E = \frac{(6.626 \times 10^{-34} J \cdot \cancel{s})(3.00 \times 10^8 \cancel{m}/\cancel{s})}{102.6\ \cancel{nm}\left(\frac{\cancel{m}}{10^9\ \cancel{nm}}\right)} = 1.937 \times 10^{-18} J \qquad 1.937 \times 10^{-18} J = -1.937 \times 10^{-18} J$$

$$-1.937 \times 10^{-18} J = -2.18 \times 10^{-18}\left(\frac{1}{1^2} - \frac{1}{n^2}\right);\ n = 3$$

For λ = 97.23 nm:

$$E = \frac{(6.626 \times 10^{-34} J \cdot \cancel{s})(3.00 \times 10^8 \cancel{m}/\cancel{s})}{97.23\ \cancel{nm}\left(\frac{\cancel{m}}{10^9\ \cancel{nm}}\right)} = 2.044 \times 10^{-18} J \qquad 2.044 \times 10^{-18} J = -2.044 \times 10^{-18} J$$

$$-2.044 \times 10^{-18} J = -2.18 \times 10^{-18}\left(\frac{1}{1^2} - \frac{1}{n^2}\right);\ n = 4$$

Check: The values obtained are all integers, which is correct. The values of *n*; 2,3,4, are reasonable. The values of *n* increase as the wavelength decreases because the two *n* levels involved are further apart and more energy is released as the electron relaxes to the *n* = 1 level.

47. **Given:** Φ = 193 kJ/mol **Find:** threshold frequency(ν)
Conceptual Plan: **Φ kJ/ mol → Φ kJ/ atom → Φ J/ atom → ν**

$$\frac{6.022 \times 10^{23}\ \text{atoms}}{\text{mol}} \qquad \frac{1000\ J}{kJ} \qquad \Phi = h\nu$$

$$\textbf{Solution: } \nu = \frac{\Phi}{h} = \frac{\left(\frac{193\ \cancel{kJ}}{\cancel{mol}}\right)\left(\frac{\cancel{mol}}{6.022 \times 10^{23}\ \text{atoms}}\right)\left(\frac{1000\ \cancel{J}}{\cancel{kJ}}\right)}{6.626 \times 10^{-34}\ \cancel{J} \cdot s} = 4.84 \times 10^{14} s^{-1}$$

Check: The units of the answer, s^{-1}, are correct. The magnitude of the answer puts the frequency in the infrared range and is a reasonable answer.

48. **Given:** m = 2 amu; v = 1 x 10^6 m/s **Find:** λ
Conceptual Plan: **m(amu) → m(g) → m(kg) and then m,v → λ**

$$\frac{1.661 \times 10^{-24} g}{amu} \qquad \frac{kg}{1000\ g} \qquad \lambda = \frac{h}{mv}$$

Solution: $\dfrac{6.626 \times 10^{-34} \dfrac{\text{kg} \cdot \text{m}^2}{\text{s}^2} \cdot \text{s}}{(2\ \text{amu})\left(\dfrac{1.661 \times 10^{-24}\ \text{g}}{\text{amu}}\right)\left(\dfrac{\text{kg}}{1000\ \text{g}}\right)\left(1 \times 10^6\ \text{m}/\text{s}\right)} = 2 \times 10^{-13}\ \text{m}$

Check: The units of the answer, m, are correct. The magnitude of the answer is reasonable since it is a smaller wavelength than for an electron and a deuteron has a much larger mass than an electron.

49. **Given:** $\nu_{low} = 30 s^{-1}$ $\nu_{hi} = 1.5 \times 10^4\ s^{-1}$; speed = 344 m/s **Find:** $\lambda_{low} - \lambda_{hi}$

Conceptual Plan: $\nu_{low} \rightarrow \lambda_{low}$ and $\nu_{hi} = \lambda_{hi}$ then $\lambda_{low} - \lambda_{hi}$

$\lambda\nu = \text{speed}$

Solution: $\lambda = \dfrac{\text{speed}}{\nu}$ $\quad \lambda_{low} = \dfrac{344\ \text{m}/\text{s}}{30\ \text{s}^{-1}} = 11\text{m}$ $\quad \lambda_{hi} = \dfrac{344\ \text{m}/\text{s}}{1.5 \times 10^4\ \text{s}^{-1}} = 0.023\text{m}$ $\quad 11\text{m} - 0.023\text{m} = 11\text{m}$

Check: The units of the answer, m, are correct. The magnitude is reasonable since the value is only determined by the low frequency value because of significant figures.

50. **Given:** $d = 1.5 \times 10^8$ km, $\nu = 1.0 \times 10^{14}\ s^{-1}$ **Find:** number of wave crests

Conceptual Plan: $\nu \rightarrow \lambda$ and then d(km) → d(m) → number of waves → number of crests

$\nu = \dfrac{c}{\lambda}$ $\qquad \dfrac{1000\text{m}}{\text{km}}$ $\quad \dfrac{d}{\lambda}$

Solution: $\dfrac{3.00 \times 10^8 \text{m}/\text{s}}{1.0 \times 10^{14}\ \text{s}^{-1}} = 3.0 \times 10^{-6}\ \text{m}$ $\qquad \dfrac{1.5 \times 10^8\ \text{km} \times \dfrac{1000\ \text{m}}{\text{km}}}{3.0 \times 10^{-6}\ \text{m}} = 5.0 \times 10^{16}$ waves

Since wavelength is measured crest to crest, the number of wave crests would be $5.0 \times 10^{13} + 1$

Check: The answer is reasonable since the wavelength is small and the distance traveled is large.

51. a. **Given:** $n = 1, n = 2, n = 3$, L = 155 pm **Find:** E_1, E_2, E_3

Conceptual Plan: $n \rightarrow$ **E**

$E_n = \dfrac{n^2 h^2}{8\ m\ L^2}$

Solution:

$$E_1 = \frac{1^2(6.626 \times 10^{-34}\ \text{J} \cdot \text{s})^2}{8(9.11 \times 10^{-31}\text{kg})(155\ \text{pm})^2\left(\dfrac{\text{m}}{10^{12}\ \text{pm}}\right)^2} = \frac{1(6.626 \times 10^{-34})^2\ \text{J}^2\text{s}^2}{8(9.11 \times 10^{-31}\text{kg})(155 \times 10^{-12})^2\text{m}^2}$$

$$= \frac{1(6.626 \times 10^{-34})^2\left(\dfrac{\text{kg} \cdot \text{m}^2}{\text{s}^2}\right)\text{J}\ \text{s}^2}{8(9.11 \times 10^{-31}\ \text{kg})(155 \times 10^{-12})^2\ \text{m}^2} = 2.51 \times 10^{-18}\ \text{J}$$

$$E_2 = \frac{2^2(6.626 \times 10^{-34}\ \text{J} \cdot \text{s})^2}{8(9.11 \times 10^{-31}\text{kg})(155\ \text{pm})^2\left(\dfrac{\text{m}}{10^{12}\ \text{pm}}\right)^2} = \frac{4(6.626 \times 10^{-34})^2\ \text{J}^2\text{s}^2}{8(9.11 \times 10^{-31}\text{kg})(155 \times 10^{-12})^2\text{m}^2}$$

$$= \frac{4(6.626 \times 10^{-34})^2\left(\dfrac{\text{kg} \cdot \text{m}^2}{\text{s}^2}\right)\text{J}\ \text{s}^2}{8(9.11 \times 10^{-31}\ \text{kg})(155 \times 10^{-12})^2\ \text{m}^2} = 1.00 \times 10^{-17}\ \text{J}$$

$$E_3 = \frac{3^2(6.626\times10^{-34}\ \text{J}\bullet\text{s})^2}{8(9.11\times10^{-31}\text{kg})(155\ \cancel{\text{pm}})^2\left(\frac{\text{m}}{10^{12}\ \cancel{\text{pm}}}\right)^2} = \frac{9(6.626\times10^{-34})^2\,\text{J}^2\text{s}^2}{8(9.11\times10^{-31}\text{kg})(155\times10^{-12})^2\text{m}^2}$$

$$= \frac{9(6.626\times10^{-34})^2\left(\frac{\cancel{\text{kg}\bullet\text{m}^2}}{\cancel{\text{s}^2}}\right)\text{J}\,\cancel{\text{s}^2}}{8(9.11\times10^{-31}\ \cancel{\text{kg}})(155\times10^{-12})^2\ \cancel{\text{m}^2}} = 2.26\times10^{-17}\ \text{J}$$

Check: The units of the answers, J, are correct. The answers seem reasonable since the energy is increasing with increasing *n* level.

b. **Given:** $n = 1 \rightarrow n = 2$ and $n = 2 \rightarrow n = 3$ **Find:** λ

Conceptual Plan: $n = 1, n = 2 \rightarrow \Delta E_{atom} \rightarrow \Delta E_{photon} \rightarrow \lambda$

$\Delta E_{atom} = E_2 - E_1$ $\quad \Delta E_{atom} \rightarrow -\Delta E_{photon}$ $\quad E=\frac{hc}{\lambda}$

Solution: Using the energies calculated in part a:

$E_2 - E_1 = (1.00 \times 10^{-17}\ \text{J} - 2.51 \times 10^{-18}\ \text{J}) = 7.49 \times 10^{-18}\ \text{J}$

$$\lambda = \frac{(6.626\times10^{-34}\ \cancel{\text{J}}\bullet\cancel{\text{s}})(3.00\times10^8\ \text{m}/\cancel{\text{s}})}{7.49\times10^{-18}\ \cancel{\text{J}}} = 2.65\times10^{-8}\ \text{m} = 26.5\ \text{nm}$$

$E_3 - E_2 = (2.26 \times 10^{-17}\ \text{J} - 1.00 \times 10^{-17}\ \text{J}) = 1.26 \times 10^{-17}\ \text{J}$

$$\lambda = \frac{(6.626\times10^{-34}\ \cancel{\text{J}}\bullet\cancel{\text{s}})(3.00\times10^8\ \text{m}/\cancel{\text{s}})}{1.26\times10^{-17}\ \cancel{\text{J}}} = 1.58\times10^{-8}\ \text{m} = 15.8\ \text{nm}$$

These wavelengths would lie in the UV region.

Check: The units of the answers, m, are correct. The magnitude of the answers is reasonable based on the energies obtained for the levels.

52. **Given:** $n = 1$, $\nu = 8.85 \times 10^{13}\text{s}^{-1}$ **Find:** E, λ

Conceptual Plan: $n,\nu \rightarrow E \rightarrow \lambda$

$E=(n+\frac{1}{2})h\nu \quad E=\frac{hc}{\lambda}$

Solution:

$$E=(1+\frac{1}{2})(6.626\times10^{-34}\ \text{J}\bullet\cancel{\text{s}})(8.85\times10^{13}\ \cancel{\text{s}^{-1}})=8.80\times10^{-20}\ \text{J} \qquad \lambda=\frac{hc}{E}=\frac{(6.626\times10^{-34}\ \cancel{\text{J}}\bullet\cancel{\text{s}})(3.00\times10^8\ \text{m}/\cancel{\text{s}})}{8.80\times10^{-20}\ \cancel{\text{J}}}=2.26\times10^{-6}\ \text{m}$$

Check: The units of the answer, J and m, are correct. The magnitude of the answer puts the vibrational frequency in the infrared region, which is reasonable.

53. For the 1s orbital: In the Excel spreadsheet, call column A: r; and column B: ψ(1s). Make the values for r column A: 0 – 200. In column B, put the equation for the wavefunction written as: =(POWER(1/3.1415,1/2))*(1/POWER(53,3/2))*(EXP(-A2/53)). Go to make chart, choose xy scatter.

e.g., sample values

r	ψ(1s)
0	0.00146224
1	0.00143491
2	0.00140809
3	0.00138177
4	0.00135594
5	0.0013306
6	0.00130573

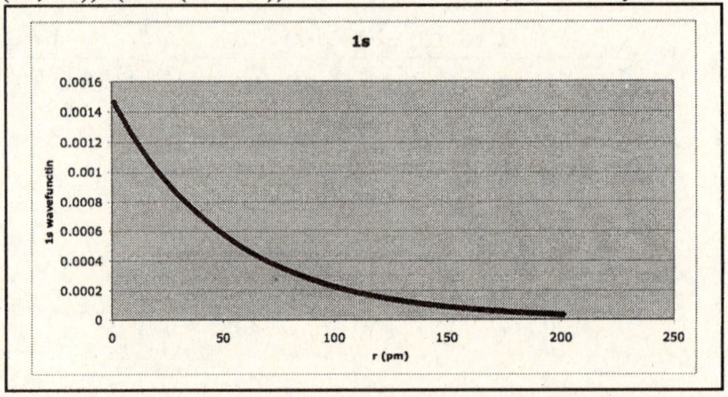

For the 2s orbital: In the same Excel spreadsheet, call column A: r; and column C: ψ(2s). Use the same values for r in column A: 0 – 200. In column C, put the equation for the wavefunction written as:

=(POWER(1/((32)*(3.1415)),1/2))*(1/POWER(53,3/2))*(2-(A2/53))*(EXP(-A2/53)). Go to make chart, choose xy scatter.

e.g., sample values

r	ψ(2s)
0	0.000516979
1	0.00050253
2	0.000488441
3	0.000474702
4	0.000461307
5	0.000448247
6	0.000435513

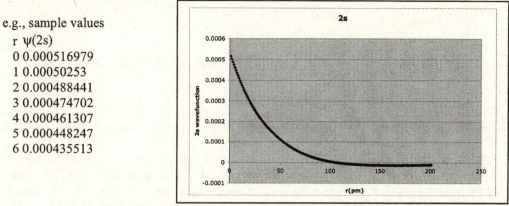

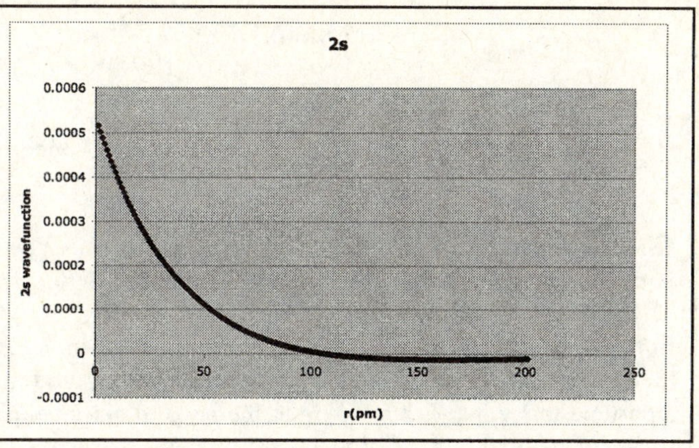

Note: The plot for the 2s orbital extends below the x axis. The x-intercept represents the radial node of the orbital.

54. **Given:** $\Delta E = E_m - E_n = -2.18\text{x}10^{-18}(1/m^2) - [-2.18 \text{ x } 10^{-18}(1/n^2)]$; $E = hc/\lambda$ **Find:** $1/\lambda = R(1/m^2 - 1/n^2)$
Conceptual Plan: $\Delta E_{atom} \rightarrow E_{photon} \rightarrow 1/\lambda$

$$\Delta E_{atom} = E_m - E_n \quad \Delta E_{atom} \rightarrow -\Delta E_{photon} \qquad E=\frac{hc}{\lambda}$$

Solution: $$\Delta E = E_m - E_n = -2.18\text{x}10^{-18}\left(\frac{1}{m^2}\right) - [-2.18 \text{ x } 10^{-18}\left(\frac{1}{n^2}\right)] = -2.18 \text{ x } 10^{-18}\left(\frac{1}{m^2}-\frac{1}{n^2}\right)$$

$$\Delta E_{atom} = -\Delta E_{photon}$$

$$E_{photon} = -(-2.18 \text{ x } 10^{-18}\left(\frac{1}{m^2}-\frac{1}{n^2}\right)) = \frac{hc}{\lambda} \qquad \frac{1}{\lambda} = \frac{2.18 \text{ x } 10^{-18}}{hc}\left(\frac{1}{m^2}-\frac{1}{n^2}\right) = 1.1\text{x}10^{7}\left(\frac{1}{m^2}-\frac{1}{n^2}\right)$$

$$\frac{1}{\lambda} = R\left(\frac{1}{m^2}-\frac{1}{n^2}\right)$$

55. **Given:** threshold frequency = $2.25 \text{ x } 10^{14} s^{-1}$; $\lambda = 5.00 \text{ x } 10^{-7}$ m **Find:** v of electron
Conceptual Plan: $\nu \rightarrow \Phi$ **and then** $\lambda \rightarrow E$ **and then** $\rightarrow KE \rightarrow v$

$$\Phi = h\nu \qquad E=\frac{hc}{\lambda} \qquad KE = E - \Phi \qquad KE = 1/2\ mv^2$$

Solution:

$$\Phi = (6.626\text{x}10^{-34}\text{ J}\cdot\text{s})(2.25\text{x}10^{14}\text{ s}^{-1}) = 1.4\underline{9}1\text{x}10^{-19}\text{ J} \qquad E = \frac{(6.626\text{x}10^{-34}\text{ J}\cdot\text{s})(3.00\text{x}10^{8}\text{ m/s})}{5.00\text{x}10^{-7}\text{ m}} = 3.9\underline{7}6\text{x}10^{-19}\text{ J}$$

$$KE = 3.9\underline{7}6\text{x}10^{-19}\text{ J} - 1.4\underline{9}1\text{x}10^{-19}\text{ J} = 2.485\text{x}10^{-19}\text{ J} \qquad v^2 = \frac{2.485\text{x}10^{-19}\ \frac{\text{kg}\cdot\text{m}^2}{\text{s}^2}}{\frac{1}{2}(9.11\text{x}10^{-31}\text{ kg})} = 5.455\text{x}10^{11}\ \frac{\text{m}^2}{\text{s}^2}$$

$v = 7.39 \text{ x } 10^5$ m/s

Check: The units of the answer, m/s, are correct. The magnitude of the answer is reasonable for the speed of an electron.

56. **Given:** $\lambda = 2.8 \text{ x } 10^{-4}$ cm; m = 2.0 g; $\Delta T = 2.0$ K **Find:** number of photons
Conceptual Plan: $\lambda(\text{cm}) \rightarrow \lambda(\text{m}) \rightarrow E_{photon}$ **and** $m, \Delta T \rightarrow q_{water}$ **and then** $\rightarrow$ **number photons**

$$\frac{\text{m}}{100\text{ cm}} \qquad E=\frac{hc}{\lambda} \qquad q = mC_s\Delta T \qquad \frac{q_{water}}{E_{photon}}$$

Solution: $$E_{photon} = \frac{(6.626 \times 10^{-34}\ J\cdot \cancel{s})(3.00 \times 10^{8}\ \cancel{m}/\cancel{s})}{(2.8 \times 10^{-4}\ \cancel{cm})\left(\frac{\cancel{m}}{100\ \cancel{cm}}\right)} = 7.1 \times 10^{-20}\ J/photon$$

$$q = (2.0\ \cancel{g})\left(4.18 \frac{J}{\cancel{g}\cdot °C}\right)\left(\frac{\cancel{°C}}{\cancel{K}}\right)(2.0\ \cancel{K}) = 16.7\ J \qquad \text{number of photons} = \frac{16.7\ \cancel{J}}{7.1 \times 10^{-20}\ \cancel{J}/photon} = 2.4 \times 10^{20}\ photons$$

Check: The units of the answer, photons, is correct. The magnitude of the answer seems reasonable because a large amount of heat energy is needed to raise the temperature of the water.

57. In the Bohr model of the atom, the electron travels in a circular orbit around the nucleus. It is a 2-dimensional model. The electron is constrained to move only from one orbit to another orbit. But, the electron is treated as a particle that behaves according to the laws of classical physics. The quantum-mechanical model of the atom is 3-dimensional. In this model, we treat the electron, an absolutely small particle, differently than we treat particles with classical physics. The electron is in an orbital, which gives us the probability of finding the electron within a volume of space.
Because the electron in the Bohr model is constrained to a circular orbit, it would theoretically be possible to know both the position and the velocity of the electron simultaneously. This contradicts the Heisenberg uncertainty principle which states that position and velocity are complementary terms which cannot both be known with precision.

58. The transition from $n = 3 \rightarrow n = 2$ would cause the photoelectric effect while the transition from $n = 4 \rightarrow n = 3$ would not. Because the n levels get closer together as n increases, the energy difference between the 4 and 3 level would be less than the energy difference between the 3 and 2 levels. Therefore, the energy of the photon emitted when the electron moves from 4 to 3 would not be above the threshold energy for the metal. The energy of the photon emitted when the electron makes the transition from $n = 3$ to $n = 2$ is larger and surpasses the threshold energy, thus causing the photoelectric effect.

59. a) Since the interference pattern is caused by single electrons interfering with themselves, the pattern remains the same even when the rate of the electrons passing through the slits is one electron per minute. It will simply take longer for the full pattern to develop.

b) When a light is placed behind the slits to determine which hole the electron passes through, the light flashes to indicate which hole the electron passed through, but the interference pattern is now absent. With the laser on, the electrons hit positions directly behind each slit, as if they were ordinary particles.

c) Diffraction occurs when a wave encounters an obstacle of a slit that is comparable in size to its wavelength. The wave bends around the slit. The diffraction of light through two slits separated by a distance comparable to the wavelength of the light results in an interference pattern. Each slit acts as a new wave source, and the two new waves interfere with each other, which result in a pattern of bright and dark lines.

d) Since the mass of the bullets and their particle size are not absolutely small, the bullets will not produce an interference pattern when they pass through the slits. The de Broglie wavelength produced by the bullets will not be sufficiently large enough to interfere with the bullet trajectory and no interference pattern will be observed.

Chapter 8
Periodic Properties of the Elements

1. a) P Phosphorus has 15 electrons. Distribute two of these into the 1*s* orbital, two into the 2*s* orbital, six into the 2*p* orbital, two into the 3s orbital, and three into the 3*p* orbital. $1s^22s^22p^63s^23p^3$

   b) C Carbon has 6 electrons. Distribute two of these into the 1*s* orbital, two into the 2*s* orbital, and two into the 2*p* orbital. $1s^22s^22p^2$

   c) Na Sodium has 11 electrons. Distribute two of these into the 1*s* orbital, two into the 2*s* orbital, six into the 2*p* orbital, and one into the 3*s* orbital. $1s^22s^22p^63s^1$

   d) Ar Argon has 18 electrons. Distribute two of these into the 1*s* orbital, two into the 2*s* orbital, six into the 2*p* orbital, two into the 3*s* orbital, and six into the 3*p* orbital. $1s^22s^22p^63s^23p^6$

2. a) O Oxygen has 8 electrons. Distribute two of these into the 1*s* orbital, two into the 2*s* orbital, and 4 into the 2*p* orbital. $1s^22s^22p^4$

   b) Si Silicon has 14 electrons. Distribute two of these into the 1*s* orbital, two into the 2*s* orbital, six into the 2*p* orbital, two into the 3*s* orbital, and two into the 3*p* orbital. $1s^22s^22p^63s^23p^2$

   c) Ne Neon has 10 electrons. Distribute two of these into the 1*s* orbital, two into the 2*s* orbital, and six into the 2*p* orbital. $1s^22s^22p^6$

   d) K Potassium has 19 electrons. Distribute two of these into the 1*s* orbital, two into the 2*s* orbital, six into the 2*p* orbital, two into the 3*s* orbital, six into the 3*p* orbital, and one into the 4s orbital. $1s^22s^22p^63s^23p^64s^1$

3. a) N Nitrogen has 7 electrons and has the electron configuration: $1s^22s^22p^3$. Draw a box for each orbital, putting the lowest energy orbital (1s) on the far left and proceeding to orbitals of higher energy to the right. Distribute the 7 electrons into the boxes representing the orbitals, allowing a maximum of two electrons per orbital and remembering Hund's rule. You can see from the diagram that nitrogen has 3 unpaired electrons.

1*s*	2*s*	2*p*
⇂↾	⇂↾	↑ | ↑ | ↑

   b) F Fluorine has 9 electrons and has the electron configuration: $1s^22s^22p^5$. Draw a box for each orbital, putting the lowest energy orbital (1s) on the far left and proceeding to orbitals of higher energy to the right. Distribute the 9 electrons into the boxes representing the orbitals, allowing a maximum of two electrons per orbital and remembering Hund's rule. You can see from the diagram that fluorine has 1 unpaired electron.

1*s*	2*s*	2*p*
⇂↾	⇂↾	⇂↾ | ⇂↾ | ↑

   c) Mg Magnesium has 12 electrons and has the electron configuration: $1s^22s^22p^63s^2$. Draw a box for each orbital, putting the lowest energy orbital (1s) on the far left and proceeding to orbitals of higher energy to the right. Distribute the 12 electrons into the boxes representing the orbitals, allowing a maximum of two electrons per orbital and remembering Hund's rule. You can see from the diagram that magnesium has no unpaired electrons.

1*s*	2*s*	2*p*	3*s*
⇂↾	⇂↾	⇂↾ | ⇂↾ | ⇂↾	⇂↾

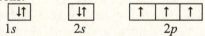

   d) Al Aluminum has 13 electrons and has the electron configuration: $1s^22s^22p^63s^23p^1$. Draw a box for each orbital, putting the lowest energy orbital (1s) on the far left and proceeding to orbitals of higher energy to the right. Distribute the 13 electrons into the boxes representing the orbitals, allowing a maximum of two electrons per orbital and remembering Hund's rule. You can see from the diagram that aluminum has 1 unpaired electron.

↓↑	↓↑	↓↑ ↓↑ ↓↑	↓↑	↑
1*s*	2*s*	2*p*	3*s*	3*p*

4. a) S Sulfur has 16 electrons and has the electron configuration: $1s^22s^22p^63s^23p^4$. Draw a box for each orbital, putting the lowest energy orbital (1s) on the far left and proceeding to orbitals of higher energy to the right. Distribute the 16 electrons into the boxes representing the orbitals, allowing a maximum of two electrons per orbital and remembering Hund's rule. You can see from the diagram that sulfur has 2 unpaired electrons.

↓↑	↓↑	↓↑ ↓↑ ↓↑	↓↑	↓↑ ↑ ↑
1*s*	2*s*	2*p*	3*s*	3*p*

b) Ca Calcium has 20 electrons and has the electron configuration: $1s^22s^22p^63s^23p^64s^2$. Draw a box for each orbital, putting the lowest energy orbital (1s) on the far left and proceeding to orbitals of higher energy to the right. Distribute the 20 electrons into the boxes representing the orbitals, allowing a maximum of two electrons per orbital and remembering Hund's rule. You can see from the diagram that nitrogen has no unpaired electrons.

↓↑	↓↑	↓↑ ↓↑ ↓↑	↓↑	↓↑ ↓↑ ↓↑	↓↑
1*s*	2*s*	2*p*	3*s*	3*p*	4*s*

c) Ne Neon has 10 electrons and has the electron configuration: $1s^22s^22p^6$. Draw a box for each orbital, putting the lowest energy orbital (1s) on the far left and proceeding to orbitals of higher energy to the right. Distribute the 10 electrons into the boxes representing the orbitals, allowing a maximum of two electrons per orbital and remembering Hund's rule. You can see from the diagram that neon has no unpaired electrons.

↓↑	↓↑	↓↑ ↓↑ ↓↑
1*s*	2*s*	2*p*

d) He Helium has 2 electrons and has the electron configuration: $1s^2$. Draw a box for each orbital, putting the lowest energy orbital (1s) on the far left and proceeding to orbitals of higher energy to the right. Distribute the 2 electrons into the boxes representing the orbitals, allowing a maximum of two electrons per orbital and remembering Hund's rule. You can see from the diagram that helium has 0 unpaired electrons.

↓↑
1*s*

5. a) P The atomic number of P is 15. The noble gas that precedes P in the periodic table is neon, so the inner electron configuration is [Ne]. Obtain the outer electron configuration by tracing the elements between Ne and P and assigning electrons to the appropriate orbitals. Begin with [Ne]. Because P is in row 3, add two 3*s* electrons. Next add three 3*p* electrons as you trace across the *p* block to P which is in the third column of the p block.

P $[Ne]3s^23p^3$

b) Ge The atomic number of Ge is 32. The noble gas that precedes Ge in the periodic table is argon, so the inner electron configuration is [Ar]. Obtain the outer electron configuration by tracing the elements between Ar and Ge and assigning electrons to the appropriate orbitals. Begin with [Ar]. Because Ge is in row 4, add two 4*s* electrons. Next, add ten 3*d* electrons as you trace across the *d* block. Finally add two 4*p* electrons as you trace across the *p* block to Ge which is in the second column of the *p* block.

Ge $[Ar]4s^23d^{10}4p^2$

c) Zr The atomic number of Zr is 40. The noble gas that precedes Zr in the periodic table is krypton, so the inner electron configuration is [Kr]. Obtain the outer electron configuration by tracing the elements between Kr and Zr and assigning electrons to the appropriate orbitals. Begin with [Kr]. Because Zr is in row 5, add two 5*s* electrons. Next, add two 4*d* electrons as you trace across the *d* block to Zr which is in the second column.

Zr $[Kr]5s^24d^2$

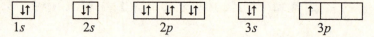

d) I The atomic number of I is 53. The noble gas that precedes I in the periodic table is krypton, so the inner electron configuration is [Kr]. Obtain the outer electron configuration by tracing the elements between Kr and I and assigning electrons to the appropriate orbitals. Begin with [Kr]. Because I is in row 5, add two 5*s* electrons. Next, add ten 4*d* electrons as you trace across the d block. Finally add five 5*p* electrons as you trace across the *p* block to I which is in the fifth column of the *p* block.

I $[Kr]5s^2 4d^{10} 5p^5$

6. a) $[Ar]\ 4s^2 3d^{10} 4p^6$ To determine the element corresponding to the electron configuration, begin with Ar then trace across the 4*s* block, the 3*d* block, and then the 4*p* block until you come to the sixth column. The element is Kr.

   b) $[Ar]\ 4s^2 3d^2$ To determine the element corresponding to the electron configuration, begin with Ar then trace across the 4*s* block, and then 3*d* block until you come to the second column. The element is Ti.

   c) $[Kr]\ 5s^2 4d^{10} 5p^2$ To determine the element corresponding to the electron configuration, begin with Kr then trace across the 5*s* block, the 4*d* block, and then the 5*p* block until you come to the second column. The element is Sn.

   d) $[Kr]\ 5s^2$ To determine the element corresponding to the electron configuration, begin with Kr then trace across the 5*s* block to the second column. The element is Sr.

7. a) Li is in period 2, and the first column in the *s* block so Li has one 2*s* electron.

   b) Cu is in period 4, and the ninth column in the *d* block (n – 1) so Cu should have nine 3*d* electrons, however, it is one of our exceptions, so it has ten 3*d* electrons.

   c) Br is in period 4, and the fifth column of the *p* block, so Br has five 4*p* electrons.

   d) Zr is in period 5, and the second column of the *d* block (n – 1), so Zr has two 4*d* electrons.

8. a) Mg is in period 3, and the second column of the *s* block, so Mg has two 3*s* electrons.

   b) Cr is in period 4, and the fourth column of the *d* block (n – 1), so Cr should have four 3*d* electrons, however, Cr is one of our exceptions, so it has five 3*d* electrons.

   c) Y is in period 5, and the first column of the *d* block (n – 1), so Y has one 4*d* electron.

   d) Pb is in period 6, and the second column of the *p* block, so Pb has two 6*p* electrons.

9. a) In period 4, an element with five valence electrons could be V or As.

   b) In period 4, an element with four 4*p* electrons would be in the fourth column of the *p* block and is Se.

   c) In period 4, an element with three 3*d* electrons would be in the third column of the *d* block (n – 1) and is V.

   d) In period 4, an element with a complete outer shell would be in the sixth column of the *p* block and is Kr.

10. a) In period 3, an element with three valence electrons would be in the first column of the *p* block and is Al.

    b) In period 3, an element with four 3*p* electrons would be in the fourth column of the *p* block and is S.

    c) In period 3, an element with six 3*p* electrons would be in the sixth column of the *p* block and is Ar.

    d) In period 3, an element with two 3*s* electrons and no 3*p* electrons would be in the second column of the *s* block and is Mg.

11. a) Ba is in column 2A, so it has two valence electrons.

    b) Cs is in column 1A, so it has one valence electron.

c) Ni is in column 8 of the *d* block, so it has 10 valence electrons (8 from the *d* block and 2 from the *s* block).

d) S is in column 6A, so it has six valence electrons.

12. a) Al is in column 3A, so it has three valence electrons. Al is a metal and will tend to lose the three valence electrons to achieve the noble gas configuration of Ne.

b) Sn is in column 4A, so it has four valence electrons. Sn is a metal and will tend to lose the valence electrons to obtain a completely filled $n = 3$ level.

c) Br is in column 7A, so it has seven valence electrons. Br is a nonmetal and will tend to gain an electron to achieve the noble gas configuration of Kr.

d) Se is in column 6A, so it has six valence electrons. Se is a nonmetal and will tend to gain electrons to achieve the noble gas configuration of Kr.

13. a) The outer electron configuration ns^2 would belong to a reactive metal in the alkaline earth family.

b) The outer electron configuration ns^2np^6 would belong to an unreactive nonmetal in the noble gas family.

c) The outer electron configuration ns^2np^5 would belong to a reactive nonmetal in the halogen family.

d) The outer electron configuration ns^2np^2 would belong to an element in the carbon family. If $n = 2$, the element is a nonmetal, if $n = 3$ or 4, the element is a metalloid, and if $n = 5$ or 6, the element is a metal.

14. a) The outer electron configuration ns^2 would belong to a metal in the alkaline earth family.

b) The outer electron configuration ns^2np^6 would belong to a nonmetal in the noble gas family.

c) The outer electron configuration ns^2np^5 would belong to a nonmetal in the halogen family.

d) The outer electron configuration ns^2np^2 would belong to an element in the carbon family. If $n = 2$, the element is a nonmetal, if $n = 3$ or 4, the element is a metalloid, and if $n = 5$ or 6, the element is a metal.

15. The valence electrons in nitrogen would experience a greater effective nuclear charge. Be has 4 protons and N has 7 protons. Both atoms have 2 core electrons that predominately contribute to the shielding while the valence electrons will contribute a slight shielding effect. So, Be has an effective nuclear charge of slightly more than 2+ and N has an effective nuclear charge of slightly more than 5+.

16. S(16) = [Ne]$3s^23p^4$ Mg(12) = [Ne]$3s^2$ Al(13) = [Ne]$3s^23p^1$ Si(14) = [Ne]$3s^23p^2$
All four atoms have the same number of core electrons that contribute to shielding. So, the effective nuclear charge will decrease with decreasing number of protons. S > Si > Al > Mg

17. a) K(19) [Ar]$4s^1$ $Z_{eff} = Z - \text{core electrons} = 19 - 18 = 1+$

b) Ca(20) [Ar]$4s^2$ $Z_{eff} = Z - \text{core electrons} = 20 - 18 = 2+$

c) O(8) [He]$2s^22p^4$ $Z_{eff} = Z - \text{core electrons} = 8 - 2 = 6+$

d) C(6) [He]$2s^22p^2$ $Z_{eff} = Z - \text{core electrons} = 6 - 2 = 4+$

18. B has an electron configuration of $1s^22s^22p^1$. To estimate the effective nuclear charge experienced by the outer electrons we need to distinguish between two different types of shielding: (1) the shielding of the outermost electrons by the core electrons and (2) the shielding of the outermost electrons by each other. The three outermost electrons in boron experience the 5+ charge of the nucleus through the shield of the two 1*s* core electrons. We can estimate that the shielding experience by any one of the outermost electrons due to the core electrons is nearly 2. For the 2*s* electrons the shielding due to the other 2*s* electron is nearly zero. For the 2*p* electron however, we would expect that the 2s electrons would contribute some shielding because although the 2*p* orbital penetrates the 2*s* orbital to some degree most of the 2*p* orbital lies outside the 2*s* orbital. So the

effective nuclear charge would be slightly greater than 3+ and the effective nuclear charge felt by the 2*s* electrons would be greater than the effective nuclear charge felt by the 2*p* electrons.

19. a) Al or In In atoms are larger than Al atoms because, as you trace the path between Al and In on the periodic table, you move down a column. Atomic size increases as you move down a column because the outermost electrons occupy orbitals with a higher principal quantum number that are therefore larger, resulting in a larger atom.

    b) Si or N Si atoms are larger than N atoms because, as you trace the path between N and Si on the periodic table, you move down a column (atomic size increases) and then to the left across a period (atomic size increases). These effects add together for an overall increase.

    c) P or Pb Pb atoms are larger than P atoms because, as you trace the path between P and Pb on the periodic table, you move down a column (atomic size increases) and then to the left across a period (atomic size increases). These effects add together for an overall increase.

    d) C or F C atoms are larger than F atoms because, as you trace the path between C and F on the periodic table, you move to the right within the same period. As you move to the right across a period, the effective nuclear charge experienced by the outermost electrons increase, which results in a smaller size.

20. a) Sn of Si Sn atoms are larger than Si atoms because, as you trace the path between Si and Sn on the periodic table, you move down a column. Atomic size increases as you move down a column because the outermost electrons occupy orbitals with a higher principal quantum number that are therefore larger, resulting in a larger atom.

    b) Br or Ga Ga atoms are larger than Br atoms because, as you trace the path between Ga and Br on the periodic table, you move to the right within the same period. As you move to the right across a period, the effective nuclear charge experienced by the outermost electrons increases, which results in a smaller size.

    c) Sn or Bi Based on periodic trends alone, you cannot tell which atom is larger, because as you trace the path between Sn and Bi, you go to the right across a period (atomic size decreases) and then down a column (atomic size increases). These effects tend to oppose each other, and it is not easy to tell which will predominate.

    d) Se or Sn Sn atoms are larger than Se atoms because, as you trace the path between Se and Sn on the periodic table, you move down a column (atomic size increases) and then to the left across a period (atomic size increases). These effects add together for an overall increase.

21. Ca, Rb, S, Si, Ge, F F is above and to the right of the other elements, so we start with F as the smallest atom. As you trace a path from F to S you move to the left (size increases) and down (size increases), then you move to Si to the left (size increases) then down to Ge (size increases) then to the left to Ca (size increases) then to the left and down to Rb (size increases). So, in order of increasing atomic radii: F < S < Si < Ge < Ca < Rb

22. Cs, Sb, S, Pb, Se Cs is below and to the left of the other elements, so we start with Cs as the largest atom. As you trace a path from Cs to Pb you move to the right in the same period (size decreases), then from Pb to Sb you move up a column and then to the right (size decreases) from Sb to Se you move up the column and then to the right (size decreases) and finally from Se to S you move up the column (size decreases). So, in order of decreasing radii: Cs > Pb > Sb > Se > S

23. a) O^{2-} Begin by writing the electron configuration of the neutral atom.
    O $1s^2 2s^2 2p^4$
    Since this ion has a 2 – charge, add two electrons to write the electron configuration of the ion.
    O^{2-} $1s^2 2s^2 2p^6$ This is isoelectronic with Ar

    b) Br^- Begin by writing the electron configuration of the neutral atom.
    Br $[Ar]4s^2 3d^{10} 4p^5$
    Since this ion has a 1 – charge, add one electron to write the electron configuration of the ion.
    Br^- $[Ar]4s^2 3d^{10} 4p^6$ This is isoelectronic with Kr

c) Sr^{2+} Begin by writing the electron configuration of the neutral atom.

Sr [Kr]$5s^2$

Since this ion has a 2+ charge, remove two electrons to write the electron configuration of the ion.

Sr^{2+} [Kr]

d) Co^{3+} Begin by writing the electron configuration of the neutral atom.

Co [Ar]$4s^2 3d^7$

Since this ion has a 3+ charge, remove three electrons to write the electron configuration of the ion. Since it is a transition metal, remove the electrons from the 4*s* orbital before removing electrons from the 3*d* orbitals.

Co^{3+} [Ar]$4s^0 3d^6$

e) Cu^{2+} Begin by writing the electron configuration of the neutral atom. Remember, Cu is one of our exceptions.

Cu [Ar]$4s^1 3d^{10}$

Since this ion has a 2+ charge, remove two electrons to write the electron configuration of the ion. Since it is a transition metal, remove the electrons from the 4*s* orbital before removing electrons from the 3*d* orbitals.

Cu^{2+} [Ar]$4s^0 3d^9$

24. a) Cl^- Begin by writing the electron configuration of the neutral atom.

Cl [Ne]$3s^2 3p^5$

Since this ion has a 1 – charge, add one electron to write the electron configuration of the ion.

Cl^- [Ne]$3s^2 3p^6$ This is isoelectronic with Ar

b) P^{3-} Begin by writing the electron configuration of the neutral atom.

P [Ne]$3s^2 3p^3$

Since this ion has a 3 – charge, add three electrons to write the electron configuration of the ion.

P^{3-} [Ne]$3s^2 3p^6$ This is isoelectronic with Ar

c) K^+ Begin by writing the electron configuration of the neutral atom.

K [Ar]$4s^1$

Since this ion has a 1+ charge, remove one electron to write the electron configuration of the ion.

K^+ [Ar]

d) Mo^{3+} Begin by writing the electron configuration of the neutral atom. Remember, Mo is one of our exceptions.

Mo [Kr]$5s^1 4d^5$

Since this ion has a 3+ charge, remove three electrons to write the electron configuration of the ion. Since it is a transition metal, remove the electrons from the 5*s* orbital before removing electrons from the 4*d* orbitals.

Mo^{3+} [Kr]$5s^0 4d^3$

e) V^{3+} Begin by writing the electron configuration of the neutral atom.

V [Ar]$4s^2 3d^3$

Since this ion has a 3+ charge, remove three electrons to write the electron configuration of the ion. Since it is a transition metal, remove the electrons from the 4*s* orbital before removing electrons from the 3*d* orbitals.

V^{3+} [Ar]$4s^0 3d^2$

25. a) V^{5+} Begin by writing the electron configuration of the neutral atom.

V [Ar]$4s^2 3d^3$

Since this ion has a 5+ charge, remove five electrons to write the electron configuration of the ion. Since it is a transition metal, remove the electrons from the 4*s* orbital before removing electrons from the 3*d* orbitals.

V^{5+} [Ar]$4s^0 3d^0$ = [Ne]$3s^2 3p^6$

[Ne] | ⇅ | | ⇅ | ⇅ | ⇅ |

3*s* 3*p*

V^{5+} is diamagnetic

b) Cr^{3+} Begin by writing the electron configuration of the neutral atom. Remember, Cr is one of our exceptions.

Cr [Ar]$4s^1 3d^5$

Since this ion has a 3+ charge, remove three electrons to write the electron configuration of the ion. Since it is a transition metal, remove the electrons from the 4*s* orbital before removing electrons from the 3*d* orbitals.

Cr^{3+} [Ar]$4s^0 3d^3$

[Ar] | | | ↑ | ↑ | ↑ | | |

4*s* 3*d*

Cr^{3+} is paramagnetic

c) Ni^{2+} Begin by writing the electron configuration of the neutral atom.

Ni [Ar]$4s^2 3d^8$

Since this ion has a 2+ charge, remove two electrons to write the electron configuration of the ion. Since it is a transition metal, remove the electrons from the 4*s* orbital before removing electrons from the 3*d* orbitals.

Ni^{2+} [Ar]$4s^0 3d^8$

[Ar] | | | ⇅ | ⇅ | ⇅ | ↑ | ↑ |

4*s* 3*d*

Ni^{2+} is paramagnetic

d) Fe^{3+} Begin by writing the electron configuration of the neutral atom.

Fe [Ar]$4s^2 3d^6$

Since this ion has a 3+ charge, remove three electrons to write the electron configuration of the ion. Since it is a transition metal, remove the electrons from the 4s orbital before removing electrons from the 3d orbitals.

Fe^{3+} [Ar]$4s^0 3d^5$

[Ar] | | | ↑ | ↑ | ↑ | ↑ | ↑ |

4s 3d

Fe^{3+} is paramagnetic

26. a) Cd^{2+} Begin by writing the electron configuration of the neutral atom.

Cd [Kr]$5s^2 4d^{10}$

Since this ion has a 2+ charge, remove two electrons to write the electron configuration of the ion. Since it is a transition metal, remove the electrons from the 5*s* orbital before removing electrons from the 4*d* orbitals.

Cd^{2+} [Kr]$5s^0 4d^{10}$

[Kr] | | | ⇅ | ⇅ | ⇅ | ⇅ | ⇅ |

5*s* 4*d*

Cd^{2+} is diamagnetic

b) Au^+ Begin by writing the electron configuration of the neutral atom. Remember Au is one of our exceptions.

Au [Xe]$6s^1 4f^{14} 5d^{10}$

Since this ion has a + charge, remove one electron to write the electron configuration of the ion. Since it is a transition metal, remove the electrons from the 6*s* orbital before removing electrons from the 5*d* or 4*f* orbitals.

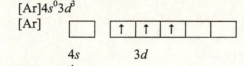

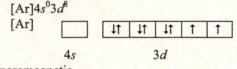

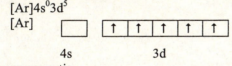

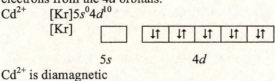

Au^+ $[Xe]6s^04f^{14}5d^{10}$

[Xe]

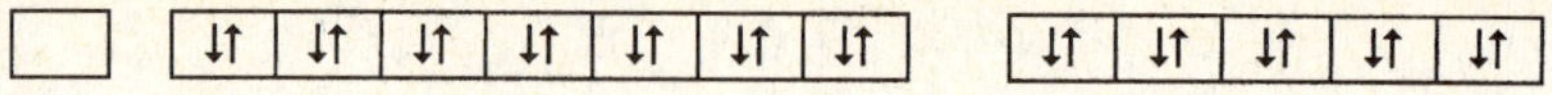

6*s* 4*f* 5*d*

Au^+ is diamagnetic

c) Mo^{3+} Begin by writing the electron configuration of the neutral atom. Remember, Mo is one of our exceptions.

Mo $[Kr]5s^14d^5$

Since this ion has a 3+ charge, remove three electrons to write the electron configuration of the ion. Since it is a transition metal, remove the electrons from the 5*s* orbital before removing electrons from the 4*d* orbitals.

Mo^{3+} $[Kr]5s^04d^3$

[Kr]

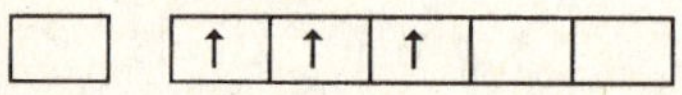

5*s* 4*d*

Mo^{3+} is paramagnetic

d) Zr^{2+} Begin by writing the electron configuration of the neutral atom.

Zr $[Kr]5s^24d^2$

Since this ion has a 2+ charge, remove two electrons to write the electron configuration of the ion. Since it is a transition metal, remove the electrons from the 5*s* orbital before removing electrons from the 4*d* orbitals.

Zr^{2+} $[Kr]5s^04d^2$

[Kr]

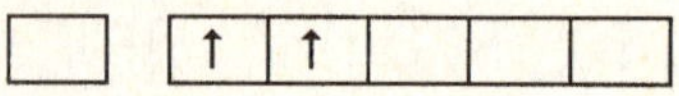

5*s* 4*d*

Zr^{2+} is paramagnetic

27. a) Li or Li^+ — A Li atom is larger than Li^+ because cations are smaller than the atoms from which they are formed.

b) I^- or Cs^+ — An I^- ion is larger than a Cs^+ ion because, although they are isoelectronic, I^- has two fewer protons than Cs^+, resulting in a lesser pull on the electrons and therefore larger radius.

c) Cr or Cr^{3+} — A Cr atom is larger than Cr^{3+} because cations are smaller than the atoms from which they are formed.

d) O or O^{2-} — An O^{2-} ion is larger than an O atom because anions are larger than the atoms from which they are formed.

28. a) Sr or Sr^{2+} — A Sr atom is larger than Sr^{2+} because cations are smaller than the atoms from which they are formed.

b) N or N^{3-} — An N^{3-} ion is larger than an N atom because anions are larger than the atoms from which they are formed.

c) Ni or Ni^{2+} — A Ni atom is larger than Ni^{2+} because cations are smaller than the atoms from which they are formed.

d) S^{2-} or Ca^{2+} — An S^{2-} ion is larger than a Ca^{2+} ion because, although they are isoelectronic, S^{2-} has four fewer protons than Ca^{2+}, resulting in a lesser pull on the electrons and therefore larger radius.

29. Since all the species are isoelectronic, the radius will depend on the number of protons in each species. The fewer the protons the larger the radius.

F: Z = 9; Ne: Z = 10; O: Z = 8; Mg: Z = 12; Na: Z = 11

So: $O^{2-} > F^- > Ne > Na^+ > Mg^{2+}$

30. Since all the species are isoelectronic, the radius will depend on the number of protons in each species. The fewer the protons, the larger the radius.
Se: Z = 34; Kr: Z = 36; Sr: Z = 38; Rb: Z = 37; Br: Z = 35
So: $Sr^{2+} < Rb^{+} < Kr < Br^{-} < Se^{2-}$

31. a) Br or Bi   Br has a higher ionization energy than Bi because, as you trace the path between Br and Bi on the periodic table, you move down a column (ionization energy decreases) and then to the left across a period (ionization energy decreases). These effects sum together for an overall decrease.

b) Na or Rb   Na has a higher ionization energy than Rb because, as you trace a path between Na and Rb on the periodic table, you move down a column. Ionization energy decreases as you go down a column because of the increasing size of orbitals with increasing n.

c) As or At   Based on periodic trends alone, it is impossible to tell which has a higher ionization energy because, as you trace the path between As and At you go to the right across a period (ionization energy increases) and then down a column (ionization energy decreases). These effects tend to oppose each other, and it is not obvious which will dominate.

d) P or Sn   P has a higher ionization energy than Sn because, as you trace the path between P and Sn on the periodic table, you move down a column (ionization energy decreases) and then to the left across a period (ionization energy decreases). These effects sum together for an overall decrease.

32. a) P or I   Based on periodic trends alone, it is impossible to tell which has a higher ionization energy because, as you trace the path between P and I you go to the right across a period (ionization energy increases) and then down a column (ionization energy decreases). These effects tend to oppose each other, and it is not obvious which will dominate.

b) Se or Cl   Cl has a higher ionization energy than Se because, as you trace the path between Cl and Se on the periodic table, you move down a column (ionization energy decreases) and then to the left across a period (ionization energy decreases). These effects sum together for an overall decrease.

c) P or Sb   P has a higher ionization energy than Sb because, as you trace a path between P and Sb on the periodic table, you move down a column. Ionization energy decreases as you go down a column because of the increasing size of orbitals with increasing n.

d) Ga or Ge   Ge has a higher ionization energy than Ga because, as you trace a path between Ga and Ge on the periodic table, you move to the right within the same period. Ionization energy increases as you go to the right because of increasing effective nuclear charge.

33. Since ionization energy increases as you move to the right across a period and increases as you move up a column, the element with the smallest first ionization energy would be the element farthest to the left and lowest down on the periodic table. So, In has the smallest ionization energy, as you trace a path to the right and up on the periodic table, the next element reached is Si, continuing up and to the right you reach N and then continuing to the right you reach F. So, in the order of increasing first ionization energy the elements are:
In < Si < N < F.

34. Since ionization energy increases as you move to the right across a period and increases as you move up a column, the element with the largest first ionization energy would be the element farthest to the right and highest up on the periodic table. So, Cl has the largest ionization energy, as you trace a path to the left on the periodic table you reach S, as you move down a column and to the left you reach Sn, moving down the column you reach Pb. So, in the order of decreasing first ionization energy the elements are: Cl > S > Sn > Pb.

35. The jump in ionization energy occurs when you change from removing a valence electron to removing a core electron. To determine where this jump occurs you need to look at the electron configuration of the atom.
a) Be $1s^2 2s^2$ The first and second ionization energies involve removing $2s$ electrons, the third ionization energy removes a core electron, so the jump will occur between the second and third ionization energies.

b) N $1s^22s^22p^3$ The first five ionization energies involve removing the $2p$ and $2s$ electrons, the sixth ionization energy removes a core electron, so the jump will occur between the fifth and sixth ionization energies.

c) O $1s^22s^22p^4$ The first six ionization energies involve removing the $2p$ and $2s$ electrons, the seventh ionization energy removes a core electron, so the jump will occur between the sixth and seventh ionization energies.

d) Li $1s^22s^1$ The first ionization energy involve removing a $2s$ electron, the second ionization energy removes a core electron, so the jump will occur between the first and second ionization energies.

36. The jump occurs between IE_3 and IE_4, so removing the first three electrons involves removing valence electrons and the fourth electron is a core electron, so the valence electron configuration would be ns^2np^1; this puts the element in column 3A and would be Al.

37. a) Na or Rb Na has a more negative electron affinity than Rb. In column 1A electron affinity becomes less negative as you go down the column.

b) B or S S has a more negative electron affinity than B. As you trace from B to S in the periodic table, you move to the right which shows the value of the electron affinity becoming more negative. Also, as you move from period 2 to period 3, the value of the electron affinity becomes more negative. Both of these trends sum together for the value of the electron affinity to become more negative.

c) C or N C has the more negative electron affinity. As you trace from C to N across the periodic table, you would normally expect N to have the more negative electron affinity. However, N has a half-filled p sublevel which lends it extra stability, therefore it is harder to add an electron.

d) Li or F F has the more negative electron affinity. As you trace from Li to F on the periodic table you move to the right in the period. As you go to the right across a period, the value of the electron affinity generally becomes more negative.

38. a) Mg or S S has the more negative electron affinity. As you trace from Mg to S on the periodic table you move to the right in the period. As you go to the right across a period, the value of the electron affinity generally becomes more negative.

b) K or Cs K has the more negative electron affinity. In column 1A as you go down the column, the electron affinity becomes less negative.

c) Si or P Si has the more negative electron affinity. As you trace from Si to P across the periodic table, you would normally expect P to have the more negative electron affinity. However, P has a half-filled p sublevel which lends extra stability, therefore it is harder to add an electron.

d) Ga or Br Br has the more negative electron affinity. As you trace from Ga to Br on the periodic table, you move to the right in the period. As you go to the right across a period, the value of the electron affinity generally becomes more negative.

39. a) Sr or Sb Sr is more metallic than Sb because, as we trace the path between Sr and Sb on the periodic table, we move to the right within the same period. Metallic character decreases as you go to the right.

b) As or Bi Bi is more metallic because, as we trace a path between As and Bi on the periodic table, we move down a column in the same family (metallic character increases).

c) Cl or O Based on periodic trends alone, we cannot tell which is more metallic because, as we trace the path between O and Cl, we go to the right across a period (metallic character decreases) and then down a column (metallic character increases). These effects tend to oppose each other, and it is not easy to tell which will predominate.

d) S or As As is more metallic than S because, as we trace the path between S and As on the periodic table, we move down a column (metallic character increases) and then to the left across a period (metallic character increases). These effects add together for an overall increase.

40. a) Sb or Pb Pb is more metallic than Sb because, as we trace the path between Sb and Pb on the periodic table, we move down a column (metallic character increases) and then to the left across a period (metallic character increases). These effects add together for an overall increase.

b) K or Ge K is more metallic than Ge because, as we trace the path between K and Ge on the periodic table, we move to the right within the same period. Metallic character decreases as you go to the right.

c) Ge or Sb Based on periodic trends alone, we cannot tell which is more metallic because as we trace the path between Ge and Sb, we go to the right across a period (metallic character decreases) and then down a column (metallic character increases). These effects tend to oppose each other, and it is not easy to tell which will predominate.

d) As or Sn Sn is more metallic than As because, as we trace the path between As and Sn on the periodic table, we move down a column (metallic character increases) and then to the left across a period (metallic character increases). These effects add together for an overall increase.

41. The order of increasing metallic character is: S < Se < Sb < In < Ba < Fr.
Metallic character decreases as you move left to right across a period and decreases as you move up a column, therefore, the element with the least metallic character will be to the top right of the periodic table. So, of these elements, S has the least metallic character. As you move down the column, the next element is Se, as you continue down and then to the right, you reach Sb, continuing to the right goes to In, going down the column and then to the right comes to Ba and then down the column and to the right is Fr.

42. The order of decreasing metallic character is: Sr > Ga > Al > Si > P > N.
Metallic character decreases as you move left to right across a period and decreases as you move up a column, therefore, the element with the greatest metallic character will be at the bottom left of the periodic table. So, of these elements, Sr has the most metallic character. As you trace up the column and then to the right across the period the next element is Ga, trace up the column to Al, then to the right to Si and then P, trace up the column to N.

43. Br: $1s^22s^22p^63s^23p^64s^23d^{10}4p^5$
Kr: $1s^22s^22p^63s^23p^64s^23d^{10}4p^6$
Krypton has a completely filled *p* sublevel giving it chemical stability. Bromine needs one electron to achieve a completely filled *p* sublevel and therefore has a highly negative electron affinity. It therefore easily takes on an electron and is reduced to the bromide ion, giving it the added stability of the filled *p* sublevel.

44. K: $1s^22s^22p^63s^23p^64s^1$
Ar: $1s^22s^22p^63s^23p^6$
Argon has a completely filled *p* sublevel giving it chemical stability. Potassium has one electron in the 4*s* sublevel and can easily lose this electron so it has a low first ionization energy. It therefore loses the 4*s* electron to achieve an argon electron configuration, giving it the added stability of the filled p sublevel.

45. Write the electron configuration of vanadium
V: [Ar] $4s^23d^3$
Since this ion has a 3+ charge, remove three electrons to write the electron configuration of the ion. Since it is a transition metal, remove the electrons from the 4*s* orbital before removing electrons from the 3*d* orbitals.
V^{3+}: [Ar] $4s^03d^2$
Both vanadium and the V^{3+} ion have unpaired electrons and are paramagnetic.

46. Begin by writing the electron configuration of the neutral atom. Remember, Cu is one of our exceptions.
Cu [Ar]$4s^13d^{10}$
Since this ion has a 1+ charge, remove one electron to write the electron configuration of the ion. Since it is a transition metal, remove the electrons from the 4*s* orbital before removing electrons from the 3*d* orbitals.
Cu^+ [Ar]$4s^03d^{10}$

Cu contains one unpaired electron in the 4*s* orbital and is paramagnetic, Cu^+ has all paired electrons in the 3*d* orbitals and is diamagnetic.

47. Since K^+ has a 1+ charge you would need a cation with a similar size and a 1+ charge. Looking at the ions in the same family, Na^+ would be too small and Rb^+ would be too large. If we then consider Ar^+ and Ca^+ we would have ions of similar size and charge. Between these two Ca^+ would be the easier to achieve because the first ionization energy of Ca is similar to that of K while the first ionization energy of Ar is much larger. However, the second ionization energy of Ca is relatively low, making it easy to lose the second electron.

48. Since Na^+ has a 1+ charge you would need a cation with a similar size and a 1+ charge. Looking at the ions in the same family, Li^+ would be too small and K^+ would be too large. If we then consider Ne^+ and Mg^+ we would have ions of similar size and charge. Between these two Mg^+ would be the easier to achieve because the first ionization energy of Mg is similar to that of Na while the first ionization energy of Ne is much larger. However, the second ionization energy of Mg is relatively low, making it easy to lose the second electron.

49. C has an outer shell electron configuration of ns^2np^2; based on this you would expect Si and Ge, which are in the same family, to be most like carbon. Ionization energies for both Si and Ge are similar and tend to be slightly lower than C but all are intermediate in the range of first ionization energies. The electron affinities of Si and Ge are close to that of C.

50. a) Si and Ga Ga would be larger than Si because, as you trace from Si to Ga on the periodic table, you move down a column (radius increases) and then to the left across the period (radius increases). The sum of these two trends would give you a larger radius for Ga.

b) Si and Ge Ge would be larger than Si because, as you trace from Si to Ge on the periodic table, you move down a column and the radius increases.

c) Si and As As would be most similar to Si in atomic radius because, as you trace from Si to As on the periodic table, you move down the column (radius increase) and then to the right across the period (radius decreases). The sum of these two trends would make As smaller than Ga and Ge and thus closer to the radius of Si.

51. a) N: $[He]2s^22p^3$ Mg: $[Ne]3s^2$ O: $[He]2s^22p^4$
F: $[He]2s^22p^5$ Al: $[Ne]3s^23p^1$

b) Mg > Al > N > O > F
c) Al < Mg < O < N < F (from the table)

d) Mg and Al would have the largest radius because they are in period $n = 3$, Al is smaller than Mg because radius decreases as you move to the right across the period. F is smaller than O is smaller than N because as you move to the right across the period, radius decreases.
The first ionization energy of Al is smaller than the first ionization energy of Mg because Al loses the electron from the 3*p* orbital which is shielded by the electrons in the 3*s* orbital, while Mg loses the electron from the filled 3*s* orbital which has added stability because it is a filled orbital. The first ionization energy of O is lower than the first ionization energy of N because N has a half-filled 2*p* orbitals which adds extra stability, thus making it harder to remove the electron and the fourth electron in the O 2*p* orbitals experiences added electron-electron repulsion because it must pair with another electron in the same 2*p* orbital, thus making it easier to remove.

52. a) P: $[Ne]3s^23p^3$ Ca: $[Ar]4s^2$ Si$[Ne]3s^23p^2$
S: $[Ne]3s^23p^4$ Ga: $[Ar]4s^23d^{10}4p^1$

b) Ca > Ga > Si > P > S
c) Ga < Ca < Si < S < P (from table)

d) Ca and Ga would have the largest radius because they are in period $n = 4$, Ga is smaller than Ca because radius increases as you move to the left across the period. S is smaller than P is smaller than Si because as you move to the right across the period, radius decreases.
The first ionization energy of Ga is smaller than the first ionization energy of Ca because Ga loses the electron from the 4*p* orbital which is shielded by the electrons in the 4*s* orbital, while Ca loses the electron from the

filled 4*s* orbital which has added stability because it is a filled orbital. The first ionization energy of S is lower than the first ionization energy of P because P has a half-filled 3*p* orbital which adds extra stability, thus making it harder to remove the electron and the fourth electron in the S 3*p* orbitals experiences added electron-electron repulsion because it must pair with another electron in the same 3*p* orbital, thus making it easier to remove.

53. As you move to the right across a row in the periodic table for the main-group elements, the effective nuclear charge (Z_{eff}) experienced by the electrons in the outermost principal energy level increases, resulting in a stronger attraction between the outermost electrons and the nucleus and therefore, smaller atomic radii.
Across the row of transition elements, the number of electrons in the outermost principal energy level (highest n value) is nearly constant. As another proton is added to the nucleus with each successive element, another electron is added, but that electron goes into an $n_{highest} - 1$ orbital (a core level). The number of outermost electrons stays constant and they experience a roughly constant effective nuclear charge, keeping the radius approximately constant after the first couple of elements in the series.

54. Across the row of transition elements, the number of electrons in the outermost principal energy level (highest n value) is nearly constant. As another proton is added to the nucleus with each successive element another electron is added, but the electron goes into an $n_{highest} - 1$ orbital. So, even though the atomic number of Cu is higher than that of V, the outermost electron experiences roughly the same effective nuclear charge and thus, the radii of the two elements are nearly the same. Since the radius of the two elements is nearly the same, the volume occupied by the element will be nearly the same, since the mass increases as the atomic number increases, the mass of Cu is greater than the mass of V, density is mass/ volume, so the density of Cu should be greater than the density of V.
We find that the densities are: Cu = 8.96 g/cm^3 V = 5.49 g/cm^3 and our prediction was correct.

55. The noble gases all have a filled outer quantum level, very high first ionization energies, and positive values for the electron affinity and are thus particularly unreactive. The lighter noble gases will not form any compounds because the ionization energies of He and Ne are both over 2000 kJ/mol. Since ionization energy decreases as you move down a column we find that the heavier noble gases, Ar, Kr, and Xe do form some compounds, they have ionization energies that are close to the ionization energy of H and can thus be forced to lose an electron.

56. The halogens will all add an electron to achieve the stability of the noble gas configuration, thus they are all powerful oxidizing agents (they are reduced). F would be the strongest because it adds the electron to the $n = 2$ level achieving the electron configuration of Ne. Since the $n = 2$ level lies lower in energy than the outermost level of the other halogens, it is more energetically favorable for F to gain the noble gas configuration than for the other halogens. This combined with the high ionization energy and relatively exothermic electron affinity makes F very reactive. As you move down the column, the *n* level of the outermost electrons increases, making it less energetically favorable for each of the successive halogens to gain an electron.

57. Group 6A: ns^2np^4 Group 7A: ns^2np^5
The electron affinity of the group 7A elements are more negative than the group 6A elements in the same period because group 7A requires only one electron to achieve the noble gas configuration ns^2np^6 while the group 6A elements require 2 electrons. Adding one electron to the group 6A element will not give them any added stability and leads to extra electron-electron repulsions so the value of the electron affinity is less negative than that for group 7A.

58. Group 5A : ns^2np^3 Group 4A: ns^2np^2
The electron affinity of group 5A elements is more positive than the group 4A elements in the same period because group 5A has a half-filled p sublevel. Adding an electron to this group adds a fourth electron into the p sublevel and increases the electron-electron repulsions. It also eliminates the stability of the half-filled sublevel. Adding an electron to a group 4A element however, adds a third electron into the p sublevel, giving it the added stability of the half-filled sublevel.

59. 35 = Br = $[Ar]4s^23d^{10}4p^5$ 53 = I = $[Kr]5s^24d^{10}5p^6$
Br and I are both halogens with an outermost electron configuration of ns^2np^5, the next element with the same outermost electron configuration is 85, At.

60. Begin by writing the electron configuration of the neutral atom. $S = 1s^22s^22p^63s^23p^4$

S^+ loses one electron: $1s^22s^22p^63s^23p^3$

1s	2s	2p	3s	3p
[⇅]	[⇅]	[⇅][⇅][⇅]	[⇅]	[↑][↑][↑]

S^{2+} loses two electrons: $1s^22s^22p^63s^23p^2$

1s	2s	2p	3s	3p
[⇅]	[⇅]	[⇅][⇅][⇅]	[⇅]	[↑][↑][ ]

S^{3+} loses three electrons: $1s^22s^22p^63s^23p^1$

1s	2s	2p	3s	3p
[⇅]	[⇅]	[⇅][⇅][⇅]	[⇅]	[↑][ ][ ]

S^{4+} loses four electrons: $1s^22s^22p^63s^23p^0$

1s	2s	2p	3s	3p
[⇅]	[⇅]	[⇅][⇅][⇅]	[⇅]	[ ][ ][ ]

S^{5+} loses five electrons: $1s^22s^22p^63s^23p^0$

1s	2s	2p	3s	3p
[⇅]	[⇅]	[⇅][⇅][⇅]	[↑]	[ ][ ][ ]

S^{6+} loses six electrons: $1s^22s^22p^63s^23p^0$

1s	2s	2p	3s	3p
[⇅]	[⇅]	[⇅][⇅][⇅]	[ ]	[ ][ ][ ]

61. a) Using Excel, make a table of radius, atomic number, and density. Using xy scatter, make a chart of radius vs. density. With an exponential trendline, estimate the density of argon and xenon. Also, make a chart or atomic number vs. density. With a linear trendline, estimate the density of argon and xenon.

element	radius(pm)	atomic number	density
He	32	2	0.18
Ne	70	10	0.90
Ar	98	18	
Kr	112	36	3.75
Xe	130	54	
Rn		86	9.73
		118	

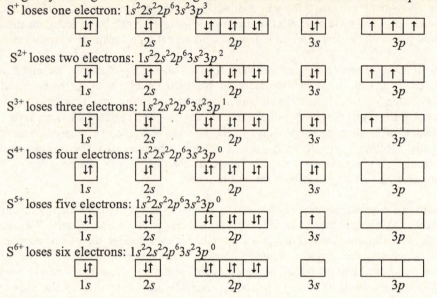

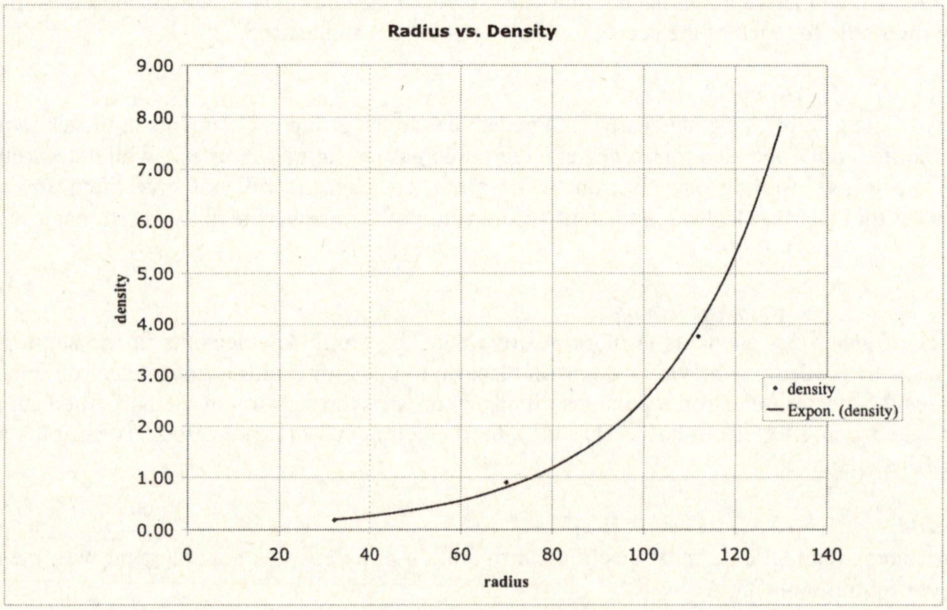

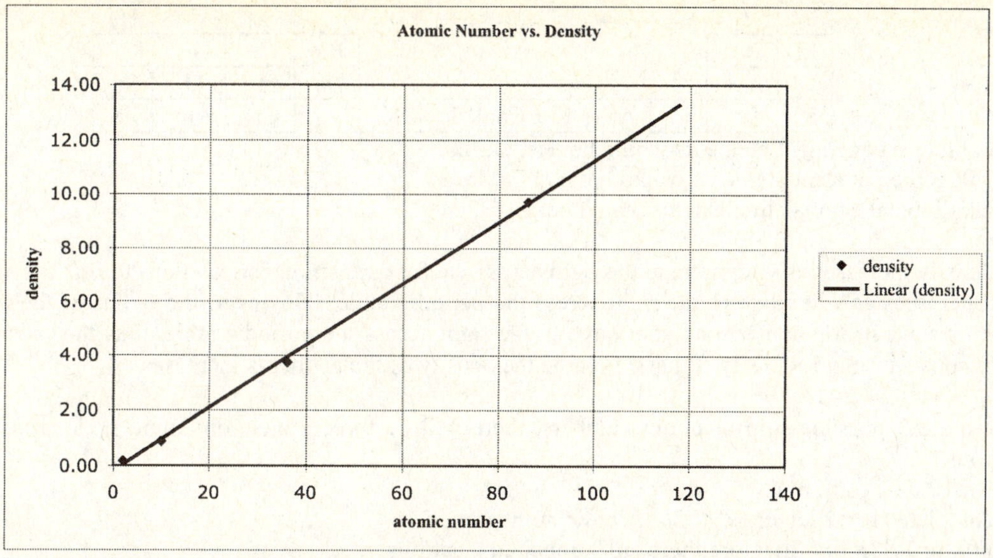

From the radius vs. density chart: Ar has a density of ~ 2 g/L and Xe has a density of ~ 7.7 g/L. From the atomic number vs. density chart: Ar has a density of ~1.8 g/L and Xe has a density of ~6 g/L.

b) Using the chart of atomic number vs. density, element 118 would be predicted to have a density of ~ 13 g/L.

c) **Given:** Ne: M = 20.18 g/mol; r = 70 pm **Find:** mass of neon; d neon
Conceptual Plan: M → m_{atom} and then r → vol_{atom} and then → d

$$\frac{6.022 \times 10^{23} \text{ atoms}}{\text{mol}} \qquad V = \frac{4}{3}\pi r^3 \qquad d = \frac{\text{mass}}{\text{vol}}$$

Solution: $\frac{20.18 \text{ g}}{\cancel{\text{mol}}} \times \frac{\cancel{\text{mol}}}{6.022 \times 10^{23} \text{ atoms}} = 3.35 \times 10^{-23}$ g/atom

$$V = \frac{4}{3} \times 3.14 \times (70 \cancel{\text{pm}})^3 \times \left(\frac{\cancel{\text{m}}}{10^{12} \cancel{\text{pm}}}\right)^3 \times \frac{\text{L}}{0.0010 \cancel{\text{m}^3}} = 1.\underline{4}4 \times 10^{-27} \text{ L}$$

$$d = \frac{3.35 \times 10^{-23} \text{ g}}{1.\underline{4}4 \times 10^{-27} \text{ L}} = 2.\underline{3}3 \times 10^4 = 2.3 \times 10^4 \text{ g/L}$$

Check: The units of the answer, g/L, are correct. This density is significantly larger than the actual density of neon gas. This suggests that a L of neon is composed of primarily empty space.

d) **Given:** Ne: M = 20.18 g/ mol, d = 0.90 g/L; Kr: M = 83.30 g/ mol, d = 3.75 g/L; Ar: M = 39.95 g/ mol
Find: d of argon in g/L
Conceptual Plan: d → mol/L → atoms/L for Kr and Ne and then atoms/L → mol/L → d for Ar

$$\text{mol} = \frac{\text{mass}}{\text{molar mass}} \qquad \frac{6.022 \times 10^{23} \text{ atoms}}{\text{mol}} \qquad \frac{\text{mol}}{6.022 \times 10^{23} \text{ atoms}} \qquad \frac{39.95 \text{ g}}{\text{mol}}$$

Solution:

for Ne: $\frac{0.90 \cancel{\text{g}}}{\text{L}} \times \frac{\cancel{\text{mol}}}{20.18 \cancel{\text{g}}} \times \frac{6.022 \times 10^{23} \text{ atoms}}{\cancel{\text{mol}}} = 2.69 \times 10^{22}$ atoms/ L

for Kr: $\frac{3.75 \cancel{\text{g}}}{\text{L}} \times \frac{\cancel{\text{mol}}}{83.80 \cancel{\text{g}}} \times \frac{6.022 \times 10^{23} \text{ atoms}}{\cancel{\text{mol}}} = 2.69 \times 10^{22}$ atoms/ L

for Ar: $\frac{2.69 \times 10^{22} \cancel{\text{atoms}}}{\text{L}} \times \frac{\cancel{\text{mol}}}{6.022 \times 10^{23} \cancel{\text{atoms}}} \times \frac{39.95 \text{ g}}{\cancel{\text{mol}}} = 1.78$ g/ L

This value is similar to the value calculated in part a. The value of the density calculated from the radius was 2 g/L and the value of the density calculated from the atomic number was 1.8 g/L.
Check: The units of the answer, g/L , are correct. The value of the answer agrees with published value.

62. If there were only two p orbitals there would only be 4 p block columns and if there were only three d orbitals there would be only 6 d block columns. So the periodic table would have 12 columns.

H											He
Li	Be							B	C	N	O
F	Ne							Na	Mg	Al	Si
P	S	Cl	Ar	K	Ca	Sc	Ti	V	Cr	Mn	Fe

The noble gas equivalent elements would be: He, O, Si, Fe
The halogen equivalent elements would be: N, Al, Mn
The alkali metal equivalent elements would be: Li, F, P

63. The density increases as you move to the right across the first transition series. For the first transition series, the mass increases as you move to the right across the periodic table. However, the radius of the transition series elements stays nearly constant as you move to the right across the periodic table, thus the volume will remain nearly constant. Since density is mass/ volume the density of the elements increases.

64. If there are 3 possible spin quantum numbers, there will be three *s* electrons, nine *p* electrons, and fifteen *d* electrons.
   a) Ne(10 e): $1s^3 2s^3 2p^4$
   b) completed n = 2 level: $1s^3 2s^3 2p^9$ Atomic number = 15
   c) F(9 e): $1s^3 2s^3 2p^3$ therefore there are 3 unpaired electrons.

65. The longest wavelength would be associated with the lowest energy state next to the ground state of carbon which has 2 unpaired electrons:
Ground state of carbon:

1*s*	2*s*	2*p*		
⇂↿	⇂↿	↑	↑	

Longest wavelength: one of the *p* electrons flipped in its orbital which requires the least amount of energy.

1*s*	2*s*	2*p*		
⇂↿	⇂↿	↑	↓	

The next wavelength would be associated with the pairing of the two *p* electrons in the same orbital because this requires energy and raises the energy.

1*s*	2*s*	2*p*		
⇂↿	⇂↿	⇂↿		

The next wavelength would be associated with the energy needed to promote one of the *s* electrons to a *p* orbital.

1*s*	2*s*	2*p*		
⇂↿	↑	↑	↑	↑

66. Element darmstadium (110) would be in the column with Ni, Pd, and Pt so it might be expected to have an electron configuration similar to Ni or Pd or Pt.
Similar to Ni: $[Rn]7s^2 5f^{14} 6d^8$
Similar to Pd: $[Rn]7s^0 5f^{14} 6d^{10}$
Similar to Pt: $[Rn]\ 7s^1 5f^{14} 6d^9$

67. The element that would fill the 8*s* and 8*p* orbitals would have atomic number 168. The element is in the noble gas family and would have the properties of noble gases. It would have the electron configuration of $[118]8s^2 5g^{18} 6f^{14} 7d^{10} 8p^6$. The outer shell electron (highest *n* level) configuration would be $8s^2 8p^6$. The element would be relatively inert, have a first ionization energy less than 1037 kJ/mol (the first ionization energy of Rn), and have a positive electron affinity. It would be difficult to form compounds with other elements but would be able to form compounds with fluorine.

68. To determine the second ionization energies, look at the electron configuration of the 1+ ions.

First write the electron configuration of the atom:		Then write the electron configuration of the 1+ ion:	
Li	$1s^2 2s^1$	Li^+	$1s^2 2s^0$
Be	$1s^2 2s^2$	Be^+	$1s^2 2s^1$
B	$1s^2 2s^2 2p^1$	B^+	$1s^2 2s^2$
C	$1s^2 2s^2 2p^2$	C^+	$1s^2 2s^2 2p^1$
N	$1s^2 2s^2 2p^3$	N^+	$1s^2 2s^2 2p^2$
O	$1s^2 2s^2 2p^4$	O^+	$1s^2 2s^2 2p^3$
F	$1s^2 2s^2 2p^5$	F^+	$1s^2 2s^2 2p^4$

Based on the electron configuration of the ions, Li^+ should have the largest second ionization energy since the removal of the second electron involves removing a core electron. The lowest second ionization energy should be Be^+ since removing the second electron takes you to the $1s^2$ (stable) configuration.
O would have the highest second ionization energy because the electron configuration of O^+ has a half-filled p orbital which is particularly stable and therefore it would require more energy to remove the second electron. N^+ would have the lowest second ionization energy because the size of N^+ would be larger than the radius of the F^+, so the attraction between the outer electron and the nucleus would be less in N^+ than in F^+, making it easier to remove the electron.

69. When you move down the column from Al to Ga, the size of the atom actually decreases because not much shielding is contributed by the 3d electrons in the Ga atom while there is a large increase in the nuclear charge, therefore, the effective nuclear charge is greater for Ga than for Al, so the ionization energy does not decrease. As you go from In to Tl, the ionization energy actually increases because the 4f electrons do not contribute to the shielding of the outermost electrons and there is a large increase in the effective nuclear charge.

70. ΔE for the reaction based on the ionization energy and the electron affinity = +147 kJ/ mol

$Na(g) \rightarrow Na^+(g) + e^-$	IE = +496 kJ/mol
$Cl(g) + e^- \rightarrow Cl^-(g)$	EA = - 349 kJ/mol
$Na(g) + Cl(g) \rightarrow Na^+(g) + Cl^-(g)$	ΔE = +147 kJ/mol

71. The second electron is added to an ion with a 1 – charge, so there is a large repulsive force that has to be overcome to add the second electron. Thus, it will require energy to add the second electron and the second electron affinity will have a positive value.

72. The diagonal relationship between some elements could be explained because the atomic size of the atoms on the diagonal would be about the same. Because on the diagonal, the radius would increase as you go down the column but will decrease as you move to the right across the period. Also, the size of the ion formed would be about the same. Therefore, you might expect the elements to have similar behavior.

73. If six electrons rather than eight electrons led to a stable configuration, the electron configuration of the stable configuration would be: ns^2np^4
    a) A noble gas would have the electron configuration ns^2np^4. This could correspond to the O atom.
    b) A reactive nonmetal would have one less electron than the stable configuration. This would have the electron configuration ns^2np^3. This could correspond to the N atom
    c) A reactive metal would have one have one more electron than the stable configuration. This would have the electron configuration of ns^1. This could correspond to the Li atom.

74. a) True: An electron in a 3*s* orbital is more shielded than an electron in a 2*s* orbital. This is true since there are more core electrons below a 3s orbital.

    b) True: An electron in a 3*s* orbital penetrates into the region occupied by the core electrons more than electrons in a 3*p* orbital. Examine Figure 8.5, the radial distribution functions for the 3*s*, 3*p* and 3*d* orbitals. You will see that the 3*s* electrons penetrate more deeply than the 3*p* electrons and more than the 3*d* electrons.

    c) False: An electron in an orbital that penetrates closer to the nucleus will experience <u>less</u> shielding than an electron in an orbital that does not penetrate as far.

    d) True: An electron in an orbital that penetrates close to the nucleus will tend to experience a higher effective nuclear charge than one that does not. Since the orbital penetrates closer to the nucleus, the electron will experience less shielding and, therefore, a higher effective nuclear charge.

75. An electron in a 5*p* orbital could have any one of the following combinations of quantum numbers.
5,1, -1,+1/2   5,1,-1,-1/2   5,1,0,+1/2   5,1,0,-1/2   5,1,1,+1/2   5,1,1,-1/2
An electron in a 6*d* orbital could have any one of the following combinations of quantum numbers.
6,2,-2,+1/2   6,2,-2,-1/2   6,2,-1,+1/2   6,2,-1,-1/2   6,2,0,+1/2   6,2,0,-1/2
6,2,1,+1/2   6,2,1,-1/2   6,2,2,+1/2   6,2,2,-1/2

76. The 4*s* electrons in calcium have relatively low ionization energies (IE_1 = 590 kJ/mol; IE_2 = 1145 kJ/mol) because they are valence electrons. The energetic cost for calcium to lose a third electron is extraordinarily high because the next electron to be lost is a core electron. Similarly, the electron affinity of fluorine to gain one electron (- 328 kJ/mol) is highly exothermic because the added electron completes fluoride's valence shell. The gain of a second electron by the negatively charged fluoride anion would not be favorable. Therefore, we would expect calcium and fluoride to combine in a 1:2 ratio.

Chapter 9
Chemical Bonding I: Lewis Theory

1. N : $1s^22s^22p^3$  The electrons included in the Lewis structure are: $2s^22p^3$.

2. Ne : $1s^22s^22p^6$  The electrons included in the Lewis structure are: $2s^22p^6$.

3. a) Al: $1s^22s^22p^63s^23p^1$

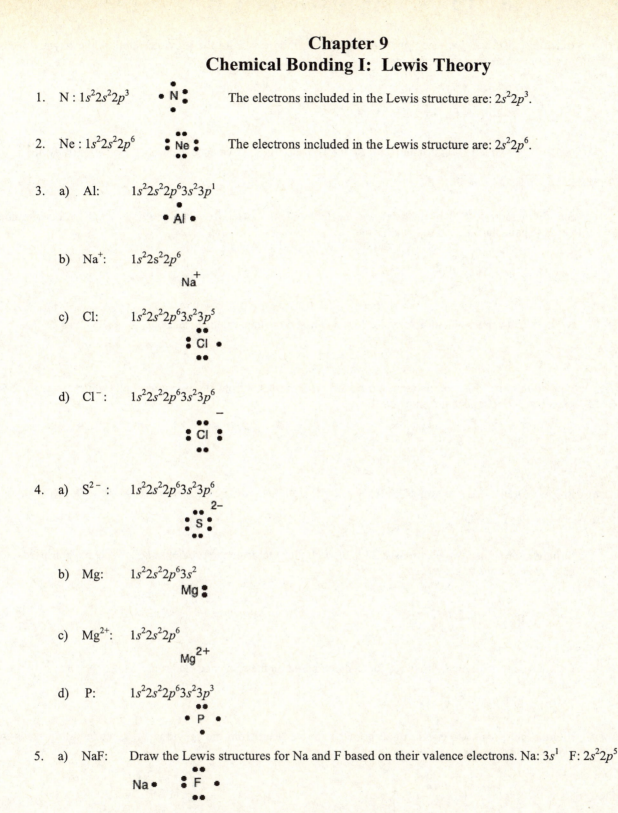

b) Na^+: $1s^22s^22p^6$

Na^+

c) Cl: $1s^22s^22p^63s^23p^5$

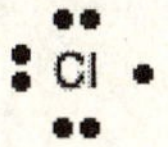

d) Cl^-: $1s^22s^22p^63s^23p^6$

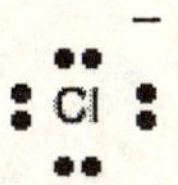

4. a) S^{2-} : $1s^22s^22p^63s^23p^6$

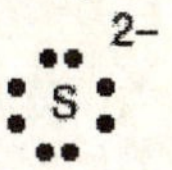

b) Mg: $1s^22s^22p^63s^2$

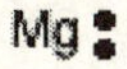

c) Mg^{2+}: $1s^22s^22p^6$

Mg^{2+}

d) P: $1s^22s^22p^63s^23p^3$

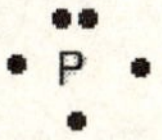

5. a) NaF: Draw the Lewis structures for Na and F based on their valence electrons. Na: $3s^1$ F: $2s^22p^5$

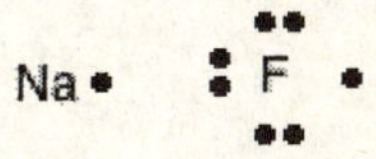

Sodium must lose one electron and be left with the octet from the previous shell, while fluorine needs to gain one electron to get an octet.

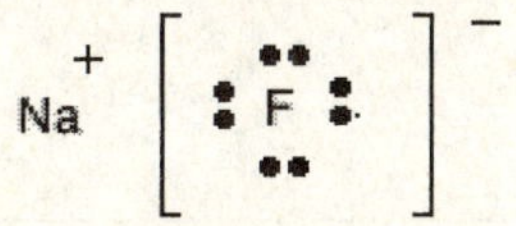

Ca $\overset{\bullet\bullet}{\underset{\bullet}{\,\cdot\,O}}\,\cdot$

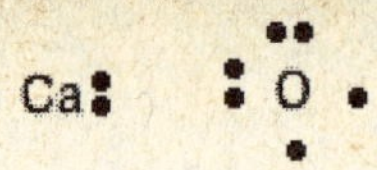

Calcium must lose two electrons and be left with the octet from the previous shell, while oxygen needs to gain two electrons to get an octet.

$$Ca^{2+} \left[\,\vdots\,\overset{\bullet\bullet}{\underset{\bullet\bullet}{O}}\,\vdots\, \right]^{2-}$$

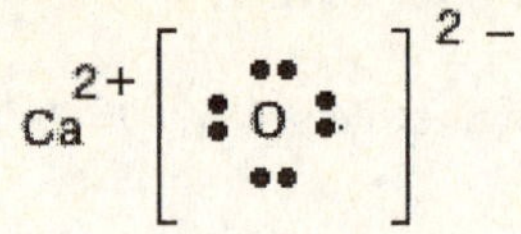

c)   SrBr$_2$:   Draw the Lewis structures for Sr and Br based on their valence electrons.   Sr: $5s^2$ Br: $4s^24p^5$

Sr$\,\vdots\quad\vdots\,\overset{\bullet\bullet}{\underset{\bullet\bullet}{Br}}\,\cdot$

Strontium must lose two electrons and be left with the octet from the previous shell, while bromine needs to gain one electron to get an octet.

$$Sr^{2+}\ 2\left[\,\vdots\,\overset{\bullet\bullet}{\underset{\bullet\bullet}{Br}}\,\vdots\, \right]^{-}$$

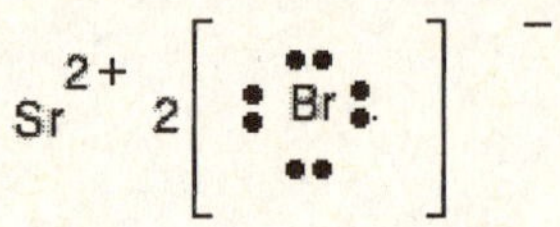

d)   K$_2$O: Draw the Lewis structures for K and O based on their valence electrons.   K: $4s^1$ O: $2s^22p^4$

K $\cdot\quad\,\vdots\,\overset{\bullet\bullet}{\underset{\bullet}{O}}\,\cdot$

Potassium must lose one electron and be left with the octet from the previous shell, while oxygen needs to gain two electrons to get an octet.

$$2K^{+}\ \left[\,\vdots\,\overset{\bullet\bullet}{\underset{\bullet\bullet}{O}}\,\vdots\, \right]^{2-}$$

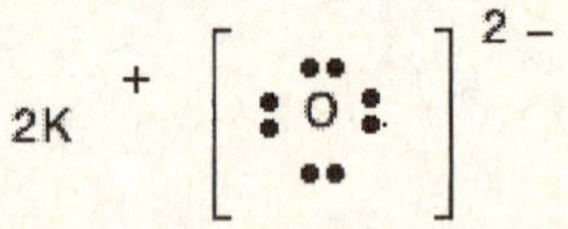

6.   a)   SrO: Draw the Lewis structures for Sr and O based on their valence electrons.   Sr: $5s^2$ O: $2s^22p^4$

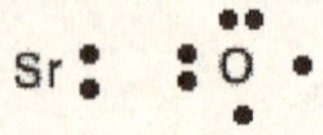

Sr$\,\vdots\quad\vdots\,\overset{\bullet\bullet}{\underset{\bullet}{O}}\,\cdot$

Strontium must lose two electrons and be left with the octet from the previous shell, while oxygen needs to gain two electrons to get an octet.

$$Sr^{2+}\ \left[\,\vdots\,\overset{\bullet\bullet}{\underset{\bullet\bullet}{O}}\,\vdots\, \right]^{2-}$$

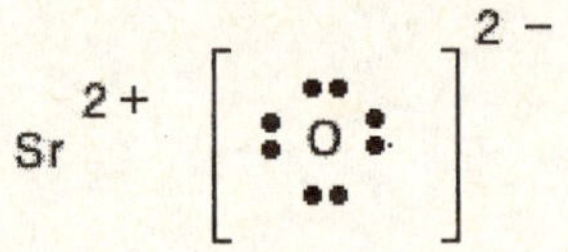

b)   Li$_2$S: Draw the Lewis structures for Li and S based on their valence electrons.   Li: $2s^1$ S: $3s^23p^4$

Li $\cdot\quad\,\vdots\,\overset{\bullet\bullet}{\underset{\bullet}{S}}\,\cdot$

Lithium must lose one electron and be left with the octet from the previous shell, while sulfur needs to gain two electrons to get an octet.

$$2\ Li^{+}\ \left[\,\vdots\,\overset{\bullet\bullet}{\underset{\bullet\bullet}{S}}\,\vdots\, \right]^{2-}$$

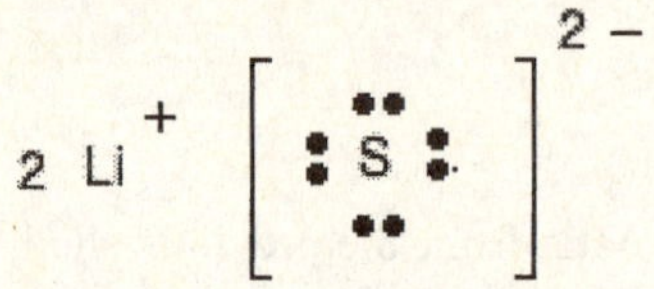

c) CaI₂: Draw the Lewis structures for Ca and I based on their valence electrons. Ca: $4s^2$ I: $5s^25p^5$

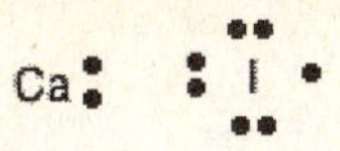

Calcium must lose two electrons and be left with the octet from the previous shell, while iodine needs to gain one electron to get an octet.

$$Ca^{2+} \; 2\left[\; :\overset{\bullet\bullet}{\underset{\bullet\bullet}{I}}: \;\right]^{-}$$

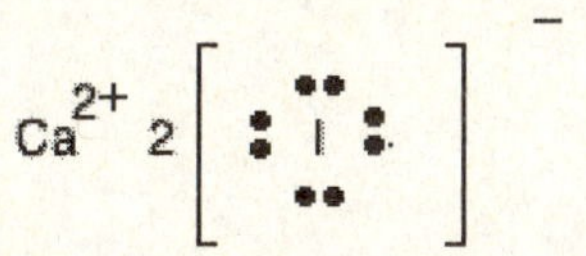

d) RbF: Draw the Lewis structures for Rb and F based on their valence electrons. Ca: $5s^1$ I: $2s^22p^5$

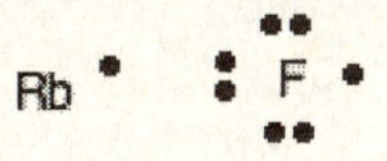

Rubidium must lose one electron and be left with the octet from the previous shell, while fluorine needs to gain one electron to get an octet.

$$Rb^{+} \left[\; :\overset{\bullet\bullet}{\underset{\bullet\bullet}{F}}: \;\right]^{-}$$

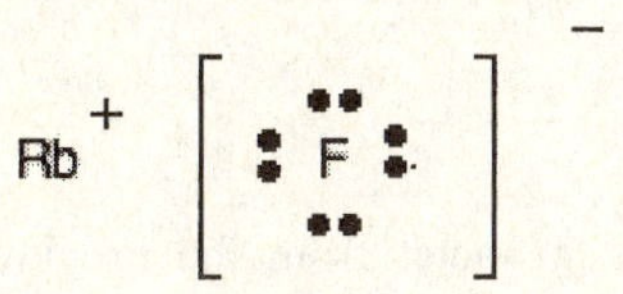

7. a) Sr and Se: Draw the Lewis structures for Sr and Se based on their valence electrons.
        Sr: $5s^2$ Se: $4s^24p^4$

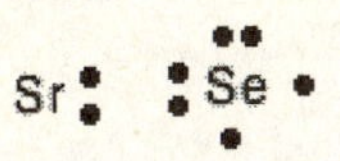

Strontium must lose two electrons and be left with the octet from the previous shell, while selenium needs to gain two electrons to get an octet.

$$Sr^{2+} \left[\; :\overset{\bullet\bullet}{\underset{\bullet\bullet}{Se}}: \;\right]^{2-}$$

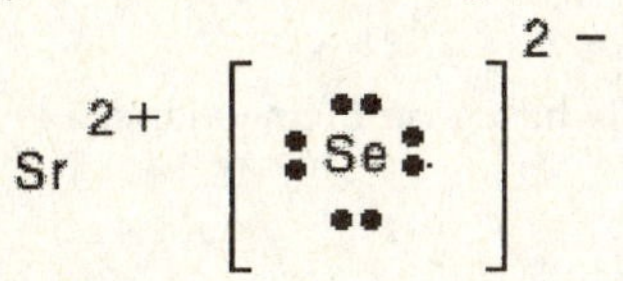

Thus, we need one Sr^{2+} and one Se^{2-}. Write the formula with subscripts (if necessary) to indicate the number of atoms.
        SrSe

b) Ba and Cl   Draw the Lewis structures for Ba and Cl based on their valence electrons.
        Ba: $6s^2$ Cl: $3s^23p^5$

$$Ba \; \overset{\bullet}{\underset{\bullet}{\;}} \quad : \overset{\bullet\bullet}{Cl} \bullet$$

Barium must lose two electrons and be left with the octet from the previous shell, while chlorine needs to gain one electron to get an octet.

$$Ba^{2+} \; 2\left[\; :\overset{\bullet\bullet}{\underset{\bullet\bullet}{Cl}}: \;\right]^{-}$$

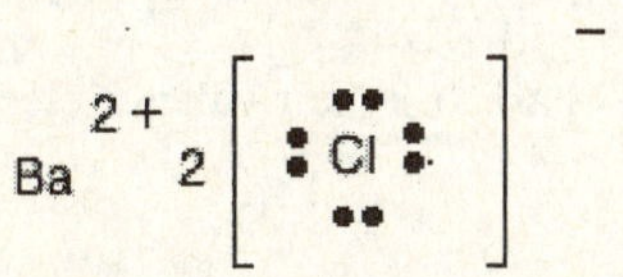

Thus, we need one Ba^{2+} and two Cl^-. Write the formula with subscripts (if necessary) to indicate the number of atoms.
        BaCl₂

c) Na and S Draw the Lewis structures for Na and S based on their valence electrons.

Na: $3s^1$ S: $3s^2 3p^4$

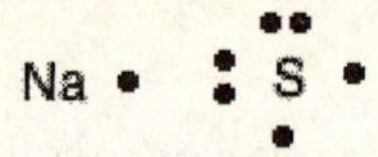

Sodium must lose one electron and be left with the octet from the previous shell, while sulfur needs to gain two electrons to get an octet.

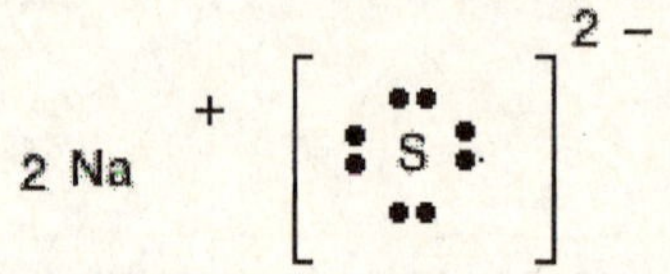

Thus, we need two Na^+ and one S^{2-}. Write the formula with subscripts (if necessary) to indicate the number of atoms.

Na_2S

d) Al and O Draw the Lewis structures for Al and O based on their valence electrons.

Al: $3s^2 3p^1$ O: $2s^2 2p^4$

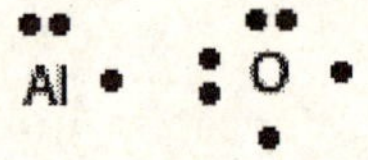

Aluminum must lose three electrons and be left with the octet from the previous shell, while oxygen needs to gain two electrons to get an octet.

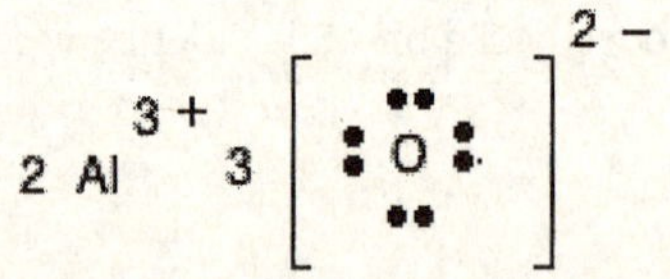

Thus, we need two Al^{3+} and three O^{2-} in order to lose and gain the same number of electrons. Write the formula with subscripts (if necessary) to indicate the number of atoms.

Al_2O_3

8. a) Ca and N Draw the Lewis structures for Ca and N based on their valence electrons.

Ca: $4s^2$ N: $2s^2 2p^3$

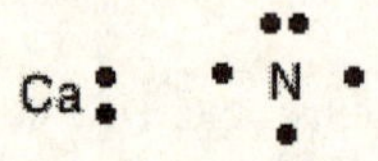

Calcium must lose two electrons and be left with the octet from the previous shell, while nitrogen needs to gain three electrons to get an octet.

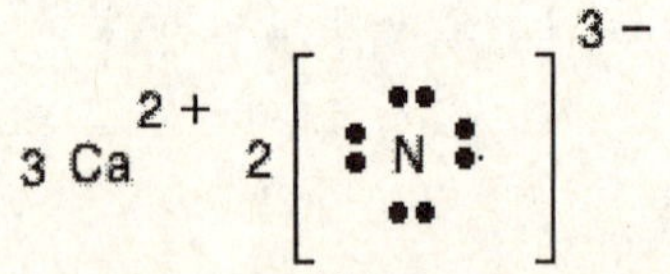

Thus, we need three Ca^{2+} and two N^{3-} in order to lose and gain the same number of electrons. Write the formula with subscripts (if necessary) to indicate the number of atoms.

Ca_3N_2

b) Mg and I Draw the Lewis structures for Mg and I based on their valence electrons.

Mg: $3s^2$ I: $5s^2 5p^5$

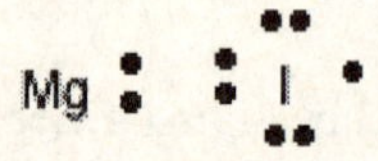

Magnesium must lose two electrons and be left with the octet from the previous shell, while iodine needs to gain one electron to get an octet.

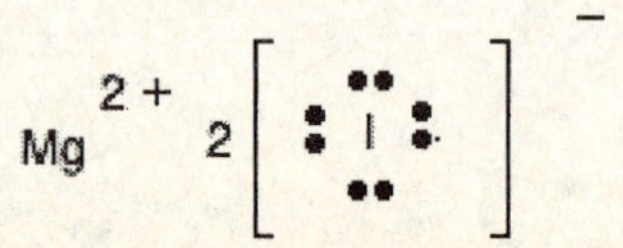

Thus, we need one Mg^{2+} and two I^-. Write the formula with subscripts (if necessary) to indicate the number of atoms.

MgI_2

c) Ca and S Draw the Lewis structures for Ca and S based on their valence electrons.

Ca: $4s^2$ S: $3s^23p^4$

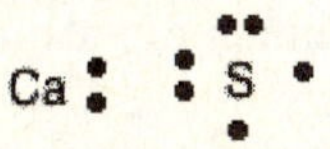

Calcium must lose two electrons and be left with the octet from the previous shell, while sulfur needs to gain two electrons to get an octet.

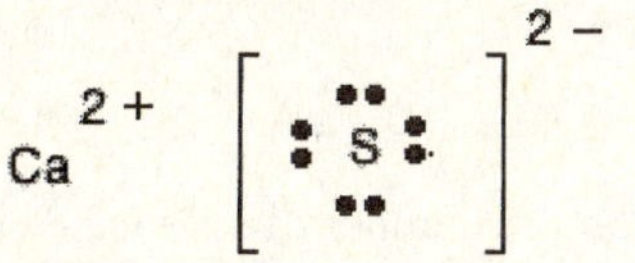

Thus, we need one Ca^{2+} and one S^{2-}. Write the formula with subscripts (if necessary) to indicate the number of atoms.

CaS

d) Cs and F Draw the Lewis structures for Cs and F based on their valence electrons.

Cs: $6s^1$ F: $2s^22p^5$

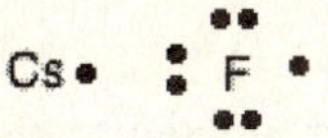

Cesium must lose one electron and be left with the octet from the previous shell, while fluorine needs to gain one electron to get an octet.

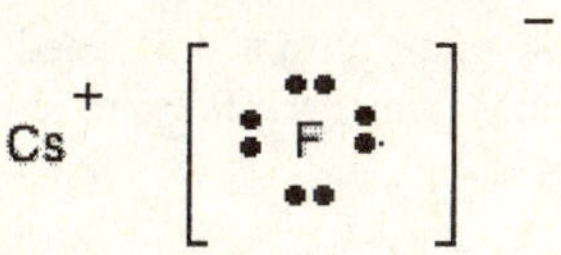

Thus, we need one Cs^+ and one F^-. Write the formula with subscripts (if necessary) to indicate the number of atoms.

CsF

9. As the size of the alkaline metal ions increases down the column, so does the distance between the metal cation and the oxide anion. Therefore, the magnitude of the lattice energy of the oxides decreases, making the formation of the oxides less exothermic and the compounds less stable. Since the ions cannot get as close to each other, they therefore do not release as much energy.

10. Rubidium is below potassium on the periodic table and iodine is below bromine on the periodic table. Therefore, both the rubidium ion and the iodide ion are larger than the potassium ion and the bromide ion. So, the rubidium ion and the iodide ion cannot get as close to each other as the potassium ion and the bromide ion, therefore, they do not release as much energy and the lattice energy of potassium bromide is more exothermic.

11. Cesium is slightly larger than barium, but oxygen is slightly larger than fluorine, so we cannot use size to explain the difference in the lattice energy. However, the charge on cesium ion is 1+ and the charge on fluoride ion is 1 −, while the charge on barium ion is 2+ and the charge on oxide ion is 2 −. The coulombic equation states that the magnitude of the potential also depends on the product of the charges. Since the product of the charges for CsF = 1 −, and the product of the charges for BaO = 4 −, the stabilization for BaO relative to CsF should be about four times greater, which is what we see in its much more exothermic lattice energy.

12. RbBr < KCl < SrO < CaO. KCl and RbBr both have a product of the charges of 1 −, while SrO and CaO have a product of the charges of 4 −. So, the lattice energies of KCl and RbBr are less than SrO and CaO. Within KCl and RbBr, rubidium ion is larger than potassium ion and bromide ion is larger than chloride ion. Therefore, the rubidium ion and the bromide ion will be farther apart, leading to a smaller lattice energy. Between SrO and CaO, the strontium ion is larger than the calcium ion, so the strontium oxide will have a smaller (less negative) lattice energy.

13. a) Hydrogen: Write the Lewis structure of each atom based on the number of valence electrons.

H• •H

When the two hydrogen atoms share their electrons, they each get a duet, which is a stable configuration for hydrogen

H —— H

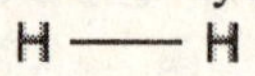

b) The halogens: Write the Lewis structure of each atom based on the number of valence electrons.

$\overset{\bullet\bullet}{\underset{\bullet\bullet}{:X}}\bullet\quad\bullet\overset{\bullet\bullet}{\underset{\bullet\bullet}{X:}}$

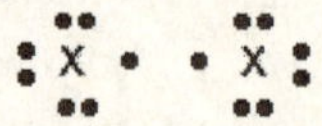

If the two halogens pair together they can each achieve an octet, which is a stable configuration. So, the halogens are predicted to exist as diatomic molecules.

$\overset{\bullet\bullet}{\underset{\bullet\bullet}{:X}} —— \overset{\bullet\bullet}{\underset{\bullet\bullet}{X:}}$

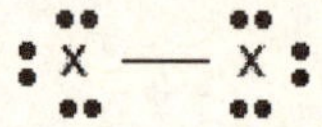

c.) Oxygen: Write the Lewis structure of each atom based on the number of valence electrons.

$\overset{\bullet\bullet}{\underset{\bullet}{:O}}\bullet\quad\bullet\overset{\bullet\bullet}{\underset{\bullet}{O:}}$

In order to achieve a stable octet on each oxygen, the oxygen atoms will need to share two electron pairs. So, oxygen is predicted to exist as a diatomic molecule with a double bond.

$\overset{\bullet\bullet}{:O} == \overset{\bullet\bullet}{O:}$

d) Nitrogen: Write the Lewis structure of each atom based on the number of valence electrons.

$\bullet\overset{\bullet\bullet}{\underset{\bullet}{N}}\bullet\quad\bullet\overset{\bullet\bullet}{\underset{\bullet}{N}}\bullet$

In order to achieve a stable octet on each nitrogen, the nitrogen atoms will need to share three electron pairs. So, nitrogen is predicted to exist as a diatomic molecule with a triple bond.

$\overset{\bullet\bullet}{N}\equiv\equiv\overset{\bullet\bullet}{N}$

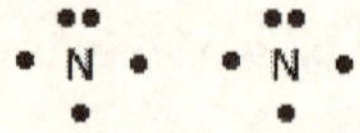

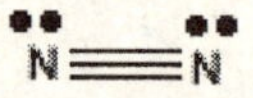

14. Write the Lewis structure for N and H based on the number of valence electrons.

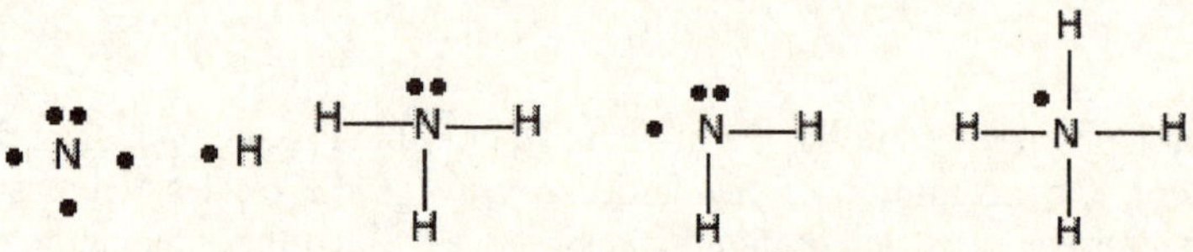

If nitrogen combines with three hydrogen atoms, the nitrogen will achieve a stable octet and each hydrogen will have a duet of electrons. This is a stable configuration. If the nitrogen were to combine with only two hydrogen atoms, the nitrogen could only achieve a seven electron configuration, which is not stable. Also, if the nitrogen were to combine with four hydrogen atoms, the nitrogen would have a nine electron configuration which is not stable. So, Lewis theory predicts that nitrogen will combine with three hydrogen atoms.

15. a) PH_3: Write the Lewis structure for each atom based on the valence electron.

$\bullet\overset{\bullet\bullet}{\underset{\bullet}{P}}\bullet\quad\bullet H$

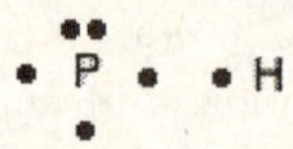

The phosphorus will share an electron pair with each hydrogen in order to achieve a stable octet.

H—$\overset{\bullet\bullet}{P}$—H
|
H

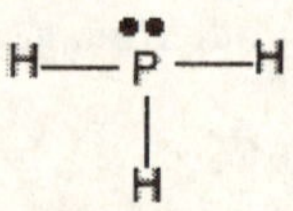

b) SCl_2: Write the Lewis structure for each atom based on the valence electron.

$\overset{\bullet\bullet}{\underset{\bullet}{:S}}\bullet\quad\overset{\bullet\bullet}{\underset{\bullet\bullet}{:Cl}}\bullet$

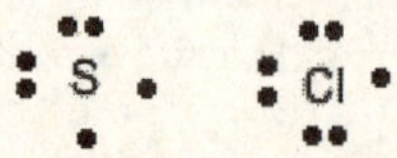

The sulfur will share an electron pair with each chlorine in order to achieve a stable octet.

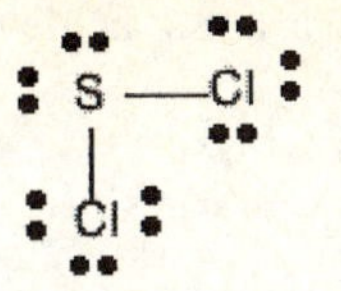

c) HI: Write the Lewis structure for each atom based on the valence electron.

The iodine will share an electron pair with hydrogen in order to achieve a stable octet.

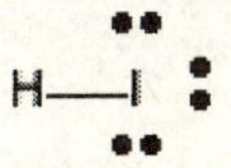

d) CH_4: Write the Lewis structure for each atom based on the valence electron.

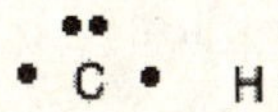

The carbon will share an electron pair with each hydrogen in order to achieve a stable octet.

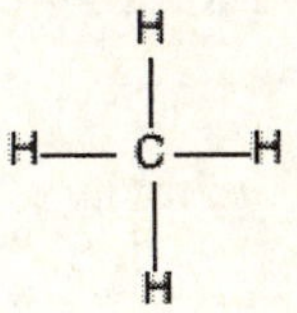

16. a) NF_3: Write the Lewis structure for each atom based on the valence electron.

The nitrogen will share an electron pair with each fluorine in order to achieve a stable octet.

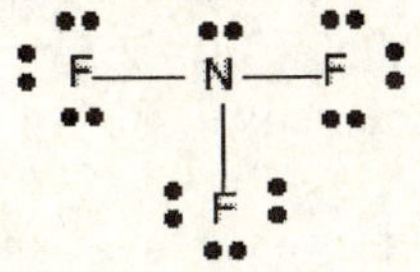

b) HBr: Write the Lewis structure for each atom based on the valence electron.

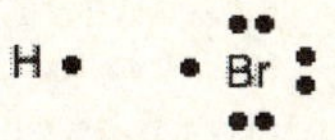

The bromine will share an electron pair with hydrogen in order to achieve a stable octet.

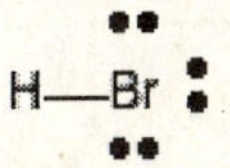

c) SBr_2: Write the Lewis structure for each atom based on the valence electron.

The sulfur will share an electron pair with each bromine in order to achieve a stable octet.

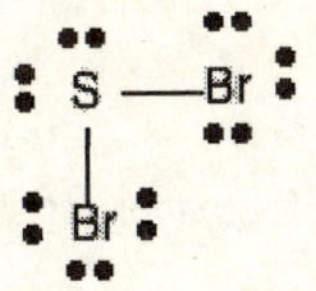

d) CCl_4: Write the Lewis structure for each atom based on the valence electron.

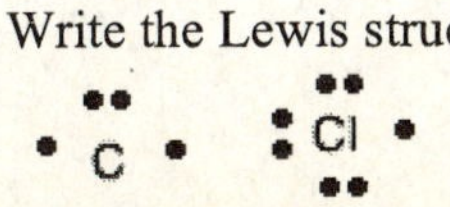

The carbon will share an electron pair with each chlorine in order to achieve a stable octet.

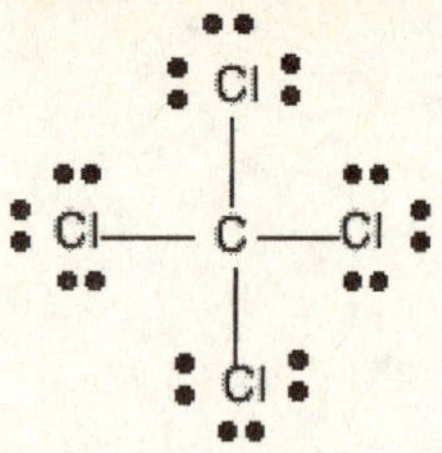

17\. a) Br and Br: pure covalent From Figure 9.9 we find the electronegativity of Br is 2.5. Since both atoms are the same, the electronegativity difference (ΔEN) = 0, and using Table 9.1 we classify this bond as pure covalent.

b) C and Cl: polar covalent From Figure 9.9 we find the electronegativity of C is 2.5 and Cl is 3.0. The electronegativity difference (ΔEN) is $\Delta EN = 3.0 - 2.5 = 0.5$. Using Table 9.1 we classify this bond as polar covalent.

c) C and S: pure covalent From Figure 9.9 we find the electronegativity of C is 2.5 and S is 2.5. The electronegativity difference (ΔEN) is $\Delta EN = 2.5 - 2.5 = 0$. Using Table 9.1 we classify this bond as pure covalent.

d) Sr and O: ionic From Figure 9.9 we find the electronegativity of Sr is 1.0 and O is 3.5. The electronegativity difference (ΔEN) is $\Delta EN = 3.5 - 1.0 = 2.5$. Using Table 9.1 we classify this bond as ionic.

18\. a) C and N: polar covalent From Figure 9.9 we find the electronegativity of C is 2.5 and N is 3.0. The electronegativity difference (ΔEN) is $\Delta EN = 3.0 - 2.5 = 0.5$. Using Table 9.1 we classify this bond as polar covalent.

b) N and S: polar covalent From Figure 9.9 we find the electronegativity of S is 2.5 and N is 3.0. The electronegativity difference (ΔEN) is $\Delta EN = 3.0 - 2.5 = 0.5$. Using Table 9.1 we classify this bond as polar covalent.

c) K and F: ionic From Figure 9.9 we find the electronegativity of K is 0.8 and F is 4.0. The electronegativity difference (ΔEN) is $\Delta EN = 4.0 - 0.8 = 3.2$. Using Table 9.1 we classify this bond as ionic.

d) N and N: pure covalent From Figure 9.9 we find the electronegativity of N is 3.0. Since both atoms are the same, the electronegativity difference (ΔEN) = 0, and using Table 9.1 we classify this bond as pure covalent.

19\. CO: Write the Lewis structure for each atom based on the valence electron.

The carbon will share three electron pairS with oxygen in order to achieve a stable octet.
The oxygen atom is more electronegative than the carbon atom, so the oxygen will have a partial negative charge and the carbon will have a partial positive charge.

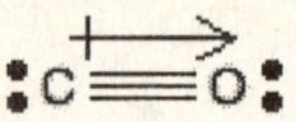

To estimate the percent ionic character, determine the difference in electronegativity between carbon and oxygen.
From figure 9.9 we find the electronegativity of C is 2.5 and O is 3.5. The electronegativity difference (ΔEN) is $\Delta EN = 3.5 - 2.5 = 1.0$.
From figure 9.11, we can estimate a percent ionic character of 25%

20. BrF: Write the Lewis structure for each atom based on the valence electron.

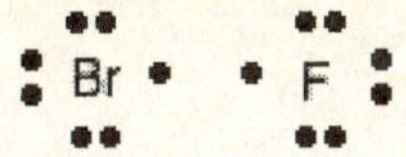

The bromine and fluorine will share an electron pair to achieve a stable octet.
The fluorine atom is more electronegative than the bromine atom, so the fluorine will have a partial negative charge and the bromine will have a partial positive charge.

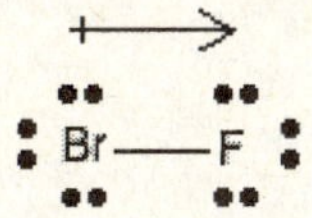

To estimate the percent ionic character, determine the difference in electronegativity between bromine and fluorine.
From figure 9.9 we find the electronegativity of Br = 2.0 and F = 4.0. The electronegativity difference (ΔEN) is $\Delta EN = 4.0 - 2.0 = 2.0$.
From figure 9.11, we can estimate a percent ionic character of 70%

21. a.) CI_4: Write the correct skeletal structure for the molecule.

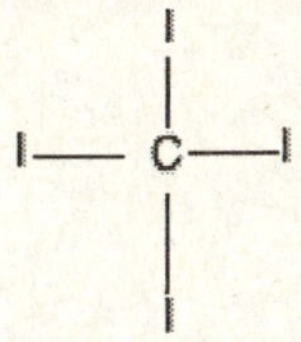

Calculate the total number of electrons for the Lewis structure by summing the valence electron of each atom in the molecule.

(number of valence e^- for C) + 4(number of valence e^- for I) = 4 + 4(7) = 32

Distribute the electrons among the atoms, giving octets (or duets for H) to as many atoms as possible. Begin with the bonding electrons, and then proceed to lone pairs on terminal atoms and finally to lone pairs of the central atom.

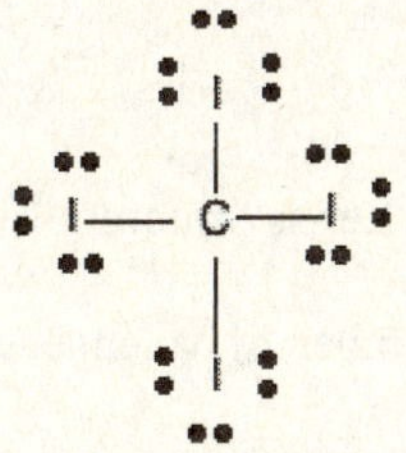

All 32 valence electrons are used
If any atom lacks an octet, form double or triple bonds as necessary to give them octets.
All atoms have octets, duets for H, structure is complete.

b) N_2O: Write the correct skeletal structure for the molecule
N is the less electronegative, so it is central

Calculate the total number of electrons for the Lewis structure by summing the valence electron of each atom in the molecule

2(number of valence e^- for N) + (number of valence e^- for O) = 2(5) + 6 = 16

Distribute the electrons among the atoms, giving octets (or duets for H) to as many atoms as possible. Begin with the bonding electrons, and then proceed to lone pairs on terminal atoms and finally to lone pairs of the central atom.

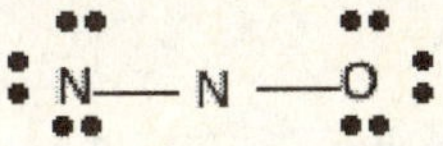

All 16 valence electrons are used
If any atom lacks an octet, form double or triple bonds as necessary to give them octets.

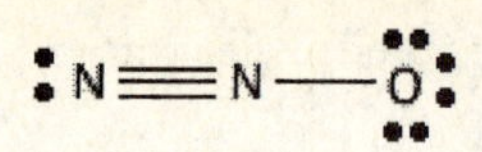

All atoms have octets, duets for H, structure is complete

c) SiH_4: Write the correct skeletal structure for the molecule
H is always terminal so Si is the central atom

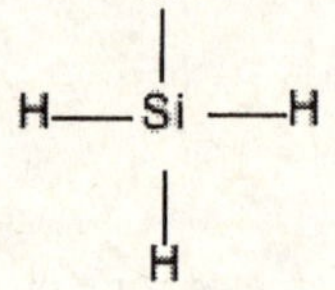

Calculate the total number of electrons for the Lewis structure by summing the valence electrons of each atom in the molecule

(number of valence e^- for Si) + 4(number of valence e^- for H) = 4 + 4(1) = 8

Distribute the electrons among the atoms, giving octets (or duets for H) to as many atoms as possible. Begin with the bonding electrons, and then proceed to lone pairs on terminal atoms and finally to lone pairs of the central atom.

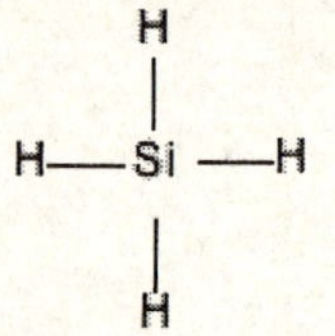

All 8 valence electrons are used.

If any atom lacks an octet, form double or triple bonds as necessary to give them octets.
All atoms have octets, duets for H, structure is complete.

d) Cl_2CO: Write the correct skeletal structure for the molecule
C is the least electronegative, so it is the central atom

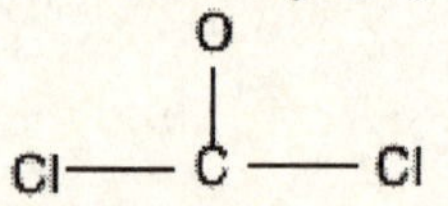

Calculate the total number of electrons for the Lewis structure by summing the valence electrons of each atom in the molecule

(number of valence e^- for C) + 2(number of valence e^- for Cl) + (number of valence e^- for O) = 4 + 2(7) + 6 = 24

Distribute the electrons among the atoms, giving octets (or duets for H) to as many atoms as possible. Begin with the bonding electrons, and then proceed to lone pairs on terminal atoms and finally to lone pairs of the central atom.

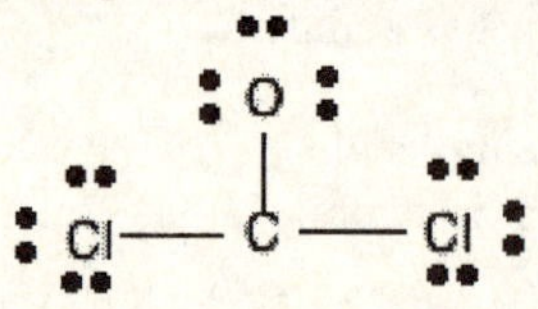

All 24 valence electrons are used.

If any atom lacks an octet, form double or triple bonds as necessary to give them octets.

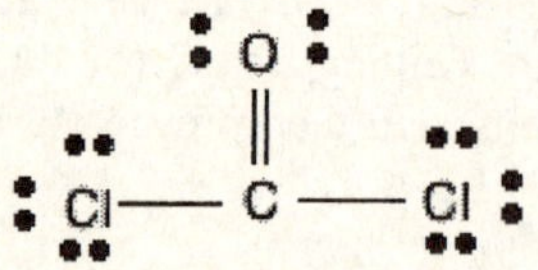

All atoms have octets, duets for H, structure is complete.

e) H_3COH: Write the correct skeletal structure for the molecule
C is less electronegative and H is terminal

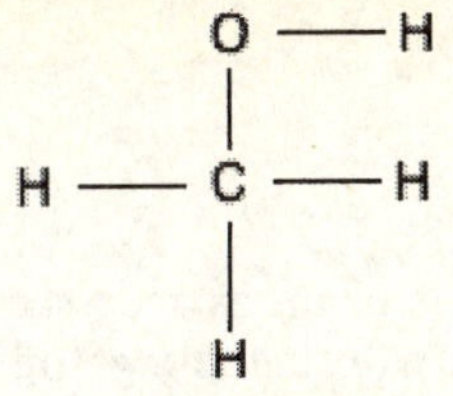

Calculate the total number of electrons for the Lewis structure by summing the valence electrons of each atom in the molecule

(number of valence e^- for C) + 4(number of valence e^- for H) + (number of valence e^- for O) = 4 + 4(1) + 6 = 14

Distribute the electrons among the atoms, giving octets (or duets for H) to as many atoms as possible. Begin with the bonding electrons, and then proceed to lone pairs on terminal atoms and finally to lone pairs of the central atom.

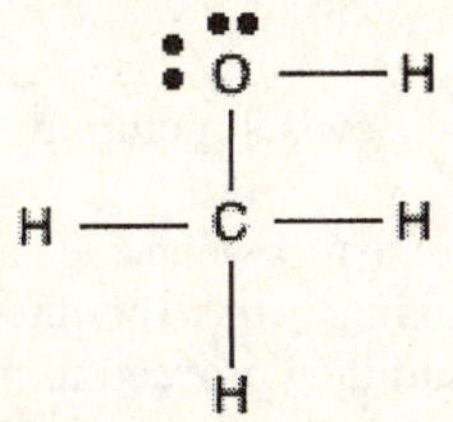

All 14 valence electrons are used.

If any atom lacks an octet, form double or triple bonds as necessary to give them octets.
All atoms have octets, duets for H, structure is complete.

f) OH^-: Write the correct skeletal structure for the ion

O — H

Calculate the total number of electrons for the Lewis structure by summing the valence electrons of each atom in the ion and adding 1 for the 1 – charge.

(number of valence e^- for O) + (number of valence e^- for H) + 1 = 6 + 1 + 1 = 8

Distribute the electrons among the atoms, giving octets (or duets for H) to as many atoms as possible. Begin with the bonding electrons, and then proceed to lone pairs on terminal atoms and finally to lone pairs of the central atom.

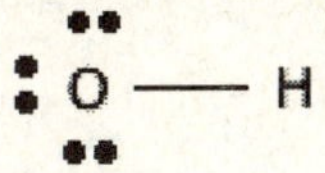

All 8 valence electrons are used.

If any atom lacks an octet, form double or triple bonds as necessary to give them octets.
Lastly, write the Lewis structure in brackets with the charge of the ion in the upper right-hand corner.

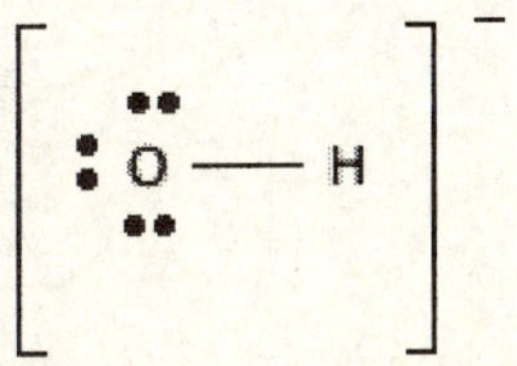

g) BrO^-: Write the correct skeletal structure for the ion

Br — O

Calculate the total number of electrons for the Lewis structure by summing the valence electrons of each atom in the ion and adding 1 for the 1 – charge.

(number of valence e^- for O) + (number of valence e^- for Br) + 1 = 6 + 7 + 1 = 14

Distribute the electrons among the atoms, giving octets (or duets for H) to as many atoms as possible. Begin with the bonding electrons, and then proceed to lone pairs on terminal atoms and finally to lone pairs of the central atom.

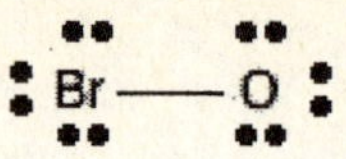

All 14 valence electrons are used.

If any atom lacks an octet, form double or triple bonds as necessary to give them octets.

Lastly, write the Lewis structure in brackets with the charge of the ion in the upper right-hand corner.

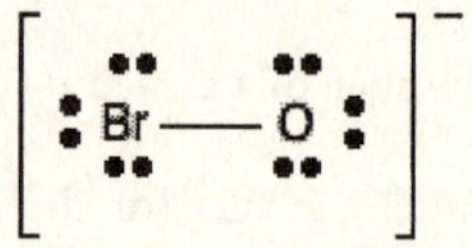

22. a) N_2H_2: Write the correct skeletal structure for the molecule

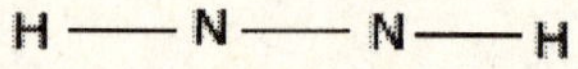

Calculate the total number of electrons for the Lewis structure by summing the valence electrons of each atom in the molecule

$2(\text{number of valence } e^- \text{ for N}) + 2(\text{number of valence } e^- \text{ for H}) = 2(5) + 2(1) = 12$

Distribute the electrons among the atoms, giving octets (or duets for H) to as many atoms as possible. Begin with the bonding electrons, and then proceed to lone pairs on terminal atoms and finally to lone pairs of the central atom.

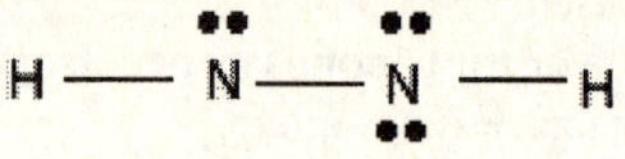

All 12 valence electrons are used.

If any atom lacks an octet, form double or triple bonds as necessary to give them octets.

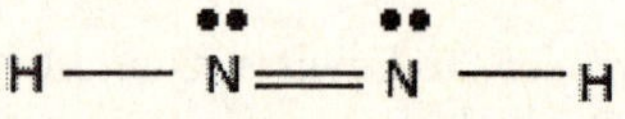

All atoms have octets, duets for H, structure is complete.

b) N_2H_4: Write the correct skeletal structure for the molecule

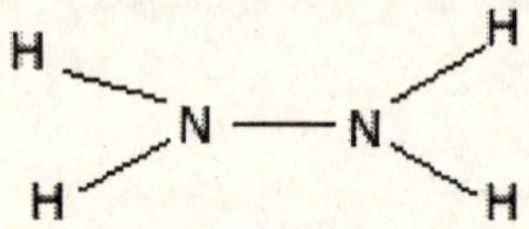

Calculate the total number of electrons for the Lewis structure by summing the valence electrons of each atom in the molecule

$2(\text{number of valence } e^- \text{ for N}) + 4(\text{number of valence } e^- \text{ for H}) = 2(5) + 4(1) = 14$

Distribute the electrons among the atoms, giving octets (or duets for H) to as many atoms as possible. Begin with the bonding electrons, and then proceed to lone pairs on terminal atoms and finally to lone pairs of the central atom.

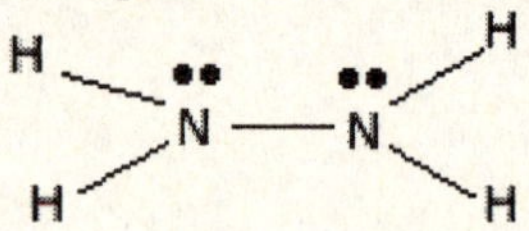

All 14 valence electrons are used.

If any atom lacks an octet, form double or triple bonds as necessary to give them octets.

All atoms have octets, duets for H, structure is complete.

c) C_2H_2: Write the correct skeletal structure for the molecule

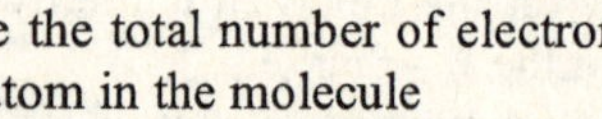

Calculate the total number of electrons for the Lewis structure by summing the valence electrons of each atom in the molecule

$2(\text{number of valence } e^- \text{ for C}) + 2(\text{number of valence } e^- \text{ for H}) = 2(4) + 2(1) = 10$

Distribute the electrons among the atoms, giving octets (or duets for H) to as many atoms as possible. Begin with the bonding electrons, and then proceed to lone pairs on terminal atoms and finally to lone pairs of the central atom.

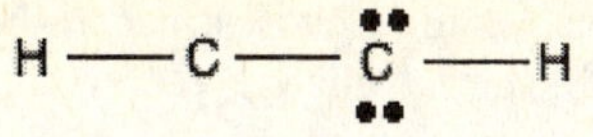

All 10 valence electrons are used.

If any atom lacks an octet, form double or triple bonds as necessary to give them octets.

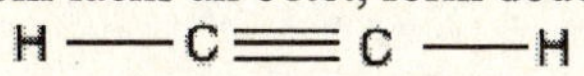

All atoms have octets, duets for H, structure is complete.

d) C_2H_4: Write the correct skeletal structure for the molecule

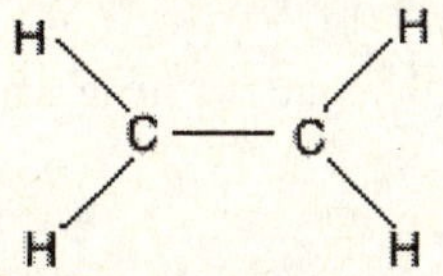

Calculate the total number of electrons for the Lewis structure by summing the valence electrons of each atom in the molecule

$$2(\text{number of valence e}^- \text{ for C}) + 4(\text{number of valence e}^- \text{ for H}) = 2(4) + 4(1) = 12$$

Distribute the electrons among the atoms, giving octets (or duets for H) to as many atoms as possible. Begin with the bonding electrons, and then proceed to lone pairs on terminal atoms and finally to lone pairs of the central atom.

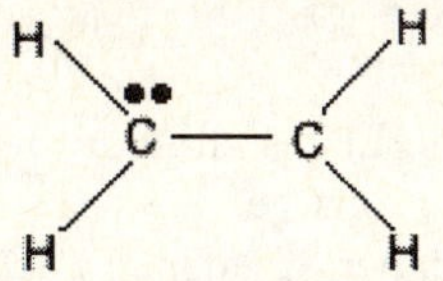

All 12 valence electrons are used.

If any atom lacks an octet, form double or triple bonds as necessary to give them octets.

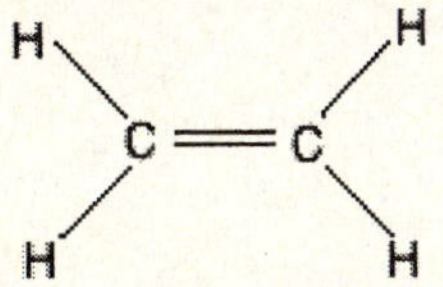

All atoms have octets, duets for H, structure is complete.

e) H_3COCH_3: Write the correct skeletal structure for the molecule

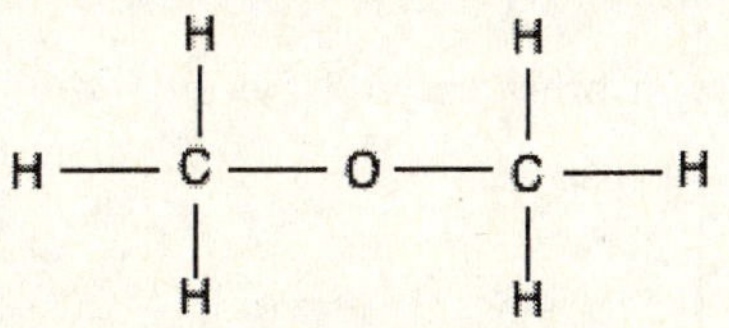

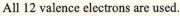

Calculate the total number of electrons for the Lewis structure by summing the valence electrons of each atom in the molecule

$$2(\text{number of valence e}^- \text{ for C}) + (\text{number of valence e}^- \text{ for O}) + 6(\text{number of valence e}^- \text{ for H}) = 2(4) + 6 + 6(1) = 20$$

Distribute the electrons among the atoms, giving octets (or duets for H) to as many atoms as possible. Begin with the bonding electrons, and then proceed to lone pairs on terminal atoms and finally to lone pairs of the central atom.

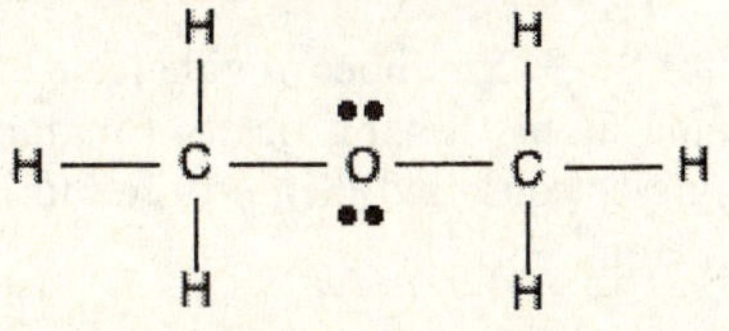

All 20 valence electrons are used.

If any atom lacks an octet, form double or triple bonds as necessary to give them octets.

All atoms have octets, duets for H, structure is complete.

f) CN^-: Write the correct skeletal structure for the ion

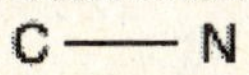

Calculate the total number of electrons for the Lewis structure by summing the valence electrons of each atom in the ion and adding 1 for the 1 – charge.

(number of valence e⁻ for C) + (number of valence e⁻ for N) + 1 = 4 + 5 + 1 = 10

Distribute the electrons among the atoms, giving octets (or duets for H) to as many atoms as possible. Begin with the bonding electrons, and then proceed to lone pairs on terminal atoms and finally to lone pairs of the central atom.

$$\overset{\bullet\bullet}{C}\text{—}\overset{\bullet\bullet}{\underset{\bullet\bullet}{N}}\colon$$

All 10 valence electrons are used.

If any atom lacks an octet, form double or triple bonds as necessary to give them octets.

$$\colon\!\! C\!\equiv\! N \colon$$

Lastly, write the Lewis structure in brackets with the charge of the ion in the upper right-hand corner.

$$\left[\colon\!\! C\!\equiv\! N \colon\right]^{-}$$

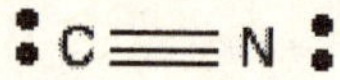

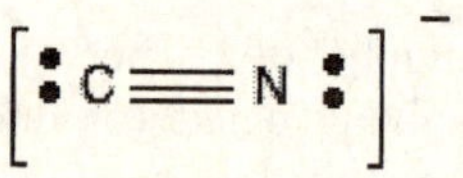

g) NO_2^-:   Write the correct skeletal structure for the ion

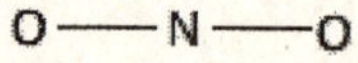

O —— N —— O

Calculate the total number of electrons for the Lewis structure by summing the valence electrons of each atom in the ion and adding 1 for the 1 – charge.

2(number of valence e⁻ for O) + (number of valence e⁻ for N) + 1 = 2(6) + 5 + 1 = 18

Distribute the electrons among the atoms, giving octets (or duets for H) to as many atoms as possible. Begin with the bonding electrons, and then proceed to lone pairs on terminal atoms and finally to lone pairs of the central atom.

$$\colon\!\overset{\bullet\bullet}{\underset{\bullet\bullet}{O}}\text{—}\overset{\bullet\bullet}{N}\text{—}\overset{\bullet\bullet}{\underset{\bullet\bullet}{O}}\!\colon$$

All 18 valence electrons are used.

If any atom lacks an octet, form double or triple bonds as necessary to give them octets.

$$\colon\!\overset{\bullet\bullet}{\underset{\bullet\bullet}{O}}\text{—}\overset{\bullet\bullet}{N}\!=\!\overset{\bullet\bullet}{O}\!\colon$$

Lastly, write the Lewis structure in brackets with the charge of the ion in the upper right-hand corner.

$$\left[\colon\!\overset{\bullet\bullet}{\underset{\bullet\bullet}{O}}\text{—}\overset{\bullet\bullet}{N}\!=\!\overset{\bullet\bullet}{O}\!\colon\right]^{-}$$

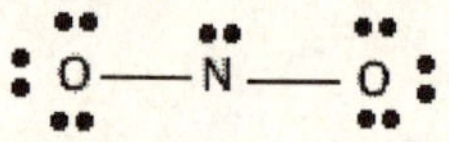

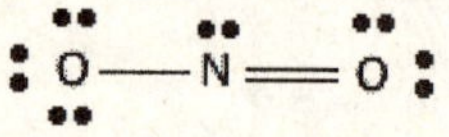

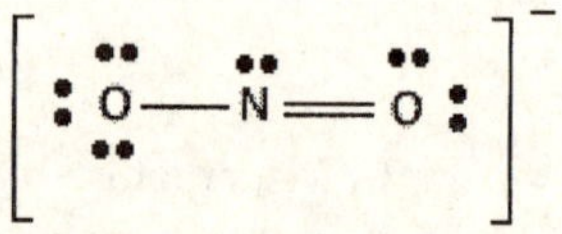

23. a) SeO_2:   Write the correct skeletal structure for the molecule
Se is the less electronegative, so it is central

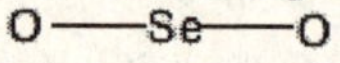

O —— Se —— O

Calculate the total number of electrons for the Lewis structure by summing the valence electrons of each atom in the molecule

(number of valence e⁻ for Se) + 2(number of valence e⁻ for O) = 6 + 2(6) = 18

Distribute the electrons among the atoms, giving octets (or duets for H) to as many atoms as possible. Begin with the bonding electrons, and then proceed to lone pairs on terminal atoms and finally to lone pairs of the central atom.

$$\colon\!\overset{\bullet\bullet}{\underset{\bullet\bullet}{O}}\text{—}\overset{\bullet\bullet}{Se}\text{—}\overset{\bullet\bullet}{\underset{\bullet\bullet}{O}}\!\colon$$

All 18 valence electrons are used.

If any atom lacks an octet, form double or triple bonds as necessary to give them octets.

$$\colon\!\overset{\bullet\bullet}{\underset{\bullet\bullet}{O}}\text{—}\overset{\bullet\bullet}{Se}\!=\!\overset{\bullet\bullet}{O}\!\colon$$

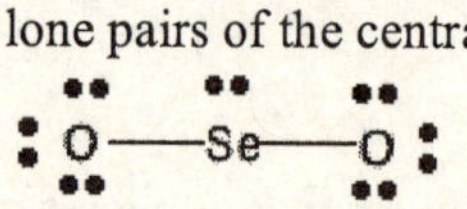

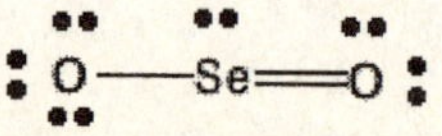

All atoms have octets, duets for H, structure is complete. However, the double bond can form from either oxygen atom, so there are two resonance forms.

Calculate the formal charge on each atom by finding the number of valence electrons and subtracting the number of lone pair electrons and one-half the number of bonding electrons.

	O	Se	O	O	Se	O
number of valence electrons	6	6	6	6	6	6
- number of lone pair electrons	6	2	4	4	2	6
- 1/2(number of bonding electrons)	1	3	2	2	3	1
Formal charge	−1	+1	0	0	+1	−1

b) CO_3^{2-}: Write the correct skeletal structure for the ion

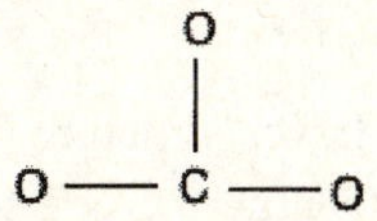

Calculate the total number of electrons for the Lewis structure by summing the valence electrons of each atom in the ion and adding 2 for the 2 – charge.

$$3(\text{number of valence e}^- \text{ for O}) + (\text{number of valence e}^- \text{ for C}) + 2 = 3(6) + 4 + 2 = 24$$

Distribute the electrons among the atoms, giving octets (or duets for H) to as many atoms as possible. Begin with the bonding electrons, and then proceed to lone pairs on terminal atoms and finally to lone pairs of the central atom.

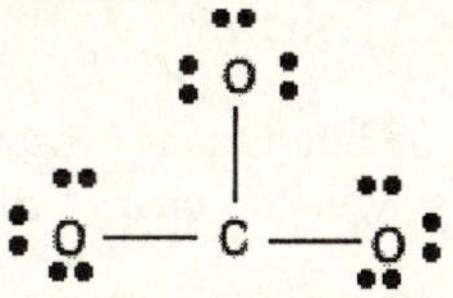

All 24 valence electrons are used.

If any atom lacks an octet, form double or triple bonds as necessary to give them octets.

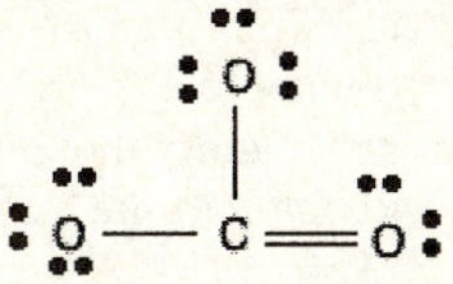

Lastly, write the Lewis structure in brackets with the charge of the ion in the upper right-hand corner.

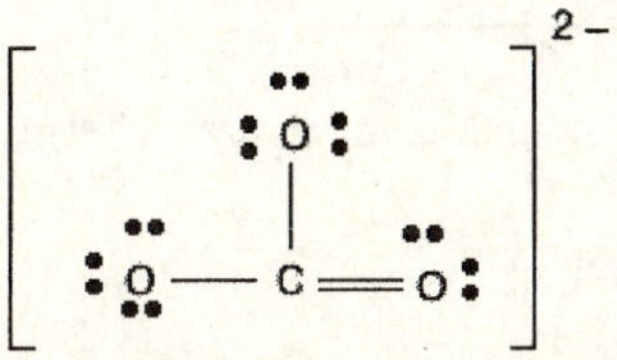

All atoms have octets, duets for H, structure is complete. However, the double bond can form from any oxygen atom, so there are three resonance forms.

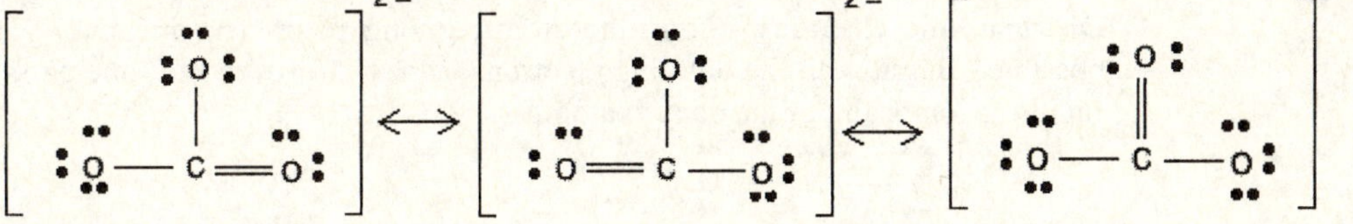

Calculate the formal charge on each atom by finding the number of valence electrons and subtracting the number of lone pair electrons and one-half the number of bonding electrons.

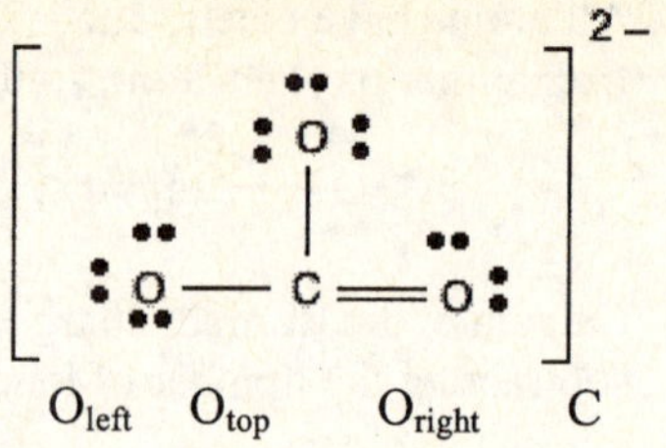

	O_{left}	O_{top}	O_{right}	C
number of valence electrons	6	6	6	4
- number of lone pair electrons	6	6	4	0
- 1/2(number of bonding electrons)	1	1	2	4
Formal charge	−1	−1	0	0

The sum of the formal charges is – 2 which is the overall charge of the ion. The other resonance forms would be the same.

c) ClO^-: Write the correct skeletal structure for the ion

Cl —— O

Calculate the total number of electrons for the Lewis structure by summing the valence electrons of each atom in the ion and adding 1 for the 1 – charge.

(number of valence e^- for O) + (number of valence e^- for Cl) + 1 = 6 + 7 + 1 = 14

Distribute the electrons among the atoms, giving octets (or duets for H) to as many atoms as possible. Begin with the bonding electrons, and then proceed to lone pairs on terminal atoms and finally to lone pairs of the central atom.

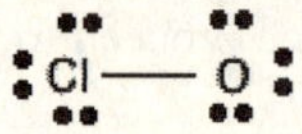

All 14 valence electrons are used.

If any atom lacks an octet, form double or triple bonds as necessary to give them octets.

Lastly, write the Lewis structure in brackets with the charge of the ion in the upper right-hand corner.

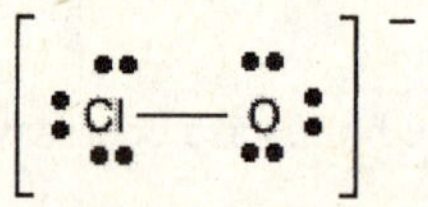

All atoms have octets, duets for H, structure is complete.

Calculate the formal charge on each atom by finding the number of valence electrons and subtracting the number of lone pair electrons and one-half the number of bonding electrons.

	Cl	O
number of valence electrons	7	6
- number of lone pair electrons	6	6
- 1/2(number of bonding electrons)	1	1
Formal charge	0	−1

The sum of the formal charges is – 1 which is the overall charge of the ion.

d) NO_2^-: Write the correct skeletal structure for the ion

O —— N —— O

Calculate the total number of electrons for the Lewis structure by summing the valence electrons of each atom in the ion and adding 1 for the 1 – charge.

2(number of valence e^- for O) + (number of valence e^- for N) + 1 = 2(6) + 5 + 1 = 18

Distribute the electrons among the atoms, giving octets (or duets for H) to as many atoms as possible. Begin with the bonding electrons, and then proceed to lone pairs on terminal atoms and finally to lone pairs of the central atom.

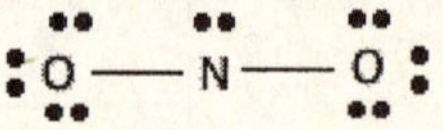

All 14 valence electrons are used.

If any atom lacks an octet, form double or triple bonds as necessary to give them octets.

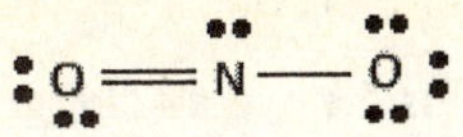

Lastly, write the Lewis structure in brackets with the charge of the ion in the upper right-hand corner.

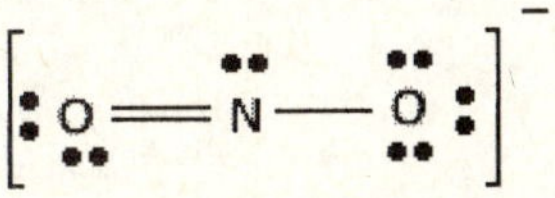

All atoms have octets, duets for H, structure is complete. However, the double bond can form from either oxygen atom, so there are two resonance forms.

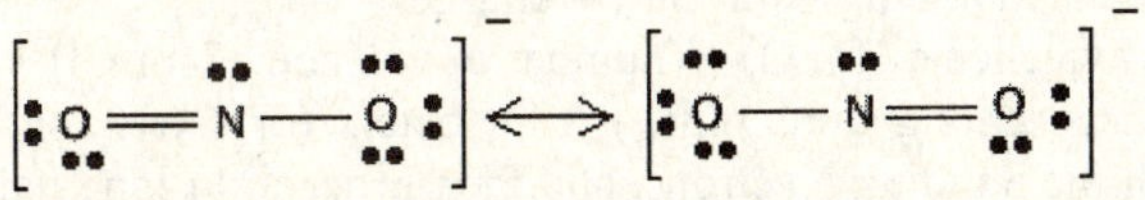

Calculate the formal charge on each atom by finding the number of valence electrons and subtracting the number of lone pair electrons and one-half the number of bonding electrons. Using the left side structure:

	O	N	O
number of valence electrons	6	5	6
- number of lone pair electrons	4	2	6
- 1/2(number of bonding electrons)	2	3	1
Formal charge	0	0	– 1

The sum of the formal charges is – 1 which is the overall charge of the ion.

24. a) ClO_3^-: Write the correct skeletal structure for the ion

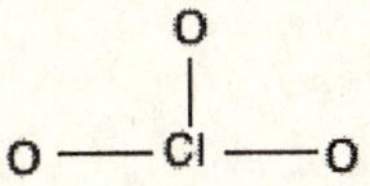

Calculate the total number of electrons for the Lewis structure by summing the valence electrons of each atom in the ion and adding 1 for the 1 – charge.

$$3(\text{number of valence e}^- \text{ for O}) + (\text{number of valence e}^- \text{ for Cl}) + 1 = 3(6) + 7 + 1 = 26$$

Distribute the electrons among the atoms, giving octets (or duets for H) to as many atoms as possible. Begin with the bonding electrons, and then proceed to lone pairs on terminal atoms and finally to lone pairs of the central atom.

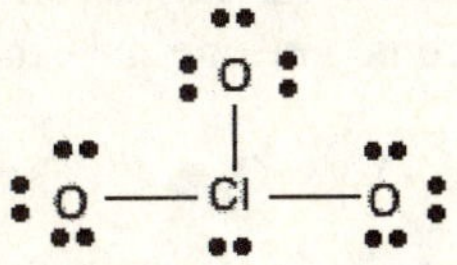

All 26 valence electrons are used.

If any atom lacks an octet, form double or triple bonds as necessary to give them octets.

Lastly, write the Lewis structure in brackets with the charge of the ion in the upper right-hand corner.

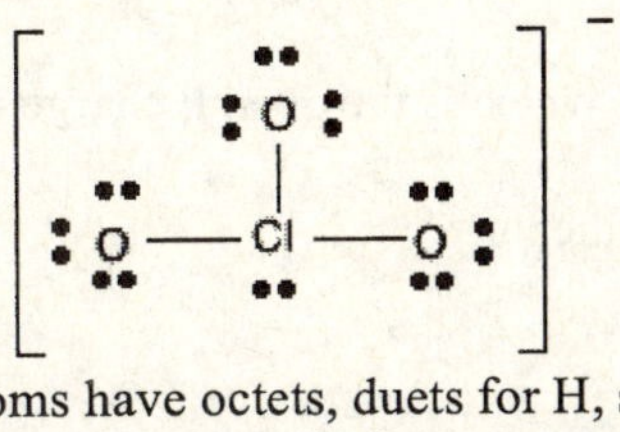

All atoms have octets, duets for H, structure is complete.

Calculate the formal charge on each atom by finding the number of valence electrons and subtracting the number of lone pair electrons and one-half the number of bonding electrons.

	O_{left}	O_{top}	O_{right}	Cl
number of valence electrons	6	6	6	7
- number of lone pair electrons	6	6	6	2
- 1/2(number of bonding electrons)	1	1	1	3
Formal charge	–1	–1	–1	+2

The sum of the formal charges is – 1 which is the overall charge of the ion.

b) ClO_4^-: Write the correct skeletal structure for the ion

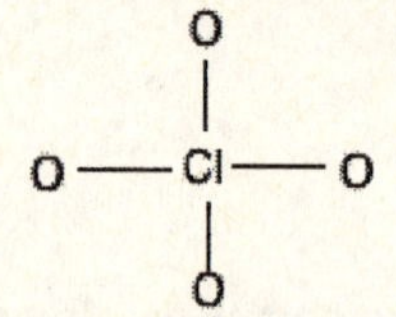

Calculate the total number of electrons for the Lewis structure by summing the valence electrons of each atom in the ion and adding 1 for the 1 – charge.

4(number of valence e^- for O) + (number of valence e^- for Cl) + 1 = 4(6) + 7 + 1 = 32

Distribute the electrons among the atoms, giving octets (or duets for H) to as many atoms as possible. Begin with the bonding electrons, and then proceed to lone pairs on terminal atoms and finally to lone pairs of the central atom.

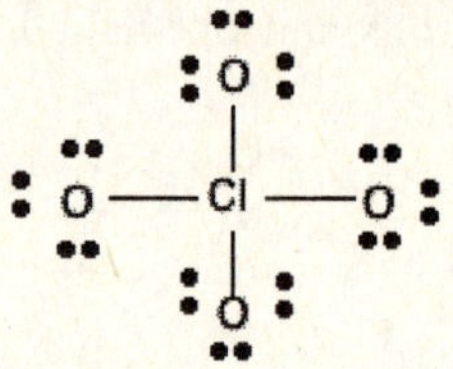

All 26 valence electrons are used.

If any atom lacks an octet, form double or triple bonds as necessary to give them octets.
Lastly, write the Lewis structure in brackets with the charge of the ion in the upper right-hand corner.

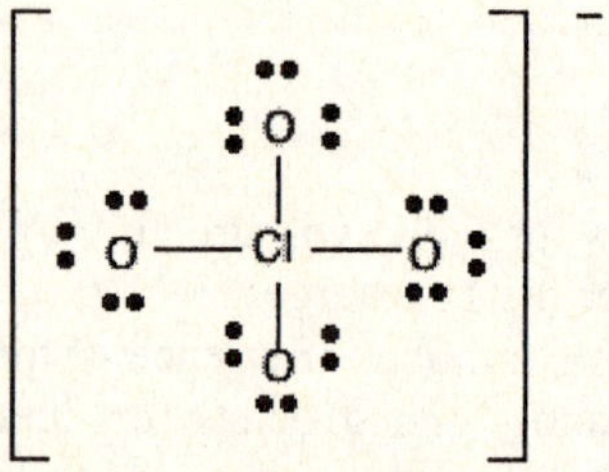

All atoms have octets, duets for H, structure is complete.
Calculate the formal charge on each atom be finding the number of valence electrons and subtracting the number of lone pair electrons and one-half the number of bonding electrons. Using the left side structure:

	O_{left}	O_{top}	O_{right}	O_{bottom}	Cl
number of valence electrons	6	6	6	6	7
- number of lone pair electrons	6	6	6	6	0
- 1/2(number of bonding electrons)	1	1	1	1	4
Formal charge	–1	–1	–1	–1	+3

The sum of the formal charges is – 1 which is the overall charge of the ion.

c) NO_3^-: Write the correct skeletal structure for the ion

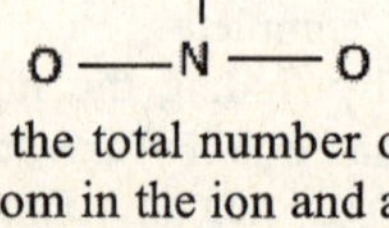

Calculate the total number of electrons for the Lewis structure by summing the valence electrons of each atom in the ion and adding 1 for the 1 – charge.

3(number of valence e^- for O) + (number of valence e^- for N) + 1 = 3(6) + 5 + 1 = 24

Distribute the electrons among the atoms, giving octets (or duets for H) to as many atoms as possible. Begin with the bonding electrons, and then proceed to lone pairs on terminal atoms and finally to lone pairs of the central atom.

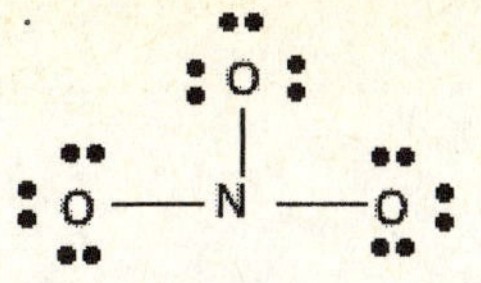

All 24 valence electrons are used.

If any atom lacks an octet, form double or triple bonds as necessary to give them octets.

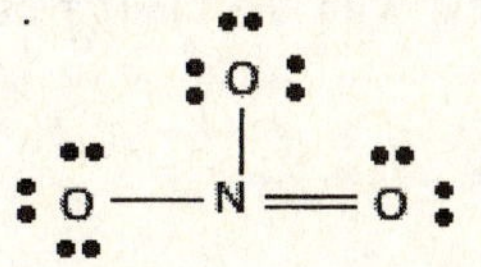

Lastly, write the Lewis structure in brackets with the charge of the ion in the upper right-hand corner.

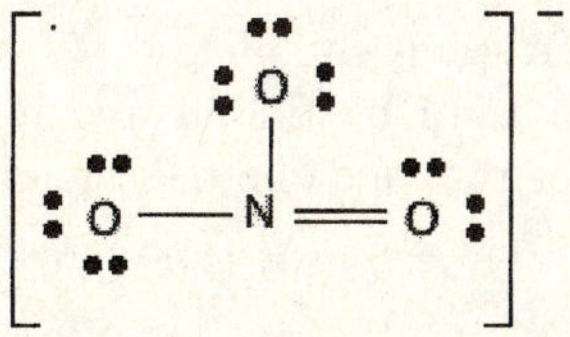

All atoms have octets, duets for H, structure is complete. However, the double bond can form from any oxygen atom, so there are three resonance forms.

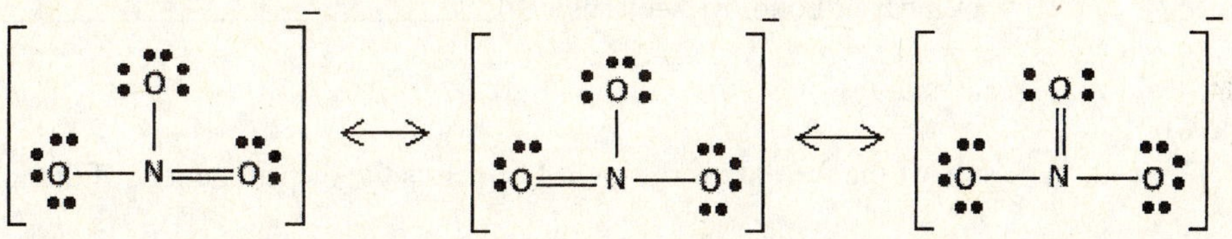

Calculate the formal charge on each atom be finding the number of valence electrons and subtracting the number of lone pair electrons and one-half the number of bonding electrons. Using the left hand structure

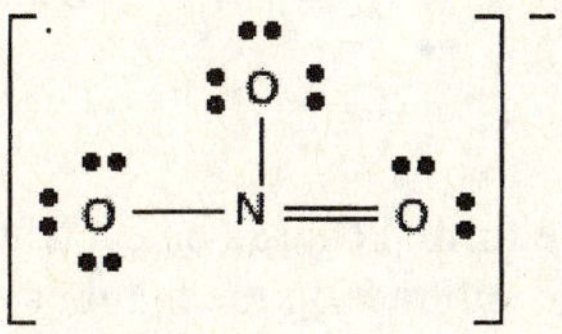

	O_{left}	O_{top}	O_{right}	N
number of valence electronss	6	6	6	5
- number of lone pair electrons	6	6	4	0
- 1/2(number of bonding electrons)	1	1	2	4
Formal charge	–1	–1	0	+1

The sum of the formal charges is – 1 which is the overall charge of the ion. The other resonance forms would be the same.

d) NH_4^+: Write the correct skeletal structure for the ion

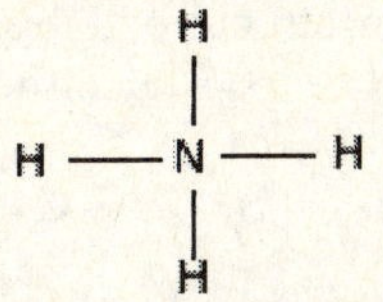

Calculate the total number of electrons for the Lewis structure by summing the valence electrons of each atom in the ion and subtracting 1 for the 1+ charge.

4(number of valence e^- for H) + (number of valence e^- for N) – 1 = 4(1) + 5 – 1 = 8

Distribute the electrons among the atoms, giving octets (or duets for H) to as many atoms as possible. Begin with the bonding electrons, and then proceed to lone pairs on terminal atoms and finally to lone pairs of the central atom.

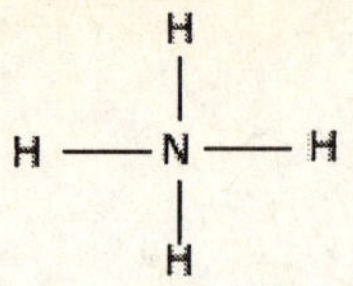

All 8 valence electrons are used.

If any atom lacks an octet, form double or triple bonds as necessary to give them octets.

Lastly, write the Lewis structure in brackets with the charge of the ion in the upper right-hand corner.

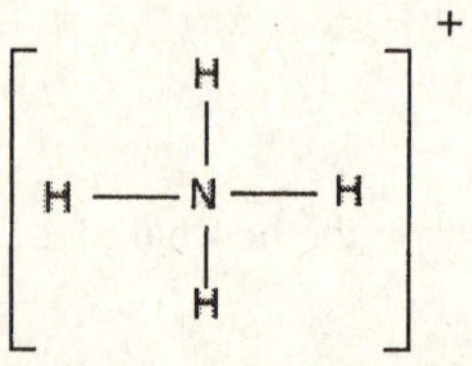

All atoms have octets, duets for H, structure is complete.

Calculate the formal charge on each atom be finding the number of valence electrons and subtracting the number of lone pair electrons and one-half the number of bonding electrons.

	H_{left}	H_{top}	H_{right}	H_{bottom}	N
number of valence electrons	1	1	1	1	5
- number of lone pair electrons	0	0	0	0	0
- 1/2(number of bonding electrons)	1	1	1	1	4
Formal charge	0	0	0	0	+1

The sum of the formal charges is + 1 which is the overall charge of the ion.

25.

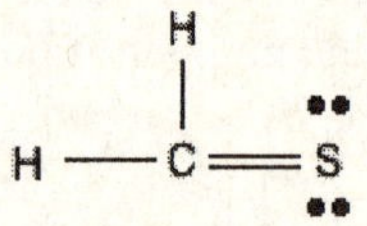

I

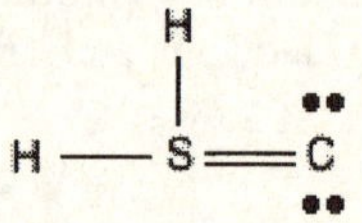

II

Calculate the formal charge on each atom in structure I by finding the number of valence electrons and subtracting the number of lone pair electrons and one-half the number of bonding electrons.

	H_{left}	H_{top}	C	S
number of valence electrons	1	1	4	6
- number of lone pair electrons	0	0	0	4
- 1/2(number of bonding electrons)	1	1	4	2
Formal charge	0	0	0	0

The sum of the formal charges is 0 which is the overall charge of the molecule.

Calculate the formal charge on each atom in structure II by finding the number of valence electrons and subtracting the number of lone pair electrons and one-half the number of bonding electrons

	H_{left}	H_{top}	S	C
number of valence electrons	1	1	6	4
- number of lone pair electrons	0	0	0	4
- 1/2(number of bonding electrons)	1	1	4	2
Formal charge	0	0	+2	– 2

The sum of the formal charges is 0 which is the overall charge of the molecule.

Structure I is the better Lewis structure because it has the least amount of formal charge on each atom.

26.

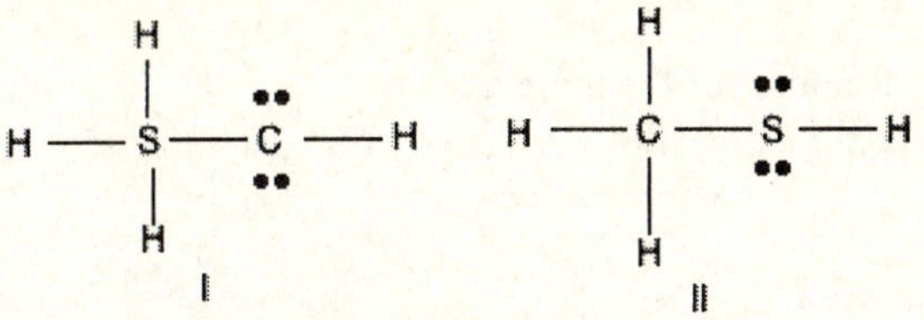

Calculate the formal charge on each atom in structure I by finding the number of valence electrons and subtracting the number of lone pair electrons and one-half the number of bonding electrons.

	H_{left}	H_{top}	H_{right}	H_{bottom}	S	C
number of valence electrons	1	1	1	1	6	4
- number of lone pair electrons	0	0	0	0	0	4
- 1/2(number of bonding electrons)	1	1	1	1	4	2
Formal charge	0	0	0	0	+2	– 2

The sum of the formal charges is 0 which is the overall charge of the molecule.
Calculate the formal charge on each atom in structure by finding the number of valence electrons and subtracting the number of lone pair electrons and one-half the number of bonding electrons.

	H_{left}	H_{top}	H_{right}	H_{bottom}	C	S
number of valence electrons	1	1	1	1	4	6
- number of lone pair electrons	0	0	0	0	0	4
- 1/2(number of bonding electrons)	1	1	1	1	4	2
Formal charge	0	0	0	0	0	0

The sum of the formal charges is 0 which is the overall charge of the molecule.
Structure II is the better Lewis structure because it has the least amount of formal charge on each atom.

27. 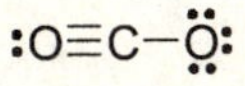 does not provide a significant contribution to the resonance hybrid as it has a +1 formal charge on a very electronegative oxygen.

	O_{left}	O_{right}	C
number of valence electrons	6	6	4
- number of lone pair electrons	2	6	0
- 1/2(number of bonding electrons)	3	1	4
Formal charge	+1	–1	0

28. Compare the two forms of each molecule with O as a central atom and a terminal atom. Determine the formal charge on the central atom for all the structures.

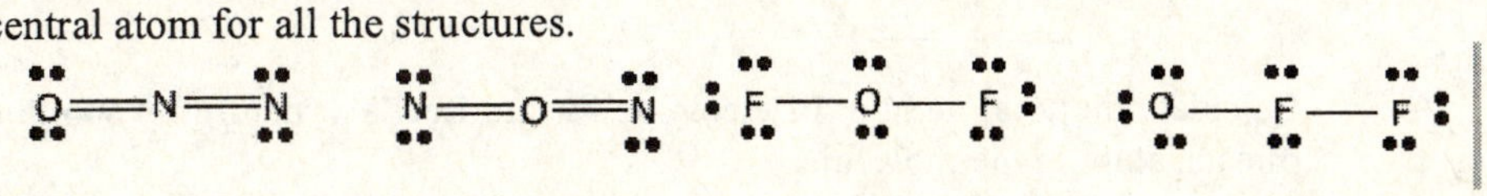

I II III IV

	I	II	III	IV
	N	O	O	F
number of valence electrons	5	6	6	7
- number of lone pair electrons	0	0	4	4
- 1/2(number of bonding electrons)	3	4	4	2
Formal charge	+1	+2	0	+1

We can see from the formal charge on the central atom, O has to be terminal for the N_2O molecule. When O is the central atom it has a +2 formal charge and it is the more electronegative atom. So, this would not be a good structure. For the OF_2 molecule, O has to be the central atom. When O is central it has a formal charge of 0, when the F is central it has a formal charge of +1. This puts a positive formal charge on the most electronegative atom, which is not acceptable.

29. a) BCl_3: Write the correct skeletal structure for the molecule.
B is the less electronegative, so it is central

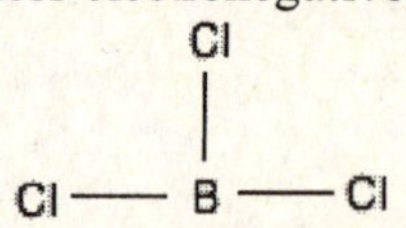

Calculate the total number of electrons for the Lewis structure by summing the valence electrons of each atom in the molecule

(number of valence e^- for B) + 3(number of valence e^- for Cl) = 3 +3(7) = 24

Distribute the electrons among the atoms, giving octets (or duets for H) to as many atoms as possible. Begin with the bonding electrons, and then proceed to lone pairs on terminal atoms and finally to lone pairs of the central atom.

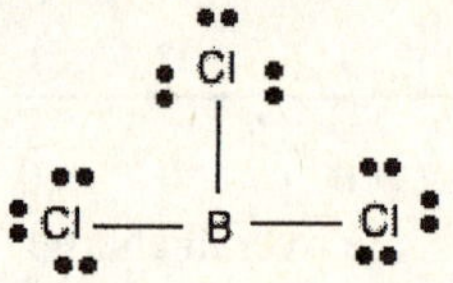

All 24 valence electrons are used.

B has an incomplete octet. If we complete the octet, there is a formal charge of – 1 on the B which is less electronegative than Cl.

b) NO_2: Write the correct skeletal structure for the molecule.
N is the less electronegative, so it is central

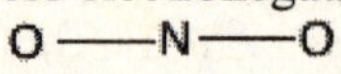

Calculate the total number of electrons for the Lewis structure by summing the valence electrons of each atom in the molecule

(number of valence e^- for N) + 2(number of valence e^- for O) = 5 +2(6) = 17

Distribute the electrons among the atoms, giving octets (or duets for H) to as many atoms as possible. Begin with the bonding electrons, and then proceed to lone pairs on terminal atoms and finally to lone pairs of the central atom.

All 17 valence electrons are used.

N has an incomplete octet. It has 7 electrons because we have an odd number of valence electrons.

c) BH_3: Write the correct skeletal structure for the molecule.
B is the less electronegative, so it is central

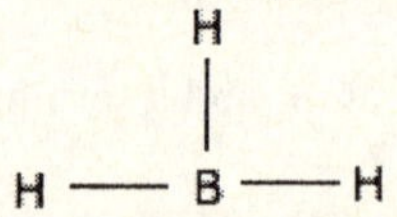

Calculate the total number of electrons for the Lewis structure by summing the valence electrons of each atom in the molecule

(number of valence e^- for B) + 3(number of valence e^- for H) = 3 +3(1) = 6

Distribute the electrons among the atoms, giving octets (or duets for H) to as many atoms as possible. Begin with the bonding electrons, and then proceed to lone pairs on terminal atoms and finally to lone pairs of the central atom.

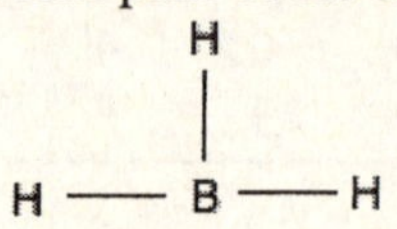

All 6 valence electrons are used.

B has an incomplete octet. H cannot double bond, so it is not possible to complete the octet on B with a double bond.

30. a) BBr_3: Write the correct skeletal structure for the molecule.
B is the less electronegative, so it is central

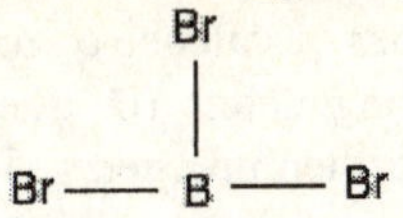

Calculate the total number of electrons for the Lewis structure by summing the valence electrons of each atom in the molecule

(number of valence e^- for B) + 3(number of valence e^- for Br) = 3 +3(7) = 24

Distribute the electrons among the atoms, giving octets (or duets for H) to as many atoms as possible. Begin with the bonding electrons, and then proceed to lone pairs on terminal atoms and finally to lone pairs of the central atom.

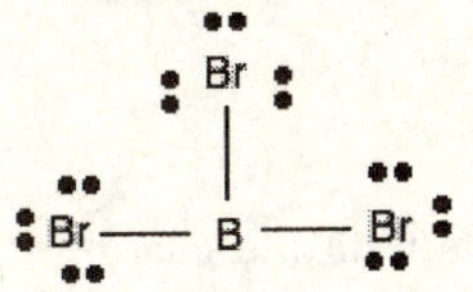

All 24 valence electrons are used.

B has an incomplete octet. If we complete the octet, there is a formal charge of – 1 on the B which is less electronegative than Br.

b) NO: Write the correct skeletal structure for the molecule.

Calculate the total number of electrons for the Lewis structure by summing the valence electrons of each atom in the molecule

(number of valence e^- for N) + (number of valence e^- for O) = 5 +6 = 11

Distribute the electrons among the atoms, giving octets (or duets for H) to as many atoms as possible. Begin with the bonding electrons, and then proceed to lone pairs on terminal atoms and finally to lone pairs of the central atom.

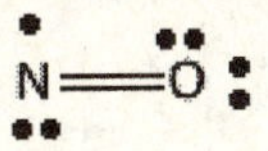

All 11 valence electrons are used.

N has an incomplete octet. It has 7 electrons because we have an odd number of valence electrons.

c) ClO_2: Write the correct skeletal structure for the molecule.
Cl is less electronegative so it is central

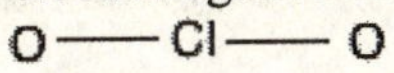

Calculate the total number of electrons for the Lewis structure by summing the valence electrons of each atom in the molecule

(number of valence e^- for Cl) + 2(number of valence e^- for O) = 5 +2(6) = 17

Distribute the electrons among the atoms, giving octets (or duets for H) to as many atoms as possible. Begin with the bonding electrons, and then proceed to lone pairs on terminal atoms and finally to lone pairs of the central atom.

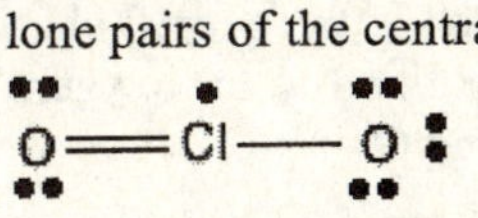

All 17 valence electrons are used.

Cl has an incomplete octet. It has 7 electrons because we have an odd number of valence electrons.

31. a) PO_4^{3-}: Write the correct skeletal structure for the ion.

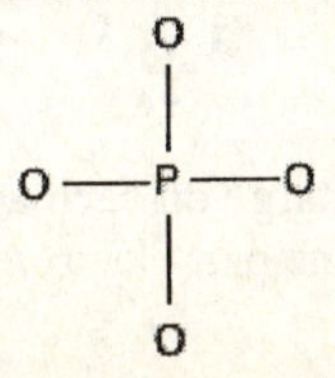

Calculate the total number of electrons for the Lewis structure by summing the valence electrons of each atom in the ion and adding 3 for the 3 – charge.

$$4(\text{number of valence } e^- \text{ for O}) + (\text{number of valence } e^- \text{ for P}) + 3 = 4(6) + 5 + 3 = 32$$

Distribute the electrons among the atoms, giving octets (or duets for H) to as many atoms as possible. Begin with the bonding electrons, and then proceed to lone pairs on terminal atoms and finally to lone pairs of the central atom.

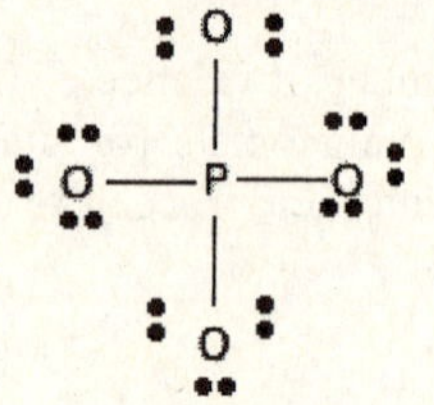

All 32 valence electrons are used.

Lastly, write the Lewis structure in brackets with the charge of the ion in the upper right-hand corner.

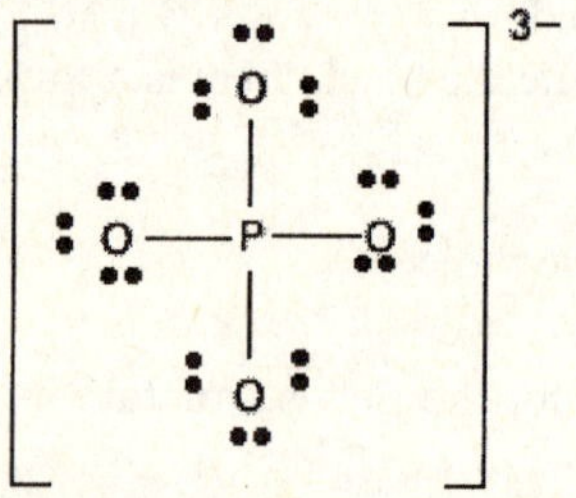

All atoms have octets, duets for H, structure is complete.

Calculate the formal charge on each atom by finding the number of valence electrons and subtracting the number of lone pair electrons and one-half the number of bonding electrons.

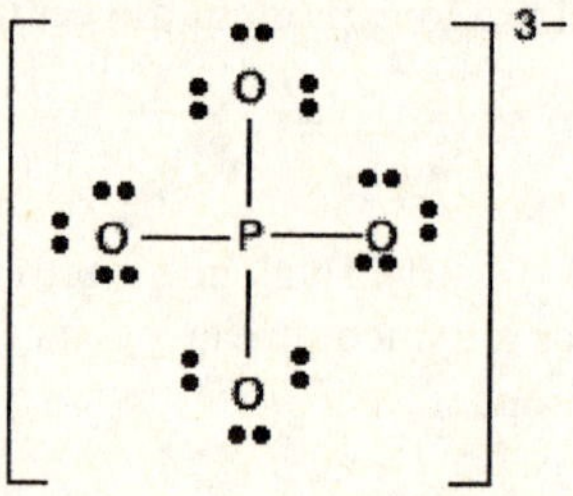

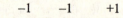

	O_{left}	O_{top}	O_{right}	O_{bottom}	P
number of valence electrons	6	6	6	6	5
- number of lone pair electrons	6	6	6	6	0
- 1/2(number of bonding electrons)	1	1	1	1	4
Formal charge	–1	–1	–1	–1	+1

The sum of the formal charges is –3, which is the overall charge of the ion. However, we can write a resonance structure with a double bond to an oxygen because P can expand its octet. This leads to lower formal charges on P and O.

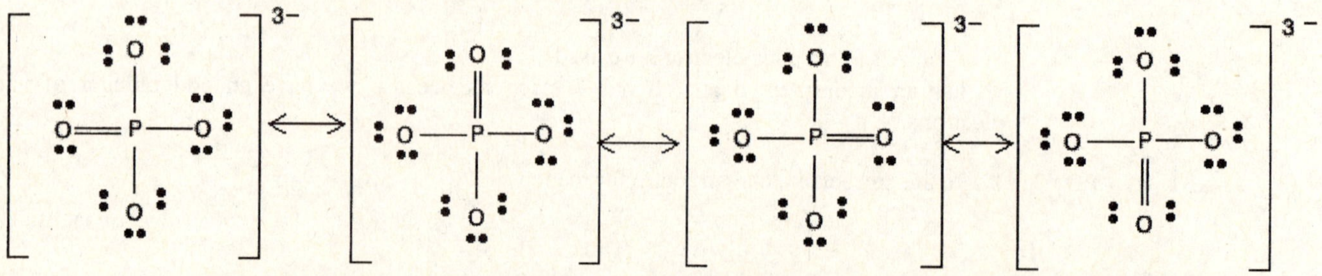

Using the leftmost structure calculate the formal charge on each atom by finding the number of valence electrons and subtracting the number of lone pair electrons and one-half the number of bonding electrons.

	O_{left}	O_{top}	O_{right}	O_{bottom}	P
number of valence electrons	6	6	6	6	5
- number of lone pair electrons	4	6	6	6	0
- 1/2(number of bonding electrons)	2	1	1	1	5
Formal charge	0	–1	–1	–1	0

The sum of the formal charges is –3, which is the overall charge of the ion. These resonance forms would all have the lower formal charges associated with the double bonded O and P.

b) CN^-: Write the correct skeletal structure for the ion.

Calculate the total number of electrons for the Lewis structure by summing the valence electrons of each atom in the ion and adding 1 for the 1 – charge.

(number of valence e^- for C) + (number of valence e^- for N) + 1 = 4 + 5 + 1 = 10

Distribute the electrons among the atoms, giving octets (or duets for H) to as many atoms as possible. Begin with the bonding electrons, and then proceed to lone pairs on terminal atoms and finally to lone pairs of the central atom.

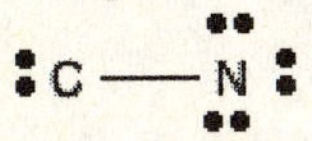

All 10 valence electrons are used.

If any atom lacks an octet, form double or triple bonds as necessary to give them octets.

Lastly, write the Lewis structure in brackets with the charge of the ion in the upper right-hand corner.

All atoms have octets, duets for H, structure is complete.

Calculate the formal charge on each atom by finding the number of valence electrons and subtracting the number of lone pair electrons and one-half the number of bonding electrons.

	C	N
number of valence electrons	4	5
- number of lone pair electrons	2	2
- 1/2(number of bonding electrons)	3	3
Formal charge	–1	0

The sum of the formal charges is – 1, which is the overall charge of the ion

c) SO_3^{2-}: Write the correct skeletal structure for the ion.

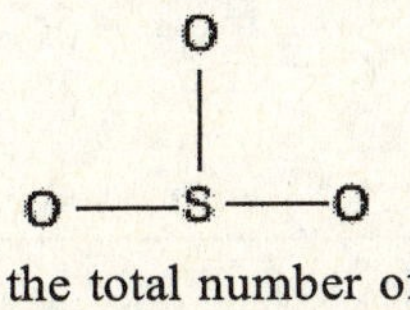

Calculate the total number of electrons for the Lewis structure by summing the valence electrons of each atom in the ion and adding 2 for the 2 – charge.

3(number of valence e^- for O) + (number of valence e^- for S) + 2 = 3(6) + 6 + 2 = 26

Distribute the electrons among the atoms, giving octets (or duets for H) to as many atoms as possible. Begin with the bonding electrons, and then proceed to lone pairs on terminal atoms and finally to lone pairs of the central atom.

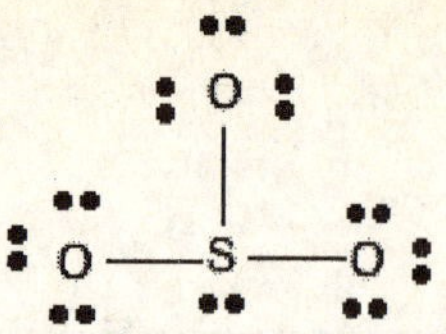

All 26 valence electrons are used.

Lastly, write the Lewis structure in brackets with the charge of the ion in the upper right-hand corner.

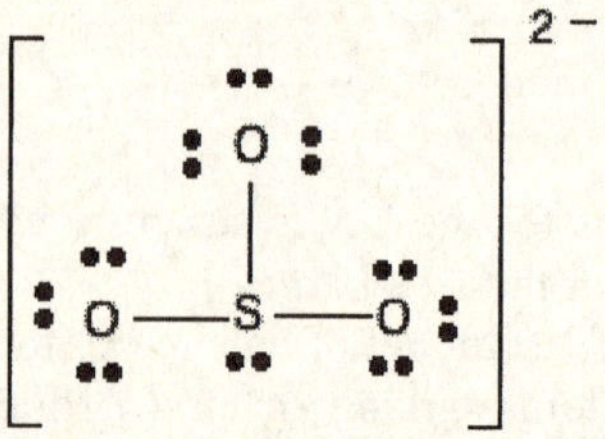

Calculate the formal charge on each atom by finding the number of valence electrons and subtracting the number of lone pair electrons and one-half the number of bonding electrons.

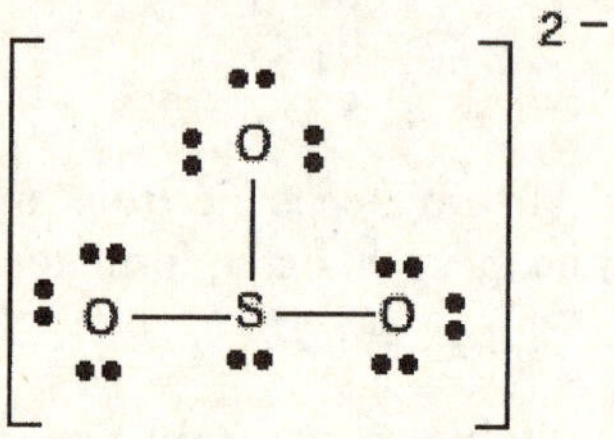

	O_{left}	O_{top}	O_{right}	S
number of valence electrons	6	6	6	6
- number of lone pair electrons	6	6	6	2
- 1/2(number of bonding electrons)	1	1	1	3
Formal charge	–1	–1	–1	+1

The sum of the formal charges is –2, which is the overall charge of the ion. However, we can write a resonance structure with a double bond to an oxygen because S can expand its octet. This leads to a lower formal charge

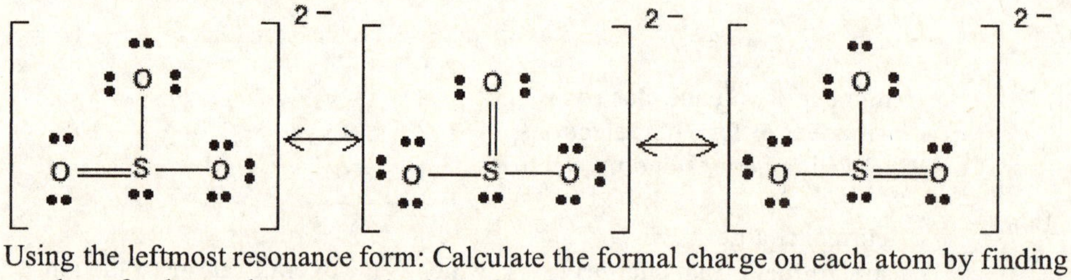

Using the leftmost resonance form: Calculate the formal charge on each atom by finding the number of valence electrons and subtracting the number of lone pair electrons and one-half the number of bonding electrons.

	O_{left}	O_{top}	O_{right}	S
number of valence electrons	6	6	6	6
- number of lone pair electrons	4	6	6	2
- 1/2(number of bonding electrons)	2	1	1	4
Formal charge	0	–1	–1	0

The sum of the formal charges is –2, which is the overall charge of the ion. These resonance forms would all have the lower formal charge on the double bonded O and S.

d) ClO_2^-: Write the correct skeletal structure for the ion.

O — Cl — O

Calculate the total number of electrons for the Lewis structure by summing the valence electrons of each atom in the ion and adding 1 for the 1 – charge.

$$2(\text{number of valence } e^- \text{ for O}) + (\text{number of valence } e^- \text{ for Cl}) + 1 = 2(6) + 7 + 1 = 20$$

Distribute the electrons among the atoms, giving octets (or duets for H) to as many atoms as possible. Begin with the bonding electrons, and then proceed to lone pairs on terminal atoms and finally to lone pairs of the central atom.

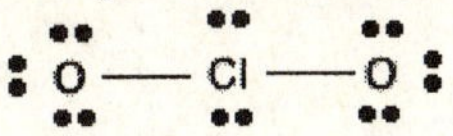

All 20 valence electrons are used.

Lastly, write the Lewis structure in brackets with the charge of the ion in the upper right-hand corner.

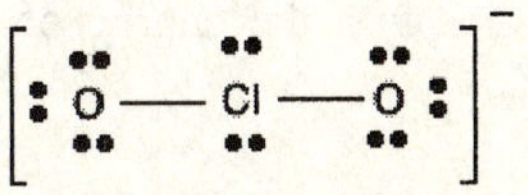

All atoms have octets, duets for H, structure is complete.

Calculate the formal charge on each atom by finding the number of valence electrons and subtracting the number of lone pair electrons and one-half the number of bonding electrons.

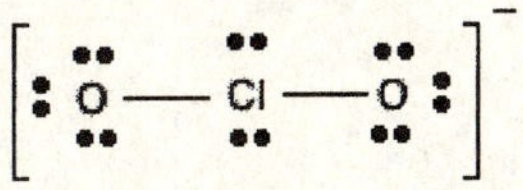

	O_{left}	O_{right}	Cl
number of valence electrons	6	6	7
- number of lone pair electrons	6	6	4
- 1/2(number of bonding electrons)	1	1	2
Formal charge	–1	–1	+1

The sum of the formal charges is –1, which is the overall charge of the ion. However, we can write a resonance structure with a double bond to an oxygen because Cl can expand its octet. This leads to a lower formal charge

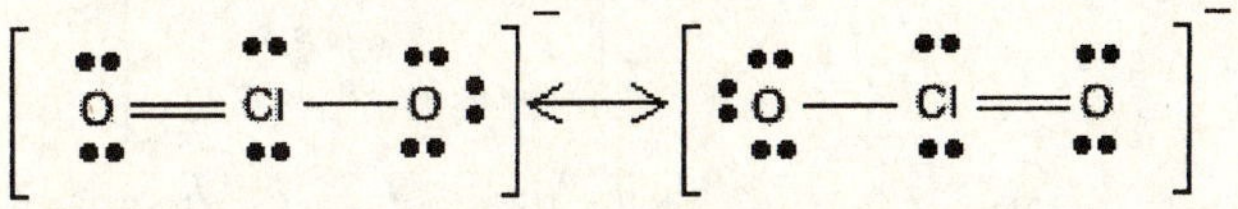

Using the leftmost resonance form: Calculate the formal charge on each atom by finding the number of valence electrons and subtracting the number of lone pair electrons and one – half the number of bonding electrons.

	O_{left}	O_{right}	Cl
number of valence electrons	6	6	7
- number of lone pair electrons	4	6	4
- 1/2(number of bonding electrons)	2	1	3
Formal charge	0	–1	0

The sum of the formal charges is –1, which is the overall charge of the ion. These resonance forms would all have the lower formal charge on the double bonded O and Cl.

32. a) SO_4^{2-}: Write the correct skeletal structure for the ion.

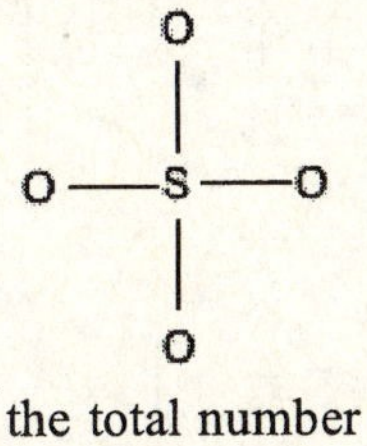

Calculate the total number of electrons for the Lewis structure by summing the valence electrons of each atom in the molecule and adding 2 for the 2 – charge.

$$4(\text{number of valence e}^- \text{ for O}) + (\text{number of valence e}^- \text{ for S}) + 2 = 4(6) + 6 + 2 = 32$$

Distribute the electrons among the atoms, giving octets (or duets for H) to as many atoms as possible. Begin with the bonding electrons, and then proceed to lone pairs on terminal atoms and finally to lone pairs of the central atom.

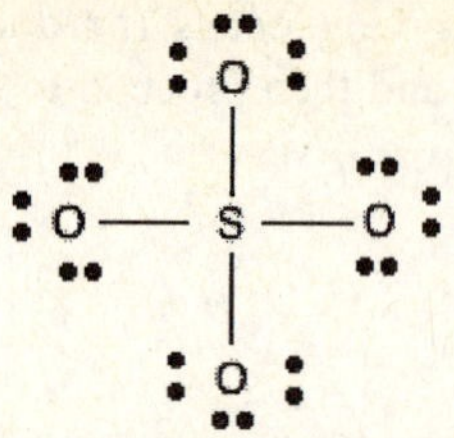

All 32 valence electrons are used.

Lastly, write the Lewis structure in brackets with the charge of the ion in the upper right-hand corner.

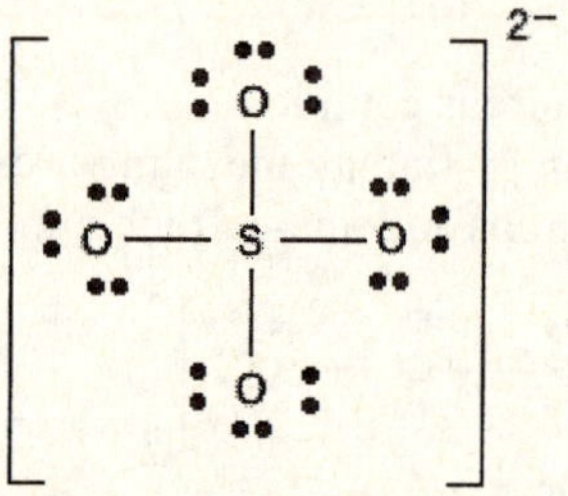

All atoms have octets, duets for H, structure is complete.

Calculate the formal charge on each atom by finding the number of valence electrons and subtracting the number of lone pair electrons and one-half the number of bonding electrons.

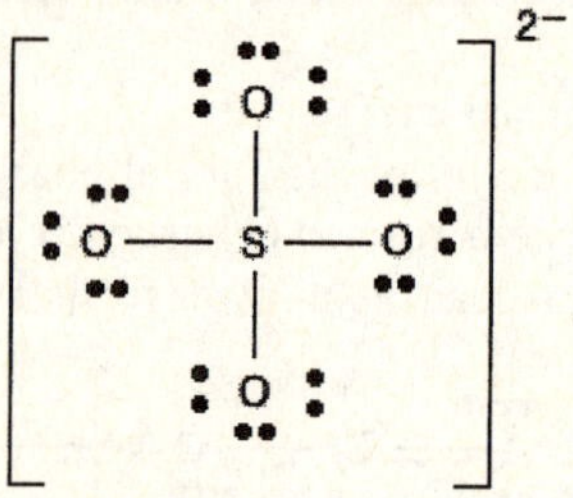

	O_{left}	O_{top}	O_{right}	O_{bottom}	S
number of valence electrons	6	6	6	6	6
- number of lone pair electrons	6	6	6	6	0
- 1/2(number of bonding electrons)	1	1	1	1	4
Formal charge	−1	−1	− 1	− 1	+2

The sum of the formal charges is – 2, which is the overall charge of the ion. However, we can write a resonance structure with double bonds to two oxygen atoms because S can expand its octet. This leads to lower formal charges.

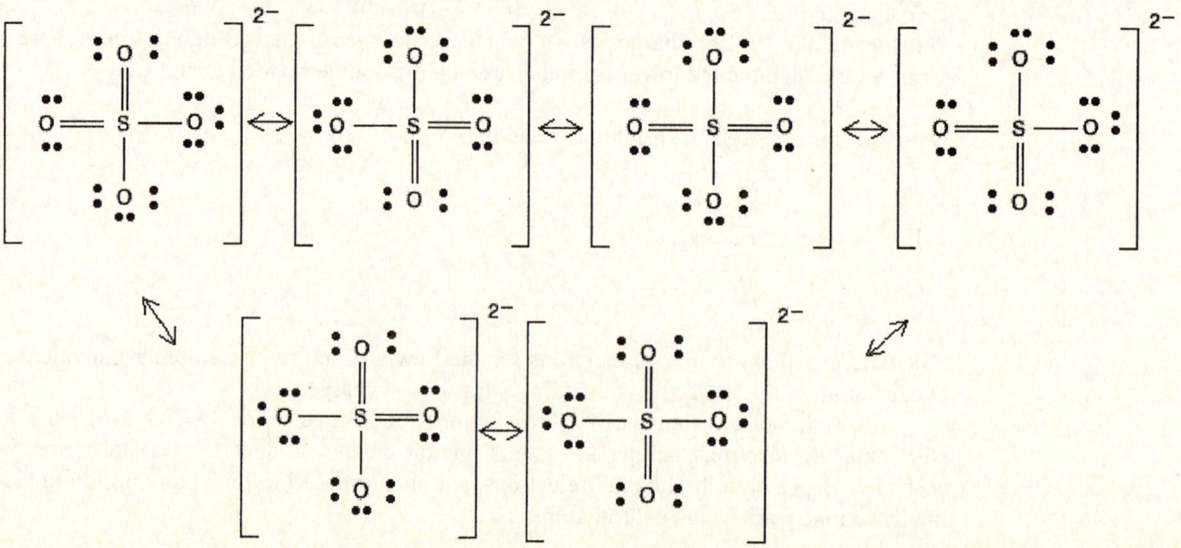

Using the leftmost resonance form: Calculate the formal charge on each atom by finding the number of valence electrons and subtracting the number of lone pair electrons and one-half the number of bonding electrons.

	O_{left}	O_{top}	O_{right}	O_{bottom}	S
number of valence electrons	6	6	6	6	6
- number of lone pair electrons	4	4	6	6	0
- 1/2(number of bonding electrons)	2	2	1	1	6
Formal charge	0	0	− 1	− 1	0

The sum of the formal charges is −2, which is the overall charge of the ion. These resonance forms would all have the lower formal charges on the double bonded O and S.

b) HSO_4^-: Write the correct skeletal structure for the ion

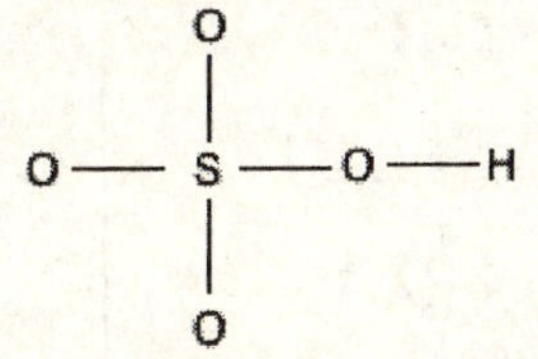

Calculate the total number of electrons for the Lewis structure by summing the valence electrons of each atom in the ion and adding 1 for the 1 – charge.

4(number of valence e^- for O) + (number of valence e^- for S) + (number of valence e^- for H) + 1 = 4(6) + 6 + 1 +1 = 32

Distribute the electrons among the atoms, giving octets (or duets for H) to as many atoms as possible. Begin with the bonding electrons, and then proceed to lone pairs on terminal atoms and finally to lone pairs of the central atom.

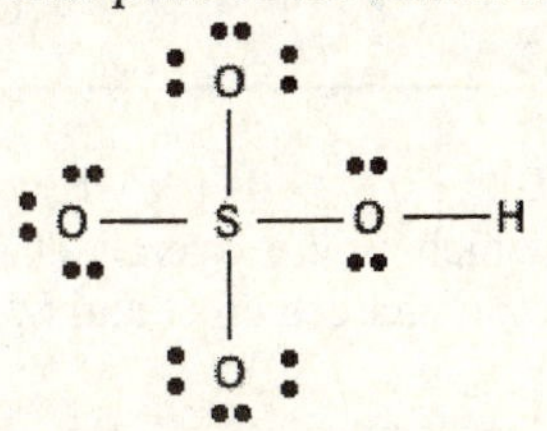

All 32 valence electrons are used.

Lastly, write the Lewis structure in brackets with the charge of the ion in the upper right-hand corner.

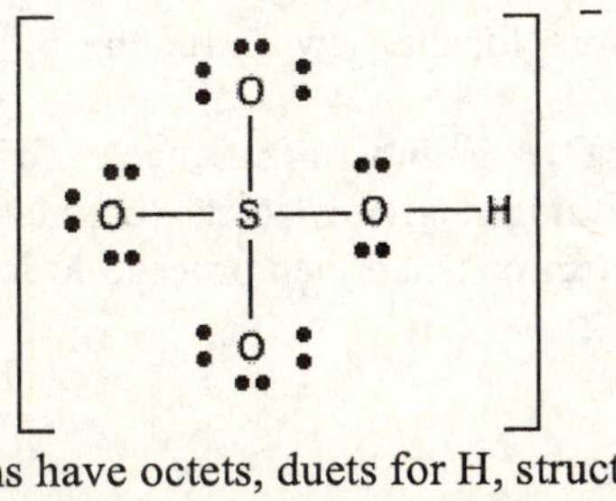

All atoms have octets, duets for H, structure is complete.

Calculate the formal charge on each atom by finding the number of valence electrons and subtracting the number of lone pair electrons and one-half the number of bonding electrons.

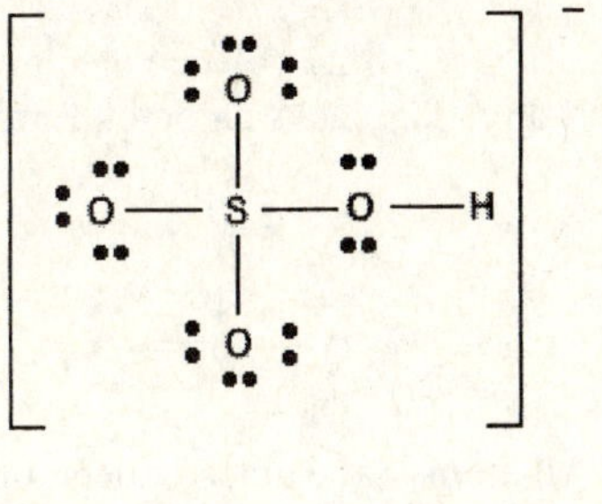

	O_{left}	O_{top}	O_{right}	O_{bottom}	S	H
number of valence electrons	6	6	6	6	6	1
- number of lone pair electrons	6	6	4	6	0	0
- 1/2(number of bonding electrons)	1	1	2	2	4	1
Formal charge	−1	−1	0	−1	+2	0

The sum of the formal charges is – 1, which is the overall charge of the ion. However, we can write a resonance structure with double bonds to two oxygen atoms because S can expand its octet. This leads to a lower formal charges.

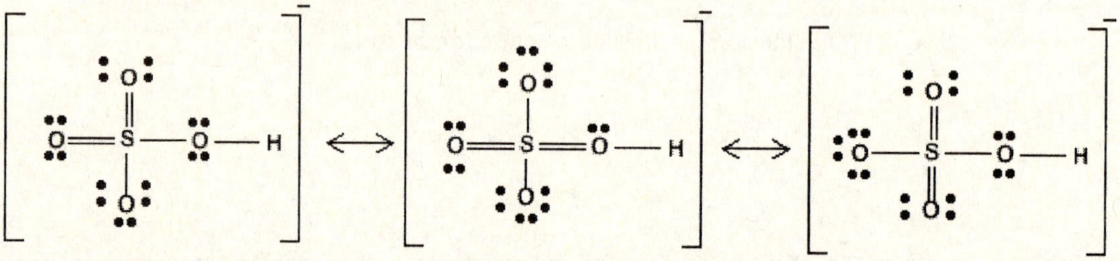

Using the leftmost resonance form: Calculate the formal charge on each atom by finding the number of valence electrons and subtracting the number of lone pair electrons and one-half the number of bonding electrons.

	O_{left}	O_{top}	O_{right}	O_{bottom}	S	H
number of valence electrons	6	6	6	6	6	1
- number of lone pair electrons	4	4	4	6	0	0
- 1/2(number of bonding electrons)	2	2	2	2	6	1
Formal charge	0	0	0	−1	0	0

The sum of the formal charges is −1, which is the overall charge of the ion. Each of these resonance forms would all have lower formal charges on O and S.

c) SO_3 : Write the correct skeletal structure for the molecule

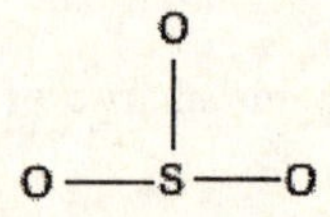

Calculate the total number of electrons for the Lewis structure by summing the valence electrons of each atom in the molecule.

$$3(\text{number of valence } e^- \text{ for O}) + (\text{number of valence } e^- \text{ for S}) = 3(6) + 6 = 24$$

Distribute the electrons among the atoms, giving octets (or duets for H) to as many atoms as possible. Begin with the bonding electrons, and then proceed to lone pairs on terminal atoms and finally to lone pairs of the central atom.

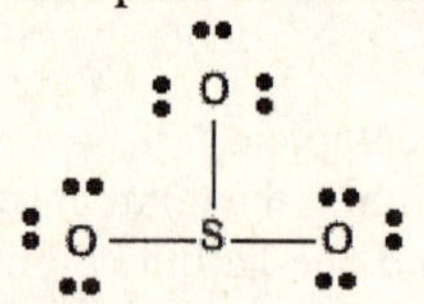

All 24 valence electrons are used.

If any atom lacks an octet, form double or triple bonds as necessary to give them octets.

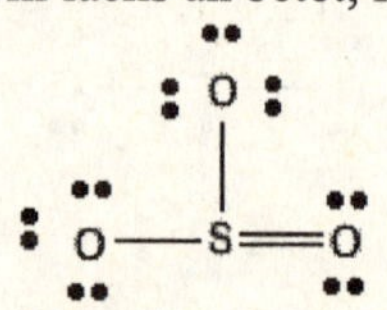

All atoms have octets, duets for H, structure is complete.

Calculate the formal charge on each atom by finding the number of valence electrons and subtracting the number of lone pair electrons and one-half the number of bonding electrons.

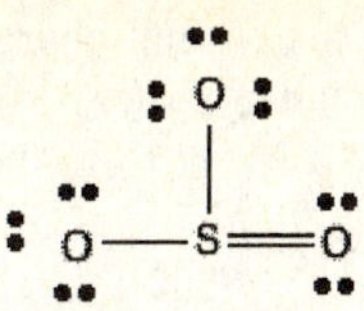

	O_{left}	O_{top}	O_{right}	S
number of valence electrons	6	6	6	6
- number of lone pair electrons	6	6	4	0
- 1/2(number of bonding electrons)	1	1	2	4
Formal charge	−1	−1	0	+2

The sum of the formal charges is 0, which is the overall charge of the molecule. However, we can write a resonance structure with a double bond to all oxygen atoms because S can expand its octet. This leads to lower formal charges.

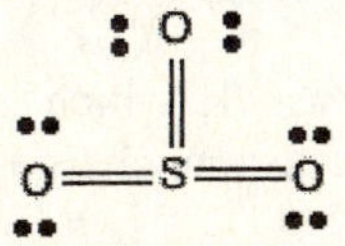

Calculate the formal charge on each atom by finding the number of valence electrons and subtracting the number of lone pair electrons and one-half the number of bonding electrons.

	O_{left}	O_{top}	O_{right}	S
number of valence electrons	6	6	6	6
- number of lone pair electrons	4	4	4	0
- 1/2(number of bonding electrons)	2	2	2	6
Formal charge	0	0	0	0

The sum of the formal charges is 0, which is the overall charge of the molecule. This resonance form would have the lower formal charges on each atom.

d) BrO_2^-: Write the correct skeletal structure for the ion

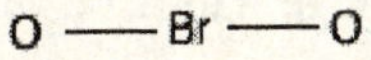

Calculate the total number of electrons for the Lewis structure by summing the valence electrons of each atom in the molecule and adding 1 for the 1 – charge.

2(number of valence e^- for O) + (number of valence e^- for Br) + 1 = 2(6) + 7 + 1 = 20

Distribute the electrons among the atoms, giving octets (or duets for H) to as many atoms as possible. Begin with the bonding electrons, and then proceed to lone pairs on terminal atoms and finally to lone pairs of the central atom.

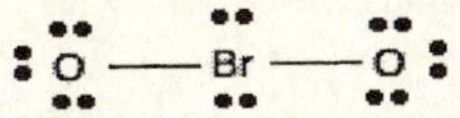

All 20 valence electrons are used.

Lastly, write the Lewis structure in brackets with the charge of the ion in the upper right-hand corner.

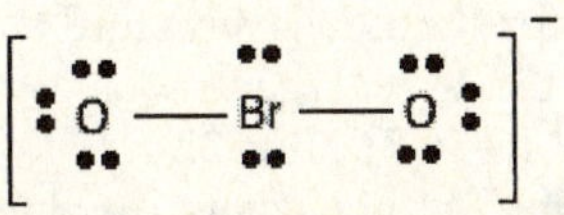

All atoms have octets, duets for H, structure is complete.

Calculate the formal charge on each atom by finding the number of valence electrons and subtracting the number of lone pair electrons and one-half the number of bonding electrons.

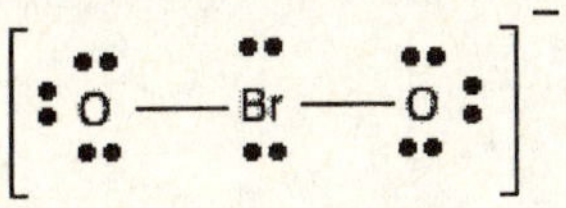

	O_{left}	O_{right}	Br
number of valence electrons	6	6	7
- number of lone pair electrons	6	6	4
- 1/2(number of bonding electrons)	1	1	2
Formal charge	−1	−1	+1

The sum of the formal charges is – 1, which is the overall charge of the ion. However, we can write a resonance structure with a double bond to an oxygen because Br can expand its octet. This leads to a lower formal charge

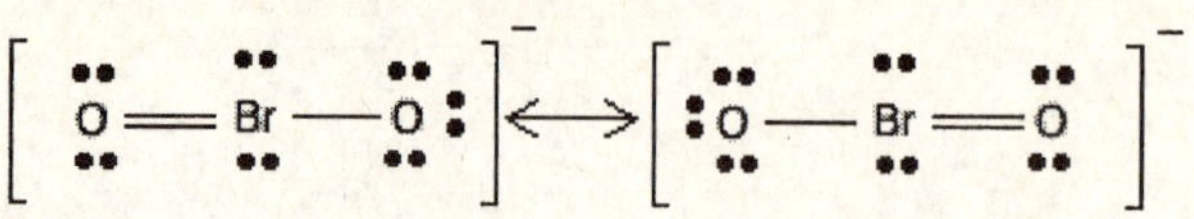

Using the leftmost resonance form: Calculate the formal charge on each atom by finding the number of valence electrons and subtracting the number of lone pair electrons and one – half the number of bonding electrons.

	O_{left}	O_{right}	Br
number of valence electron	6	6	7
- number of lone pair electrons	4	6	4
- 1/2(number of bonding electrons)	2	1	3
Formal charge	0	–1	0

The sum of the formal charges is –1, which is the overall charge of the ion. These resonance forms would both have the lower formal charges on the double bonded O and Br.

33. a) PF_5: Write the correct skeletal structure for the molecule

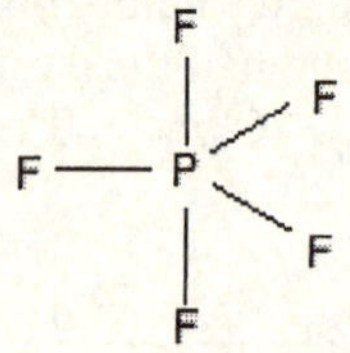

Calculate the total number of electrons for the Lewis structure by summing the valence electrons of each atom in the molecule

(number of valence e^- for P) + 5(number of valence e^- for F) = 5 +5(7) = 40

Distribute the electrons among the atoms, giving octets (or duets for H) to as many atoms as possible. Begin with the bonding electrons, and then proceed to lone pairs on terminal atoms and finally to lone pairs of the central atom. Arrange additional electrons around the central atom, giving it an expanded octet of up to 12 electrons.

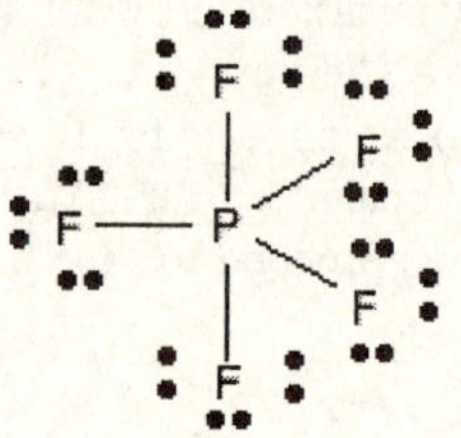

b) I_3^-: Write the correct skeletal structure for the ion

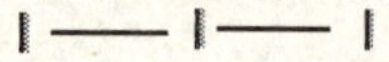

Calculate the total number of electrons for the Lewis structure by summing the valence electrons of each atom in the ion and adding 1 for the 1 – charge.

3(number of valence e^- for I) + 1 = 3(7) + 1 = 22

Distribute the electrons among the atoms, giving octets (or duets for H) to as many atoms as possible. Begin with the bonding electrons, and then proceed to lone pairs on terminal atoms and finally to lone pairs of the central atom. Arrange additional electrons around the central atom, giving it an expanded octet of up to 12 electrons.

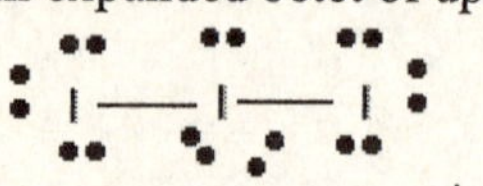

Lastly, write the Lewis structure in brackets with the charge of the ion in the upper right-hand corner.

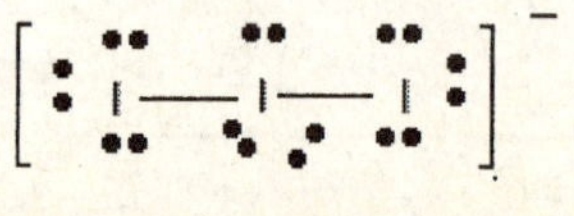

c) SF_4: Write the correct skeletal structure for the molecule

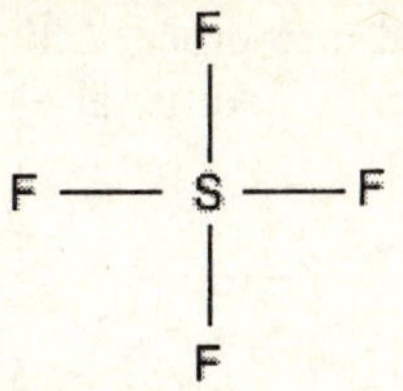

Calculate the total number of electrons for the Lewis structure by summing the valence electrons of each atom in the molecule

(number of valence e^- for S) + 4(number of valence e^- for F) = 6 + 4(7) = 34

Distribute the electrons among the atoms, giving octets (or duets for H) to as many atoms as possible. Begin with the bonding electrons, and then proceed to lone pairs on terminal atoms and finally to lone pairs of the central atom. Arrange additional electrons around the central atom, giving it an expanded octet of up to 12 electrons.

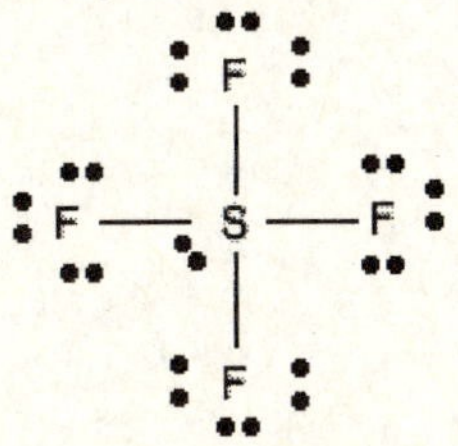

d) GeF_4: Write the correct skeletal structure for the molecule

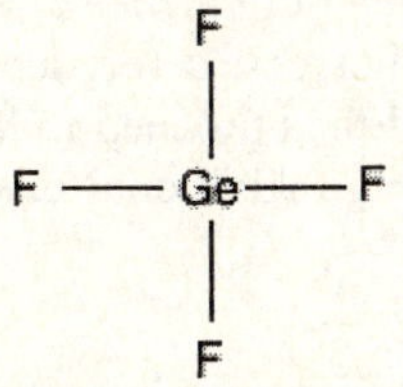

Calculate the total number of electrons for the Lewis structure by summing the valence electrons of each atom in the molecule

(number of valence e^- for Ge) + 4(number of valence e^- for F) = 4 + 4(7) = 32

Distribute the electrons among the atoms, giving octets (or duets for H) to as many atoms as possible. Begin with the bonding electrons, and then proceed to lone pairs on terminal atoms and finally to lone pairs of the central atom. Arrange additional electrons around the central atom, giving it an expanded octet of up to 12 electrons.

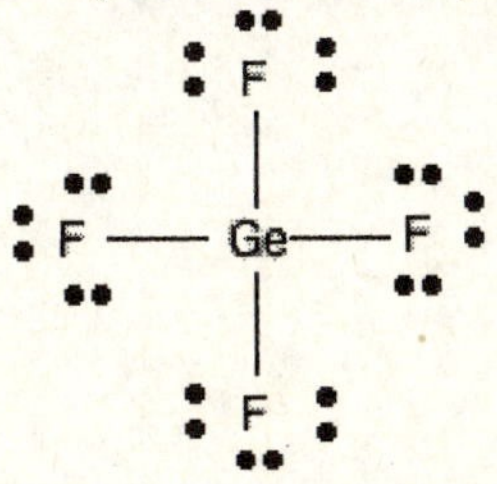

34. a) ClF_5: Write the correct skeletal structure for the molecule

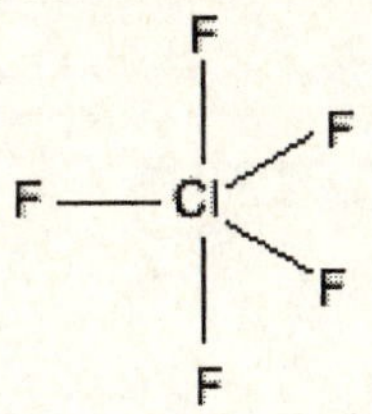

Calculate the total number of electrons for the Lewis structure by summing the valence electrons of each atom in the molecule

(number of valence e^- for Cl) + 5(number of valence e^- for F) = 7 +5(7) = 42

Distribute the electrons among the atoms, giving octets (or duets for H) to as many atoms as possible. Begin with the bonding electrons, and then proceed to lone pairs on terminal atoms and finally to lone pairs of the central atom. Arrange additional electrons around the central atom, giving it an expanded octet of up to 12 electrons.

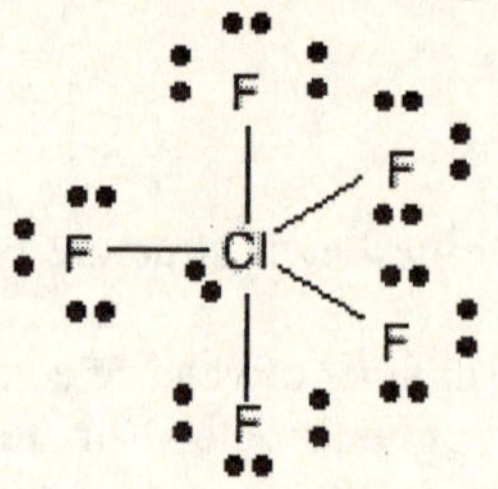

b) AsF_6^- : Write the correct skeletal structure for the ion.

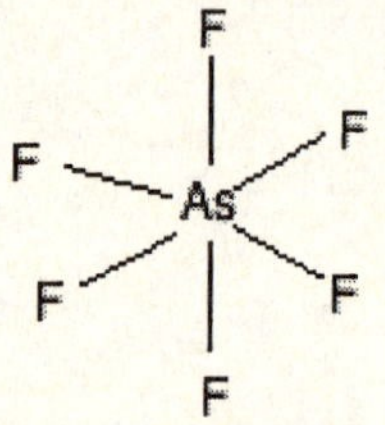

Calculate the total number of electrons for the Lewis structure by summing the valence electrons of each atom in the ion and adding one for the 1 – charge.

(number of valence e^- for As) + 6(number of valence e^- for F) + 1 = 5 +6(7) + 1 = 48

Distribute the electrons among the atoms, giving octets (or duets for H) to as many atoms as possible. Begin with the bonding electrons, and then proceed to lone pairs on terminal atoms and finally to lone pairs of the central atom. Arrange additional electrons around the central atom, giving it an expanded octet of up to 12 electrons.

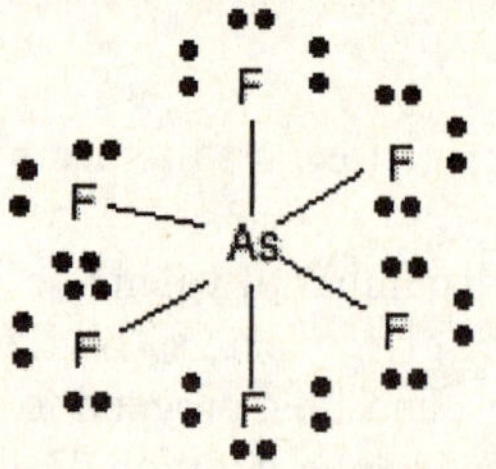

Lastly, write the Lewis structure in brackets with the charge of the ion in the upper right-hand corner.

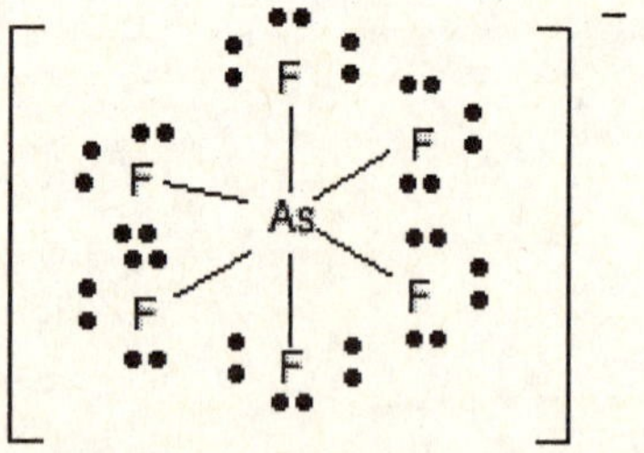

c) Cl_3PO: Write the correct skeletal structure for the molecule.

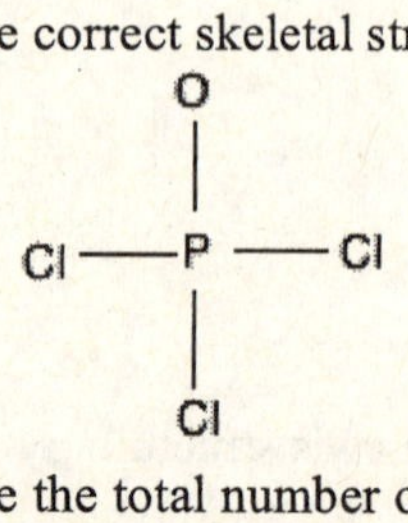

Calculate the total number of electrons for the Lewis structure by summing the valence electron of each atom in the molecule

(number of valence e^- for P) + (number of valence e^- for O) + 3(number of valence e^- for Cl)= 5 +6 + 3(7) = 32

Distribute the electrons among the atoms, giving octets (or duets for H) to as many atoms as possible. Begin with the bonding electrons, and then proceed to lone pairs on terminal atoms and finally to lone pairs of the central atom. Arrange additional electrons around the central atom, giving it an expanded octet of up to 12 electrons.

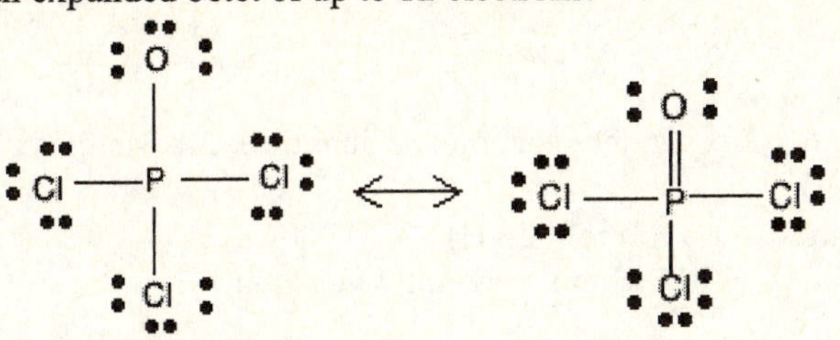

d) IF_5: Write the correct skeletal structure for the molecule

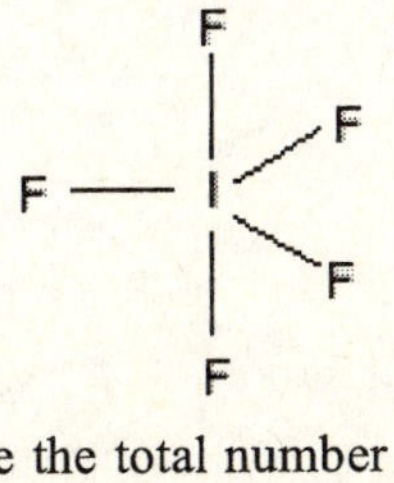

Calculate the total number of electrons for the Lewis structure by summing the valence electrons of each atom in the molecule

(number of valence e^- for I) + 5(number of valence e^- for F) = 7 +5(7) = 42

Distribute the electrons among the atoms, giving octets (or duets for H) to as many atoms as possible. Begin with the bonding electrons, and then proceed to lone pairs on terminal atoms and finally to lone pairs of the central atom. Arrange additional electrons around the central atom, giving it an expanded octet of up to 12 electrons.

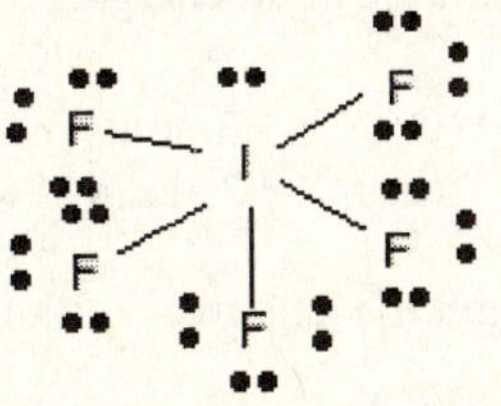

35. Bond strength: $H_3CCH_3 < H_2CCH_2 < HCCH$
Bond length: $H_3CCH_3 > H_2CCH_2 > HCCH$
Write the Lewis structures for the three compounds. Compare the C – C bonds. Triple bonds are stronger than double bonds which are stronger than single bonds. Also, single bonds are longer than double bonds are longer than triple bonds.

HCCH (10 e^-) H_2CCH_2 (12 e^-) H_3CCH_3(14 e^-)

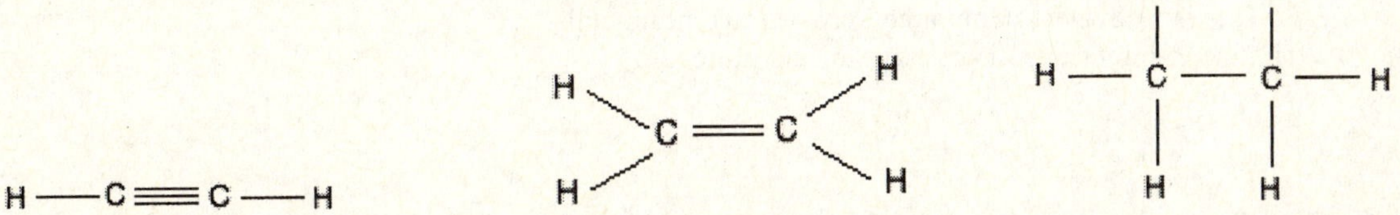

36. Strongest bond: HNNH Shortest bond: HNNH
Write the Lewis structures for the three compounds. Compare the N – N bonds. Triple bonds are stronger than double bonds which are stronger than single bonds. Also, single bonds are longer than double bonds are longer than triple bonds.

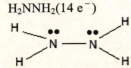

H_2NNH_2(14 e^-) HNNH(12 e^-)

37. Rewrite the reaction using the Lewis structures of the molecules involved.

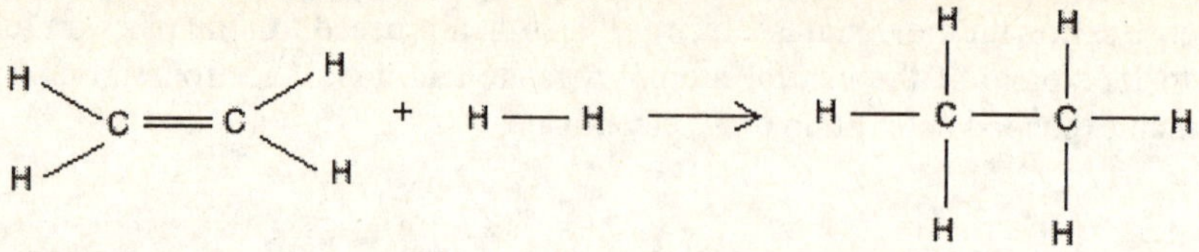

Determine which bonds are broken in the reaction and sum the bond energies of these

$\Sigma(\Delta H\text{'s bonds broken})$
$= 4mol(C - H) + 1mol(C = C) + 1mol(H - H)$
$= 4mol(414\ kJ/mol) + 1mol(611\ kJ/mol) + 1mol(436\ kJ/mol)$
$= 2703\ kJ/mol$

Determine which bonds are formed in the reaction and sum the negatives of the bond energies of these

$\Sigma(-\Delta H\text{'s of bonds formed})$
$= - 6mol(C - H) - 1mol(C - C)$
$= - 6mol(414\ kJ/mol) - 1mol\ (347\ kJ/mol)$
$= -2831\ kJ/mol$

Find ΔH_{rxn} by summing the results of the two steps.

$\Delta H_{rxn} = \Sigma(\Delta H\text{'s bonds broken}) + \Sigma(-\Delta H\text{'s of bonds formed})$
$= 2703\ kJ/mol - 2831\ kJ/mol$
$= - 128\ kJ/mol$

38. Rewrite the reaction using the Lewis structures of the molecules involved.

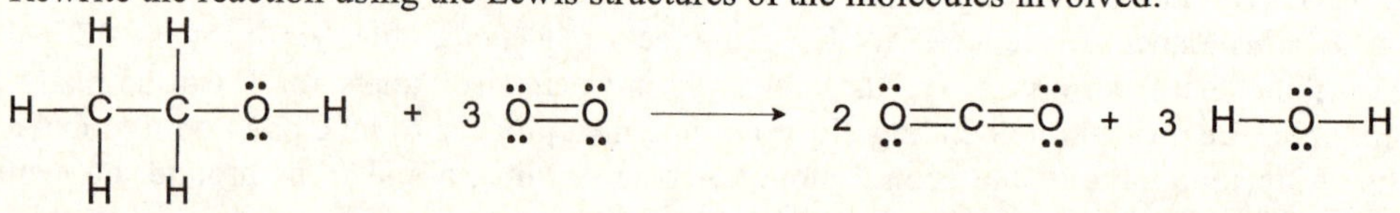

Determine which bonds are broken in the reaction and sum the bond energies of these

$\Sigma(\Delta H\text{'s bonds broken})$
$= 5mol(C - H) + 1mol\ (C - C) + 1mol(C - O) + 1mol\ (O - H) + 3mol(O = O)$
$= 5mol(414\ kJ/mol) + 1mol\ (347\ kJ/mol) + 1mol(350\ kJ/mol) + 1mol(464\ kJ/mol) + 3mol(498kJ/mol)$
$= 4735\ kJ/mol$

Determine which bonds are formed in the reaction and sum the negatives of the bond energies of these

$\Sigma(-\Delta H\text{'s of bonds formed})$
$= - 4mol(C = O) - 6mol(O - H)$
$= - 4mol(799\ kJ/mol) - 6mol(464\ kJ/mol)$
$= -5980\ kJ/mol$

Find ΔH_{rxn} by summing the results of the two steps.

$\Delta H_{rxn} = \Sigma(\Delta H\text{'s bonds broken}) + \Sigma(-\Delta H\text{'s of bonds formed})$
$= 4735\ kJ/mol - 5980\ kJ/mol$
$= - 1245\ kJ/mol$

39. a) BI_3: This is a covalent compound between two nonmetals.
Write the correct skeletal structure for the molecule

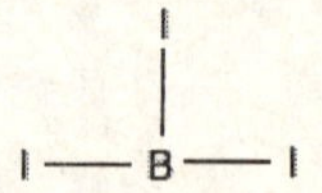

Calculate the total number of electrons for the Lewis structure by summing the valence electrons of each atom in the molecule

(number of valence e^- for B) + (number of valence e^- for I) = 3 +3(7) = 24

Distribute the electrons among the atoms, giving octets (or duets for H) to as many atoms as possible. Begin with the bonding electrons, and then proceed to lone pairs on terminal atoms and finally to lone pairs of the central atom.

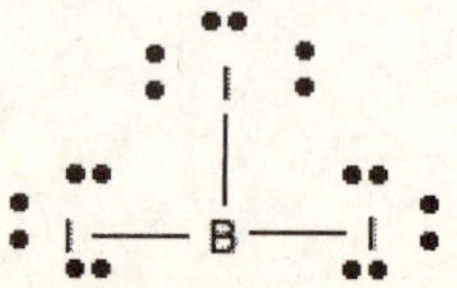

b) K_2S: This is an ionic compound between a metal and nonmetal
Draw the Lewis structures for K and S based on their valence electrons. K: $4s^1$ S: $3s^23p^4$

Potassium must lose one electron and be left with the octet from the previous shell, while sulfur needs to gain two electrons to get an octet.

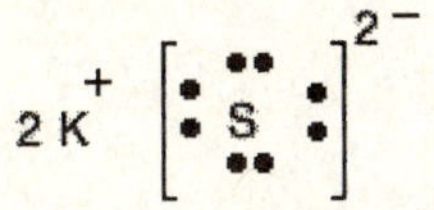

c) HCFO: This is a covalent compound between nonmetals.
Write the correct skeletal structure for the molecule

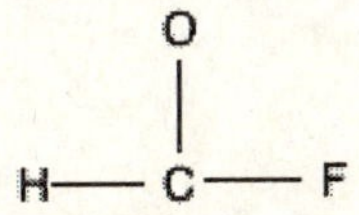

Calculate the total number of electrons for the Lewis structure by summing the valence electrons of each atom in the molecule

(number of valence e^- for H)+(number of valence e^- for C)+(number of valence e^- for F)+(number of valence e^- for O) = 1 + 4 + 7 +6 = 18

Distribute the electrons among the atoms, giving octets (or duets for H) to as many atoms as possible. Begin with the bonding electrons, and then proceed to lone pairs on terminal atoms and finally to lone pairs of the central atom.

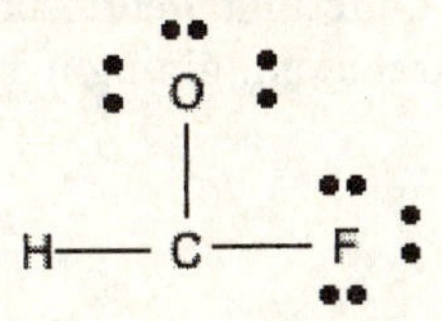

If any atom lacks an octet, form double or triple bonds as necessary to give them octets.

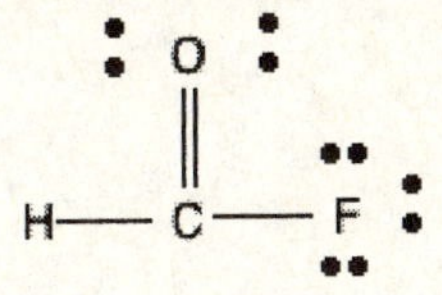

d) PBr_3: This is a covalent compound between two nonmetals.
Write the correct skeletal structure for the molecule

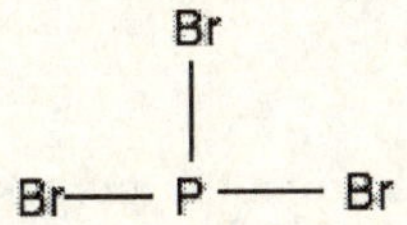

Calculate the total number of electrons for the Lewis structure by summing the valence electrons of each atom in the molecule

(number of valence e^- for P) + 3(number of valence e^- for Br) = 5 + 3(7) = 26

Distribute the electrons among the atoms, giving octets (or duets for H) to as many atoms as possible. Begin with the bonding electrons, and then proceed to lone pairs on terminal atoms and finally to lone pairs of the central atom.

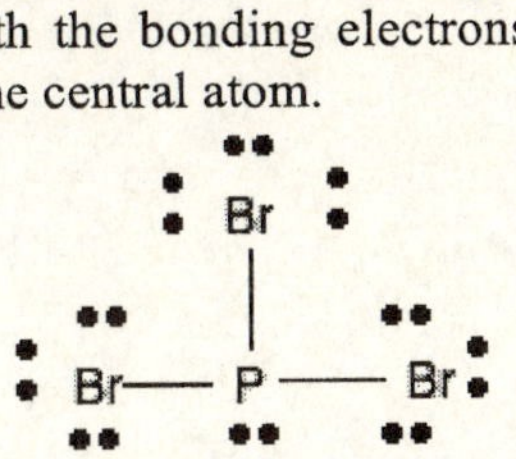

40. a) Al_2O_3: This is an ionic compound between a metal and nonmetal
Draw the Lewis structures for Al and O based on their valence electrons. Al: $3s^23p^1$ O: $2s^22p^4$

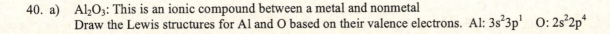

Aluminum must lose three electrons and be left with the octet from the previous shell, while oxygen needs to gain two electrons to get an octet.

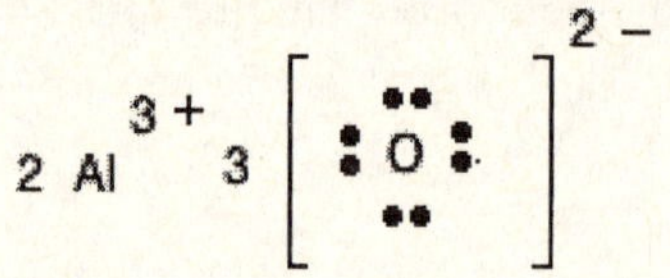

b) ClF_5: This is a covalent compound between two nonmetals.
Write the correct skeletal structure for the molecule

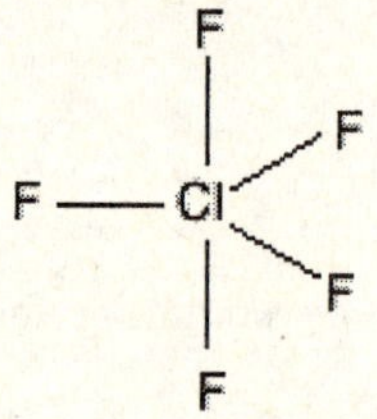

Calculate the total number of electrons for the Lewis structure by summing the valence electrons of each atom in the molecule

(number of valence e^- for Cl) + 5(number of valence e^- for F) = 7 +5(7) = 42

Distribute the electrons among the atoms, giving octets (or duets for H) to as many atoms as possible. Begin with the bonding electrons, and then proceed to lone pairs on terminal atoms and finally to lone pairs of the central atom. Arrange additional electrons around the central atom, giving it an expanded octet of up to 12 electrons.

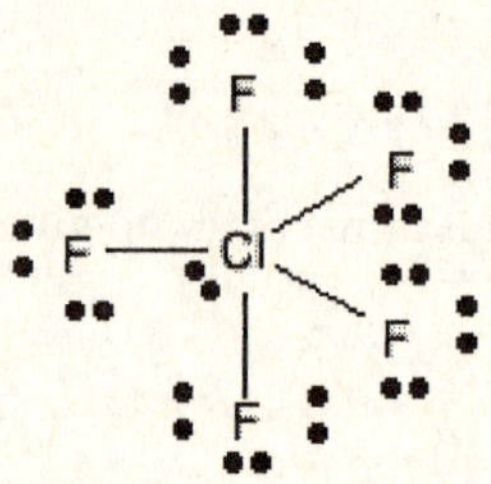

c) MgI_2: This is an ionic compound between a metal and nonmetal
Draw the Lewis structures for Mg and I based on their valence electrons. Mg: $3s^2$ I: $5s^25p^5$

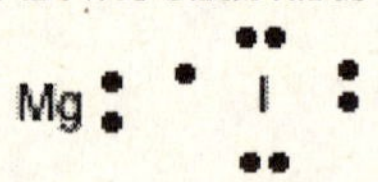

Magnesium must lose two electrons and be left with the octet from the previous shell, while iodine needs to gain one electron to get an octet.

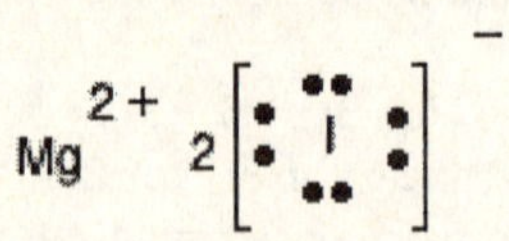

d) XeO_4: This is a covalent compound between two nonmetals.
Write the correct skeletal structure for the molecule

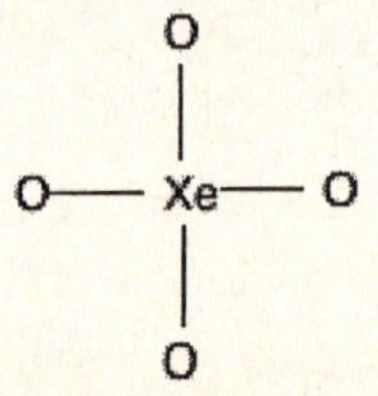

Calculate the total number of electrons for the Lewis structure by summing the valence electrons of each atom in the molecule (number of valence e^- for Xe) + 4(number of valence e^- for O) = 8 + 4(6) = 32

Distribute the electrons among the atoms, giving octets (or duets for H) to as many atoms as possible. Begin with the bonding electrons, and then proceed to lone pairs on terminal atoms and finally to lone pairs of the central atom.

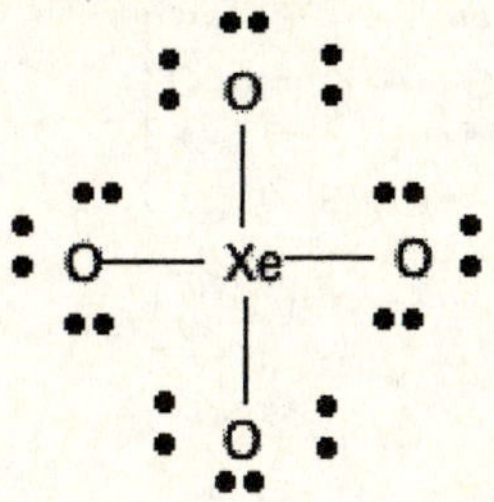

41. a) $BaCO_3$: Ba^{2+}

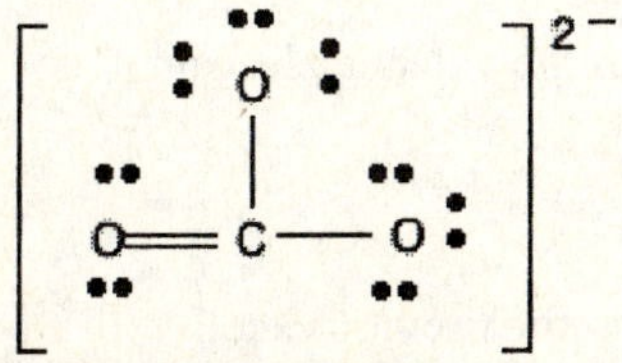

Determine the cation and anion

Ba^{2+} CO_3^{2-}

Write the Lewis structure for the barium cation based on the valence electrons

Ba $5s^2$ Ba^{2+} $5s^0$

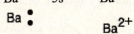

Ba must lose two electrons and be left with the octet from the previous shell

Write the Lewis structure for the covalent anion.

Write the correct skeletal structure for the ion

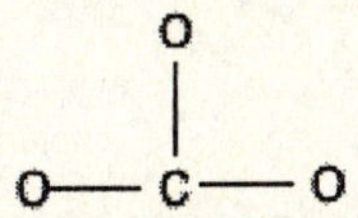

Calculate the total number of electrons for the Lewis structure by summing the valence electrons of each atom in the ion and adding two for the 2 – charge.

(number of valence e^- for C) + 3(number of valence e^- for O) = 4 + 3(6) + 2 = 24

Distribute the electrons among the atoms, giving octets (or duets for H) to as many atoms as possible. Begin with the bonding electrons, and then proceed to lone pairs on terminal atoms and finally to lone pairs of the central atom.

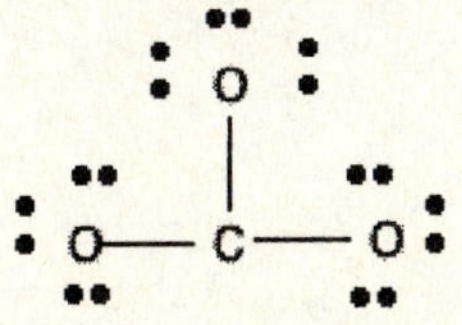

If any atom lacks an octet, form double or triple bonds as necessary to give them octets.

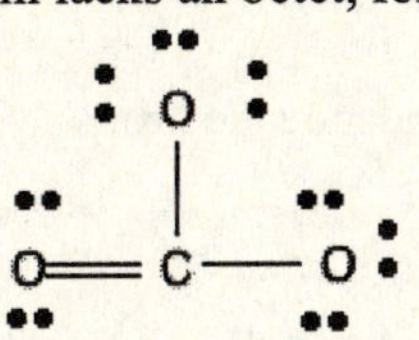

Lastly, write the Lewis structure in brackets with the charge of the ion in the upper right-hand corner.

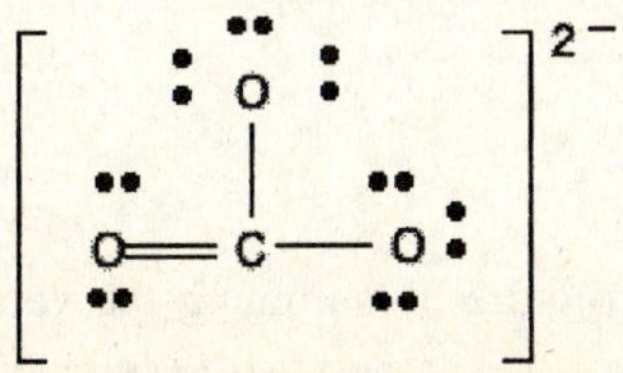

The double bond can be between the C and any of the oxygen atoms, so there are resonance structures

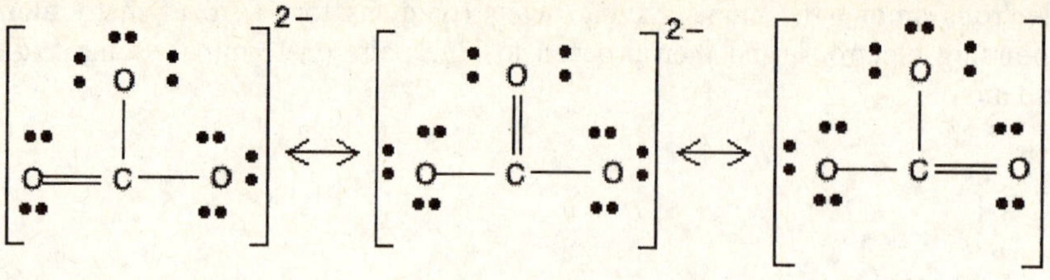

b) $Ca(OH)_2$: Ca^{2+}

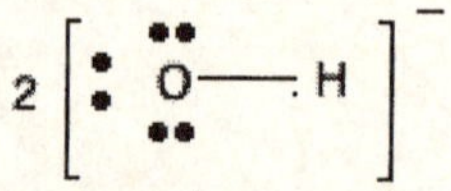

Determine the cation and anion

Ca^{2+} OH^-

Write the Lewis structure for the calcium cation based on the valence electrons

Ca $4s^2$ Ca^{2+} $4s^0$

Ca :

Ca^{2+}

Ca must lose two electrons and be left with the octet from the previous shell

Write the Lewis structure for the covalent anion.

Write the correct skeletal structure for the ion

O—H

Calculate the total number of electrons for the Lewis structure by summing the valence electrons of each atom in the ion and adding one for the 1 – charge.

(number of valence e^- for H) + (number of valence e^- for O) +1 = 1 + 6 +1 = 8

Distribute the electrons among the atoms, giving octets (or duets for H) to as many atoms as possible. Begin with the bonding electrons, and then proceed to lone pairs on terminal atoms and finally to lone pairs of the central atom.

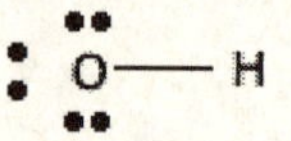

Lastly, write the Lewis structure in brackets with the charge of the ion in the upper right-hand corner.

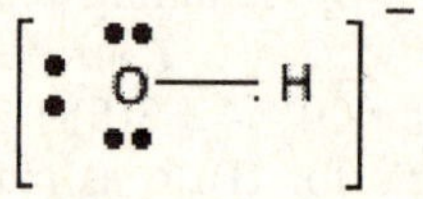

c) KNO_3: K^+

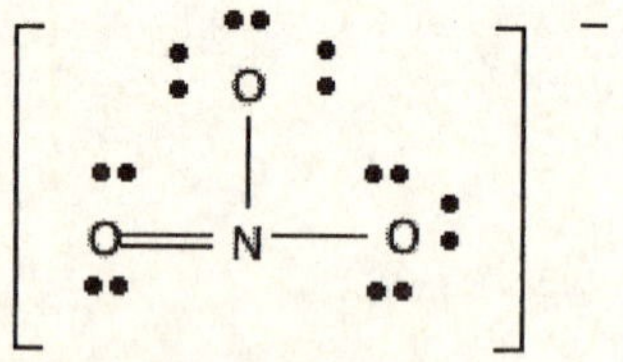

Determine the cation and anion

K^+ NO_3^-

Write the Lewis structure for the potassium cation based on the valence electrons

K $4s^1$ K^+ $4s^0$

K•

K^+

K must lose one electron and be left with the octet from the previous shell

Write the Lewis structure for the covalent anion.

Write the correct skeletal structure for the ion

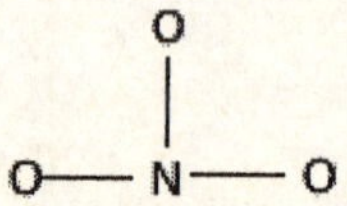

Calculate the total number of electrons for the Lewis structure by summing the valence electrons of each atom in the ion and adding one for the 1 – charge.

(number of valence e^- for N) + (number of valence e^- for O) = 5 + 3(6) +1 = 24

Distribute the electrons among the atoms, giving octets (or duets for H) to as many atoms as possible. Begin with the bonding electrons, and then proceed to lone pairs on terminal atoms and finally to lone pairs of the central atom.

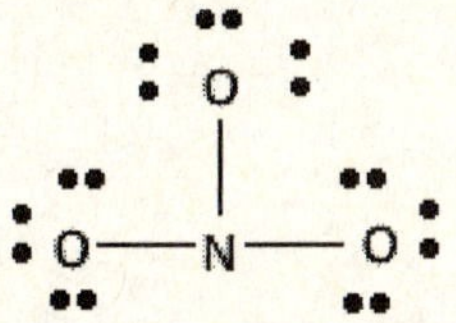

If any atom lacks an octet, form double or triple bonds as necessary to give them octets.

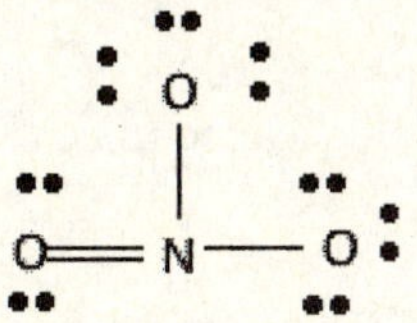

Lastly, write the Lewis structure in brackets with the charge of the ion in the upper right-hand corner.

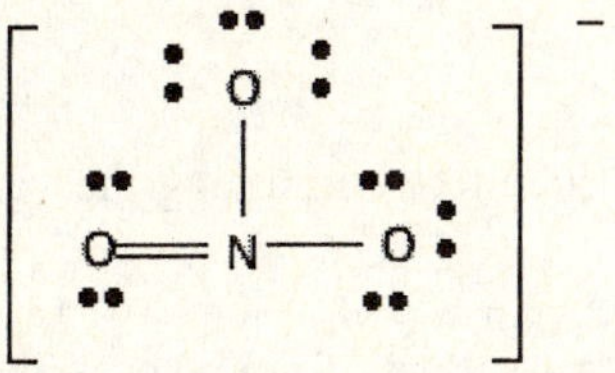

The double bond can be between the N and any of the oxygen atoms, so there are resonance structures

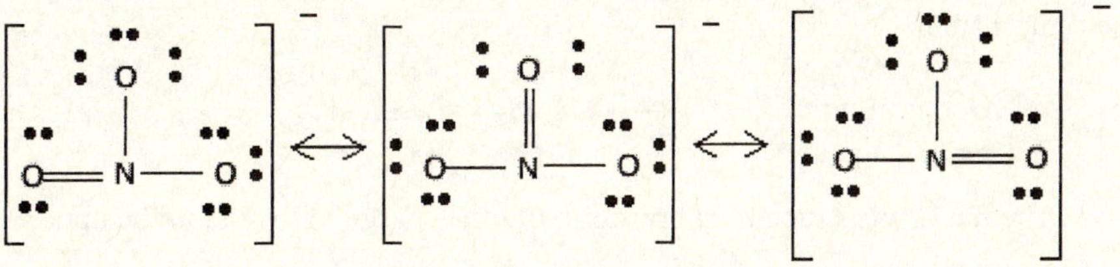

d) LiIO: Li^+

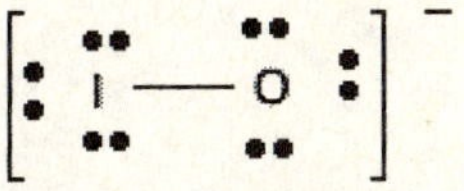

Determine the cation and anion

Li^+ IO^-

Write the Lewis structure for the lithium cation based on the valence electrons

Li $2s^1$ Li^+ $2s^0$

Li•

Li^+

Li must lose one electrons and be left with the octet from the previous shell

Write the Lewis structure for the covalent anion.

Write the correct skeletal structure for the ion

I —— O

Calculate the total number of electrons for the Lewis structure by summing the valence electrons of each atom in the ion and adding one for the 1 – charge.

(number of valence e^- for I) + (number of valence e^- for O) = 7 +6 + 1 = 14

Distribute the electrons among the atoms, giving octets (or duets for H) to as many atoms as possible. Begin with the bonding electrons, and then proceed to lone pairs on terminal atoms and finally to lone pairs of the central atom.

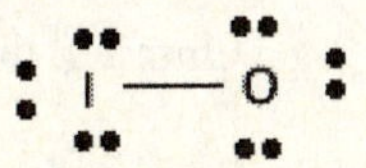

Lastly, write the Lewis structure in brackets with the charge of the ion in the upper right-hand corner.

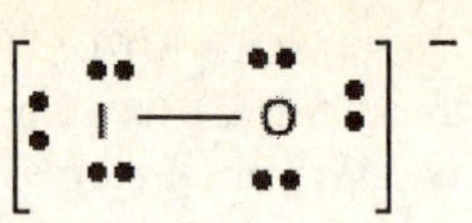

$$\left[\, : \ddot{I} - \ddot{O} : \,\right]^{-}$$

42. a) RbIO$_2$: Rb$^+$

$$\left[\, : \ddot{O} - \ddot{I} - \ddot{O} : \,\right]^{-}$$

Determine the cation and anion
      Rb$^+$ IO$_2^-$
Write the Lewis structure for the Rubidium cation based on the valence electrons
      Rb 5s^1 Rb$^+$ 5s^0
      Rb•
                          Rb$^+$
Rb must lose one electrons and be left with the octet from the previous shell
Write the Lewis structure for the covalent anion.
Write the correct skeletal structure for the ion

Calculate the total number of electrons for the Lewis structure by summing the valence electrons of each atom in the ion and adding one for the 1 – charge.
      (number of valence e$^-$ for I) +2 (number of valence e$^-$ for O) = 7 + 2(6) +1 = 20
Distribute the electrons among the atoms, giving octets (or duets for H) to as many atoms as possible. Begin with the bonding electrons, and then proceed to lone pairs on terminal atoms and finally to lone pairs of the central atom.

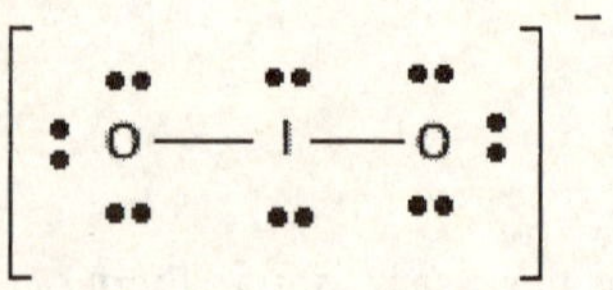

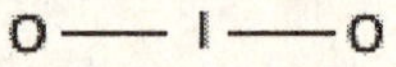

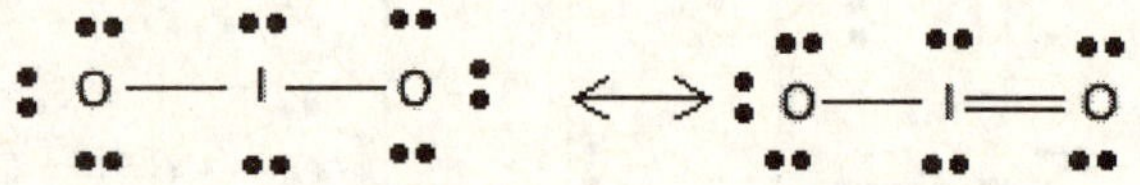

Lastly, write the Lewis structure in brackets with the charge of the ion in the upper right-hand corner.

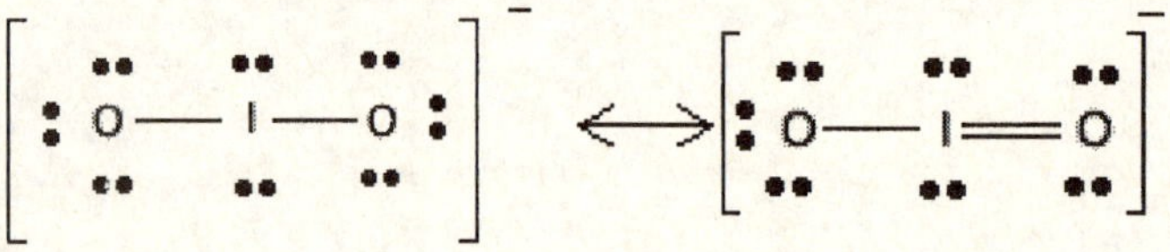

$$\left[\, : \ddot{O} - \ddot{I} - \ddot{O} : \,\right]^{-} \longleftrightarrow \left[\, : \ddot{O} - \ddot{I} = \ddot{O} \,\right]^{-}$$

b) NH$_4$Cl:

$$\left[\begin{array}{c} H \\ | \\ H - N - H \\ | \\ H \end{array}\right]^{+} \qquad \left[\, : \ddot{C}l : \,\right]^{-}$$

Determine the cation and anion
      NH$_4^+$ Cl$^-$
Write the Lewis structure for the covalent cation.
Write the correct skeletal structure for the ion

Calculate the total number of electrons for the Lewis structure by summing the valence electrons of each atom in the ion and subtracting one for the 1 + charge.
      (number of valence e$^-$ for N) + 4(number of valence e$^-$ for H) – 1 = 5 + 4(1) – 1 = 8

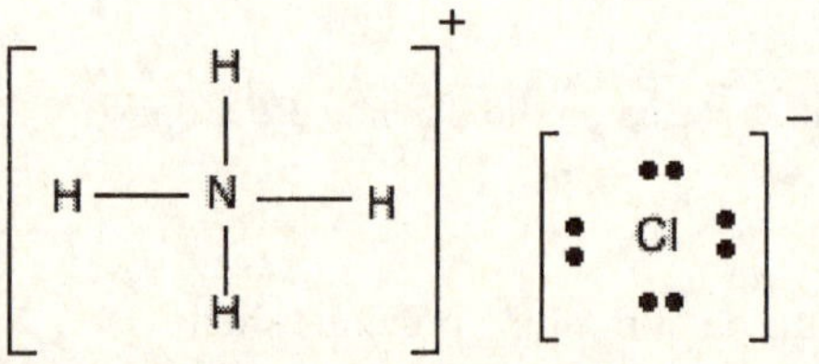

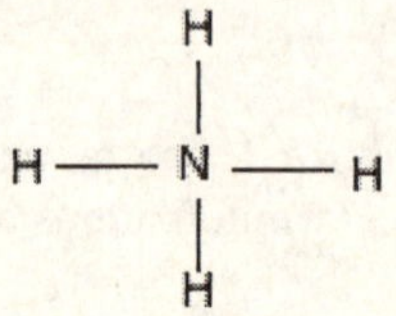

Distribute the electrons among the atoms, giving octets (or duets for H) to as many atoms as possible. Begin with the bonding electrons, and then proceed to lone pairs on terminal atoms and finally to lone pairs of the central atom.

Lastly, write the Lewis structure in brackets with the charge of the ion in the upper right-hand corner.

$$\left[\begin{array}{c} H \\ | \\ H - N - H \\ | \\ H \end{array} \right]^{+}$$

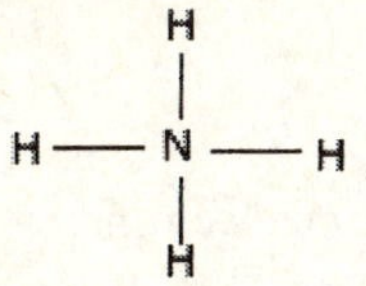

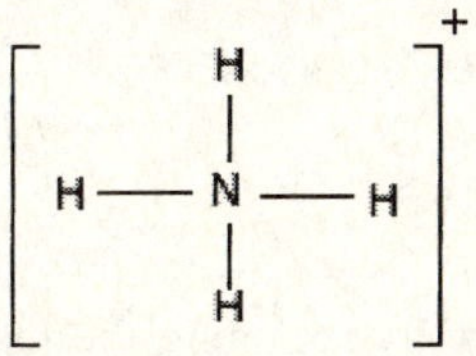

Write the Lewis structure for the chlorine anion based on the valence electrons
    Cl $3s^2 3p^5$

$$\cdot \ \ddot{\underset{\cdot\cdot}{Cl}} \vdots$$

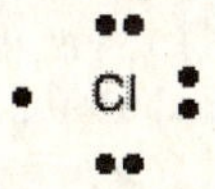

Cl must gain one electron to complete its octet.

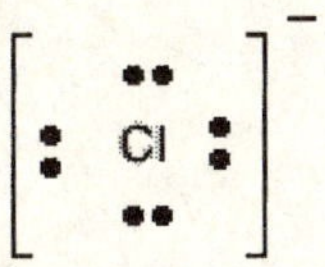

$$\left[\ \vdots \ddot{\underset{\cdot\cdot}{Cl}} \vdots \ \right]^{-}$$

c) KOH: K^+

$$\left[\ \vdots \ddot{\underset{\cdot\cdot}{O}} - \cdot H \ \right]^{-}$$

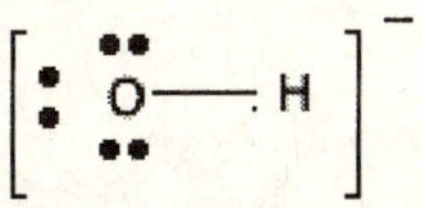

Determine the cation and anion
    K^+ OH^-
Write the Lewis structure for the potassium cation based on the valence electrons
    K $4s^1$ K^+ $4s^0$
    K •
            K^+
K must lose one electron and be left with the octet from the previous shell
Write the Lewis structure for the covalent anion.
Write the correct skeletal structure for the ion
        O — H

Calculate the total number of electrons for the Lewis structure by summing the valences electrons of each atom in the ion and adding one for the 1 – charge.
    (number of valence e⁻ for H) + (number of valence e⁻ for O) +1 = 1 + 6 +1 = 8
Distribute the electrons among the atoms, giving octets (or duets for H) to as many atoms as possible. Begin with the bonding electrons, and then proceed to lone pairs on terminal atoms and finally to lone pairs of the central atom.

$$\vdots \ddot{\underset{\cdot\cdot}{O}} - H$$

Lastly, write the Lewis structure in brackets with the charge of the ion in the upper right-hand corner.

$$\left[\ \vdots \ddot{\underset{\cdot\cdot}{O}} - \cdot H \ \right]^{-}$$

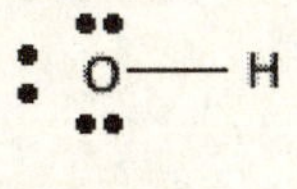

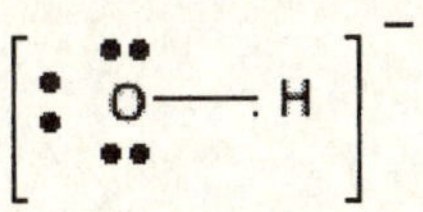

d) $Sr(CN)_2$: Sr^+

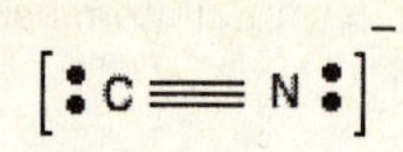

Determine the cation and anion

Sr^+ CN^-

Write the Lewis structure for the strontium cation based on the valence electrons

Sr $5s^1$ Sr^+ $5s^0$

Sr•

Sr^+

Sr must lose one electron and be left with the octet from the previous shell

Write the Lewis structure for the covalent anion.

Write the correct skeletal structure for the ion

Calculate the total number of electrons for the Lewis structure by summing the valence electrons of each atom in the ion and adding one for the 1 – charge.

(number of valence e^- for N) + (number of valence e^- for C) = 5 + 4 + 1 = 10

Distribute the electrons among the atoms, giving octets (or duets for H) to as many atoms as possible. Begin with the bonding electrons, and then proceed to lone pairs on terminal atoms and finally to lone pairs of the central atom.

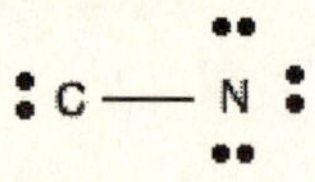

Complete octets on both atoms by forming a triple bond

Lastly, write the Lewis structure in brackets with the charge of the ion in the upper right-hand corner.

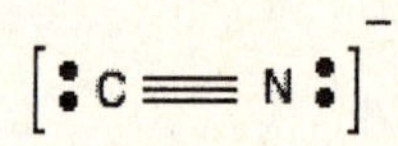

43. a) C_4H_8: Write the correct skeletal structure for the molecule

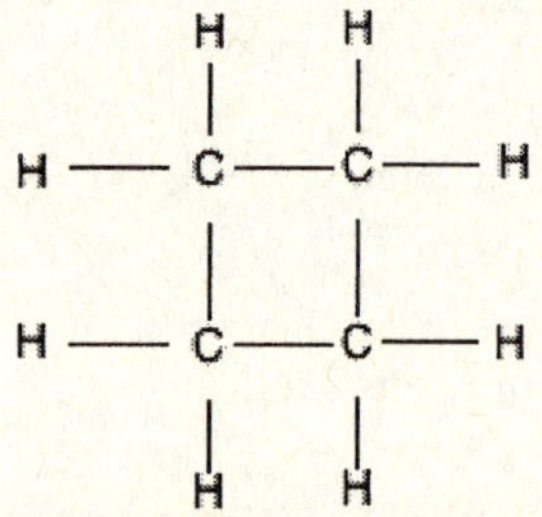

Calculate the total number of electrons for the Lewis structure by summing the valence electrons of each atom in the molecule.

4 (number of valence e^- for C) + 8(number of valence e^- for H) = 4(4) + 8(1) = 24

Distribute the electrons among the atoms, giving octets (or duets for H) to as many atoms as possible.

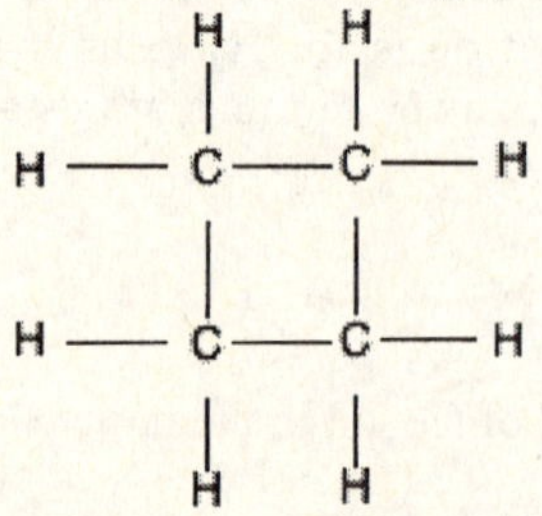

All atoms have octets or duets for H,

b) C_4H_4: Write the correct skeletal structure for the molecule

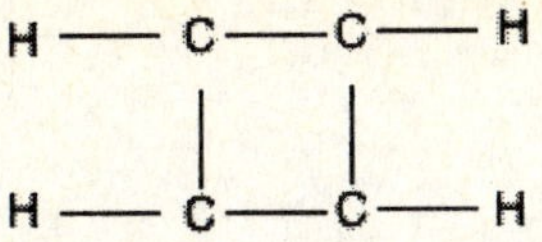

Calculate the total number of electrons for the Lewis structure by summing the valence electrons of each atom in the molecule.

4 (number of valence e^- for C) + 4(number of valence e^- for H) = 4(4) + 4(1) = 20

Distribute the electrons among the atoms, giving octets (or duets for H) to as many atoms as possible.

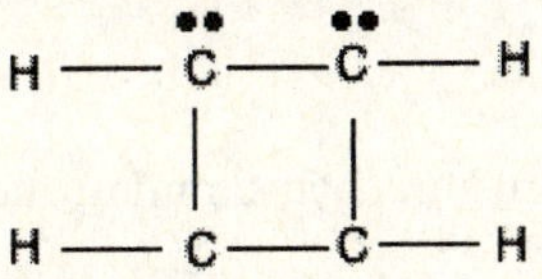

Complete octets by forming double bonds on alternating carbons, draw resonance structures.

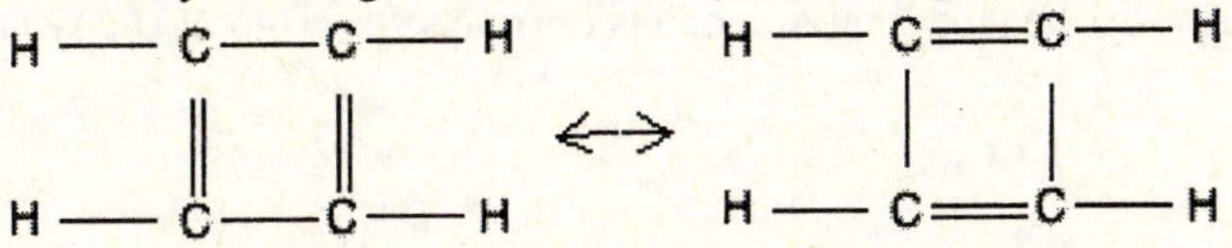

c) C_6H_{12}: Write the correct skeletal structure for the molecule.

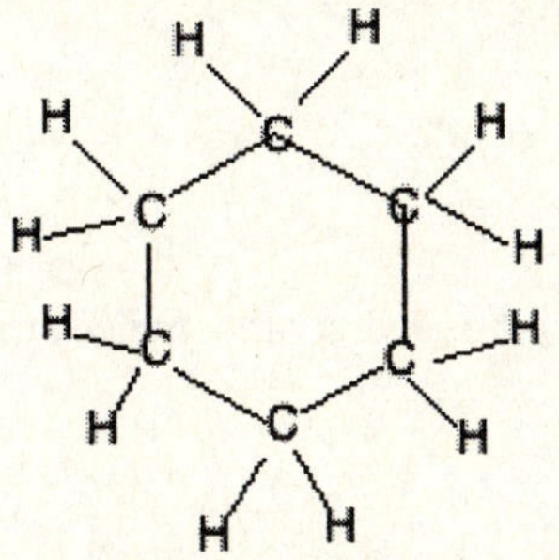

Calculate the total number of electrons for the Lewis structure by summing the valence electrons of each atom in the molecule.

6 (number of valence e^- for C) + 12(number of valence e^- for H) = 6(4) + 12(1) = 36

Distribute the electrons among the atoms, giving octets (or duets for H) to as many atoms as possible. Begin with the bonding electrons, and then proceed to lone pairs on terminal atoms and finally to lone pairs of the central atom.

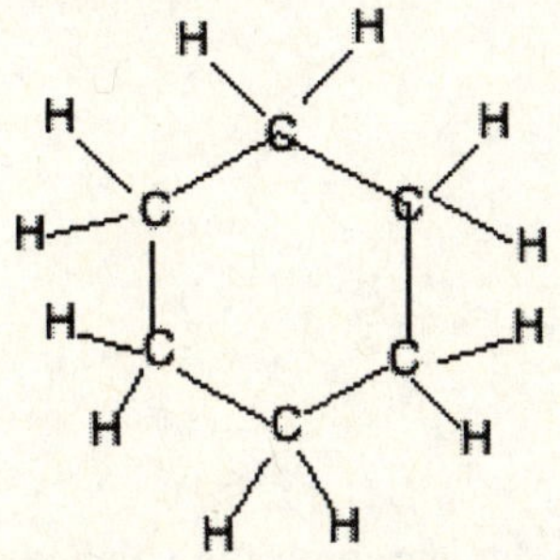

All 36 electrons are used and all atoms have octets or duets for H.

d) C_6H_6: Write the correct skeletal structure for the molecule.

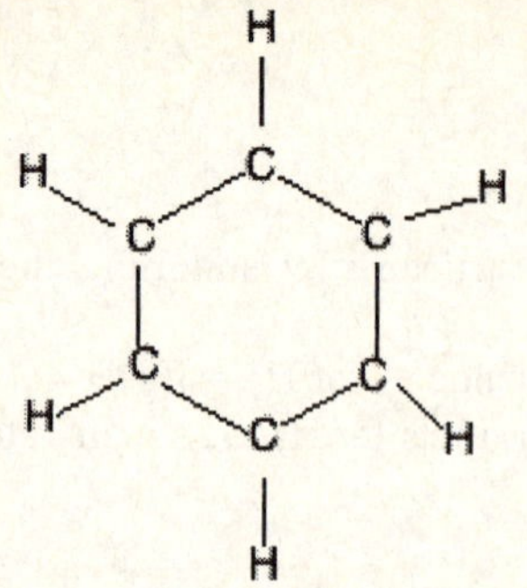

Calculate the total number of electrons for the Lewis structure by summing the valence electrons of each atom in the molecule.

6 (number of valence e^- for C) + 6(number of valence e^- for H) = 6(4) + 6(1) = 30

Distribute the electrons among the atoms, giving octets (or duets for H) to as many atoms as possible.

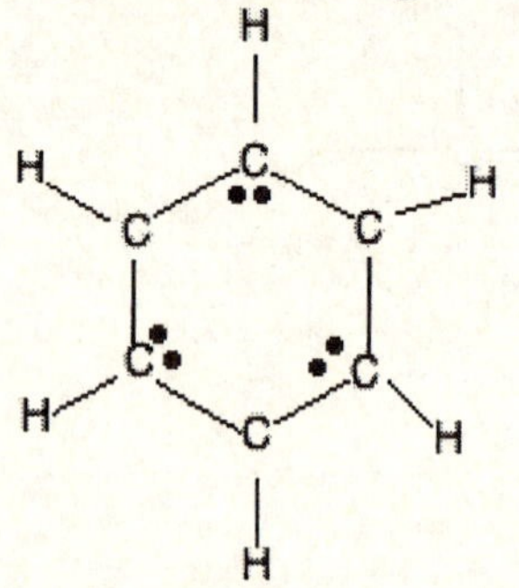

Complete octets by forming double bonds on alternating carbons, draw resonance structures.

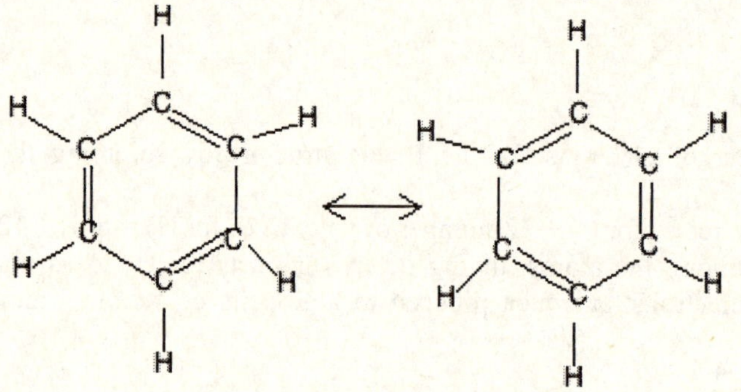

44. H_2NCH_2COOH

Write the correct skeletal structure for the molecule

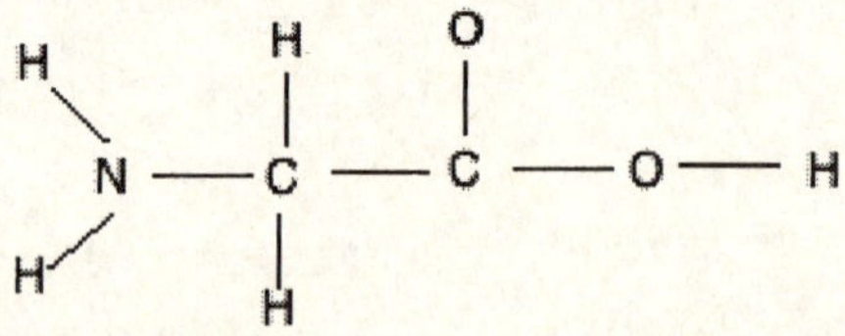

Calculate the total number of electrons for the Lewis structure by summing the valence electrons of each atom in the molecule.

2 (number of valence e^- for C) + 2(number of valence e^- for O) + (number of valence e^- for N) + 5(number of valence e^- for H) = 2(4) + 2(6) + 5 + 5(1) = 30

Distribute the electrons among the atoms, giving octets (or duets for H) to as many atoms as possible. Begin with the bonding electrons, and then proceed to lone pairs on terminal atoms and finally to lone pairs of the central atom.

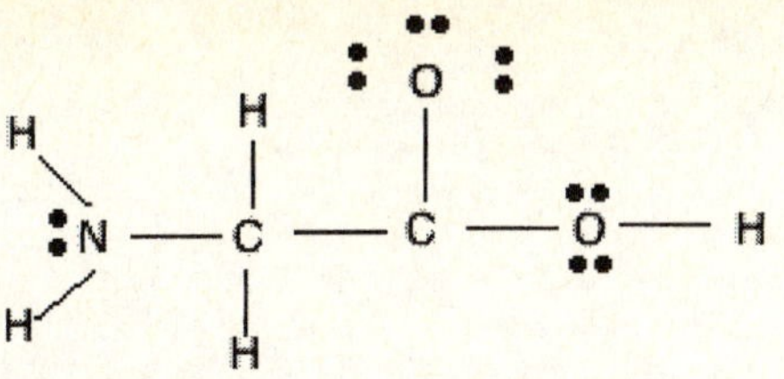

All 30 electrons are used.
Draw a double bond to satisfy the octet on C

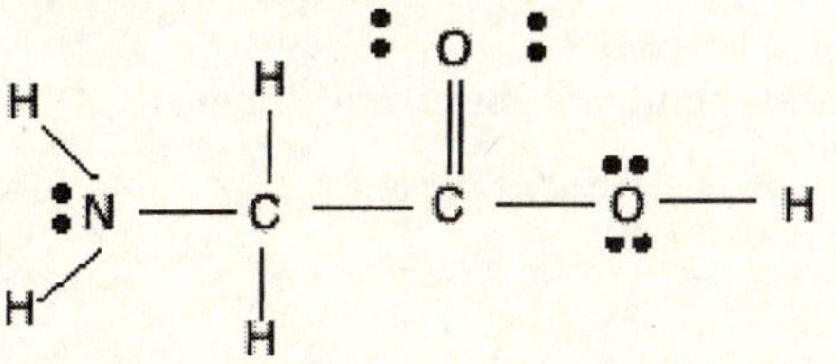

45. **Given:** 26.01% C; 4.38 % H; 69.52 % O; molar mass = 46.02 g/mol
Find: molecular formula and Lewis structure
Conceptual Plan:
convert mass to mol of each element → pseudoformula → empirical formula

$$\frac{1 \text{ mol C}}{12.01 \text{ g C}} \quad \frac{1 \text{ mol H}}{1.008 \text{ g H}} \quad \frac{1 \text{ mol O}}{16.00 \text{ g O}} \qquad \text{divide by smallest number}$$

→ molecular formula→ Lewis structure
empirical formula x n

Solution:

$$26.01 \cancel{\text{g C}} \times \frac{1 \text{ mol C}}{12.01 \cancel{\text{g C}}} = 2.166 \text{ mol C}$$

$$4.38 \cancel{\text{g H}} \times \frac{1 \text{ mol H}}{1.008 \cancel{\text{g H}}} = 4.3\underline{4}5 \text{ mol H}$$

$$69.52 \cancel{\text{g O}} \times \frac{1 \text{ mol O}}{16.00 \cancel{\text{g O}}} = 4.345 \text{ mol O}$$

$C_{2.166}H_{4.345}O_{4.345}$

$$C_{\frac{2.166}{2.166}}H_{\frac{4.345}{2.166}}O_{\frac{4.345}{2.166}} \rightarrow CH_2O_2$$

The correct empirical formula is CH_2O_2
empirical formula mass = (12.01 g/mol) + 2(1.008 g/mol) + 2(16.00 g/mol) = 46.03 g/mol

$$n = \frac{\text{molar mass}}{\text{formula molar mass}} = \frac{46.02 \text{ g/mol}}{46.03 \text{ g/mol}} = 1$$

molecular formula $= CH_2O_2 \times 1$
$= CH_2O_2$

Write the correct skeletal structure for the molecule

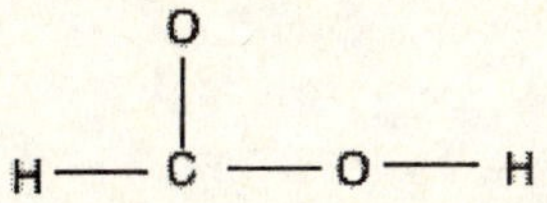

Calculate the total number of electrons for the Lewis structure by summing the valence electrons of each atom in the molecule.

(number of valence e^- for C) + 2(number of valence e^- for O) + 2(number of valence e^- for H)
= 4 + 2(6) + 2(1) = 18

Distribute the electrons among the atoms, giving octets (or duets for H) to as many atoms as possible. Begin with the bonding electrons, and then proceed to lone pairs on terminal atoms and finally to lone pairs of the central atom.

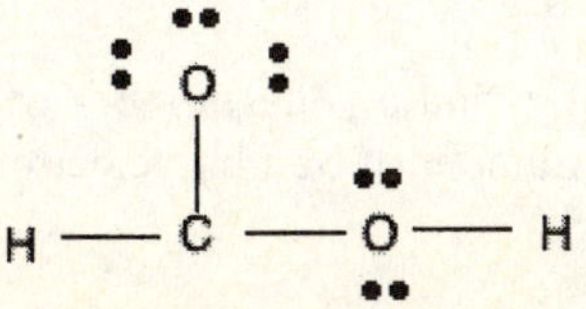

Complete the octet on C by forming a double bond

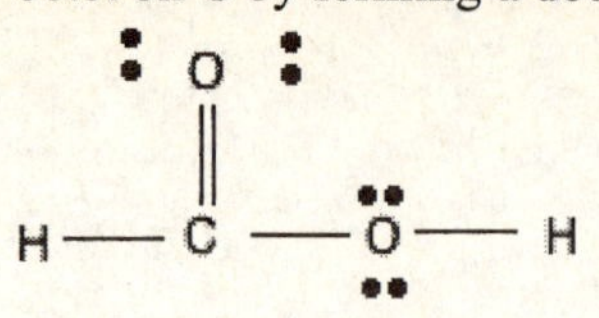

46. **Given:** 28.57% C; 4.80 % H; 66.64 % N; molar mass = 42.04 g/mol
Find: molecular formula and Lewis structure
Conceptual Plan:
convert mass to mol of each element → pseudoformula → empirical formula

$$\frac{1 \text{ mol C}}{12.01 \text{ g C}} \quad \frac{1 \text{ mol H}}{1.008 \text{ g H}} \quad \frac{1 \text{ mol N}}{14.00 \text{ g N}} \qquad \text{divide by smallest number}$$

→ molecular formula→ Lewis structure
empirical formula x n

Solution:

$$28.57 \text{ g C} \times \frac{1 \text{ mol C}}{12.01 \text{ g C}} = 2.380 \text{ mol C}$$

$$4.80 \text{ g H} \times \frac{1 \text{ mol H}}{1.008 \text{ g H}} = 4.7\underline{6}2 \text{ mol H}$$

$$66.64 \text{ g N} \times \frac{1 \text{ mol N}}{14.00 \text{ g N}} = 4.760 \text{ mol N}$$

$$C_{2.380}H_{4.762}N_{4.762}$$

$$C_{\frac{2.380}{2.380}}H_{\frac{4.762}{2.380}}N_{\frac{4.762}{2.380}} \rightarrow CH_2N_2$$

The correct empirical formula is CH_2N_2
empirical formula mass = (12.01 g/mol) + 2(1.008 g/mol) + 2(14.00 g/mol) = 42.03 g/mol

$$n = \frac{\text{molar mass}}{\text{formula molar mass}} = \frac{42.04 \text{ g/mol}}{42.03 \text{ g/mol}} = 1$$

molecular formula $= CH_2N_2 \times 1 = CH_2N_2$

Write the correct skeletal structure for the molecule

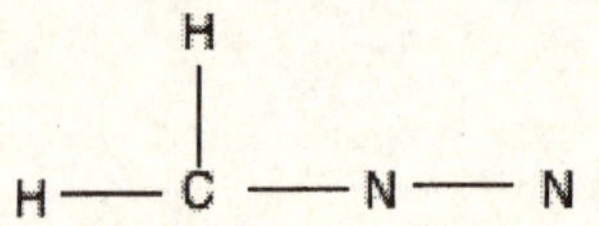

Calculate the total number of electrons for the Lewis structure by summing the valence electrons of each atom in the molecule.

(number of valence e^- for C) + 2(number of valence e^- for N) + 2(number of valence e^- for H)
= 4 + 2(5) + 2(1) = 16

Distribute the electrons among the atoms, giving octets (or duets for H) to as many atoms as possible. Begin with the bonding electrons, and then proceed to lone pairs on terminal atoms and finally to lone pairs of the central atom.

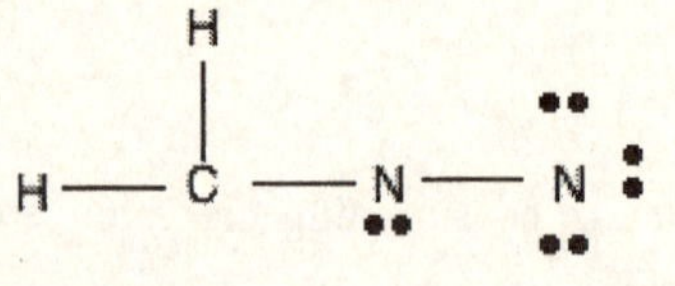

Complete the octet on C and N by forming a double bonds

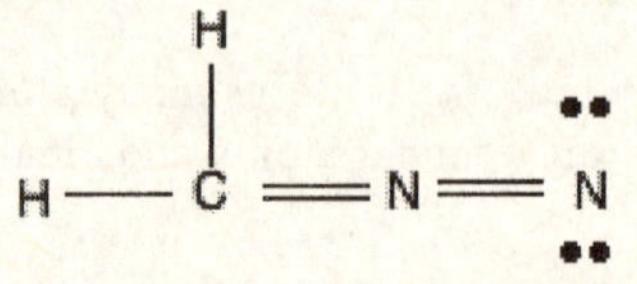

Calculate the formal charge on each atom in the structure by finding the number of valence electrons and subtracting the number of lone pair electrons and one-half the number of bonding electrons.

	H_{left}	H_{top}	C	N_{left}	N_{right}
number of valence electron	1	1	4	5	5
- number of lone pair electrons	0	0	0	0	4
- 1/2(number of bonding electrons)	1	1	4	4	2
Formal charge	0	0	0	+1	- 1

The diazomethane molecule has nitrogen atoms next to each other with a +1 and a – 1 charge. Nitrogen is more electronegative than C, which has a 0 formal charge. The nitrogen with the +1 charge is not a very stable configuration for nitrogen, particularly next to the 0 formal charge C atom.

47. The lattice energy of Al_2O_3 is –15,916 kJ/mol. The thermite reaction is exothermic due to the energy released when the Al_2O_3 lattice forms. The lattice energy of Al_2O_3 is much more negative than the lattice energy of Fe_2O_3.

48. For NaCl, E is proportional to (1+)(1–) = –1. While for XY, E is proportional to (3+)(3–) = –9. So, the relative stabilization for XY relative to NaCl should be roughly nine times greater. $\Delta H_{lattice}$ (XY) = 9 x $\Delta H_{lattice}$ (NaCl) = 9(–787kJ/mol) = –7083kJ/mol

49. HNO_3 Write the correct skeletal structure for the molecule

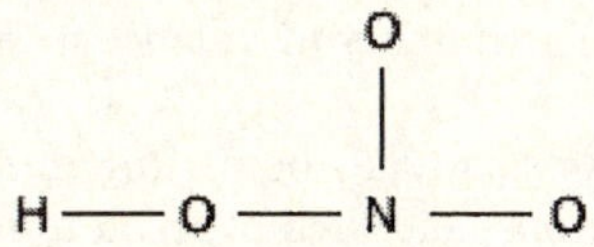

Calculate the total number of electrons for the Lewis structure by summing the valence electrons of each atom in the molecule.

3(number of valence e^- for O) + (number of valence e^- for N) + (number of valence e^- for H) = 3(6) + 5 +1 = 24

Distribute the electrons among the atoms, giving octets (or duets for H) to as many atoms as possible. Begin with the bonding electrons, and then proceed to lone pairs on terminal atoms and finally to lone pairs of the central atom.

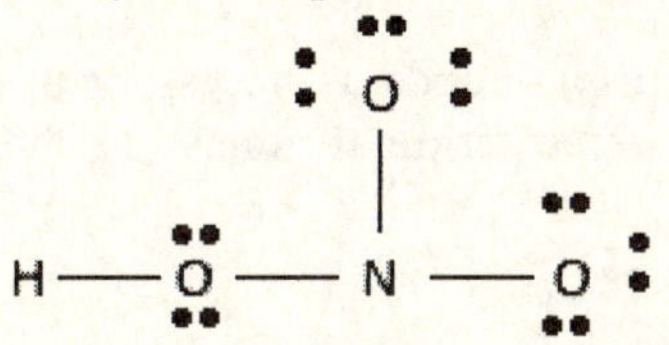

All 24 valence electrons are used

If any atom lacks an octet, form double or triple bonds as necessary to give them octets. The double bond can be formed to any of the three oxygen atoms, so there are three resonance forms.

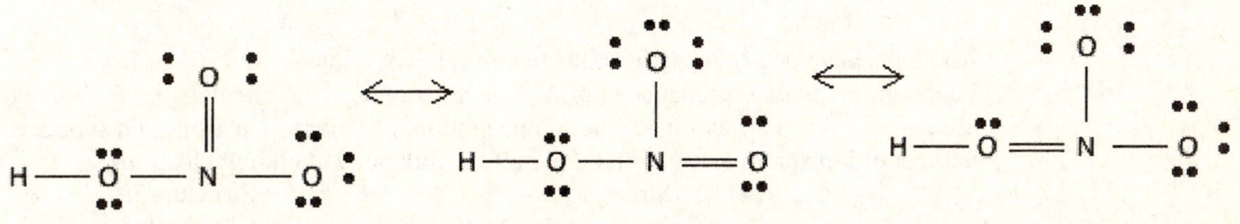

I II III

All atoms have octets, duets for H, structures is complete.

To determine which resonance hybrid(s) is most important, calculate the formal charge on each atom in each structure by finding the number of valence electrons and subtracting the number of lone pair electrons and one-half the number of bonding electrons.

	Structure I					Structure II				
	O_{left}	O_{top}	O_{right}	N	H	O_{left}	O_{top}	O_{right}	N	H
number of valence electrons	6	6	6	5	1	6	6	6	5	1
- number of lone pair electrons	4	4	6	0	0	4	6	4	0	0
- 1/2(number of bonding electrons)	2	2	1	4	1	2	1	2	4	1
Formal charge	0	0	–1	+1	0	0	–1	0	+1	0

Structure III

	O_{left}	O_{top}	O_{right}	N	H
number of valence electrons	6	6	6	5	1
- number of lone pair electrons	2	6	6	0	0
- 1/2(number of bonding electrons)	3	1	1	4	1
Formal charge	+1	−1	−1	+1	0

The sum of the formal charges is 0 for each structure, which is the overall charge of the molecule. However, in structures I and II, the individual formal charges are lower and these two forms would contribute equally to the structure of HNO_3 and structure III would be less important since the individual formal charges are higher.

50. Cl_2CO Write the correct skeletal structure for the molecule

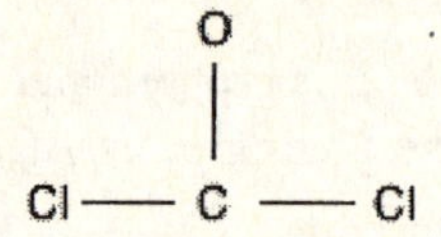

Calculate the total number of electrons for the Lewis structure by summing the valence electron of each atom in the molecule.

1(number of valence e^- for O) + (number of valence e^- for C) + 2(number of valence e^- for Cl) = 6 + 4 + 2(7) = 24

Distribute the electrons among the atoms, giving octets (or duets for H) to as many atoms as possible. Begin with the bonding electrons, and then proceed to lone pairs on terminal atoms and finally to lone pairs of the central atom.

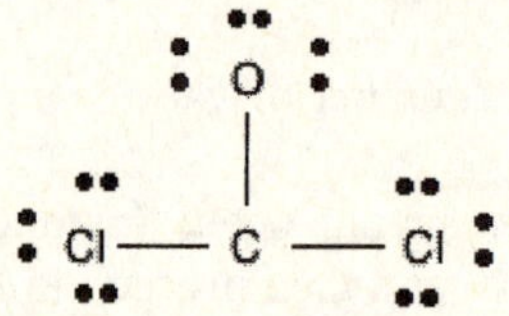

All 24 valence electrons are used

If any atom lacks an octet, form double or triple bonds as necessary to give them octets. The double bond can be formed to any of the three terminal atoms, so there are three resonance forms.

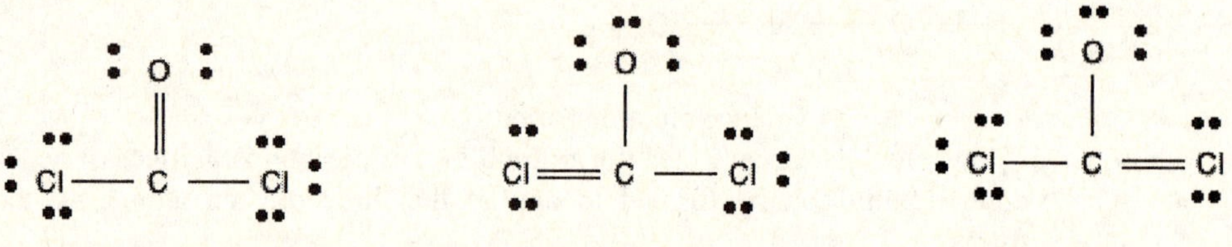

I II III

All atoms have octets, duets for H, structures are complete.

To determine which resonance hybrids(s) is most important, calculate the formal charge on each atom in each structure by finding the number of valence electrons and subtracting the number of lone pair electrons and one-half the number of bonding electrons.

	Structure I Cl_{left}	Cl_{right}	O	C	Structure II Cl_{left}	Cl_{right}	O	C
number of valence electrons	7	7	6	4	7	7	6	4
- number of lone pair electrons	6	6	4	0	4	6	6	0
- 1/2(number of bonding electrons)	1	1	2	4	2	1	1	4
Formal charge	0	0	0	0	+1	0	−1	0

Structure III

	Cl_{left}	Cl_{right}	O	C
number of valence electrons	7	7	6	4
- number of lone pair electrons	6	4	6	0
- 1/2(number of bonding electrons)	1	2	1	4
Formal charge	0	+1	−1	0

The sum of the formal charges is 0 for each structure, which is the overall charge of the molecule. However, in structures I the individual formal charges are lower and this form would be more important to Cl_2CO than structure II and structure III.

51. CNO^- Write the skeletal structure:
$$C - N - O$$
Determine the number of valence electrons.
(valence e^- from C) + (valence e^- from N) + (valence e^- from O) +1(from the negative charge)
$$4 + 5 + 6 + 1 = 16$$

$$:\overset{\bullet\bullet}{\underset{\bullet\bullet}{C}} \!-\! N \!-\! \overset{\bullet\bullet}{\underset{\bullet\bullet}{O}}:$$

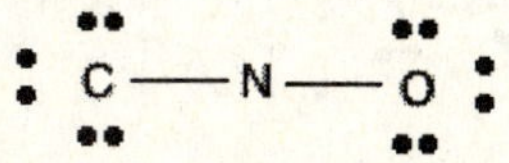

Distribute the electrons to complete octets if possible.

$$\left[\overset{\bullet\bullet}{\underset{\bullet\bullet}{C}} \!=\! N \!=\! \overset{\bullet\bullet}{\underset{\bullet\bullet}{O}} \right]^- \longleftrightarrow \left[:\!C \!\equiv\! N \!-\! \overset{\bullet\bullet}{\underset{\bullet\bullet}{O}}: \right]^- \longleftrightarrow \left[\overset{\bullet\bullet}{\underset{\bullet\bullet}{:\!C}} \!-\! N \!\equiv\! O\overset{\bullet}{\underset{\bullet}{:}} \right]^-$$

I   II   III

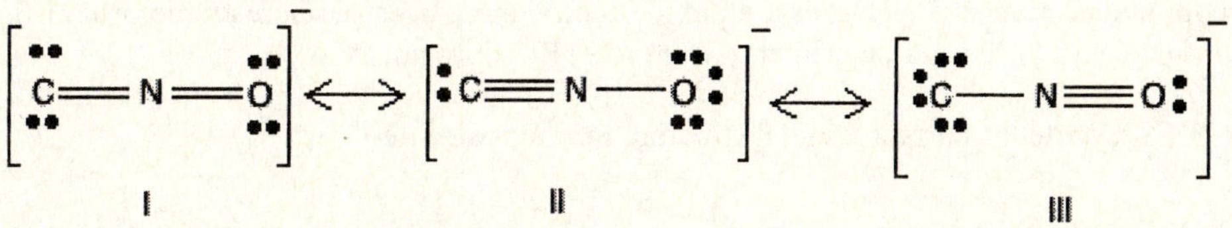

Determine the formal charge on each atom for each structure

	Structure I			Structure II		
	C	N	O	C	N	O
number of valence electrons	4	5	6	4	5	6
- number of lone pair electrons	4	0	4	2	0	6
- 1/2(number of bonding electrons)	2	4	2	3	4	1
Formal charge	−2	+1	0	−1	+1	−1

	Structure III		
	C	N	O
number of valence electrons	4	5	6
- number of lone pair electrons	6	0	2
- 1/2(number of bonding electrons)	1	4	3
Formal charge	−3	+1	+1

Structures I, II, and III all follow the octet rule but have varying degrees of negative formal charge on carbon, which is the least electronegative atom. Also the amount of formal charge is very high in all three resonance forms. Therefore, none of these resonance forms contribute to the stability of the fulminate ion and the ion is not very stable.

52. The Lewis structures for the three ions will be similar. So, we can write the Lewis structure for one ion and use it to determine the other two.
Br_3^-: Write the correct skeletal structure for the ion
     Br —— Br —— Br

Calculate the total number of electrons for the Lewis structure by summing the valence electrons of each atom in the ion and adding 1 for the 1 − charge.
     3(number of valence e^- for Br) + 1 = 3(7) +1 = 22
Distribute the electrons among the atoms, giving octets to as many atoms as possible. Begin with the bonding electrons, and then proceed to lone pairs on terminal atoms and finally to lone pairs of the central atom. Assign electrons above 8 to the central atom.

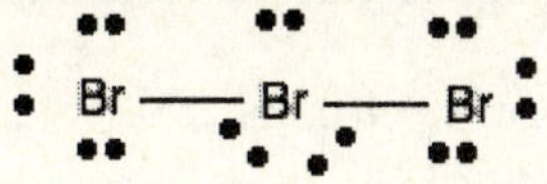

Lastly, place the ion in brackets and place the charge in the upper right hand corner.

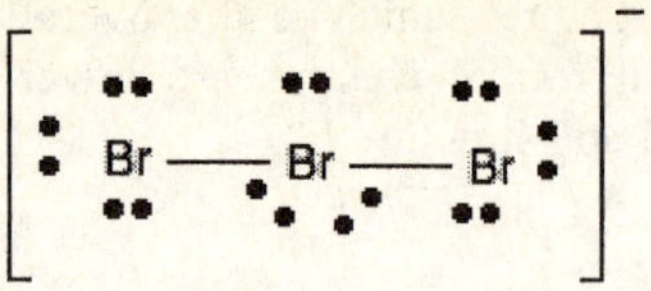

Since Br, I, and F are all in the same family, each of these ions would have 22 electrons and should have the same Lewis structure

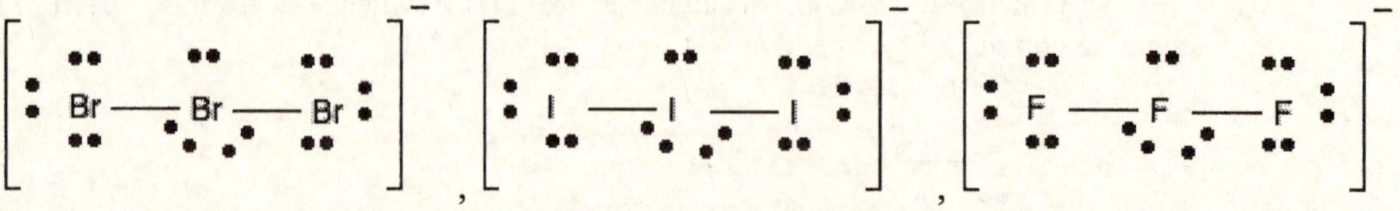

All three ions are written with 5 electron groups around the central atom. Bromine and iodine can accommodate 10 electrons around the central atom. However, fluorine cannot. Fluorine is in period 2 and can accommodate at most 8 electrons around the central atom because there are no orbitals low enough in energy to hybridize with the 2s and 2p orbitals. Therefore, F_3^- does not exist.

53. $HCSNH_2$: Write the correct skeletal structure for the molecule.

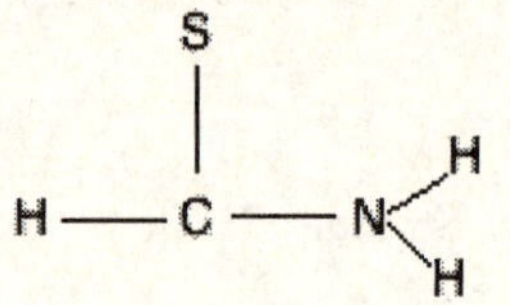

Calculate the total number of electrons for the Lewis structure by summing the valence electrons of each atom in the molecule.

(number of valence e^- for N) + (number of valence e^- for S) + (number of valence e^- for C) + 3(number of valence e^- for H) = 5 + 6 + 4 + 3(1) = 18

Distribute the electrons among the atoms, giving octets (or duets for H) to as many atoms as possible. Begin with the bonding electrons, and then proceed to lone pairs on terminal atoms and finally to lone pairs of the central atom.

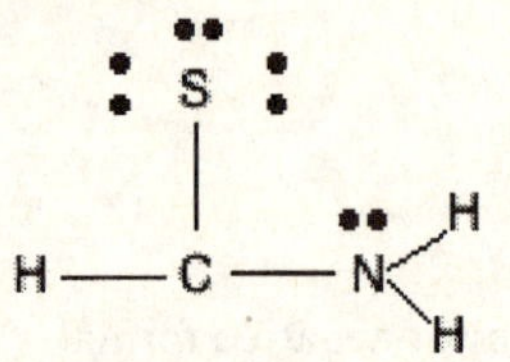

Complete the octet on C by forming a double bond.

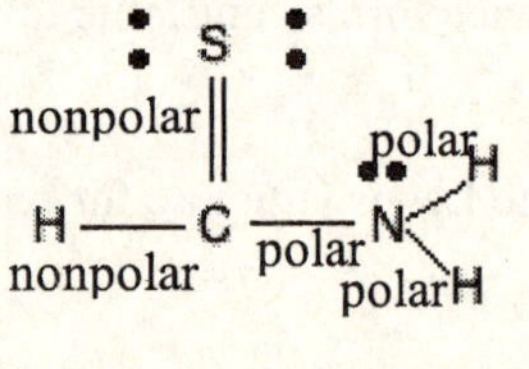

54. H_2NCONH_2 Write the correct skeletal structure for the molecule.

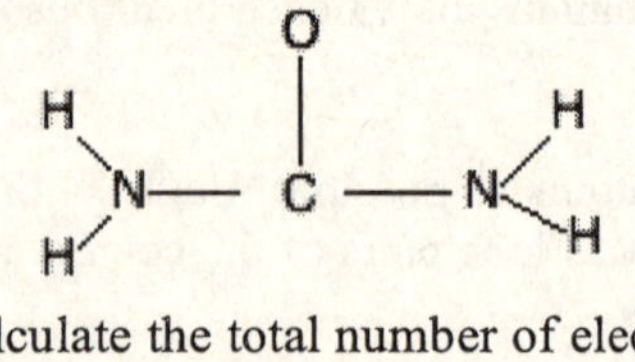

Calculate the total number of electrons for the Lewis structure by summing the valence electron of each atom in the molecule.

2(number of valence e^- for N) + (number of valence e^- for O) + 4(number of valence e^- for H) + (number of valence e^- for C)
= 2(5) + 6 + 2(1) + 4 = 24

Distribute the electrons among the atoms, giving octets (or duets for H) to as many atoms as possible.
Begin with the bonding electrons, and then proceed to lone pairs on terminal atoms and finally to lone pairs of the central atom.

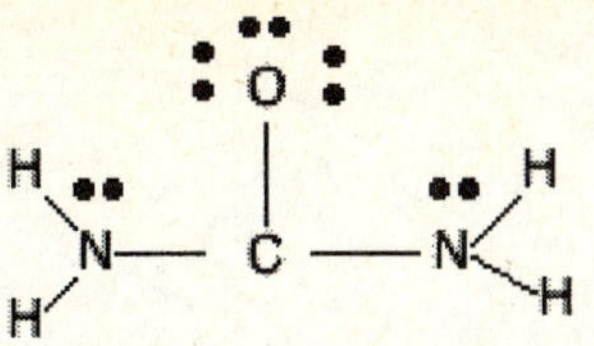

Complete the octet on C with a double bond.

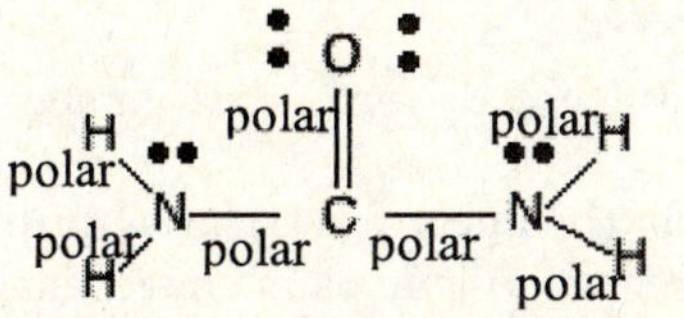

The C – O bond would be the most polar because it has the greater difference in electronegativity.

ΔEN C— O = 1.0 ΔEN C – N = 0.5 ΔEN N – H = 0.9

55. a) O_2^-: Write the correct skeletal structure for the radical

Calculate the total number of electrons for the Lewis structure by summing the valence electrons of each atom in the radical and adding 1 for the 1 – charge.

2(number of valence e^- for O) + 1 = 2(6) + 1 = 13

Distribute the electrons among the atoms, giving octets to as many atoms as possible. Begin with the bonding electrons, and then proceed to lone pairs on terminal atoms and finally to lone pairs of the central atom.

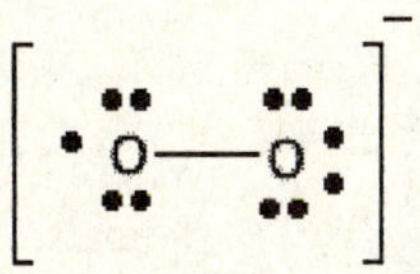

All 13 valence electrons are used.

O has an incomplete octet. It has 7 electrons because we have an odd number of valence electrons.

b) O^-: Write the Lewis structure based on the valence electrons: $2s^22p^5$.

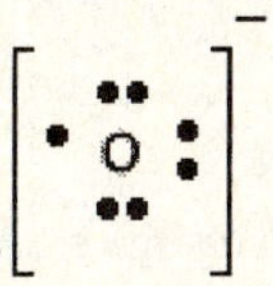

c) OH: Write the correct skeletal structure for the molecule

Calculate the total number of electrons for the Lewis structure by summing the valence electrons of each atom in the molecule

(number of valence e^- for O) + (number of valence e^- for H) = 6 + 1 = 7

Distribute the electrons among the atoms, giving octets (or duets for H) to as many atoms as possible. Begin with the bonding electrons, and then proceed to lone pairs on terminal atoms and finally to lone pairs of the central atom.

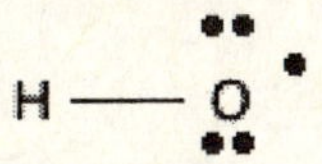

All 7 valence electrons are used.

O has an incomplete octet. It has 7 electrons because we have an odd number of valence electrons.

d) CH_3OO: Write the correct skeletal structure for the radical

C is the less electronegative, so it is central

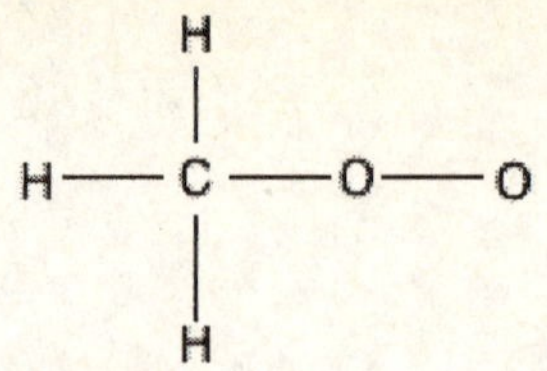

Calculate the total number of electrons for the Lewis structure by summing the valence electrons of each atom in the molecule

3(number of valence e^- for H) + (number of valence e^- for C) + 2(number of valence e^- for O) = 3(1) + 4 +2(6) = 19

Distribute the electrons among the atoms, giving octets (or duets for H) to as many atoms as possible. Begin with the bonding electrons, and then proceed to lone pairs on terminal atoms and finally to lone pairs of the central atom.

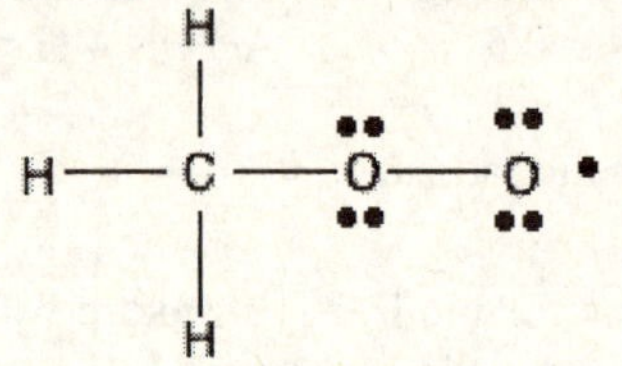

All 19 valence electrons are used.

O has an incomplete octet. It has 7 electrons because we have an odd number of valence electrons.

56.

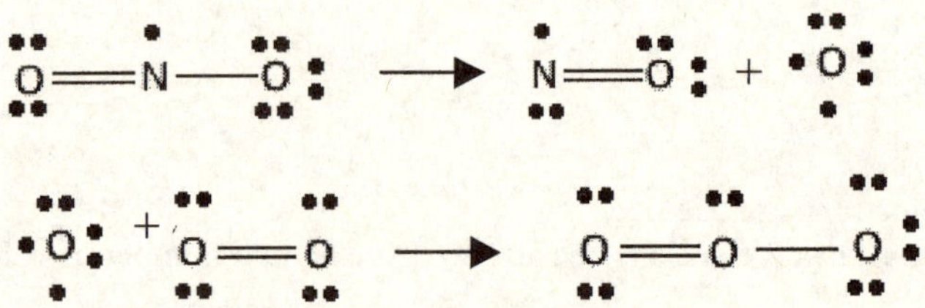

Where NO_2 and NO are the free radicals.

57. Rewrite the reaction using the Lewis structures of the molecules involved.

$$H-H\,(g) + 1/2\ O=O\,(g) \rightarrow H-O-H$$

Determine which bonds are broken in the reaction and sum the bond energies of these

$\Sigma(\Delta H\text{'s bonds broken})$
$= 1\text{mol}(H-H) + 1/2\text{mol}(O=O)$
$= 1\text{mol}(436\text{ kJ/mol}) + 1/2\text{mol}(498)$
$= 685\text{ kJ/mol}$

Determine which bonds are formed in the reaction and sum the negatives of the bond energies of these

$\Sigma(-\Delta H\text{'s of bonds formed})$
$= -\,2\text{mol}(O-H)$
$= -\,2\text{mol}(464\text{ kJ/mol})$
$= -928\text{ kJ/mol}$

Find ΔH_{rxn} by summing the results of the two steps.

$\Delta H_{rxn} = \Sigma(\Delta H\text{'s bonds broken}) + \Sigma(-\Delta H\text{'s of bonds formed})$
$= 685\text{ kJ/mol} - 928\text{ kJ/mol}$
$= -\,243\text{ kJ/mol}$

$CH_4(g) + 2O_2(g) \rightarrow CO_2(g) + 2H_2O(g)$

Rewrite the reaction using the Lewis structures of the molecules involved.

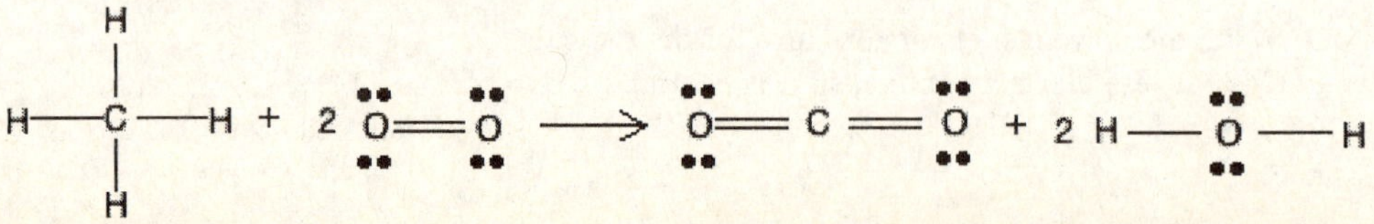

Determine which bonds are broken in the reaction and sum the bond energies of these

$\Sigma(\Delta H\text{'s bonds broken})$
$= 4\text{mol}(C - H) + 2\text{mol}(O = O)$
$= 4\text{mol}(414\ \text{kJ/mol}) + 2\text{mol}(498)$
$= 2652\ \text{kJ/mol}$

Determine which bonds are formed in the reaction and sum the negatives of the bond energies of these

$\Sigma(-\Delta H\text{'s of bonds formed})$
$= -\ 2\text{mol}(C = O) - 4\text{mol}(O - H)$
$= -\ 2\text{mol}(799\ \text{kJ/mol}) - 4\text{mol}(464\ \text{kJ/mol})$
$= -3454\ \text{kJ/mol}$

Find ΔH_{rxn} by summing the results of the two steps.

$\Delta H_{rxn} = \Sigma(\Delta H\text{'s bonds broken}) + \Sigma(-\Delta H\text{'s of bonds formed})$
$= 2653\ \text{kJ/mol} - 3454\ \text{kJ/mol}$
$= -\ 802\ \text{kJ/mol}$

Compare

	kJ/mol	kJ/g
H_2	–243	–120
CH_4	–802	–50.1

So, methane yields more energy per mole but hydrogen yields more energy per gram.

58. octane = C_8H_{18}

$C_8H_{18}(l) + 25/2O_2(g) \rightarrow 8CO_2(g) + 9H_2O(g)$

Rewrite the reaction using the Lewis structures of the molecules involved.

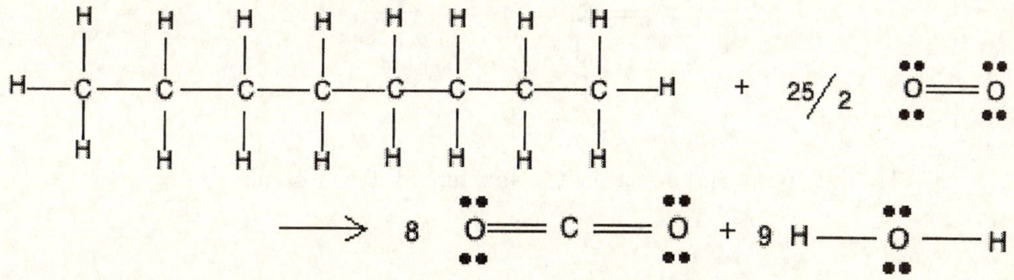

Determine which bonds are broken in the reaction and sum the bond energies of these

$\Sigma(\Delta H\text{'s bonds broken})$
$= 18\text{mol}(C - H) + 25/2\ \text{mol}(O = O) + 7\text{mol}(C - C)$
$= 18\text{mol}(414\ \text{kJ/mol}) + 25/2\text{mol}(498) + 7\text{mol}(347)$
$= 16106\ \text{kJ/mol}$

Determine which bonds are formed in the reaction and sum the negatives of the bond energies of these

$\Sigma(-\Delta H\text{'s of bonds formed})$
$= -\ 16\text{mol}(C = O) - 18\text{mol}(O - H)$
$= -\ 16\text{mol}(799\ \text{kJ/mol}) - 18\text{mol}(464\ \text{kJ/mol})$
$= -21136\ \text{kJ/mol}$

Find ΔH_{rxn} by summing the results of the two steps.

$\Delta H_{rxn} = \Sigma(\Delta H\text{'s bonds broken}) + \Sigma(-\Delta H\text{'s of bonds formed})$
$= 16106\ \text{kJ/mol} - 21136\ \text{kJ/mol}$
$= -\ 5030\ \text{kJ/mol}$

$\Delta H_{rxn} = \Sigma(\Delta H_f(\text{products})) - \Sigma(\Delta H_f(\text{reactants}))$
$= [8\text{mol}\Delta H_f(CO_2(g)) + 9\text{mol}\Delta H_f(H_2O(g))] - [1\text{mol}\Delta H_f(C_8H_{18}(l)) + 25/2\text{mol}\Delta H_f(O_2(g))]$
$= [8\text{mol}(-393.5\ \text{kJ/mol}) + 9\text{mol}(-241.8\ \text{kJ/mol})] - [1\text{mol}(-250\ \text{kJ/mol}) + 25/2\text{mol}(0)]$
$= -5074\ \text{kJ/mol}$

$$\%\ \text{difference} = \frac{-5074\ \text{kJ} - (-5030\ \text{kJ})}{-5074\ \text{kJ}} \times 100 = 0.8672\%$$

You would expect the value calculated from the heats of formation to be more accurate. The bond energy values are average values, not values for a specific molecule. The heats of formation are for specific compounds.

59. a) Cl_2O_7: Write the correct skeletal structure for the molecule

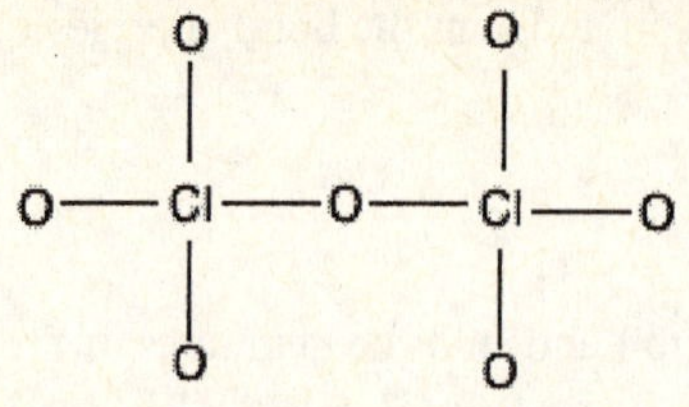

Calculate the total number of electrons for the Lewis structure by summing the valence electrons of each atom in the molecule

2(number of valence e^- for Cl) + 7(number of valence e^- for O) = 2(7)5 + 7(6) = 56

Distribute the electrons among the atoms, giving octets (or duets for H) to as many atoms as possible. Begin with the bonding electrons, and then proceed to lone pairs on terminal atoms and finally to lone pairs of the central atom.

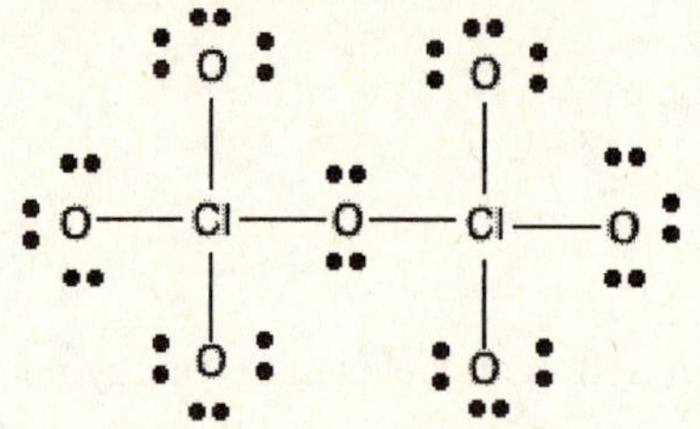

Form double bonds to minimize formal charge.

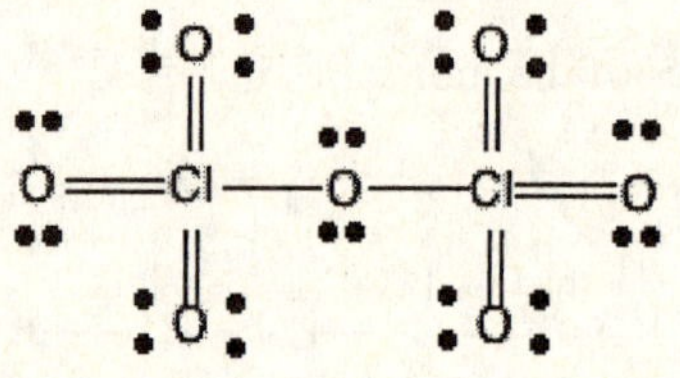

b) H_3PO_3: Write the correct skeletal structure for the molecule

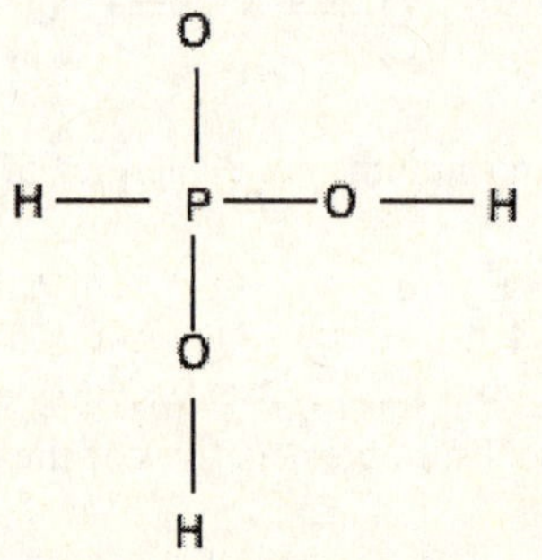

Calculate the total number of electrons for the Lewis structure by summing the valence electrons of each atom in the molecule

(number of valence e^- for P) + 3(number of valence e^- for O) + 3(number of valence e^- for H) = 5 + 3(6) + 3(1) = 26

Distribute the electrons among the atoms, giving octets (or duets for H) to as many atoms as possible. Begin with the bonding electrons, and then proceed to lone pairs on terminal atoms and finally to lone pairs of the central atom.

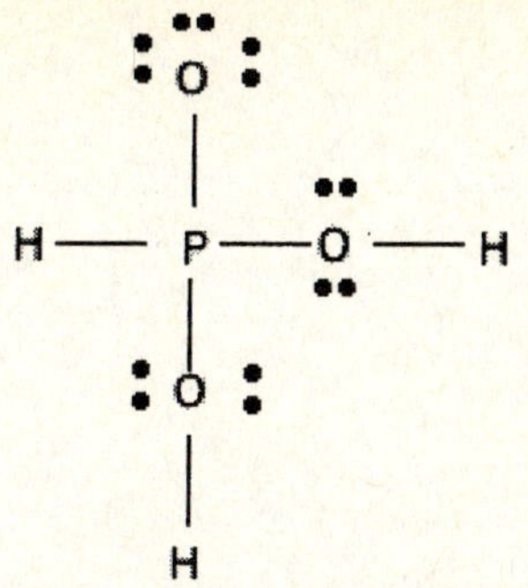

Form double bond to minimize formal charge.

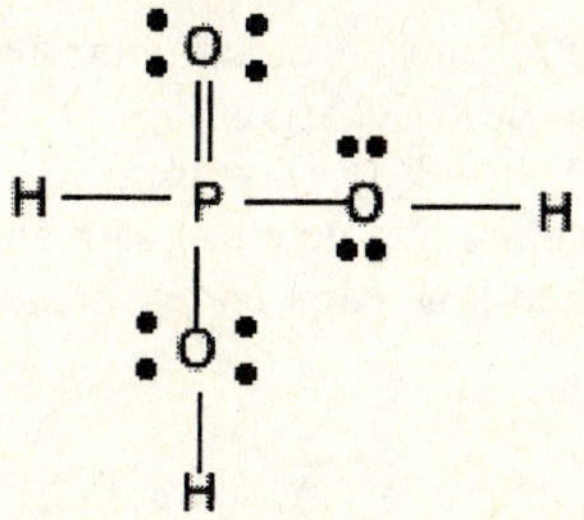

c) H_3AsO_4: Write the correct skeletal structure for the radical

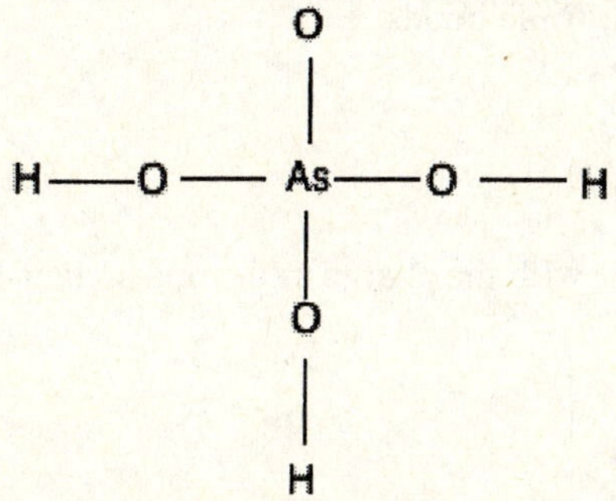

Calculate the total number of electrons for the Lewis structure by summing the valence electrons of each atom in the molecule

(number of valence e^- for As) + 4(number of valence e^- for O) + 3(number of valence e^- for H) = 5 + 4(6) + 3(1) = 32

Distribute the electrons among the atoms, giving octets (or duets for H) to as many atoms as possible. Begin with the bonding electrons, and then proceed to lone pairs on terminal atoms and finally to lone pairs of the central atom.

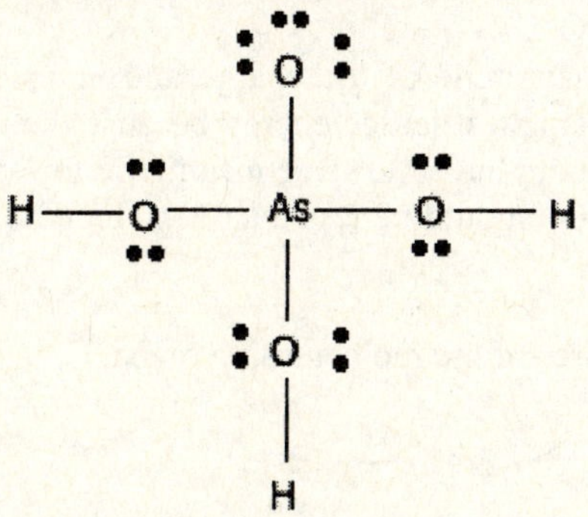

Form double bond to minimize formal charge.

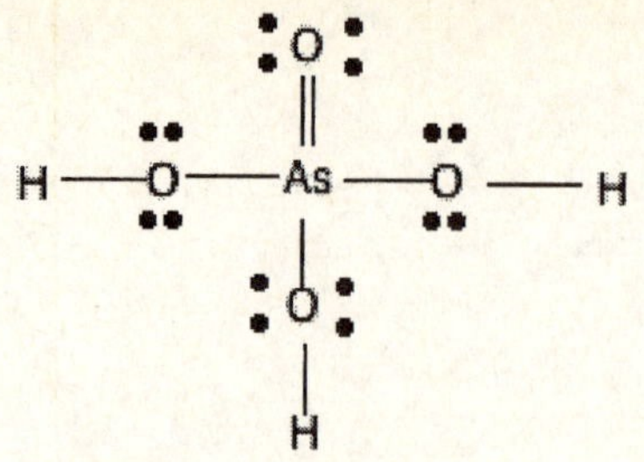

60. N_3^-: Write the correct skeletal structure for the ion

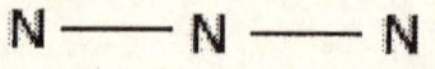

Calculate the total number of electrons for the Lewis structure by summing the valence electrons of each atom in the ion and adding 1 for the 1 – charge.

3(number of valence e^- for N) +1 = 3(5) + 1 = 16

Distribute the electrons among the atoms, giving octets to as many atoms as possible. Begin with the bonding electrons, and then proceed to lone pairs on terminal atoms and finally to lone pairs of the central atom.

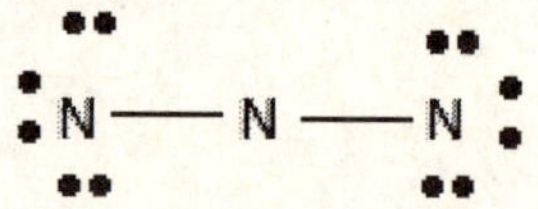

Complete octets with double or triple bonds.

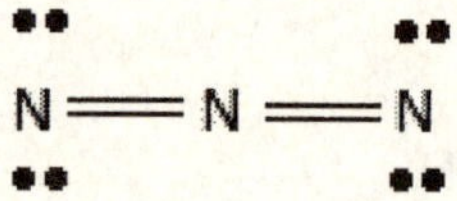

Lastly, write the ion in brackets with the charge in the upper right-hand corner.

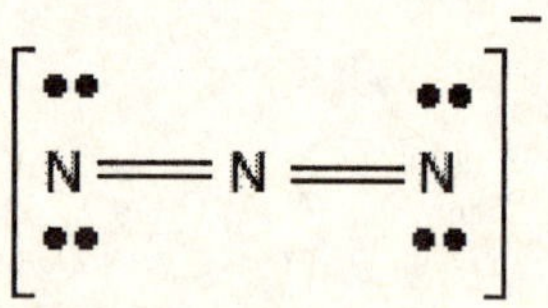

Write the resonance forms.

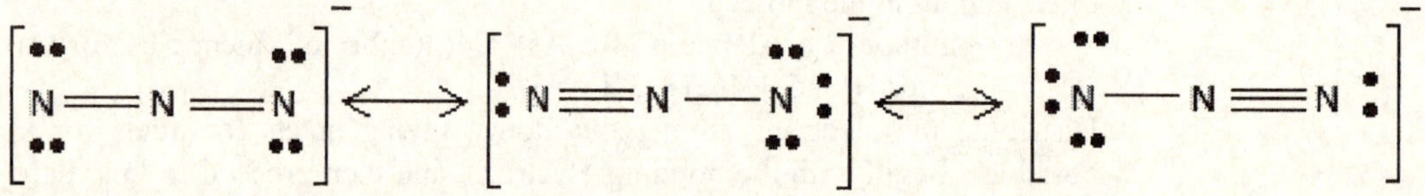

61. $Na^+F^- < Na^+O^{2-} < Mg^{2+}F^- < Mg^{2+}O^{2-} < Al^{3+}O^{2-}$

The lattice energy is proportional to the magnitude of the charge and inversely proportional to the distance between the atoms. Na^+F^- would have the smallest lattice energy because the magnitude of the charges on Na and F are the smallest. $Mg^{2+}F^-$ and Na^+O^{2-} both have the same magnitude formal charge, the O^{2-} is larger than F^- in size, Na^+ is larger than Mg^{2+}, so Na^+O^{2-} should be less than $Mg^{2+}F^-$. The magnitude of the charge makes $Mg^{2+}O^{2-} < Al^{3+}O^{2-}$.

62. Rewrite the reaction using the Lewis structures of the molecules involved.

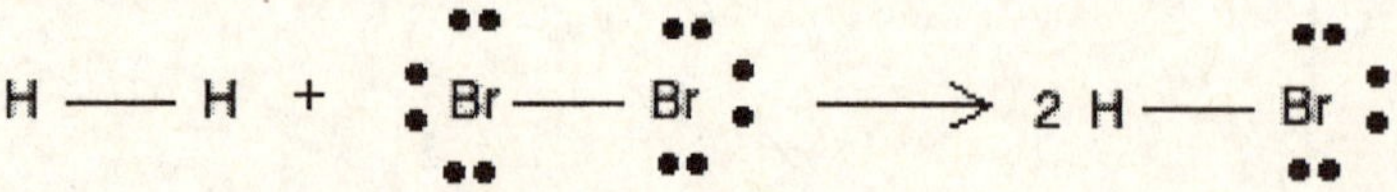

Determine which bonds are broken in the reaction and sum the bond energies of these

$\Sigma(\Delta H\text{'s bonds broken})$

$= 1\text{mol (H – H)} + 1\text{mol(Br – Br)}$

$= 1\text{mol(436 kJ/mol)} + 1\text{mol(193 kJ/mol)}$

$= 629\text{ kJ/mol}$

Determine which bonds are formed in the reaction and sum the negatives of the bond energies of these

Σ(-ΔH's of bonds formed)
= – 2mol(H – Br)
= – 2mol(364 kJ/mol)
= – 728 kJ/mol

Find ΔH_{rxn} by summing the results of the two steps.

ΔH_{rxn} = Σ(ΔH's bonds broken) + Σ(-ΔH's of bonds formed)
= 629 kJ/mol – 728 kJ/mol
= – 99 kJ/mol

ΔH_f° from the table = –36.3 kJ/mol

The value for ΔH_f° would be expected to be 1/2 the value calculated from the bond energies but it is not. ΔH_f° is for the formation of HBr from elements in the standard state. The standard state of Br is Br_2(l) while we used bond energies for Br_2(g). If you include the value for the formation of Br_2(g), (–30.9kJ/mol), we obtain a value of – 51.8 kJ/mol for the reaction. This is still not ½ the value of ΔH_f° for the formation of HBr. Since it is not, we can account for the difference by looking at the types of bonds broken and formed. The H_2 and Br_2 bonds are pure covalent bonds while the HBr bond formed will be polar covalent. The distribution of the electron density will be unequal.

63.

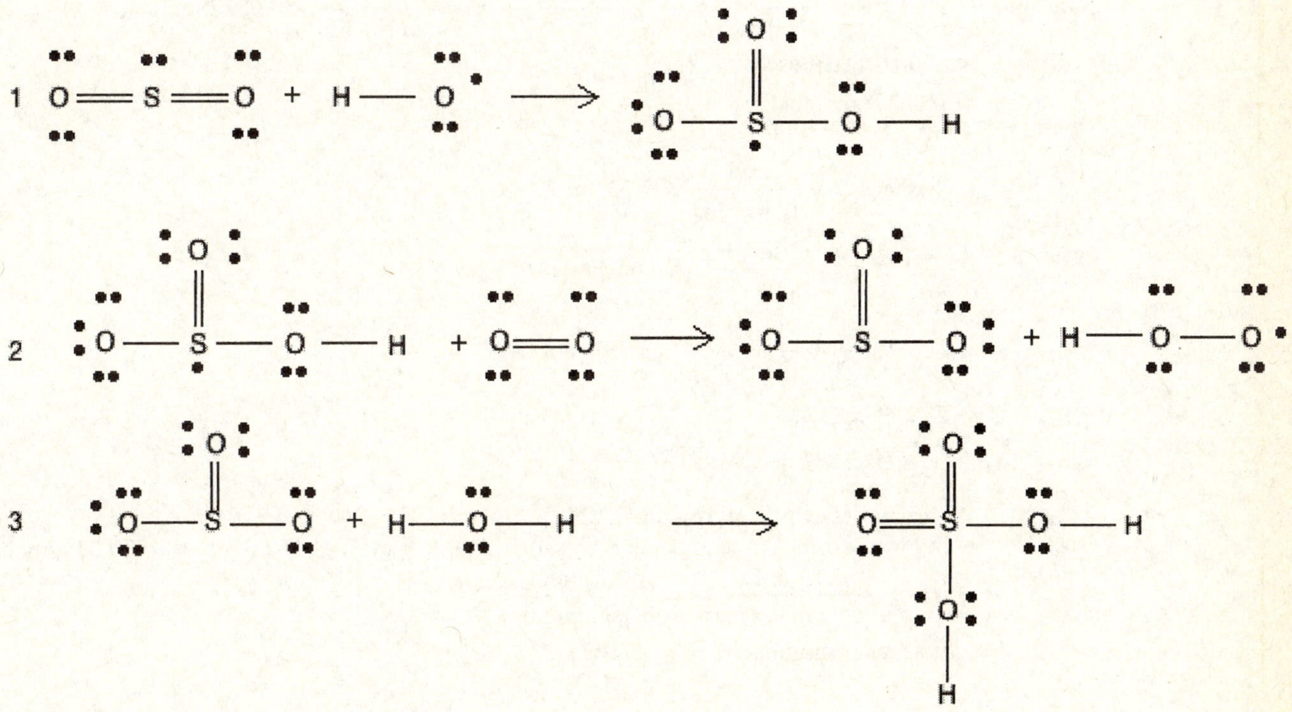

Step 1:

Bonds broken: 2mol(S = O) +1mol(H – O) = 2mol(523kJ/mol) + 1mol(464kJ/mol) = 1510 kJ/mol

Bonds formed: –2mol(S – O) –1mol(S = O) –1mol(O – H) =
–2mol(265 kJ/mol) –1mol(523 kJ/mol) –1mol(464 kJ/mol) = –1517 kJ/mol

ΔH_{step} = –7 kJ/mol

Step 2:

Bonds broken: 2mol(S – O) + 1mol(S = O) + 1mol(O – H) + 1mol(O = O) =
2mol(265 kJ/mol)+1mol(523 kJ/mol)+1mol(464 kJ/mol)+1mol(498 kJ/mol) =
2015 kJ/mol

Bonds formed: : –2mol(S – O) –1mol(S = O) –1mol(O – H) –1mol(O – O) =
–2mol(265 kJ/mol) –1mol(523 kJ/mol) –1mol(464 kJ/mol) –1mol(142 kJ/mol)=
–1659 kJ/mol

ΔH_{step}= +356 kJ/mol

Step 3:

Bonds broken: 2mol(S – O) + 1mol(S = O) + 2mol(O – H) =
2mol(265 kJ/mol)+1mol(523 kJ/mol)+2mol(464 kJ/mol) = 1981 kJ/mol

Bonds formed: $-2mol(S-O) - 2mol(S=O) - 2mol(O-H) =$
$-2mol(265\ kJ/mol) + -2mol(523\ kJ/mol) + -2mol(464\ kJ/mol) = -2504\ kJ/mol$
$\Delta H_{step} = -523\ kJ/mol$

Hess's law states that ΔH for the reaction is the sum of ΔH of the steps:
$\Delta H_{rxn} = (-7\ kJ/mol) + (+356\ kJ/mol) + (-523\ kJ/mol) = -174\ kJ/mol$

64. **Given:** 0.167 g acid; 27.8 mL 0.100 M NaOH; 40.00% C; 6.71 % H; 53.29 % O
Find: molar mass, molecular formula, Lewis structure
Conceptual Plan: mL → L → mol NaOH → mol acid → molar mass and then:

$\frac{L}{1000\ mL}$ mol = VM mol acid = mol base $\frac{mass}{mol}$

convert mass to mol of each element → pseudoformula → empirical formula → molecular formula

$\frac{1\ mol\ C}{12.01\ g\ C}$ $\frac{1\ mol\ H}{1.008\ g\ H}$ $\frac{1\ mol\ N}{14.00\ g\ N}$ divide by smallest number empirical formula x n

→Lewis structure

Solution: $27.8\ mL \times \frac{1\ L}{1000\ mL} \times \frac{0.100\ mol\ NaOH}{L} \times \frac{1\ mol\ acid}{1\ mol\ NaOH} = 0.00278\ mol\ acid$

$$\frac{0.167\ g\ acid}{0.00278\ mol\ acid} = 60.1\ g/mol$$

$$40.00\ \cancel{g\,C} \times \frac{1\ mol\ C}{12.01\ \cancel{g\,C}} = 3.331\ mol\ C$$

$$6.71\ \cancel{g\,H} \times \frac{1\ mol\ H}{1.008\ \cancel{g\,H}} = 6.6\underline{5}7\ mol\ H$$

$$53.29\ \cancel{g\,O} \times \frac{1\ mol\ O}{16.00\ \cancel{g\,O}} = 3.331\ mol\ O$$

$C_{3.331}H_{6.657}O_{3.331}$

$C_{\frac{3.331}{3.331}}H_{\frac{6.657}{3.331}}O_{\frac{3.331}{3.331}} \rightarrow CH_2O$

The correct empirical formula is CH_2O

empirical formula mass = (12.01 g/mol) + 2(1.008 g/mol) + (16.00 g/mol) = 30.03 g/mol

$$n = \frac{\text{molar mass}}{\text{formula molar mass}} = \frac{60.1\ g/mol}{30.03\ g/mol} = 2$$

molecular formula $= CH_2O \times 2$
$= C_2H_4O_2$

Write the correct skeletal structure for the molecule

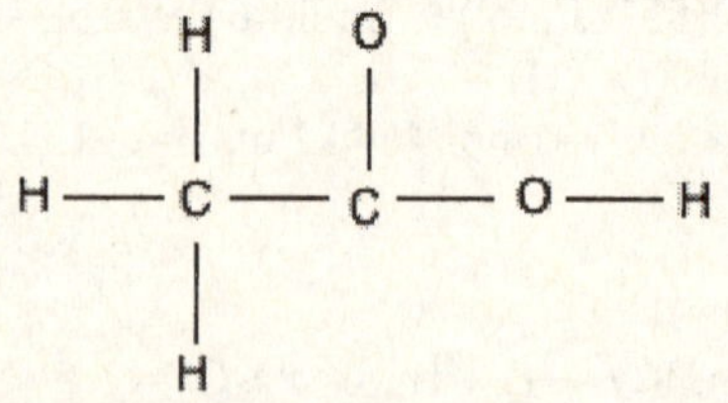

Calculate the total number of electrons for the Lewis structure by summing the valence electrons of each atom in the molecule.

2(number of valence e^- for C) + 2(number of valence e^- for O) + 4(number of valence e^- for H) = 2(4) + 2(6) + 4(1) = 24

Distribute the electrons among the atoms, giving octets (or duets for H) to as many atoms as possible. Begin with the bonding electrons, and then proceed to lone pairs on terminal atoms and finally to lone pairs of the central atom.

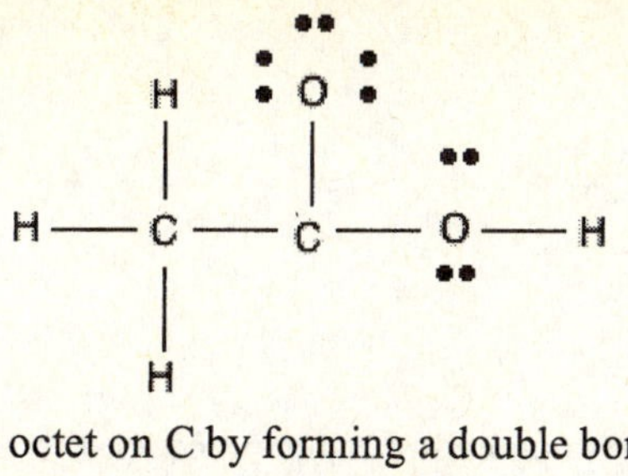

Complete the octet on C by forming a double bond

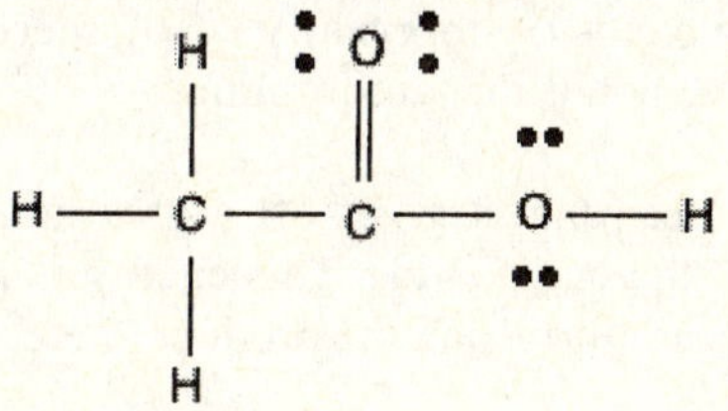

65. **Given:** μ = 1.08 D HCl, 20% ionic and μ = 1.82 D HF, 45% ionic **Find:** r
Conceptual Plan: μ → μ_{calc} → r

$$\%\text{ ionic character} = \frac{\mu}{\mu_{calc}} \qquad \mu_{calc} = qr$$

Solution: For HCl $\quad \mu_{calc} = \dfrac{1.08}{0.20} = 5.4\,D \qquad \dfrac{5.4\,D \times \dfrac{3.34 \times 10^{-30}\ C\cdot m}{D} \times \dfrac{10^{12}\ pm}{m}}{1.6 \times 10^{-19}\ C} = 113\ pm$

For HF $\quad \mu_{calc} = \dfrac{1.82}{0.45} = 4.\underline{0}4\,D \qquad \dfrac{4.\underline{0}4\,D \times \dfrac{3.34 \times 10^{-30}\ C\cdot m}{D} \times \dfrac{10^{12}\ pm}{m}}{1.6 \times 10^{-19}\ C} = 84\ pm$

From Table 9.4, the bond length of HCl = 127 pm, and HF = 92 pm. Both of these values are slightly higher than the calculated values.

66. Formation reaction: $6C(s) + 3H_2(g) \rightarrow C_6H_6(g)$ $\Delta H_f^\circ = 82.9$ kJ/mol
Using bond energies we would have the reaction: $6C(g) + 3H_2(g) \rightarrow C_6H_6(g)$ so we have to include in the bond energy calculation the energy needed to convert C(s) → C(g) (718.4 kJ/mol)
6molC(s) → 6molC(g) 6mol(718.4 kJ/mol) = 4310.4 kJ/mol
Rewrite the reaction with the Lewis structures.

6C(g) + 3 H – H(g) →

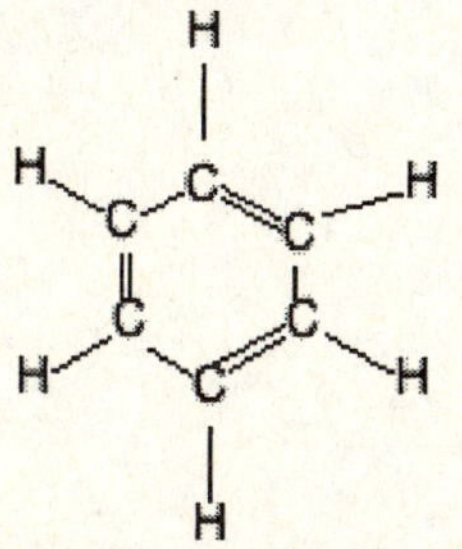

(g)

bonds broken: 3mol(H – H) = 3mol(436 kJ/mol) = 1308 kJ/mol
bonds formed: –3mol(C = C) – 3mol(C – C) – 6mol(C – H) =
–3mol(611kJ/mol) – 3mol(347 kJ/mol) – 6mol(414 kJ/mol) = –5358 kJ/mol
ΔH from bond energies = +(4310 kJ) + (1308 kJ) – 5358 kJ = +260 kJ/mol
The difference between the value calculated from bond energies (260 kJ/mol) and ΔH_f° = 82.9 kJ/mol for benzene leads us to conclude there is a great deal of stabilization from the two resonance forms and that they contribute much to the formation of benzene.

67. In order for the four P atoms to be equivalent, they must all be in the same electronic environment. That is, they must all see the same number of bonds and lone pair electrons. The only way to achieve this is with a tetrahedral configuration where the P atoms are at the four points of the tetrahedron.

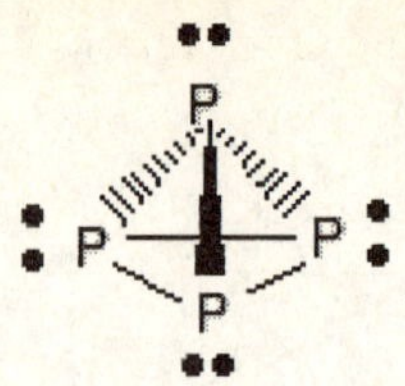

68. a) is true: Strong bonds break and weak bonds form. In an endothermic reaction, the energy required to break the bonds is greater than the energy given off when the bonds are formed ($\Delta H > 0$), therefore, in an endothermic reaction the bonds that are breaking are stronger than the bonds that are forming.

69. When we say that a compound is "energy rich" we mean that it gives off a great amount of energy when it reacts. It means that there is a lot of energy stored in the compound. This energy is released when the weak bonds in the compound break and much stronger bonds are formed in the product, thereby releasing energy.

70. In solid covalent compounds, the electrons in the bonds are shared directly between the atoms involved in the molecule. Each molecule is a distinct unit. Ionic compounds, on the other hand, are not distinct units. Rather, they are composed of alternating positive and negative ions in a three-dimensional crystalline array.

71. Lewis theory is successful because it allows us to understand and predict many chemical observations. We can use it to determine the formula of ionic compounds, to account for low melting points and boiling points of molecular compounds compared to ionic compounds. Lewis theory allows us to predict what molecules or ions will be stable, which will be more reactive, and which will not exist. Lewis theory, however, does not really tell us anything about how the bonds in the molecules and ions form. It does not give us a way to account for the paramagnetism of oxygen. And, by itself, Lewis theory does not really tell us anything about the shape of the molecule or ion.

Chapter 10
Chemical Bonding II: Molecular Shapes, Valence Bond Theory, and Molecular Orbital Theory

1. 4 electron pairs: A trigonal pyramidal molecular geometry has three bonding groups and one lone pair of electrons, so there are four electron pairs on atom A.

2. 3 electron pairs: A trigonal planar molecular geometry has three bonding groups and no lone pairs of electrons so there are three electron pairs on atom A.

3. a) 4 total electron groups, 4 bonding groups, 0 lone pair
   A tetrahedral molecular geometry has four bonding groups and no lone pair. So, there are four total electron groups, four bonding groups, and no lone pair.

   b) 5 total electron groups, 3 bonding groups, 2 lone pairs
   A T-shaped molecular geometry has three bonding groups and two lone pairs. So, there are five total electron groups, three bonding groups, and two lone pairs.

   c) 6 total electron groups, 5 bonding groups, 1 lone pair
   A square pyramidal molecular geometry has five bonding groups and one lone pair. So, there are six total electron groups, five bonding groups, and one lone pair.

4. a) 6 total electron groups, 6 bonding groups, 0 lone pair
   An octahedral molecular geometry has six bonding groups and no lone pair. So, there are six total electron groups, six bonding groups, and no lone pairs.

   b) 6 electron groups, 4 bonding groups, 2 lone pairs
   A square planar molecular geometry has four bonding groups and two lone pairs. So, there are six total electron groups, four bonding groups, and two lone pairs.

   c) 5 electron groups, 4 bonding groups, 1 lone pair
   A seesaw molecular geometry has four bonding groups and one lone pair. So, there are five total electron groups, four bonding groups, and one lone pair.

5. a) PF_3: Electron geometry – tetrahedral; molecular geometry – trigonal pyramidal; bond angle = 109.5°
   Because of the lone pair, the bond angle will be less than 109.5°.
   Draw a Lewis structure for the molecule:
   PF_3 has 26 valence electrons

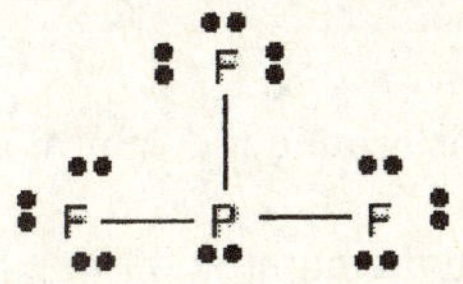

   Determine the total number of electron groups around the central atom:
   There are four electron groups on P

   Determine the number of bonding groups and the number of lone pairs around the central atom:
   There are three bonding groups and one lone pair

   Use Table 10.1 to determine the electron geometry and molecular geometry and bond angles:
   Four electron groups is tetrahedral electron geometry, three bonding groups and one lone pair is trigonal pyramidal molecular geometry, the idealized bond angles for tetrahedral are 109.5°; however, the lone pair will make the bond angle less than idealized.

   b) SBr_2: Electron geometry – tetrahedral; molecular geometry – bent; bond angle = 109.5°
   Because of the lone pairs, the bond angle will be less than 109.5°.
   Draw a Lewis structure for the molecule:
   SBr_3 has 20 valence electrons

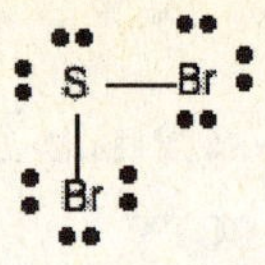

Determine the total number of electron groups around the central atom:
There are four electron groups on S
Determine the number of bonding groups and the number of lone pairs around the central atom:
There are two bonding groups and two lone pair
Use Table 10.1 to determine the electron geometry and molecular geometry and bond angles:
Four electron groups is tetrahedral electron geometry, two bonding groups and two lone pair is a bent molecular geometry, the idealized bond angles for tetrahedral are 109.5°; however, the lone pairs will make the bond angle less than idealized.

c) $CHCl_3$: Electron geometry – tetrahedral; molecular geometry – tetrahedral; bond angle = 109.5°
Because there are no lone pairs, the bond angle will be 109.5°.
Draw a Lewis structure for the molecule:
$CHCl_3$ has 26 valence electrons

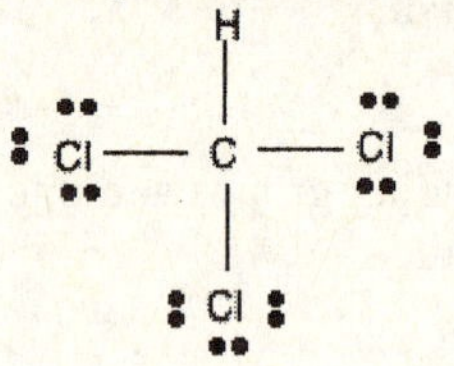

Determine the total number of electron groups around the central atom:
There are four electron groups on C
Determine the number of bonding groups and the number of lone pairs around the central atom:
There are four bonding groups and no lone pair
Use Table 10.1 to determine the electron geometry and molecular geometry and bond angles:
Four electron groups is tetrahedral electron geometry, four bonding groups and no lone pair is a tetrahedral molecular geometry, the idealized bond angles for tetrahedral are 109.5°; however, because the attached atoms have different electronegativities the bond angles are less than idealized.

d) CS_2: Electron geometry – linear; molecular geometry – linear; bond angle = 180°
Because there are no lone pairs, the bond angle will 180°.
Draw a Lewis structure for the molecule:
CS_2 has 16 valence electrons

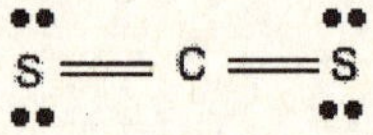

Determine the total number of electron groups around the central atom:
There are two electron groups on C
Determine the number of bonding groups and the number of lone pairs around the central atom:
There are two bonding groups and no lone pair
Use Table 10.1 to determine the electron geometry and molecular geometry and bond angles:
Two electron groups is linear geometry, two bonding groups and no lone pair is linear molecular geometry, the idealized bond angle is 180°

6. a) CF_4: Electron geometry – tetrahedral; molecular geometry – tetrahedral; bond angle = 109.5°
Draw a Lewis structure for the molecule:
CF_4 has 32 valence electrons

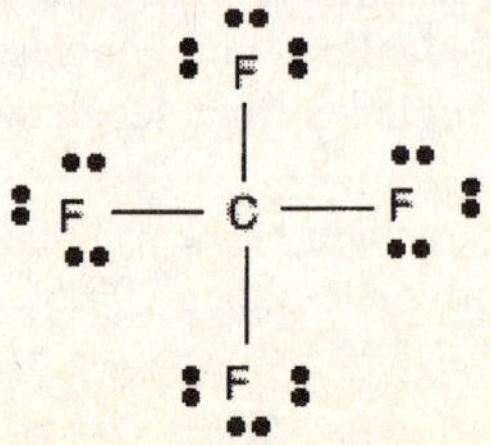

Determine the total number of electron groups around the central atom:
There are four electron groups on C
Determine the number of bonding groups and the number of lone pairs around the central atom:
There are four bonding groups and no lone pairs.
Use Table 10.1 to determine the electron geometry and molecular geometry and bond angles:
Four electron groups is tetrahedral electron geometry, four bonding groups and no lone pairs is a tetrahedral molecular geometry, the idealized bond angles for tetrahedral are 109.5°

b) NF_3: Electron geometry – tetrahedral; molecular geometry – trigonal pyramidal; bond angle = 109.5°
Because of the lone pair, the bond angle will be less than 109.5°.
Draw a Lewis structure for the molecule:
NF_3 has 26 valence electrons

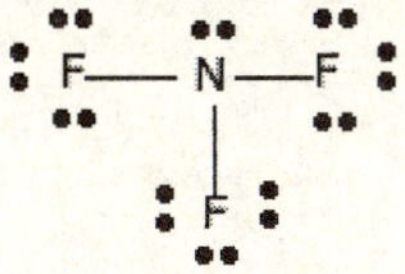

Determine the total number of electron groups around the central atom:
There are four electron groups on N
Determine the number of bonding groups and the number of lone pairs around the central atom:
There are three bonding groups and one lone pair
Use Table 10.1 to determine the electron geometry and molecular geometry and bond angles:
Four electron groups is tetrahedral electron geometry, three bonding groups and one lone pair is a trigonal pyramidal molecular geometry, the idealized bond angles for tetrahedral are 109.5°; however, the lone pair will make the bond angle less than idealized.

c) OF_2: Electron geometry – tetrahedral; molecular geometry – bent; bond angle = 109.5°
Because of the lone pairs, the bond angle will be less than 109.5°.
Draw a Lewis structure for the molecule:
OF_2 has 20 valence electrons

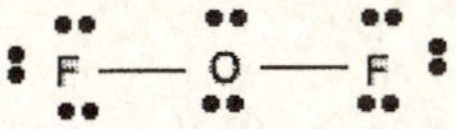

Determine the total number of electron groups around the central atom:
There are four electron groups on O
Determine the number of bonding groups and the number of lone pairs around the central atom:
There are two bonding groups and two lone pairs.
Use Table 10.1 to determine the electron geometry and molecular geometry and bond angles:
Four electron groups is tetrahedral electron geometry, two bonding groups and two lone pairs is a bent molecular geometry, the idealized bond angles for tetrahedral are 109.5°, however, the lone pairs will make the bond angle less than idealized.

d) H_2S: Electron geometry – tetrahedral; molecular geometry – bent; bond angle = 109.5°
Because of the lone pair, the bond angle will be less than 109.5°.
Draw a Lewis structure for the molecule:
H_2S has 8 valence electrons

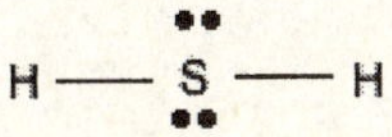

Determine the total number of electron groups around the central atom:
There are four electron groups on S
Determine the number of bonding groups and the number of lone pairs around the central atom:
There are two bonding groups and two lone pairs
Use Table 10.1 to determine the electron geometry and molecular geometry and bond angles:
Four electron groups is tetrahedral electron geometry, two bonding groups and two lone pairs is a bent molecular geometry, the idealized bond angles for tetrahedral are 109.5°; however, the lone pairs will make the bond angle less than idealized.

7. H_2O will have the smaller bond angle because lone pair–lone pair repulsions are greater than lone pair–bonding pair repulsions.

Draw the Lewis structures for both structures:

H_3O^+ has 8 valence electrons $\qquad$ H_2O has 8 valence electrons

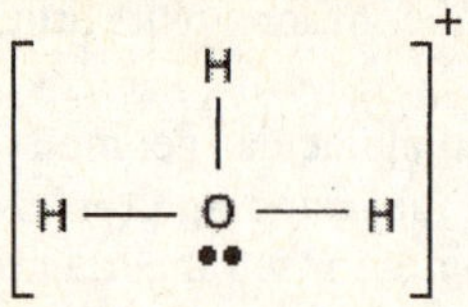

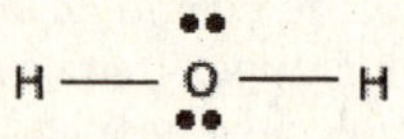

3 bonding groups and 1 lone pair $\qquad$ 2 bonding groups and 2 lone pairs

Both have 4 electron groups, but the 2 lone pairs in H_2O will cause the bond angle to be smaller because of the lone pair–lone pair repulsions.

8. ClO_3^- will have the smaller bond angle because lone pair–bonding pair repulsions are greater than bonding pair–bonding pair repulsions.

Draw the Lewis structures for both structures:

ClO_3^- has 26 valence electrons $\qquad$ ClO_4^- has 32 valence electrons

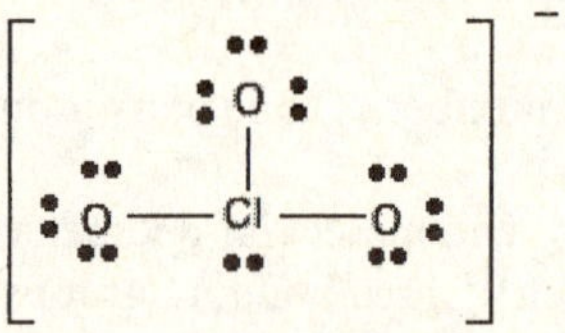

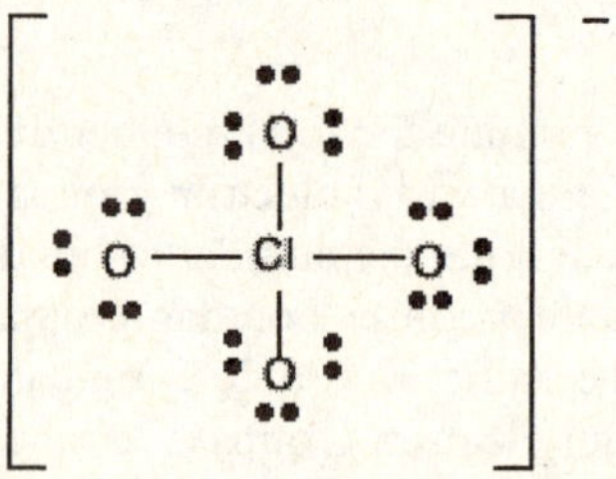

3 bonding groups and 1 lone pair $\qquad$ 4 bonding groups and no lone pairs

Both have 4 electron groups, but the lone pair in ClO_3^- will cause the bond angle to be smaller because of the lone pair–bonding pair repulsions.

9. a) SF_4 $\qquad$ Draw a Lewis structure for the molecule:

SF_4 has 34 valence electrons

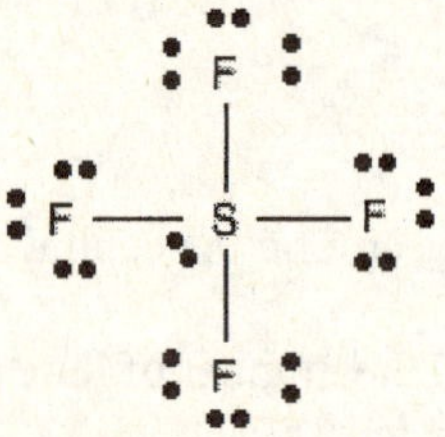

Determine the total number of electron groups around the central atom:

There are five electron groups on S

Determine the number of bonding groups and the number of lone pairs around the central atom:

There are four bonding groups and one lone pair

Use Table 10.1 to determine the electron geometry and molecular geometry:

The electron geometry is trigonal bipyramidal so the molecular geometry is seesaw

Sketch the molecule:

b) ClF_3 $\qquad$ Draw a Lewis structure for the molecule:

ClF_3 has 28 valence electrons

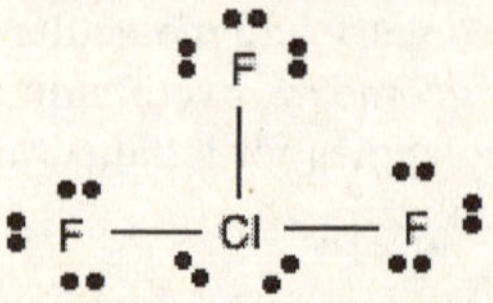

Determine the total number of electron groups around the central atom:

There are five electron groups on Cl

Determine the number of bonding groups and the number of lone pairs around the central atom:
There are three bonding groups and two lone pairs.
Use Table 10.1 to determine the electron geometry and molecular geometry:
The electron geometry is trigonal bipyramidal so the molecular geometry is T-shape
Sketch the molecule:

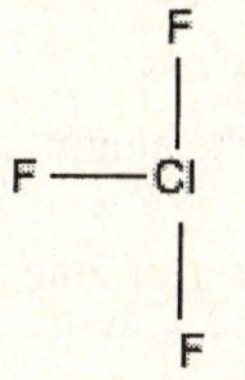

c) IF_2^- Draw a Lewis structure for the ion:
IF_2^- has 22 valence electrons

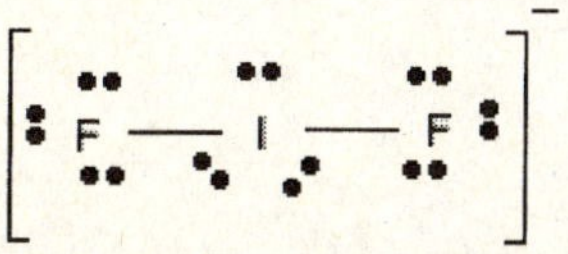

Determine the total number of electron groups around the central atom:
There are five electron groups on I
Determine the number of bonding groups and the number of lone pairs around the central atom:
There are two bonding groups and three lone pairs.
Use Table 10.1 to determine the electron geometry and molecular geometry:
The electron geometry is trigonal bipyramidal so the molecular geometry is linear
Sketch the ion:

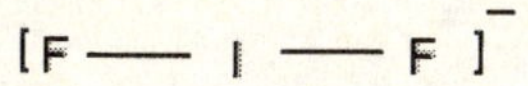

d) IBr_4^- Draw a Lewis structure for the ion:
IBr_4^- has 36 valence electrons

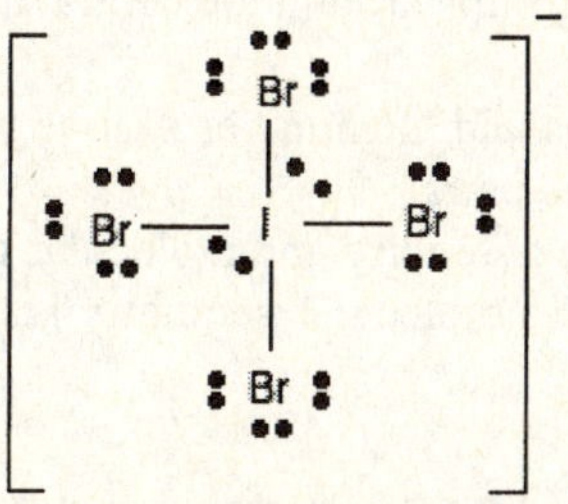

Determine the total number of electron groups around the central atom:
There are six electron groups on I
Determine the number of bonding groups and the number of lone pairs around the central atom:
There are four bonding groups and two lone pairs.
Use Table 10.1 to determine the electron geometry and molecular geometry:
The electron geometry is octahedral so the molecular geometry is square planar
Sketch the ion:

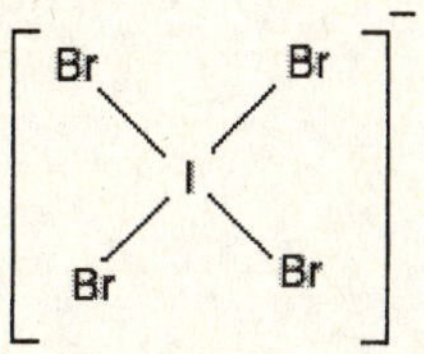

10\. a) BrF_5 Draw a Lewis structure for the molecule:
BrF_5 has 42 valence electrons

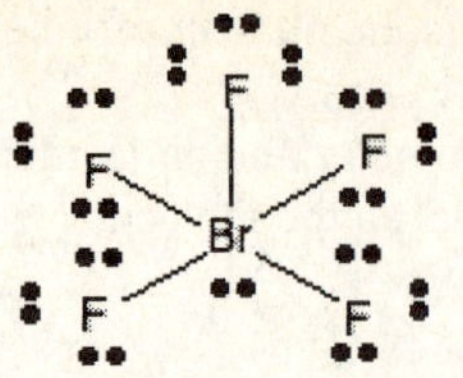

Determine the total number of electron groups around the central atom:
There are six electron groups on Br
Determine the number of bonding groups and the number of lone pairs around the central atom:
There are five bonding groups and one lone pair
Use Table 10.1 to determine the electron geometry and molecular geometry:
The electron geometry is octahedral so the molecular geometry is square pyramidal
Sketch the molecule:

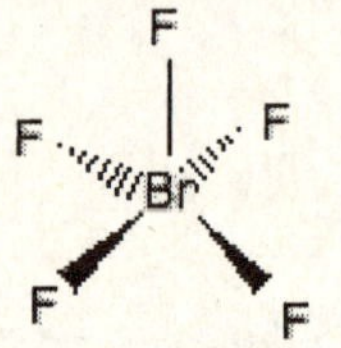

b) SCl_6 Draw a Lewis structure for the molecule:
SCl_6 has 48 valence electrons

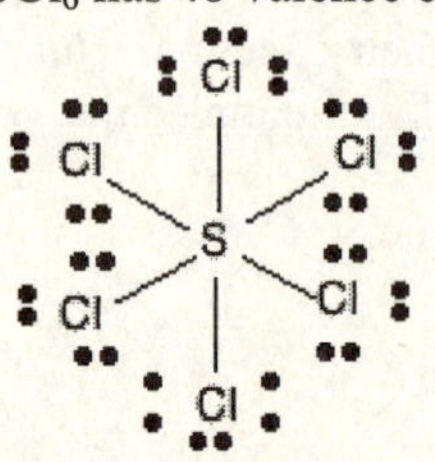

Determine the total number of electron groups around the central atom:
There are six electron groups on S
Determine the number of bonding groups and the number of lone pairs around the central atom:
There are six bonding groups and no lone pairs.
Use Table 10.1 to determine the electron geometry and molecular geometry:
The electron geometry is octahedral so the molecular geometry is octahedral
Sketch the molecule:

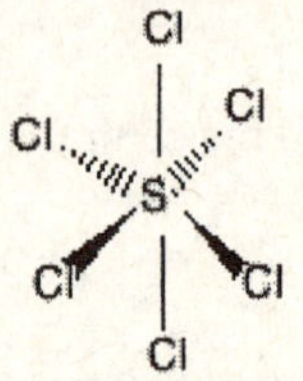

c) PF_5 Draw a Lewis structure for the molecule:
PF_5 has 40 valence electrons

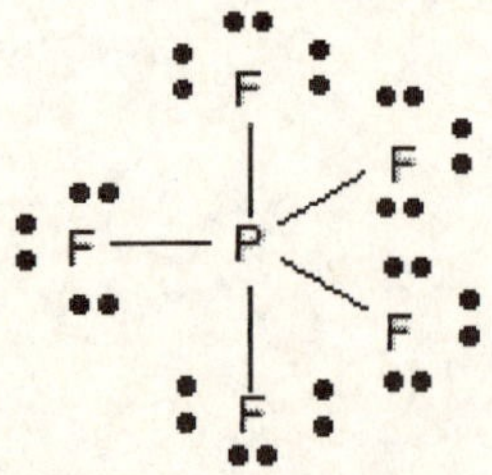

Determine the total number of electron groups around the central atom:
There are five electron groups on P
Determine the number of bonding groups and the number of lone pairs around the central atom:
There are five bonding groups and no lone pairs.

Use Table 10.1 to determine the electron geometry and molecular geometry:
The electron geometry is trigonal bipyramidal so the molecular geometry is trigonal bipyramidal
Sketch the molecule:

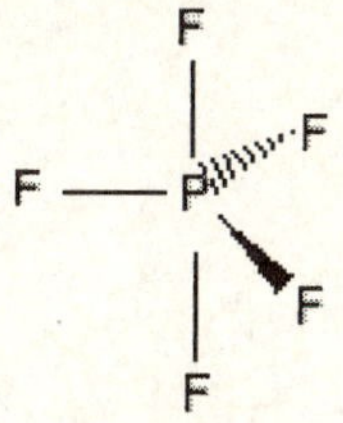

d) IF_4^+ Draw a Lewis structure for the ion:

IF_4^+ has 34 valence electrons

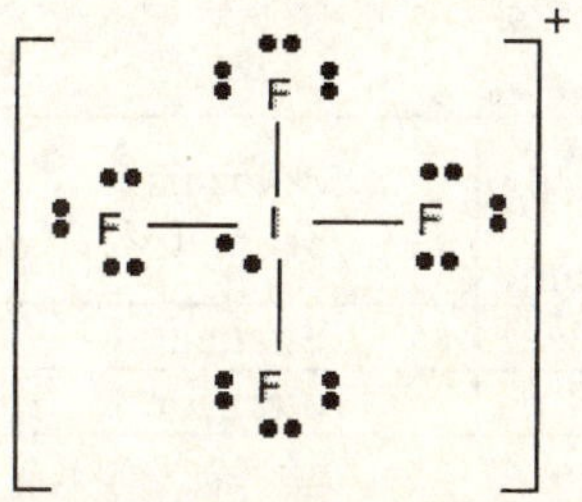

Determine the total number of electron groups around the central atom:
There are five electron groups on I
Determine the number of bonding groups and the number of lone pairs around the central atom:
There are four bonding groups and one lone pair
Use Table 10.1 to determine the electron geometry and molecular geometry:
The electron geometry is trigonal bipyramidal so the molecular geometry is seesaw
Sketch the ion:

11. a) C_2H_2 Draw the Lewis structure:

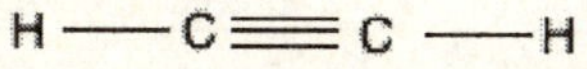

Atom	Number of Electron Groups	Number of Lone Pairs	Molecular Geometry
Left C	2	0	Linear
Right C	2	0	Linear

Sketch the molecule:

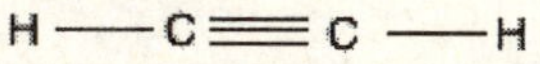

b) C_2H_4 Draw the Lewis structure:

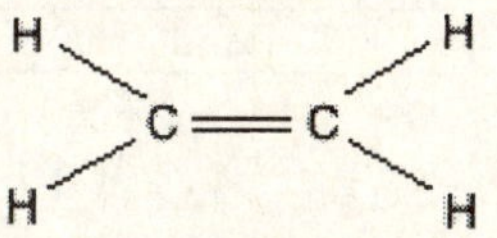

Atom	Number of Electron Groups	Number of Lone Pairs	Molecular Geometry
Left C	3	0	Trigonal planar
Right C	3	0	Trigonal planar

Sketch the molecule:

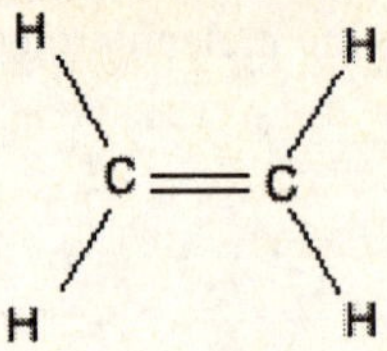

c) C_2H_6 Draw the Lewis structure:

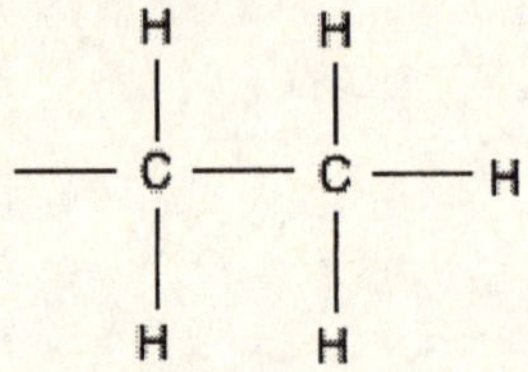

Atom	Number of Electron Groups	Number of Lone Pairs	Molecular Geometry
Left C	4	0	Tetrahedral
Right C	4	0	Tetrahedral

Sketch the molecule:

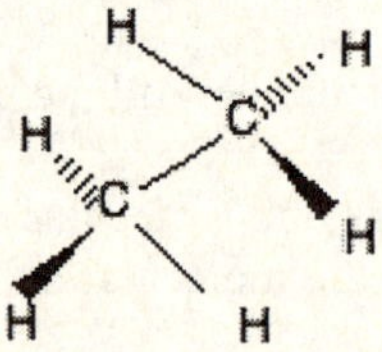

12. a) N_2 Draw the Lewis structure:

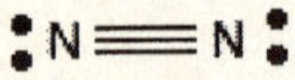

Atom	Number of Electron Groups	Number of Lone Pairs	Molecular Geometry
Left N	2	1	Linear
Right N	2	1	Linear

Sketch the molecule:

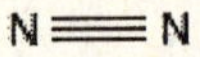

b) N_2H_2 Draw the Lewis structure:

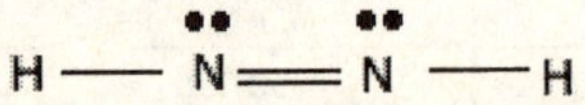

Atom	Number of Electron Groups	Number of Lone Pairs	Molecular Geometry
Left N	3	1	Bent
Right N	3	1	Bent

Sketch the molecule:

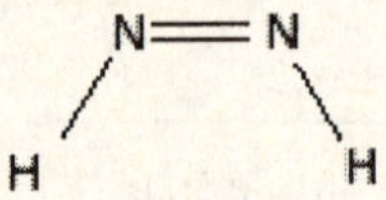

c) N_2H_4 Draw the Lewis structure:

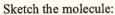

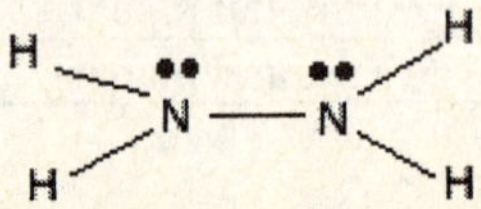

Atom	Number of Electron Groups	Number of Lone Pairs	Molecular Geometry
Left N	4	1	Trigonal pyramidal
Right C	4	1	Trigonal pyramidal

Sketch the molecule:

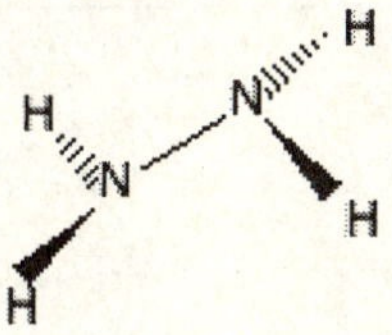

13. a) Four pairs of electrons gives a tetrahedral electron geometry, the lone pair would cause lone pair–bonded pair repulsions and would have a trigonal pyramidal molecular geometry.

    b) Five pairs of electrons gives a trigonal bipyramidal electron geometry, the lone pair occupies an equatorial position in order to minimize lone pair–bonded pair repulsions and the molecule would have a seesaw molecular geometry.

    c) Six pairs of electrons gives an octahedral electron geometry, the two lone pair would occupy opposite position in order to minimize lone pair–lone pair repulsions. The molecular geometry would be square planar.

14. a) Four pairs of electrons gives a tetrahedral electron geometry, the two lone pair would cause repulsions that would lead to a bent molecular geometry.

    b) Five pairs of electrons gives a trigonal bipyramidal geometry. The three lone pair would occupy equatorial positions in order to minimize the lone pair–lone pair repulsions. This would give a linear molecular geometry.

    c) Six pairs of electrons gives an octahedral electron geometry. The lone pair would occupy a position to minimize the lone pair–bonded pair repulsions and gives a square pyramidal molecular geometry.

15. a) CH_3OH Draw the Lewis structure, determine the geometry about each interior atom:

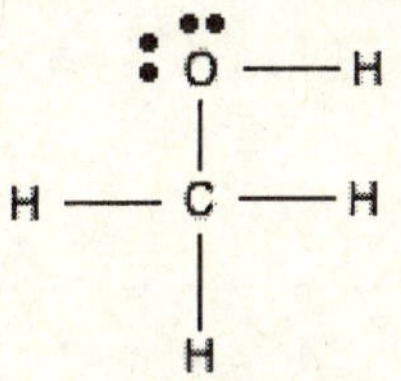

Atom	Number of Electron Groups	Number of Lone Pairs	Molecular Geometry
C	4	0	Tetrahedral
O	4	2	Bent

Sketch the molecule:

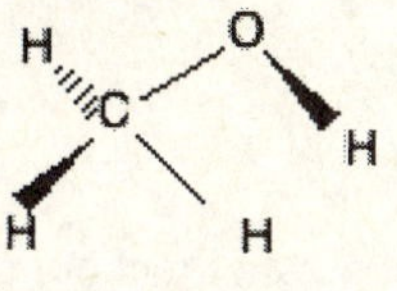

    b) CH_3OCH_3 Draw the Lewis structure, determine the geometry about each interior atom:

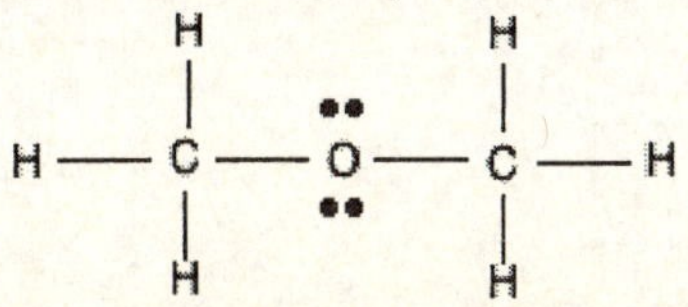

Atom	Number of Electron Groups	Number of Lone Pairs	Molecular Geometry
C	4	0	Tetrahedral
O	4	2	Bent
C	4	0	Tetrahedral

Sketch the molecule:

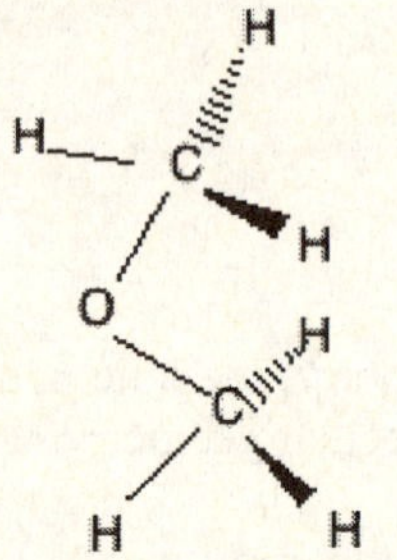

c) H_2O_2 Draw the Lewis structure, determine the geometry about each interior atom:

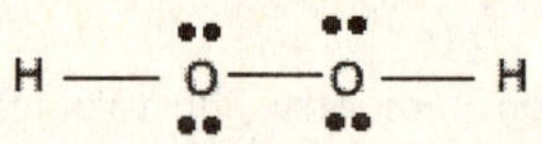

Atom	Number of Electron Groups	Number of Lone Pairs	Molecular Geometry
O	4	2	Bent
O	4	2	Bent

Sketch the molecule:

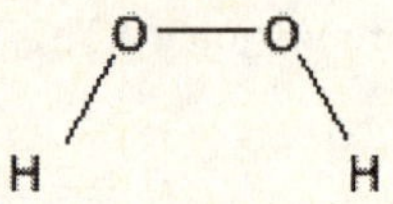

16. a) CH_3NH_2 Draw the Lewis structure, determine the geometry about each interior atom:

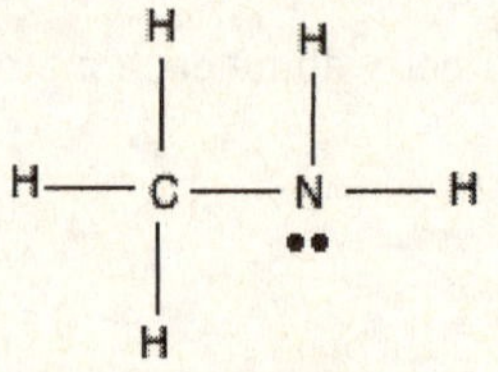

Atom	Number of Electron Groups	Number of Lone Pairs	Molecular Geometry
C	4	0	Tetrahedral
N	4	1	Trigonal Pyramidal

Sketch the molecule:

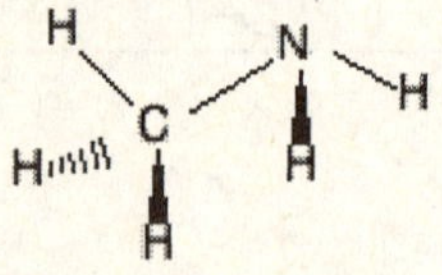

b) $CH_3CO_2CH_3$ Draw the Lewis structure, determine the geometry about each interior atom:

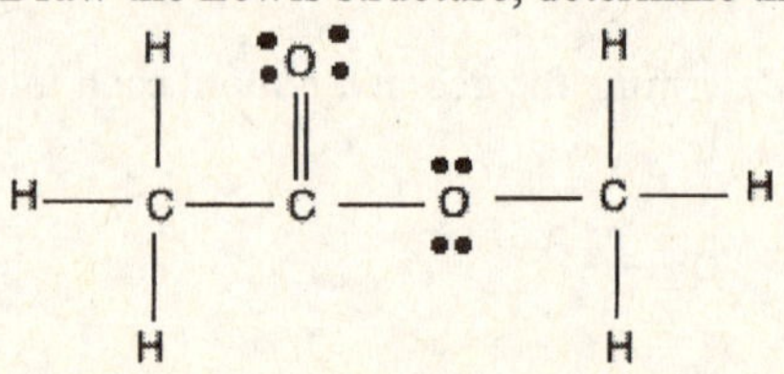

Atom	Number of Electron Groups	Number of Lone Pairs	Molecular Geometry
Left C	4	0	Tetrahedral
Center C	3	0	Trigonal Planar
O	4	2	Bent
Right C	4	0	Tetrahedral

Sketch the molecule:

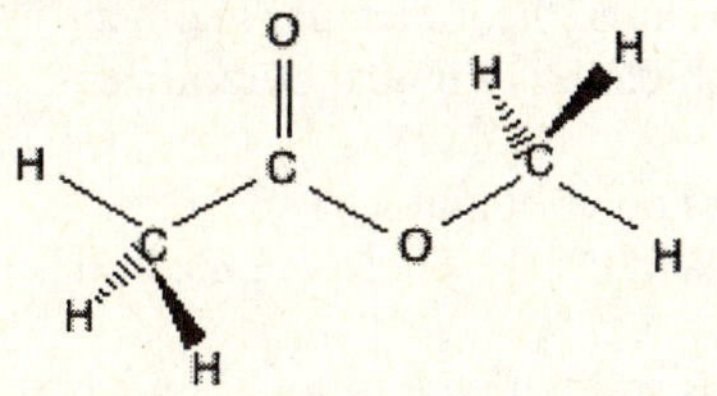

c) NH_2CO_2H Draw the Lewis structure, determine the geometry about each interior atom

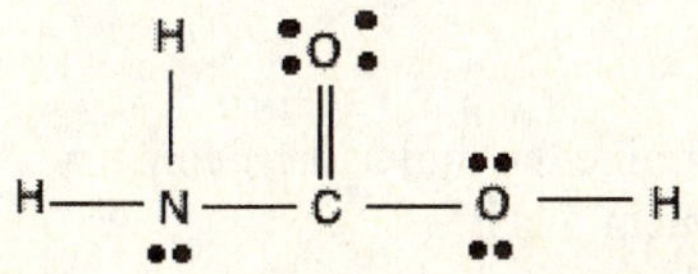

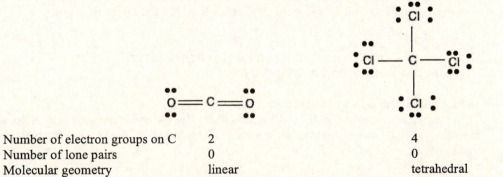

Atom	Number of Electron Groups	Number of Lone Pairs	Molecular Geometry
N	4	1	Trigonal Pyramidal
C	3	0	Trigonal Planar
O	4	2	Bent

Sketch the molecule:

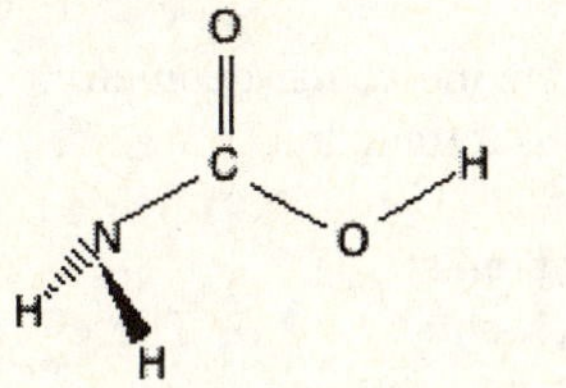

17. Draw the Lewis structure for CO_2 and CCl_4 and determine the molecular geometry.

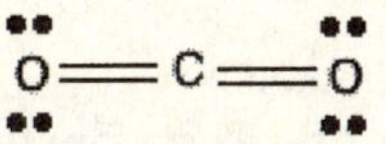

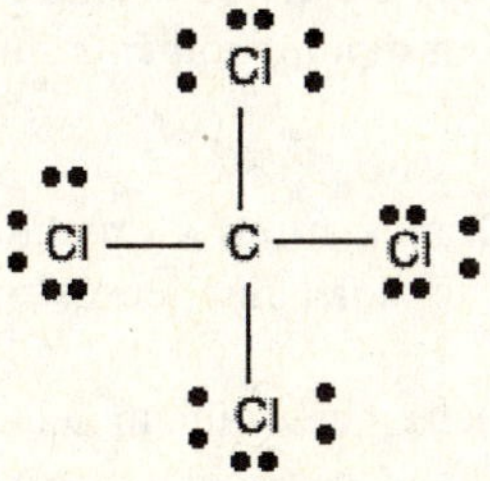

Number of electron groups on C	2	4
Number of lone pairs	0	0
Molecular geometry	linear	tetrahedral

Even though each molecule contains polar bonds, the sum of the bond dipoles gives a net dipole of zero for each molecule.

18. Draw the Lewis structure of CH_3F and determine the molecular geometry.

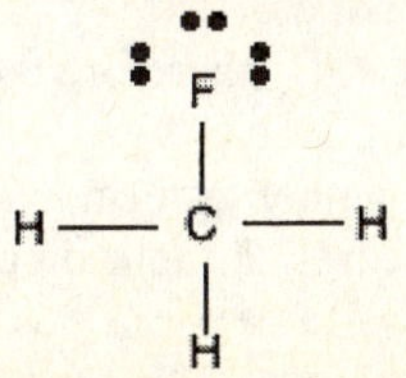

Number of electron groups on C	4
Number of lone pairs	0
Molecular geometry	tetrahedral

The molecule is tetrahedral but is polar because the C – H bond dipoles are different from the C – F bond dipole. Because the bond dipoles are different, the sum of the bond dipoles is NOT zero. Therefore, the molecule is polar.

19. a) PF_3 – polar

Draw the Lewis structure and determine the molecular geometry:
The molecular geometry from problem 5 is trigonal pyramidal.

Determine if the molecule contains polar bonds:
The electronegativities of P = 2.1 and F = 4. Therefore the bonds are polar.

Determine whether the polar bonds add together to form a net dipole:
Because the molecule is trigonal pyramidal, the three dipole moments sum to a nonzero net dipole moment. The molecule is polar.

b) SBr_2 – polar

Draw the Lewis structure and determine the molecular geometry:
The molecular geometry from problem 5 is bent.

Determine if the molecule contains polar bonds:
The electronegativities of S = 2.5 and Br = 2.0. Therefore the bonds are polar.

Determine whether the polar bonds add together to form a net dipole:
Because the molecule is bent, the two dipole moments sum to a nonzero net dipole moment. The molecule is polar.

c) $CHCl_3$ – polar

Draw the Lewis structure and determine the molecular geometry:
The molecular geometry from problem 5 is tetrahedral.

Determine if the molecule contains polar bonds:
The electronegativities of C = 2.5, H = 2.1, and Cl = 3.0. Therefore the bonds are polar.

Determine whether the polar bonds add together to form a net dipole:
Because the bonds have different dipole moments because of the different atoms involved, the four dipole moments sum to a nonzero net dipole moment. The molecule is polar.

d) CS_2 – nonpolar

Draw the Lewis structure and determine the molecular geometry:
The molecular geometry from problem 5 is linear.

Determine if the molecule contains polar bonds:
The electronegativities of C = 2.5 and S = 2.5. Therefore the bonds are nonpolar. Also, the molecule is linear, which would result in a zero net dipole even if the bonds were polar.
The molecule is nonpolar.

20. a) CF_4 – nonpolar

Draw the Lewis structure and determine the molecular geometry:
The molecular geometry from problem 6 is tetrahedral.

Determine if the molecule contains polar bonds:
The electronegativities of C = 2.5 and F = 4.0. Therefore the bonds are polar.

Determine whether the polar bonds add together to form a net dipole:
Because the molecular geometry is tetrahedral, the four dipole moments sum to a zero net dipole moment. The molecule is nonpolar.

b) NF_3 – polar

Draw the Lewis structure and determine the molecular geometry:
The molecular geometry from problem 6 is trigonal pyramidal.

Determine if the molecule contains polar bonds:
The electronegativities of N = 3.0 and F = 4.0. Therefore the bonds are polar.

Determine whether the polar bonds add together to form a net dipole:
Because the molecular geometry is trigonal pyramidal, the three dipole moments sum to a nonzero net dipole moment. The molecule is polar.

c) OF_2 – polar

Draw the Lewis structure and determine the molecular geometry:
The molecular geometry from problem 6 is bent

Determine if the molecule contains polar bonds:
The electronegativities of O = 3.5 and F = 4.0. Therefore the bonds are polar.

Determine whether the polar bonds add together to form a net dipole:
Because the molecular geometry is bent, the two dipole moments sum to a nonzero net dipole moment. The molecule is polar.

d) H_2S – polar

Draw the Lewis structure and determine the molecular geometry:
The molecular geometry from problem 6 is bent

Determine if the molecule contains polar bonds:
The electronegativities of H = 2.1 and S = 2.5. Therefore the bonds are polar.

Determine whether the polar bonds add together to form a net dipole:
Because the molecular geometry is bent, the two dipole moments sum to a nonzero net dipole moment. The molecule is polar.

21. a) ClO_3^- – polar

Draw the Lewis structure and determine the molecular geometry:

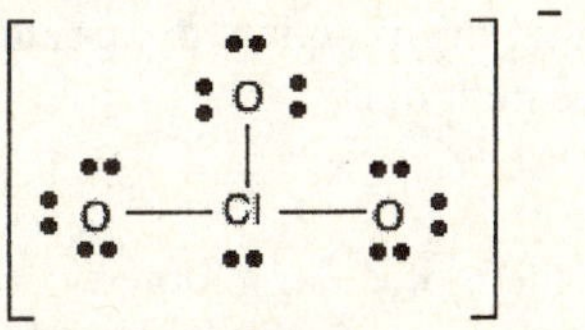

Four electron pairs, with one lone pair gives a trigonal pyramidal molecular geometry

Determine if the molecule contains polar bonds:
The electronegativities of Cl = 3.0 and O = 3.5. Therefore the bonds are polar.

Determine whether the polar bonds add together to form a net dipole:
Because the molecular geometry is trigonal pyramidal, the three dipole moments sum to a nonzero net dipole moment. The molecule is polar.

b) SCl_2 – polar

Draw the Lewis structure and determine the molecular geometry:

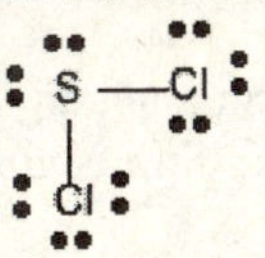

Four electron pairs with two lone pairs gives a bent molecular geometry

Determine if the molecule contains polar bonds:

The electronegativities of S = 2.5 and Cl = 3.0. Therefore the bonds are polar.

Determine whether the polar bonds add together to form a net dipole:
Because the molecular geometry is bent, the two dipole moments sum to a nonzero net dipole moment. The molecule is polar.

c) SCl_4 – polar
Draw the Lewis structure and determine the molecular geometry:

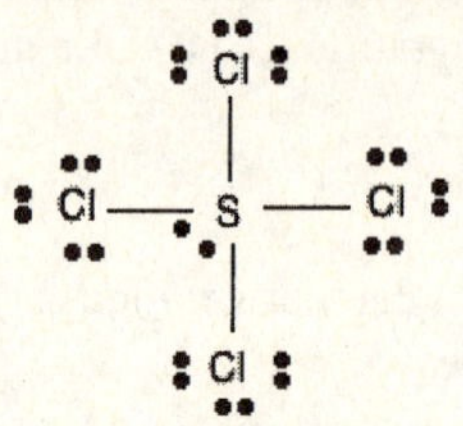

Five electron pairs with one lone pair give a seesaw molecular geometry

Determine if the molecule contains polar bonds:
The electronegativities of S = 2.5 and Cl = 3.0. Therefore the bonds are polar.

Determine whether the polar bonds add together to form a net dipole:
Because the molecular geometry is seesaw, the four dipole moments sum to a nonzero net dipole moment. The molecule is polar.

d) $BrCl_5$ – nonpolar
Draw the Lewis structure and determine the molecular geometry:

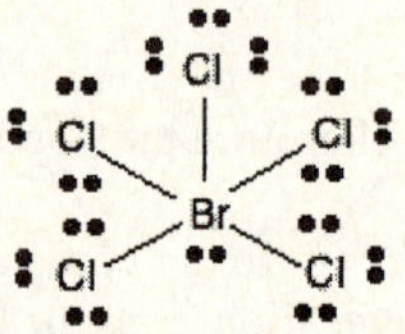

Six electron pairs with one lone pair give square pyramidal molecular geometry

Determine if the molecule contains polar bonds:
The electronegativity of Br = 2.8 and Cl = 3.0. The difference is only 0.2, therefore the bonds are nonpolar. Even though the molecular geometry is square pyramidal, the five bonds are nonpolar so there is no net dipole. The molecule is nonpolar.

22. a) $SiCl_4$ – nonpolar
Draw the Lewis structure and determine the molecular geometry:

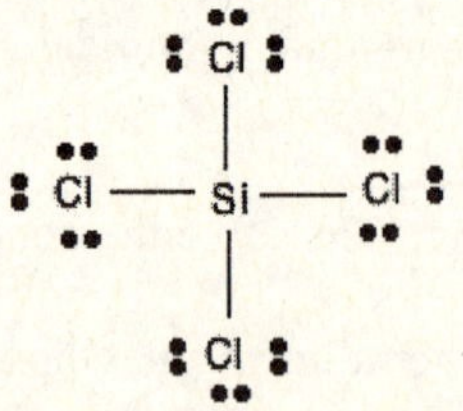

Four electron pairs with no lone pairs gives a tetrahedral molecular geometry

Determine if the molecule contains polar bonds:
The electronegativities of Cl = 3.0 and Si = 1.8. Therefore the bonds are polar.

Determine whether the polar bonds add together to form a net dipole:
Because the molecular geometry is tetrahedral, the four dipole moments sum to a zero net dipole moment. The molecule is nonpolar.

b) CF_2Cl_2 – polar
Draw the Lewis structure and determine the molecular geometry:

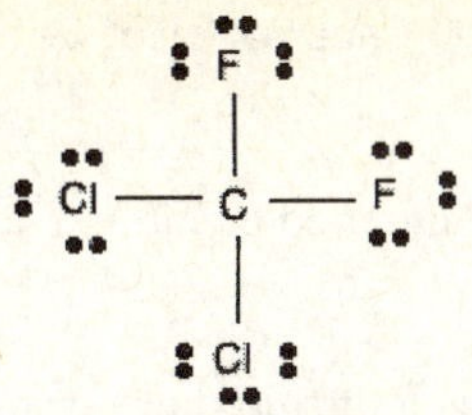

Four electron pairs with no lone pair gives a tetrahedral molecular geometry

Determine if the molecule contains polar bonds:
The electronegativities of C = 2.5, F = 4.0, and Cl = 3.0. Therefore the bonds are polar.

Determine whether the polar bond add together to form a net dipole:
Even though the molecular geometry is tetrahedral which normally yields a nonpolar molecule, the four dipole moments sum to a nonzero net dipole moment because of the different electronegativities of Cl and F. The molecule is polar.

c) SeF_6 – nonpolar
Draw the Lewis structure and determine the molecular geometry:

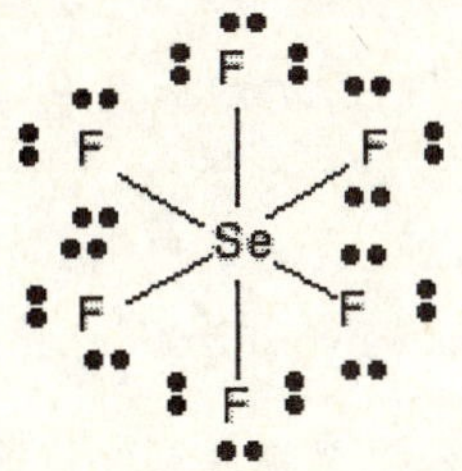

Six electron pairs with no lone pair gives an octahedral molecular geometry

Determine if the molecule contains polar bonds:
The electronegativities of Se = 3.0 and F = 4.0. Therefore the bonds are polar.

Determine whether the polar bonds add together to form a net dipole:
Because the molecular geometry is octahedral, the six dipole moments sum to a zero net dipole moment. The molecule is nonpolar.

d) IF_5 – polar
Draw the Lewis structure and determine the molecular geometry:

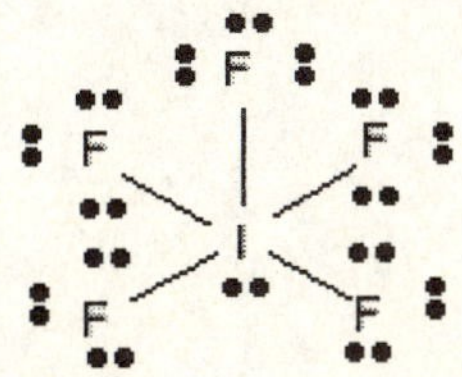

Six electron pairs with one lone pair give square pyramidal molecular geometry

Determine if the molecule contains polar bonds:
The electronegativities of I = 2.0 and F = 4.0. Therefore the bonds are polar.

Determine whether the polar bond add together to form a net dipole:
Because the molecular geometry is square pyramidal, the five dipole moments sum to a nonzero net dipole moment. The molecule is polar.

23. a) Be $2s^2$ — 0 bonds can form. Beryllium contains no unpaired electrons, so no bonds can form without hybridization.

b) P $3s^23p^3$ — 3 bonds can form. Phosphorus contains 3 unpaired electrons, so 3 bonds can form without hybridization.

c) F $2s^2 2p^5$ — 1 bond can form. Fluorine contains 1 unpaired electron, so 1 bond can form without hybridization.

24. a) B $2s^2 2p^1$ — 1 bond can form. Boron contains 1 unpaired electron, so 1 bond can form without hybridization.

b) N $2s^2 2p^3$ — 3 bonds can form. Nitrogen contains 3 unpaired electrons, so 3 bonds can form without hybridization.

c) O $2s^2 2p^4$ — 2 bonds can form. Oxygen contains 2 unpaired electrons, so 2 bonds can form without hybridization.

25. PH_3

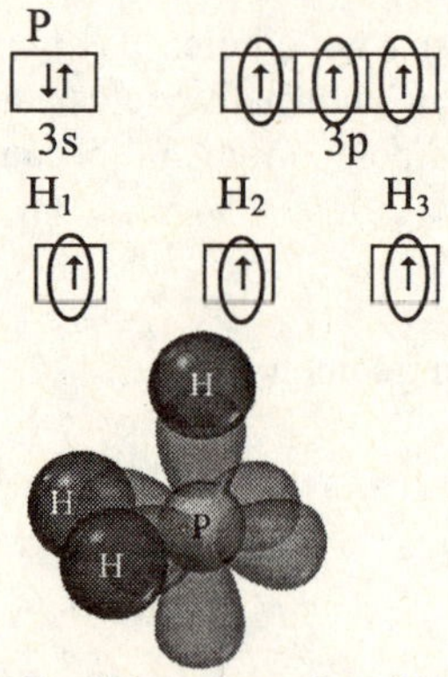

The unhybridized bond angles should be 90°. So, without hybridization, there is good agreement between valence bond theory and the actual bond angle of 93.3°

26. SF_2

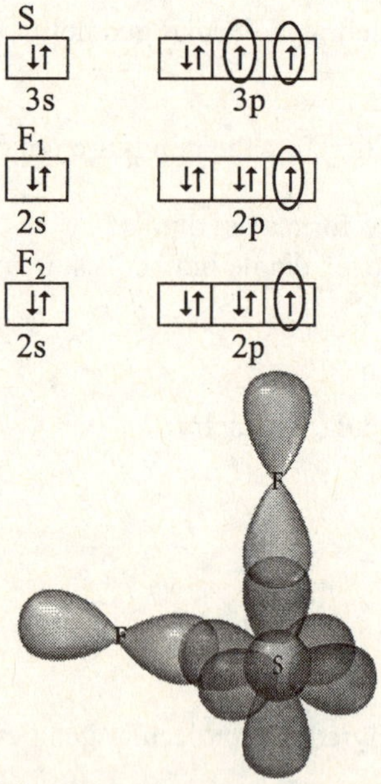

The unhybridized bond angles should be 90°. So, without hybridization, there is not very good agreement between valence bond theory and the actual bond angle of 98.2°

27. C $2s^2 2p^2$

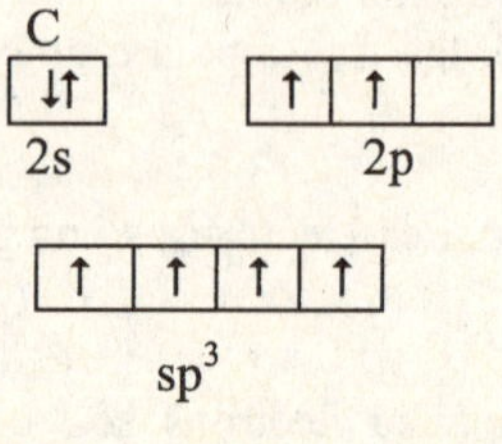

28. C $2s^2 2p^2$

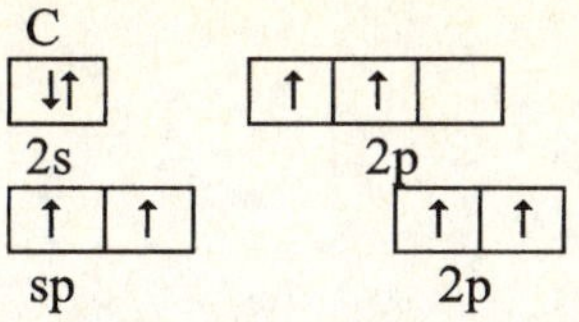

29. sp^2 Only sp^2 hybridization of this set of orbitals has a remaining p orbital to form a π bond.
sp^3 hybridization utilizes all 3 p orbitals
sp^3d^2 hybridization utilizes all 3 p orbitals and 2 d orbitals

30. sp^3d sp^3d hybridization utilizes an s orbital, 3 p orbitals, and d orbital. Since 5 orbitals are used, 5 hybrid orbitals are formed and 5 bonds can form.
sp^3 hybridization utilizes an s orbital and 3 p orbitals. 4 orbitals are used, so 4 hybrid orbitals form and 4 bonds can form.
sp^2 hybridization utilizes an s orbital and 2 p orbitals. 3 orbitals are used, so 3 hybrid orbitals form. This allows 3 σ and 1 π bond to form for a total of 4 bonds formed.

31. a) CCl_4 Write the Lewis structure for the molecule:

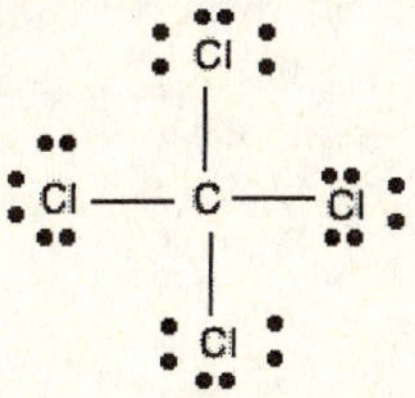

Use VSEPR to predict the electron geometry:
Four electron groups around the central atom give tetrahedral electron geometry

Select the correct hybridization for the central atom based on the electron geometry:
Tetrahedral electron geometry has sp^3 hybridization

Sketch the molecule and label the bonds:

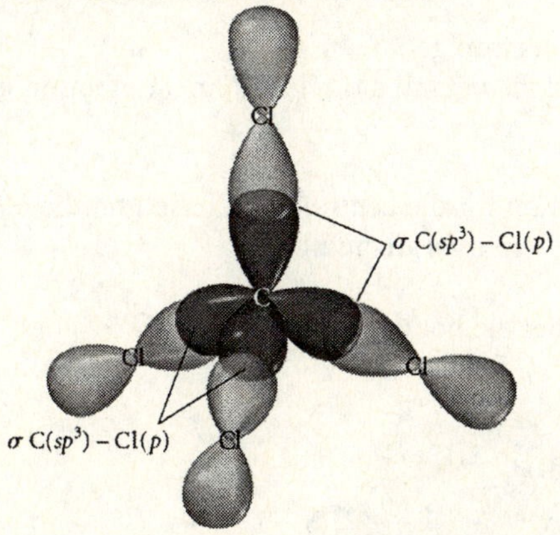

b) NH_3 Write the Lewis structure for the molecule:

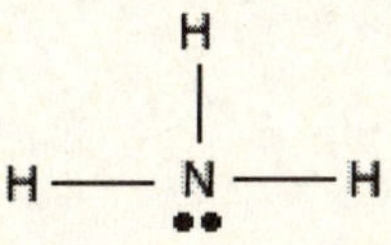

Use VSEPR to predict the electron geometry:
Four electron groups around the central atom give tetrahedral electron geometry

Select the correct hybridization for the central atom based on the electron geometry:
Tetrahedral electron geometry has sp^3 hybridization

Sketch the molecule and label the bonds:

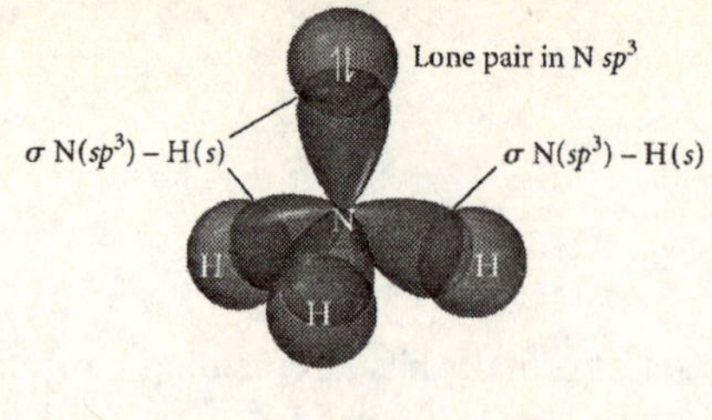

c) OF_2 Write the Lewis structure for the molecule:

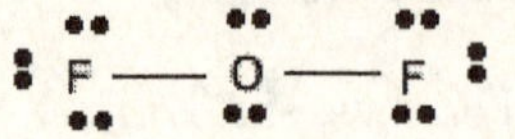

Use VSEPR to predict the electron geometry:
Four electron groups around the central atom give tetrahedral electron geometry

Select the correct hybridization for the central atom based on the electron geometry:
Tetrahedral electron geometry has sp^3 hybridization

Sketch the molecule and label the bonds:

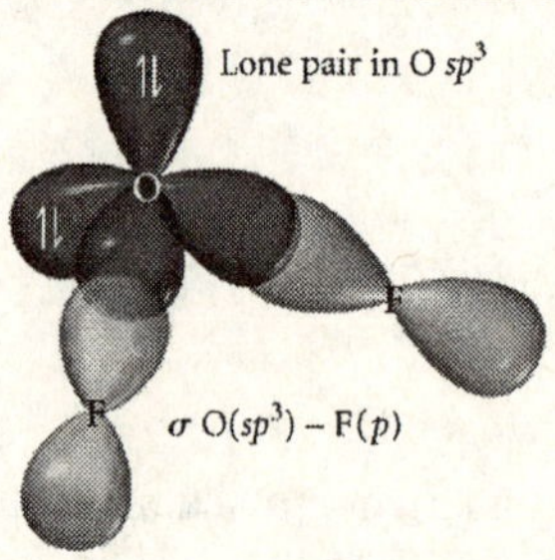

d) CO_2 Write the Lewis structure for the molecule:

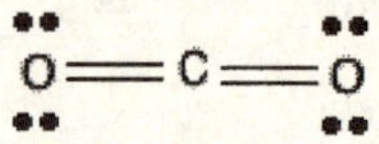

Use VSEPR to predict the electron geometry:
Two electron groups around the central atom gives linear electron geometry

Select the correct hybridization for the central atom based on the electron geometry:
Linear electron geometry has sp hybridization

Sketch the molecule and label the bonds:

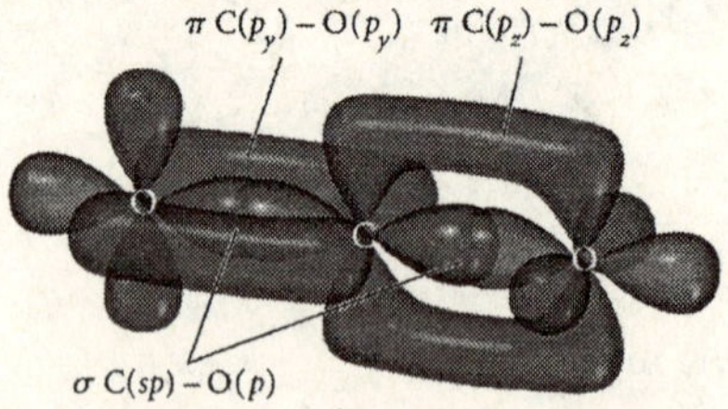

32. a) CH_2Br_2 Write the Lewis structure for the molecule:

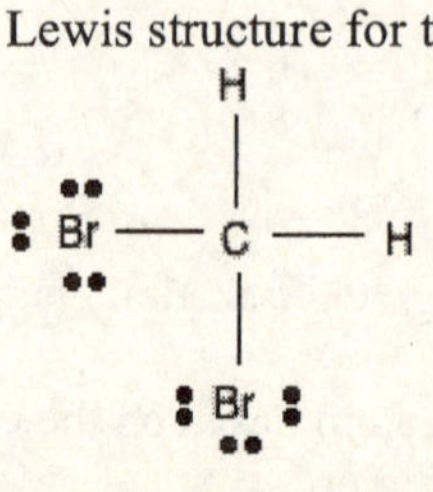

Use VSEPR to predict the electron geometry:
Four electron groups around the central atom give tetrahedral electron geometry

Select the correct hybridization for the central atom based on the electron geometry:
Tetrahedral electron geometry has sp^3 hybridization

Sketch the molecule and label the bonds:

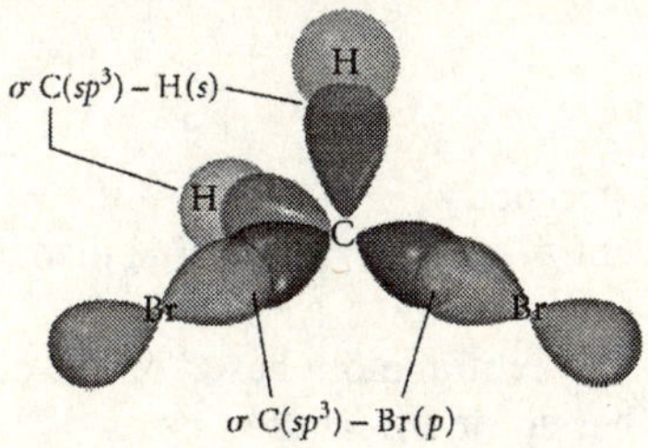

b) SO_2 Write the Lewis structure for the molecule:

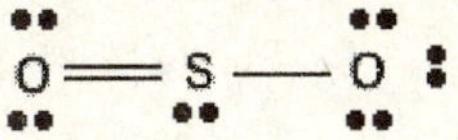

Use VSEPR to predict the electron geometry:
Three electron groups around the central atom gives trigonal planar electron geometry

Select the correct hybridization for the central atom based on the electron geometry:
Trigonal planar electron geometry has sp^2 hybridization

Sketch the molecule and label the bonds:

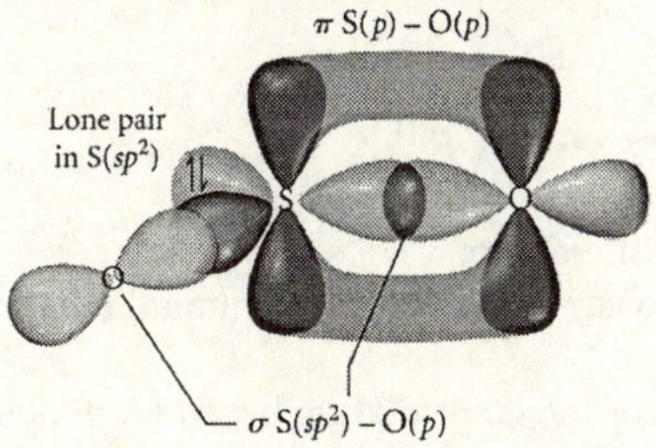

c) NF_3 Write the Lewis structure for the molecule:

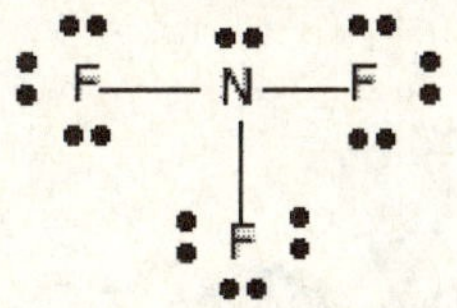

Use VSEPR to predict the electron geometry:
Four electron groups around the central atom give tetrahedral electron geometry

Select the correct hybridization for the central atom based on the electron geometry:
Tetrahedral electron geometry has sp^3 hybridization

Sketch the molecule and label the bonds:

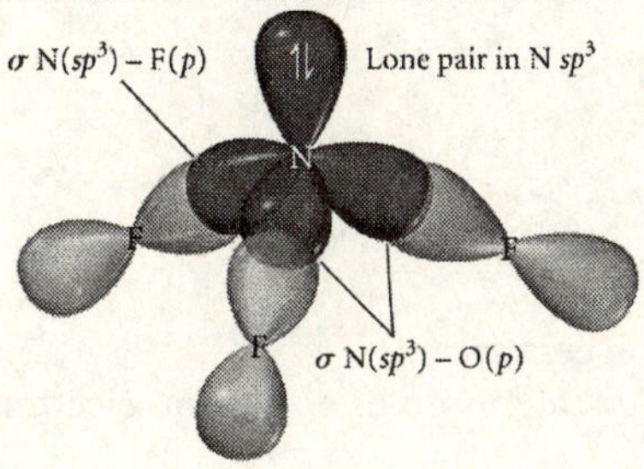

d) BF_3 Write the Lewis structure for the molecule:

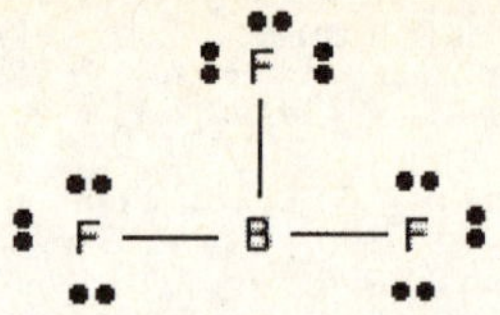

Use VSEPR to predict the electron geometry:
Three electron groups around the central atom gives trigonal planar electron geometry

Select the correct hybridization for the central atom based on the electron geometry:
Trigonal planar electron geometry has sp^2 hybridization

Sketch the molecule and label the bonds:

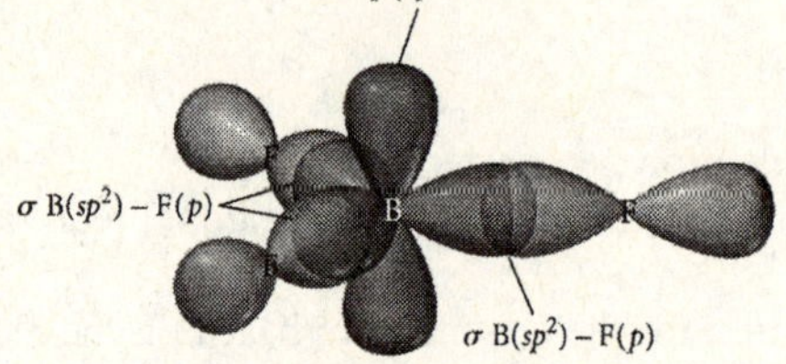

33. a) $COCl_2$ Write the Lewis structure for the molecule:

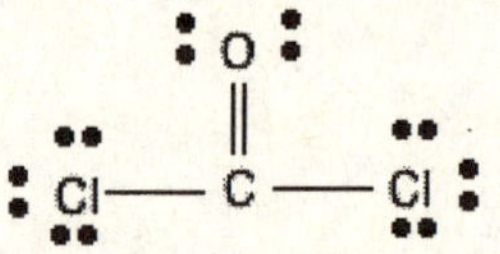

Use VSEPR to predict the electron geometry:
Three electron groups around the central atom gives trigonal planar electron geometry

Select the correct hybridization for the central atom based on the electron geometry:
Trigonal planar electron geometry has sp^2 hybridization

Sketch the molecule and label the bonds:

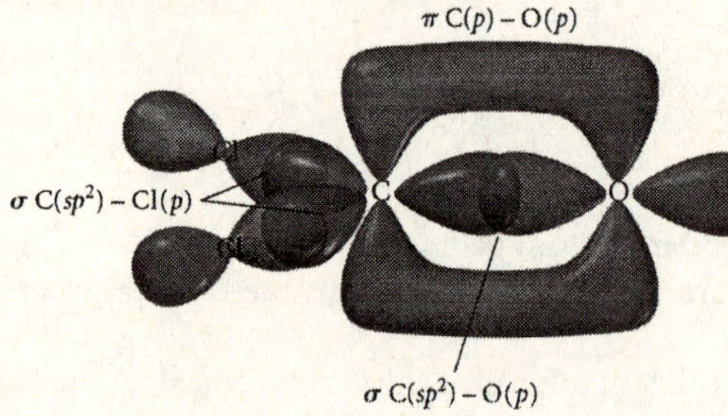

b) BrF_5 Write the Lewis structure for the molecule:

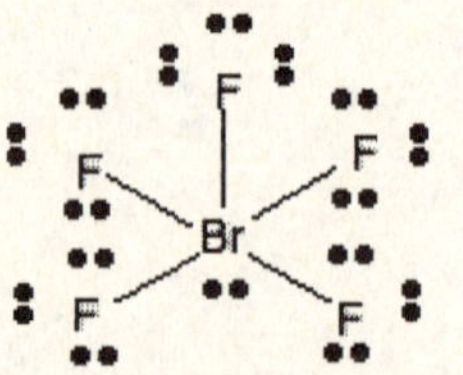

Use VSEPR to predict the electron geometry:
Six electron pairs around the central atoms gives octahedral electron geometry

Select the correct hybridization for the central atom based on the electron geometry:
Octahedral electron geometry has sp^3d^2 hybridization

Sketch the molecule and label the bonds:

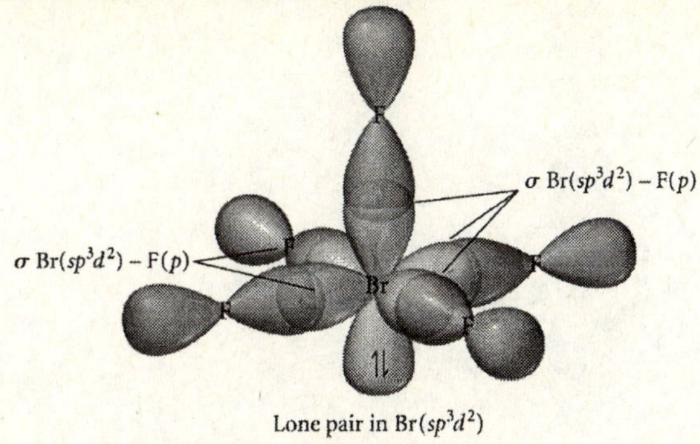

c) XeF_2 Write the Lewis structure for the molecule:

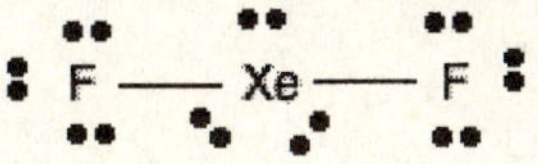

Use VSEPR to predict the electron geometry:
Five electron groups around the central atom gives trigonal bipyramidal geometry

Select the correct hybridization for the central atom based on the electron geometry:
Trigonal bipyramidal geometry has sp^3d hybridization

Sketch the molecule and label the bonds:

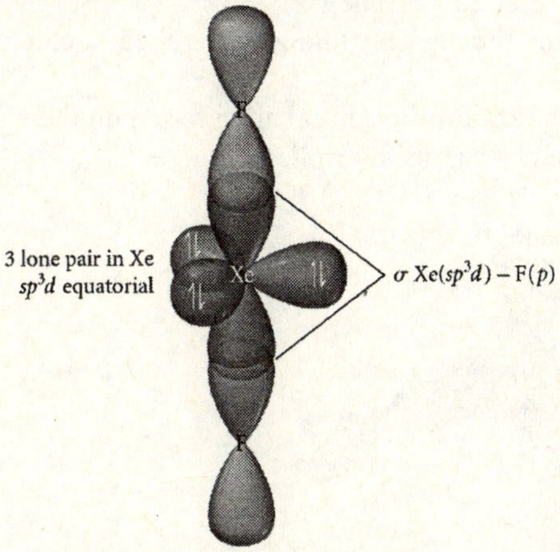

d) I_3^- Write the Lewis structure for the molecule:

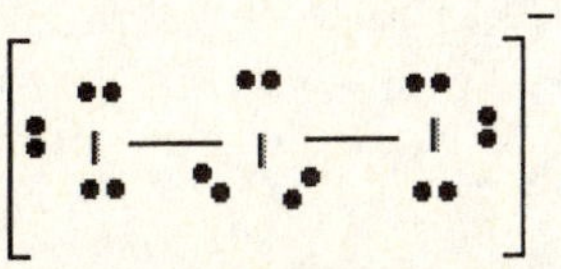

Use VSEPR to predict the electron geometry:
Five electron groups around the central atom gives trigonal bipyramidal geometry

Select the correct hybridization for the central atom based on the electron geometry:
Trigonal bipyramidal geometry has sp^3d hybridization

Sketch the molecule and label the bonds:

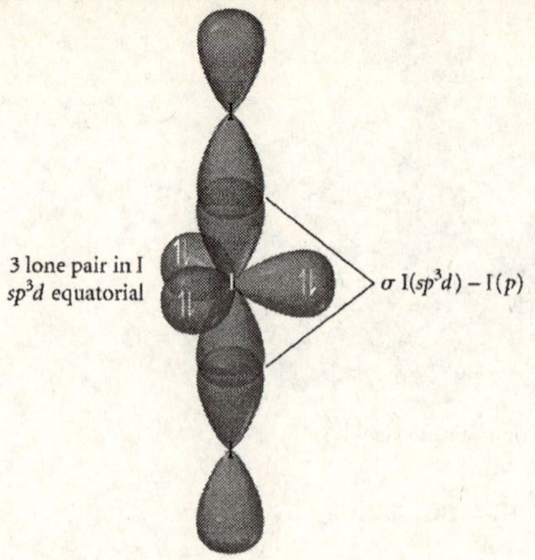

34. a) SO_3^{2-} Write the Lewis structure for the ion:

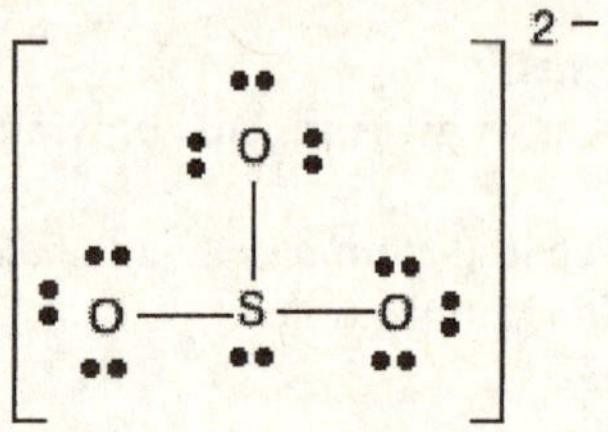

Use VSEPR to predict the electron geometry:
Four electron groups around the central atom give tetrahedral electron geometry

Select the correct hybridization for the central atom based on the electron geometry:
Tetrahedral electron geometry has sp^3 hybridization

Sketch the molecule and label the bonds:

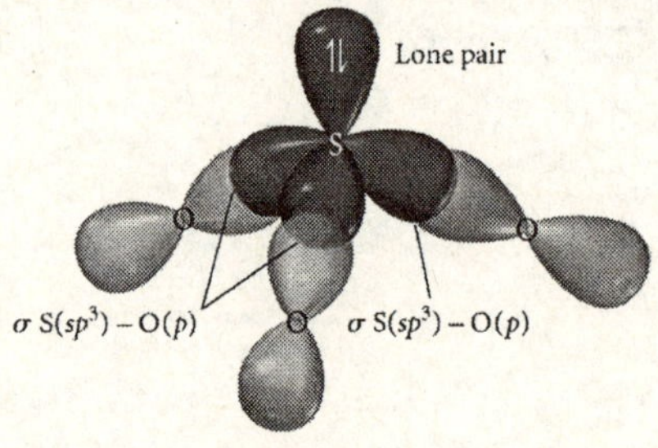

b) PF_6^- Write the Lewis structure for the ion:

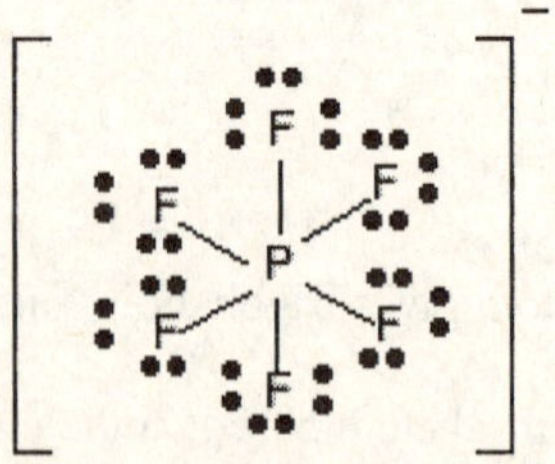

Use VSEPR to predict the electron geometry:
Six electron pairs around the central atoms gives octahedral electron geometry

Select the correct hybridization for the central atom based on the electron geometry:
Octahedral electron geometry has sp^3d^2 hybridization

Sketch the molecule and label the bonds:

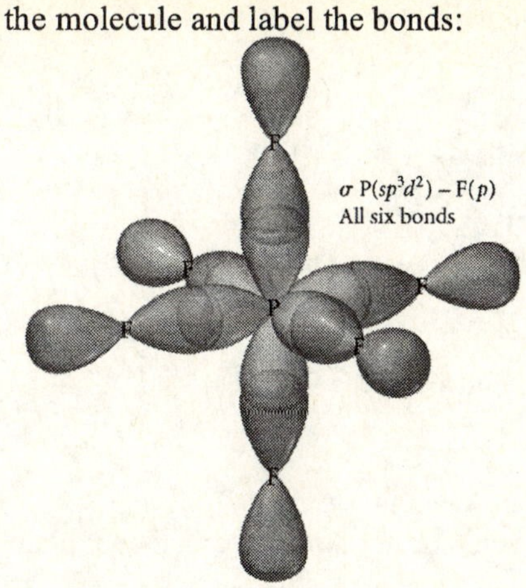

c) BrF_3 Write the Lewis structure for the molecule:

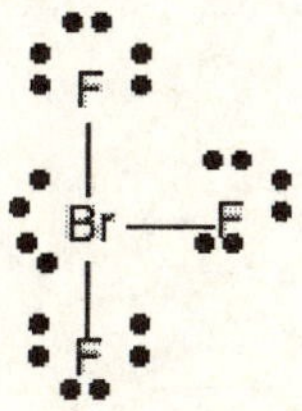

Use VSEPR to predict the electron geometry:
Five electron groups around the central atom gives trigonal bipyramidal geometry

Select the correct hybridization for the central atom based on the electron geometry:
Trigonal bipyramidal geometry has sp^3d hybridization

Sketch the molecule and label the bonds:

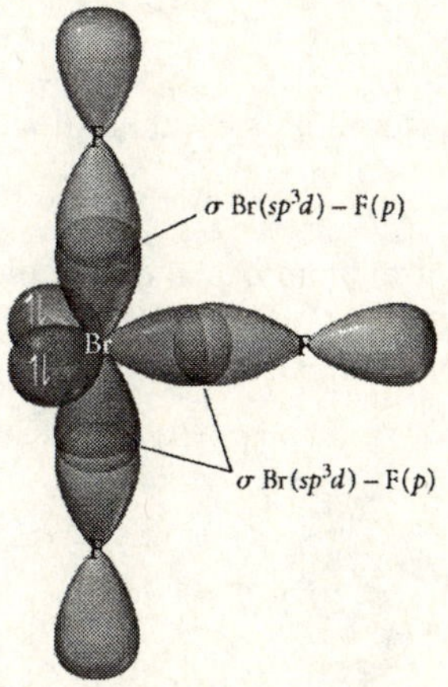

d) HCN Write the Lewis structure for the molecule:

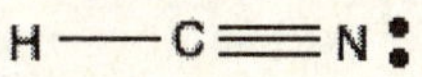

Use VSEPR to predict the electron geometry:
Two electron groups around the central atom gives linear electron geometry

Select the correct hybridization for the central atom based on the electron geometry:
Linear electron geometry has sp hybridization

Sketch the molecule and label the bonds:

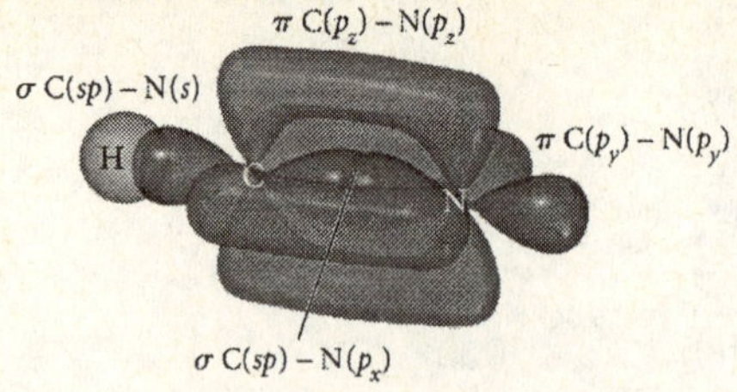

35. a) N_2H_2 Write the Lewis structure for the molecule:

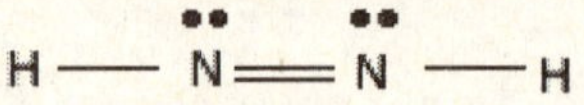

Use VSEPR to predict the electron geometry:
Three electron groups around each interior atom gives trigonal planar electron geometry

Select the correct hybridization for the central atom based on the electron geometry:
Trigonal planar electron geometry has sp^2 hybridization

Sketch the molecule and label the bonds:

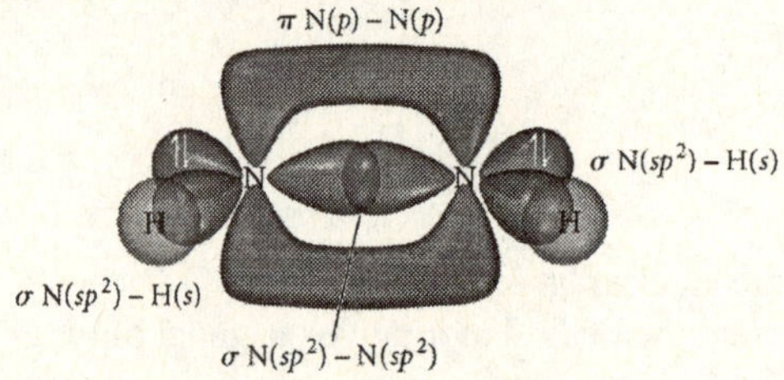

b) N_2H_4 Write the Lewis structure for the molecule:

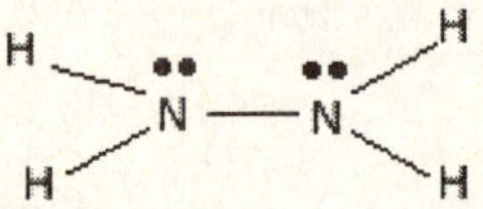

Use VSEPR to predict the electron geometry:
Four electron groups around each interior atom gives tetrahedral electron geometry

Select the correct hybridization for the central atom based on the electron geometry:
Tetrahedral electron geometry has sp^3 hybridization

Sketch the molecule and label the bonds:

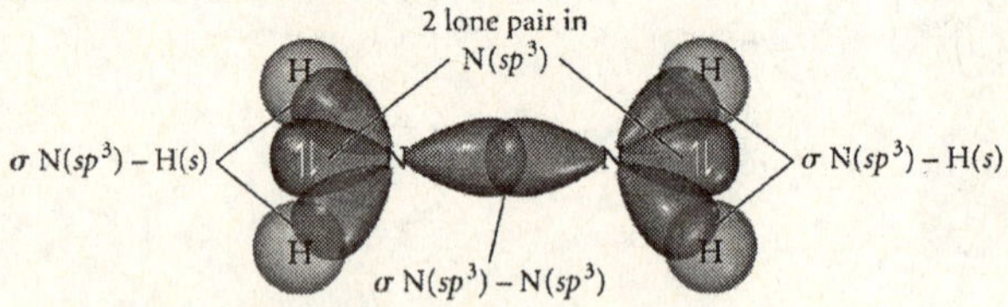

c) CH_3NH_2 Write the Lewis structure for the molecule:

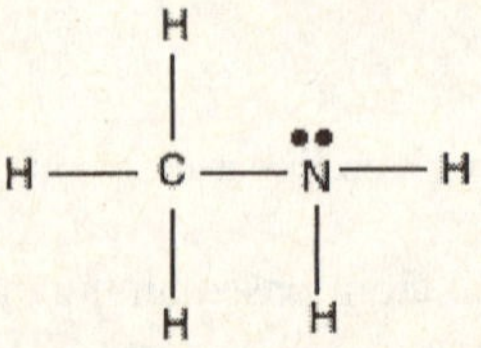

Use VSEPR to predict the electron geometry:
Four electron groups around the C gives tetrahedral electron geometry, four electron groups around the N gives tetrahedral geometry.

Select the correct hybridization for the central atom based on the electron geometry:
Tetrahedral electron geometry has sp^3 hybridization of both C and N

Sketch the molecule and label the bonds:

Lone pair in N(sp^3)

σ C(sp^3) – H(s)

σ N(sp^3) – H(s)

σ C(sp^3) – N(sp^3)

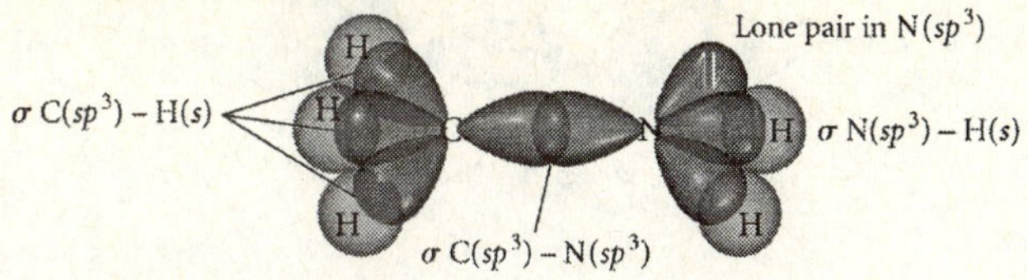

36. a) C_2H_2 Write the Lewis structure for the molecule:

H —— C ≡≡≡ C —— H

Use VSEPR to predict the electron geometry:
Two electron groups around each interior atom gives linear electron geometry

Select the correct hybridization for the central atom based on the electron geometry:
Linear electron geometry has sp hybridization

Sketch the molecule and label the bonds:

π C(p) – C(p)
2 bonds formed

σ C(sp) – H(s)

σ C(sp) – H(s)

σ C(sp) – C(sp)

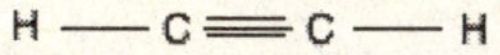

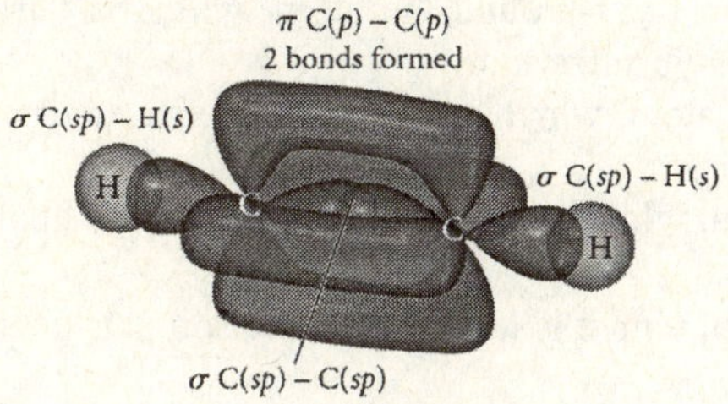

b) C_2H_4 Write the Lewis structure for the molecule:

Use VSEPR to predict the electron geometry:
Three electron groups around each interior atom gives trigonal planar electron geometry

Select the correct hybridization for the central atom based on the electron geometry:
Trigonal planar electron geometry has sp^2 hybridization

Sketch the molecule and label the bonds:

π C(p) – C(p)

σ C(sp^2) – H(s)

σ C(sp^2) – H(s)

σ C(sp^2) – C(sp^2)

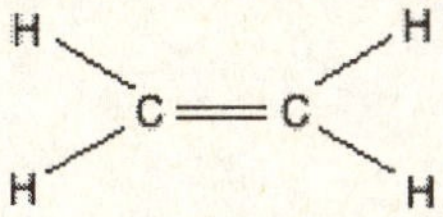

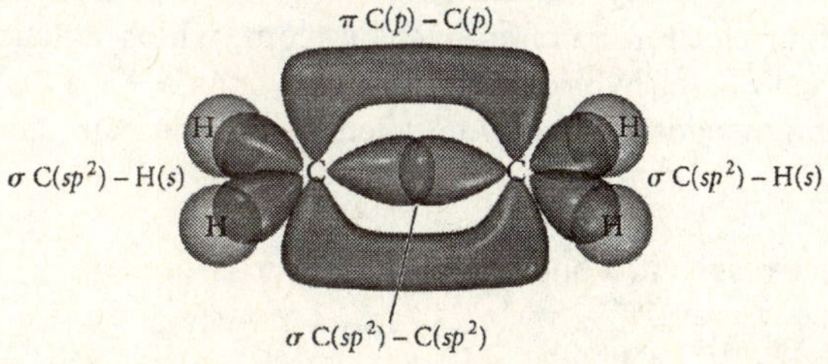

c) C_2H_6 Write the Lewis structure for the molecule:

Use VSEPR to predict the electron geometry:
Four electron groups around each interior atom gives tetrahedral electron geometry

Select the correct hybridization for the central atom based on the electron geometry:
Tetrahedral electron geometry has sp^3 hybridization

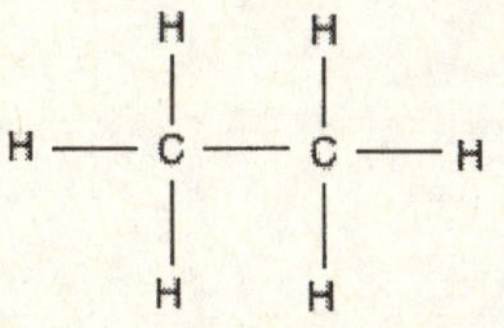

Sketch the molecule and label the bonds:

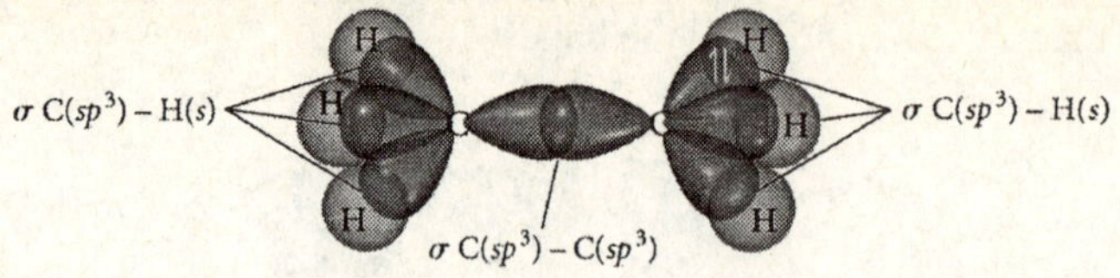

37. 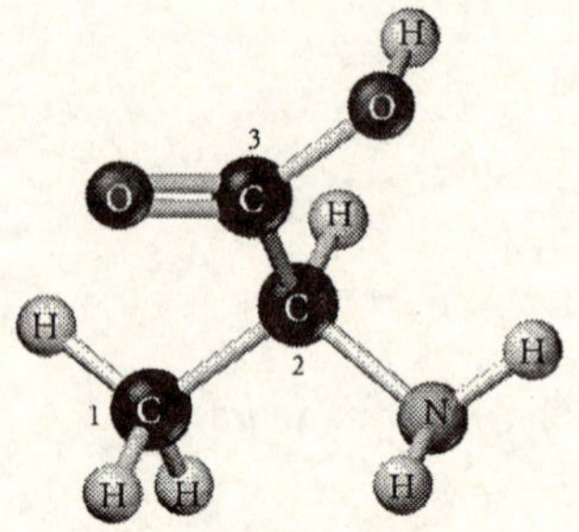

C – 1 and C – 2 each have four electron pairs around the atom, which is tetrahedral electron pair geometry. Tetrahedral electron pair geometry is sp^3 hybridization.
C – 3 has three electron pairs around the atom, which is trigonal planar electron pair geometry. Trigonal planar electron pair geometry is sp^2 hybridization.
O has four electron pairs around the atom, which is tetrahedral electron pair geometry. Tetrahedral electron pair geometry is sp^3 hybridization.
N has four electron pairs around the atom, which is tetrahedral electron pair geometry. Tetrahedral electron pair geometry is sp^3 hybridization.

38. 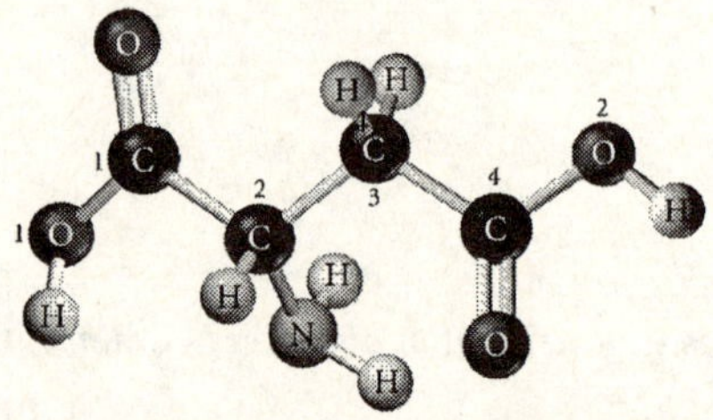

C – 1 and C – 4 each have three electron pairs around the atom, which is trigonal planar electron pair geometry. Trigonal planar electron pair geometry is sp^2 hybridization.
C – 2 and C – 3 each have four electron pairs around the atom, which is tetrahedral electron pair geometry. Tetrahedral electron pair geometry is sp^3 hybridization.
O – 1 and O – 2 each have four electron pairs around the atom, which is tetrahedral electron pair geometry. Tetrahedral electron pair geometry is sp^3 hybridization.
N has four electron pairs around the atom, which is tetrahedral electron pair geometry. Tetrahedral electron pair geometry is sp^3 hybridization.

39. 1s + 1s constructive interference results in a bonding orbital

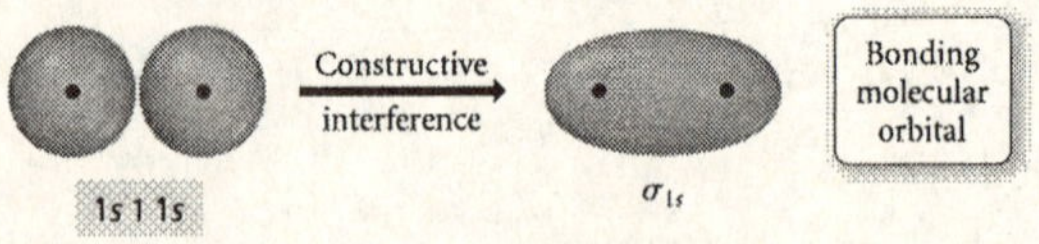

40. 1s – 1s destructive interference results in an antibonding orbital

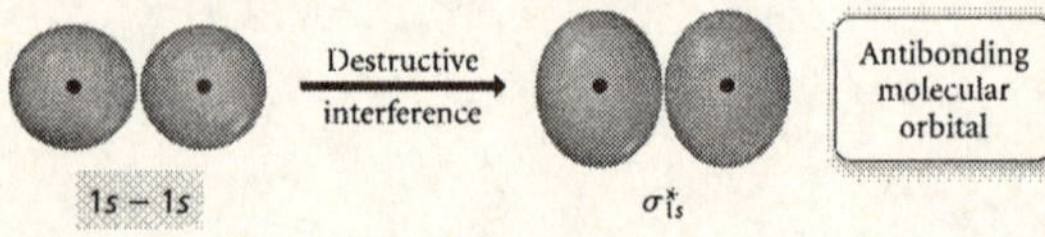

41. Be_2^+ has 7 electrons

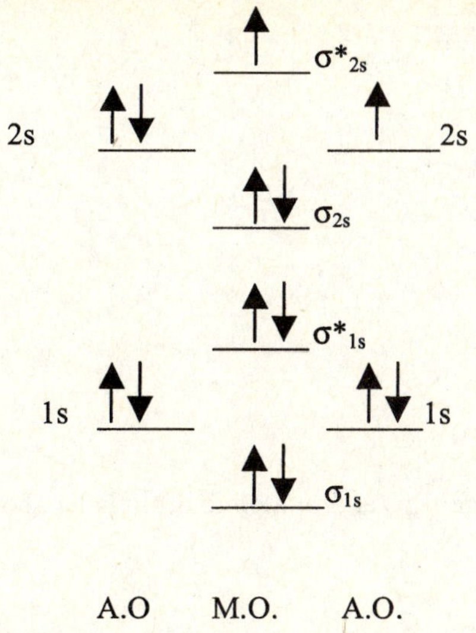

Be_2^- has 9 electrons

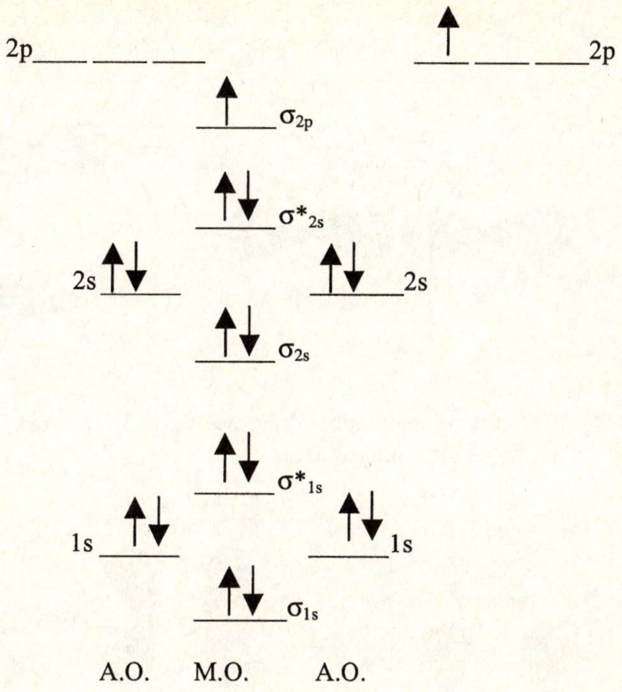

A.O.= Atomic Orbital; M.O. = Molecular Orbital

Bond order = $\frac{4-3}{2}=\frac{1}{2}$ stable

Bond order = $\frac{5-4}{2}=\frac{1}{2}$ stable

42. Li_2^+ has 5 electrons

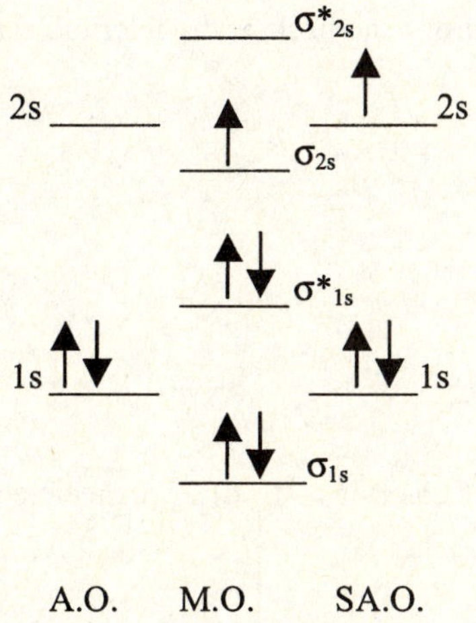

Li_2^- has 7 electrons

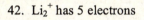

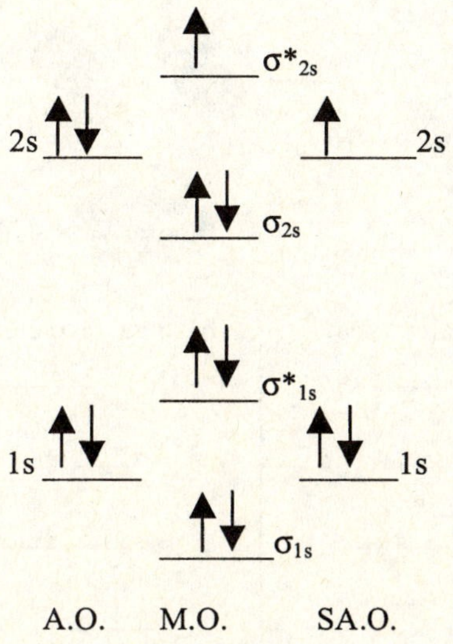

A.O.= Atomic Orbital; M.O. = Molecular Orbital

Bond order = $\frac{3-2}{2}=\frac{1}{2}$ stable

Bond order = $\frac{4-3}{2}=\frac{1}{2}$ stable

43. The bonding and antibonding molecular orbitals from the combination of p_x and p_x atomic orbitals lie along the internuclear axis.

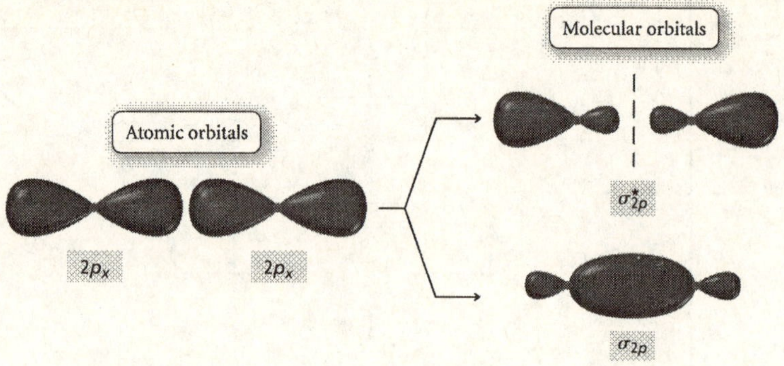

44. The bonding and antibonding molecular orbital from the combination of p_y and p_y atomic orbitals lie above and below the internuclear axis.

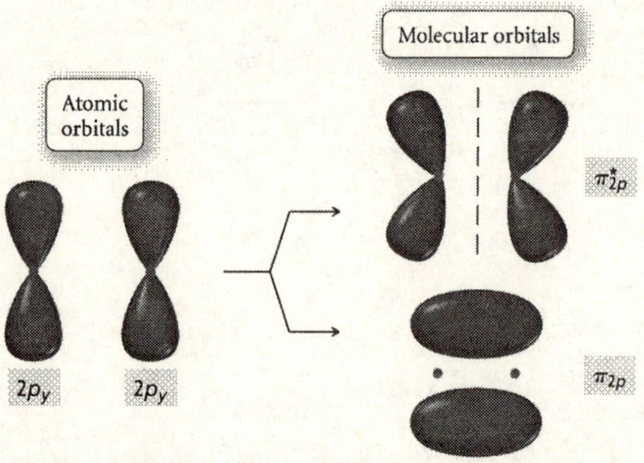

When the p_z and the p_z orbitals combine, similar bonding and antibonding molecular orbitals form. The only difference between the resulting MOs is a rotation about the internuclear axis. The energies and the names of the bonding and antibonding MOs obtained from the combination of the p_z atomic orbitals are identical to those obtained from the combination of the p_y atomic orbitals which lie in front and back of the internuclear axis.

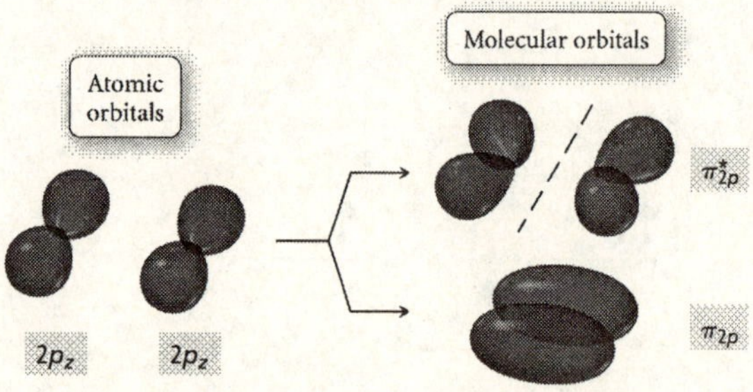

45. a) 4 valence electrons | b) 6 valence electrons | c) 8 valence electrons | d) 9 valence electrons

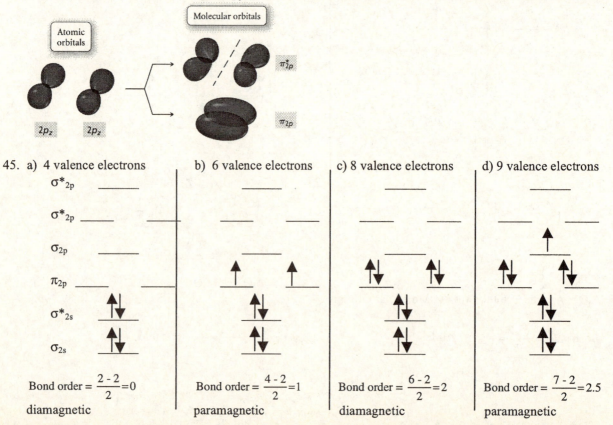

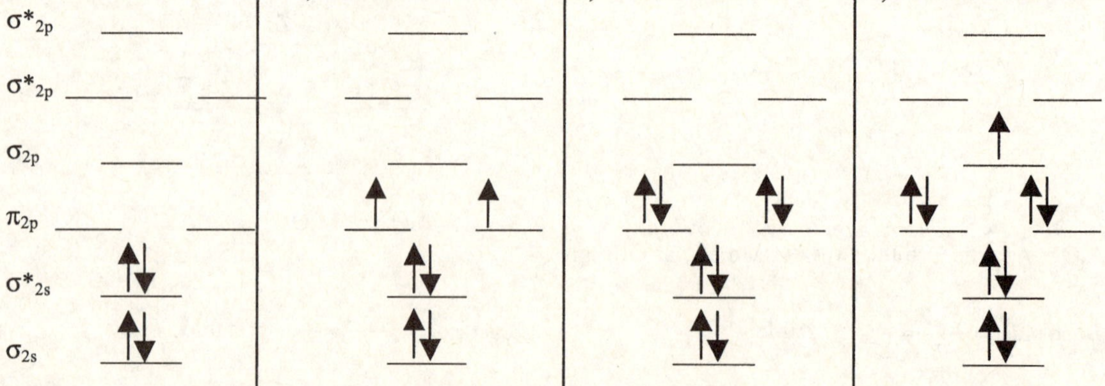

a) Bond order = $\frac{2-2}{2}$=0
diamagnetic

b) Bond order = $\frac{4-2}{2}$=1
paramagnetic

c) Bond order = $\frac{6-2}{2}$=2
diamagnetic

d) Bond order = $\frac{7-2}{2}$=2.5
paramagnetic

46. a) 10 valence electrons b) 12 valence electrons c) 13 valence electrons d) 14 valence electrons

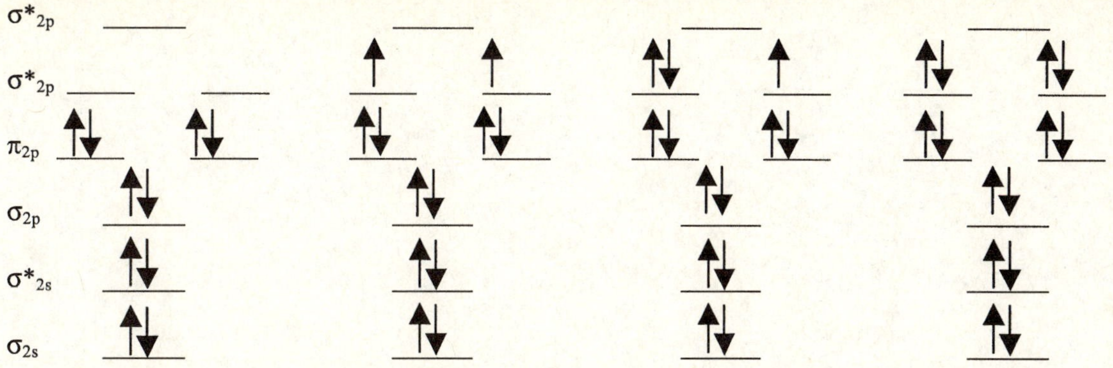

a) Bond order = $\frac{8-2}{2}=3$
diamagnetic

b) Bond order = $\frac{8-4}{2}=2$
paramagnetic

c) Bond order = $\frac{8-5}{2}=1.5$
paramagnetic

d) Bond order = $\frac{8-6}{2}=1$
diamagnetic

47. a) Write an energy level diagram for the molecular orbitals in H_2^{2-}. The ion has 4 valence electrons. Assign the electrons to the molecular orbitals beginning with the lowest energy orbitals and following Hund's rule.

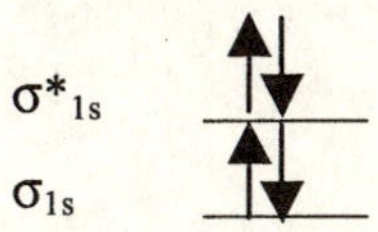

Bond order = $\frac{2-2}{2}=0$ With a bond order of 0, the ion will not exist.

b) Write an energy level diagram for the molecular orbitals in Ne_2. The molecule has 16 valence electrons. Assign the electrons to the molecular orbitals beginning with the lowest energy orbitals and following Hund's rule.

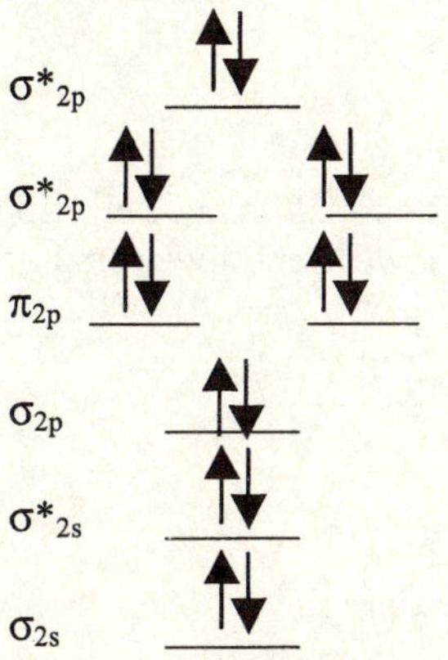

Bond order = $\frac{8-8}{2}=0$ With a bond order of 0, the molecule will not exist.

c) Write an energy level diagram for the molecular orbitals in He_2^{2+}. The ion has 2 valence electrons. Assign the electrons to the molecular orbitals beginning with the lowest energy orbitals and following Hund's rule.

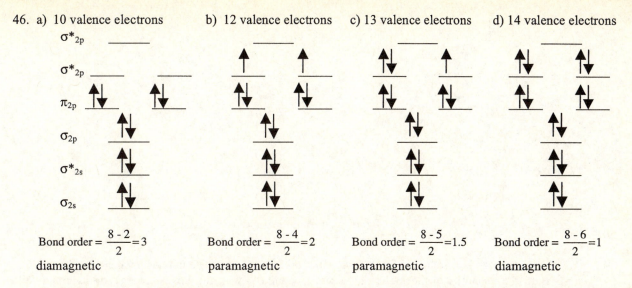

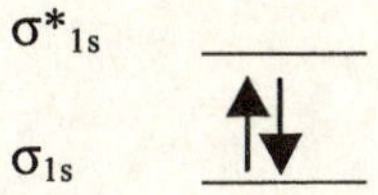

Bond order = $\frac{2-0}{2}=1$ With a bond order of 1, the ion will exist.

d) Write an energy level diagram for the molecular orbitals in F_2^{2-}. The molecule has 16 valence electrons. Assign the electrons to the molecular orbitals beginning with the lowest energy orbitals and following Hund's rule.

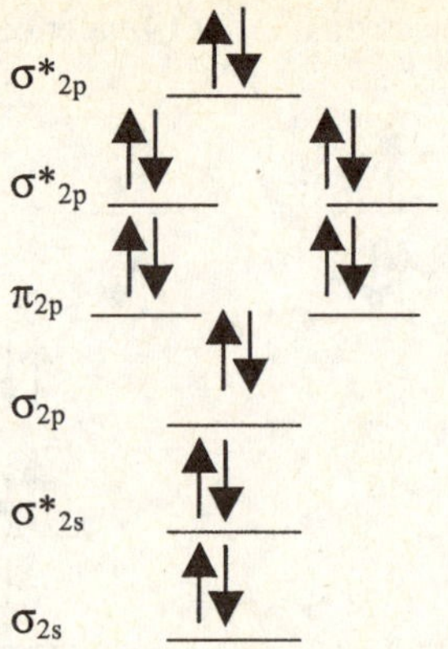

Bond order = $\frac{8-8}{2}=0$ With a bond order of 0, the ion will not exist.

48. a) Write an energy level diagram for the molecular orbitals in C_2^{2+}. The ion has 6 valence electrons. Assign the electrons to the molecular orbitals beginning with the lowest energy orbitals and following Hund's rule.

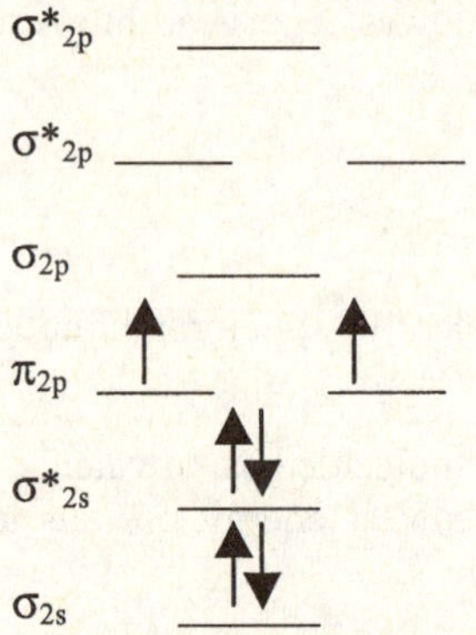

Bond order = $\frac{4-2}{2}=1$ With a bond order of 1, the ion will exist.

b) Write an energy level diagram for the molecular orbitals in Li_2. The ion has 2 valence electrons. Assign the electrons to the molecular orbitals beginning with the lowest energy orbitals and following Hund's rule.

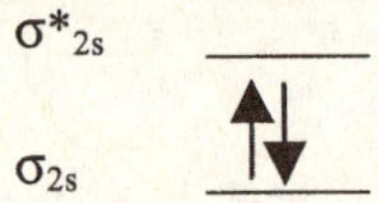

Bond order = $\frac{2-0}{2}=1$ With a bond order of 1, the molecule will exist.

c) Write an energy level diagram for the molecular orbitals in Be_2^{2+}. The ion has 2 valence electrons. Assign the electrons to the molecular orbitals beginning with the lowest energy orbitals and following Hund's rule.

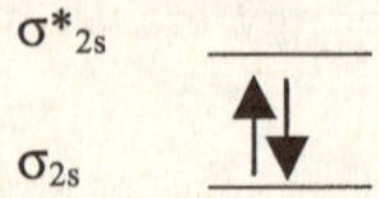

Bond order = $\frac{2-0}{2}=1$ With a bond order of 1, the ion will exist.

d) Write an energy level diagram for the molecular orbitals in Li_2^{2-}. The ion has 4 valence electrons. Assign the electrons to the molecular orbitals beginning with the lowest energy orbitals and following Hund's rule.

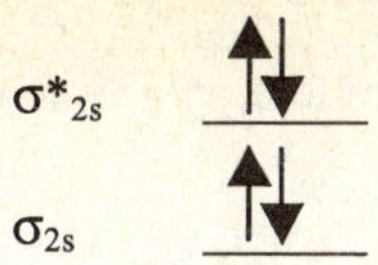

Bond order = $\frac{2-2}{2}$=0 With a bond order of 0, the ion will not exist.

49. C_2^- has the highest bond order, the highest bond energy, and the shortest bond.
Write an energy level diagram for the molecular orbitals in C_2.
Assign the electrons to the molecular orbitals beginning with the lowest energy orbitals and following Hund's rule for each of the species.
C_2 (8 valence electrons); C_2^+ (7 valence electrons): C_2^- (9 valence electrons)

C_2 C_2^+ C_2^-

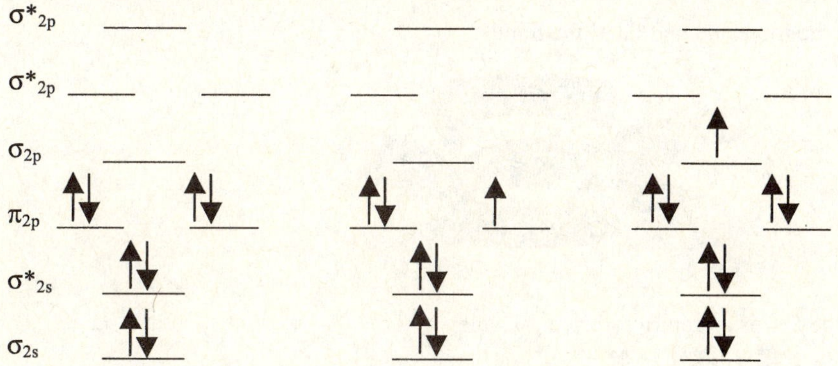

Bond order = $\frac{6-2}{2}$=2 Bond order = $\frac{5-2}{2}$=1.5 Bond order = $\frac{7-2}{2}$=2.5

C_2^- has the highest bond order at 2.5. Bond order is directly related to bond energy, so C_2^- has the largest bond energy and bond order is inversely related to bond length, so C_2^- has the shortest bond length.

50. O_2 has the highest bond order, the highest bond energy, and the shortest bond.
Write an energy level diagram for the molecular orbitals in O_2.
Assign the electrons to the molecular orbitals beginning with the lowest energy orbitals and following Hund's rule for each of the species.
O_2 (12 valence electrons); O_2^- (13 valence electrons): O_2^{2-} (14 valence electrons)

O_2 (12 valence electrons) O_2^- (13 valence electrons) O_2^{2-} (14 valence electrons)

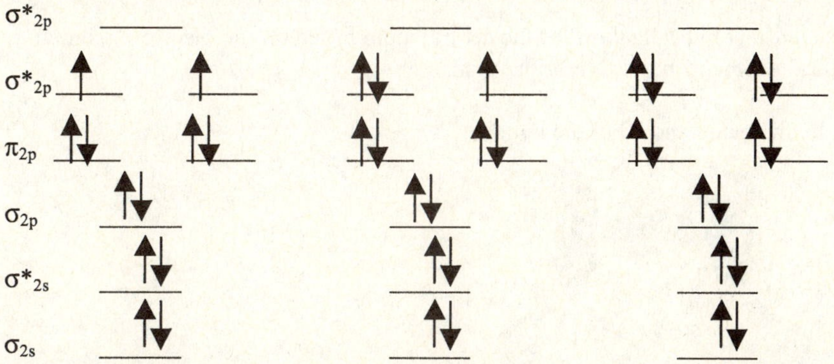

Bond order = $\frac{8-4}{2}$=2 Bond order = $\frac{8-5}{2}$=1.5 Bond order = $\frac{8-6}{2}$=1

O_2 has the highest bond order at 2. Bond order is directly related to bond energy, so O_2 has the largest bond energy and bond order is inversely related to bond length, so O_2 has the shortest bond length.

51. a) COF_2 Write the Lewis structure for the molecule:

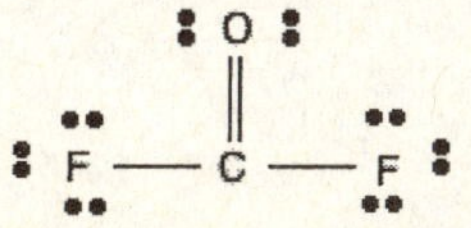

Use VSEPR to predict the electron geometry:

Three electron groups around the central atom gives trigonal planar electron geometry. Three bonding pairs of electrons give trigonal planar molecular geometry.

Determine if the molecule contains polar bonds:
The electronegativities of C = 2.5, O = 3.5, and F = 4.0 Therefore the bonds are polar.

Determine whether the polar bonds add together to form a net dipole:
Even though a trigonal planar molecular geometry normally is nonpolar, because the bonds have different dipole moments, the sum of the dipole moments is not zero. The molecule is polar.

Select the correct hybridization for the central atom based on the electron geometry:
Trigonal planar geometry has sp^2 hybridization.

Sketch the molecule and label the bonds:

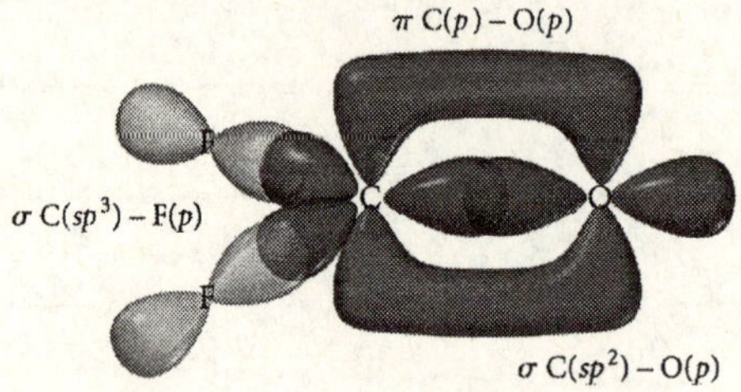

b) S_2Cl_2 Write the Lewis structure for the molecule:

Use VSEPR to predict the electron geometry:
Four electron groups around the central atom gives tetrahedral electron geometry. Two bonding pairs and two lone pairs of electrons gives bent molecular geometry.

Determine if the molecule contains polar bonds:
The electronegativities of S = 2.5 and Cl = 3.0. Therefore the bonds are polar.

Determine whether the polar bonds add together to form a net dipole:
In a bent molecular geometry the sum of the dipole moments is not zero. The molecule is polar.

Select the correct hybridization for the central atom based on the electron geometry:
Tetrahedral geometry has sp^3 hybridization.

Sketch the molecule and label the bonds:

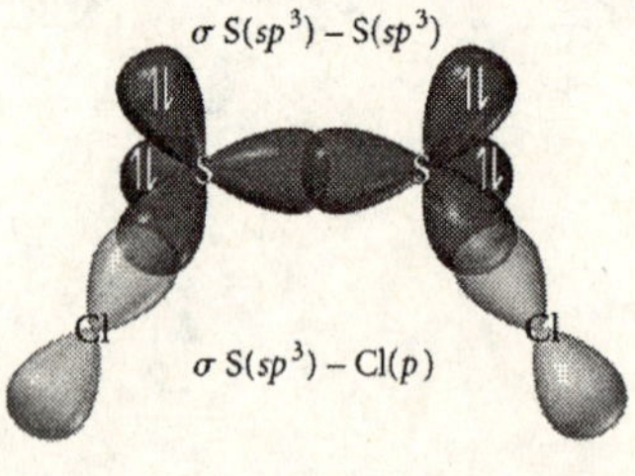

c) SF_4 Write the Lewis structure for the molecule:

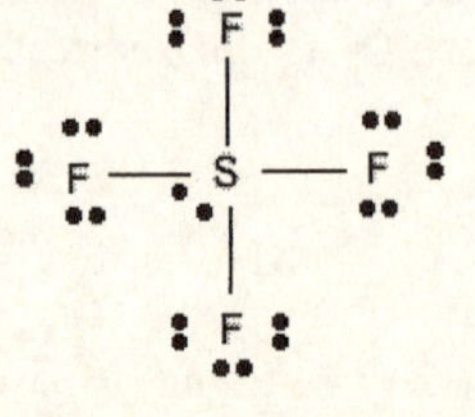

Use VSEPR to predict the electron geometry:

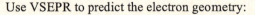

Five electron groups around the central atom gives trigonal bipyramidal electron geometry. Four bonding pairs and one lone pair of electrons gives seesaw molecular geometry.

Determine if the molecule contains polar bonds:
The electronegativities of S = 2.5 and F = 4.0. Therefore the bonds are polar.

Determine whether the polar bonds add together to form a net dipole:
In a seesaw molecular geometry the sum of the dipole moments is not zero. The molecule is polar.

Select the correct hybridization for the central atom based on the electron geometry:
Trigonal bipyramidal electron geometry has sp^3d hybridization.

Sketch the molecule and label the bonds:

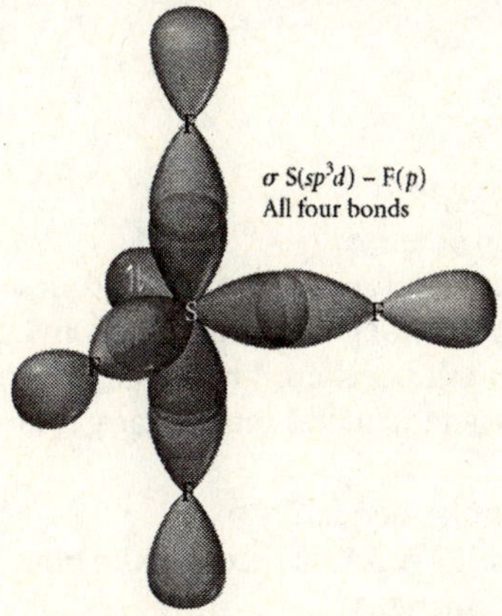

52. a) IF_5 Write the Lewis structure for the molecule:

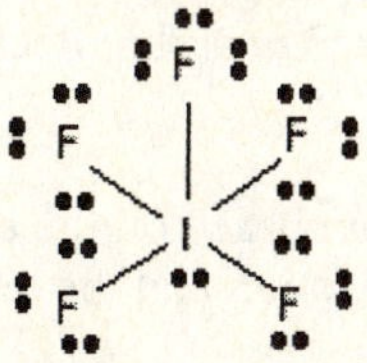

Use VSEPR to predict the electron geometry:
Six electron groups around the central atom gives octahedral electron geometry. Five bonding pairs and one lone pair of electrons gives square pyramidal molecular geometry.

Determine if the molecule contains polar bonds:
The electronegativities of I = 2.5 and F = 4.0. Therefore the bonds are polar.

Determine whether the polar bonds add together to form a net dipole:
In a square pyramidal molecular geometry the sum of the dipole moments is not zero. The molecule is polar.

Select the correct hybridization for the central atom based on the electron geometry:
Octahedral electron geometry has sp^3d^2 hybridization

Sketch the molecule and label the bonds:

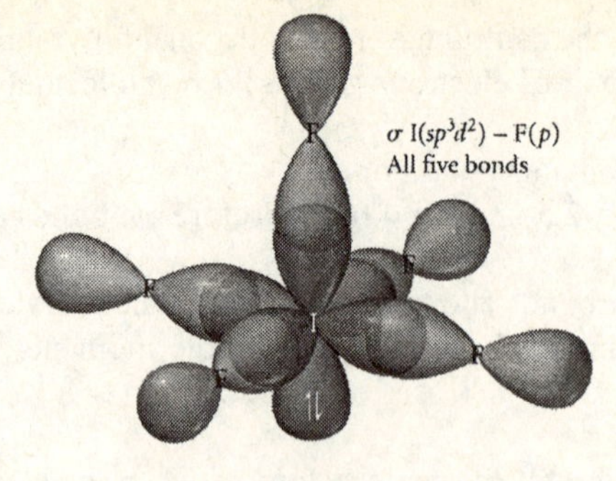

b) CH_2CHCH_3

Write the Lewis structure for the molecule:

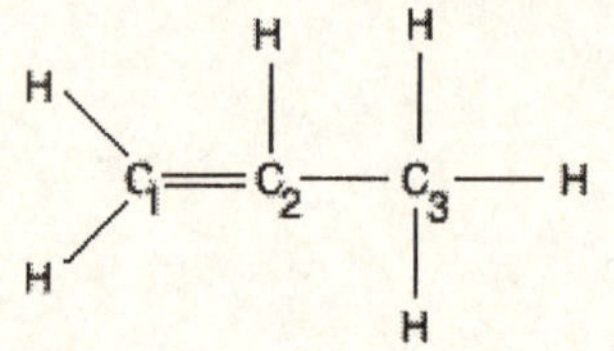

Use VSEPR to predict the electron geometry:
C – 1 and C – 2 each have three electron groups around the atom, this gives trigonal planar electron geometry. Three bonding pairs of electrons give trigonal planar molecular geometry. C – 3 has four electron groups around the C atom, four electron groups gives tetrahedral electron geometry. Four bonding groups gives tetrahedral molecular geometry.

Determine if the molecule contains polar bonds:
The electronegativities of C = 2.5 and H = 2.1. Therefore the bonds are slightly polar because the difference in electronegativity is less than 0.5.

Determine whether the polar bond add together to form a net dipole:
The trigonal planar molecular geometry and the tetrahedral molecular geometry give a net dipole moment of zero. The molecule is nonpolar.

Select the correct hybridization for the central atom based on the electron geometry:
Trigonal planar electron geometry has sp^2 hybridization and the tetrahedral electron geometry has sp^3 hybridization.

Sketch the molecule and label the bonds:

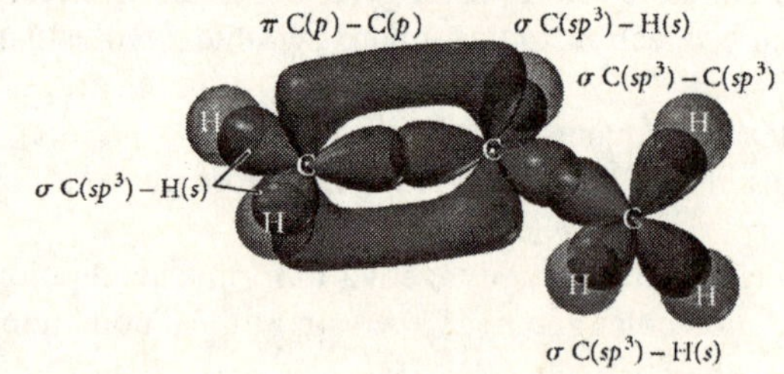

c) CH_3SH Write the Lewis structure for the molecule:

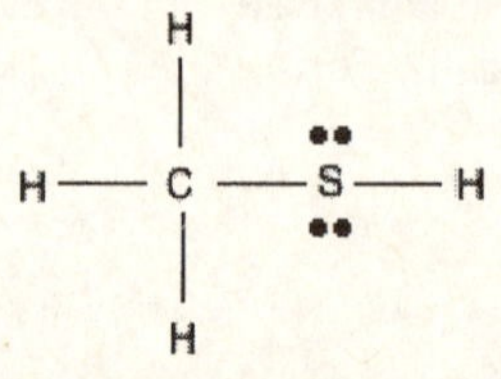

Use VSEPR to predict the electron geometry:
Four electron groups around the C atom and the S atom gives tetrahedral electron geometry. Four bonding pairs of electrons around the C gives tetrahedral molecular geometry and two bonding groups and two lone pairs around the S gives bent molecular geometry.

Determine if the molecule contains polar bonds:
The electronegativities of C = 2.5, S = 2.5, and H = 2.1 The C – H bonds and the S – H bond will be slightly polar, the C – S bond will be nonpolar.

Determine whether the polar bond add together to form a net dipole:
In both molecular geometries the sum of the dipole moments is not zero. The molecule is polar.

Select the correct hybridization for the central atom based on the electron geometry:
Tetrahedral electron geometry has sp^3 hybridization on both C and S.

Sketch the molecule and label the bonds:

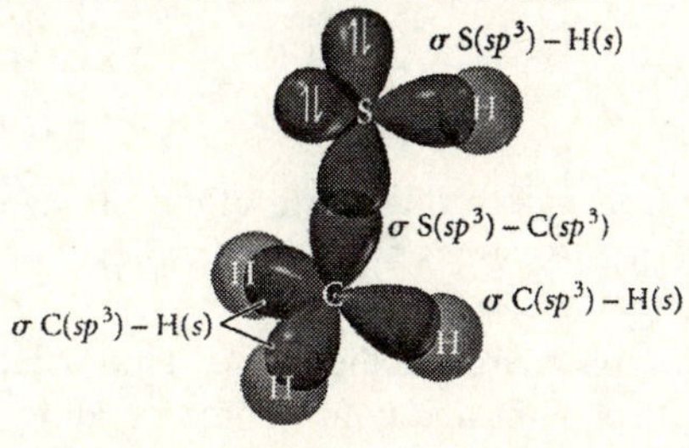

53. a) serine

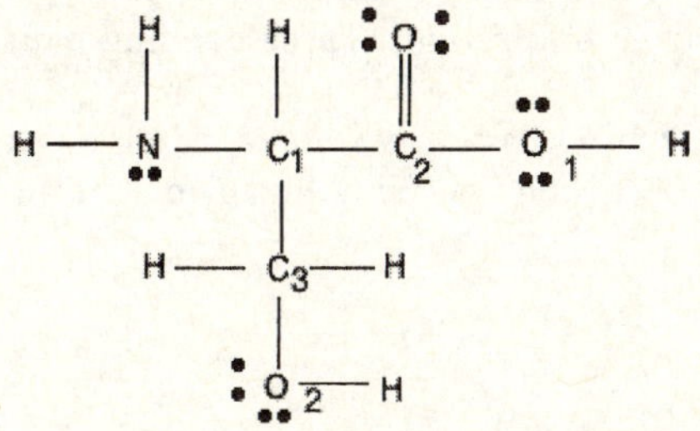

C – 1 and C – 3 each have four electron pairs around the atom. Four electron pairs gives tetrahedral electron geometry, tetrahedral electron geometry has sp^3 hybridization. Four bonding pairs and zero lone pairs gives tetrahedral molecular geometry.
C – 2 has three electron pairs around the atom. Three electron pairs gives trigonal planar geometry, trigonal planar geometry has sp^2 hybridization. Three bonding pairs and zero lone pair gives trigonal planar molecular geometry.
N has four electron pairs around the atom. Four electron pairs gives tetrahedral electron geometry, tetrahedral electron geometry has sp^3 hybridization. Three bonding pairs and one lone pair gives trigonal pyramidal molecular geometry.
O – 1 and O – 2 each has four electron pairs around the atom. Four electron pairs gives tetrahedral electron geometry, tetrahedral electron geometry has sp^3 hybridization. Two bonding pairs and two lone pairs gives bent molecular geometry.

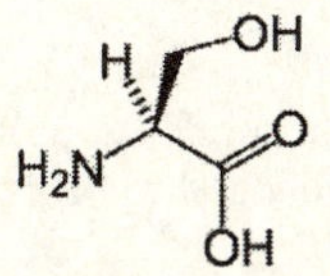

b) asparagine

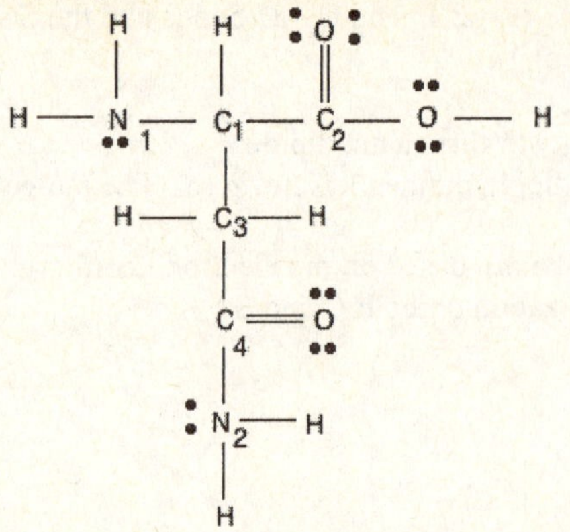

C – 1 and C – 3 each have four electron pairs around the atom. Four electron pairs gives tetrahedral electron geometry, tetrahedral electron geometry has sp^3 hybridization. Four bonding pairs and zero lone pairs gives tetrahedral molecular geometry.
C – 2 and C – 4 each have three electron pairs around the atom. Three electron pairs gives trigonal planar geometry, trigonal planar geometry has sp^2 hybridization. Three bonding pairs and zero lone pairs gives trigonal planar molecular geometry.
N – 1 and N – 2 each have four electron pairs around the atom. Four electron pairs gives tetrahedral electron geometry, tetrahedral electron geometry has sp^3 hybridization. Three bonding pairs and one lone pair gives trigonal pyramidal molecular geometry.
O has four electron pairs around the atom. Four electron pairs gives tetrahedral electron geometry, tetrahedral electron geometry has sp^3 hybridization. Two bonding pairs and two lone pairs gives bent molecular geometry.

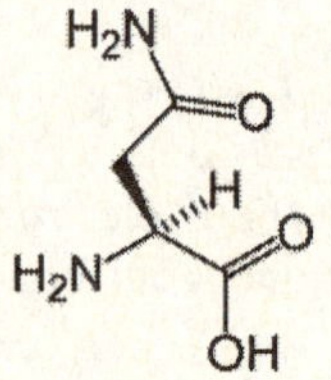

c) cysteine

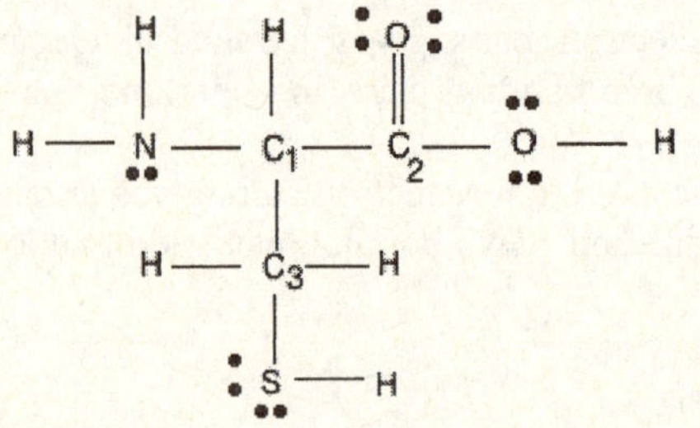

C – 1 and C – 3 each have four electron pairs around the atom. Four electron pairs gives tetrahedral electron geometry, tetrahedral electron geometry has sp^3 hybridization. Four bonding pairs and zero lone pairs gives tetrahedral molecular geometry.
C – 2 has three electron pairs around the atom. Three electron pairs gives trigonal planar geometry, trigonal planar geometry has sp^2 hybridization. Three bonding pairs and zero lone pairs gives trigonal planar molecular geometry.
N has four electron pairs around the atom. Four electron pairs gives tetrahedral electron geometry, tetrahedral electron geometry has sp^3 hybridization. Three bonding pairs and one lone pair gives trigonal pyramidal molecular geometry.
O and S have four electron pairs around the atom. Four electron pairs gives tetrahedral electron geometry, tetrahedral electron geometry has sp^3 hybridization. Two bonding pairs and two lone pairs gives bent molecular geometry.

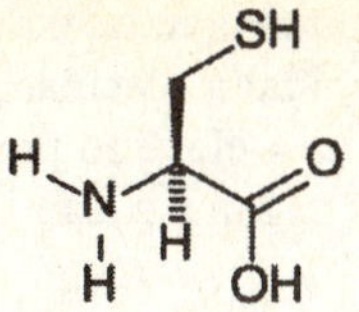

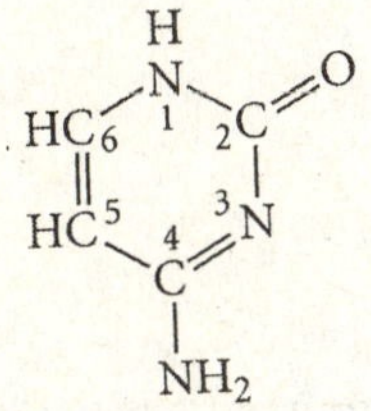

(a)

54. a) cytosine

N – 1 has 3 bonding pairs of electrons and 1 lone pair, 4 electron pairs is tetrahedral electron geometry and sp^3 hybridization. 3 bonding pairs of electrons gives trigonal pyramidal molecular geometry.
C – 2 has 3 bonding pairs of electrons and 0 lone pairs, 3 electron pairs is trigonal planar geometry and sp^2 hybridization. 3 bonding pairs of electrons gives trigonal planar molecular geometry.
N – 3 has 2 bonding pairs of electrons and 1 lone pair, 3 electron pairs is trigonal planar electron geometry and sp^2 hybridization. 2 bonding pairs and 1 lone pair gives bent molecular geometry.
C – 4 has 3 bonding pairs of electrons and 0 lone pairs, 3 electron pairs is trigonal planar geometry and sp^2 hybridization. 3 bonding pairs of electrons gives trigonal planar molecular geometry.
C – 5 has 3 bonding pairs of electrons and 0 lone pairs, 3 electron pairs is trigonal planar geometry and sp^2 hybridization. 3 bonding pairs of electrons gives trigonal planar molecular geometry.
C – 6 has 3 bonding pairs of electrons and 0 lone pairs, 3 electron pairs is trigonal planar geometry and sp^2 hybridization. 3 bonding pairs of electrons gives trigonal planar molecular geometry.
N outside the ring has 3 bonding pairs of electrons and 1 lone pair, 4 electron pairs is tetrahedral electron geometry and sp^3 hybridization. 3 bonding pairs of electrons gives trigonal pyramidal molecular geometry.

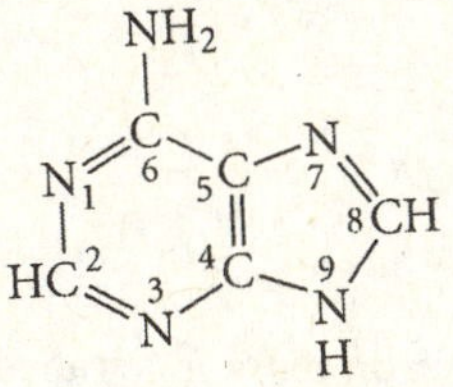

(b)

b) adenine

N – 1 has 2 bonding pairs of electrons and 1 lone pair, 3 electron pairs is trigonal planar electron geometry and sp^2 hybridization. 2 bonding pairs and 1 lone pair gives bent molecular geometry.
C – 2 has 3 bonding pairs of electrons and 0 lone pairs, 3 electron pairs is trigonal planar geometry and sp^2 hybridization. 3 bonding pairs of electrons gives trigonal planar molecular geometry.
N – 3 has 2 bonding pairs of electrons and 1 lone pair, 3 electron pairs is trigonal planar electron geometry and sp^2 hybridization. 2 bonding pairs and 1 lone pair gives bent molecular geometry.
C – 4 has 3 bonding pairs of electrons and 0 lone pairs, 3 electron pairs is trigonal planar geometry and sp^2 hybridization. 3 bonding pairs of electrons gives trigonal planar molecular geometry.
C – 5 has 3 bonding pairs of electrons and 0 lone pairs, 3 electron pairs is trigonal planar geometry and sp^2 hybridization. 3 bonding pairs of electrons gives trigonal planar molecular geometry.
C – 6 has 3 bonding pairs of electrons and 0 lone pairs, 3 electron pairs is trigonal planar geometry and sp^2 hybridization. 3 bonding pairs of electrons gives trigonal planar molecular geometry.
N – 7 has 2 bonding pairs of electrons and 1 lone pair, 3 electron pairs is trigonal planar electron geometry and sp^2 hybridization. 2 bonding pairs and 1 lone pair gives bent molecular geometry
C – 8 has 3 bonding pairs of electrons and 0 lone pairs, 3 electron pairs is trigonal planar geometry and sp^2 hybridization. 3 bonding pairs of electrons gives trigonal planar molecular geometry.

N – 9 has 3 bonding pairs of electrons and 1 lone pair, 4 electron pairs is tetrahedral electron geometry and sp^3 hybridization. 3 bonding pairs of electrons gives trigonal pyramidal molecular geometry.
N outside the ring has 3 bonding pairs of electrons and 1 lone pair, 4 electron pairs is tetrahedral electron geometry and sp^3 hybridization. 3 bonding pairs of electrons gives trigonal pyramidal molecular geometry.

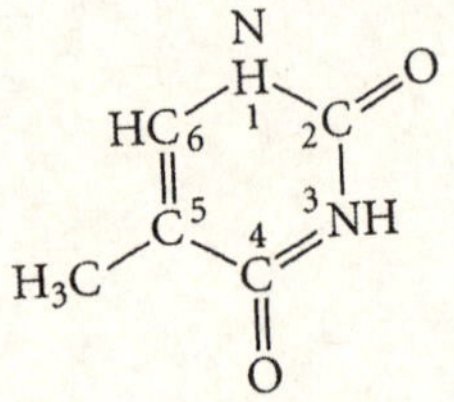

(c)

c) thymine
N – 1 has 3 bonding pairs of electrons and 1 lone pair, 4 electron pairs is tetrahedral electron geometry and sp^3 hybridization. 3 bonding pairs of electrons gives trigonal pyramidal molecular geometry.
C – 2 has 3 bonding pairs of electrons and 0 lone pairs, 3 electron pairs is trigonal planar geometry and sp^2 hybridization. 3 bonding pairs of electrons gives trigonal planar molecular geometry.
N – 3 has 3 bonding pairs of electrons and 1 lone pair, 4 electron pais is tetrahedral electron geometry and sp^3 hybridization. 3 bonding pairs and 1 lone pair gives trigonal pyramidal molecular geometry.
C – 4 has 3 bonding pairs of electrons and 0 lone pairs, 3 electron pairs is trigonal planar geometry and sp^2 hybridization. 3 bonding pairs of electrons gives trigonal planar molecular geometry.
C – 5 has 3 bonding pairs of electrons and 0 lone pairs, 3 electron pairs is trigonal planar geometry and sp^2 hybridization. 3 bonding pairs of electrons gives trigonal planar molecular geometry.
C – 6 has 3 bonding pairs of electrons and 0 lone pairs, 3 electron pairs is trigonal planar geometry and sp^2 hybridization. 3 bonding pairs of electrons gives trigonal planar molecular geometry.
C outside the ring has 4 bonding pairs of electrons and 0 lone pairs, 4 electron pairs is tetrahedral electron geometry and sp^3 hybridization. 4 bonding pairs of electrons gives tetrahedral molecular geometry.

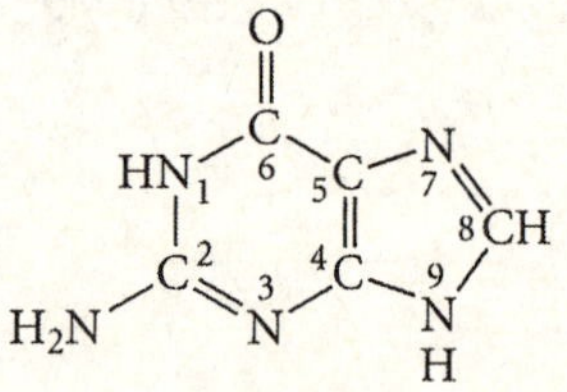

(d)

d) guanine
N – 1 has 3 bonding pairs of electrons and 1 lone pair, 4 electron pairs is tetrahedral electron geometry and sp^3 hybridization. 3 bonding pairs of electrons gives trigonal pyramidal molecular geometry.
C – 2 has 3 bonding pairs of electrons and 0 lone pairs, 3 electron pairs is trigonal planar geometry and sp^2 hybridization. 3 bonding pairs of electrons gives trigonal planar molecular geometry.
N – 3 has 2 bonding pairs of electrons and 1 lone pair, 3 electron pairs is trigonal planar electron geometry and sp^2 hybridization. 2 bonding pairs and 1 lone pair gives bent molecular geometry.
C – 4 has 3 bonding pairs of electrons and 0 lone pairs, 3 electron pairs is trigonal planar geometry and sp^2 hybridization. 3 bonding pairs of electrons gives trigonal planar molecular geometry.
C – 5 has 3 bonding pairs of electrons and 0 lone pairs, 3 electron pairs is trigonal planar geometry and sp^2 hybridization. 3 bonding pairs of electrons gives trigonal planar molecular geometry.
C – 6 has 3 bonding pairs of electrons and 0 lone pairs, 3 electron pairs is trigonal planar geometry and sp^2 hybridization. 3 bonding pairs of electrons gives trigonal planar molecular geometry.
N – 7 has 2 bonding pairs of electrons and 1 lone pair, 3 electron pairs is trigonal planar electron geometry and sp^2 hybridization. 2 bonding pairs and 1 lone pair gives bent molecular geometry.
C – 8 has 3 bonding pair of electrons and 0 lone pair, 3 electron pairs is trigonal planar geometry and sp^2 hybridization. 3 bonding pairs of electrons gives trigonal planar molecular geometry.
N – 9 has 3 bonding pairs of electrons and 1 lone pair, 4 electron pairs is tetrahedral electron geometry and sp^3 hybridization. 3 bonding pairs of electrons gives trigonal pyramidal molecular geometry.
N outside the ring has 3 bonding pairs of electrons and 1 lone pair, 4 electron pairs is tetrahedral electron geometry and sp^3 hybridization. 3 bonding pairs of electrons gives trigonal pyramidal molecular geometry.

55. 4 π bonds; 25 σ bonds; the lone pair on the O's and N – 2 occupy sp^2 orbitals, the lone pairs on N – 1, N – 3, and N – 4 occupy sp^3 orbitals.

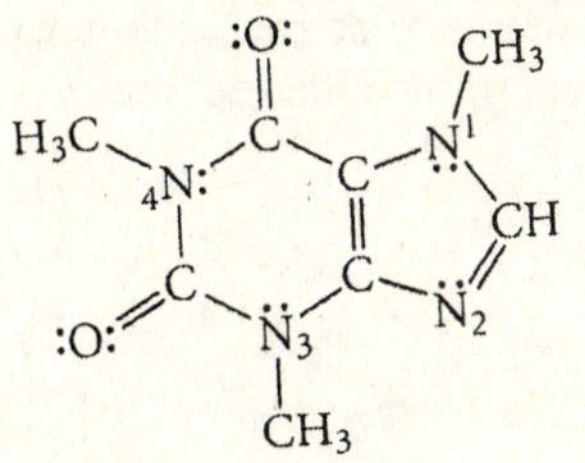

56. 5 π bonds; 21 σ bonds;

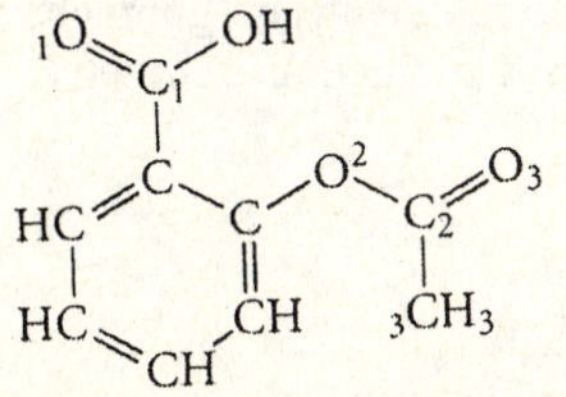

There is rotation around the bond from C – 1 to the ring and from C – 1 to OH bond. There is rotation around the O – 2 to the ring bond and around the O – 2 to C – 2 bond. There is rotation around the C – 2 to C – 3 bond. The C – 1 to O – 1 bond is rigid, the ring structure is rigid and the C – 2 to O – 3 bond is rigid.

57. a) water soluble – the 4 C – OH bonds, the C = O bond and the C – O bonds in the ring, make the molecule polar. Because of the large electronegativity difference between the C and O, each of the bonds will have a dipole moment. The sum of the dipole moments does NOT give a net zero dipole moment, so the molecule is polar. Since it is polar, it will be water soluble.

b) fat soluble – There is only one C – O bond in the molecule. The dipole moment from this bond is not enough to make the molecule polar because of all of the nonpolar components of the molecule. The C – H bonds in the structure lead to a net dipole of zero for most of the sites in the molecule. Since the molecule is nonpolar, it is fat soluble.

c) water soluble – the carboxylic acid function (COOH group) along with the N atom in the ring make the molecule polar. Because of the electronegativity difference between the C and O and the C and N atoms, the bonds will have a dipole moment and the net dipole moment of the molecule is NOT zero, so the molecule is polar. Since the molecule is polar, it is water soluble.

d) fat soluble – The two O atoms in the structure contribute a very small amount to the net dipole moment of this molecule. The majority of the molecule is nonpolar because there is no net dipole moment at the interior C atoms. Because the molecule is nonpolar it is fat soluble.

58. The soap molecule has a nonpolar hydrocarbon end and an anionic end when it is dissolved in water. In order to dissolve in water, the sodium stearate congregates to form small spheres (called micelles) with the nonpolar ends on the insides and the anionic ends on the surface. The anionic end interacts with the polar water molecules, while the nonpolar hydrocarbon end can attract and interact with the nonpolar grease. This allows the soapy water to remove the grease.

59. BrF (14 valence electrons)

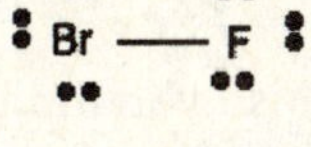

no central atom, no hybridization, no electron structure

BrF_2^- (22 valence electrons)

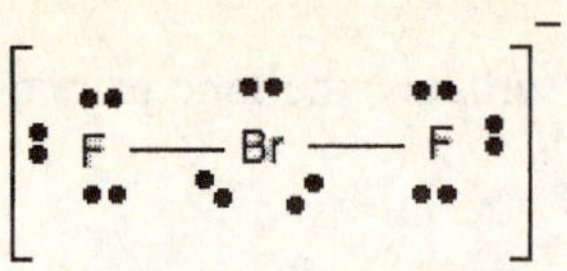

Five electron pairs on the central atom, electron geometry is trigonal bipyramidal, two bonding pairs and three lone pairs gives a linear molecular geometry. An electron geometry of trigonal bipyramidal has sp^3d hybridization.

BrF_3 (28 valence electrons)

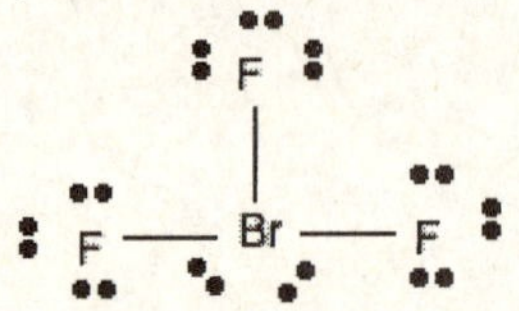

Five electron pairs on the central atom, electron geometry is trigonal bipyramidal, three bonding pairs and two lone pairs gives a T-shaped molecular geometry. An electron geometry of trigonal bipyramidal has sp^3d hybridization.

BrF_4^- (36 valence electrons)

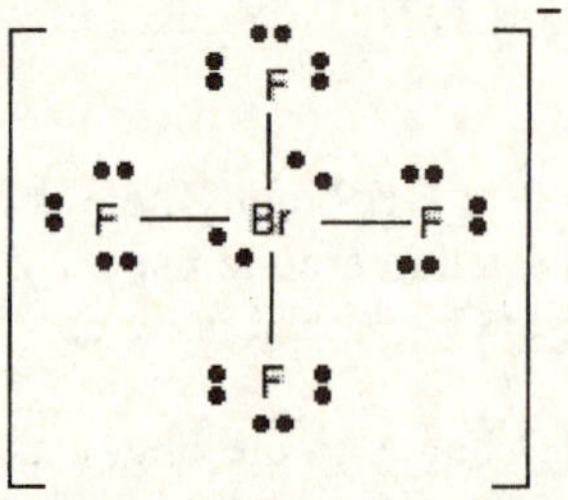

Six electron pairs on the central atom, electron geometry is octahedral, four bonding pairs and two lone pairs gives a square planar molecular geometry. An electron geometry of octahedral has sp^3d^2 hybridization.

BrF_5 (42 valence electrons)

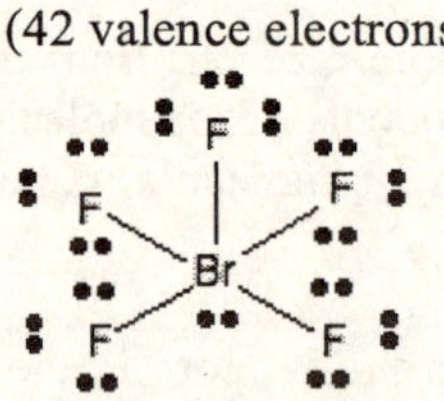

Six electron pairs on the central atom, electron geometry is octahedral, five bonding pairs and one lone pair gives a square pyramidal molecular geometry. An electron geometry of octahedral has sp^3d^2 hybridization.

60. Write the Lewis structure:

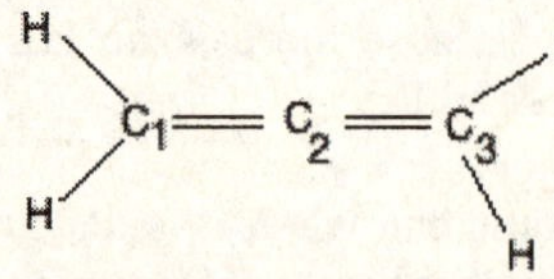

C – 1 and C – 3 each have 3 pairs of electrons and a trigonal planar structure giving sp^2 hybridization on the C with a p orbital left for the π bond. C – 2 has 2 pairs of electrons and is linear which gives sp hybridization and 2 p orbitals left for the π bonds. According to valence bond theory, the π bonds are formed by the sideways overlap of p orbitals. Since the remaining p orbitals on C – 2 are perpendicular to each other, the π bonds formed between C – 1 and C – 2 and between C – 2 and C – 3 must also be perpendicular to each other. Therefore, the two trigonal planar structures at C –1 and C – 3 will be perpendicular to each other.

61. According to valence bond theory, CH_4, NH_3, and H_2O are all sp^3 hybridized. This hybridization results in a tetrahedral electron group configuration with a 109.5° bond angle. NH_3 and H_2O deviate from this idealized bond angle because their lone electron pairs exist in their own sp^3 orbitals. The presence of lone pairs lowers the tendency for the central atom's orbitals to hybridize. As a result, as lone pairs are added, the bond angle moves further from the 109.5° hybrid angle to the 90° unhybridized angle.

62. a) In the isomerization, you need to break the C – C π bond but not the σ bond. So the energy needed would be the difference between the bond energy of a C = C bond and a C – C bond.

C = C 611 kJ/mol

C – C 347 kJ/mol

Therefore, the energy needed to break the π bond would be 264 kJ/mol = 2.64 x 10^5 J/mol

$$\frac{264\ \text{kJ}}{\text{mol}} \times \frac{1000\ \text{J}}{\text{kJ}} \times \frac{\text{mol}}{6.022 \times 10^{23}\ \text{molecules}} = 4.38 \times 10^{-19}\ \frac{\text{J}}{\text{molecule}}$$

b) **Given:** 4.38 x 10^{-19} J/ molecule **Find:** ν and part of the spectrum

Conceptual Plan: E → ν

E = hν

Solution: $\nu = \frac{E}{h} = \frac{4.38 \times 10^{-19}\ \text{J}}{6.62 \times 10^{-34}\ \text{J}\cdot\text{s}} = 6.62 \times 10^{14}\ \text{s}^{-1}$

This frequency is right near the border of the ultraviolet–visible region of the electromagnetic spectrum.

63. Write the Lewis structure for :

Determine electron pair geometry around each central atom:

Determine the molecular geometry, determine idealized bond angles and predict actual bond angles:

NO_2

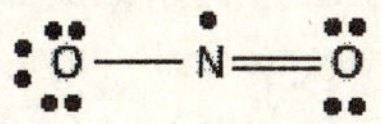

2 bonding pairs and a lone electron gives trigonal planar electron geometry, the molecular geometry will be bent. Trigonal planar electron geometry has idealized bond angles of 120°. The bond angle is expected to be slightly less than 120° because of the lone electron occupying the third sp^2 orbital.

NO_2^+

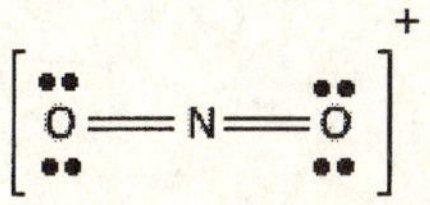

Two bonding pairs of electrons and no lone pairs gives linear electron geometry and molecular geometry. Linear electron geometry has a bond angle of 180°.

NO_2^-

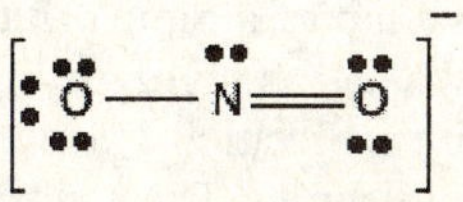

Two bonding pairs of electrons and one lone pair gives trigonal planar electron geometry, the molecular geometry will be bent.

Trigonal planar electron geometry has idealized bond angles of 120°. The bond angle is expected to be less than 120° because of the lone pair electrons occupying the third sp^2 orbital. Further, the bond angle should be less than the bond angle in NO_2 because the presence of lone pairs lowers the tendency for the central atom's orbitals to hybridize. As a result, as lone pairs are added, the bond angle moves further from the 120° hybrid angle to the 90° unhybridized angle and the two electrons will increase this tendency.

64. As you move down the column from F to Cl to Br to I, the atomic radius of the atoms increases. Because of this, the larger atoms cannot be accommodated with the smaller bond angle. The attached atoms themselves would begin to overlap their orbitals. So, as the size of the attached atom increases, the bond angle becomes larger, approaching the hybridized 109.5° angle.

65. Statement a is the best statement.

Statement b neglects the lowering of potential energy that arises from the interaction of the lone pair electrons with the bonding electrons.

Statement c neglects the interaction of the electrons altogether. The bonds form to accommodate the electrons, not the other way around.

66. A molecule with four bond groups and one lone pair would need 5 equivalent positions around the central atom. In two dimensions, this could be accommodated with a pentagon shape around the central atom. The idealized bond angles would be 108°; however, because of the lone pair occupying one of the positions, the bond angles would be less than 108°

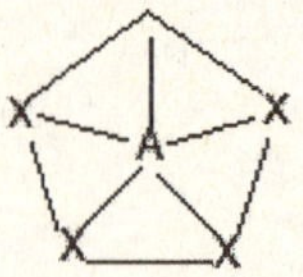

67. In Lewis theory, a covalent bond comes from the sharing of electrons.

A single bond shares 2 electrons (one pair)
A double bond shares 4 electrons (two pairs)
A triple bond shares 6 electrons (three pairs)

In valence bond theory, a covalent bond forms when orbitals overlap. The orbitals can be unhybridized or hybridized orbitals.

A single bond forms when a σ bond is formed from the overlap of an s orbital with an s orbital, an s orbital with a p orbital or a p orbital and a p orbital overlapping end to end.
A double bond is a combination of a σ bond and a π bond. The π bond forms from the sideways overlap of a p orbital on each of the atoms involved in the bond. The p orbitals must have the same orientation.
A triple bond is a combination of a σ bond and a 2π bonds. The π bonds form from the sideways overlap of a p orbital on each of the atoms involved in the bond. The p orbitals must have the same orientation so each π bond is formed from a different set of p orbitals.

In molecular orbital theory, molecular orbitals form. These are combinations of the atomic orbitals of the atoms involved in the bond. The bonds form when the valence electrons occupy more bonding molecular orbitals than antibonding molecular orbitals. This is calculated by the bond order.

A single bond, has a bond order of 1.
A double bond, has a bond order of 2.
A triple bond, has a bond order of 3.

All three models show the formation of bonds between two atoms. All three models show the formation of the same number of bonds between the atoms involved. The Lewis theory tells us only about the number of bonds formed and combined with VSEPR theory allows us to predict the shape of the molecule. It does not, however, tell us anything about how the bonds are formed. Valence bond theory addresses the formation of the different types of bonds, sigma and pi. In valence bond theory the bonds form from the overlap of atomic orbitals on the individual atoms involved in the bonds and the atoms are localized between the two atoms involved in the bond. Molecular orbital theory approaches the formation of bonds by looking at the entire molecule. The electrons are not restricted to any two individual atoms, but treated as belonging to the whole molecule. The electrons reside in molecular orbitals that are part of the entire molecule rather than being restricted to individual atoms. Each model gives us information about the molecule. The amount of information that we need determines the model that we choose to use.

68. In period 2, the atoms are smaller and they do not have d orbitals available to hybridize, so they cannot accommodate as many atoms around the central atom. In order to complete the octet of electrons multiple bonds must form. In the period 3 and higher atoms, more atoms can surround the central atom and the d orbitals can hybridize with the s and p orbitals. There are now more orbitals available to overlap, and there is space for them to do so. The central atom, therefore, can attain a stable configuration of eight or more electrons without having to multiple bond.

Chapter 11
Liquids, Solids, and Intermolecular Forces

1. a) dispersion forces

   b) dispersion forces and dipole–dipole forces

   c) dispersion forces

   d) dispersion forces, dipole–dipole forces, and hydrogen bonding

   e) dispersion forces

   f) dispersion forces, dipole–dipole forces and hydrogen bonding

   g) dispersion forces and dipole–dipole forces

   h) dispersion forces

2. a) dispersion forces and dipole–dipole forces

   b) dispersion forces, dipole–dipole forces and hydrogen bonding

   c) dispersion forces

   d) dispersion forces

   e) dispersion forces and dipole–dipole forces

   f) dispersion forces and dipole–dipole forces

   g) dispersion forces, dipole–dipole forces and hydrogen bonding

   h) dispersion forces

3. a) CH_4 < b) CH_3CH_3 < c) CH_3CH_2Cl < d) CH_3CH_2OH. The first two molecules only exhibit dispersion forces, so the boiling point increases with increasing molar mass. The third molecule also exhibits dipole–dipole forces, which are stronger than dispersion forces. The last molecule exhibits hydrogen bonding. Since these are the strongest intermolecular forces in this group, the last molecule has the highest boiling point.

4. a) H_2S < b) H_2Se < c) H_2O The first two molecules only exhibit dispersion forces and dipole–dipole forces, so the boiling point increases with increasing molar mass. The third molecule also exhibits hydrogen bonding. Since these are the strongest intermolecular forces in this group, the last molecule has the highest boiling point.

5. a) CH_3OH has the higher boiling point since it exhibits hydrogen bonding.

   b) CH_3CH_2OH has the higher boiling point since it exhibits hydrogen bonding.

   c) CH_3CH_3 has the higher boiling point since it has the larger molar mass.

6. a) NH_3 has the higher boiling point since it exhibits hydrogen bonding.

   b) CS_2 has the higher boiling point since it has the larger molar mass.

   c) NO_2 has the higher boiling point since it exhibits dipole–dipole forces.

7. a) Br_2 has the higher vapor pressure since it has the smaller molar mass.

   b) H_2S has the higher vapor pressure since it does not exhibit hydrogen bonding.

c) PH_3 has the higher vapor pressure since it does not exhibit hydrogen bonding.

8. a) CH_4 has the higher vapor pressure since it has the smaller molar mass and it does not exhibit dipole–dipole forces.
   b) CH_3OH has the higher vapor pressure since it has the smaller molar mass, and both exhibit hydrogen bonding.
   c) H_2CO has the higher vapor pressure since it has the smaller molar mass and it does not exhibit hydrogen bonding.

9. a) This will not form a homogeneous solution, since one is polar and one is nonpolar.

   b) This will form a homogeneous solution. There will be ion–dipole interactions between the K^+ and Cl^- ions and the water molecules. There will also be dispersion forces, dipole–dipole forces, and hydrogen bonding between the water molecules.

   c) This will form a homogeneous solution. There will be dispersion forces present.

   d) This will form a homogeneous solution. There will be dispersion forces, dipole–dipole forces, and hydrogen bonding.

10. a) This will form a homogeneous solution. There will be dispersion forces present.

    b) This will not form a homogeneous solution, since one is polar and one is nonpolar.

    c) This will form a homogeneous solution. There will be ion–dipole interactions between the Li^+ and NO_3^- ions and the water molecules. There will also be dispersion forces, dipole–dipole forces, and hydrogen bonding between the water molecules.

    d) This will not form a homogeneous solution, since one is polar and one is nonpolar.

11. Water will have the higher surface tension since it exhibits hydrogen bonding, a strong intermolecular force. Acetone cannot form hydrogen bonds.

12. a) Water "wets" surfaces that are capable of dipole–dipole interactions. The water will form strong adhesive forces with the surface when these dipole–dipole forces are present and so the water will spread to cover as much of the surface as possible. Water does not experience strong intermolecular forces with oil and other nonpolar surfaces. The water will bead up, maximizing the cohesive interactions, which involve strong hydrogen bonds. So water will bead up on surfaces that can only exhibit dispersion forces.

    b) Mercury will bead up on surfaces since it is not capable of forming strong intermolecular interactions (only dispersion forces).

13. Compound A will have the higher viscosity since it can interact with other molecules along the entire molecule, not just at a single point. Also the molecule is very flexible and the molecules can get tangled with each other.

14. Multigrade oils contain polymers (long molecules made up of repeating structural units) that coil at low temperatures but unwind at high temperatures. At low temperatures, the coiled polymers — because of their compact shape — do not contribute very much to the viscosity of the oil. As the temperature increases, however, the molecules unwind and their long shape results in intermolecular forces and molecular entanglements that prevent the viscosity from decreasing as much as it would normally. The result is an oil whose viscosity is less temperature-dependent than it would be otherwise, allowing the same oil to be used over a wider range of temperatures.

15. In a clean glass tube the water can generate strong adhesive interactions with the glass (due to the dipoles at the surface of the glass). Water experiences adhesive forces with glass that are stronger than its cohesive forces, causing it to climb the surface of a glass tube. When grease or oil coats the glass this interferes with the formation of these adhesive interactions with the glass, since oils are nonpolar and cannot interact strongly with the dipoles in the water. Without experiencing these strong intermolecular forces with oil, the water's cohesive forces will be greater and it will be drawn away from the surface of the tube.

16. Water can generate strong adhesive interactions with the glass (due to the dipoles at the surface of the glass), but hexane is nonpolar and cannot interact strongly with the glass surface.

17. The water in the 12 cm diameter beaker will evaporate more quickly because there is more surface area for the molecules to evaporate from. The vapor pressure will be the same in the two containers because the vapor pressure is the pressure of the gas when it is in dynamic equilibrium with the liquid (evaporation rate = condensation rate). The vapor pressure is dependent only on the substance and the temperature. The 12 cm diameter container will reach this dynamic equilibrium faster.

18. The acetone will evaporate more quickly since it is not capable of forming hydrogen bonds, so the intermolecular forces are much weaker. This will also result in a larger vapor pressure at the same temperature as the water.

19. The boiling point and higher heat of vaporization of oil are much higher than that of water, so it will not vaporize as quickly as the water. The evaporation of water cools your skin because evaporation is an endothermic process.

20. Water molecules have a lower kinetic energy at room temperature than at 100 °C. The heat of vaporization is the energy difference between the molecules in the liquid phase and the gas phase. Since the energy of the liquid is lower at room temperature, then the energy difference that must be overcome to become steam is greater, so the heat of vaporization is greater.

21. **Given:** 955 kJ from candy bar, water d = 1.00 g/ml **Find:** L(H_2O) vaporized at 100.0 °C
Other: ΔH°_{vap} = 40.7 kJ/mol
Conceptual Plan: $q \rightarrow$ **mol H_2O** $\rightarrow$ **g H_2O** $\rightarrow$ **mL H_2O** $\rightarrow$ **L H_2O**

$$\frac{1\text{ mol}}{40.7\text{ kJ}} \quad \frac{18.01\text{ g}}{1\text{ mol}} \quad \frac{1.00\text{ mL}}{1.00\text{ g}} \quad \frac{1\text{ L}}{1000\text{ mL}}$$

Solution: $$955\text{ kJ} \times \frac{1\text{ mol}}{40.7\text{ kJ}} \times \frac{18.01\text{ g}}{1\text{ mol}} \times \frac{1.00\text{ mL}}{1\text{ g}} \times \frac{1\text{ L}}{1000\text{ mL}} = 0.423\text{ L }H_2O$$

Check: The units (L) are correct. The magnitude of the answer (< 1 L) makes physical sense because we are vaporizing about 20 moles of water.

22. **Given:** 55.0 mL water, d = 1.00 g/ml, heated to 100.0 °C **Find:** heat (kJ) to vaporize at 100.0 °C
Other: ΔH°_{vap} = 40.7 kJ/mol
Conceptual Plan: **mL H_2O** $\rightarrow$ **g H_2O** $\rightarrow$ **mol H_2O** $\rightarrow$ q

$$\frac{1.00\text{ g}}{1.00\text{ mL}} \quad \frac{1\text{ mol}}{18.01\text{ g}} \quad \frac{40.7\text{ kJ}}{1\text{ mol}}$$

Solution: $$55.0\text{ mL} \times \frac{1.00\text{ g}}{1.00\text{ mL}} \times \frac{1\text{ mol}}{18.01\text{ g}} \times \frac{40.7\text{ kJ}}{1\text{ mol}} = 124\text{ kJ}$$

Check: The units (kJ) are correct. The magnitude of the answer (124 kJ) makes physical sense because we are vaporizing about 3 moles of water.

23. **Given:** 0.88 g water condenses on iron block 75.0 g at T_i = 22 °C **Find:** T_f(iron block)
Other: ΔH°_{vap} = 44.0 kJ/mol; C_{Fe} = 0.449 J/g · °C from text
Conceptual Plan: **g H_2O** $\rightarrow$ **mol H_2O** $\rightarrow$ q_{H2O} **(kJ)** $\rightarrow$ q_{H2O} **(J)** $\rightarrow$ q_{Fe} **then** q_{Fe}, m_{Fe}, $T_i \rightarrow T_f$

$$\frac{1\text{ mol}}{18.01\text{ g}} \quad \frac{-44.0\text{ kJ}}{1\text{ mol}} \quad \frac{1000\text{ J}}{1\text{ kJ}} \quad -q_{H2O} = q_{Fe} \quad q = m\,C_s\left(T_f - T_i\right)$$

Solution: $$0.88\text{ g} \times \frac{1\text{ mol}}{18.01\text{ g}} \times \frac{-44.0\text{ kJ}}{1\text{ mol}} \times \frac{1000\text{ J}}{1\text{ kJ}} = -2\underline{1}49.92\text{ J then } -q_{H2O} = q_{Fe} = 2\underline{1}49.92\text{ J then}$$

$q = m\,C_s\left(T_f - T_i\right)$ Rearrange to solve for T_f.

$$T_f = \frac{m\,C_s\,T_i + q}{m\,C_s} = \frac{\left(75.0\text{ g} \times 0.449\frac{\text{J}}{\text{g}\cdot{}^\circ\text{C}} \times 22\,{}^\circ\text{C}\right) + 2\underline{1}49.92\text{ J}}{75.0\text{ g} \times 0.449\frac{\text{J}}{\text{g}\cdot{}^\circ\text{C}}} = 86\,{}^\circ\text{C}$$

Check: The units (°C) are correct. The temperature rose, which is consistent with heat being added to the block. The magnitude of the answer (86 °C) makes physical sense because even though we have ~ 1/20 th of a mole, the energy involved in condensation is very large.

24. **Given:** 1.02 g rubbing alcohol (C_3H_8O) evaporated from aluminum block 55.0 g at T_i = 25 °C
Find: T_f(aluminum block) **Other:** $\Delta H°_{vap}$ = 45.4 kJ/mol; C_{Al} = 0.903 J/g · °C from text
Conceptual Plan:
g C_3H_8O → mol C_3H_8O → q_{C3H8O} (kJ) → q_{C3H8O} (J) → q_{Al} then q_{Al}, m_{Fe}, T_i → T_f

$$\frac{1\ \text{mol}}{60.09\ \text{g}} \quad \frac{-45.4\ \text{kJ}}{1\ \text{mol}} \quad \frac{1000\ \text{J}}{1\ \text{kJ}} \quad -q_{H2O} = q_{Al} \quad q = m\,C_S\left(T_f - T_i\right)$$

Solution: $1.02\ \text{g} \times \frac{1\ \text{mol}}{60.09\ \text{g}} \times \frac{45.4\ \text{kJ}}{1\ \text{mol}} \times \frac{1000\ \text{J}}{1\ \text{kJ}} = 77\underline{0}.644\ \text{J}$ then $-q_{H2O} = q_{Al} = -77\underline{0}.644\ \text{J}$ then

$q = m\,C_s\left(T_f - T_i\right)$ Rearrange to solve for T_f.

$$T_f = \frac{m\,C_s\,T_i + q}{m\,C_s} = \frac{\left(55.0\ \text{g} \times 0.903\ \frac{\text{J}}{\text{g}\cdot°\text{C}} \times 25\ °\text{C}\right) - 77\underline{0}.644\ \text{J}}{55.0\ \text{g} \times 0.903\ \frac{\text{J}}{\text{g}\cdot°\text{C}}} = 9.5\ °\text{C}$$

Check: The units (°C) are correct. The temperature dropped, which is consistent with heat being removed from the block. The magnitude of the answer (9.5 °C) makes physical sense because even though we have only a fraction of a mole, the energy involved in vaporization is very large.

25. **Given:**

Temperature (K)	Vapor Pressure (torr)
200	65.3
210	134.3
220	255.7
230	456.0
235	597.0

Find: $\Delta H°_{vap}$ (NH_3) and normal boiling point

Conceptual Plan: To find the heat of vaporization, use Excel or similar software to make a plot of the natural log of vapor pressure (ln *P*) as a function of the inverse of the temperature in K (1/*T*). Then fit the points to a line and determine the slope of the line. Since the slope = – $\Delta H_{vap}/R$, we find the heat of vaporization as follows:
slope = – $\Delta H_{vap}/R$ → ΔH_{vap} = – slope × *R* then J → kJ.

$$\frac{1\ \text{kJ}}{1000\ \text{J}}$$

For the normal boiling point, use the equation of the best fit line, substitute 760 torr for the pressure and calculate the temperature.
Solution: Data was plotted in Excel.
The slope of the best fitting line is – 29$\underline{6}$9.9 K.

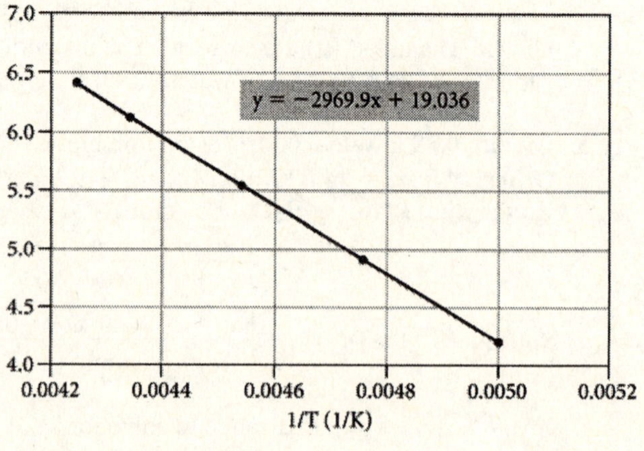

$$\Delta H_{vap} = -\,slope \times R = --\,29\underline{6}9.9\ \text{K} \times \frac{8.314\ \text{J}}{\text{K mol}} =$$

$$= \frac{2.4\underline{6}917 \times 10^4\ \text{J}}{\text{mol}} \times \frac{1\ \text{kJ}}{1000\ \text{J}} = 24.7\ \frac{\text{kJ}}{\text{mol}}$$

$$\ln P = -\,29\underline{6}9.9\text{K}\left(\frac{1}{T}\right) + 19.03\underline{6} \rightarrow$$

$$\ln 760 = -\,29\underline{6}9.9\text{K}\left(\frac{1}{T}\right) + 19.03\underline{6} \rightarrow$$

$$29\underline{6}9.9\text{K}\left(\frac{1}{T}\right) = 19.03\underline{6} - 6.6\underline{3}332 \rightarrow$$

$$T = \frac{29\underline{6}9.9\text{K}}{12.40\underline{2}68} = 239\ \text{K}$$

Check: The units (kJ/mol) are correct. The magnitude of the answer (25) is consistent with other values in the text.

26. **Given:**

Temperature (K)	Vapor Pressure (torr)
65	130.5
70	289.5
75	570.8
80	1028
85	1718

Find: ΔH°_{vap} (N_2) and normal boiling point

Conceptual Plan: To find the heat of vaporization, use Excel or similar software to make a plot of the natural log of vapor pressure (ln *P*) as a function of the inverse of the temperature in K (1/*T*). Then fit the points to a line and determine the slope of the line. Since the slope = – $\Delta H_{vap}/R$, we find the heat of vaporization as follows:

slope = – $\Delta H_{vap}/R$ → ΔH_{vap} = – slope × *R* then J → kJ.

$$\frac{1\ \text{kJ}}{1000\ \text{J}}$$

For the normal boiling point, use the equation of the best fit line, substitute 760 torr for the pressure and calculate the temperature.

Solution: Data was plotted in Excel.

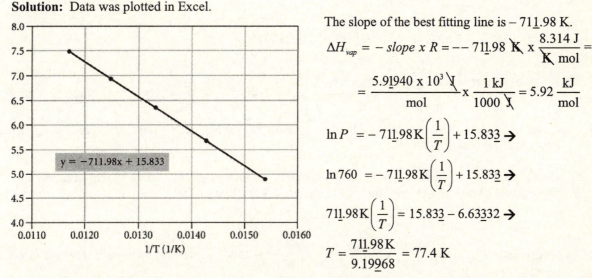

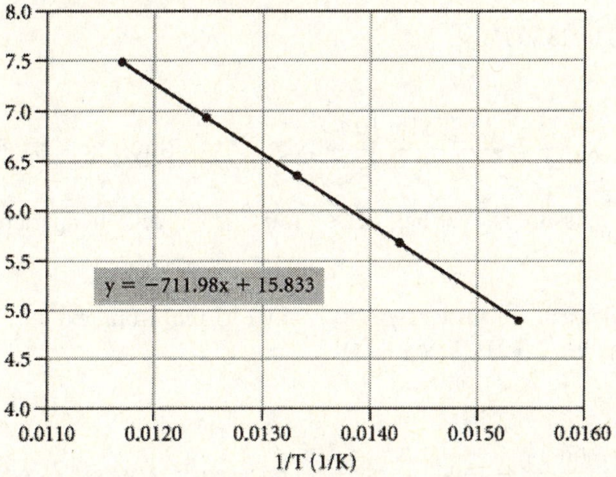

The slope of the best fitting line is – 711.98 K.

$$\Delta H_{vap} = -\ slope\ x\ R = --\ 711.98\ \text{K}\ \text{x}\ \frac{8.314\ \text{J}}{\text{K mol}} =$$

$$= \frac{5.91940 \text{ x } 10^3\ \text{J}}{\text{mol}} \text{ x } \frac{1\ \text{kJ}}{1000\ \text{J}} = 5.92\ \frac{\text{kJ}}{\text{mol}}$$

$$\ln P = -\ 711.98\text{K}\left(\frac{1}{T}\right) + 15.833 \rightarrow$$

$$\ln 760 = -\ 711.98\text{K}\left(\frac{1}{T}\right) + 15.833 \rightarrow$$

$$711.98\text{K}\left(\frac{1}{T}\right) = 15.833 - 6.63332 \rightarrow$$

$$T = \frac{711.98\text{K}}{9.19968} = 77.4\ \text{K}$$

Check: The units (kJ/mol) are correct. The magnitude of the answer (6) is lower than other values quoted in the text. This is consistent with the fact that nitrogen boils at such a low temperature.

27. **Given:** ethanol, ΔH°_{vap} = 38.56 kJ/mol; normal boiling point = 78.4 °C **Find:** P_{Ethanol} at 15 °C

Conceptual Plan: °C → K and kJ → J then $\Delta H^\circ_{vap}, T_1, P_1, T_2 \rightarrow P_2$

$$\text{K} = {}^\circ\text{C} + 273.15 \qquad \frac{1000\ \text{J}}{1\ \text{kJ}} \qquad \ln\frac{P_2}{P_1} = \frac{-\Delta H_{vap}}{R}\left(\frac{1}{T_2} - \frac{1}{T_1}\right)$$

Solution: T_1 = 78.4 °C + 273.15 = 351.6 K; T_2 = 15 °C + 273.15 = 288 K;

$$\frac{38.56\ \text{kJ}}{\text{mol}} \text{ x } \frac{1000\ \text{J}}{1\ \text{kJ}} = 3.856 \text{ x } 10^4\ \frac{\text{J}}{\text{mol}} \quad P_1 = 760\ \text{torr} \quad \ln\frac{P_2}{P_1} = \frac{-\Delta H_{vap}}{R}\left(\frac{1}{T_2} - \frac{1}{T_1}\right)$$ Substitute values in equation.

$$\ln\frac{P_2}{760\ \text{torr}} = \frac{-\ 3.856 \text{ x } 10^4\ \frac{\text{J}}{\text{mol}}}{8.314\ \frac{\text{J}}{\text{K}\cdot\text{mol}}}\left(\frac{1}{288\ \text{K}} - \frac{1}{351.6\ \text{K}}\right) = -\ 2.91302 \rightarrow$$

$$\frac{P_2}{760\ \text{torr}} = e^{-2.91302} = 0.054311 \rightarrow P_2 = 0.054311 \text{ x } 760\ \text{torr} = 41\ \text{torr}$$

Check: The units (torr) are correct. Since 15 °C is significantly below the boiling point, we expect the answer to be much less than 760 torr.

28. **Given:** benzene, $\Delta H°_{vap}$ = 30.72 kJ/mol; normal boiling point = 80.1 °C; P_2 = 445 torr **Find:** T_2
Conceptual Plan: **°C → K and kJ → J then $\Delta H°_{vap}, T_1, P_1, T_2 \rightarrow P_2$**

$$K = °C + 273.15 \qquad \frac{1000\ J}{1\ kJ} \qquad \ln\frac{P_2}{P_1} = \frac{-\Delta H_{vap}}{R}\left(\frac{1}{T_2} - \frac{1}{T_1}\right)$$

Solution: T_1 = 80.1 °C + 273.15 = 353.3 K; $\frac{30.72\ kJ}{mol} \times \frac{1000\ J}{1\ kJ} = 3.072 \times 10^4\ \frac{J}{mol}$; P_1 = 760 torr;

P_2 = 445 torr $\ln\frac{P_2}{P_1} = \frac{-\Delta H_{vap}}{R}\left(\frac{1}{T_2} - \frac{1}{T_1}\right)$ Substitute values in equation.

$$\ln\frac{445\ torr}{760\ torr} = \frac{-3.072 \times 10^4\ \frac{J}{mol}}{8.314\ \frac{J}{K \cdot mol}}\left(\frac{1}{T_2} - \frac{1}{353.3\ K}\right) \rightarrow$$

$$-0.535244 = -3.694972 \times 10^3\ K\left(\frac{1}{T_2} - 0.002830456\right)$$

$$\rightarrow \frac{1.44857 \times 10^{-4}}{K} = \left(\frac{1}{T_2} - \frac{0.002830456}{K}\right) \rightarrow \frac{1}{T_2} = \frac{2.97531 \times 10^{-3}}{K} \rightarrow T_2 = 336.0990\ K = 63\ °C$$

Check: The units (°C) are correct. Since the pressure is over half of 760 torr, we expect a temperature a little lower than the boiling point.

29. **Given:** 47.5 g water freezes **Find:** energy released **Other:** $\Delta H°_{fus}$ = 6.02 kJ/mol from text
Conceptual Plan: **g H_2O → mol H_2O → q_{H2O} (kJ) → q_{H2O} (J)**

$$\frac{1\ mol}{18.01\ g} \qquad \frac{-6.02\ kJ}{1\ mol} \qquad \frac{1000\ J}{1\ kJ}$$

Solution: $47.5\ g \times \frac{1\ mol}{18.01\ g} \times \frac{-6.02\ kJ}{1\ mol} \times \frac{1000\ J}{1\ kJ} = -15900$ J or 15900 J released or 15.9 kJ released

Check: The units (J) are correct. The magnitude (15900 J) makes sense since we are freezing about 3 moles of water. Freezing is exothermic, so heat is released.

30. **Given:** 25.0 g dry ice(CO_2) sublimation **Find:** heat required **Other:** $\Delta H°_{sub}$ = 32.3 kJ/mol
Conceptual Plan: **g dry ice → mol dry ice → $q_{dry\ ice}$ (kJ) → $q_{dry\ ice}$ (J)**

$$\frac{1\ mol}{44.01\ g} \qquad \frac{32.3\ kJ}{1\ mol} \qquad \frac{1000\ J}{1\ kJ}$$

Solution: $25.0\ g \times \frac{1\ mol}{44.01\ g} \times \frac{32.3\ kJ}{1\ mol} \times \frac{1000\ J}{1\ kJ} =$ 18300 J required or 18.3 kJ absorbed

Check: The units (J) are correct. The magnitude (18300 J) makes sense since we are subliming just over 0.5 moles of dry ice. Sublimation is endothermic, so heat is required.

31. **Given:** 8.5 g ice; 255 g water **Find:** ΔT of water
Other: $\Delta H°_{fus}$ = 6.0 kJ/mol; C_{H2O} = 4.18 J/g · °C from text
Conceptual Plan: The first step is to calculate how much heat is removed from the water to melt the ice.
$q_{ice} = -q_{water}$ so g (ice) → mol (ice) → q_{fus}(kJ) → q_{fus} (J) → q_{water} (J) then $q, m, C_s \rightarrow \Delta T_1$

$$\frac{1\ mol}{18.01 g} \qquad \frac{6.0\ kJ}{1\ mol} \qquad \frac{1000\ J}{1\ kJ} \qquad q_{water} = -q_{ice} \qquad q = m C_S \Delta T_1$$

Now we have slightly cooled water (at a temperature of T_1) in contact with 0.0 °C water, and we can calculate a second temperature drop of the water due to mixing of the water that was ice with the initially room temperature water. so $q_{ice} = -q_{water}$ with $m, C_s \rightarrow \Delta T_2$ with $\Delta T_1\ \Delta T_2 \rightarrow \Delta T_{Total}$

$$q = m\ C_S \Delta T_2\ \ then\ set\ q_{ice} = -q_{H2O} \qquad \Delta T_{Total} = \Delta T_1 + \Delta T_2$$

Solution: $8.5 \cancel{g} \times \dfrac{1 \cancel{mol}}{18.01 \cancel{g}} \times \dfrac{6.0 \cancel{kJ}}{1 \cancel{mol}} \times \dfrac{1000\ J}{1 \cancel{kJ}} = 2.\underline{8}3176 \times 10^3\ J$, $\quad q_{water} = -\ q_{ice} = -\ 2.\underline{8}3176 \times 10^3\ J$

$q = mC_S\Delta T$ Rearrange to solve for ΔT. $\Delta T_1 = \dfrac{q}{mC_S} = \dfrac{-\ 2.\underline{8}3176 \times 10^3\ \cancel{J}}{255\ \cancel{g} \times 4.18\ \dfrac{\cancel{J}}{\cancel{g} \cdot {}^\circ C}} = -\ 2.\underline{6}567\ {}^\circ C$.

$q = mC_S\Delta T$ substitute in values and set $q_{ice} = -\ q_{water}$.

$q_{ice} = m_{ice}\ C_{ice}\left(T_f - T_{icei}\right) = 8.5\ \cancel{g} \times 4.18\ \dfrac{J}{\cancel{g} \cdot {}^\circ\cancel{C}} \times \left(T_f - 0.0\ {}^\circ C\right) =$

$\rightarrow$

$-\ q_{water} = -\ m_{water}\ C_{water}\ \Delta T_{water2} = -\ 255\ \cancel{g} \times 4.18\ \dfrac{J}{\cancel{g} \cdot {}^\circ\cancel{C}} \times \Delta T_{water2}$

$8.5\ T_f = -\ 255\ \Delta T_{water2} = -\ 255\ (T_f - T_{f1})$ Rearrange to solve for T_f. $8.5\ T_f + 255\ T_f = 255\ T_{f1}$ $\rightarrow$

$263.5\ T_f = 255\ T_{f1}$ $\rightarrow$ $T_f = 0.96\underline{7}74\ T_{f1}$ but $\Delta T_1 = (T_{f1} - T_{i1}) = -\ 2.\underline{6}567\ {}^\circ C$ which says that

$T_{f1} = T_{i1} - 2.\underline{6}567\ {}^\circ C$ and $\Delta T_{Total} = (T_f - T_{i1})$ so

$\Delta T_{Total} = 0.96\underline{7}74\ T_{f1} - T_{i1} = 0.96\underline{7}74\left(T_{i1} - 2.\underline{6}567\ {}^\circ C\right) - T_{i1} = -\ 2.\underline{6}567\ {}^\circ C - 0.03\underline{2}26 T_{i1}$.

This implies that the larger the initial temperature of the water, the larger the temperature drop. If the initial temperature was 90 °C, the temperature drop would be 5.6 °C. If the initial temperature was 25 °C, the temperature drop would be 3.5°C. If the initial temperature was 5 °C, the temperature drop would be 2.8 °C. This makes physical sense because the lower the initial temperature of the water, the less kinetic energy it initially has and the smaller the heat transfer from the water to the melted ice will be.

Check: The units (°C) are correct. The temperature drop from the melting of the ice is only 2.7 °C because the mass of the water is so much larger than the ice.

32. **Given:** 352 mL water, $T_i = 25\ {}^\circ C$, $T_f = 5\ {}^\circ C$, $d = 1.0\ g/mL$; ice $T_i = 0\ {}^\circ C$, $T_f = 5\ {}^\circ C$ **Find:** g (ice)
    Other: $\Delta H^\circ_{fus} = 6.02\ kJ/mol$; $C_{H2O} = 4.18\ J/g \cdot {}^\circ C$ from text
    Conceptual Plan:
    mL $\rightarrow$ g then m, C_s, T_i, T_f $\rightarrow$ q_{water} (J) $\rightarrow$ q_{ice} (J) then $q_{ice}, \Delta H^\circ_{fus}, C_{H2O}, T_i, T_f$ $\rightarrow$ g (ice)

    $\dfrac{1.0\ g}{1.0\ mL}$ $\qquad\qquad q = mC_S(T_f - T_i)$ $\quad q_{ice} = -\ q_{water}$ $\quad q_{ice} = mC_S\left(T_f - T_i\right) + m \times \dfrac{1\ mol}{18.01 g} \times \dfrac{6.02\ kJ}{1\ mol} \times \dfrac{1000\ J}{1\ kJ}$

    Solution: $352\ \cancel{mL} \times \dfrac{1.0\ g}{1.0\ \cancel{mL}} = 3\underline{5}2\ g$

    $q_{water} = m_{water}\ C_{water}\left(T_f - T_i\right) = 3\underline{5}2\ \cancel{g} \times 4.18\ \dfrac{J}{\cancel{g} \cdot {}^\circ\cancel{C}} \times (5\ \cancel{{}^\circ C} - 25\ \cancel{{}^\circ C}) = -\ 2.\underline{9}4272 \times 10^4\ J$,

    $q_{ice} = -\ q_{water} = 2.\underline{9}4272 \times 10^4\ J$ then $q_{ice} = mC_S\left(T_f - T_i\right) + m \times \dfrac{1\ \cancel{mol}}{18.01 g} \times \dfrac{6.02\ \cancel{kJ}}{1\ \cancel{mol}} \times \dfrac{1000\ J}{1\ \cancel{kJ}}$ substitute

    values. $2.\underline{9}4272 \times 10^4\ J = m \times 4.18\ \dfrac{J}{g \cdot {}^\circ\cancel{C}} \times \left(5\ \cancel{{}^\circ C} - 0\ \cancel{{}^\circ C}\right) + m \times \dfrac{33\underline{4}.26\ J}{1 g}$ Rearrange to solve for m.

    $m = \dfrac{2.\underline{9}4272 \times 10^4\ \cancel{J}}{\dfrac{35\underline{5}.16\ \cancel{J}}{1 g}} = 83\ g$

    Check: The units (g) are correct. The temperature drop is large and so is the amount of water we want to cool, so the mass seems reasonable.

33. **Given:** 10.0 g ice $T_i = -10.0\ {}^\circ C$ to steam at $T_f = 110.0\ {}^\circ C$ **Find:** heat required (kJ)
    Other: $\Delta H^\circ_{fus} = 6.02\ kJ/mol$; $\Delta H^\circ_{vap} = 40.7\ kJ/mol$; $C_{ice} = 2.09\ J/g \cdot {}^\circ C$; $C_{H2O} = 4.18\ J/g \cdot {}^\circ C$; $C_{steam} = 2.01\ J/g \cdot {}^\circ C$
    Conceptual Plan: Follow the heating curve in Figure 11.36. $q_{Total} = q_1 + q_2 + q_3 + q_4 + q_5$ where q_1, q_3, and q_5 are heating of a single phase then J $\rightarrow$ kJ and q_2 and q_4 are phase transitions.

    $q = mC_S(T_f - T_i)$ $\qquad\qquad \dfrac{1\ kJ}{1000\ J}$ $\qquad q = m \times \dfrac{1\ mol}{18.01 g} \times \dfrac{\Delta H}{1\ mol}$

Solution:

$q_1 = m_{ice}\, C_{ice}(T_{icef} - T_{icei}) = 10.0\ \text{g} \times 2.09\ \frac{\text{J}}{\text{g}\cdot{}^\circ\text{C}} \times (0.0\ ^\circ\text{C} - -10.0\ ^\circ\text{C}) = 209\ \text{J} \times \frac{1\ \text{kJ}}{1000\ \text{J}} = 0.209\ \text{kJ}$,

$q_2 = m \times \frac{1\ \text{mol}}{18.01\text{g}} \times \frac{\Delta H_{fus}}{1\ \text{mol}} = 10.0\ \text{g} \times \frac{1\ \text{mol}}{18.01\ \text{g}} \times \frac{6.02\ \text{kJ}}{1\ \text{mol}} = 3.3\underline{4}3\ \text{kJ}$,

$q_3 = m_{water}\, C_{water}(T_{waterf} - T_{wateri}) = 10.0\ \text{g} \times 4.18\ \frac{\text{J}}{\text{g}\cdot{}^\circ\text{C}} \times (100.0\ ^\circ\text{C} - 0.0\ ^\circ\text{C}) = 4180\ \text{J} \times \frac{1\ \text{kJ}}{1000\ \text{J}} = 4.18\ \text{kJ}$,

$q_4 = m \times \frac{1\ \text{mol}}{18.01\text{g}} \times \frac{\Delta H_{vap}}{1\ \text{mol}} = 10.0\ \text{g} \times \frac{1\ \text{mol}}{18.01\ \text{g}} \times \frac{40.7\ \text{kJ}}{1\ \text{mol}} = 22.\underline{5}99\ \text{kJ}$,

$q_5 = m_{steam}\, C_{steam}(T_{steamf} - T_{steami}) = 10.0\ \text{g} \times 2.01\ \frac{\text{J}}{\text{g}\cdot{}^\circ\text{C}} \times (110.0\ ^\circ\text{C} - 100.0\ ^\circ\text{C})$

$= 201\ \text{J} \times \frac{1\ \text{kJ}}{1000\ \text{J}} = 0.201\ \text{kJ}.$

$q_{Total} = q_1 + q_2 + q_3 + q_4 + q_5 = 0.209\ \text{kJ} + 3.3\underline{4}3\ \text{kJ} + 4.18\ \text{kJ} + 22.\underline{5}99\ \text{kJ} + 0.201\ \text{kJ} = 30.5\ \text{kJ}$

Check: The units (kJ) are correct. The amount of heat is dominated the vaporization step. Since we have less than 1 mole we expect less than 41 kJ.

34. **Given:** 1.00 mole ice $T_i = 145.0\ ^\circ\text{C}$ to ice at $T_f = -50.0\ ^\circ\text{C}$ **Find:** heat evolved (kJ) **Other:** $\Delta H^\circ_{fus} = 6.02$ kJ/mol; $\Delta H^\circ_{vap} = 40.7$ kJ/mol; $C_{ice} = 2.09$ J/g · °C; $C_{H2O} = 4.18$ J/g · °C; $C_{steam} = 2.01$ J/g · °C
Conceptual Plan: mol → g Follow the heating curve in Figure 11.36, only in reverse.

$\frac{18.01\ \text{g}}{1\ \text{mol}}$

$q_{Total} = q_1 + q_2 + q_3 + q_4 + q_5$ **where q_1, q_3, and q_5 are heating of a single phase then J → kJ**

$q = m\,C_S(T_f - T_i)$ $\qquad \frac{1\ \text{kJ}}{1000\ \text{J}}$

and q_2 and q_4 are phase transitions.

$q = mol \times \frac{\Delta H}{1\ \text{mol}}$

Solution: $1.00\ \text{mol} \times \frac{18.01\ \text{g}}{1\ \text{mol}} = 18.\underline{0}1\ \text{g}$

$q_1 = m_{steam}\, C_{steam}(T_{steamf} - T_{steami}) = 18.\underline{0}1\ \text{g} \times 2.01\ \frac{\text{J}}{\text{g}\cdot{}^\circ\text{C}} \times (100.0\ ^\circ\text{C} - 145.0\ ^\circ\text{C}) = -16\underline{2}9.00\ \text{J} \times \frac{1\ \text{kJ}}{1000\ \text{J}}$

$= -1.6\underline{2}9\ \text{kJ}$

$q_2 = mol \times \frac{-\Delta H_{vap}}{1\text{mol}} = 1.00\ \text{mol} \times \frac{-40.7\ \text{kJ}}{1\ \text{mol}} = -40.7\ \text{kJ}$,

$q_3 = m_{water}\, C_{water}(T_{waterf} - T_{wateri}) = 18.\underline{0}1\ \text{g} \times 4.18\ \frac{\text{J}}{\text{g}\cdot{}^\circ\text{C}} \times (0.0\ ^\circ\text{C} - 100.0\ ^\circ\text{C}) = -75\underline{2}8.2\ \text{J} \times \frac{1\ \text{kJ}}{1000\ \text{J}}$

$= -7.5\underline{2}82\ \text{kJ}$,

$q_4 = mol \times \frac{-\Delta H_{fus}}{1\text{mol}} = 1.00\ \text{mol} \times \frac{-6.02\ \text{kJ}}{1\ \text{mol}} = -6.02\ \text{kJ}$,

$q_5 = m_{ice}\, C_{ice}(T_{icef} - T_{icei}) = 18.\underline{0}1\ \text{g} \times 2.09\ \frac{\text{J}}{\text{g}\cdot{}^\circ\text{C}} \times (-50.0\ ^\circ\text{C} - 0.0\ ^\circ\text{C}) = -18\underline{8}2.0\ \text{J} \times \frac{1\ \text{kJ}}{1000\ \text{J}}$

$= -1.8\underline{8}20\ \text{kJ}$

$q_{Total} = q_1 + q_2 + q_3 + q_4 + q_5 = -1.6\underline{2}9\ \text{kJ} - 40.7\ \text{kJ} - 7.5\underline{2}82\ \text{kJ} - 6.02\ \text{kJ} - 1.8\underline{8}20\ \text{kJ}$

$= -57.8\ \text{kJ}$ or 57.8 kJ released.

Check: The units (kJ) are correct. The amount of heat is dominated the vaporization step. Since we have less than 1 mole we expect less than 41 kJ.

35. a) solid

b) liquid

c) gas

d) supercritical fluid

e) solid/liquid equilibrium

f) liquid/gas equilibrium

g) solid/liquid/gas equilibrium

36. a) 184.4 °C

b) 113.6 °C

c) solid

d) gas

37. **Given:** nitrogen, normal boiling point = 77.3 K, normal melting point = 63.1 K, critical temperature = 126.2 K, critical pressure = 2.55×10^4 torr, triple point at 63.1 K and 94.0 torr
Find: Sketch phase diagram. Does nitrogen have a stable liquid phase at 1 atm?

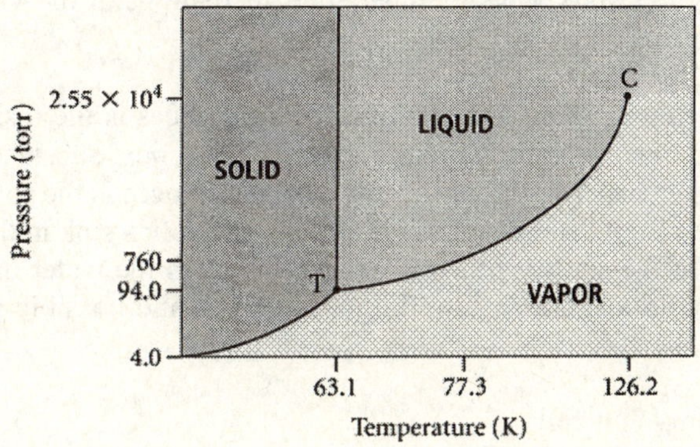

Nitrogen has a stable liquid phase at 1 atm.

38. **Given:** argon, normal boiling point = 87.2 K, normal melting point = 84.1 K, critical temperature = 150.8 K, critical pressure = 48.3 atm, triple point at 83.7 K and 0.68 atm
Find: Sketch phase diagram. Which has the greater density, solid or liquid argon?

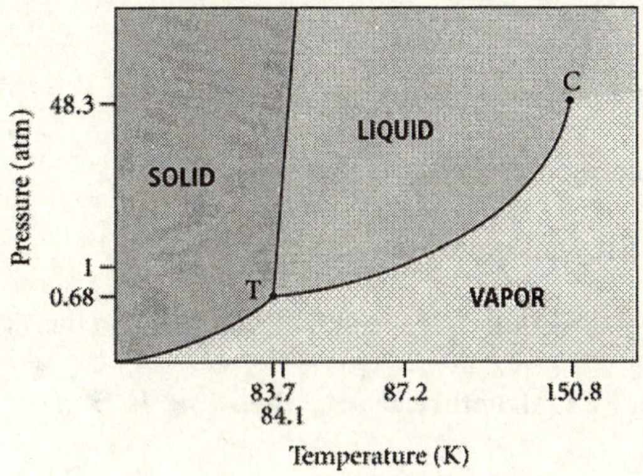

The solid has the higher density because the slope of the solid/liquid equilibrium line is positive. If we start in the liquid and increase the pressure, we will cross into the solid phase, the dense phase.

39. a) 0.027 mmHg, the higher of the two triple points

b) The Rhombic phase is denser because if we start in the Monoclinic phase at 100 °C and increase the pressure, we will cross into the Rhombic phase.

40. The triple point marked "O" shows the equilibrium of Ice II, Ice III, and Ice V. Ice II is denser than Ice I because you can generate Ice II from Ice I by increasing the pressure (and pushing the molecules closer together). Ice III would sink in liquid water. Note that the slope of the Ice III/liquid line has the typical positive slope.

41. Water has a low molar mass (18.01 g/mol), yet it is a liquid at room temperature. Water's high boiling point for its molar mass can be understood by examining the structure of the water molecule. The bent geometry of the water molecule and the highly polar nature of the O–H bonds result in a molecule with a significant dipole moment. Water's two O–H bonds (hydrogen directly bonded to oxygen) allow a water molecule to form strong hydrogen bonds with four other water molecules, resulting in a relatively high boiling point.

42. Water's high polarity also allows it to dissolve many other polar and ionic compounds, and even a number of nonpolar gases such as oxygen and carbon dioxide (by inducing a dipole moment in their molecules). Consequently, water is the main solvent within living organisms, transporting nutrients and other important compounds throughout the body. Water is also the main solvent of the environment, allowing aquatic animals, for example, to survive by breathing dissolved oxygen and aquatic plants to survive by using dissolved carbon dioxide for photosynthesis.

43. Water has an exceptionally high specific heat capacity, which has a moderating effect on the climate of coastal cities. Also, its high ΔH_{vap} causes water evaporation and condensation to have a strong effect on temperature. A tremendous amount of heat can be stored in large bodies of water. The heat will be absorbed or released from the large bodies of water preferentially over the land around it. In some cities, such as San Francisco, for example, the daily fluctuation in temperature can be less than 10 °C. This same moderating effect occurs over the entire planet, two-thirds of which is covered by water. In other words, without water, the daily temperature fluctuations on our planet might be more like those on Mars, where temperature fluctuations of 63 °C (113 °F) have been measured between midday and early morning.

44. One significant difference between the phase diagram of water and that of other substances is the fusion curve for water which has a negative slope. The fusion curve within the phase diagrams for most substances has a positive slope because increasing pressure favors the denser phase, which for most substances is the solid phase. This negative slope means that ice is less dense than liquid water and so ice floats. The solids sink in the liquids of most other substances. The frozen layer of ice at the surface of a winter lake insulates the water in the lake from further freezing. If this ice layer sank, it would kill bottom-dwelling aquatic life and possibly allow the lake to freeze solid, eliminating virtually all life in the lake.

45. a) 8 corner atoms x (1/8 atom / unit cell) = 1 atom / unit cell

b) 8 corner atoms x (1/8 atom / unit cell) + 1 atom in center = (1 + 1) atoms / unit cell = 2 atoms / unit cell

c) 8 corner atoms x (1/8 atom / unit cell) + 6 face-centered atoms x (1/2 atom / unit cell) = (1 + 3) atoms / unit cell = 4 atoms / unit cell

46. a) coordination number of 12 since this is a face-centered cubic structure

b) coordination number of 12 since this is a hexagonal closest packed structure

c) coordination number of 8 since this is a body-centered cubic structure

47. **Given:** platinum, face-centered cubic structure, r = 139 pm **Find:** edge length of unit cell and density (g/cm^3)
Conceptual Plan:
$r \rightarrow l$ **and** $l \rightarrow V(\text{pm}^3) \rightarrow V(\text{cm}^3)$ **and** $\mathfrak{M}$**, FCC structure** $\rightarrow m$ **then** $m, V \rightarrow d$

$l = 2\sqrt{2}\,r$ $\quad V = l^3$ $\quad \dfrac{(1\text{ cm})^3}{(10^{10}\text{ pm})^3}$ $\quad m = \dfrac{4\text{ atoms}}{\text{unit cell}} \times \dfrac{\mathfrak{M}}{N_A}$ $\quad d = m/V$

Solution: $l = 2\sqrt{2}\,r = 2\sqrt{2} \times 139\text{ pm} = 39\underline{3}.151\text{ pm} = 393\text{ pm}$ and

$$V = l^3 = (393.151 \text{ pm})^3 \times \frac{(1 \text{ cm})^3}{(10^{10} \text{ pm})^3} = 6.07682 \times 10^{-23} \text{ cm}^3$$ and

$$m = \frac{4 \text{ atoms}}{\text{unit cell}} \times \frac{\mathfrak{M}}{N_A} = \frac{4 \text{ atoms}}{\text{unit cell}} \times \frac{195.09 \text{ g}}{1 \text{ mol}} \times \frac{1 \text{ mol}}{6.022 \times 10^{23} \text{ atoms}} = 1.295848 \times 10^{-21} \frac{\text{g}}{\text{unit cell}}$$ then

$$d = \frac{m}{V} = \frac{1.295848 \times 10^{-21} \frac{\text{g}}{\text{unit cell}}}{6.07682 \times 10^{-23} \frac{\text{cm}^3}{\text{unit cell}}} = 21.3 \frac{\text{g}}{\text{cm}^3}$$

Check: The units (pm and g/cm^3) are correct. The magnitude (393 pm) makes sense because it must be larger than the radius of an atom. The magnitude (21 g/ cm^3) is consistent for Pt from Chapter 1.

48. **Given:** molybdenum, body-centered cubic structure, r = 136 pm **Find:** edge length of unit cell and density (g/cm^3)

Conceptual Plan:

$r \rightarrow l$ and $l \rightarrow V(\text{pm}^3) \rightarrow V(\text{cm}^3)$ and $\mathfrak{M}$, BCC structure $\rightarrow m$ then $m, V \rightarrow d$

$l = \frac{4r}{\sqrt{3}}$ $\quad V = l^3 \quad \frac{(1 \text{ cm})^3}{(10^{10} \text{ pm})^3} \quad m = \frac{2 \text{ atoms}}{\text{unit cell}} x \frac{\mathfrak{M}}{N_A} \quad d = m/V$

Solution: $l = \frac{4r}{\sqrt{3}} = \frac{4 \times 136 \text{ pm}}{\sqrt{3}} = 314.079 \text{ pm} = 314 \text{ pm}$ and

$$V = l^3 = (314.079 \text{ pm})^3 \times \frac{(1 \text{ cm})^3}{(10^{10} \text{ pm})^3} = 3.09823 \times 10^{-23} \text{ cm}^3$$ and

$$m = \frac{2 \text{ atoms}}{\text{unit cell}} x \frac{\mathfrak{M}}{N_A} = \frac{2 \text{ atoms}}{\text{unit cell}} \times \frac{95.94 \text{ g}}{1 \text{ mol}} \times \frac{1 \text{ mol}}{6.022 \times 10^{23} \text{ atoms}} = 3.186317 \times 10^{-22} \frac{\text{g}}{\text{unit cell}}$$ then

$$d = \frac{m}{V} = \frac{3.186317 \times 10^{-22} \frac{\text{g}}{\text{unit cell}}}{3.09823 \times 10^{-23} \frac{\text{cm}^3}{\text{unit cell}}} = 10.3 \frac{\text{g}}{\text{cm}^3}$$

Check: The units (pm and g/cm^3) are correct. The magnitude (314 pm) makes sense because it must be larger than the radius of an atom. The magnitude (10 g/ cm^3) is reasonable for a metal density.

49. **Given:** rhodium, face-centered cubic structure, d = 12.41 g/cm^3 **Find:** r (Rh)

Conceptual Plan:

$\mathfrak{M}$, FCC structure $\rightarrow m$ then $m, V \rightarrow d$ then $V(\text{cm}^3) \rightarrow l(\text{cm}) \rightarrow l(\text{pm})$ then $l \rightarrow r$

$m = \frac{4 \text{ atoms}}{\text{unit cell}} \times \frac{\mathfrak{M}}{N_A} \quad d = m/V \quad V = l^3 \quad \frac{10^{10} \text{ pm}}{1 \text{ cm}} \quad l = 2\sqrt{2}\,r$

Solution: $$m = \frac{4 \text{ atoms}}{\text{unit cell}} \times \frac{\mathfrak{M}}{N_A} = \frac{4 \text{ atoms}}{\text{unit cell}} \times \frac{102.905 \text{ g}}{1 \text{ mol}} \times \frac{1 \text{ mol}}{6.022 \times 10^{23} \text{ atoms}} = 6.835271 \times 10^{-22} \frac{\text{g}}{\text{unit cell}}$$

then $d = \frac{m}{V}$ Rearrange to solve for V. $$V = \frac{m}{d} = \frac{6.835271 \times 10^{-22} \frac{\text{g}}{\text{unit cell}}}{12.41 \frac{\text{g}}{\text{cm}^3}} = 5.507873 \times 10^{-23} \frac{\text{cm}^3}{\text{unit cell}}$$ then

$V = l^3$ Rearrange to solve for l.

$$l = \sqrt[3]{V} = \sqrt[3]{5.507873 \times 10^{-23} \text{ cm}^3} = 3.804831 \times 10^{-8} \text{ cm} \times \frac{10^{10} \text{ pm}}{1 \text{ cm}} = 380.4831 \text{ pm}$$ then $l = 2\sqrt{2}\,r$ Rearrange

to solve for r. $$r = \frac{l}{2\sqrt{2}} = \frac{380.4831 \text{ pm}}{2\sqrt{2}} = 134.5 \text{ pm}$$

Check: The units (pm) are correct. The magnitude (135 pm) is consistent with atom diameters.

50. **Given:** barium, body-centered cubic structure, $d = 3.59\ \text{g/cm}^3$ **Find:** r (Ba)
Conceptual Plan:
$\mathfrak{M}$, BCC structure → m then m, V → d then $V(\text{cm}^3)$ → l (cm) → l (pm) then l → r

$$m = \frac{2\ \text{atoms}}{\text{unit cell}} \times \frac{\mathfrak{M}}{N_A} \qquad d = m/V \qquad V = l^3 \qquad \frac{10^{10}\ \text{pm}}{1\ \text{cm}} \qquad l = \frac{4r}{\sqrt{3}}$$

Solution: $m = \frac{2\ \text{atoms}}{\text{unit cell}} \times \frac{\mathfrak{M}}{N_A} = \frac{2\ \text{atoms}}{\text{unit cell}} \times \frac{137.34\ \text{g}}{1\ \text{mol}} \times \frac{1\ \text{mol}}{6.022 \times 10^{23}\ \text{atoms}} = 4.56\underline{1}275 \times 10^{-22}\ \frac{\text{g}}{\text{unit cell}}$ then

$d = \frac{m}{V}$ Rearrange to solve for V. $V = \frac{m}{d} = \frac{4.56\underline{1}275 \times 10^{-22}\ \frac{\text{g}}{\text{unit cell}}}{3.59\ \frac{\text{g}}{\text{cm}^3}} = 1.27\underline{0}550 \times 10^{-22}\ \frac{\text{cm}^3}{\text{unit cell}}$ then $V = l^3$

Rearrange to solve for l. $l = \sqrt[3]{V} = \sqrt[3]{1.27\underline{0}550 \times 10^{-22}\ \text{cm}^3} = 5.02\underline{7}336 \times 10^{-8}\ \text{cm} \times \frac{10^{10}\ \text{pm}}{1\ \text{cm}} = 502.\underline{7}336\ \text{pm}$

then $l = \frac{4r}{\sqrt{3}}$ Rearrange to solve for r. $r = \frac{l\sqrt{3}}{4} = \frac{502.\underline{7}336\ \text{pm} \times \sqrt{3}}{4} = 217.7\ \text{pm}$

Check: The units (pm) are correct. The magnitude (218 pm) is consistent with atom diameters.

51. **Given:** polonium, simple cubic structure, $d = 9.3\ \text{g/cm}^3$; $r = 167$ pm; $\mathfrak{M} = 209$ g/mol **Find:** estimate N_A
Conceptual Plan:
r → l and l → $V(\text{pm}^3)$ → $V(\text{cm}^3)$ then d, V → m then $\mathfrak{M}$, SC structure → m

$$l = 2r \qquad V = l^3 \qquad \frac{(1\ \text{cm})^3}{(10^{10}\ \text{pm})^3} \qquad d = m/V \qquad m = \frac{1\ \text{atom}}{\text{unit cell}} \times \frac{\mathfrak{M}}{N_A}$$

Solution: $l = 2r = 2 \times 167\ \text{pm} = 334\ \text{pm}$ and $V = l^3 = (334\ \text{pm})^3 \times \frac{(1\ \text{cm})^3}{(10^{10}\ \text{pm})^3} = 3.7\underline{2}597 \times 10^{-23}\ \text{cm}^3$ then

$d = \frac{m}{V}$ Rearrange to solve for m. $m = dV = 9.3\ \frac{\text{g}}{\text{cm}^3} \times \frac{3.7\underline{2}597 \times 10^{-23}\ \text{cm}^3}{\text{unit cell}} = 3.4\underline{6}515 \times 10^{-22}\ \frac{\text{g}}{\text{unit cell}}$

then $m = \frac{1\ \text{atom}}{\text{unit cell}} \times \frac{\mathfrak{M}}{N_A}$ Rearrange to solve for N_A.

$$N_A = \frac{1\ \text{atom}}{\text{unit cell}} \times \frac{\mathfrak{M}}{m} = \frac{1\ \text{atom}}{\text{unit cell}} \times \frac{209\ \text{g}}{1\ \text{mol}} \times \frac{1\ \text{unit cell}}{3.4\underline{6}515 \times 10^{-22}\ \text{g}} = 6.03 \times 10^{23}\ \frac{\text{atom}}{\text{mol}}$$

Check: The units (atoms/mol) are correct. The magnitude (6×10^{23}) is consistent with Avogadro's number.

52. **Given:** palladium, face-centered cubic structure, $d = 12.0\ \text{g/cm}^3$; $r = 138$ pm; $\mathfrak{M} = 106.42$ g/mol
Find: estimate N_A
Conceptual Plan:
r → l and l → $V(\text{pm}^3)$ → $V(\text{cm}^3)$ then d, V → m then $\mathfrak{M}$, FCC structure → m

$$l = 2\sqrt{2}\,r \qquad V = l^3 \qquad \frac{(1\ \text{cm})^3}{(10^{10}\ \text{pm})^3} \qquad d = m/V \qquad m = \frac{4\ \text{atoms}}{\text{unit cell}} \times \frac{\mathfrak{M}}{N_A}$$

Solution: $l = 2\sqrt{2}\,r = 2\sqrt{2} \times 138\ \text{pm} = 39\underline{0}.323\ \text{pm}$ and

$V = l^3 = (39\underline{0}.323\ \text{pm})^3 \times \frac{(1\ \text{cm})^3}{(10^{10}\ \text{pm})^3} = 5.9\underline{4}665 \times 10^{-23}\ \text{cm}^3$ then $d = \frac{m}{V}$ Rearrange to solve for m.

$m = dV = 12.0\ \frac{\text{g}}{\text{cm}^3} \times \frac{5.9\underline{4}665 \times 10^{-23}\ \text{cm}^3}{\text{unit cell}} = 7.1\underline{3}598 \times 10^{-22}\ \frac{\text{g}}{\text{unit cell}}$ then $m = \frac{4\ \text{atoms}}{\text{unit cell}} \times \frac{\mathfrak{M}}{N_A}$ Rearrange to solve for N_A. $N_A = \frac{4\ \text{atoms}}{\text{unit cell}} \times \frac{\mathfrak{M}}{m} = \frac{4\ \text{atoms}}{\text{unit cell}} \times \frac{106.42\ \text{g}}{1\ \text{mol}} \times \frac{1\ \text{unit cell}}{7.1\underline{3}598 \times 10^{-22}\ \text{g}} = 5.97 \times 10^{23}\ \frac{\text{atom}}{\text{mol}}$

Check: The units (atoms/mol) are correct. The magnitude (6×10^{23}) is consistent with Avogadro's number.

53. a) atomic, since Ar is an atom

b) molecular, since water is a molecule

c) ionic, since K_2O is an ionic solid

d) atomic, since iron is an atom

54. a) ionic, since $CaCl_2$ is an ionic solid

b) molecular, since CO_2 is a molecule

c) atomic, since nickle (Ni) is an atom

d) molecular, since I_2 is a molecule

55. LiCl has the highest melting point since it is the only ionic solid in the group. The other three solids are held together by intermolecular forces while LiCl is held together by stronger coulombic interactions between the cations and anions of the crystal lattice.

56. C (diamond) has the highest melting point (3800 °C). Both covalent network solids and ionic solids have high melting points. NaCl has a melting point of 801 °C. In diamond (Figure 11.57a), each carbon atom forms four covalent bonds to four other carbon atoms in a tetrahedral geometry. This structure extends throughout the entire crystal, so that a diamond crystal can be thought of as a giant molecule, held together by these covalent bonds. Since covalent bonds are very strong, covalent atomic solids have high melting points.

57. a) TiO_2 because it is an ionic solid

b) $SiCl_4$ because it has a higher molar mass, and therefore, has stronger dispersion forces

c) Xe because it has a higher molar mass, and therefore, has stronger dispersion forces

d) CaO because the ions have greater charge, stronger dipole–dipole interactions

58. a) Fe because it is an atomic solid held together by metallic bonding

b) KCl because it is an ionic solid

c) Ti because it is an atomic solid held together by metallic bonding

d) H_2O because it is capable of hydrogen bonding

59. The Ti atoms occupy the corner positions and the center of the unit cell – 8 corner atoms x (1/8 atom / unit cell) + 1 atom in center = (1 + 1) Ti atoms / unit cell = 2 Ti atoms / unit cell. The O atoms occupy four positions on the top and bottom faces and two positions inside the unit cell – 4 face-centered atoms x (1/2 atom / unit cell) + 2 atoms in the interior = (2 + 2) O atoms / unit cell = 4 O atoms / unit cell. Therefore there are 2 Ti atoms / unit cell and 4 O atoms / unit cell, so the ratio Ti:O is 2:4 or 1:2. The formula for the compound is TiO_2.

60. The Re atoms occupy the corner positions and the center of the unit cell – 8 corner atoms x (1/8 atom / unit cell) = 1 Re atom / unit cell. The O atoms occupy twelve edge positions – 12 edge atoms x (1/4 atom / unit cell) = 3 O atoms / unit cell. Therefore there are 1 Re atom / unit cell and 3 O atoms / unit cell, so the ratio Re:O is 1:3. The formula for the compound is ReO_3.

61. In CsCl: The Cs atoms occupy the center of the unit cell – 1 atom in center = 1 Cs atom / unit cell. The Cl atoms occupy corner positions of the unit cell – 8 corner atoms x (1/8 atom / unit cell) = 1 Cl atom / unit cell. Therefore there are 1 Cl atom / unit cell and 1 Cl atom / unit cell, so the ratio Cs:Cl is 1:1. The formula for the compound is CsCl, as expected.

In $BaCl_2$: The Ba atoms occupy the corner positions and the face-centered positions of the unit cell – 8 corner atoms x (1/8 atom / unit cell) + 6 face-centered atoms x (1/2 atom / unit cell) = (1 + 3) Ba atoms / unit cell = 4

Ba atoms / unit cell. The Cl atoms occupy eight positions inside the unit cell – 8 Cl atoms / unit cell. Therefore there are 4 Ba atoms / unit cell and 8 Cl atoms / unit cell, so the ratio Ba:Cl is 4:8 or 1:2. The formula for the compound is $BaCl_2$, as expected.

62. In Li_2O: The Li atoms occupy eight positions inside the unit cell – 8 Li atoms / unit cell. The O atoms occupy the corner positions and the face-centered positions of the unit cell – 8 corner atoms x (1/8 atom / unit cell) + 6 face-centered atoms x (1/2 atom / unit cell) = (1 + 3) Li atoms / unit cell = 4 O atoms / unit cell. Therefore there are 4 O atoms / unit cell and 8 Li atoms / unit cell, so the ratio Li:O is 8:4 or 2:1. The formula for the compound is Li_2O, as expected.

In AgI: The Ag atoms occupy four positions inside the unit cell – 4 Ag atoms / unit cell. The I atoms occupy the corner positions and the face-centered positions of the unit cell – 8 corner atoms x (1/8 atom / unit cell) + 6 face-centered atoms x (1/2 atom / unit cell) = (1+ 3) I atoms / unit cell = 4 I atoms / unit cell. Therefore there are 4 I atoms / unit cell and 4 Ag atoms / unit cell, so the ratio Ag:I is 4:4 or 1:1. The formula for the compound is AgI, as expected.

63. a) Zn should have little or no band gap because it is the only metal in the group.

64. **Given:** 5.45 g sodium crystal **Find:** number of molecular orbitals in the valence band
Conceptual Plan:
g → mol → Na_N → number of valence electrons → number of molecular orbitals

$$\frac{1\text{ mol}}{22.99\text{ g}} \quad \frac{6.022\times10^{23}\text{ atoms}}{1\text{ mol}} \quad \frac{1\text{ 3s electron}}{1\text{ Na atom}} \quad \frac{1\text{ molecular orbital}}{1\text{ 3s electron}}$$

Solution: $5.45\text{ g Na} \times \dfrac{1\text{ mol Na}}{22.99\text{ g Na}} \times \dfrac{6.022\times10^{23}\text{ Na atoms}}{1\text{ mol Na}} \times \dfrac{1\text{ 3s electron}}{1\text{ Na atom}} \times \dfrac{1\text{ molecular orbital}}{1\text{ 3s electron}}$

$= 1.43 \times 10^{23}$ molecular orbitals

Check: The units (number of molecular orbitals) are correct. The magnitude (10^{23}) is expected since there are so many orbitals because we have about a ¼ mole of atoms. Metals can conduct electricity because of these large numbers of orbitals.

65. The general trend is that melting point increases with increasing mass. This is due to the fact that the electrons of the larger molecules are held more loosely and a stronger dipole moment can be induced more easily. HF is the exception to the rule. It has a relatively high melting point due to hydrogen bonding.

66. The general trend is that boiling point increases with increasing mass. This is due to the fact that the electrons of the larger molecules are held more loosely and a stronger dipole moment can be induced more easily. H_2O is the exception to the rule. It has a relatively high boiling point due to hydrogen bonding.

67. **Given:** $P_{H2O} = 23.76$ torr at 25 °C; 1.25 g water in 1.5 L container **Find:** m (H_2O) as liquid
Conceptual Plan:
°C → K and torr → atm then *P, V, T* → mol (*g*) → g (*g*) then g (*g*), g (l)$_i$ → g (l)$_f$

$$K = °C + 273.15 \quad \frac{1\text{ atm}}{760\text{ torr}} \quad PV = nRT \quad \frac{18.01\text{ g}}{1\text{ mol}} \quad g\,(l)_f = g\,(l)_i - g\,(g)$$

Solution: $T = 25\text{ °C} + 273.15 = 298\text{ K}$, $23.76\text{ torr} \times \dfrac{1\text{ atm}}{760\text{ torr}} = 0.031\underline{2}632\text{ atm}$ then $PV = nRT$

Rearrange to solve for n. $n = \dfrac{PV}{RT} = \dfrac{0.031\underline{2}632\text{ atm} \times 1.5\text{ L}}{0.08206\dfrac{\text{L}\cdot\text{atm}}{\text{K}\cdot\text{mol}} \times 298\text{ K}} = 0.001\underline{9}1768\text{ mol}$ then

$0.001\underline{9}1768\text{ mol} \times \dfrac{18.01\text{ g}}{1\text{ mol}} = 0.03\underline{4}5375\text{ g}$ in gas phase then

$g\,(l)_f = g\,(l)_i - g\,(g) = 1.25\text{ g} - 0.03\underline{4}5375\text{ g} = 1.22\text{ g}$ remaining as liquid Yes, there is 1.22 g of liquid.

Check: The units (g) are correct. The magnitude (1.2 g) is expected since very little material is expected to be in the gas phase.

68. **Given:** P_{CCl3F} = 856 torr at 300 K; 11.5 g CCl_3F in 1.0 L container **Find:** m (CCl_3F) as liquid
Conceptual Plan: torr → atm then P, V, T → mol (g) → g (g) then g (g), g (l)$_i$ → g (l)$_f$

$$\frac{1\text{ atm}}{760\text{ torr}} \qquad PV = nRT \qquad \frac{137.36\text{ g}}{1\text{ mol}} \qquad g(l)_f = g(l)_i - g(g)$$

Solution: $856\text{ torr} \times \frac{1\text{ atm}}{760\text{ torr}} = 1.12\underline{6}32\text{ atm}$ then $PV = nRT$ Rearrange to solve for n.

$$n = \frac{PV}{RT} = \frac{1.12\underline{6}32\text{ atm} \times 1.0\text{ L}}{0.08206\frac{\text{L}\cdot\text{atm}}{\text{K}\cdot\text{mol}} \times 300\text{ K}} = 0.04\underline{5}7517\text{ mol then}$$

$$0.04\underline{5}7517\text{ mol} \times \frac{137.36\text{ g}}{1\text{ mol}} = 6.\underline{2}845\text{ g in gas phase then}$$

$g(l)_f = g(l)_i - g(g) = 11.5\ g - 6.\underline{2}845\text{ g} = 5.2\text{ g}$ remaining as liquid Yes, there is 5.2 g of liquid.

Check: The units (g) are correct. The magnitude (5 g) is expected since even at moderate pressures little material is expected to be in the gas phase.

69. Since we are starting at a temperature that is higher and a pressure that is lower than the triple point, the phase transitions will be gas → liquid → solid or condensation followed by freezing.

70. The solid is denser than the liquid. Since the triple point temperature is lower than the normal melting point, the slope of the fusion curve must be positive. This means that as you start in the liquid phase and increase the pressure you will eventually cross into the solid phase. As pressure increases, the phases get denser, as the atoms, molecules, or ions are pushed closer and closer together.

71. **Given:** Ice: T_1 = 0 °C exactly, m = 53.5 g; Water: T_1 = 75 °C, m = 115 g **Find:** T_f
Other: $\Delta H°_{fus}$ = 6.0 kJ/mol; C_{H2O} = 4.18 J/g·°C
Conceptual Plan: $q_{ice} = -q_{water}$ **so g (ice) → mol (ice) →** q_{fus}**(kJ) →** q_{fus} **(J) →** q_{water} **(J) then**

$$\frac{1\text{ mol}}{18.01\text{g}} \qquad \frac{6.02\text{ kJ}}{1\text{ mol}} \qquad \frac{1000\text{ J}}{1\text{ kJ}} \qquad q_{water} = -q_{ice}$$

q, m, C_s → ΔT then $T_i, \Delta T$ → T_2 now we have slightly cooled water in contact with 0.0 °C water

$$q = mC_S\Delta T \qquad \Delta T = T_2 - T_i$$

so $q_{ice} = -q_{water}$ **with m, C_s, T_i → T_f**

$$q = m\ C_s\left(T_f - T_i\right) \text{ then set } q_{ice} = -q_{water}$$

Solution: $53.5\text{ g} \times \frac{1\text{ mol}}{18.01\text{ g}} \times \frac{6.02\text{ kJ}}{1\text{ mol}} \times \frac{1000\text{ J}}{1\text{ kJ}} = 1.7\underline{8}828 \times 10^4\text{ J}$, $q_{water} = -q_{ice} = -1.7\underline{8}828 \times 10^4$ J

$q = mC_S\Delta T$ Rearrange to solve for ΔT. $\Delta T = \frac{q}{mC_S} = \frac{-1.7\underline{8}828 \times 10^4\text{ J}}{115\text{ g} \times 4.18\frac{\text{J}}{\text{g}\cdot°\text{C}}} = -37.\underline{2}017\ °\text{C}$ then

$\Delta T = T_2 - T_i$. Rearrange to solve for T_2. $T_2 = \Delta T + T_i = -37.\underline{2}017\ °\text{C} + 75\ °\text{C} = 3\underline{7}.798\ °\text{C}$

$q = m\ C_s\left(T_f - T_i\right)$ substitute in values and set $q_{ice} = -q_{water}$.

$$q_{ice} = m_{ice}\ C_{ice}\left(T_f - T_{icei}\right) = 53.5\text{ g} \times 4.18\frac{\text{J}}{\text{g}\cdot°\text{C}} \times\left(T_f - 0.0\ °\text{C}\right) =$$

$$-q_{water} = -m_{water}\ C_{water}\left(T_f - T_{water2}\right) = -115\text{ g} \times 4.18\frac{\text{J}}{\text{g}\cdot°\text{C}} \times\left(T_f - 3\underline{7}.798°\text{C}\right)$$

Rearrange to solve for T_f.

$$53.5\,T_f = -115\left(T_f - 3\underline{7}.798°\text{C}\right) \rightarrow 53.5\,T_f = -115\,T_f + 4\underline{3}46.8°\text{C} \rightarrow -4\underline{3}46.8\ °\text{C} = -16\underline{8}.5\,T_f$$

$$\rightarrow T_f = \frac{-4\underline{3}46.8\ °\text{C}}{-16\underline{8}.5} = 2\underline{5}.8\ °\text{C} = 26\ °\text{C}$$

Check: The units (°C) are correct. The temperature is between the two initial temperatures. Since the ice mass is about half the water mass, we are not surprised that the temperature is closer to the original ice temperature.

72. **Given:** Steam: $T_i = 100$ °C, $m = 0.552$ g; Water: $T_i = 5.0$ °C, $m = 4.25$ g **Find:** T_f
Other: $\Delta H°_{vap} = 40.7$ kJ/mol; $C_{H2O} = 4.18$ J/g·°C
Conceptual Plan:

$q_{steam} = -q_{water}$ **so g (steam) → mol (steam) →** q_{vap}**(kJ) →** q_{fvap} **(J) →** q_{water} **(J) then**

$\frac{1\text{ mol}}{18.01\text{g}}$ $\quad \frac{-40.7\text{ kJ}}{1\text{ mol}}$ $\quad \frac{1000\text{ J}}{1\text{ kJ}}$ $\quad q_{steam} = -q_{ice}$

q, m, C_s → ΔT then T_i, ΔT → T_2 now we have slightly warmed water in contact with 100.0 °C water

$q = mC_S\Delta T$ $\quad \Delta T = T_2 - T_i$

so $q_{steam} = -q_{water}$ **with** **m, C_s, T_i → T_f**

$q = m\,C_s\left(T_f - T_i\right)$ *then set* $q_{steam} = -q_{water}$

Solution: $0.552\ \cancel{g} \times \frac{1\ \cancel{mol}}{18.01\ \cancel{g}} \times \frac{-40.7\ \cancel{kJ}}{1\ \cancel{mol}} \times \frac{1000\text{ J}}{1\ \cancel{kJ}} = -1.2\underline{4}744 \times 10^3\text{ J}$,

$q_{water} = -q_{steam} = -(-1.2\underline{4}744 \times 10^3\text{ J})$ $\quad q = mC_S\Delta T$ Rearrange to solve for ΔT.

$$\Delta T = \frac{q}{mC_S} = \frac{1.2\underline{4}744 \times 10^3\ \cancel{J}}{4.25\ \cancel{g} \times 4.18\ \frac{\cancel{J}}{\cancel{g}\cdot°C}} = 70.\underline{2}190\text{ °C} \quad \text{then } \Delta T = T_2 - T_i.$$

Rearrange to solve for T_2. $T_2 = \Delta T + T_i = 70.\underline{2}190\text{ °C} + 5.0\text{ °C} = 75.\underline{2}190\text{ °C}$ $\quad q = m\,C_s\left(T_f - T_i\right)$ substitute in values and set $q_{steam} = -q_{water}$.

$$q_{steam} = m_{steam}\,C_{steam}\left(T_f - T_{steami}\right) = 0.552\ \cancel{g} \times 4.18\ \frac{\text{J}}{\cancel{g\cdot°C}} \times \left(T_f - 100.0\text{ °C}\right) =$$

$$-q_{water} = -m_{water}\,C_{water}\left(T_f - T_{water2}\right) = -4.25\ \cancel{g} \times 4.18\ \frac{\text{J}}{\cancel{g\cdot°C}} \times \left(T_f - 75.\underline{2}190\text{ °C}\right)$$

Rearrange to solve for T_f.

$$0.552\left(T_f - 100.0\text{ °C}\right) = -4.25\left(T_f - 75.\underline{2}190\text{°C}\right) \rightarrow 0.552\,T_f - 55.2\text{°C} = -4.25\,T_f + 31\underline{9}.681\text{°C} \rightarrow$$

$$-37\underline{4}.881\text{ °C} = -4.8\underline{0}2\,T_f \rightarrow T_f = \frac{-37\underline{4}.881\text{ °C}}{-4.8\underline{0}2} = 78.1\text{ °C}$$

Check: The units (°C) are correct. The temperature is between the two initial temperatures. Since there is so much heat involved in the vaporization process, we are not surprised that the temperature is closer to the original steam temperature.

73. **Given:** Home: 6.0 m x 10.0 m x 2.2 m; $T = 30$ °C, $P_{H2O} = 85$ % of $P°_{H2O}$ **Find:** m (H_2O) removed
Other: $P°_{H2O} = 31.86$ mm Hg from text
Conceptual Plan:

l, w, h → V(m³) → V(cm³) → V(L) and $P°_{H2O}$ → P_{H2O} (mm Hg) → P_{H2O} (atm) and

$V = l\,w\,h$ $\quad \frac{(100\text{ cm})^3}{(1\text{ m})^3}$ $\quad \frac{1\text{ L}}{1000\text{ cm}^3}$ $\quad P_{H2O} = 0.85\,P^{\circ}_{H2O}$ $\quad \frac{1\text{ atm}}{760\text{ mmHg}}$

°C → K then P, V, T → mol (H_2O) → g (H_2O)

$\text{K} = \text{°C} + 273.15$ $\quad PV = nRT$ $\quad \frac{18.01\text{g}}{1\text{ mol}}$

Solution: $V = l\,wh = 6.0\text{ m} \times 10.0\text{ m} \times 2.2\text{ m} = 1\underline{3}2\ \cancel{m^3} \times \frac{(100\ \cancel{cm})^3}{(1\ \cancel{m})^3} \times \frac{1\text{ L}}{1000\ \cancel{cm^3}} = 1.\underline{3}2 \times 10^5\text{ L}$,

$P_{H2O} = 0.85\,P^0_{H2O} = 0.85 \times 31.86\ \cancel{mmHg} \times \frac{1\text{ atm}}{760\ \cancel{mmHg}} = 0.03\underline{5}633\text{ atm}$, $T = 30\text{ °C} + 273.15 = 3\underline{0}3$ K,

then $PV = nRT$ Rearrange to solve for n.

$$n = \frac{PV}{RT} = \frac{0.035633 \text{ atm} \times 1.32 \times 10^5 \text{ L}}{0.08206 \frac{\text{L}\cdot\text{atm}}{\text{K}\cdot\text{mol}} \times 303 \text{ K}} = 189.17 \text{ mol} \text{ then } 189.17 \text{ mol} \times \frac{18.01 \text{ g}}{1 \text{ mol}} = 3400 \text{ g to remove}$$

Check: The units (g) are correct. The magnitude of the answer (3400 g) makes sense since the volume of the house is so large. We are removing almost 200 moles of water.

74. **Given:** Flask with 0.55 g water at $T = 28\ °C$, $P°_{H2O} = 28.36$ mm Hg
Find: minimum V of flask for all vapor
Conceptual Plan:
g (H_2O) → mol (H_2O) and $P°_{H2O}$ (mm Hg) → P_{H2O} (atm) and °C → K then P, n, T → V

$\frac{1 \text{ mol}}{18.01 \text{g}}$ $\frac{1 \text{ atm}}{760 \text{ mm Hg}}$ $K = °C + 273.15$ $PV = nRT$

Solution: $0.55 \text{ g} \times \frac{1 \text{ mol}}{18.01 \text{ g}} = 0.030539 \text{ mol}$, $P_{H2O} = 28.63 \text{ mm Hg} \times \frac{1 \text{ atm}}{760 \text{ mm Hg}} = 0.03767105 \text{ atm}$,

$T = 28\ °C + 273.15 = 301$ K, then $PV = nRT$ Rearrange to solve for V.

$$V = \frac{nRT}{P} = \frac{0.030539 \text{ mol} \times 0.08206 \frac{\text{L}\cdot\text{atm}}{\text{K}\cdot\text{mol}} \times 301 \text{ K}}{0.03767105 \text{ atm}} = 20.02 \text{ L} = 2.0 \times 10^1 \text{ L}$$

Check: The units (L) are correct. The magnitude of the answer (20 L) makes sense since we have about 1/30 mole and a pressure of about 1/30 atm and at STP one mole of a gas occupies 22 L.

75. CsCl has a higher melting point than AgI because of its higher coordination number. In CsCl, one anion bonds to 8 cations (and vice versa) while in AgI, one anion bonds only to 4 cations.

76. KCl has a higher melting point than copper iodide because of its higher coordination number. In KCl, one anion bonds to 6 cations (and vice versa) while in copper iodide, one anion bonds only to 4 cations.

77. a) atoms are connected across the face diagonal (c), so $c = 4r$

b) From the Pythagorean Theorem $c^2 = a^2 + b^2$, from part a) $c = 4r$ and for a cubic structure $a = l, b = l$ so $(4r)^2 = l^2 + l^2$ → $16r^2 = 2l^2$ → $8r^2 = l^2$ → $l = \sqrt{8r^2}$ → $l = 2\sqrt{2}\,r$

78. a) atoms are connected across the cube diagonal (c), so $c = 4r$

b) Since b forms the diagonal of the face, where each edge length $a = l$ and combining this with the Pythagorean Theorem $b^2 = l^2 + l^2$ → $b^2 = 2l^2$ → $b = \sqrt{2}\,l$

c) From the Pythagorean Theorem $c^2 = a^2 + b^2$, from part a) $c = 4r$ and from part b) $b = \sqrt{2}\,l$ and for a cubic structure $a = l$ so $(4r)^2 = l^2 + (\sqrt{2}\,l)^2$ → $16r^2 = l^2 + 2l^2$ → $16r^2 = 3l^2$ → $4r = \sqrt{3}\,l$ → $l = \frac{4r}{\sqrt{3}}$

79. **Given:** diamond, V (unit cell) = 0.0454 nm^3; d = 3.52 g/cm^3 **Find:** number of carbon atoms / unit cell
Conceptual Plan: **V(nm^3) → V(cm^3) then d, V → m → mol → atoms**

$\frac{(1 \text{ cm})^3}{(10^7 \text{ nm})^3}$ $d = m/V$ $\frac{1 \text{ mol}}{12.01 \text{ g}}$ $\frac{6.022 \times 10^{23} \text{ atoms}}{1 \text{ mol}}$

Solution: $0.0454 \text{ nm}^3 \times \frac{(1 \text{ cm})^3}{(10^7 \text{ nm})^3} = 4.54 \times 10^{-23} \text{ cm}^3$ then $d = \frac{m}{V}$ Rearrange to solve for m.

$m = dV = 3.52 \frac{\text{g}}{\text{cm}^3} \times 4.54 \times 10^{-23} \text{ cm}^3 = 1.59808 \times 10^{-22}$ g then

$$\frac{1.59808 \times 10^{-22}\ \cancel{g}}{\text{unit cell}} \times \frac{1\ \cancel{mol}}{12.01\ \cancel{g}} \times \frac{6.022 \times 10^{23}\ \text{atoms}}{1\ \cancel{mol}} = 8.01\ \frac{\text{C atoms}}{\text{unit cell}} = 8\ \frac{\text{C atoms}}{\text{unit cell}}$$

Check: The units (atoms) are correct. The magnitude (8) makes sense because it is a fairly small number and our answer within our calculation error of an integer.

80. **Given:** metal, $d = 12.3\ \text{g/cm}^3$; $r = 0.134$ nm, face-centered cubic lattice **Find:** $\mathfrak{M}$

Conceptual Plan:

$r \rightarrow l \rightarrow V(\text{nm}^3) \rightarrow V(\text{cm}^3)$ **then** $d, V \rightarrow m$ **then** m, **FCC structure** $\rightarrow \mathfrak{M}$

$l = 2\sqrt{2}\, r$ $\quad V = l^3 \quad \frac{(1\ \text{cm})^3}{(10^7\ \text{nm})^3} \quad d = m/V \quad m = \frac{4\ \text{atoms}}{\text{unit cell}} \times \frac{\mathfrak{M}}{N_A}$

Solution: $l = 2\sqrt{2}\, r = 2\sqrt{2} \times 0.134\ \text{nm} = 0.379009\ \text{nm}$,

$V = l^3 = (0.379009\ \text{nm})^3 = 0.0544439\ \cancel{\text{nm}^3} \times \frac{(1\ \text{cm})^3}{(10^7\ \cancel{\text{nm}})^3} = 5.44439 \times 10^{-23}\ \text{cm}^3$ then $d = \frac{m}{V}$ Rearrange to solve for m. $m = d\,V = 12.3\ \frac{\text{g}}{\cancel{\text{cm}^3}} \times 5.44439 \times 10^{-23}\ \cancel{\text{cm}^3} = 6.69660 \times 10^{-22}\ \text{g}$ then $m = \frac{4\ \text{atoms}}{\text{unit cell}} \times \frac{\mathfrak{M}}{N_A}$

Rearrange to solve for $\mathfrak{M}$.

$$\mathfrak{M} = \frac{\text{unit cell}}{4\ \text{atoms}} \times N_A \times m = \frac{\cancel{\text{unit cell}}}{4\ \cancel{\text{atoms}}} \times \frac{6.022 \times 10^{23}\ \cancel{\text{atoms}}}{1\ \text{mol}} \times \frac{6.69660 \times 10^{-22}\ \text{g}}{\cancel{\text{unit cell}}} = 101\ \frac{\text{g}}{\text{mol}}\ \text{Ruthenium}$$

Check: The units (g/mol) are correct. The magnitude (101) makes sense because it is a reasonable atomic mass for a metal and it is close to Ruthenium.

81. a) $CO_2\,(s) \rightarrow CO_2\,(g)$ at 194.7 K

b) $CO_2\,(s) \rightarrow$ triple point at 216.5 K $\rightarrow CO_2\,(g)$ just above 216.5 K

c) $CO_2\,(s) \rightarrow CO_2\,(l)$ at somewhat above 216 K $\rightarrow CO_2\,(g)$ at around 250 K

d) CO_2 (s) $\rightarrow CO_2$ above the critical point where there is no distinction between liquid and gas. This change occurs at about 300 K.

82. If atmospheric pressure is 2500 mm Hg, water would still be a liquid. At a higher pressure atmospheric pressure, water would remain a liquid to a lower temperature than 0 °C, this could reduce the damage done to organisms that are exposed to cold temperatures. At a higher atmospheric pressure there would be more molecules in the gas phase and the atmosphere would not behave as ideally, water might condense more readily, lowering the vapor pressure of water. This could have an adverse effect on living organisms. At higher atmospheric pressures, cell walls would need to be stronger to withstand higher pressures. This would most likely make the cell walls less permeable and affect many biological systems.

83. **Given:** KCl, rock salt structure **Find:** density (g/cm^3) **Other:** $r\,(K^+) = 133$ pm; $r\,(Cl^-) = 181$ pm from Chapter 8

Conceptual Plan: Rock salt structure is a face-centered cubic structure with anions at the lattice points and cations in the holes between lattice sites → assume $r = r(Cl^-)$, but $\mathfrak{M} = \mathfrak{M}(\text{KCl})$

$r(K^+), r(Cl^-) \rightarrow l$ **and** $l \rightarrow V(\text{pm}^3) \rightarrow V(\text{cm}^3)$ **and , FCC structure** $\rightarrow m$ **then** $m, V \rightarrow d$

from Figure 11.52 $l = 2r(Cl^-) + 2r(K^+)$ $\quad V = l^3 \quad \frac{(1\ \text{cm})^3}{(10^{10}\ \text{pm})^3} \quad m = \frac{4\ \text{formula units}}{\text{unit cell}} \times \frac{\mathfrak{M}}{N_A} \quad d = m/V$

Solution: $l = 2r(Cl^-) + 2r(K^+) = 2(181\ \text{pm}) + 2(133\ \text{pm}) = 628\ \text{pm}$ and

$V = l^3 = (628\ \cancel{\text{pm}})^3 \times \frac{(1\ \text{cm})^3}{(10^{10}\ \cancel{\text{pm}})^3} = 2.47673 \times 10^{-22}\ \text{cm}^3$ and

$$m = \frac{4 \text{ formula units}}{\text{unit cell}} \times \frac{\mathfrak{M}}{N_A} = \frac{4 \text{ formula units}}{\text{unit cell}} \times \frac{74.55 \text{ g}}{1 \text{ mol}} \times \frac{1 \text{ mol}}{6.022 \times 10^{23} \text{ formula units}}$$

$$= 4.95\underline{1}976 \times 10^{-22} \frac{\text{g}}{\text{unit cell}}$$

then $$d = \frac{m}{V} = \frac{4.95\underline{1}976 \times 10^{-22} \frac{\text{g}}{\text{unit cell}}}{2.4\underline{7}673 \times 10^{-22} \frac{\text{cm}^3}{\text{unit cell}}} = 1.9\underline{9}940 \frac{\text{g}}{\text{cm}^3} = 2.00 \frac{\text{g}}{\text{cm}^3}$$

Check: The units (g/cm^3) are correct. The magnitude (2 g/ cm^3) is reasonable for a salt density. The published value is 1.98 g/cm^3. This method of estimating the density gives a value that is close to the experimentally measured density.

84. **Given:** butane (C_4H_{10}), $\Delta H°_{vap} = 22.44$ kJ/mol; normal boiling point = – 0.4 °C, 0.55 g; 250 mL flask
Find: amount of butane present as a liquid at – 22 °C, and at 25 °C
Conceptual Plan:
at each temperature: °C → K and kJ → J then $\Delta H°_{vap}, T_1, P_1, T_2 \rightarrow P_2$

$K = °C + 273.15$ $\quad \frac{1000 \text{ J}}{1 \text{ kJ}} \quad \ln \frac{P_2}{P_1} = \frac{-\Delta H_{vap}}{R}\left(\frac{1}{T_2} - \frac{1}{T_1}\right)$

and mL → L then P_2, V, T_2 → mol → g (g) → g (l)

$\frac{1 \text{ L}}{1000 \text{ mL}} \quad PV = nRT \quad \frac{58.12 \text{ g}}{1 \text{ mol}} \quad g\,(l) = g_{Total} - g\,(g)$

Solution: $T_1 = -0.4 °C + 273.15 = 272.8$ K; $T_2 = -22 °C + 273.15 = 251$ K;

$$\frac{22.44 \text{ kJ}}{\text{mol}} \times \frac{1000 \text{ J}}{1 \text{ kJ}} = 2.244 \times 10^4 \frac{\text{J}}{\text{mol}} \quad P_1 = 1 \text{ atm} \quad \ln \frac{P_2}{P_1} = \frac{-\Delta H_{vap}}{R}\left(\frac{1}{T_2} - \frac{1}{T_1}\right)$$

Substitute values in equation.

$$\ln \frac{P_2}{1 \text{ atm}} = \frac{-2.244 \times 10^4 \frac{\text{J}}{\text{mol}}}{8.314 \frac{\text{J}}{\text{K} \cdot \text{mol}}}\left(\frac{1}{251 \text{ K}} - \frac{1}{272.8 \text{ K}}\right) = -0.85\underline{9}312 \rightarrow \frac{P_2}{1 \text{ atm}} = e^{-0.85\underline{9}312} = 0.42\underline{3}45 \rightarrow$$

$P_2 = 0.42\underline{3}45 \times 1 \text{ atm} = 0.42\underline{3}45$ atm and $250 \text{ mL} \times \frac{1 \text{ L}}{1000 \text{ mL}} = 0.25$ L then $PV = nRT$

Rearrange to solve for n. $$n = \frac{PV}{RT} = \frac{0.42\underline{3}45 \text{ atm} \times 0.25 \text{ L}}{0.08206 \frac{\text{L} \cdot \text{atm}}{\text{K} \cdot \text{mol}} \times 251 \text{ K}} = 0.005\underline{1}397 \text{ mol}$$ then

$$0.005\underline{1}397 \text{ mol} \times \frac{58.12 \text{ g}}{1 \text{ mol}} = 0.2\underline{9}872 \text{ g in gas phase}$$ then

$g\,(l) = g_{Total} - g\,(g) = 0.55 \text{ g} - 0.2\underline{9}872 \text{ g} = 0.25$ g as liquid at – 22 °C

$T_1 = -0.4 °C + 273.15 = 272.8$ K; $T_2 = 25 °C + 273.15 = 298$ K; $\frac{22.44 \text{ kJ}}{\text{mol}} \times \frac{1000 \text{ J}}{1 \text{ kJ}} = 2.244 \times 10^4 \frac{\text{J}}{\text{mol}}$

$P_1 = 1$ atm $\ln \frac{P_2}{P_1} = \frac{-\Delta H_{vap}}{R}\left(\frac{1}{T_2} - \frac{1}{T_1}\right)$ Substitute values in equation.

$$\ln \frac{P_2}{1 \text{ atm}} = \frac{-2.244 \times 10^4 \frac{\text{J}}{\text{mol}}}{8.314 \frac{\text{J}}{\text{K} \cdot \text{mol}}}\left(\frac{1}{298 \text{ K}} - \frac{1}{272.8 \text{ K}}\right) = 0.83\underline{6}667 \rightarrow \frac{P_2}{1 \text{ atm}} = e^{0.83\underline{6}667} = 2.3\underline{0}866 \rightarrow$$

P_2 = 2.30866 x 1 atm = 2.30866 atm and 250 ~~mL~~ x $\frac{1\text{ L}}{1000\text{ mL}}$ = 0.25 L then $PV = nRT$

Rearrange to solve for *n*. $n = \frac{PV}{RT} = \frac{2.30866\text{ atm} \times 0.25\text{ L}}{0.08206 \frac{\text{L}\cdot\text{atm}}{\text{K}\cdot\text{mol}} \times 298\text{ K}}$ = 0.023602 mol then

0.023602 ~~mol~~ x $\frac{58.12\text{ g}}{1\text{ mol}}$ = 1.3718 g in gas phase since the amount that can be put in the gas phase is greater than the available amount, there is no liquid present at 25 °C.

Check: The units (g) are correct. The magnitude of the answers (0.25 g and 0 g) makes sense since we expect less to be in the liquid phase at a higher temperature. The second temperature is over the boiling point, so we expect a lot in the gas phase.

85. Decreasing the pressure will decrease the temperature of liquid nitrogen. Because the nitrogen is boiling, its temperature must be constant at a given pressure. As the pressure decreases, the boiling point decreases, and therefore so does the temperature. Remember that vaporization is an endothermic process, so as the nitrogen vaporizes it will remove heat from the liquid, dropping its temperature. If the pressure drops below the pressure of the triple point, the phase change will shift from vaporization to sublimation and the liquid nitrogen will become solid.

86. **Given:** cubic closest packing structure **Find:** fraction of empty space to 5 significant figures
Conceptual Plan: cubic closest packing is the same as face-centered cubic structure so calculate the volume of the atoms and the volume of the unit cell then $V_{atoms}, V_{unit\ cell} \rightarrow \%\ V_{empty}$

$$\frac{4\ atoms}{unit\ cell} \text{ and } V_{atom} = \frac{\frac{4}{3}\pi r^3}{atom} \qquad V_{unit\ cell} = l^3 \text{ and } l = 2\sqrt{2}\,r \qquad \%\ V_{empty} = \frac{V_{unit\ cell} - V_{atoms}}{V_{unit\ cell}} \times 100\ \%$$

Solution: $V_{atoms} = \frac{4\ atoms}{unit\ cell} \times \frac{\frac{4}{3}\pi r^3}{atom} = \frac{\frac{16}{3}\pi r^3}{unit\ cell}$ and $V_{unit\ cell} = l^3 = \left(2\sqrt{2}\,r\right)^3$ so

$$\%\ V_{empty} = \frac{V_{unit\ cell} - V_{atoms}}{V_{unit\ cell}} \times 100\ \% \text{ or}$$

$$\%\ V_{empty} = \frac{\left(2\sqrt{2}\,r\right)^3 - \frac{16}{3}\pi r^3}{\left(2\sqrt{2}\,r\right)^3} \times 100\ \% = \frac{2^3\, 2^{3/2}\, r^3 - \frac{2^4}{3}\pi r^3}{2^3\, 2^{3/2}\, r^3} \times 100\ \% = \frac{\sqrt{2} - \frac{\pi}{3}}{\sqrt{2}} \times 100\ \% = 25.952\ \%$$

Check: The units (%) are correct. The magnitude (26 %) is consistent with what is stated in the text.

87. **Given:** cubic closest packing structure = cube with touching spheres of radius = *r* on alternating corners of a cube **Find:** body diagonal of cube and radius of tetrahedral hole
Solution: The cell edge length = l and $l^2 + l^2 = (2r)^2 \rightarrow 2\,l^2 = 4\,r^2 \rightarrow l^2 = 2\,r^2$. Since body diagonal = *BD* is the hypotenuse of the right triangle formed by the face diagonal and the cell edge we have $(BD)^2 = l^2 + (2r)^2 = 2\,r^2 + 4r^2 = 6\,r^2 \rightarrow BD = \sqrt{6}\,r$. The radius of the tetrahedral hole = r_T is half the body diagonal minus the radius of the sphere or

$$r_T = \frac{BD}{2} - r = \frac{\sqrt{6}\,r}{2} - r = \left(\frac{\sqrt{6}}{2} - 1\right)r = \left(\frac{\sqrt{6}-2}{2}\right)r = \left(\frac{\sqrt{3}\sqrt{2} - \sqrt{2}\sqrt{2}}{\sqrt{2}\sqrt{2}}\right)r = \left(\frac{\sqrt{3}-\sqrt{2}}{\sqrt{2}}\right)r \approx 0.22474\,r$$

88. **Given:** $\Delta H°_{fus}$ = – 6.02 kJ/mol @ 0.0 °C; C_s (liquid water) = 75.2 J/mol·K; C_s (ice) = 37.7 J/mol·K
Find: ΔH_{fus} at –10.0 °C
Conceptual Plan: Assume exactly 1 mol H_2O for all calculations (report answer as kJ/mol). Since K = °C =273.13, ΔT (°C) = ΔT (K). Construct the following path:

	#2		
Liquid @ 0.0 °C	→	**solid @ 0.0 °C**	**According to Hess' Law $\Delta H_1 + \Delta H_2 + \Delta H_3 = \Delta H_4$**
↑ **#1**		↓ **#3**	**= $\Delta H°_{fus}$ @ –10.0 °C**
Liquid @ –10.0 °C	→	**solid @ –10.0 °C**	**at constant *P*, $\Delta H = q$**
	#4		**for steps #1 and #3 $q = nC_s\Delta T$**

for steps #1 and #3 J → kJ

$\frac{1 \text{ kJ}}{1000 \text{ J}}$

Solution:

$$\Delta H_1 = q_1 = nC_S\Delta T = 1 \text{ mol} \times 75.2 \frac{\text{J}}{\text{mol} \cdot {}^\circ\text{C}} \times (0.00\,{}^\circ\text{C} - -10.0\,{}^\circ\text{C}) = +752 \text{ J} \times \frac{1 \text{ kJ}}{1000 \text{ J}} = +0.752 \text{ kJ},$$

$$\Delta H_2 = q_2 = n\Delta H = 1 \text{ mol} \times -6.02 \frac{\text{kJ}}{\text{mol}} = -6.02 \text{ kJ},$$

$$\Delta H_3 = q_3 = nC_S\Delta T = 1 \text{ mol} \times 37.7 \frac{\text{J}}{\text{mol} \cdot {}^\circ\text{C}} \times (-10.00\,{}^\circ\text{C} - 0.0\,{}^\circ\text{C}) = -377 \text{ J} \times \frac{1 \text{ kJ}}{1000 \text{ J}} = -0.377 \text{ kJ} \quad \text{so}$$

$$\Delta H_4 = \Delta H_1 + \Delta H_2 + \Delta H_3 = 0.752 \text{ kJ} - 6.02 \text{ kJ} - 0.377 \text{ kJ} = -5.65 \text{ kJ}$$

Check: The units (kJ/mol) are correct. We expect freezing to release less energy at –10 °C because we are below the normal freezing point.

89. Melting of an ice cube in a glass of water will not raise or lower the level of the liquid in the glass as long as the ice is always floating in the liquid. This is because the ice will displace a volume of water based on its mass. By the same logic, melting floating icebergs will not raise the ocean levels (assuming that the dissolved solids content, and thus the density, will not change when the icebergs melt). Dissolving ice formations that are supported by land will raise the ocean levels, just as pouring more water into the glass will raise the liquid level in the glass.

90. The water, a container with a larger surface area will evaporate more quickly because there is more surface area for the molecules to evaporate from. Vapor pressure is the pressure of the gas when it is in dynamic equilibrium with the liquid (evaporation rate = condensation rate). The vapor pressure is dependent only on the substance and the temperature. The larger the surface area, the more quickly it will reach this equilibrium state.

91. Substance A will have the larger change in vapor pressure with the same temperature change. To understand this consider the Clausius–Clapeyron Equation: $\ln \frac{P_2}{P_1} = \frac{-\Delta H_{vap}}{R}\left(\frac{1}{T_2} - \frac{1}{T_1}\right)$, if we use the same temperatures we see that $\frac{P_2}{P_1} \alpha\, e^{-\Delta H_{vap}}$. So the smaller the heat of vaporization, the larger the final vapor pressure. We can also consider that the lower the heat of vaporization, the easier it is to convert the substance from a liquid to a gas. This again leads to Substance A having the larger change in vapor pressure.

92. The triple point will be at a lower temperature since the fusion equilibrium line has a positive slope. This means that we will be increasing both temperature and pressure as we travel from the triple point to the normal melting point.

93. $\Delta H_{sub} = \Delta H_{fus} + \Delta H_{vap}$ as long as the heats of fusion and vaporization are measured at the same temperatures.

94. The liquid segment will have the least steep slope because it takes the most kJ/mol to raise the temperature of the phase.

95. Water has an exceptionally high specific heat capacity, which has a moderating effect on the temperature of the root cellar. A large amount of heat can be stored in a large vat of water. The heat will be absorbed or released from the large bodies of water preferentially over the area around it. As the temperature of the air drops, the water will release heat, keeping the temperature more constant. If the temperature of the cellar falls enough to begin to freeze the water, the heat given off during the freezing will further protect the food in the cellar.

Chapter 12
Solutions

1. a) hexane, toluene, or CCl_4; dispersion forces

   b) water, methanol, acetone; dispersion, dipole–dipole, hydrogen bonding

   c) hexane, toluene, or CCl_4; dispersion forces

   d) water, acetone, methanol, ethanol; dispersion, ion–dipole

2. a) water, methanol, ethanol; dispersion, dipole–dipole, hydrogen bonding

   b) water, acetone, methanol, ethanol; dispersion, ion–dipole

   c) hexane, toluene, or CCl_4; dispersion forces

   d) water, acetone, methanol, ethanol; dispersion, ion–dipole

3. $HOCH_2CH_2CH_2OH$ would be more soluble in water because it has –OH groups on both ends of the molecule, so it can hydrogen bond on both ends, not just one end.

4. CH_2Cl_2 would be more soluble in water because it is a polar molecule and can exhibit dipole–dipole interactions with the water molecules. CCl_4 is a nonpolar molecule.

5. a) water; dispersion, dipole–dipole, hydrogen bonding

   b) hexane; dispersion forces

   c) water; dispersion, dipole–dipole

   d) water; dispersion, dipole–dipole, hydrogen bonding

6. a) hexane; dispersion forces

   b) water; dispersion, dipole–dipole, hydrogen bonding

   c) hexane; dispersion forces

   d) water; dispersion, dipole–dipole, hydrogen bonding

7. a) endothermic

   b) The lattice energy is greater in magnitude than the heat of hydration.

   c)

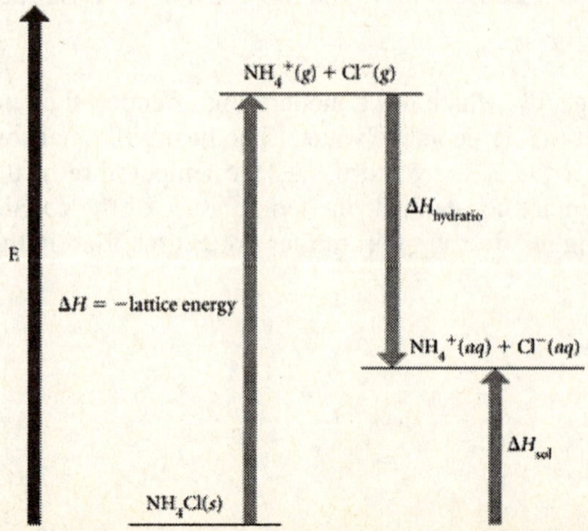

d) The solution forms because chemical systems tend towards greater entropy.

8. a) exothermic

b) The lattice energy is smaller in magnitude than the heat of hydration

c)

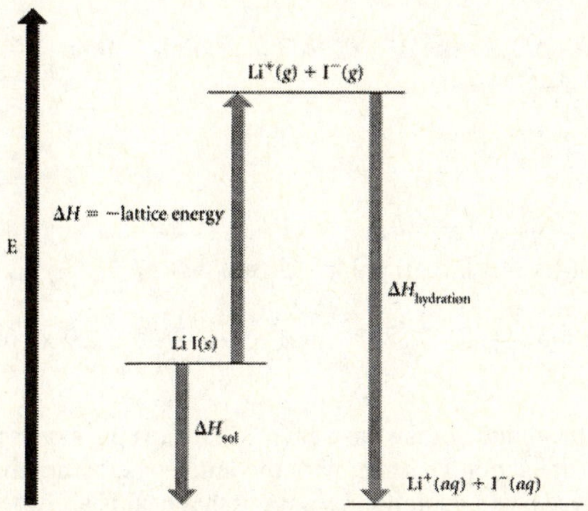

d) The solution forms because chemical systems tend towards lower energy and greater entropy.

9. **Given:** $AgNO_3$: Lattice Energy = – 820. kJ/mol, ΔH_{soln} = + 22.6 kJ/mol **Find:** $\Delta H_{hydration}$
Conceptual Plan: Lattice Energy, ΔH_{soln} → $\Delta H_{hydration}$

$\Delta H_{soln} = \Delta H_{solute} + \Delta H_{hydration}$ where $\Delta H_{solute} = -\Delta H_{lattice}$

Solution: $\Delta H_{soln} = \Delta H_{solute} + \Delta H_{hydration}$ where $\Delta H_{solute} = -\Delta H_{lattice}$ so $\Delta H_{hydration} = \Delta H_{soln} + \Delta H_{lattice}$

$\Delta H_{hydration}$ = 22.6 kJ/mol – 820. kJ/mol = – 797 kJ/mol

Check: The units (kJ/mol) are correct. The magnitude of the answer (– 800) makes physical sense because the lattice energy is so negative and so it dominates the calculation.

10. **Given:** LiCl: Lattice Energy = – 834 kJ/mol, ΔH_{soln} = – 37.0 kJ/mol; NaCl: Lattice Energy = – 769 kJ/mol, ΔH_{soln} = + 3.88 kJ/mol; **Find:** $\Delta H_{hydration}$ and which has stronger ion–dipole interactions
Conceptual Plan: Lattice Energy, ΔH_{soln} → $\Delta H_{hydration}$ then compare values

$\Delta H_{soln} = \Delta H_{solute} + \Delta H_{hydration}$ where $\Delta H_{solute} = -\Delta H_{lattice}$

Solution: $\Delta H_{soln} = \Delta H_{solute} + \Delta H_{hydration}$ where $\Delta H_{solute} = -\Delta H_{lattice}$ so $\Delta H_{hydration} = \Delta H_{soln} + \Delta H_{lattice}$

LiCl: $\Delta H_{hydration}$ = – 37.0 kJ/mol – 834 kJ/mol = – 871 kJ/mol

NaCl: $\Delta H_{hydration}$ = + 3.88 kJ/mol – 769 kJ/mol = – 765 kJ/mol

Since $\Delta H_{hydration}$ of LiCl is more negative than for NaCl, LiCl has the stronger ion–dipole interactions
Check: The units (kJ/mol) are correct. The magnitude of the answer (– 900 and – 800) makes physical sense because the lattice energies are so negative and so they dominate the calculation. We expect stronger interactions with LiCl since the Li^+ ion is smaller than the Na^+ ion and so its charge density is higher and will interact more strongly with the dipoles of the water molecules.

11. **Given:** LiI: Lattice Energy = – 7.3 x 10^2 kJ/mol, $\Delta H_{hydration}$ = – 793 kJ/mol; 15.0 g LiI
Find: ΔH_{soln} and heat evolved
Conceptual Plan: Lattice Energy, $\Delta H_{hydration}$ → ΔH_{soln} and g → mol then mol, ΔH_{soln} → q

$\Delta H_{soln} = \Delta H_{solute} + \Delta H_{hydration}$ where $\Delta H_{solute} = -\Delta H_{lattice}$ $\quad \frac{1 \text{ mol}}{133.843 \text{ g}}$ $\quad q = n\,\Delta H_{soln}$

Solution: $\Delta H_{soln} = \Delta H_{solute} + \Delta H_{hydration}$ where $\Delta H_{solute} = -\Delta H_{lattice}$ so $\Delta H_{soln} = \Delta H_{hydration} - \Delta H_{lattice}$

ΔH_{soln} = – 793 kJ/mol – (– 7$\underline{3}$0 kJ/mol) = – $\underline{6}$0 kJ/mol = – 6 x 10^1 kJ/mol and

$$15.0 \text{ g} \times \frac{1 \text{ mol}}{133.843 \text{ g}} = 0.11\underline{2}072 \text{ mol} \quad \text{then}$$

$$q = n\,\Delta H_{soln} = 0.11\underline{2}072 \text{ mol} \times -6 \times 10^1 \frac{\text{kJ}}{\text{mol}} = -7 \text{ kJ or } 7 \text{ kJ released}$$

Check: The units (kJ/mol and kJ) are correct. The magnitude of the answer (– 60) makes physical sense because the lattice energy and the heat of hydration are about the same. The magnitude of the heat (7) makes physical sense since 15 g is much less than a mole, and so the amount of heat released is going to be small.

12. **Given:** KNO_3: Lattice Energy = – 163.8 kcal/mol, $\Delta H_{\text{hydration}}$ = – 155.5 kcal/mol; 1.00 x 10^2 kJ absorbed
Find: ΔH_{soln} and m (KNO_3)
Conceptual Plan: Lattice Energy, $\Delta H_{\text{hydration}}$ → ΔH_{soln} (kcal) → ΔH_{soln} (kcal) then
$\Delta H_{soln} = \Delta H_{solute} + \Delta H_{hydration}$ where $\Delta H_{solute} = -\Delta H_{lattice}$
mol, q → ΔH_{soln} then mol → g
$q = n\,\Delta H_{soln}$ $\qquad \dfrac{101.11\text{ g}}{1\text{ mol}}$

Solution: $\Delta H_{soln} = \Delta H_{solute} + \Delta H_{hydration}$ where $\Delta H_{solute} = -\Delta H_{lattice}$ so $\Delta H_{soln} = \Delta H_{hydration} - \Delta H_{lattice}$
$\Delta H_{soln} = [-163.8\text{ kcal/mol} + (-155.5\text{ kcal/mol})](4.184\text{ kJ/kcal}) = 34.7\text{ kJ/mol}$ then $q = n\,\Delta H_{soln}$ Rearrange to solve for n. $n = \dfrac{q}{\Delta H_{soln}} = \dfrac{1.00\times10^2\text{ kJ}}{34.7\,\dfrac{\text{kJ}}{\text{mol}}} = 2.8818\text{ mol}$ then $2.8818\text{ mol} \times \dfrac{101.11\text{ g}}{1\text{ mol}} = 2.9 \times 10^2\text{ g}$

Check: The units (kJ/mol and g) are correct. The magnitude of the answer (+ 35) makes physical sense because the lattice energy and the heat of hydration are about the same, with the lattice energy dominating, so the answer is positive. The problem hints at a positive heat of solution by saying that heat is absorbed. The magnitude of the mass (290) makes physical sense since 100 kJ will require over 2 moles of salt.

13. The solution is unsaturated since we are dissolving 25g of NaCl per 100 g of water and the solubility from the figure is ~ 35 g NaCl per 100 g of water.

14. The solution is almost saturated since we are dissolving 32 g of KNO_3 per 100 g of water and the solubility from the figure is ~ 36 g KNO_3 per 100 g of water.

15. At 40 °C the solution has 45 g of KNO_3 per 100 g of water and it can contain up to 63 g of KNO_3 per 100 g of water. At 0 °C the solubility from the figure is ~ 14 g KNO_3 per 100 g of water, so ~ 31 g KNO_3 per 100 g of water will precipitate out of solution.

16. At 60 °C the solution has 42 g of KCl per 100 g of water and it can contain up to 45 g of KCl per 100 g of water. At 0 °C the solubility from the figure is ~ 26 g KCl per 100 g of water, so ~ 16 g KCl per 100 g of water will precipitate out of solution.

17. Since the solubility of gases decrease as the temperature increased, dissolved oxygen will be removed from the solution.

18. Since the solubility of gases decrease as the temperature increased, dissolved oxygen was removed from the solution and so there was no oxygen in the water for the fish to breathe.

19. Henry's Law says that as pressure increases, nitrogen will more easily dissolve in blood. To reverse this process, divers should ascend to lower pressures.

20. Henry's Law says that as pressure increases, oxygen will more easily dissolve in blood. To reverse this process, divers should ascend to lower pressures or breathe special gas mixtures with lower oxygen levels.

21. **Given:** room temperature, 80.0 L aquarium, P_{Total} = 1.0 atm; χ_{N2} = 0.78 **Find:** m (N_2)
Other: $k_H(N_2) = 6.1 \times 10^{-4}$ M/L at 25 °C
Conceptual Plan: $P_{\text{Total}}, \chi_{N2}$ → P_{N2} then $P_{N2}, k_H(N_2)$ → S_{N2} then L → mol → g
$P_{N2} = \chi_{N2}\,P_{Total}$ $\qquad S_{N2} = k_H(N_2)P_{N2}$ $\qquad S_{N2}$ $\qquad \dfrac{28.01\text{ g}}{1\text{ mol}}$

Solution: $P_{N2} = \chi_{N2}\,P_{Total} = 0.78 \times 1.0\text{ atm} = 0.78\text{ atm}$ then

$S_{N2} = k_H(N_2)P_{N2} = 6.1 \times 10^{-4}\,\dfrac{\text{M}}{\text{atm}} \times 0.78\text{ atm} = 4.758 \times 10^{-4}\text{ M}$ then

$$80.0\ \text{L} \times 4.758 \times 10^{-4}\ \frac{\text{mol}}{\text{L}} \times \frac{28.01\ \text{g}}{1\ \text{mol}} = 1.1\ \text{g}$$

Check: The units (g) are correct. The magnitude of the answer (1) seems reasonable since we have 80 L of water and expect much less than a mole of nitrogen.

22. **Given:** Helium, 25 °C, P_{He} = 1.0 atm **Find:** S_{He} (M) **Other:** k_H(He) = 3.7 x 10^{-4} M/L at 25 °C
Conceptual Plan: $P_{He}, k_H(He) \rightarrow S_{He}$

$$S_{He} = k_H(He)P_{He}$$

Solution: $S_{He} = k_H(He)P_{He} = 3.7 \times 10^{-4}\ \frac{\text{M}}{\text{atm}} \times 1.0\ \text{atm} = 3.7 \times 10^{-4}\ \text{M}$

Check: The units (M) are correct. The magnitude of the answer (10^{-4}) seems reasonable since this is the value of k_H.

23. **Given:** NaCl and water; 133 g NaCl in 1.00 L solution **Find:** M, *m*, and mass percent
Other: d = 1.08 g/mL
Conceptual Plan: $g_{NaCl} \rightarrow mol$ and $L \rightarrow mL \rightarrow g_{soln}$ and $g_{soln}\ g_{NaCl} \rightarrow g_{H2O} \rightarrow kg_{H2O}$ then

$$\frac{1\ \text{mol NaCl}}{58.44\ \text{g NaCl}} \qquad \frac{1000\ \text{mL}}{1\ \text{L}}\ \frac{1.08\ \text{g}}{1\ \text{mL}} \qquad g_{H2O} = g_{soln} - g_{NaCl} \qquad \frac{1\ \text{kg}}{1000\ \text{g}}$$

mol, $V \rightarrow$ M and **mol, $kg_{H2O} \rightarrow m$** and **$g_{soln}\ g_{NaCl} \rightarrow$ mass percent**

$$\text{M} = \frac{\text{amount solute (moles)}}{\text{volume solution (L)}} \qquad m = \frac{\text{amount solute (moles)}}{\text{mass solvent (kg)}} \qquad mass\ percent = \frac{mass\ solute}{mass\ solution} \times 100\ \%$$

Solution: $133\ \text{g NaCl} \times \frac{1\ \text{mol NaCl}}{58.44\ \text{g NaCl}} = 2.27584\ \text{mol NaCl}$ and

$1.00\ \text{L} \times \frac{1000\ \text{mL}}{1\ \text{L}} \times \frac{1.08\ \text{g}}{1\ \text{mL}} = 1080\ \text{g soln}$ and

$g_{H2O} = g_{soln} - g_{NaCl} = 1080\ \text{g} - 133\ \text{g} = 947\ \text{g H}_2\text{O} \times \frac{1\ \text{kg}}{1000\ \text{g}} = 0.947\ \text{kg H}_2\text{O}$ then

$$\text{M} = \frac{\text{amount solute (moles)}}{\text{volume solution (L)}} = \frac{2.27584\ \text{mol NaCl}}{1.00\ \text{L soln}} = 2.28\ \text{M} \quad \text{and}$$

$$m = \frac{\text{amount solute (moles)}}{\text{mass solvent (kg)}} = \frac{2.27584\ \text{mol NaCl}}{0.947\ \text{kg H}_2\text{O}} = 2.4\ m \quad \text{and}$$

$$mass\ percent = \frac{mass\ solute}{mass\ solution} \times 100\ \% = \frac{133\ \text{g NaCl}}{1080\ \text{g soln}} \times 100\ \% = 12.3\ \%\ \text{by mass}.$$

Check: The units (M, m, and percent by mass) are correct. The magnitude of the answer (2.28 M) seems reasonable since we have 133 g NaCl, which is a couple of moles and we have 1 L. The magnitude of the answer (2.4 *m*) seems reasonable since it is a little higher than the molarity, which we expect since we only use the solvent weight in the denominator. The magnitude of the answer (12 %) seems reasonable since we have 133 g NaCl and just over 1000 g of solution.

24. **Given:** KNO_3 and water; 88.4 g KNO_3 in 1.50 L solution **Find:** M, *m*, and mass percent
Other: d = 1.05 g/mL
Conceptual Plan: $g_{KNO3} \rightarrow mol$ and $L \rightarrow mL \rightarrow g_{soln}$ and $g_{soln}\ g_{KNO3} \rightarrow g_{H2O} \rightarrow kg_{H2O}$ then

$$\frac{1\ \text{mol KNO}_3}{101.11\ \text{g KNO}_3} \qquad \frac{1000\ \text{mL}}{1\ \text{L}}\ \frac{1.05\ \text{g}}{1\ \text{mL}} \qquad g_{H2O} = g_{soln} - g_{KNO3} \qquad \frac{1\ \text{kg}}{1000\ \text{g}}$$

mol, $V \rightarrow$ M and **mol, $kg_{H2O} \rightarrow m$** and **$g_{soln}\ g_{KNO3} \rightarrow$ mass percent**

$$\text{M} = \frac{\text{amount solute (moles)}}{\text{volume solution (L)}} \qquad m = \frac{\text{amount solute (moles)}}{\text{mass solvent (kg)}} \qquad mass\ percent = \frac{mass\ solute}{mass\ solution} \times 100\ \%$$

Solution: $88.4 \text{ g NaCl} \times \dfrac{1 \text{ mol } KNO_3}{101.11 \text{ g } KNO_3} = 0.87\underline{4}295 \text{ mol } KNO_3$ and

$1.50 \text{ L} \times \dfrac{1000 \text{ mL}}{1 \text{ L}} \times \dfrac{1.05 \text{ g}}{1 \text{ mL}} = 15\underline{7}5 \text{ g soln}$ and

$g_{H2O} = g_{soln} - g_{NaCl} = 15\underline{7}5 \text{ g} - 88.4 \text{ g} = 14\underline{8}6.6 \text{ g } H_2O \times \dfrac{1 \text{ kg}}{1000 \text{ g}} = 1.4\underline{8}66 \text{ kg } H_2O$ then

$$M = \frac{\text{amount solute (moles)}}{\text{volume solution (L)}} = \frac{0.87\underline{4}295 \text{ mol } KNO_3}{1.50 \text{ L soln}} = 0.583 \text{ M} \text{ and}$$

$$m = \frac{\text{amount solute (moles)}}{\text{mass solvent (kg)}} = \frac{0.87\underline{4}295 \text{ mol } KNO_3}{1.4\underline{8}66 \text{ kg } H_2O} = 0.588\ m \text{ and}$$

$$\textit{mass percent} = \frac{\textit{mass solute}}{\textit{mass solution}} \times 100\,\% = \frac{88.4 \text{ g } KNO_3}{15\underline{7}5 \text{ g soln}} \times 100\,\% = 5.61\ \% \text{ by mass}.$$

Check: The units (M, m, and percent by mass) are correct. The magnitude of the answer (0.58 M) seems reasonable since we have 88.4 g KNO_3, which is less than a mole and we have 1.5 L. The magnitude of the answer (0.59 *m*) seems reasonable since it is a little higher than the molarity, which we expect since we only use the solvent weight in the denominator. The magnitude of the answer (6 %) seems reasonable since we have 88 g KNO_3 and ~ 1500 g of solution.

25. **Given:** initial solution: 50.0 mL of 5.00 M KI; final solution contains: 3.25 g KI in 25.0 mL
Find: final volume to dilute initial solution to
Conceptual Plan:
final solution: g_{KI} → mol and mL → L then mol, V → M_2 then M_1, V_1, M_2 → V_2

$\dfrac{1 \text{ mol KI}}{166.006 \text{ g KI}}$ $\dfrac{1 \text{ L}}{1000 \text{ mL}}$ $M = \dfrac{\text{amount solute (moles)}}{\text{volume solution (L)}}$ $M_1 V_1 = M_2 V_2$

Solution: $3.25 \text{ g KI} \times \dfrac{1 \text{ mol KI}}{166.006 \text{ g KI}} = 0.019\underline{5}776 \text{ mol KI}$ and $25.0 \text{ mL} \times \dfrac{1 \text{ L}}{1000 \text{ mL}} = 0.0250 \text{ mL}$

then $M = \dfrac{\text{amount solute (moles)}}{\text{volume solution (L)}} = \dfrac{0.019\underline{5}776 \text{ mol KI}}{0.0250 \text{ L soln}} = 0.78\underline{3}104 \text{ M}$ then $M_1 V_1 = M_2 V_2$.

Rearrange to solve for V_2. $V_2 = \dfrac{M_1}{M_2} \times V_1 = \dfrac{5.00 \text{ M}}{0.78\underline{3}104 \text{ M}} \times 50.0 \text{ mL} = 319 \text{ mL diluted volume}$

Check: The units (mL) are correct. The magnitude of the answer (319 mL) seems reasonable since we are starting with a concentration of 5 M and ending with a concentration of less than 1 M.

26. **Given:** initial solution: 125 mL of 8.00 M $CuCl_2$; final solution contains: 5.9 g $CuCl_2$ in 50.0 mL
Find: final volume to dilute initial solution to
Conceptual Plan:
final solution: g_{CuCl2} → mol and mL → L then mol, V → M_2 then M_1, V_1, M_2 → V_2

$\dfrac{1 \text{ mol } CuCl_2}{134.45 \text{ g } CuCl_2}$ $\dfrac{1 \text{ L}}{1000 \text{ mL}}$ $M = \dfrac{\text{amount solute (moles)}}{\text{volume solution (L)}}$ $M_1 V_1 = M_2 V_2$

Solution: $5.9 \text{ g } CuCl_2 \times \dfrac{1 \text{ mol } CuCl_2}{134.45 \text{ g } CuCl_2} = 0.04\underline{3}884 \text{ mol } CuCl_2$ and $50.0 \text{ mL} \times \dfrac{1 \text{ L}}{1000 \text{ mL}} = 0.0500 \text{ mL}$

then $M = \dfrac{\text{amount solute (moles)}}{\text{volume solution (L)}} = \dfrac{0.04\underline{3}884 \text{ mol } CuCl_2}{0.0500 \text{ L soln}} = 0.8\underline{7}768 \text{ M}$ then $M_1 V_1 = M_2 V_2$. Rearrange to

solve for V_2. $V_2 = \dfrac{M_1}{M_2} \times V_1 = \dfrac{8.00 \text{ M}}{0.8\underline{7}768 \text{ M}} \times 125 \text{ mL} = 1100 \text{ mL diluted volume}$

Check: The units (mL) are correct. The magnitude of the answer (1100 mL) seems reasonable since we are starting with a concentration of 8 M and ending with a concentration of less than 1 M.

27. **Given:** $AgNO_3$ and water; 3.4 % Ag by mass, 4.8 L solution **Find:** m (Ag) **Other:** d = 1.01 g/mL
Conceptual Plan: **L → mL → g_{soln} → g_{Ag}**

$$\frac{1000\text{ mL}}{1\text{ L}} \quad \frac{1.01\text{ g}}{1\text{ mL}} \quad \frac{3.4\text{ g Ag}}{100\text{ g soln}}$$

Solution: $4.8\text{ L} \times \frac{1000\text{ mL}}{1\text{ L}} \times \frac{1.01\text{ g}}{1\text{ mL}} = 4\underline{8}48\text{ g soln}$ then

$$4\underline{8}48\text{ g soln} \times \frac{3.4\text{ g Ag}}{100\text{ g soln}} = 160\text{ g Ag} = 1.6 \times 10^2\text{ g Ag}.$$

Check: The units (g) are correct. The magnitude of the answer (160 g) seems reasonable since we have almost 5000 g solution.

28. **Given:** dioxin and water; 0.085 % dioxin by mass, 2.5 L solution **Find:** m (dioxin) **Other:** d = 1.00 g/mL
Conceptual Plan: **L → mL → g_{soln} → g_{dioxin}**

$$\frac{1000\text{ mL}}{1\text{ L}} \quad \frac{1.01\text{ g}}{1\text{ mL}} \quad \frac{0.085\text{ g dioxin}}{100\text{ g soln}}$$

Solution: $2.5\text{ L} \times \frac{1000\text{ mL}}{1\text{ L}} \times \frac{1.00\text{ g}}{1\text{ mL}} = 2\underline{5}00\text{ g soln}$ then $2\underline{5}00\text{ g soln} \times \frac{0.085\text{ g dioxin}}{100\text{ g soln}} = 2.1\text{ g dioxin}$.

Check: The units (g) are correct. The magnitude of the answer (2 g) seems reasonable since we have 2500 g solution and a low concentration.

29. **Given:** Ca^{2+} and water; 0.0085 % Ca^{2+} by mass, 1.2 g Ca **Find:** m (water)
Conceptual Plan: **g_{Ca} → g_{soln} → g_{H2O}**

$$\frac{100\text{ g soln}}{0.0085\text{ g Ca}} \qquad g_{H2O} = g_{soln} - g_{Ca}$$

Solution: $1.2\text{ g Ca} \times \frac{100\text{ g soln}}{0.0085\text{ g Ca}} = 14\underline{1}18\text{ g soln}$ then

$$g_{H2O} = g_{soln} - g_{Ca} = 14\underline{1}18\text{ g} - 1.2\text{ g} = 1.4 \times 10^4\text{ g water}.$$

Check: The units (g) are correct. The magnitude of the answer (10^4 g) seems reasonable since we have such a low concentration of Ca.

30. **Given:** Pb and water; 0.0011 % Pb by mass, 150 mg Pb **Find:** V (mL) **Other:** d = 1.00 g/mL
Conceptual Plan: **mg_{Pb} → g_{Pb} → g_{soln} → mL**

$$\frac{1000\text{ g}}{1\text{ mg}} \quad \frac{100\text{ g soln}}{0.0011\text{ g Pb}} \quad \frac{1\text{ mL}}{1.00\text{ g}}$$

Solution: $150\text{ mg Pb} \times \frac{1\text{ g Pb}}{1000\text{ mg Pb}} \times \frac{100\text{ g soln}}{0.0011\text{ g Pb}} \times \frac{1\text{ mL}}{1.00\text{ g}} = 1.4 \times 10^4\text{ mL}$

Check: The units (mL) are correct. The magnitude of the answer (10^4 g) seems reasonable since we have such a low concentration of Pb.

31. **Given:** concentrated HNO_3: 70.3 % HNO_3 by mass, d = 1.41 g/mL; final solution: 1.15 L of 0.100 M HNO_3
Find: describe final solution preparation
Conceptual Plan:
M_2, V_2 → mol_{HNO3} → g_{HNO3} → $g_{conc\ acid}$ → $mL_{conc\ acid}$ then describe method

$$mol = M\,V \quad \frac{63.02\text{ g HNO}_3}{1\text{ mol HNO}_3} \quad \frac{100\text{ g conc acid}}{70.3\text{ g HNO}_3} \quad \frac{1\text{ mL}}{1.41\text{ g}}$$

Solution: $mol = M\,V = 0.100\frac{\text{mol HNO}_3}{1\text{ L soln}} \times 1.15\text{ L soln} = 0.115\text{ mol HNO}_3$ then

$0.115 \cancel{\text{mol HNO}_3} \times \frac{63.02 \cancel{\text{ g HNO}_3}}{1 \cancel{\text{ mol HNO}_3}} \times \frac{100 \cancel{\text{ g conc acid}}}{70.3 \cancel{\text{ g HNO}_3}} \times \frac{1 \text{ mL conc acid}}{1.41 \cancel{\text{ g conc acid}}} = 7.31 \text{ mL conc acid}$. Prepare the solution by putting 1.00 L of distilled water in a container. Carefully pour in the 7.31 mL of the concentrated acid, mix the solution, and allow it to cool. Finally add enough water to generate a total volume of solution (1.15 L). It is important to add acid to water, and not the reverse, since there is such a large amount of heat released upon mixing.

Check: The units (mL) are correct. The magnitude of the answer (7 g) seems reasonable since we are starting with such a very concentrated solution and diluting it to a low concentration.

32. **Given:** concentrated HCl: 37.0 % HCl by mass, $d = 1.20$ g/mL; final solution: 2.85 L of 0.500 M HCl
Find: describe final solution preparation
Conceptual Plan: $M_2, V_2 \rightarrow \text{mol}_{HCl} \rightarrow g_{HCl} \rightarrow g_{\text{conc acid}} \rightarrow \text{mL}_{\text{conc acid}}$ **then describe method**

$$mol = M\,V \qquad \frac{36.46 \text{ g HCl}}{1 \text{ mol HCl}} \qquad \frac{100 \text{ g conc acid}}{37.0 \text{ g HCl}} \qquad \frac{1 \text{ mL}}{1.20 \text{ g}}$$

Solution: $mol = M\,V = 0.500 \frac{\text{mol HCl}}{1 \cancel{\text{ L soln}}} \times 2.85 \cancel{\text{ L soln}} = 1.4\underline{2}5 \text{ mol HCl}$ then

$$1.4\underline{2}5 \cancel{\text{ mol HCl}} \times \frac{36.46 \cancel{\text{ g HCl}}}{1 \cancel{\text{ mol HCl}}} \times \frac{100 \cancel{\text{ g conc acid}}}{37.0 \cancel{\text{ g HCl}}} \times \frac{1 \text{ mL conc acid}}{1.20 \cancel{\text{ g conc acid}}} = 117 \text{ mL conc acid.}$$

Prepare the solution by putting 2.5 L of distilled water in a container. Carefully pour in the 117 mL of the concentrated acid, mix the solution, and allow it to cool. Finally add enough water to generate a total volume of solution. It is important to add acid to water, and not the reverse, since there is such a large amount of heat released upon mixing.

Check: The units (mL) are correct. The magnitude of the answer (117 g) seems reasonable since we are starting with such a concentrated solution and diluting it to a low concentration.

33. a) **Given:** 1.00×10^2 mL of 0.500 M KCl **Find:** describe final solution preparation
Conceptual Plan: **mL → L then** $M, V \rightarrow \text{mol}_{KCl} \rightarrow g_{KCl}$ **then describe method**

$$\frac{1 \text{ L}}{1000 \text{ mL}} \qquad mol = M\,V \qquad \frac{74.56 \text{ g KCl}}{1 \text{ mol KCl}}$$

Solution: $1.00 \times 10^2 \cancel{\text{ mL}} \times \frac{1 \text{L}}{1000 \cancel{\text{ mL}}} = 0.100 \text{ L}$

$$mol = M\,V = 0.500 \frac{\text{mol KCl}}{1 \cancel{\text{ L soln}}} \times 0.100 \cancel{\text{ L soln}} = 0.0500 \text{ mol KCl}$$

then $0.0500 \cancel{\text{ mol KCl}} \times \frac{74.56 \text{g KCl}}{1 \cancel{\text{ mol KCl}}} = 3.73 \text{ g KCl}$.

Prepare the solution by carefully adding 3.73 g KCl to a 100-mL volumetric flask. Add ~ 75 mL of distilled water and agitate the solution until the salt dissolves completely. Finally add enough water to generate a total volume of solution (add water to the mark on the flask).

Check: The units (g) are correct. The magnitude of the answer (4 g) seems reasonable since we are making a small volume of solution and the formula weight of KCl is ~ 75 g/mol.

b) **Given:** 1.00×10^2 g of 0.500 m KCl **Find:** describe final solution preparation
Conceptual Plan:
$m \rightarrow \text{mol}_{KCl}$ **/1 kg solvent** $\rightarrow g_{KCl}$**/1 kg solvent then** g_{KCl}**/1 kg solvent,** $g_{soln} \rightarrow g_{KCl}, g_{H2O}$

$$m = \frac{\text{amount solute (moles)}}{\text{mass solvent (kg)}} \qquad \frac{74.56 \text{ g KCl}}{1 \text{ mol KCl}} \qquad g_{soln} = g_{KCl} + g_{H2O}$$

then describe method

Solution: $m = \frac{\text{amount solute (moles)}}{\text{mass solvent (kg)}}$ so $0.500\ m = \frac{0.500 \text{ mol KCl}}{1 \text{ kg H}_2\text{O}}$ so

$$\frac{0.500 \cancel{\text{ mol KCl}}}{1 \text{ kg H}_2\text{O}} \times \frac{74.56 \text{g KCl}}{1 \cancel{\text{ mol KCl}}} = \frac{37.\underline{2}8 \text{ g KCl}}{1000 \text{ g H}_2\text{O}} \qquad g_{soln} = g_{KCl} + g_{H2O} \text{ so } g_{soln} - g_{KCl} = g_{H2O}$$

substitute into ratio $\frac{0.037\underline{2}8 \text{ g KCl}}{1 \text{ g H}_2\text{O}} = \frac{x \text{ g KCl}}{100 \text{ g soln} - x \text{ g KCl}}$ Rearrange and solve for x g KCl

$0.037\underline{2}8\,(100 \text{ g soln} - x \text{ g KCl}) = x \text{ g KCl} \rightarrow 3.7\underline{2}8 - 0.037\underline{2}8\,(x \text{ g KCl}) = x \text{ g KCl} \rightarrow$

$3.7\underline{2}8 = 1.037\underline{2}8\,(x \text{ g KCl}) \rightarrow \frac{3.7\underline{2}8}{1.037\underline{2}8} = x \text{ g KCl} = 3.59 \text{ g KCl}$ then

$g_{H2O} = g_{soln} - g_{KCl} = 100.\text{ g} - 3.59 \text{ g} = 96.41 \text{ g H}_2\text{O}$.

Prepare the solution by carefully adding 3.59 g KCl to a container with 96.41 g of distilled water and agitate the solution until the salt dissolves completely.

Check: The units (g) are correct. The magnitude of the answer (3.6 g) seems reasonable since we are making a small volume of solution and the formula weight of KCl is ~ 75 g/mol.

c) **Given:** 1.00×10^2 g of 5.0 % KCl by mass **Find:** describe final solution preparation

Conceptual Plan: $\mathbf{g_{soln} \rightarrow g_{KCl}}$ **then** $\mathbf{g_{KCl}, g_{soln} \rightarrow g_{H2O}}$

$\frac{5.0 \text{ g KCl}}{100 \text{ g soln}}$ $\qquad g_{soln} = g_{KCl} + g_{H2O}$

then describe method

Solution: $1.00 \times 10^2 \text{ g soln} \times \frac{5.0 \text{ g KCl}}{100 \text{ g soln}} = 5.0 \text{ g KCl}$ then $g_{soln} = g_{KCl} + g_{H2O}$.

So $g_{H2O} = g_{soln} - g_{KCl} = 100.\text{ g} - 5.0 \text{ g} = 95 \text{ g H}_2\text{O}$

Prepare the solution by carefully adding 5.0 g KCl to a container with 95 g of distilled water and agitate the solution until the salt dissolves completely.

Check: The units (g) are correct. The magnitude of the answer (5 g) seems reasonable since we are making a small volume of solution and the solution is 5 % by mass KCl.

34. a) **Given:** 125 mL of 0.100 M $NaNO_3$ **Find:** describe final solution preparation

Conceptual Plan: **mL → L then M, V → $\mathbf{mol_{NaNO3}}$ → $\mathbf{g_{NaNO3}}$ then describe method**

$\frac{1 \text{ L}}{1000 \text{ mL}}$ $\qquad mol = M\,V$ $\qquad \frac{85.00 \text{ g NaNO}_3}{1 \text{ mol NaNO}_3}$

Solution: $125 \text{ mL} \times \frac{1 \text{ L}}{1000 \text{ mL}} = 0.125 \text{ L}$

$mol = M\,V = 0.100 \frac{\text{mol NaNO}_3}{1 \text{ L soln}} \times 0.125 \text{ L soln} = 0.0125 \text{ mol NaNO}_3$ then

$0.0125 \text{ mol NaNO}_3 \times \frac{85.00 \text{ g NaNO}_3}{1 \text{ mol NaNO}_3} = 1.06 \text{ g NaNO}_3$.

Prepare the solution by carefully adding 1.06 g $NaNO_3$ to a container. Add ~ 100 mL of distilled water and agitate the solution until the salt dissolves completely. Finally add enough water to generate a total volume of solution (125 mL).

Check: The units (g) are correct. The magnitude of the answer (7 g) seems reasonable since we are making a small volume of solution and the formula weight of KCl is ~ 75 g/mol.

b) **Given:** 125 g of 0.100 m $NaNO_3$ **Find:** describe final solution preparation

Conceptual Plan: $\boldsymbol{m}$ **→ $\mathbf{mol_{NaNO3}}$ /1 kg solvent → $\mathbf{g_{NaNO3}}$/1 kg solvent then**

$m = \frac{\text{amount solute (moles)}}{\text{mass solvent (kg)}}$ $\qquad \frac{85.00 \text{ g NaNO}_3}{1 \text{ mol NaNO}_3}$

$\mathbf{g_{NaNO3}}$/1 kg solvent, $\mathbf{g_{soln}}$ → $\mathbf{g_{NaNO3}, g_{H2O}}$ then describe method

$g_{soln} = g_{NaNO3} + g_{H2O}$

Solution: $m = \frac{\text{amount solute (moles)}}{\text{mass solvent (kg)}}$ so $0.100\ m = \frac{0.500 \text{ mol NaNO}_3}{1 \text{ kg H}_2\text{O}}$ so

$$\frac{0.100 \text{ mol NaNO}_3}{1 \text{ kg H}_2\text{O}} \times \frac{85.00 \text{ g NaNO}_3}{1 \text{ mol NaNO}_3} = \frac{8.5\underline{0} \text{ g NaNO}_3}{1000 \text{ g H}_2\text{O}}$$ then $g_{soln} = g_{NaNO3} + g_{H2O}$ so

$g_{soln} - g_{NaNO3} = g_{H2O}$ substitute into ratio $\frac{0.0085\underline{0} \text{ g NaNO}_3}{1 \text{ g H}_2\text{O}} = \frac{x \text{ g NaNO}_3}{125 \text{ g soln} - x \text{ g NaNO}_3}$

Rearrange and solve for x g $NaNO_3$ $0.00850\,(125 \text{ g soln} - x \text{ g NaNO}_3) = x \text{ g NaNO}_3$ →

$1.0\underline{6}25 - 0.00850\,(x \text{ g NaNO}_3) = x \text{ g NaNO}_3$ → $1.0\underline{6}25 = 1.00850\,(x \text{ g NaNO}_3)$ →

$\frac{1.0\underline{6}25}{1.00850} = x \text{ g NaNO}_3 = 1.05 \text{ g NaNO}_3$ then $g_{H2O} = g_{soln} - g_{KCl} = 125 \text{ g} - 1.05 \text{ g} = 124 \text{ g H}_2\text{O}$.

Prepare the solution by carefully adding 1.05 g $NaNO_3$ to a container with 124 g of distilled water and agitate the solution until the salt dissolves completely.

Check: The units (g) are correct. The magnitude of the answer (1 g) seems reasonable since we are making a small volume of solution and the formula weight of $NaNO_3$ is 85 g/mol.

c) **Given:** 125 g of 1.0 % $NaNO_3$ by mass **Find:** describe final solution preparation

Conceptual Plan: $g_{soln} \rightarrow g_{NaNO3}$ then $g_{NaNO3}, g_{soln} \rightarrow g_{H2O}$

$\frac{1.0 \text{ g NaNO}_3}{100 \text{ g soln}}$ $g_{soln} = g_{NaNO3} + g_{H2O}$

then describe method

Solution: $125 \text{ g soln} \times \frac{1.0 \text{ g NaNO}_3}{100 \text{ g soln}} = 1.\underline{2}5 \text{ g NaNO}_3$ then $g_{soln} = g_{NaNO3} + g_{H2O}$. So

$g_{H2O} = g_{soln} - g_{KCl} = 125 \text{ g} - 1.25 \text{ g} = 124 \text{ g H}_2\text{O}$

Prepare the solution by carefully adding 1.3 g $NaNO_3$ to a container with 124 g of distilled water and agitate the solution until the salt dissolves completely.

Check: The units (g) are correct. The magnitude of the answer (1 g) seems reasonable since we are making a small volume of solution and the solution is 1 % by mass $NaNO_3$.

35. a) **Given:** 28.4 g of glucose ($C_6H_{12}O_6$) in 355 g water; final volume = 378 mL **Find:** molarity

Conceptual Plan: mL → L and $g_{C6H12O6} \rightarrow mol_{C6H12O6}$ then $mol_{C6H12O6}, V \rightarrow M$

$\frac{1 \text{ L}}{1000 \text{ mL}}$ $\frac{1 \text{ mol C}_6\text{H}_{12}\text{O}_6}{180.16 \text{ g C}_6\text{H}_{12}\text{O}_6}$ $M = \frac{\text{amount solute (moles)}}{\text{volume solution (L)}}$

Solution: $378 \text{ mL} \times \frac{1 \text{ L}}{1000 \text{ mL}} = 0.378 \text{ L}$ and

$$28.4 \text{ g C}_6\text{H}_{12}\text{O}_6 \times \frac{1 \text{ mol C}_6\text{H}_{12}\text{O}_6}{180.16 \text{ g C}_6\text{H}_{12}\text{O}_6} = 0.15\underline{7}638 \text{ mol C}_6\text{H}_{12}\text{O}_6$$

$$M = \frac{\text{amount solute (moles)}}{\text{volume solution (L)}} = \frac{0.15\underline{7}638 \text{ mol C}_6\text{H}_{12}\text{O}_6}{0.378 \text{ L}} = 0.417 \text{ M}$$

Check: The units (M) are correct. The magnitude of the answer (0.4 M) seems reasonable since we have 1/8 mole in about 1/3 L.

b) **Given:** 28.4 g of glucose ($C_6H_{12}O_6$) in 355 g water; final volume = 378 mL **Find:** molality

Conceptual Plan: $g_{H2O} \rightarrow kg_{H2O}$ and $g_{C6H12O6} \rightarrow mol_{C6H12O6}$ then $mol_{C6H12O6}, kg_{H2O} \rightarrow m$

$\frac{1 \text{ kg}}{1000 \text{ g}}$ $\frac{1 \text{ mol C}_6\text{H}_{12}\text{O}_6}{180.16 \text{ g C}_6\text{H}_{12}\text{O}_6}$ $m = \frac{\text{amount solute (moles)}}{\text{mass solvent (kg)}}$

Solution: $355 \text{ g} \times \frac{1 \text{ kg}}{1000 \text{ g}} = 0.355 \text{ kg}$ and

$$28.4 \text{ g C}_6\text{H}_{12}\text{O}_6 \times \frac{1 \text{ mol C}_6\text{H}_{12}\text{O}_6}{180.16 \text{ g C}_6\text{H}_{12}\text{O}_6} = 0.15\underline{7}638 \text{ mol C}_6\text{H}_{12}\text{O}_6$$

$$m = \frac{\text{amount solute (moles)}}{\text{mass solvent (kg)}} = \frac{0.15\underline{7}638 \text{ mol } C_6H_{12}O_6}{0.355 \text{ kg}} = 0.444\ m$$

Check: The units (m) are correct. The magnitude of the answer (0.4 m) seems reasonable since we have 1/8 mole in about 1/3 kg.

c) **Given:** 28.4 g of glucose ($C_6H_{12}O_6$) in 355 g water; final volume = 378 mL **Find:** percent by mass
Conceptual Plan: $g_{C6H12O6}, g_{H2O} \rightarrow g_{soln}$ **then** $g_{C6H12O6}, g_{soln} \rightarrow$ **percent by mass**

$$g_{soln} = g_{C6H12O6} + g_{H2O} \qquad mass\ percent = \frac{mass\ solute}{mass\ solution} \times 100\ \%$$

Solution: $g_{soln} = g_{C6H12O6} + g_{H2O} = 28.4\text{ g} + 355\text{ g} = 38\underline{3}.4\text{ g soln}$ then

$$mass\ percent = \frac{mass\ solute}{mass\ solution} \times 100\ \% = \frac{28.4\text{ g } C_6H_{12}O_6}{38\underline{3}.4\text{ g soln}} \times 100\ \% = 7.41 \text{ percent by mass}$$

Check: The units (percent by mass) are correct. The magnitude of the answer (7 %) seems reasonable since we are dissolving 28 g in 355 g.

d) **Given:** 28.4 g of glucose ($C_6H_{12}O_6$) in 355 g water; final volume = 378 mL **Find:** mole fraction
Conceptual Plan:
$g_{C6H12O6} \rightarrow mol_{C6H12O6}$ **and** $g_{H2O} \rightarrow mol_{H2O}$ **then** $mol_{C6H12O6}, mol_{H2O} \rightarrow \chi_{C6H12O6}$

$$\frac{1 \text{ mol } C_6H_{12}O_6}{180.16 \text{ g } C_6H_{12}O_6} \qquad \frac{1 \text{ mol } H_2O}{18.02 \text{ g } H_2O} \qquad \chi = \frac{\text{amount solute (in moles)}}{\text{total amount of solute and solvent (in moles)}}$$

Solution: $28.4 \text{ g } C_6H_{12}O_6 \times \frac{1 \text{ mol } C_6H_{12}O_6}{180.16 \text{ g } C_6H_{12}O_6} = 0.15\underline{7}638 \text{ mol } C_6H_{12}O_6$ and

$355 \text{ g } H_2O \times \frac{1 \text{ mol } H_2O}{18.02 \text{ g } H_2O} = 19.\underline{7}003 \text{ mol } H_2O$ then

$$\chi = \frac{\text{amount solute (in moles)}}{\text{total amount of solute and solvent (in moles)}} = \frac{0.15\underline{7}638 \text{ mol}}{0.15\underline{7}638 \text{ mol} + 19.\underline{7}003 \text{ mol}} = 0.00794$$

Check: The units (none) are correct. The magnitude of the answer (0.008) seems reasonable since we have many more grams of water and water has a much lower molecular weight.

e) **Given:** 28.4 g of glucose ($C_6H_{12}O_6$) in 355 g water; final volume = 378 mL **Find:** mole percent
Conceptual Plan: **use answer from part d)** **then** $\chi_{C6H12O6} \rightarrow$ **mole percent**

$$\chi \times 100\ \%$$

Solution: *mole percent* = $\chi \times 100\ \% = 0.00794 \times 100\ \% = 0.794$ mole percent

Check: The units (%) are correct. The magnitude of the answer (0.8) seems reasonable since we have many more grams of water and water has a much lower molecular weight and we are just increasing the answer from part d) by a factor of 100.

36. a) **Given:** 20.2 mL of methanol (CH_3OH) in 100.0 mL water; final volume = 118 mL **Find:** molarity
Other: $d(CH_3OH) = 0.782$ g/mL; $d(H_2O) = 1.00$ g/mL
Conceptual Plan: **mL → L and** $mL_{CH3OH} \rightarrow g_{CH3OH} \rightarrow mol_{CH3OH}$ **then** $mol_{CH3OH}, V \rightarrow$ **M**

$$\frac{1\text{L}}{1000 \text{ mL}} \qquad \frac{0.782 \text{ g}}{1 \text{ mL}} \quad \frac{1 \text{ mol } CH_3OH}{32.04 \text{ g } CH_3OH} \qquad M = \frac{\text{amount solute (moles)}}{\text{volume solution (L)}}$$

Solution: $118 \text{ mL} \times \frac{1\text{L}}{1000 \text{ mL}} = 0.118 \text{ L}$ and

$$20.2 \text{ mL } CH_3OH \times \frac{0.782 \text{ g } CH_3OH}{1 \text{ mL } CH_3OH} \times \frac{1 \text{ mol } CH_3OH}{32.04 \text{ g } CH_3OH} = 0.49\underline{3}021 \text{ mol } CH_3OH$$

$$M = \frac{\text{amount solute (moles)}}{\text{volume solution (L)}} = \frac{0.49\underline{3}021 \text{ mol } CH_3OH}{0.118 \text{ L}} = 4.18 \text{ M}$$

Check: The units (M) are correct. The magnitude of the answer (4 M) seems reasonable since we have 1/2 mole in about 1/8 L.

b) **Given:** 20.2 mL of methanol (CH_3OH) in 100.0 mL water; final volume = 118 mL **Find:** molality
Other: $d(CH_3OH) = 0.782$ g/mL; $d(H_2O) = 1.00$ g/mL
Conceptual Plan:
$\mathbf{mL_{CH3OH} \rightarrow g_{CH3OH} \rightarrow mol_{CH3OH}}$ **and** $\mathbf{mL_{H2O} \rightarrow g_{H2O} \rightarrow kg_{H2O}}$ **then** $\mathbf{mol_{CH3OH}, kg_{H2O} \rightarrow \textit{m}}$

$$\frac{0.782\text{ g}}{1\text{ mL}} \quad \frac{1\text{ mol CH}_3\text{OH}}{32.04\text{ g CH}_3\text{OH}} \qquad \frac{1.00\text{ g}}{1\text{ mL}} \quad \frac{1\text{kg}}{1000\text{ g}} \qquad m = \frac{\text{amount solute (moles)}}{\text{mass solvent (kg)}}$$

Solution: $$20.2\text{ mL CH}_3\text{OH} \times \frac{0.782\text{ g CH}_3\text{OH}}{1\text{ mL CH}_3\text{OH}} \times \frac{1\text{ mol CH}_3\text{OH}}{32.04\text{ g CH}_3\text{OH}} = 0.49\underline{3}021\text{ mol CH}_3\text{OH} \text{ and}$$

$$100.0\text{ mL} \times \frac{1.00\text{ g}}{1\text{ mL}} \times \frac{1\text{kg}}{1000\text{ g}} = 0.1000\text{ kg} \text{ then}$$

$$m = \frac{\text{amount solute (moles)}}{\text{mass solvent (kg)}} = \frac{0.49\underline{3}021\text{ mol CH}_3\text{OH}}{0.1000\text{ kg}} = 4.93\ m$$

Check: The units (m) are correct. The magnitude of the answer (5 m) seems reasonable since we have 1/2 mole in 1/10 kg.

c) **Given:** 20.2 mL of methanol (CH_3OH) in 100.0 mL water; final volume = 118 mL
Find: percent by mass
Other: $d(CH_3OH) = 0.782$ g/mL; $d(H_2O) = 1.00$ g/mL
Conceptual Plan: $\mathbf{mL_{CH3OH} \rightarrow g_{CH3OH}}$ **and** $\mathbf{mL_{H2O} \rightarrow g_{H2O}}$ **then** $\mathbf{g_{CH3OH}, g_{H2O} \rightarrow g_{soln}}$ **then**

$$\frac{0.782\text{ g}}{1\text{ mL}} \qquad \frac{1.00\text{ g}}{1\text{ mL}} \qquad g_{soln} = g_{C6H12O6} + g_{H2O}$$

$\mathbf{g_{CH3OH}, g_{soln} \rightarrow}$ **percent by mass**

$$mass\ percent = \frac{mass\ solute}{mass\ solution} \times 100\ \%$$

Solution: $$20.2\text{ mL CH}_3\text{OH} \times \frac{0.782\text{ g CH}_3\text{OH}}{1\text{ mL CH}_3\text{OH}} = 15.\underline{7}964\text{ g CH}_3\text{OH} \text{ and}$$

$$100.0\text{ mL} \times \frac{1.00\text{g}}{1\text{ mL}} = 100.0\text{ g} \quad g_{soln} = g_{CH3OH} + g_{H2O} = 15.\underline{7}964\text{ g} + 100.0\text{ g} = 115.\underline{7}964\text{ g soln} \text{ then}$$

$$mass\ percent = \frac{mass\ solute}{mass\ solution} \times 100\ \% = \frac{15.\underline{7}964\text{ g CH}_3\text{OH}}{115.\underline{7}964\text{ g soln}} \times 100\ \% = 13.6 \text{ percent by mass}$$

Check: The units (percent by mass) are correct. The magnitude of the answer (14 %) seems reasonable since we are dissolving 16 g in 100 g.

d) **Given:** 20.2 mL of methanol (CH_3OH) in 100.0 mL water; final volume = 118 mL
Find: mole fraction
Other: $d(CH_3OH) = 0.782$ g/mL; $d(H_2O) = 1.00$ g/mL
Conceptual Plan: $\mathbf{mL_{CH3OH} \rightarrow g_{CH3OH} \rightarrow mol_{CH3OH}}$ **and** $\mathbf{mL_{H2O} \rightarrow g_{H2O} \rightarrow mol_{H2O}}$ **then**

$$\frac{0.782\text{ g}}{1\text{ mL}} \quad \frac{1\text{ mol CH}_3\text{OH}}{32.04\text{ g CH}_3\text{OH}} \qquad \frac{1.00\text{ g}}{1\text{ mL}} \quad \frac{1\text{ mol H}_2\text{O}}{18.01\text{ g H}_2\text{O}}$$

$\mathbf{mol_{CH3OH}, mol_{H2O} \rightarrow \chi_{CH3OH}}$

$$\chi = \frac{\text{amount solute (in moles)}}{\text{total amount of solute and solvent (in moles)}}$$

Solution: $$20.2\text{ mL CH}_3\text{OH} \times \frac{0.782\text{ g CH}_3\text{OH}}{1\text{ mL CH}_3\text{OH}} \times \frac{1\text{ mol CH}_3\text{OH}}{32.04\text{ g CH}_3\text{OH}} = 0.49\underline{3}021\text{ mol CH}_3\text{OH} \text{ and}$$

$$100.0\ \text{mL} \times \frac{1.00\ \text{g}}{1\ \text{mL}} \times \frac{1\ \text{mol}\,H_2O}{18.01\ \text{g}\,H_2O} = 5.55\underline{2}471\ \text{mol}\,H_2O$$ then

$$\chi = \frac{\text{amount solute (in moles)}}{\text{total amount of solute and solvent (in moles)}} = \frac{0.49\underline{3}021\ \text{mol}}{0.49\underline{3}021\ \text{mol} + 5.55\underline{2}471\ \text{mol}} = 0.0815$$

Check: The units (none) are correct. The magnitude of the answer (0.08) seems reasonable since we have many more grams of water and water has a lower molecular weight.

e) **Given:** 20.2 mL of methanol (CH_3OH) in 100.0 mL water; final volume = 118 mL
Find: mole percent
Other: $d\,(CH_3OH) = 0.782$ g/mL; $d\,(H_2O) = 1.00$ g/mL
Conceptual Plan: use answer from part d) then $\chi_{C6H12O6}$ **→ mole percent**
$\chi \times 100\ \%$

Solution: *mole percent* = $\chi \times 100\ \% = 0.0815 \times 100\ \% = 8.15$ mole percent

Check: The units (%) are correct. The magnitude of the answer (8) seems reasonable since we have many more grams of water and water has a lower molecular weight and we are just increasing the answer from part d) by a factor of 100.

37. **Given:** 3.0 % H_2O_2 by mass, $d = 1.01$ g/mL **Find:** molarity
Conceptual Plan:
Assume exactly 100 g of solution; $g_{Solution} \rightarrow g_{H2O2} \rightarrow mol_{H2O2}$ **and** $g_{Solution} \rightarrow mL_{Solution} \rightarrow L_{Solution}$

$$\frac{3.0\ \text{g}\ H_2O_2}{100\ \text{g Solution}} \quad \frac{1\ \text{mol}\ H_2O_2}{34.02\ \text{g}\ H_2O_2} \quad \frac{1\ \text{mL}}{1.01\ \text{g}} \quad \frac{1\ \text{L}}{1000\ \text{mL}}$$

then $mol_{H2O2}, L_{Solution} \rightarrow M$

$$M = \frac{\text{amount solute (moles)}}{\text{volume solution (L)}}$$

Solution: $$100\ \text{g Solution} \times \frac{3.0\ \text{g}\ H_2O_2}{100\ \text{g Solution}} \times \frac{1\ \text{mol}\ H_2O_2}{34.02\ \text{g}\ H_2O_2} = 0.08\underline{8}1834\ \text{mol}\ H_2O_2 \text{ and}$$

$$100\ \text{g Solution} \times \frac{1\ \text{mL Solution}}{1.01\ \text{g Solution}} \times \frac{1\ \text{L Solution}}{1000\ \text{mL Solution}} = 0.099\underline{0}099\ \text{L Solution}$$ then

$$M = \frac{\text{amount solute (moles)}}{\text{volume solution (L)}} = \frac{0.08\underline{8}1834\ \text{mol}\ H_2O_2}{0.099\underline{0}099\ \text{L Solution}} = 0.89\ M\ H_2O_2.$$

Check: The units (M) are correct. The magnitude of the answer (1) seems reasonable since we are starting with a low concentration solution and pure water is ~ 55.5 M.

38. **Given:** 4.55 % NaOCl by mass, $d = 1.02$ g/mL **Find:** molarity
Conceptual Plan:
Assume exactly 100 g of solution; $g_{Solution} \rightarrow g_{NaOCl} \rightarrow mol_{NaOCl}$ **and** $g_{Solution} \rightarrow mL_{Solution} \rightarrow L_{Solution}$

$$\frac{4.55\ \text{g NaOCl}}{100\ \text{g Solution}} \quad \frac{1\ \text{mol NaOCl}}{74.44\ \text{g NaOCl}} \quad \frac{1\ \text{mL}}{1.02\ \text{g}} \quad \frac{1\ \text{L}}{1000\ \text{mL}}$$

then $mol_{H2O2}, L_{Solution} \rightarrow M$

$$M = \frac{\text{amount solute (moles)}}{\text{volume solution (L)}}$$

Solution: $$100\ \text{g Solution} \times \frac{4.55\ \text{g NaOCl}}{100\ \text{g Solution}} \times \frac{1\ \text{mol NaOCl}}{74.44\ \text{g NaOCl}} = 0.061\underline{1}2305\ \text{mol NaOCl and}$$

$$100\ \text{g Solution} \times \frac{1\ \text{mL Solution}}{1.02\ \text{g Solution}} \times \frac{1\ \text{L Solution}}{1000\ \text{mL Solution}} = 0.098\underline{0}392\ \text{L Solution}$$ then

$$M = \frac{\text{amount solute (moles)}}{\text{volume solution (L)}} = \frac{0.061\underline{1}2305\ \text{mol NaOCl}}{0.098\underline{0}392\ \text{L Solution}} = 0.623\ M\ \text{NaOCl}.$$

Check: The units (M) are correct. The magnitude of the answer (1) seems reasonable since we are starting with a low concentration solution and pure water is ~ 55.5 M.

39. **Given:** 36 % HCl by mass **Find:** molality and mole fraction
Conceptual Plan:
Assume exactly 100 g of solution; $g_{Solution} \rightarrow g_{HCl} \rightarrow mol_{HCl}$ **and** $g_{HCl}, g_{Solution} \rightarrow g_{Solvent} \rightarrow kg_{Solvent}$

$$\frac{36\text{ g HCl}}{100\text{ g Solution}} \quad \frac{1\text{ mol HCl}}{36.46\text{ g HCl}} \quad g_{soln} = g_{HCl} + g_{H2O} \quad \frac{1\text{ kg}}{1000\text{ g}}$$

then $mol_{HCl}, kg_{Solvent} \rightarrow m$ **and** $g_{Solvent} \rightarrow mol_{Solvent}$ **then** $mol_{HCl}, mol_{Solvent} \rightarrow \chi_{HCl}$

$$m = \frac{\text{amount solute (moles)}}{\text{mass solvent (kg)}} \quad \frac{1\text{ mol H}_2\text{O}}{18.02\text{ g H}_2\text{O}} \quad \chi = \frac{\text{amount solute (in moles)}}{\text{total amount of solute and solvent (in moles)}}$$

Solution: $$100\text{ g Solution} \times \frac{36\text{ g HCl}}{100\text{ g Solution}} = 36\text{ g HCl} \times \frac{1\text{ mol HCl}}{36.46\text{ g HCl}} = 0.9\underline{8}7383\text{ mol HCl}$$ and

$g_{soln} = g_{HCl} + g_{H2O}$ Rearrange to solve for $g_{Solvent}$. $g_{H2O} = g_{soln} - g_{HCl} = 100\text{ g} - 36\text{ g} = 64\text{ g H}_2\text{O}$

$$64\text{ g H}_2\text{O} \times \frac{1\text{ kg H}_2\text{O}}{1000\text{ g H}_2\text{O}} = 0.064\text{ kg H}_2\text{O}$$ then

$$m = \frac{\text{amount solute (moles)}}{\text{mass solvent (kg)}} = \frac{0.9\underline{8}7383\text{ mol HCl}}{0.064\text{ kg}} = 15\ m\text{ HCl}$$ and

$$64\text{ g H}_2\text{O} \times \frac{1\text{ mol H}_2\text{O}}{18.02\text{ g H}_2\text{O}} = 3.\underline{5}5161\text{ mol H}_2\text{O}$$ then

$$\chi = \frac{\text{amount solvent (in moles)}}{\text{total amount of solute and solvent (in moles)}} = \frac{0.9\underline{8}7383\text{ mol}}{0.9\underline{8}7383\text{ mol} + 3.\underline{5}5161\text{ mol}} = 0.22.$$

Check: The units (m and unitless) are correct. The magnitudes of the answers (15 and 0.2) seem reasonable since we are starting with a high concentration solution and the molar mass of water is much less than that of HCl.

40. **Given:** 5.0 % NaCl by mass **Find:** molality and mole fraction
Conceptual Plan:
Assume exactly 100 g of solution; $g_{Solution} \rightarrow g_{HCl} \rightarrow mol_{NaCl}$ **and** $g_{NaCl}, g_{Solution} \rightarrow g_{Solvent} \rightarrow kg_{Solvent}$

$$\frac{5.0\text{ g NaCl}}{100\text{ g Solution}} \quad \frac{1\text{ mol NaCl}}{58.44\text{ g NaCl}} \quad g_{soln} = g_{HCl} + g_{H2O} \quad \frac{1\text{ kg}}{1000\text{ g}}$$

then $mol_{HCl}, kg_{Solvent} \rightarrow m$ **and** $g_{Solvent} \rightarrow mol_{Solvent}$ **then** $mol_{HCl}, mol_{Solvent} \rightarrow \chi_{HCl}$

$$m = \frac{\text{amount solute (moles)}}{\text{mass solvent (kg)}} \quad \frac{1\text{ mol H}_2\text{O}}{18.02\text{ g H}_2\text{O}} \quad \chi = \frac{\text{amount solute (in moles)}}{\text{total amount of solute and solvent (in moles)}}$$

Solution: $$100\text{ g Solution} \times \frac{5.0\text{ g NaCl}}{100\text{ g Solution}} = 5.0\text{ g HCl} \times \frac{1\text{ mol NaCl}}{58.44\text{ g NaCl}} = 0.08\underline{5}5578\text{ mol NaCl}$$ and

$g_{soln} = g_{HCl} + g_{H2O}$ Rearrange to solve for $g_{Solvent}$. $g_{H2O} = g_{soln} - g_{HCl} = 100.0\text{ g} - 5.0\text{ g} = 95.0\text{ g H}_2\text{O}$

$$95.0\text{ g H}_2\text{O} \times \frac{1\text{ kg H}_2\text{O}}{1000\text{ g H}_2\text{O}} = 0.0950\text{ kg H}_2\text{O}$$ then

$$m = \frac{\text{amount solute (moles)}}{\text{mass solvent (kg)}} = \frac{0.08\underline{5}5578\text{ mol NaCl}}{0.0950\text{ kg}} = 0.901\ m\text{ NaCl}$$ and

$$95.0\text{ g H}_2\text{O} \times \frac{1\text{ mol H}_2\text{O}}{18.02\text{ g H}_2\text{O}} = 5.\underline{2}7192\text{ mol H}_2\text{O}$$ then

$$\chi = \frac{\text{amount solvent (in moles)}}{\text{total amount of solute and solvent (in moles)}} = \frac{0.08\underline{5}5578 \ \cancel{\text{mol}}}{0.08\underline{5}5578 \ \cancel{\text{mol}} + 5.\underline{2}7192 \ \cancel{\text{mol}}} = 0.016 .$$

Check: The units (m and unitless) are correct. The magnitudes of the answers (1 and 0.02) seem reasonable since we are starting with a low concentration solution and the molar mass of water is much less than that of NaCl.

41. The level has decreased more in the beaker filled with pure water. The dissolved salt in the seawater decreases the vapor pressure and subsequently lowers the rate of vaporization.

42. b) Assume that the solutions obey Raoult's Law ($P_{solution} = \chi_{solvent} \ P^o_{solvent}$). Each of the solutions has the same amount of solvent, so we need to compare the number of moles of particles in the solvent. Without doing any calculations we can see that b) will have the lower number of particles than a) because b) has the higher molecular weight. Potassium acetate will generate ~ 2 moles of particles per mole of the salt, so it will generate more than the other two. So solution a) will have the highest vapor pressure.

43. **Given:** 28.5 g of glycerin ($C_3H_8O_3$) in 125 mL water at 30 °C; $P°_{H2O}$ = 31.8 torr **Find:** P_{H2O}
Other: $d(H_2O)$ = 1.00 g/mL; glycerin is not ionic solid
Conceptual Plan:
$\mathbf{g_{C3H8O3} \rightarrow mol_{C3H8O3}}$ **and** $\mathbf{mL_{H2O} \rightarrow g_{H2O} \rightarrow mol_{H2O}}$ **then** $\mathbf{mol_{C3H8O3}, mol_{H2O} \rightarrow \chi_{H2O}}$

$$\frac{1 \ \text{mol} \ C_3H_8O_3}{92.09 \ \text{g} \ C_3H_8O_3} \qquad \frac{1.00 \ \text{g}}{1 \ \text{mL}} \quad \frac{1 \ \text{mol} \ H_2O}{18.01 \ \text{g} \ H_2O} \qquad \chi = \frac{\text{amount solute (in moles)}}{\text{total amount of solute and solvent (in moles)}}$$

then $\mathbf{\chi_{H2O}, P°_{H2O} \rightarrow P_{H2O}}$

$$P_{solution} = \chi_{solvent} \ P^o_{solvent}$$

Solution: $28.5 \ \cancel{\text{g} \ C_3H_8O_3} \times \frac{1 \ \text{mol} \ C_3H_8O_3}{92.09 \ \cancel{\text{g} \ C_3H_8O_3}} = 0.30\underline{9}466 \ \text{mol} \ C_3H_8O_3$ and

$$125 \ \cancel{\text{mL}} \times \frac{1.00 \ \cancel{\text{g}}}{1 \ \cancel{\text{mL}}} \times \frac{1 \ \text{mol} \ H_2O}{18.01 \ \cancel{\text{g} \ H_2O}} = 6.9\underline{4}059 \ \text{mol} \ H_2O \qquad \text{then}$$

$$\chi = \frac{\text{amount solvent (in moles)}}{\text{total amount of solute and solvent (in moles)}} = \frac{6.9\underline{4}059 \ \cancel{\text{mol}}}{0.30\underline{9}466 \ \cancel{\text{mol}} + 6.9\underline{4}059 \ \cancel{\text{mol}}} = 0.95\underline{7}316 \ \text{then}$$

$$P_{solution} = \chi_{solvent} \ P^o_{solvent} = 0.95\underline{7}316 \times 31.8 \ \text{torr} = 30.4 \ \text{torr}$$

Check: The units (torr) are correct. The magnitude of the answer (30 torr) seems reasonable since it is a drop from the pure vapor pressure. Very few moles of glycerin are added, so the pressure will not drop much.

44. **Given:** 10.85 % naphthalene($C_{10}H_8$) by mass in hexane(C_6H_{14}) at 25 °C; $P°_{C6H14}$ = 151 torr **Find:** P_{C6H14}
Conceptual Plan: % naphthalene($\mathbf{C_{10}H_8}$) by mass $\rightarrow$ $\mathbf{g_{C10H8}, g_{C6H14}}$ **then** $\mathbf{g_{C10H8} \rightarrow mol_{C10H8}}$ **and**

$$\frac{10.85 \ \text{g} \ C_{10}H_8}{100 \ \text{g} \ (C_{10}H_8 + C_6H_{14})} \qquad \frac{1 \ \text{mol} \ C_{10}H_8}{128.16 \ \text{g} \ C_{10}H_8}$$

$\mathbf{g_{C6H14} \rightarrow mol_{C6H14}}$ **then** $\mathbf{mol_{C10H8}, mol_{C6H14} \rightarrow \chi_{C6H14}}$ **then** $\mathbf{\chi_{C6H14}, P°_{C6H14} \rightarrow P_{C6H14}}$

$$\frac{1 \ \text{mol} \ C_6H_{14}}{86.17 \ \text{g} \ C_6H_{14}} \qquad \chi = \frac{\text{amount solute (in moles)}}{\text{total amount of solute and solvent (in moles)}} \qquad P_{solution} = \chi_{solvent} \ P^o_{solvent}$$

Solution: $\frac{10.85 \ \text{g} \ C_{10}H_8}{100 \ \text{g} \ (C_{10}H_8 + C_6H_{14})}$ means 10.85 g $C_{10}H_8$ and (100 g – 10.85 g) = 89.15 g C_6H_{14} then

$$10.85 \ \cancel{\text{g} \ C_{10}H_8} \times \frac{1 \ \text{mol} \ C_{10}H_8}{128.16 \ \cancel{\text{g} \ C_{10}H_8}} = 0.0846\underline{5}980 \ \text{mol} \ C_{10}H_8 \ \text{and}$$

$$89.15 \ \text{g} \ C_6H_{14} \times \frac{1 \ \text{mol} \ C_6H_{14}}{86.17 \ \cancel{\text{g} \ C_6H_{14}}} = 1.03\underline{4}583 \ \text{mol} \ C_6H_{14} \qquad \text{then}$$

$$\chi = \frac{\text{amount solvent (in moles)}}{\text{total amount of solute and solvent (in moles)}} = \frac{1.034583 \text{ mol}}{0.08465980 \text{ mol} + 1.034583 \text{ mol}} = 0.9243599$$

then $P_{solution} = \chi_{solvent} P^{o}_{solvent} = 0.9243599 \text{ x } 151 \text{ torr} = 140. \text{ torr } C_6H_{14}$

Check: The units (torr) are correct. The magnitude of the answer (140 torr) seems reasonable since it is a drop from the pure vapor pressure. Only a fraction of a mole of naphthalene is added, so the pressure will not drop much.

45. **Given:** 5.50 % NaCl by mass in water at 25 °C; **Find:** P_{H2O} **Other:** $P°_{H2O} = 23.78$ torr, $i_{NaCl} = 1.9$

Conceptual Plan: % NaCl by mass → g_{NaCl}, g_{H2O} then g_{NaCl} → mol_{NaCl} and

$$\frac{5.50 \text{ g NaCl}}{100 \text{ g (NaCl + H}_2\text{O)}} \qquad \frac{1 \text{ mol NaCl}}{58.44 \text{ g NaCl}}$$

g_{H2O} → mol_{H2O} then mol_{NaCl}, mol_{H2O} → χ_{H2O} then χ_{H2O}, $P°_{H2O}$ → P_{H2O}

$$\frac{1 \text{ mol H}_2\text{O}}{18.01 \text{ g H}_2\text{O}} \qquad \chi = \frac{\text{amount solute (in moles)}}{\text{total amount of solute and solvent (in moles)}} \qquad P_{solution} = \chi_{solvent} P^{o}_{solvent}$$

Solution: $\frac{5.50 \text{ g NaCl}}{100 \text{ g (NaCl + H}_2\text{O)}}$ means 5.50 g NaCl and (100 g – 5.50 g) = 94.5 g H_2O then

$$5.50 \text{ g NaCl x } \frac{1 \text{ mol NaCl}}{58.44 \text{ g NaCl}} = 0.0941136 \text{ mol NaCl and } 94.5 \text{ g H}_2\text{O x } \frac{1 \text{ mol H}_2\text{O}}{18.01 \text{ g H}_2\text{O}} = 5.24708 \text{ mol H}_2\text{O}$$

the number of moles of solute = i_{NaCl} x n_{NaCl} so

$$\chi_{solv} = \frac{\text{amount solvent (in moles)}}{\text{total amount solute and solvent particles (in moles)}} = \frac{5.24708 \text{ mol}}{5.24708 \text{ mol} + 1.9(0.0941136 \text{ mol})} = 0.96704$$

then $P_{soln} = \chi_{solv} P^{o}_{solv} = 0.96704 \times 23.78 \text{ torr} = 23.0 \text{ torr}$

Check: The units (torr) are correct. The magnitude of the answer (23 torr) seems reasonable since it is a drop from the pure vapor pressure. Only a fraction of a mole of NaCl is added, so the pressure will not drop much.

46. **Given:** $CaCl_2$ and water; $P°_{H2O} = 92.6$ mm Hg, $P°_{H2O} = 81.6$ mm Hg at 50 °C; **Find:** mass percent $CaCl_2$

Conceptual Plan: P_{H2O}, $P°_{H2O}$ → χ_{H2O} then assume 1 mol water, χ_{H2O} → mol_{CaCl2} then

$$P_{solution} = \chi_{solvent} P^{o}_{solvent} \qquad \chi_{H2O} = \frac{\text{mol H}_2\text{O}}{\text{mol H}_2\text{O} + 3 \text{ x mol CaCl}_2}$$

mol_{CaCl2} → g_{CaCl2} and mol_{H2O} → g_{H2O} then g_{CaCl2}, g_{H2O} → percent by mass $CaCl_2$

$$\frac{110.99 \text{ g CaCl}_2}{1 \text{ mol CaCl}_2} \qquad \frac{18.01 \text{ g H}_2\text{O}}{1 \text{ mol H}_2\text{O}} \qquad \textit{mass percent} = \frac{\textit{mass solute}}{\textit{mass solution}} \times 100 \%$$

Solution: $P_{solution} = \chi_{solvent} P^{o}_{solvent}$ Rearrange to solve for χ_{H2O}.

$$\chi_{solvent} = \frac{P_{solution}}{P^{o}_{solvent}} = \frac{81.6 \text{ mm Hg}}{92.6 \text{ mm Hg}} = 0.881210 \quad \text{then assume 1 mol water}$$

$$\chi_{H2O} = \frac{\text{mol H}_2\text{O}}{\text{mol H}_2\text{O} + 3 \text{ x mol CaCl}_2} \quad \text{so} \quad 0.881210 = \frac{1 \text{ mol}}{1 \text{ mol} + 3 \text{ x mol CaCl}_2}$$

Rearrange to solve for mol_{CaCl2}.

$$0.881210\,(1 \text{ mol} + 3 \text{ x mol CaCl}_2) = 1 \text{ mol} \rightarrow \text{mol CaCl}_2 = \frac{1 \text{ mol} - 0.881210 \text{ mol}}{2.64363} = 0.0449345 \text{ mol CaCl}_2$$

$$\text{then} \quad 0.0449345 \text{ mol CaCl}_2 \text{ x} \frac{110.99 \text{ g CaCl}_2}{1 \text{ mol CaCl}_2} = 4.98727 \text{ g CaCl}_2 \quad \text{and} \quad 1 \text{ mol H}_2\text{O} = 18.01 \text{ g H}_2\text{O} \quad \text{then}$$

$$\textit{mass percent} = \frac{\textit{mass solute}}{\textit{mass solution}} \times 100\% = \frac{4.98727 \text{ g CaCl}_2}{4.98727 \text{ g} + 18.01 \text{ g}} \text{ x } 100 \% = 21.69 \text{ percent by mass CaCl}_2$$

Check: The units (%) are correct. The magnitude of the answer (22 %) seems reasonable since there is a significant drop of the pure vapor pressure and the formula mass of the salt is much larger than water's molar mass.

47. **Given:** 50.0 g of heptane (C_7H_{16}) and 50.0 g of octane (C_8H_{18}) at 25 °C; $P°_{C7H16} = 45.8$ torr; $P°_{C8H18} = 10.9$ torr

a) **Find:** P_{C7H16}, P_{C8H18}

Conceptual Plan:

$g_{C7H16} \rightarrow mol_{C7H16}$ and $g_{C8H18} \rightarrow mol_{C8H18}$ then $mol_{C7H16}, mol_{C8H18} \rightarrow \chi_{C7H16}, \chi_{C8H18}$

$$\frac{1\text{ mol }C_7H_{16}}{100.20\text{ g }C_7H_{16}} \qquad \frac{1\text{ mol }C_8H_{18}}{114.22\text{ g }C_8H_{18}} \qquad \chi_{C7H16} = \frac{\text{amount }C_7H_{16}\text{ (in moles)}}{\text{total amount (in moles)}} \qquad \chi_{C8H18} = 1 - \chi_{C7H16}$$

then $\chi_{C7H16}, P°_{C7H16} \rightarrow P_{C7H16}$ and $\chi_{C8H18}, P°_{C8H18} \rightarrow P_{C8H18}$

$$P_{C7H16} = \chi_{C7H16}\,P^o_{C7H16} \qquad P_{C8H18} = \chi_{C8H18}\,P^o_{C8H18}$$

Solution: $50.0\ \cancel{g\,C_7H_{16}} \times \frac{1\text{ mol}C_7H_{16}}{100.20\ \cancel{g\,C_7H_{16}}} = 0.49\underline{9}002\text{ mol}C_7H_{16}$ and

$50.0\ \cancel{g\,C_8H_{18}} \times \frac{1\text{ mol}C_8H_{18}}{114.22\ \cancel{g\,C_8H_{18}}} = 0.43\underline{7}752\text{ mol}C_8H_{18}$ then

$$\chi_{C7H16} = \frac{\text{amount }C_7H_{16}\text{ (in moles)}}{\text{total amount (in moles)}} = \frac{0.49\underline{9}002\ \cancel{mol}}{0.49\underline{9}002\ \cancel{mol} + 0.43\underline{7}752\ \cancel{mol}} = 0.53\underline{2}693 \text{ and}$$

$\chi_{C8H18} = 1 - \chi_{C7H16} = 1 - 0.53\underline{2}693 = 0.46\underline{7}307$ then

$P_{C7H16} = \chi_{C7H16}\,P^o_{C7H16} = 0.53\underline{2}693 \times 45.8\text{ torr} = 24.4\text{ torr}$ and

$P_{C8H18} = \chi_{C8H18}\,P^o_{C8H18} = 0.46\underline{7}307 \times 10.9\text{ torr} = 5.09\text{ torr}$

Check: The units (torr) are correct. The magnitude of the answer (24 and 5 torr) seems reasonable since it we expect a drop in half from the pure vapor pressures since we have roughly a 50:50 mole ratio of the two components.

b) **Find:** P_{Total}

Conceptual Plan: $P_{C7H16}, P_{C8H18} \rightarrow P_{Total}$

$$P_{Total} = P_{C7H16} + P_{C8H18}$$

Solution: $P_{Total} = P_{C7H16} + P_{C8H18} = 24.4\text{ torr} + 5.09\text{ torr} = 29.5\text{ torr}$

Check: The units (torr) are correct. The magnitude of the answer (30 torr) seems reasonable considering the two pressures.

c) **Find:** mass percent composition of the gas phase

Conceptual Plan: since $n \propto P$ and we are calculating a mass percent, which is a ratio of masses, we can simply convert 1 torr to 1 mole so

$P_{C7H16}, P_{C8H18} \rightarrow n_{C7H16}, n_{C8H18}$ then $mol_{C7H16} \rightarrow g_{C7H16}$ and $mol_{C8H18} \rightarrow g_{C8H18}$

$$\frac{100.20\text{ g }C_7H_{16}}{1\text{ mol }C_7H_{16}} \qquad \frac{114.22\text{ g }C_8H_{18}}{1\text{ mol }C_8H_{18}}$$

then $g_{C7H16}, g_{C8H18} \rightarrow$ mass percents

$$mass\ percent = \frac{mass\ solute}{mass\ solution} \times 100\ \%$$

Solution: so $n_{C7H16} = 24.4$ mol and $n_{C8H18} = 5.09$ mol then

$24.4\ \cancel{mol\,C_7H_{16}} \times \frac{100.20\text{ g }C_7H_{16}}{1\ \cancel{mol\,C_7H_{16}}} = 24\underline{4}4.88\text{ g }C_7H_{16}$ and

$5.09\ \cancel{mol\,C_8H_{18}} \times \frac{114.22\text{ g }C_8H_{18}}{1\ \cancel{mol\,C_8H_{18}}} = 58\underline{1}.380\text{ g }C_8H_{18}$ then

$$mass\ percent = \frac{mass\ solute}{mass\ solution} \times 100\,\% = \frac{24\underline{4}4.88\ \text{g}\ \cancel{C_7H_{16}}}{24\underline{4}4.88\ \text{g}\ \cancel{C_7H_{16}} + 58\underline{1}.380\ \text{g}\ \cancel{C_8H_{18}}} \times 100\,\% =$$

= 80.8 percent by mass C_7H_{16}

then 100 % – 80.8 % = 19.2 percent by mass C_8H_{18}
Check: The units (%) are correct. The magnitudes of the answers (81 % and 19 %) seem reasonable considering the two pressures.

d) The two mass percents are different because the vapor is richer in the more volatile component (the lighter molecule).

48. **Given:** pentane(C_5H_{12}) and hexane(C_6H_{14}) P_{Total} = 258 torr; $P°_{C5H12}$ = 425 torr; $P°_{C6H14}$ = 151 torr at 25°C
Find: χ_{C5H12}, χ_{C6H14}
Conceptual Plan: $P_{Total} = P_{C5H12} + P_{C6H14}$ **where** $P_{C5H12} = \chi_{C5H12}\, P^o_{C5H12}$ **and** $P_{C6H14} = \chi_{C6H14}\, P^o_{C6H14}$ **but** $\chi_{C6H14} = 1 - \chi_{C5H12}$ **so** $P_{Total} = \chi_{C5H12}\, P^o_{C5H12} + (1-\chi_{C5H12})P^o_{C6H14}$ **substitute in values and solve for** χ_{C5H12}, χ_{C6H14}

Solution: $P_{Total} = \chi_{C5H12}\, P^o_{C5H12} + (1-\chi_{C5H12})P^o_{C6H14}$ so $258 \text{ torr} = \chi_{C5H12}\, 425 \text{ torr} + (1-\chi_{C5H12})\, 151 \text{ torr} \rightarrow$

$$258 - 151 = \chi_{C5H12}(425 - 151) \rightarrow \chi_{C5H12} = \frac{107}{274} = 0.391 \text{ and } \chi_{C6H14} = 1 - \chi_{C5H12} = 1 - 0.391 = 0.609$$

Check: The units (none) are correct. The magnitudes of the answers (0.4 and 0.6) seem reasonable since the total vapor pressure is closer to the vapor pressure of hexane, so we expect there to be more hexane in the liquid.

49. **Given:** 55.8 g of glucose ($C_6H_{12}O_6$) in 455 g water **Find:** T_f and T_b
Other: K_f = 1.86 °C/*m*; K_b = 0.512 °C/*m*;
Conceptual Plan: $g_{H2O} \rightarrow kg_{H2O}$ **and** $g_{C6H12O6} \rightarrow mol_{C6H12O6}$ **then** $mol_{C6H12O6}, kg_{H2O} \rightarrow m$

$$\frac{1\text{kg}}{1000\text{ g}} \qquad \frac{1 \text{ mol } C_6H_{12}O_6}{180.16 \text{ g } C_6H_{12}O_6} \qquad m = \frac{\text{amount solute (moles)}}{\text{mass solvent (kg)}}$$

$m, K_f \rightarrow \Delta T_f \rightarrow T_f$ **and** $m, K_b \rightarrow \Delta T_b \rightarrow T_b$

$$\Delta T_f = K_f m \quad T_f = T_f^\circ - \Delta T_f \qquad \Delta T_b = K_b m \quad \Delta T_b = T_b - T_b^\circ$$

Solution: $455 \text{ g} \times \frac{1\text{kg}}{1000 \text{ g}} = 0.455 \text{ kg}$ and $55.8 \text{ g } C_6H_{12}O_6 \times \frac{1 \text{ mol } C_6H_{12}O_6}{180.16 \text{ g } C_6H_{12}O_6} = 0.30\underline{9}725 \text{ mol } C_6H_{12}O_6$

then $m = \frac{\text{amount solute (moles)}}{\text{mass solvent (kg)}} = \frac{0.30\underline{9}725 \text{ mol } C_6H_{12}O_6}{0.455 \text{ kg}} = 0.68\underline{0}714\ m$ then

$\Delta T_f = K_f m = 1.86 \frac{°C}{m} \times 0.68\underline{0}714\ m = 1.27\ °C$ then $T_f = T_f^\circ - \Delta T_f = 0.00\ °C - 1.27\ °C = -1.27\ °C$ and

$\Delta T_b = K_b m = 0.512 \frac{°C}{m} \times 0.68\underline{0}714\ m = 0.349\ °C$ and $\Delta T_b = T_b - T_b^\circ$ so

$T_b = T_b^\circ + \Delta T_b = 100.000°C + 0.349\ °C = 100.349\ °C$

Check: The units (**°C**) are correct. The magnitudes of the answers seem reasonable since the molality is ~ 2/3. The shift in boiling point is less than the shift in freezing point because the constant is smaller for freezing.

50. **Given:** 21.2 g of ethylene glycol ($C_2H_6O_2$) in 85.4 g water **Find:** T_f and T_b
Other: K_f = 1.86 °C/*m*; K_b = 0.512 °C/*m*;
Conceptual Plan: $g_{H2O} \rightarrow kg_{H2O}$ **and** $g_{C2H6O2} \rightarrow mol_{C2H6O2}$ **then** $mol_{C2H6O2}, kg_{H2O} \rightarrow m$

$$\frac{1\text{kg}}{1000\text{ g}} \qquad \frac{1 \text{ mol } C_2H_6O_2}{62.07 \text{ g } C_2H_6O_2} \qquad m = \frac{\text{amount solute (moles)}}{\text{mass solvent (kg)}}$$

$m, K_f \rightarrow \Delta T_f \rightarrow T_f$ **and** $m, K_b \rightarrow \Delta T_b \rightarrow T_b$

$$\Delta T_f = K_f m \quad T_f = T_f^\circ - \Delta T_f \quad \Delta T_b = K_b m \quad \Delta T_b = T_b - T_b^\circ$$

Solution: $84.5 \text{ g} \times \frac{1\text{kg}}{1000 \text{ g}} = 0.0845 \text{ kg}$ and $21.2 \text{ g } C_2H_6O_2 \times \frac{1 \text{ mol } C_2H_6O_2}{62.07 \text{ g } C_2H_6O_2} = 0.34\underline{1}561 \text{ mol } C_2H_6O_2$

then $m = \frac{\text{amount solute (moles)}}{\text{mass solvent (kg)}} = \frac{0.34\underline{1}561 \text{ mol } C_2H_6O_2}{0.0845 \text{ kg}} = 4.0\underline{4}214\ m$ then

$\Delta T_f = K_f m = 1.86 \frac{°C}{m} \times 4.04214\ m = 7.52\ °C$ then $T_f = T_f^o - \Delta T_f = 0.00\ °C - 7.52\ °C = -7.52\ °C$ and

$\Delta T_b = K_b m = 0.512 \frac{°C}{m} \times 4.04214\ m = 2.07\ °C$ and $\Delta T_b = T_b - T_b^o$ so

$T_b = T_b^o + \Delta T_b = 100.00°C + 2.07\ °C = 102.07\ °C$

Check: The units (°C) are correct. The magnitudes of the answers seem reasonable since the molality is ~ 4. The shift in boiling point is less than the shift in freezing point because the constant is smaller for freezing.

51. **Given:** 17.5 g of unknown nonelectrolyte in 100.0 g water, $T_f = -1.8\ °C$ **Find:** $\mathfrak{M}$
Other: $K_f = 1.86\ °C/m$
Conceptual Plan: $g_{H2O} \rightarrow kg_{H2O}$ and $T_f \rightarrow \Delta T_f$ then $\Delta T_f, K_f \rightarrow m$ then $m, kg_{H2O} \rightarrow mol_{Unk}$

$\frac{1\ kg}{1000\ g}$ $T_f = T_f^o - \Delta T_f$ $\Delta T_f = K_f m$ $m = \frac{\text{amount solute (moles)}}{\text{mass solvent (kg)}}$

then $g_{Unk}, mol_{Unk} \rightarrow \mathfrak{M}$

$\mathfrak{M} = \frac{g_{Unk}}{mol_{Unk}}$

Solution: $100.0\ g \times \frac{1\ kg}{1000\ g} = 0.1000\ kg$ and $T_f = T_f^o - \Delta T_f$ so

$\Delta T_f = T_f^o - T_f = 0.00\ °C - (-1.8\ °C) = +1.8\ °C$ $\Delta T_f = K_f m$ Rearrange to solve for m.

$m = \frac{\Delta T_f}{K_f} = \frac{1.8\ °C}{1.86 \frac{°C}{m}} = 0.96774\ m$ then $m = \frac{\text{amount solute (moles)}}{\text{mass solvent (kg)}}$ so

$mol_{Unk} = m_{Unk} \times kg_{H2O} = 0.96774 \frac{\text{mol Unk}}{kg} \times 0.1000\ kg = 0.096774\ \text{mol Unk}$ then

$\mathfrak{M} = \frac{g_{Unk}}{mol_{Unk}} = \frac{17.5\ g}{0.096774\ mol} = 180 \frac{g}{mol} = 1.8 \times 10^2 \frac{g}{mol}$

Check: The units (g/mol) are correct. The magnitude of the answer (180 g/mol) seems reasonable since the molality is ~ 0.1 and we have ~18 g. It is a reasonable molecular weight for a solid or liquid.

52. **Given:** 35.9 g of unknown nonelectrolyte in 150.0 g water, $T_f = -1.3\ °C$ **Find:** $\mathfrak{M}$
Other: $K_f = 1.86\ °C/m$;
Conceptual Plan:
$g_{H2O} \rightarrow kg_{H2O}$ and $T_f \rightarrow \Delta T_f$ then $\Delta T_f, K_f \rightarrow m$ then $m, kg_{H2O} \rightarrow mol_{Unk}$

$\frac{1\ kg}{1000\ g}$ $T_f = T_f^o - \Delta T_f$ $\Delta T_f = K_f m$ $m = \frac{\text{amount solute (moles)}}{\text{mass solvent (kg)}}$

then $g_{Unk}, mol_{Unk} \rightarrow \mathfrak{M}$

$\mathfrak{M} = \frac{g_{Unk}}{mol_{Unk}}$

Solution: $150.0\ g \times \frac{1\ kg}{1000\ g} = 0.1500\ kg$ and $T_f = T_f^o - \Delta T_f$ so

$\Delta T_f = T_f^o - T_f = 0.00\ °C - -1.3\ °C = +1.3\ °C$ $\Delta T_f = K_f m$ Rearrange to solve for m.

$m = \frac{\Delta T_f}{K_f} = \frac{1.3\ °C}{1.86 \frac{°C}{m}} = 0.69892\ m$ then $m = \frac{\text{amount solute (moles)}}{\text{mass solvent (kg)}}$ so

$mol_{Unk} = m_{Unk} \times kg_{H2O} = 0.69892 \frac{\text{mol Unk}}{kg} \times 0.1500\ kg = 0.104839\ \text{mol Unk}$ then

$$\mathfrak{M} = \frac{g_{Unk}}{mol_{Unk}} = \frac{35.9\text{ g}}{0.104839\text{ mol}} = 340\ \frac{\text{g}}{\text{mol}}$$

Check: The units (g/mol) are correct. The magnitude of the answer (340 g/mol) seems reasonable since the molality is ~ 0.7 and we have ~36 g. It is a reasonable molecular weight for a solid or liquid.

53. **Given:** 24.6 g of glycerin ($C_3H_8O_3$) in 250.0 mL of solution at 298 K **Find:** Π

Conceptual Plan: $\textbf{mL}_{\textbf{soln}} \rightarrow \textbf{L}_{\textbf{soln}}$ **and** $\textbf{g}_{C3H8O3} \rightarrow \textbf{mol}_{C3H8O3}$ **then** $\textbf{mol}_{C3H8O3}, \textbf{L}_{\textbf{soln}} \rightarrow \textbf{M}$ **then**

$$\frac{1\text{ L}}{1000\text{ mL}} \qquad \frac{1\text{ mol } C_3H_8O_3}{92.09\text{ g } C_3H_8O_3} \qquad M = \frac{\text{amount solute (moles)}}{\text{volume solution (L)}}$$

M, *T* → Π

$\Pi = MRT$

Solution:

$$250.0\text{ mL} \times \frac{1\text{ L}}{1000\text{ mL}} = 0.2500\text{ L} \text{ and } 24.6\text{ g } C_3H_8O_3 \times \frac{1\text{ mol } C_3H_8O_3}{92.09\text{ g } C_3H_8O_3} = 0.267130\text{ mol } C_3H_8O_3 \text{ then}$$

$$M = \frac{\text{amount solute (moles)}}{\text{volume solution (L)}} = \frac{0.267130\text{ mol } C_3H_8O_3}{0.2500\text{ L}} = 1.06852\text{ M then}$$

$$\Pi = MRT = 1.06852\ \frac{\text{mol}}{\text{L}} \times 0.08206\ \frac{\text{L}\cdot\text{atm}}{\text{K}\cdot\text{mol}} \times 298\text{ K} = 26.1\text{ atm}$$

Check: The units (atm) are correct. The magnitude of the answer (26 atm) seems reasonable since the molarity is ~ 1.

54. **Given:** sucrose ($C_{12}H_{22}O_{11}$) in 5.00 x 10^2 g water; Π = 8.55 atm at 298 K **Find:** *m* ($C_{12}H_{22}O_{11}$)

Other: d = 1.0 g/mL

Conceptual Plan: **Π, *T* → M** **then** $\textbf{g}_{H2O}, d, \mathfrak{M} \rightarrow \textbf{g}_{C12H22O11}$

$$\Pi = MRT \qquad M = \frac{\text{amount solute (moles)}}{\text{volume solution (L)}} \text{ with } \frac{1\text{ L}}{1000\text{ mL}}, \frac{342.30\text{ g } C_{12}H_{22}O_{11}}{1\text{ mol } C_{12}H_{22}O_{11}} \text{ and } \frac{1.0\text{ mL}}{1.0\text{ g}}$$

Solution: $\Pi = MRT$ for M. $$M = \frac{\Pi}{RT} = \frac{8.55\text{ atm}}{0.08206\ \frac{\text{L}\cdot\text{atm}}{\text{K}\cdot\text{mol}} \times 298\text{ K}} = 0.349638\ \frac{\text{mol}}{\text{L}}$$

Substitute quantities into the definition of M.

$$M = \frac{\text{amount solute (moles)}}{\text{volume solution (L)}} = \frac{(x\text{ g } C_{12}H_{22}O_{11})\left(\frac{1\text{ mol } C_{12}H_{22}O_{11}}{342.30\text{ g } C_{12}H_{22}O_{11}}\right)}{(5.00 \times 10^2\text{ g } H_2O + x\text{ g } C_{12}H_{22}O_{11}) \times \frac{1.0\text{ mL}}{1.0\text{ g}} \times \frac{1\text{ L}}{1000\text{ mL}}} = 0.349638\ \frac{\text{mol}}{\text{L}}$$

Rearrange to solve for $x\text{ g } C_{12}H_{22}O_{11}$. $x\text{ g } C_{12}H_{22}O_{11} = 0.119681 \times (5.00 \times 10^2\text{ g } H_2O + x\text{ g } C_{12}H_{22}O_{11}) \rightarrow$

$$0.880312 \times (x\text{ g } C_{12}H_{22}O_{11}) = 59.8405 \rightarrow x\text{ g } C_{12}H_{22}O_{11} = \frac{59.8405}{0.880312} = 68.0\text{ g } C_{12}H_{22}O_{11}$$

Check: The units (g) are correct. The magnitude of the answer (68 g) seems reasonable since the molarity is ~ 1/3 and we have 0.5 L of water.

55. **Given:** 27.55 mg unknown protein in 25.0 mL solution; Π = 3.22 torr at 25 °C **Find:** $\mathfrak{M}_{\text{unknown protein}}$

Conceptual Plan:

°C → K **and** **torr → atm** **then** **Π, *T* → M** **then** $\textbf{mL}_{\textbf{soln}} \rightarrow \textbf{L}_{\textbf{soln}}$ **then**

$$K = °C + 273.15 \qquad \frac{1\text{ atm}}{760\text{ torr}} \qquad \Pi = MRT \qquad \frac{1\text{ L}}{1000\text{ mL}}$$

L_{soln}, M → $mol_{unknown\ protein}$ and mg → g then $g_{unknown\ protein}$, $mol_{unknown\ protein}$ → $\mathfrak{M}_{unknown\ protein}$

$$M = \frac{\text{amount solute (moles)}}{\text{volume solution (L)}} \qquad \frac{1\text{g}}{1000\text{ mg}} \qquad \mathfrak{M} = \frac{g_{unknown\ protein}}{mol_{unknown\ protein}}$$

Solution: 25 °C + 273.15 = 298 K and $3.22\ \text{torr} \times \frac{1\ \text{atm}}{760\ \text{torr}} = 0.00423684\ \text{atm}$ $\Pi = MRT$ for M.

$$M = \frac{\Pi}{RT} = \frac{0.00423684\ \text{atm}}{0.08206 \frac{\text{L}\cdot\text{atm}}{\text{K}\cdot\text{mol}} \times 298\ \text{K}} = 1.73258 \times 10^{-4}\ \frac{\text{mol}}{\text{L}} \text{ then } 25.0\ \text{mL} \times \frac{1\text{L}}{1000\ \text{mL}} = 0.0250\ \text{L}$$

then $M = \frac{\text{amount solute (moles)}}{\text{volume solution (L)}}$ Rearrange to solve for $mol_{unknown\ protein}$.

$$mol_{unknown\ protein} = M \times L = 1.73258 \times 10^{-4} \frac{\text{mol}}{\text{L}} \times 0.0250\ \text{L} = 4.33146 \times 10^{-6}\ \text{mol} \text{ and}$$

$$27.55\ \text{mg} \times \frac{1\text{g}}{1000\ \text{mg}} = 0.02755\ \text{g} \text{ then}$$

$$\mathfrak{M} = \frac{g_{unknown\ protein}}{mol_{unknown\ protein}} = \frac{0.02755\ \text{g}}{4.33146 \times 10^{-6}\ \text{mol}} = 6.36 \times 10^{3}\ \frac{\text{g}}{\text{mol}}$$

Check: The units (g/mol) are correct. The magnitude of the answer (6400 g/mol) seems reasonable for a large biological molecule. A small amount of material is put into 0.025 L, so the concentration is very small and the molecular weight is large.

56. **Given:** 18.75 mg of hemoglobin in 15.0 mL of solution at 25 °C, $\mathfrak{M}_{hemoglobin} = 6.5 \times 10^4$ g/mol **Find:** Π
Conceptual Plan: mL_{soln} → L_{soln} **and** mg_H → g_H → mol_H **then** mol_H, L_{soln} → **M then**

$$\frac{1\text{L}}{1000\text{ mL}} \qquad \frac{1\text{g}}{1000\text{ mg}} \qquad \frac{1\text{ mol H}}{6.5 \times 10^4\text{ g H}} \qquad M = \frac{\text{amount solute (moles)}}{\text{volume solution (L)}}$$

M, *T* → Π
$\Pi = MRT$

Solution:

$$15.0\ \text{mL} \times \frac{1\text{L}}{1000\ \text{mL}} = 0.0150\ \text{L} \text{ and } 18.75\ \text{mg H} \times \frac{1\text{g}}{1000\ \text{mg}} \times \frac{1\ \text{mol H}}{6.5 \times 10^4\ \text{g H}} = 2.88465 \times 10^{-7}\ \text{mol H}$$

then $M = \frac{\text{amount solute (moles)}}{\text{volume solution (L)}} = \frac{2.88465 \times 10^{-7}\ \text{mol H}}{0.0150\ \text{L}} = 1.9231 \times 10^{-5}\ \text{M}$ then

$$\Pi = MRT = 1.9231 \times 10^{-5} \frac{\text{mol}}{\text{L}} \times 0.08206 \frac{\text{L}\cdot\text{atm}}{\text{K}\cdot\text{mol}} \times 298\ \text{K} = 4.7 \times 10^{-4}\ \text{atm} = 0.36\ \text{torr}$$

Check: The units (atm) are correct. The magnitude of the answer (10^{-4} atm) seems reasonable since the molarity is so small.

57. a) **Given:** 0.100 *m* of K_2S, completely dissociated **Find:** T_f, T_b
Other: K_f = 1.86 °C/*m*; K_b = 0.512 °C/*m*;
Conceptual Plan: ***m*, *i*, K_f** → ΔT_f **then** ΔT_f → T_f **and** ***m*, *i*, K_b** → ΔT_b **then** ΔT_b → T_b

$$\Delta T_f = K_f i m \quad i = 3 \qquad T_f = T_f^{\circ} - \Delta T_f \qquad \Delta T_b = K_b i m \quad i = 3 \qquad T_b = T_b^{\circ} + \Delta T_b$$

Solution: $\Delta T_f = K_f\, i\, m = 1.86 \frac{°\text{C}}{m} \times 3 \times 0.100\ m = 0.558\ °\text{C}$ then

$T_f = T_f^o - \Delta T_f = 0.000°\text{C} - 0.558\ °\text{C} = -0.558\ °\text{C}$ and

$\Delta T_b = K_b\, i\, m = 0.512 \frac{°\text{C}}{m} \times 3 \times 0.100\ m = 0.154\ °\text{C}$ then

$T_b = T_b^{\circ} - \Delta T_b = 100.000°\text{C} + 0.154\ °\text{C} = 100.154\ °\text{C}$.

Check: The units (°C) are correct. The magnitude of the answer (– 0.6 °C and 100.2 °C) seems reasonable since the molality of the particles is 0.3. The shift in boiling point is less than the shift in freezing point because the constant is smaller for freezing.

b) **Given:** 21.5 g $CuCl_2$ in 4.50 x 10^2 g water, completely dissociated **Find:** T_f, T_b
Other: K_f = 1.86 °C/*m*; K_b = 0.512 °C/*m*;
Conceptual Plan: $g_{H2O} \rightarrow kg_{H2O}$ **and** $g_{CuCl2} \rightarrow mol_{CuCl2}$ **then** $mol_{CuCl2}, kg_{H2O} \rightarrow m$

$$\frac{1\text{kg}}{1000\text{ g}} \qquad \frac{1\text{ mol CuCl}_2}{134.46\text{ g CuCl}_2} \qquad m = \frac{\text{amount solute (moles)}}{\text{mass solvent (kg)}}$$

$m, i, K_f \rightarrow \Delta T_f \rightarrow T_f$ **and** $m, i, K_b \rightarrow \Delta T_b \rightarrow T_b$

$\Delta T_f = K_f\, i\, m \quad i = 3 \quad T_f = T_f^{\circ} - \Delta T_f \qquad \Delta T_b = K_b\, i\, m \quad i = 3 \quad T_b = T_b^{\circ} + \Delta T_b$

Solution:

$$4.5 \times 10^2\ \text{g} \times \frac{1\text{kg}}{1000\ \text{g}} = 0.450\text{ kg} \text{ and } 21.5\ \text{g CuCl}_2 \times \frac{1\text{ mol CuCl}_2}{134.46\ \text{g CuCl}_2} = 0.15\underline{9}904\text{ mol CuCl}_2 \text{ then}$$

$$m = \frac{\text{amount solute (moles)}}{\text{mass solvent (kg)}} = \frac{0.15\underline{9}904\text{ mol CuCl}_2}{0.450\text{ kg}} = 0.35\underline{5}341\ m \text{ then}$$

$$\Delta T_f = K_f\, i\, m = 1.86 \frac{°\text{C}}{m} \times 3 \times 0.35\underline{5}341\ m = 1.98\ °\text{C} \text{ then}$$

$$T_f = T_f^{\circ} - \Delta T_f = 0.000°\text{C} - 1.98\ °\text{C} = -1.98\ °\text{C} \text{ and}$$

$$\Delta T_b = K_b\, i\, m = 0.512 \frac{°\text{C}}{m} \times 3 \times 0.35\underline{5}341\ m = 0.546\ °\text{C} \text{ then}$$

$T_b = T_b^{\circ} - \Delta T_b = 100.000°\text{C} + 0.546\ °\text{C} = 100.546\ °\text{C}$.

Check: The units (°C) are correct. The magnitude of the answer (– 2 °C and 100.5 °C) seems reasonable since the molality of the particles is ~ 1. The shift in boiling point is less than the shift in freezing point because the constant is smaller for freezing.

c) **Given:** 5.5 % by mass $NaNO_3$, completely dissociated **Find:** T_f, T_b
Other: K_f = 1.86 °C/*m*; K_b = 0.512 °C/*m*;
Conceptual Plan:
percent by mass $\rightarrow g_{NaNO3}, g_{H2O}$ **then** $g_{H2O} \rightarrow kg_{H2O}$ **and** $g_{NaNO3} \rightarrow mol_{NaNO3}$ **then**

$$mass\ percent = \frac{mass\ solute}{mass\ solution} \times 100\ \% \qquad \frac{1\text{kg}}{1000\text{ g}} \qquad \frac{1\text{ mol NaNO}_3}{84.99\text{ g NaNO}_3}$$

$mol_{NaNO3}, kg_{H2O} \rightarrow m$ **then** $m, i, K_f \rightarrow \Delta T_f \rightarrow T_f$ **and** $m, i, K_b \rightarrow \Delta T_b \rightarrow T_b$

$$m = \frac{\text{amount solute (moles)}}{\text{mass solvent (kg)}} \qquad \Delta T_f = K_f\, i\, m \quad i = 2 \quad T_f = T_f^{\circ} - \Delta T_f \qquad \Delta T_b = K_b\, i\, m \quad i = 2 \quad T_b = T_b^{\circ} + \Delta T_b$$

Solution: $mass\ percent = \frac{mass\ solute}{mass\ solution} \times 100\ \%$ so 5.5 % by mass $NaNO_3$ means 5.5 g $NaNO_3$ and 100.0 g – 5.5 g = 94.5 g water. Then $94.5\ \text{g} \times \frac{1\text{kg}}{1000\ \text{g}} = 0.0945\text{ kg}$ and

$$5.5\ \text{g NaNO}_3 \times \frac{1\text{ mol NaNO}_3}{84.99\ \text{g NaNO}_3} = 0.06\underline{4}713\text{ mol NaNO}_3 \text{ then}$$

$$m = \frac{\text{amount solute (moles)}}{\text{mass solvent (kg)}} = \frac{0.06\underline{4}713\text{ mol NaNO}_3}{0.0945\text{ kg}} = 0.6\underline{8}480\ m \text{ then}$$

$$\Delta T_f = K_f\, i\, m = 1.86 \frac{°\text{C}}{m} \times 2 \times 0.6\underline{8}480\ m = 2.5\ °\text{C} \text{ then}$$

$T_f = T_f^o - \Delta T_f = 0.000\,°C - 2.3\,°C = -2.5\,°C$ and

$\Delta T_b = K_b\, i\, m = 0.512 \frac{°C}{m} \times 2 \times 0.68480\, m = 0.70\,°C$ then

$T_b = T_b^o - \Delta T_b = 100.000\,°C + 0.64\,°C = 100.70\,°C$.

Check: The units (°C) are correct. The magnitude of the answer (– 2.5 °C and 100.7 °C) seems reasonable since the molality of the particles is ~ 1. The shift in boiling point is less than the shift in freezing
point because the constant is smaller for freezing.

58. a) **Given:** 10.5 g $FeCl_3$ in 1.50 x 10^2 g water, completely dissociated **Find:** T_f, T_b
Other: $K_f = 1.86$ °C/*m*; $K_b = 0.512$ °C/*m*;
Conceptual Plan: $g_{H2O} \rightarrow kg_{H2O}$ and $g_{FeCl3} \rightarrow mol_{FeCl3}$ then $mol_{FeCl3}, kg_{H2O} \rightarrow m$

$\frac{1\,kg}{1000\,g}$ $\frac{1\,mol\,FeCl_3}{162.21\,g\,FeCl_3}$ $m = \frac{\text{amount solute (moles)}}{\text{mass solvent (kg)}}$

$m, i, K_f \rightarrow \Delta T_f \rightarrow T_f$ and $m, i, K_b \rightarrow \Delta T_b \rightarrow T_b$

$\Delta T_f = K_f\, i\, m$ $i = 4$ $T_f = T_f^o - \Delta T_f$ $\Delta T_b = K_b\, i\, m$ $i = 4$ $T_b = T_b^o + \Delta T_b$

Solution: $4.5 \times 10^2\, g \times \frac{1\,kg}{1000\,g} = 0.450\,kg$ and

$10.5\, g\,FeCl_3 \times \frac{1\,mol\,FeCl_3}{162.21\,g\,FeCl_3} = 0.0647325\,mol\,FeCl_3$ then

$m = \frac{\text{amount solute (moles)}}{\text{mass solvent (kg)}} = \frac{0.0647325\,mol\,FeCl_3}{0.150\,kg} = 0.431550\, m$ then

$\Delta T_f = K_f\, i\, m = 1.86 \frac{°C}{m} \times 4 \times 0.431550\, m = 3.21\,°C$ then

$T_f = T_f^o - \Delta T_f = 0.000\,°C - 3.21\,°C = -3.21\,°C$ and

$\Delta T_b = K_b\, i\, m = 0.512 \frac{°C}{m} \times 4 \times 0.431550\, m = 0.884\,°C$ then

$T_b = T_b^o - \Delta T_b = 100.000\,°C + 0.884\,°C = 100.884\,°C$.

Check: The units (°C) are correct. The magnitude of the answer (– 3 °C and 100.9 °C) seems reasonable since the molality of the particles is ~ 2. The shift in boiling point is less than the shift in freezing point because the constant is smaller for freezing.

b) **Given:** 3.5 % by mass KCl, completely dissociated **Find:** T_f, T_b
Other: $K_f = 1.86$ °C/*m*; $K_b = 0.512$ °C/*m*;
Conceptual Plan:
percent by mass $\rightarrow g_{KCl}, g_{H2O}$ **then** $g_{H2O} \rightarrow kg_{H2O}$ **and** $g_{KCl} \rightarrow mol_{KCl}$ **then**

$\text{mass percent} = \frac{\text{mass solute}}{\text{mass solution}} \times 100\,\%$ $\frac{1\,kg}{1000\,g}$ $\frac{1\,mol\,KCl}{74.55\,g\,KCl}$

$mol_{NaNO3}, kg_{H2O} \rightarrow m$ then $m, i, K_f \rightarrow \Delta T_f \rightarrow T_f$ and $m, i, K_b \rightarrow \Delta T_b \rightarrow T_b$

$m = \frac{\text{amount solute (moles)}}{\text{mass solvent (kg)}}$ $\Delta T_f = K_f\, i\, m$ $i = 2$ $T_f = T_f^o - \Delta T_f$ $\Delta T_b = K_b\, i\, m$ $i = 2$ $T_b = T_b^o + \Delta T_b$

Solution: $\text{mass percent} = \frac{\text{mass solute}}{\text{mass solution}} \times 100\,\%$ so 3.5 % by mass KCl means

3.5 g KCl and 100.0 g – 3.5 g = 96.5 g water. Then $96.5\, g \times \frac{1\,kg}{1000\,g} = 0.0965\,kg$ and

$$3.5\ \cancel{g\,KCl} \times \frac{1\ \text{mol KCl}}{74.55\ \cancel{g\,KCl}} = 0.04\underline{6}948\ \text{mol KCl}$$

then $$m = \frac{\text{amount solute (moles)}}{\text{mass solvent (kg)}} = \frac{0.04\underline{6}948\ \text{mol KCl}}{0.0965\ \text{kg}} = 0.4\underline{8}651\ m$$ then

$\Delta T_f = K_f\, i\, m = 1.86 \frac{°C}{\cancel{m}} \times 2 \times 0.4\underline{8}651\ \cancel{m} = 1.8\ °C$ then $T_f = T_f^o - \Delta T_f = 0.000°C - 1.8\ °C = -1.8\ °C$

and $\Delta T_b = K_b\, i\, m = 0.512 \frac{°C}{\cancel{m}} \times 2 \times 0.4\underline{8}651\ \cancel{m} = 0.50\ °C$ then

$T_b = T_b^o - \Delta T_b = 100.000°C + 0.50\ °C = 100.50\ °C$.

Check: The units (°C) are correct. The magnitude of the answer (– 2 °C and 100.5 °C) seems reasonable since the molality of the particles is ~ 1. The shift in boiling point is less than the shift in freezing point because the constant is smaller for freezing.

c) **Given:** 0.150 m of MgF_2, completely dissociated **Find:** T_f, T_b
Other: $K_f = 1.86$ °C/m; $K_b = 0.512$ °C/m;
Conceptual Plan:
$m, i, K_f \rightarrow \Delta T_f$ **then** $\Delta T_f \rightarrow T_f$ **and** $m, i, K_b \rightarrow \Delta T_b$ **then** $\Delta T_b \rightarrow T_b$
$\Delta T_f = K_f\, i\, m\ \ i = 3$ $\quad T_f = T_f^o - \Delta T_f$ $\quad \Delta T_b = K_b\, i\, m\ \ i = 3$ $\quad T_b = T_b^o + \Delta T_b$

Solution: $\Delta T_f = K_f\, i\, m = 1.86 \frac{°C}{\cancel{m}} \times 3 \times 0.150\ \cancel{m} = 0.837\ °C$ then

$T_f = T_f^o - \Delta T_f = 0.000°C - 0.837\ °C = -0.837\ °C$ and

$\Delta T_b = K_b\, i\, m = 0.512 \frac{°C}{\cancel{m}} \times 3 \times 0.150\ \cancel{m} = 0.230\ °C$ then

$T_b = T_b^o - \Delta T_b = 100.000°C + 0.230\ °C = 100.230\ °C$.

Check: The units (°C) are correct. The magnitude of the answer (– 0.8 °C and 100.2 °C) seems reasonable since the molality of the particles is 0.5. The shift in boiling point is less than the shift in freezing point because the constant is smaller for freezing.

59. a) **Given:** 0.100 m of $FeCl_3$ **Find:** T_f **Other:** $K_f = 1.86$ °C/m; $i_{measured} = 3.4$
Conceptual Plan: $m, i, K_f \rightarrow \Delta T_f$ **then** $\Delta T_f \rightarrow T_f$
$\Delta T_f = K_f\, i\, m$ $\quad T_f = T_f^o - \Delta T_f$

Solution: $\Delta T_f = K_f\, i\, m = 1.86 \frac{°C}{\cancel{m}} \times 3.4 \times 0.100\ \cancel{m} = 0.632\ °C$ then

$T_f = T_f^o - \Delta T_f = 0.000°C - 0.632\ °C = -0.632\ °C$.

Check: The units (°C) are correct. The magnitude of the answer (– 0.6 °C) seems reasonable since the molality of the particles is 0.3.

b) **Given:** 0.085 M of K_2SO_4 at 298 K **Find:** Π **Other:** $i_{measured} = 2.6$
Conceptual Plan: **M,** i**,** $T \rightarrow \Pi$
$\Pi = i\,M\,R\,T$

Solution: $\Pi = i\,M\,R\,T = 2.6 \times 0.085 \frac{\cancel{mol}}{\cancel{L}} \times 0.08206 \frac{\cancel{L} \cdot atm}{\cancel{K} \cdot \cancel{mol}} \times 298\ \cancel{K} = 5.4\ atm$

Check: The units (atm) are correct. The magnitude of the answer (5 atm) seems reasonable since the molarity is ~ 0.2.

c) **Given:** 1.22 % by mass $MgCl_2$ **Find:** T_b **Other:** $K_b = 0.512$ °C/m; $i_{measured} = 2.7$

Conceptual Plan:

percent by mass → g_{MgCl2}, g_{H2O} then g_{H2O} → kg_{H2O} and g_{MgCl2} → mol_{MgCl2} then

$$mass\ percent = \frac{mass\ solute}{mass\ solution} \times 100\ \% \qquad \frac{1\,kg}{1000\ g} \qquad \frac{1\ mol\,MgCl_2}{95.22\ g\ MgCl_2}$$

mol_{MgCl2}, kg_{H2O} → m then m, i, K_b → ΔT_b → T_b

$$m = \frac{\text{amount solute (moles)}}{\text{mass solvent (kg)}} \qquad \Delta T_b = K_b\, i\, m \quad T_b = T_b^o + \Delta T_b$$

Solution: $mass\ percent = \dfrac{mass\ solute}{mass\ solution} \times 100\,\%$ so 1.22 % by mass $MgCl_2$ means 1.22 g $MgCl_2$ and

100.00 g – 1.22 g = 98.78 g water. Then $98.78\ \cancel{g} \times \dfrac{1\,kg}{1000\ \cancel{g}} = 0.09878$ kg and

$1.22\ \cancel{g\,MgCl_2} \times \dfrac{1\ mol\,MgCl_2}{95.22\ \cancel{g\ MgCl_2}} = 0.0128124\ mol\ MgCl_2$ then

$m = \dfrac{\text{amount solute (moles)}}{\text{mass solvent (kg)}} = \dfrac{0.0128124\ mol\ MgCl_2}{0.09878\ kg} = 0.129706\ m$ then

$\Delta T_b = K_b\, i\, m = 0.512 \dfrac{°C}{\cancel{m}} \times 2.7 \times 0.129706\ \cancel{m} = 0.18\ °C$ then

$T_b = T_b^o - \Delta T_b = 100.000°C + 0.18\ °C = 100.18\ °C$.

Check: The units (°C) are correct. The magnitude of the answer (100.2 °C) seems reasonable since the molality of the particles is ~ 1/3.

60. a) **Given:** NaCl; 1.50 x 10^2 g water and $T_f = -1.0$ °C **Find:** m(NaCl)

Other: $K_f = 1.86$ °C/m; $i_{measured} = 1.9$

Conceptual Plan:

T_f → ΔT_f then ΔT_f, i, K_f → m then g_{H2O} → kg_{H2O} then m, kg_{H2O} → mol_{NaCl}

$$T_f = T_f^o - \Delta T_f \qquad \Delta T_f = K_f\, i m \qquad \frac{1\,kg}{1000\ g} \qquad m = \frac{\text{amount solute (moles)}}{\text{mass solvent (kg)}}$$

then mol_{NaCl} → g_{NaCl}

$$\frac{58.44\ g\ NaCl}{1\ mol\ NaCl}$$

Solution: $T_f = T_f^o - \Delta T_f$ so $\Delta T_f = T_f^o - T_f = 0.0°C - 1.0\ °C = -1.0\ °C$ then $\Delta T_f = K_f\, i\, m$

Rearrange to solve for m. $m = \dfrac{\Delta T_f}{K_f\, i} = \dfrac{1.0\ \cancel{°C}}{1.86 \dfrac{\cancel{°C}}{m} \times 1.9} = 0.28297\,m$ NaCl then

$1.50 \times 10^2\ \cancel{g} \times \dfrac{1\,kg}{1000\ \cancel{g}} = 0.150$ kg then $m = \dfrac{\text{amount solute (moles)}}{\text{mass solvent (kg)}}$ so

$mol_{NaCl} = m \times kg_{H2O} = 0.28297 \dfrac{mol\ NaCl}{\cancel{kg_{H2O}}} \times 0.150\ \cancel{kg_{H2O}} = 0.042445\ mol\ NaCl$ then

$0.042445\ \cancel{mol\ NaCl} \times \dfrac{58.44\ g\,NaCl}{1\ mol\ \cancel{NaCl}} = 2.5\ g\,NaCl$

Check: The units (g) are correct. The magnitude of the answer (2.5 g) seems reasonable since the temperature change is moderate and NaCl has a low formula mass.

b) **Given:** $MgSO_4$; 2.50×10^2 mL solution and $\Pi = 3.82$ atm at 298 K **Find:** $m(MgSO_4)$
Other: $i_{measured} = 1.3$
Conceptual Plan:
°C → K then $i, \Pi, T \rightarrow$ M then $mL_{soln} \rightarrow L_{soln}$ then L_{soln}, M → $mol_{MgSO4} \rightarrow g_{MgSO4}$

$K = °C + 273.15$ $\quad \Pi = iMRT$ $\quad \dfrac{1\,L}{1000\,mL}$ $\quad M = \dfrac{\text{amount solute (moles)}}{\text{volume solution (L)}}$ $\quad \dfrac{120.38\text{ g }MgSO_4}{1\text{ mol }MgSO_4}$

Solution: $\Pi = iMRT$ Rearrange to solve for M.

$$M = \frac{\Pi}{i\,RT} = \frac{3.82\ \text{atm}}{1.3 \times 0.08206 \frac{L\cdot \text{atm}}{K\cdot mol} \times 298\ K} = 0.1\underline{2}016\ \frac{mol\,MgSO_4}{L} \quad \text{then}$$

$$2.50 \times 10^2\ mL \times \frac{1L}{1000\ mL} = 0.250\ L \quad \text{then} \quad M = \frac{\text{amount solute (moles)}}{\text{volume solution (L)}} \quad \text{so}$$

$$mol_{MgSO4} = M \times L_{soln} = 0.1\underline{2}016\ \frac{mol\,MgSO_4}{L} \times 0.250\ L = 0.03\underline{0}041\ \text{mol } MgSO_4 \quad \text{finally}$$

$$0.03\underline{0}041\ \text{mol } MgSO_4 \times \frac{120.38\text{ g }MgSO_4}{1\text{ mol }MgSO_4} = 3.6\text{ g }MgSO_4$$

Check: The units (g) are correct. The magnitude of the answer (4 g) seems reasonable since the pressure is moderate and $MgSO_4$ has a low formula mass.

c) **Given:** $FeCl_3$; 2.50×10^2 g water and $T_b = 102$ °C **Find:** $m(FeCl_3)$ **Other:** $K_b = 0.512$ °C/m; $i_{measured} = 3.4$
Conceptual Plan:
$T_b \rightarrow \Delta T_b$ then $\Delta T_b, i, K_b \rightarrow m$ then $g_{H2O} \rightarrow kg_{H2O}$ then $m, kg_{H2O} \rightarrow mol_{FeCl3}$

$T_b = T_b^\circ + \Delta T_b$ $\quad \Delta T_b = K_b\,i\,m$ $\quad \dfrac{1\,kg}{1000\,g}$ $\quad m = \dfrac{\text{amount solute (moles)}}{\text{mass solvent (kg)}}$

then $mol_{FeCl3} \rightarrow g_{FeCl3}$

$\dfrac{162.21\text{ g }FeCl_3}{1\text{ mol }FeCl_3}$

Solution: $T_b = T_b^\circ + \Delta T_b$ so $\Delta T_b = T_b - T_b^\circ = 102°C - 100\ °C = 2\ °C$ then $\Delta T_b = K_b\,i\,m$

Rearrange to solve for m. $$m = \frac{\Delta T_b}{K_b\,i} = \frac{2\ °C}{0.512\ \frac{°C}{m} \times 3.4} = \underline{1}.149\ m\ FeCl_3 \quad \text{then}$$

$$2.50 \times 10^2\ g \times \frac{1kg}{1000\ g} = 0.250\ kg \quad \text{then} \quad m = \frac{\text{amount solute (moles)}}{\text{mass solvent (kg)}} \quad \text{so}$$

$$mol_{FeCl3} = m \times kg_{H2O} = \underline{1}.149\ \frac{mol\ FeCl_3}{kg_{H2O}} \times 0.250\ kg_{H2O} = 0.\underline{2}872\ \text{mol } FeCl_3 \quad \text{then}$$

$$0.\underline{2}872\ \text{mol } FeCl_3 \times \frac{162.21\text{ g }FeCl_3}{1\text{ mol }FeCl_3} = \underline{4}7\text{ g }FeCl_3 = 50\text{ g }FeCl_3$$

Check: The units (g) are correct. The magnitude of the answer (50 g) seems reasonable since the temperature change is significant and we are making 0.25 L of solution.

61. **Given:** 0.100 M of ionic solution, $\Pi = 8.3$ atm at 25 °C **Find:** $i_{measured}$
Conceptual Plan: °C → K then Π, M, $T \rightarrow i$

$K = °C + 273.15$ $\quad \Pi = iMRT$

Solution: 25 °C + 273.15 = 298 K then $\Pi = iMRT$ Rearrange to solve for i.

$$i = \frac{\Pi}{MRT} = \frac{8.3\ \text{atm}}{0.100\ \frac{mol}{L} \times 0.08206\ \frac{L\cdot \text{atm}}{K\cdot mol} \times 298\ K} = 3.4$$

Check: The units (none) are correct. The magnitude of the answer (3) seems reasonable for an ionic solution with a high osmotic pressure.

62. **Given:** 8.92 g of KBr in 500.0 mL solution, $\Pi = 6.97$ atm at 25 °C **Find:** i_{measured}
Conceptual Plan:
°C → K and mL_{soln} → L_{soln} and g_{KBr} → mol_{KBr} then mol_{KBr}, L_{soln} → M then

$K = °C + 273.15$ $\quad \frac{1\,L}{1000\,mL}$ $\quad \frac{1\,mol\,KBr}{119.01\,g\,KBr}$ $\quad M = \frac{\text{amount solute (moles)}}{\text{volume solution (L)}}$

Π, M, *T* → *i*
$\Pi = i\,M\,R\,T$

Solution: 25 °C + 273.15 = 298 K and $500.0\,mL \times \frac{1\,L}{1000\,mL} = 0.5000\,L$ and

$8.92\,g\,KBr \times \frac{1\,mol\,KBr}{119.01\,g\,KBr} = 0.07\underline{4}9517\,mol\,KBr$ then

$M = \frac{\text{amount solute (moles)}}{\text{volume solution (L)}} = \frac{0.07\underline{4}9517\,mol\,KBr}{0.5000\,L} = 0.14\underline{9}903\,\frac{mol\,KBr}{L}$ then $\Pi = i\,M\,R\,T$ Rearrange to

solve for *i*. $i = \frac{\Pi}{MRT} = \frac{6.97\,atm}{0.14\underline{9}903\,\frac{mol}{L} \times 0.08206\,\frac{L \cdot atm}{K \cdot mol} \times 298\,K} = 1.90$

Check: The units (none) are correct. The magnitude of the answer (1.9) seems reasonable for KBr since we expect *i* to be 2 if it completely dissociates. Since both ions are large and have only one charge each, we expect *i* to be close to the theoretical value.

63. Chloroform is polar and has stronger solute–solvent interactions than nonpolar carbon tetrachloride.

64. Each molecule has one –OH group that is capable of hydrogen bonding. Since phenol is a smaller molecule than naphthol, the –OH group has a bigger impact on the overall polarity of the molecule.

65. **Given:** $KClO_4$: Lattice Energy = – 599 kJ/mol, $\Delta H_{\text{hydration}}$ = – 548 kJ/mol; 10.0 g $KClO_4$ in 100.00 mL solution **Find:** ΔH_{soln} and ΔT **Other:** C_s = 4.05 J/g °C; d = 1.05 g/mL
Conceptual Plan:
Lattice Energy, $\Delta H_{\text{hydration}}$ → ΔH_{soln} and g → mol then mol, ΔH_{soln} → *q*(kJ) → *q*(J)

$\Delta H_{soln} = \Delta H_{solute} + \Delta H_{hydration}$ where $\Delta H_{solute} = -\Delta H_{lattice}$ $\quad \frac{1\,mol}{138.56\,g}$ $\quad q = n\,\Delta H_{soln}$ $\quad \frac{1000\,J}{1\,kJ}$

then mL_{soln} → g_{soln} then *q*, g_{soln}, C_s → ΔT

$\frac{1.05\,g}{1\,mL}$ $\quad q = m\,C_s\,\Delta T$

Solution: $\Delta H_{soln} = \Delta H_{solute} + \Delta H_{hydration}$ where $\Delta H_{solute} = -\Delta H_{lattice}$ so $\Delta H_{soln} = \Delta H_{hydration} - \Delta H_{lattice}$

$\Delta H_{soln} = -548\,kJ/mol - (-599\,kJ/mol) = +51\,kJ/mol$ and $10.0\,g \times \frac{1\,mol}{138.56\,g} = 0.07\underline{2}1709\,mol$ then

$q = n\,\Delta H_{soln} = 0.07\underline{2}1709\,mol \times 51\,\frac{kJ}{mol} = +3.\underline{6}807\,kJ \times \frac{1000\,J}{1\,kJ} = +3\underline{6}80.7\,J$ absorbed then

$100.0\,mL \times \frac{1.05\,g}{1\,mL} = 105\,g$ Since heat is absorbed when $KClO_4$ dissolves, the temperature will drop or $q = -3\underline{6}80.7\,J$ and $q = m\,C_s\,\Delta T$ Rearrange to solve for ΔT.

$\Delta T = \frac{q}{m\,C_s} = \frac{-3\underline{6}80.7\,J}{105\,g \times 4.05\,\frac{J}{g \cdot °C}} = -8.7\,°C$

Check: The units (kJ/mol and °C) are correct. The magnitude of the answer (51 kJ/mol) makes physical sense because the lattice energy is larger than the heat of hydration. The magnitude of the temperature change (– 9 °C) makes physical sense since heat is absorbed and the heat of solution is fairly small.

66. **Given:** NaOH: Lattice Energy = – 887 kJ/mol, $\Delta H_{\text{hydration}}$ = – 932 kJ/mol; 25.0 g NaOH in solution, T_i = 25.0 °C; T_f = 100.0 °C **Find:** ΔH_{soln} and m (solution) **Other:** C_s = 4.01 J/g °C; d = 1.05 g/mL

Conceptual Plan:

Lattice Energy, $\Delta H_{\text{hydration}}$ → ΔH_{soln} and g → mol then mol, ΔH_{soln} → q(kJ) → q(J)

$\Delta H_{soln} = \Delta H_{solute} + \Delta H_{hydration}$ where $\Delta H_{solute} = -\Delta H_{lattice}$ $\quad \frac{1\text{ mol}}{40.00\text{ g}} \quad q = n\,\Delta H_{soln} \quad \frac{1000\text{ J}}{1\text{ kJ}}$

and T_i, T_f → ΔT then q, ΔT, C_s → g_{soln} then g_{soln} → mL_{soln}

$\Delta T = T_f - T_i \quad q = m\,C_s\,\Delta T \quad \frac{1\text{ mL}}{1.05\text{ g}}$

Solution: $\Delta H_{soln} = \Delta H_{solute} + \Delta H_{hydration}$ where $\Delta H_{solute} = -\Delta H_{lattice}$ so $\Delta H_{soln} = \Delta H_{hydration} - \Delta H_{lattice}$

ΔH_{soln} = – 932 kJ/mol – – 887 kJ/mol = – 45 kJ/mol and $25.0\text{ g} \times \frac{1\text{ mol}}{40.00\text{ g}} = 0.625$ mol then

$q = n\,\Delta H_{soln} = 0.625\text{ mol} \times -45\frac{\text{kJ}}{\text{mol}} = -28.125\text{ kJ} \times \frac{1000\text{ J}}{1\text{ kJ}} = -28125$ J released then

$\Delta T = T_f - T_i = 100.0°\text{C} - 25.0°\text{C} = 75.0\ °\text{C}$ Since heat is released when NaOH dissolves the temperature will rise or $q = +28125$ J and $q = m\,C_s\,\Delta T$ Rearrange to solve for m.

$$m = \frac{q}{C_s\,\Delta T} = \frac{+28125\text{ J}}{4.01\frac{\text{J}}{\text{g}\cdot°\text{C}} \times 75.0\ °\text{C}} = 93.526\text{ g soln then } 93.526\text{ g} \times \frac{1\text{ mL}}{1.05\text{ g}} = 89\text{ mL soln}$$

Check: The units (kJ/mol and mL) are correct. The magnitude of the answer (– 45 kJ/mol) makes physical sense because the lattice energy is smaller than the heat of hydration. The magnitude of the solution volume (90 mL) makes physical sense since we have 2/3 mole of NaOH and large temperature change. NaOH is a strong base and so we expect heat to be released and need to take precautions in the lab.

67. **Given:** Argon, 0.0537 L; 25 °C, P_{Ar} = 1.0 atm to make 1.0 L saturated solution **Find:** k_H(Ar)

Conceptual Plan: °C → K and P_{Ar}, V, T → mol_{He} then mol_{He}, V_{soln}, P_{Ar} → $k_H(He)$

$\text{K} = °\text{C} + 273.15 \quad PV = nRT \quad S_{Ar} = k_H(Ar)P_{Ar}$ with $S_{Ar} = \frac{mol_{Ar}}{L_{soln}}$

Solution: 25 °C + 273.15 = 298 K and $PV = nRT$ Rearrange to solve for n.

$$n = \frac{PV}{RT} = \frac{1.0\text{ atm} \times 0.0537\text{ L}}{0.08206\frac{\text{L}\cdot\text{atm}}{\text{K}\cdot\text{mol}} \times 298\text{ K}} = 0.00219597\text{ mol then } S_{Ar} = k_H(Ar)P_{Ar} \text{ with } S_{Ar} = \frac{mol_{Ar}}{L_{soln}}$$

Substitute in values and rearrange to solve for k_H.

$$k_H(Ar) = \frac{mol_{Ar}}{L_{soln}P_{He}} = \frac{0.00219597\text{ mol}}{1.0\text{ L}_{soln} \times 1.0\text{ atm}} = 2.2 \times 10^{-3}\ \frac{\text{M}}{\text{atm}}$$

Check: The units (M/atm) are correct. The magnitude of the answer (10^{-3}) seems reasonable since it is consistent with other values in the text.

68. **Given:** gas: 1.65 L; 25 °C, P = 725 torr; and k_H = 0.112 M/atm **Find:** volume of saturated solution

Conceptual Plan: °C → K and torr → atm P, V, T → mol_{gas} then mol_{He}, k_H, P → V_{soln}

$\text{K} = °\text{C} + 273.15 \quad \frac{1\text{ atm}}{760\text{ torr}} \quad PV = nRT \quad S_{gas} = k_H(gas)P_{gas}$ with $S_{gas} = \frac{mol_{gas}}{L_{soln}}$

Solution: 25 °C + 273.15 = 298 K and $725\text{ torr} \times \frac{1\text{ atm}}{760\text{ torr}} = 0.953947$ atm then $PV = nRT$ Rearrange to solve for n.

$$n = \frac{PV}{RT} = \frac{0.953947\text{ atm} \times 1.65\text{ L}}{0.08206\frac{\text{L}\cdot\text{atm}}{\text{K}\cdot\text{mol}} \times 298\text{ K}} = 0.0643666\text{ mol then } S_{gas} = k_H(gas)P_{gas}$$

with $S_{gas} = \frac{mol_{gas}}{L_{soln}}$ Substitute in values and rearrange to solve for V_{soln}.

$$L_{soln} = \frac{mol_{gas}}{k_H(gas)P_{gas}} = \frac{0.064\underline{3}666\ \cancel{mol}}{0.112\ \frac{\cancel{mol}}{L \cdot \cancel{atm}} \times 0.95\underline{3}947\ \cancel{atm}} = 0.602\ \text{L soln}$$

Check: The units (L) are correct. The magnitude of the answer (0.6 L) seems reasonable since the Henry's Law constant is so large.

69. **Given:** 0.0020 ppm by mass Hg = legal limit; 0.0040 ppm by mass Hg = contaminated water; 50.0 mg Hg ingested **Find:** volume of contaminated water

Conceptual Plan: $mg_{Hg} \rightarrow g_{Hg} \rightarrow g_{H2O} \rightarrow mL_{H2O} \rightarrow L_{H2O}$

$$\frac{1\text{ g}}{1000\text{ mg}} \quad \frac{10^6\text{ g water}}{0.0040\text{ g Hg}} \quad \frac{1\text{ mL}}{1.00\text{ g}} \quad \frac{1\text{ L}}{1000\text{ mL}}$$

Solution:

$$50.0\ \cancel{mg\,Hg} \times \frac{1\ \cancel{g\,Hg}}{1000\ \cancel{mg\,Hg}} \times \frac{10^6\ \cancel{g\ water}}{0.0040\ \cancel{g\,Hg}} \times \frac{1\ \cancel{mL\ water}}{1.00\ \cancel{g\ water}} \times \frac{1\text{ L water}}{1000\ \cancel{mL\ water}} = 1.3 \times 10^4\text{ L water}.$$

Check: The units (L) are correct. The magnitude of the answer (10^4 L) seems reasonable since the concentration is so low.

70. **Given:** 2.4 g Na ingested / day; 0.050 % Na by mass in water; d = 1.0 g/mL **Find:** volume of water

Conceptual Plan: $g_{Na} \rightarrow g_{H2O} \rightarrow mL_{H2O} \rightarrow L_{H2O}$

$$\frac{100.000\text{ g water}}{0.050\text{ g Na}} \quad \frac{1\text{ mL}}{1.0\text{ g}} \quad \frac{1\text{ L}}{1000\text{ mL}}$$

Solution: $$2.4\ \cancel{g\,Na} \times \frac{100.000\ \cancel{g\ solution}}{0.050\ \cancel{g\,Na}} \times \frac{1\ \cancel{mL\ water}}{1.0\ \cancel{g\ solution}} \times \frac{1\text{ L water}}{1000\ \cancel{mL\ water}} = 4.8\text{ L water}$$

Check: The units (L) are correct. The magnitude of the answer (5 L) seems reasonable since the concentration is low, but not extremely low.

71. **Given:** 12.5 % NaCl by mass in water at 55 °C; 2.5 L vapor **Find:** g_{H2O} in vapor
Other: $P°_{H2O}$ = 118 torr, i_{NaCl} = 2.0 (complete dissociation)

Conceptual Plan: % NaCl by mass → g_{NaCl}, g_{H2O} then $g_{NaCl} \rightarrow mol_{NaCl}$ and

$$\frac{12.5\text{ g NaCl}}{100\text{ g }(NaCl + H_2O)} \qquad \frac{1\text{ mol NaCl}}{58.44\text{ g NaCl}}$$

$g_{H2O} \rightarrow mol_{H2O}$ then $mol_{NaCl}, mol_{H2O} \rightarrow \chi_{NaCl} \rightarrow \chi_{H2O}$ then $\chi_{H2O}, P°_{H2O} \rightarrow P_{H2O}$

$$\frac{1\text{ mol }H_2O}{18.02\text{ g }H_2O} \quad \chi = \frac{\text{amount solute (in moles)}}{\text{total amount of solute and solvent (in moles)}} \quad \chi_{H2O} = 1 - i_{NaCl}\chi_{NaCl} \quad P_{solution} = \chi_{solvent} P^o_{solvent}$$

then torr → atm and °C → K $P, V, T \rightarrow mol_{H2O} \rightarrow g_{H2O}$

$$\frac{1\text{ atm}}{760\text{ torr}} \quad K = °C + 273.15 \quad PV = nRT \quad \frac{18.02\text{ g }H_2O}{1\text{ mol }H_2O}$$

Solution: $\frac{12.5\text{ g NaCl}}{100\text{ g }(NaCl + H_2O)}$ means 12.5 g NaCl and (100 g – 12.5 g) = 87.5 g H_2O then

$$12.5\ \cancel{g\,NaCl} \times \frac{1\text{ mol NaCl}}{58.44\ \cancel{g\,NaCl}} = 0.21\underline{3}895\text{ mol NaCl} \text{ and } 87.5\ \cancel{g\,H_2O} \times \frac{1\text{ mol }H_2O}{18.02\ \cancel{g\,H_2O}} = 4.8\underline{5}572\text{ mol }H_2O$$

then $$\chi = \frac{\text{amount solute (in moles)}}{\text{total amount of solute and solvent (in moles)}} = \frac{0.21\underline{3}895\ \cancel{mol}}{0.21\underline{3}895\ \cancel{mol} + 4.8\underline{5}572\ \cancel{mol}} = 0.042\underline{1}916$$

then $\chi_{H2O} = 1 - i_{NaCl}\chi_{NaCl} = 1 - (2.0 \times 0.042\underline{1}916) = 0.91\underline{5}617$ then

$P_{solution} = \chi_{solvent} P^o_{solvent} = 0.91\underline{5}617 \times 118\text{ torr} = 10\underline{8}.043\text{ torr }H_2O$ then

$$10\underline{8}.043\ \cancel{torr\ H_2O} \times \frac{1\text{ atm}}{760\ \cancel{torr}} = 0.14\underline{2}162\text{ atm}$$

and 55 °C + 273.15 = 328 K then $PV = nRT$ Rearrange to solve for n.

$$n = \frac{PV}{RT} = \frac{0.142162\ \text{atm} \times 2.5\ \text{L}}{0.08206 \frac{\text{L} \cdot \text{atm}}{\text{K} \cdot \text{mol}} \times 328\ \text{K}} = 0.013204\ \text{mol}$$ then

$$0.013204\ \text{mol}\,H_2O \times \frac{18.02\ \text{g}\,H_2O}{1\ \text{mol}\,H_2O} = 0.24\ \text{g}\,H_2O$$

Check: The units (g) are correct. The magnitude of the answer (0.2 g) seems reasonable since there is very little mass in a vapor.

72. **Given:** 19.5 mg water in 1 L vapor at 25 °C **Find:** mole percent solute in solution **Other:** $P°_{H2O} = 23.78$ torr
Conceptual Plan:
mg $_{H2O}$ → g $_{H2O}$ → mol $_{H2O}$ and °C → K then V, mol$_{H2O}$, T → P_{H2O} then

$\frac{1\ \text{g}}{1000\ \text{mg}}$ $\frac{1\ \text{mol}\,H_2O}{18.01\ \text{g}\,H_2O}$ $K = °C + 273.15$ $PV = nRT$

then atm → torr then P_{H2O}, $P°_{H2O}$ → χ_{H2O} → χ_{Solute} → mole percent solute

$\frac{760\ \text{torr}}{1\ \text{atm}}$ $P_{H2O} = \chi_{H2O}\,P^o_{H2O}$ $\chi_{Solute} = 1 - \chi_{H2O}$ $\textit{mole percent solute} = \chi_{Solute} \times 100\ \%$

Solution: $19.5\ \text{mg}\,H_2O \times \frac{1\ \text{g}\,H_2O}{1000\ \text{mg}\,H_2O} \times \frac{1\ \text{mol}\,H_2O}{18.01\ \text{g}\,H_2O} = 0.00108273\ \text{mol}\,H_2O$ and

55 °C + 273.15 = 328 K then $PV = nRT$ Rearrange to solve for P.

$$P = \frac{nRT}{V} = \frac{0.00108273\ \text{mol} \times 0.08206 \frac{\text{L} \cdot \text{atm}}{\text{K} \cdot \text{mol}} \times 298\ \text{K}}{1.00\ \text{L}} = 0.0264770\ \text{atm}$$

$0.0264770\ \text{atm} \times \frac{760\ \text{torr}}{1\ \text{atm}} = 20.1225\ \text{torr}$ and $P_{H2O} = \chi_{H2O}\,P^o_{H2O}$ Rearrange to solve for χ_{H2O}.

$\chi_{H2O} = \frac{P_{H2O}}{P^o_{H2O}} = \frac{20.1225\ \text{torr}}{23.78\ \text{torr}} = 0.846194$ then $\chi_{Solute} = 1 - \chi_{H2O} = 1 - 0.846194 = 0.153806$ then

$\textit{mole percent solute} = \chi_{Solute} \times 100\ \% = 0.153806 \times 100\ \% = 15.4$ mole percent

Check: The units (mole percent) are correct. The magnitude of the answer (15 mole percent) seems reasonable since we expect there to be more water than solute.

73. **Given:** $T_b = 106.5$ °C aqueous solution **Find:** T_f **Other:** $K_f = 1.86$ °C/m; $K_b = 0.512$ °C/m
Conceptual Plan: $T_b \rightarrow \Delta T_b$ **then** $\Delta T_b, K_b \rightarrow m$ **then** $m, K_f \rightarrow \Delta T_f \rightarrow T_f$

$T_b = T_b^o + \Delta T_b$ $\Delta T_b = K_b m$ $\Delta T_f = K_f m$ $T_f = T_f^o - \Delta T_f$

Solution: $T_b = T_b^o + \Delta T_b$ so $\Delta T_b = T_b - T_b^o = 106.5°C - 100.0\ °C = 6.5\ °C$ then $\Delta T_b = K_b m$

Rearrange to solve for m. $m = \frac{\Delta T_b}{K_b} = \frac{6.5\ °C}{0.512 \frac{°C}{m}} = 12.695\ m$ then

$\Delta T_f = K_f m = 1.86 \frac{°C}{m} \times 12.695\ m = 23.6\ °C$ then $T_f = T_f^o - \Delta T_f = 0.000°C - 23.6\ °C\ °C = -\ 24\ °C$.

Check: The units (°C) are correct. The magnitude of the answer (– 24 °C) seems reasonable since the shift in boiling point is less than the shift in freezing point because the constant is smaller for freezing.

74. **Given:** $P_{H2O} = 20.5$ torr at 25 °C aqueous solution **Find:** T_b **Other:** $P°_{H2O} = 23.78$ torr; $K_b = 0.512$ °C/m
Conceptual Plan: P_{H2O}, $P°_{H2O}$ → χ_{H2O} **assume 1kg water kg $_{H2O}$ → mol$_{H2O}$ then**

$P_{H2O} = \chi_{H2O}\,P^o_{H2O}$ $\frac{1\ \text{mol}\,H_2O}{18.01\ \text{g}\,H_2O}$

$mol_{H2O}, \chi_{H2O} \rightarrow mol_{Solute}$ then $mol_{Solute}, kg_{H2O} \rightarrow m_{Solute}$ then $m, K_b \rightarrow \Delta T_b \rightarrow T_b$

$$\chi_{H2O} = \frac{\text{moles } H_2O}{\text{moles } H_2O + \text{moles solute}} \qquad m = \frac{\text{amount solute (moles)}}{\text{mass solvent (kg)}} \qquad \Delta T_b = K_b m \quad T_b = T_b^o + \Delta T_b$$

Solution: $P_{H2O} = \chi_{H2O} P^o_{H2O}$ Rearrange to solve for χ_{H2O}. $\chi_{H2O} = \frac{P_{H2O}}{P^o_{H2O}} = \frac{20.5 \text{ torr}}{23.78 \text{ torr}} = 0.86\underline{2}069$ then

$$1000 \text{ g } H_2O \times \frac{1 \text{ mol } H_2O}{18.01 \text{ g } H_2O} = 55.5\underline{2}470 \text{ mol } H_2O \text{ then}$$

$$\chi_{H2O} = \frac{\text{moles } H_2O}{\text{moles } H_2O + \text{moles solute}} = \frac{55.5\underline{2}470 \text{ mol}}{55.5\underline{2}470 \text{ mol} + x \text{ mol}} = 0.86\underline{2}069 \text{ Solve for x moles of solute.}$$

$$55.5\underline{2}470 \text{ mol} = 0.86\underline{2}069\,(55.5\underline{2}470 \text{ mol} + x \text{ mol}) \rightarrow x = \frac{(55.5\underline{2}470 - 47.\underline{8}661) \text{ mol}}{0.86\underline{2}069} = 8.\underline{8}8395 \text{ mol}$$

$$\text{then } m = \frac{\text{amount solute (moles)}}{\text{mass solvent (kg)}} = \frac{8.\underline{8}8395 \text{ mol}}{1 \text{ kg}} = 8.\underline{8}8395\, m \text{ then}$$

$$\Delta T_b = K_b\, m = 0.512 \frac{°C}{m} \times 8.\underline{8}8395\, m = 4.5 \text{ °C then } T_b = T_b^o + \Delta T_b = 100.0 \text{ °C} + 4.5 \text{ °C} = 104.5 \text{ °C}.$$

Check: The units (°C) are correct. The magnitude of the answer (4.5 °C) seems reasonable since there is a significant lowering of the vapor pressure.

75. a) **Given:** 0.90 % NaCl by mass per volume; isotonic aqueous solution at 25 °C; KCl; $i = 1.9$
Find: % KCl by mass per volume
Conceptual Plan: Isotonic solutions will have the same number of particles. Since *i* is the same, $\frac{1 \text{ mol KCl}}{1 \text{ mol NaCl}}$

The new % mass per volume will be the mass ratio of the two salts.

$$\text{percent by mass per volume} = \frac{\text{mass solute}}{V} \times 100\,\% \qquad \frac{1 \text{ mol NaCl}}{58.44 \text{ g NaCl}} \text{ and } \frac{74.56 \text{ g KCl}}{1 \text{ mol KCl}}$$

Solution:

$$\text{percent by mass per volume} = \frac{\text{mass solute}}{V} \times 100\,\% =$$

$$= \frac{0.0090 \text{ g NaCl}}{V} \times \frac{1 \text{ mol NaCl}}{58.44 \text{ g NaCl}} \times \frac{1 \text{ mol KCl}}{1 \text{ mol NaCl}} \times \frac{74.56 \text{ g KCl}}{1 \text{ mol KCl}} \times 100\,\% = 1.1\,\% \text{ KCl by mass per volume}$$

Check: The units (% KCl by mass per volume) are correct. The magnitude of the answer (1.1) seems reasonable since the molar mass of KCl is larger than the molar mass of NaCl.

b) **Given:** 0.90 % NaCl by mass per volume; isotonic aqueous solution at 25 °C; NaBr; $i = 1.9$
Find: % NaBr by mass per volume
Conceptual Plan: Isotonic solutions will have the same number of particles. Since *i* is the same, $\frac{1 \text{ mol NaBr}}{1 \text{ mol NaCl}}$

the new % mass per volume will be the mass ratio of the two salts.

$$\text{percent by mass per volume} = \frac{\text{mass solute}}{V} \times 100\,\% \qquad \frac{1 \text{ mol NaCl}}{58.44 \text{ g NaCl}} \text{ and } \frac{102.90 \text{ g NaBr}}{1 \text{ mol NaBr}}$$

Solution:

$$\text{percent by mass per volume} = \frac{\text{mass solute}}{V} \times 100\,\% =$$

$$= \frac{0.0090 \text{ g NaCl}}{V} \times \frac{1 \text{ mol NaCl}}{58.44 \text{ g NaCl}} \times \frac{1 \text{ mol NaBr}}{1 \text{ mol NaCl}} \times \frac{102.90 \text{ g NaBr}}{1 \text{ mol NaBr}} \times 100\,\%$$

$= 1.6\,\%$ NaBr by mass per volume

Check: The units (% NaBr by mass per volume) are correct. The magnitude of the answer (1.6) seems reasonable since the molar mass of NaBr is larger than the molar mass of NaCl.

c) **Given:** 0.90 % NaCl by mass per volume; isotonic aqueous solution at 25 °C; glucose ($C_6H_{12}O_6$); $i = 1.9$
Find: % glucose by mass per volume
Conceptual Plan: Isotonic solutions will have the same number of particles. Since glucose is a nonelectrolyte, the *i* is not the same, then use the mass ratio of the two compounds.

$\frac{1.9 \text{ mol } C_6H_{12}O_6}{1 \text{ mol NaCl}}$ $\quad$ $percent\ by\ mass\ per\ volume = \frac{mass\ solute}{V} \times 100\ \%$ $\quad$ $\frac{1 \text{ mol NaCl}}{58.44 \text{ g NaCl}}$ and $\frac{180.16 \text{ g } C_6H_{12}O_6}{1 \text{ mol } C_6H_{12}O_6}$

Solution:

$$percent\ by\ mass\ per\ volume = \frac{mass\ solute}{V} \times 100\ \% =$$

$$= \frac{0.0090 \text{ g NaCl}}{V} \times \frac{1 \text{ mol NaCl}}{58.44 \text{ g NaCl}} \times \frac{1.9 \text{ mol } C_6H_{12}O_6}{1 \text{ mol NaCl}} \times \frac{180.16 \text{ g } C_6H_{12}O_6}{1 \text{ mol } C_6H_{12}O_6} \times 100\ \% =$$

$= 5.3\ \%\ C_6H_{12}O_6$ by mass per volume

Check: The units (% $C_6H_{12}O_6$ by mass per volume) are correct. The magnitude of the answer (1.6) seems reasonable since the molar mass of $C_6H_{12}O_6$ is larger than the molar mass of NaCl and we need more moles of $C_6H_{12}O_6$ since it is a nonelectrolyte.

76. **Given:** 28.5 g of magnesium citrate ($Mg_3(C_6H_5O_3)_2$) in 235 mL of solution at 37 °C, complete dissociation
Find: Π
Conceptual Plan:
$mL_{soln} \rightarrow L_{soln}$ and $g_{Mg3(C6H5O3)2} \rightarrow mol_{Mg3(C6H5O3)2}$ then $mol_{Mg3(C6H5O3)2}, L_{soln} \rightarrow M$

$\frac{1 \text{ L}}{1000 \text{ mL}}$ $\quad$ $\frac{1 \text{ mol } Mg_3(C_6H_5O_3)_2}{323.14 \text{ g } Mg_3(C_6H_5O_3)_2}$ $\quad$ $M = \frac{\text{amount solute (moles)}}{\text{volume solution (L)}}$

then $M, i, T \rightarrow \Pi$
$\Pi = iMRT$ where $i = 5$

Solution: $235 \text{ mL} \times \frac{1 \text{ L}}{1000 \text{ mL}} = 0.235 \text{ L}$ and

$$28.5 \text{ g } Mg_3(C_6H_5O_3)_2 \times \frac{1 \text{ mol } Mg_3(C_6H_5O_3)_2}{323.14 \text{ g } Mg_3(C_6H_5O_3)_2} = 0.08\underline{8}1982 \text{ mol } Mg_3(C_6H_5O_3)_2 \text{ then}$$

$$M = \frac{\text{amount solute (moles)}}{\text{volume solution (L)}} = \frac{0.08\underline{8}1982 \text{ mol } Mg_3(C_6H_5O_3)_2}{0.235 \text{ L}} = 0.37\underline{5}311 \text{ M then}$$

$$\Pi = iMRT = 5 \times 0.37\underline{5}311 \frac{\text{mol}}{\text{L}} \times 0.08206 \frac{\text{L} \cdot \text{atm}}{\text{K} \cdot \text{mol}} \times 310.\ \text{K} = 47.7 \text{ atm}$$

Check: The units (atm) are correct. The magnitude of the answer (48 atm) seems reasonable since the molarity is ~ 1.5.

77. **Given:** 4.5701 g of $MgCl_2$ and 43.238 g water, $P_{soln} = 0.3624$ atm, $P°_{soln} = 0.3804$ atm at 348.0 K
Find: $i_{measured}$
Conceptual Plan: $g_{MgCl2} \rightarrow mol_{MgCl2}$ and $g_{H2O} \rightarrow mol_{H2O}$ then $P_{soln}, P°_{soln}, \rightarrow \chi_{MgCl2}$

$\frac{1 \text{ mol } MgCl_2}{95.218 \text{ g } MgCl_2}$ $\quad$ $\frac{1 \text{ mol } H_2O}{18.015 \text{ g } H_2O}$ $\quad$ $P_{Soln} = (1 - \chi_{MgCl2})\, P^o_{H2O}$

then $mol_{MgCl2}, mol_{H2O}, \chi_{MgCl2} \rightarrow i$

$$\chi_{MgCl2} = \frac{i\ (\text{moles } MgCl_2)}{\text{moles } H_2O + i\ (\text{moles } MgCl_2)}$$

Solution: $4.5701 \text{ g } MgCl_2 \times \frac{1 \text{ mol } MgCl_2}{95.218 \text{ g } MgCl_2} = 0.04799\underline{6}177 \text{ mol } MgCl_2$ and

$$43.238 \text{ g H}_2\text{O} \times \frac{1 \text{ mol H}_2\text{O}}{18.015 \text{ g H}_2\text{O}} = 2.4001110 \text{ mol H}_2\text{O} \quad \text{then} \quad P_{Soln} = (1 - \chi_{MgCl2})\, P^o_{H2O} \quad \text{so}$$

$$\chi_{MgCl2} = 1 - \frac{P_{Soln}}{P^o_{H2O}} = 1 - \frac{0.3624 \text{ atm}}{0.3804 \text{ atm}} = 0.04731861 \quad \text{Solve for } i.$$

$$i\,(0.047996177) = 0.04731861\,(2.4001110 + i\,(0.047996177)) \rightarrow$$

$$i\,(0.047996177 - 0.002271112) = 0.1135699 \rightarrow i = \frac{0.1135699}{0.04572506} = 2.484.$$

Check: The units (none) are correct. The magnitude of the answer (2.5) seems reasonable for $MgCl_2$ since we expect i to be 3 if it completely dissociates. Since Mg is small and doubly charges, we expect a significant drop from 3.

78. **Given:** 7.050 g of HNO_2 and 1.000 kg of water, $T_f = -0.2929$ **°C** **Find:** fraction dissociated
Other: $K_f = 1.86$ °C/m;
Conceptual Plan:
g $_{HNO2}$ **→ mol** $_{HNO2}$ **then mol** $_{HNO2}$, **kg** $_{H2O}$ **→** m **then** $m, \Delta T_{f\,actual}, K_f$ **→** i_{actual} **then**

$$\frac{1 \text{ mol HNO}_2}{47.02 \text{ g HNO}_2} \qquad m = \frac{\text{amount solute (moles)}}{\text{mass solvent (kg)}} \qquad \Delta T_{f\,actual} = i_{actual} m K_f$$

i_{actual} **→ fraction dissociated**

fraction dissociated = $i_{actual} - 1$

Solution: $7.050 \text{ g HNO}_2 \times \frac{1 \text{ mol HNO}_2}{47.02 \text{ g HNO}_2} = 0.1499425 \text{ mol HNO}_2$ then

$$m = \frac{\text{amount solute (moles)}}{\text{mass solvent (kg)}} = \frac{0.1499425 \text{ mol HNO}_2}{1.000 \text{ kg}} = 0.1499425\ m \quad \text{then } \Delta T_{f\,actual} = i_{actual} m K_f \rightarrow$$

$$0.2929 = i\,(0.1499425\ m)\left(\frac{1.86\ °\text{C}}{m}\right) \rightarrow i = 1.0502 \text{ then fraction dissociated} = i_{actual} - 1 = 1.0502 - 1 = 0.050$$

Check: The units (none) are correct. The magnitude of the answer (0.05) seems reasonable since weak acids do not fully dissociate.

79. **Given:** $T_b = 375.5$ K aqueous solution **Find:** P_{H2O} **Other:** $P°_{H2O} = 0.2467$ atm; $K_b = 0.512$ °C/m
Conceptual Plan: T_b **→** ΔT_b **then** $\Delta T_b, K_b$ **→** m **assume 1kg water kg** $_{H2O}$ **→ mol** $_{H2O}$ **then**

$$T_b = T_b^o + \Delta T_b \qquad \Delta T_b = K_b m \qquad \frac{1 \text{ mol H}_2\text{O}}{18.02 \text{ g H}_2\text{O}}$$

m **→ mol** $_{Solute}$ **then mol** $_{H2O}$, **mol** $_{Solute}$ **→** χ_{H2O} **then** $\chi_{H2O}, P°_{H2O}$ **→** P_{H2O}

$$m = \frac{\text{amount solute (moles)}}{\text{mass solvent (kg)}} \qquad \chi_{H2O} = \frac{\text{moles H}_2\text{O}}{\text{moles H}_2\text{O} + \text{moles solute}} \qquad P_{H2O} = \chi_{H2O} P^o_{H2O}$$

Solution: $T_b = T_b^o + \Delta T_b$ so $\Delta T_b = T_b - T_b^o = 375.5 \text{ K} - 373.15 \text{ K} = 2.4 \text{ K} = 2.4\ °\text{C}$ then

$$\Delta T_b = K_b\, m \quad \text{Rearrange to solve for } m. \quad m = \frac{\Delta T_b}{K_b} = \frac{2.4\ °\text{C}}{0.512 \frac{°\text{C}}{m}} = 4.6875\ m \quad \text{then}$$

$$1000 \text{ g H}_2\text{O} \times \frac{1 \text{ mol H}_2\text{O}}{18.02 \text{ g H}_2\text{O}} = 55.49390 \text{ mol H}_2\text{O} \quad \text{then}$$

$$m = \frac{\text{amount solute (moles)}}{\text{mass solvent (kg)}} = \frac{x \text{ mol}}{1 \text{ kg}} = 4.6875\ m \quad \text{so}$$

$$x = 4.6875 \text{ mol then } \chi_{H2O} = \frac{\text{moles H}_2\text{O}}{\text{moles H}_2\text{O} + \text{moles solute}} = \frac{55.49390 \text{ mol}}{55.49390 \text{ mol} + 4.6875 \text{ mol}} = 0.9221106$$

then $P_{H2O} = \chi_{H2O} P^o_{H2O} = 0.9221106 \times 0.2467 \text{ atm} = 0.227 \text{ atm}$

Check: The units (atm) are correct. The magnitude of the answer (0.227 atm) seems reasonable since the mole fraction is lowered by ~ 8%.

80. **Given:** 0.438 M K_2CrO_4 aqueous solution; $d = 1.063$ g/mL at 298 K; complete dissociation   **Find:** P_{Soln}
    Other: $P°_{H2O} = 0.0313$ atm
    Conceptual Plan:
    assume 1 L solution, so we have 0.438 mol_{K2CrO4} $\rightarrow$ **g_{K2CrO4}** **then** **mL_{soln}** $\rightarrow$ **g_{soln}** **then**

$$\frac{194.20 \text{ g } K_2CrO_4}{1 \text{ mol } K_2CrO_4} \qquad\qquad \frac{1.063 \text{ g}}{1 \text{ mL}}$$

g_{K2CrO4}, g_{soln} $\rightarrow$ **g_{H2O}** $\rightarrow$ **mol_{H2O}** **then** **mol_{H2O}, mol_{Solute}** $\rightarrow$ **χ_{H2O}** **then**   **χ_{H2O}, $P°_{H2O}$** $\rightarrow$ **P_{H2O}**

$$g \text{ H}_2\text{O} = g \text{ soln} - g \text{ K}_2\text{CrO}_4 \quad \frac{1 \text{ mol H}_2\text{O}}{18.01 \text{ g H}_2\text{O}} \qquad \chi_{H2O} = \frac{\text{moles H}_2\text{O}}{\text{moles H}_2\text{O} + \text{moles solute}} \qquad P_{Soln} = (1 - i\chi_{K_2CrO_4}) \, P^o_{H2O} \quad i = 3$$

Solution: $0.438 \ \overline{\text{mol K}_2\text{CrO}_4} \ \times \ \dfrac{194.20 \text{ g K}_2\text{CrO}_4}{1 \ \overline{\text{mol K}_2\text{CrO}_4}} = 85.\underline{0}596 \text{ g K}_2\text{CrO}_4$ **then**

$1000 \ \overline{\text{mL}} \ \times \ \dfrac{1.063 \text{ g}}{1 \ \overline{\text{mL}}} = 1063 \text{ g soln}$ **then**

$g \text{ H}_2\text{O} = g \text{ soln} - g \text{ K}_2\text{CrO}_4 = 1063 \text{ g soln} - 85.\underline{0}596 \text{ g K}_2\text{CrO}_4 = 977.\underline{9}404 \text{ g H}_2\text{O}$ **then**

$977.\underline{9}404 \ \overline{\text{g H}_2\text{O}} \ \times \ \dfrac{1 \text{ mol H}_2\text{O}}{18.01 \ \overline{\text{g H}_2\text{O}}} = 54.\underline{2}999 \text{ mol H}_2\text{O}$ **then**

$$\chi_{K_2CrO_4} = \frac{\text{moles K}_2\text{CrO}_4}{\text{moles H}_2\text{O} + \text{moles K}_2\text{CrO}_4} = \frac{0.438 \ \overline{\text{mol}}}{54.\underline{2}999 \ \overline{\text{mol}} + 0.438 \ \overline{\text{mol}}} = 0.00800\underline{1}77$$ **then**

$P_{Soln} = (1 - i\chi_{K_2CrO_4}) \, P^o_{H2O} = (1 - 3 \times 0.00800\underline{1}77) \, 0.0313 \text{ atm} = 0.0305 \text{ atm}$

Check: The units (atm) are correct. The magnitude of the answer (0.305 atm) seems reasonable since the mole fraction is lowered by < 1 %.

81. **Given:** equal masses of carbon tetrachloride (CCl_4) and chloroform ($CHCl_3$) at 316 K; $P°_{CCl4} = 0.354$ atm; $P°_{CHCl3} = 0.526$ atm **Find:** χ_{CCl4}, χ_{CHCl3} in vapor; and P_{CHCl3} in flask of condensed vapor
    Conceptual Plan: assume 100 grams of each g_{CCl4} $\rightarrow$ **mol_{CCl4}** **and** **g_{CHCl3}** $\rightarrow$ **mol_{CHCl3}** **then**

$$\frac{1 \text{ mol CCl}_4}{153.82 \text{ g CCl}_4} \qquad\qquad \frac{1 \text{ mol CHCl}_3}{119.38 \text{ g CHCl}_3}$$

mol_{CCl4}, mol_{CHCl3} $\rightarrow$ **χ_{CCl4}, χ_{CHCl3} then χ_{CCl4}, $P°_{CCl4}$** $\rightarrow$ **P_{CCl4} and χ_{CHCl3}, $P°_{CHCl3}$** $\rightarrow$ **P_{CHCl3} then**

$$\chi_{CCl4} = \frac{\text{amount CCl}_4 \text{ (in moles)}}{\text{total amount (in moles)}} \quad \chi_{CHCl3} = 1 - \chi_{CCl4} \qquad P_{CCl4} = \chi_{CCl4} \, P^o_{CCl4} \qquad\qquad P_{CHCl3} = \chi_{CHCl3} \, P^o_{CHCl3}$$

P_{CCl4}, P_{CHCl3} $\rightarrow$ **P_{Total} then since $n \, \alpha \, P$ and we are calculating a mass percent, which is a ratio of masses,**
$$P_{Total} = P_{CCl4} + P_{CHCl3}$$
we can simply convert 1 atm to 1 mole so P_{CCl4}, P_{CHCl3} $\rightarrow$ **n_{CCl4}, n_{CHCl3} then**

$$\chi_{CCl4} = \frac{\text{amount CCl}_4 \text{ (in moles)}}{\text{total amount (in moles)}}$$

mol_{CCl4}, mol_{CHCl3} $\rightarrow$ **χ_{CCl4}, χ_{CHCl3} then for the second vapor χ_{CHCl3}, $P°_{CHCl3}$** $\rightarrow$ **P_{CHCl3}**

$$\chi_{CHCl3} = 1 - \chi_{CCl4} \qquad\qquad P_{CHCl3} = \chi_{CHCl3} \, P^o_{CHCl3}$$

Solution: $100.00 \ \overline{\text{g CCl}_4} \ \times \ \dfrac{1 \text{ mol CCl}_4}{153.82 \ \overline{\text{g CCl}_4}} = 0.6501\underline{1}051 \text{ mol CCl}_4$ **and**

$100.00 \ \overline{\text{g CHCl}_3} \ \times \ \dfrac{1 \text{ mol CHCl}_3}{119.38 \ \overline{\text{g CHCl}_3}} = 0.8376\underline{6}125 \text{ mol CHCl}_3$ **then**

$$\chi_{CCl4} = \frac{\text{amount CCl}_4 \text{ (in moles)}}{\text{total amount (in moles)}} = \frac{0.6501\underline{1}051 \ \overline{\text{mol}}}{0.6501\underline{1}051 \ \overline{\text{mol}} + 0.8376\underline{6}125 \ \overline{\text{mol}}} = 0.4369\underline{6}925$$ **and**

$\chi_{CHCl3} = 1 - \chi_{CCl4} = 1 - 0.43696925 = 0.56303075$ then

$P_{CCl4} = \chi_{CCl4}\, P^o_{CCl4} = 0.43696925 \text{ x } 0.354 \text{ atm} = 0.154687 \text{ atm}$ and

$P_{CHCl3} = \chi_{CHCl3}\, P^o_{CHCl3} = 0.56303075 \text{ x } 0.526 \text{ atm} = 0.296154 \text{ atm}$ then

$P_{Total} = P_{CCl4} + P_{CHCl3} = 0.154687 \text{ atm} + 0.296154 \text{ atm} = 0.450841 \text{ atm}$ then

$\text{mol}_{CCl4} = 0.154687 \text{ mol}$ and $\text{mol}_{CHCl3} = 0.296154 \text{ mol}$ then

$$\chi_{CCl4} = \frac{\text{amount } CCl_4 \text{ (in moles)}}{\text{total amount (in moles)}} = \frac{0.154687 \cancel{\text{mol}}}{0.154687 \cancel{\text{mol}} + 0.296154 \cancel{\text{mol}}} = 0.343108 = 0.343 \text{ in the first vapor}$$

and $\chi_{CHCl3} = 1 - \chi_{CCl4} = 1 - 0.343108 = 0.656892 = 0.657$ in the first vapor then in the second vapor

$P_{CHCl3} = \chi_{CHCl3}\, P^o_{CHCl3} = 0.656892 \text{ x } 0.526 \text{ atm} = 0.345525 \text{ atm} = 0.346 \text{ atm}$

Check: The units (none and atm) are correct. The magnitudes of the answers seem reasonable since it we expect the lighter component to be found preferentially in the vapor phase. This effect is magnified in the second vapor.

82. In the previous problem we saw that the original liquid has the $\chi_{CHCl3} = 0.563$, and when the second vapor is condensed it rose to 0.657. Continue the calculation scheme from above.

$P_{CCl4} = \chi_{CCl4}\, P^o_{CCl4} = 0.343108 \text{ x } 0.354 \text{ atm} = 0.121460 \text{ atm}$ converting to moles; $\text{mol}_{CCl4} = 0.121460 \text{ mol}$ and $\text{mol}_{CHCl3} = 0.345525 \text{ mol}$ then

$$\chi_{CCl4} = \frac{\text{amount } CCl_4 \text{ (in moles)}}{\text{total amount (in moles)}} = \frac{0.121460 \cancel{\text{mol}}}{0.121460 \cancel{\text{mol}} + 0.345525 \cancel{\text{mol}}} = 0.260094 = 0.260 \text{ in the second vapor}$$

and $\chi_{CHCl3} = 1 - \chi_{CCl4} = 1 - 0.260094 = 0.739906 = 0.740$ in the second vapor then in the third vapor

$P_{CHCl3} = \chi_{CHCl3}\, P^o_{CHCl3} = 0.739906 \text{ x } 0.526 \text{ atm} = 0.389191 \text{ atm}$ and

$P_{CCl4} = \chi_{CCl4}\, P^o_{CCl4} = 0.260094 \text{ x } 0.354 \text{ atm} = 0.0920732 \text{ atm}$ converting to moles

$\text{mol}_{CCl4} = 0.0920732 \text{ mol}$ and $\text{mol}_{CHCl3} = 0.389191 \text{ mol}$ then

$$\chi_{CCl4} = \frac{\text{amount } CCl_4 \text{ (in moles)}}{\text{total amount (in moles)}} = \frac{0.0920732 \cancel{\text{mol}}}{0.0920732 \cancel{\text{mol}} + 0.389191 \cancel{\text{mol}}} = 0.191315 = 0.191 \text{ in the third vapor}$$

and $\chi_{CHCl3} = 1 - \chi_{CCl4} = 1 - 0.191315 = 0.808685 = 0.809$ in the third vapor then in the fourth vapor

$P_{CHCl3} = \chi_{CHCl3}\, P^o_{CHCl3} = 0.808685 \text{ x } 0.526 \text{ atm} = 0.425368 \text{ atm}$ and

$P_{CCl4} = \chi_{CCl4}\, P^o_{CCl4} = 0.191315 \text{ x } 0.354 \text{ atm} = 0.0677255 \text{ atm}$ converting to moles

$\text{mol}_{CCl4} = 0.0677255 \text{ mol}$ and $\text{mol}_{CHCl3} = 0.425368 \text{ mol}$ then

$$\chi_{CCl4} = \frac{\text{amount } CCl_4 \text{ (in moles)}}{\text{total amount (in moles)}} = \frac{0.0677255 \cancel{\text{mol}}}{0.0677255 \cancel{\text{mol}} + 0.425368 \cancel{\text{mol}}} = 0.137348 = 0.137 \text{ in the fourth vapor}$$

and $\chi_{CHCl3} = 1 - \chi_{CCl4} = 1 - 0.137348 = 0.862652 = 0.863$ in the fourth vapor. The concentration of the lighter component in the gas phase, or distillate, increases with each step.

83. **Given:** N_2: $k_H(N_2) = 6.1 \text{ x } 10^{-4}$ M/L at 25 °C; 14.6 mg/L at 50 °C and 1.00 atm; $P_{N2} = 0.78$ atm; O_2: $k_H(O_2) = 1.3 \text{ x } 10^{-3}$ M/L at 25 °C; 27.8 mg/L at 50 °C and 1.00 atm; $P_{O2} = 0.21$ atm; and 1.5 L water
Find: $V(N_2)$ and $V(O_2)$
Conceptual Plan: at 25 °C: $P_{Total}, \chi_{N2} \rightarrow P_{N2}$ then $P_{N2}, k_H(N_2) \rightarrow S_{N2}$ then L → mol

$P_{N2} = \chi_{N2}\, P_{Total}$ $\quad S_{N2} = k_H(N_2) P_{N2}$ $\quad S_{N2}$

at 50 °C: L → mL → mg → g → mol then mol$_{25\,°C}$, mol$_{25\,°C}$ → mol$_{removed}$ then °C → K

$\frac{1000 \text{ mL}}{1 \text{ L}}$ $\frac{14.6 \text{ mg}}{1 \text{ L}}$ $\frac{1 \text{ g}}{1000 \text{ mg}}$ $\frac{1 \text{ mol}}{28.01 \text{ g}}$ $\quad mol_{removed} = mol_{25°C} - mol_{50°C}$ $\quad K = °C + 273.15$

then $P, n, T \rightarrow V$

$PV = nRT$

at 25 °C: $P_{Total}, \chi_{O2} \rightarrow P_{O2}$ then $P_{O2}, k_H(O_2) \rightarrow S_{O2}$ then L → mol

$P_{O2} = \chi_{O2}\, P_{Total}$ $\quad S_{O2} = k_H(O_2) P_{O2}$ $\quad S_{O2}$

at 50 °C: L → mL → mg → g → mol then mol$_{25\,°C}$, mol$_{25\,°C}$ → mol$_{removed}$ then °C → K

$\frac{1000 \text{ mL}}{1 \text{ L}}$ $\frac{27.8 \text{ mg}}{1 \text{ L}}$ $\frac{1 \text{ g}}{1000 \text{ mg}}$ $\frac{1 \text{ mol}}{32.00 \text{ g}}$ $mol_{removed} = mol_{25°C} - mol_{50°C}$ $K = °C + 273.15$

then *P, n, T* → *V*

$PV = nRT$

Solution:

at 25 °C: $P_{N2} = \chi_{N2}\, P_{Total} = 0.78 \text{ x } 1.0 \text{ atm} = 0.78 \text{ atm}$ then

$S_{N2} = k_H(N_2)P_{N2} = 6.1 \text{ x } 10^{-4} \frac{\text{M}}{\text{atm}} \text{ x } 0.78 \text{ atm} = 4.\underline{7}58 \text{ x } 10^{-4} \text{ M}$ then

$1.5 \text{ L x } 4.\underline{7}58 \text{ x } 10^{-4} \frac{\text{mol}}{\text{L}} = 0.00071\underline{3}71 \text{ mol}$

at 50 °C: $1.5 \text{ L x } \frac{14.6 \text{ mg}}{1 \text{ L} \cdot \text{atm}} \text{ x } 0.78 \text{ atm x } \frac{1 \text{ g}}{1000 \text{ mg}} \text{ x } \frac{1 \text{ mol}}{28.01 \text{ g}} = 0.00060\underline{9}85 \text{ mol}$ then

$mol_{removed} = mol_{25°C} - mol_{50°C} = 0.00071\underline{3}71 \text{ mol} - 0.00060\underline{9}85 \text{ mol} = 1.\underline{0}39 \text{ x } 10^{-4} \text{ mol}$.

then $50 \text{ °C} + 273.15 = 323 \text{ K}$ then $PV = nRT$ Rearrange to solve for *V*.

$$V = \frac{nRT}{P} = \frac{1.\underline{0}39 \text{ x } 10^{-4} \text{ mol x } 0.08206 \frac{\text{L} \cdot \text{atm}}{\text{K} \cdot \text{mol}} \text{ x } 323 \text{ K}}{1.00 \text{ atm}} = 0.002\underline{7}539 \text{ L}$$

at 25 °C: $P_{O2} = \chi_{O2}\, P_{Total} = 0.21 \text{ x } 1.0 \text{ atm} = 0.21 \text{ atm}$ then

$S_{O2} = k_H(O_2)P_{O2} = 1.3 \text{ x } 10^{-3} \frac{\text{M}}{\text{atm}} \text{ x } 0.21 \text{ atm} = 2.\underline{7}3 \text{ x } 10^{-4} \text{ M}$ then

$1.5 \text{ L x } 2.\underline{7}3 \text{ x } 10^{-4} \frac{\text{mol}}{\text{L}} = 0.0004\underline{0}95 \text{ mol}$

at 50 °C: $1.5 \text{ L x } \frac{27.8 \text{ mg}}{1 \text{ L} \cdot \text{atm}} \text{ x } 0.21 \text{ atm x } \frac{1 \text{ g}}{1000 \text{ mg}} \text{ x } \frac{1 \text{ mol}}{32.00 \text{ g}} = 0.0002\underline{7}366 \text{ g}$ then

$mol_{removed} = mol_{25°C} - mol_{50°C} = 0.0004\underline{0}95 \text{ mol} - 0.0002\underline{7}366 \text{ mol} = 1.\underline{3}58 \text{ x } 10^{-4} \text{ mol}$

then $50 \text{ °C} + 273.15 = 323 \text{ K}$ then $PV = nRT$ Rearrange to solve for *V*.

$$V = \frac{nRT}{P} = \frac{1.\underline{3}58 \text{ x } 10^{-4} \text{ mol x } 0.08206 \frac{\text{L} \cdot \text{atm}}{\text{K} \cdot \text{mol}} \text{ x } 323 \text{ K}}{1.00 \text{ atm}} = 0.003\underline{5}994 \text{ L}$$ finally

$V_{Total} = V_{N2} + V_{O2} = 0.002\underline{7}526 \text{ L} + 0.003\underline{5}994 \text{ L} = 0.0064 \text{ L}$

Check: The units (L) are correct. The magnitude of the answer (0.006 L) seems reasonable since we have so little dissolved gas at room temperature and most is still soluble at 50 °C.

84. **Given:** pentane (C_5H_{12}) and hexane (C_6H_{14}): 35.5 percent by mass C_5H_{12} in vapor at 25 °C; $P°_{C5H12} = 425$ torr; $P°_{C6H14} = 151$ torr **Find:** percent by mass C_5H_{12} and percent by mass C_6H_{14} in solution

Conceptual Plan:

mass percents → g$_{C5H12}$, g$_{C6H14}$ then g$_{C5H12}$ → mol$_{C5H12}$ and g$_{C6H14}$ → mol$_{C6H14}$ then

$mass\ percent = \frac{mass\ solute}{mass\ solution} \times 100\ \%$ $\frac{1 \text{ mol } C_5H_{12}}{72.15 \text{ g } C_5H_{12}}$ $\frac{1 \text{ mol } C_6H_{14}}{86.17 \text{ g } C_6H_{14}}$

mol$_{C5H12}$, mol$_{C6H14}$ → $\chi_{C5H12vapor}$ then $\chi_{C5H12vapor}$, $P°_{C5H12}$, $P°_{C6H14}$ → $\chi_{C5H12soln}$ then assume

$\chi_{C5H12} = \frac{\text{amount } C_5H_{12} \text{ (in moles)}}{\text{total amount (in moles)}}$ $\chi_{C5H12vapor} = \frac{P_{C5H12}}{P_{Total}} = \frac{\chi_{C5H12soln} P^o_{C5H12}}{\chi_{C5H12soln} P^o_{C5H12} + (1 - \chi_{C5H12soln}) P^o_{C6H14}}$

1 total mole of solution → mol$_{C5H12}$, mol$_{C6H14}$ then mol$_{C5H12}$ → g$_{C5H12}$ and mol$_{C6H14}$ → g$_{C6H14}$

$\chi_{C6H14soln} = 1 - \chi_{C5H12soln}$ $\frac{72.15 \text{ g } C_5H_{12}}{1 \text{ mol } C_5H_{12}}$ $\frac{86.17 \text{ g } C_6H_{14}}{1 \text{ mol } C_6H_{14}}$

finally g_{C5H12}, g_{C6H14} → **mass percents**

$$mass\ percent = \frac{mass\ solute}{mass\ solution} \times 100\ \%$$

Solution: $mass\ percent = \dfrac{mass\ solute}{mass\ solution} \times 100\ \%$ means that 35.5 g C_5H_{12} and 100.0 g – 35.5 g = 64.5 g C_6H_{14}

then $35.5\ \text{g}\ C_5H_{12} \times \dfrac{1\ \text{mol}\ C_5H_{12}}{72.15\ \text{g}\ C_5H_{12}} = 0.49\underline{2}030\ \text{mol}\ C_5H_{12}$ and

$64.5\ \text{g}\ C_6H_{14} \times \dfrac{1\ \text{mol}\ C_6H_{14}}{86.17\ \text{g}\ C_6H_{14}} = 0.74\underline{8}520\ \text{mol}\ C_6H_{14}$ then

$$\chi_{C5H12} = \frac{\text{amount } C_5H_{12} \text{ (in moles)}}{\text{total amount (in moles)}} = \frac{0.49\underline{2}030\ \text{mol}}{0.49\underline{2}030\ \text{mol} + 0.74\underline{8}520\ \text{mol}} = 0.39\underline{6}622 \text{ then}$$

$$\chi_{C5H12vapor} = \frac{P_{C5H12}}{P_{Total}} = \frac{\chi_{C5H12soln} P^o_{C5H12}}{\chi_{C5H12soln} P^o_{C5H12} + (1 - \chi_{C5H12soln}) P^o_{C6H14}}$$ Substitute in values and solve for $\chi_{C5H12soln}$.

$$0.39\underline{6}622 = \frac{\chi_{C5H12soln} \times 425\ \text{torr}}{\chi_{C5H12soln} \times 425\ \text{torr} + (1 - \chi_{C5H12soln}) \times 151\ \text{torr}} \rightarrow$$

$$0.39\underline{6}622\,(425\chi_{C5H12soln} + 151(1 - \chi_{C5H12soln})) = 425\ \chi_{C5H12soln} \rightarrow$$

$$16\underline{8}.564\chi_{C5H12soln} + 59.\underline{8}900 - 59.\underline{8}900\ \chi_{C5H12soln} = 425\ \chi_{C5H12soln} \rightarrow 31\underline{6}.326\chi_{C5H12soln} = 59.\underline{8}900 \rightarrow$$

$$\chi_{C5H12soln} = \frac{59.\underline{8}900}{31\underline{6}.326} = 0.18\underline{9}330 \text{ then } \chi_{C6H14soln} = 1 - \chi_{C5H12soln} = 1 - 0.18\underline{9}330 = 0.81\underline{0}670 \text{ so}$$

mol $_{C5H12}$ = 0.18$\underline{9}$330 mol and mol $_{C6H14}$ = 0.81$\underline{0}$670 mol then

$0.18\underline{9}330\ \text{mol}\ C_5H_{12} \times \dfrac{72.15\ \text{g}\ C_5H_{12}}{1\ \text{mol}\ C_5H_{12}} = 13.\underline{6}602\ \text{g}\ C_5H_{12}$ and

$0.81\underline{0}670\ \text{mol}\ C_6H_{14} \times \dfrac{86.17\ \text{g}\ C_6H_{14}}{1\ \text{mol}\ C_6H_{14}} = 69.\underline{8}554\ \text{g}\ C_6H_{14}$ finally

$$mass\ percent = \frac{mass\ solute}{mass\ solution} \times 100\ \% = \frac{13.\underline{6}602\ \text{g}\ C_5H_{12}}{13.\underline{6}602\ \text{g}\ C_5H_{12} + 69.\underline{8}554\ \text{g}\ C_6H_{14}} \times 100\ \%$$

= 16.4 percent by mass C_5H_{12}

and 100.0 – 16.4 = 83.6 mass percent C_6H_{14}

Check: The units (mass percent) are correct. We expect the mass percent of C_6H_{14} to be much higher than the mass percent of C_5H_{12} because the vapor is richer in this component and it is the less volatile phase.

85. **Given:** 1.10 g glucose ($C_6H_{12}O_6$) and sucrose ($C_{12}H_{22}O_{11}$) mixture in 25.0 mL solution and Π = 3.78 atm at 298 K **Find:** percent composition of mixture

Conceptual Plan: **Π, *T* → M then mL$_{soln}$ → L$_{soln}$ then L$_{soln}$, M → mol$_{mixture}$ then**

$$\Pi = MRT \qquad \frac{1\ \text{L}}{1000\ \text{mL}} \qquad M = \frac{\text{amount solute (moles)}}{\text{volume solution (L)}}$$

mol$_{mixture}$, g$_{mixture}$ → mol$_{C6H12O6}$, mol$_{C12H22O11}$ then

$$g_{mixture} = mol\ C_6H_{12}O_6 \times \frac{180.16\ g\ C_6H_{12}O_6}{1\ mol\ C_6H_{12}O_6} + mol\ C_{12}H_{22}O_{11} \times \frac{342.30\ g\ C_{12}H_{22}O_{11}}{1\ mol\ C_{12}H_{22}O_{11}} \text{ with } mol_{mixture} = mol\ C_6H_{12}O_6 + mol\ C_{12}H_{22}O_{11}$$

mol$_{C6H12O6}$ → g$_{C6H12O6}$ and mol$_{C12H22O11}$ → g$_{C12H22O11}$ and g$_{C6H12O6}$, g$_{C12H22O11}$ → mass percents

$$\frac{180.16\ \text{g}\ C_6H_{12}O_6}{1\ \text{mol}\ C_6H_{12}O_6} \qquad \frac{342.30\ \text{g}\ C_{12}H_{22}O_{11}}{1\ \text{mol}\ C_{12}H_{22}O_{11}} \qquad mass\ percent = \frac{mass\ solute}{mass\ solution} \times 100\ \%$$

Solution: $\Pi = MRT$ Rearrange to solve for M.

$$M = \frac{\Pi}{RT} = \frac{3.78\ \text{atm}}{0.08206\ \frac{\text{L}\cdot\text{atm}}{\text{K}\cdot\text{mol}} \times 298\ \text{K}} = 0.15\underline{4}577\ \frac{\text{mol mixture}}{\text{L}}$$ then $$25.0\ \text{mL} \times \frac{1\ \text{L}}{1000\ \text{mL}} = 0.0250\ \text{L}$$

then $M = \frac{\text{amount solute (moles)}}{\text{volume solution (L)}}$ so

$$mol_{mixture} = M \times L_{soln} = 0.15\underline{4}577\ \frac{\text{mol mixture}}{\text{L}} \times 0.0250\ \text{L} = 0.0038\underline{6}442\ \text{mol mixture}$$ then

$$g_{mixture} = \text{mol } C_6H_{12}O_6 \times \frac{180.16\ \text{g } C_6H_{12}O_6}{1\ \text{mol } C_6H_{12}O_6} + \text{mol } C_{12}H_{22}O_{11} \times \frac{342.30\ \text{g } C_{12}H_{22}O_{11}}{1\ \text{mol } C_{12}H_{22}O_{11}}$$ with

$mol_{mixture} = mol\ C_6H_{12}O_6 + mol\ C_{12}H_{22}O_{11}$ so

$$1.10\ \text{g} = \text{mol } C_6H_{12}O_6 \times \frac{180.16\ \text{g } C_6H_{12}O_6}{1\ \text{mol } C_6H_{12}O_6} + (0.0038\underline{6}442\ \text{mol} - \text{mol } C_6H_{12}O_6) \times \frac{342.30\ \text{g } C_{12}H_{22}O_{11}}{1\ \text{mol } C_{12}H_{22}O_{11}} \rightarrow$$

$$1.10 = 180.16 \times \text{mol } C_6H_{12}O_6 + 1.3\underline{2}228 - 342.30 \times \text{mol } C_6H_{12}O_6 \rightarrow 162.14\ x\ \text{mol } C_6H_{12}O_6 = 0.2\underline{2}228 \rightarrow$$

$$x\ \text{mol } C_6H_{12}O_6 = \frac{0.2\underline{2}228}{162.14} = 0.001\underline{3}7091\ \text{mol } C_6H_{12}O_6$$ then

$$mol\ C_{12}H_{22}O_{11} = mol_{mixture} - mol\ C_6H_{12}O_6 = 0.0038\underline{6}442\ \text{mol} - 0.001\underline{3}7091\ \text{mol} = 0.002\underline{4}935\ \text{mol } C_{12}H_{22}O_{11}$$ then

$$0.001\underline{3}7091\ \text{mol } C_6H_{12}O_6 \times \frac{180.16\ \text{g } C_6H_{12}O_6}{1\ \text{mol } C_6H_{12}O_6} = 0.2\underline{4}698\ \text{g } C_6H_{12}O_6$$ and

$$0.002\underline{4}935\ \text{mol } C_{12}H_{22}O_{11} \times \frac{342.30\ \text{g } C_{12}H_{22}O_{11}}{1\ \text{mol } C_{12}H_{22}O_{11}} = 0.8\underline{5}353\ \text{g } C_{12}H_{22}O_{11}$$ finally

$$mass\ percent = \frac{mass\ solute}{mass\ solution} \times 100\,\% = \frac{0.2\underline{4}698\ \text{g } C_6H_{12}O_6}{0.2\underline{4}698\ \text{g } C_6H_{12}O_6 + 0.8\underline{5}353\ \text{g } C_{12}H_{22}O_{11}} \times 100\,\%$$

$= 2\underline{2}.44\ \%\ C_6H_{12}O_6$ by mass

and $100.00\,\% - 2\underline{2}.44\,\% = 7\underline{7}.56\ \%\ C_{12}H_{22}O_{11}$ by mass

Check: The units (% by mass) are correct. We expect the percent by $C_6H_{12}O_6$ to be larger than that for $C_{12}H_{22}O_{11}$ since the $g_{mixture}$ /$mol_{mixture}$ = 285 g/mol, which is closer to $C_{12}H_{22}O_{11}$ than $C_6H_{12}O_6$ and the molar mass of $C_{12}H_{22}O_{11}$ is larger than the molar mass of $C_6H_{12}O_6$.

86. **Given:** 631 mL methanol (CH_3OH) and 501 mL water; solution = 14.29 M CH_3OH, d (CH_3OH) = 0.792 g/mL **Find:** volume change on mixing

Conceptual Plan: $V_{CH3OH}, V_{H2O} \rightarrow V_{before\ mixing}$ **then** $mL_{CH3OH} \rightarrow g_{CH3OH} \rightarrow mol_{CH3OH}$ **then**

$V_{before\ mixing} = V_{CH3OH} + V_{H2O}$ $\qquad \frac{0.792\ \text{g}}{1\ \text{mL}}$ $\qquad \frac{1\ \text{mol } CH_3OH}{32.04\ \text{g } CH_3OH}$

mol_{CH3OH}, **M** $\rightarrow$ L_{soln} $\rightarrow$ mL_{soln} **then** $V_{before\ mixing}$, $L_{soln} \rightarrow \Delta V_{mixing}$

$M = \frac{\text{amount solute (moles)}}{\text{volume solution (L)}}$ $\qquad \frac{1000\ \text{mL}}{1\ \text{L}}$ $\qquad \Delta V_{mixing} = V_{before\ mixing} - V_{soln}$

Solution: $V_{before\,mixing} = V_{CH3OH} + V_{H2O} = 631\ \text{mL} + 501\ \text{mL} = 1132\ \text{mL}$ then

$$631\ \text{mL} \times \frac{0.792\ \text{g}}{1\ \text{mL}} \times \frac{1\ \text{mol } CH_3OH}{32.04\ \text{g } CH_3OH} = 15.\underline{5}978\ \text{mol } CH_3OH$$ then $M = \frac{\text{amount solute (moles)}}{\text{volume solution (L)}}$ so

$$L_{soln} = \frac{mol_{CH3OH}}{M} = \frac{15.\underline{5}978\ \text{mol } CH_3OH}{14.29\ \frac{\text{mol } CH_3OH}{\text{L}}} = 1.0\underline{9}152\ \text{L}$$ then $$1.0\underline{9}152\ \text{L} \times \frac{1000\ \text{mL}}{1\ \text{L}} = 10\underline{9}1.52\ \text{mL}$$ then

$$\Delta V_{mixing} = V_{before\,mixing} - V_{soln} = 1132\ \text{mL} - 10\underline{9}1.52\ \text{mL} = \underline{4}0.48\ \text{mL} = 4 \times 10^1\ \text{mL}$$

Check: The units (mL) are correct. Since the intermolecular forces between a methanol molecule and a water molecule are different than between two water molecules or between two methanol molecules, the spacing between molecules changes and thus the volume changes. The amount of the change (~ 4 %) is reasonable.

87. **Given:** isopropyl alcohol ($(CH_3)_2CHOH$) and propyl alcohol ($CH_3CH_2CH_2OH$) at 313 K; solution 2/3 by mass isopropyl alcohol $P_{2/3}$ = 0.110 atm; solution 1/3 by mass isopropyl alcohol $P_{1/3}$ = 0.089 atm; **Find:** $P°_{iso}$ and $P°_{pro}$ and explain why they are different
Conceptual Plan: since these are isomers, they have the same molar mass and so the fraction by mass is the same as the mole fraction so mole fractions, P_{soln}'s → $P°$'s

$$\chi_{iso} = \frac{\text{amount iso (in moles)}}{\text{total amount (in moles)}} \quad \chi_{pro} = 1 - \chi_{iso} \quad P_{iso} = \chi_{iso} P^{\circ}_{iso} \quad P_{pro} = \chi_{pro} P^{\circ}_{pro} \text{ and } P_{soln} = P_{iso} + P_{pro}$$

Solution:

Solution 1: $\chi_{iso} = 2/3$ and $\chi_{iso} = 1/3$ $P_{soln} = P_{iso} + P_{pro}$ so $0.110 \text{ atm} = 2/3 P^{\circ}_{iso} + 1/3 P^{\circ}_{pro}$ and

Solution 2: $\chi_{iso} = 1/3$ and $\chi_{iso} = 2/3$ $P_{soln} = P_{iso} + P_{pro}$ so $0.089 \text{ atm} = 1/3 P^{\circ}_{iso} + 2/3 P^{\circ}_{pro}$. We now have 2 equations and 2 unknowns and a number of ways to solve this. One way is to rearrange the first equation for $P°_{iso}$ and then substitute into the other equation. Thus, $P^{\circ}_{iso} = 3/2\,(0.110 \text{ atm} - 1/3 P^{\circ}_{pro})$ and

$$0.089 \text{ atm} = \frac{1}{3}\frac{3}{2}(0.110 \text{ atm} - 1/3 P^{\circ}_{pro}) + \frac{2}{3}P^{\circ}_{pro} \rightarrow 0.089 \text{ atm} = 0.0550 \text{ atm} - \frac{1}{6}P^{\circ}_{pro} + \frac{2}{3}P^{\circ}_{pro} \rightarrow$$

$$\frac{1}{2}P^{\circ}_{pro} = 0.03\underline{4}0 \text{ atm} \rightarrow P^{\circ}_{pro} = 0.06\underline{8}0 \text{ atm} = 0.068 \text{ atm} \text{ and then}$$

$$P^{\circ}_{iso} = 3/2\,(0.110 \text{ atm} - 1/3 P^{\circ}_{pro}) = 3/2\,(0.110 \text{ atm} - 1/3(0.06\underline{8}0 \text{ atm})) = 0.131 \text{ atm}$$

The major intermolecular attractions are between the OH groups. The OH group at the end of the chain in propyl alcohol is more accessible than the one in the middle of the chain in isopropyl alcohol. In addition, the molecular shape of propyl alcohol is a straight chain of carbon atoms, while that of isopropyl alcohol has a branched chain and is more like a ball. The contact area between two ball-like objects is smaller than that of two chain-like objects. The smaller contact area in isopropyl alcohol means the molecules do not attract each other as strongly as do those of propyl alcohol. As a result of both of these factors, the vapor pressure of isopropyl alcohol is higher.
Check: The units (atm) are correct. The magnitude of the answers seems reasonable since the solution partial pressures are both ~0.1 atm.

88. **Given:** metal, M, of atomic weight 96 forms MF_x salt; 9.18 g of MF_x completely dissociates in 100.0 g water, T_b = 374.38 K; **Find:** x and formula unit **Other:** K_b = 0.512 °C/*m*
Conceptual Plan:
g_{H2O} → kg_{H2O} and T_b → ΔT_b then ΔT_f, *i*, K_f → *m* then *m*, kg_{H2O} → mol_{Unk}

$$\frac{1\text{kg}}{1000\text{ g}} \quad T_b = T_b^{o} + \Delta T_b \quad \Delta T_b = K_b\, i\, m \text{ where } i = 1 + x \quad m = \frac{\text{amount solute (moles)}}{\text{mass solvent (kg)}}$$

then g_{MFx}, mol_{MFx} → $\mathfrak{M}$ → x

$$\mathfrak{M} = \frac{g_{MFx}}{mol_{MFx}} \quad \frac{1 \text{ mol } MF_x}{(96 + 19x)\text{ g } MF_x} \quad \frac{(96 + 19x)\text{ g } MF_x}{1 \text{ mol } MF_x}$$

Solution: $100.0 \text{ g} \times \frac{1\text{kg}}{1000 \text{ g}} = 0.1000 \text{ kg}$ and $T_b = T_b^{o} + \Delta T_b$ so

$\Delta T_b = T_b - T_b^{o} = 374.38 \text{ K} - 373.15 \text{ K} = +1.23 \text{ K} = +1.23\ °\text{C}$ then $\Delta T_b = K_b\, i\, m$ where $i = 1 + x$ so

$\Delta T_b = K_b(1+x)m$ Rearrange to solve for *m*. $m = \frac{\Delta T_b}{K_b(1+x)} = \frac{1.23\ °\text{C}}{0.512 \frac{°\text{C}}{m}(1+x)} = \frac{2.4\underline{0}234}{(1+x)}\, m$ then

$$m = \frac{\text{amount solute (moles)}}{\text{mass solvent (kg)}} \text{ so}$$

$$mol_{MFx} = m_{MFx} \times kg_{H2O} = \frac{2.40234 \text{ mol MF}_x}{(1+x) \text{ kg}} \times 0.1000 \text{ kg} = \frac{0.240234}{(1+x)} \text{ mol MF}_x \quad \text{then}$$

$$\mathfrak{M} = \frac{g_{MFx}}{mol_{MFx}} = \frac{9.18 \text{ g}}{\frac{0.240234}{(1+x)} \text{ mol MF}_x} = \frac{1 \text{ mol MF}_x}{(96+19x) \text{ g MF}_x} \quad \text{Rearrange and solve for x.}$$

$$(9.18 \text{ g})(1 \text{ mol MF}_x) = (96+19x)(\text{g MF}_x)\left(\frac{0.240234}{(1+x)} \text{ mol MF}_x\right) \rightarrow 9.18(1+x) = 0.240234(96+19x)$$

$$\rightarrow 9.18x - 4.5645x = 23.062 - 9.18 \rightarrow 4.6155x = 13.882 \rightarrow x = \frac{13.882}{4.6155} = 3.0 \text{ so the salt is MF}_3.$$

Since molybdenum has an atomic mass ~ 96 amu, the salt is MoF_3.
Check: The units (none) are correct. The answer (3.0) seems reasonable since it is a small integer.

89. a) The two substances mix because their intermolecular forces between themselves are roughly equal to the forces between each other and there is a pervasive tendency to increase randomness, which happens when the two substances mix.

b) $\Delta H_{soln} \approx 0$, since the intermolecular forces between themselves are roughly equal to the forces between each other.

c) ΔH_{solute} and $\Delta H_{solvent}$ are positive, ΔH_{mix} is negative and equals the sum of ΔH_{solute} and $\Delta H_{solvent}$.

90. The warm coolant water should not be put directly into the river without cooling, because it will raise the temperature of the water. When water is warmed, there is less dissolved oxygen in the water and this will be detrimental to aquatic life that depends on this dissolved oxygen.

91. d) More solute particles are found in an ionic solution, since the solute breaks apart into its ions. The vapor pressure is lowered the more solute particle are present.

92. b) NaCl. If all of the substances have the same cost per kilogram, we need to determine which substance will generate the largest number of particles per kilogram (or gram). $HOCH_2CH_2OH$ generates 1 mol particle / 62.07 g; NaCl generates 2 mol particles / 58.44 g; KCl generates 2 mol particles / 74.56 g; $MgCl_2$ generates 3 mol particles / 95.22 g; and $SrCl_2$ generates 3 particles / 158.53 g. So NaCl will generate 1 mole of particles for each 29 g.

93. The balloon not only loses He, it also takes in N_2 and O_2 from the air surrounding the balloon (due to the tendency for mixing), increasing the density of the balloon.

Chapter 13
Chemical Kinetics

1. a) $\text{Rate} = -\frac{1}{2}\frac{\Delta[\text{HBr}]}{\Delta t} = \frac{\Delta[H_2]}{\Delta t} = \frac{\Delta[Br_2]}{\Delta t}$

b) **Given:** first 15.0 s; 0.500 M to 0.455 M **Find:** average rate

Conceptual Plan: $t_1, t_2, [\text{HBr}]_1, [\text{HBr}]_2 \rightarrow$ **average rate**

$$\text{Rate} = -\frac{1}{2}\frac{\Delta[\text{HBr}]}{\Delta t}$$

Solution: $\text{Rate} = -\frac{1}{2}\frac{[\text{HBr}]_{t_2} - [\text{HBr}]_{t_1}}{t_2 - t_1} = -\frac{1}{2}\frac{0.455\text{ M} - 0.500\text{ M}}{15.0\text{s} - 0.0\text{ s}} = 1.5 \times 10^{-3}\text{ M s}^{-1}$

Check: The units (M s^{-1}) are correct. The magnitude of the answer (10^{-3}M s^{-1}) makes physical sense because rates are always positive and we are not changing the concentration much in 15 s.

c) **Given:** 0.500 L vessel and part b) data **Find:** mol_{Br2} formed

Conceptual Plan: **average rate, t_1, t_2, → $\Delta[Br_2]$ then $\Delta[Br_2]$, L → mol_{Br2} formed**

$$\text{Rate} = \frac{\Delta[Br_2]}{\Delta t} \qquad M = \frac{mol_{Br2}}{L}$$

Solution: $\text{Rate} = 1.5 \times 10^{-3}\text{ M s}^{-1} = \frac{\Delta[Br_2]}{\Delta t} = \frac{\Delta[Br_2]}{15.0\text{s} - 0.0\text{ s}}$ Rearrange to solve for $\Delta[Br_2]$.

$\Delta[Br_2] = 1.5 \times 10^{-3}\frac{\text{M}}{\text{s}} \times 15.0\text{ s} = 0.02\underline{2}5\text{ M}$ then $M = \frac{mol_{Br2}}{L}$. Rearrange to solve for mol_{Br2}.

$0.2\underline{2}5\frac{\text{mol Br}_2}{\text{L}} \times 0.500\text{ L} = 0.011\text{ mol Br}_2$.

Check: The units (mol) are correct. The magnitude of the answer (0.01 mol) makes physical sense because the times are the same as in part b) and we need to divide by 2 twice because of the stoichiometric coefficient difference and the volume of the vessel.

2. a) $\text{Rate} = -\frac{1}{2}\frac{\Delta[N_2O]}{\Delta t} = \frac{1}{2}\frac{\Delta[N_2]}{\Delta t} = \frac{\Delta[O_2]}{\Delta t}$

b) **Given:** first 10.0 s; 0.018 mol O_2 in 0.250 L **Find:** average rate

Conceptual Plan: mol_{O2}, **L → M then $t_1, t_2, [O_2]_1, [O_2]_2 \rightarrow$ average rate**

$$M = \frac{mol_{Br2}}{L} \qquad \text{Rate} = \frac{\Delta[O_2]}{\Delta t}$$

Solution: at $t_1 = 0$ s we have no O_2. $M = \frac{mol_{O2}}{L} = \frac{0.018\text{ mol}O_2}{0.250\text{ L}} = 0.072\text{ M}$ then

$\text{Rate} = \frac{[O_2]_{t_2} - [O_2]_{t_1}}{t_2 - t_1} = \frac{0.072\text{ M} - 0.000\text{ M}}{10.0\text{s} - 0.0\text{ s}} = 7.2 \times 10^{-3}\text{ M s}^{-1}$

Check: The units (M s^{-1}) are correct. The magnitude of the answer (10^{-3} M s^{-1}) makes physical sense because rates are always positive and we are not changing the concentration much in 10 s.

c) **Given:** part b) data **Find:** $\frac{\Delta[N_2O]}{\Delta t}$

Conceptual Plan: average rate → $\frac{\Delta[N_2O]}{\Delta t}$

$$\text{Rate} = -\frac{1}{2}\frac{\Delta[N_2O]}{\Delta t} = \frac{\Delta[O_2]}{\Delta t}$$

Solution: $\text{Rate} = -\frac{1}{2}\frac{\Delta[N_2O]}{\Delta t} = \frac{\Delta[O_2]}{\Delta t}$ Rearrange to solve for $\frac{\Delta[N_2O]}{\Delta t}$.

$\frac{\Delta[N_2O]}{\Delta t} = -2\frac{\Delta[O_2]}{\Delta t} = -2 \times 7.2 \times 10^{-3}\text{ M s}^{-1} = -0.014\text{ M s}^{-1}$

Check: The units ($M\ s^{-1}$) are correct. The magnitude of the answer ($-\ 0.01\ M\ s^{-1}$) makes physical sense because multiply by 2 because of the stoichiometric coefficient difference and the change in concentration with time is negative since this is a reactant.

3. a) $\text{Rate} = -\frac{1}{2}\frac{\Delta[A]}{\Delta t} = -\frac{\Delta[B]}{\Delta t} = \frac{1}{3}\frac{\Delta[C]}{\Delta t}$

b) **Given:** $\frac{\Delta[A]}{\Delta t} = -0.100\ M/s$ **Find:** $\frac{\Delta[B]}{\Delta t}$, and $\frac{\Delta[C]}{\Delta t}$

Conceptual Plan: $\frac{\Delta[A]}{\Delta t} \rightarrow \frac{\Delta[B]}{\Delta t}$**, and** $\frac{\Delta[C]}{\Delta t}$

$$\text{Rate} = -\frac{1}{2}\frac{\Delta[A]}{\Delta t} = -\frac{\Delta[B]}{\Delta t} = \frac{1}{3}\frac{\Delta[C]}{\Delta t}$$

Solution: $\text{Rate} = -\frac{1}{2}\frac{\Delta[A]}{\Delta t} = -\frac{\Delta[B]}{\Delta t} = \frac{1}{3}\frac{\Delta[C]}{\Delta t}$ Substitute in value and solve for the two desired values.

$-\frac{1}{2}\frac{-0.100\ M}{s} = -\frac{\Delta[B]}{\Delta t}$ so $\frac{\Delta[B]}{\Delta t} = -\ 0.0500\ M s^{-1}$ and $-\frac{1}{2}\frac{-0.100\ M}{s} = \frac{1}{3}\frac{\Delta[C]}{\Delta t}$ so $\frac{\Delta[C]}{\Delta t} = 0.150\ M s^{-1}$

Check: The units ($M\ s^{-1}$) are correct. The magnitude of the answer ($-\ 0.05\ M\ s^{-1}$) makes physical sense because fewer moles of B are reacting for every mole of A and the change in concentration with time is negative since this is a reactant. The magnitude of the answer ($0.15\ M\ s^{-1}$) makes physical sense because more moles of C are being formed for every mole of A reacting and the change in concentration with time is positive since this is a product.

4. a) $\text{Rate} = -\frac{\Delta[A]}{\Delta t} = -2\frac{\Delta[B]}{\Delta t} = \frac{1}{2}\frac{\Delta[C]}{\Delta t}$

b) **Given:** $\frac{\Delta[C]}{\Delta t} = 0.025\ M/s$ **Find:** $\frac{\Delta[B]}{\Delta t}$, and $\frac{\Delta[A]}{\Delta t}$

Conceptual Plan: $\frac{\Delta[C]}{\Delta t} \rightarrow \frac{\Delta[B]}{\Delta t}$**, and** $\frac{\Delta[A]}{\Delta t}$

$$\text{Rate} = -\frac{\Delta[A]}{\Delta t} = -2\frac{\Delta[B]}{\Delta t} = \frac{1}{2}\frac{\Delta[C]}{\Delta t}$$

Solution: $\text{Rate} = -\frac{\Delta[A]}{\Delta t} = -2\frac{\Delta[B]}{\Delta t} = \frac{1}{2}\frac{\Delta[C]}{\Delta t}$ Substitute in value and solve for the two desired values.

$-2\frac{\Delta[B]}{\Delta t} = \frac{1}{2}\frac{0.025\ M}{s}$ so $\frac{\Delta[B]}{\Delta t} = -\ 0.0063\ M s^{-1}$ and $-\frac{\Delta[A]}{\Delta t} = \frac{1}{2}\frac{0.025\ M}{s}$ so $\frac{\Delta[A]}{\Delta t} = -\ 0.013\ M s^{-1}$

Check: The units ($M\ s^{-1}$) are correct. The magnitude of the answer ($-\ 0.006\ M\ s^{-1}$) makes physical sense because fewer moles of B are reacting for every mole of C being formed and the change in concentration with time is negative since this is a reactant. The magnitude of the answer ($-\ 0.01\ M\ s^{-1}$) makes physical sense because fewer moles of A are reacting for every mole of C being formed and the change in concentration with time is negative since this is a reactant. B has the smallest stoichiometric coefficient and so its rate of change has the smallest magnitude.

5. a) **Given:** $[C_4H_8]$ versus time data **Find:** average rate between 0 and 10 s, and between 40 and 50 s

Conceptual Plan: $\mathbf{t_1, t_2, [C_4H_8]_1, [C_4H_8]_2 \rightarrow}$ **average rate**

$$\text{Rate} = -\frac{\Delta[C_4H_8]}{\Delta t}$$

Solution: for 0 to 10 s: $\text{Rate} = -\frac{[C_4H_8]_{t_2} - [C_4H_8]_{t_1}}{t_2 - t_1} = -\frac{0.913\ M - 1.000\ M}{10.\ s - 0.\ s} = 8.7 \times 10^{-3}\ M\ s^{-1}$ and

for 40 to 50 s: $\text{Rate} = -\frac{[C_4H_8]_{t_2} - [C_4H_8]_{t_1}}{t_2 - t_1} = -\frac{0.637\ M - 0.697\ M}{50.\ s - 40.\ s} = 6.0 \times 10^{-3}\ M\ s^{-1}$

Check: The units ($M\ s^{-1}$) are correct. The magnitude of the answer ($10^{-3}\ M\ s^{-1}$) makes physical sense because rates are always positive and we are not changing the concentration much in 10 s. Also reactions slow as they proceed because the concentration of the reactants is decreasing.

b) **Given:** [C_4H_8] versus time data **Find:** $\frac{\Delta[C_2H_4]}{\Delta t}$ between 20 and 30 s

Conceptual Plan: $\mathbf{t_1, t_2, [C_4H_8]_1, [C_4H_8]_2} \rightarrow \frac{\Delta[C_2H_4]}{\Delta t}$

$$\text{Rate} = -\frac{\Delta[C_4H_8]}{\Delta t} = \frac{1}{2}\frac{\Delta[C_2H_4]}{\Delta t}$$

Solution: $\text{Rate} = -\frac{[C_4H_8]_{t_2} - [C_4H_8]_{t_1}}{t_2 - t_1} = -\frac{0.763\ M - 0.835\ M}{30.\ s - 20.\ s} = 7.2 \times 10^{-3}\ M\ s^{-1} = \frac{1}{2}\frac{\Delta[C_2H_4]}{\Delta t}$

Rearrange to solve for $\frac{\Delta[C_2H_4]}{\Delta t}$. So $\frac{\Delta[C_2H_4]}{\Delta t} = 2(7.2 \times 10^{-3}\ M\ s^{-1}) = 1.4 \times 10^{-2}\ M\ s^{-1}$.

Check: The units ($M\ s^{-1}$) are correct. The magnitude of the answer ($10^{-2}\ M\ s^{-1}$) makes physical sense because rate of product formation is always positive and we are not changing the concentration much in 10 s. The rate of change of the product is faster than the decline of the reactant because of the stoichiometric coefficients.

6. a) **Given:** [NO_2] versus time data **Find:** average rate between 10 and 20 s, and between 50 and 60 s
Conceptual Plan: $\mathbf{t_1, t_2, [NO_2]_1, [NO_2]_2} \rightarrow$ **average rate**

$$\text{Rate} = -\frac{\Delta[NO_2]}{\Delta t}$$

Solution: for 10 to 20 s: $\text{Rate} = -\frac{[NO_2]_{t_2} - [NO_2]_{t_1}}{t_2 - t_1} = -\frac{0.904\ M - 0.951\ M}{20.\ s - 10.\ s} = 4.7 \times 10^{-3}\ M\ s^{-1}$ and

for 50 to 60 s: $\text{Rate} = -\frac{[NO_2]_{t_2} - [NO_2]_{t_1}}{t_2 - t_1} = -\frac{0.740\ M - 0.778\ M}{60.\ s - 50.\ s} = 3.8 \times 10^{-3}\ M\ s^{-1}$

Check: The units ($M\ s^{-1}$) are correct. The magnitude of the answer ($10^{-3}\ M\ s^{-1}$) makes physical sense because rates are always positive and we are not changing the concentration much in 10 s. Also reactions slow as they proceed because the concentration of the reactants is decreasing.

b) **Given:** [NO_2] versus time data **Find:** $\frac{\Delta[O_2]}{\Delta t}$ between 50 and 60 s

Conceptual Plan: **average rate from part a)** $\rightarrow \frac{\Delta[O_2]}{\Delta t}$

$$\text{Rate} = -\frac{\Delta[NO_2]}{\Delta t} = 2\frac{\Delta[O_2]}{\Delta t}$$

Solution: $\text{Rate} = -\frac{\Delta[NO_2]}{\Delta t} = 2\frac{\Delta[O_2]}{\Delta t}$ Substitute in value and solve for the desired value.

$3.8 \times 10^{-3}\ M\ s^{-1} = 2\frac{\Delta[O_2]}{\Delta t}$ so $\frac{\Delta[O_2]}{\Delta t} = 1.9 \times 10^{-3}\ M\ s^{-1}$

Check: The units ($M\ s^{-1}$) are correct. The magnitude of the answer ($10^{-3}\ M\ s^{-1}$) makes physical sense because rate of product formation is always positive and we are not changing the concentration much in 10 s. The rate of change of the product is slower than the decline of the reactant because of the stoichiometric coefficients.

7. a) **Given:** [Br_2] versus time plot
Find: (i) average rate between 0 and 25 s; (ii) instantaneous rate at 25 s; and (iii) instantaneous rate of HBr formation at 50 s
Conceptual Plan: **(i)** $\mathbf{t_1, t_2, [Br_2]_1, [Br_2]_2} \rightarrow$ **average rate** **then**

$$\text{Rate} = -\frac{\Delta[Br_2]}{\Delta t}$$

(ii) draw tangent at 25 s and determine slope → instantaneous rate then

$$\text{Rate} = -\frac{\Delta[Br_2]}{\Delta t}$$

(iii) draw tangent at 50 s and determine slope → instantaneous rate → $\dfrac{\Delta[HBr]}{\Delta t}$

$$\text{Rate} = -\frac{\Delta[Br_2]}{\Delta t} \qquad \text{Rate} = \frac{1}{2}\frac{\Delta[HBr]}{\Delta t}$$

Solution:

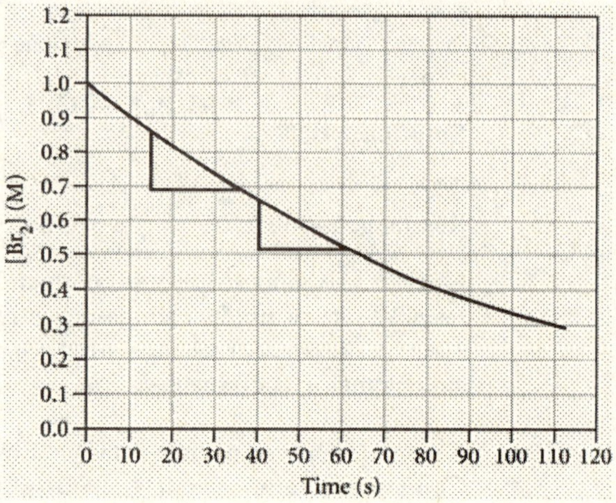

(i) $\text{Rate} = -\dfrac{[Br_2]_{t_2} - [Br_2]_{t_1}}{t_2 - t_1} = -\dfrac{0.75\text{ M} - 1.00\text{ M}}{25\text{ s} - 0.\text{ s}} =$

$= 1.0 \times 10^{-2}\text{ M s}^{-1}$

and (ii) at 25 s:

$\text{Slope} = \dfrac{\Delta y}{\Delta x} = \dfrac{0.68\text{ M} - 0.85\text{ M}}{35\text{ s} - 15\text{ s}} = -8.5 \times 10^{-3}\text{ M s}^{-1}$

since the slope $= \dfrac{\Delta[Br_2]}{\Delta t}$ and $\text{Rate} = -\dfrac{\Delta[Br_2]}{\Delta t}$,

then Rate $= -(-8.5 \times 10^{-3}\text{ M s}^{-1}) = 8.5 \times 10^{-3}\text{ M s}^{-1}$

(iii) at 50 s:

$\text{Slope} = \dfrac{\Delta y}{\Delta x} = \dfrac{0.53\text{ M} - 0.66\text{ M}}{60.\text{ s} - 40.\text{ s}} = -6.5 \times 10^{-3}\text{ M s}^{-1}$

since the slope $= \dfrac{\Delta[Br_2]}{\Delta t}$ and $\text{Rate} = -\dfrac{\Delta[Br_2]}{\Delta t} = \dfrac{1}{2}\dfrac{\Delta[HBr]}{\Delta t}$ then

$\dfrac{\Delta[HBr]}{\Delta t} = -2\dfrac{\Delta[Br_2]}{\Delta t} = -2(-6.5 \times 10^{-3}\text{ M s}^{-1}) = 1.3 \times 10^{-2}\text{ M s}^{-1}$

Check: The units (M s^{-1}) are correct. The magnitude of the first answer is larger than the second answer because the rate is slowing down and the first answer includes the initial portion of the data. The magnitudes of the answers (10^{-3} M s^{-1}) make physical sense because rates are always positive and we are not changing the concentration much.

b) **Given:** $[Br_2]$ versus time data; and $[HBr]_0 = 0$ M
Find: plot [HBr] with time

Conceptual Plan: Since $\text{Rate} = -\dfrac{\Delta[Br_2]}{\Delta t} = \dfrac{1}{2}\dfrac{\Delta[HBr]}{\Delta t}$. **The rate of change of [HBr] will be twice that of $[Br_2]$. The plot will start at the origin.**

Solution:

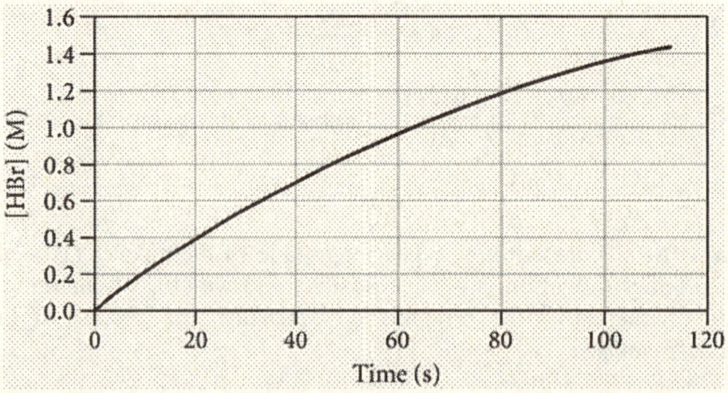

Check: The units (M versus s) are correct. The plot makes sense because the plot has the same general shape of the original plot, only we are increasing instead of decreasing and our concentration axis has changed by a factor of two (to account for the difference in stoichiometric coefficients).

8. **Given:** $[H_2O_2]$ versus time plot; and 1.5 L H_2O_2 initially **Find:** (a) average rate between 10 and 20 s; (b) instantaneous rate at 30 s; (c) instantaneous rate of O_2 formation at 50 s; and (d) mol_{O_2} formed in first 50 s

Conceptual Plan: **(a)** **t_1, t_2, $[H_2O_2]_1$, $[H_2O_2]_2$ → average rate** **then**

$$\text{Rate} = -\frac{1}{2}\frac{\Delta[H_2O_2]}{\Delta t}$$

(b) draw tangent at 30 s and determine slope → instantaneous rate **then**

$$\text{Rate} = -\frac{1}{2}\frac{\Delta[H_2O_2]}{\Delta t}$$

(c) draw tangent at 50 s and determine slope → instantaneous rate → $\frac{\Delta[O_2]}{\Delta t}$

$$\text{Rate} = -\frac{1}{2}\frac{\Delta[H_2O_2]}{\Delta t} \qquad \text{Rate} = \frac{\Delta[O_2]}{\Delta t}$$

(d) $[H_2O_2]_{0\,s}$, $[H_2O_2]_{50\,s}$ → Δ $[H_2O_2]$ → Δ $[O_2]$ then Δ $[O_2]$, V → mol_{O2}

$$\Delta[H_2O_2] = [H_2O_2]_{50\,s} - [H_2O_2]_{0\,s} \qquad \text{Rate} = -\frac{1}{2}\frac{\Delta[H_2O_2]}{\Delta t} = \frac{\Delta[O_2]}{\Delta t} \qquad M = \frac{mol_{O2}}{L}$$

Solution:

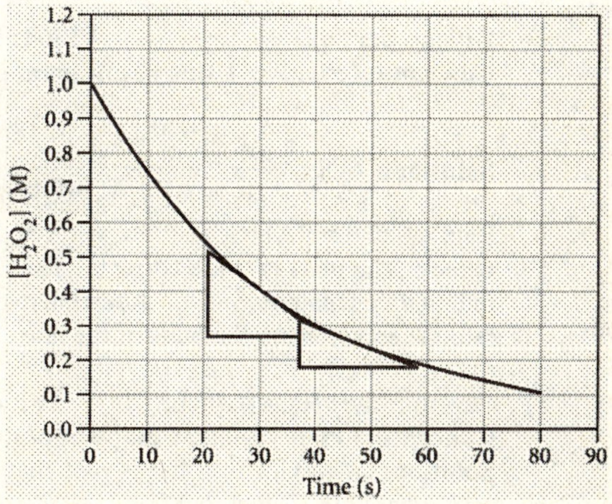

(a) $\text{Rate} = -\frac{[H_2O_2]_{t_2} - [H_2O_2]_{t_1}}{t_2 - t_1} = -\frac{0.55\text{ M} - 0.75\text{ M}}{20.\text{ s} - 10.\text{ s}}$ $= 2.0 \times 10^{-2}\text{ M s}^{-1}$ and

(b) at 30 s:

$\text{Slope} = \frac{\Delta y}{\Delta x} = \frac{0.28\text{ M} - 0.52\text{ M}}{40.\text{ s} - 20.\text{ s}} = -1.2 \times 10^{-2}\text{ M s}^{-1}$ since

the slope $= \frac{\Delta[H_2O_2]}{\Delta t}$ and $\text{Rate} = -\frac{1}{2}\frac{\Delta[H_2O_2]}{\Delta t}$, then

$\text{Rate} = (-0.5)(-1.2 \times 10^{-2}\text{ M s}^{-1}) = 6.0 \times 10^{-3}\text{ M s}^{-1}$.

(c) at 50 s:

$\text{Slope} = \frac{\Delta y}{\Delta x} = \frac{0.15\text{ M} - 0.28\text{ M}}{60.\text{ s} - 40.\text{ s}} = -6.5 \times 10^{-3}\text{ M s}^{-1}$ since the slope $= \frac{\Delta[H_2O_2]}{\Delta t}$ and

$\text{Rate} = -\frac{1}{2}\frac{\Delta[H_2O_2]}{\Delta t} = \frac{\Delta[O_2]}{\Delta t}$ then $\frac{\Delta[O_2]}{\Delta t} = (-0.5)(-6.5 \times 10^{-3}\text{ M s}^{-1}) = 3.3 \times 10^{-3}\text{ M s}^{-1}$

(d) $\Delta[H_2O_2] = [H_2O_2]_{50\,s} - [H_2O_2]_{0\,s} = 0.23\text{ M} - 1.00\text{ M} = 0.77\text{ M}$ since

$\Delta[O_2] = -\frac{1}{2}\Delta[H_2O_2] = (-0.5)(0.77\text{ M}) = 0.3\underline{8}5\text{ M}$ then $M = \frac{mol_{O2}}{L}$ so

$mol_{O2} = ML = 0.3\underline{8}5\frac{\text{mol }O_2}{\cancel{L}} \times 1.5\ \cancel{L} = 0.58\text{ mol }O_2$

Check: (a) The units (M s^{-1}) are correct. The magnitude of the first answer is reasonable considering the concentrations and times involved (1M / 100 s). (b) The units (M s^{-1}) are correct. We expect the answer in this part to be less than in the first part because the rate is decreasing as the reaction proceeds. (c) The units (M s^{-1}) are correct. We expect the answer in this part to be less than in the first part because the rate is decreasing as the reaction proceeds. (d) The units (mol) are correct. We expect an answer less than 1 mol since the drop in reactant concentration is less than 1 M, we have 1.5 L and only half as much O_2 is generated as hydrogen peroxide is consumed.

9. a) **Given:** Rate versus [A] plot **Find:** reaction order
Conceptual Plan: Look at shape of plot and match to possibilities.
Solution: The plot is a linear plot, so Rate α [A] or the reaction is first order.
Check: The order of the reaction is a common reaction order.

b) **Given:** part a) **Find:** sketch plot of [A] versus time
Conceptual Plan: **Using result from part a), shape plot of [A] versus time should be curved with [A] decreasing. Use 1.0 M as initial concentration.**

Solution:

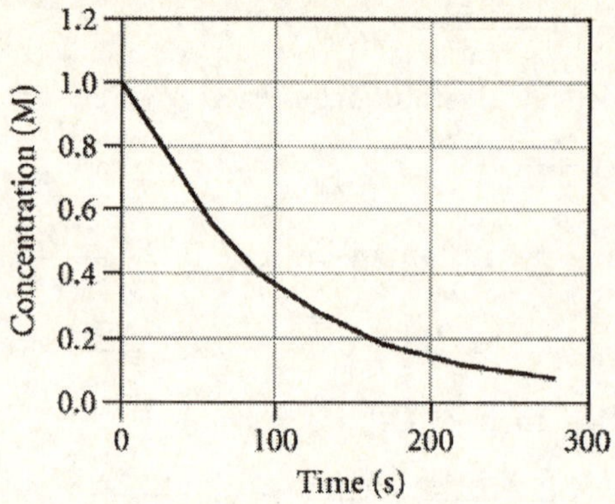

Check: The plot has a shape that matches the one in the text for first order plots.

c) **Given: part a)** **Find:** write a rate law and estimate k
Conceptual Plan: Using result from part a), the slope of the plot is the rate constant.

Solution: Slope $= \frac{\Delta y}{\Delta x} = \frac{0.010 \frac{\cancel{M}}{s} - 0.00 \frac{\cancel{M}}{s}}{1.0\ \cancel{M} - 0.0\ \cancel{M}} = 0.010\ s^{-1}$ so Rate $= k\,[A]^1$ or Rate $= k\,[A]$ or Rate $= 0.010\ s^{-1}\,[A]$

Check: The units (s^{-1}) are correct. The magnitude of the answer ($10^{-2}\ s^{-1}$) makes physical sense because, rate and concentration data. Remember that concentration is in units of M, so plugging the rate constant into the equation has the units of the rate as $M\ s^{-1}$, which is correct.

10. a) **Given:** Rate versus [A] plot **Find:** reaction order
Conceptual Plan: Look at shape of plot and match to possibilities.
Solution: The plot is a linear plot that is horizontal, so Rate is independent of [A] or the reaction is zero order with respect to A.
Check: The order of the reaction is a common reaction order.

b) **Given:** part a) **Find:** sketch plot of [A] versus time
Conceptual Plan: Using result from part a), shape plot of [A] versus time should be a straight line with [A] decreasing. Use 1.0 M as initial concentration.
Solution:

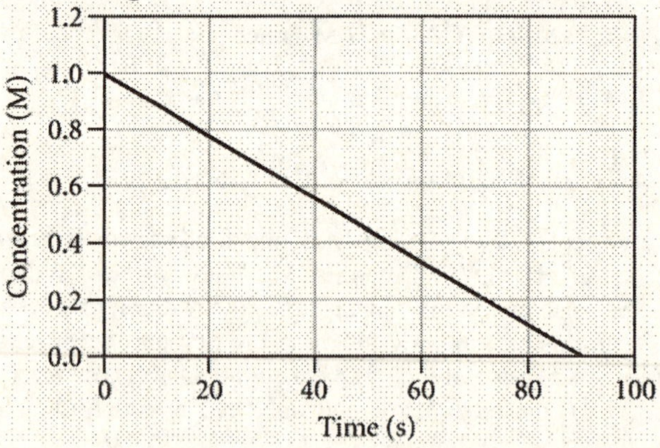

Check: The plot has a shape that matches the one in the text for zero order plots.

c) **Given: part a)** **Find:** write a rate law and estimate k
Conceptual Plan: Using result from part a), the rate is equal to the rate constant.
Solution: Rate $= k\,[A]^0$ or Rate $= k$ or Rate $= 0.011\ M\ s^{-1}$
Check: The units (Ms^{-1}) are correct. The magnitude of the answer ($10^{-2}\ Ms^{-1}$) makes physical sense because of the rate and concentration data. Plugging the rate constant into the equation, the rate has the units of $M\ s^{-1}$, which is correct.

11. **Given:** reaction order: a) first-order; b) second-order; and c) zero-order **Find:** units of k

Conceptual Plan: **Using rate law, rearrange to solve for k.**

$$\text{Rate} = k\,[A]^n \text{, where n = reaction order}$$

Solution: for all cases Rate has units of M s^{-1} and [A] has units of M

a) $\text{Rate} = k\,[A]^1 = k\,[A]$ so $k = \dfrac{\text{Rate}}{[A]} = \dfrac{\frac{\cancel{M}}{s}}{\cancel{M}} = s^{-1}$;

b) $\text{Rate} = k\,[A]^2$ so $k = \dfrac{\text{Rate}}{[A]^2} = \dfrac{\frac{\cancel{M}}{s}}{\cancel{M}M} = M^{-1}\,s^{-1}$;

c) $\text{Rate} = k\,[A]^0 = k = M\,s^{-1}$.

Check: The units (s^{-1}, $M^{-1}s^{-1}$ and Ms^{-1}) are correct. The units for k change with the reaction order so that the units on the rate remain as M s^{-1}.

12. **Given:** k = 0.053/s and $[N_2O_5]$ = 0.055 M; reaction order: a) first-order; b) second-order and zero-order (change units on k as necessary) **Find:** rate
Conceptual Plan: **Using rate law, substitute in values to solve for Rate.**

$$\text{Rate} = k\,[N_2O_5]^n \text{, where n = reaction order}$$

Solution: for all cases Rate has units of M s^{-1} and [A] has units of M. Use the results from problem #35 to choose the appropriate units for k.

a) $\text{Rate} = k\,[N_2O_5]^1 = k\,[N_2O_5] = \dfrac{0.053}{s} \times 0.055\text{ M} = 2.9 \times 10^{-3}\,\dfrac{M}{s}$;

b) $\text{Rate} = k\,[N_2O_5]^2 = k\,[N_2O_5] = \dfrac{0.053}{M\,s} \times (0.055\text{ M})^2 = 1.6 \times 10^{-4}\,\dfrac{M}{s}$; and

$\text{Rate} = k\,[N_2O_5]^0 = k = 5.3 \times 10^{-2}\,\dfrac{M}{s}$

Check: The units (Ms^{-1}) are correct. The magnitude of the rate changes as the order of the reaction changes, because we are multiplying by the concentration a different number of times in each case. The higher the order the lower the rate since the concentration is less than 1 M.

13. **Given:** A, B, and C react to form products. Reaction is first order in A, second order in B, and zero order in C **Find:** a) rate law; b) overall order of reaction; c) factor change in rate if [A] doubled; d) factor change in rate if [B] doubled; e) factor change in rate if [C] doubled; and f) factor change in rate if [A], [B], and [C] doubled.
Conceptual Plan:
a) Using general rate law form, substitute in values for orders.

$$\text{Rate} = k\,[A]^m\,[B]^n\,[C]^p \text{, where m, n, and p = reaction orders}$$

b) Using rate law in part a) add up all reaction orders.

$$\textit{overall reaction order} = m + n + p$$

c) through f) Using rate law from part a) substitute in concentration changes.

$$\frac{\text{Rate 2}}{\text{Rate 1}} = \frac{k\,[A]_2^1\,[B]_2^2}{k\,[A]_1^1\,[B]_1^2}$$

Solution:

a) m = 1, n = 2, and p = 0 so $\text{Rate} = k\,[A]^1[B]^2[C]^0$ or $\text{Rate} = k\,[A][B]^2$.

b) $\textit{overall reaction order} = m + n + p = 1 + 2 + 0 = 3$ so it is a third order reaction overall.

c) $\dfrac{\text{Rate 2}}{\text{Rate 1}} = \dfrac{k\,[A]_2^1\,[B]_2^2}{k\,[A]_1^1\,[B]_1^2}$ and $[A]_2 = 2\,[A]_1$, $[B]_2 = [B]_1$, $[C]_2 = [C]_1$, so $\dfrac{\text{Rate 2}}{\text{Rate 1}} = \dfrac{\cancel{k}(2\,\cancel{[A]_1})^1\,\cancel{[B]_1^2}}{\cancel{k}\,\cancel{[A]_1^1}\,\cancel{[B]_1^2}} = 2$ so the reaction rate doubles (factor of 2).

d) $\frac{\text{Rate 2}}{\text{Rate 1}} = \frac{k\,[A]_2^1\,[B]_2^2}{k\,[A]_1^1\,[B]_1^2}$ and $[A]_2 = [A]_1$, $[B]_2 = 2\,[B]_1$, $[C]_2 = [C]_1$, so $\frac{\text{Rate 2}}{\text{Rate 1}} = \frac{\cancel{k}\,\cancel{[A]_1}\,(2\,\cancel{[B]_1})^2}{\cancel{k}\,\cancel{[A]_1^1}\,\cancel{[B]_1^2}} = 2^2 = 4$

so the reaction rate quadruples (factor of 4).

e) $\frac{\text{Rate 2}}{\text{Rate 1}} = \frac{k\,[A]_2^1\,[B]_2^2}{k\,[A]_1^1\,[B]_1^2}$ and $[A]_2 = [A]_1$, $[B]_2 = [B]_1$, $[C]_2 = 2\,[C]_1$, so $\frac{\text{Rate 2}}{\text{Rate 1}} = \frac{\cancel{k}\,\cancel{[A]_1^1}\,\cancel{[B]_1^2}}{\cancel{k}\,\cancel{[A]_1^1}\,\cancel{[B]_1^2}} = 1$ so the reaction rate is unchanged (factor of 1).

f) $\frac{\text{Rate 2}}{\text{Rate 1}} = \frac{k\,[A]_2^1\,[B]_2^2}{k\,[A]_1^1\,[B]_1^2}$ and $[A]_2 = 2\,[A]_1$, $[B]_2 = 2\,[B]_1$, $[C]_2 = 2\,[C]_1$, so

$\frac{\text{Rate 2}}{\text{Rate 1}} = \frac{\cancel{k}\,(2\,\cancel{[A]_1})^1\,(2\,\cancel{[B]_1})^2}{\cancel{k}\,\cancel{[A]_1^1}\,\cancel{[B]_1^2}} = 2 \times 2^2 = 8$ so the reaction rate goes up by a factor of 8.

Check: The units (none) are correct. The rate law is consistent with the orders given and the overall order is larger that any of the individual orders. The factors are consistent with the reaction orders. The larger the order, the larger the factor. When all concentrations are changed the rate changes the most. If a reactant is not in the rate law, then changing its concentration has no effect on the reaction rate.

14. **Given:** A, B, and C react to form products. Reaction is zero order in A, one-half order in B, and second order in C **Find:** a) rate law; b) overall order of reaction; c) factor change in rate if [A] doubled; d) factor change in rate if [B] doubled; e) factor change in rate if [C] doubled; and f) factor change in rate if [A], [B], and [C] doubled.

Conceptual Plan:

a) Using general rate law form, substitute in values for orders.

$$\text{Rate} = k\,[A]^m\,[B]^n\,[C]^p \text{ , where m, n, and p = reaction orders}$$

b) Using rate law in part a) add up all reaction orders.

$$\textit{overall reaction order} = m + n + p$$

c) through f) Using rate law from part a) substitute in concentration changes.

$$\frac{\text{Rate 2}}{\text{Rate 1}} = \frac{k\,[B]_2^{1/2}\,[C]_2^2}{k\,[B]_1^{1/2}\,[C]_1^2}$$

Solution:

a) m = 0, n = 1/2, and p = 2 so $\text{Rate} = k\,[A]^0[B]^{1/2}[C]^2$ or $\text{Rate} = k[B]^{1/2}[C]^2$.

b) *overall reaction order* $= m + n + p = 0 + 1/2 + 2 = 5/2 = 2.5$ so it is a two and a half order reaction overall.

c) $\frac{\text{Rate 2}}{\text{Rate 1}} = \frac{k\,[B]_2^{1/2}\,[C]_2^2}{k\,[B]_1^{1/2}\,[C]_1^2}$ and $[A]_2 = 2\,[A]_1$, $[B]_2 = [B]_1$, $[C]_2 = [C]_1$, so $\frac{\text{Rate 2}}{\text{Rate 1}} = \frac{\cancel{k}\,\cancel{[B]_1^{1/2}}\,\cancel{[C]_1^2}}{\cancel{k}\,\cancel{[B]_1^{1/2}}\,\cancel{[C]_1^2}} = 1$ so the reaction rate is unchanged (factor of 1).

d) $\frac{\text{Rate 2}}{\text{Rate 1}} = \frac{k\,[B]_2^{1/2}\,[C]_2^2}{k\,[B]_1^{1/2}\,[C]_1^2}$ and $[A]_2 = [A]_1$, $[B]_2 = 2\,[B]_1$, $[C]_2 = [C]_1$, so $\frac{\text{Rate 2}}{\text{Rate 1}} = \frac{\cancel{k}(2\,\cancel{[B]_1})^{1/2}\,\cancel{[C]_1^2}}{\cancel{k}\,\cancel{[B]_1^{1/2}}\,\cancel{[C]_1^2}} = 2^{1/2}$

so the reaction rate increases by a factor of $2^{1/2}$ or $\sqrt{2}$.

e) $\frac{\text{Rate 2}}{\text{Rate 1}} = \frac{k\,[B]_2^{1/2}\,[C]_2^2}{k\,[B]_1^{1/2}\,[C]_1^2}$ and $[A]_2 = [A]_1$, $[B]_2 = [B]_1$, $[C]_2 = 2\,[C]_1$, so

$\frac{\text{Rate 2}}{\text{Rate 1}} = \frac{\cancel{k}\,\cancel{[B]_1^{1/2}}\,(2\,\cancel{[C]_1})^2}{\cancel{k}\,\cancel{[B]_1^{1/2}}\,\cancel{[C]_1^2}} = 2^2 = 4$ so the reaction rate quadruples (factor of 4).

f) $\frac{\text{Rate 2}}{\text{Rate 1}} = \frac{k\,[B]_2^{1/2}\,[C]_2^2}{k\,[B]_1^{1/2}\,[C]_1^2}$ and $[A]_2 = 2\,[A]_1$, $[B]_2 = 2\,[B]_1$, $[C]_2 = 2\,[C]_1$, so

$$\frac{\text{Rate 2}}{\text{Rate 1}} = \frac{k\,(2\,[B]_1)^{1/2}\,(2\,[C]_1)^2}{k\,[B]_1^{1/2}\,[C]_1^2} = 2^{1/2} \times 2^2 = 2^{5/2}$$ so the reaction rate goes up by a factor of $2^{5/2}$ or $4\sqrt{2}$.

Check: The units (none) are correct. The rate law is consistent with the orders given and the overall order is larger that any of the individual orders. The factors are consistent with the reaction orders. The larger the order, the larger the factor. When all concentrations are changed the rate changes the most. If a reactant is not in the rate law, then changing its concentration has no effect on the reaction rate.

15. **Given:** table of [A] versus initial rate **Find:** rate law and k

Conceptual Plan: Using general rate law form, compare rate ratios to determine reaction order.

$$\frac{\text{Rate 2}}{\text{Rate 1}} = \frac{k\,[A]_2^n}{k\,[A]_1^n}$$

Then use one of the concentration/ initial rate pairs to determine *k*.

$$\text{Rate} = k[A]^n$$

Solution: $\frac{\text{Rate 2}}{\text{Rate 1}} = \frac{k\,[A]_2^n}{k\,[A]_1^n}$ Comparing the first two sets of data: $\frac{0.210\ \text{M/s}}{0.053\ \text{M/s}} = \frac{k\,(0.200\ \text{M})^n}{k\,(0.100\ \text{M})^n}$ and $3.9623 = 2^n$ so n = 2. If we compare the first and the last data sets: $\frac{0.473\ \text{M/s}}{0.053\ \text{M/s}} = \frac{k\,(0.300\ \text{M})^n}{k\,(0.100\ \text{M})^n}$ and $8.9245 = 3^n$ so n = 2. This second comparison is not necessary, but it increases our confidence in the reaction order. So $\text{Rate} = k[A]^2$. Selecting the second data set and rearranging the rate equation

$$k = \frac{\text{Rate}}{[A]^2} = \frac{0.210\frac{\text{M}}{\text{s}}}{(0.200\,\text{M})^2} = 5.25\ \text{M}^{-1}\ \text{s}^{-1}$$ so $\text{Rate} = 5.25\ \text{M}^{-1}\ \text{s}^{-1}[A]^2$

Check: The units (none and $\text{M}^{-1}\ \text{s}^{-1}$) are correct. The rate law is a common form. The rate is changing more rapidly than the concentration, so second order is consistent. The rate constant is consistent with the units necessary to get rate as M/s and the magnitude is reasonable since we have a second order reaction.

16. **Given:** table of [A] versus initial rate **Find:** rate law and k

Conceptual Plan: Using general rate law form, compare rate ratios to determine reaction order.

$$\frac{\text{Rate 2}}{\text{Rate 1}} = \frac{k\,[A]_2^n}{k\,[A]_1^n}$$

Then use one of the concentration/ initial rate pairs to determine *k*.

$$\text{Rate} = k[A]^n$$

Solution: $\frac{\text{Rate 2}}{\text{Rate 1}} = \frac{k\,[A]_2^n}{k\,[A]_1^n}$ Comparing the first two sets of data: $\frac{0.016\ \text{M/s}}{0.008\ \text{M/s}} = \frac{k\,(0.30\ \text{M})^n}{k\,(0.15\ \text{M})^n}$ and $2 = 2^n$ so n = 1. If we compare the first and the last data sets: $\frac{0.032\ \text{M/s}}{0.008\ \text{M/s}} = \frac{k\,(0.032\ \text{M})^n}{k\,(0.008\ \text{M})^n}$ and $4 = 4^n$ so n = 1. This second comparison is not necessary, but it increases our confidence in the reaction order. So $\text{Rate} = k[A]$.

Selecting the second data set and rearranging the rate equation $k = \frac{\text{Rate}}{[A]} = \frac{0.016\frac{\text{M}}{\text{s}}}{0.30\ \text{M}} = 5.3 \times 10^{-2}\ \text{s}^{-1}$ so

$\text{Rate} = 5.3 \times 10^{-2}\ \text{s}^{-1}[A]$.

Check: The units (none and s^{-1}) are correct. The rate law is a common form. The rate is changing as rapidly as the concentration is consistent with first order. The rate constant is consistent with the units necessary to get rate as M/s and the magnitude is reasonable since we have a first order reaction.

17. **Given:** table of $[NO_2]$ and $[F_2]$ versus initial rate **Find:** rate law, k, and overall order

Conceptual Plan: Using general rate law form, compare rate ratios to determine reaction order of each reactant. Be sure to choose data that changes only one concentration at a time.

$$\frac{\text{Rate 2}}{\text{Rate 1}} = \frac{k\,[NO_2]_2^m\,[F_2]_2^n}{k\,[NO_2]_1^m\,[F_2]_1^n}$$

Then use one of the concentration/ initial rate pairs to determine *k*.

$$\text{Rate} = k\,[NO_2]^m\,[F_2]^n$$

Solution: $\frac{\text{Rate 2}}{\text{Rate 1}} = \frac{k\,[NO_2]_2^m\,[F_2]_2^n}{k\,[NO_2]_1^m\,[F_2]_1^n}$ Comparing the first two sets of data:

$\frac{0.051\ \cancel{M/s}}{0.026\ \cancel{M/s}} = \frac{\cancel{k}\,(0.200\ \cancel{M})^m\,(\cancel{0.100\ M})^n}{\cancel{k}\,(0.100\ \cancel{M})]^m\,(\cancel{0.100\ M})^n}$ and $1.9615 = 2^m$ so m = 1. If we compare the second and the third data sets: $\frac{0.103\ \cancel{M/s}}{0.051\ \cancel{M/s}} = \frac{\cancel{k}\,(\cancel{0.200\ M})^m\,(0.200\ \cancel{M})^n}{\cancel{k}\,(\cancel{0.200\ M})^m\,(0.100\ \cancel{M})^n}$ and $2 = 2^n$ so n = 1. Other comparisons can be made, but are not necessary. They should reinforce these values of the reaction orders. So $\text{Rate} = k\,[NO_2]\,[F_2]$.

Selecting the last data set and rearranging the rate equation $k = \frac{\text{Rate}}{[NO_2]\,[F_2]} = \frac{0.411\frac{\cancel{M}}{s}}{(0.400\ \cancel{M})\,(0.400\ M)} = 2.57\ M^{-1}\,s^{-1}$

so $\text{Rate} = 2.57\ M^{-1}\,s^{-1}[NO_2]\,[F_2]$ and the reaction is second order overall.

Check: The units (none and $M^{-1}s^{-1}$) are correct. The rate law is a common form. The rate is changing as rapidly as each concentration is changing, which is consistent with first order in each reactant. The rate constant is consistent with the units necessary to get rate as M/s and the magnitude is reasonable since we have a second order reaction.

18. **Given:** table of $[CH_3Cl]$ and $[Cl_2]$ versus initial rate **Find:** rate law, *k*, and overall order
Conceptual Plan: Using general rate law form, compare rate ratios to determine reaction order of each reactant. Be sure to choose data that changes only one concentration at a time.

$$\frac{\text{Rate 2}}{\text{Rate 1}} = \frac{k\,[CH_3Cl]_2^m\,[Cl_2]_2^n}{k\,[CH_3Cl]_1^m\,[Cl_2]_1^n}$$

Then use one of the concentration/ initial rate pairs to determine *k*.

$$\text{Rate} = k\,[CH_3Cl]^m\,[Cl_2]^n$$

Solution: $\frac{\text{Rate 2}}{\text{Rate 1}} = \frac{k\,[CH_3Cl]_2^m\,[Cl_2]_2^n}{k\,[CH_3Cl]_1^m\,[Cl_2]_1^n}$ Comparing the first two sets of data:

$\frac{0.029\ \cancel{M/s}}{0.014\ \cancel{M/s}} = \frac{\cancel{k}\,(0.100\ \cancel{M})^m\,(\cancel{0.050\ M})^n}{\cancel{k}\,(0.050\ \cancel{M})^m\,(\cancel{0.050\ M})^n}$ and $2.0714 = 2^m$ so m = 1. If we compare the second and the third data sets: $\frac{0.041\ \cancel{M/s}}{0.029\ \cancel{M/s}} = \frac{\cancel{k}\,(\cancel{0.100\ M})^m\,(0.100\ \cancel{M})^n}{\cancel{k}\,(\cancel{0.100\ M})^m\,(0.050\ \cancel{M})^n}$ and $1.414 = 2^n$ so n = 1/2. Other comparisons can be made, but are not necessary. They should reinforce these values of the reaction orders. So $\text{Rate} = k\,[CH_3Cl]\,[Cl_2]^{1/2}$. Selecting the last data set and rearranging the rate equation

$k = \frac{\text{Rate}}{[CH_3Cl]\,[Cl_2]^{1/2}} = \frac{0.115\frac{\cancel{M}}{s}}{(0.200\ \cancel{M})\,(0.200\ M)^{1/2}} = 1.29\ M^{-1/2}\,s^{-1}$ so $\text{Rate} = 1.29\ M^{-1/2}\,s^{-1}[CH_3Cl]\,[Cl_2]^{1/2}$ and the reaction is one and a half order overall.

Check: The units (none and $M^{-1/2}s^{-1}$) are correct. The rate law is not as common as others, but is reasonable. The rate is changing as rapidly as the CH_3Cl concentration is changing, which is consistent with first order in this reactant. The rate is changing a bit more slowly than the Cl_2 concentration, which is consistent with half order in this reactant. The rate constant is consistent with the units necessary to get rate as M/s and the magnitude is reasonable since we have a one and a half order reaction.

19. a) The reaction is zero order. Since the slope of the plot is independent of the concentration, there is no dependence of the concentration of the reactant in the rate law.

b) The reaction is first order. The expression for the half-life of a first order reaction is $t_{1/2} = \frac{0.693}{k}$, which is independent of the reactant concentration.

c) The reaction is second order. The integrated rate expression for a second order reaction is $\frac{1}{[A]_t} = kt + \frac{1}{[A]_0}$, which is linear when the inverse of the concentration is plotted versus time.

20. a) The reaction is second order. The expression for the half-life of a second order reaction, $t_{1/2} = \frac{1}{k[A]_0}$, which shows that the half-life decreases as concentration increases.

b) The reaction is first order. The integrated rate expression for a first order reaction is $\ln[A]_t = -kt + \ln[A]_0$ which is linear when the natural log of the concentration is plotted versus time.

c) The reaction is zero order. The expression for the half-life of a zero order reaction, $t_{1/2} = \frac{[A]_0}{2k}$, which shows that the half-life increases as concentration increases.

21. **Given:** table of [AB] versus time **Find:** reaction order, *k*, and [AB] at 25 s
Conceptual Plan: Look at the data and see if any common reaction orders can be eliminated. If the data does not show an equal concentration drop with time, then zero order can be eliminated. Look for changes in the half-life (compare time for concentration to drop to one half of any value). If the half-life is not constant, then the first order can be eliminated. If the half-life is getting longer as the concentration drops, this might suggest second order. Plot the data as indicated by the appropriate rate law. Determine *k* from the slope of the plot. Finally calculate the [AB] at 25 s by using the appropriate integrated rate expression.
Solution: By the above logic, we can eliminate both the zero order and the first order reactions. (Alternatively, you could make all three plots and only one should be linear.) This suggests that we should have a second order reaction. Plot 1/[AB] versus time. Since $\frac{1}{[AB]_t} = kt + \frac{1}{[AB]_0}$, the slope will be the rate constant. The slope can be determined by measuring Δy/Δx on the plot or by using functions, such as "add trendline" in Excel. Thus the rate constant is 0.0225 M^{-1} s^{-1} and the rate law is Rate = $0.0225\ M^{-1}s^{-1}[AB]^2$.

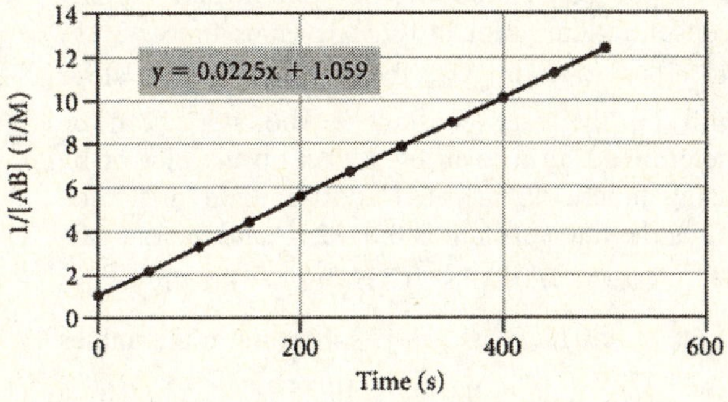

Finally, use $\frac{1}{[AB]_t} = kt + \frac{1}{[AB]_0}$, substitute in the values of $[AB]_0$, 25 s and *k* and rearrange to solve for [AB] at 25 s.

$$[AB]_t = \frac{1}{kt + \frac{1}{[AB]_0}} = \frac{1}{(0.0225\ M^{-1}\ \cancel{s^{-1}})(25\ \cancel{s}) + \left(\frac{1}{0.950\ M}\right)} = 0.619\ M$$

Check: The units (none, $M^{-1}s^{-1}$, and M) are correct. The rate law is a common form. The plot was extremely linear, confirming second order kinetics. The rate constant is consistent with the units necessary to get rate as M/s and the magnitude is reasonable since we have a second order reaction. The [AB] at 25 s is in between the values at 0 s and 50 s.

22. **Given:** table of $[N_2O_5]$ versus time **Find:** reaction order, *k*, and $[N_2O_5]$ at 250 s
Conceptual Plan: Look at the data and see if any common reaction orders can be eliminated. If the data does not show an equal concentration drop with time, then zero order can be eliminated. Look for changes in the half-life (compare time for concentration to drop to one half of any value). If the half-life is not constant, then the first order can be eliminated. If the half-life is getting longer as the concentration drops, this might suggest second order. Plot the data as indicated by the appropriate rate

law. Determine k from the slope of the plot. Finally calculate the $[N_2O_5]$ at 250 s by using the appropriate integrated rate expression.

Solution: By the above logic, we can see that the reaction is most likely first order. It takes just under 75 s for the concentration to be cut in half for any concentration. Plot ln $[N_2O_5]$ versus time. Since $\ln[A]_t = -kt + \ln[A]_0$, the negative of the slope will be the rate constant. The slope can be determined by measuring Δy/Δx on the plot or by using functions, such as "add trendline" in Excel.

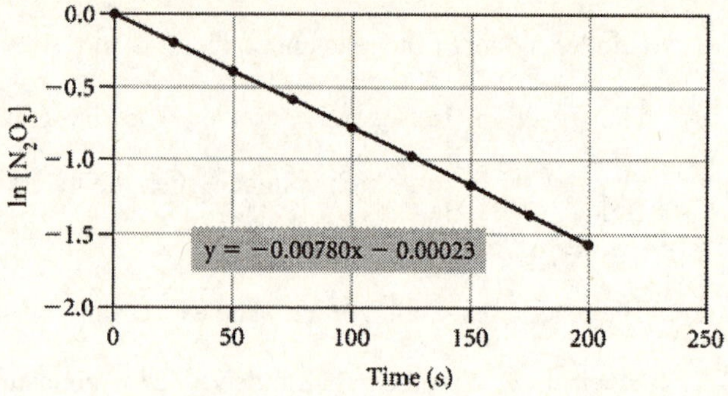

Thus the rate constant is 0.00780 s^{-1} and the rate law is Rate = $0.00780\ s^{-1}[N_2O_5]$.

Finally, use $\ln[N_2O_5]_t = -kt + \ln[N_2O_5]_0$, substitute in the values of $[N_2O_5]_0$, 250 s and k and rearrange to solve for $[N_2O_5]$ at 250 s. $\ln[N_2O_5]_{250\,s} = -(0.00780\ s^{-1})(250\ s) + \ln[1.000]_0$ then

$[N_2O_5]_{250\,s} = e^{-1.95} = 0.142\ M$.

Check: The units (none, s^{-1}, and M) are correct. The rate law is a common form. The plot was extremely linear, confirming first order kinetics. The rate constant is consistent with the units necessary to get rate as M/s and the magnitude is reasonable since we have a first order reaction. The $[N_2O_5]$ at 250 s is less than the value at 200 s.

23. **Given:** table of $[C_4H_8]$ versus time **Find:** reaction order, k, and reaction rate when $[C_4H_8]$ = 0.25 M

Conceptual Plan: Look at the data and see if any common reaction orders can be eliminated. If the data does not show an equal concentration drop with time, then zero order can be eliminated. Look for changes in half-life (compare time for concentration to drop to one half of any value). If the half-life is not constant, then the first order can be eliminated. If the half-life is getting longer as the concentration drops, this might suggest second order. Plot the data as indicated by the appropriate rate law. Determine k from the slope of the plot. Finally calculate the reaction rate when $[C_4H_8]$ = 0.25 M by using the rate law.

Solution: By the above logic, we can see that the reaction is most likely first order. It takes about 60 s for the concentration to be cut in half for any concentration. Plot ln $[C_4H_8]$ versus time. Since $\ln[A]_t = -kt + \ln[A]_0$, the negative of the slope will be the rate constant. The slope can be determined by measuring Δy/Δx on the plot or by using functions, such as "add trendline" in Excel. Thus the rate constant is 0.0112 s^{-1} and the rate law is Rate = $0.0112\ s^{-1}[C_4H_8]$. Finally, use Rate = $0.0112\ s^{-1}[C_4H_8]$, substitute in the values of $[C_4H_8]$. Rate = $0.0112\ s^{-1}[0.25\ M] = 2.8 \times 10^{-3}\ M\ s^{-1}$.

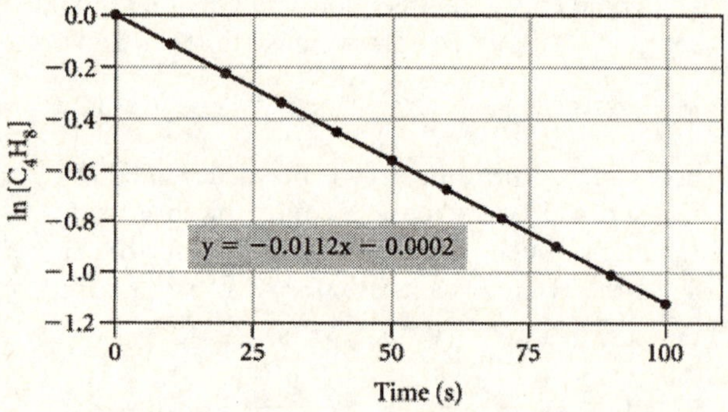

Check: The units (none, s^{-1}, and M s^{-1}) are correct. The rate law is a common form. The plot was extremely linear, confirming first order kinetics. The rate constant is consistent with the units necessary to get rate as M/s and the magnitude is reasonable since we have a first order reaction. The rate when $[C_4H_8]$ = 0. 25 M is consistent with the average rate using 90 s and 100 s.

24. **Given:** table of [A] versus time **Find:** reaction order, k, and reaction rate when [A] = 0.10 M

Conceptual Plan: Look at the data and see if any common reaction orders can be eliminated. If the data does not show an equal concentration drop with time, then zero order can be eliminated. Look for changes in half-life (compare time for concentration to drop to one half of any value). If the half-life is not constant, then the first order can be eliminated. If the half-life is getting longer as the concentration drops, this might suggest second order. Plot the data as indicated by the appropriate rate law. Determine k from the slope of the plot. Finally calculate the reaction rate when [A] = 0.10 M by using the rate law.

Solution: By the above logic, we can see that the reaction is most likely zero order. There is a difference of about 0.085 M between each data point, so the rate is independent of the [A]. Plot [A] versus time. Since

$[A]_t = -kt + [A]_0$, the negative of the slope will be the rate constant. The slope can be determined by measuring Δy/Δx on the plot or by using functions, such as "add trendline" in Excel. Thus the rate constant is 3.41 x 10^{-3} M s^{-1} and the rate law is $\text{Rate} = 3.41 \times 10^{-3}\ \text{M s}^{-1}$. Finally, since the rate is independent of concentration, $\text{Rate} = 3.41 \times 10^{-3}\ \text{M s}^{-1}$ at 0.10 M and all other concentrations. NOTE: A plot is not necessary since the kinetics are so simple.

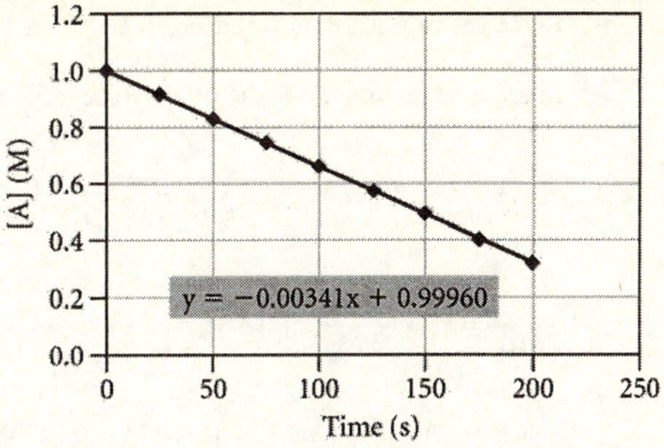

Check: The units (none, M s^{-1}, and M s^{-1}) are correct. The rate law is a common form. The plot was extremely linear, confirming zero order kinetics. The rate constant is consistent with the units necessary to get rate as M/s and the magnitude is reasonable since we have a zero order reaction. The rate is the same as any average rate that can be calculated using the data.

25. **Given:** plot of ln [A] versus time has slope = – 0.0045/s; $[A]_0$ = 0.250 M
Find: a) *k*; b) rate law; c) $t_{1/2}$; and d) [A] after 225 s
Conceptual Plan:
a) **A plot of ln [A] versus time is linear for a first order reaction. Using** $\ln[A]_t = -kt + \ln[A]_0$**, the rate constant is the negative of the slope.**
b) **Rate law is first order. Add rate constant from part a).**
c) **For a first order reaction,** $t_{1/2} = \frac{0.693}{k}$**. Substitute in k from part a).**
d) **Use the integrated rate law,** $\ln[A]_t = -kt + \ln[A]_0$**, and substitute in k and the initial concentration.**
Solution:
a) Since the rate constant is the negative of the slope, $k = 4.5 \times 10^{-3}\ s^{-1}$.
b) Since the reaction is first order, Rate = 4.5 x 10^{-3} s^{-1} [A].
c) $t_{1/2} = \frac{0.693}{k} = \frac{0.693}{0.0045/s} = 1.5 \times 10^2\ s$.
d) $\ln[A]_t = -kt + \ln[A]_0$, and substitute in *k* and the initial concentration. So
$\ln[A]_t = -(0.0045/\cancel{s})\ (225\ \cancel{s}) + \ln 0.250\ \text{M} = -2.39879$ and $[A]_{250s} = e^{-2.39879} = 0.0908\ \text{M}$
Check: The units (s^{-1}, none, s, and M) are correct. The rate law is a common form. The rate constant is consistent with value of the slope. The half-life is consistent with a small value of *k*. The concentration at 225 s is consistent with being between one and two half-lives.

26. **Given:** plot of 1/[AB] versus time has slope = 0.055/M s; $[A]_0$ = 0.250 M
Find: a) *k*; b) rate law; c) $t_{1/2}$ when $[AB]_0$ = 0.55 M; and d) [A] and [B] after 75 s
Conceptual Plan:
a) **A plot of 1/[AB] versus time is linear for a second order reaction. Using** $\frac{1}{[AB]_t} = kt + \frac{1}{[AB]_0}$**, the rate constant is the slope.**
b) **Rate law is second order. Add rate constant from part a).**
c) **For a second order reaction,** $t_{1/2} = \frac{1}{k[AB]_0}$**. Substitute in *k* from part a).**
d) **Use the integrated rate law,** $\frac{1}{[AB]_t} = kt + \frac{1}{[AB]_0}$**, and substitute in *k*, t, and the initial concentration to get the [AB]. Then $[AB]_0$, [AB] → [A], [B].**

$$\Delta[AB] = [AB]_{0s} - [AB]_{75s} \quad \text{with} \quad \frac{1\ \text{mol A}}{1\ \text{mol AB}} \quad \text{and} \quad \frac{1\ \text{mol B}}{1\ \text{mol AB}}$$

Solution:

a) Since the rate constant is the slope, $k = 5.5 \times 10^{-2}\ M^{-1}\ s^{-1}$.

b) Since the reaction is second order, Rate $= 5.5 \times 10^{-2}\ M^{-1}\ s^{-1}\ [AB]^2$.

c) $t_{1/2} = \dfrac{1}{k[AB]_0}$ so $t_{1/2} = \dfrac{1}{(5.5 \times 10^{-2}\ \cancel{M^{-1}}\ s^{-1})(0.550\ \cancel{M})} = 33\ s$.

d) $\dfrac{1}{[AB]_t} = kt + \dfrac{1}{[AB]_0}$ so $[AB]_t = \dfrac{1}{kt + \dfrac{1}{[AB]_0}} = \dfrac{1}{(5.5 \times 10^{-2}\ M^{-1}\ \cancel{s^{-1}})(75\ \cancel{s}) + \left(\dfrac{1}{0.250\ M}\right)} = 0.12\underline{3}08\ M$ then

$\Delta[AB] = [AB]_{0s} - [AB]_{75s} = 0.250\ M - 0.12\underline{3}08\ M = 0.12\underline{6}92\ M\ AB$ so

$0.12\underline{6}92 \dfrac{\cancel{mol\ AB}}{L} \times \dfrac{1\ mol\ A}{1\ \cancel{mol\ AB}} = 0.13\ M\ A$ and $0.12\underline{6}92 \dfrac{\cancel{mol\ AB}}{L} \times \dfrac{1\ mol\ B}{1\ \cancel{mol\ AB}} = 0.13\ M\ B$.

Check: The units ($M^{-1}\ s^{-1}$, none, s, and M) are correct. The rate law is a common form. The rate constant is consistent with value of the slope. The half-life is consistent with a small value of k. The concentration at 75 s is consistent with being about one half-life.

27. **Given:** decomposition of SO_2Cl_2, first order; $k = 1.42 \times 10^{-4}\ s^{-1}$
Find: a) $t_{1/2}$; b) t to decrease to 25 % of $[SO_2Cl_2]_0$; c) t to 0.78 M when $[SO_2Cl_2]_0 = 1.00$ M; and d) $[SO_2Cl_2]$ after 2.00×10^2 s and 5.00×10^2 s when $[SO_2Cl_2]_0 = 0.150$ M
Conceptual Plan:

a) $k \rightarrow t_{1/2}$

$$t_{1/2} = \frac{0.693}{k}$$

b) $[SO_2Cl_2]_0$**, 25 % of** $[SO_2Cl_2]_0$**,** $k \rightarrow t$

$$\ln[A]_t = -kt + \ln[A]_0$$

c) $[SO_2Cl_2]_0$**,** $[SO_2Cl_2]_t$**,** $k \rightarrow t$

$$\ln[A]_t = -kt + \ln[A]_0$$

d) $[SO_2Cl_2]_0$**, t,** $k \rightarrow [SO_2Cl_2]_t$

$$\ln[A]_t = -kt + \ln[A]_0$$

Solution:

a) $t_{1/2} = \dfrac{0.693}{k} = \dfrac{0.693}{1.42 \times 10^{-4}\ s^{-1}} = 4.88 \times 10^3\ s$

b) $[SO_2Cl_2]_t = 0.25\ [SO_2Cl_2]_0$. Since $\ln[SO_2Cl_2]_t = -kt + \ln[SO_2Cl_2]_0$ rearrange to solve for t

$$t = -\frac{1}{k}\ln\frac{[SO_2Cl_2]_t}{[SO_2Cl_2]_0} = -\frac{1}{1.42 \times 10^{-4}\ s^{-1}}\ln\frac{0.25\ \cancel{[SO_2Cl_2]_0}}{\cancel{[SO_2Cl_2]_0}} = 9.8 \times 10^3\ s$$

c) $[SO_2Cl_2]_t = 0.78$ M; $[SO_2Cl_2]_0 = 1.00$ M. Since $\ln[SO_2Cl_2]_t = -kt + \ln[SO_2Cl_2]_0$ rearrange to solve for t

$$t = -\frac{1}{k}\ln\frac{[SO_2Cl_2]_t}{[SO_2Cl_2]_0} = -\frac{1}{1.42 \times 10^{-4}\ s^{-1}}\ln\frac{0.78\ \cancel{M}}{1.00\ \cancel{M}} = 1.7 \times 10^3\ s$$

d) $[SO_2Cl_2]_0 = 0.150$ M and 2.00×10^2 s in

$\ln[SO_2Cl_2]_t = -(1.42 \times 10^{-4}\ \cancel{s^{-1}})(2.00 \times 10^2\ \cancel{s}) + \ln 0.150\ M = -1.9\underline{2}552 \rightarrow$

$[SO_2Cl_2]_t = e^{-1.92552} = 0.146\ M$ and $[SO_2Cl_2]_0 = 0.150$ M and 5.00×10^2 s in

$\ln[SO_2Cl_2]_t = -(1.42 \times 10^{-4}\ \cancel{s^{-1}})(5.00 \times 10^2\ \cancel{s}) + \ln 0.150\ M = -1.9\underline{6}812 \rightarrow [SO_2Cl_2]_t = e^{-1.96812} = 0.140\ M$

Check: The units (s, s, s, and M) are correct. The rate law is a common form. The half-life is consistent with a small value of *k*. The time to 25% is consistent with two half-lives. The time to 0.78 M is consistent with being less than one half-life. The final concentrations are consistent with the time being less than one half-life.

28. **Given:** decomposition of XY, second order in XY; $k = 7.02 \times 10^{-3}\ M^{-1}\ s^{-1}$
Find: a) $t_{1/2}$ when $[XY]_0 = 0.100$ M; b) t to decrease to 12.5 % of $[XY]_0 = 0.100$ M and 0.200 M; c) t to 0.062 M when $[XY]_0 = 0.150$ M; and d) [XY] after 5.0×10^1 s and 5.50×10^2 s when $[XY]_0 = 0.050$ M

Conceptual Plan:

a) $[XY]_0, k \rightarrow t_{1/2}$

$$t_{1/2} = \frac{1}{k[A]_0}$$

b) $[XY]_0$, 12.5 % of $[XY]_0, k \rightarrow t$

$$\frac{1}{[A]_t} = kt + \frac{1}{[A]_0}$$

c) $[XY]_0, [XY]_t, k \rightarrow t$

$$\frac{1}{[A]_t} = kt + \frac{1}{[A]_0}$$

d) $[XY]_0, t, k \rightarrow [XY]_t$

$$\frac{1}{[A]_t} = kt + \frac{1}{[A]_0}$$

Solution:

a) $$t_{1/2} = \frac{1}{k[XY]_0} = \frac{1}{(7.02 \times 10^{-3}\ \cancel{M^{-1}}\ s^{-1})(0.100\ \cancel{M})} = 1.42 \times 10^3\ s$$

b) $[XY]_t = 0.125\ [XY]_0 = 0.125 \times 0.100\ M = 0.0125$ M. Since $\frac{1}{[XY]_t} = kt + \frac{1}{[XY]_0}$ rearrange to solve

for t $$t = \frac{1}{k}\left(\frac{1}{[XY]_t} - \frac{1}{[XY]_0}\right) = \frac{1}{(7.02 \times 10^{-3}\ \cancel{M^{-1}}\ s^{-1})}\left(\frac{1}{0.0125\ \cancel{M}} - \frac{1}{0.100\ \cancel{M}}\right) = 9.97 \times 10^3\ s$$ and

$[XY]_t = 0.125\ [XY]_0 = 0.125 \times 0.200\ M = 0.0250$ M. Since $\frac{1}{[XY]_t} = kt + \frac{1}{[XY]_0}$ rearrange to solve

for t $$t = \frac{1}{k}\left(\frac{1}{[XY]_t} - \frac{1}{[XY]_0}\right) = \frac{1}{(7.02 \times 10^{-3}\ \cancel{M^{-1}}\ s^{-1})}\left(\frac{1}{0.0250\ \cancel{M}} - \frac{1}{0.200\ \cancel{M}}\right) = 4.99 \times 10^3\ s$$

c) $[XY]_t = 0.062$; $[XY]_0 = 0.150$ M. Since $\frac{1}{[XY]_t} = kt + \frac{1}{[XY]_0}$ rearrange to solve for t

$$t = \frac{1}{k}\left(\frac{1}{[XY]_t} - \frac{1}{[XY]_0}\right) = \frac{1}{(7.02 \times 10^{-3}\ \cancel{M^{-1}}\ s^{-1})}\left(\frac{1}{0.062\ \cancel{M}} - \frac{1}{0.150\ \cancel{M}}\right) = 1.3 \times 10^3\ s$$

d) $[XY]_0 = 0.050$ M and 5.0×10^1 s in $\frac{1}{[XY]_t} = kt + \frac{1}{[XY]_0} \rightarrow$

$$\frac{1}{[XY]_t} = (7.02 \times 10^{-3}\ M^{-1}\ \cancel{s^{-1}})(5.0 \times 10^1\ \cancel{s}) + \frac{1}{0.050\ M} = \frac{20.351}{M}$$ so [XY] = 0.049 M and

$[XY]_0 = 0.050$ M and 5.50×10^2 s in $$\frac{1}{[XY]_t} = (7.02 \times 10^{-3}\ M^{-1}\ \cancel{s^{-1}})(5.50 \times 10^2\ \cancel{s}) + \frac{1}{0.050\ M} = \frac{23.861}{M}$$

so [XY] = 0.042 M.

Check: The units (s, s, s, s, M, and M) are correct. The rate law is a common form. The half-life is consistent with a small value of *k*. The time to 12.5 % is consistent with three half-lives, where the half-life time is increasing. The next time (5000 s) is shorter because the initial concentration is higher. The last time is the shortest because it is less than a half-life with an intermediate concentration. The final concentrations are consistent with the time being much less than one half-life.

29. **Given:** $t_{1/2}$ for radioactive decay of U-238 = 4.5 billion years and independent of $[U\text{-}238]_0$ **Find:** t to decrease by 10 %; number U-238 atoms today, when 1.5×10^{18} atoms formed 13.8 billion years ago

Conceptual Plan: $t_{1/2}$ independent of concentration implies first order kinetics, $t_{1/2} \rightarrow k$ then

$$t_{1/2} = \frac{0.693}{k}$$

90 % of $[U\text{-}238]_0$, $k \rightarrow$ t and $[U\text{-}238]_0$, t, $k \rightarrow [U\text{-}238]_t$

$\ln[A]_t = -kt + \ln[A]_0$ $\qquad$ $\ln[A]_t = -kt + \ln[A]_0$

Solution: $t_{1/2} = \frac{0.693}{k}$ rearrange to solve for k. $k = \frac{0.693}{t_{1/2}} = \frac{0.693}{4.5 \times 10^9 \text{ yr}} = 1.\underline{5}4 \times 10^{-10} \text{ yr}^{-1}$ then

$[U\text{-}238]_t = 0.10\ [U\text{-}238]_0$. Since $\ln[U\text{-}238]_t = -kt + \ln[U\text{-}238]_0$ rearrange to solve for t

$t = -\frac{1}{k}\ln\frac{[U\text{-}238]_t}{[U\text{-}238]_0} = -\frac{1}{1.\underline{5}4 \times 10^{-10} \text{ yr}^{-1}}\ln\frac{0.90\ \cancel{[U\text{-}238]_0}}{\cancel{[U\text{-}238]_0}} = 6.8 \times 10^8$ yr and $[U\text{-}238]_0 = 1.5 \times 10^{18}$ atoms;

$t = 13.8 \times 10^9$ yr

$\ln[U\text{-}238]_t = -kt + \ln[U\text{-}238]_0 = -(1.\underline{5}4 \times 10^{-10}\ \cancel{\text{yr}^{-1}})(13.8 \times 10^9\ \cancel{\text{yr}}) + \ln(1.5 \times 10^{18} \text{ atoms}) = 39.7\underline{2}6797 \rightarrow$

$[U\text{-}238]_t = e^{39.7\underline{2}6797} = 1.8 \times 10^{17}$ atoms

Check: The units (yr and atoms) are correct. The time to 10 % decay is consistent with less than one half-life. The final concentration is consistent with the time being about three half-lives.

30. **Given:** $t_{1/2}$ for radioactive decay of C-14 = 5730 years
Find: t to decrease by 25 %; mmol C-14 atoms left, after 2255 yr in sample initially contains 1.5 mmol C-14
Conceptual Plan: radioactive decay implies first order kinetics, $t_{1/2} \rightarrow k$ then

$$t_{1/2} = \frac{0.693}{k}$$

75 % of $[C\text{-}14]_0$, $k \rightarrow$ t and $[C\text{-}14]_0$, t, $k \rightarrow [C\text{-}14]_t$

$\ln[A]_t = -kt + \ln[A]_0$ $\qquad$ $\ln[A]_t = -kt + \ln[A]_0$

Solution: $t_{1/2} = \frac{0.693}{k}$ rearrange to solve for k. $k = \frac{0.693}{t_{1/2}} = \frac{0.693}{5730 \text{ yr}} = 1.2\underline{0}942 \times 10^{-4} \text{ yr}^{-1}$ then

$[C\text{-}14]_t = 0.75\ [C\text{-}14]_0$. Since $\ln[C\text{-}14]_t = -kt + \ln[C\text{-}14]_0$ rearrange to solve for t

$t = -\frac{1}{k}\ln\frac{[C\text{-}14]_t}{[C\text{-}14]_0} = -\frac{1}{1.2\underline{0}942 \times 10^{-4} \text{ yr}^{-1}}\ln\frac{0.75\ \cancel{[C\text{-}14]_0}}{\cancel{[C\text{-}14]_0}} = 2.4 \times 10^3$ yr and $[C\text{-}14]_0 = 1.5$ mmol;

t = 2255 yr $\ln[C\text{-}14]_t = -kt + \ln[C\text{-}14]_0 = -(1.2\underline{0}942 \times 10^{-4}\ \cancel{\text{yr}^{-1}})(2255\ \cancel{\text{yr}}) + \ln(1.5 \text{ mmol}) = 0.13\underline{2}741 \rightarrow$

$[C\text{-}14]_t = e^{0.13\underline{2}741} = 1.1$ mmol

Check: The units (yr and mmol) are correct. The time to 25 % decay is consistent with less than one half-life. The final concentration is consistent with the time being less than one half-life.

31.

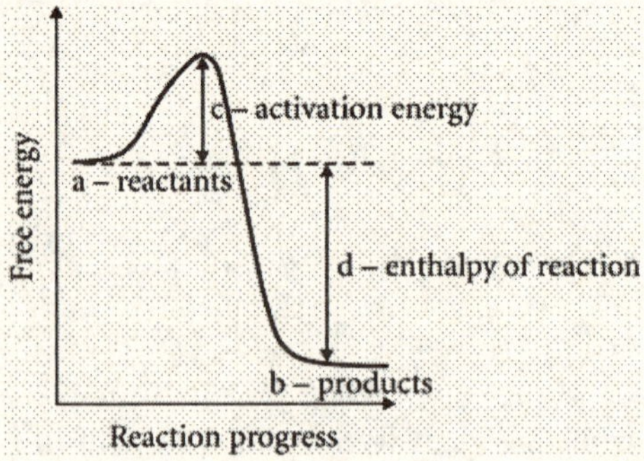

32.

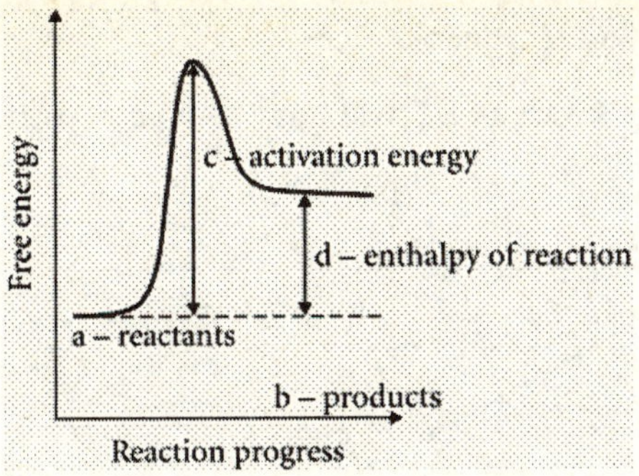

33. **Given:** activation energy = 56.8 kJ/mol, frequency factor = 1.5×10^{11} /s, 25 °C **Find:** rate constant

Conceptual Plan: **°C → K and kJ/mol → J/mol then E_a, T, A → k**

$K = °C + 273.15 \qquad \frac{1000\ J}{1\ kJ} \qquad k = A\,e^{-E_a/RT}$

Solution: $T = 25°C + 273.15 = 298$ K and $\frac{56.8\ kJ}{mol} \times \frac{1000\ J}{1\ kJ} = 5.68 \times 10^4\ \frac{J}{mol}$ then

$$k = A\,e^{-E_a/RT} = (1.5 \times 10^{11}\ s^{-1})\,e^{\frac{-5.68 \times 10^4 \frac{J}{mol}}{\left(8.314 \frac{J}{K\ mol}\right)298\ K}} = 17\ s^{-1}$$

Check: The units (s^{-1}) are correct. The rate constant is consistent with a large activation energy and a large frequency factor.

34. **Given:** 32 °C, rate constant = 0.055/s, and frequency factor = 1.2×10^{13} /s **Find:** activation energy

Conceptual Plan: **°C → K then E_a, T, A → k then J/mol → kJ/mol**

$K = °C + 273.15 \qquad k = A\,e^{-E_a/RT} \qquad \frac{1\ kJ}{1000\ J}$

Solution: $T = 32°C + 273.15 = 305$ K then $k = A\,e^{-E_a/RT}$ Rearrange to solve for E_a.

$$E_a = -R\,T \ln\left(\frac{k}{A}\right) = -8.314 \frac{J}{K\ mol} \times 305\ K \times \ln\left(\frac{0.055\ s^{-1}}{1.2 \times 10^{13}\ s^{-1}}\right) = 8.37 \times 10^4 \frac{J}{mol} \times \frac{1\ kJ}{1000\ J} = 83.7 \frac{kJ}{mol}.$$

Check: The units (kJ/mol) are correct. The activation energy is consistent with a modest rate constant and a large frequency factor.

35. **Given:** table of rate constant versus T **Find:** E_a, and A

Conceptual Plan: **Since** $\ln k = \frac{-E_a}{R}\left(\frac{1}{T}\right) + \ln A$ **a plot of ln k versus $1/T$ will have a slope = $-E_a/R$ and an intercept = ln A.**

Solution: The slope can be determined by measuring Δy/Δx on the plot or by using functions, such as "add trendline" in Excel. Since the slope = – 30189 K = $-E_a/R$ then

$$E_a = -(slope)R = -(-30189\ K)\left(8.314 \frac{J}{K\ mol}\right)\left(\frac{1\ kJ}{1000\ J}\right) = 251 \frac{kJ}{mol}.$$

and intercept = 27.399 = ln A then $A = e^{intercept} = e^{27.399} = 7.93 \times 10^{11}\ s^{-1}$.

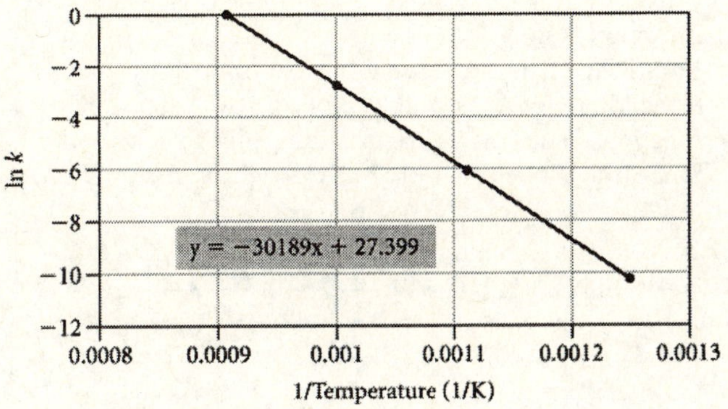

Check: The units (kJ/mol and s^{-1}) are correct. The plot was extremely linear, confirming Arrhenius behavior. The activation and frequency factor are typical for many reactions.

36. **Given:** table of rate constant versus T **Find:** E_a, and A

Conceptual Plan: **Since** $\ln k = \frac{-E_a}{R}\left(\frac{1}{T}\right) + \ln A$ **a plot of ln k versus $1/T$ will have a slope = $-E_a/R$ and an intercept = ln A.**

Solution: The slope can be determined by measuring Δy/Δx on the plot or by using functions, such as "add trendline" in Excel. Since the slope = – 10283 K = $-E_a/R$ then $E_a = -(slope)R =$

$$= -(-10283\ \text{K})\left(8.314\frac{\text{J}}{\text{K mol}}\right)\left(\frac{1\ \text{kJ}}{1000\ \text{J}}\right)$$

$$= 85.5\frac{\text{kJ}}{\text{mol}}$$

and intercept = 29.967 = ln A then $A = e^{intercept} = e^{29.967} = 1.03 \times 10^{13}\ \text{s}^{-1}$.

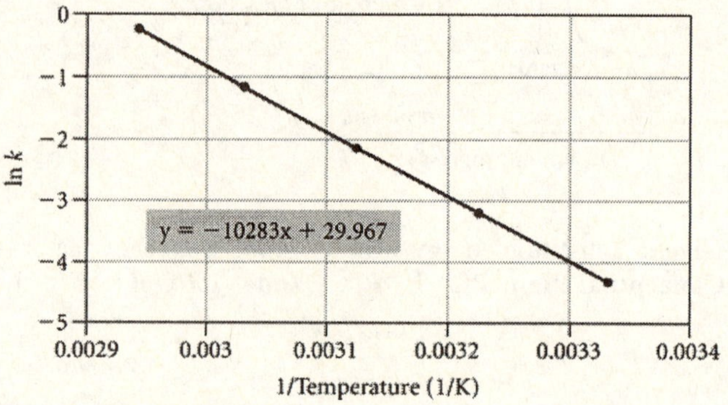

Check: The units (kJ/mol and s^{-1}) are correct. The plot was extremely linear, confirming Arrhenius behavior. The activation and frequency factor are typical for many reactions.

37. **Given:** table of rate constant versus T **Find:** E_a, and A

Conceptual Plan: **Since** $\ln k = \frac{-E_a}{R}\left(\frac{1}{T}\right) + \ln A$ **a plot of ln k versus $1/T$ will have a slope = $-E_a/R$ and an intercept = ln A.**

Solution: The slope can be determined by measuring Δy/Δx on the plot or by using functions, such as "add trendline" in Excel. Since the slope = – 2767.2 K = $-E_a/R$ then $E_a = -(slope)R =$

$$= -(-2767.2\ \text{K})\left(8.314\frac{\text{J}}{\text{K mol}}\right)\left(\frac{1\ \text{kJ}}{1000\ \text{J}}\right) =$$

$$= 23.0\frac{\text{kJ}}{\text{mol}}$$

and intercept = 25.112 = ln A then $A = e^{intercept} = e^{25.112} = 8.05 \times 10^{10}\ \text{s}^{-1}$.

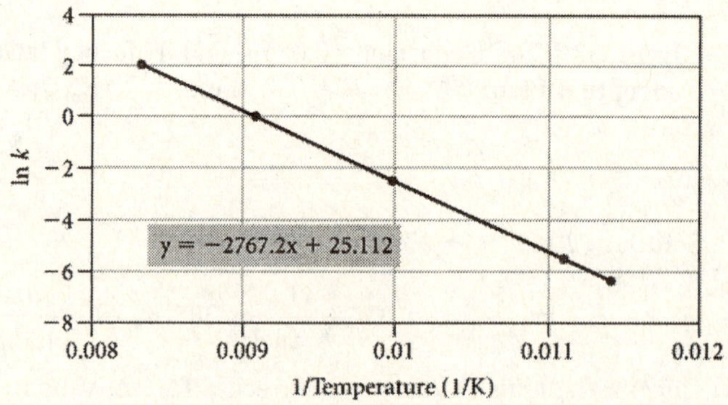

Check: The units (kJ/mol and s^{-1}) are correct. The plot was extremely linear, confirming Arrhenius behavior. The activation and frequency factor are typical for many reactions.

38. **Given:** table of rate constant versus T **Find:** E_a, and A

Conceptual Plan: **Since** $\ln k = \frac{-E_a}{R}\left(\frac{1}{T}\right) + \ln A$ **a plot of ln k versus $1/T$ will have a slope = $-E_a/R$ and an intercept = ln A.**

Solution: The slope can be determined by measuring Δy/Δx on the plot or by using functions, such as "add trendline" in Excel. Since the slope = – 11624 K = $-E_a/R$ then $E_a = -(slope)R =$

$$= -(-11624\ \text{K})\left(8.314\frac{\text{J}}{\text{K mol}}\right)\left(\frac{1\ \text{kJ}}{1000\ \text{J}}\right)$$

$$= 96.6\frac{\text{kJ}}{\text{mol}}$$

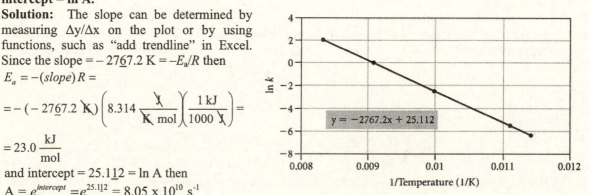

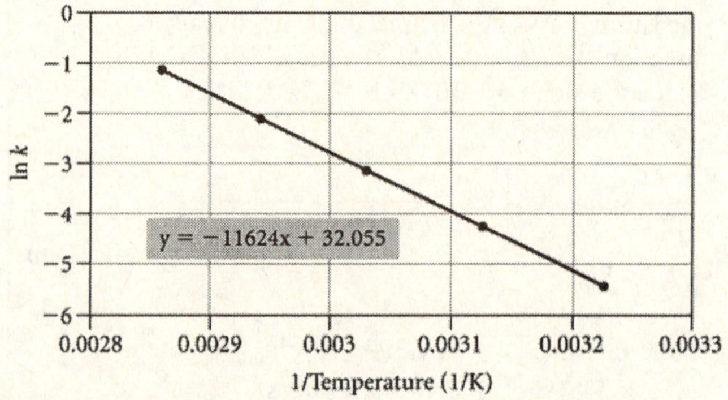

and intercept = 32.055 = ln A then $A = e^{intercept} = e^{32.055} = 8.34 \times 10^{13}\ \text{s}^{-1}$.

Check: The units (kJ/mol and s^{-1}) are correct. The plot was extremely linear, confirming Arrhenius behavior. The activation and frequency factor are typical for many reactions.

39. **Given:** rate constant = 0.0117/s at 400. K, and 0.689/s at 450. K **Find:** a) E_a and b) rate constant at 425 K
Conceptual Plan: a) $k_1, T_1, k_2, T_2 \rightarrow E_a$ then J/mol $\rightarrow$ kJ/mol b) $E_a, k_1, T_1, T_2 \rightarrow k_2$

$$\ln\left(\frac{k_2}{k_1}\right)=\frac{E_a}{R}\left(\frac{1}{T_1}-\frac{1}{T_2}\right) \qquad \frac{1\text{ kJ}}{1000\text{ J}} \qquad \ln\left(\frac{k_2}{k_1}\right)=\frac{E_a}{R}\left(\frac{1}{T_1}-\frac{1}{T_2}\right)$$

Solution: a) $\ln\left(\frac{k_2}{k_1}\right)=\frac{E_a}{R}\left(\frac{1}{T_1}-\frac{1}{T_2}\right)$ Rearrange to solve for E_a.

$$E_a=\frac{R\ln\left(\frac{k_2}{k_1}\right)}{\left(\frac{1}{T_1}-\frac{1}{T_2}\right)}=\frac{8.314\frac{\text{J}}{\text{K mol}}\ln\left(\frac{0.689\text{ s}^{-1}}{0.0117\text{ s}^{-1}}\right)}{\left(\frac{1}{400.\text{ K}}-\frac{1}{450.\text{ K}}\right)}=1.22\times10^5\frac{\text{J}}{\text{mol}}\times\frac{1\text{ kJ}}{1000\text{ J}}=122\frac{\text{kJ}}{\text{mol}}\text{ and}$$

b) $\ln\left(\frac{k_2}{k_1}\right)=\frac{E_a}{R}\left(\frac{1}{T_1}-\frac{1}{T_2}\right)$ with k_1, = 0.0117/s, T_1 = 400. K, T_2 = 425 K Rearrange to solve for k_2.

$$\ln k_2=\frac{E_a}{R}\left(\frac{1}{T_1}-\frac{1}{T_2}\right)+\ln k_1=\frac{1.22\times10^5\frac{\text{J}}{\text{mol}}}{8.314\frac{\text{J}}{\text{K mol}}}\left(\frac{1}{400.\text{ K}}-\frac{1}{425\text{ K}}\right)+\ln 0.0117\text{ s}^{-1}=-2.29\underline{0}2 \rightarrow$$

$k_2=e^{-2.2902}=0.101\text{ s}^{-1}$

Check: The units (kJ/mol and s^{-1}) are correct. The activation energy is typical for a reaction. The rate constant at 425 K is in between the values given at 400 K and 450 K.

40. **Given:** rate constant = 0.000122/s at 27 °C, and 0.228/s at 77 °C **Find:** a) E_a and b) rate constant at 17 °C
Conceptual Plan:
a) °C $\rightarrow$ K then $k_1, T_1, k_2, T_2 \rightarrow E_a$ then J/mol $\rightarrow$ kJ/mol;

$$\text{K}=°\text{C}+273.15 \qquad \ln\left(\frac{k_2}{k_1}\right)=\frac{E_a}{R}\left(\frac{1}{T_1}-\frac{1}{T_2}\right) \qquad \frac{1\text{ kJ}}{1000\text{ J}}$$

b) °C $\rightarrow$ K then $E_a, k_1, T_1, T_2 \rightarrow k_2$

$$\text{K}=°\text{C}+273.15 \qquad \ln\left(\frac{k_2}{k_1}\right)=\frac{E_a}{R}\left(\frac{1}{T_1}-\frac{1}{T_2}\right)$$

Solution: T_1= 27°C + 273.15 = 300. K and T_2 = 77°C + 273.15 = 350. K then $\ln\left(\frac{k_2}{k_1}\right)=\frac{E_a}{R}\left(\frac{1}{T_1}-\frac{1}{T_2}\right)$

a) Rearrange to solve for E_a.

$$E_a=\frac{R\ln\left(\frac{k_2}{k_1}\right)}{\left(\frac{1}{T_1}-\frac{1}{T_2}\right)}=\frac{8.314\frac{\text{J}}{\text{K mol}}\ln\left(\frac{0.228\text{ s}^{-1}}{0.000122\text{ s}^{-1}}\right)}{\left(\frac{1}{300.\text{ K}}-\frac{1}{350.\text{ K}}\right)}=1.32\times10^5\frac{\text{J}}{\text{mol}}\times\frac{1\text{ kJ}}{1000\text{ J}}=132\frac{\text{kJ}}{\text{mol}}\text{ and}$$

b) $\ln\left(\frac{k_2}{k_1}\right)=\frac{E_a}{R}\left(\frac{1}{T_1}-\frac{1}{T_2}\right)$ with k_1, = 0.000122/s, T_1 = 300. K, T_2 = 17°C + 273.15 = 290. K

Rearrange to solve for k_2.

$$\ln k_2=\frac{E_a}{R}\left(\frac{1}{T_1}-\frac{1}{T_2}\right)+\ln k_1=\frac{1.32\times10^5\frac{\text{J}}{\text{mol}}}{8.314\frac{\text{J}}{\text{K mol}}}\left(\frac{1}{300.\text{ K}}-\frac{1}{290.\text{ K}}\right)+\ln 0.000122\text{ s}^{-1}=-10.8\underline{2}98 \rightarrow$$

$k_2=e^{-10.8298}=0.0000198\text{ s}^{-1}=1.98\times10^{-5}\text{ s}^{-1}$

Check: The units (kJ/mol and s^{-1}) are correct. The activation energy is typical for a reaction. The rate constant at 17 °C is smaller than the values given at 27 °C.

41. **Given:** rate constant doubles from at 10.0 °C to 20.0 °C **Find:** E_a
Conceptual Plan: °C $\rightarrow$ K then $k_1, T_1, k_2, T_2 \rightarrow E_a$ then J/mol $\rightarrow$ kJ/mol;

$$K = °C + 273.15 \qquad \ln\left(\frac{k_2}{k_1}\right) = \frac{E_a}{R}\left(\frac{1}{T_1} - \frac{1}{T_2}\right) \qquad \frac{1\ \text{kJ}}{1000\ \text{J}}$$

Solution: $T_1 = 10.0\ °C + 273.15 = 283.2$ K and $T_2 = 20.0\ °C + 273.15 = 293.2$ K and $k_2 = 2\,k_1$ then

$\ln\left(\frac{k_2}{k_1}\right) = \frac{E_a}{R}\left(\frac{1}{T_1} - \frac{1}{T_2}\right)$ Rearrange to solve for E_a.

$$E_a = \frac{R \ln\left(\frac{k_2}{k_1}\right)}{\left(\frac{1}{T_1} - \frac{1}{T_2}\right)} = \frac{8.314 \frac{\text{J}}{\cancel{\text{K}}\ \text{mol}} \ln\left(\frac{2\,\cancel{k_1}}{\cancel{k_1}}\right)}{\left(\frac{1}{283.2\ \cancel{\text{K}}} - \frac{1}{293.2\ \cancel{\text{K}}}\right)} = 4.785 \times 10^4 \frac{\cancel{\text{J}}}{\text{mol}} \times \frac{1\ \text{kJ}}{1000\ \cancel{\text{J}}} = 47.85 \frac{\text{kJ}}{\text{mol}}$$

Check: The units (kJ/mol) are correct. The activation energy is typical for a reaction.

42. **Given:** rate constant triples from at 20.0 °C to 35.0 °C **Find:** E_a
Conceptual Plan: **°C → K then k_1, T_1, k_2, T_2 → E_a then J/mol → kJ/mol;**

$$K = °C + 273.15 \qquad \ln\left(\frac{k_2}{k_1}\right) = \frac{E_a}{R}\left(\frac{1}{T_1} - \frac{1}{T_2}\right) \qquad \frac{1\ \text{kJ}}{1000\ \text{J}}$$

Solution: $T_1 = 20.0\ °C + 273.15 = 293.2$ K and $T_2 = 35.0\ °C + 273.15 = 808.2$ K and $k_2 = 3\,k_1$ then

$\ln\left(\frac{k_2}{k_1}\right) = \frac{E_a}{R}\left(\frac{1}{T_1} - \frac{1}{T_2}\right)$ Rearrange to solve for E_a.

$$E_a = \frac{R \ln\left(\frac{k_2}{k_1}\right)}{\left(\frac{1}{T_1} - \frac{1}{T_2}\right)} = \frac{8.314 \frac{\text{J}}{\cancel{\text{K}}\ \text{mol}} \ln\left(\frac{3\,\cancel{k_1}}{\cancel{k_1}}\right)}{\left(\frac{1}{293.2\ \cancel{\text{K}}} - \frac{1}{308.2\ \cancel{\text{K}}}\right)} = 5.502 \times 10^4 \frac{\cancel{\text{J}}}{\text{mol}} \times \frac{1\ \text{kJ}}{1000\ \cancel{\text{J}}} = 55.02 \frac{\text{kJ}}{\text{mol}}$$

Check: The units (kJ/mol) are correct. The activation energy is typical for a reaction.

43. Reaction a. would have the faster rate because the orientation factor, *p*, would be larger for this reaction since the reactants are symmetrical.

44. Reaction b. would have the smaller orientation factor, because we are reacting an asymmetric molecule with a homonuclear diatomic molecule (symmetrical) and so the orientation is important. In reaction a both reacting species are symmetrical and so orientation is unimportant.

45. Since the first reaction is the slow step, it is the rate determining step. Using this first step to determine the rate law, Rate = $k_1\,[AB]^2$. Since this is the observed rate law, this mechanism is consistent with the experimental data.

46. a) The reaction cannot occur in a single step in which X and Y collide, because the rate law would be Rate = k [X] [Y]. This is not consistent with the stated rate law of Rate = $k\,[X]^2\,[Y]$.

b) Since the second step is the rate determining step, Rate = $k_3\,[X_2]\,[Y]$. X_2 is an intermediate, so its concentration cannot appear in the rate law. Using the fast equilibrium in the first step, we see that $k_1[X]^2 = k_2\,[X_2]$ or $[X_2] = \frac{k_1}{k_2}[X]^2$. Substituting this into the first rate expression we get that Rate = $\frac{k_3\,k_1}{k_2}\,[X]^2\,[Y]$. Simplifying this expression we see Rate = $k\,[X]^2\,[Y]$, which is consistent with the experimentally derived rate law.

47. a) The overall reaction is the sum of the steps in the mechanism:

$$Cl_2\,(g) \underset{k_2}{\overset{k_1}{\rightleftharpoons}} \cancel{2\,Cl\,(g)}$$

$$\cancel{Cl\,(g)} + CHCl_3\,(g) \xrightarrow{k_3} HCl\,(g) + \cancel{CCl_3\,(g)}$$

$$\cancel{Cl\,(g)} + \cancel{CCl_3\,(g)} \xrightarrow{k_4} CCl_4\,(g)$$

$$Cl_2\,(g) + CHCl_3\,(g) \rightarrow HCl\,(g) + CCl_4\,(g)$$

b) The intermediates are the species that are generated by one step and consumed by other steps. These are a Cl (*g*) and CCl_3 (*g*).

c) Since the second step is the rate determining step, Rate = k_3 [Cl] [$CHCl_3$]. Since Cl is an intermediate, its concentration cannot appear in the rate law. Using the fast equilibrium in the first step, we see that $k_1[Cl_2] = k_2\,[Cl]^2$ or $[Cl] = \sqrt{\frac{k_1}{k_2}[Cl_2]}$. Substituting this into the first rate expression we get that Rate = $k_3\sqrt{\frac{k_1}{k_2}}[Cl_2]^{1/2}[CHCl_3]$. Simplifying this expression we see Rate = $k\,[Cl_2]^{1/2}\,[CHCl_3]$.

48. a) The overall reaction is the sum of the steps in the mechanism:

$$NO_2\,(g) + Cl_2\,(g) \xrightarrow[k_1]{} ClNO_2\,(g) + \cancel{Cl(g)}$$

$$NO_2\,(g) + \cancel{Cl(g)} \xrightarrow[k_2]{} ClNO_2\,(g)$$

$$2\,NO_2\,(g) + Cl_2\,(g) \rightarrow 2\,ClNO_2\,(g)$$

b) The intermediates are the species that are generated by one step and consumed by other steps. This is Cl (*g*).

c) Since the first step is the rate determining step, Rate = k_1 [NO_2] [Cl_2]. Since both of these species are reactants, this is the predicted rate law.

49. Heterogeneous catalysts require a large surface area because catalysis can only happen at the active sites on the surface. A greater surface area means greater opportunity for the substrate to react, which results in a speedier reaction.

50. The initial and final energies (reactants and products) remain the same. The activation energy drops, from 75 kJ/mol to a smaller value, for example 30 kJ/mol. There are usually more steps in the reaction progress diagram.

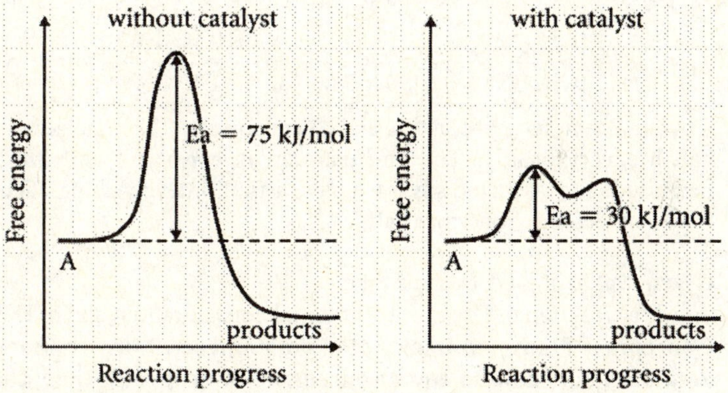

51. Assume Rate ratio α *k* ratio (since concentration terms will cancel each other) and $k = A\,e^{-E_a/RT}$.
$T = 25\,°C + 273.15 = 298\,K$, $E_{a1} = 1.25 \times 10^5$ J/mol, and $E_{a2} = 5.5 \times 10^4$ J/mol. Ratio of rates will be

$$\frac{k_2}{k_1} = \frac{\cancel{A}\,e^{-E_{a2}/RT}}{\cancel{A}\,e^{-E_{a1}/RT}} = \frac{e^{\frac{-5.5\times10^4\,\frac{\cancel{J}}{\cancel{mol}}}{\left(8.314\,\frac{\cancel{J}}{\cancel{K}\,\cancel{mol}}\right)298\,\cancel{K}}}}{e^{\frac{-1.25\times10^5\,\frac{\cancel{J}}{\cancel{mol}}}{\left(8.314\,\frac{\cancel{J}}{\cancel{K}\,\cancel{mol}}\right)298\,\cancel{K}}}} = \frac{e^{-22.199}}{e^{-50.453}} = 10^{12}.$$

52. Assume Rate ratio α *k* ratio (since concentration terms will cancel each other) and $k = A\,e^{-E_a/RT}$.
$T = 25\,°C + 273.15 = 298\,K$. $E_{a1} = 1.25 \times 10^5$ J/mol and $E_{a2} = 5.5 \times 10^{54}$ J/mol. Ratio of rates will be

$$\frac{k_2}{k_1} = 10^6 = \frac{A\, e^{-E_{a2}/RT}}{A\, e^{-E_{a1}/RT}} = \frac{e^{\frac{-E_{a2}}{\left(8.314 \frac{\text{J}}{\text{K mol}}\right)298\text{ K}}}}{e^{\frac{-1.08 \times 10^5 \frac{\text{J}}{\text{mol}}}{\left(8.314 \frac{\text{J}}{\text{K mol}}\right)298\text{ K}}}} = \frac{e^{\frac{-E_{a2}}{2.47756 \times 10^3 \frac{\text{J}}{\text{mol}}}}}{1.17 \times 10^{-19}} \rightarrow e^{\frac{-E_{a2}}{2.47756 \times 10^3 \frac{\text{J}}{\text{mol}}}} = 1.17 \times 10^{-13} \rightarrow$$

$$\frac{-E_{a2}}{2.47756 \times 10^3 \frac{\text{J}}{\text{mol}}} = \ln(1.17 \times 10^{-13}) = -29.7766 \rightarrow E_{a2} = 7.38 \times 10^4 \frac{\text{J}}{\text{mol}} = 73.8 \frac{\text{kJ}}{\text{mol}}.$$

53. **Given:** table of $[CH_3CN]$ versus time **Find:** a) reaction order, k; b) $t_{1/2}$; and c) t for 90 % conversion
Conceptual Plan: a) and b) Look at the data and see if any common reaction orders can be eliminated. If the data does not show an equal concentration drop with time, then zero order can be eliminated. Look for changes in the half-life (compare time for concentration to drop to one half of any value). If the half-life is not constant, then the first order can be eliminated. If the half-life is getting longer as the concentration drops, this might suggest second order. Plot the data as indicated by the appropriate rate law or if it is first order and there is an obvious half-life in the date, a plot is not necessary. Determine *k* from the slope of the plot (or using half-life equation for first order). c) Finally calculate the time to 90 % conversion using the appropriate integrated rate equation.
Solution: a and b) By the above logic, we can see that the reaction is first order. It takes 15.0 h for the concentration to be cut in half for any concentration (1.000 M to 0.501 M; 0.794 M to 0.398 M; and 0.631 M to 0.316 M), so $t_{1/2}$ = 15.0 h. Then use $t_{1/2} = \frac{0.693}{k}$ and rearrange to solve for k.

$$k = \frac{0.693}{t_{1/2}} = \frac{0.693}{15.0\text{ h}} = 0.0462\text{ h}^{-1}.$$

c) $[CH_3CN]_t$ = 0.10 $[CH_3CN]_0$. Since $\ln[CH_3CN]_t = -kt + \ln[CH_3CN]_0$ rearrange to solve for t

$$t = -\frac{1}{k}\ln\frac{[CH_3CN]_t}{[CH_3CN]_0} = -\frac{1}{0.0462\text{ hr}^{-1}}\ln\frac{0.10\,[CH_3CN]_0}{[CH_3CN]_0} = 49.8\text{ hr}.$$

Check: The units (none, h^{-1}, h, and h) are correct. The rate law is a common form. The data showed a constant half-life very clearly. The rate constant is consistent with the units necessary to get rate as M/s and the magnitude is reasonable since we have a first order reaction. The time to 90 % conversion is consistent with a time between three and four half-lives.

54. **Given:** table of $[X_2Y]$ versus time
Find: a) reaction order, k; b) $t_{1/2}$ at initial concentration; and c) [X] at 10.0 h
Conceptual Plan: a) Look at the data and see if any common reaction orders can be eliminated. If the data does not show an equal concentration drop with time, then zero order can be eliminated. Look for changes in the half-life (compare time for concentration to drop to one half of any value). If the half-life is not constant, then the first order can be eliminated. If the half-life is getting longer as the concentration drops, this might suggest second order. Plot the data as indicated by the appropriate rate law. Determine *k* from the slope of the plot. b) Calculate the half-life with the appropriate equation. c) Finally calculate the $[X_2Y]$ at 10.0 h using the appropriate integrated rate expression and then convert this to a change in $[X_2Y]$ and then to [X] using the reaction stoichiometry.
Solution:

a) By the above logic, we can eliminate both the zero order and the first order reactions. (Alternatively, you could make all three plots and only one should be linear.) This suggests that we should have a second order reaction. Plot 1/[X_2Y] versus time. Since $\frac{1}{[X_2Y]_t} = kt + \frac{1}{[X_2Y]_0}$, the slope will be the rate constant. The slope can be determined by measuring Δy/Δx on the plot or by using functions, such as "add trendline" in Excel. Thus the rate constant is 1.6827 M^{-1} h^{-1} and the rate law is Rate = $1.68\text{ M}^{-1}\text{h}^{-1}[X_2Y]^2$.

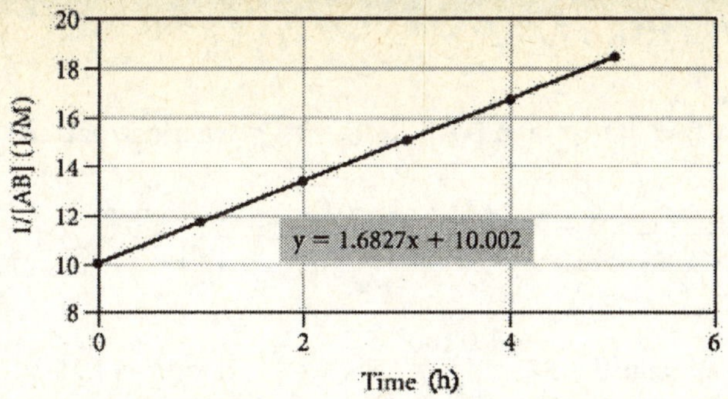

b) $t_{1/2} = \dfrac{1}{k[X_2Y]_0}$ so $t_{1/2} = \dfrac{1}{(1.6\underline{8}27\ \cancel{M^{-1}}\ h^{-1})\ (0.100\ \cancel{M})} = 5.94\ h$.

c) Finally, use $\dfrac{1}{[X_2Y]_t} = kt + \dfrac{1}{[X_2Y]_0}$, substitute in the values of $[X_2Y]_0$, 10.0 h, and k and rearrange to solve for $[X_2Y]$ at 10.0 h.

$$[X_2Y]_t = \frac{1}{kt + \dfrac{1}{[X_2Y]_0}} = \frac{1}{(1.6\underline{8}27\ M^{-1}\ \cancel{h^{-1}})\ (10.0\ \cancel{h}) + \left(\dfrac{1}{0.100\ M}\right)} = 0.037\underline{2}759\ M$$ then

$$\Delta[X_2Y] = [X_2Y]_0 - [X_2Y]_{10.0\,hr} = 0.100\ M - 0.037\underline{2}759\ M = 0.06\underline{2}724\ M$$ then

$$\frac{0.06\underline{2}724\ \cancel{mol\ X_2Y}}{L} \times \frac{2\ mol\ X}{1\ \cancel{mol\ X_2Y}} = 0.13\ M\ X\ .$$

Check: The units (none, $M^{-1}h^{-1}$, h, and M) are correct. The rate law is a common form. The plot was extremely linear, confirming second order kinetics. The rate constant is consistent with the units necessary to get rate as M/s and the magnitude is reasonable since we have a second order reaction. The half-life is consistent with the data table, which indicates that the half-life is a little over 5 h. The [X] at 10 h s is consistent with the changes that we see in the data table through 5 h.

55. **Given:** $\text{Rate} = k\dfrac{[A][C]^2}{[B]^{1/2}}$ = 0.0115 M/s at certain initial concentrations of A, B and C; double A and C concentration and triple B concentration **Find:** reaction rate

Conceptual Plan: $[A]_1$, $[B]_1$, $[C]_1$, Rate 1, $[A]_2$, $[B]_2$, $[C]_2$ → **Rate 2**

$$\frac{\text{Rate 2}}{\text{Rate 1}} = \frac{k\dfrac{[A]_2[C]_2^2}{[B]_2^{1/2}}}{k\dfrac{[A]_1[C]_1^2}{[B]_1^{1/2}}}$$

Solution: $\dfrac{\text{Rate 2}}{\text{Rate 1}} = \dfrac{k\dfrac{[A]_2[C]_2^2}{[B]_2^{1/2}}}{k\dfrac{[A]_1[C]_1^2}{[B]_1^{1/2}}}$ Rearrange to solve for Rate 2. $\text{Rate 2} = \dfrac{k\dfrac{[A]_2[C]_2^2}{[B]_2^{1/2}}}{k\dfrac{[A]_1[C]_1^2}{[B]_1^{1/2}}}\text{Rate 1}$ $[A]_2 = 2[A]_1$, $[B]_2 = 3[B]_1$, $[C]_2 = 2[C]_1$ and Rate 1 = 0.0115 M/s so

$$\text{Rate 2} = \frac{\cancel{k}\dfrac{2\cancel{[A]_1}\,(2\cancel{[C]_1})^2}{(3\cancel{[B]_1})^{1/2}}}{\cancel{k}\dfrac{\cancel{[A]_1}\,\cancel{[C]_1^2}}{\cancel{[B]_1^{1/2}}}}\ 0.0115\ \frac{M}{s} = \frac{2^3}{3^{1/2}}\ 0.0115\ \frac{M}{s} = 0.0531\ \frac{M}{s}$$

Check: The units ($M\ s^{-1}$) are correct. The should increase because we have a factor of 8 (2^3) divided by the square root of three.

56. **Given:** $\text{Rate} = k\dfrac{[O_3]^2}{[O_2]}$ initially 1.0 mol O_3, and 1.0 mol O_2, in 1.0 L

Find: fraction O_3 reacted when reaction rate is cut in half

Conceptual Plan:

mol, L → M then $[O_3]_1$, $[O_2]_1$, Rate 1, Rate 2 → $[O_3]_2$ then $[O_3]_1$, $[O_3]_2$ → O_3 fraction reacted

$$M = \frac{mol}{L} \qquad \frac{\text{Rate 2}}{\text{Rate 1}} = \frac{k\dfrac{[O_3]_2^2}{[O_2]_2}}{k\dfrac{[O_3]_1^2}{[O_2]_1}} \qquad O_3\ \textit{fraction reacted} = \frac{[O_3]_1 - [O_3]_2}{[O_3]_1}$$

Solution: $M = \dfrac{mol}{L}$ so $[O_3]_1 = \dfrac{1.0\text{ mol}}{1.0\text{ L}} = 1.0\text{ M}$ and $[O_2]_1 = \dfrac{1.0\text{ mol}}{1.0\text{ L}} = 1.0\text{ M}$. Rate 1 = 2 Rate 2.

Let x = $\Delta[O_3]$ so $[O_3]_2 = [O_3]_1 - x$ and $[O_2]_2 = [O_2]_1 + 3/2\ x$. Substitute values into $\dfrac{\text{Rate 2}}{\text{Rate 1}} = \dfrac{k\dfrac{[O_3]_2^2}{[O_2]_2}}{k\dfrac{[O_3]_1^2}{[O_2]_1}}$ and

rearrange to solve for x. $\dfrac{\cancel{\text{Rate 2}}}{2\ \cancel{\text{Rate 2}}} = \dfrac{\cancel{k}\dfrac{(1.0\text{ M} - x)^2}{(1.0\text{ M} + 3/2\ x)}}{\cancel{k}\dfrac{(1.0\text{ M})^{\cancel{2}}}{\cancel{(1.0\text{ M})}}}$ → $0.50\text{ M }(1.0\text{ M} + 3/2\ x) = (1.0\text{ M} - x)^2$ →

$0.50 + 0.75\ x = 1.0 - 2.0x + x^2$ → $0 = x^2 - 2.\underline{7}5x + 0.\underline{5}0$ solve with quadratic equation

$\left(x = \dfrac{-b \pm \sqrt{b^2 - 4ac}}{2a}\right)$. So $x = \dfrac{2.\underline{7}5 \pm \sqrt{(-2.\underline{7}5)^2 - (4)(1.0)(0.50)}}{2(1.0)} = \dfrac{2.\underline{7}5 \pm \sqrt{5.\underline{5}625}}{2.0} = \dfrac{2.\underline{7}5 \pm 2.\underline{3}585}{2.0} =$

$= 0.\underline{1}9575$ M or $2.\underline{5}5$ M The answer must be $0.\underline{1}9575$ M because the other answer is larger than our initial concentration (and is, therefore, impossible).

$$O_3\ \textit{fraction reacted} = \frac{[O_3]_1 - [O_3]_2}{[O_3]_1} = \frac{x}{[O_3]_1} = \frac{0.\underline{1}9575\ \cancel{\text{M}}}{1.0\ \cancel{\text{M}}} = 0.2$$

Check: The units (M) are correct. The concentration is reasonable since there are two forces slowing down the reaction: 1) the decrease in the reactant and 2) the increase of the product (which appears in the rate law). The calculation can be double checked by substituting in the value of x and the resulting rate = 0.5 *k*.

57. **Given:** N_2O_5 decomposes to NO_2 and O_2, first order in $[N_2O_5]$; $t_{1/2}$ = 2.81 h at 25 °C; *V* = 1.5 L, $P°_{N2O5}$ = 745 torr **Find:** P_{O2} after 215 minutes

Conceptual Plan:

Write a balanced reaction. then $t_{1/2}$ → *k* then °C → K and torr → atm then

$N_2O_5 \rightarrow 2\ NO_2 + ½\ O_2 \qquad t_{1/2} = \dfrac{0.693}{k} \qquad K = °C + 273.15 \qquad \dfrac{1\text{ atm}}{760\text{ torr}}$

$P°_{N2O5}$, *V*, *T* → *n*/*V* then min → h $[N_2O_5]_0$, t, *k* → $[N_2O_5]_t$ then $[N_2O_5]_0$, $[N_2O_5]_t$ → $[O_2]_t$

$PV = nRT \qquad \dfrac{1\text{ hr}}{60\text{ min}} \qquad \ln[A]_t = -kt + \ln[A]_0 \qquad [O_2]_t = ([N_2O_5]_0 - [N_2O_5]_t) \times \dfrac{1/2\text{ mol }O_2}{1\text{ mol }N_2O_5}$

then $[O_2]_t$, *V*, *T* → $P°_{O2}$ then finally atm → torr

$PV = nRT \qquad \dfrac{760\text{ torr}}{1\text{ atm}}$

Solution: $t_{1/2} = \dfrac{0.693}{k}$ and rearrange to solve for *k*. $k = \dfrac{0.693}{t_{1/2}} = \dfrac{0.693}{2.81\text{hr}} = 0.24\underline{6}619\text{ hr}^{-1}$. Then

$T = 25\text{ °C} + 273.15 = 298\text{ K}$. $745\ \cancel{\text{torr}} \times \dfrac{1\text{ atm}}{760\ \cancel{\text{torr}}} = 0.98\underline{0}263\text{ atm}$ then $PV = nRT$ Rearrange to solve for

n/*V*. $\dfrac{n}{V} = \dfrac{P}{RT} = \dfrac{0.98\underline{0}263\ \cancel{\text{atm}}}{0.08206\dfrac{\text{L}\cdot\cancel{\text{atm}}}{\cancel{\text{K}}\cdot\text{mol}} \times 298\ \cancel{\text{K}}} = 0.040\underline{0}862\text{ M}$ then $215\ \cancel{\text{min}} \times \dfrac{1\text{ hr}}{60\ \cancel{\text{min}}} = 3.5\underline{8}333\text{ hr}$.

Since $\ln[N_2O_5]_t = -kt + \ln[N_2O_5]_0 = -(0.24\underline{6}619\text{ hr}^{-1})(3.5\underline{8}333\text{ hr}) + \ln(0.040\underline{0}862\text{ M}) = -4.1\underline{0}044$ →
$[N_2O_5]_t = e^{-4.1\underline{0}044} = 0.016\underline{5}653\text{ M}$ then

$$[O_2]_t = ([N_2O_5]_0 - [N_2O_5]_t) \times \frac{1/2 \text{ mol } O_2}{1 \text{ mol } N_2O_5} = (0.040\underline{0}862 \frac{\text{mol } N_2O_5}{L} - 0.016\underline{5}653 \frac{\text{mol } N_2O_5}{L}) \times \frac{1/2 \text{ mol } O_2}{1 \text{ mol } N_2O_5} =$$

$$= 0.011\underline{7}605 \text{ M } O_2$$

then finally $PV = nRT$ rearrange to solve for P.

$$P = \frac{n}{V}RT = 0.011\underline{7}605 \frac{\text{mol}}{L} \times 0.08206 \frac{L \text{ atm}}{K \text{ mol}} \times 298 \text{ K} = 0.287589 \text{ atm} \times \frac{760 \text{ torr}}{1 \text{ atm}} = 219 \text{ torr}$$

Check: The units (torr) are correct. The pressure is reasonable because it must be less than one half of the original pressure.

58. **Given:** Cyclopropane (C_3H_6) reacts, first order in $[C_3H_6]$; $k = 5.87 \times 10^{-4}$ /s at 485 °C; $V = 2.5$ L, $P°_{C3H6} = 722$ torr   **Find:** t to $P_{C3H6} = 100.0$ torr

Conceptual Plan: Since $P \, \alpha \, M$ we do not need to convert P to M. $P°_{C3H6}, P_{C3H6}, k \rightarrow t$
$$\ln[A]_t = -kt + \ln[A]_0$$

Solution: $\ln[C_3H_6]_t = -kt + \ln[C_3H_6]_0$ rearrange to solve for t

$$t = -\frac{1}{k}\ln\frac{[C_3H_6]_t}{[C_3H_6]_0} = -\frac{1}{k}\ln\frac{P_{C3H6}}{P^0_{C3H60}} = -\frac{1}{5.87 \times 10^{-4} \text{ s}^{-1}}\ln\frac{100.\text{ torr}}{722 \text{ torr}} = 3.37 \times 10^3 \text{ s} = 56.1 \text{ min}$$

Check: The units (s or min) are correct. The time is reasonable because it is about three half-lives (pressure dropped to 14 % of original pressure).

59. **Given:** I_2 formation from I atoms, second order in I; $k = 1.5 \times 10^{10}$ M^{-1} s^{-1}, $[I]_0 = 0.0100$ M
Find: t to decrease by 95 %
Conceptual Plan: $[I]_0, [I]_t, k \rightarrow t$
$$\frac{1}{[A]_t} = kt + \frac{1}{[A]_0}$$

Solution: $[I]_t = 0.05 \, [I]_0 = 0.05 \times 0.0100$ M $= 0.0005$ M. Since $\frac{1}{[I]_t} = kt + \frac{1}{[I]_0}$ rearrange to solve for t

$$t = \frac{1}{k}\left(\frac{1}{[I]_t} - \frac{1}{[I]_0}\right) = \frac{1}{(1.5 \times 10^{10} \text{ M}^{-1}\text{s}^{-1})}\left(\frac{1}{0.0005 \text{ M}} - \frac{1}{0.0100 \text{ M}}\right) = 1.\underline{2}67 \times 10^{-7} \text{ s} = 1 \times 10^{-7} \text{ s}.$$

Check: The units (s) are correct. We expect the time to be extremely small because the rate constant is so large.

60. **Given:** sucrose hydrolysis, first order in $[C_{12}H_{22}O_{11}]$; $k = 1.8 \times 10^{-4}$ s^{-1} at 25 °C; $V = 2.55$ L, $[C_{12}H_{22}O_{11}]_0 = 0.150$ M, and 195 min   **Find:** $m(C_{12}H_{22}O_{11})$ hydrolyzed
Conceptual Plan: min $\rightarrow$ s   then $[C_{12}H_{22}O_{11}]_0, t, k \rightarrow [C_{12}H_{22}O_{11}]_t$ then
$$\frac{60 \text{ s}}{1 \text{ min}} \qquad\qquad \ln[A]_t = -kt + \ln[A]_0$$

$V, [C_{12}H_{22}O_{11}]_0, [C_{12}H_{22}O_{11}]_t \rightarrow$ mol $C_{12}H_{22}O_{11}$ hydrolyzed $\rightarrow$ g $C_{12}H_{22}O_{11}$ hydrolyzed
$$mol \, C_{12}H_{22}O_{11} = ([C_{12}H_{22}O_{11}]_0 - [C_{12}H_{22}O_{11}]_t) \times V \qquad \frac{342.30 \text{ g } C_{12}H_{22}O_{11}}{1 \text{ mol } C_{12}H_{22}O_{11}}$$

Solution: $195 \text{ min} \times \frac{60 \text{ sec}}{1 \text{ min}} = 11\underline{7}00 \text{ s}$. Since

$$\ln[C_{12}H_{22}O_{11}]_t = -kt + \ln[C_{12}H_{22}O_{11}]_0 = -(1.8 \times 10^{-4} \text{ s}^{-1})(11\underline{7}00 \text{ s}) + \ln(0.150 \text{ M}) = -4.0\underline{0}312 \rightarrow$$

$$[C_{12}H_{22}O_{11}]_t = e^{-4.0\underline{0}312} = 0.018\underline{2}585 \text{ M} \quad \text{then}$$

$$mol \, C_{12}H_{22}O_{11} = ([C_{12}H_{22}O_{11}]_0 - [C_{12}H_{22}O_{11}]_t) \times V =$$

$$\left(0.150 \frac{\text{mol } C_{12}H_{22}O_{11}}{L} - 0.018\underline{2}585 \frac{\text{mol } C_{12}H_{22}O_{11}}{L}\right) \times 2.55 \text{ L} = 0.33\underline{5}941 \text{ mol } C_{12}H_{22}O_{11}$$

$$0.33\underline{5}941 \text{ mol } C_{12}H_{22}O_{11} \times \frac{342.30 \text{ g } C_{12}H_{22}O_{11}}{1 \text{ mol } C_{12}H_{22}O_{11}} = 115 \text{ g } C_{12}H_{22}O_{11}.$$

Check: The units (g) are correct. The mass is reasonable because it must be less than the original amount in solution (131 g). The amount is close to the original amount in solution because the final sucrose concentration is so low, because we have gone over three half-lives.

61. a) There are two elementary steps in the reaction mechanism because there are two peaks in the reaction progress diagram.

b)

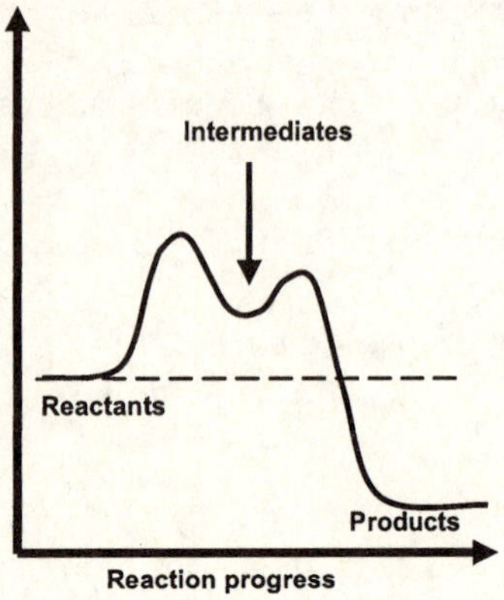

c) The first step is the rate limiting step because it has the higher activation energy.

d) The overall reaction is exothermic because the products are at a lower energy than the reactants.

62. a) The first step is the rate limiting step because it has the higher activation energy.

b) Since the first step is the rate determining step, Rate = k_1 [HCl] [$H_2C{=}CH_2$]. The reaction will be second order overall.

c) The overall reaction is exothermic because the products are at a lower energy than the reactants.

63. **Given:** n-butane desorption from single crystal aluminum oxide, first order; $k = 0.128\ s^{-1}$ at 150 K; initially completely covered
Find: a) $t_{½}$; b) t for 25 % and for 50 % to desorb; c) fraction remaining after 10 s and 20 s
Conceptual Plan: a) $k \rightarrow t_{½}$ b) $[C_4H_{10}]_0$, $[C_4H_{10}]_t$, $k \rightarrow t$ c) $[C_4H_{10}]_0$, t, $k \rightarrow [C_4H_{10}]_t$

$t_{1/2} = \frac{0.693}{k}$ $\qquad \ln[A]_t = -kt + \ln[A]_0$ $\qquad \ln[A]_t = -kt + \ln[A]_0$

Solution:

a) $t_{1/2} = \frac{0.693}{k} = \frac{0.693}{0.128\ s^{-1}} = 5.41\ s$.

b) $\ln[C_4H_{10}]_t = -kt + \ln[C_4H_{10}]_0$ Rearrange to solve for t. For 25 % desorbed $[C_4H_{10}]_t = 0.75\ [C_4H_{10}]_0$

and $t = -\frac{1}{k}\ln\frac{[C_4H_{10}]_t}{[C_4H_{10}]_0} = -\frac{1}{0.128\ s^{-1}}\ln\frac{0.75\ [C_4H_{10}]_0}{[C_4H_{10}]_0} = 2.2\ s$. For 50 % desorbed

$[C_4H_{10}]_t = 0.50\ [C_4H_{10}]_0$ and $t = -\frac{1}{k}\ln\frac{[C_4H_{10}]_t}{[C_4H_{10}]_0} = -\frac{1}{0.128\ s^{-1}}\ln\frac{0.50\ [C_4H_{10}]_0}{[C_4H_{10}]_0} = 5.4\ s$.

c) For 10 s $\ln[C_4H_{10}]_t = -kt + \ln[C_4H_{10}]_0 = -(0.128\ s^{-1})(10\ s) + \ln(1.00) = -1.\underline{2}8 \rightarrow$

$[C_4H_{10}]_t = e^{-1.28} = 0.28$ = fraction covered and for 20 s

$\ln[C_4H_{10}]_t = -kt + \ln[C_4H_{10}]_0 = -(0.128\ s^{-1})(20\ s) + \ln(1.00) = -2.\underline{5}6 \rightarrow$

$[C_4H_{10}]_t = e^{-2.56} = 0.077$ = fraction covered.

Check: The units (s, s, s, none, and none) are correct. The half-life is reasonable considering the size of the rate constant. The time to 25 % desorbed is less than one half-life. The time to 50 % desorbed is the half-life.

The fraction at 10 s is consistent with about two half-lives. The fraction covered at 20 s is consistent with about four half-lives.

64. **Given:** 120 nm film n-pentane evaporation from single crystal aluminum oxide, zero order; $k = 1.92 \times 10^{13}$ molecules / cm^2 s at 120 K; initially coverage = 8.9×10^{16} molecules / cm^2
Find: a) $t_{1/2}$; b) fraction remaining after 10 s
Conceptual Plan:
a) $[C_5H_{12}]_0, k \rightarrow t_{1/2}$

$$t_{1/2} = \frac{[A]_0}{2k}$$

b) $[C_5H_{12}]_0, t, k \rightarrow [C_5H_{12}]_t$ then $[C_5H_{12}]_0, [C_5H_{12}]_t \rightarrow$ fraction remaining

$$[A]_t = -kt + [A]_0 \qquad \textit{fraction remaining} = \frac{[C_5H_{12}]_t}{[C_5H_{12}]_0}$$

Solution:

a) $$t_{1/2} = \frac{[C_5H_{12}]_0}{2k} = \frac{8.9 \times 10^{16}\ \frac{\text{molecules}}{\text{cm}^2}}{2 \times 1.92 \times 10^{13}\ \frac{\text{molecules}}{\text{cm}^2\ \text{s}}} = 2.3 \times 10^3\ \text{s}$$

b) $$[C_5H_{12}]_t = -kt + [C_5H_{12}]_0 = -\left(1.92 \times 10^{13}\ \frac{\text{molecules}}{\text{cm}^2\ \text{s}}\right)(10.\ \text{s}) + 8.9 \times 10^{16}\ \frac{\text{molecules}}{\text{cm}^2} = 8.\underline{8}808 \times 10^{16}\ \frac{\text{molecules}}{\text{cm}^2}$$

$$\textit{fraction remaining} = \frac{[C_5H_{12}]_t}{[C_5H_{12}]_0} = \frac{8.\underline{8}808 \times 10^{16}\ \frac{\text{molecules}}{\text{cm}^2}}{8.9 \times 10^{16}\ \frac{\text{molecules}}{\text{cm}^2}} = 0.9\underline{9}784 = 1.0$$ so within experimental error, all are remaining on the surface.

Check: The units (s and none) are correct. The half-life is reasonable considering the size of the rate constant. The fraction at 10 s is the fact that the time is very, very small compared to the half-life.

65. a) **Given:** table of rate constant versus T **Find:** E_a, and A
Conceptual Plan: **First convert temperature data into Kelvins (°C + 273.15 = K). Since** $\ln k = \frac{-E_a}{R}\left(\frac{1}{T}\right) + \ln A$ **a plot of ln k versus $1/T$ will have a slope = $-E_a/R$ and an intercept = ln A.**
Solution: The slope can be determined by measuring Δy/Δx on the plot or by using functions, such as "add trendline" in Excel.
Since the slope = $-$ 10$\underline{7}$59 K = $-$ E_a/R then
$E_a = -(slope) R =$

$$= -(-10\underline{7}59\ \text{K})\left(8.314\ \frac{\text{J}}{\text{K mol}}\right)\left(\frac{1\ \text{kJ}}{1000\ \text{J}}\right) \text{ an}$$

$$= 89.5\ \frac{\text{kJ}}{\text{mol}}$$

d intercept = 26.7$\underline{6}$9 = ln A then
$A = e^{intercept} = e^{26.7\underline{6}9} = 4.22 \times 10^{11}\ \text{s}^{-1}$.

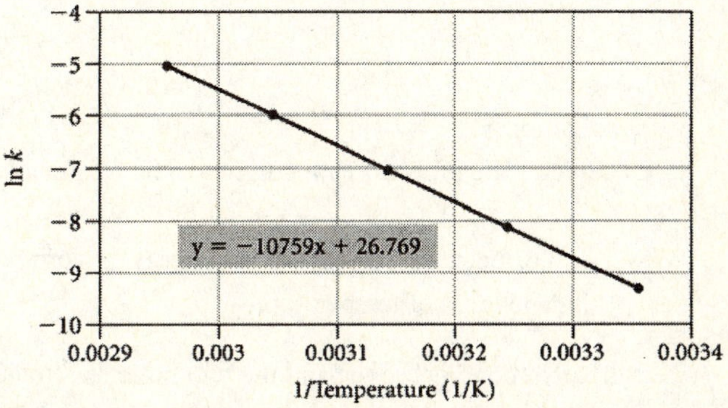

Check: The units (kJ/mol and s^{-1}) are correct.
The plot was extremely linear, confirming Arrhenius behavior. The activation and frequency factor are typical for many reactions.

b) **Given:** part a) results **Find:** k at 15 °C
Conceptual Plan: **°C → K then T, E_a, A → k**

$$°\text{C} + 273.15 = \text{K} \qquad \ln k = \frac{-E_a}{R}\left(\frac{1}{T}\right) + \ln A$$

Solution: 15 °C + 273.15 = 288 K then

$$\ln k = \frac{-E_a}{R}\left(\frac{1}{T}\right) + \ln A = \frac{-89.5\ \frac{\text{kJ}}{\text{mol}} \times \frac{1000\ \text{J}}{1\ \text{kJ}} E_a}{8.314\ \frac{\text{J}}{\text{K mol}}}\left(\frac{1}{288\ \text{K}}\right) + \ln(4.22 \times 10^{11}\ s^{-1}) = -10.610 \rightarrow$$

$k = e^{-10.610} = 2.5 \times 10^{-5}\ M^{-1}\ s^{-1}$.

Check: The units ($M^{-1}\ s^{-1}$) are correct. The value of the rate constant is less than the value at 25 °C.

c) **Given:** part a) results, 0.155 M C_2H_5Br and 0.250 M OH^- at 75 °C **Find:** initial reaction rate
Conceptual Plan:
°C → K then T, E_a, A → k then k, $[C_2H_5Br]$, $[OH^-]$ → initial reaction rate

$°C + 273.15 = K$ $\quad \ln k = \frac{-E_a}{R}\left(\frac{1}{T}\right) + \ln A$ $\quad \text{Rate} = k\,[C_2H_5Br]\,[OH^-]$

Solution: 75 °C + 273.15 = 348 K then

$$\ln k = \frac{-E_a}{R}\left(\frac{1}{T}\right) + \ln A = \frac{-89.5\ \frac{\text{kJ}}{\text{mol}} \times \frac{1000\ \text{J}}{1\ \text{kJ}} E_a}{8.314\ \frac{\text{J}}{\text{K mol}}}\left(\frac{1}{348\ \text{K}}\right) + \ln(4.22 \times 10^{11}\ s^{-1}) = -4.1656 \rightarrow$$

$k = e^{-4.1656} = 1.5521 \times 10^{-2}\ M^{-1}\ s^{-1}$.

$\text{Rate} = k\,[C_2H_5Br]\,[OH^-] = (1.5521 \times 10^{-2}\ M^{-1}\ s^{-1})\,(0.155\ M)\,(0.250\ M) = 6.0 \times 10^{-4}\ M\ s^{-1}$

Check: The units ($M\ s^{-1}$) are correct. The value of the rate is reasonable considering the value of the rate constant (larger than in the table) and the fact that the concentrations are less than 1 M.

66. **Given:** $k = 2.35 \times 10^{-4}\ s^{-1}$ at 293 K and $k = 9.15 \times 10^{-4}\ s^{-1}$ at 303 K **Find:** A
Conceptual Plan: °C → K then k_1, T_1, k_2, T_2 → E_a then k_2, T_2, E_a → A

$K = °C + 273.15$ $\quad \ln k = \frac{-E_a}{R}\left(\frac{1}{T}\right) + \ln A$ $\quad k = A\,e^{-E_a/RT}$

Solution: $T_1 = 293$ K, $k_1 = 2.35 \times 10^{-4}\ s^{-1}$; and $T_2 = 303$ K and $k_2 = 9.15 \times 10^{-4}\ s^{-1}$ then

$\ln\left(\frac{k_2}{k_1}\right) = \frac{E_a}{R}\left(\frac{1}{T_1} - \frac{1}{T_2}\right)$ Rearrange to solve for E_a.

$$E_a = \frac{R \ln\left(\frac{k_2}{k_1}\right)}{\left(\frac{1}{T_1} - \frac{1}{T_2}\right)} = \frac{8.314\ \frac{\text{J}}{\text{K mol}} \ln\left(\frac{9.15 \times 10^{-4}\ s^{-1}}{2.35 \times 10^{-4}\ s^{-1}}\right)}{\left(\frac{1}{293\ \text{K}} - \frac{1}{303\ \text{K}}\right)} = 1.00334 \times 10^5\ \frac{\text{J}}{\text{mol}}$$

. Since $k = A\,e^{-E_a/RT}$ rearrange to solve for A.

$$A = k\,e^{E_a/RT} = 9.15 \times 10^{-4}\ s^{-1}\ e^{\frac{1.00334 \times 10^5\ \frac{\text{J}}{\text{mol}}}{\left(8.314\ \frac{\text{J}}{\text{K mol}}\right)303\ \text{K}}} = 1.8 \times 10^{14}\ s^{-1}.$$

Check: The units (s^{-1}) are correct. The frequency factor is typical for a reaction.

67. a) No, because the activation energy is zero. This means that the rate constant ($k = A\,e^{-E_a/RT}$) will be independent of temperature.

b) No bond is broken and the two radicals attract each other.

d) Formation of diatomic gases from atomic gases.

68. a) Nitrogen has a triple bond, so it will take more energy to break the N≡N bond than the H–H bond.

b) **Given:** E_a = 315 kJ/mol for reaction 1 and E_a = 23 kJ/mol for reaction 2; frequency factor similar; and 25 °C **Find:** ratio of rate constants

Conceptual Plan: °C → K then T, E_as, A → k_1/k_2

$$°C + 273.15 = K \qquad \frac{k_1}{k_2} = \frac{A\,e^{-E_{a1}/RT}}{A\,e^{-E_{a2}/RT}}$$

Solution: T = 25 °C + 273.15 = 298 K. E_{a1} = 315 kJ/mol and E_{a2} = 23 kJ/mol. Ratio of rate constants will be

$$\frac{k_1}{k_2} = \frac{A\,e^{-E_{a1}/RT}}{A\,e^{-E_{a2}/RT}} = \frac{e^{\frac{-315\frac{kJ}{mol}\times\frac{1000\,J}{1\,kJ}}{\left(8.314\frac{J}{K\,mol}\right)298\,K}}}{e^{\frac{-23\frac{kJ}{mol}\times\frac{1000\,J}{1\,kJ}}{\left(8.314\frac{J}{K\,mol}\right)298\,K}}} = 6.5 \times 10^{-52}.$$

Check: The units (none) are correct. Since there is a large difference between the activation energies, we expect a large difference in the rate constants.

69. **Given:** $t_{1/2}$ for radioactive decay of C-14 = 5730 years; bone has 19.5 % C-14 in living bone
Find: age of bone
Conceptual Plan:
radioactive decay implies first order kinetics, $t_{1/2} \rightarrow k$ then 19.5 % of [C-14]$_0$, $k \rightarrow$ t

$$t_{1/2} = \frac{0.693}{k} \qquad ln[A]_t = -kt + ln[A]_0$$

Solution: $t_{1/2} = \frac{0.693}{k}$ rearrange to solve for k. $k = \frac{0.693}{t_{1/2}} = \frac{0.693}{5730\text{ yr}} = 1.2\underline{0}942 \times 10^{-4}\text{ yr}^{-1}$ then

[C-14]$_t$ = 0.195 [C-14]$_0$. Since $ln[\text{C-14}]_t = -kt + ln[\text{C-14}]_0$ rearrange to solve for t

$$t = -\frac{1}{k}\ln\frac{[\text{C-14}]_t}{[\text{C-14}]_0} = -\frac{1}{1.2\underline{0}942 \times 10^{-4}\text{ yr}^{-1}}\ln\frac{0.195\,[\text{C-14}]_0}{[\text{C-14}]_0} = 1.35 \times 10^4\text{ yr}.$$

Check: The units (yr) are correct. The time to 19.5 % decay is consistent with the time being between two and three half-lives.

70. **Given:** $t_{1/2}$ for radioactive decay of U-238 = 4.5 billion years; rock has 83.2 % of original U-238
Find: age of rock
Conceptual Plan:
radioactive decay implies first order kinetics, $t_{1/2} \rightarrow k$ then 82.3 % of [U-238]$_0$, $k \rightarrow$ t

$$t_{1/2} = \frac{0.693}{k} \qquad ln[A]_t = -kt + ln[A]_0$$

Solution: $t_{1/2} = \frac{0.693}{k}$ rearrange to solve for k. $k = \frac{0.693}{t_{1/2}} = \frac{0.693}{4.5 \times 10^9\text{ yr}} = 1.54 \times 10^{-10}\text{ yr}^{-1}$ then

[U-238]$_t$ = 0.823 [U-238]$_0$. Since $\ln[\text{U-238}]_t = -kt + \ln[\text{U-238}]_0$ rearrange to solve for t

$$t = -\frac{1}{k}\ln\frac{[\text{U-238}]_t}{[\text{U-238}]_0} = -\frac{1}{1.54 \times 10^{-10}\text{ yr}^{-1}}\ln\frac{0.832\,[\text{U-238}]_0}{[\text{U-238}]_0} = 1.19 \times 10^9\text{ yr}.$$

Check: The units (yr) are correct. The time to 82.3 % decay is consistent with the time being less than one half-life.

71. a) For each, check that all steps sum to overall reaction and that the predicted rate law is consistent with experimental data (Rate = k [H_2] [I_2]).
For the first mechanism, the single step is the overall reaction. The rate law is determined by the stoichiometry, so Rate = k [H_2] [I_2] and the mechanism is valid.
For the second mechanism, the overall reaction is the sum of the steps in the mechanism:

$$I_2(g) \underset{k_2}{\overset{k_1}{\rightleftharpoons}} \cancel{2\,I(g)}$$

$$H_2(g) + \cancel{2\,I(g)} \xrightarrow{k_3} 2\,HI(g)$$ So the sum matches the overall reaction.

$$H_2(g) + I_2(g) \rightarrow 2\,HI(g)$$

Since the second step is the rate determining step, Rate = k_3 [H_2] [I]^2. Since I is an intermediate, its concentration cannot appear in the rate law. Using the fast equilibrium in the first step, we see that $k_1[I_2] = k_2 [I]^2$ or $[I]^2 = \frac{k_1}{k_2}[I_2]$. Substituting this into the first rate expression, we get that

Rate = $k_3 \frac{k_1}{k_2} [H_2] [I_2]$ and the mechanism is valid.

For the third mechanism, the overall reaction is the sum of the steps in the mechanism:

$$I_2\ (g) \underset{k_2}{\overset{k_1}{\rightleftharpoons}} \cancel{2\,I(g)}$$

$$H_2\ (g) + \cancel{I(g)} \underset{k_4}{\overset{k_3}{\rightleftharpoons}} \cancel{H_2I(g)}$$

$$\cancel{H_2I(g)} + \cancel{I(g)} \xrightarrow[k_5]{} 2\ HI\ (g)$$

$$H_2\ (g) + I_2\ (g) \rightarrow 2\ HI\ (g)$$

So the sum matches the overall reaction.

Since the third step is the rate determining step, Rate = k_5 [H_2I] [I]. Since H_2I and I are intermediates, their concentrations cannot appear in the rate law. Using the fast equilibrium in the first step, we see that $k_1[I_2] = k_2 [I]^2$ or $[I] = \sqrt{\frac{k_1}{k_2}[I_2]}$. Using the fast equilibrium in the second step, we see that $k_3[H_2]$ [I] $= k_4$ [H_2I] or $[H_2I] = \frac{k_3}{k_4} [H_2] [I]$. Substituting both of these expressions for the intermediates into the first rate expression, we get that Rate = $k_5 [H_2I] [I] = k_5 \frac{k_3}{k_4} [H_2] [I] \sqrt{\frac{k_1}{k_2}[I_2]}$. Substituting again for [I] we get Rate = $k_5 \frac{k_3}{k_4} [H_2] \sqrt{\frac{k_1}{k_2}[I_2]} \sqrt{\frac{k_1}{k_2}[I_2]}$. Simplifying this expression we see

Rate = $k_5 \frac{k_3}{k_4} \frac{k_1}{k_2} [H_2] [I_2]$ and the mechanism is valid.

b) To distinguish between mechanisms, you could look for the buildup of I(*g*) and/or H_2I(*g*), the intermediates.

72. a) The overall reaction is the sum of the steps in the mechanism:

$$NH_3\ (aq) + OCl^-\ (aq) \underset{k_2}{\overset{k_1}{\rightleftharpoons}} \cancel{NH_2Cl\ (aq)} + \cancel{OH^-\ (aq)}$$

$$\cancel{NH_2Cl\ (aq)} + NH_3\ (aq) \xrightarrow[k_3]{} \cancel{N_2H_5^+\ (aq)} + Cl^-\ (aq)$$

$$\cancel{N_2H_5^+\ (aq)} + \cancel{OH^-\ (aq)} \xrightarrow[k_4]{} N_2H_4\ (aq) + H_2O\ (l)$$

$$2NH_3\ (aq) + OCl^-\ (aq) \rightarrow N_2H_4\ (aq) + H_2O\ (l) + Cl^-\ (aq)$$

So the sum matches the overall reaction.

b) Since the second step is the rate determining step, Rate = k_3 [NH_2Cl] [NH_3]. Since NH_2Cl is an intermediate, its concentration cannot appear in the rate law. Using the fast equilibrium in the first step, we see that $k_1[NH_3] [OCl^-] = k_2 [NH_2Cl] [OH^-]$ or $[NH_2Cl] = \frac{k_1}{k_2}\frac{[NH_3]\,[OCl^-]}{[OH^-]}$. Substituting this into the first rate expression, we get that Rate = $k_3 \frac{k_1}{k_2}\frac{[NH_3]\,[OCl^-]}{[OH^-]} [NH_3]$ or

Rate = $k_3 \frac{k_1}{k_2}\frac{[NH_3]^2\,[OCl^-]}{[OH^-]}$. In a pH neutral solution, $[OH^-] = 10^{-7}$ (see Chapter 15), so that the rate law can be expressed as Rate = $k[NH_3]_2\ [OCl^-]$.

73. a) For a zero order reaction, the rate is independent of the concentration. If the first half goes in the first 100 minutes, the second half will go in the second 100 minutes. This means that there will be none or 0 % left at 200 minutes.

b) For a first order reaction, the half-life is independent of concentration. This means that if half of the reactant decomposes in the first 100 minutes, then half of this (or another 25 % of the original amount) will decompose in the second 100 minutes. This means that at 200 minutes 50 % + 25 % = 75 % has decomposed or 25 % remains.

c) For a second order reaction, $t_{1/2} = \dfrac{1}{k[\mathrm{A}]_0} = 100 \text{ min}$ and the integrated rate expression is

$\dfrac{1}{[\mathrm{A}]_t} = kt + \dfrac{1}{[\mathrm{A}]_0}$. We can rearrange the first expression to solve for k as $k = \dfrac{1}{100 \text{ min } [\mathrm{A}]_0}$.

Substituting this and 200 minutes into the integrated rate expression, we get $\dfrac{1}{[\mathrm{A}]_t} = \dfrac{200 \cancel{\text{min}}}{100 \cancel{\text{min}} [\mathrm{A}]_0} + \dfrac{1}{[\mathrm{A}]_0}$

$\rightarrow \dfrac{1}{[\mathrm{A}]_t} = \dfrac{3}{[\mathrm{A}]_0} \rightarrow \dfrac{[\mathrm{A}]_t}{[\mathrm{A}]_0} = \dfrac{1}{3}$ or 33 % remains.

74. **Given:** $t_{1/2}$ for radioactive decay of Pu-239 = 24,000 years; 1 mole initially to 1 atom **Find:** t
Conceptual Plan:
radioactive decay implies first order kinetics, $t_{1/2} \rightarrow k$ then [Pu-239]$_0$, [Pu-239]$_t$, $k \rightarrow$ t

$$t_{1/2} = \frac{0.693}{k} \qquad \ln[\mathrm{A}]_t = -kt + \ln[\mathrm{A}]_0$$

Solution: $t_{1/2} = \dfrac{0.693}{k}$ rearrange to solve for k. $k = \dfrac{0.693}{t_{1/2}} = \dfrac{0.693}{24000 \text{ yr}} = 2.8875 \times 10^{-5} \text{ yr}^{-1}$ then

[Pu-239]$_t$ = 1 and [Pu-239]$_0$ = 6.022 x 10^{23}. Since $\ln[\text{Pu-239}]_t = -kt + \ln[\text{Pu-239}]_0$ rearrange to solve for t,

$$t = -\frac{1}{k}\ln\frac{[\text{Pu-239}]_t}{[\text{Pu-239}]_0} = -\frac{1}{2.8875 \times 10^{-5} \text{ yr}^{-1}}\ln\frac{1 \cancel{\text{ atom Pu-239}}}{6.022 \times 10^{23} \cancel{\text{ atom Pu-239}}} = 1.9 \times 10^{6} \text{ yr}.$$

Check: The units (yr) are correct. The time to decay 1 atom is consistent with the time being 79 half-lives, which makes sense since $2^{-79} = 6.09 \times 10^{23}$.

75. Using the energy diagram show and using Hess' Law, we can see that the activation energy for the decomposition is equal to the activation energy for the formation reaction plus the heat of formation of 2 moles of HI or
$E_{a\,formation} = E_{a\,decomposition} + 2\,\Delta H_f^0(\mathrm{HI})$. So
$E_{a\,formation} = 185 \text{ kJ} + 2 \text{mol}(-5.65 \text{ kJ/mol}) = 174 \text{ kJ}$.
Check: Since the reaction is endothermic, we expect the activation energy in the reverse direction to be less in the forward direction.

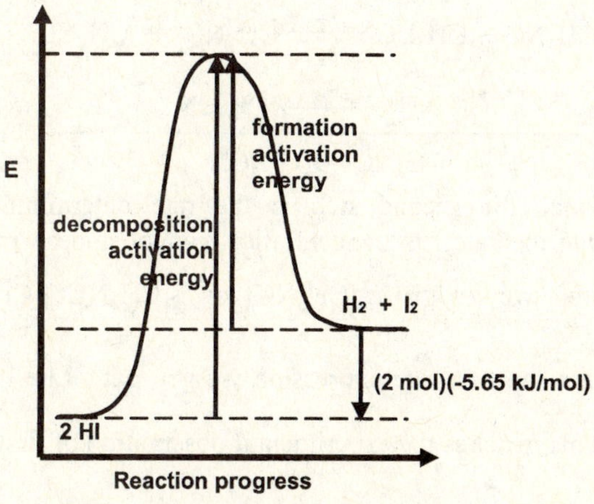

Note: energy axis is not to scale.

76. **Given:** first order reaction, E_a = 249 kJ/mol, A = 1.6 x 10^{14} s^{-1}, and 710 K
Find: k, fraction decomposed in 15 min and T for double the reaction rate
Conceptual Plan:
T, A, $E_a \rightarrow k$ then min $\rightarrow$ s then k, t $\rightarrow$ fraction decomposed then k_2/k_1, T_1, $E_a \rightarrow T_2$

$$k = Ae^{-E_a/RT} \qquad \frac{60 \text{ s}}{1 \text{ min}} \qquad \ln[\mathrm{A}]_t = -kt + \ln[\mathrm{A}]_0 \qquad \ln\left(\frac{k_2}{k_1}\right) = \frac{E_a}{R}\left(\frac{1}{T_1} - \frac{1}{T_2}\right)$$

Solution: Since $k = Ae^{-E_a/RT} = 1.6 \times 10^{14}\ s^{-1}\ e^{\frac{-249 \frac{kJ}{mol} \times \frac{1000\ J}{1\ kJ}}{\left(8.314 \frac{J}{K\ mol}\right) 710\ K}} = 7.\underline{6}657 \times 10^{-5}\ s^{-1} = 7.7 \times 10^{-5}\ s^{-1}$ then $15\ min \times \frac{60\ s}{1\ min} = 9\underline{0}0\ s$ in $\ln[C_2H_5Cl]_t = -kt + \ln[C_2H_5Cl]_0$ Rearrange to solve for fraction remaining $\rightarrow \frac{[C_2H_5Cl]_t}{[C_2H_5Cl]_0} = e^{-kt} = e^{-(7.\underline{6}657 \times 10^{-5}\ s^{-1})(9\underline{0}0\ s)} = 0.9\underline{3}333$ thus fraction decomposed $= 1 - 0.9\underline{3}333 = 0.0\underline{6}667 =$ 0.07. $T_1 = 710$ K, $k_2 / k_1 = 2$ and $\ln\left(\frac{k_2}{k_1}\right) = \frac{E_a}{R}\left(\frac{1}{T_1} - \frac{1}{T_2}\right)$ Rearrange to solve for T_2.

$$T_2 = \frac{\frac{E_a}{R}}{\frac{E_a}{RT_1} - \ln\left(\frac{k_2}{k_1}\right)} = \frac{\frac{249 \frac{kJ}{mol} \times \frac{1000\ J}{1\ kJ}}{8.314 \frac{J}{K\ mol}}}{\left(\frac{249 \frac{kJ}{mol} \times \frac{1000\ J}{1\ kJ}}{8.314 \frac{J}{K\ mol} \times 710\ K}\right) - \ln(2)} = 7\underline{2}1.86\ K = 720\ K$$

Check: The units (s^{-1}, none, and K) are correct. The reaction rate is reasonable considering the frequency factor, activation energy, and T. The fraction decomposed is reasonable since 900 s is a small fraction of the half-life. The temperature is reasonable since many reactions double their rate with an increase in temp of 10 K.

77. a) Since the rate determining step involves the collision of two molecules, the expected reaction order would be second order.

b) The proposed mechanism is:

$$CH_3NC + \cancel{CH_3NC} \underset{k_2}{\overset{k_1}{\rightleftharpoons}} \cancel{CH_3NC^*} + \cancel{CH_3NC} \quad \text{(fast)}$$

$$\cancel{CH_3NC^*} \xrightarrow{k_3} CH_3CN \quad \text{(slow)}$$

So the sum matches the overall reaction.

$$CH_3NC \rightarrow CH_3CN$$

Since the second step is the rate determining step, Rate = $k_3[CH_3NC^*]$. Since CH_3NC^* is an intermediate, its concentration cannot appear in the rate law. Using the fast equilibrium in the first step, we see that $k_1[CH_3NC]^2 = k_2[CH_3NC^*][CH_3NC]$ or $[CH_3NC^*] = \frac{k_1}{k_2}[CH_3NC]$. Substituting this into the first rate expression we get that Rate $= k_3 \frac{k_1}{k_2}[CH_3NC]$, which simplifies to Rate $= k\,[CH_3NC]$. This matches the experimental observation of first order and the mechanism is valid.

78. a) Rate $= k\,[A]^{1/2}$ and Rate $= -\frac{d[A]}{dt}$ so $\frac{d[A]}{dt} = -k\,[A]^{1/2}$ moving the A terms to the left and the t and constants to the right we have $\frac{d[A]}{[A]^{1/2}} = -k\,dt$. Integrating we get $\int_{[A]_0}^{[A]} \frac{d[A]}{[A]^{1/2}} = -\int_0^t k\,dt$. When we evaluate this integral

$$2[A]^{1/2}\Big|_{[A]_0}^{[A]} = -k\,t\Big|_0^t \rightarrow 2[A]_t^{1/2} - 2[A]_0^{1/2} = -k\,t \rightarrow 2[A]_t^{1/2} = -k\,t + 2[A]_0^{1/2}$$ the integrated rate law.

b) To derive the half-life, set $[A]_t = 1/2[A]_0$. Substituting this into $2[A]_t^{1/2} = -k\,t + 2[A]_0^{1/2}$ we get

$$2(1/2[A]_0)^{1/2} = -k\,t_{1/2} + 2[A]_0^{1/2} \rightarrow t_{1/2} = \frac{2[A]_0^{1/2} - 2(1/2[A]_0)^{1/2}}{k} \rightarrow t_{1/2} = \frac{(2-\sqrt{2})[A]_0^{1/2}}{k}.$$

79. a) **Given:** N_2O_5 decomposes to NO_2 and O_2, first order in $[N_2O_5]$; $k = 7.48 \times 10^{-3}\ s^{-1}$; $P°_{N2O5} = 0.100$ atm
Find: t to $P_{Total} = 0.145$ atm
Conceptual Plan: **Write a balanced reaction. then**
$N_2O_5 \rightarrow 2\ NO_2 + \frac{1}{2} O_2$
Write expression for P_{Total} in terms of amount reacted then $x, P°_{N2O5}, k \rightarrow t$
let $x = P_{N2O5\,reacted}$ $P_{Total} = P_{N2O5} + P_{NO2} + P_{O2}$ $\qquad \ln[A]_t = -kt + \ln[A]_0$

Solution: $P_{Total} = P_{N2O5} + P_{NO2} + P_{O2} = (0.100\,\text{atm} - x) + (2x) + (1/2x) = 0.100\ \text{atm} + 1.5\,x$ Set $P_{Total} = 0.145\ \text{atm} = 0.100\ \text{atm} + 1.5\,x$ and solve for x. $x = \dfrac{0.145\ \text{atm} - 0.100\ \text{atm}}{1.5} = 0.030\ \text{atm}$ then $P_{N2O5} = 0.100\ \text{atm} - x = 0.100\ \text{atm} - 0.030\ \text{atm} = 0.070\ \text{atm}$. Since $P \propto n/V$ or M
$\ln[N_2O_5]_t = -kt + \ln[N_2O_5]_0$ rearrange to solve for t.

$$t = -\frac{\ln\dfrac{[N_2O_5]_t}{[N_2O_5]_0}}{k} = -\frac{\ln\left(\dfrac{0.070\ \cancel{\text{atm}}}{0.100\ \cancel{\text{atm}}}\right)}{7.48 \times 10^{-3}\ s^{-1}} = 4\underline{7}.684\ s = 48\ s$$

Check: The units (s) are correct. The time is reasonable because it is less than one half-life and the amount decomposing is less that 50 %.

b) **Given:** N_2O_5 decomposes to NO_2 and O_2, first order in $[N_2O_5]$; $k = 7.48 \times 10^{-3}\ s^{-1}$; $P°_{N2O5} = 0.100$ atm
Find: t to $P_{Total} = 0.200$ atm
Conceptual Plan: **Write a balanced reaction. then**
$N_2O_5 \rightarrow 2\ NO_2 + \frac{1}{2} O_2$
Write expression for P_{Total} in terms of amount reacted then $x, P°_{N2O5}, k \rightarrow t$
let $x = P_{N2O5\,reacted}$ $P_{Total} = P_{N2O5} + P_{NO2} + P_{O2}$ $\qquad \ln[A]_t = -kt + \ln[A]_0$

Solution: $P_{Total} = P_{N2O5} + P_{NO2} + P_{O2} = (0.100\,\text{atm} - x) + (2x) + (1/2x) = 0.100\ \text{atm} + 1.5\,x$ Set $P_{Total} = 0.200\ \text{atm} = 0.100\ \text{atm} + 1.5\,x$ and solve for x. $x = \dfrac{0.200\ \text{atm} - 0.100\ \text{atm}}{1.5} = 0.06\underline{6}667\ \text{atm}$ then $P_{N2O5} = 0.100\ \text{atm} - x = 0.100\ \text{atm} - 0.06\underline{6}667\ \text{atm} = 0.03\underline{3}333\ \text{atm}$. Since $P \propto n/V$ or M
$\ln[N_2O_5]_t = -kt + \ln[N_2O_5]_0$ rearrange to solve for t.

$$t = -\frac{\ln\dfrac{[N_2O_5]_t}{[N_2O_5]_0}}{k} = -\frac{\ln\left(\dfrac{0.03\underline{3}333\ \cancel{\text{atm}}}{0.100\ \cancel{\text{atm}}}\right)}{7.48 \times 10^{-3}\ s^{-1}} = 1\underline{4}6.873\ s = 150\ s = 1.5 \times 10^2\ s.$$

Check: The units (s) are correct. The time is reasonable because it is between one and two half-lives and the amount decomposing is 67 %.

c) **Given:** N_2O_5 decomposes to NO_2 and O_2, first order in $[N_2O_5]$; $k = 7.48 \times 10^{-3}\ s^{-1}$; $P°_{N2O5} = 0.100$ atm
Find: P_{Total} after 100 s
Conceptual Plan:
$P°_{N2O5}, k, t \rightarrow P_{N2O5}$ **then Write a balanced reaction. then** $P°_{N2O5}, P_{N2O5} \rightarrow x$ **then**
$\ln[A]_t = -kt + \ln[A]_0$ $\qquad N_2O_5 \rightarrow 2\ NO_2 + \frac{1}{2} O_2$ $\qquad x = P^0_{N2O5} - P_{N2O5}$
Write expression for P_{Total} in terms of amount reacted.
let $x = P_{N2O5\,reacted}$ $P_{Total} = P_{N2O5} + P_{NO2} + P_{O2}$

Solution: Since $P \propto n/V$ or M
$\ln[N_2O_5]_t = -kt + \ln[N_2O_5]_0 = -(7.48 \times 10^{-3}\ s^{-1})(100\ s) + \ln(0.100\ atm) = -3.0\underline{5}059$
$P_{N2O5} = e^{-3.0\underline{5}059} = 0.047\underline{3}312$ atm so $x = P^0_{N2O5} - P_{N2O5} = 0.100\ \text{atm} - 0.047\underline{3}312\ \text{atm} = 0.05\underline{2}669$ atm
finally $P_{Total} = P_{N2O5} + P_{NO2} + P_{O2} = (0.100\,\text{atm} - x) + (2x) + (1/2x) = 0.100\ \text{atm} + 1.5\,x =$
$= 0.100\ \text{atm} + 1.5(0.05\underline{2}669\ \text{atm}) = 0.179$ atm

Check: The units (atm) are correct. The pressure is reasonable because the time is between those for parts a and b.

80. For this mechanism, the overall reaction is the sum of the steps in the mechanism:

$$Cl_2\ (g) \underset{k_2}{\overset{k_1}{\rightleftharpoons}} 2\ Cl\ (g)$$

$$\cancel{Cl\ (g)} + CO\ (g) \underset{k_4}{\overset{k_3}{\rightleftharpoons}} \cancel{ClCO\ (g)}$$

$$\cancel{ClCO\ (g)} + Cl_2\ (g) \xrightarrow[k_5]{} Cl_2CO\ (g) + \cancel{Cl\ (g)}$$

$$CO\ (g) + 2\ Cl_2\ (g) \rightarrow Cl_2CO\ (g) + 2\ Cl\ (g)$$

There is no overall reaction given. Since the third step is the rate determining step, Rate = k_5 [ClCO] [Cl_2]. Since ClCO is an intermediate, its concentration cannot appear in the rate law. Using the fast equilibrium in the second step, we see that k_3[Cl] [CO] = k_4 [ClCO] or [ClCO] = $\frac{k_3}{k_4}$ [Cl] [CO]. Substituting this expression into the first rate expression we get Rate = $k_5\frac{k_3}{k_4}$ [Cl] [CO] [Cl_2]. Since Cl is an intermediate, its concentration cannot appear in the rate law. Using the fast equilibrium in the first step, we see that $k_1[Cl_2] = k_2\ [Cl]^2$ or $[Cl] = \sqrt{\frac{k_1}{k_2}[Cl_2]}$. Substituting this expression into last first rate expression we get that Rate = k_5 $\frac{k_3}{k_4}\sqrt{\frac{k_1}{k_2}[Cl_2]}$ [CO] [Cl_2] = $k_5\frac{k_3}{k_4}\sqrt{\frac{k_1}{k_2}}$ [CO] $[Cl_2]^{3/2}$. Simplifying this expression we see Rate = k[CO] $[Cl_2]^{3/2}$.

81. **Given:** N_2O_3 decomposes to NO_2 and NO, first order in [N_2O_3], table of [NO_2] versus time, at 50,000 s = [N_2O_3] = 0 **Find:** k

Conceptual Plan: Write a balanced reaction. then write expression for P_{NO2} in terms P_{N2O3}, then

$N_2O_3 \rightarrow NO_2 + NO$ $\qquad$ $P_{N2O3} = 0.784\ atm - P_{NO2}$

Plot ln P_{N2O3} versus time (this will have the same slope as ln [N_2O_3] versus time). Since $\ln[A]_t = -kt + \ln[A]_0$, the negative of the slope will be the rate constant.

Solution: $P_{N2O3} = 0.784\ atm - P_{NO2}$ Plot ln P_{N2O3} versus time (this will have the same slope as ln [N_2O_3] versus time). Since $\ln[A]_t = -kt + \ln[A]_0$, the negative of the slope will be the rate constant. The slope can be determined by measuring Δy/Δx on the plot or by using functions, such as "add trendline" in Excel. The last point (50,000 s) cannot be plotted because the concentration is 0 and the ln 0 is undefined,

Thus the rate constant is $3.20 \times 10^{-4}\ s^{-1}$ and the rate law is Rate = $3.20\ 10^{-4}\ s^{-1}[N_2O_3]$.

Check: The units (s^{-1}) are correct. The rate constant is typical for a reaction.

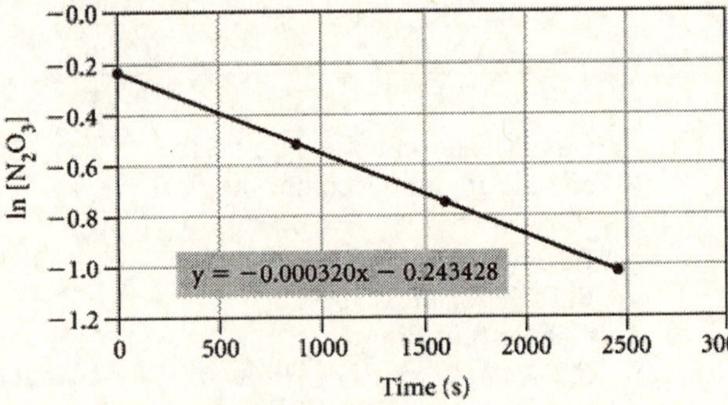

82. Since Rate = k [$CHCl_3$] $[Cl_2]^{1/2}$, Rate a = $k\ (3)(3)^{1/2} = 5.2\ k$, Rate b = $k\ (4)(2)^{1/2} = 5.7\ k$, and Rate c = $k\ (2)(4)^{1/2}$ = 4 k. So b has the fastest rate.

83. Reaction A is second order. A plot of 1/[A] versus time will be linear $\left(\frac{1}{[A]_t} = kt + \frac{1}{[A]_0}\right)$. Reaction B is first order. A plot of ln [A] versus time will be linear ($\ln[A]_t = -kt + \ln[A]_0$). Reactant concentrations drop more quickly for first order reactions than for second order reactions.

84. A reaction that slows down as the reaction proceeds and has a half-life that is dependent on the concentration is a second order reaction. Statement (a) will be false because this describes a first order reaction. Statement (b) is true because it describes a second order reaction. Statements (c) and (d) are false because they describe a zero order reaction.

Chapter 14
Chemical Equilibrium

1. The equilibrium constant is defined as the concentrations of the products raised to their stoichiometric coefficients divided by the concentrations of the reactants raised to their stoichiometric coefficients.

   a) $K = \dfrac{[SbCl_3][Cl_2]}{[SbCl_5]}$

   b) $K = \dfrac{[NO]^2[Br_2]}{[BrNO]^2}$

   c) $K = \dfrac{[CS_2][H_2]^4}{[CH_4][H_2S]^2}$

   d) $K = \dfrac{[CO_2]^2}{[CO]^2[O_2]}$

2. a) The equilibrium constant is defined as the concentrations of the products **raised to their stoichiometric coefficients** divided by the concentrations of the reactants **raised to their stoichiometric coefficients.**

   $K = \dfrac{[H_2]^2[S_2]}{[H_2S]^2}$

   b) The equilibrium constant is defined as the **concentrations of the products** raised to their stoichiometric coefficients **divided by the concentrations of the reactants** raised to their stoichiometric coefficients.

   $K = \dfrac{[COCl_2]}{[CO][Cl_2]}$

3. With an equilibrium constant of 1.4×10^{-5}, the value of the equilibrium constant is small, therefore, the concentration of reactants will be greater than the concentration of products. This is independent of the initial concentration of the reactants and products.

4. Figure a at equilibrium has 8 $C_2H_4Cl_2$; 2 Cl_2; and 2 C_2H_4.
   Figure b at equilibrium has 6 $C_2H_4Br_2$; 4 Br_2; and 4 C_2H_4.
   Figure c at equilibrium has 3 $C_2H_4I_2$; 7 I_2; and 7 C_2H_4.
   Since the equilibrium constant is concentration of products/ concentration of reactants, the equilibrium situation which has the largest concentration of products will have the largest equilibrium constant. Therefore, $K_{Cl2} > K_{Br2} > K_{I2}$.

5. (i) has 10 H_2 and 10 I_2
   (ii) has 7 H_2 and 7 I_2 and 6 HI
   (iii) has 5 H_2 and 5 I_2 and 10 HI
   (iv) has 4 H_2 and 4 I_2 and 12 HI
   (v) has 3 H_2 and 3 I_2 and 14 HI
   (vi) has 3 H_2 and 3 I_2 and 14 HI

   a) Concentration of (v) and (vi) are the same so the system reached equilibrium at (v).

   b) If a catalyst was added to the system, the system would reach the conditions at (v) sooner since a catalyst speeds up the reaction but does not change the equilibrium conditions.

   c) The final figure (vi) would have the same amount of reactants and products since a catalyst speeds up the reaction, but does not change the equilibrium concentrations.

6. The equilibrium constant gives us the ratio of products to reactants at equilibrium, it does not say how long it takes to reach equilibrium. So, after 15 minutes, if the smaller equilibrium constant has more products, then the kinetics of that reaction are faster.

7. a) If you reverse the reaction, invert the equilibrium constant. So, $K' = \frac{1}{K_p} = \frac{1}{2.26x10^4} = 4.42 \text{ x } 10^{-5}$. The reactants will be favored.

b) If you multiply the coefficients in the equation by a factor, raise the equilibrium constant to the same factor.
So, $K' = (K_p)^{1/2} = (2.26 \text{ x } 10^4)^{1/2} = 1.50 \text{ x } 10^2$. The products will be favored.

c) Begin with the reverse of the reaction and invert the equilibrium constant.
$K_{reverse} = \frac{1}{K_p} = \frac{1}{2.26x10^4} = 4.42 \text{ x } 10^{-5}$
Then, multiply the reaction by 2 and raise the value of $K_{reverse}$ to the 2 power.
$K' = (K_{reverse})^2 = (4.42 \text{ x } 10^{-5})^2 = 1.96 \text{ x } 10^{-9}$. The reactants will be favored.

8. a) The reaction is multiplied by 1/2, so raise the value of the equilibrium constant to 1/2.
$K' = (K_p)^{1/2} = (2.2 \text{ x } 10^6)^{1/2} = 1.5 \text{ x } 10^3$. The products will be favored.

b) The reaction is multiplied by 3, so raise the value of the equilibrium constant to 3.
$K' = (K_p)^3 = (2.2 \text{ x } 10^6)^3 = 1.1 \text{ x } 10^{19}$. The products will be favored.

c) Begin with the reverse of the reaction and invert the equilibrium constant.
$K_{reverse} = \frac{1}{K_p} = \frac{1}{2.2x10^6} = 4.5 \text{ x } 10^{-7}$
Then, multiply the reaction by 2 and raise the value of $K_{reverse}$ to the 2 power.
$K' = (K_{reverse})^2 = (4.5 \text{ x } 10^{-7})^2 = 2.1 \text{ x } 10^{-13}$. The reactants will be favored.

9. To find the equilibrium constant for reaction 3, you need to combine reactions 1 and 2 to get reaction 3. Begin by reversing reaction 2, then multiply reaction 1 by 2 and add the two new reactions. When you add reactions you multiply the values of K.

$N_2(g) + O_2(g) \rightleftharpoons \cancel{2NO}(g)$	$K_1 = \frac{1}{K_p} = \frac{1}{2.1 \text{ x } 10^{30}} = 4.\underline{7}6 \text{ x } 10^{-31}$
$\cancel{2NO}(g) + Br_2\ (g) \rightleftharpoons 2\ NOBr(g)$	$K_2 = (K_p)^2 = (5.3)^2 = 2\underline{8}.09$
$N_2(g) + O_2(g) + Br_2\ (g) \rightleftharpoons 2\ NOBr(g)$	$K_3 = K_1K_2 = (4.\underline{7}6 \text{ x } 10^{-31})(2\underline{8}.09) = 1.3 \text{ x } 10^{-29}$

10. To find the equilibrium constant for reaction 3, you need to combine reactions 1 and 2 to get reaction 3. Begin by multiplying reaction 1 by 2 and then reverse reaction 2 and add the two new reactions. When you add reactions you multiply the values of K.

$2\ A(s) \rightleftharpoons \cancel{B}(g) + \cancel{2C}(g)$	$K_1 = (K_p)^2 = (0.0334)^2 = 1.1 \text{ x } 10^{-3}$
$\cancel{B}(g) + \cancel{2C}(g) \rightleftharpoons 3\ D(g)$	$K_2 = \frac{1}{K_p} = \frac{1}{2.35} = 0.425$
$2\ A(s) \rightleftharpoons 3\ D(g)$	$K' = K_1K_2 = (1.1 \text{ x } 10^{-3})(0.425) = 4.68 \text{ x } 10^{-4}$

11. a) **Given:** $K_p = 6.26 \text{ x } 10^{-22}$ $T = 298K$ **Find:** K_c
Conceptual Plan: $K_p \rightarrow K_c$
$K_p = K_c(RT)^{\Delta n}$

Solution: Δn = mol product gas – mol reactant gas = 2 – 1 = 1

$$K_c = \frac{K_p}{(RT)^{\Delta n}} = \frac{6.26 \text{ x } 10^{-22}}{\left(0.08206 \frac{L \cdot atm}{mol \cdot \cancel{K}} \text{ x } 298\ \cancel{K}\right)^1} = 2.56 \text{ x } 10^{-23}$$

Check: Substitute into the equation and confirm that you get the original value of K_p.

$$K_p = K_c(RT)^{\Delta n} = (2.56 \times 10^{-23})(0.08206 \frac{\text{L}\bullet\text{atm}}{\text{mol}\bullet\text{K}} \times 298)^1 = 6.26 \times 10^{-22}$$

b) **Given:** $K_p = 7.7 \times 10^{24}$ T = 298K **Find:** K_c
Conceptual Plan: $K_p \rightarrow K_c$

$K_p = K_c(RT)^{\Delta n}$

Solution: Δn = mol product gas – mol reactant gas = 4 – 2 = 2

$$K_c = \frac{K_p}{(RT)^{\Delta n}} = \frac{7.7 \times 10^{24}}{(0.08206 \frac{\text{L}\bullet\text{ atm}}{\text{mol}\bullet\text{ K}} \times 298 \text{ K})^2} = 1.3 \times 10^{22}$$

Check: Substitute into the equation and confirm that you get the original value of K_p.

$$K_p = K_c(RT)^{\Delta n} = (1.3 \times 10^{22})(0.08206 \frac{\text{L}\bullet\text{atm}}{\text{mol}\bullet\text{K}} \times 298)^2 = 7.7 \times 10^{24}$$

c) **Given:** $K_p = 81.9$ T = 298K **Find:** K_c
Conceptual Plan: $K_p \rightarrow K_c$

$K_p = K_c(RT)^{\Delta n}$

Solution: Δn = mol product gas – mol reactant gas = 2 – 2 = 0

$$K_c = \frac{K_p}{(RT)^{\Delta n}} = \frac{81.9}{(0.08206 \frac{\text{L}\bullet\text{ atm}}{\text{mol}\bullet\text{ K}} \times 298 \text{ K})^0} = 81.9$$

Check: Substitute into the equation and confirm that you get the original value of K_p.

$$K_p = K_c(RT)^{\Delta n} = (81.9)(0.08206 \frac{\text{L}\bullet\text{atm}}{\text{mol}\bullet\text{K}} \times 298)^0 = 81.9$$

12. a) **Given:** $K_c = 5.9 \times 10^{-3}$ T = 298K **Find:** K_p
Conceptual Plan: $K_p \rightarrow K_c$

$K_p = K_c(RT)^{\Delta n}$

Solution: Δn = mol product gas – mol reactant gas = 2 – 1 = 1

$$K_p = K_c(RT)^{\Delta n} = 5.9 \times 10^{-3}(0.08206 \frac{\text{L}\bullet\text{ atm}}{\text{mol}\bullet\text{ K}} \times 298 \text{ K})^1 = 1.4 \times 10^{-1}$$

Check: Substitute into the equation and confirm that you get the original value of K_p.

$$K_c = \frac{K_p}{(RT)^{\Delta n}} = \frac{1.4 \times 10^{-1}}{(0.08206 \frac{\text{L}\bullet\text{ atm}}{\text{mol}\bullet\text{ K}} \times 298 \text{ K})^1} = 0.0059$$

b) **Given:** $K_c = 3.7 \times 10^{8}$ T = 298K **Find:** K_p
Conceptual Plan: $K_p \rightarrow K_c$

$K_p = K_c(RT)^{\Delta n}$

Solution: Δn = mol product gas – mol reactant gas = 2 – 4 = – 2

$$K_p = K_c(RT)^{\Delta n} = 3.7 \times 10^{8}(0.08206 \frac{\text{L}\bullet\text{ atm}}{\text{mol}\bullet\text{ K}} \times 298 \text{ K})^{-2} = 6.2 \times 10^{5}$$

Check: Substitute into the equation and confirm that you get the original value of K_p.

$$K_c = \frac{K_p}{(RT)^{\Delta n}} = \frac{6.2 \times 10^{5}}{(0.08206 \frac{\text{L}\bullet\text{ atm}}{\text{mol}\bullet\text{ K}} \times 298 \text{ K})^{-2}} = 3.7 \times 10^{8}$$

c) **Given:** $K_c = 4.10 \times 10^{-31}$ T = 298K **Find:** K_p
Conceptual Plan: $K_p \rightarrow K_c$

$K_p = K_c(RT)^{\Delta n}$

Solution: Δn = mol product gas – mol reactant gas = 2 – 2 = 0

$$K_p = K_c(RT)^{\Delta n} = 4.10 \times 10^{-31} (0.08206 \frac{\text{L}\bullet\text{ atm}}{\text{mol}\bullet\text{ K}} \times 298 \text{ K})^0 = 4.10 \times 10^{-31}$$

Check: Substitute into the equation and confirm that you get the original value of K_p.

$$K_c = \frac{K_p}{(RT)^{\Delta n}} = \frac{4.10 \times 10^{-31}}{(0.08206 \frac{L \cdot atm}{mol \cdot \cancel{K}} \times 298 \cancel{K})^0} = 4.10 \times 10^{-31}$$

13. a) Since H_2O is a liquid, it is omitted from the equilibrium expression. $K_{eq} = \frac{[HCO_3^-][OH^-]}{[CO_3^{2-}]}$

b) Since $KClO_3$ and KCl are both solids, they are omitted from the equilibrium expression. $K_{eq} = [O_2]^3$

c) Since H_2O is a liquid, it is omitted from the equilibrium expression. $K_{eq} = \frac{[H_3O^+][F^-]}{[HF]}$

d) Since H_2O is a liquid, it is omitted from the equilibrium expression. $K_{eq} = \frac{[NH_4^+][OH^-]}{[NH_3]}$

14. Since PCl_3 is a liquid, it is omitted from the equilibrium expression. $K_{eq} = \frac{[Cl_2]}{[PCl_5]}$

15. **Given:** At equilibrium: [CO] = 0.105M, $[H_2]$ = 0.114M, $[CH_3OH]$ = 0.185M **Find:** K_c
Conceptual Plan: Balanced reaction → equilibrium expression → K_c
Solution: $K_c = \frac{[CH_3OH]}{[CO][H_2]^2} = \frac{(0.185)}{(0.105)(0.114)^2} = 136$

Check: The answer is reasonable since the concentration of products is greater than the concentration of reactants and the equilibrium constant should be greater than 1.

16. **Given:** At equilibrium: $[NH_3]$ = 0.278M, $[H_2S]$ = 0.355M **Find:** K_c
Conceptual Plan: Balanced reaction → equilibrium expression → K_c
Solution: Since NH_4HS is a solid, it is omitted from the equilibrium expression:
$K_c = [NH_3][H_2S] = (0.278)(0.355) = 0.0987$

Check: The answer is reasonable since the concentration of products is less than 1M, the equilibrium constant is less than 1.

17. At 500K: **Given:** At equilibrium: $[N_2]$ = 0.115M, $[H_2]$ = 0.105M, and $[NH_3]$ = 0.439 **Find:** K_c
Conceptual Plan: Balanced reaction → equilibrium expression → K_c
Solution: $K_c = \frac{[NH_3]^2}{[N_2][H_2]^3} = \frac{(0.439)^2}{(0.115)(0.105)^3} = 1.45 \times 10^3$

Check: The value is reasonable since the concentration of products is greater than the concentration of reactants.

At 575K: **Given:** At equilibrium: $[N_2]$ = 0.110M, $[NH_3]$ = 0.128, K_c = 9.6 **Find:** $[H_2]$
Conceptual Plan: Balanced reaction → equilibrium expression → $[H_2]$
Solution: $K_c = \frac{[NH_3]^2}{[N_2][H_2]^3}$ $9.6 = \frac{(0.128)^2}{(0.110)(x)^3}$ $x = 0.249$

Check: Plug the value for x back into the equilibrium expression, and check the value.

$$9.6 = \frac{(0.128)^2}{(0.110)(0.249)^3}$$

At 775K: **Given:** At equilibrium: $[N_2]$ = 0.120M, $[H_2]$ = 0.140, K_c = 0.0584 **Find:** $[NH_3]$
Conceptual Plan: Balanced reaction → equilibrium expression → $[NH_3]$
Solution: $K_c = \frac{[NH_3]^2}{[N_2][H_2]^3}$ $0.0584 = \frac{(x)^2}{(0.120)(0.140)^3}$ $x = 0.00439$

Check: Plug the value for x back into the equilibrium expression, and check the value.

$$0.0584 = \frac{(0.00439)^2}{(0.120)(0.140)^3}$$

18. At 25°C : **Given:** At equilibrium: $[H_2] = 0.0355M$, $[I_2] = 0.0388M$, and $[HI] = 0.922$ **Find:** K_c
Conceptual Plan: Balanced reaction → equilibrium expression → K_c

Solution: $K_c = \frac{[HI]^2}{[H_2][I_2]} = \frac{(0.922)^2}{(0.0355)(0.0388)} = 617$

Check: The value is reasonable since the concentration of products is greater than the concentration of reactants.

At 340°C: **Given:** At equilibrium: $[I_2] = 0.0455$, and $[HI] = 0.387$, $K_c = 9.6$ **Find:** $[H_2]$
Conceptual Plan: Balanced reaction → equilibrium expression → $[H_2]$

Solution: $K_c = \frac{[HI]^2}{[H_2][I_2]}$ $\quad 9.6 = \frac{(0.387)^2}{(x)(0.0455)}$ $\quad x = 0.343$

Check: Plug the value for x back into the equilibrium expression, and check the value.

$$9.6 = \frac{(0.387)^2}{(0.343)(0.0455)}$$

At 445°C: **Given:** At equilibrium: $[H_2] = 0.0485M$, $[I_2] = 0.0468$, $K_c = 50.2$ **Find:** [HI]
Conceptual Plan: Balanced reaction → equilibrium expression → [HI]

Solution: $K_c = \frac{[HI]^2}{[H_2][I_2]}$ $\quad 50.2 = \frac{(x)^2}{(0.0485)(0.0468)}$ $\quad x = 0.338$

Check: Plug the value for x back into the equilibrium expression, and check the value.

$$50.2 = \frac{(0.338)^2}{(0.0485)(0.0468)}$$

19. **Given:** $P_{NO} = 118$ torr; $P_{Br2} = 176$ torr, $K_p = 28.4$ **Find:** P_{NOBr}
Conceptual Plan: **torr → atm and then balanced reation → equilibrium expression → P_{NOBr}**

$$\frac{1\text{ atm}}{760\text{ torr}}$$

Solution: $P_{NO} = 118 \text{ torr x } \frac{1\text{ atm}}{760\text{ torr}} = 0.15\underline{5}3\text{ atm}$ $\quad P_{Br_2} = 176\text{ torr x }\frac{1\text{ atm}}{760\text{ torr}} = 0.23\underline{1}6$

$K_p = \frac{P_{NOBr}^2}{P_{NO}^2 P_{Br_2}}$ $\quad 28.4 = \frac{x^2}{(0.15\underline{5}3)^2(0.23\underline{1}6)}$ $\quad x = 0.398\text{ atm} = 303\text{ torr}$

Check: Plug the value for x back into the equilibrium expression, and check the value.

$$28.4 = \frac{(0.398)^2}{(0.15\underline{5}3)^2(0.23\underline{1}6)}$$

20. **Given:** $P_{SO2} = 117$ torr; $P_{Cl2} = 255$ torr, $K_p = 2.91 \times 10^3$ **Find:** P_{SO2Cl2}
Conceptual Plan: **torr → atm and then balanced reation → equilibrium expression → P_{SO2Cl2}**

$$\frac{1\text{ atm}}{760\text{ torr}}$$

Solution: $P_{SO_2} = 117\text{ torr x }\frac{1\text{ atm}}{760\text{ torr}} = 0.15\underline{3}9\text{ atm}$ $\quad P_{Cl_2} = 255\text{ torr x }\frac{1\text{ atm}}{760\text{ torr}} = 0.33\underline{5}5$

$K_p = \frac{P_{SO_2}P_{Cl_2}}{P_{SO_2Cl_2}}$ $\quad 2.91 \times 10^3 = \frac{(0.15\underline{3}9)(0.33\underline{5}5)}{(x)}$ $\quad x = 1.77 \times 10^{-5}\text{ atm} = 0.0135\text{ torr}$

Check: Plug the value for x back into the equilibrium expression, and check the value.

$$2.91 \times 10^3 = \frac{(0.15\underline{3}9)(0.33\underline{5}5)}{(1.77 \times 10^{-5})}$$

21. **Given:** $[Fe^{3+}]_{initial} = 1.0 \times 10^{-3}$ M; $[SCN^-]_{initial} = 8.0 \times 10^{-4}$ M; $[FeSCN^{2+}]_{eq} = 1.7 \times 10^{-4}$M **Find:** K_c

Conceptual Plan:
1. **Prepare ICE table**
2. **Calculate concentration change for known value**
3. **Calculate concentration changes for other reactants/products**
4. **Determine equilibrium concentration**
5. **Write the equilibrium expression and determine K_c**

Solution: $Fe^{3+}(aq) + SCN^-(aq) \leftrightarrows FeSCN^{3+}(aq)$

	$[Fe^{3+}]$	$[SCN^-]$	$[FeSCN^{3+}]$
I	1.0×10^{-3}	8.0×10^{-4}	0.00
C	-1.7×10^{-4}	-1.7×10^{-4}	$+1.7\times10^{-4}$
E	8.3×10^{-4}	6.3×10^{-4}	1.7×10^{-4}

$$K_c = \frac{[FeSCN^{2+}]}{[Fe^{3+}][SCN^-]} = \frac{(1.7 \times 10^{-4})}{(8.3 \times 10^{-4})(6.3 \times 10^{-4})} = 3.3 \times 10^2$$

22. **Given:** $[SO_2Cl_2]_{initial} = 0.020$ M; $[Cl_2]_{eq} = 1.2 \times 10^{-2}$M **Find:** K_c

Conceptual Plan:
1. **Prepare ICE table**
2. **Calculate concentration change for known value**
3. **Calculate concentration changes for other reactants/products**
4. **Determine equilibrium concentration**
5. **Write the equilibrium expression and determine K_c**

Solution: $SO_2Cl_2(g) \leftrightarrows SO_2(g) + Cl_2(g)$

	$[SO_2Cl_2]$	$[SO_2]$	$[Cl_2]$
I	0.020	0.00	0.00
C	-1.2×10^{-2}	$+1.2\times10^{-2}$	$+1.2\times10^{-2}$
E	0.0080	1.2×10^{-2}	1.2×10^{-2}

$$K_c = \frac{[SO_2][Cl_2]}{[SO_2Cl_2]} = \frac{(1.2 \times 10^{-2})(1.2 \times 10^{-2})}{(0.0080)} = 0.018$$

23. **Given:** 3.67 L flask; 0.763 g H_2 initial; 96.9 g I_2 initial; 90.4 g HI equilibrium **Find:** K_c

Conceptual Plan: **g → mol → M and then**

$$n = \frac{g}{\text{molar mass}} \qquad M = \frac{n}{V}$$

1. **Prepare ICE table**
2. **Calculate concentration change for known value**
3. **Calculate concentration changes for other reactants/products**
4. **Determine equilibrium concentration**
5. **Write the equilibrium expression and determine K_c**

Solution: $0.763 \text{ g } H_2 \times \frac{1 \text{ mol } H_2}{2.016 \text{ g } H_2} = 0.37\underline{8}5 \text{ mol } H_2$ $\qquad \frac{0.37\underline{8}5 \text{ mol } H_2}{3.67 \text{ L}} = 0.103$ M

$96.9 \text{ g } I_2 \times \frac{1 \text{ mol } I_2}{253.8 \text{ g } I_2} = 0.38\underline{1}8 \text{ mol } I_2$ $\qquad \frac{0.38\underline{1}8 \text{ mol } I_2}{3.67 \text{ L}} = 0.104$ M

$90.4 \text{ g HI} \times \frac{1 \text{ mol HI}}{127.9 \text{ g } I_2} = 0.70\underline{6}8 \text{ mol HI}$ $\qquad \frac{0.70\underline{6}8 \text{ mol HI}}{3.67 \text{ L}} = 0.193$ M

$H_2(g) + I_2(g) \leftrightarrows 2HI(g)$

	$[H_2]$	$[I_2]$	$[HI]$
I	0.103	0.104	0.00
C	-0.0965	-0.0965	+0.193
E	0.0065	0.0075	0.193

$$K_c = \frac{[HI]^2}{[H_2][I_2]} = \frac{(0.193)^2}{(0.0065)(0.0075)} = 764$$

24. **Given:** 5.19 L flask; 26.9 g CO initial; 2.34 g H_2 initial; 8.65 g CH_3OH equilibrium **Find:** K_c

Conceptual Plan: **g → mol → M and then**

$$n = \frac{g}{\text{molar mass}} \qquad M = \frac{n}{V}$$

1. **Prepare ICE table**
2. **Calculate concentration change for known value**
3. **Calculate concentration changes for other reactants/products**
4. **Determine equilibrium concentration**
5. **Write the equilibrium expression and determine K_c**

Solution: $26.9 \text{ g } CO_2 \times \frac{1 \text{ mol CO}}{28.01 \text{ g CO}} = 0.96\underline{0}4 \text{ mol CO}$ $\frac{0.96\underline{0}4 \text{ mol CO}}{5.19 \text{ L}} = 0.185 \text{ M}$

$2.34 \text{ g } H_2 \times \frac{1 \text{ mol } H_2}{2.016 \text{ g } H_2} = 1.1\underline{6}1 \text{ mol } H_2$ $\frac{1.1\underline{6}1 \text{ mol } H_2}{5.19 \text{ L}} = 0.224 \text{ M}$

$8.65 \text{ g } CH_3OH \times \frac{1 \text{ mol } CH_3OH}{32.04 \text{ g } CH_3OH} = 0.27\underline{0}0 \text{ mol } CH_3OH$ $\frac{0.27\underline{0}0 \text{ mol } CH_3OH}{5.19 \text{ L}} = 0.0520 \text{ M}$

	$CO(g)$ +	$2H_2(g)$	⇆ $CH_3OH(g)$
	[CO]	[H_2]	[CH_3OH]
I	0.185	0.224	0.00
C	- 0.0520	- 0.104	+0.0520
E	0.133	0.120	0.0520

$$K_c = \frac{[CH_3OH]}{[CO][H_2]} = \frac{(0.0520)}{(0.133)(0.120)} = 27.2$$

25. **Given:** $K_c = 8.5 \times 10^{-3}$; $[NH_3] = 0.166M$; $[H_2S] = 0.166M$ **Find:** Will solid form or decompose?
Conceptual Plan: Calculate Q → compare Q and K_c
Solution: $Q = [NH_3][H_2S] = (0.166)(0.166) = 0.0276$
$Q = 0.0276$ and $K_c = 8.5 \times 10^{-3}$ so $Q > K_c$ and the reaction will shift to the left, so more solid will form.

26. **Given:** $K_p = 2.4 \times 10^{-4}$; $P_{H2} = 0.112\text{atm}$; $P_{S2} = 0.055\text{atm}$; $P_{H2S} = 0.445\text{atm}$ **Find:** Is the reaction at equilibrium?
Conceptual Plan: Calculate Q → compare Q and K_p
Solution: $Q = \frac{P_{H_2}^2 P_{S_2}}{P_{H_2S}^2} = \frac{(0.112)^2(0.055)}{(0.445)^2} = 3.48 \times 10^{-3}$ $Q = 3.48 \times 10^{-3}$ and $K_p = 2.4 \times 10^{-4}$
so $Q > K_p$ and the system is not at equilibrium and the reaction will shift to the left to reach equilibrium.

27. **Given:** 6.55 g Ag_2SO_4, 1.5 L solution, $K_c = 1.1 \times 10^{-5}$ **Find:** will more solid dissolve
Conceptual Plan:
g Ag_2SO_4 → mol Ag_2SO_4 → [Ag_2SO_4] → [Ag^+],[SO_4^{-2}] → calculate Q and compare to K_c.
$\frac{1 \text{ mol } Ag_2SO_4}{311.81 \text{ g}}$ $[\,] = \frac{\text{mol } Ag_2SO_4}{\text{vol solution}}$ $Q = [Ag^+]^2[SO_4^{2-}]$

Solution: $6.55 \text{ g } Ag_2SO_4 \left(\frac{1 \text{ mol } Ag_2SO_4}{311.81 \text{ g } Ag_2SO_4}\right) = 0.0210 \text{ mol } Ag_2SO_4$

$\frac{0.0210 \text{ mol } Ag_2SO_4}{1.5 \text{ L solution}} = 0.01\underline{4}0 \text{ M } Ag_2SO_4$

$[Ag^+] = 2[Ag_2SO_4] = 2(0.01\underline{4}0 \text{ M}) = 0.02\underline{8}0 \text{ M}$ $[SO_4^{2-}] = [Ag_2SO_4] = 0.01\underline{4}0 \text{ M}$

$Q = [Ag^+]^2[SO_4^{2-}] = (0.02\underline{8}0)^2(0.01\underline{4}0) = 1.1 \times 10^{-5}$

$Q = K_c$ so the system is at equilibrium and is a saturated solution. Therefore, if more solid is added it will not dissolve.

28. **Given:** $K_p = 6.7$ at 298 K; 2.25 L flask; $NO_2 = 0.055$ mol; $N_2O_4 = 0.082$ mol **Find:** Is the reaction at equilibrium?

Conceptual Plan: K_p → K_c and mol → M and then calculate Q and compare to K_c.

$K_p = K_c(RT)^{\Delta n}$ $M = \frac{\text{mol}}{V}$

Solution: $K_c = \frac{K_p}{(RT)^{\Delta n}} = \frac{(6.7)}{[(.0821\frac{L \cdot atm}{mol \cdot K})(298K)]^{-1}} = 1\underline{6}4$

$$Q = \frac{[N_2O_4]}{[NO]^2} = \frac{\left(\frac{0.082 \text{ mol}}{2.25 \text{ L}}\right)}{\left(\frac{0.055 \text{ mol}}{2.25 \text{ L}}\right)^2} = 61$$

$Q < K_c$, the reaction is not at equilibrium and will shift to the right.

29. a) **Given:** [A] = 1.0 M, [B] = 0.0 $K_c = 2.0$; $a = 1$, $b = 1$ **Find:** [A], [B] at equilibrium
Conceptual Plan: Prepare an ICE table; and then calculate Q; compare Q and K_c and predict the direction of the reaction; represent the change with x; sum the table and determine the equilibrium values; put the equilibrium values in the equilibrium expression and solve for x. Determine [A] and [B].
Solution:

	A(g) ⇆	B(g)
	[A]	[B]
I	1.0	0.00
C	- x	x
E	1 – x	x

$Q = \frac{[B]}{[A]} = \frac{0}{1} = 0$ $\quad$ $Q < K$ therefore, the reaction will proceed to the right by x

$K_c = \frac{[B]}{[A]} = \frac{(x)}{(1 - x)} = 2.0$ $\quad$ x = 0.67

[A] = 1 - 0.67 = 0.33M $\quad$ [B] = 0.67M

Check: Plug the values into the equilibrium expression: $K_c = \frac{0.67}{0.33} = 2.0$

b) **Given:** [A] = 1.0 M, [B] = 0.0 $K_c = 2.0$; $a = 2$, $b = 2$ **Find:** [A], [B] at equilibrium
Conceptual Plan: Prepare an ICE table; and then calculate Q; compare Q and K_c and predict the direction of the reaction; represent the change with x; sum the table and determine the equilibrium values; put the equilibrium values in the equilibrium expression and solve for x. Determine [A] and [B].
Solution:

	2A(g) ⇆	2B(g)
	[A]	[B]
I	1.0	0.00
C	- 2x	2x
E	1 – 2x	2x

$Q = \frac{[B]^2}{[A]^2} = \frac{0}{1} = 0$ $\quad$ $Q < K$ therefore, the reaction will proceed to the right by x

$K_c = \frac{[B]^2}{[A]^2} = \frac{(2x)^2}{(1 - 2x)^2} = 2.0$ $\quad$ x = 0.2$\underline{9}$3

[A] = 1 - 2(0.2$\underline{9}$3) = 0.414 = 0.41M $\quad$ [B] = 2(0.2$\underline{9}$3) = 0.586 = 0.59M

Check: Plug the values into the equilibrium expression: $K_c = \frac{(0.59)^2}{(0.41)^2} = 2.0$

c) **Given:** [A] = 1.0 M, [B] = 0.0 $K_c = 2.0$; $a = 1$, $b = 2$ **Find:** [A], [B] at equilibrium
Conceptual Plan: Prepare an ICE table; and then calculate Q; compare Q and K_c and predict the direction of the reaction; represent the change with x; sum the table and determine the equilibrium values; put the equilibrium values in the equilibrium expression and solve for x. Determine [A] and [B].

Solution:

	$A(g) \leftrightarrows$	$2B(g)$
	[A]	[B]
I	1.0	0.00
C	- x	2x
E	1 – x	2x

$$Q = \frac{[B]^2}{[A]} = \frac{0}{1} = 0$$ $Q < K$ therefore, the reaction will proceed to the right by x

$$K_c = \frac{[B]^2}{[A]} = \frac{(2x)^2}{(1 - x)} = 2.0 \qquad 4x^2 + 2x - 2 = 0$$

$(4x - 2)(x + 1) = 0$; $x = -1$ or $x = 0.5$, therefore, $x = 0.5$

$[A] = 1 - 0.50 = 0.50M \qquad [B] = 2x = 2(0.50) = 1.0M$

Check: Plug the values into the equilibrium expression: $K_c = \frac{(1.0)^2}{0.50} = 2.0$

30. a) **Given:** [A] = 1.0 M, [B] = 1.0, [C] = 0.0 $K_c = 4.0$; $a = 1$, $b = 1$, c = 2 **Find:**[A],[B],[C] at equilibrium
Conceptual Plan: Prepare an ICE table; and then calculate Q; compare Q and K_c and predict the direction of the reaction; represent the change with x; sum the table and determine the equilibrium values; put the equilibrium values in the equilibrium expression and solve for x. Determine [A], [B], and [C].
Solution:

	$A(g)$	$+B(g)$	$\leftrightarrows$	$2C(g)$
	[A]	[B]		[C]
I	1.0	1.0		0.0
C	- x	-x		2x
E	1 – x	1 – x		2x

$$Q = \frac{[C]^2}{[A][B]} = \frac{0}{(1)(1)} = 0$$ $Q < K$ therefore, the reaction will proceed to the right by x

$$K_c = \frac{[C]^2}{[A][B]} = \frac{(2x)^2}{(1 - x)(1 - x)} = 4.0 \qquad x = 0.50$$

$[A] = [B] = 1 - 0.50 = 0.50\ M \qquad [C] = 2x = 1.0M$

Check: Plug the values into the equilibrium expression: $K_c = \frac{(1.0)^2}{(0.50)(0.50)} = 4.0$

b) **Given:** [A] = 1.0 M, [B] = 1.0, [C] = 0.0 $K_c = 4.0$; $a = 1$, $b = 1$, c = 1 **Find:**[A],[B],[C] at equilibrium
Conceptual Plan: Prepare an ICE table; and then calculate Q; compare Q and K_c and predict the direction of the reaction; represent the change with x; sum the table and determine the equilibrium values; put the equilibrium values in the equilibrium expression and solve for x. Determine [A], [B], and [C].
Solution:

	$A(g)$	$+B(g)$	$\leftrightarrows$	$C(g)$
	[A]	[B]		[C]
I	1.0	1.0		0.0
C	- x	-x		x
E	1 – x	1 – x		x

$$Q = \frac{[C]}{[A][B]} = \frac{0}{(1)(1)} = 0$$ $Q < K$ therefore, the reaction will proceed to the right by x

$$K_c = \frac{[C]}{[A][B]} = \frac{(x)}{(1 - x)(1 - x)} = 4.0 \qquad 4x^2 - 9x + 4 = 0$$

$x = 1.64$ or $x = 0.61$, therefore, $x = 0.61$

$[A] = [B] = 1 - 0.61 = 0.39\ M \qquad [C] = x = 0.61M$

Check: Plug the values into the equilibrium expression: $K_c = \frac{(0.61)}{(0.39)(0.39)} = 4.0$

c) **Given:** [A] = 1.0 M, [B] = 1.0, [C] = 0.0 K_c = 4.0; *a* = 2, *b* = 1, c = 1 **Find:**[A],[B],[C] at equilibrium
Conceptual Plan: Prepare an ICE table; and then calculate *Q*; compare *Q* and K_c and predict the direction of the reaction; represent the change with x; sum the table and determine the equilibrium values; put the equilibrium values in the equilibrium expression and solve for x. Determine [A], [B], and [C].
Solution:

	2A(g)	+B(g) ⇆	C(g)
	[A]	[B]	[C]
I	1.0	1.0	0.0
C	- 2x	-x	x
E	1 – 2x	1 – x	x

$Q = \frac{[C]}{[A]^2[B]} = \frac{0}{(1)(1)} = 0$ $\quad Q < K$ therefore, the reaction will proceed to the right by x

$K_c = \frac{[C]}{[A]^2[B]} = \frac{(x)}{(1 - 2x)^2(1 - x)} = 4.0$

$- 16x^3 + 32x^2 - 21x + 4.0 = 0$

x = 0.326

[A] = 1 – 2x = 0.348M; [B] = 1 – x = 0.674M; [C] = x = 0.326M

Check: Plug the values into the equilibrium expression: $K_c = \frac{(0.326)}{(0.348)^2(0.674)} = 4.0$

31. **Given:** [N_2O_4] = 0.0500 M, [NO_2] = 0.0 K_c = 0.513 **Find:** [N_2O_4], [NO_2] at equilibrium
Conceptual Plan: Prepare an ICE table; and then calculate *Q*; compare *Q* and K_c and predict the direction of the reaction; represent the change with x; sum the table and determine the equilibrium values; put the equilibrium values in the equilibrium expression and solve for x. Determine [N_2O_4] and [NO_2].
Solution:

	N_2O_4 (g) ⇆	2 NO_2(g)
	[N_2O_4]	[NO_2]
I	0.0500	0.00
C	- x	2x
E	0.0500 – x	2x

$Q = \frac{[NO_2]^2}{[N_2O_4]} = \frac{0}{0.0500} = 0$ $\quad Q < K$ therefore, the reaction will proceed to the right by x

$K_c = \frac{[NO_2]^2}{[N_2O_4]} = \frac{(2x)^2}{(0.0500 - x)} = 0.513 \qquad 4x^2 + 0.513x - 0.02565 = 0$

$\frac{-b \pm \sqrt{b^2 - 4ac}}{2a} = \frac{-0.513 \pm \sqrt{(0.513)^2 - 4(4)(-0.02565)}}{2(4)} = \frac{-0.513 \pm \sqrt{0.6735}}{2(4)}$

x = - 0.1667 or x = 0.0385, therefore, x = 0.0385

[A] = 0.0500 - 0.03850 = 0.0115M $\qquad$ [B] = 2x = 2(0.03850) = 0.0770M

Check: Plug the values into the equilibrium expression: $K_c = \frac{(0.0770)^2}{0.0115} = 0.515$

32. **Given:** [CO] = 0.1500 M, [Cl_2] = 0.175, [$COCl_2$] = 0.0 K_c = 255 **Find:** [CO], [Cl_2], [$COCl_2$] at equilibrium
Conceptual Plan: Prepare an ICE table; and then calculate *Q*; compare *Q* and K_c and predict the direction of the reaction; represent the change with x; sum the table and determine the equilibrium values; put the equilibrium values in the equilibrium expression and solve for x. Determine [CO], [Cl_2]. and [$COCl_2$].
Solution:

	CO (g)	+Cl_2 (g)	⇆$COCl_2$ (g)
	[CO]	[Cl_2]	[$COCl_2$]
I	0.1500	0.175	0.0
C	- x	-x	x
E	0.1500–x	0.175 – x	x

$$Q = \frac{[COCl_2]}{[CO][Cl_2]} = \frac{0}{(0.1500)(0.175)} = 0 \qquad Q < K \text{ therefore, the reaction will proceed to the right by x}$$

$$K_c = \frac{[COCl_2]}{[CO][Cl_2]} = \frac{x}{(0.1500 - x)(0.175 - x)} = 255 \qquad 255x^2 - 83.875x + 6.69375 = 0$$

$$\frac{-b \pm \sqrt{b^2 - 4ac}}{2a} = \frac{-(-83.875) \pm \sqrt{(-83.875)^2 - 4(255)(6.69375)}}{2(255)}$$

$x = 0.19\underline{2}7$ or $x = 0.13\underline{6}2$, therefore, $x = 0.13\underline{6}2$

$[CO] = 0.1500 - 0.13\underline{6}2 = 0.01\underline{3}77M$

$[Cl_2] = 0.175 - 0.13\underline{6}2 = 0.03\underline{8}78M$

$[COCl_2] = x = 0.13\underline{6}2M$

Check: Plug the values into the equilibrium expression: $K_c = \frac{(0.1362)}{(0.01377)(0.03878)} = 255$

33. **Given:** $[CO] = 0.10$ M, $[CO_2] = 0.0$ $K_c = 4.0 \times 10^3$ **Find:** $[CO_2]$ at equilibrium

Conceptual Plan: Prepare an ICE table; and then calculate Q; compare Q and K_c and predict the direction of the reaction; represent the change with x; sum the table and determine the equilibrium values; put the equilibrium values in the equilibrium expression and solve for x. Determine [CO], $[Cl_2]$, and $[COCl_2]$.

Solution:

	NiO(s) +CO (g)	⇆ Ni(s) + CO_2 (g)
	[CO]	$[CO_2]$
I	0.10	0.0
C	- x	x
E	0.10 –x	x

$$Q = \frac{[CO_2]}{[CO]} = \frac{0}{(0.10)} = 0 \qquad Q < K \text{ therefore, the reaction will proceed to the right by x}$$

$$K_c = \frac{[CO_2]}{[CO]} = \frac{x}{(0.10 - x)} = 4.0 \times 10^3 \qquad 4.0 \times 10^3(0.10 - x) = x$$

$x = 0.10$

$[CO_2] = 0.10$ M

Check: since the equilibrium constant is so large, the reaction goes essentially to completion, therefore, it is reasonable that the concentration of product is 0.10 M

34. **Given:** $[CO] = 0.125M$ M, $[H_2O] = 0.125M[CO_2] = 0.0$, $[H_2] = 0.0$ $K_c = 102$

Find: [CO], $[H_2O]$, $[CO_2]$, $[H_2]$ at equilibrium

Conceptual Plan: Prepare an ICE table; and then calculate Q; compare Q and K_c and predict the direction of the reaction; represent the change with x; sum the table and determine the equilibrium values; put the equilibrium values in the equilibrium expression and solve for x. Determine [CO], $[Cl_2]$, and $[COCl_2]$.

Solution:

	CO (g)	$+H_2O$(g) ⇆	CO_2 (g) +	H_2(g)
	[CO]	$[H_2O]$	$[CO_2]$	$[H_2]$
I	0.125	0.125	0.0	0.0
C	- x	-x	x	x
E	0.125–x	0.125 – x	x	x

$$Q = \frac{[CO_2][H_2]}{[CO][H_2O]} = \frac{0}{(0.125)(0.125)} = 0 \qquad Q < K \text{ therefore, the reaction will proceed to the right by x}$$

$$K_c = \frac{[CO_2][H_2]}{[CO][H_2O]} = \frac{(x)(x)}{(0.125 - x)(0.125 - x)} = 102$$

$$\sqrt{\frac{(x)(x)}{(0.125 - x)(0.125 - x)}} = \sqrt{102}$$

$$\frac{x}{0.125 - x} = \pm 10.0995$$

$\pm 10.0995(0.125 - x) = x$

$x = 0.11373$ or $x = -0.1387$, therefore, $x = 0.1137$

$[CO] = [H_2O] = 0.125 - 0.1137 = 0.0113 = 0.011M$

$[H_2] = [CO_2] = x = 0.11373 = 0.114M$

Check: Plug the values into the equilibrium expression: $K_c = \frac{(0.1137)(0.1137)}{(0.0113)(0.0113)} = 101.2 = 1.0 \times 10^2$; within 1 significant figure of true value, answers are valid.

35. **Given:** $[HC_2H_3O_2] = 0.210M$ M, $[H_3O^+] = 0.0$, $[C_2H_3O_2^-] = 0.0$, $K_c = 1.8 \times 10^{-5}$
Find: $[HC_2H_3O2]$, $[H_2O^+]$, $[C_2H_3O_2^-]$ at equilibrium
Conceptual Plan: Prepare an ICE table; and then calculate Q; compare Q and K_c and predict the direction of the reaction; represent the change with x; sum the table and determine the equilibrium values; put the equilibrium values in the equilibrium expression and solve for x. Determine [CO], $[Cl_2]$, and $[COCl_2]$.
Solution:

	$HC_2H_3O_2$ (aq)	$+H_2O(l)$	⇆	H_3O^+ (aq) +	$C_2H_3O_2^-$(aq)
	$[HC_2H_3O_2]$	$[H_2O]$		$[H_3O^+]$	$[C_2H_3O_2^-]$
I	0.210			0.0	0.0
C	- x			x	x
E	0.210–x			x	x

$$Q = \frac{[H_3O^+][C_2H_3O_2^-]}{[HC_2H_3O_2]} = \frac{0}{(0.210)} = 0$$ $Q < K$ therefore, the reaction will proceed to the right by x

$$K_c = \frac{[H_3O^+][C_2H_3O_2^-]}{[HC_2H_3O_2]} = \frac{(x)(x)}{(0.210 - x)} = 1.8 \times 10^{-5}$$

assume x is small compared to 0.210

$x^2 = 0.210(1.8 \times 10^{-5})$

$x = 0.00194$ check assumption: $\frac{0.00194}{0.210} \times 100 = 0.92\ \%$; assumption valid

$[H_3O^+] = [C_2H_3O_2^-] = 0.00194M$

$[HC_2H_3O_2] = 0.210 - 0.00194 = 0.2081 = 0.208M$

Check: Plug the values into the equilibrium expression: $K_c = \frac{(0.00194)(0.00194)}{(0.208)} = 1.81 \times 10^{-5}$; within 1 significant figure of true value, answers are valid.

36. **Given:** $[SO_2Cl_2] = 0.175M$ M, $[SO_2] = 0.0$, $[Cl_2] = 0.0$ $K_c = 2.99 \times 10^{-7}$
Find: $[SO_2Cl_2]$, $[SO_2]$, $[Cl_2]$ at equilibrium
Conceptual Plan: Prepare an ICE table; and then calculate Q; compare Q and K_c and predict the direction of the reaction; represent the change with x; sum the table and determine the equilibrium values; put the equilibrium values in the equilibrium expression and solve for x. Determine $[SO_2Cl_2]$, $[SO_2]$, and $[Cl_2]$.
Solution:

	SO_2Cl_2 (g)	⇆	SO_2 (g)	+ Cl_2(g)
	$[SO_2Cl_2]$		$[SO_2]$	$[Cl_2]$
I	0.175		0.0	0.0
C	- x		x	x
E	0.175–x		x	x

$$Q = \frac{[SO_2][Cl_2]}{[SO_2Cl_2]} = \frac{0}{(0.175)} = 0$$ $Q < K$ therefore, the reaction will proceed to the right by x

$$K_c = \frac{[SO_2][Cl_2]}{[SO_2Cl_2]} = \frac{(x)(x)}{(0.175 - x)} = 2.99 \times 10^{-7}$$

assume x is small compared to 0.175

$$x^2 = 2.99 \times 10^{-7}(0.175)$$

$$x = 2.2\underline{8}7 \times 10^{-4} \qquad \text{check assumption: } \frac{2.2\underline{8}7 \times 10^{-4}}{0.175} \times 100 = 0.13\%; \text{ assumption valid.}$$

$$[SO_2] = [Cl_2] = x = 2.2\underline{8}7 \times 10^{-4} = 2.29 \times 10^{-4} M$$

$$[SO_2Cl_2] = 0.175 - 2.2\underline{8}7 \times 10^{-4} = 0.17\underline{4}7 = 0.175 M$$

Check: Plug the values into the equilibrium expression:

$$K_c = \frac{(0.2.2\underline{8}7 \times 10^{-4})(2.2\underline{8}7 \times 10^{-4})}{(0.175)} = 2.996 \times 10^{-7} = 3.00 \times 10^{-7};$$

within 1 significant figure of the true value, therefore, answers are valid.

37. **Given:** $P_{Br2} = 755$ torr, $P_{Cl2} = 735$ torr, $P_{BrCl} = 0.0$, $K_p = 1.11 \times 10^{-4}$ **Find:** P_{BrCl} at equilibrium
Conceptual Plan: Torr → atm and then: Prepare an ICE table; and then calculate *Q*; compare *Q* and K_c and predict the direction of the reaction; represent the change with x; sum the table and determine the equilibrium values; put the equilibrium values in the equilibrium expression and solve for x. Determine P_{BrCl}
Solution:

$$P_{Br_2} = 755 \text{ torr} \times \frac{1 \text{ atm}}{760 \text{ torr}} = 0.99\underline{3}4 \text{ atm} \qquad P_{Cl_2} = 735 \text{ torr} \times \frac{1 \text{ atm}}{760 \text{ torr}} = 0.96\underline{7}1 \text{ atm}$$

	Br_2 (g)	$+Cl_2$(g)	⇆ 2BrCl (g)
	P_{Br2}	P_{Cl2}	P_{BrCl}
I	0.9934	0.9671	0.0
C	- x	-x	2x
E	0.9934–x	0.9671– x	2x

$$Q = \frac{P^2_{BrCl}}{P_{Br_2}P_{Cl_2}} = \frac{0}{(0.9934)(0.9671)} = 0 \qquad Q < K \text{ therefore, the reaction will proceed to the right by x}$$

$$K_p = \frac{P^2_{BrCl}}{P_{Br_2}P_{Cl_2}} = \frac{(2x)^2}{(0.9934 - x)(0.9671 - x)} = 1.11 \times 10^{-4}$$

asusme x is small compared to 0.9934 and 0.9671

$$\frac{(2x)^2}{(0.9934)(0.9671)} = 1.11 \times 10^{-4} \qquad 4x^2 = 1.066 \times 10^{-4}$$

$$x = 0.00516 = 3.92 \text{ torr}$$

$$P_{BrCl} = 2x = 2(3.92 \text{ torr}) = 7.84 \text{ torr}$$

Check: Plug the values into the equilibrium expression:

$$K_c = \frac{(2(0.00516))^2}{(0.9934 - 0.00516)(0.99671 - 0.00516)} = 1.065 \times 10^{-4} = 1.12 \times 10^{-4};$$

within 1 significant figure of the true value, therefore, answers are valid.

38. **Given:** $P_{CO} = 1344$ torr, $P_{H2O} = 1766$ torr, $P_{CO2} = 0.0$, $P_{H2} = 0.0$, $K_p = 0.0611$ **Find:** P_{CO2}, P_{H2} at equilibrium
Conceptual Plan: Torr → atm and then: Prepare an ICE table; and then calculate *Q*; compare *Q* and K_c and predict the direction of the reaction; represent the change with x; sum the table and determine the equilibrium values; put the equilibrium values in the equilibrium expression and solve for x. Determine P_{CO2} and P_{H2}

Solution: $P_{CO} = 1344 \text{ torr} \times \frac{1 \text{ atm}}{760 \text{ torr}} = 1.76\underline{8}4 \text{ atm} \qquad P_{H_2O} = 1766 \text{ torr} \times \frac{1 \text{ atm}}{760 \text{ torr}} = 2.32\underline{3}7 \text{ atm}$

	CO (g)	$+H_2O$(g)	⇆CO_2 (g) +	H_2(g)
	P_{CO}	P_{H2O}	P_{CO2}	P_{H2}
I	1.76$\underline{8}$4	2.32$\underline{3}$7	0.0	0.0
C	- x	-x	x	x
E	1.7684–x	2.3237 – x	x	x

$$Q = \frac{P_{CO_2}P_{H_2}}{P_{CO}P_{H_2O}} = \frac{0}{(1.7684)(2.3237)} = 0 \qquad Q < K \text{ therefore, the reaction will proceed to the right by x}$$

$$K_p = \frac{P_{CO_2}P_{H_2}}{P_{CO}P_{H_2O}} = \frac{(x)(x)}{(1.7684 - x)(2.3237 - x)} = 0.0611$$

$x^2 = (0.0611)(4.1092 - 4.0921x + x^2)$

$x^2 = (0.25107 - 0.2500x + 0.0611x^2)$

$0.9389x^2 + 0.2500\,x - 0.25107 = 0$

$$\frac{-b \pm \sqrt{b^2 - 4ac}}{2a} = \frac{-0.2500 \pm \sqrt{(0.2500)^2 - 4(0.9389)(-0.25107)}}{2(0.9389)}$$

$x = 0.4008$ or $- 0.6671$ so: $x = 0.4008$

$P_{CO_2} = P_{H_2} = x = 0.4008$ atm $= 305$ torr

Check: Plug the values into the equilibrium expression:

$$K_p = \frac{(0.4008)^2}{(1.7684 - 0.4008)(2.3237 - 0.4008)} = 0.06108 = 0.061;$$

within 1 significant figure of the true value, therefore, answers are valid.

39. a) **Given:** $[A] = 1.0$ M, $[B] = [C] = 0.0$, $K_c = 1.0$ **Find:** [A], [B], [C] at equilibrium

Conceptual Plan: Prepare an ICE table; and then calculate Q; compare Q and K_c and predict the direction of the reaction; represent the change with x; sum the table and determine the equilibrium values; put the equilibrium values in the equilibrium expression and solve for x. Determine [A], [B], and [C].

Solution:

	A (g)	⇆	B (g) +	C(g)
	[A]		[B]	[C]
I	1.0		0.0	0.0
C	- x		x	x
E	1.0–x		x	x

$$Q = \frac{[B][C]}{[A]} = \frac{0}{(1.0)} = 0$$ $Q < K$ therefore, the reaction will proceed to the right by x

$$K_c = \frac{[B][C]}{[A]} = \frac{(x)(x)}{(1.0 - x)} = 1.0$$

$x^2 = 1.0(1.0 - x)$

$x^2 + x - 1 = 0$

$$\frac{-b \pm \sqrt{b^2 - 4ac}}{2a} \quad \frac{-1 \pm \sqrt{1^2 - 4(1)(-1)}}{2(1)}$$

$x = 0.6\underline{1}8$ or $x = - 1.618$, therefore, $x = 0.6\underline{1}8$

$[B] = [C] = x = 0.6\underline{1}8 = 0.62$M

$[A] = 1.0 - 0.6\underline{1}8 = 0.382 = 0.38$M

Check: Plug the values into the equilibrium expression:

$$K_c = \frac{(0.62)(0.62)}{(0.38)} = 1.01 = 1.0;$$ which is the equilibrium constant so the values are correct.

b) **Given:** $[A] = 1.0$ M, $[B] = [C] = 0.0$, $K_c = 0.010$ **Find:** [A], [B], [C] at equilibrium

Conceptual Plan: Prepare an ICE table; and then calculate Q; compare Q and K_c and predict the direction of the reaction; represent the change with x; sum the table and determine the equilibrium values; put the equilibrium values in the equilibrium expression and solve for x. Determine [A], [B], and [C].

Solution:

	A (g)	⇆	B (g) +	C(g)
	[A]		[B]	[C]
I	1.0		0.0	0.0
C	- x		x	x
E	1.0–x		x	x

$$Q = \frac{[B][C]}{[A]} = \frac{0}{(1.0)} = 0 \qquad Q < K \text{ therefore, the reaction will proceed to the right by x}$$

$$K_c = \frac{[B][C]}{[A]} = \frac{(x)(x)}{(1.0 - x)} = 0.010$$

$x^2 = 0.010(1.0 - x)$

$x^2 + 0.010x - 0.010 = 0$

$$\frac{-b \pm \sqrt{b^2 - 4ac}}{2a} \qquad \frac{-(0.010) \pm \sqrt{(0.010)^2 - 4(1)(-0.010)}}{2(1)}$$

$x = 0.09\underline{5}12$ or $x = -0.1015$, therefore, $x = 0.09\underline{5}12$

$[B] = [C] = x = 0.09\underline{5}12 = 0.095M$

$[A] = 1.0 - 0.09\underline{5}12 = 0.90488 = 0.90M$

Check: Plug the values into the equilibrium expression:

$$K_c = \frac{(0.095)(0.095)}{(0.90)} = 0.01002 = 0.010;$$ which is the equilibrium constant so the values are correct.

c) **Given:** $[A] = 1.0$ M, $[B] = [C] = 0.0$, $K_c = 1.0 \times 10^{-5}$ **Find:** [A], [B], [C] at equilibrium

Conceptual Plan: Prepare an ICE table; and then calculate Q; compare Q and K_c and predict the direction of the reaction; represent the change with x; sum the table and determine the equilibrium values; put the equilibrium values in the equilibrium expression and solve for x. Determine [A], [B], and [C].

Solution:

	A (g) ⇆	B (g) +	C(g)
	[A]	[B]	[C]
I	1.0	0.0	0.0
C	- x	x	x
E	1.0–x	x	x

$$Q = \frac{[B][C]}{[A]} = \frac{0}{(1.0)} = 0 \qquad Q < K \text{ therefore, the reaction will proceed to the right by x}$$

$$K_c = \frac{[B][C]}{[A]} = \frac{(x)(x)}{(1.0 - x)} = 1.0 \times 10^{-5}$$

assume x is small compared to 1.0

$x^2 = 1.0(1.0 \times 10^{-5})$

$x = 0.003\underline{1}6$ check assumption: $\frac{0.003\underline{1}6}{1.0} \times 100 = 0.32\%$, assumption valid

$[B] = [C] = x = 0.003\underline{1}6 = 0.0032M$

$[A] = 1.0 - 0.003\underline{1}6 = 0.9968 = 1.0M$

Check: Plug the values into the equilibrium expression:

$$K_c = \frac{(0.0032)(0.0032)}{(1.0)} = 1.024 \times 10^{-5} = 1.0 \times 10^{-5};$$

which is the equilibrium constant so the values are correct

40. a) **Given:** $P_B = 1.0$ atm, $P_A = 00$ atm, $P_{BrCl} = 0.0$, $K_p = 1.0$ **Find:** P_B, P_A at equilibrium

Conceptual Plan: Prepare an ICE table; and then calculate Q; compare Q and K_p and predict the direction of the reaction; represent the change with x; sum the table and determine the equilibrium values; put the equilibrium values in the equilibrium expression and solve for x. Determine P_A, P_B

Solution:

	A (g) ⇆	2B (g)
	P_A	P_B
I	0.0	1.0
C	x	-2x
E	x	1.0 - 2x

$Q = \frac{P_B^2}{P_A}$ = since there is no A, the reaction shifts to the left

$K_p = \frac{P_B^2}{P_A} = \frac{(1.0 - 2x)^2}{(x)} = 1.0$

$1.0 - 4x + 4x^2 = 1.0x$

$4x^2 - 5x + 1.0 = 0$

$\frac{-b \pm \sqrt{b^2 - 4ac}}{2a} \quad \frac{-(-5) \pm \sqrt{(-5)^2 - 4(4)(1)}}{2(4)}$

x = 0.25 or x = 1.0, therefore, x = 0.25

$P_A = x = 0.25\text{atm}; P_B = 1.0 - 2x = 1.0 - 0.50 = 0.50\text{atm}$

Check: Plug the values into the equilibrium expression:

$K_p = \frac{(0.50)^2}{(0.25)} = 1.0$; which is the equilibrium constant

b) **Given:** $P_B = 1.0$ atm, $P_A = 00$ atm, $P_{BrCl} = 0.0$, $K_p = 1.0 \times 10^{-4}$ **Find:** P_B, P_A at equilibrium
Conceptual Plan: Prepare an ICE table; and then calculate Q; compare Q and K_p and predict the direction of the reaction; represent the change with x; sum the table and determine the equilibrium values; put the equilibrium values in the equilibrium expression and solve for x. Determine P_A, P_B
Solution:

	A (g)	⇆	2B (g)
	P_A		P_B
I	0.0		1.0
C	x		-2x
E	x		1.0 - 2x

$Q = \frac{P_B^2}{P_A}$ = since there is no A, the reaction shifts to the left

$K_p = \frac{P_B^2}{P_A} = \frac{(1.0 - 2x)^2}{(x)} = 1.0 \times 10^{-4}$

$1.0 - 4x + 4x^2 = (1.0 \times 10^{-4})x$

$4x^2 - 4.0001x + 1.0 = 0$

$\frac{-b \pm \sqrt{b^2 - 4ac}}{2a} = \frac{-(-4.0001) \pm \sqrt{(-4.0001)^2 - 4(4)(1)}}{2(4)} = \frac{-(-4.0001) \pm \sqrt{8.0 \times 10^{-4}}}{2(4)}$

$x = 0.5\underline{0}35$ or $x = 0.4\underline{9}65$, therefore, $x = 0.4\underline{9}65$

$P_A = x = 0.4\underline{9}65 = 0.50\text{atm}; P_B = 1.0 - 2x = 1.0 - 2(0.4\underline{9}65) = 0.007\text{atm}$

Check: Plug the values into the equilibrium expression:

$K_p = \frac{(0.007)^2}{(0.4\underline{9}65)} = 9.867 \times 10^{-5} = 1.0 \times 10^{-4}$; which is the equilibrium constant

c) **Given:** $P_B = 1.0$ atm, $P_A = 0.0$ atm, $P_{BrCl} = 0.0$, $K_p = 1.0 \times 10^{5}$ **Find:** P_B, P_A at equilibrium
Conceptual Plan: Prepare an ICE table; and then calculate Q; compare Q and K_p and predict the direction of the reaction; represent the change with x; sum the table and determine the equilibrium values; put the equilibrium values in the equilibrium expression and solve for x. Determine P_A, P_B.
Solution:

	A (g)	⇆	2B (g)
	P_A		P_B
I	0.0		1.0
C	x		-2x
E	x		1.0 - 2x

$$Q = \frac{P_B^2}{P_A} = \text{since there is no A, the reaction shifts to the left}$$

$$K_p = \frac{P_B^2}{P_A} = \frac{(1.0 - 2x)^2}{(x)} = 1.0 \times 10^5$$

assume 2x is small compared to 1.0

$$\frac{1.0}{x} = 1.0 \times 10^5$$

$x = 1.0 \times 10^{-5}$ check assumption: $\frac{2(1.0 \times 10^{-5})}{1.0} \times 100\% = 0.002\ \%$; assumption valid

$P_A = x = 1.0 \times 10^{-5}$atm; $P_B = 1.0 - 2x = 1.0 - 2(1.0 \times 10^{-5}) = 0.99998$ atm

Check: Plug the values into the equilibrium expression:

$$K_p = \frac{(0.99998)^2}{(1.0 \times 10^{-5})} = 9.9996 \times 10^4 = 1.0 \times 10^5$$;which is the equilibrium constant

41. **Given:** $CO(g) + Cl_2(g) \leftrightarrows COCl_2(g)$ at equilibrium **Find:** what is the effect of each of the following
    a) $COCl_2$ is added to the reaction mixture: Adding $COCl_2$ increases the concentration of $COCl_2$ and causes the reaction to shift to the left.

    b) Cl_2 is added to the reaction mixture: Adding Cl_2 increases the concentration of Cl_2 and causes the reaction to shift to the right.

    c) $COCl_2$ is removed from the reaction mixture: removing the $COCl_2$ decreases the concentration of $COCl_2$ and causes the reaction to shift to the right.

42. **Given:** $2BrNO(g) \leftrightarrows 2NO(g) + Br_2(g)$ at equilibrium **Find:** what is the effect of each of the following
    a) NO is added to the reaction mixture: Adding NO increases the concentration of NO and causes the reaction to shift to the left.

    b) BrNO is added to the reaction mixture: Adding BrNO increases the concentration of BrNO and causes the reaction to shift to the right.

    c) Br_2 is removed from the reaction mixture: Removing Br_2 decreases the concentration of Br_2 and causes the reaction to shift to the right.

43. **Given:** $2KClO_3(s) \leftrightarrows 2KCl(s) + 3O_2(g)$ at equilibrium **Find:** what is the effect of each of the following
    a) O_2 is removed from the reaction mixture: Removing the O_2 decreases the concentration of O_2 and causes the reaction to shift to the right.

    b) KCl is added to the reaction mixture: Adding KCl does not cause any change in the reaction. KCl is a solid and the concentration remains constant so the addition of more solid does not change the equilibrium concentration.

    c) $KClO_3$ is added to the reaction mixture: Adding $KClO_3$ does not cause any change in the reaction. $KClO_3$ is a solid and the concentration remains constant so the addition of more solid does not change the equilibrium concentration.

    d) O_2 is added to the reaction mixture: Adding O_2 increases the concentration of O_2 and causes the reaction to shift to the left.

44. **Given:** $C(s) + H_2O(g) \leftrightarrows CO(g) + H_2(g)$ **Find:** what is the effect of each of the following
    a) C is added to the reaction mixture: Adding C does not cause any change in the reaction. C is a solid and the concentration remains constant so the addition of more solid does not change the equilibrium concentration.

    b) H_2O is condensed and removed from the reaction mixture: Removing the H_2O decreases the concentration of H_2O and causes the reaction to shift to the left.

c) CO is added to the reaction mixture: Adding CO increases the concentration of CO and causes the reaction to shift to the left.

d) H_2 is removed from the reaction mixture: Removing the H_2 decreases the concentration of H_2 and causes the reaction to shift to the right.

45. a) **Given:** $I_2(g) \leftrightarrows 2I(g)$ at equilibrium **Find:** the effect of increasing the volume
The chemical equation has 2 moles of gas on the right and 1 mole of gas on the left. Increasing the volume of the reaction mixture decreases the pressure and causes the reaction to shift to the right (toward the side with more moles of gas particles).

b) **Given:** $2H_2S(g) \leftrightarrows 2H_2(g) + S_2(g)$ **Find:** the effect of decreasing the volume
The chemical equation has 3 moles of gas on the right and 2 moles of gas on the left. Decreasing the volume of the reaction mixture increases the pressure and causes the reaction to shift to the left (toward the side with fewer moles of gas particles).

c) **Given:** $I_2(g) + Cl_2(g) \leftrightarrows 2ICl(g)$ **Find:** the effect of decreasing the volume
The chemical equation has 2 moles of gas on the right and 2 moles of gas on the left. Decreasing the volume of the reaction mixture increases the pressure but causes no shift in the reaction because the moles are equal on both sides.

46. a) **Given:** $CO(g) + H_2O(g) \leftrightarrows CO_2(g) + H_2(g)$ **Find:** the effect of decreasing the volume
The chemical equation has 2 moles of gas on the right and 2 moles of gas on the left. Decreasing the volume of the reaction mixture increases the pressure but causes no shift in the reaction because the moles are equal on both sides

b) **Given:** $PCl_3(g) + Cl_2(g) \leftrightarrows PCl_5(g)$ **Find:** the effect of increasing the volume
The chemical equation has 2 moles of gas on the left and 1 mole of gas on the right. Increasing the volume of the reaction mixture decreases the pressure and causes the reaction to shift to the left (toward the side with more moles of gas particles).

c) **Given:** $CaCO_3(s) \leftrightarrows CaO(s) + CO_2(g)$ **Find:** the effect of increasing the volume
The chemical equation has 1 mole of gas on the right and 0 moles of gas on the left. Increasing the volume of the reaction mixture decreases the pressure and causes the reaction to shift to the right (toward the side with more moles of gas particles).

47. **Given:** $C(s) + CO_2(g) \leftrightarrows 2CO(g)$ is endothermic **Find:** the effect of increasing the temperature
Since the reaction is endothermic we can think of the heat as a reactant, increasing the temperature is equivalent to adding a reactant causing the reaction to shift to the right. This will cause an increase in the concentration of products and a decrease in the concentration of reactant, therefore, the value of K will increase.
Find: the effect of decreasing the temperature
Since the reaction is endothermic we can think of the heat as a reactant, decreasing the temperature is equivalent to removing a reactant causing the reaction to shift to the left. This will cause a decrease in the concentration of products and an increase in the concentration of reactants, therefore, the value of K will decrease.

48. **Given:** $C_6H_{12}O_6(s) + 6\,O_2(g) \leftrightarrows 6CO_2(g) + 6\,H_2O(g)$ is exothermic
Find: the effect of increasing the temperature
Since the reaction is exothermic we can think of the heat as a product, increasing the temperature is equivalent to adding a product causing the reaction to shift to the left. This will cause a decrease in the concentration of products and an increase in the concentration of reactant, therefore, the value of K will decrease.

Find: the effect of decreasing the temperature
Since the reaction is exothermic we can think of the heat as a product, decreasing the temperature is equivalent to removing a product causing the reaction to shift to the right. This will cause an increase in the concentration of products and a decrease in the concentration of reactants, therefore, the value of K will increase.

49. **Given:** $C(s) + 2H_2(g) \leftrightarrows CH_4(g)$ is exothermic **Find:** determine which will favor CH_4
a) adding more C to the reaction mixture: does NOT favor CH_4. Adding C does not cause any change in the reaction. C is a solid and the concentration remains constant so the addition of more solid does not change the equilibrium concentration.

b) adding more H_2 to the reaction mixture: favors CH_4. Adding H_2 increases the concentration of H_2 causing the reaction to shift to the right.

c) raising the temperature of the reaction mixture: does NOT favor CH_4. Since the reaction is exothermic we can think of heat as a product, raising the temperature is equivalent to adding a product causing the reaction to shift to the left.

d) lowering the volume of the reaction mixture: favors CH_4. The chemical equation has 1 mole of gas on the right and 2 moles of gas on the left. Decreasing the volume of the reaction mixture increases the pressure and causes the reaction to shift to the right (toward the side with fewer moles of gas particles).

e) adding a catalyst to the reaction mixture: does NOT favor CH_4. A catalyst added to the reaction mixture only speeds up the reaction, it does not change the equilibrium concentration.

f) adding neon gas to the reaction mixture: does NOT favor CH_4. Adding an inert gas to a reaction mixture at a fixed volume has no effect on the equilibrium

50. **Given:** $C(s) + H_2O(g) \leftrightarrows CO(g) + H_2(g)$ is endothermic **Find:** Determine which will favor H_2

a) adding more C to the reaction mixture: does NOT favor H_2. Adding C does not cause any change in the reaction. C is a solid and the concentration remains constant so the addition of more solid does not change the equilibrium concentration.

b) adding more H_2O to the reaction mixture: favors H_2. Adding H_2O increases the concentration and causes the reaction to shift to the right.

c) raising the temperature of the reaction mixture: favors H_2. Since the reaction is endothermic we can think of heat as a reactant, raising the temperature is equivalent to adding a reactant causing the reaction to shift to the right.

d) increasing the volume of the reaction mixture: favors H_2. The chemical equation has 2 moles of gas on the right and 1 mole of gas on the left. Increasing the volume of the reaction mixture decreases the pressure and causes the reaction to shift to the right (toward the side with more moles of gas particles).

e) adding a catalyst to the reaction mixture: does NOT favor H_2. A catalyst added to the reaction mixture only speeds up the reaction, it does not change the equilibrium concentration.

f) adding an inert gas to the reaction mixture: does NOT favor H_2. Adding an inert gas to a reaction mixture at a fixed volume has no effect on the equilibrium.

51. a) To find the value of K for the new equation, combine the two given equations to yield the new equation. Reverse equation 1, and use $1/K_1$ and then add to equation 2. To find K for equation 3 use $(1/K_1)(K_2)$.

$HbO_2(aq)$	$\leftrightarrows$	~~Hb(aq)~~ + $O_2(aq)$	$K_1 = 1/1.8$
~~Hb(aq)~~ + $CO(aq)$	$\leftrightarrows$	$HbCO(aq)$	$K_2 = 306$
$HbO_2(aq) + CO(aq)$	$\leftrightarrows$	$HbCO(aq) + O_2(aq)$	$K_3 = K_1K_2 = (1/1.8)(306) = 170$

b) **Given:** $O_2 = 20\%$, $CO = 0.10\%$ **Find:** The ratio $\frac{[HbCO]}{[HbO_2]}$

Conceptual Plan: Determine the equilibrium expression and then determine $\frac{[HbCO]}{[HbO_2]}$

Solution: $K = \frac{[HbCO][O_2]}{[HbO_2][CO]}$ $\quad 170 = \frac{[HbCO](20.0)}{[HbO_2](0.10)} \quad \frac{[HbCO]}{[HbO_2]} = 170\left(\frac{0.10}{20.0}\right) = \frac{0.85}{1.0}$

Since the ratio is almost 1:1, 0.10% CO will replace about 50% of the O_2 in the blood. The CO blocks the uptake of O_2 by the blood and is therefore highly toxic.

52. **Given:** P = 1 atm, T = 298 K, $N_2 = 78\%$, $O_2 = 21\%$, $K_p = 4.1 \times 10^{-31}$ **Find:** [NO] in molecules/cm^3

Conceptual Plan:

%vol → n → M and then K_p → K_c and then: Prepare an ICE table; represent the change with

$PV = nRT \qquad \frac{n}{1\text{ L air}} \qquad K_p = K_c(RT)^{\Delta n}$

x; sum the table and determine the equilibrium values; put the equilibrium values in the equilibrium expression and solve for x. Determine [NO] in mol/L → molecules/cm³

$$\frac{6.022 \times 10^{23} \text{ molecules}}{\text{mole}} \quad \frac{1\text{ L}}{1000\text{ mL}} \quad \frac{\text{mL}}{\text{cm}^3}$$

Solution: Assume 1 L of air.

$$n_{N_2} = \frac{(1\text{ atm})(0.78\text{ L})}{\left(0.0821\frac{\text{L}\cdot\text{atm}}{\text{mol}\cdot\text{K}}\right)(298\text{ K})} = 0.03\underline{1}88\text{ mol} \qquad n_{O_2} = \frac{(1\text{ atm})(0.21\text{ L})}{\left(0.0821\frac{\text{L}\cdot\text{atm}}{\text{mol}\cdot\text{K}}\right)(298\text{ K})} = 0.008\underline{5}8\text{ mol}$$

$[N_2] = 0.03\underline{1}88$ mol/L $\qquad [O_2] = 0.008\underline{5}8$ mol/L

$$K_p = K_c(RT)^{\Delta n} \qquad K_c = \frac{K_p}{(RT)^{\Delta n}} = \frac{4.1 \times 10^{-31}}{\left(0.0821\frac{\text{L}\cdot\text{atm}}{\text{mol}\cdot\text{K}})(298\text{K}\right)^0} = 4.1 \times 10^{-31}$$

	N_2 (g)	$+O_2$ (g)	⇆ 2 NO (g)
	$[N_2]$	$[O_2]$	[NO]
I	0.03$\underline{1}$9	0.008$\underline{5}$8	0.0
C	- x	-x	2x
E	0.03$\underline{1}$9–x	0.008$\underline{5}$8– x	2x

The reaction will proceed to the right by x

$$K_C = \frac{[NO]^2}{[N_2][O_2]} = \frac{(2x)^2}{(0.0319 - x)(0.00858 - x)} = 4.1 \times 10^{-31}$$

Assume x is small compared to 0.0085

$x = 5.\underline{2}9 \times 10^{-18}$ $\qquad$ check assumption: $\frac{5.\underline{2}9 \times 10^{-18}}{0.00858} \times 100\% = 6.2 \times 10^{-14}\%$

$[NO] = 2x = 2(5.\underline{2}9 \times 10^{-18}) = 1.\underline{0}6 \times 10^{-17}$M

$$1.\underline{0}6 \times 10^{-17}\frac{\text{mol}}{\text{L}} \times \frac{6.022 \times 10^{23}\text{ molecules}}{\text{mol}} \times \frac{\text{L}}{1000\text{ mL}} \times \frac{\text{mL}}{\text{cm}^3} = 6.38 \times 10^3\frac{\text{molecules}}{\text{cm}^3}$$

$$= 6.4 \times 10^3\frac{\text{molecules}}{\text{cm}^3}$$

Check: The answer is reasonable since the reaction has a small equilibrium constant so you would not expect to produce much product.

The reaction to produce NO is endothermic, so we can think of heat as a reactant. So, raising the temperature (as in an automobile engine) shifts the reaction to the right, producing more NO.

53. **Given:** $C_2H_4(g) + Cl_2(g)$ ⇆ $C_2H_4Cl_2(g)$ is exothermic
Find: Which of the following will maximize $C_2H_4Cl_2$
a) increasing the reaction volume. Will not maximize $C_2H_4Cl_2$. The chemical equation has 1 mole of gas on the right and 2 moles of gas on the left. Increasing the volume of the reaction mixture decreases the pressure and causes the reaction to shift to the left (toward the side with more moles of gas particles).

b) removing $C_2H_4Cl_2$ as it forms. Will maximize $C_2H_4Cl_2$. Removing the $C_2H_4Cl_2$ will decrease the concentration of $C_2H_4Cl_2$ and will cause the reaction to shift to the right, producing more $C_2H_4Cl_2$.

c) lowering the reaction temperature. Will maximize $C_2H_4Cl_2$. The reaction is exothermic so we can think of heat as a product. Lowering the temperature will cause the reaction to shift to the right, producing more $C_2H_4Cl_2$.

d) adding Cl_2. Will maximize $C_2H_4Cl_2$. Adding Cl_2 increases the concentration of Cl_2 so the reaction shifts to the right, which will produce more $C_2H_4Cl_2$.

54. **Given:** $C_2H_4(g) + I_2(g)$ ⇆ $C_2H_4I_2(g)$ is endothermic
Find: Which of the following will maximize $C_2H_4I_2$
a) decreasing the reaction volume: Will maximize $C_2H_4I_2$. The chemical equation has 1 mole of gas on the right and 2 moles of gas on the left. Decreasing the volume of the reaction mixture increases the pressure and causes the reaction to shift to the right (toward the side with less moles of gas particles).

b) removing I_2 from the reaction mixture. Will not maximize $C_2H_4I_2$. Removing I_2 decreases the concentration of I_2 and the reaction will shift to the left to produce more I_2.

c) raising the temperature of the reaction. Will maximize $C_2H_4I_2$. The reaction is endothermic so we can think of heat as a reactant. Raising the temperature will cause the reaction to shift to the right, producing more $C_2H_4I_2$.

d) adding C_2H_4 to the reaction mixture. Will maximize $C_2H_4I_2$. Adding C_2H_4 will increase the concentration of C_2H_4 and cause the reaction to shift to the right to produce more $C_2H_4I_2$.

55. **Given:** Reaction 1 at equilibrium: $P_{H2} = 0.958$ atm; $P_{I2} = 0.877$ atm; $P_{HI} = 0.0200$ atm: reaction 2: $P_{H2} = P_{I2} = 0.621$ atm; $P_{HI} = 0.101$ atm **Find:** Is reaction 2 at equilibrium, if not, what is the P_{HI} at equilibrium.
Conceptual Plan: Use equilibrium partial pressures to determine K_p. Use K_p to determine if reaction 2 is at equilibrium. Prepare an ICE table; and then calculate Q; compare Q and K_p and predict the direction of the reaction; represent the change with x; sum the table and determine the equilibrium values; put the equilibrium values in the equilibrium expression and solve for x. Determine P_{HI}.
Solution: $H_2(g) + I_2(g) \leftrightarrows 2HI(g)$

	P_{H2}	P_{I2}	P_{HI}
Reaction 1:	0.958	0.877	0.020

$$K_p = \frac{P_{HI}^2}{P_{H_2}P_{I_2}} = \frac{(0.020)^2}{(0.958)(0.877)} = 4.7\underline{6}10 \times 10^{-4}$$

	$H_2(g)$ +	$I_2(g) \leftrightarrows$	$2HI(g)$
Reaction 2:	P_{H2}	P_{I2}	P_{HI}
I	0.621	0.621	0.101
C	x	x	-2x
E	0.621+x	0.621+x	0.101 -2x

$$Q = \frac{P_{HI}^2}{P_{H_2}P_{I_2}} = \frac{0.101^2}{(0.621)(0.621)} = 0.0264\text{: } Q > K \text{ so the reaction shifts to the left}$$

$$K_p = \frac{P_{HI}^2}{P_{H_2}P_{I_2}} = \frac{(0.101 - x)^2}{(0.621+x)(0.621+x)} = 4.7\underline{6}10 \times 10^{-4}$$

$$\sqrt{\frac{(0.101 - 2x)^2}{(0.621+x)(0.621+x)}} = \sqrt{4.7\underline{6}10 \times 10^{-4}}$$

$$\frac{(0.101 - 2x)}{(0.621+x)} = 2.1\underline{8}2 \times 10^{-2}$$

$x = 0.04325 = 0.0433$

$P_{H_2} = P_{I_2} = 0.621 + x = 0.621 + 0.0433 = 0.664$ atm; $P_{HI} = 0.101 - 2x = 0.101 - 2(0.0433) = 0.0144$ atm

Check: Plug the values into the equilibrium expression:

$$K_p = \frac{(0.0144)^2}{(0.664)} = 4.703 \times 10^{-4} = 4.70 \times 10^{-4}$$; this value is close to the original equlbirium constant

56. **Given:** Reaction 1 initial: $H_2S = 0.500$ M; $SO_2 = 0.500$M; H_2O at equilibrium = 0.0011 M: reaction 2: $H_2S = 0.250$ M; $SO_2 = 0.325$M; **Find:** $[H_2O]$ at equilibrium in reaction 2.
Conceptual Plan: Prepare an ICE table. Determine the equilibrium concentrations in reaction 1 and determine K_c. Use K_c to determine the equilibrium concentrations for reaction 2. Prepare an ICE table; and then calculate Q; compare Q and K_c and predict the direction of the reaction; represent the change with x; sum the table and determine the equilibrium values; put the equilibrium values in the equilibrium expression and solve for x. Determine $[H_2O]$.
Solution: $2H_2S(g) + SO_2(g) \leftrightarrows 3S(s) + 2H_2O(g)$

Reaction 1:	$[H_2S]$	$[SO_2]$	[S]	$[H_2O]$
I	0.500	0.500	constant	0
C	-2x	-x		2x
E	0.500-x	0.500-x		0.0011

$2x = 0.0011$, $x = 5.5 \times 10^{-4}$

$[H_2S] = 0.500 - 0.0011 = 0.4989$ $[SO_2] = 0.500 - 5.5 \times 10^{-4} = 0.49945$

$$K_c = \frac{[H_2O]^2}{[H_2S]^2[SO_2]} = \frac{(0.0011)^2}{(0.4989)^2(0.49945)} = 9.7\underline{3}3 \times 10^{-6}$$

$2H_2S(g) + SO_2(g) \leftrightarrows 3S(s) + 2H_2O(g)$

Reaction 2:	$[H_2S]$	$[SO_2]$	$[S]$	$[H_2O]$
I	0.250	0.325	constant	
C	-2x	-x		2x
E	0.250-x	0.325-x		2x

Reaction shifts to the right

$$K_c = \frac{[H_2O]^2}{[H_2S]^2[SO_2]} = \frac{(2x)^2}{(0.250\text{-}x)^2(0.325\text{-}x)} = 9.7\underline{3}3 \times 10^{-6}$$

Assume x is small compared to 0.250 and 0.325

$4x^2 = 1.9\underline{7}70 \times 10^{-7}$

$x^2 = 4.9\underline{4}25 \times 10^{-8}$

$x = 2.2\underline{2}3 \times 10^{-4}$ check assumption: $\frac{2.22 \times 10^{-4}}{0.250} = 8.88 \times 10^{-4} \times 100\% = 0.089\%$; assumption valid

$[H_2O] = 2x = 2(2.22 \times 10^{-4}) = 4.44 \times 10^{-4} M$

$[H_2S] = 0.250 - 2(2.22 \times 10^{-4}) = 0.250 M$

$[SO_2] = 0.325 - 2.22 \times 10^{-4} = 0.325 M$

Check: Plug the values into the equilibrium expression:

$K_c = \frac{(4.44 \times 10^{-4})^2}{(0.250)^2(0.325)} = 9.71 \times 10^{-6}$; this value is close to the original equilibrium constant

57. **Given:** 200.0 L container; 1.27 kg N_2; 0.310 kg H_2; 725K; $K_p = 5.3 \times 10^{-5}$
Find: mass in g of NH_3 and % yield
Conceptual Plan:

$K_p \rightarrow K_c$ and then kg → g → mol → M and then prepare an ICE table; represent the change

$K_p = K_c(RT)^{\Delta n}$ $\quad \frac{1000 \text{ g}}{\text{kg}}$ $\quad \frac{\text{g}}{\text{molar mass}}$ $\quad \frac{\text{mol}}{\text{vol}}$

with x; sum the table and determine the equilibrium values; put the equilibrium values in the equilibrium expression and solve for x. Determine $[NH_3]$. Then M → mol → g and then determine

M x vol $\quad$ mol x molar mass

theoretical yield NH_3 → % yield

determine limiting reactant $\quad \frac{\text{actual yield}}{\text{theoretical yield}}$

Solution:

$$K_p = K_c(RT)^{\Delta n} \qquad K_c = \frac{K_p}{(RT)^{\Delta n}} = \frac{5.3 \times 10^{-5}}{\left((0.0821 \frac{\text{L}\cdot\text{atm}}{\text{mol}\cdot\text{K}})(725\text{K})\right)^{-2}} = 0.1\underline{8}77$$

$$n_{N_2} = 1.27 \text{ kg } N_2 \times \frac{1000 \text{ g}}{\text{kg}} \times \frac{1 \text{ mol } N_2}{28.00 \text{ g } N_2} = 45.\underline{3}57 \text{ mol } N_2 \qquad [N_2] = \frac{45.\underline{3}57 \text{ mol}}{200.0 \text{ L}} = 0.22\underline{6}78$$

$$n_{H_2} = 0.310 \text{ kg } H_2 \times \frac{1000 \text{ g}}{\text{kg}} \times \frac{1 \text{ mol } H_2}{2.016 \text{ g } H_2} = 15\underline{3}.77 \text{ mol } H_2 \qquad [H_2] = \frac{15\underline{3}.77 \text{ mol}}{200.0 \text{ L}} = 0.76\underline{8}85$$

$N_2(g) + 3H_2(g) \leftrightarrows 2NH_3(g)$

Reaction 1:	$[N_2]$	$[H_2]$	$[NH_3]$
I	0.22\underline{6}8	0.76\underline{8}9	0.0
C	-x	-3x	2x
E	0.2268-x	0.7689-3x	2x

Reaction shifts to the right

$$K_c = \frac{[NH_3]^2}{[N_2][H_2]^3} = \frac{(2x)^2}{(0.2268\text{-}x)(0.7689\text{-}3x)^3} = 0.1877$$

Assume x is small compared to 0.2268 and 3x is small compared to 0.7689

$$\frac{(2x)^2}{(0.2268)(0.7689)^3} = 0.1877 \qquad x = 0.06956$$

check assumptions: $\frac{0.06956}{0.2268} \times 100\% = 30.7$ not valid and $\frac{3(0.06956)}{0.7689} \times 100\% = 27.1\%$

Use method of successive substitution to solve for x. This yields

$x = 0.0461$

$[NH_3] = 2x = 2(0.0461) = 0.0922M$

Check: Plug the values into the equilibrium expression:

$$K_c = \frac{(0.0922)^2}{(0.2268 - 0.0461)(0.7689 - 3(0.0461))^3} = 0.1876 \;;$$

this value is close to the original equilibrium constant

Determine grams NH_3: $0.0922 \frac{\text{mol } NH_3}{\text{L}} \times \frac{200.0 \text{ L}}{} \times \frac{17.02 \text{ g } NH_3}{\text{mol } NH_3} = 313.8 \text{ g} = 3.1 \times 10^2 \text{ g}$

Determine the theoretical yield: Determine the limiting reactant:

$$1.27 \text{ kg } N_2 \times \frac{1000 \text{ g}}{\text{kg}} \times \frac{1 \text{ mol } N_2}{28.0 \text{ g } N_2} \times \frac{2 \text{ mol } NH_3}{1 \text{ mol } N_2} \times \frac{17.02 \text{ g } NH_3}{\text{mol } NH_3} = 1544 \text{ g } NH_3$$

$$0.310 \text{ kg } H_2 \times \frac{1000 \text{ g}}{\text{kg}} \times \frac{1 \text{ mol } H_2}{2.016 \text{ g } H_2} \times \frac{2 \text{ mol } NH_3}{3 \text{ mol } H_2} \times \frac{17.02 \text{ g } NH_3}{\text{mol } NH_3} = 1745 \text{ g } NH_3$$

N_2 produces the least amount of NH_3 therefore, it is the limiting reactant and the theoretical yield is 1.54×10^3 g NH_3.

$$\% \text{ yield} = \frac{3.1 \times 10^2 \text{ g}}{1.54 \times 10^3 \text{ g}} \times 100 = 20.\%$$

58. **Given:** V = 85.0 L, 22.3 kg CH_4, 55.4 kg CO_2 T = 825 K, $K_p = 4.5 \times 10^2$ **Find:** g H_2 at equilibrium
Conceptual Plan: kg → g → mol → P and then prepare an ICE table; represent the change

$\frac{1000 \text{ g}}{\text{kg}}$ $\quad \frac{\text{g}}{\text{molar mass}}$ $\quad PV = nRT$

with x; sum the table and determine the equilibrium values; put the equilibrium values in the equilibrium expression and solve for x. Determine P(H_2). Then P → mol → g and then determine

PV = nRT $\quad$ mol x molar mass

theoretical yield H_2 → % yield

determine limiting reactant $\quad \frac{\text{actual yield}}{\text{theoretical yield}}$

Solution:

$$n_{CH_4} = 23.3 \text{ kg} \times \frac{1000 \text{ g}}{\text{kg}} \times \frac{1 \text{ mol}}{16.0 \text{ g}} = 1456.2 \text{ mol}$$

$$P_{CH_4} = \frac{nRT}{V} = \frac{(1456 \text{ mol})(0.0821 \frac{\text{L} \cdot \text{atm}}{\text{mol} \cdot \text{K}})(825 \text{ K})}{85.0 \text{ L}} = 1160 \text{ atm}$$

$$n_{CO_2} = 55.4 \text{ kg} \times \frac{1000 \text{ g}}{\text{kg}} \times \frac{1 \text{ mol}}{44.0 \text{ g}} = 1259 \text{ mol}$$

$$P_{CO_2} = \frac{nRT}{V} = \frac{(1259 \text{ mol})(0.0821 \frac{\text{L} \cdot \text{atm}}{\text{mol} \cdot \text{K}})(825 \text{ K})}{85.0 \text{ L}} = 1003 \text{ atm}$$

	$CH_4(g)$ +	$CO_2(g)$ ⇆	$2CO(g)$ +	$2H_2(g)$
	P_{CH4}	P_{CO2}	P_{CO}	P_{H2}
I	1160	1003	0.00	0.00
C	-x	-x	2x	2x
E	1160-x	1003-x	2x	2x

$$K_p = \frac{P_{CO}^2 P_{H_2}}{P_{CH_2} P_{CO_2}} = \frac{(2x)^2(2x)^2}{(1160 - x)(1003 - x)} = 4.5 \times 10^2$$

Assume x is small compared to 1003 and 1160

$x^4 = 3.27 \times 10^7$ $\quad x = 75.6$

check assumptions: $\frac{75.6}{1003} \times 100\% = 7.54\%$ and $\frac{75.6}{1160} \times 100\% = 6.52\%$ assumptions not valid

Use method of successive substitutions which yields

x = 73.0

Check:

Plug into equilibrium expression: $K_p = \frac{P_{CO}^2 P_{H_2}}{P_{CH_2} P_{CO_2}} = \frac{(2(73.0))^2(2(73.0))^2}{(1160 - 73.0)(1003 - 73.0)} = 4.49 \times 10^2 = 4.5 \times 10^2$;

which is the equilibrium constant

Determine the grams of H_2.

$P_{H_2} = 2(73.0) = 146$ atm $\quad n_{H_2} = \frac{PV}{RT} = \frac{(146 \text{ atm})(85.0 \text{ L})}{(0.0821 \frac{\text{L} \cdot \text{atm}}{\text{mol} \cdot \text{K}})(825 \text{ K})} = 18\underline{3}.2$ mol

$18\underline{3}.2 \text{ mol} \times \frac{2.016 \text{ g } H_2}{\text{mol}} = 369.4 \text{ g} = 369 \text{ g } H_2$

Determine % yield

$1456 \text{ mol } CH_4 \times \frac{2 \text{ mol } H_2}{\text{mol } CH_4} = 2912 \text{ mol } H_2$

$1259 \text{ mol } CO_2 \times \frac{2 \text{ mol } H_2}{\text{mol } CO_2} = 2518 \text{ mol } H_2$

CO_2 is limiting reactant

$\frac{18\underline{3}.2 \text{ mol } H_2}{2518 \text{ mol } H_2} \times 100 = 7.28\%$ yield

59. **Given:** At equilibrium: $P_{CO} = 0.30$ atm; $P_{Cl2} = 0.10$ atm; $P_{COCl2} = 0.60$ atm, add 0.40 atm Cl_2
Find: P_{CO} when system returns to equilibrium.
Conceptual Plan: Use equilibrium partial pressures to determine K_p. For the new conditions: prepare an ICE table; represent the change with x; sum the table and determine the equilibrium values; put the equilibrium values in the equilibrium expression and solve for x. Determine P_{CO}.
Solution: $CO(g) + Cl_2(g) \leftrightarrows COCl_2(g)$

Condition 1:	P_{CO}	P_{Cl2}	P_{COCl2}
	0.30	0.10	0.60

$$K_p = \frac{P_{COCl_2}}{P_{CO} P_{Cl_2}} = \frac{(0.60)}{(0.30)(0.10)} = 20.$$

$CO(g) + Cl_2(g) \leftrightarrows COCl_2(g)$

Condition 2:	P_{CO}	P_{Cl2}	P_{COCl2}
I	0.30	0.10+0.40	0.60
C	-x	-x	+x
E	0.30-x	0.50-x	0.60 + x

Reaction shifts to the right because the concentration of Cl_2 was increased

$$K_p = \frac{P_{COCl_2}}{P_{CO} P_{Cl_2}} = \frac{(0.60+x)}{(0.30 - x)(0.50 - x)} = 20$$

$20x^2 - 17x + 2.4 = 0$

$$\frac{-b \pm \sqrt{b^2 - 4ac}}{2a} = \frac{-(-17) \pm \sqrt{(-17)^2 - 4(20)(2.4)}}{2(20)}$$

$x = 0.67$ or 0.18 So, $x = 0.18$

$P_{CO} = 0.30 - 0.18 = 0.12$ atm; $P_{Cl_2} = 0.50 - 0.18 = 0.32$; $P_{COCl_2} = 0.60 + 0.18 = 0.78$ atm

Check: Plug the values into the equilibrium expression:

$$K_p = \frac{(0.78)}{(0.12)(0.32)} = 20.3 = 20.$$; this is the same as the original equilibrium constant.

60. **Given:** SO_2 P = 3.00 atm; O_2 P = 1.00 atm; at equilibrium, $P_{total} = 3.75$ atm; T = 27°C **Find:** K_c

Conceptual Plan: Prepare an ICE table; represent the change with x; sum the table and determine the equilibrium values; use the total pressure and solve for x. Determine partial pressure of each at equilibrium. Determine $K_p \rightarrow K_c$.

XXX $K_p = K_c(RT)^{\Delta n}$

Solution: $2SO_2(g) + O_2(g) \leftrightarrows 2SO_3(g)$

	P_{SO2}	P_{O2}	P_{SO3}
I	3.00	1.00	0.00
C	-2x	-x	+2x
E	3.00-2x	1.00-x	2x

$P_{Total} = P_{SO_2} + P_{O_2} + P_{SO_3}$ $\quad 3.75 = 3.00 - 2x + (1.00 - x) + 2x$

$x = 0.25$ $\quad P_{SO_2} = (3.00 - 2(0.25) = 2.50$ atm; $P_{O_2} = (1.00 - 0.25) = 0.75$ atm; $P_{SO_3} = 2(0.25) = 0.50$ atm

$$K_p = \frac{P^2_{SO_3}}{P^2_{SO_2}P_{O_2}} = \frac{(0.50)^2}{(2.50)^2(0.75)} = 0.05\underline{33} = 0.053$$

Check: The value of the pressure of SO_3 is small compared to the pressures of SO_2 and O_2, therefore, you would expect K_p to be less than 1.

$$K_p = K_c(RT)^{\Delta n} \qquad 0.05\underline{33} = K_c((0.0821\frac{L \cdot atm}{mol \cdot K})(27 + 273K))^{-1}$$

$K_c = 1.\underline{3}12 = 1.3$

61. **Given:** $K_p = 0.76$; P_{total} at equilibrium = 1.00 **Find:** $P_{initial}$ CCl_4

Conceptual Plan: Prepare an ICE table; represent the P(CCl_4) with A and the change with x; sum the table and determine the equilibrium values; use the total pressure and solve for A in terms of x. Determine partial pressure of each at equilibrium, use the equilibrium expression to determine x, and then determine A.

Solution: $CCl_4(g) \leftrightarrows C(s) + 2Cl_2(g)$

	P_{CCl4}	P_C	P_{Cl2}
I	A	constant	0.00
C	-x		+2x
E	A-x		2x

$P_{Total} = P_{CCl_4} + P_{Cl_2}$ $\quad 1.0 = A - x + 2x \quad A = 1 - x$

$P_{CCl_4} = (A - x) = (1-x) - x = 1 - 2x$; $P_{Cl_2} = (2x)$

$$K_p = \frac{P^2_{Cl_2}}{P_{CCl_4}} = \frac{(2x)^2}{(1-2x)} = 0.76$$

$4x^2 + 1.52x - 0.76 = 0$ $\quad x = 0.285$ or -0.665 so $x = 0.2\underline{8}5$

$A = 1 - x = 1.0 - 0.285 = 0.715 = 0.72$ atm

Check: Plug values into equilibrium expression:

$$K_p = \frac{P^2_{Cl_2}}{P_{CCl_4}} = \frac{(2x)^2}{(A-x)} = \frac{(2(0.285))^2}{(0.715-0.285)} = 0.755 = 0.76;$$ the original equilibrium constant

62. **Given:** $K = 3.0$; $SO_2 = 2.4$ mol initial; $SO_3 = 1.2$ mol equilibrium **Find:** mol NO_2 initial

Conceptual Plan: Assume 1.0 L, prepare ICE table; represent the change with x; sum the table and determine the equilibrium values; determine initial values.

Solution: $SO_2(g) + NO_2(g) \leftrightarrows SO_3(g) + NO(g)$

	$[SO_2]$	$[NO_2]$	$[SO_3]$	$[NO]$
I	2.4	y	0.0	0.0
C	-x	-x	x	x
E	1.2	y-1.2	1.2	1.2

x = 1.2, fill in table

$$K = \frac{[SO_3][NO]}{[SO_2][NO_2]} \quad 3.0 = \frac{(1.2)(1.2)}{(1.2)(y - 1.2)} \quad y = 1.6$$

mol NO_2 initial = 1.6 mol

63. **Given:** V = 0.654 L, T = 1000 K, $K_p = 3.9 \times 10^{-2}$ **Find:** mass CaO as equilibrium

Conceptual Plan: $K_p \rightarrow P_{CO_2} \rightarrow n(CO_2) \rightarrow n(CaO) \rightarrow g$

PV = nRT stoichiometry g = n(molar mass)

Solution: Since $CaCO_3$ and CaO are solids, they are not included in the equilibrium expression.

$$K_p = P_{CO_2} = 3.9 \times 10^{-2} \qquad n = \frac{PV}{RT} = \frac{(3.9 \times 10^{-2}\ \text{atm})(0.654\ \text{L})}{\left(0.0821\ \frac{\text{L}\cdot\text{atm}}{\text{mol}\cdot\text{K}}\right)(1000\ \text{K})} = 3.\underline{1}06 \times 10^{-4}\ \text{mol CO}_2$$

$$3.\underline{1}06 \times 10^{-4}\ \text{mol CO}_2 \times \frac{1\ \text{mol CaO}}{1\ \text{mol CO}_2} \times \frac{56.1\ \text{g CaO}}{1\ \text{mol CaO}} = 0.0174\ \text{g} = 0.017\ \text{g CaO}$$

Check: The small value of K would give a small amount of products, so we would not expect to have a large mass of CaO formed.

64. **Given:** at equilibrium: N_2O_4, P = 0.28 atm; NO_2, P = 1.1 atm; T = 350 K **Find:** equilibrium pressures when volume doubles.

Conceptual Plan: P(N_2O_4), P(NO_2) → K_p and then P when volume doubles and then prepare ICE table; represent the change with x; sum the table and determine the equilibrium values.

Solution: $K_p = \frac{P^2_{NO_2}}{P_{N_2O_4}} = \frac{(1.1)^2}{0.28} = 4.\underline{3}21$

When the volume is doubled, the partial pressure of each gas will decrease by half.

$N_2O_2(g) \leftrightarrows 2NO_2(g)$

	P_{N2O4}	P_{NO2}	
I	0.28/2	1.1/2	
C	-x	+2x	Reaction shifts to the side with more moles, to the right
E	0.14-x	0.55+2x	

$$K_p = \frac{P^2_{NO_2}}{P_{N_2O_4}} = \frac{(0.55 + 2x)^2}{(0.14 - x)} = 4.\underline{3}21$$

$$4x^2 + 6.521x - 0.3024 = 0$$

$$\frac{-b \pm \sqrt{b^2 - 4ac}}{2a} = \frac{-(6.521) \pm \sqrt{(6.521)^2 - 4(4)(0.3024)}}{2(4)}$$

x = 0.0451 or - 1.675 So, x = 0.0451

$P_{N_2O_4} = 0.14 - 0.0451 = 0.0949$ atm = 0.095 atm; $P_{NO_2} = 0.55 + 2(0.0451) = 0.6402 = 0.64$ atm

Check: Plug the values into the equilibrium expression:

$$K_p = \frac{(0.6402)^2}{(0.0949)} = 4.31 = 4.3.$$; this is the same as the original equilibrium constant.

65. a) **Given:** NO (P = 522 torr), O_2 (P = 421 torr); at equilibrium, P_{total} = 748 torr **Find:** K_p

Conceptual Plan: Prepare an ICE table; represent the change with x; sum the table and determine the equilibrium values; use the total pressure and solve for x. torr → atm → K_p

Solution: $2NO(g) + O_2(g) \leftrightarrows 2NO_2(g)$

	P_{NO}	P_{O2}	P_{NO2}
I	522 torr	421 torr	0.00
C	-2x	-x	+2x
E	522-2x	421-x	2x

$P_{Total} = P_{NO} + P_{O_2} + P_{NO_2}$ $748 = 522 - 2x + (421 - x) + 2x$

$x = 195$ torr $P_{NO} = (522 - 2(195) = 132$ torr; $P_{O_2} = (421 - 195) = 226$ torr; $P_{NO_2} = 2(195) = 390$ torr

$P_{NO} = 132 \text{ torr x } \frac{1 \text{ atm}}{760 \text{ torr}} = 0.17\underline{3}7 \text{ atm};$ $P_{O_2} = 226 \text{ torr x } \frac{1 \text{ atm}}{760 \text{ torr}} = 0.29\underline{7}4 \text{ atm};$

$P_{NO_2} = 390 \text{ torr x } \frac{1 \text{ atm}}{760 \text{ torr}} = 0.51\underline{3}2 \text{ atm}$

$$K_p = \frac{P^2_{NO_2}}{P^2_{NO}P_{O_2}} = \frac{(0.51\underline{3}2)^2}{(0.17\underline{3}7)^2(0.29\underline{7}4)} = 29.34 = 29.3$$

b) **Given:** NO (P = 255 torr), O_2 (P = 185 torr), $K_p = 29.3$ **Find:** equilibrium P_{NO2}
Conceptual Plan:

torr → atm and then prepare an ICE table; represent the change with x; sum the table and

$\frac{\text{atm}}{760 \text{ torr}}$

determine the equilibrium values; put the equilibrium values in the equilibrium expression and solve for x. Determine P N_2O_4.

Solution: $P_{NO} = 255 \text{ torr x } \frac{1 \text{ atm}}{760 \text{ torr}} = 0.33\underline{5}5 \text{ atm}$ $P_{O_2} = 185 \text{ torr x } \frac{1 \text{ atm}}{760 \text{ torr}} = 0.24\underline{3}4 \text{ atm}$

	$2NO(g)$ +	$O_2(g)$ ⇆	$2NO_2(g)$
	P_{NO}	P_{O2}	P_{NO2}
I	0.3355	0.2434	0.00
C	-2x	-x	+2x
E	0.3355-2x	0.2434-x	2x

$$K_p = \frac{P^2_{NO_2}}{P^2_{NO}P_{O_2}} = \frac{(2x)^2}{(0.3355 - 2x)^2(0.2434\text{-}x)} = 29.3$$

$-117.2x^3 + 63.847x^2 - 12.869x + 0.80272 = 0$

$x = 0.11\underline{1}2$ $P_{NO_2} = 2x = 2(0.11\underline{1}2) = 0.22\underline{2}4$ atm

$0.22\underline{2}4 \text{ atm x } \frac{760 \text{ torr}}{1 \text{ atm}} = 169.1 \text{ torr} = 169 \text{ torr}$

Check: Plug the values into the equilibrium expression:

$K_p = \frac{(0.2224)^2}{(0.1131)^2(0.1322)} = 29.249 = 29.2$; this is within 1 significant figure of the original equilibrium constant.

66. **Given:** 2.75 L, 950 K, 0.100 mol SO_2, 0.100 mol O_2; K_p =0.355 **Find:** P_{total} at equilibrium
Conceptual Plan:

n → P and then prepare an ICE table; represent the change with x; sum the table and

PV = nRT

determine the equilibrium values; put the equilibrium values in the equilibrium expression and solve for x and then determine P for each reactant and product → P_{total}

$P_{total} = P_{SO_2} + P_{O_2} + P_{SO_3}$

Solution: $P = \frac{nRT}{V}$ $P_{O_2} = P_{SO_2} = \frac{(0.100 \cancel{\text{mol}})(0.0821 \frac{\cancel{\text{L}} \cdot \text{atm}}{\cancel{\text{mol}} \cdot \cancel{\text{K}}})(950 \cancel{\text{K}})}{2.75 \cancel{\text{L}}} = 2.8\underline{3}6 \text{ atm}$

	$2SO_2(g)$ +	$O_2(g)$ ⇆	$2SO_3(g)$
	P_{SO2}	P_{O2}	P_{SO3}
I	2.836	2.836	0.00
C	-2x	-x	+2x
E	2.836-2x	2.836-x	2x

$$K_p = \frac{P^2_{SO_3}}{P^2_{SO_2}P_{O_2}} = \frac{(2x)^2}{(2.836 - 2x)^2(2.836 - x)} = 0.355$$

$$-1.42x^3 + 8.054x^2 - 14.275x + 8.096 = 0$$

$$x = 0.6629 = 0.663$$

$$P_{SO_3} = 2x = 2(0.663) = 1.3\underline{2}6 = 1.33 \text{ atm}$$

$$P_{SO_2} = (2.836 - 2x) = (2.836 - 2(0.663)) = 1.5\underline{1}0 = 1.51 \text{ atm}$$

$$P_{O_2} = (2.836 - x) = (2.836 - 0.663) = 2.1\underline{7}3 = 2.17 \text{ atm}$$

$$P_{Total} = 1.33 + 1.51 + 2.17 = 5.01 \text{ atm}$$

Check: Plug the values into the equilibrium expression:

$$K_p = \frac{P^2_{SO_3}}{P^2_{SO_2}P_{O_2}} = \frac{(1.33)^2}{(1.51)^2(2.17)} = 0.3575;$$

this is within 3 digits of the significant figures of the original equilibrium constant

67. **Given:** P_{NOCl} at equilibrium = 115 torr; $K_p = 0.27$, T = 700 K, **Find:** initial pressure NO, Cl_2

Conceptual Plan:

torr → atm and then prepare an ICE table; represent the change with x; sum the table and

$\frac{\text{atm}}{\text{760 torr}}$

determine the equilibrium values; put the equilibrium values in the equilibrium expression and determine initial pressure.

Solution: $115 \text{ torr} \times \frac{1 \text{ atm}}{760 \text{ torr}} = 0.15\underline{1}3 \text{ atm}$

	$2NO(g) +$	$Cl_2(g) \leftrightarrows$	$2NOCl(g)$	
	P_{NO}	P_{Cl2}	P_{NOCl}	Let A = initial pressure of NO and Cl_2
I	A	A	0.00	
C	-2x	-x	+2x	
E	A-2x	A-x	0.151	2x = 0.151, so, x = 0.0756
	A-0.151	A-0.0756		

$$K_p = \frac{P^2_{NOCL}}{P^2_{NO}P_{Cl_2}} = \frac{(0.151)^2}{(A - 0.151)^2(A - 0.0756)} = 0.27$$

$$0.27A^3 - 0.1019A^2 + 0.01231A - 0.022336 = 0$$

$$A = 0.565$$

$$P_{NO} = P_{Cl_2} = A = 0.565\text{atm} = 429 \text{ torr}$$

Check: Plug the values into the equilibrium expression:

$$K_p = \frac{P^2_{NOCL}}{P^2_{NO}P_{Cl_2}} = \frac{(0.151)^2}{(0.565 - 0.151)^2(0.565 - 0.0756)} \quad 0.2\underline{7}2 \;;$$

this is within 1 digit of the significant figure of the original equilibrium constant

68. **Given:** $P(N_2O_4) = 1$ atm, K_p reaction 1 = 1×10^4, K_p reaction 2 = 0.10 **Find:** which component will have $P > 0.2$atm

Conceptual Plan: Use reaction 2 to determine $P(NO_2)$. Then, prepare an ICE table; represent the change with x; sum the table and determine the equilibrium values; put the equilibrium values in the equilibrium expression and solve for x

Solution:

	$2NO(g) \leftrightarrows$	$N_2O_4(g)$
	P_{NO}	P_{N2O4}
I	0.00	1.00
C	+2x	-x
E	+2x	1.00 - x

$$K_p = \frac{P_{N_2O_4}}{P^2_{NO}} = \frac{(1.00 - x)}{(2x)^2} = 0.10$$

$$0.40x^2 + x - 1.00 = 0$$

$$\frac{-b \pm \sqrt{b^2 - 4ac}}{2a} = \frac{-1 \pm \sqrt{1^2 - 4(0.40)(-1.00)}}{2(0.40)}$$

$x = 0.7655 = 0.77$

$P_{NO_2} = 2x = 1.54$

Because K_p for reaction 1 is so large, essentially all of the material is products, so the P(NO) and P(O_2) in reaction 1 will be negligible.

69. **Given:** P= 0.750 atm, density = 0.520 g/L, T = 337°C **Find:** K_c

Conceptual Plan: Prepare an ICE table; represent the P(CCl_4) with A and the change with x; sum the table and determine the equilibrium values; use the total pressure and solve for A in terms of x. Determine partial pressure of each at equilibrium in terms of x and then use the density to determine

$$d = \frac{PM}{RT}$$

the apparent molar mass and then use the mole fraction (in terms of P) and the molar mass of each

$$\chi_A = \frac{P_A}{P_{Total}}$$

gas to determine x.

Solution: $2NO_2(g) \leftrightarrows 2NO(g) + O_2(g)$

	P_{NO2}	P_{NO}	P_{O2}
I	A	0.00	0.00
C	-2x	+2x	+x
E	A-2x	2x	x

$P_{Total} = P_{NO_2} + P_{NO} + P_{O_2}$ 0.750 = A -2x + 2x + x A = 0.750 - x

$P_{NO_2} = (A - 2x) = (0.750\text{-}x) - 2x = (0.750 - 3x)$; $P_{NO} = (2x)$; $P_{O_2} = x$

$$d = \frac{PM}{RT} \quad M = \frac{dRT}{P} = \frac{(0.520\frac{g}{L})(0.0821\frac{L \cdot atm}{mol \cdot K})(610\ K)}{0.750\ atm} = 34.72\ g/mol$$

$$M = \chi_{NO_2}M_{NO_2} + \chi_{NO}M_{NO} + \chi_{O_2}M_{O_2} = \frac{P_{NO_2}}{P_{total}}M_{NO_2} + \frac{P_{NO}}{P_{total}}M_{NO} + \frac{P_{O_2}}{P_{total}}M_{O_2}$$

$$P_{Total}M = P_{NO_2}M_{NO_2} + P_{NO}M_{NO} + P_{O_2}M_{O_2}$$

$(0.750)(34.7) = (0.750 - 3x)(46.0) + 2x(30.0) + x(32.0)$

$x = 0.184$

$$K_p = \frac{P_{NO}^2 P_{O_2}}{P_{NO_2}^2} = \frac{(2x)^2(x)}{(0.750\text{-}3x)} = \frac{(2(0.184))^2(0.184)}{(0.750\text{-}3(0.184))^2} = 0.63\underline{5}6$$

$$K_p = K_c(RT)^{\Delta n} \qquad 0.63\underline{5}6 = K_c((0.0821\frac{L \cdot atm}{mol \cdot K})(610K))^1$$

$K_c = 1.27 \times 10^{-2}$

70. **Given:** Reaction 1 K_c = 7.75, Reaction 2 K_c = 4.00, $[N_2O_5]_{initial}$ = 4.00M, $[O_2]_{equil}$ = 4.50M

Find: concentration of other species at equilibrium

Conceptual Plan: Combine reaction 1 and reaction 2 to get the overall reaction and then prepare an ICE table; represent the change with x; sum the table and determine the equilibrium values; use the total pressure and solve for x, use x to determine equilibrium concentration.

Solution:

$N_2O_5(g) \leftrightarrows \cancel{N_2O_3(g)} + O_2(g)$ $K_1 = 7.75$

$\cancel{N_2O_3(g)} \leftrightarrows N_2O(g) + O_2(g)$ $K_2 = 4.00$

$N_2O_5(g) \leftrightarrows N_2O(g) + 2O_2(g)$ $K = K_1K_2 = 3\underline{1}.0$

	N_2O_5	N_2O	O_2
I	4.00	0.00	0.00
C	-x	x	2x
E	4.00 – x	x	4.50

$$K_c = \frac{[N_2O][O_2]^2}{[N_2O_5]} = 31.0 = \frac{(x)(4.50)^2}{(4.00 - x)}$$

$x = 2.42$

$[N_2O_5] = 4.00 - x = 4.00 - 2.42 = 1.58M$

$[N_2O] = x = 2.42M$

$[O_2] = 4.50M$

Use either reaction 1 or reaction 2 and solve for $[N_2O_3]$

$$K_c = \frac{[N_2O_3][O_2]}{[N_2O_5]} = 7.75 = \frac{(y)(4.50)}{(1.58)}$$

$y = 2.72$ $\quad [N_2O_3] = y = 2.72M$

Check: Plug the values into the equilibrium expression into any of the equilibrium expressions: e.g., For the overall reaction:

$$K_c = \frac{[N_2O][O_2]^2}{[N_2O_5]} = \frac{(2.42)(4.50)^2}{(1.58)} = 31.0\text{; the equilibrium constant}$$

e.g., For reaction 2:

$$K_c = \frac{[N_2O][O_2]}{[N_2O_3]} = \frac{(2.42)(4.50)}{(2.72)} = 4.00$$

71. Yes, the direction will depend on the volume. If the initial moles of A and B are equal, the initial concentrations of A and B are equal regardless of the volume. Since $K_c = \frac{[B]^2}{[A]} = 1$, if the volume is such that the $[A] = [B] < 1.0$, then $Q < K$ and the reaction goes to the right to reach equilibrium, but, if the volume is such that the $[A] = [B] > 1.0$, then $Q > K$ and the reaction goes to the left to reach equilibrium.

72. $K_p = 0.50$ means that P(products) < P(reactants). If the reactants and products are in their standard states, then P of each reactant and product = 1.0. So $Q > K$ and to reach equilibrium the reaction will have to shift to the left.

73. An examination of the data shows when $P_A = 1.0$ then $P_B = 1.0$, therefore, $K_p = \frac{P_B^b}{P_A^a} = \frac{(1.0)^b}{(1.0)^a} = 1.0$.

Therefore, the value of the numerator and denominator must be equal. We see from the data that $P_B = \sqrt{P_A}$, so $P_B^2 = P_A$. Since, the stoichiometric coefficients become exponents in the equilibrium expression, $a = 1$ and $b = 2$.

Chapter 15
Acids and Bases

1. a) acid $HNO_3(aq) \rightarrow H^+(aq) + NO_3^-(aq)$

   b) acid $NH_4^+(aq) \rightarrow H^+(aq) + NH_3(aq)$

   c) base $KOH(aq) \rightarrow K^+(aq) + OH^-(aq)$

   d) acid $HC_2H_3O_2(aq) \rightarrow H^+(aq) + C_2H_3O_2^-(aq)$

2. a) base $NaOH(aq) \rightarrow Na^+(aq) + OH^-(aq)$

   b) acid $H_2SO_4(aq) \rightarrow 2H^+(aq) + SO_4^{2-}(aq)$

   c) acid $HBr(aq) \rightarrow H^+(aq) + Br^-(aq)$

   d) base $Sr(OH)_2 \rightarrow Sr^{2+}(aq) + 2OH^-(aq)$

3. a) Since H_2CO_3 donates a proton to H_2O, it is the acid. After H_2CO_3 donates the proton, it becomes HCO_3^-, the conjugate base. Since H_2O accepts a proton, it is the base. After H_2O accepts the proton, it becomes H_3O^+, the conjugate acid.

   b) Since H_2O donates a proton to NH_3, it is the acid. After H_2O donates the proton, it becomes OH^-, the conjugate base. Since NH_3 accepts a proton, it is the base. After NH_3 accepts the proton, it becomes NH_4^+, the conjugate acid.

   c) Since HNO_3 donates a proton to H_2O, it is the acid. After HNO_3 donates the proton, it becomes NO_3^-, the conjugate base. Since H_2O accepts a proton, it is the base. After H_2O accepts the proton, it becomes H_3O^+, the conjugate acid.

   d) Since H_2O donates a proton to C_5H_5N, it is the acid. After H_2O donates the proton, it becomes OH^-, the conjugate base. Since C_5H_5N accepts a proton, it is the base. After C_5H_5N accepts the proton, it becomes $C_5H_5NH^+$, the conjugate acid.

4. a) Since HI donates a proton to H_2O, it is the acid. After HI donates the proton, it becomes I^-, the conjugate base. Since H_2O accepts a proton, it is the base. After H_2O accepts the proton, it becomes H_3O^+, the conjugate acid.

   b) Since H_2O donates a proton to CH_3NH_2, it is the acid. After H_2O donates the proton, it becomes OH^-, the conjugate base. Since CH_3NH_2 accepts a proton, it is the base. After CH_3NH_2 accepts the proton, it becomes $CH_3NH_3^+$, the conjugate acid.

   c) Since H_2O donates a proton to CO_3^{2-}, it is the acid. After H_2O donates the proton, it becomes OH^-, the conjugate base. Since CO_3^{2-} accepts a proton, it is the base. After CO_3^{2-} accepts the proton, it becomes HCO_3^-, the conjugate acid.

   d) Since HBr donates a proton to H_2O, it is the acid. After HBr donates the proton, it becomes Br^-, the conjugate base. Since H_2O accepts a proton, it is the base. After H_2O accepts the proton, it becomes H_3O^+, the conjugate acid.

5. a) Cl^- $HCl(aq) + H_2O(l) \rightarrow H_3O^+(aq) + Cl^-(aq)$

   b) HSO_3^- $H_2SO_3(aq) + H_2O(l) \leftrightarrows H_3O^+(aq) + HSO_3^-(aq)$

   c) CHO_2^- $HCHO_2(aq) + H_2O(l) \leftrightarrows H_3O^+(aq) + CHO_2^-(aq)$

   d) F^- $HF(aq) + H_2O(l) \leftrightarrows H_3O^+(aq) + F^-(aq)$

6. a) NH_4^+ $NH_3(aq) + H_2O(l) \leftrightarrows NH_4^+(aq) + OH^-(aq)$

b) $HClO_4$ $HClO_4(aq) + H_2O(l) \rightarrow H_3O^+(aq) + ClO_4^-(aq)$

c) H_2SO_4 $H_2SO_4(aq) + H_2O(l) \rightarrow H_3O^+(aq) + HSO_4^-(aq)$

d) HCO_3^- $HCO_3^-(aq) + H_2O(l) \leftrightarrows H_2CO_3(aq) + OH^-(aq)$

7. $H_2PO_4^- + H_2O(l) \leftrightarrows H_3O^+(aq) + HPO_4^{2-}(aq)$
$H_2PO_4^- + H_2O(l) \leftrightarrows H_3PO_4(aq) + OH^-(aq)$

8. $HCO_3^-(aq) + H_2O(l) \leftrightarrows H_3O^+(aq) + CO_3^{2-}(aq)$
$HCO_3^-(aq) + H_2O(l) \leftrightarrows H_2CO_3(aq) + OH^-(aq)$

9. a) HNO_3 is a strong acid

b) HCl is a strong acid

c) HBr is a strong acid

d) H_2SO_3 is a weak acid $H_2SO_3(aq) + H_2O(l) \leftrightarrows H_3O^+(aq) + HSO_3^-(aq)$

$$K_a = \frac{[H_3O^+][HSO_3^-]}{[H_2SO_3]}$$

10. a) HF is a weak acid $HF(aq) + H_2O(l) \leftrightarrows H_3O^+(aq) + F^-(aq)$

$$K_a = \frac{[H_3O^+][F^-]}{[HF]}$$

b) $HCHO_2$ is a weak acid $HCHO_2(aq) + H_2O(l) \leftrightarrows H_3O^+(aq) + CHO_2^-(aq)$

$$K_a = \frac{[H_3O^+][CHO_2^-]}{[HCHO_2]}$$

c) H_2SO_4 is a strong acid

d) H_2CO_3 is a weak acid $H_2CO_3(aq) + H_2O(l) \leftrightarrows H_3O^+(aq) + HCO_3^-(aq)$

$$K_{a_1} = \frac{[H_3O^+][HCO_3^-]}{[H_2CO_3]}$$

11. a) contains no HA, 10 H^+, and 10 A^-

b) contains 3 HA, 3 H^+, and 7 A^-

c) contains 9 HA, 1 H^+, and 1 A^-

So, solution a > solution b > solution c

12. HCl is a strong acid, $K_a(HF) = 3.5 \times 10^{-4}$, $K_a(HClO) = 2.9 \times 10^{-8}$, $K_a(HC_6H_5O) = 1.3 \times 10^{-10}$
The larger the value of K_a the stronger the acid and the greater the $[H_3O^+]$.
The order of decreasing $[H_3O^+]$ is HCl > HF > HClO > HC_6H_5O

13. a) F^- is a stronger base than Cl^-.
F^- is the conjugate base of HF (a weak acid), Cl^- is the conjugate base of HCl (a strong acid), the weaker the acid, the stronger the conjugate base.

b) NO_2^- is a stronger base than NO_3^-.
NO_2^- is the conjugate base of HNO_2 (a weak acid), NO_3^- is the conjugate base of HNO_3 (a strong acid), the weaker the acid, the stronger the conjugate base.

c) ClO^- is a stronger base than F^-.
F^- is the conjugate base of HF ($K_a = 3.5 \times 10^{-4}$), ClO^- is the conjugate base of HClO ($K_a = 2.9 \times 10^{-8}$)
HClO is the weaker acid, the weaker the acid, the stronger the conjugate base.

14. a) ClO_2^- is a stronger base than ClO_4^-.
ClO_2^- is the conjugate base of $HClO_2$ (a weak acid), ClO_4^- is the conjugate base of $HClO_4$(a strong acid), the weaker the acid, the stronger the conjugate base.

b) H_2O is a stronger base than Cl^-.
H_2O is the conjugate base of H_3O^+, Cl^- is the conjugate base of HCl (a strong acid), the weaker the acid, the stronger the conjugate base.

c) CN^- is stronger base than ClO^-.
CN^- is the conjugate base of HCN ($K_a = 4.9 \times 10^{-10}$), ClO^- is the conjugate base of HClO ($K_a = 2.9 \times 10^{-8}$), the weaker the acid, the stronger the conjugate base.

15. a) **Given:** $K_w = 1.0 \times 10^{-14}$, $[H_3O^+] = 9.7 \times 10^{-9}$ M **Find:** $[OH^-]$
Conceptual Plan: $[H_3O^+] \rightarrow [OH^-]$

$K_w = 1 \times 10^{-14} = [H_3O^+][OH^-]$

Solution:

$K_w = 1 \times 10^{-14} = (9.7 \times 10^{-9})[OH^-]$

$[OH^-] = 1.0 \times 10^{-6}$M

$[OH^-] > [H_3O^+]$ so the solution is basic

b) **Given:** $K_w = 1.0 \times 10^{-14}$, $[H_3O^+] = 2.2 \times 10^{-6}$ M **Find:** $[OH^-]$
Conceptual Plan: $[H_3O^+] \rightarrow [OH^-]$

$K_w = 1 \times 10^{-14} = [H_3O^+][OH^-]$

Solution:

$K_w = 1 \times 10^{-14} = (2.2 \times 10^{-6})[OH^-]$

$[OH^-] = 4.5 \times 10^{-9}$M

$[H_3O^+] > [OH^-]$ so the solution is acidic

c) **Given:** $K_w = 1.0 \times 10^{-14}$, $[H_3O^+] = 1.2 \times 10^{-9}$ M **Find:** $[OH^-]$
Conceptual Plan: $[H_3O^+] \rightarrow [OH^-]$

$K_w = 1 \times 10^{-14} = [H_3O^+][OH^-]$

Solution:

$K_w = 1 \times 10^{-14} = (1.2 \times 10^{-9})[OH^-]$

$[OH^-] = 8.3 \times 10^{-6}$M

$[OH^-] > [H_3O^+]$ so the solution is basic

16. a) **Given:** $K_w = 1.0 \times 10^{-14}$, $[OH^-] = 5.1 \times 10^{-4}$ M **Find:** $[H_3O^+]$
Conceptual Plan: $[OH^-] \rightarrow [H_3O^+]$

$K_w = 1 \times 10^{-14} = [H_3O^+][OH^-]$

Solution:

$K_w = 1 \times 10^{-14} = [H_3O^+](5.1 \times 10^{-4})$

$[H_3O^+] = 2.0 \times 10^{-11}$M

$[OH^-] > [H_3O^+]$ so the solution is basic

b) **Given:** $K_w = 1.0 \times 10^{-14}$, $[OH^-] = 1.7 \times 10^{-12}$ M **Find:** $[H_3O^+]$
Conceptual Plan: $[OH^-] \rightarrow [H_3O^+]$

$K_w = 1 \times 10^{-14} = [H_3O^+][OH^-]$

Solution:

$K_w = 1 \times 10^{-14} = [H_3O^+](1.7 \times 10^{-12})$

$[H_3O^+] = 5.9 \times 10^{-3}$M

$[H_3O^+] > [OH^-]$ so the solution is acidic

c) **Given:** $K_w = 1.0 \times 10^{-14}$, $[OH^-] = 2.8 \times 10^{-2}$ M **Find:** $[H_3O^+]$
Conceptual Plan: $[OH^-] \rightarrow [H_3O^+]$

$K_w = 1 \times 10^{-14} = [H_3O^+][OH^-]$

Solution:

$K_w = 1 \times 10^{-14} = [H_3O^+](2.8 \times 10^{-2})$

$[H_3O^+] = 3.6 \times 10^{-13}$M

$[OH^-] > [H_3O^+]$ so the solution is basic

17. a) **Given:** $[H_3O^+] = 1.7 \times 10^{-8}$ M **Find:** pH and pOH
Conceptual Plan: $[H_3O^+] \rightarrow$ pH $\rightarrow$ pOH

pH = -log$[H_3O^+]$ pH + pOH = 14

Solution: pH = -log(1.7×10^{-8}) = 7.77 pOH = 14.00 - 7.77 = 6.23

pH > 7 so the solution is basic.

b) **Given:** $[H_3O^+] = 1. \times 10^{-7}$ M **Find:** pH and pOH
Conceptual Plan: $[H_3O^+] \rightarrow$ pH $\rightarrow$ pOH

pH = -log$[H_3O^+]$ pH + pOH = 14

Solution: pH = -log(1.0×10^{-7}) = 7.00 pOH = 14.00 - 7.00 = 7.00

pH = 7 so the solution is neutral.

c) **Given:** $[H_3O^+] = 2.2 \times 10^{-6}$ M **Find:** pH and pOH
Conceptual Plan: $[H_3O^+] \rightarrow$ pH $\rightarrow$ pOH

pH = -log$[H_3O^+]$ pH + pOH = 14

Solution: pH = -log(2.2×10^{-6}) = 5.66 pOH = 14.00 - 5.66 = 8.34

pH < 7 so the solution is acidic.

18. a) **Given:** pH = 8.55 **Find:** $[H_3O^+]$, $[OH^-]$
Conceptual Plan: pH $\rightarrow [H_3O^+] \rightarrow [OH^-]$

pH = -log$[H_3O^+]$ *XXX* $K_w = 1 \times 10^{-14} = [H_3O^+][OH^-]$

Solution: pH = -log$[H_3O^+]$ 8.55 = -log$[H_3O^+]$

$-8.55 = \log[H_3O^+]$ $10^{-8.55} = 10^{\log[H_3O^+]}$

$10^{-8.55} = [H_3O^+]$ $[H_3O^+] = 2.8 \times 10^{-9}$

$K_w = 1 \times 10^{-14} = (2.8 \times 10^{-9})[OH^-]$

$[OH^-] = 3.6 \times 10^{-6}$M

b) **Given:** pH = 11.23 **Find:** $[H_3O^+]$, $[OH^-]$
Conceptual Plan: pH $\rightarrow [H_3O^+] \rightarrow [OH^-]$

pH = -log$[H_3O^+]$ *XXX* $K_w = 1 \times 10^{-14} = [H_3O^+][OH^-]$

Solution: pH = -log$[H_3O^+]$ 11.23 = -log$[H_3O^+]$

$-11.23 = \log[H_3O^+]$ $10^{-11.23} = 10^{\log[H_3O^+]}$

$10^{-11.23} = [H_3O^+]$ $[H_3O^+] = 5.9 \times 10^{-12}$

$K_w = 1 \times 10^{-14} = (5.9 \times 10^{-12})[OH^-]$

$[OH^-] = 1.7 \times 10^{-3}$M

c) **Given:** pH = 2.87 **Find:** $[H_3O^+]$, $[OH^-]$
Conceptual Plan: pH $\rightarrow [H_3O^+] \rightarrow [OH^-]$

pH = -log$[H_3O^+]$ *XXX* $K_w = 1 \times 10^{-14} = [H_3O^+][OH^-]$

Solution: pH = -log$[H_3O^+]$ 2.87 = -log$[H_3O^+]$

$-2.87 = \log[H_3O^+]$ $10^{-2.87} = 10^{\log[H_3O^+]}$

$10^{-2.87} = [H_3O^+]$ $[H_3O^+] = 1.3 \times 10^{-3}$

$K_w = 1 \times 10^{-14} = (1.3 \times 10^{-3})[OH^-]$

$[OH^-] = 7.4 \times 10^{-12} M$

19. $pH = -\log[H_3O^+]$ $K_w = 1 \times 10^{-14} = [H_3O^+][OH^-]$

$[H_3O^+]$	$[OH^-]$	pH	Acidic or basic
7.1×10^{-4}	1.4×10^{-11}	**3.15**	acidic
3.7×10^{-9}	2.7×10^{-6}	8.43	basic
8×10^{-12}	1×10^{-3}	**11.1**	basic
6.2×10^{-4}	**1.6×10^{-11}**	3.20	acidic

$[H_3O^+] = 10^{-3.15} = 7.1 \times 10^{-4}$ $\quad [OH^-] = \dfrac{1 \times 10^{-14}}{7.1 \times 10^{-4}} = 1.4 \times 10^{-11}$

$[OH^-] = \dfrac{1 \times 10^{-14}}{3.7 \times 10^{-9}} = 2.7 \times 10^{-6}$ $\quad pH = -\log(3.7 \times 10^{-9}) = 8.43$

$[H_3O^+] = 10^{-11.1} = 8 \times 10^{-12}$ $\quad [OH^-] = \dfrac{1 \times 10^{-14}}{8 \times 10^{-12}} = 1 \times 10^{-3}$

$[H_3O^+] = \dfrac{1 \times 10^{-14}}{1.6 \times 10^{-11}} = 6.2 \times 10^{-4}$ $\quad pH = -\log(6.2 \times 10^{-4}) = 3.20$

20. $pH = -\log[H_3O^+]$ $K_w = 1 \times 10^{-14} = [H_3O^+][OH^-]$

$[H_3O^+]$	$[OH^-]$	pH	Acidic or basic
3.5×10^{-3}	2.9×10^{-12}	2.46	acidic
2.6×10^{-8}	**3.8×10^{-7}**	7.58	basic
1.8×10^{-9}	5.6×10^{-6}	8.74	basic
7.1×10^{-8}	1.4×10^{-7}	**7.15**	basic

$[OH^-] = \dfrac{1 \times 10^{-14}}{3.5 \times 10^{-3}} = 2.9 \times 10^{-12}$ $\quad pH = -\log(3.5 \times 10^{-3}) = 2.46$

$[H_3O^+] = \dfrac{1 \times 10^{-14}}{3.8 \times 10^{-7}} = 2.6 \times 10^{-8}$ $\quad pH = -\log(2.6 \times 10^{-8}) = 7.58$

$[OH^-] = \dfrac{1 \times 10^{-14}}{1.8 \times 10^{-9}} = 5.6 \times 10^{-6}$ $\quad pH = -\log(1.8 \times 10^{-9}) = 8.74$

$[H_3O^+] = 10^{-7.15} = 7.1 \times 10^{-8}$ $\quad [OH^-] = \dfrac{1 \times 10^{-14}}{7.1 \times 10^{-8}} = 1.4 \times 10^{-7}$

21. **Given:** $K_w = 2.4 \times 10^{-14}$ at 37°C **Find:** $[H_3O^+]$, pH

Conceptual Plan: $K_w \rightarrow [H_3O^+] \rightarrow pH$

$K_w = [H_3O^+][OH^-]$ $\quad pH = -\log[H_3O^+]$

Solution: $H_2O(l) + H_2O(l) \leftrightarrows H_3O^+(aq) + OH^-(aq)$

$K_w = [H_3O^+][OH^-]$

$[H_3O^+] = [OH^-] = \sqrt{K_w} = \sqrt{2.4 \times 10^{-14}} = 1.5 \times 10^{-7}$

$pH = -\log[H_3O^+] = -\log(1.5 \times 10^{-7}) = 6.81$

Check: The value of K_w increased indicating more products formed, so the $[H_3O^+]$ increases and the pH decreases from the values at 25°C

22. The increasing value of K_w indicates more products are formed as the temperature increases. According to Le Chatelier, this means the heat is a reactant. Therefore, the autoionization of water is endothermic.

23. a) **Given:** 0.15M HCl (strong acid) **Find:** $[H_3O^+]$, $[OH^-]$, pH

Conceptual Plan: $[HCl] \rightarrow [H_3O^+] \rightarrow$ **pH and then** $[H_3O^+] \rightarrow [OH^-]$

$[HCl] \rightarrow [H_3O^+]$ $\quad pH = -\log[H_3O^+]$ $\quad [H_3O^+][OH^-] = 1 \times 10^{-14}$

Solution: $0.15 \text{ M HCl} = 0.15 \text{ M } H_3O^+$ $\quad pH = -\log(0.15) = 0.82$

$[OH^-] = 1 \times 10^{-14} / 0.15 \text{ M} = 6.67 \times 10^{-14}$

Check: HCl is a strong acid with a relatively high concentration, so we expect the pH to be low and the $[OH^-]$ to be small.

b) **Given:** 0.025M HNO_3 (strong acid) **Find:** $[H_3O^+]$,$[OH^-]$, pH
Conceptual Plan: $[HNO_3] \rightarrow [H_3O^+] \rightarrow$ **pH and then** $[H_3O^+] \rightarrow [OH^-]$
$[HNO_3] \rightarrow [H_3O^+]$ pH = -log$[H_3O^+]$ $[H_3O^+][OH^-] = 1 \times 10^{-14}$
Solution: 0.025 M HNO_3 = 0.025 M H_3O^+ pH = –log(0.025) = 1.60
$[OH^-] = 1.0 \times 10^{-14} / 0.025\ M = 4.0 \times 10^{-13}$
Check: HNO_3 is a strong acid, so we expect the pH to be low and the $[OH^-]$ to be small.

c) **Given:** 0.072M HBr and 0.015M HNO_3 (strong acids) **Find:** $[H_3O^+]$,$[OH^-]$, pH
Conceptual Plan: $[HBr] + [HNO_3] \rightarrow [H_3O^+] \rightarrow$ **pH and then** $[H_3O^+] \rightarrow [OH^-]$
$[HBr] + [HNO_3] \rightarrow [H_3O^+]$ pH = -log$[H_3O^+]$ $[H_3O^+][OH^-] = 1 \times 10^{-14}$
Solution: 0.072 M HBr = 0.072 M H_3O^+ and 0.015M HNO_3 = 0.015M H_3O^+
Total H_3O^+ = 0.072M + 0.015M = 0.087M pH = –log(0.087) = 1.06
$[OH^-] = 1.0 \times 10^{-14} / 0.087\ M = 1.1 \times 10^{-13}$
Check: HBr and HNO_3 are both strong acids and completely dissociate, this gives a relatively high concentration, so we expect the pH to be low and the $[OH^-]$ to be small.

d) **Given:** HNO_3 = 0.855% by mass, $d_{solution}$ = 1.01 g/mL **Find:** $[H_3O^+]$,$[OH^-]$, pH
Conceptual Plan:
% mass $HNO_3 \rightarrow$ g $HNO_3 \rightarrow$ mol HNO_3 and then g soln $\rightarrow$ mL soln $\rightarrow$ L soln $\rightarrow$ M HNO_3
$\frac{\%}{100}$ $\frac{\text{mol } HNO_3}{63.018\text{ g } HNO_3}$ $\frac{1.01\text{ g soln}}{\text{mL soln}}$ $\frac{1000\text{ mL soln}}{\text{L soln}}$ $\frac{\text{mol } HNO_3}{\text{L soln}}$
$\rightarrow$ **M $H_3O^+ \rightarrow$ pH and then $[H_3O^+] \rightarrow [OH^-]$**
$[HNO_3] \rightarrow [H_3O^+]$ pH = -log$[H_3O^+]$ $[H_3O^+][OH^-] = 1 \times 10^{-14}$

Solution: $\frac{0.855\text{ g } HNO_3}{100\text{ g soln}} \times \frac{1\text{ mol } HNO_3}{63.018\text{ g } HNO_3} \times \frac{1.01\text{ g soln}}{\text{mL soln}} \times \frac{1000\text{ mL soln}}{\text{L soln}} = 0.137\text{ M } HNO_3$
0.137 M HNO_3 = 0.137 M H_3O^+ pH = –log(0.137) = 0.863
$[OH^-] = 1.00 \times 10^{-14} / 0.137\ M = 7.30 \times 10^{-14}$
Check: HNO_3 is a strong acid and completely dissociates, this gives a relatively high concentration, so we expect the pH to be low and the $[OH^-]$ to be small.

24. a) **Given:** 0.028M HI (strong acid) **Find:** $[H_3O^+]$,$[OH^-]$, pH
Conceptual Plan: $[HI] \rightarrow [H_3O^+] \rightarrow$ **pH and then** $[H_3O^+] \rightarrow [OH^-]$
$[HI] \rightarrow [H_3O^+]$ pH = -log$[H_3O^+]$ $[H_3O^+][OH^-] = 1 \times 10^{-14}$
Solution: 0.028 M HI = 0.028 M H_3O^+ pH = –log(0.028) = 1.55
$[OH^-] = 1.0 \times 10^{-14} / 0.028\ M = 3.6 \times 10^{-13}$
Check: HI is a strong acid with a relatively high concentration, so we expect the pH to be low and the $[OH^-]$ to be small.

b) **Given:** 0.115M $HClO_4$ (strong acid) **Find:** $[H_3O^+]$,$[OH^-]$, pH
Conceptual Plan: $[HClO_4] \rightarrow [H_3O^+] \rightarrow$ **pH and then** $[H_3O^+] \rightarrow [OH^-]$
$[HClO_4] \rightarrow [H_3O^+]$ pH = -log$[H_3O^+]$ $[H_3O^+][OH^-] = 1 \times 10^{-14}$
Solution: 0.115 M $HClO_4$ = 0.115 M H_3O^+ pH = –log(0.115) = 0.939
$[OH^-] = 1.0 \times 10^{-14} / 0.115\ M = 8.7 \times 10^{-14}$
Check: $HClO_4$ is a strong acid, so we expect the pH to be low and the $[OH^-]$ to be small.

c) **Given:** 0.055M $HClO_4$ and 0.028M HCl (strong acids) **Find:** $[H_3O^+]$,$[OH^-]$, pH
Conceptual Plan: $[HCl] + [HClO_4] \rightarrow [H_3O^+] \rightarrow$ **pH and then** $[H_3O^+] \rightarrow [OH^-]$
$[HCl] + [HClO_4] \rightarrow [H_3O^+]$ pH = -log$[H_3O^+]$ $[H_3O^+][OH^-] = 1 \times 10^{-14}$
Solution: 0.055 M $HClO_4$ = 0.055 M H_3O^+ and 0.028M HNO_3 = 0.028M H_3O^+
Total H_3O^+ = 0.055M + 0.028M = 0.083M pH = –log(0.083) = 1.08
$[OH^-] = 1.0 \times 10^{-14} / 0.083\ M = 1.2 \times 10^{-13}$
Check: $HClO_4$ and HNO_3 are both strong acids and completely dissociate, this gives a relatively high concentration, so we expect the pH to be low and the $[OH^-]$ to be small.

d) **Given:** HCl = 1.85% by mass, $d_{solution}$ = 1.01 g/mL **Find:** $[H_3O^+]$,$[OH^-]$, pH
Conceptual Plan:
% mass HCl $\rightarrow$ g HCl $\rightarrow$ mol HCl and then g soln $\rightarrow$ mL soln $\rightarrow$ L soln $\rightarrow$ M HCl
$\frac{\%}{100}$ $\frac{\text{mol HCl}}{35.458\text{ g HCl}}$ $\frac{1.01\text{ g soln}}{\text{mL soln}}$ $\frac{1000\text{ mL soln}}{\text{L soln}}$ $\frac{\text{mol HCl}}{\text{L soln}}$

→M H_3O^+ → pH and then $[H_3O^+]$ → $[OH^-]$

$[HCl] \rightarrow [H_3O^+]$ $pH = -\log[H_3O^+]$ $[H_3O^+][OH^-] = 1 \times 10^{-14}$

Solution: $\frac{1.85 \text{ g HCl}}{100 \text{ g soln}} \times \frac{1 \text{ mol HCl}}{35.458 \text{ g HCl}} \times \frac{1.01 \text{ g soln}}{\text{mL soln}} \times \frac{1000 \text{ mL soln}}{\text{L soln}} = 0.527 \text{ M HCl}$

$0.527 \text{ M HCl} = 0.527 \text{ M } H_3O^+$ $pH = -\log(0.137) = 0.278$

$[OH^-] = 1.0 \times 10^{-14} / 0.527 \text{ M} = 1.9 \times 10^{-14}$

Check: HCl is a strong acid and completely dissociates, this gives a relatively high concentration, so we expect the pH to be low and the $[OH^-]$ to be small.

25. a) **Given:** pH = 1.25, 0.250 L **Find:** g HI

Conceptual Plan: pH → $[H_3O^+]$ → [HI] → mol HI → g HI

$pH = -\log[H_3O^+]$ $[H_3O^+] \rightarrow [HI]$ mol = MV g = mol(127.9 g/mol)

Solution: $[H_3O^+] = 10^{-1.25} = 0.056\text{M} = [HI]$ $\frac{0.056 \text{ mol HI}}{\text{L}} \times 0.250 \text{ L} \times \frac{127.90 \text{ g HI}}{\text{mol HI}} = 1.8 \text{ g HI}$

b) **Given:** pH = 1.75, 0.250 L **Find:** g HI

Conceptual Plan: pH → $[H_3O^+]$ → [HI] → mol HI → g HI

$pH = -\log[H_3O^+]$ $[H_3O^+] \rightarrow [HI]$ mol = MV g = mol(127.9 g/mol)

Solution $[H_3O^+] = 10^{-1.75} = 0.01\underline{7}8\text{M} = [HI]$ $\frac{0.01\underline{7}8 \text{ mol HI}}{\text{L}} \times 0.250 \text{ L} \times \frac{127.90 \text{ g HI}}{\text{mol HI}} = 0.57 \text{ g HI}$

c) **Given:** pH = 2.85, 0.250 L **Find:** g HI

Conceptual Plan: pH → $[H_3O^+]$ → [HI] → mol HI → g HI

$pH = -\log[H_3O^+]$ $[H_3O^+] \rightarrow [HI]$ mol = MV g = mol(127.9 g/mol)

Solution $[H_3O^+] = 10^{-2.85} = 0.0014\text{M} = [HI]$ $\frac{0.0014 \text{ mol HI}}{\text{L}} \times 0.250 \text{ L} \times \frac{127.90 \text{ g HI}}{\text{mol HI}} = 0.045 \text{ g HI}$

26. a) **Given:** pH = 2.50, 0.500 L **Find:** g $HClO_4$

Conceptual Plan: pH → $[H_3O^+]$ → $[HClO_4]$ → mol $HClO_4$ → g $HClO_4$

$pH = -\log[H_3O^+]$ $[H_3O^+] \rightarrow [HClO_4]$ mol = MV g = mol(100.46 g/mol)

Solution:

$[H_3O^+] = 10^{-2.50} = 0.003\underline{1}6\text{M} = [HClO_4]$

$\frac{0.003\underline{1}6 \text{ mol } HClO_4}{\text{L}} \times 0.500 \text{ L} \times \frac{100.46 \text{ g } HClO_4}{\text{mol } HClO_4} = 0.16 \text{ g } HClO_4$

b) **Given:** pH = 1.50, 0.500 L **Find:** g $HClO_4$

Conceptual Plan: pH → $[H_3O^+]$ → $[HClO_4]$ → mol $HClO_4$ → g $HClO_4$

$pH = -\log[H_3O^+]$ $[H_3O^+] \rightarrow [HClO_4]$ mol = MV g = mol(100.46 g/mol)

Solution:

$[H_3O^+] = 10^{-1.50} = 0.03\underline{1}6\text{M} = [HClO_4]$

$\frac{0.03\underline{1}6 \text{ mol } HClO_4}{\text{L}} \times 0.500 \text{ L} \times \frac{100.46 \text{ g } HClO_4}{\text{mol } HClO_4} = 1.6 \text{ g } HClO_4$

c) **Given:** pH = 0.50, 0.500 L **Find:** g $HClO_4$

Conceptual Plan: pH → $[H_3O^+]$ → $[HClO_4]$ → mol $HClO_4$ → g $HClO_4$

$pH = -\log[H_3O^+]$ $[H_3O^+] \rightarrow [HClO_4]$ mol = MV g = mol(100.46 g/mol)

Solution:

$[H_3O^+] = 10^{-0.50} = 0.3\underline{1}6\text{M} = [HClO_4]$

$\frac{0.3\underline{1}6 \text{ mol } HClO_4}{\text{L}} \times 0.500 \text{ L} \times \frac{100.46 \text{ g } HClO_4}{\text{mol } HClO_4} = 16 \text{ g } HClO_4$

27. **Given:** 0.100M benzoic acid. $K_a = 6.5 \times 10^{-5}$ **Find:** $[H_3O^+]$, pH

Conceptual Plan: Write a balanced reaction. Prepare an ICE table; represent the change with x; sum the table and determine the equilibrium values; put the equilibrium values in the equilibrium expression and solve for x. Determine $[H_3O^+]$ and pH

Solution: $HC_7H_5O_2(aq) + H_2O(l) \leftrightarrows H_3O^+(aq) + C_7H_5O_2^-(aq)$

I	0.100M	0.0	0.0
C	- x	x	x
E	0.100 – x	x	x

$$K_a = \frac{[H_3O^+][C_7H_5O_2^-]}{[HC_7H_5O_2]} = \frac{(x)(x)}{(0.100 - x)} = 6.5 \times 10^{-5}$$

Assume x is small compared to 0.100

$x^2 = (6.5 \times 10^{-5})(0.100)$ $\quad$ $x = 2.5 \times 10^{-3}\ M = [H_3O^+]$

check assumption: $\frac{2.5 \times 10^{-3}}{0.100} \times 100\% = 2.5\%$; assumption valid

$pH = -\log(2.5 \times 10^{-3}) = 2.60$

28. **Given:** 0.200M formic acid. $K_a = 1.8 \times 10^{-4}$ **Find:** $[H_3O^+]$, pH
Conceptual Plan: Write a balanced reaction. Prepare an ICE table; represent the change with x; sum the table and determine the equilibrium values; put the equilibrium values in the equilibrium expression and solve for x. Determine $[H_3O^+]$ and pH
Solution: $HCH_2O(aq) + H_2O(l) \leftrightarrows H_3O^+(aq) + CH_2O^-(aq)$

I	0.200M	0.0	0.0
C	- x	x	x
E	0.200 – x	x	x

$$K_a = \frac{[H_3O^+][CH_2O^-]}{[HCH_2O]} = \frac{(x)(x)}{(0.200 - x)} = 1.8 \times 10^{-4}$$

Assume x is small compared to 0.200

$x^2 = (1.8 \times 10^{-4})(0.200)$ $\quad$ $x = 6.0 \times 10^{-3}\ M = [H_3O^+]$

check assumption: $\frac{6.0 \times 10^{-3}}{0.200} \times 100\% = 3.0\%$, assumption valid

$pH = -\log(6.0 \times 10^{-3}) = 2.22$

29. a) **Given:** 0.500M HNO_2. $K_a = 4.6 \times 10^{-4}$ **Find:** pH
Conceptual Plan: Write a balanced reaction. Prepare an ICE table; represent the change with x; sum the table and determine the equilibrium values; put the equilibrium values in the equilibrium expression and solve for x. Determine $[H_3O^+]$ and pH
Solution: $HNO_2(aq) + H_2O(l) \leftrightarrows H_3O^+(aq) + NO_2^-(aq)$

I	0.500M	0.0	0.0
C	- x	x	x
E	0.500 – x	x	x

$$K_a = \frac{[H_3O^+][NO_2^-]}{[HNO_2]} = \frac{(x)(x)}{(0.500 - x)} = 4.6 \times 10^{-4}$$

Assume x is small compared to 0.500

$x^2 = (4.6 \times 10^{-4})(0.500)$ $\quad$ $x = 0.015M = [H_3O^+]$

Check assumption: $\frac{0.015}{0.500} \times 100\% = 3.0\%$ assumption valid

$pH = -\log(0.015) = 1.82$

b) **Given:** 0.100M HNO_2. $K_a = 4.6 \times 10^{-4}$ **Find:** pH
Conceptual Plan: Write a balanced reaction. Prepare an ICE table; represent the change with x; sum the table and determine the equilibrium values; put the equilibrium values in the equilibrium expression and solve for x. Determine $[H_3O^+]$ and pH
Solution: $HNO_2(aq) + H_2O(l) \leftrightarrows H_3O^+(aq) + NO_2^-(aq)$

I	0.100M	0.0	0.0
C	- x	x	x
E	0.100 – x	x	x

$$K_a = \frac{[H_3O^+][NO_2^-]}{[HNO_2]} = \frac{(x)(x)}{(0.100 - x)} = 4.6 \times 10^{-4}$$

Assume x is small compared to 0.100

$x^2 = (4.6 \times 10^{-4})(0.100)$ $\quad x = 0.0068M = [H_3O^+]$

Check assumption: $\frac{0.0068}{0.100} \times 100\% = 6.8\%$ assumption not valid

$x^2 = (4.6 \times 10^{-4})(0.100 - x)$ $\quad x^2 + 4.6 \times 10^{-4}x - 4.6 \times 10^{-5} = 0$

$x = 0.00656$

$pH = -\log(0.00656) = 2.18$

c) **Given:** 0.100M HNO_2. $K_a = 4.6 \times 10^{-4}$ **Find:** pH
Conceptual Plan: Write a balanced reaction. Prepare an ICE table; represent the change with x; sum the table and determine the equilibrium values; put the equilibrium values in the equilibrium expression and solve for x. Determine $[H_3O^+]$ and pH
Solution: $HNO_2(aq) + H_2O(l) \leftrightarrows H_3O^+(aq) + NO_2^-(aq)$

	HNO_2	H_3O^+	NO_2^-
I	0.0100M	0.0	0.0
C	- x	x	x
E	0.0100 – x	x	x

$$K_a = \frac{[H_3O^+][NO_2^-]}{[HNO_2]} = \frac{(x)(x)}{(0.0100 - x)} = 4.6 \times 10^{-4}$$

Assume x is small compared to 0.100

$x^2 = (4.6 \times 10^{-4})(0.0100)$ $\quad x = 0.0021M = [H_3O^+]$

Check assumption: $\frac{0.0021}{0.0100} \times 100\% = 21\%$ assumption not valid

$x^2 = (4.6 \times 10^{-4})(0.0100 - x)$ $\quad x^2 + 4.6 \times 10^{-4}x - 4.6 \times 10^{-6} = 0$

$x = 0.0019$

$pH = -\log(0.0019) = 2.72$

30. a) **Given:** 0.250 M HF $K_a = 3.5 \times 10^{-4}$ **Find:** pH
Conceptual Plan: Write a balanced reaction. Prepare an ICE table; represent the change with x; sum the table and determine the equilibrium values; put the equilibrium values in the equilibrium expression and solve for x. Determine $[H_3O^+]$ and pH
Solution: $HF(aq) + H_2O(l) \leftrightarrows H_3O^+(aq) + F^-(aq)$

	HF	H_3O^+	F^-
I	0.250M	0.0	0.0
C	- x	x	x
E	0.250 – x	x	x

$$K_a = \frac{[H_3O^+][F^-]}{[HF]} = \frac{(x)(x)}{(0.250 - x)} = 3.5 \times 10^{-4}$$

Assume x is small compared to 0.250

$x^2 = (3.5 \times 10^{-4})(0.250)$ $\quad x = 0.009\underline{3}5M = [H_3O^+]$

Check assumption: $\frac{0.009\underline{3}5}{0.250} \times 100\% = 3.7\%$ assumption valid

$pH = -\log(0.009\underline{3}5) = 2.03$

b) **Given:** 0.0500 M HF $K_a = 3.5 \times 10^{-4}$ **Find:** pH
Conceptual Plan: Write a balanced reaction. Prepare an ICE table; represent the change with x; sum the table and determine the equilibrium values; put the equilibrium values in the equilibrium expression and solve for x. Determine $[H_3O^+]$ and pH
Solution: $HF(aq) + H_2O(l) \leftrightarrows H_3O^+(aq) + F^-(aq)$

	HF	H_3O^+	F^-
I	0.0500M	0.0	0.0
C	- x	x	x
E	0.0500 – x	x	x

$$K_a = \frac{[H_3O^+][F^-]}{[HF]} = \frac{(x)(x)}{(0.0500 - x)} = 3.5 \times 10^{-4}$$

Assume x is small compared to 0.0500

$x^2 = (3.5 \times 10^{-4})(0.0500)$ $\qquad x = 0.00418M = [H_3O^+]$

Check assumption: $\frac{0.00418}{0.050} \times 100\% = 8.3\%$ assumption not valid

$x^2 + 3.5 \times 10^{-4} x - 1.75 \times 10^{-5} = 0$ $\qquad x = 0.0040 = [H_3O^+]$

pH = -log(0.0040) = 2.40

c) **Given:** 0.0250 M HF $K_a = 3.5 \times 10^{-4}$ $\qquad\qquad$ **Find:** pH
    Conceptual Plan: Write a balanced reaction. Prepare an ICE table; represent the change with x; sum the table and determine the equilibrium values; put the equilibrium values in the equilibrium expression and solve for x. Determine [H$_3$O$^+$] and pH
    Solution: HF(aq) + H$_2$O(l) $\leftrightarrows$ $\quad$ H$_3$O$^+$(aq) + F$^-$(aq)
    I $\qquad$ 0.0250M $\qquad\qquad\qquad$ 0.0 $\qquad$ 0.0
    C $\qquad$ - x $\qquad\qquad\qquad\qquad$ x $\qquad$ x
    E $\qquad$ 0.0250 – x $\qquad\qquad\quad$ x $\qquad$ x

$$K_a = \frac{[H_3O^+][F^-]}{[HF]} = \frac{(x)(x)}{(0.0250 - x)} = 3.5 \times 10^{-4}$$

Assume x is small compared to 0.0250

$x^2 = (3.5 \times 10^{-4})(0.0250)$ $\qquad x = 0.00295M = [H_3O^+]$

Check assumption: $\frac{0.00295}{0.0250} \times 100\% = 11.8\%$ assumption not valid

$x^2 + 3.5 \times 10^{-4} x - 8.75 \times 10^{-6} = 0$ $\qquad x = 0.00279 = [H_3O^+]$

pH = -log(0.00279) = 2.55

31. **Given:** 15.0 mL glacial acetic, d = 1.05 g/mL, dilute to 1.50 L, $K_a = 1.8 \times 10^{-5}$ $\qquad$ **Find:** pH
    Conceptual Plan: mL acetic acid → g acetic acid → mol acetic acid → M and then write a balanced reaction. $\qquad XXX \frac{1.05 \text{ g}}{mL}$ $XXX \frac{\text{mol acetic acid}}{60.05 \text{ g}}$ $\qquad XXX$ M = $\frac{mol}{L}$

Prepare an ICE table; represent the change with x; sum the table and determine the equilibrium values; put the equilibrium values in the equilibrium expression and solve for x. Determine [H$_3$O$^+$] and pH

Solution: 15.0 mL x $\frac{1.05 \text{ g}}{mL}$ x $\frac{1 \text{ mol}}{60.05 \text{ g}}$ x $\frac{1}{1.50 \text{ L}}$ = 0.1748 M

HC$_2$H$_3$O$_2$(aq) + H$_2$O(l) $\leftrightarrows$ H$_3$O$^+$(aq) + C$_2$H$_3$O$_2^-$ (aq)
I $\qquad$ 0.1748M $\qquad\qquad\qquad$ 0.0 $\qquad$ 0.0
C $\qquad$ - x $\qquad\qquad\qquad\qquad$ x $\qquad$ x
E $\qquad$ 0.1748 – x $\qquad\qquad\quad$ x $\qquad$ x

$$K_a = \frac{[H_3O^+][\ C_2H_3O_2^-]}{[HC_2H_3O_2]} = \frac{(x)(x)}{(0.1748 - x)} = 1.8 \times 10^{-5}$$

Assume x is small compared to 0.1748

$x^2 = (1.8 \times 10^{-5})(0.1748)$ $\qquad x = 0.00177M = [H_3O^+]$

Check assumption: $\frac{0.00177}{0.1748} \times 1005 = 1.0\%$ assumption valid

pH = -log(0.00177) = 2.75

32. **Given:** 1.35% formic acid, d = 1.01 g/mL, $K_a = 1.8 \times 10^{-4}$ $\qquad\qquad$ **Find:** pH
    Conceptual Plan: % formic acid → g formic acid → mol and g soln → mL soln → L soln and then M
    $\qquad\qquad\qquad\qquad\qquad\qquad XXX \frac{1.01 \text{ g soln}}{mL \text{ soln}}$ $\quad XXX \frac{mol}{46.03 \text{ g}}$ $\qquad XXX \frac{1000 \text{ mL soln}}{L \text{ soln}}$

Write a balanced reaction. Prepare an ICE table; represent the change with x; sum the table and determine the equilibrium values; put the equilibrium values in the equilibrium expression and solve for x. Determine [H$_3$O$^+$] and pH

Solution: $\frac{1.35 \text{ g HCHO}}{100 \text{ g soln}} \times \frac{\text{mol HCHO}}{46.03 \text{ g}} \times \frac{1.01 \text{ g soln}}{\text{mL soln}} \times \frac{1000 \text{ mL soln}}{\text{L soln}} = 0.29\underline{6}2\text{M}$

	$HCHO_2(aq) + H_2O(l) \leftrightarrows$	$H_3O^+(aq) +$	$CHO_2^-(aq)$
I	0.2962M	0.0	0.0
C	- x	x	x
E	0.2962 – x	x	x

$$K_a = \frac{[H_3O^+][CHO_2^-]}{[HCHO_2]} = \frac{(x)(x)}{(0.2962 - x)} = 1.8 \times 10^{-4}$$

Assume x is small compared to 0.2962

$x^2 = (1.8 \times 10^{-4})(0.2962)$ $\quad$ $x = 0.007\underline{3}0\text{M} = [H_3O^+]$

Check assumption: $\frac{0.007\underline{3}0}{0.2962} \times 100\% = 2.5\%$ assumption valid

$pH = -\log(0.007\underline{3}0) = 2.14$

33. **Given:** 0.185 M HA, pH = 2.95 **Find:** K_a

Conceptual Plan: pH → $[H_3O^+]$ and then write a balanced reaction, prepare an ICE table, calculate equilibrium concentrations, and then plug into the equilibrium expression to solve for K_a.

Solution: $[H_3O^+] = 10^{-2.95} = 0.0012 \text{ M} = [A^-]$

	$HA(aq) + H_2O(l) \leftrightarrows$	$H_3O^+(aq) +$	$A^-(aq)$
I	0.185M	0.0	0.0
C	- x	x	x
E	0.185 – 0.0012	0.0012	0.0012

$$K_a = \frac{[H_3O^+][A^-]}{[HA]} = \frac{(0.0012)(0.0012)}{(0.185 - 0.0012)} = 7.8 \times 10^{-6}$$

34. **Given:** 0.115 M HA, pH = 3.29 **Find:** K_a

Conceptual Plan: pH → $[H_3O^+]$ and then write a balanced reaction, prepare an ICE table, calculate equilibrium concentrations, and then plug into the equilibrium expression to solve for K_a.

Solution: $[H_3O^+] = 10^{-3.29} = 5.\underline{1}3 \times 10^{-4} \text{ M} = [A^-]$

	$HA(aq) + H_2O(l) \leftrightarrows$	$H_3O^+(aq) +$	$A^-(aq)$
I	0.185M	0.0	0.0
C	- x	x	x
E	$0.115 - 5.\underline{1}3 \times 10^{-4}$	$5.\underline{1}3 \times 10^{-4}$	$5.\underline{1}3 \times 10^{-4}$

$$K_a = \frac{[H_3O^+][A^-]}{[HA]} = \frac{(5.\underline{1}3 \times 10^{-4})(5.\underline{1}3 \times 10^{-4})}{(0.115 - 5.\underline{1}3 \times 10^{-4})} = 2.3 \times 10^{-6}$$

35. **Given:** 0.125 HCN $K_a = 4.9 \times 10^{-10}$ **Find:** % ionization

Conceptual Plan: Write a balanced reaction. Prepare an ICE table; represent the change with x; sum the table and determine the equilibrium values; put the equilibrium values in the equilibrium expression and solve for x and then x → % ionization

$$\% \text{ ionization} = \frac{x}{[HCN]_{original}} \times 100$$

Solution:

	$HCN(aq) + H_2O(l) \leftrightarrows$	$H_3O^+(aq) +$	$CN^-(aq)$
I	0.125M	0.0	0.0
C	- x	x	x
E	0.125 – x	x	x

$$K_a = \frac{[H_3O^+][CN^-]}{[HCN]} = \frac{(x)(x)}{(0.125 - x)} = 4.9 \times 10^{-10}$$

Assume x is small compared to 0.125

$x^2 = (4.9 \times 10^{-10})(0.125)$ $\quad$ $x = 7.\underline{8}3 \times 10^{-6}$

$\% \text{ ionization} = \frac{7.\underline{8}3 \times 10^{-6}}{0.125} \times 100 = 0.0063\%$ ionized

36. **Given:** 0.225 $HC_7H_5O_2$ $K_a = 6.5 \times 10^{-5}$ **Find:** % ionization
Conceptual Plan: Write a balanced reaction. Prepare an ICE table; represent the change with x; sum the table and determine the equilibrium values; put the equilibrium values in the equilibrium expression and solve for x and then x → % ionization

$$\% \text{ ionization} = \frac{x}{[HC_7H_5O_2]_{original}} \times 100$$

Solution: $HC_7H_5O_2\,(aq) + H_2O(l) \leftrightarrows H_3O^+(aq) + C_7H_5O_2^-\,(aq)$

I	0.125M	0.0	0.0
C	- x	x	x
E	0.125 – x	x	x

$$K_a = \frac{[H_3O^+][C_7H_5O_2^-]}{[HC_7H_5O_2]} = \frac{(x)(x)}{(0.225 - x)} = 6.5 \times 10^{-5}$$

Assume x is small compared to 0.225

$x^2 = (6.5 \times 10^{-5})(0.1225)$ $x = 0.00382$

$$\% \text{ ionization} = \frac{0.00382}{0.225} \times 100 = 1.7\% \text{ ionized}$$

37. a) **Given:** 1.00M $HC_2H_3O_2$ $K_a = 1.8 \times 10^{-5}$ **Find:** % ionization
Conceptual Plan: Write a balanced reaction. Prepare an ICE table; represent the change with x; sum the table and determine the equilibrium values; put the equilibrium values in the equilibrium expression and solve for x and then x → % ionization

$$\% \text{ ionization} = \frac{x}{[HC_2H_3O_2]_{original}} \times 100$$

Solution: $HC_2H_3O_2\,(aq) + H_2O(l) \leftrightarrows H_3O^+(aq) + C_2H_3O_2^-\,(aq)$

I	1.0 M	0.0	0.0
C	- x	x	x
E	1.00 – x	x	x

$$K_a = \frac{[H_3O^+][C_2H_3O_2^-]}{[HC_2H_3O_2]} = \frac{(x)(x)}{(1.00 - x)} = 1.8 \times 10^{-5}$$

Assume x is small compared to1.00

$x^2 = (1.8 \times 10^{-5})(1.00)$ $x = 0.00424$

$$\% \text{ ionization} = \frac{0.00424}{1.00} \times 100 = 0.42\% \text{ ionized}$$

b) **Given:** 0.500 M $HC_2H_3O_2$ $K_a = 1.8 \times 10^{-5}$ **Find:** % ionization
Conceptual Plan: Write a balanced reaction. Prepare an ICE table; represent the change with x; sum the table and determine the equilibrium values; put the equilibrium values in the equilibrium expression and solve for x and then x → % ionization

$$\% \text{ ionization} = \frac{x}{[HC_2H_3O_2]_{original}} \times 100$$

Solution: $HC_2H_3O_2\,(aq) + H_2O(l) \leftrightarrows H_3O^+(aq) + C_2H_3O_2^-\,(aq)$

I	0.500 M	0.0	0.0
C	- x	x	x
E	0.500 – x	x	x

$$K_a = \frac{[H_3O^+][C_2H_3O_2^-]}{[HC_2H_3O_2]} = \frac{(x)(x)}{(0.500 - x)} = 1.8 \times 10^{-5}$$

Assume x is small compared to 0.500

$x^2 = (1.8 \times 10^{-5})(0.500)$ $x = 0.00300$

$$\% \text{ ionization} = \frac{0.00300}{0.500} \times 100 = 0.60\% \text{ ionized}$$

c) **Given:** 0.100M $HC_2H_3O_2$ $K_a = 1.8 \times 10^{-5}$ **Find:** % ionization

Conceptual Plan: Write a balanced reaction. Prepare an ICE table; represent the change with x; sum the table and determine the equilibrium values; put the equilibrium values in the equilibrium expression and solve for x and then x → % ionization

$$\% \text{ ionization} = \frac{x}{[HC_2H_3O_2]_{original}} \times 100$$

Solution: $HC_2H_3O_2\,(aq) + H_2O(l) \leftrightarrows H_3O^+(aq) + C_2H_3O_2^-\,(aq)$

I	0.100 M	0.0	0.0
C	- x	x	x
E	0.100 – x	x	x

$$K_a = \frac{[H_3O^+][C_2H_3O_2^-]}{[HC_2H_3O_2]} = \frac{(x)(x)}{(0.100 - x)} = 1.8 \times 10^{-5}$$

Assume x is small compared to 1.00

$x^2 = (1.8 \times 10^{-5})(0.100)$ $x = 0.00134$

$$\% \text{ ionization} = \frac{0.00134}{0.100} \times 100 = 1.3\% \text{ ionized}$$

d) **Given:** 0.0500M $HC_2H_3O_2$ $K_a = 1.8 \times 10^{-5}$ **Find:** % ionization

Conceptual Plan: Write a balanced reaction. Prepare an ICE table; represent the change with x; sum the table and determine the equilibrium values; put the equilibrium values in the equilibrium expression and solve for x and then x → % ionization

$$\% \text{ ionization} = \frac{x}{[HC_2H_3O_2]_{original}} \times 100$$

Solution: $HC_2H_3O_2\,(aq) + H_2O(l) \leftrightarrows$ $H_3O^+(aq) +$ $C_2H_3O_2^-\,(aq)$

I	0.0500 M	0.0	0.0
C	- x	x	x
E	0.0500 – x	x	x

$$K_a = \frac{[H_3O^+][C_2H_3O_2^-]}{[HC_2H_3O_2]} = \frac{(x)(x)}{(0.0500 - x)} = 1.8 \times 10^{-5}$$

Assume x is small compared to 0.0500

$x^2 = (1.8 \times 10^{-5})(0.0500)$ $x = 9.49 \times 10^{-4}$

$$\% \text{ ionization} = \frac{9.49 \times 10^{-4}}{0.0500} \times 100 = 1.9\% \text{ ionized}$$

38. a) **Given:** 1.00M $HCHO_2$ $K_a = 1.8 \times 10^{-4}$ **Find:** % ionization

Conceptual Plan: Write a balanced reaction. Prepare an ICE table; represent the change with x; sum the table and determine the equilibrium values; put the equilibrium values in the equilibrium expression and solve for x and then x → % ionization

$$\% \text{ ionization} = \frac{x}{[HCHO_2]_{original}} \times 100$$

Solution: $HCHO_2\,(aq) + H_2O(l) \leftrightarrows H_3O^+(aq) + CHO_2^-\,(aq)$

I	1.0 M	0.0	0.0
C	- x	x	x
E	1.00 – x	x	x

$$K_a = \frac{[H_3O^+][CHO_2^-]}{[HCHO_2]} = \frac{(x)(x)}{(1.00 - x)} = 1.8 \times 10^{-4}$$

Assume x is small compared to1.00

$x^2 = (1.8 \times 10^{-4})(1.00)$ $x = 0.0134$

$$\% \text{ ionization} = \frac{0.0134}{1.00} \times 100 = 1.3\% \text{ ionized}$$

b) **Given:** 0.500 M $HCHO_2$ $K_a = 1.8 \times 10^{-4}$ **Find:** % ionization
Conceptual Plan: Write a balanced reaction. Prepare an ICE table; represent the change with x; sum the table and determine the equilibrium values; put the equilibrium values in the equilibrium expression and solve for x and then x → % ionization

$$\% \text{ ionization} = \frac{x}{[HCHO_2]_{original}} \times 100$$

Solution: $HCHO_2\,(aq) + H_2O(l) \leftrightarrows H_3O^+(aq) + CHO_2^-(aq)$

I	0.500 M	0.0	0.0
C	- x	x	x
E	0.500 – x	x	x

$$K_a = \frac{[H_3O^+][CHO_2^-]}{[HCHO_2]} = \frac{(x)(x)}{(0.500 - x)} = 1.8 \times 10^{-4}$$

Assume x is small compared to 1.00

$x^2 = (1.8 \times 10^{-4})(0.500)$ $x = 0.009\underline{4}9$

$$\% \text{ ionization} = \frac{0.009\underline{4}9}{0.500} \times 100 = 1.9\% \text{ ionized}$$

c) **Given:** 0.100M $HCHO_2$ $K_a = 1.8 \times 10^{-4}$ **Find:** % ionization
Conceptual Plan: Write a balanced reaction. Prepare an ICE table; represent the change with x; sum the table and determine the equilibrium values; put the equilibrium values in the equilibrium expression and solve for x and then x → % ionization

$$\% \text{ ionization} = \frac{x}{[HCHO_2]_{original}} \times 100$$

Solution: $HCHO_2\,(aq) + H_2O(l) \leftrightarrows H_3O^+(aq) + CHO_2^-(aq)$

I	0.100 M	0.0	0.0
C	- x	x	x
E	0.100 – x	x	x

$$K_a = \frac{[H_3O^+][CHO_2^-]}{[HCHO_2]} = \frac{(x)(x)}{(0.100 - x)} = 1.8 \times 10^{-4}$$

Assume x is small compared to 0.100

$x^2 = (1.8 \times 10^{-4})(0.100)$ $x = 0.004\underline{2}4$

$$\% \text{ ionization} = \frac{0.004\underline{2}4}{0.100} \times 100 = 4.2\% \text{ ionized}$$

d) **Given:** 0.0500 M $HCHO_2$ $K_a = 1.8 \times 10^{-4}$ **Find:** % ionization
Conceptual Plan: Write a balanced reaction. Prepare an ICE table; represent the change with x; sum the table and determine the equilibrium values; put the equilibrium values in the equilibrium expression and solve for x and then x → % ionization

$$\% \text{ ionization} = \frac{x}{[HCHO_2]_{original}} \times 100$$

Solution: $HCHO_2\,(aq) + H_2O(l) \leftrightarrows H_3O^+(aq) + CHO_2^-(aq)$

I	0.0500 M	0.0	0.0
C	- x	x	x
E	0.0500 – x	x	x

$$K_a = \frac{[H_3O^+][CHO_2^-]}{[HCHO_2]} = \frac{(x)(x)}{(0.0500 - x)} = 1.8 \times 10^{-4}$$

Assume x is small compared to 0.0500

$x^2 = (1.8 \times 10^{-4})(0.0500)$ $x = 0.003\underline{0}3$

$$\% \text{ ionization} = \frac{0.003\underline{0}0}{0.0500} \times 100\% = 6.0\% \text{ ionized}$$

39. **Given:** 0.148 M HA 1.55% dissociation **Find:** K_a
Conceptual Plan: M → $[H_3O^+]$ → K_a and then write a balanced reaction, determine equilibrium concentration and plug into the equilibrium expression.

Solution: $(0.148\ M\ HA)(0.0155) = 0.002294\ [H_3O^+] = [A^-]$

	$HA(aq) + H_2O(l) \leftrightarrows$	$H_3O^+(aq) + A^-(aq)$	
I	0. 148M	0.0	0.0
C	- x	x	x
E	0.148 – 0.002294	0.002294	0.002294

$$K_a = \frac{[H_3O^+][A^-]}{[HA]} = \frac{(0.002294)(0.002294)}{(0.148 - 0.002294)} = 3.61 \times 10^{-5}$$

40. **Given:** 0.085 M HA 0.59% dissociation **Find:** K_a

Conceptual Plan: M → $[H_3O^+]$ → K_a and then write a balanced reaction, determine equilibrium concentration and plug into the equilibrium expression.

Solution: $(0.085\ M\ HA)(0.0059) = 5.02 \times 10^{-4}\ [H_3O^+] = [A^-]$

	$HA(aq) + H_2O(l) \leftrightarrows$	$H_3O^+(aq) +$	$A^-(aq)$
I	0. 085M	0.0	0.0
C	- x	x	x
E	$0.085 - 5.02 \times 10^{-4}$	5.02×10^{-4}	5.02×10^{-4}

$$K_a = \frac{[H_3O^+][A^-]}{[HA]} = \frac{(5.02 \times 10^{-4})(5.02 \times 10^{-4})}{(0.085 - 5.02 \times 10^{-4})} = 3.0 \times 10^{-6}$$

41. a) **Given:** 0.250 M HF $K_a = 3.5 \times 10^{-4}$ **Find:** pH, % dissociation

Conceptual Plan: Write a balanced reaction. Prepare an ICE table; represent the change with x; sum the table and determine the equilibrium values; put the equilibrium values in the equilibrium expression and solve for x and then x → % ionization

$$\%\ \text{ionization} = \frac{x}{[HF]_{original}} \times 100$$

Solution: $HF(aq) + H_2O(l) \leftrightarrows H_3O^+(aq) + F^-(aq)$

I	0.250M	0.0	0.0
C	- x	x	x
E	0.250 – x	x	x

$$K_a = \frac{[H_3O^+][F^-]}{[HF]} = \frac{(x)(x)}{(0.250 - x)} = 3.5 \times 10^{-4}$$

Assume x is small compared to 0.250

$x^2 = (3.5 \times 10^{-4})(0.250)$ $x = 0.00935M = [H_3O^+]$

$$\frac{0.00935}{0.250} \times 100\% = 3.7\%$$

$pH = -\log(0.00935) = 2.03$

b) **Given:** 0.100 M HF $K_a = 3.5 \times 10^{-4}$ **Find:** pH, % dissociation

Conceptual Plan: Write a balanced reaction. Prepare an ICE table; represent the change with x; sum the table and determine the equilibrium values; put the equilibrium values in the equilibrium expression and solve for x and then x → % ionization

$$\%\ \text{ionization} = \frac{x}{[HF]_{original}} \times 100$$

Solution: $HF(aq) + H_2O(l) \leftrightarrows H_3O^+(aq) + F^-(aq)$

I	0.1000M	0.0	0.0
C	- x	x	x
E	0.100 – x	x	x

$$K_a = \frac{[H_3O^+][F^-]}{[HF]} = \frac{(x)(x)}{(0.100 - x)} = 3.5 \times 10^{-4}$$

Assume x is small compared to 0.100

$x^2 = (3.5 \times 10^{-4})(0.100)$ $x = 0.00592M = [H_3O^+]$

$$\frac{0.00592}{0.100} \times 100\% = 5.9\%$$

$x > 5.0$ % therefore, assumption invalid

$x^2 + 3.5 \times 10^{-4}x - 3.5 \times 10^{-5} = 0$

$x = 0.00574$ or -0.00609

$pH = -\log(0.00574) = 2.24$

$$\% \text{ dissociation} = \frac{0.00574}{0.100} \times 100 = 5.7\%$$

c) **Given:** 0.050 M HF $K_a = 3.5 \times 10^{-4}$ **Find:** pH, % dissociation

Conceptual Plan: Write a balanced reaction. Prepare an ICE table; represent the change with x; sum the table and determine the equilibrium values; put the equilibrium values in the equilibrium expression and solve for x and then x → % ionization

$$\% \text{ ionization} = \frac{x}{[HF]_{original}} \times 100$$

Solution: $HF(aq) + H_2O(l) \leftrightarrows H_3O^+(aq) + F^-(aq)$

I	0.050M	0.0	0.0
C	- x	x	x
E	0.050 – x	x	x

$$K_a = \frac{[H_3O^+][F^-]}{[HF]} = \frac{(x)(x)}{(0.050 - x)} = 3.5 \times 10^{-4}$$

Assume x is small compared to 0.050

$x^2 = (3.5 \times 10^{-4})(0.050)$ $x = 0.00418M = [H_3O^+]$

$$\frac{0.00418}{0.050} \times 100\% = 8.4\%$$

Assumption invalid.

$x^2 + 3.5 \times 10^{-4}x - 1.75 \times 10^{-5} = 0$

$x = 0.00401$ or -0.00436

$pH = -\log(0.00401) = 2.40$

$$\% \text{ dissociation} = \frac{0.00401}{0.050} \times 100\% = 8.0\%$$

42. a) **Given:** 0.100 M HA, $K_a = 1.0 \times 10^{-5}$ **Find:** pH, % dissociation

Conceptual Plan: Write a balanced reaction. Prepare an ICE table; represent the change with x; sum the table and determine the equilibrium values; put the equilibrium values in the equilibrium expression and solve for x and then x → % ionization

$$\% \text{ ionization} = \frac{x}{[HA]_{original}} \times 100$$

Solution: $HA(aq) + H_2O(l) \leftrightarrows H_3O^+(aq) + A^-(aq)$

I	0.100M	0.0	0.0
C	- x	x	x
E	0.100 – x	x	x

$$K_a = \frac{[H_3O^+][A^-]}{[HA]} = \frac{(x)(x)}{(0.100 - x)} = 1.0 \times 10^{-5}$$

Assume x is small compared to 0.100

$x^2 = (1.0 \times 10^{-5})(0.100)$ $x = 0.00100M = [H_3O^+]$

$$\frac{0.00100}{0.100} \times 100\% = 1.0\%$$

$pH = -\log(0.00100) = 3.00$

b) **Given:** 0.100 M HA, $K_a = 1.0 \times 10^{-3}$ **Find:** pH, % dissociation
Conceptual Plan: Write a balanced reaction. Prepare an ICE table; represent the change with x; sum the table and determine the equilibrium values; put the equilibrium values in the equilibrium expression and solve for x and then x → % ionization

$$\% \text{ ionization} = \frac{x}{[HA]_{original}} \times 100$$

Solution: $HA(aq) + H_2O(l) \leftrightarrows H_3O^+(aq) + A^-(aq)$

I	0.100M	0.0	0.0
C	- x	x	x
E	0.100 – x	x	x

$$K_a = \frac{[H_3O^+][A^-]}{[HA]} = \frac{(x)(x)}{(0.100 - x)} = 1.0 \times 10^{-3}$$

$x^2 = (1.0 \times 10^{-3} - x)(0.100)$ $\qquad x^2 + 1 \times 10^{-3}x - 1 \times 10^{-4} = 0$

$x = 0.009\underline{5} = [H_3O^+]$

$$\frac{0.009\underline{5}}{0.100} \times 100\% = 9.5\%$$

$pH = -\log(0.009\underline{5}) = 2.02$

c) **Given:** 0.100 M HA, $K_a = 1.0 \times 10^{-1}$ **Find:** pH, % dissociation
Conceptual Plan: Write a balanced reaction. Prepare an ICE table; represent the change with x; sum the table and determine the equilibrium values; put the equilibrium values in the equilibrium expression and solve for x and then x → % ionization

$$\% \text{ ionization} = \frac{x}{[HA]_{original}} \times 100$$

Solution: $HA(aq) + H_2O(l) \leftrightarrows H_3O^+(aq) + A^-(aq)$

I	0.100M	0.0	0.0
C	- x	x	x
E	0.100 – x	x	x

$$K_a = \frac{[H_3O^+][A^-]}{[HA]} = \frac{(x)(x)}{(0.100 - x)} = 1.0 \times 10^{-1}$$

$x^2 = (1.0 \times 10^{-1} - x)(0.100)$ $\qquad x^2 + 0.10x - 0.01 = 0$

$x = 0.06\underline{1}6 = [H_3O^+]$

$$\frac{0.06\underline{1}6}{0.100} \times 100\% = 61.8\%$$

$pH = -\log(0.06\underline{1}6) = 1.21$

43. $H_3PO_4(aq) + H_2O(l) \leftrightarrows H_3O^+(aq) + H_2PO_4^-(aq)$ $\qquad K_{a1} = \frac{[H_3O^+][H_2PO_4^-]}{[H_3PO_4]}$

$H_2PO_4^-(aq) + H_2O(l) \leftrightarrows H_3O^+(aq) + HPO_4^{2-}(aq)$ $\qquad K_{a2} = \frac{[H_3O^+][HPO_4^{2-}]}{[H_2PO_4^-]}$

$HPO_4^{2-}(aq) + H_2O(l) \leftrightarrows H_3O^+(aq) + PO_4^{3-}(aq)$ $\qquad K_{a3} = \frac{[H_3O^+][PO_4^{3-}]}{[HPO_4^{2-}]}$

44. $H_2CO_3(aq) + H_2O(l) \leftrightarrows H_3O^+(aq) + HCO_3^-(aq)$ $\qquad K_{a1} = \frac{[H_3O^+][HCO_3^-]}{[H_2CO_3]}$

$HCO_3^-(aq) + H_2O(l) \leftrightarrows H_3O^+(aq) + CO_3^{2-}(aq)$ $\qquad K_{a2} = \frac{[H_3O^+][CO_3^{2-}]}{[HCO_3^-]}$

45. a) **Given:** 0.350 M H_3PO_4 $\quad K_{a1} = 7.5 \times 10^{-3}$, $K_{a2} = 6.2 \times 10^{-8}$ **Find:** $[H_3O^+]$, pH
Conceptual Plan: K_{a1} much larger than K_{a2}, so use K_{a1} to calculate $[H_3O^+]$. Write a balanced reaction. Prepare an ICE table; represent the change with x; sum the table and determine the equilibrium values; put the equilibrium values in the equilibrium expression and solve for x.
Solution: $H_3PO_4(aq) + H_2O(l) \leftrightarrows H_3O^+(aq) + H_2PO_4^-(aq)$

I	0.350M	0.0	0.0
C	- x	x	x
E	0.350 – x	x	x

$$K_a = \frac{[H_3O^+][H_2PO_4^-]}{[H_3PO_4]} = \frac{(x)(x)}{(0.350 - x)} = 7.3 \times 10^{-3}$$

Assume x is small compared to 0.350

$x^2 = (7.3 \times 10^{-3})(0.350)$ $\quad x = 0.0512M = [H_3O^+]$

Check assumption: $\frac{0.0512}{0.350} \times 100\% = 14.6\%$ assumption not valid

$x^2 + 7.5 \times 10^{-3}\ x - 0.002625 = 0$ $\quad x = 0.04\underline{7}62 = [H_3O^+]$

$pH = -\log(0.04\underline{7}62) = 1.32$

b) **Given:** 0.350 M $H_2C_2O_4$ $K_{a1} = 6.0 \times 10^{-2}$, $K_{a2} = 6.0 \times 10^{-5}$ **Find:** $[H_3O^+]$, pH
Conceptual Plan: K_{a1} much larger than K_{a2}, so use K_{a1} to calculate $[H_3O^+]$. Write a balanced reaction. Prepare an ICE table; represent the change with x; sum the table and determine the equilibrium values; put the equilibrium values in the equilibrium expression and solve for x.
Solution: $H_2C_2O_4(aq) + H_2O(l) \leftrightarrows H_3O^+(aq) + HC_2O_4^-(aq)$

I	0.350M	0.0	0.0
C	- x	x	x
E	0.350 – x	x	x

$$K_a = \frac{[H_3O^+][HC_2O_4^-]}{[H_2C_2O_4]} = \frac{(x)(x)}{(0.350 - x)} = 6.0 \times 10^{-2}$$

$x^2 + 6.0 \times 10^{-2}\ x - 0.021 = 0$ $\quad x = 0.1\underline{1}79 = 0.12\ M\ [H_3O^+]$

$pH = -\log(0.1\underline{1}79) = 0.93$

46 a) **Given:** 0.125 M H_2CO_3 $K_{a1} = 4.3 \times 10^{-7}$, $K_{a2} = 5.6 \times 10^{-11}$ **Find:** $[H_3O^+]$, pH
Conceptual Plan: K_{a1} much larger than K_{a2}, so use K_{a1} to calculate $[H_3O^+]$. Write a balanced reaction. Prepare an ICE table; represent the change with x; sum the table and determine the equilibrium values; put the equilibrium values in the equilibrium expression and solve for x.
Solution: $H_2CO_3(aq) + H_2O(l) \leftrightarrows H_3O^+(aq) + HCO_3^-(aq)$

I	0.125M	0.0	0.0
C	- x	x	x
E	0.125 – x	x	x

$$K_a = \frac{[H_3O^+][HCO_3^-]}{[H_2CO_3]} = \frac{(x)(x)}{(0.125 - x)} = 4.3 \times 10^{-7}$$

Assume x is small

$x^2 = (4.3 \times 10^{-7})(0.125)$ $\quad x = 2.\underline{3}2 \times 10^{-4} = [H_3O^+]$

$\frac{2.\underline{3}2 \times 10^{-4}}{0.125} \times 100\% = 1.9\%$; assumption valid

$pH = -\log(2.\underline{3}2 \times 10^{-4}) = 3.63$

b) **Given:** 0.125 M $H_3C_6H_5O_3$ $K_{a1} = 7.4 \times 10^{-4}$, $K_{a2} = 1.7 \times 10^{-5}$, $_{a3} = 4.0 \times 10^{-7}$
Find: $[H_3O^+]$, pH
Conceptual Plan: K_{a1} and K_{a2} are only 10^{-1} apart, so use both to calculate $[H_3O^+]$. Write a balanced reaction. Prepare an ICE table; represent the change with x; sum the table and determine the equilibrium values; put the equilibrium values in the equilibrium expression and solve for x.
Solution: $H_3C_6H_5O_3(aq) + H_2O(l) \leftrightarrows H_3O^+(aq) + H_2C_6H_5O_3^-(aq)$

I	0.125M	0.0	0.0
C	- x	x	x
E	0.125 – x	x	x

$$K_{a_1} = \frac{[H_3O^+][H_2C_6H_5O_3^-]}{[H_3C_6H_5O_3]} = \frac{(x)(x)}{(0.125 - x)} = 7.4 \times 10^{-4}$$

Assume x is small

$x^2 = (7.4 \times 10^{-4})(0.125)$ $x = 9.\underline{6}2 \times 10^{-3} = [H_3O^+]$

$\frac{9.\underline{6}2 \times 10^{-3}}{0.125} \times 100\% = 7.8\%$; assumption not valid

$x^2 + 7.4 \times 10^{-4}\, x - 9.25 \times 10^{-5} = 0$ $x = 0.009\underline{2}55 = [H_3O^+] = [H_2C_6H_5O_3^-]$

and then:

	$H_2C_6H_5O_3^-$ (aq) + H_2O(l) $\leftrightarrows$	H_3O^+(aq) +	$HC_6H_5O_3^{2-}$(aq)
I	0.009255M	0.009255	0.0
C	- y	y	y
E	0.009255 – y	0.009255 + y	y

$$K_a = \frac{[H_3O^+][HC_6H_5O_3^{2-}]}{[H_2C_6H_5O_3^-]} = \frac{(0.009255 + y)(y)}{(0.009255 - y)} = 1.7 \times 10^{-5}$$

Assume y is small   Then, $y = 1.7 \times 10^{-5}$

$\frac{1.7 \times 10^{-5}}{0.009255} \times 100\% = 1.8\%$; assumption valid

$[H_3O^+] = 1.7 \times 10^{-5}$ (from second ionization)

$[H_3O^+] = 0.009\underline{2}55 + 1.7 \times 10^{-5} = 0.009\underline{2}7$ M $pH = -\log(0.009\underline{2}7) = 2.03$

47. a) **Given:** 0.15 M NaOH **Find:** $[OH^-]$, $[H_3O^+]$, pH, pOH

Conceptual Plan: $[NaOH] \rightarrow [OH^-] \rightarrow [H_3O^+] \rightarrow pH \rightarrow pOH$

$K_w = [H_3O^+][OH^-]$ $pH = -\log[H_3O^+]$ $pH + pOH = 14$

Solution: $[OH^-] = [NaOH] = 0.15M$

$$[H_3O^+] = \frac{K_w}{[OH^-]} = \frac{1 \times 10^{-14}}{0.15\ M} = 6.7 \times 10^{-14} M$$

$pH = -\log(6.7 \times 10^{-14}) = 13.18$

$pOH = 14.00 - 13.18 = 0.82$

b) **Given:** 1.5×10^{-3} M $Ca(OH)_2$ **Find:** $[OH^-]$, $[H_3O^+]$, pH, pOH

Conceptual Plan: $[Ca(OH)_2] \rightarrow [OH^-] \rightarrow [H_3O^+] \rightarrow pH \rightarrow pOH$

$K_w = [H_3O^+][OH^-]$ $pH = -\log[H_3O^+]$ $pH + pOH = 14$

Solution: $[OH^-] = 2[Ca(OH)_2] = 2(1.5 \times 10^{-3}) = 0.0030$ M

$$[H_3O^+] = \frac{K_w}{[OH^-]} = \frac{1 \times 10^{-14}}{0.0030\ M} = 3.\underline{3}3 \times 10^{-12} M$$

$pH = -\log(3.\underline{3}3 \times 10^{-12}) = 11.48$

$pOH = 14.00 - 11.48 = 2.52$

c) **Given:** 4.8×10^{-4} M $Sr(OH)_2$ **Find:** $[OH^-]$, $[H_3O^+]$, pH, pOH

Conceptual Plan: $[Sr(OH)_2] \rightarrow [OH^-] \rightarrow [H_3O^+] \rightarrow pH \rightarrow pOH$

$K_w = [H_3O^+][OH^-]$ $pH = -\log[H_3O^+]$ $pH + pOH = 14$

Solution: $[OH^-] = [Sr(OH)_2] = 2(4.8 \times 10^{-4}) = 9.6 \times 10^{-4}$ M

$$[H_3O^+] = \frac{K_w}{[OH^-]} = \frac{1 \times 10^{-14}}{9.6 \times 10^{-4}\ M} = 1.\underline{0}4 \times 10^{-11} M$$

$pH = -\log(1.\underline{0}4 \times 10^{-11}) = 10.98$

$pOH = 14.00 - 10.98 = 3.02$

d) **Given:** 8.7×10^{-5} M KOH **Find:** $[OH^-]$, $[H_3O^+]$, pH, pOH

Conceptual Plan: $[KOH] \rightarrow [OH^-] \rightarrow [H_3O^+] \rightarrow pH \rightarrow pOH$

$K_w = [H_3O^+][OH^-]$ $pH = -\log[H_3O^+]$ $pH + pOH = 14$

Solution: $[OH^-] = [KOH] = 8.7 \times 10^{-5}$ M

$$[H_3O^+] = \frac{K_w}{[OH^-]} = \frac{1 \times 10^{-14}}{8.7 \times 10^{-5} M} = 1.\underline{1} \times 10^{-10} M$$

$pH = -\log(1.\underline{1} \times 10^{-10}) = 9.94$

$pOH = 14.00 - 9.94 = 4.06$

48. a) **Given:** 8.77×10^{-3} M LiOH **Find:** $[OH^-]$, $[H_3O^+]$, pH, pOH

Conceptual Plan: **$[LiOH] \rightarrow [OH^-] \rightarrow [H_3O^+] \rightarrow pH \rightarrow pOH$**

$K_w = [H_3O^+][OH^-]$ $pH = -\log[H_3O^+]$ $pH + pOH = 14$

Solution: $[OH^-] = [LiOH] = 8.77 \times 10^{-3}$ M

$$[H_3O^+] = \frac{K_w}{[OH^-]} = \frac{1 \times 10^{-14}}{8.77 \times 10^{-3} M} = 1.1\underline{4}0 \times 10^{-12} M$$

$pH = -\log(1.1\underline{4}0 \times 10^{-12}) = 11.943$

$pOH = 14.00 - 11.943 = 2.057$

b) **Given:** 0.0112 M $Ba(OH)_2$ **Find:** $[OH^-]$, $[H_3O^+]$, pH, pOH

Conceptual Plan: **$[Ba(OH)_2] \rightarrow [OH^-] \rightarrow [H_3O^+] \rightarrow pH \rightarrow pOH$**

$K_w = [H_3O^+][OH^-]$ $pH = -\log[H_3O^+]$ $pH + pOH = 14$

Solution: $[OH^-] = 2[Ba(OH)_2] = 2(0.0112) = 0.0224$ M

$$[H_3O^+] = \frac{K_w}{[OH^-]} = \frac{1 \times 10^{-14}}{0.0224\ M} = 4.4\underline{6}4 \times 10^{-13} M$$

$pH = -\log(4.4\underline{6}4 \times 10^{-13}) = 12.350$

$pOH = 14.000 - 12.350 = 1.650$

c) **Given:** 1.9×10^{-4} M KOH **Find:** $[OH^-]$, $[H_3O^+]$, pH, pOH

Conceptual Plan: **$[KOH] \rightarrow [OH^-] \rightarrow [H_3O^+] \rightarrow pH \rightarrow pOH$**

$K_w = [H_3O^+][OH^-]$ $pH = -\log[H_3O^+]$ $pH + pOH = 14$

Solution: $[OH^-] = [KOH] = 1.9 \times 10^{-4}$ M

$$[H_3O^+] = \frac{K_w}{[OH^-]} = \frac{1 \times 10^{-14}}{1.9 \times 10^{-4} M} = 5.\underline{2}6 \times 10^{-11} M$$

$pH = -\log(5.\underline{2}6 \times 10^{-11}) = 10.28$

$pOH = 14.00 - 10.28 = 3.72$

d) **Given:** 5.0×10^{-4} M $Ca(OH)_2$ **Find:** $[OH^-]$, $[H_3O^+]$, pH, pOH

Conceptual Plan: **$[Ca(OH)_2] \rightarrow [OH^-] \rightarrow [H_3O^+] \rightarrow pH \rightarrow pOH$**

$K_w = [H_3O^+][OH^-]$ $pH = -\log[H_3O^+]$ $pH + pOH = 14$

Solution: $[OH^-] = [Ca(OH)_2] = 2(5.0 \times 10^{-4}) = 0.0010$ M

$$[H_3O^+] = \frac{K_w}{[OH^-]} = \frac{1 \times 10^{-14}}{0.0010\ M} = 1.\underline{0}0 \times 10^{-11} M$$

$pH = -\log(1.\underline{0}0 \times 10^{-11}) = 11.00$

$pOH = 14.00 - 10.90 = 3.00$

49. **Given:** 3.85% KOH by mass, d = 1.01 g/mL **Find:** pH

Conceptual Plan:

% mass → g KOH → mol KOH and mass soln → mL soln → L soln → M KOH → $[OH^{-1}]$

$\frac{1 \text{ mol KOH}}{56.01 \text{ g KOH}}$ $\frac{1.01 \text{ g soln}}{\text{mL soln}}$ $\frac{1000 \text{ mL soln}}{\text{L soln}}$ $\frac{\text{mol KOH}}{\text{L soln}}$

→ pOH → pH

$pOH = -\log[OH^-]$ $pH + pOH = 14$

Solution: $$\frac{3.85 \text{ g KOH}}{100.0 \text{ g soln}} \times \frac{1 \text{ mol KOH}}{56.01 \text{ g KOH}} \times \frac{1.01 \text{ g soln}}{\text{mL soln}} \times \frac{1000 \text{ mL soln}}{\text{L soln}} = 0.69\underline{4}2 \text{ M KOH}$$

$[OH^-] = [KOH] = 0.69\underline{4}2$ M $pOH = -\log(0.69\underline{4}2) = 0.159$ $pH = 14.000 - 0.159 = 13.841$

50. **Given:** 1.55% NaOH by mass, d = 1.01 g/mL **Find:** pH

Conceptual Plan:

% mass → g NaOH → mol NaOH and mass soln → mL soln → L soln → MNaOH → $[OH^{-1}]$

$\frac{1 \text{ mol NaOH}}{40.01 \text{ g NaOH}}$ $\frac{1.01 \text{ g soln}}{\text{mL soln}}$ $\frac{1000 \text{ mL soln}}{\text{L soln}}$ $\frac{\text{mol NaOH}}{\text{L soln}}$

→ pOH → pH

$pOH = -\log[OH^-]$ $pH + pOH = 14$

Solution: $\frac{1.55 \text{ g NaOH}}{100.0 \text{ g soln}} \times \frac{1 \text{ mol NaOH}}{40.01 \text{ g KOH}} \times \frac{1.01 \text{ g soln}}{\text{mL soln}} \times \frac{1000 \text{ mL soln}}{\text{L soln}} = 0.39\underline{1}3 \text{ M KOH}$

$[OH^-] = [NaOH] = 0.39\underline{1}3$ M pOH = -log(0.39$\underline{1}$3) = 0.407 pH = 14.000 - 0.407 = 13.593

51. a) $NH_3(aq) + H_2O(l) \leftrightarrows NH_4^+(aq) + OH^-(aq)$ $K_b = \frac{[NH_4^+][OH^-]}{[NH_3]}$

b) $HCO_3^-(aq) + H_2O(l) \leftrightarrows H_2CO_3(aq) + OH^-(aq)$ $K_b = \frac{[H_2CO_3][OH^-]}{[HCO_3^-]}$

c) $CH_3NH_2(aq) + H_2O(l) \leftrightarrows CH_3NH_3^+(aq) + OH^-(aq)$ $K_b = \frac{[CH_3NH_3^+][OH^-]}{[CH_3NH_2]}$

52. a) $CO_3^{2-} + H_2O(l) \leftrightarrows HCO_3^-$ aq) + $OH^-(aq)$ $K_b = \frac{[HCO_3^-][OH^-]}{[CO_3^{2-}]}$

b) $C_6H_5NH_2 + H_2O(l) \leftrightarrows C_6H_5NH_3^+(aq) + OH^-(aq)$ $K_b = \frac{[C_6H_5NH_3^+][OH^-]}{[C_6H_5NH_2]}$

c) $C_2H_5NH_2 + H_2O(l) \leftrightarrows C_2H_5NH_3^+(aq) + OH^-(aq)$ $K_b = \frac{[C_2H_5NH_3^+][OH^+]}{[C_2H_5NH_2]}$

53. **Given:** 0.15M NH_3 $K_b = 1.76 \times 10^{-5}$ **Find:** $[OH^-]$,pH, pOH

Conceptual Plan: Write a balanced reaction. Prepare an ICE table; represent the change with x; sum the table and determine the equilibrium values; put the equilibrium values in the equilibrium expression and solve for x.

$x = [OH^-]$ → pOH → pH

$pOH = -\log[OH^-]$ $pH + pOH = 14$

Solution: $NH_3(aq) + H_2O(l) \leftrightarrows NH_4^+(aq) + OH^-(aq)$

	NH_3	NH_4^+	OH^-
I	0.15	0.0	0.0
C	-x	x	x
E	0.15 – x	x	x

$K_b = \frac{[NH_4^+][OH^-]}{[NH_3]} = \frac{(x)(x)}{(0.15 - x)} = 1.76 \times 10^{-5}$

Assume x is small

$x^2 = (1.76 \times 10^{-5})(0.15)$ $x = [OH^-] = 0.001\underline{6}2$ M

pOH = -log(0.001$\underline{6}$2) = 2.79

pH = 14.00 - 2.79 = 11.21

54. **Given:** 0.125M CO_3^{2-} $K_b = 1.8 \times 10^{-4}$ **Find:** $[OH^-]$,pH, pOH

Conceptual Plan: Write a balanced reaction. Prepare an ICE table; represent the change with x; sum the table and determine the equilibrium values; put the equilibrium values in the equilibrium expression and solve for x.

$x = [OH^-]$ → pOH → pH

$pOH = -\log[OH^-]$ $pH + pOH = 14$

Solution: $CO_3^{2-} + H_2O(l) \leftrightarrows HCO_3^-(aq) + OH^-(aq)$

	CO_3^{2-}	HCO_3^-	OH^-
I	0.125	0.0	0.0
C	-x	x	x
E	0.125 – x	x	x

$$K_b = \frac{[HCO_3^-][OH^-]}{[CO_3^-]} = \frac{(x)(x)}{(0.125 - x)} = 1.8 \times 10^{-5}$$

Assume x is small

$x^2 = (1.8 \times 10^{-4})(0.125)$ $\quad x = [OH^-] = 0.004\underline{7}4$ M

$$\frac{0.004\underline{7}4}{0.125} \times 100\% = 3.8\%; \text{ assumption is valid}$$

$pOH = -\log(0.004\underline{7}4) = 2.32$

$pH = 14.00 - 2.32 = 11.68$

55. **Given:** $pK_b = 10.4$, 455 mg/L caffeine **Find:** pH

Conceptual Plan: $pK_b \rightarrow K_b$ **and then mg/L** $\rightarrow$ **g/L** $\rightarrow$ **mol/L and then write a balanced reaction. Prepare an ICE table; represent the change with x; sum the table and determine the equilibrium values; put the equilibrium values in the equilibrium expression and solve for x.**

$x = [OH^-] \rightarrow pOH \rightarrow pH$

$pOH = -\log[OH^-]$ $\quad pH + pOH = 14$

Solution: $K_b = 10^{-10.4} = \underline{3}.98 \times 10^{-11}$

$$\frac{455 \text{ mg caffeine}}{\text{L soln}} \times \frac{\text{g caffeine}}{1000 \text{ mg caffeine}} \times \frac{1 \text{ mol caffeine}}{194.19 \text{ g}} = 0.0023\underline{4}2 \text{ M caffeine}$$

$C_8H_{10}N_4O_2(aq) + H_2O(l) \leftrightarrows HC_8H_{10}N_4O_2^+(aq) + OH^-(aq)$

	$C_8H_{10}N_4O_2$	$HC_8H_{10}N_4O_2^+$	OH^-
I	0.0023\underline{4}2	0.0	0.0
C	-x	x	x
E	0.0023\underline{4}2– x	x	x

$$K_b = \frac{[HC_8H_{10}N_4O_2^+][OH^-]}{[C_8H_{10}N_4O_2]} = \frac{(x)(x)}{(0.0023\underline{4}2 - x)} = \underline{3}.98 \times 10^{-11}$$

Assume x is small

$x^2 = (\underline{3}.98 \times 10^{-11})(0.0023\underline{4}2)$ $\quad x = [OH^-] = \underline{3}.05 \times 10^{-7}$ M

$$\frac{\underline{3}.05 \times 10^{-7} \text{ M}}{0.0023\underline{4}2} \times 100\% = 0.013\%; \text{ assumption is valid}$$

$pOH = -\log(\underline{3}.05 \times 10^{-7}) = 6.5$

$pH = 14.00 - 6.5 = 7.5$

56. **Given:** $pK_b = 4.2$, 225 mg/L amphetamine **Find:** pH

Conceptual Plan: $pK_b \rightarrow K_b$ **and then mg/L** $\rightarrow$ **g/L** $\rightarrow$ **mol/L and then write a balanced reaction. Prepare an ICE table; represent the change with x; sum the table and determine the equilibrium values; put the equilibrium values in the equilibrium expression and solve for x.**

$x = [OH^-] \rightarrow pOH \rightarrow pH$

$pOH = -\log[OH^-]$ $\quad pH + pOH = 14$

Solution: $K_b = 10^{-4.2} = \underline{6}.31 \times 10^{-5}$

$$\frac{225 \text{ mg amphetamine}}{\text{L soln}} \times \frac{\text{g amphetamine}}{1000 \text{ mg amphetamine}} \times \frac{1 \text{ mol amphetamine}}{135.21 \text{ g}} = 0.0016\underline{6}4 \text{ M amphetamine}$$

$C_9H_{13}N(aq) + H_2O(l) \leftrightarrows C_9H_{13}NH^+(aq) + OH^-(aq)$

	$C_9H_{13}N$	$C_9H_{13}NH^+$	OH^-
I	0.0016\underline{6}4	0.0	0.0
C	-x	x	x
E	0.0016\underline{6}4– x	x	x

$$K_b = \frac{[C_9H_{13}NH^+][OH^-]}{[C_9H_{13}N]} = \frac{(x)(x)}{(0.0016\underline{6}4 - x)} = \underline{6}.31 \times 10^{-5}$$

Assume x is small

$x^2 = (\underline{6}.31 \times 10^{-5})(0.0016\underline{6}4)$ $\qquad$ $x = [OH^-] = \underline{3}.24 \times 10^{-4}$ M

$$\frac{\underline{3}.24 \times 10^{-4}}{0.0016\underline{6}4} \times 100\% = 19.4\%;\text{ assumption not valid}$$

$x^2 + \underline{6}.31 \times 10^{-5} x - 1.05 \times 10^{-7} = 0$

$x = [OH^-] = \underline{2}.94 \times 10^{-4}$M

$pOH = -\log(\underline{2}.94 \times 10^{-4}) = 3.5$

$pH = 14.00 - 3.5 = 10.5$

57. **Given:** 0.150 M morphine, pH = 10.5 **Find:** K_b
Conceptual Plan:
pH → pOH → [OH⁻] and then write a balanced equation, prepare an ICE table, determine
pH = pOH = 14 $\quad$ pOH = -log[OH⁻]
equilibrium concentrations → K_b
Solution: pOH = 14.0 – 10.5 = 3.5 $\quad$ $[OH^-] = 10^{-3.5} = \underline{3}.16 \times 10^{-4} = [Hmorphine^+]$

	morphine(aq) + H_2O(l) ⇆	$Hmorphine^+$(aq)+	OH^-(aq)
I	0.150	0.0	0.0
C	-x	x	x
E	0.150– x	3.16×10^{-4}	3.16×10^{-4}

$$K_b = \frac{[Hmorphine^+][OH^-]}{[morphine]} = \frac{(3.16 \times 10^{-4})(3.16 \times 10^{-4})}{(0.150 - 3.16 \times 10^{-4})} = \underline{6}.68 \times 10^{-7} = 7 \times 10^{-7}$$

58. **Given:** 0.135 M base, pH = 11.23 **Find:** K_b
Conceptual Plan:
pH → pOH → [OH⁻] and then write a balanced equation, prepare an ICE table, determine
pH = pOH = 14 $\quad$ pOH = -log[OH⁻]
equilibrium concentrations → K_b
Solution: pOH = 14.00 – 11.23 = 2.77 $\quad$ $[OH^-] = 10^{-2.77} = 1.\underline{6}98 \times 10^{-3} = [HB^+]$

	B(aq) + H_2O(l) ⇆	HB^+(aq)+	OH^-(aq)
I	0.135	0.0	0.0
C	-x	x	x
E	0.135– x	$1.\underline{6}98 \times 10^{-3}$	$1.\underline{6}98 \times 10^{-3}$

$$K_b = \frac{[HB^+][OH^-]}{[B]} = \frac{(1.\underline{6}98 \times 10^{-3})(1.\underline{6}98 \times 10^{-3})}{(0.135 - 1.\underline{6}98 \times 10^{-3})} = 2.1\underline{6} \times 10^{-5} = 2.2 \times 10^{-5}$$

59. a) pH neutral: Br^- is the conjugate base of a strong acid, therefore, it is pH neutral.

b) weak base: ClO^- is the conjugate base of a weak acid, therefore, it is a weak base.
$ClO^-(aq) + H_2O(l) \leftrightharpoons HClO(aq) + OH^-(aq)$

c) weak base: CN^- is the conjugate base of a weak acid, therefore, it is a weak base.
$CN^-(aq) + H_2O(l) \leftrightharpoons HCN(aq) + OH^-(aq)$

d) pH neutral: Cl^- is the conjugate base of a strong acid, therefore, it is pH neutral.

60. a) weak base: $C_7H_5O_2^-$ is the conjugate base of a weak acid, therefore, it is a weak base.
$C_7H_5O_2^-(aq) + H_2O(l) \leftrightharpoons H\,C_7H_5O_2(aq) + OH^-(aq)$

b) pH neutral: I^- is the conjugate base of a strong acid, therefore, it is pH neutral.

c) pH neutral: NO_3^- is the conjugate base of a strong acid, therefore, it is pH neutral.

d) weak base: F^- is the conjugate base of a weak acid, therefore, it is a weak base.
$F^-(aq) + H_2O(l) \leftrightharpoons HF(aq) + OH^-(aq)$

61. **Given:** $[F^-] = 0.140$ M, $K_a(HF) = 3.5 \times 10^{-4}$ **Find:** $[OH^-]$, pOH

Conceptual Plan:

Determine K_b. Write a balanced reaction. Prepare an ICE table; represent the change with x;

$$K_b = \frac{K_w}{K_a}$$

sum the table and determine the equilibrium values; put the equilibrium values in the equilibrium expression and solve for x. Determine $[OH^-] \rightarrow$ pOH $\rightarrow$ pH

$$pOH = -\log[OH^-] \quad pH + pOH = 14$$

Solution: $F^-(aq) + H_2O(l) \leftrightarrows HF(aq) + OH^-(aq)$

	F^-	HF	OH^-
I	0.140	0.0	0.0
C	-x	x	x
E	0.140– x	x	x

$$K_b = \frac{K_w}{K_a} = \frac{1 \times 10^{-14}}{3.5 \times 10^{-4}} = \frac{(x)(x)}{(0.140 - x)}$$

Assume x is small

$x = 2.0 \times 10^{-6} = [OH^-]$ $\quad$ $pOH = -\log(2.0 \times 10^{-6}) = 5.70$

$pH = 14.00 - 5.70 = 8.30$

62. **Given:** $[HCO_3^-] = 0.250$ M, $K_a(H_2CO_3) = 4.3 \times 10^{-7}$ **Find:** $[OH^-]$, pOH

Conceptual Plan:

Determine K_b. Write a balanced reaction. Prepare an ICE table; represent the change with x;

$$K_b = \frac{K_w}{K_a}$$

sum the table and determine the equilibrium values; put the equilibrium values in the equilibrium expression and solve for x. Determine $[OH^-] \rightarrow$ pOH $\rightarrow$ pH

$$pOH = -\log[OH^-] \quad pH + pOH = 14$$

Solution: $HCO_3^-(aq) + H_2O(l) \leftrightarrows H_2CO_3(aq) + OH^-(aq)$

	HCO_3^-	H_2CO_3	OH^-
I	0.250	0.0	0.0
C	-x	x	x
E	0.250– x	x	x

$$K_b = \frac{K_w}{K_a} = \frac{1 \times 10^{-14}}{4.3 \times 10^{-7}} = \frac{(x)(x)}{(0.250 - x)}$$

Assume x is small

$x = 7.6 \times 10^{-5} = [OH^-]$ $\quad$ $pOH = -\log(7.6 \times 10^{-5}) = 4.12$

$pH = 14.00 - 4.12 = 9.88$

63. a) weak acid: NH_4^+ is the conjugate acid of a weak base, therefore, it is a weak acid.
$NH_4^+(aq) + H_2O(l) \leftrightarrows H_3O^+(aq) + NH_3(aq)$

b) pH neutral: Na^+ is the counterion of a strong base, therefore, it is pH neutral.

c) weak acid: The Co^{3+} cation is a small, highly charged metal cation, therefore, it is a weak acid.
$Co(H_2O)_6^{3+}(aq) + H_2O(l) \leftrightarrows Co(H_2O)_5(OH)^{2+}(aq) + H_3O^+(aq)$

d) weak acid: $CH_2NH_3^+$ is the conjugate acid of a weak base, therefore, it is a weak acid.
$CH_2NH_3^+(aq) + H_2O(l) \leftrightarrows H_3O^+(aq) + CH_2NH_2(aq)$

64. a) pH neutral: Sr^{2+} is the counterion of a strong base, therefore, it is pH neutral.

b) weak acid: The Mn^{3+} cation is a small, highly charged metal cation, therefore, it is a weak acid.
$Mn(H_2O)_6^{3+}(aq) + H_2O(l) \leftrightarrows Mn(H_2O)_5(OH)^{2+}(aq) + H_3O^+(aq)$

c) weak acid: $C_5H_5NH^+$ is the conjugate acid of a weak base, therefore, it is a weak acid.
$C_5H_5NH^+(aq) + H_2O(l) \leftrightarrows H_3O^+(aq) + C_5H_5NH(aq)$

d) pH neutral: Li^+ is the counterion of a strong base, therefore, it is pH neutral.

65. a) acidic: $FeCl_3$ Fe^{3+} is a small, highly charged metal cation and is therefore, acidic. Cl^- is the conjugate base of a strong acid, therefore, it is pH neutral.

b) basic: NaF Na^+ is the counterion of a strong base, therefore, it is pH neutral. F^- is the conjugate base of a weak acid, therefore, it is basic.

c) pH neutral: $CaBr_2$ Ca^{2+} is the counterion of a strong base, therefore, it is pH neutral. Br^- is the conjugate base of a strong acid, therefore, it is pH neutral.

d) acidic: NH_4Br NH_4^+ is the conjugate acid of a weak base, therefore, it is acidic. Br^- is the conjugate base of a strong acid, therefore, it is pH neutral.

e) acidic: $C_6H_5NH_3NO_2$ $C_6H_5NH_3^+$ is the conjugate acid of a weak base, therefore, it is a weak acid. NO_2^- is the conjugate base of a weak acid, therefore, it is a weak base. To determine pH, compare K values. $K_a\ (C_6H_5NH_3^+) = \dfrac{1 \times 10^{-14}}{3.9 \times 10^{-10}} = 2.6 \times 10^{-5}$ $K_b(NO_2^-) = \dfrac{1 \times 10^{-14}}{4.6 \times 10^{-4}} = 2.2 \times 10^{-11}$

$K_a > K_b$ therefore, the solution is acidic.

66. a) acidic: $Al(NO_3)_3$ Al^{3+} is a small, highly charged metal cation and is therefore, acidic. NO_3^- is the conjugate base of a strong acid, therefore, it is pH neutral.

b) acidic: $C_2H_5NH_3NO_3$ $C_2H_5NH_3^+$ is the conjugate acid of a weak base, therefore, it is acidic. NO_3^- is the conjugate base of a strong acid, therefore, it is pH neutral.

c) basic: K_2CO_3 K^+ is the counterion of a strong base, therefore, it is pH neutral. CO_3^{2-} is the conjugate base of a weak acid, therefore, it is basic.

d) pH neutral: RbI Rb^+ is the counterion of a strong base, therefore, it is pH neutral. I^- is the conjugate base of a strong acid, therefore, it is pH neutral.

e) basic NH_4ClO NH_4^+ is the conjugate acid of a weak base, therefore, it is a weak acid. ClO^- is the conjugate base of a weak acid, therefore, it is a weak base. To determine pH, compare K values.

$K_a\ (NH_4^+) = \dfrac{1 \times 10^{-14}}{1.8 \times 10^{-5}} = 5.6 \times 10^{-10}$ $K_b(ClO^-) = \dfrac{1 \times 10^{-14}}{2.9 \times 10^{-8}} = 3.4 \times 10^{-7}$

$K_b > K_a$ therefore, the solution is basic.

67. Identify each species and determine acid, base, neutral.
NaCl pH neutral: Na^+ is the counterion of a strong base, therefore, it is pH neutral. Cl^- is the conjugate base of a strong acid, therefore, it is pH neutral.
NH_4Cl acidic: NH_4^+ is the conjugate acid of a weak base, therefore, it is acidic. Cl^- is the conjugate base of a strong acid, therefore, it is pH neutral.
$NaHCO_3$ basic: Na^+ is the counterion of a strong base, therefore, it is pH neutral. HCO_3^- is the conjugate base of a weak acid, therefore, it is basic.
NH_4ClO_2 acidic: NH_4ClO_2 NH_4^+ is the conjugate acid of a weak base, therefore, it is a weak acid. ClO_2^- is the conjugate base of a weak acid, therefore, it is a weak base. $K_a(NH_4^+) = 5.6 \times 10^{-10}$ $K_b(ClO_2^-) = 9.1 \times 10^{-13}$
NaOH strong base
Increasing acidity: $NaOH < NaHCO_3 < NaCl < NH_4ClO_2 < NH_4Cl$

68. Identify each species and determine acid, base, neutral.
CH_3NH_3Br acidic: $CH_3NH_3^+$ is the conjugate acid of a weak base, therefore, it is acidic. Br^- is the conjugate base of a strong acid, therefore, it is pH neutral.
KOH strong base
KBr pH neutral: K^+ is the counterion of a strong base, therefore, it is pH neutral. Br^- is the conjugate base of a strong acid, therefore, it is pH neutral.
KCN basic: K^+ is the counterion of a strong base, therefore, it is pH neutral. CN^- is the conjugate base of a weak acid, therefore, it is basic.
$C_5H_5NHNO_2$ acidic: $C_5H_5NH^+$ is the conjugate acid of a weak base, therefore, it is acidic. NO_2^- is the conjugate base of a weak acid, therefore, it is basic. $K_a(C_5H_5NH^+) = 5.9 \times 10^{-6}$; $K_b(NO_2^-) = 2.2 \times 10^{-11}$
Increasing basicity: $CH_3NH_3Br < C_5H_5NHNO_2 < KBr < KCN < KOH$

69. a) **Given:** 0.10 M NH_4Cl **Find:** pH

Conceptual Plan: Identify each species and determine which will contribute to pH. Write a balanced reaction. Prepare an ICE table; represent the change with x; sum the table and determine the equilibrium values; put the equilibrium values in the equilibrium expression and solve for x. Determine $[H_3O^+]$ → pH

Solution: NH_4^+ is the conjugate acid of a weak base, therefore, it is acidic. Cl^- is the conjugate base of a strong acid, therefore, it is pH neutral.

	NH_4^+ (aq) + $H_2O(l)$ ⇆	NH_3 (aq) +	H_3O^+ (aq)
I	0.10	0.0	0.0
C	-x	x	x
E	0.10– x	x	x

$$K_a = \frac{K_w}{K_b} = \frac{1 \times 10^{-14}}{1.8 \times 10^{-5}} = 5.\underline{5}6 \times 10^{-10} = \frac{(x)(x)}{(0.10 - x)}$$

Assume x is small

$x = 7.\underline{4}5 \times 10^{-6} = [H_3O^+]$ $pH = -\log(7.\underline{4}5 \times 10^{-6}) = 5.13$

b) **Given:** 0.10 M $NaC_2H_3O_2$ **Find:** pH

Conceptual Plan: Identify each species and determine which will contribute to pH. Write a balanced reaction. Prepare an ICE table; represent the change with x; sum the table and determine the equilibrium values; put the equilibrium values in the equilibrium expression and solve for x. Determine $[OH^-]$ → pOH → pH

$pOH = -\log[OH^-]$ $pH + pOH = 14$

Solution: Na^+ is the counterion of a strong base, therefore, it is pH neutral. $C_2H_3O_2^-$ is the conjugate base of a weak acid, therefore, it is basic.

	$C_2H_3O_2^-$ (aq) + $H_2O(l)$ ⇆	$HC_2H_3O_2$ (aq)+	OH^- (aq)
I	0.10	0.0	0.0
C	-x	x	x
E	0.10– x	x	x

$$K_b = \frac{K_w}{K_a} = \frac{1 \times 10^{-14}}{1.8 \times 10^{-5}} = 5.\underline{5}6 \times 10^{-10} = \frac{(x)(x)}{(0.10 - x)}$$

Assume x is small

$x = 7.\underline{4}5 \times 10^{-6} = [OH^-]$ $pOH = -\log(7.\underline{4}5 \times 10^{-6}) = 5.13$

$pH = 14.00 - 5.13 = 8.87$

c) **Given:** 0.10 M NaCl **Find:** pH

Conceptual Plan: Identify each species and determine which will contribute to pH.

Solution: Na^+ is the counterion of a strong base, therefore, it is pH neutral. Cl^- is the conjugate base of a strong acid, therefore, it is pH neutral.

pH = 7.0

70. a) **Given:** 0.20 M $NaCHO_2$ **Find:** pH

Conceptual Plan: Identify each species and determine which will contribute to pH. Write a balanced reaction. Prepare an ICE table; represent the change with x; sum the table and determine the equilibrium values; put the equilibrium values in the equilibrium expression and solve for x. Determine $[OH^-]$ → pOH → pH

$pOH = -\log[OH^-]$ $pH + pOH = 14$

Solution: Na^+ is the counterion of a strong base, therefore, is pH neutral. CHO_2^- is the conjugate base of a weak acid, therefore, it is basic.

	CHO_2^- (aq) + $H_2O(l)$ ⇆	$HCHO_2$ (aq)+	OH^- (aq)
I	0.20	0.0	0.0
C	-x	x	x
E	0.20– x	x	x

$$K_b = \frac{K_w}{K_a} = \frac{1 \times 10^{-14}}{1.8 \times 10^{-4}} = 5.\underline{5}6 \times 10^{-11} = \frac{(x)(x)}{(0.20 - x)}$$

Assume x is small

$x = 3.\underline{3}3 \times 10^{-6} = [OH^-]$ $pOH = -\log(3.\underline{3}3 \times 10^{-6}) = 5.48$

$pH = 14.00 - 5.48 = 8.52$

b) **Given:** 0.20 M CH_3NH_3I **Find:** pH
Conceptual Plan: Identify each species and determine which will contribute to pH. Write a balanced reaction. Prepare an ICE table; represent the change with x; sum the table and determine the equilibrium values; put the equilibrium values in the equilibrium expression and solve for x. Determine $[H_3O^+] \rightarrow pH$
Solution: $CH_3NH_4^+$ is the conjugate acid of a weak base, therefore, it is acidic. Cl^- is the conjugate base of a strong acid, therefore, it is pH neutral.

	$CH_3NH_3^+(aq) + H_2O(l) \leftrightarrows$	$CH_3NH_2(aq)$ +	$H_3O^+(aq)$
I	0.20	0.0	0.0
C	-x	x	x
E	0.20– x	x	x

$$K_a = \frac{K_w}{K_b} = \frac{1 \times 10^{-14}}{4.4 \times 10^{-4}} = 2.\underline{2}7 \times 10^{-11} = \frac{(x)(x)}{(0.20 - x)}$$

Assume x is small

$x = 2.\underline{1}3 \times 10^{-6} = [H_3O^+]$ $pH = -\log(2.\underline{1}3 \times 10^{-6}) = 5.67$

c) **Given:** 0.20 M KI **Find:** pH
Conceptual Plan: Identify each species and determine which will contribute to pH.
Solution: K^+ is the counterion of a strong base, therefore, it is pH neutral. I^- is the conjugate base of a strong acid, therefore, it is pH neutral.
pH = 7.0

71. **Given:** 0.15 M KF **Find:** concentration of all species
Conceptual Plan: Identify each species and determine which will contribute to pH. Write a balanced reaction. Prepare an ICE table; represent the change with x; sum the table and determine the equilibrium values; put the equilibrium values in the equilibrium expression and solve for x. And then, determine $[OH^-] \rightarrow [H_3O^+]$
$K_w = [H_3O^+][OH^-]$
Solution: K^+ is the counterion of a strong base, therefore, is pH neutral. F^- is the conjugate base of a weak acid, therefore, it is basic.

	$F^-(aq) + H_2O(l) \leftrightarrows$	$HF(aq)+$	$OH^-(aq)$
I	0.15	0.0	0.0
C	-x	x	x
E	0.15– x	x	x

$$K_b = \frac{K_w}{K_a} = \frac{1 \times 10^{-14}}{3.5 \times 10^{-4}} = \frac{(x)(x)}{(0.15 - x)}$$

Assume x is small

$x = 2.1 \times 10^{-6} = [OH^-] = [HF]$ $[H_3O^+] = \dfrac{K_w}{[OH^-]} = \dfrac{1 \times 10^{-14}}{2.1 \times 10^{-6}} = 4.8 \times 10^{-9}$

$[K^+] = 0.15$ M
$[F^-] = (0.15 - 2.1 \times 10^{-6}) = 0.15M$
$[HF] = 2.1 \times 10^{-6}$
$[OH^-] = 2.1 \times 10^{-6}$
$[H_3O^+] = 4.8 \times 10^{-9}$

72. **Given:** 0.225 M $C_6H_5NH_3Cl$ **Find:** concentration of all species
Conceptual Plan: Identify each species and determine which will contribute to pH. Write a balanced reaction. Prepare an ICE table; represent the change with x; sum the table and determine the equilibrium values; put the equilibrium values in the equilibrium expression and solve for x. And then, determine $[H_3O^+] \rightarrow [OH^-]$
$K_w = [H_3O^+][OH^-]$
Solution: $C_6H_5NH_3^+$ is the conjugate acid of a weak base, therefore, it is a weak acid. Cl^- is the conjugate base of a strong acid, therefore, it is pH neutral.

	$C_6H_5NH_3^+(aq)$ + $H_2O(l)$ ⇆	$C_6H_5NH_2(aq)$ +	$H_3O^+(aq)$
I	0.225	0.0	0.0
C	-x	x	x
E	0.225– x	x	x

$$K_a = \frac{K_w}{K_b} = \frac{1 \times 10^{-14}}{3.9 \times 10^{-10}} = 2.\underline{5}6 \times 10^{-5} = \frac{(x)(x)}{(0.225 - x)}$$

Assume x is small

$$x = 0.0024 = [H_3O^+] = [CH_5NH_2] \qquad [OH^-] = \frac{K_w}{[H_3O^+]} = \frac{1 \times 10^{-14}}{0.0024} = 4.2 \times 10^{-12}$$

$[C_6H_5NH_3^+] = 0.225 - 0.0024 = 0.223$ M
$[Cl^-] = 0.225$ M
$[C_6H_5NH_2] = 0.0024$ M
$[H_3O^+] = 0.0024$ M
$[OH^-] = 4.2 \times 10^{-12}$ M

73. a) HCl is the stronger acid. HCl is the weaker bond, therefore, it is more acidic.

    b) HF is the stronger acid. F is more electronegative than O, so the bond is more polar and more acidic.

    c) H_2Se is the stronger acid. The H – Se bond is weaker, therefore, it is more acidic.

74. Increasing acid strength: NaH < H_2S < H_2Te < HI
H_2Te is a stronger acid than H_2S because the H – Te bond is weaker. HI is a stronger acid than H_2Te because I is more electronegative than Te. NaH is not acidic because H is more electronegative than Na.

75. a) H_2SO_4 is the stronger acid because it has more oxygen atoms.

    b) $HClO_2$ is the stronger acid because it has more oxygen atoms.

    c) HClO is the stronger acid because Cl is more electronegative than Br.

    d) CCl_3COOH is the stronger acid because Cl is more electronegative than H.

76. Increasing acid strength: HIO_3 < $HBrO_3$ < $HClO_3$.
Cl is more electronegative than Br which is more electronegative than I.

77. S^{2-} is the stronger base. Base strength is determined from the corresponding acid. The weaker the acid, the stronger the base. H_2S is the weaker acid because it has a stronger bond.

78. AsO_4^{3-} is the stronger base. Base strength is determined from the corresponding acid. The weaker the acid, the stonger the base. H_3AsO_4 is the weaker acid because P is more electronegative than As.

79. a) Lewis acid: Fe^{3+} has an empty d orbital and can accept lone pair electrons.

    b) Lewis acid: BH_3 has an empty p orbital to accept a lone pair of electrons.

    c) Lewis base: NH_3 has a lone pair of electrons to donate.

    d) Lewis base: F^- has lone pair electrons to donate.

80. a) Lewis acid: $BeCl_2$ has empty p orbitals to accept lone pair electrons.

    b) Lewis base: OH^- has lone pair electrons to donate.

    c) Lewis acid: $B(OH)_3$ has an empty p orbital to accept a lone pair of electrons.

    d) Lewis base: CN^- has lone pair electrons to donate.

81. a) Fe^{3+} accepts electron pair from H_2O, so Fe^{3+} is the Lewis acid and H_2O is the Lewis base.

b) Zn^{2+} accepts electron pair from NH_3, so Zn^{2+} is the Lewis acid and NH_3 is the Lewis base.

c) The empty p orbital on B accepts an electron pair from $(CH_3)_3N$, so BF_3 is the Lewis acid and $(CH_3)_3N$ is the Lewis base.

82. a) Ag^+ accepts electron pair from NH_3, so Ag^+ is the Lewis acid and NH_3 is the Lewis base.

b) The empty p orbital on Al accepts an electron pair from NH_3, $AlBr_3$ is the Lewis acid and NH_3 is the Lewis base.

c) The empty p orbital on B accepts an electron pair from F^-, so BF_3 is the Lewis acid and F^- is the Lewis base.

83. a) weak acid. The beaker contains 10 HF molecule, $2H_3O^+$ ions, and $2F^-$ ions. Since both the molecule and the ions exist in solution, the acid is a weak acid.

b) strong acid. The beaker contains $12H_3O^+$ ions and $2OH^-$ ions. Since the molecule is completely ionized in solution, the acid is a strong acid.

c) weak acid. The beaker contains 10 $HCHO_2$ molecules, $2H_3O^+$ ions, and $2CHO_2^-$ ions. Since both the molecule and the ions exist in solution, the acid is a weak acid.

d) strong acid. The beaker contains $12H_3O^+$ ions and $2NO_3^-$ ions. Since the molecule is completely ionized in solution, the acid is a strong acid.

84. a) weak base. The beaker contains 11 NH_3 molecules, 2 NH_4^+ ions, and $2OH^-$ ions. Since both the molecule and the ions exist in solution, the acid is a weak acid.

b) strong base. The beaker contains 12 Na^+ ions and $12OH^-$ ions. Since the molecule is completely ionized in solution, it is a strong base.

c) weak base. The beaker contains 12 Na^+, 10 H_2CO_3 molecules, $2HCO_3^-$ ions, and $2OH^-$ ions. The beaker contains the HCO_3^- ion from the $NaHCO_3$ and H_2CO_3 molecules from the reaction of the salt.

d) strong base. The beaker contains 12 Sr^{2+} ions and 24 OH^- ions. Since the molecule is completely ionized in solution, it is a strong base.

85. $HbH^+(aq) + O_2(aq) \leftrightarrows HbO_2(aq) + H^+(aq)$
Using LeChatelier's Principle, if the $[H^+]$ increases, the reaction will shift left, if the $[H^+]$ decreases, the reaction will shift right. So, if the pH of blood is too acidic (low pH; $[H^+]$ increased), the reaction will shift to the left. This will cause less of the HbO_2 in the blood and decrease the oxygen-carrying capacity of the hemoglobin in the blood.

86. $CO_2(g) + H_2O(l) \leftrightarrows H_2CO_3(aq)$
$H_2CO_3(aq) + H_2O(l) \leftrightarrows HCO_3^-(aq) + H_3O^+(aq)$
As the concentration of CO_2 in the atmosphere increases, more will dissolve in H_2O and form H_2CO_3. The H_2CO_3 will then act as a weak acid with H_2O and form HCO_3^- and H_3O^+, causing the water (oceans) to be more acidic. If the pH of the oceans decreases, the added H_3O^+ will react with the CO_3^{2-} ion in the $CaCO_3$ in the limestone structures and decompose the $CaCO_3$.
$2H_3O^+(aq) + CaCO_3(s) \rightarrow Ca^{2+}(aq) + H_2CO_3(aq) + 2H_2O \rightarrow H_2O(l) + CO_2(g)$

87. **Given:** 4.00×10^2 mg $Mg(OH)_2$, 2.00×10^2 mL HCl solution, pH = 1.3
Find: volume neutralized, % neutralized.
Conceptual Plan:
mg $Mg(OH)_2$ → g $Mg(OH)_2$ → mol $Mg(OH)_2$ and then pH → $[H_3O^+]$ and then mol $Mg(OH)_2$

$$\frac{\text{g Mg(OH)}_2}{1000\text{ mg}} \qquad \frac{\text{mol Mg(OH)}_2}{58.326\text{ g}} \qquad \text{pH} = -\log[H_3O^+]$$

mol OH^- → mol H_3O^+ → vol H_3O^+ → % neutralized.

$$\frac{2\ OH^-}{Mg(OH)_2} \quad \frac{H_3O^+}{OH^-} \quad \frac{mol\ H_3O^+}{M(H_3O^+)} \quad \frac{vol\ HCl\ neutralized}{total\ vol\ HCl} \times 100$$

Solution: $[H_3O^+] = 10^{-1.3} = 0.05011\ M = 0.05\ M$

$$4.00 \times 10^2\ mg\ Mg(OH)_2 \times \frac{g\ Mg(OH)_2}{1000\ mg\ Mg(OH)_2} \times \frac{1\ mol\ Mg(OH)_2}{58.326\ g\ Mg(OH)_2} \times \frac{2\ mol\ OH^-}{1\ mol\ Mg(OH)_2}$$

$$\times \frac{1\ mol\ H_3O^+}{1\ mol\ OH^-} \times \frac{1\ L}{0.0501\ mol\ H_3O^+} \times \frac{1000\ mL}{L} = 27\underline{3}.7\ mL = 274\ mL\ neutralized$$

Stomach contains 2.00×10^2 mL HCl at pH = 1.3 and 4.00×10^2 mg will neutralize 274 mL of pH 1.3 HCl, so all of the stomach acid will be neutralized.

88. **Given:** 4.3 billion L, pH = 5.5 **Find:** mass in kg $CaCO_3$
Conceptual Plan:
pH → $[H_3O^+]$ and then vol lake → mol $[H_3O^+]$ → mol $CaCO_3$ → g $CaCO_3$ → kg $CaCO_3$

$$pH = -\log[H_3O^+] \quad mol = vol \times M \quad \frac{mol\ CaCO_3}{2\ mol\ H_3O^+} \quad \frac{100.09\ g\ CaCO_3}{mol\ CaCO_3} \quad \frac{kg\ CaCO_3}{1000\ g\ CaCO_3}$$

Solution: $[H_3O^+] = 10^{-5.5} = 3.\underline{1}6 \times 10^{-6}\ M$

$$4.3 \times 10^9\ L \times \frac{3.\underline{1}6 \times 10^{-6}\ mol\ H_3O^+}{L} \times \frac{1\ mol\ CaCO_3}{2\ mol\ H_3O^+} \times \frac{100.09\ g\ CaCO_3}{mol\ CaCO_3} \times \frac{1\ kg}{1000\ g}$$

$= 680.01\ kg = 6.8 \times 10^2\ kg\ CaCO_3$

89. **Given:** pH of Great Lakes acid rain = 4.5, West Coast = 5.4
Find: $[H_3O^+]$ and ratio of Great Lakes/ West Coast
Conceptual Plan: pH → $[H_3O^+]$, and then ratio of $[H_3O^+]$ Great Lakes to West Coast.
$pH = -\log[H_3O^+]$
Solution: Great Lakes: $[H_3O^+] = 10^{-4.5} = 3.16 \times 10^{-5} M$ West Coast: $[H_3O^+] = 10^{-5.4} = 3.98 \times 10^{-6} M$

$$\frac{Great\ Lakes}{West\ Coast} = \frac{\underline{3}.16 \times 10^{-5} M}{\underline{3}.98 \times 10^{-6} M} = 7.94 = 8\ times\ more\ acidic$$

90. **Given:** pH of Sauvignon Blanc = 3.23, Cabernet Sauvignon = 3.64
Find: $[H_3O^+]$ and ratio of Sauvignon Blanc to Cabernet Sauvignon
Conceptual Plan: pH → $[H_3O^+]$, and then ratio of $[H_3O^+]$ Sauvignon Blanc to Cabernet Sauvignon.
$pH = -\log[H_3O^+]$
Solution:

Sauvignon Blanc: $[H_3O^+] = 10^{-3.23} = 5.8\underline{8}8 \times 10^{-4} M$
Cabernet Sauvignon: $[H_3O^+] = 10^{-3.64} = 2.2\underline{9}1 \times 10^{-4} M$

$$\frac{Sauvignon\ Blanc}{Cabernet\ Sauvignon} = \frac{5.8\underline{8}8 \times 10^{-4} M}{2.2\underline{9}1 \times 10^{-4} M} = 2.570 = 2.57\ times\ more\ acidic$$

91. **Given:** 6.5×10^2 mg aspirin, 8 ounces water, $pK_a = 3.5$ **Find:** pH of solution
Conceptual Plan:
mg aspirin → g aspirin → mol aspirin and ounces → quart → L and then [aspirin] and pK_a

$$\frac{g\ aspirin}{1000\ mg} \quad \frac{mol\ aspirin}{180.16\ g} \quad \frac{1\ qt}{32\ ounces} \quad \frac{1\ L}{1.0567\ qt} \quad \frac{mol\ aspirin}{L\ soln} \quad pK_a = -\log K_a$$

→ K_a and then: Write a balanced reaction. Prepare an ICE table; represent the change with x; sum the table and determine the equilibrium values; put the equilibrium values in the equilibrium expression and solve for x. Determine $[H_3O^+]$ → pH
$pH = -\log[H_3O^+]$

Solution: $$\frac{6.5 \times 10^2\ mg\ aspirin}{8\ ounces} \times \frac{1\ gram\ aspirin}{1000\ mg\ aspirin} \times \frac{mol\ aspirin}{180.16\ g} \times \frac{32\ ounces}{qt} \times \frac{1.0567\ qt}{1\ L} = 0.01\underline{5}2\ M$$

$K_a = 10^{-3.5} = 3.\underline{1}6 \times 10^{-4}$

$aspirin(aq) + H_2O(l) \rightleftarrows H_3O^+(aq) + aspirin^-(aq)$

I	0.0152M	0.0	0.0
C	- x	x	x
E	0.0152– x	x	x

$$K_a = \frac{[H_3O^+][aspirin^-]}{[aspirin]} = \frac{(x)(x)}{(0.0152 - x)} = 3.16 \times 10^{-4}$$

$x^2 + 3.16 \times 10^{-4}x - 4.80 \times 10^{-6} = 0$ $\quad x = 2.04 \times 10^{-3}$ M $= [H_3O^+]$

pH = $-\log(2.04 \times 10^{-3}) = 2.69 = 2.7$

92. **Given:** 565 mg/L ddC, $pK_b = 9.8$ **Find:** % protonated
Conceptual Plan: mg → g → mol → M and then pK_b → K_b and then: Write a balanced reaction. Prepare an ICE table; represent the change with x; sum the table and determine the equilibrium values; put the equilibrium values in the equilibrium expression and solve for x.

Solution: $\frac{565 \text{ mg ddC}}{L} \times \frac{\text{g ddC}}{1000 \text{ mg ddC}} \times \frac{\text{mol ddC}}{224.22 \text{ g ddC}} = 0.0025199$ M

$K_b = 10^{-9.8} = 1.585 \times 10^{-10}$

	ddC(aq) + H_2O(l) ⇆	$HddC^+$(aq)+	OH^-(aq)
I	0.0025199	0.0	0.0
C	-x	x	x
E	0.0025199– x	x	x

$$K_b = \frac{[HddC^+][OH^-]}{[ddC]} = \frac{(x)(x)}{(0.0025199 - x)} = 1.585 \times 10^{-10}$$

Assume x is small

$x^2 = (1.585 \times 10^{-10})(0.0025199)$ $\quad x = [OH^-] = [HddC^+] = 6.32 \times 10^{-7}$ M

$$\% \text{ protonated} = \frac{[HddC^+]_{equilibrium}}{[ddC]_{original}} \times 100\% = \frac{(6.32 \times 10^{-7})}{(0.0025199)} \times 100\% = 0.025\ \%$$

93. a) **Given:** 0.0100M $HClO_4$ **Find:** pH
Conceptual Plan: $[HClO_4]$ → $[H_3O^+]$ → **pH**
$[HClO_4]$ → $[H_3O^+]$ $\quad$ pH = $-\log[H_3O^+]$
Solution: 0.0100 M $HClO_4$ = 0.0100 M H_3O^+ $\quad$ pH = $-\log(0.0100) = 2.000$

b) **Given:** 0.115M $HClO_2$, $K_a = 1.1 \times 10^{-2}$ **Find:** pH
Conceptual Plan: Write a balanced reaction. Prepare an ICE table; represent the change with x; sum the table and determine the equilibrium values; put the equilibrium values in the equilibrium expression and solve for x. Determine $[H_3O^+]$ and pH
Solution: $HClO_2$ (aq) + H_2O(l) ⇆ H_3O^+(aq) + ClO_2^- (aq)

I	0.115M	0.0	0.0
C	- x	x	x
E	0.115 – x	x	x

$$K_a = \frac{[H_3O^+][ClO_2^-]}{[HClO_2]} = \frac{(x)(x)}{(0.115 - x)} = 1.1 \times 10^{-2}$$

Assume x is small compared to 0.115

$x^2 = (1.1 \times 10^{-2})(0.115)$ $\quad x = 0.0356$M $= [H_3O^+]$

Check assumption: $\frac{0.0356}{0.115} \times 100 = 30.9\%$ assumption not valid

$x^2 + 1.1 \times 10^{-2}x - 0.001265 = 0$

x = 0.03049 or -0.0415

pH = $-\log(0.03049) = 1.52$

c) **Given:** 0.045M $Sr(OH)_2$ **Find:** pH
Conceptual Plan: $[Sr(OH)_2]$ → $[OH^-]$ → $[H_3O^+]$ → **pH**
$K_w = [H_3O^+][OH^-]$ $\quad$ pH = $-\log[H_3O^+]$ $\quad$ pH + pOH =14
Solution: $[OH^-] = [Sr(OH)_2] = 2(0.045) = 0.090$ M

$$[H_3O^+] = \frac{K_w}{[OH^-]} = \frac{1 \times 10^{-14}}{0.090\ M} = 1.\underline{1}1 \times 10^{-13} M$$

$$pH = -\log(1.\underline{1}1 \times 10^{-13}) = 12.95$$

d) **Given:** 0.0852 KC_6H_5O, K_a (HC_6H_5O)= 4.9×10^{-10} **Find:** pH

Conceptual Plan: Identify each species and determine which will contribute to pH. Write a balanced reaction. Prepare an ICE table; represent the change with x; sum the table and determine the equilibrium values; put the equilibrium values in the equilibrium expression and solve for x. Determine $[OH^-]$ → pOH → pH

$pOH = -\log[OH^-]$ $pH + pOH = 14$

Solution: K^+ is the counterion of a strong base, therefore, is pH neutral. CN^- is the conjugate base of a weak acid, therefore, it is basic.

	$CN^-(aq) + H_2O(l) \leftrightarrows$	$HCN(aq) +$	$OH^-(aq)$
I	0.0852	0.0	0.0
C	-x	x	x
E	0.0852 – x	x	x

$$K_b = \frac{K_w}{K_a} = \frac{1 \times 10^{-14}}{4.9 \times 10^{-10}} = 2.\underline{0}4 \times 10^{-5} = \frac{(x)(x)}{(0.0852 - x)}$$

Assume x is small

$x = 1.\underline{3}2 \times 10^{-3} = [OH^-]$ $pOH = -\log(1.\underline{3}2 \times 10^{-3}) = 2.88$

$pH = 14.00 - 2.88 = 11.12$

e) **Given:** 0.155 NH_4Cl, K_b (NH_3)= 1.8×10^{-5} **Find:** pH

Conceptual Plan: Identify each species and determine which will contribute to pH. Write a balanced reaction. Prepare an ICE table; represent the change with x; sum the table and determine the equilibrium values; put the equilibrium values in the equilibrium expression and solve for x. Determine $[H_3O^+]$ → pH

Solution: NH_4^+ is the conjugate acid of a weak base, therefore, it is acidic. Cl^- is the conjugate base of a strong acid, therefore, it is pH neutral.

	$NH_4^+(aq) + H_2O(l) \leftrightarrows$	$NH_3(aq) +$	$H_3O^+(aq)$
I	0.155	0.0	0.0
C	-x	x	x
E	0.155– x	x	x

$$K_a = \frac{K_w}{K_b} = \frac{1 \times 10^{-14}}{1.8 \times 10^{-5}} = 5.\underline{5}6 \times 10^{-10} = \frac{(x)(x)}{(0.155 - x)}$$

Assume x is small

$x = 9.\underline{2}8 \times 10^{-6} = [H_3O^+]$ $pH = -\log(9.\underline{2}8 \times 10^{-6}) = 5.03$

94. a) **Given:** 0.0650M HNO_3 **Find:** pH

Conceptual Plan: **$[HNO_3]$ → $[H_3O^+]$ → pH**

$[HNO_3] \rightarrow [H_3O^+]$ $pH = -\log[H_3O^+]$

Solution: 0.0650M HNO_3= 0.0650M H_3O^+ $pH = -\log(0.0650) = 1.187$

b) **Given:** 0.150M HNO_2 ,$K_a = 4.6 \times 10^{-4}$ **Find:** pH

Conceptual Plan: Write a balanced reaction. Prepare an ICE table; represent the change with x; sum the table and determine the equilibrium values; put the equilibrium values in the equilibrium expression and solve for x. Determine $[H_3O^+]$ and pH

Solution: $HNO_2(aq) + H_2O(l) \leftrightarrows H_3O^+(aq) + NO_2^-(aq)$

	HNO_2	H_3O^+	NO_2^-
I	0.150M	0.0	0.0
C	- x	x	x
E	0.150 – x	x	x

$$K_a = \frac{[H_3O^+][NO_2^-]}{[HNO_2]} = \frac{(x)(x)}{(0.150 - x)} = 4.6 \times 10^{-4}$$

Assume x is small compared to 0.150

$x^2 = (4.6 \times 10^{-4})(0.150)$ $x = 0.00\underline{8}31M = [H_3O^+]$

Check assumption: $\dfrac{0.008\underline{3}1}{0.150}$ x 100 = 5.54% assumption not valid

$x + 4.6 \times 10^{-4} - 6.9 \times 10^{-5} = 0$

$x = 0.008\underline{0}8M = [H_3O^+]$

$pH = -log(0.008\underline{0}8) = 2.09$

c) **Given:** 0.0195M KOH   **Find:** pH

Conceptual Plan: [KOH] → [OH⁻] → [H₃O⁺] → pH

$K_w = [H_3O^+][OH^-]$ $pH = -log[H_3O^+]$ $pH + pOH = 14$

Solution: $[OH^-] = [KOH] = (0.0195) = 0.0195 \text{ M}$

$$[H_3O^+] = \frac{K_w}{[OH^-]} = \frac{1 \times 10^{-14}}{0.0195 \text{ M}} = 5.1\underline{2}8 \times 10^{-13}M$$

$pH = -log(5.1\underline{2}8 \times 10^{-13}) = 12.290$

d) **Given:** 0.245 CH_3NH_3I, K_b (CH_3NH_3)= 4.4×10^{-4} **Find:** pH

Conceptual Plan: **Identify each species and determine which will contribute to pH. Write a balanced reaction. Prepare an ICE table; represent the change with x; sum the table and determine the equilibrium values; put the equilibrium values in the equilibrium expression and solve for x. Determine [H₃O⁺] → pH**

Solution: $CH_3NH_3^+$ is the conjugate acid of a weak base, therefore, it is acidic. I^- is the conjugate base of a strong acid, therefore, it is pH neutral.

$CH_3NH_4^+ \text{ (aq)} + H_2O\text{(l)} \leftrightarrows CH_3NH_2\text{ (aq)} + H_3O^+\text{ (aq)}$

I	0.245	0.0	0.0
C	-x	x	x
E	0.245– x	x	x

$$K_a = \frac{K_w}{K_b} = \frac{1 \times 10^{-14}}{4.4 \times 10^{-4}} = 2.\underline{2}72 \times 10^{-11} = \frac{(x)(x)}{(0.245 - x)}$$

Assume x is small

$x = 2.\underline{3}6 \times 10^{-6} = [H_3O^+]$ $pH = -log(2.\underline{3}6 \times 10^{-6}) = 5.63$

e) **Given:** 0.312 KC_6H_5O, K_a (HC_6H_5O)= 1.3×10^{-10} **Find:** pH

Conceptual Plan: **Identify each species and determine which will contribute to pH. Write a balanced reaction. Prepare an ICE table; represent the change with x; sum the table and determine the equilibrium values; put the equilibrium values in the equilibrium expression and solve for x. Determine [OH⁻] → pOH → pH**

$pOH = -log[OH^-]$ $pH + pOH = 14$

Solution: K^+ is the counterion of a strong base, therefore, it is pH neutral. $C_6H_5O^-$ is the conjugate base of a weak acid, therefore, it is basic.

$C_6H_5O^- \text{ (aq)} + H_2O\text{(l)} \leftrightarrows HC_6H_5O\text{ (aq)} + OH^-\text{ (aq)}$

I	0. 312	0.0	0.0
C	-x	x	x
E	0. 312– x	x	x

$$K_b = \frac{K_w}{K_a} = \frac{1 \times 10^{-14}}{1.3 \times 10^{-10}} = 7.\underline{6}9 \times 10^{-5} = \frac{(x)(x)}{(0.312 - x)}$$

Assume x is small

$x = 4.\underline{9}0 \times 10^{-3} = [OH^-]$ $pOH = -log(4.\underline{9}0 \times 10^{-3}) = 2.31$

$pH = 14.00 - 2.31 = 11.69$

95. a) sodium cyanide = NaCN   nitric acid = HNO_3

$H^+\text{(aq)} + CN^-\text{ (aq)} \leftrightarrows HCN\text{(aq)}$

b) ammonium chloride = NH_4Cl sodium hydroxide = NaOH

$NH_4^+\text{(aq)} + OH^-\text{ (aq)} \leftrightarrows NH_3\text{(aq)} + H_2O\text{(l)}$

---

c) sodium cyanide = NaCN ammonium bromide = NH_4Br
$NH_4^+(aq) + CN^-(aq) \leftrightarrows NH_3(aq) + HCN(aq)$

d) potassium hydrogen sulfate = $KHSO_4$ lithium acetate = $LiC_2H_3O_2$
$HSO_4^-(aq) + C_2H_3O_2^-(aq) \leftrightarrows SO_4^{2-}(aq) + HC_2H_3O_2(aq)$

e) sodium hypochlorite = NaClO ammonia = NH_3
No reaction, both are bases

96. **Given:** 0.682 g opium, 8.92 mL of 0.0116 M H_2SO_4, **Find:** % morphine
Conceptual Plan:
vol H_2SO_4 → mol H_2SO_4 → mol H_3O^+ → mol morphine → g morphine → % morphine

$$\text{mol} = VM \qquad \frac{\text{mol } H_3O^+}{H_2SO_4} \qquad \frac{\text{mol morphine}}{\text{mol } H_3O^+} \qquad \frac{285\ 4 \text{ g morphine}}{\text{mol morphine}} \qquad \frac{\text{g morphine}}{\text{g opium}} \times 100$$

Solution:

$$8.92 \text{ mL } H_2SO_4 \times \frac{1 \text{ L}}{1000 \text{ mL}} \times \frac{0.0116 \text{ mol } H_2SO_4}{\text{L}} \times \frac{2 \text{ mol } H_3O^+}{\text{mol } H_2SO_4} \times \frac{\text{mol morphine}}{\text{mol } H_3O^+} \times \frac{285.4 \text{ g morphine}}{\text{mol morphine}}$$

$= 0.059\underline{0}6$ g morphine

$$\frac{0.059\underline{0}6 \text{ g morphine}}{0.682 \text{ g opium}} \times 100 = 8.66\ \%$$

97. **Given:** 1.0M urea, pH = 7.050 **Find:** K_a $Hurea^+$
Conceptual Plan: pH → pOH → [OH^-] → K_b(urea) → K_a($Hurea^+$). Write a balanced reaction.
pH + pOH = 14 pOH = -log[OH^-]
Prepare an ICE table; represent the change with x; sum the table and determine the equilibrium values; put the equilibrium values in the equilibrium expression and determine K_b.
Solution: pOH = 14.000 – 7.050 = 6.950 $[OH^{-1}] = 10^{-6.950} = 1.1\underline{2}2 \times 10^{-7}$M

	urea (aq) + H_2O(l) ⇆	$Hurea^+$ (aq)+	OH^-(aq)
I	1.0	0.0	0.0
C	-x	x	x
E	1.0– $1.1\underline{2}2 \times 10^{-7}$	$1.1\underline{2}2 \times 10^{-7}$	$1.1\underline{2}2 \times 10^{-7}$

$$K_b = \frac{[Hurea^+][OH^-]}{[urea]} = \frac{(1.1\underline{2}2 \times 10^{-7})(1.1\underline{2}2 \times 10^{-7})}{(1.0 - 1.1\underline{2}2 \times 10^{-7})} = 1.2\underline{5}89 \times 10^{-14}$$

$$K_a = \frac{K_w}{K_b} = \frac{1.00 \times 10^{-14}}{1.2\underline{5}89 \times 10^{-14}} = 0.79\underline{4}3 = 0.794$$

98. **Given:** 0.1 M $HC_2H_3O_2$, $K_a = 1.8 \times 10^{-5}$; 0.10 M NH_4Cl, $K_b = 1.8 \times 10^{-5}$ **Find:** [NH_3]
Conceptual Plan: Use $HC_2H_3O_2$ to determine [H_3O^+], then use NH_4Cl to determine [NH_3]. Write a balanced reaction. Prepare an ICE table; represent the change with x; sum the table and determine the equilibrium values.
Solution: $HC_2H_3O_2(aq) + H_2O(l) \leftrightarrows H_3O^+(aq) + C_2H_3O_2^-(aq)$

	$HC_2H_3O_2$	H_3O^+	$C_2H_3O_2^-$
I	0.10 M	0.0	0.0
C	- x	x	x
E	0.10– x	x	x

$$K_a = \frac{[H_3O^+][C_2H_3O_2^-]}{[HC_2H_3O_2]} = \frac{(x)(x)}{(0.10 - x)} = 1.8 \times 10^{-5}$$

Assume x is small compared to 0.10

$x^2 = (1.8 \times 10^{-5})(0.10)$

$x = [H_3O^+] = 0.00134$ M

	NH_4^+ (aq) + H_2O(l) ⇆	NH_3 (aq) +	H_3O^+(aq)
I	0.10 M	0.0	0.00134
C	-y	y	y
E	0.10 – y	y	0.00134+ y

$$K_a = \frac{K_w}{K_b} = \frac{1 \times 10^{-14}}{1.8 \times 10^{-5}} = 5.\underline{5}6 \times 10^{-10} = \frac{(y)(0.00134 + y)}{(0.10 - y)}$$

Assume y is small compared to 0.10

$y^2 + 0.00134\,y - 5.56 \times 10^{-11}$ $\quad y = [NH_3] = 4.1 \times 10^{-8}M$

99. The calculation is incorrect because it neglects the contribution from the autoionization of water.
HI is a strong acid, so $[H_3O^+]$ from HI = 1.0×10^{-7}

	$H_2O(l) + H_2O(l) \leftrightarrows$	$H_3O^+(aq)$	+ $OH^-(aq)$
I		1×10^{-7}	0.0
C	-x	x	x
E		$1 \times 10^{-7} + x$	x

$K_w = [H_3O^+][OH^-] = 1.0 \times 10^{-14}$

$(1 \times 10^{-7} + x)(x) = 1.0 \times 10^{-14}$ $\quad x^2 + 1 \times 10^{-7}\,x - 1.0 \times 10^{-14} = 0$

$x = 6.18 \times 10^{-8}$

$[H_3O^+] = (1 \times 10^{-7} + x) = (1 \times 10^{-7} + 6.18 \times 10^{-8}) = 1.618 \times 10^{-7}$

$pH = -\log(1.618 \times 10^{-7}) = 6.79$

100. **Given:** 2.55 g HA, molar mass = 85.0 g/mol, 250.0 g H_2O, FP = - 0.257 °C **Find:** K_a
Conceptual Plan:
g HA → mol HA → m HA → *i* → % ionization and then mol HA → M HA and then: write a

$$\text{mol HA} = \frac{\text{g HA}}{\text{molar mass}} \quad \frac{\text{mol HA}}{\text{kg } H_2O} \quad \Delta T = iK_f m$$

balanced reaction, prepare an ICE table, calculate equilibrium concentrations, and then plug into the equilibrium expression to solve for K_a.

Solution: $2.55 \text{ g HA} \times \frac{1 \text{ mol}}{85.0 \text{ g HA}} = 0.0300 \text{ mol HA}$ $\quad \frac{0.0300 \text{ mol HA}}{\left(250.0 \text{ g } H_2O \times \frac{\text{kg}}{1000 \text{ g}}\right)} = 0.120\ m$

$\Delta T = iK_f m$ $\quad \Delta T = 0.000°C - (-0.257°C) = 0.257°C$

$(0.257\ \cancel{°C}) = i(1.86 \frac{\cancel{°C}}{\cancel{m}})(0.120\ \cancel{m})$ $\quad i = 1.151$ = moles of particles in solution/mol HA

$HA(aq) + H_2O(l) \leftrightarrows$	$H_3O^+(aq) + A^-(aq)$	
1 – y	y	y

So: (1 – y) + y + y = 1.151 $\quad$ y = 0.151 = fraction of HA dissociated

$$\text{M HA} = \frac{0.0300 \text{ mol HA}}{(250.0 \text{ g} + 2.55 \text{ g})\left(\frac{1.00 \text{ mL}}{1.00 \text{ g}}\right)\left(\frac{\text{L}}{1000 \text{ mL}}\right)} = 0.1188 \text{ M}$$

$(0.1188 \text{ M})(0.151) = 0.01794 \text{ M} = [H_3O^+] = [A^-]$ at equilibrium

	$HA(aq) + H_2O(l) \leftrightarrows$	$H_3O^+(aq)$	+ $A^-(aq)$
I	0.1188M	0.0	0.0
C	- x	x	x
E	0. 1188 – 0.01794	0.01794	0.01794

$$K_a = \frac{[H_3O^+][A^-]}{[HA]} = \frac{(0.01794)(0.01794)}{(0.1188 - 0.01794)} = 0.003191 = 3.2 \times 10^{-3}$$

101. **Given:** 0.00115M HCl, 0.01000M $HClO_2$ $K_a = 1.1 \times 10^{-2}$ **Find:** pH
Conceptual Plan: Use HCl to determine $[H_3O^+]$, use $[H_3O^+]$ and $HClO_2$ to determine dissociation of $HClO_2$. Write a balanced reaction, prepare an ICE table, calculate equilibrium concentrations, and then plug into the equilibrium expression.
Solution: 0.00115M HCl = 0.00115M H_3O^+

	$HClO_2(aq) + H_2O(l) \leftrightarrows$	$H_3O^+(aq)$ +	$ClO_2^-(aq)$
I	0.0100M	0.00115	0.0
C	- x	x	x
E	0.01000 – x	0.00115 + x	x

$$K_a = \frac{[H_3O^+][ClO_2^-]}{[HClO_2]} = \frac{(0.00115 + x)(x)}{(0.0100 - x)} = 1.1 \times 10^{-2}$$

$x^2 + 0.01215x - 1.1 \times 10^{-4} = 0$ $\quad x = 0.006045$

$[H_3O^+] = 0.00115 + 0.006045 = 0.0071\underline{9}5$

$pH = -\log(0.0071\underline{9}5) = 2.14$

102. Volume should be increased to 4 L.
$HA(aq) + H_2O(l) \leftrightarrows H_3O^+(aq) + A^-(aq)$
Initial equilibrium conditions: $[HA] = A$, $[H_3O^+] = [A^-] = x$

$$K_a = \frac{[H_3O^+][A^-]}{[HA]} = \frac{(x)(x)}{(A)}$$

Second equilibrium conditions: $[HA] = A/y$, $[H_3O^+] = [A^-] = x/2$

$$K_a = \frac{[H_3O^+][A^-]}{[HA]} = \frac{(x)(x)}{(A)} = \frac{\left(\frac{x}{2}\right)\left(\frac{x}{2}\right)}{\left(\frac{A}{y}\right)}$$

$\frac{x^2A}{y} = \frac{x^2A}{4}$ $\quad$ y = 4, so for the concentration of $[H^+]$ to be halved,

the [A] needs to decrease to 1/4 th the original concentration.

So, the volume has to increase to 4 L

103. **Given:** 1.0M HA, $K_a = 1.0 \times 10^{-8}$ $K = 4.0$ for reaction 2 **Find:** $[H^+]$, $[A^-]$, $[HA_2^-]$
Conceptual Plan: Combine reaction 1 and reaction 2 and determine equilibrium expression and the value of *K*. Prepare an ICE table, calculate equilibrium concentrations, and then plug into the equilibrium expression
Solution:

	$HA(aq)$	$\leftrightarrows$	$H^+(aq) +$	$A^-(aq)$	$K = 1.0 \times 10^{-8}$
	$HA(aq) + A^-(aq)$	$\leftrightarrows$	$HA_2^+(aq)$		$K = 4.0$
	$2HA(aq)$	$\leftrightarrows$	$H^+(aq) +$	$HA_2^+(aq)$	$K = 4.0 \times 10^{-8}$
I	1.0		0.0	0.0	
C	- 2x		x	x	
E	1.0 – 2x		x	x	

$$K = \frac{[H^+][HA_2^-]}{[HA]^2} = 4.0 \times 10^{-8} = \frac{(x)(x)}{(1.0 - 2x)^2}$$

$x^2 + 16 \times 10^{-8}x - 4 \times 10^{-8} = 0$ $\quad x = 1.\underline{9}992 \times 10^{-4}$

	$HA(aq)$	$\leftrightarrows$	$H^+(aq) +$	$A^-(aq)$	$K = 1.0 \times 10^{-8}$
I	1.0		1.9992×10^{-4}	0.0	
C	- 2y		y	y	
E	1.0 – 2y		$1.9992 \times 10^{-4} + y$	y	

$$K = \frac{[H^+][A^-]}{[HA]} = 1.0 \times 10^{-8} = \frac{(1.9992 \times 10^{-4} + y)(y)}{(1.0 - y)}$$

$y^2 + 1.9992 \times 10^{-4}x - 1 \times 10^{-8} = 0$ $\quad y = 4.\underline{2}9 \times 10^{-5}$

$[H^+] = x + y = 1.\underline{9}992 \times 10^{-4} + 4.\underline{2}9 \times 10^{-5} = 2.4 \times 10^{-4}$

$[A^-] = y = 4.29 \times 10^{-5}$

$[HA_2^-] = x = 2.0 \times 10^{-4}$

104. In the gas phase, $(CH_3)_3N$ is a stronger Lewis base than CH_3NH_2 because the polar C – H bond repels the H^+ that needs to be added. In the liquid phase, the steric hindrance from the size of the CH_3 groups becomes more pronounced, therefore, it is harder to add the H^+ ion.

105. Solution b would be most acidic.
a) 0.0100 M HCl (strong acid) and 0.0100 M KOH(strong base), since the concentrations are equal, the acid and base will completely neutralize each other and the resulting solution will be pH neutral.

b) 0.0100 M HF (weak acid) and 0.0100 KBr (salt). K_a (HF) = 3.5 x 10^{-4}. The weak acid will produce an acidic solution. K^+ is the counterion of a strong base and is pH neutral. Br^- is the conjugate base of a strong acid and is pH neutral.

c) 0.0100 M NH_4Cl (salt) and 0.100 M CH_3NH_3Br. $K_b(NH_3)$ = 1.8 x 10^{-5}, $K_b(CH_3NH_2)$ = 4.4 x 10^{-4}. NH_4^+ is the conjugate acid of a weak base and $CH_3NH_3^+$ is the conjugate acid of a weak base. Cl^- and Br^- are the conjugate base of a strong acid and will be pH neutral. The solution will be acidic, however, K_a for the conjugate acids is much smaller K_a for HF, so the solution will be acidic, but not as acidic as HF.

d) 0.100 M NaCN (salt) and 0.100 M $CaCl_2$. Na^+ and Ca^{2+} ion are the counterion of a strong base, therefore, they are pH neutral. Cl^- is the conjugate base of a strong acid and is pH neutral. CN^- is the conjugate base of a weak acid and will produce a basic solution.

106. Solution a would be most basic.

a) 0.100 M NaClO(salt) and 0.100 M NaF(salt). Na^+ is the counterion of a strong base and is pH neutral. ClO^- is the conjugate base of a weak acid (HClO, K_a = 2.9 x 10^{-8}). F^- is the conjugate base of a weak acid (HF, K_a = 3.5 x 10^{-4}), the solution will be basic and ClO^- is a stronger base than F^-.

b) 0.0100 M KCl(salt) and 0.0100 M $KClO_2$(salt). K^+ is the counterion of a strong base and is pH neutral. Cl^- is the conjugate base of a strong acid and is pH neutral. ClO_2^- is the conjugate base of a weak acid ($HClO_2$, K_a = 1.1 x 10^{-2}) and will produce a basic solution, however, ClO_2^- is a weaker base than ClO^-

c) 0.0100 M HNO_3 (strong acid) and 0.0100 M NaOH(strong base), since the concentrations are equal, the acid and base will completely neutralize each other and the resulting solution will be pH neutral.

d) 0.0100 M NH_4Cl(salt) and 0.0100 M HCN(weak acid). NH_4^+ is the conjugate acid of a weak base and will be acidic. HCN is a weak acid and will be acidic. Cl^- is the conjugate base of a strong acid and is pH neutral.

107. $CH_3COOH < CH_2ClCOOH < CHCl_2COOH < CCl_3COOH$

Since Cl is more electronegative than H, as you add Cl you increase the amount of electronegativity and pull electron density away from the O – H bond, making it more acidic.

Chapter 16
Aqueous Ionic Equilibrium

1. The only solution that HNO_2 will ionize less in is d) 0.10 M $NaNO_2$. It is the only solution that generates a common ion (NO_2^-) with nitrous acid.

2. Formic acid is $HCHO_2$, which dissociates to H+ and CHO_2^-. The only solution that generates a common ion (CHO_2^-) with formic acid is c) $NaCHO_2$.

3. a) **Given:** 0.15 M $HCHO_2$ and 0.10 M $NaCHO_2$ **Find:** pH **Other:** K_a ($HCHO_2$) = 1.8×10^{-4}

Conceptual Plan:

M $NaCHO_2$ → M CHO_2^- then M $HCHO_2$, M CHO_2^- → $[H_3O^+]$ → pH

$NaCHO_2\,(aq) \rightarrow Na^+\,(aq) + CHO_2^-\,(aq)$ ICE Chart $pH = -\log[H_3O^+]$

Solution: Since 1 CHO_2^- ion is generated for each $NaCHO_2$, $[CHO_2^-] = 0.10$ M CHO_2^-.

$$HCHO_2\,(aq) + H_2O\,(l) \rightleftharpoons H_3O^+\,(aq) + CHO_2^-\,(aq)$$

	$[HCHO_2]$	$[H_3O^+]$	$[CHO_2^-]$
Initial	0.15	≈ 0.00	0.10
Change	$-x$	$+x$	$+x$
Equil	$0.15 - x$	$+x$	$0.10 + x$

$K_a = \dfrac{[H_3O^+][CHO_2^-]}{[HCHO_2]} = 1.8 \times 10^{-4} = \dfrac{x(0.10 + x)}{0.15 - x}$ Assume x is small ($x << 0.10 < 0.15$) so

$\dfrac{x(0.10 + \cancel{x})}{0.15 - \cancel{x}} = 1.8 \times 10^{-4} = \dfrac{x(0.10)}{0.15}$ and $x = 2.7 \times 10^{-4}$ M = $[H_3O^+]$. Confirm that assumption is valid

$\dfrac{2.7 \times 10^{-4}}{0.10} \times 100\,\% = 0.27\,\%$ so assumption is valid. Finally,

$pH = -\log[H_3O^+] = -\log(2.7 \times 10^{-4}) = 3.57$

Check: The units (none) are correct. The magnitude of the answer makes physical sense because pH should be greater than $-\log(0.15) = 0.82$ because this is a weak acid and there is a common ion effect.

b) **Given:** 0.12 M NH_3 and 0.18 M NH_4Cl **Find:** pH **Other:** K_b (NH_3) = 1.79×10^{-5}

Conceptual Plan:

M NH_4Cl → M NH_4^+ then M NH_3, M NH_4^+ → $[OH^-]$ → $[H_3O^+]$ → pH

$NH_4Cl\,(aq) \rightarrow NH_4^+\,(aq) + Cl^-\,(aq)$ ICE Chart $K_w = [H_3O^+][OH^-]$ $pH = -\log[H_3O^+]$

Solution: Since 1 NH_4^+ ion is generated for each NH_4Cl, $[NH_4^+] = 0.18$ M NH_4^+.

$$NH_3\,(aq) + H_2O\,(l) \rightleftharpoons NH_4^+\,(aq) + OH^-\,(aq)$$

	$[NH_3]$	$[NH_4^+]$	$[OH^-]$
Initial	0.12	0.18	≈ 0.00
Change	$-x$	$+x$	$+x$
Equil	$0.12 - x$	$0.18 + x$	$+x$

$K_b = \dfrac{[NH_4^+][OH^-]}{[NH_3]} = 1.79 \times 10^{-5} = \dfrac{(0.18 + x)x}{0.12 - x}$

Assume x is small ($x << 0.12 < 0.18$) so $\dfrac{(0.18 + \cancel{x})x}{0.12 - \cancel{x}} = 1.79 \times 10^{-5} = \dfrac{(0.18)x}{0.12}$ and $x = 1.\underline{1}9333 \times 10^{-5}$ M = $[OH^-]$. Confirm that assumption is valid $\dfrac{1.\underline{1}9333 \times 10^{-5}}{0.12} \times 100\,\% = 9.9 \times 10^{-3}\,\%$ so assumption is valid.

$K_w = [H_3O^+][OH^-]$ so $[H_3O^+] = \dfrac{K_w}{[OH^-]} = \dfrac{1.0 \times 10^{-14}}{1.\underline{1}9333 \times 10^{-5}} = 8.\underline{3}799 \times 10^{-10}$ M. Finally,

$pH = -\log[H_3O^+] = -\log(8.\underline{3}799 \times 10^{-10}) = 9.08$

Check: The units (none) are correct. The magnitude of the answer makes physical sense because pH should be less than $14 + \log(0.12) = 13.1$ because this is a weak base and there is a common ion effect.

4. a) **Given:** 0.175 M $HC_2H_3O_2$ and 0.110 M $KC_2H_3O_2$ **Find:** pH
Other: K_a ($HC_2H_3O_2$) = 1.8×10^{-5}
Conceptual Plan:
M $KC_2H_3O_2$ → M $C_2H_3O_2^-$ then M $HC_2H_3O_2$, M $C_2H_3O_2^-$ → $[H_3O^+]$ → pH

$KC_2H_3O_2\,(aq) \rightarrow K^+\,(aq) + C_2H_3O_2^-\,(aq)$ ICE Chart $pH = -\log [H_3O^+]$

Solution: Since 1 $C_2H_3O_2^-$ ion is generated for each $KC_2H_3O_2$, $[C_2H_3O_2^-] = 0.110$ M $C_2H_3O_2^-$.

$$HC_2H_3O_2\,(aq) + H_2O\,(l) \rightleftharpoons H_3O^+\,(aq) + C_2H_3O_2^-\,(aq)$$

	$[HC_2H_3O_2]$	$[H_3O^+]$	$[C_2H_3O_2^-]$
Initial	0.175	≈ 0.00	0.110
Change	$-x$	$+x$	$+x$
Equil	$0.175 - x$	$+x$	$0.110 + x$

$$K_a = \frac{[H_3O^+][HC_2H_3O_2^-]}{[HC_2H_3O_2]} = 1.8 \times 10^{-5} = \frac{x(0.110 + x)}{0.175 - x}$$ Assume x is small ($x << 0.110 < 0.175$) so

$$\frac{x(0.110 + \cancel{x})}{0.175 - \cancel{x}} = 1.8 \times 10^{-5} = \frac{x(0.110)}{0.175}$$ and $x = 2.\underline{8}636 \times 10^{-5}$ M = $[H_3O^+]$. Confirm that assumption is valid

$$\frac{2.\underline{8}636 \times 10^{-5}}{0.110} \times 100\,\% = 0.026\,\%$$ so assumption is valid. Finally,

$pH = -\log [H_3O^+] = -\log (2.\underline{8}636 \times 10^{-5}) = 4.54$

Check: The units (none) are correct. The magnitude of the answer makes physical sense because pH should be greater than $-\log (0.110) = 0.96$ because this is a weak acid and there is a common ion effect.

b) **Given:** 0.195 M CH_3NH_2 and 0.105 M CH_3NH_3Br **Find:** pH
Other: K_b (CH_3NH_2) = 4.4×10^{-4}
Conceptual Plan: M CH_3NH_3Br → M $CH_3NH_3^+$ then

$CH_3NH_3Br\,(aq) \rightarrow CH_3NH_3^+\,(aq) + Br^-\,(aq)$

M CH_3NH_3, M $CH_3NH_3^+$ → $[OH^-]$ → $[H_3O^+]$ → pH

ICE Chart $K_w = [H_3O^+][OH^-]$ $pH = -\log [H_3O^+]$

Solution: Since 1 $CH_3NH_3^+$ ion is generated for each CH_3NH_3Br, $[CH_3NH_3^+] = 0.105$ M $CH_3NH_3^+$.

$$CH_3NH_2\,(aq) + H_2O\,(l) \rightleftharpoons CH_3NH_3^+\,(aq) + OH^-\,(aq)$$

	$[CH_3NH_2]$	$[CH_3NH_3^+]$	$[OH^-]$
Initial	0.195	0.105	≈ 0.00
Change	$-x$	$+x$	$+x$
Equil	$0.195 - x$	$0.105 + x$	$+x$

$$K_b = \frac{[CH_3NH_3^+][OH^-]}{[CH_3NH_2]} = 4.4 \times 10^{-4} = \frac{(0.105 + x)x}{0.195 - x}$$

Assume x is small ($x << 0.105 < 0.195$) so $\frac{(0.105 + \cancel{x})x}{0.195 - \cancel{x}} = 4.4 \times 10^{-4} = \frac{(0.105)x}{0.195}$ and $x = 8.\underline{1}7143 \times 10^{-4}$ M

$= [OH^-]$. Confirm that assumption is valid $\frac{8.\underline{1}7143 \times 10^{-4}}{0.105} \times 100\,\% = 0.78\,\%$ so assumption is valid.

$$K_w = [H_3O^+][OH^-] \text{ so } [H_3O^+] = \frac{K_w}{[OH^-]} = \frac{1.0 \times 10^{-14}}{8.\underline{1}7143 \times 10^{-4}} = 1.\underline{2}2378 \times 10^{-11} \text{ M}.$$

Finally, $pH = -\log [H_3O^+] = -\log (1.\underline{2}2378 \times 10^{-11}) = 10.91$.

Check: The units (none) are correct. The magnitude of the answer makes physical sense because pH should be less than $14 + \log (0.195) = 13.3$ because this is a weak base and there is a common ion effect.

5. **Given:** 0.15 M $HC_7H_5O_2$ in pure water and in 0.10 M $NaC_7H_5O_2$
Find: % ionization in both solutions **Other:** K_a ($HC_7H_5O_2$) = 6.5×10^{-5}

Conceptual Plan:
pure water: M $HC_7H_5O_2$ → $[H_3O^+]$ → % ionization then in $NaC_7H_5O_2$ solution:

ICE Chart $\quad \% \ ionization = \frac{[H_3O^+]_{equil}}{[HC_7H_5O_2]_0} \times 100\ \%$

M $NaC_7H_5O_2$ → M $C_7H_5O_2^-$ then M H $C_7H_5O_2$, M $C_7H_5O_2^-$ → $[H_3O^+]$ → % ionization

$NaC_7H_5O_2\ (aq) \rightarrow Na^+\ (aq) + C_7H_5O_2^-\ (aq)$ $\quad$ ICE Chart $\quad \% \ ionization = \frac{[H_3O^+]_{equil}}{[HC_7H_5O_2]_0} \times 100\ \%$

Solution: in pure water:

$$HC_7H_5O_2\,(aq) + H_2O(l) \rightleftharpoons H_3O^+\,(aq) + C_7H_5O_2^-\,(aq)$$

	$[HC_7H_5O_2]$	$[H_3O^+]$	$[C_7H_5O_2^-]$
Initial	0.15	≈ 0.00	0.00
Change	$-x$	$+x$	$+x$
Equil	$0.15 - x$	$+x$	$+x$

$$K_a = \frac{[H_3O^+][C_7H_5O_2^-]}{[HC_7H_5O_2]} = 6.5 \times 10^{-5} = \frac{x^2}{0.15 - x}$$

Assume x is small ($x << 0.10$) so $\frac{x^2}{0.15 - \cancel{x}} = 6.5 \times 10^{-5} = \frac{x^2}{0.15}$ and $x = 3.\underline{1}225 \times 10^{-3}$ M $= [H_3O^+]$. Then

$\% \ ionization = \frac{[H_3O^+]_{equil}}{[HC_7H_5O_2]_0} \times 100\ \% = \frac{3.\underline{1}225 \times 10^{-3}}{0.15} \times 100\ \% = 2.1\ \%$, which also confirms that the assumption is valid (since it is less than 5 %). In $NaC_7H_5O_2$ solution: Since 1 $C_7H_5O_2^-$ ion is generated for each $NaC_7H_5O_2$, $[C_7H_5O_2^-] = 0.10$ M $C_7H_5O_2^-$.

$$HC_7H_5O_2\,(aq) + H_2O(l) \rightleftharpoons H_3O^+\,(aq) + C_7H_5O_2^-\,(aq)$$

	$[HC_7H_5O_2]$	$[H_3O^+]$	$[C_7H_5O_2^-]$
Initial	0.15	≈ 0.00	0.10
Change	$-x$	$+x$	$+x$
Equil	$0.15 - x$	$+x$	$0.10 + x$

$K_a = \frac{[H_3O^+][C_7H_5O_2^-]}{[HC_7H_5O_2]} = 6.5 \times 10^{-5} = \frac{x(0.10 + x)}{0.15 - x}$ Assume x is small ($x << 0.10 < 0.15$) so

$\frac{x(0.10 + \cancel{x})}{0.15 - \cancel{x}} = 6.5 \times 10^{-5} = \frac{x(0.10)}{0.15}$ and $x = 9.\underline{7}5 \times 10^{-5}$ M $= [H_3O^+]$, then

$\% \ ionization = \frac{[H_3O^+]_{equil}}{[HC_7H_5O_2]_0} \times 100\ \% = \frac{9.\underline{7}5 \times 10^{-5}}{0.15} \times 100\ \% = 0.065\ \%$, which also confirms that the assumption is valid (since it is less than 5 %). The percent ionization in the sodium benzoate solution is less than in pure water because of the common ion effect. An increase in one of the products (benzoate ion) shifts the equilibrium to the left, so less acid dissociates.

Check: The units (%) are correct. The magnitude of the answer makes physical sense because the acid is weak and so the percent ionization is low. With a common ion present, the percent ionization decreases.

6. **Given:** 0.13 M $HCHO_2$ in pure water and in 0.11 M $KCHO_2$ **Find:** % ionization in both solutions
Other: K_a ($HCHO_2$) = 1.8×10^{-4}
Conceptual Plan: pure water: M $HCHO_2$ → $[H_3O^+]$ → % ionization then in $KCHO_2$ solution:

ICE Chart $\quad \% \ ionization = \frac{[H_3O^+]_{equil}}{[HCHO_2]_0} \times 100\ \%$

M $KCHO_2$ → M CHO_2^- then M $HCHO_2$, M CHO_2^- → $[H_3O^+]$ → % ionization

$KCHO_2\ (aq) \rightarrow K^+\ (aq) + CHO_2^-\ (aq)$ $\quad$ *ICE Chart* $\quad \% \ ionization = \frac{[H_3O^+]_{equil}}{[HCHO_2]_0} \times 100\ \%$

Solution: in pure water:

$$HCHO_2\,(aq) + H_2O\,(l) \rightleftharpoons H_3O^+\,(aq) + CHO_2^-\,(aq)$$

	$[HCHO_2]$	$[H_3O^+]$	$[CHO_2^-]$
Initial	0.13	≈ 0.00	0.00
Change	$-x$	$+x$	$+x$
Equil	$0.13 - x$	$+x$	$+x$

$$K_a = \frac{[H_3O^+][CHO_2^-]}{[HCHO_2]} = 1.8 \times 10^{-4} = \frac{x^2}{0.13 - x}$$

Assume x is small ($x << 0.10$) so $\dfrac{x^2}{0.13 - \cancel{x}} = 1.8 \times 10^{-4} = \dfrac{x^2}{0.13}$ and $x = 4.\underline{8}374 \times 10^{-3}$ M $= [H_3O^+]$. Then

$\% \ ionization = \dfrac{[H_3O^+]_{equil}}{[HCHO_2]_0} \times 100\ \% = , \dfrac{4.\underline{8}374 \times 10^{-3}}{0.13} \times 100\ \% = 3.7\ \%$ which also confirms that the assumption is valid (since it is less than 5 %).

In $KCHO_2$ solution: Since 1 CHO_2^- ion is generated for each $KCHO_2$, $[CHO_2^-] = 0.11$ M CHO_2^-.

$$HCHO_2\,(aq) + H_2O\,(l) \rightleftharpoons H_3O^+\,(aq) + CHO_2^-\,(aq)$$

	$[HCHO_2]$	$[H_3O^+]$	$[CHO_2^-]$
Initial	0.13	≈ 0.00	0.10
Change	$-x$	$+x$	$+x$
Equil	$0.13 - x$	$+x$	$0.10 + x$

$$K_a = \frac{[H_3O^+][CHO_2^-]}{[HCHO_2]} = 1.8 \times 10^{-4} = \frac{x(0.11 + x)}{0.13 - x}$$

Assume x is small ($x << 0.11 < 0.13$) so $\dfrac{x(0.11 + \cancel{x})}{0.13 - \cancel{x}} = 1.8 \times 10^{-4} = \dfrac{x(0.11)}{0.13}$ and $x = 2.\underline{1}273 \times 10^{-4}$ M $= [H_3O^+]$.

Then $\% \ ionization = \dfrac{[H_3O^+]_{equil}}{[HCHO_2]_0} \times 100\ \% = \dfrac{2.\underline{1}273 \times 10^{-4}}{0.13} \times 100\ \% = 0.16\ \%$, which also confirms that the assumption is valid (since it is less than 5 %). The percent ionization in the potassium formate solution is less than in pure water because of the common ion effect. An increase in one of the products (formate ion) shifts the equilibrium to the left, so less acid dissociates.

Check: The units (%) are correct. The magnitude of the answer makes physical sense because the acid is weak and so the percent ionization is low. With a common ion present, the percent ionization decreases.

7. a) **Given:** 0.15 M HF **Find:** pH **Other:** K_a (HF) = 3.5×10^{-4}

Conceptual Plan: M HF → $[H_3O^+]$ → pH

ICE Chart; $pH = -\log [H_3O^+]$

Solution:

$$HF\,(aq) + H_2O\,(l) \rightleftharpoons H_3O^+\,(aq) + F^-\,(aq)$$

	$[HF]$	$[H_3O^+]$	$[F^-]$
Initial	0.15	≈ 0.00	0.00
Change	$-x$	$+x$	$+x$
Equil	$0.15 - x$	$+x$	$+x$

$$K_a = \frac{[H_3O^+][F^-]}{[HF]} = 3.5 \times 10^{-4} = \frac{x^2}{0.15 - x}$$

Assume x is small ($x << 0.15$) so $\dfrac{x^2}{0.15 - \cancel{x}} = 3.5 \times 10^{-4} = \dfrac{x^2}{0.15}$ and $x = 7.\underline{2}457 \times 10^{-3}$ M $= [H_3O^+]$. Confirm that assumption is valid $\dfrac{7.\underline{2}457 \times 10^{-3}}{0.15} \times 100\ \% = 4.8\ \% < 5\ \%$ so assumption is valid.

Finally, $pH = -\log [H_3O^+] = -\log (7.\underline{2}457 \times 10^{-3}) = 2.14$.

Check: The units (none) are correct. The magnitude of the answer makes physical sense because pH should be greater than $-\log (0.15) = 0.82$ because this is a weak acid.

b) **Given:** 0.15 M NaF **Find:** pH **Other:** K_a (HF) = 3.5×10^{-4}

Conceptual Plan:

M NaF → M F^- and $K_a \rightarrow K_b$ then M F^- → $[OH^-]$ → $[H_3O^+]$ → pH

$NaF\,(aq) \rightarrow Na^+\,(aq) + F^-\,(aq)$; $K_w = K_a\,K_b$; *ICE Chart*; $K_w = [H_3O^+][OH^-]$; $pH = -\log [H_3O^+]$

Solution: Since 1 F^- ion is generated for each NaF, $[F^-] = 0.15$ M F^-. Since $K_w = K_a K_b$, rearrange to solve for K_b. $K_b = \frac{K_w}{K_a} = \frac{1.00 \times 10^{-14}}{3.5 \times 10^{-4}} = 2.\underline{8}571 \times 10^{-11}$

$$F^-(aq) + H_2O(l) \rightleftharpoons HF(aq) + OH^-(aq)$$

	$[F^-]$	[HF]	$[OH^-]$
Initial	0.15	0.00	≈ 0.00
Change	$-x$	$+x$	$+x$
Equil	$0.15 - x$	$+x$	$+x$

$$K_b = \frac{[HF][OH^-]}{[F^-]} = 2.\underline{8}571 \times 10^{-11} = \frac{x^2}{0.15 - x}$$

Assume x is small ($x << 0.15$) so $\frac{x^2}{0.15 - \cancel{x}} = 2.\underline{8}571 \times 10^{-11} = \frac{x^2}{0.15}$ and $x = 2.\underline{0}702 \times 10^{-6}$ M = $[OH^-]$.

Confirm that assumption is valid $\frac{2.\underline{0}702 \times 10^{-6}}{0.15} \times 100\ \% = 0.0014\ \% < 5\ \%$ so assumption is valid.

$K_w = [H_3O^+][OH^-]$ so $[H_3O^+] = \frac{K_w}{[OH^-]} = \frac{1.0 \times 10^{-14}}{2.\underline{0}702 \times 10^{-6}} = 4.\underline{8}305 \times 10^{-9}$ M. Finally,

$pH = -\log[H_3O^+] = -\log(4.\underline{8}305 \times 10^{-9}) = 8.32$.

Check: The units (none) are correct. The magnitude of the answer makes physical sense because pH should be slightly basic, since the fluoride ion is a very weak base.

c) **Given:** 0.15 M HF and 0.15 M NaF **Find:** pH **Other:** K_a (HF) = 3.5×10^{-4}

Conceptual Plan: M NaF → M F^- then M HF, M F^- → $[H_3O^+]$ → pH

NaF (aq) → Na^+ (aq) + F^- (aq) ICE Chart $pH = -\log[H_3O^+]$

Solution: Since 1 F^- ion is generated for each NaF, $[F^-] = 0.15$ M F^-.

$$HF(aq) + H_2O(l) \rightleftharpoons H_3O^+(aq) + F^-(aq)$$

	[HF]	$[H_3O^+]$	$[F^-]$
Initial	0.15	≈ 0.00	0.15
Change	$-x$	$+x$	$+x$
Equil	$0.15 - x$	$+x$	$0.15 + x$

$$K_a = \frac{[H_3O^+][F^-]}{[HF]} = 3.5 \times 10^{-4} = \frac{x(0.15 + x)}{0.15 - x}$$

Assume x is small ($x << 0.15$) so $\frac{x(0.15 + \cancel{x})}{0.15 - \cancel{x}} = 3.5 \times 10^{-4} = \frac{x(\cancel{0.15})}{\cancel{0.15}}$ and $x = 3.5 \times 10^{-4}$ M = $[H_3O^+]$.

Confirm that assumption is valid $\frac{3.5 \times 10^{-4}}{0.15} \times 100\ \% = 0.23\ \% < 5\ \%$ so assumption is valid.

Finally, $pH = -\log[H_3O^+] = -\log(3.5 \times 10^{-4}) = 3.46$.

Check: The units (none) are correct. The magnitude of the answer makes physical sense because pH should be greater than in part a (2.14) because of the common ion effect suppressing the dissociation of the weak acid.

8. a) **Given:** 0.18 M CH_3NH_2 **Find:** pH **Other:** K_b (CH_3NH_2) = 4.4×10^{-4}

Conceptual Plan: M CH_3NH_3 → $[OH^-]$ → $[H_3O^+]$ → pH

ICE Chart $K_w = [H_3O^+][OH^-]$ $pH = -\log[H_3O^+]$

Solution:

$$CH_3NH_2(aq) + H_2O(l) \rightleftharpoons CH_3NH_3^+(aq) + OH^-(aq)$$

	$[CH_3NH_2]$	$[CH_3NH_3^+]$	$[OH^-]$
Initial	0.18	0.00	≈ 0.00
Change	$-x$	$+x$	$+x$
Equil	$0.18 - x$	$+x$	$+x$

$$K_b = \frac{[CH_3NH_3^+][OH^-]}{[CH_3NH_2]} = 4.4 \times 10^{-4} = \frac{x^2}{0.18 - x}$$

Assume x is small ($x << 0.18$) so $\frac{x^2}{0.18 - \cancel{x}} = 4.4 \times 10^{-4} = \frac{x^2}{0.18}$ and $x = 8.\underline{8}994 \times 10^{-3}$ M = $[OH^-]$.

Confirm that assumption is valid $\frac{8.8994 \times 10^{-3}}{0.18} \times 100\% = 4.9\% < 5\%$ so assumption is valid.

$$K_w = [H_3O^+][OH^-] \text{ so } [H_3O^+] = \frac{K_w}{[OH^-]} = \frac{1.0 \times 10^{-14}}{8.8994 \times 10^{-3}} = 1.1237 \times 10^{-12}\text{ M}.$$

Finally, $pH = -\log[H_3O^+] = -\log(1.1237 \times 10^{-12}) = 11.95$.

Check: The units (none) are correct. The magnitude of the answer makes physical sense because pH should be less than 14 + log (0.18) = 13.3 because this is a weak base.

b) **Given:** 0.18 M CH_3NH_3Cl **Find:** pH **Other:** K_b (CH_3NH_2) = 4.4×10^{-4}

Conceptual Plan:

M CH_3NH_3Cl → M $CH_3NH_3^+$ and K_b → K_a then M $CH_3NH_3^+$ → $[H_3O^+]$ → pH

$CH_3NH_3Cl\ (aq) \rightarrow CH_3NH_3^+\ (aq) + Cl^-\ (aq)$ $K_w = K_a K_b$ *ICE Chart* $pH = -\log[H_3O^+]$

Solution: Since 1 $CH_3NH_3^+$ ion is generated for each CH_3NH_3Cl, $[CH_3NH_3^+]$ = 0.18 M $CH_3NH_3^+$. Since $K_w = K_a K_b$, rearrange to solve for K_b. $K_b = \frac{K_w}{K_a} = \frac{1.00 \times 10^{-14}}{4.4 \times 10^{-4}} = 2.2727 \times 10^{-11}$

$$CH_3NH_3^+(aq) + H_2O(l) \rightleftharpoons H_3O^+(aq) + CH_3NH_2(aq)$$

	$[CH_3NH_3^+]$	$[H_3O^+]$	$[CH_3NH_2]$
Initial	0.18	≈ 0.00	0.00
Change	$-x$	$+x$	$+x$
Equil	$0.18 - x$	$+x$	$+x$

$$K_b = \frac{[H_3O^+][CH_3NH_2]}{[CH_3NH_3^+]} = 2.2727 \times 10^{-11} = \frac{x^2}{0.18 - x}$$ Assume x is small ($x << 0.18$) so

$$\frac{x^2}{0.18 - \cancel{x}} = 2.2727 \times 10^{-11} = \frac{x^2}{0.18}$$ and $x = 2.0226 \times 10^{-6}\text{ M} = [H_3O^+]$.

Confirm that assumption is valid $\frac{2.0226 \times 10^{-6}}{0.18} \times 100\% = 0.0012\% < 5\%$ so assumption is valid.

Finally, $pH = -\log[H_3O^+] = -\log(2.0226 \times 10^{-6}) = 5.69$.

Check: The units (none) are correct. The magnitude of the answer makes physical sense because pH should be slightly acidic, since the methylammonium cation is a very weak acid.

c) **Given:** 0.18 M CH_3NH_2 and 0.18 M CH_3NH_3Cl **Find:** pH **Other:** K_b (CH_3NH_2) = 4.4×10^{-4}

Conceptual Plan: M CH_3NH_3Cl → M $CH_3NH_3^+$ then

$CH_3NH_3Cl\ (aq) \rightarrow CH_3NH_3^+\ (aq) + Cl^-\ (aq)$

M CH_3NH_2, M $CH_3NH_3^+$ → $[OH^-]$ → $[H_3O^+]$ → pH

ICE Chart $K_w = [H_3O^+][OH^-]$ $pH = -\log[H_3O^+]$

Solution: Since 1 $CH_3NH_3^+$ ion is generated for each CH_3NH_3Cl, $[CH_3NH_3^+]$ = 0.105 M $CH_3NH_3^+$.

$$CH_3NH_2(aq) + H_2O(l) \rightleftharpoons CH_3NH_3^+(aq) + OH^-(aq)$$

	$[CH_3NH_2]$	$[CH_3NH_3^+]$	$[OH^-]$
Initial	0.18	0.18	≈ 0.00
Change	$-x$	$+x$	$+x$
Equil	$0.18 - x$	$0.18 + x$	$+x$

$$K_b = \frac{[CH_3NH_3^+][OH^-]}{[CH_3NH_2]} = 4.4 \times 10^{-4} = \frac{(0.18 + x)x}{0.18 - x}$$

Assume x is small ($x << 0.18$) so $\frac{(0.18 + \cancel{x})x}{0.18 - \cancel{x}} = 4.4 \times 10^{-4} = \frac{(\cancel{0.18})x}{\cancel{0.18}}$ and $x = 4.4 \times 10^{-4}\text{ M} = [OH^-]$.

Confirm that assumption is valid $\frac{4.4 \times 10^{-4}}{0.18} \times 100\% = 0.24\% < 5\%$ so assumption is valid.

$$K_w = [H_3O^+][OH^-] \text{ so } [H_3O^+] = \frac{K_w}{[OH^-]} = \frac{1.0 \times 10^{-14}}{4.4 \times 10^{-4}} = 2.2727 \times 10^{-11}\text{ M}.$$ Finally,

$pH = -\log[H_3O^+] = -\log(2.2727 \times 10^{-11}) = 10.64$

Check: The units (none) are correct. The magnitude of the answer makes physical sense because pH should be less than 14 + log (0.18) = 13.3 because this is a weak base and there is a common ion effect.

9. When an acid (such as HCl) is added it will react with the conjugate base of the buffer system as follows: $HCl + NaC_2H_3O_2 \rightarrow HC_2H_3O_2 + NaCl$. When a base (such as NaOH) is added it will react with the weak acid of the buffer system as follows: $NaOH + HC_2H_3O_2 \rightarrow H_2O + NaC_2H_3O_2$. The reaction generates the other buffer system component.

10. When an acid (such as HCl) is added it will react with the conjugate base of the buffer system as follows: $HCl + NH_3 \rightarrow NH_4Cl$. When a base (such as NaOH) is added it will react with the weak acid of the buffer system as follows: $NaOH + NH_4Cl \rightarrow H_2O + NH_3 + NaCl$. The reaction generates the other buffer system component.

11. a) **Given:** 0.15 M $HCHO_2$ and 0.10 M $NaCHO_2$ **Find:** pH **Other:** K_a ($HCHO_2$) = 1.8×10^{-4}
Conceptual Plan: identify acid and base components then M $NaCHO_2$ → M CHO_2^- then
acid = $HCHO_2$ base = CHO_2^- $NaCHO_2\ (aq) \rightarrow Na^+\ (aq) + CHO_2^-\ (aq)$
K_a, M $HCHO_2$, M CHO_2^- → pH

$$pH = pK_a + \log\frac{[base]}{[acid]}$$

Solution: Acid = $HCHO_2$, so [acid] = [$HCHO_2$] = 0.15 M. Base = CHO_2^-. Since 1 CHO_2^- ion is generated for each $NaCHO_2$, [CHO_2^-] = 0.10 M CHO_2^- = [base]. Then

$$pH = pK_a + \log\frac{[base]}{[acid]} = -\log(1.8 \times 10^{-4}) + \log\frac{0.10\ \cancel{M}}{0.15\ \cancel{M}} = 3.57\,.$$

Note that in order to use the Henderson–Hasselbalch Equation, the assumption that x is small must be valid. This was confirmed in problem #3.
Check: The units (none) are correct. The magnitude of the answer makes physical sense because pH should be less than the pK_a of the acid because there is more acid than base. The answer agrees with problem #3.

b) **Given:** 0.12 M NH_3 and 0.18 M NH_4Cl **Find:** pH **Other:** K_b (NH_3) = 1.79×10^{-5}
Conceptual Plan:
identify acid and base components then M NH_4Cl → M NH_4^+ and K_b → pK_b → pK_a
acid = NH_4^+ base = NH_3 $NH_4Cl\ (aq) \rightarrow NH_4^+\ (aq) + Cl^-\ (aq)$ $pK_b = -\log K_b$ $14 = pK_a + pK_b$
then pK_a, M NH_3, M NH_4^+ → pH

$$pH = pK_a + \log\frac{[base]}{[acid]}$$

Solution: Base = NH_3, [base] = [NH_3] = 0.12 M Acid = NH_4^+. Since 1 NH_4^+ ion is generated for each NH_4Cl, [NH_4^+] = 0.18 M NH_4^+ = [acid].
Since K_b (NH_3) = 1.79×10^{-5}, $pK_b = -\log K_b = -\log(1.79 \times 10^{-5}) = 4.75$. Since $14 = pK_a + pK_b$,

$$pK_a = 14 - pK_b = 14 - 4.75 = 9.25 \text{ then } pH = pK_a + \log\frac{[base]}{[acid]} = 9.25 + \log\frac{0.12\ \cancel{M}}{0.18\ \cancel{M}} = 9.07$$

Note that in order to use the Henderson–Hasselbalch Equation, the assumption that x is small must be valid. This was confirmed in problem #3.
Check: The units (none) are correct. The magnitude of the answer makes physical sense because pH should be less than the pK_a of the acid because there is more acid than base. The answer agrees with problem #3, within the error of the value.

12. a) **Given:** 0.175 M $HC_2H_3O_2$ and 0.110 M $KC_2H_3O_2$ **Find:** pH **Other:** K_a ($HC_2H_3O_2$) = 1.8×10^{-5}
Conceptual Plan: identify acid and base components then M $KC_2H_3O_2$ → M $C_2H_3O_2^-$ then
acid = $HC_2H_3O_2$ base = $C_2H_3O_2^-$ $KC_2H_3O_2\ (aq) \rightarrow K^+\ (aq) + C_2H_3O_2^-\ (aq)$
M $HC_2H_3O_2$, M $C_2H_3O_2^-$ → pH

$$pH = pK_a + \log\frac{[base]}{[acid]}$$

Solution: Acid = $HC_2H_3O_2$, so [acid] = [$HC_2H_3O_2$] = 0.175 M. Base = $C_2H_3O_2^-$. Since 1 $C_2H_3O_2^-$ ion is generated for each $KC_2H_3O_2$, [$C_2H_3O_2^-$] = 0.110 M $C_2H_3O_2^-$ = [base]. Then

$$pH = pK_a + \log\frac{[base]}{[acid]} = -\log(1.8 \times 10^{-5}) + \log\frac{0.110\ \cancel{M}}{0.175\ \cancel{M}} = 4.54\,.$$

Note that in order to use the Henderson–Hasselbalch Equation, the assumption that x is small must be valid. This was confirmed in problem #4.

Check: The units (none) are correct. The magnitude of the answer makes physical sense because pH should be less than the pK_a of the acid because there is more acid than base. The answer agrees with problem #4.

b) **Given:** 0.195 M CH_3NH_2 and 0.105 M CH_3NH_3Br **Find:** pH **Other:** K_b (CH_3NH_2) = 4.4 x 10^{-4}
Conceptual Plan: identify acid and base components then M CH_3NH_3Br → M $CH_3NH_3^+$ and
acid = $CH_3NH_3^+$ base = CH_3NH_2 $CH_3NH_3Br\ (aq) \rightarrow CH_3NH_3^+\ (aq) + Br^-\ (aq)$
K_b → pK_b → pK_a then pK_a, M NH_3, M NH_4^+ → pH

$pK_b = -\log K_b \quad 14 = pK_a + pK_b \qquad pH = pK_a + \log \frac{[base]}{[acid]}$

Solution: Base = CH_3NH_2, [base] = [CH_3NH_2] = 0.195 M Acid = NH_3^+. Since 1 $CH_3NH_3^+$ ion is generated for each CH_3NH_3Br, [$CH_3NH_3^+$] = 0.105 M $CH_3NH_3^+$ = [acid].
Since K_b (CH_3NH_2) = 4.4 x 10^{-4}, $pK_b = -\log K_b = -\log(4.4 \times 10^{-4}) = 3.36$. Since $14 = pK_a + pK_b$,

$$pK_a = 14 - pK_b = 14 - 3.36 = 10.64 \text{ then } pH = pK_a + \log\frac{[base]}{[acid]} = 10.64 + \log\frac{0.195\ \cancel{M}}{0.105\ \cancel{M}} = 10.91.$$

Note that in order to use the Henderson–Hasselbalch Equation, the assumption that x is small must be valid. This was confirmed in problem #4.
Check: The units (none) are correct. The magnitude of the answer makes physical sense because pH should be greater than the pK_a of the acid because there is more base than acid. The answer agrees with problem #4.

13. a) **Given:** 0.125 M HClO and 0.150 M KClO **Find:** pH **Other:** K_a (HClO) = 2.9 x 10^{-8}
Conceptual Plan: identify acid and base components then M KClO → M ClO^- then
acid = HClO base = ClO^- $KClO\ (aq) \rightarrow K^+\ (aq) + ClO^-\ (aq)$
M HClO, M ClO^- → pH

$pH = pK_a + \log\frac{[base]}{[acid]}$

Solution: Acid = HClO, so [acid] = [HClO] = 0.125 M. Base = ClO^-. Since 1 ClO^- ion is generated for each KClO, [ClO^-] = 0.150 M ClO^- = [base]. Then

$$pH = pK_a + \log\frac{[base]}{[acid]} = -\log(2.9 \times 10^{-8}) + \log\frac{0.150\ \cancel{M}}{0.125\ \cancel{M}} = 7.62.$$

Check: The units (none) are correct. The magnitude of the answer makes physical sense because pH should be greater than the pK_a of the acid because there is more base than acid.

b) **Given:** 0.175 M $C_2H_5NH_2$ and 0.150 M $C_2H_5NH_3Br$ **Find:** pH **Other:** K_b ($C_2H_5NH_2$) = 5.6 x 10^{-4}
Conceptual Plan: identify acid and base components then M $C_2H_5NH_3Br$ → M $C_2H_5NH_3^+$ and
acid = $C_2H_5NH_3^+$ base = $C_2H_5NH_2$ $C_2H_5NH_3Br\ (aq) \rightarrow C_2H_5NH_3^+\ (aq) + Br^-\ (aq)$
K_b → pK_b → pK_a then pK_a, M $C_2H_5NH_2$, M $C_2H_5NH_3^+$ → pH

$pK_b = -\log K_b \quad 14 = pK_a + pK_b \qquad pH = pK_a + \log\frac{[base]}{[acid]}$

Solution: Base = $C_2H_5NH_2$, [base] = [$C_2H_5NH_2$] = 0.175 M Acid = $C_2H_5NH_3^+$. Since 1 $C_2H_5NH_3^+$ ion is generated for each $C_2H_5NH_3Br$, [$C_2H_5NH_3^+$] = 0.150 M $C_2H_5NH_3^+$ = [acid]. Since K_b ($C_2H_5NH_2$) = 5.6 x 10^{-4}, $pK_b = -\log K_b = -\log(5.6 \times 10^{-4}) = 3.25$. Since $14 = pK_a + pK_b$,

$$pK_a = 14 - pK_b = 14 - 3.25 = 10.75 \text{ then } pH = pK_a + \log\frac{[base]}{[acid]} = 10.75 + \log\frac{0.175\ \cancel{M}}{0.150\ \cancel{M}} = 10.82.$$

Check: The units (none) are correct. The magnitude of the answer makes physical sense because pH should be greater than the pK_a of the acid because there is more base than acid.

c) **Given:** 10.0 g $HC_2H_3O_2$ and 10.0 g $NaC_2H_3O_2$ in 150.0 mL solution **Find:** pH
Other: K_a ($HC_2H_3O_2$) = 1.8 x 10^{-5}
Conceptual Plan:
identify acid and base components then mL → L and g $HC_2H_3O_2$ → mol $HC_2H_3O_2$

acid = $HC_2H_3O_2$ base = $C_2H_3O_2^-$ $\quad \frac{1\ L}{1000\ mL} \quad \frac{1\ mol\ HC_2H_3O_2}{60.05\ g\ HC_2H_3O_2}$

then mol $HC_2H_3O_2$, L → M $HC_2H_3O_2$ and g $NaC_2H_3O_2$ → mol $NaC_2H_3O_2$ then

$$M = \frac{mol}{L} \qquad \frac{1 \text{ mol } NaC_2H_3O_2}{82.04 \text{ g } NaC_2H_3O_2}$$

mol $NaC_2H_3O_2$, L → M $NaC_2H_3O_2$ → M $C_2H_3O_2^-$ then M $HC_2H_3O_2$, M $C_2H_3O_2^-$ → pH

$$M = \frac{mol}{L} \qquad NaC_2H_3O_2\,(aq) \rightarrow Na^+\,(aq) + C_2H_3O_2^-\,(aq) \qquad pH = pK_a + \log\frac{[base]}{[acid]}$$

Solution: $150.0 \text{ mL} \times \frac{1 \text{ L}}{1000 \text{ mL}} = 0.1500 \text{ L}$ and

$$10.0 \text{ g } HC_2H_3O_2 \times \frac{1 \text{ mol } HC_2H_3O_2}{60.05 \text{ g } HC_2H_3O_2} = 0.16\underline{6}528 \text{ mol } HC_2H_3O_2$$

then $M = \frac{mol}{L} = \frac{0.16\underline{6}528 \text{ mol } HC_2H_3O_2}{0.1500 \text{ L}} = 1.1\underline{1}019 \text{ M } HC_2H_3O_2$ and

$$10.0 \text{ g } NaC_2H_3O_2 \times \frac{1 \text{ mol } NaC_2H_3O_2}{82.04 \text{ g } NaC_2H_3O_2} = 0.12\underline{1}892 \text{ mol } NaC_2H_3O_2 \text{ then}$$

$M = \frac{mol}{L} = \frac{0.12\underline{1}892 \text{ mol } NaC_2H_3O_2}{0.1500 \text{ L}} = 0.81\underline{2}612 \text{ M } NaC_2H_3O_2$. So Acid = $HC_2H_3O_2$, so [acid] = $[HC_2H_3O_2]$ = 1.1$\underline{1}$019 M and Base = $C_2H_3O_2^-$. Since 1 $C_2H_3O_2^-$ ion is generated for each $NaC_2H_3O_2$, $[C_2H_3O_2^-]$ = 0.81$\underline{2}$612 M $C_2H_3O_2^-$ = [base]. Then

$$pH = pK_a + \log\frac{[base]}{[acid]} = -\log(1.8 \times 10^{-5}) + \log\frac{0.81\underline{2}612 \text{ M}}{1.1\underline{1}019 \text{ M}} = 4.61.$$

Check: The units (none) are correct. The magnitude of the answer makes physical sense because pH should be less than the pK_a of the acid because there is more acid than base.

14. a) **Given:** 0.155 M $HC_3H_5O_2$ (propanoic acid) and 0.110 M $KC_3H_5O_2$ (potassium propanoate) **Find:** pH **Other:** K_a ($HC_3H_5O_2$) = 1.3×10^{-5}

Conceptual Plan:

identify acid and base components then M $KC_3H_5O_2$ → M $C_3H_5O_2^-$ then

acid = $HC_3H_5O_2$ base = $C_3H_5O_2^-$ $KC_3H_5O_2\,(aq) \rightarrow K^+\,(aq) + C_3H_5O_2^-\,(aq)$

M $HC_3H_5O_2$, M $C_3H_5O_2^-$ → pH

$$pH = pK_a + \log\frac{[base]}{[acid]}$$

Solution: Acid = $HC_3H_5O_2$, so [acid] = $[HC_3H_5O_2]$ = 0.155 M. Base = $C_3H_5O_2^-$. Since 1 $C_3H_5O_2^-$ ion is generated for each $KC_3H_5O_2$, $[C_3H_5O_2^-]$ = 0.110 M $C_3H_5O_2^-$ = [base]. Then

$$pH = pK_a + \log\frac{[base]}{[acid]} = -\log(1.3 \times 10^{-5}) + \log\frac{0.110 \text{ M}}{0.155 \text{ M}} = 4.74.$$

Check: The units (none) are correct. The magnitude of the answer makes physical sense because pH should be less than the pK_a of the acid because there is more acid than base.

b) **Given:** 0.15 M C_5H_5N and 0.10 M C_5H_5NHCl **Find:** pH **Other:** K_b (C_5H_5N) = 1.7×10^{-9}

Conceptual Plan:

identify acid and base components then M CH_3NH_3Cl → M $CH_3NH_3^+$ and

acid = $C_5H_5NH^+$ base = C_5H_5N $C_5H_5NHCl\,(aq) \rightarrow C_5H_5NH^+\,(aq) + Cl^-\,(aq)$

K_b → pK_b → pK_a then pK_a, M C_5H_5N, M $C_5H_5NH^+$ → pH

$$pK_b = -\log K_b \quad 14 = pK_a + pK_b \qquad pH = pK_a + \log\frac{[base]}{[acid]}$$

Solution: Base = C_5H_5N, [base] = $[C_5H_5NH]$ = 0.15 M Acid = $C_5H_5NH^+$. Since 1 $C_5H_5NH^+$ ion is generated for each C_5H_5NHCl, $[C_5H_5NH^+]$ = 0.10 M $C_5H_5NH^+$ = [acid]. Since K_b (C_5H_5N) = = 1.7×10^{-9}, $pK_b = -\log K_b = -\log(1.7 \times 10^{-9}) = 8.77$. Since $14 = pK_a + pK_b$,

$$pK_a = 14 - pK_b = 14 - 8.77 = 5.23 \text{ then } pH = pK_a + \log\frac{[base]}{[acid]} = 5.23 + \log\frac{0.15 \text{ M}}{0.10 \text{ M}} = 5.41.$$

Check: The units (none) are correct. The magnitude of the answer makes physical sense because pH should be greater than the pK_a of the acid because there is more base than acid.

c) **Given:** 15.0 g HF and 25.0 g NaF in 125 mL solution **Find:** pH **Other:** K_a (HF) = 3.5 x 10^{-4}
Conceptual Plan:
identify acid and base components then mL → L and g HF → mol HF

acid = HF base = F^- $\quad \frac{1\ L}{1000\ mL} \quad \frac{1\ mol\ HF}{20.01\ g\ HF}$

then mol HF, L → M HF and g NaF → mol NaF then

$M = \frac{mol}{L} \quad \frac{1\ mol\ NaF}{41.99\ g\ NaF}$

mol NaF, L → M NaF → M F^- then M HF, M F^- → pH

$M = \frac{mol}{L} \quad NaF\ (aq) \rightarrow Na^+\ (aq) + F^-\ (aq) \quad pH = pK_a + \log\frac{[base]}{[acid]}$

Solution: $125\ mL \times \frac{1\ L}{1000\ mL} = 0.125\ L$ and $15.0\ g\ HF \times \frac{1\ mol\ HF}{20.01\ g\ HF} = 0.749625\ mol\ HF$ then

$M = \frac{mol}{L} = \frac{0.749625\ mol\ HF}{0.125\ L} = 5.997\ M\ HF$ and $25.0\ g\ NaF \times \frac{1\ mol\ NaF}{41.99\ g\ NaF} = 0.595380\ mol\ NaF$ then

$M = \frac{mol}{L} = \frac{0.595380\ mol\ NaF}{0.125\ L} = 4.76304\ M\ NaF$. So Acid = HF, so [acid] = [HF] = 5.997 M and Base = F^-. Since 1F^- ion is generated for each NaF, $[F^-]$ = 4.76304 M F^- = [base]. Then

$pH = pK_a + \log\frac{[base]}{[acid]} = -\log(3.5 \times 10^{-4}) + \log\frac{4.76304\ M}{5.997\ M} = 3.36$.

Check: The units (none) are correct. The magnitude of the answer makes physical sense because pH should be less than the pK_a of the acid because there is more acid than base.

15. a) **Given:** 50.0 mL of 0.15 M $HCHO_2$ and 75.0 mL of 0.13 M $NaCHO_2$ **Find:** pH
Other: K_a ($HCHO_2$) = 1.8 x 10^{-4}
Conceptual Plan:
identify acid and base components then mL $HCHO_2$, mL $NaCHO_2$ → total mL then

acid = $HCHO_2$ base = CHO_2^- $\quad$ total mL = mL $HCHO_2$ + mL $NaCHO_2$

mL $HCHO_2$, M $HCHO_2$, total mL → buffer M $HCHO_2$ and

$M_1 V_1 = M_2 V_2$

mL $NaCHO_2$, M $NaCHO_2$, total mL → buffer M $NaCHO_2$ → buffer M CHO_2^- then

$M_1 V_1 = M_2 V_2 \quad NaCHO_2\ (aq) \rightarrow Na^+\ (aq) + CHO_2^-\ (aq)$

K_a, M $HCHO_2$, M CHO_2^- → pH

$pH = pK_a + \log\frac{[base]}{[acid]}$

Solution: total mL = mL $HCHO_2$ + mL $NaCHO_2$ = 50.0 mL + 75.0 mL = 125.0 mL. Then

since $M_1 V_1 = M_2 V_2$ rearrange to solve for M_2. $M_2 = \frac{M_1 V_1}{V_2} = \frac{(0.15\ M)(50.0\ mL)}{125.0\ mL} = 0.060\ M\ HCHO_2$ and

$M_2 = \frac{M_1 V_1}{V_2} = \frac{(0.13\ M)(75.0\ mL)}{125.0\ mL} = 0.078\ M\ NaCHO_2$. Acid = $HCHO_2$, so [acid] = [$HCHO_2$] = 0.060 M. Base = CHO_2^-. Since 1 CHO_2^- ion is generated for each $NaCHO_2$, [CHO_2^-] = 0.078 M CHO_2^- = [base]. Then $pH = pK_a + \log\frac{[base]}{[acid]} = -\log(1.8 \times 10^{-4}) + \log\frac{0.078\ M}{0.060\ M} = 3.86$.

Check: The units (none) are correct. The magnitude of the answer makes physical sense because pH should be greater than the pK_a of the acid because there is more base than acid.

b) **Given:** 125.0 mL of 0.10 M NH_3 and 250.0 mL of 0.10 M NH_4Cl **Find:** pH
Other: K_b (NH_3) = 1.79 x 10^{-5}
Conceptual Plan:
identify acid and base components then mL NH_3, mL NH_4Cl → total mL then

acid = NH_4^+ base = NH_3 $\quad$ total mL = mL NH_3 + mL NH_4Cl

mL NH_3, M NH_3, total mL → buffer M NH_3 and

$M_1 V_1 = M_2 V_2$

mL NH_4Cl, M NH_4Cl, total mL → buffer M NH_4Cl → buffer M NH_4^+ and K_b → pK_b → pK_a then

$M_1 V_1 = M_2 V_2$ $NH_4Cl\,(aq) \rightarrow NH_4^+\,(aq) + Cl^-\,(aq)$ $pK_b = -\log K_b$ $14 = pK_a + pK_b$

then pK_a, M NH_3, M NH_4^+ → pH

$$pH = pK_a + \log \frac{[base]}{[acid]}$$

Solution: total mL = mL NH_3 + mL NH_4Cl = 125.0 mL + 250.0 mL = 375.0 mL. Then

since $M_1 V_1 = M_2 V_2$ rearrange to solve for M_2. $M_2 = \frac{M_1 V_1}{V_2} = \frac{(0.10\text{ M})(125.0\text{ mL})}{375.0\text{ mL}} = 0.03\underline{3}333\text{ M } NH_3$ and

$M_2 = \frac{M_1 V_1}{V_2} = \frac{(0.10\text{ M})(250.0\text{ mL})}{375.0\text{ mL}} = 0.06\underline{6}667\text{ M } NH_4Cl$. Base = NH_3, [base] = [NH_3] = 0.03$\underline{3}$333 M

Acid = NH_4^+. Since 1 NH_4^+ ion is generated for each NH_4Cl, [NH_4^+] = 0.06$\underline{6}$6667 M NH_4^+ = [acid].
Since K_b (NH_3) = 1.79×10^{-5}, $pK_b = -\log K_b = -\log(1.79 \times 10^{-5}) = 4.75$. Since $14 = pK_a + pK_b$,

$pK_a = 14 - pK_b = 14 - 4.75 = 9.25$ then $pH = pK_a + \log \frac{[base]}{[acid]} = 9.25 + \log \frac{0.03\underline{3}333\text{ M}}{0.06\underline{6}667\text{ M}} = 8.95$.

Check: The units (none) are correct. The magnitude of the answer makes physical sense because pH should be less than the pK_a of the acid because there is more acid than base.

16\. a) **Given:** 150.0 mL of 0.25 M HF and 225.0 mL of 0.30 M NaF **Find:** pH
Other: K_a (HF) = 3.5×10^{-4}
Conceptual Plan:
identify acid and base components then mL $HCHO_2$, mL $NaCHO_2$ → total mL then

acid = HF base = F^- total mL = mL HF + mL NaF

mL HF, M HF, total mL → buffer M HF and

$M_1 V_1 = M_2 V_2$

mL NaF, M NaF, total mL → buffer M NaF → buffer M F^-

$M_1 V_1 = M_2 V_2$ $NaF\,(aq) \rightarrow Na^+\,(aq) + F^-\,(aq)$

then K_a, M $HCHO_2$, M CHO_2^- → pH

$$pH = pK_a + \log \frac{[base]}{[acid]}$$

Solution: total mL = mL HF + mL NaF = 150.0 mL + 225.0 mL = 375.0 mL. Then

since $M_1 V_1 = M_2 V_2$ rearrange to solve for M_2. $M_2 = \frac{M_1 V_1}{V_2} = \frac{(0.25\text{ M})(150.0\text{ mL})}{375.0\text{ mL}} = 0.10\text{ M HF}$ and

$M_2 = \frac{M_1 V_1}{V_2} = \frac{(0.30\text{ M})(225.0\text{ mL})}{375.0\text{ mL}} = 0.18\text{ M NaF}$. Acid = HF, so [acid] = [HF] = 0.10 M. Base = F^-.

Since 1 F^- ion is generated for each NaF, [F^-] = 0.18 M F^- = [base]. Then

$pH = pK_a + \log \frac{[base]}{[acid]} = -\log(3.5 \times 10^{-4}) + \log \frac{0.18\text{ M}}{0.10\text{ M}} = 3.71$.

Check: The units (none) are correct. The magnitude of the answer makes physical sense because pH should be greater than the pK_a of the acid because there is more base than acid.

b) **Given:** 175.0 mL of 0.10 M $C_2H_5NH_2$ and 275.0 mL of 0.20 M $C_2H_5NH_3Cl$ **Find:** pH
Other: K_b ($C_2H_5NH_2$) = 5.6×10^{-4}
Conceptual Plan:
identify acid and base components then mL $C_2H_5NH_2$, mL $C_2H_5NH_3Cl$ → total mL then

acid = $C_2H_5NH_3^+$ base = $C_2H_5NH_2$ total mL = mL $C_2H_5NH_2$ + mL $C_2H_5NH_3Cl$

mL NH_3, M $C_2H_5NH_2$, total mL → buffer M $C_2H_5NH_2$ and

$M_1 V_1 = M_2 V_2$

mL NH_4Cl, M $C_2H_5NH_3Cl$, total mL → buffer M $C_2H_5NH_3Cl$ → buffer M $C_2H_5NH_3^+$ and

$M_1 V_1 = M_2 V_2$ $C_2H_5NH_3Cl\,(aq) \rightarrow C_2H_5NH_3^+\,(aq) + Cl^-\,(aq)$

$K_b \rightarrow pK_b \rightarrow pK_a$ then pK_a, M $C_2H_5NH_2$, M $C_2H_5NH_3^+ \rightarrow$ pH

$$pK_b = -\log K_b \quad 14 = pK_a + pK_b \qquad\qquad pH = pK_a + \log \frac{[base]}{[acid]}$$

Solution: total mL = mL $C_2H_5NH_2$ + mL $C_2H_5NH_3Cl$ = 175.0 mL + 275.0 mL = 450.0 mL. Then since $M_1V_1 = M_2V_2$ rearrange to solve for M_2.

$$M_2 = \frac{M_1V_1}{V_2} = \frac{(0.10\ M)(175.0\ \cancel{mL})}{450.0\ \cancel{mL}} = 0.03\underline{8}889\ M\ C_2H_5NH_2 \text{ and}$$

$$M_2 = \frac{M_1V_1}{V_2} = \frac{(0.20\ M)(275.0\ \cancel{mL})}{450.0\ \cancel{mL}} = 0.1\underline{2}222\ M\ C_2H_5NH_3Cl.$$ Base = $C_2H_5NH_2$, [base] = $[C_2H_5NH_2]$ =

$0.03\underline{8}889\ M$ Acid = $C_2H_5NH_3^+$. Since 1 $C_2H_5NH_3^+$ ion is generated for each $C_2H_5NH_3Cl$, $[C_2H_5NH_3^+]$ = $0.1\underline{2}222\ M\ C_2H_5NH_3^+$ = [acid]. Since K_b ($C_2H_5NH_2$) = 5.6×10^{-4},

$pK_b = -\log K_b = -\log(5.6 \times 10^{-4}) = 3.25$. Since $14 = pK_a + pK_b$,

$$pK_a = 14 - pK_b = 14 - 3.25 = 10.75 \text{ then } pH = pK_a + \log\frac{[base]}{[acid]} = 10.75 + \log\frac{0.03\underline{8}889\ \cancel{M}}{0.1\underline{2}222\ \cancel{M}} = 10.25.$$

Check: The units (none) are correct. The magnitude of the answer makes physical sense because pH should be less than the pK_a of the acid because there is more acid than base.

17. **Given:** NaF / HF buffer at pH = 4.00   **Find:** [NaF] / [HF]   **Other:** K_a (HF) = 3.5×10^{-4}

     Conceptual Plan: identify acid and base components   then   pH, K_a $\rightarrow$ [NaF] / [HF]

$$acid = HF \quad base = F^- \qquad\qquad pH = pK_a + \log\frac{[base]}{[acid]}$$

Solution: $pH = pK_a + \log\dfrac{[base]}{[acid]} = -\log(3.5 \times 10^{-4}) + \log\dfrac{[NaF]}{[HF]} = 4.00$. Solve for [NaF] / [HF].

$$\log\frac{[NaF]}{[HF]} = 4.00 - 3.46 = 0.54 \quad\rightarrow\quad \frac{[NaF]}{[HF]} = 10^{0.54} = 3.5.$$

Check: The units (none) are correct. The magnitude of the answer makes physical sense because the pH is greater than the pK_a of the acid, so there needs to be more base than acid.

18. **Given:** CH_3NH_2 / CH_3NH_3Cl buffer at pH = 10.24   **Find:** $[CH_3NH_2]$ / $[CH_3NH_3Cl]$

     Other: K_b (CH_3NH_2) = 4.4×10^{-4}

     Conceptual Plan: identify acid and base components   and K_b $\rightarrow$ pK_b $\rightarrow$ pK_a then

$$acid = CH_3NH_3^+ \quad base = CH_3N \qquad pK_b = -\log K_b \quad 14 = pK_a + pK_b$$

pH, K_a $\rightarrow$ $[CH_3NH_2]$ / $[CH_3NH_3Cl]$

$$pH = pK_a + \log\frac{[base]}{[acid]}$$

Solution: Since K_b (CH_3NH_2) = 4.4×10^{-4}, $pK_b = -\log K_b = -\log(4.4 \times 10^{-4}) = 3.36$. Since $14 = pK_a + pK_b$, $pK_a = 14 - pK_b = 14 - 3.36 = 10.64$ then

$$pH = pK_a + \log\frac{[base]}{[acid]} = 10.64 + \log\frac{[CH_3NH_2]}{[CH_3NH_3Cl]} = 10.24.$$ Solve for $[CH_3NH_2]$ / $[CH_3NH_3Cl]$.

$$\log\frac{[CH_3NH_2]}{[CH_3NH_3Cl]} = 10.24 - 10.64 = -0.40 \quad\rightarrow\quad \frac{[CH_3NH_2]}{[CH_3NH_3Cl]} = 10^{-0.40} = 0.39.$$

Check: The units (none) are correct. The magnitude of the answer makes physical sense because the pH is less than the pK_a of the acid, so there needs to be less base than acid.

19. **Given:** 150.0 mL buffer of 0.15 M benzoic acid at pH = 4.25   **Find:** mass sodium benzoate

     Other: K_a ($HC_7H_5O_2$) = 6.5×10^{-5}

     Conceptual Plan: identify acid and base components   then   pH, K_a, $[HC_7H_5O_2]$ $\rightarrow$ $[NaC_7H_5O_2]$

$$acid = HC_7H_5O_2 \quad base = C_7H_5O_2^- \qquad\qquad pH = pK_a + \log\frac{[base]}{[acid]}$$

mL → L then $[NaC_7H_5O_2]$, L → mol $NaC_7H_5O_2$ → g $NaC_7H_5O_2$

$\frac{1\text{ L}}{1000\text{ mL}}$ $\quad M = \frac{\text{mol}}{\text{L}}$ $\quad \frac{144.11\text{ g NaC}_7\text{H}_5\text{O}_2}{1\text{ mol NaC}_7\text{H}_5\text{O}_2}$

Solution: $\text{pH} = \text{p}K_a + \log\frac{[\text{base}]}{[\text{acid}]} = -\log(6.5 \times 10^{-5}) + \log\frac{[\text{NaC}_7\text{H}_5\text{O}_2]}{0.15\text{ M}} = 4.25$. Solve for $[NaC_7H_5O_2]$.

$\log\frac{[\text{NaC}_7\text{H}_5\text{O}_2]}{0.15\text{ M}} = 4.25 - 4.19 = 0.0\underline{6}291 \rightarrow \frac{[\text{NaC}_7\text{H}_5\text{O}_2]}{0.15\text{ M}} = 10^{0.0\underline{6}291} = 1.\underline{1}559 \rightarrow [\text{NaC}_7\text{H}_5\text{O}_2] = 0.1\underline{7}338\text{ M}$.

Convert to moles using $M = \frac{\text{mol}}{\text{L}}$.

$\frac{0.1\underline{7}338\text{ mol NaC}_7\text{H}_5\text{O}_2}{1\text{ L}} \times 0.150\text{ L} = 0.02\underline{6}007\text{ mol NaC}_7\text{H}_5\text{O}_2 \times \frac{144.11\text{ g NaC}_7\text{H}_5\text{O}_2}{1\text{ mol NaC}_7\text{H}_5\text{O}_2} = 3.7\text{g NaC}_7\text{H}_5\text{O}_2$.

Check: The units (g) are correct. The magnitude of the answer makes physical sense because the volume of solution is small and the concentration is low, so much less than a mole is needed.

20. **Given:** 2.55 L buffer of 0.155 M NH_3 at pH = 9.55 **Find:** mass ammonium chloride
Other: K_b (NH_3) = 1.79×10^{-5}
Conceptual Plan: identify acid and base components then K_b → pK_b → pK_a then

acid = NH_4^+ base = NH_3 $\quad pK_b = -\log K_b \quad 14 = pK_a + pK_b$

pH, K_a, $[NH_3]$ → $[NH_4Cl]$ then $[NH_4Cl]$, L → mol NH_4Cl → g NH_4Cl

$\text{pH} = \text{p}K_a + \log\frac{[\text{base}]}{[\text{acid}]}$ $\quad M = \frac{\text{mol}}{\text{L}}$ $\quad \frac{53.49\text{ g NH}_4\text{Cl}}{1\text{ mol NH}_4\text{Cl}}$

Solution: Since K_b (NH_3) = 1.79×10^{-5}, $pK_b = -\log K_b = -\log(1.79 \times 10^{-5}) = 4.75$. Since $14 = pK_a + pK_b$,

$pK_a = 14 - pK_b = 14 - 4.75 = 9.25$ then $\text{pH} = \text{p}K_a + \log\frac{[\text{base}]}{[\text{acid}]} = 9.25 + \log\frac{0.155\text{ M}}{[\text{NH}_4\text{Cl}]} = 9.55$.

Solve for $[NH_4Cl]$. $\log\frac{0.155\text{ M}}{[\text{NH}_4\text{Cl}]} = 9.55 - 9.25 = 0.30 \rightarrow \frac{0.155\text{ M}}{[\text{NH}_4\text{Cl}]} = 10^{0.30} = 1.9\underline{9}526 \rightarrow$

$[\text{NH}_4\text{Cl}] = 0.07\underline{7}6841\text{ M}$. Convert to moles using $M = \frac{\text{mol}}{\text{L}}$.

$\frac{0.07\underline{7}6841\text{ mol NH}_4\text{Cl}}{1\text{ L}} \times 2.55\text{ L} = 0.19\underline{8}094\text{ mol NH}_4\text{Cl} \times \frac{53.49\text{ g NH}_4\text{Cl}}{1\text{ mol NH}_4\text{Cl}} = 10.6\text{ g NH}_4\text{Cl}$.

Check: The units (g) are correct. The magnitude of the answer makes physical sense because the volume of solution is large and the concentration is low, so less than a mole is needed.

21. a) **Given:** 250.0 mL buffer 0.250 M $HC_2H_3O_2$ and 0.250 M $NaC_2H_3O_2$ **Find:** initial pH
Other: K_a ($HC_2H_3O_2$) = 1.8×10^{-5}
Conceptual Plan: identify acid and base components then M $NaC_2H_3O_2$ → M $C_2H_3O_2^-$ then

acid = $HC_2H_3O_2$ base = $C_2H_3O_2^-$ $\quad NaC_2H_3O_2\,(aq) \rightarrow Na^+\,(aq) + C_2H_3O_2^-\,(aq)$

M $HC_2H_3O_2$, M $C_2H_3O_2^-$ → pH

$\text{pH} = \text{p}K_a + \log\frac{[\text{base}]}{[\text{acid}]}$

Solution: Acid = $HC_2H_3O_2$, so [acid] = $[HC_2H_3O_2]$ = 0.250 M. Base = $C_2H_3O_2^-$. Since 1 $C_2H_3O_2^-$ ion is generated for each $NaC_2H_3O_2$, $[C_2H_3O_2^-]$ = 0.250 M $C_2H_3O_2^-$ = [base]. Then

$\text{pH} = \text{p}K_a + \log\frac{[\text{base}]}{[\text{acid}]} = -\log(1.8 \times 10^{-5}) + \log\frac{0.250\text{ M}}{0.250\text{ M}} = 4.74$.

Check: The units (none) are correct. The magnitude of the answer makes physical sense because pH is equal to the pK_a of the acid because there are equal amounts of acid and base.

b) **Given:** 250.0 mL buffer 0.250 M $HC_2H_3O_2$ and 0.250 M $NaC_2H_3O_2$, add 0.0050 mol HCl
Find: pH **Other:** K_a ($HC_2H_3O_2$) = 1.8×10^{-5}

Conceptual Plan: Part I: Stoichiometry:

mL → L then $[NaC_2H_3O_2]$, L → mol $NaC_2H_3O_2$ and $[HC_2H_3O_2]$, L → mol $HC_2H_3O_2$

$\frac{1\text{ L}}{1000\text{ mL}}$ $\qquad M = \frac{\text{mol}}{\text{L}}$ $\qquad M = \frac{\text{mol}}{\text{L}}$

write balanced equation then

$HCl + NaC_2H_3O_2 \rightarrow HC_2H_3O_2 + NaCl$

mol $NaC_2H_3O_2$, mol $HC_2H_3O_2$, mol HCl → mol $NaC_2H_3O_2$, mol $HC_2H_3O_2$ then

set up stoichiometry table

Part II: Equilibrium:

mol $NaC_2H_3O_2$, mol $HC_2H_3O_2$, L, K_a → pH

$$pH = pK_a + \log\frac{[\text{base}]}{[\text{acid}]}$$

Solution: $250.0\ \cancel{\text{mL}} \times \frac{1\text{ L}}{1000\ \cancel{\text{mL}}} = 0.2500\text{ L}$ then

$\frac{0.250\text{ mol } HC_2H_3O_2}{1\ \cancel{\text{L}}} \times 0.250\ \cancel{\text{L}} = 0.0625\text{ mol } HC_2H_3O_2$ and

$\frac{0.250\text{ mol } NaC_2H_3O_2}{1\ \cancel{\text{L}}} \times 0.250\ \cancel{\text{L}} = 0.0625\text{ mol } NaC_2H_3O_2$ set up table to track changes:

	HCl (*aq*)	+ $NaC_2H_3O_2$ (*aq*) →	$HC_2H_3O_2$ (*aq*)	+ NaCl (*aq*)
Before addition	≈ 0.00 mol	0.0625 mol	0.0625 mol	0.00 mol
Addition	0.0050 mol	–	–	–
After addition	≈ 0.00 mol	0.0575 mol	0.0675 mol	0.0050 mol

Since the amount of HCl is small, there are still significant amounts of both buffer components, so the Henderson–Hasselbalch Equation can be used to calculate the new pH.

$$pH = pK_a + \log\frac{[\text{base}]}{[\text{acid}]} = -\log(1.8 \times 10^{-5}) + \log\frac{\frac{0.0575\ \cancel{\text{mol}}}{\cancel{0.250\text{ L}}}}{\frac{0.0675\ \cancel{\text{mol}}}{\cancel{0.250\text{ L}}}} = 4.68\,.$$

Check: The units (none) are correct. The magnitude of the answer makes physical sense because the pH dropped slightly when acid was added.

c) **Given:** 250.0 mL buffer 0.250 M $HC_2H_3O_2$ and 0.250 M $NaC_2H_3O_2$, add 0.0050 mol NaOH
Find: pH **Other:** K_a ($HC_2H_3O_2$) = 1.8×10^{-5}

Conceptual Plan: Part I: Stoichiometry:

mL → L then $[NaC_2H_3O_2]$, L → mol $NaC_2H_3O_2$ and $[HC_2H_3O_2]$, L → mol $HC_2H_3O_2$

$\frac{1\text{ L}}{1000\text{ mL}}$ $\qquad M = \frac{\text{mol}}{\text{L}}$ $\qquad M = \frac{\text{mol}}{\text{L}}$

write balanced equation then

$NaOH + HC_2H_3O_2 \rightarrow H_2O + NaC_2H_3O_2$

mol $NaC_2H_3O_2$, mol $HC_2H_3O_2$, mol NaOH → mol $NaC_2H_3O_2$, mol $HC_2H_3O_2$ then

set up stoichiometry table

Part II: Equilibrium:

mol $NaC_2H_3O_2$, mol $HC_2H_3O_2$, L, K_a → pH

$$pH = pK_a + \log\frac{[\text{base}]}{[\text{acid}]}$$

Solution: $250.0\ \cancel{\text{mL}} \times \frac{1\text{ L}}{1000\ \cancel{\text{mL}}} = 0.2500\text{ L}$ then

$\frac{0.250\text{ mol } HC_2H_3O_2}{1\ \cancel{\text{L}}} \times 0.2500\ \cancel{\text{L}} = 0.0625\text{ mol } HC_2H_3O_2$ and

$\frac{0.250\text{ mol } NaC_2H_3O_2}{1\ \cancel{\text{L}}} \times 0.2500\ \cancel{\text{L}} = 0.0625\text{ mol } NaC_2H_3O_2$ set up table to track changes:

	$NaOH\ (aq)$ +	$HC_2H_3O_2\ (aq)$ →	$NaC_2H_3O_2\ (aq)$ +	$H_2O\ (l)$
Before addition	≈ 0.00 mol	0.0625 mol	0.0625 mol	–
Addition	0.0050 mol	–	–	–
After addition	≈ 0.00 mol	0.0575 mol	0.0675 mol	–

Since the amount of NaOH is small, there are still significant amounts of both buffer components, so the Henderson–Hasselbalch Equation can be used to calculate the new pH.

$$pH = pK_a + \log\frac{[base]}{[acid]} = -\log(1.8 \times 10^{-5}) + \log\frac{\frac{0.0675\ \cancel{mol}}{\cancel{0.2500\ L}}}{\frac{0.0575\ \cancel{mol}}{\cancel{0.2500\ L}}} = 4.81.$$

Check: The units (none) are correct. The magnitude of the answer makes physical sense because the pH rose slightly when base was added.

22. a) **Given:** 100.0 mL buffer 0.175 M HClO and 0.150 M NaClO **Find:** initial pH
Other: K_a (HClO) = 2.9×10^{-8}
Conceptual Plan: identify acid and base components then M NaClO → M ClO⁻ then
acid = HClO base = ClO^- ; $NaClO\ (aq) \rightarrow Na^+\ (aq) + ClO^-\ (aq)$
M HClO, M ClO^- → pH

$$pH = pK_a + \log\frac{[base]}{[acid]}$$

Solution: Acid = HClO, so [acid] = [HClO] = 0.175 M. Base = ClO^-. Since 1 ClO^- ion is generated for each NaClO, $[ClO^-]$ = 0.150 M ClO^- = [base].

$$\text{Then } pH = pK_a + \log\frac{[base]}{[acid]} = -\log(2.9 \times 10^{-8}) + \log\frac{0.150\ \cancel{M}}{0.175\ \cancel{M}} = 7.47.$$

Check: The units (none) are correct. The magnitude of the answer makes physical sense because pH is less than the pK_a of the acid because there is more acid than base.

b) **Given:** 100.0 mL buffer 0.175 M HClO and 0.150 M NaClO, add 150.0 mg HBr **Find:** pH
Other: K_a (HClO) = 2.9×10^{-8}
Conceptual Plan: Part I: Stoichiometry:
mL → L then [NaClO], L → mol NaClO and [HClO], L → mol HClO and
$\frac{1\ L}{1000\ mL}$; $M = \frac{mol}{L}$; $M = \frac{mol}{L}$
mg HBr → g HBr → mol HBr write balanced equation then
$\frac{1\ g\ HBr}{1000\ mg\ HBr}$; $\frac{1\ mol\ HBr}{71.91\ g\ HBr}$; $HBr + NaClO \rightarrow HClO + NaBr$
mol NaClO, mol HClO, mol HBr → mol NaClO, mol HClO then
set up stoichiometry table
Part II: Equilibrium:
mol NaClO, mol HClO, L, K_a → pH

$$pH = pK_a + \log\frac{[base]}{[acid]}$$

Solution: $100.0\ \cancel{mL} \times \frac{1\ L}{1000\ \cancel{mL}} = 0.1000\ L$ then $\frac{0.175\ mol\ HClO}{1\ \cancel{L}} \times 0.1000\ \cancel{L} = 0.0175\ mol\ HClO$

and $\frac{0.150\ mol\ NaClO}{1\ \cancel{L}} \times 0.1000\ \cancel{L} = 0.0150\ mol\ NaClO$ and

$150.0\ \cancel{mg\ HBr} \times \frac{1\ \cancel{g\ HBr}}{1000\ \cancel{mg\ HBr}} \times \frac{1\ mol\ HBr}{71.91\ \cancel{g\ HBr}} = 0.002086\ mol\ HBr$ then set up table to track changes:

	$HBr\ (aq)$ +	$NaClO\ (aq)$ →	$HClO\ (aq)$ +	$NaBr\ (aq)$
Before addition	≈ 0.00 mol	0.0150 mol	0.0175 mol	0.00 mol
Addition	0.002086 mol	–	–	–
After addition	≈ 0.00 mol	0.0129 mol	0.0196 mol	0.002086 mol

Since the amount of HBr is small, there are still significant amounts of both buffer components, so the Henderson–Hasselbalch Equation can be used to calculate the new pH.

$$pH = pK_a + \log\frac{[\text{base}]}{[\text{acid}]} = -\log(2.9 \times 10^{-8}) + \log\frac{\dfrac{0.0129\ \cancel{\text{mol}}}{\cancel{0.1000\ \text{L}}}}{\dfrac{0.0196\ \cancel{\text{mol}}}{\cancel{0.1000\ \text{L}}}} = 7.36\ .$$

Check: The units (none) are correct. The magnitude of the answer makes physical sense because the pH dropped slightly when acid was added. The pH is closer to the pK_a of the acid than at the start.

c) **Given:** 100.0 mL buffer 0.175 M HClO and 0.150 M NaClO, add 85.0 mg NaOH **Find:** pH
Other: K_a (HClO) = 2.9×10^{-8}
Conceptual Plan: Part I: Stoichiometry:
mL → L then [NaClO], L → mol NaClO and [HClO], L → mol HClO and

$\frac{1\ \text{L}}{1000\ \text{mL}}$ $M = \frac{\text{mol}}{\text{L}}$ $M = \frac{\text{mol}}{\text{L}}$

mg NaOH → g NaOH → mol NaOH the write balanced equation then

$\frac{1\ \text{g NaOH}}{1000\ \text{mg NaOH}}$ $\frac{1\ \text{mol NaOH}}{40.00\ \text{g NaOH}}$ $NaOH + HClO \rightarrow H_2O + NaClO$

mol NaClO, mol HClO, mol NaOH → mol NaClO, mol HClO then

set up stoichiometry table

Part II: Equilibrium:
mol NaClO, mol HClO, L, K_a → pH

$$pH = pK_a + \log\frac{[\text{base}]}{[\text{acid}]}$$

Solution: $100.0\ \cancel{\text{mL}} \times \frac{1\ \text{L}}{1000\ \cancel{\text{mL}}} = 0.1000\ \text{L}$ then $\frac{0.175\ \text{mol HClO}}{1\ \cancel{\text{L}}} \times 0.1000\ \cancel{\text{L}} = 0.0175\ \text{mol HClO}$

and $\frac{0.150\ \text{mol NaClO}}{1\ \cancel{\text{L}}} \times 0.1000\ \cancel{\text{L}} = 0.0150\ \text{mol NaClO}$ and

$85.0\ \cancel{\text{mg NaOH}} \times \frac{1\ \cancel{\text{g NaOH}}}{1000\ \cancel{\text{mg NaOH}}} \times \frac{1\ \text{mol NaOH}}{40.00\ \cancel{\text{g NaOH}}} = 0.00213\ \text{mol NaOH}$ then set up table to track changes:

	NaOH (*aq*) +	HClO (*aq*) →	NaClO (*aq*) +	H_2O (*l*)
Before addition	≈ 0.00 mol	0.0175 mol	0.0150 mol	–
Addition	0.00213 mol	–	–	–
After addition	≈ 0.00 mol	0.0154 mol	0.0171 mol	–

Since the amount of NaOH is small, there are still significant amounts of both buffer components, so the Henderson–Hasselbalch Equation can be used to calculate the new pH.

$$pH = pK_a + \log\frac{[\text{base}]}{[\text{acid}]} = -\log(2.9 \times 10^{-8}) + \log\frac{\dfrac{0.0171\ \cancel{\text{mol}}}{\cancel{0.1000\ \text{L}}}}{\dfrac{0.0154\ \cancel{\text{mol}}}{\cancel{0.1000\ \text{L}}}} = 7.58\ .$$

Check: The units (none) are correct. The magnitude of the answer makes physical sense because the pH rose slightly when base was added.

23. a) **Given:** 500.0 mL pure water **Find:** initial pH and after adding 0.010 mol HCl
Conceptual Plan:
pure water has a pH of 7.00 then mL → L then mol HCl, L → $[H_3O^+]$ → pH

$\frac{1\ \text{L}}{1000\ \text{mL}}$ $M = \frac{\text{mol}}{\text{L}}$ $pH = -\log[H_3O^+]$

Solution: pure water has a pH of 7.00 so initial pH = 7.00 then $500.0\ \cancel{\text{mL}} \times \frac{1\ \text{L}}{1000\ \cancel{\text{mL}}} = 0.5000\ \text{L}$

then $M = \frac{\text{mol}}{\text{L}} = \frac{0.010 \text{ mol HCl}}{0.5000 \text{ L}} = 0.020 \text{ M HCl}$. Since HCl is a strong acid, it dissociates completely,

so $pH = -\log[H_3O^+] = -\log(0.020) = 1.70$.

Check: The units (none) are correct. The magnitudes of the answers make physical sense because the pH starts neutral and then dropps significantly when acid is added and there is no buffer present.

b) **Given:** 500.0 mL buffer 0.125 M $HC_2H_3O_2$ and 0.115 M $NaC_2H_3O_2$
Find: initial pH and after adding 0.010 mol HCl **Other:** $K_a(HC_2H_3O_2) = 1.8 \times 10^{-5}$
Conceptual Plan: initial pH:
identify acid and base components then M $NaC_2H_3O_2$ → M $C_2H_3O_2^-$ then
acid = $HC_2H_3O_2$ base = $C_2H_3O_2^-$ $NaC_2H_3O_2\ (aq) \rightarrow Na^+\ (aq) + C_2H_3O_2^-\ (aq)$
M $HC_2H_3O_2$, M $C_2H_3O_2^-$ → pH

$$pH = pK_a + \log\frac{[\text{base}]}{[\text{acid}]}$$

pH after HCl addition: Part I: Stoichiometry:
mL → L then [$NaC_2H_3O_2$], L → mol $NaC_2H_3O_2$ and [$HC_2H_3O_2$], L → mol $HC_2H_3O_2$

$$\frac{1 \text{ L}}{1000 \text{ mL}} \qquad M = \frac{\text{mol}}{\text{L}} \qquad M = \frac{\text{mol}}{\text{L}}$$

write balanced equation then
$HCl + NaC_2H_3O_2 \rightarrow HC_2H_3O_2 + NaCl$
mol $NaC_2H_3O_2$, mol $HC_2H_3O_2$, mol HCl → mol $NaC_2H_3O_2$, mol $HC_2H_3O_2$ then
set up stoichiometry table
Part II: Equilibrium:
mol $NaC_2H_3O_2$, mol $HC_2H_3O_2$, L, K_a → pH

$$pH = pK_a + \log\frac{[\text{base}]}{[\text{acid}]}$$

Solution: initial pH: Acid = $HC_2H_3O_2$, so [acid] = [$HC_2H_3O_2$] = 0.125 M. Base = $C_2H_3O_2^-$. Since 1 $C_2H_3O_2^-$ ion is generated for each $NaC_2H_3O_2$, [$C_2H_3O_2^-$] = 0.115 M $C_2H_3O_2^-$ = [base]. Then

$$pH = pK_a + \log\frac{[\text{base}]}{[\text{acid}]} = -\log(1.8 \times 10^{-5}) + \log\frac{0.115 \text{ M}}{0.125 \text{ M}} = 4.71.$$

pH after HCl addition:

$$500.0 \text{ mL} \times \frac{1 \text{ L}}{1000 \text{ mL}} = 0.5000 \text{ L} \text{ then } \frac{0.125 \text{ mol } HC_2H_3O_2}{1 \text{ L}} \times 0.5000 \text{ L} = 0.0625 \text{ mol } HC_2H_3O_2 \text{ and}$$

$$\frac{0.115 \text{ mol } NaC_2H_3O_2}{1 \text{ L}} \times 0.5000 \text{ L} = 0.0575 \text{ mol } NaC_2H_3O_2$$ set up table to track changes:

	$HCl\ (aq)$ +	$NaC_2H_3O_2\ (aq)$ →	$HC_2H_3O_2\ (aq)$ +	$NaCl\ (aq)$
Before addition	≈ 0.00 mol	0.0575 mol	0.0625 mol	0.00 mol
Addition	0.010 mol	–	–	–
After addition	≈ 0.00 mol	0.0475 mol	0.0725 mol	0.010 mol

Since the amount of HCl is small, there are still significant amounts of both buffer components, so the Henderson–Hasselbalch Equation can be used to calculate the new pH.

$$pH = pK_a + \log\frac{[\text{base}]}{[\text{acid}]} = -\log(1.8 \times 10^{-5}) + \log\frac{\frac{0.0475 \text{ mol}}{0.5000 \text{ L}}}{\frac{0.0725 \text{ mol}}{0.5000 \text{ L}}} = 4.56.$$

Check: The units (none) are correct. The magnitudes of the answers make physical sense because the pH started below the pK_a of the acid and it dropped slightly when acid was added.

c) **Given:** 500.0 mL buffer 0.155 M $CH_3CH_2NH_2$ and 0.145 M $CH_3CH_2NH_3Cl$
Find: initial pH and after adding 0.010 mol HCl **Other:** $K_b(CH_3CH_2NH_2) = 5.6 \times 10^{-4}$
Conceptual Plan: initial pH:
identify acid and base components then M $CH_3CH_2NH_3Cl$ → M $CH_3CH_2NH_3^+$
acid = $C_2H_5NH_3^+$ base = $C_2H_5NH_2$ $C_2H_5NH_3Cl\ (aq) \rightarrow C_2H_5NH_3^+\ (aq) + Cl^-\ (aq)$

and K_b → pK_b → pK_a then pK_a, M $CH_3CH_2NH_2$, M $CH_3CH_2NH_3^+$ → pH

$pK_b = -\log K_b \quad 14 = pK_a + pK_b \qquad \text{pH} = pK_a + \log\frac{[\text{base}]}{[\text{acid}]}$

pH after HCl addition: Part I: Stoichiometry:
mL → L then [$CH_3CH_2NH_2$], L → mol $CH_3CH_2NH_2$ and

$\frac{1\text{ L}}{1000\text{ mL}} \qquad M = \frac{\text{mol}}{\text{L}}$

[$CH_3CH_2NH_3Cl$], L → mol $CH_3CH_2NH_3Cl$ write balanced equation then

$M = \frac{\text{mol}}{\text{L}} \qquad HCl + CH_3CH_2NH_2 \rightarrow CH_3CH_2NH_3Cl$

mol $CH_3CH_2NH_2$, mol $CH_3CH_2NH_3Cl$, mol HCl → mol $CH_3CH_2NH_2$, mol $CH_3CH_2NH_3Cl$ then
set up stoichiometry table

Part II: Equilibrium:
mol $CH_3CH_2NH_2$, mol $CH_3CH_2NH_3Cl$, L, K_a → pH

$$\text{pH} = pK_a + \log\frac{[\text{base}]}{[\text{acid}]}$$

Solution: Base = $CH_3CH_2NH_2$, [base] = [$CH_3CH_2NH_2$] = 0.155 M Acid = $CH_3CH_2NH_3^+$. Since 1 $CH_3CH_2NH_3^+$ ion is generated for each $CH_3CH_2NH_3Cl$, [$CH_3CH_2NH_3^+$] = 0.145 M $CH_3CH_2NH_3^+$ = [acid]. Since K_b ($CH_3CH_2NH_2$) = 5.6×10^{-4}, $pK_b = -\log K_b = -\log(5.6 \times 10^{-4}) = 3.25$. Since $14 = pK_a + pK_b$, $pK_a = 14 - pK_b = 14 - 3.25 = 10.75$ then

$$\text{pH} = pK_a + \log\frac{[\text{base}]}{[\text{acid}]} = 10.75 + \log\frac{0.155\ \cancel{M}}{0.145\ \cancel{M}} = 10.78\,.$$

pH after HCl addition: $500.0\ \cancel{\text{mL}} \times \frac{1\text{ L}}{1000\ \cancel{\text{mL}}} = 0.5000\text{ L}$ then

$\frac{0.155\text{ mol }CH_3CH_2NH_2}{1\ \cancel{\text{L}}} \times 0.5000\ \cancel{\text{L}} = 0.0775\text{ mol }CH_3CH_2NH_2$ and

$\frac{0.145\text{ mol }CH_3CH_2NH_3Cl}{1\ \cancel{\text{L}}} \times 0.5000\ \cancel{\text{L}} = 0.0725\text{ mol }CH_3CH_2NH_3Cl$ set up table to track changes:

	$HCl\ (aq)$ +	$CH_3CH_2NH_2\ (aq)$ →	$CH_3CH_2NH_3Cl\ (aq)$
Before addition	≈ 0.00 mol	0.0775 mol	0.0725 mol
Addition	0.010 mol	–	–
After addition	≈ 0.00 mol	0.0675 mol	0.0825 mol

Since the amount of HCl is small, there are still significant amounts of both buffer components, so the Henderson–Hasselbalch Equation can be used to calculate the new pH.

$$\text{pH} = pK_a + \log\frac{[\text{base}]}{[\text{acid}]} = 10.75 + \log\frac{\frac{0.0675\ \cancel{\text{mol}}}{\cancel{0.5000\text{ L}}}}{\frac{0.0825\ \cancel{\text{mol}}}{\cancel{0.5000\text{ L}}}} = 10.66\,.$$

Check: The units (none) are correct. The magnitudes of the answers make physical sense because the initial pH should be greater than the pK_a of the acid because there is more base than acid and the pH drops slightly when acid is added.

24\. a) **Given:** 250.0 mL pure water **Find:** initial pH and after adding 0.010 mol NaOH
Conceptual Plan:
pure water has a pH of 7.00 then mL→ L then mol NaOH, L→ [OH^-] → [H_3O^+] → pH

$\frac{1\text{ L}}{1000\text{ mL}} \qquad M = \frac{\text{mol}}{\text{L}} \quad K_w = [H_3O^+][OH^-] \quad pH = -\log[H_3O^+]$

Solution: pure water has a pH of 7.00 so initial pH = 7.00 then $250.0\ \cancel{\text{mL}} \times \frac{1\text{ L}}{1000\ \cancel{\text{mL}}} = 0.2500\text{ L}$ then

$M = \frac{\text{mol}}{\text{L}} = \frac{0.010\text{ mol NaOH}}{0.2500\text{ L}} = 0.040\text{ M NaOH}$. Since NaOH is a strong base, it dissociates completely, so $[OH^-] == 0.020$ M. $K_w = [H_3O^+][OH^-]$ so $[H_3O^+] = \frac{K_w}{[OH^-]} = \frac{1.0 \times 10^{-14}}{0.040} = 2.5 \times 10^{-13}$ M and

$pH = -\log[H_3O^+] = -\log(2.5 \times 10^{-13}) = 12.60$.

Check: The units (none) are correct. The magnitudes of the answers make physical sense because the pH starts neutral and then rose significantly when base is added and there is no buffer present.

b) Given: 250.0 mL buffer 0.195 M $HCHO_2$ and 0.275 M $KCHO_2$
Find: initial pH and after adding 0.010 mol NaOH **Other:** K_a ($HCHO_2$) = 1.8×10^{-4}
Conceptual Plan: initial pH:
identify acid and base components then M $KCHO_2$ → M CHO_2^- then
acid = $HCHO_2$ base = CHO_2^- $KCHO_2\ (aq) \rightarrow K^+\ (aq) + CHO_2^-\ (aq)$
M $HCHO_2$, M CHO_2^- → pH

$$pH = pK_a + \log\frac{[base]}{[acid]}$$

pH after NaOH addition: Part I: Stoichiometry:
mL → L then [$KCHO_2$], L → mol $KCHO_2$ and [$HCHO_2$], L → mol $HCHO_2$

$\frac{1\ L}{1000\ mL}$ $M = \frac{mol}{L}$ $M = \frac{mol}{L}$

write balanced equation then
$NaOH + HCHO_2 \rightarrow NaCHO_2 + H_2O$
mol CHO_2^-, mol $HCHO_2$, mol NaOH → mol $NaCHO_2$, mol $HCHO_2$ then
set up stoichiometry table
Part II: Equilibrium:
mol $NaC_2H_3O_2$, mol $HC_2H_3O_2$, L, K_a → pH

$$pH = pK_a + \log\frac{[base]}{[acid]}$$

Solution: initial pH: Acid = $HCHO_2$, so [acid] = [$HCHO_2$] = 0.195 M. Base = CHO_2^-.
Since 1 CHO_2^- ion is generated for each $KCHO_2$, [CHO_2^-] = 0.275 M CHO_2^- = [base]. Then

$$pH = pK_a + \log\frac{[base]}{[acid]} = -\log(1.8 \times 10^{-4}) + \log\frac{0.275\ \cancel{M}}{0.195\ \cancel{M}} = 3.89\,.$$

pH after NaOH addition: $250.0\ \cancel{mL} \times \frac{1\ L}{1000\ \cancel{mL}} = 0.2500\ L$ then

$\frac{0.195\ mol\ HCHO_2}{1\ \cancel{L}} \times 0.2500\ \cancel{L} = 0.04875\ mol\ HCHO_2$ and

$\frac{0.275\ mol\ KCHO_2}{1\ \cancel{L}} \times 0.2500\ \cancel{L} = 0.06875\ mol\ KCHO_2$

set up table to track changes:

	NaOH (*aq*)	+ $HCHO_2$ (*aq*)	→ $NaCHO_2$ (*aq*)	+ H_2O (*l*)
Before addition	≈ 0.00 mol	0.04875 mol	0.06875 mol	–
Addition	0.010 mol	–	–	–
After addition	≈ 0.00 mol	0.03875 mol	0.07875 mol	–

Since the amount of NaOH is small, there are still significant amounts of both buffer components, so the Henderson–Hasselbalch Equation can be used to calculate the new pH.

$$pH = pK_a + \log\frac{[base]}{[acid]} = -\log(1.8 \times 10^{-4}) + \log\frac{\frac{0.07875\ \cancel{mol}}{\cancel{0.2500\ L}}}{\frac{0.03875\ \cancel{mol}}{\cancel{0.2500\ L}}} = 4.05\,.$$

Check: The units (none) are correct. The magnitudes of the answers make physical sense because the pH started above the pK_a of the acid and it rose slightly when base was added.

c) Given: 250.0 mL buffer 0.255 M $CH_3CH_2NH_2$ and 0.235 M $CH_3CH_2NH_3Cl$
Find: initial pH and after adding 0.010 mol NaOH **Other:** K_b ($CH_3CH_2NH_2$) = 5.6×10^{-4}

Conceptual Plan: initial pH:
identify acid and base components then M $CH_3CH_2NH_3Cl$ → M $CH_3CH_2NH_3^+$

acid = $C_2H_5NH_3^+$ base = $C_2H_5NH_2$ $C_2H_5NH_3Cl$ (aq) → $C_2H_5NH_3^+$ (aq) + Cl^- (aq)

and K_b → pK_b → pK_a then pK_a, M $CH_3CH_2NH_2$, M $CH_3CH_2NH_3^+$ → pH

$pK_b = -\log K_b$ $14 = pK_a + pK_b$ $\mathrm{pH} = \mathrm{p}K_a + \log\frac{[\text{base}]}{[\text{acid}]}$

pH after NaOH addition: Part I: Stoichiometry:
mL → L then [$CH_3CH_2NH_2$], L → mol $CH_3CH_2NH_2$ and

$\frac{1\text{ L}}{1000\text{ mL}}$ $M = \frac{\text{mol}}{\text{L}}$

[$CH_3CH_2NH_3Cl$], L → mol $CH_3CH_2NH_3Cl$

$M = \frac{\text{mol}}{\text{L}}$

write balanced equation then

$NaOH + CH_3CH_2NH_3Cl \rightarrow CH_3CH_2NH_2 + NaCl + H_2O$

mol $CH_3CH_2NH_2$, mol $CH_3CH_2NH_3Cl$, mol NaOH → mol $CH_3CH_2NH_2$, mol $CH_3CH_2NH_3Cl$

set up stoichiometry table

Part II: Equilibrium:
mol $CH_3CH_2NH_2$, mol $CH_3CH_2NH_3Cl$, L, K_a → pH

$$\mathrm{pH} = \mathrm{p}K_a + \log\frac{[\text{base}]}{[\text{acid}]}$$

Solution: Base = $CH_3CH_2NH_2$, [base] = [$CH_3CH_2NH_2$] = 0.255 M Acid = $CH_3CH_2NH_3^+$. Since 1 $CH_3CH_2NH_3^+$ ion is generated for each $CH_3CH_2NH_3Cl$, [$CH_3CH_2NH_3^+$] = 0.235 M $CH_3CH_2NH_3^+$ = [acid]. Since K_b ($CH_3CH_2NH_2$) = 5.6×10^{-4}, $pK_b = -\log K_b = -\log(5.6 \times 10^{-4}) = 3.25$. Since $14 = pK_a + pK_b$,

$$pK_a = 14 - pK_b = 14 - 3.25 = 10.75 \text{ then } \mathrm{pH} = \mathrm{p}K_a + \log\frac{[\text{base}]}{[\text{acid}]} = 10.75 + \log\frac{0.255\ \cancel{\text{M}}}{0.235\ \cancel{\text{M}}} = 10.79.$$

pH after HCl addition: $250.0\ \cancel{\text{mL}} \times \frac{1\text{ L}}{1000\ \cancel{\text{mL}}} = 0.2500\text{ L}$ then

$$\frac{0.255\text{ mol } CH_3CH_2NH_2}{1\ \cancel{\text{L}}} \times 0.2500\ \cancel{\text{L}} = 0.06375\text{ mol } CH_3CH_2NH_2 \text{ and}$$

$$\frac{0.235\text{ mol } CH_3CH_2NH_3Cl}{1\ \cancel{\text{L}}} \times 0.2500\ \cancel{\text{L}} = 0.05875\text{ mol } CH_3CH_2NH_3Cl$$ set up table to track changes:

	NaOH (*aq*) +	$CH_3CH_2NH_3Cl$ (*aq*) →	$CH_3CH_2NH_2$ (*aq*) +	NaCl (*aq*) +	H_2O (*aq*)
Before addition	≈ 0.00 mol	0.05875 mol	0.06375 mol	0.00 mol	–
Addition	0.010 mol	–	–	–	–
After addition	≈ 0.00 mol	0.04875 mol	0.07375 mol		

Since the amount of NaOH is small, there are still significant amounts of both buffer components, so the Henderson–Hasselbalch Equation can be used to calculate the new pH.

$$\mathrm{pH} = \mathrm{p}K_a + \log\frac{[\text{base}]}{[\text{acid}]} = 10.75 + \log\frac{\frac{0.07375\ \cancel{\text{mol}}}{\cancel{0.2500\text{ L}}}}{\frac{0.04875\ \cancel{\text{mol}}}{\cancel{0.2500\text{ L}}}} = 10.93.$$

Check: The units (none) are correct. The magnitudes of the answers make physical sense because the initial pH should be greater than the pK_a of the acid because there is more base than acid and the pH rises slightly when base is added.

25. **Given:** 350.00 mL 0.150 M HF and 0.150 M NaF buffer
Find: mass NaOH to raise pH to 4.00 and mass NaOH to raise pH to 4.00 with buffer concentrations raised to 0.350 M
Other: K_a (HF) = 3.5×10^{-4}
Conceptual Plan: identify acid and base components Since [NaF] = [HF] then initial pH = pK_a.

acid = HF base = F^- pH = pK_a

final pH, pK_a → [NaF]/[HF] and mL → L then [HF], L → mol HF and [NaF], L → mol NaF

$pH = pK_a + \log\frac{[base]}{[acid]}$ $\frac{1\ L}{1000\ mL}$ $M = \frac{mol}{L}$ $M = \frac{mol}{L}$

then write balanced equation then

$NaOH + HF \rightarrow NaF + H_2O$

mol HF, mol NaF, [NaF]/[HF] → mol NaOH → g NaOH

set up stoichiometry table $\frac{40.00\ g\ NaOH}{1\ mol\ NaOH}$

Finally, when the buffer concentrations are raised to 0.350 M, simply multiply the g NaOH by ratio of concentrations (0.350 M / 0.150 M).

Solution: initial $pH = pK_a = -\log(3.5 \times 10^{-4}) = 3.46$ then

$pH = pK_a + \log\frac{[base]}{[acid]} = -\log(3.5 \times 10^{-4}) + \log\frac{[NaF]}{[HF]} = 4.00$. Solve for [NaF] / [HF].

$\log\frac{[NaF]}{[HF]} = 4.00 - 3.46 = 0.54$ → $\frac{[NaF]}{[HF]} = 10^{0.54} = 3.5$. $350.0\ \cancel{mL} \times \frac{1\ L}{1000\ \cancel{mL}} = 0.3500\ L$ then

$\frac{0.150\ mol\ HF}{1\ \cancel{L}} \times 0.3500\ \cancel{L} = 0.0525\ mol\ HF$ and $\frac{0.150\ mol\ NaF}{1\ \cancel{L}} \times 0.3500\ \cancel{L} = 0.0525\ mol\ NaF$

set up table to track changes:

	NaOH (*aq*) +	HF (*aq*) →	NaF (*aq*) +	H_2O (*aq*)
Before addition	0.00 mol	0.0525 mol	0.0525 mol	–
Addition	x	–	–	–
After addition	≈ 0.00 mol	$(0.0525 - x)$ mol	$(0.0525 + x)$ mol	–

Since $\frac{[NaF]}{[HF]} = 3.5 = \frac{(0.0525 + x)\ \cancel{mol}}{(0.0525 - x)\ \cancel{mol}}$, solve for x. Note that the ratio of moles is the same as the ratio of concentrations, since the volume for both terms is the same. $3.5\ (0.0525 - x) = (0.0525 + x)$ → $0.18\underline{3}75 - 3.5x = 0.0525 + x$ → $0.13\underline{1}25 = 4.5x$ → $x = 0.02\underline{9}167$ mol NaOH then

$0.02\underline{9}167\ \cancel{mol\ NaOH} \times \frac{40.00\ g\ NaOH}{1\ \cancel{mol\ NaOH}} = 1.\underline{1}667\ g\ NaOH = 1.2\ g\ NaOH$. Finally multiply the NaOH mass by the ratio of concentrations $1.1667\ g\ NaOH \times \frac{0.350\ \cancel{M}}{0.150\ \cancel{M}} = 2.7\ g\ NaOH$.

Check: The units (g) are correct. The magnitudes of the answers make physical sense because there is much less than a mole of each of the buffer components, so there must be much less than a mole of NaOH. The higher the buffer concentrations, the higher the buffer capacity and the mass of NaOH it can neutralize.

26. **Given:** 100.00 mL 0.100 M NH_3 and 0.125 M NH_4Br buffer
Find: mass HCl to lower pH to 9.00 and mass HCl to lower pH to 9.00 with buffer concentrations raised to 0.250 M NH_3 and 0.400 M NH_4Br
Other: K_b (NH_3) = 1.79×10^{-5}
Conceptual Plan: identify acid and base components then K_b → pK_b → pK_a then

acid = NH_4^+ base = NH_3 $pK_b = -\log K_b$ $14 = pK_a + pK_b$

final pH, pK_a → [NH_3]/[NH_4^+] and mL → L then [NH_3], L → mol NH_3 and

$pH = pK_a + \log\frac{[base]}{[acid]}$ $\frac{1\ L}{1000\ mL}$ $M = \frac{mol}{L}$

[NH_4^+], L → mol NH_4^+ then write balanced equation then

$M = \frac{mol}{L}$ $H^+ + NH_3 \rightarrow NH_4^+$

mol NH_3, mol NH_4^+, [NH_3]/[NH_4^+] → mol HCl → g HCl

set up stoichiometry table $\frac{36.46\ g\ HCl}{1\ mol\ HCl}$

Solution: Since K_b (NH_3) = 1.79×10^{-5}, $pK_b = -\log K_b = -\log(1.79 \times 10^{-5}) = 4.75$. Since $14 = pK_a + pK_b$,

$pK_a = 14 - pK_b = 14 - 4.75 = 9.25$ then $\text{pH} = pK_a + \log\frac{[\text{base}]}{[\text{acid}]} = 9.25 + \log\frac{[NH_3]}{[NH_4^+]} = 9.00$.

Solve for $[NH_4Br]$. $\log\frac{[NH_3]}{[NH_4^+]} = 9.00 - 9.25 = -0.25 \rightarrow \frac{[NH_3]}{[NH_4^+]} = 10^{-0.25} = 0.5\underline{6}2341$.

$100.0 \text{ mL} \times \frac{1 \text{ L}}{1000 \text{ mL}} = 0.1000 \text{ L}$ then $\frac{0.100 \text{ mol } NH_3}{1 \text{ L}} \times 0.1000 \text{ L} = 0.0100 \text{ mol } NH_3$ and

$\frac{0.125 \text{ mol } NH_4Br}{1 \text{ L}} \times 0.1000 \text{ L} = 0.0125 \text{ mol } NH_4Br = 0.0125 \text{ mol } NH_4^+$ Since HCl is a strong acid,

$[HCl] = [H^+]$, and set up table to track changes:

	$H^+ (aq)$	+ $NH_3 (aq)$	$\rightarrow$ $NH_4^+ (aq)$
Before addition	≈ 0.00 mol	0.0100 mol	0.0125 mol
Addition	x	–	–
After addition	≈ 0.00 mol	(0.0100 – x) mol	(0.0125 + x) mol

Since $\frac{[NH_3]}{[NH_4^+]} = 0.5\underline{6}2341 = \frac{(0.0100 - x) \text{ mol}}{(0.0125 + x) \text{ mol}}$, solve for x. Note that the ratio of moles is the same as the ratio of concentrations, since the volume for both terms is the same. $0.5\underline{6}2341(0.0125 + x) = (0.0100 - x) \rightarrow$ $0.007\underline{0}2926 + 0.5\underline{6}2341x = 0.0100 - x \rightarrow 1.5\underline{6}2341x = 0.00\underline{2}9707 \rightarrow x = 0.00\underline{1}9015 \text{ mol HCl}$ then

$0.00\underline{1}9015 \text{ mol HCl} \times \frac{36.46 \text{ g HCl}}{1 \text{ mol HCl}} = 0.0\underline{6}932 \text{ g HCl} = 0.07 \text{ g HCl}$. For the higher concentration buffer, repeat part of the above calculations. $\frac{0.250 \text{ mol } NH_3}{1 \text{ L}} \times 0.1000 \text{ L} = 0.0250 \text{ mol } NH_3$ and

$\frac{0.400 \text{ mol } NH_4Br}{1 \text{ L}} \times 0.1000 \text{ L} = 0.0400 \text{ mol } NH_4Br = 0.0400 \text{ mol } NH_4^+$ Since HCl is a strong acid,

$[HCl] = [H^+]$, and set up table to track changes:

	$H^+ (aq)$	+ $NH_3 (aq)$	$\rightarrow$ $NH_4^+ (aq)$
Before addition	≈ 0.00 mol	0.0250 mol	0.0400 mol
Addition	x	–	–
After addition	≈ 0.00 mol	(0.0250 – x) mol	(0.0400 + x) mol

Since $\frac{[NH_3]}{[NH_4^+]} = 0.5\underline{6}2341 = \frac{(0.0250 - x) \text{ mol}}{(0.0400 + x) \text{ mol}}$, solve for x. Note that the ratio of moles is the same as the ratio of concentrations, since the volume for both terms is the same. $0.5\underline{6}2341(0.0400 + x) = (0.0250 - x) \rightarrow$ $0.02\underline{2}4936 + 0.5\underline{6}2341x = 0.0250 - x \rightarrow 1.5\underline{6}2341x = 0.00\underline{2}50636 \rightarrow x = 0.00\underline{1}6042 \text{ mol HCl}$ then

$0.00\underline{1}6042 \text{ mol HCl} \times \frac{36.46 \text{ g HCl}}{1 \text{ mol HCl}} = 0.0\underline{5}8490 \text{ g HCl} = 0.06 \text{ g HCl}$.

Check: The units (g) are correct. The magnitudes of the answers make physical sense because there is much less than a mole of each of the buffer components, so there must be much less than a mole of HCl. Also, the pH of the initial buffer is less than the pK_a of the acid, so even less acid is necessary to drop the pH below 9.00. The higher concentration buffer solution requires about the same amount of acid because the initial pH of this buffer is closer to 9.00 than the low concentration buffer.

27. a) Yes, this will be a buffer because NH_3 is a weak base and NH_4^+ is its conjugate acid. The ratio of base to acid is 0.10/0.15 = 0.67, so the pH will be within 1 pH unit of the pK_a.

b) No, this will not be a buffer solution because HCl is a strong acid and NaOH is a strong base.

c) Yes, this will be a buffer because HF is a weak acid and the NaOH will convert 20.0/50.0 = 40 % of the acid to its conjugate base.

d) No, this will not be a buffer solution because both components are bases.

e) No, this will not be a buffer solution because both components are bases.

28. a) Yes, this will be a buffer because HF is a weak acid and F^- is its conjugate base. The ratio of base to acid is (55.0 x 0.15)/(75.0 x 0.10) = 1.1, so the pH will be within 1 pH unit of the pK_a.

b) No, this will not be a buffer solution because both components are acids.

c) Yes, this will be a buffer because HF is a weak acid and the KOH will convert (135.0 x 0.050)/(165.0 x 0.10) = 41 % of the acid to its conjugate base.

d) Yes, this will be a buffer because CH_3NH_2 is a weak base and $CH_3NH_3^+$ is its conjugate acid. The ratio of base to acid is (125.0 x 0.15)/(120.0 x 0.25) = 0.63, so the pH will be within 1 pH unit of the pK_a.

e) Yes, this will be a buffer because CH_3NH_2 is a weak base and the HCl will convert (95.0 x 0.10)/(105.0 x 0.15) = 60 % of the base to its conjugate acid.

29. a) **Given:** blood buffer 0.024 M HCO_3^- and 0.0012 M H_2CO_3, pK_a = 6.1 **Find:** initial pH
Conceptual Plan: identify acid and base components then M HCO_3^-, M H_2CO_3 → pH

acid = H_2CO_3 base = HCO_3^- $\quad$ $\text{pH} = \text{p}K_a + \log\frac{[\text{base}]}{[\text{acid}]}$

Solution: Acid = H_2CO_3, so [acid] = [H_2CO_3] = 0.0012 M. Base = HCO_3^-, so [base] = [HCO_3^-] = 0.024 M HCO_3^-. Then $\text{pH} = \text{p}K_a + \log\frac{[\text{base}]}{[\text{acid}]} = 6.1 + \log\frac{0.024\ \cancel{M}}{0.0012\ \cancel{M}} = 7.4$.

Check: The units (none) are correct. The magnitude of the answer makes physical sense because pH is greater than the pK_a of the acid because there is more base than acid.

b) **Given:** 5.0 L of blood buffer **Find:** mass HCl to lower pH to 7.0
Conceptual Plan: final pH, pK_a → [HCO_3^-]/[H_2CO_3] then [HCO_3^-], L → mol HCO_3^- and

$\text{pH} = \text{p}K_a + \log\frac{[\text{base}]}{[\text{acid}]}$ $\quad$ $M = \frac{\text{mol}}{\text{L}}$

[H_2CO_3], L → mol H_2CO_3 then write balanced equation then

$M = \frac{\text{mol}}{\text{L}}$ $\quad$ $H^+ + HCO_3^- \rightarrow H_2CO_3$

mol HCO_3^-, mol H_2CO_3, [HCO_3^-]/[H_2CO_3] → mol HCl → g HCl

set up stoichiometry table $\quad$ $\frac{36.46\text{ g HCl}}{1\text{ mol HCl}}$

Solution: $\text{pH} = \text{p}K_a + \log\frac{[\text{base}]}{[\text{acid}]} = 6.1 + \log\frac{[HCO_3^-]}{[H_2CO_3]} = 7.0$. Solve for [$HCO_3^-$]/[$H_2CO_3$].

$\log\frac{[HCO_3^-]}{[H_2CO_3]} = 7.0 - 6.1 = 0.9 \rightarrow \frac{[HCO_3^-]}{[H_2CO_3]} = 10^{0.9} = \underline{7}.9433$. Then

$\frac{0.024\text{ mol }HCO_3^-}{1\ \cancel{L}} \times 5.0\ \cancel{L} = 0.12\text{ mol }HCO_3^-$ and $\frac{0.0012\text{ mol }H_2CO_3}{1\ \cancel{L}} \times 5.0\ \cancel{L} = 0.0060\text{ mol }H_2CO_3$

Since HCl is a strong acid, [HCl] = [H^+], and set up table to track changes:

	H^+ (*aq*) +	HCO_3^- (*aq*) →	H_2CO_3 (*aq*)
Before addition	≈ 0.00 mol	0.12 mol	0.0060 mol
Addition	*x*	–	–
After addition	≈ 0.00 mol	(0.12 – x) mol	(0.0060 + x) mol

Since $\frac{[HCO_3^-]}{[H_2CO_3]} = \underline{7}.9433 = \frac{(0.12 - x)\ \cancel{\text{mol}}}{(0.0060 + x)\ \cancel{\text{mol}}}$, solve for *x*. Note that the ratio of moles is the same as the ratio of concentrations, since the volume for both terms is the same. $\underline{7}.9433(0.0060 + x) = (0.12 - x) \rightarrow$ $0.04\underline{7}6598 + \underline{7}.9433x = 0.12 - x \rightarrow \underline{8}.9433x = 0.0\underline{7}234 \rightarrow x = 0.00\underline{8}0888$ mol HCl then

$0.00\underline{8}0888\ \cancel{\text{mol HCl}} \times \frac{36.46\text{ g HCl}}{1\ \cancel{\text{mol HCl}}} = 0.\underline{2}9492\text{ g HCl} = 0.3\text{ g HCl}$.

Check: The units (g) are correct. The amount of acid needed is small because the concentrations of the buffer components are very low and the buffer starts only 0.4 pH units above the final pH.

c) **Given:** 5.0 L of blood buffer **Find:** mass NaOH to raise pH to 7.8

Conceptual Plan: final pH, pK_a → $[HCO_3^-]/[H_2CO_3]$ then $[HCO_3^-]$, L → mol HCO_3^- and

$$pH = pK_a + \log\frac{[base]}{[acid]} \qquad M = \frac{mol}{L}$$

$[H_2CO_3]$, L → mol H_2CO_3 then write balanced equation then

$$M = \frac{mol}{L} \qquad OH^- + H_2CO_3 \rightarrow HCO_3^- + H_2O$$

mol HCO_3^-, mol H_2CO_3, $[HCO_3^-]/[H_2CO_3]$ → mol NaOH → g NaOH

set up stoichiometry table $\quad \frac{40.00 \text{ g NaOH}}{1 \text{ mol NaOH}}$

Solution: $pH = pK_a + \log\frac{[base]}{[acid]} = 6.1 + \log\frac{[HCO_3^-]}{[H_2CO_3]} = 7.8$.

Solve for $[HCO_3^-]/[H_2CO_3]$. $\log\frac{[HCO_3^-]}{[H_2CO_3]} = 7.8 - 6.1 = 1.7 \rightarrow \frac{[HCO_3^-]}{[H_2CO_3]} = 10^{1.7} = 50.\underline{1}1872$. Then

$\frac{0.024 \text{ mol } HCO_3^-}{1 \cancel{L}} \times 5.0 \cancel{L} = 0.12 \text{ mol } HCO_3^-$ and $\frac{0.0012 \text{ mol } H_2CO_3}{1 \cancel{L}} \times 5.0 \cancel{L} = 0.0060 \text{ mol } H_2CO_3$

Since NaOH is a strong base, [NaOH] = [OH^-], and set up table to track changes:

	OH^- (*aq*)	+ H_2CO_3 (*aq*)	→ HCO_3^- (*aq*)	+ H_2O (*l*)
Before addition	≈ 0.00 mol	0.0060 mol	0.12 mol	–
Addition	*x*	–	–	
After addition	≈ 0.00 mol	(0.0060 – *x*) mol	(0.12 + *x*) mol	

Since $\frac{[HCO_3^-]}{[H_2CO_3]} = 50.\underline{1}1872 = \frac{(0.12 + x) \cancel{mol}}{(0.0060 - x) \cancel{mol}}$, solve for *x*. Note that the ratio of moles is the same as the ratio of concentrations, since the volume for both terms is the same. $50.\underline{1}1872(0.0060 - x) = (0.12 + x)$ → $0.3\underline{0}071 - 50.\underline{1}1872x = 0.12 + x$ → $51.\underline{1}1872x = 0.1\underline{8}071$ → $x = 0.003\underline{5}351$ mol NaOH

then $0.003\underline{5}351 \cancel{\text{mol NaOH}} \times \frac{40.00 \text{ g NaOH}}{1 \cancel{\text{mol NaOH}}} = 0.1\underline{4}141 \text{ g NaOH} = 0.14 \text{ g NaOH}$.

Check: The units (g) are correct. The amount of base needed is small because the concentrations of the buffer components are very low.

30. a) **Given:** $HPO_4^{2-}/H_2PO_4^-$ buffer at pH = 7.1 **Find:** $[HPO_4^{2-}]/[H_2PO_4^-]$

Other: K_{a2} (H_3PO_4) = 6.2×10^{-8}

Conceptual Plan: identify acid and base components then pH, K_a → $[HPO_4^{2-}]/[H_2PO_4^-]$

acid = $H_2PO_4^-$ base = HPO_4^{2-} $\qquad pH = pK_a + \log\frac{[base]}{[acid]}$

Solution: $pH = pK_a + \log\frac{[base]}{[acid]} = -\log(6.2 \times 10^{-8}) + \log\frac{[HPO_4^{2-}]}{[H_2PO_4^-]} = 7.1$. Solve for $[HPO_4^{2-}]/[H_2PO_4^-]$.

$\log\frac{[HPO_4^{2-}]}{[H_2PO_4^-]} = 7.1 - 7.2 = -0.1 \rightarrow \frac{[HPO_4^{2-}]}{[H_2PO_4^-]} = 10^{-0.1} = 0.\underline{7}9433 = 0.8$.

Check: The units (none) are correct. The magnitude of the answer makes physical sense because the pH is very close, but less than, the pK_a of the acid.

b) No, H_3PO_4 and $H_2PO_4^-$ cannot be used as a buffer in the cell because the K_{a1} (H_3PO_4) = 7.5×10^{-3} and so the pK_{a1} = 2.1. In order to have an effective buffer the pK_a should be within in 1 pH unit of the desired pH (not 5.0 pH units).

31. **Given:** $HC_2H_3O_2/KC_2H_3O_2$, $HClO_2/KClO_2$, NH_3/NH_4Cl, and HClO/KClO potential buffer systems to create buffer at pH = 7.20 **Find:** best buffer system and ratio of component masses
Other: $K_a\,(HC_2H_3O_2) = 1.8 \times 10^{-5}$, $K_a\,(HClO_2) = 1.8 \times 10^{-4}$, $K_b\,(NH_3) = 1.79 \times 10^{-5}$, $K_a\,(HClO) = 2.9 \times 10^{-8}$
Conceptual Plan:
calculate p K_a of all potential buffer acids for the base $K_b \rightarrow pK_b \rightarrow pK_a$ then

$$pK_a = -\log K_a \qquad pK_b = -\log K_b \quad 14 = pK_a + pK_b$$

and choose the pK_a that is closest to 7.20. Then pH, K_a → [base]/[acid] → mass base/mass acid

$$\text{pH} = \text{p}K_a + \log\frac{[\text{base}]}{[\text{acid}]} \qquad \frac{\mathfrak{M}\,(\text{base})}{\mathfrak{M}\,(\text{acid})}$$

Solution: for $HC_2H_3O_2/KC_2H_3O_2$: $pK_a = -\log K_a = -\log(1.8 \times 10^{-5}) = 4.74$; for $HClO_2/KClO_2$:
$pK_a = -\log K_a = -\log(1.8 \times 10^{-4}) = 3.74$; for NH_3/NH_4Cl: $pK_b = -\log K_b = -\log(1.79 \times 10^{-5}) = 4.75$.
Since $14 = pK_a + pK_b$, $pK_a = 14 - pK_b = 14 - 4.75 = 9.25$; and for HClO/KClO:
$pK_a = -\log K_a = -\log(2.9 \times 10^{-8}) = 7.54$. So the HClO/KClO buffer system has the pK_a that is the closest to 7.20. So, $\text{pH} = \text{p}K_a + \log\frac{[\text{base}]}{[\text{acid}]} = 7.54 + \log\frac{[\text{KClO}]}{[\text{HClO}]} = 7.1$. Solve for [KClO]/[HClO].

$$\log\frac{[\text{KClO}]}{[\text{HClO}]} = 7.20 - 7.54 = -0.34 \rightarrow \frac{[\text{KClO}]}{[\text{HClO}]} = 10^{-0.34} = 0.4\underline{5}7088$$

Then convert to mass ratio using

$$\frac{\mathfrak{M}\,(\text{base})}{\mathfrak{M}\,(\text{acid})},\quad 0.4\underline{5}7088\,\frac{\cancel{\text{KClO mol}}/\cancel{\text{L}}}{\cancel{\text{HClO mol}}/\cancel{\text{L}}} \times \frac{\dfrac{90.55\text{ g KClO}}{\cancel{\text{mol KClO}}}}{\dfrac{52.46\text{ g HClO}}{\cancel{\text{mol HClO}}}} = 0.79\,\frac{\text{g KClO}}{\text{g HClO}}$$

Check: The units (none and g base/g acid)) are correct. The buffer system with the K_a closest to 10^{-7} is the best choice. The magnitude of the answer makes physical sense because the buffer needs more acid than base (and this fact is not overcome by the heavier molar mass of the base).

32. **Given:** HF/KF, HNO_2/KNO_2, NH_3/NH_4Cl, and HClO/KClO potential buffer systems to create buffer at pH = 9.00 **Find:** best buffer system and ratio of component masses
Other: $K_a\,(HF) = 3.5 \times 10^{-4}$, $K_a\,(HNO_2) = 4.6 \times 10^{-4}$, $K_b\,(NH_3) = 1.79 \times 10^{-5}$, $K_a\,(HClO) = 2.9 \times 10^{-8}$
Conceptual Plan:
calculate p K_a of all potential buffer acids for the base $K_b \rightarrow pK_b \rightarrow pK_a$ then

$$pK_a = -\log K_a \qquad pK_b = -\log K_b \quad 14 = pK_a + pK_b$$

and choose the pK_a that is closest to 9.00. Then pH, K_a → [base]/[acid] → mass base/mass acid

$$\text{pH} = \text{p}K_a + \log\frac{[\text{base}]}{[\text{acid}]} \qquad \frac{\mathfrak{M}\,(\text{base})}{\mathfrak{M}\,(\text{acid})}$$

Solution: for HF/KF: $pK_a = -\log K_a = -\log(3.5 \times 10^{-4}) = 3.46$; for HNO_2/KNO_2 :
$pK_a = -\log K_a = -\log(4.6 \times 10^{-4}) = 3.34$; for NH_3/NH_4Cl: $pK_b = -\log K_b = -\log(1.79 \times 10^{-5}) = 4.75$.
Since $14 = pK_a + pK_b$, $pK_a = 14 - pK_b = 14 - 4.75 = 9.25$; and for HClO/KClO:
$pK_a = -\log K_a = -\log(2.9 \times 10^{-8}) = 7.54$. So the NH_3/NH_4Cl buffer system has the pK_a that is the closest to 9.00. So, $\text{pH} = \text{p}K_a + \log\frac{[\text{base}]}{[\text{acid}]} = 9.25 + \log\frac{[NH_3]}{[NH_4Cl]} = 9.00$. Solve for $[NH_3]/[NH_4Cl]$.

$$\log\frac{[NH_3]}{[NH_4Cl]} = 9.00 - 9.25 = -0.25 \rightarrow \frac{[NH_3]}{[NH_4Cl]} = 10^{-0.25} = 0.5\underline{6}2341$$

Then convert to mass ratio using

$$\frac{\mathfrak{M}\,(\text{base})}{\mathfrak{M}\,(\text{acid})},\quad 0.5\underline{6}2341\,\frac{\cancel{NH_3\text{ mol}}/\cancel{\text{L}}}{\cancel{NH_4Cl\text{ mol}}/\cancel{\text{L}}} \times \frac{\dfrac{17.03\text{ g }NH_3}{\cancel{\text{mol }NH_3}}}{\dfrac{55.49\text{ g }NH_4Cl}{\cancel{\text{mol }NH_4Cl}}} = 0.17\,\frac{\text{g }NH_3}{\text{g }NH_4Cl}$$

Check: The units (none and g base/g acid)) are correct. The buffer system with the K_a closest to 10^{-9} is the best choice. The magnitude of the answer makes physical sense because the buffer needs more acid than base and the acid component has a heavier molar mass, so the mass ratio is small.

33. **Given:** 500.0 mL of 0.100 M HNO_2 / 0.150 M KNO_2 buffer and a) 250 mg NaOH, b) 350 mg KOH, c) 1.25 g HBr and d) 1.35 g HI **Find:** if buffer capacity is exceeded

Conceptual Plan:

mL → L then [HNO_2], L → mol HNO_2 and [KNO_2], L → mol KNO_2 then

$\frac{1\text{ L}}{1000\text{ mL}}$ $\quad M = \frac{\text{mol}}{\text{L}}$ $\quad M = \frac{\text{mol}}{\text{L}}$

then calculate moles of acid or base to be added to the buffer mg → g → mol then

$\frac{1\text{ g}}{1000\text{ mg}}$ $\quad \mathfrak{M}$

compare the added amount to the buffer amount of the opposite component. Ratio of base/acid must be between 0.1 and 10 to maintain the buffer integrity.

Solution: $500.00\text{ mL} \times \frac{1\text{ L}}{1000\text{ mL}} = 0.5000\text{ L}$ then $\frac{0.100\text{ mol } HNO_2}{1\text{ L}} \times 0.5000\text{ L} = 0.0500\text{ mol } HNO_2$ and

$\frac{0.150\text{ mol } KNO_2}{1\text{ L}} \times 0.5000\text{ L} = 0.0750\text{ mol } KNO_2$

a) for NaOH: $250\text{ mg NaOH} \times \frac{1\text{ g NaOH}}{1000\text{ mg NaOH}} \times \frac{1\text{ mol NaOH}}{40.00\text{ g NaOH}} = 0.00625\text{ mol NaOH}$. Since the buffer contains 0.0500 mol acid, the amount of acid is reduced by 0.00625/0.0500 = 12.5 % and the ratio of base/acid is still between 0.1and 10. The buffer capacity is not exceeded.

b) for KOH: $350\text{ mg KOH} \times \frac{1\text{ g KOH}}{1000\text{ mg KOH}} \times \frac{1\text{ mol KOH}}{56.11\text{ g KOH}} = 0.00624\text{ mol KOH}$. Since the buffer contains 0.0500 mol acid, the amount of acid is reduced by 0.00624/0.0500 = 12.5 % and the ratio of base/acid is still between 0.1 and 10. The buffer capacity is not exceeded.

c) for HBr: $1.25\text{ g HBr} \times \frac{1\text{ mol HBr}}{80.91\text{ g HBr}} = 0.0154496\text{ mol HBr}$. Since the buffer contains 0.0750 mol base, the amount of acid is reduced by 0.0154/0.0750 = 20.6 % and the ratio of base/acid is still between 0.1 and 10. The buffer capacity is not exceeded.

d) for HI: $1.35\text{ g HI} \times \frac{1\text{ mol HI}}{127.91\text{ g HI}} = 0.0105545\text{ mol HI}$. Since the buffer contains 0.0750 mol base, the amount of acid is reduced by 0.0106/0.0750 = 14.1 % and the ratio of base/acid is still between 0.1 and 10. The buffer capacity is not exceeded.

34. **Given:** 1.0 L of 0.125 M HNO_2 / 0.145 M $NaNO_2$ buffer and a) 1.5 g HCl, b) 1.5 g NaOH, c) 1.5 g HI
Find: [HNO_2] and [$NaNO_2$] after addition **Other:** K_a (HNO_2) = 4.6×10^{-4}
Conceptual Plan: [HNO_2], L → mol HNO_2 and [$NaNO_2$], L → mol $NaNO_2$ (= mol NO_2^-)

$M = \frac{\text{mol}}{\text{L}}$ $\quad M = \frac{\text{mol}}{\text{L}}$

then calculate moles of acid or base to be added to the buffer g → mol

$\mathfrak{M}$

then write balanced equation then mol HNO_2, mol NO_2^-, mol added species → mol HNO_2, mol NO_2^-

$H^+ + NO_2^- \rightarrow HNO_2$ $\quad$ set up stoichiometry table

$OH^- + HNO_2 \rightarrow NO_2^- + H_2O$

mol HNO_2, L → [HNO_2] and mol NO_2^- (= mol $NaNO_2$), L → [$NaNO_2$]

$M = \frac{\text{mol}}{\text{L}}$ $\quad M = \frac{\text{mol}}{\text{L}}$

Solution: $\frac{0.125\text{ mol } HNO_2}{1\text{ L}} \times 1.0\text{ L} = 0.125\text{ mol } HNO_2$ and $\frac{0.145\text{ mol } KNO_2}{1\text{ L}} \times 1.0\text{ L} = 0.145\text{ mol } KNO_2$

a) for HCl: $1.5\text{ g HCl} \times \frac{1\text{ mol HCl}}{36.46\text{ g HCl}} = 0.041141\text{ mol HCl}$. Since HCl is a strong acid, [HCl] = [H^+], and set up table to track changes:

	H^+ (*aq*)	+ NO_2^- (*aq*)	→ HNO_2 (*aq*)
Before addition	≈ 0.00 mol	0.145 mol	0.125 mol
Addition	0.041141 mol	–	–
After addition	≈ 0.00 mol	0.104 mol	0.166 mol

Because the concentrations of the acid and base components have not changed much, the buffer is still able to do its job. Finally, since there is 1.0 L of solution, $[HNO_2]$ = 0.17 M and $[NaNO_2]$ = 0.10 M.

b) for NaOH: $1.5\ \text{g NaOH} \times \frac{1\ \text{mol NaOH}}{40.00\ \text{g NaOH}} = 0.0375\ \text{mol NaOH}$. Since NaOH is a strong base, [NaOH] = $[OH^-]$, and set up table to track changes:

	OH^- (*aq*)	+ HNO_2 (*aq*)	→ NO_2^- (*aq*)	+ H_2O (*l*)
Before addition	≈ 0.00 mol	0.125 mol	0.145 mol	–
Addition	0.0375 mol	–	–	–
After addition	≈ 0.00 mol	0.0875 mol	0.1825 mol	–

Because the concentrations of the acid and base components have not changed much, the buffer is still able to do its job. Finally, since there is 1.0 L of solution, $[HNO_2]$ = 0.09 M and $[NaNO_2]$ = 0.18 M.

c) for HI: $1.5\ \text{g HI} \times \frac{1\ \text{mol HI}}{127.91\ \text{g HI}} = 0.011727\ \text{mol HI}$. Since HI is a strong acid, [HI] = $[H^+]$, and set up table to track changes:

	H^+ (*aq*)	+ NO_2^- (*aq*)	→ HNO_2 (*aq*)
Before addition	≈ 0.00 mol	0.145 mol	0.125 mol
Addition	0.011727 mol	–	–
After addition	≈ 0.00 mol	0.133 mol	0.137 mol

Because the concentrations of the acid and base components have not changed much, the buffer is still able to do its job. Finally, since there is 1.0 L of solution, $[HNO_2]$ = 0.14 M and $[NaNO_2]$ = 0.13 M.

Check: The units (M) are correct. Since the number of moles added is small compared to the buffer components, the buffer still remains active. Adding acid increases the amount of the conjugate base. Adding base increases the amount of the weak acid.

35. (i) The equivalence point of a titration is where the pH rises sharply as base is added. The pH as the equivalence point is the midpoint of the sharp rise at ~ 50 mL added base. For (a) the pH = ~ 8 and for (b) the pH = ~ 7.

(ii) Graph (a) represents a weak acid and graph (b) represents a strong acid. A strong acid titration starts at a lower pH, has a flatter initial region and a sharper rise at the equivalence point than a weak acid. The pH at the equivalence point of a strong acid is neutral, while the pH at the equivalence point of a weak acid is basic.

36. **Given:** 25.0 mL 0.100 M HCl and 0.100 M HF titrated with 0.200 M KOH

a) **Find:** volume of base to reach equivalence point

Conceptual Plan: The answer for both titrations will be the same since the initial concentration and volumes of the acids are the same and both acids are monoprotic. Write balanced equation

$HCl + KOH \rightarrow KCl + H_2O$ and $HF + KOH \rightarrow KF + H_2O$

then mL → L then [acid], L → mol acid then set mol acid = mol base and

$\frac{1\ \text{L}}{1000\ \text{mL}}$ $\quad M = \frac{\text{mol}}{\text{L}}$ $\quad$ balanced equation has 1:1 stoichiometry

[KOH], mol KOH → L KOH → mL KOH

$M = \frac{\text{mol}}{\text{L}}$ $\quad \frac{1000\ \text{mL}}{1\ \text{L}}$

Solution: $25.0\ \text{mL acid} \times \frac{1\ \text{L}}{1000\ \text{mL}} = 0.0250\ \text{L acid}$ then

$$\frac{0.100 \text{ mol acid}}{1 \text{ L}} \times 0.0250 \text{ L} = 0.00250 \text{ mol acid}$$. So mol acid = 0.00250 mol = mol KOH then

$$0.00250 \text{ mol KOH} \times \frac{1 \text{ L KOH}}{0.200 \text{ mol KOH}} = 0.0125 \text{ L KOH} \times \frac{1000 \text{ mL}}{1 \text{ L}} = 12.5 \text{ mL KOH}$$ for both titrations.

Check: The units (mL) are correct. The volume of base is half the volume of acids because the concentration of the base is twice that of the acids. The answer for both titrations is the same because the stoichiometry is the same for both titration reactions.

b) The pH at the equivalence point will be neutral for HCl (since it is a strong acid) and it will be basic for HF (since it is a weak acid).

c) The initial pH will be lower for HCl (since it is a strong acid) and so it dissociates completely. The HF (since it is a weak acid) will only partially dissociate and not drop the pH as low as HCl at the same acid concentration.

d) The titration curves will look like:

HCl:

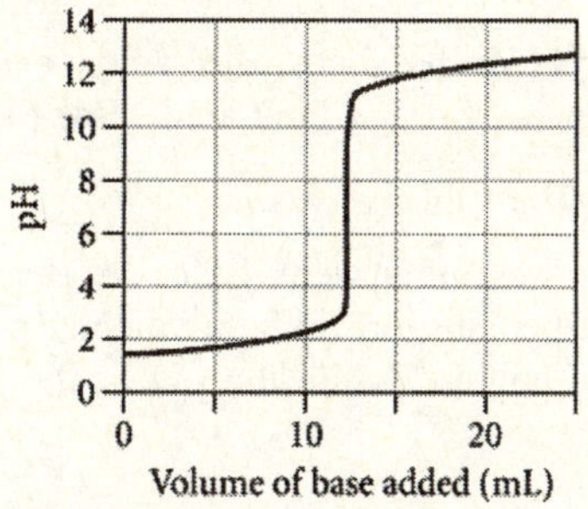

HF:

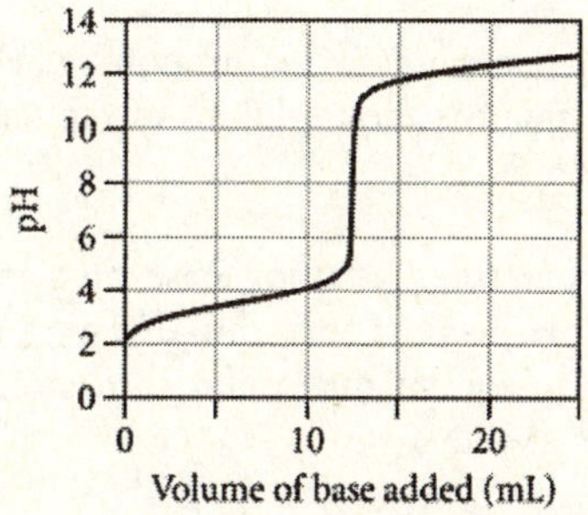

Important features to include are a low initial pH (if strong acid pH is 1 and higher for a weak acid), flat initial region (very flat for strong acid, not as flat for weak acid where pH halfway to equivalence point is the pK_a of the acid), sharp rise at equivalence point, pH at equivalence point (neutral for strong acid and higher for weak acid), and then flatten out at high pH.

37. **Given:** 20.0 mL 0.200 M KOH and 0.200 M CH_3NH_2 titrated with 0.100 M HI

a) **Find:** volume of base to reach equivalence point

Conceptual Plan: The answer for both titrations will be the same since the initial concentration and volumes of the bases are the same. Write balanced equation then mL → L then

$HI + KOH \rightarrow KI + H_2O$ and $HI + KOH \rightarrow CH_3NH_3I$ $\qquad \frac{1 \text{ L}}{1000 \text{ mL}}$

[base], L → mol base then set mol base = mol acid and [HI], mol HI → L HI → mL HI

$M = \frac{\text{mol}}{\text{L}}$ $\qquad$ balanced equation has 1:1 stoichiometry $\qquad$ $M = \frac{\text{mol}}{\text{L}}$ $\quad \frac{1000 \text{ mL}}{1 \text{ L}}$

Solution: $20.0 \text{ mL base} \times \frac{1 \text{ L}}{1000 \text{ mL}} = 0.0200 \text{ L base}$ then

$$\frac{0.200 \text{ mol base}}{1 \text{ L}} \times 0.0200 \text{ L} = 0.00400 \text{ mol base}$$. So mol base = 0.00400 mol = mol HI then

$$0.00400 \text{ mol HI} \times \frac{1 \text{ L HI}}{0.100 \text{ mol HI}} = 0.0400 \text{ L HI} \times \frac{1000 \text{ mL}}{1 \text{ L}} = 40.0 \text{ mL HI}$$ for both titrations.

Check: The units (mL) are correct. The volume of acid is twice the volume of bases because the concentration of the base is twice that of the acid in each case. The answer for both titrations is the same because the stoichiometry is the same for both titration reactions.

b) The pH at the equivalence point will be neutral for KOH (since it is a strong base) and it will be acidic for CH_3NH_2 (since it is a weak base).

c) The initial pH will be lower for CH_3NH_2 (since it is a weak base and will only partially dissociate and not raise the pH as high as KOH (since it is a strong base and so it dissociates completely) at the same base concentration.

d) The titration curves will look like:

KOH:

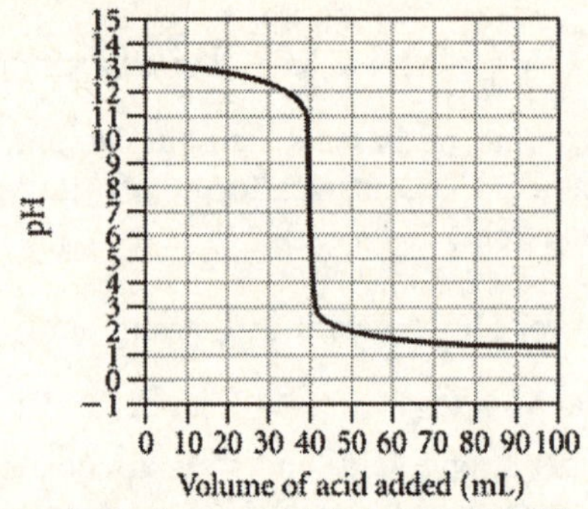

CH_3NH_2:

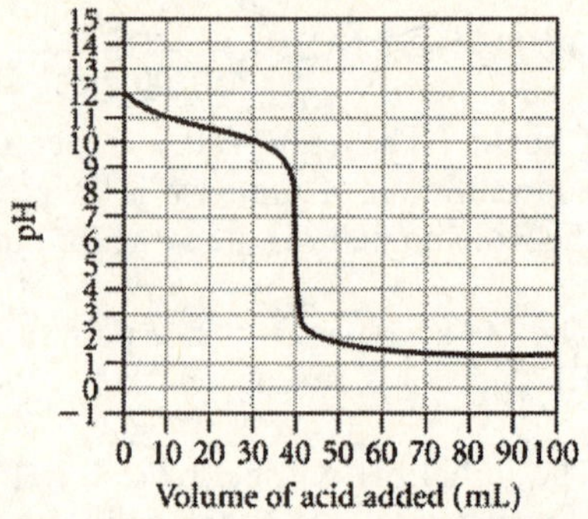

Important features to include are a high initial pH (if strong base pH is over 13 and lower for a weak base), flat initial region (very flat for strong base, not as flat for weak base where pH halfway to equivalence point is the pK_b of the base), sharp drop at equivalence point, pH at equivalence point (neutral for strong base and lower for weak base), and then flatten out at low pH.

38. (i) The equivalence point of a titration is where the pH drops sharply as acid is added. The pH at the equivalence point is the midpoint of the sharp drop at ~ 25 mL added acid. For (a) the pH = ~ 7 and for (b) the pH = ~ 5.

(ii) Graph (a) represents a strong base and graph (b) represents a weak base. A strong base titration starts at a higher pH, has a flatter initial region and a sharper drop at the equivalence point than a weak base. The pH at the equivalence point of a strong base is neutral, while the pH at the equivalence point of a weak base is acidic.

39. a) The equivalence point of a titration is where the pH rises sharply as base is added. The volume at the equivalence point is ~ 30 mL. The pH as the equivalence point is the midpoint of the sharp rise at ~ 30 mL added base, which is a pH = ~ 9.

b) At 0 mL the pH is calculated by doing an equilibrium calculation of a weak acid in water (as done in Chapter 15).

c) The pH one-half way to the equivalence point is equal to the pK_a of the acid, or ~ 15 mL.

d) The pH at the equivalence point, or ~ 30 mL, is calculated by doing an equilibrium problem with the K_b of the acid. At the equivalence point, all of the acid has been converted to its conjugate base.

e) Beyond the equivalence point, or ~ 30 mL, there is excess base. All of the acid has been converted to its conjugate base and so the pH is calculated by focusing on this excess base concentration.

40. a) The equivalence point of a titration is where the pH drops sharply as acid is added. The volume at the equivalence point is ~ 25 mL. The pH as the equivalence point is the midpoint of the sharp drop at ~ 25 mL added acid, which is a pH = ~ 5.

b) At 0 mL the pH is calculated by doing an equilibrium calculation of a weak base in water (as done in Chapter 15).

c) The pH one-half way to the equivalence point is equal to the 14 – pK_b = pK_a of the base, or ~ 12 mL.

d) The pH at the equivalence point, or ~ 25 mL, is calculated by doing an equilibrium problem with the K_a of the base. At the equivalence point, all of the base has been converted to its conjugate acid.

e) Beyond the equivalence point, or ~ 25 mL, there is excess acid. All of the base has been converted to its conjugate acid and so the pH is calculated by focusing on this excess acid concentration.

41. **Given:** 35.0 mL of 0.175 M HBr titrated with 0.200 M KOH

a) **Find:** initial pH

Conceptual Plan: Since HBr is a strong acid, it will dissociate completely, so initial pH = – log $[H_3O^+]$ = – log [HBr].
Solution: pH = – log [HBr] = – log 0.175 = 0.757

Check: The units (none) are correct. The pH is reasonable since the concentration is greater than 0.1 M and the acid dissociates completely, the pH is less than 1.

b) **Find:** volume of base to reach equivalence point
Conceptual Plan: Write balanced equation then mL → L then [HBr], L → mol HBr then

$HBr + KOH \rightarrow KBr + H_2O$ $\quad \frac{1\ L}{1000\ mL} \quad M = \frac{mol}{L}$

set mol acid (HBr) = mol base (KOH) and [KOH], mol KOH → L KOH → mL KOH

balanced equation has 1:1 stoichiometry $\quad M = \frac{mol}{L} \quad \frac{1000\ mL}{1\ L}$

Solution: $35.0\ \text{mL HBr} \times \frac{1\ L}{1000\ mL} = 0.0350\ L\ HBr$ then

$\frac{0.175\ mol\ HBr}{1\ L} \times 0.0350\ L = 0.0061\underline{2}5\ mol\ HBr$.

So mol acid = mol HBr = $0.0061\underline{2}5$ mol = mol KOH then

$0.0061\underline{2}5\ \text{mol KOH} \times \frac{1\ L}{0.200\ \text{mol KOH}} = 0.030\underline{6}25\ \text{L KOH} \times \frac{1000\ mL}{1\ L} = 30.6\ mL\ KOH$.

Check: The units (mL) are correct. The volume of base is a little less than the volume of acid because the concentration of the base is a little greater than that of the acid.

c) **Find:** pH after adding 10.0 mL of base
Conceptual Plan:
Use calculations from part b. Then mL → L then [KOH], L → mol KOH then

$\frac{1\ L}{1000\ mL} \quad M = \frac{mol}{L}$

mol HBr, mol KOH → mol excess HBr and L HBr, L KOH → total L then

set up stoichiometry table $\quad$ L HBr + L KOH = total L

mol excess HBr, L → [HBr] → pH

$M = \frac{mol}{L} \quad pH = -\log [HBr]$

Solution: $10.0\ \text{mL KOH} \times \frac{1\ L}{1000\ mL} = 0.0100\ L\ KOH$ then

$\frac{0.200\ mol\ KOH}{1\ L} \times 0.0100\ L = 0.00200\ mol\ KOH$.

Since KOH is a strong base, [KOH] = $[OH^-]$, and set up table to track changes:

	$KOH\ (aq)$	$+\ HBr\ (aq)$	$\rightarrow\ KBr\ (aq)$	$+\ H_2O\ (l)$
Before addition	0.00 mol	$0.0061\underline{2}5$ mol	0.00 mol	–
Addition	0.00200 mol	–	–	–
After addition	≈ 0.00 mol	$0.0041\underline{2}5$ mol	0.00200 mol	–

Then 0.0350 L HBr + 0.0100 L KOH = 0.0450 L total volume. So mol excess acid = mol HBr = $0.0041\underline{2}5$ mol in 0.0450 L so $[HBr] = \frac{0.0041\underline{2}5\ mol\ HBr}{0.0450\ L} = 0.091\underline{6}667\ M$ and

$pH = -\log [HBr] = -\log 0.091\underline{6}667 = 1.038$.

Check: The units (none) are correct. The pH is a little higher than the initial pH, which is expected since this is a strong acid.

d) **Find:** pH at equivalence point
Solution: Since this is a strong acid–strong base titration, the pH at the equivalence point is neutral or 7.

e) **Find:** pH after adding 5.0 mL of base beyond the equivalence point

Conceptual Plan: Use calculations from parts b & c.   Then   the pH is only dependent on the amount

of excess base and the total solution volumes.

mL excess $\rightarrow$ L excess   then   [KOH],   L excess $\rightarrow$ mol KOH excess

$$\frac{1\ L}{1000\ mL} \qquad\qquad\qquad\qquad M = \frac{mol}{L}$$

then   L HBr, L KOH to equivalence point, L KOH excess $\rightarrow$ total L   then

$$L\ HBr + L\ KOH\ to\ equivalence\ point + L\ KOH\ excess = total\ L$$

mol excess KOH, total L $\rightarrow$ [KOH] = [OH$^-$] $\rightarrow$ [H$_3$O$^+$] $\rightarrow$ pH

$$M = \frac{mol}{L} \qquad\qquad K_w = [H_3O^+][OH^-] \qquad pH = -\log[H_3O^+]$$

Solution: $5.0\ \cancel{mL\ KOH} \times \dfrac{1\ L}{1000\ \cancel{mL}} = 0.0050\ L\ KOH\ excess$ then

$\dfrac{0.200\ mol\ KOH}{1\ \cancel{L}} \times 0.0050\ \cancel{L} = 0.0010\ mol\ KOH\ excess.$ Then $0.0350\ L\ HBr + 0.0306\ L\ KOH + 0.0050\ L$

$KOH = 0.0706\ L$ total volume. $[KOH\ excess] = \dfrac{0.0010\ mol\ KOH\ excess}{0.0706\ L} = 0.01\underline{4}164\ M\ KOH\ excess$

Since KOH is a strong base, [KOH] excess = [OH$^-$]. $K_w = [H_3O^+][OH^-]$ so

$[H_3O^+] = \dfrac{K_w}{[OH^-]} = \dfrac{1.0 \times 10^{-14}}{0.01\underline{4}164} = 7.\underline{0}6 \times 10^{-13}\ M$. Finally, $pH = -\log[H_3O^+] = -\log(7.\underline{0}6 \times 10^{-13}) = 12.15$.

Check: The units (none) are correct. The pH is rising sharply at the equivalence point, so the pH after 5 mL past the equivalence point should be quite basic.

42. **Given:** 20.0 mL of 0.125 M HNO_3 titrated with 0.150 M NaOH

Find: pH at 5 different points and plot titration curve

Conceptual Plan: Choose points to calculate: (i) initial pH; (ii) pH after 5.0 mL; (iii) pH after 10.0 mL; (iv) pH at equivalence point; and pH at 25.0 mL. Points should be on both sides of the equivalence point.

(i) Since HNO$_3$ is a strong acid, it will dissociate completely, so initial pH $= -\log[H_3O^+] = -\log[HNO_3]$.

Solution: $pH = -\log[HNO_3] = -\log 0.125 = 0.903$

Check: The units (none) are correct. The pH is reasonable since the concentration is greater than 0.1 M and the acid dissociates completely, the pH is less than 1.

(ii) **Find:** pH after adding 5.0 mL of base

Conceptual Plan:

Write balanced equation   then   mL $\rightarrow$ L then [HNO$_3$], L $\rightarrow$ mol HNO$_3$ then

$$HNO_3 + NaOH \rightarrow NaNO_3 + H_2O \qquad \frac{1\ L}{1000\ mL} \qquad\qquad M = \frac{mol}{L}$$

mL $\rightarrow$ L   then   [NaOH], L $\rightarrow$ mol NaOH   then   mol HNO$_3$, mol NaOH $\rightarrow$ mol excess HNO$_3$

$$\frac{1\ L}{1000\ mL} \qquad\qquad M = \frac{mol}{L} \qquad\qquad\qquad set\ up\ stoichiometry\ table$$

and   L HNO$_3$, L NaOH $\rightarrow$ total L   then   mol excess HNO$_3$, L $\rightarrow$ [HNO$_3$] $\rightarrow$ pH

$$L\ HNO_3 + L\ NaOH = total\ L \qquad\qquad\qquad M = \frac{mol}{L} \qquad pH = -\log[HNO_3]$$

Solution: $20.0\ \cancel{mL\ HNO_3} \times \dfrac{1\ L}{1000\ \cancel{mL}} = 0.0200\ L\ HNO_3$ then

$\dfrac{0.125\ mol\ HNO_3}{1\ \cancel{L}} \times 0.0200\ \cancel{L} = 0.00250\ mol\ HNO_3$ and $5.0\ \cancel{mL\ NaOH} \times \dfrac{1\ L}{1000\ \cancel{mL}} = 0.0050\ L\ NaOH$

then $\dfrac{0.150\ mol\ NaOH}{1\ \cancel{L}} \times 0.0050\ \cancel{L} = 0.00075\ mol\ NaOH$.

This is a strong acid – strong base titration, so set up table to track changes:

	$NaOH\ (aq)$	$+\ HNO_3\ (aq)$	$\rightarrow\ NaNO_3\ (aq)$	$+\ H_2O\ (l)$
Before addition	0.00 mol	0.00250 mol	0.00 mol	–
Addition	0.00075 mol	–	–	–
After addition	≈ 0.00 mol	0.00175 mol	0.00075 mol	–

Then 0.0200 L HNO_3 + 0.0050 L NaOH = 0.0250 L total volume. So mol excess acid = mol HNO_3 = 0.00175 mol in 0.0250 L, so $[HNO_3] = \frac{0.00175 \text{ mol } HNO_3}{0.0250 \text{ L}} = 0.0700 \text{ M}$ and

$pH = -\log [HNO_3] = -\log 0.0700 = 1.155$.

Check: The units (none) are correct. The pH remains very low in a strong acid–strong base titration before the equivalence point.

(iii) Find: pH after adding 10.0 mL of base

Conceptual Plan:

Use calculations for point (ii) then mL → L then [NaOH], L → mol NaOH then

$$\frac{1 \text{ L}}{1000 \text{ mL}} \qquad M = \frac{\text{mol}}{\text{L}}$$

mol HNO_3, mol NaOH → mol excess HNO_3 and L HNO_3, L NaOH → total L then

set up stoichiometry table; L HNO_3 + L NaOH = total L

mol excess HNO_3, L → [HNO_3] → pH

$$M = \frac{\text{mol}}{\text{L}} \qquad pH = -\log [HNO_3]$$

Solution: $10.0 \text{ mL NaOH} \times \frac{1 \text{ L}}{1000 \text{ mL}} = 0.0100 \text{ L NaOH}$ then

$\frac{0.150 \text{ mol NaOH}}{1 \text{ L}} \times 0.0100 \text{ L} = 0.00150 \text{ mol NaOH}$. Set up table to track changes:

	$NaOH\ (aq)$	$+\ HNO_3\ (aq)$	$\rightarrow\ NaNO_3\ (aq)$	$+\ H_2O\ (l)$
Before addition	0.00 mol	0.00250 mol	0.00 mol	–
Addition	0.00150 mol	–	–	–
After addition	≈ 0.00 mol	0.00100 mol	0.00150 mol	–

Then 0.0200 L HNO_3 + 0.0100 L NaOH = 0.0300 L total volume. So mol excess acid = mol HNO_3 = 0.00100 mol in 0.0300 L, so $[HNO_3] = \frac{0.00100 \text{ mol } HNO_3}{0.0300 \text{ L}} = 0.0333333 \text{ M}$ and

$pH = -\log [HNO_3] = -\log 0.0333333 = 1.477$.

Check: The units (none) are correct. The pH remains very low in a strong acid–strong base titration before the equivalence point.

(iv) Find: pH at equivalence point and volume of base to reach equivalence point

Conceptual Plan: Since this is a strong acid–strong base titration, the pH at the equivalence point is neutral or 7. Use calculations for point (ii) then

set mol acid (HNO_3) = mol base (NaOH) and

balanced equation has 1:1 stoichiometry

[NaOH], mol NaOH → L NaOH → mL NaOH

$$M = \frac{\text{mol}}{\text{L}} \qquad \frac{1000 \text{ mL}}{1 \text{ L}}$$

Solution: Since this is a strong acid – strong base titration, the pH at the equivalence point is neutral or 7. So mol acid = mol HNO_3 = 0.00250 mol = mol NaOH then

$$0.00250 \text{ mol NaOH} \times \frac{1 \text{ L}}{0.150 \text{ mol NaOH}} = 0.0166667 \text{ L NaOH} \times \frac{1000 \text{ mL}}{1 \text{ L}} = 16.7 \text{ mL NaOH}.$$

Check: The units (none and mL) are correct. The equivalence point pH of a strong acid–strong base titration is neutral. The volume of base is a little less than the volume of acid because the concentration of the base is a little greater than that of the acid.

(v) Find: pH after adding 25.0 mL of base

Conceptual Plan:
Use calculations for point (ii) then mL → L then [NaOH], L → mol NaOH then

$\frac{1\text{ L}}{1000\text{ mL}}$ $M = \frac{\text{mol}}{\text{L}}$

mol HNO_3, mol NaOH → mol excess HNO_3 and L HNO_3, L NaOH → total L then

set up stoichiometry table L HNO_3 + L NaOH = total L

mol excess NaOH, total L → [NaOH] = [OH^-] → [H_3O^+] → pH

$M = \frac{\text{mol}}{\text{L}}$ $K_w = [H_3O^+][OH^-]$ $pH = -\log[H_3O^+]$

Solution: $25.0\ \text{mL NaOH} \times \frac{1\text{ L}}{1000\text{ mL}} = 0.0250\text{ L NaOH}$ then

$\frac{0.150\text{ mol NaOH}}{1\text{ L}} \times 0.0250\text{ L} = 0.00375\text{ mol NaOH}$. Set up table to track changes:

	NaOH (*aq*)	+ HNO_3 (*aq*)	→ $NaNO_3$ (*aq*)	+ H_2O (*l*)
Before addition	0.00 mol	0.00250 mol	0.00 mol	–
Addition	0.00375 mol	–	–	–
After addition	0.00125 mol	≈ 0.00 mol	0.00375 mol	–

Then 0.0200 L HNO_3 + 0.0250 L NaOH = 0.0450 L total volume. So mol excess acid = mol NaOH = 0.00125 mol in 0.0450 L, so. $[\text{NaOH excess}] = \frac{0.00125\text{ mol NaOH excess}}{0.0450\text{ L}} = 0.02\underline{7}7778\text{ M NaOH excess}$

Since NaOH is a strong base, [NaOH] excess = [OH^-]. $K_w = [H_3O^+][OH^-]$ so

$[H_3O^+] = \frac{K_w}{[OH^-]} = \frac{1.0 \times 10^{-14}}{0.02\underline{7}7778} = 3.6 \times 10^{-13}\text{ M}$. Finally, $pH = -\log[H_3O^+] = -\log(3.6 \times 10^{-13}) = 12.44$.

Check: The units (none) are correct. The pH is rising sharply at the equivalence point, so the pH over 5 mL past the equivalence point should be quite basic.

Finally plotting these 5 points, the titration curve looks like:

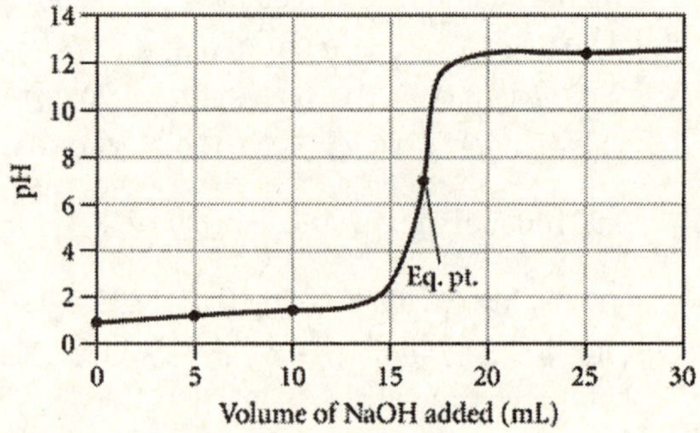

43. **Given:** 25.0 mL of 0.115 M RbOH titrated with 0.100 M HCl

a) **Find:** initial pH

Conceptual Plan: Since RbOH is a strong base, it will dissociate completely, so [RbOH] = [OH^-] → [H_3O^+] → pH

$K_w = [H_3O^+][OH^-]$ $pH = -\log[H_3O^+]$

Solution: Since RbOH is a strong base, [RbOH] excess = [OH^-]. $K_w = [H_3O^+][OH^-]$ so

$[H_3O^+] = \frac{K_w}{[OH^-]} = \frac{1.0 \times 10^{-14}}{0.115} = 8.\underline{6}9565 \times 10^{-14}\text{ M}$ and $pH = -\log[H_3O^+] = -\log(8.\underline{6}9565 \times 10^{-14}) = 13.06$

Check: The units (none) are correct. The pH is reasonable since the concentration is greater than 0.1 M and the base dissociates completely, the pH is greater than 13.

b) **Find:** volume of acid to reach equivalence point

Conceptual Plan: Write balanced equation then mL → L then [RbOH], L → mol RbOH then

HCl + RbOH → RbCl + H_2O $\frac{1\text{ L}}{1000\text{ mL}}$ $M = \frac{\text{mol}}{\text{L}}$

set mol base (RbOH) = mol acid (HCl) and [HCl], mol HCl → L HCl → mL HCl

balanced equation has 1:1 stoichiometry $\quad M = \frac{\text{mol}}{\text{L}} \quad \frac{1000 \text{ mL}}{1 \text{ L}}$

Solution: $25.0 \cancel{\text{mL RbOH}} \times \frac{1 \text{ L}}{1000 \cancel{\text{mL}}} = 0.0250 \text{ L RbOH}$ then

$\frac{0.115 \text{ mol RbOH}}{1 \cancel{\text{L}}} \times 0.0250 \cancel{\text{L}} = 0.0028\underline{7}5 \text{ mol RbOH}$. So mol base = mol RbOH = 0.0028$\underline{7}$5 mol = mol HCl then $0.0028\underline{7}5 \cancel{\text{mol HCl}} \times \frac{1 \text{ L}}{0.100 \cancel{\text{mol HCl}}} = 0.0287\underline{5} \cancel{\text{L HCl}} \times \frac{1000 \text{ mL}}{1 \cancel{\text{L}}} = 28.8 \text{ mL HCl}$.

Check: The units (mL) are correct. The volume of acid is greater than the volume of base because the concentration of the base is a little greater than that of the acid.

c) **Find:** pH after adding 5.0 mL of acid
Conceptual Plan: Use calculations from part b. Then mL → L then [HCl], L → mol HCl then

$\frac{1 \text{ L}}{1000 \text{ mL}} \quad M = \frac{\text{mol}}{\text{L}}$

mol RbOH, mol HCl → mol excess RbOH and L RbOH, L HCl → total L then

set up stoichiometry table $\quad$ L RbOH + L HCl = total L

mol excess RbOH, L → [RbOH] = [OH⁻] → [H₃O⁺] → pH

$M = \frac{\text{mol}}{\text{L}} \quad K_w = [H_3O^+][OH^-] \quad pH = -\log [H_3O^+]$

Solution: $5.0 \cancel{\text{mL HCl}} \times \frac{1 \text{ L}}{1000 \cancel{\text{mL}}} = 0.0050 \text{ L HCl}$ then $\frac{0.100 \text{ mol HCl}}{1 \cancel{\text{L}}} \times 0.0050 \cancel{\text{L}} = 0.00050 \text{ mol HCl}$.

Since HCl is a strong acid, [HCl] = [H_3O^+], and set up table to track changes:

	HCl (*aq*)	+ RbOH (*aq*)	→ RbCl (*aq*)	+ H_2O (*l*)
Before addition	0.00 mol	0.0028$\underline{7}$5 mol	0.00 mol	–
Addition	0.00050 mol	–	–	–
After addition	≈ 0.00 mol	0.0023$\underline{7}$5 mol	0.00050 mol	–

Then 0.0250 L RbOH + 0.0050 L HCl = 0.0300 L total volume. So mol excess base = mol RbOH = 0.0023$\underline{7}$5 mol in 0.0300 L so $[\text{RbOH}] = \frac{0.0023\underline{7}5 \text{ mol RbOH}}{0.0300 \text{ L}} = 0.079\underline{1}667 \text{ M}$. Since RbOH is a strong base, [RbOH] excess = [OH^-]. $K_w = [H_3O^+][OH^-]$ so $[H_3O^+] = \frac{K_w}{[OH^-]} = \frac{1.0 \times 10^{-14}}{0.079\underline{1}667} = 1.\underline{2}6316 \times 10^{-13} \text{ M}$ and $pH = -\log [H_3O^+] = -\log (1.\underline{2}6316 \times 10^{-13}) = 12.90$.

Check: The units (none) are correct. The pH is a little lower than the initial pH, which is expected since this is a strong base.

d) **Find:** pH at equivalence point
Solution: Since this is a strong acid–strong base titration, the pH at the equivalence point is neutral or 7.

e) **Find:** pH after adding 5.0 mL of acid beyond the equivalence point
Conceptual Plan: Use calculations from parts b & c. Then the pH is only dependent on the amount of excess acid and the total solution volumes. Then
mL excess → L excess then [HCl], L excess → mol HCl excess

$\frac{1 \text{ L}}{1000 \text{ mL}} \quad M = \frac{\text{mol}}{\text{L}}$

then L RbOH, L HCl to equivalence point, L HCl excess → total L then

L RbOH + L HCl to equivalence point + L HCl excess = total L

mol excess HCl, total L → [HCl] = [H₃O⁺] → pH

$M = \frac{\text{mol}}{\text{L}} \quad pH = -\log [H_3O^+]$

Solution: $5.0 \cancel{\text{mL HCl}} \times \frac{1 \text{ L}}{1000 \cancel{\text{mL}}} = 0.0050 \text{ L HCl excess}$ then

$\frac{0.100 \text{ mol HCl}}{1 \text{ L}} \times 0.0050 \text{ L} = 0.00050 \text{ mol HCl excess}$. Then 0.0250 L RbOH + 0.0288 L HCl + 0.0050 L HCl = 0.0588 L total volume. $[\text{HCl excess}] = \frac{0.00050 \text{ mol HCl excess}}{0.0588 \text{ L}} = 0.0085034 \text{ M HCl excess}$

Since HCl is a strong acid, [HCl] excess = $[H_3O^+]$. Finally, $pH = -\log [H_3O^+] = -\log (0.0085034) = 2.07$.

Check: The units (none) are correct. The pH is dropping sharply at the equivalence point, so the pH after 5 mL past the equivalence point should be quite acidic.

44. **Given:** 15.0 mL of 0.100 M $Ba(OH)_2$ titrated with 0.125 M HCl
Find: pH at 5 different points and plot titration curve
Conceptual Plan: Choose points to calculate: (i) initial pH; (ii) pH after 10.0 mL; (iii) pH after 20.0 mL; (iv) pH at equivalence point; and pH at 30.0 mL. Points should be on both sides of the equivalence point.

(i) **Find:** initial pH

Conceptual Plan: Since $Ba(OH)_2$ is a strong base, it will dissociate completely, bearing in mind that $Ba(OH)_2 \rightarrow Ba^{2+} + 2\ OH^-$ so 2 hydroxide ions are generated for each barium hydroxide and $2\ [Ba(OH)_2] = [OH^-]$ → $[H_3O^+]$ → pH

$K_w = [H_3O^+][OH^-]$ $\quad pH = -\log [H_3O^+]$

Solution: Since $Ba(OH)_2$ is a strong base, $2\ [Ba(OH)_2] = [OH^-] = 2 \times 0.100 \text{ M} = 0.200 \text{ M}$.

$K_w = [H_3O^+][OH^-]$ so $[H_3O^+] = \frac{K_w}{[OH^-]} = \frac{1.0 \times 10^{-14}}{0.200} = 5.0 \times 10^{-14} \text{ M}$ and

$pH = -\log [H_3O^+] = -\log (5.0 \times 10^{-14}) = 13.30$

Check: The units (none) are correct. The pH is reasonable since the concentration is greater than 0.1 M and the base dissociates completely, the pH is greater than 13.

(ii) **Find:** pH after adding 10.0 mL of acid

Conceptual Plan:

Write balanced equation then mL → L then $[Ba(OH)_2]$, L → mol $Ba(OH)_2$ then

$2\ HCl + Ba(OH)_2 \rightarrow BaCl_2 + 2\ H_2O$ $\quad \frac{1 \text{ L}}{1000 \text{ mL}}$ $\quad M = \frac{\text{mol}}{\text{L}}$

mL → L then [HCl], L → mol HCl then mol $Ba(OH)_2$, mol HCl → mol excess $Ba(OH)_2$ and

$\frac{1 \text{ L}}{1000 \text{ mL}}$ $\quad M = \frac{\text{mol}}{\text{L}}$ $\quad$ set up stoichiometry table

L $Ba(OH)_2$, L HCl → total L then

$M = \frac{\text{mol}}{\text{L}}$

mol excess $Ba(OH)_2$, L → $2\ [Ba(OH)_2] = [OH^-]$ → $[H_3O^+]$ → pH

L $Ba(OH)_2$ + L HCl = total L $\quad K_w = [H_3O^+][OH^-]$ $\quad pH = -\log [H_3O^+]$

Solution: $15.0 \text{ mL } Ba(OH)_2 \times \frac{1 \text{ L}}{1000 \text{ mL}} = 0.0150 \text{ L } Ba(OH)_2$ then

$\frac{0.100 \text{ mol } Ba(OH)_2}{1 \text{ L}} \times 0.0150 \text{ L} = 0.00150 \text{ mol } Ba(OH)_2$ and $10.0 \text{ mL HCl} \times \frac{1 \text{ L}}{1000 \text{ mL}} = 0.0100 \text{ L HCl}$

then $\frac{0.125 \text{ mol HCl}}{1 \text{ L}} \times 0.0100 \text{ L} = 0.00125 \text{ mol HCl}$. Since HCl is a strong acid, $[HCl] = [H_3O^+]$,

and set up table to track changes:

	2 HCl (*aq*)	+ $Ba(OH)_2$ (*aq*)	→ $BaCl_2$ (*aq*)	+ 2 H_2O (*l*)
Before addition	0.00 mol	0.00150 mol	0.00 mol	–
Addition	0.00125 mol	–	–	–
After addition	≈ 0.00 mol	0.000875 mol	0.000625 mol	–

Then 0.0150 L $Ba(OH)_2$ + 0.0100 L HCl = 0.0250 L total volume. So mol excess base = mol $Ba(OH)_2$ = 0.000875 mol in 0.0250 L so $[Ba(OH)_2] = \frac{0.000875 \text{ mol } Ba(OH)_2}{0.0250 \text{ L}} = 0.035 \text{ M}$ Since $Ba(OH)_2$ is a strong base, $2\ [Ba(OH)_2] = [OH^-] = 2 \times 0.035 \text{ M} = 0.070 \text{ M}$. $K_w = [H_3O^+][OH^-]$ so

$$[H_3O^+] = \frac{K_w}{[OH^-]} = \frac{1.0 \times 10^{-14}}{0.070} = 1.\underline{4}286 \times 10^{-13} \text{ M and}$$

$pH = -\log[H_3O^+] = -\log(1.\underline{4}286 \times 10^{-13}) = 12.85$.

Check: The units (none) are correct. The pH is a little lower than the initial pH, which is expected since this is a strong base.

(iii) Find: pH after adding 20.0 mL of acid

Conceptual Plan:

Use calculations from part (ii) then mL → L then [HCl], L → mol HCl then

$\frac{1\text{ L}}{1000\text{ mL}}$ $\quad M = \frac{\text{mol}}{\text{L}}$

mol $Ba(OH)_2$, mol HCl → mol excess $Ba(OH)_2$ and L $Ba(OH)_2$, L HCl → total L then

set up stoichiometry table $\quad$ L $Ba(OH)_2$ + L HCl = total L

mol excess $Ba(OH)_2$, L → 2 [$Ba(OH)_2$] = [OH^-] → [H_3O^+] → pH

$M = \frac{\text{mol}}{\text{L}}$ $\quad K_w = [H_3O^+][OH^-]$ $\quad pH = -\log[H_3O^+]$

Solution: $20.0 \text{ mL HCl} \times \frac{1\text{ L}}{1000\text{ mL}} = 0.0200 \text{ L HCl}$ then $\frac{0.125 \text{ mol HCl}}{1\text{ L}} \times 0.0200\text{ L} = 0.00250 \text{ mol HCl}$.

Since HCl is a strong acid, [HCl] = [H_3O^+], and set up table to track changes:

	$2\,HCl\,(aq)$ +	$Ba(OH)_2\,(aq)$ →	$BaCl_2\,(aq)$ +	$2\,H_2O\,(l)$
Before addition	0.00 mol	0.00150 mol	0.00 mol	–
Addition	0.00250 mol	–	–	–
After addition	≈ 0.00 mol	0.00025 mol	0.00125 mol	–

then

0.0150 L $Ba(OH)_2$ + 0.0200 L HCl = 0.0350 L total volume. So mol excess base = mol $Ba(OH)_2$ = 0.00025 mol in 0.0350 L so $[Ba(OH)_2] = \frac{0.00025 \text{ mol } Ba(OH)_2}{0.0350\text{ L}} = 0.007\underline{1}429\text{ M}$ Since $Ba(OH)_2$ is a strong base, 2 [$Ba(OH)_2$] = [OH^-] = 2 x 0.007$\underline{1}$429 M = 0.014286 M. $K_w = [H_3O^+][OH^-]$ so

$$[H_3O^+] = \frac{K_w}{[OH^-]} = \frac{1.0 \times 10^{-14}}{0.01\underline{4}286} = 6.\underline{9}9986 \times 10^{-13}\text{ M and } pH = -\log[H_3O^+] = -\log(6.\underline{9}9986 \times 10^{-13}) = 12.15.$$

Check: The units (none) are correct. The pH is a little lower than the initial pH, which is expected since this is a strong base.

(iv) Find: pH at equivalence point and volume of acid to reach equivalence point

Solution: Since this is a strong acid–strong base titration, the pH at the equivalence point is neutral or 7.

Conceptual Plan:

Use calculations from part (ii) then set 2 mol base ($Ba(OH)_2$) = mol acid (HCl) and

balanced equation has 1:2 stoichiometry

[HCl], mol HCl → L HCl → mL HCl

$M = \frac{\text{mol}}{\text{L}}$ $\quad \frac{1000\text{ mL}}{1\text{ L}}$

Solution: So $0.00150 \text{ mol } Ba(OH)_2 \times \frac{2 \text{ mol HCl}}{1 \text{ mol } Ba(OH)_2} = 0.00300 \text{ mol HCl}$ then

$$0.00300 \text{ mol HCl} \times \frac{1\text{ L}}{0.125 \text{ mol HCl}} = 0.0240 \text{ L HCl} \times \frac{1000\text{ mL}}{1\text{ L}} = 24.0 \text{ mL HCl}.$$

Check: The units (mL) are correct. The volume of acid is a greater less than the volume of base because two moles of acid are needed for each mole of the base.

(v) Find: pH after adding 30.0 mL of acid

Conceptual Plan: Use calculations from earlier parts. then the pH is only dependent on the amount of excess acid and the total solution volumes.

mL added, mL at equiv. pt. → mL excess → L excess then

mL excess = mL added – mL at equiv. pt. $\quad \frac{1\text{ L}}{1000\text{ mL}}$

[HCl], L excess → mol HCl excess then L $Ba(OH)_2$, L HCl → total L then

$M = \frac{mol}{L}$ $\qquad$ L $Ba(OH)_2$ + L HCl = total L

mol excess HCl, total L → [HCl] = $[H_3O^+]$ → pH

$M = \frac{mol}{L}$ $\qquad$ $pH = -\log[H_3O^+]$

Solution: mL HCl excess = mL added – mL to equiv. pt. = 30.0 mL – 24.0 mL = 6.0 mL.

$6.0 \cancel{\text{mL HCl}} \times \frac{1\text{ L}}{1000\ \cancel{\text{mL}}} = 0.0060\text{ L HCl excess}$ then

$\frac{0.125\text{ mol HCl}}{1\ \cancel{\text{L}}} \times 0.0060\ \cancel{\text{L}} = 0.00075\text{ mol HCl excess}$. Then 0.0150 L $Ba(OH)_2$ + 0.0300 L HCl =

0.0450 L total volume. $[\text{HCl excess}] = \frac{0.00075\text{ mol HCl excess}}{0.0450\text{ L}} = 0.01\underline{6}667\text{ M HCl excess}$. Since HCl is a strong acid, [HCl] excess $= [H_3O^+]$. Finally, $pH = -\log[H_3O^+] = -\log(0.01\underline{6}667) = 1.78$.

Check: The units (none) are correct. The pH is dropping sharply at the equivalence point, so the pH after 5 mL past the equivalence point should be quite acidic.

Finally plotting these 5 points, the titration curve looks like:

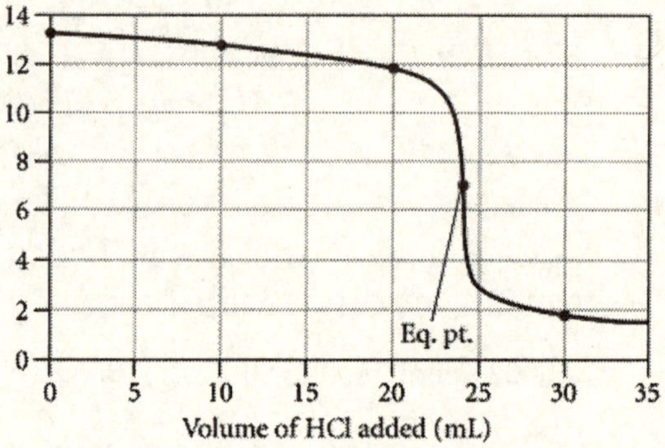

45. **Given:** 20.0 mL of 0.105 M $HC_2H_3O_2$ titrated with 0.125 M NaOH **Other:** K_a ($HC_2H_3O_2$) = 1.8×10^{-5}

a) **Find:** initial pH

Conceptual Plan:

Since $HC_2H_3O_2$ is a weak acid, set up an equilibrium problem using the initial concentration.

So M $HC_2H_3O_2$ → $[H_3O^+]$ → pH

ICE Chart $\qquad$ $pH = -\log[H_3O^+]$

Solution:

$$HC_2H_3O_2(aq) + H_2O(l) \rightleftharpoons H_3O^+(aq) + C_2H_3O_2^-(aq)$$

	$[HC_2H_3O_2]$	$[H_3O^+]$	$[C_2H_3O_2^-]$
Initial	0.105	≈ 0.00	0.00
Change	$-x$	$+x$	$+x$
Equil	$0.105 - x$	$+x$	$+x$

$K_a = \frac{[H_3O^+][C_2H_3O_2^-]}{[HC_2H_3O_2]} = 1.8 \times 10^{-5} = \frac{x^2}{0.105 - x}$ Assume x is small ($x \ll 0.105$) so

$\frac{x^2}{0.105 - \cancel{x}} = 1.8 \times 10^{-5} = \frac{x^2}{0.105}$ and $x = 1.\underline{3}748 \times 10^{-3}$ M= $[H_3O^+]$. Confirm that assumption is valid

$\frac{1.\underline{3}748 \times 10^{-3}}{0.105} \times 100\,\% = 1.3\,\% < 5\,\%$ so assumption is valid. Finally,

$pH = -\log[H_3O^+] = -\log(1.\underline{3}748 \times 10^{-3}) = 2.86$

Check: The units (none) are correct. The magnitude of the answer makes physical sense because pH should be greater than $-\log(0.105) = 0.98$ because this is a weak acid.

b) **Find:** volume of base to reach equivalence point

Conceptual Plan:

Write balanced equation then mL → L then [$HC_2H_3O_2$], L → mol $HC_2H_3O_2$ then

$HC_2H_3O_2 + NaOH \rightarrow NaC_2H_3O_2 + H_2O$ $\quad \frac{1\ L}{1000\ mL}$ $\quad M = \frac{mol}{L}$

set mol acid($HC_2H_3O_2$) = mol base(NaOH) and [NaOH], mol NaOH → L NaOH → mL NaOH

balanced equation has 1:1 stoichiometry $\quad M = \frac{mol}{L}$ $\quad \frac{1000\ mL}{1\ L}$

Solution: $20.0\ \text{mL}\ HC_2H_3O_2 \times \frac{1\ L}{1000\ mL} = 0.0200\ L\ HC_2H_3O_2$ then

$\frac{0.105\ mol\ HC_2H_3O_2}{1\ L} \times 0.0200\ L = 0.00210\ mol\ HC_2H_3O_2$. So mol acid = mol $HC_2H_3O_2$ = 0.00210 mol =

mol NaOH then $0.00210\ mol\ NaOH \times \frac{1\ L}{0.125\ mol\ NaOH} = 0.0168\ L\ NaOH \times \frac{1000\ mL}{1\ L} = 16.8\ mL\ NaOH$.

Check: The units (mL) are correct. The volume of base is a little less than the volume of acid because the concentration of the base is a little greater than that of the acid.

c) **Find:** pH after adding 5.0 mL of base

Conceptual Plan:

Use calculations from part b. Then mL → L then [NaOH], L → mol NaOH then

$\frac{1\ L}{1000\ mL}$ $\quad M = \frac{mol}{L}$

mol $HC_2H_3O_2$, mol NaOH → mol excess $HC_2H_3O_2$, mol $C_2H_3O_2^-$ and

set up stoichiometry table

L $HC_2H_3O_2$, L NaOH → total L then

L $HC_2H_3O_2$ + L NaOH = total L

then mol excess $HC_2H_3O_2$, L → [$HC_2H_3O_2$] and mol excess $C_2H_3O_2^-$, L → [$C_2H_3O_2^-$] then

$M = \frac{mol}{L}$ $\quad M = \frac{mol}{L}$

M $HC_2H_3O_2$, M $C_2H_3O_2^-$ → [H_3O^+] → pH

ICE Chart $\quad pH = -\log[H_3O^+]$

Solution: $5.0\ \text{mL NaOH} \times \frac{1\ L}{1000\ mL} = 0.0050\ L\ NaOH$ then

$\frac{0.125\ mol\ NaOH}{1\ L} \times 0.0050\ L = 0.00062\underline{5}\ mol\ NaOH$. Set up table to track changes:

	$NaOH\ (aq)$	$+\ HC_2H_3O_2\ (aq)$	$\rightarrow\ NaC_2H_3O_2\ (aq)$	$+\ H_2O\ (l)$
Before addition	0.00 mol	0.00210 mol	0.00 mol	–
Addition	0.00062̲5 mol	–	–	–
After addition	≈ 0.00 mol	0.0014̲75 mol	0.00062̲5 mol	–

Then 0.0200 L H $C_2H_3O_2$ + 0.0050 L NaOH = 0.0250 L total volume. Then

$[HC_2H_3O_2] = \frac{0.0014\underline{7}5\ mol\ HC_2H_3O_2}{0.0250\ L} = 0.0590\ M$ and

$[NaC_2H_3O_2] = \frac{0.00062\underline{5}\ mol\ C_2H_3O_2^-}{0.0250\ L} = 0.025\ M$. Since 1 $C_2H_3O_2^-$ ion is generated for each $NaC_2H_3O_2$,

$[C_2H_3O_2^-] = 0.025\ M\ C_2H_3O_2^-$.

$$HC_2H_3O_2\,(aq) + H_2O\,(l) \rightleftharpoons H_3O^+\,(aq) + C_2H_3O_2^-\,(aq)$$

	$[HC_2H_3O_2]$	$[H_3O^+]$	$[C_2H_3O_2^-]$
Initial	0.0590	≈ 0.00	0.025
Change	$-x$	$+x$	$+x$
Equil	$0.0590 - x$	$+x$	$0.025 + x$

$K_a = \frac{[H_3O^+][C_2H_3O_2^-]}{[HC_2H_3O_2]} = 1.8 \times 10^{-5} = \frac{x(0.025 + x)}{0.0590 - x}$ Assume x is small ($x << 0.025 < 0.0590$) so

$$\frac{x(0.025 + \cancel{x})}{0.0590 - \cancel{x}} = 1.8 \times 10^{-5} = \frac{x(0.025)}{0.0590}$$ and $x = 4.\underline{2}48 \times 10^{-5}$ M $= [H_3O^+]$. Confirm that assumption is valid

$$\frac{4.\underline{2}48 \times 10^{-5}}{0.025} \times 100\,\% = 0.17\,\% < 5\,\%$$ so assumption is valid. Finally,

$pH = -\log [H_3O^+] = -\log (4.\underline{2}48 \times 10^{-5}) = 4.37$.

Check: The units (none) are correct. The pH is a little higher than the initial pH, which is expected since some of the acid has been neutralized.

d) **Find:** pH at one-half of the equivalence point
Conceptual Plan: Since this is a weak acid–strong base titration, the pH at one-half the equivalence point is the pK_a of the weak acid.
Solution: $pH = pK_a = -\log K_a = -\log (1.8 \times 10^{-5}) = 4.74$.

Check: The units (none) are correct. Since this is a weak acid–strong base titration, the pH at one-half the equivalence point is the pK_a of the weak acid, so it should be a little below 5.

e) **Find:** pH at equivalence point
Conceptual Plan: Use calculations from parts b. Then since all of the weak acid has been converted to its conjugate base, the pH is only dependent on the hydrolysis reaction of the conjugate base. The mol $C_2H_3O_2^-$ = initial mol $HC_2H_3O_2$ and L $HC_2H_3O_2$, L NaOH to equivalence point → total L then

L $HC_2H_3O_2$ + L NaOH = total L

mol excess $C_2H_3O_2^-$, L → $[C_2H_3O_2^-]$ and K_a → K_b then do equilibrium calculation:

$M = \frac{mol}{L}$ $\qquad$ $K_w = K_a K_b$

$[C_2H_3O_2^-]$, K_b → $[OH^-]$ → $[H_3O^+]$ → pH

set up ICE table $\quad K_w = [H_3O^+][OH^-] \quad pH = -\log [H_3O^+]$

Solution: mol $C_2H_3O_2^-$ = initial mol $HC_2H_3O_2$ = 0.00210 mol and total volume = L $HC_2H_3O_2$ + L NaOH =

0.020 L + 0.0168 L = 0.0368 L then $[C_2H_3O_2^-] = \frac{0.00210 \text{ mol } C_2H_3O_2^-}{0.0368 \text{ L}} = 0.057\underline{0}652$ M and

$K_w = K_a K_b$. Rearrange to solve for K_b. $K_b = \frac{K_w}{K_a} = \frac{1.0 \times 10^{-14}}{1.8 \times 10^{-5}} = 5.\underline{5}556 \times 10^{-10}$. Set up ICE table

$$C_2H_3O_2^-(aq) + H_2O(l) \rightleftharpoons HC_2H_3O_2(aq) + OH^-(aq)$$

	$[C_2H_3O_2^-]$	$[HC_2H_3O_2]$	$[OH^-]$
Initial	0.057\underline{0}652	≈ 0.00	≈ 0.00
Change	$-x$	$+x$	$+x$
Equil	$0.057\underline{0}652 - x$	$+x$	$+x$

$$K_b = \frac{[HC_2H_3O_2][OH^-]}{[C_2H_3O_2^-]} = 5.\underline{5}556 \times 10^{-10} = \frac{x^2}{0.057\underline{0}652 - x}$$ Assume x is small ($x << 0.057$) so

$$\frac{x^2}{0.057\underline{0}652 - \cancel{x}} = 5.\underline{5}556 \times 10^{-10} = \frac{x^2}{0.057\underline{0}652}$$ and $x = 5.\underline{6}305 \times 10^{-6}$ M $= [OH^-]$.

Confirm that assumption is valid $\frac{5.\underline{6}305 \times 10^{-6}}{0.057\underline{0}652} \times 100\,\% = 0.0099\,\% < 5\,\%$ so assumption is valid.

$K_w = [H_3O^+][OH^-]$ so $[H_3O^+] = \frac{K_w}{[OH^-]} = \frac{1.0 \times 10^{-14}}{5.\underline{6}305 \times 10^{-6}} = 1.\underline{7}760 \times 10^{-9}$ M. Finally,

$pH = -\log [H_3O^+] = -\log (1.\underline{7}760 \times 10^{-9}) = 8.75$.

Check: The units (none) are correct. Since this is a weak acid–strong base titration, the pH at the equivalence point is basic.

f) **Find:** pH after adding 5.0 mL of base beyond the equivalence point
Conceptual Plan: Use calculations from parts b & c. Then the pH is only dependent on the amount of excess base and the total solution volumes.

mL excess → L excess then [NaOH], L excess → mol NaOH excess

$\frac{1\text{ L}}{1000\text{ mL}}$ $M = \frac{\text{mol}}{\text{L}}$

then L $HC_2H_3O_2$, L NaOH to equivalence point, L NaOH excess → total L then

L $HC_2H_3O_2$ + L NaOH to equivalence point + L NaOH excess = total L

mol excess NaOH, total L → [NaOH] = [OH^-] → [H_3O^+] → pH

$M = \frac{\text{mol}}{\text{L}}$ $K_w = [H_3O^+][OH^-]$ $pH = -\log[H_3O^+]$

Solution: $5.0\text{ mL NaOH} \times \frac{1\text{ L}}{1000\text{ mL}} = 0.0050\text{ L NaOH excess}$ then

$\frac{0.125\text{ mol NaOH}}{1\text{ L}} \times 0.0050\text{ L} = 0.000625$ mol NaOH excess. Then 0.0200 L $HC_2H_3O_2$ + 0.0168 L NaOH + 0.0050 L NaOH = 0.0418 L total volume.

$[\text{NaOH excess}] = \frac{0.000625\text{ mol NaOH excess}}{0.0418\text{ L}} = 0.01495\underline{2}2$ M NaOH excess Since NaOH is a strong base, [NaOH] excess = [OH^-]. The strong base overwhelms the weak base and is insignificant in the calculation. $K_w = [H_3O^+][OH^-]$ so $[H_3O^+] = \frac{K_w}{[OH^-]} = \frac{1.0 \times 10^{-14}}{0.01495\underline{2}2} = 6.\underline{6}88 \times 10^{-13}$ M . Finally,

$pH = -\log[H_3O^+] = -\log(6.\underline{6}88 \times 10^{-13}) = 12.17$.

Check: The units (none) are correct. The pH is rising sharply at the equivalence point, so the pH after 5 mL past the equivalence point should be quite basic.

46. **Given:** 30.0 mL of 0.165 M $HC_3H_5O_2$ titrated with 0.300 M KOH **Other:** K_a ($HC_3H_5O_2$) = 1.3×10^{-5}
Find: initial pH
Conceptual Plan: Since $HC_3H_5O_2$ is a weak acid, set up an equilibrium problem using the initial concentration. So M $HC_3H_5O_2$ → [H_3O^+] → pH

ICE Chart $pH = -\log[H_3O^+]$

Solution:

$$HC_3H_5O_2(aq) + H_2O(l) \rightleftharpoons H_3O^+(aq) + C_3H_5O_2^-(aq)$$

	$[HC_3H_5O_2]$	$[H_3O^+]$	$[C_3H_5O_2^-]$
Initial	0.165	≈ 0.00	0.00
Change	$-x$	$+x$	$+x$
Equil	$0.165 - x$	$+x$	$+x$

$K_a = \frac{[H_3O^+][C_3H_5O_2^-]}{[HC_3H_5O_2]} = 1.3 \times 10^{-5} = \frac{x^2}{0.165 - x}$ Assume x is small ($x \ll 0.165$) so

$\frac{x^2}{0.165 - x} = 1.3 \times 10^{-5} = \frac{x^2}{0.165}$ and $x = 1.\underline{4}646 \times 10^{-3}$ M= [H_3O^+]. Confirm that assumption is valid

$\frac{1.\underline{4}646 \times 10^{-3}}{0.165} \times 100\% = 0.89\% < 5\%$ so assumption is valid. Finally,

$pH = -\log[H_3O^+] = -\log(1.\underline{4}646 \times 10^{-3}) = 2.83$

Check: The units (none) are correct. The magnitude of the answer makes physical sense because pH should be greater than – log (0.165) = 0.78 because this is a weak acid.

Find: pH after adding 5.0 mL of base
Conceptual Plan:
Write balanced equation then mL → L then [$HC_3H_5O_2$], L → mol $HC_3H_5O_2$ then

$HC_3H_5O_2 + KOH \rightarrow KC_3H_5O_2 + H_2O$ $\frac{1\text{ L}}{1000\text{ mL}}$ $M = \frac{\text{mol}}{\text{L}}$

mL → L then [KOH], L → mol KOH then

$\frac{1\text{ L}}{1000\text{ mL}}$ $M = \frac{\text{mol}}{\text{L}}$

mol $HC_3H_5O_2$, mol KOH → mol excess $HC_3H_5O_2$, mol $C_3H_5O_2^-$
set up stoichiometry table

Since there are significant concentrations of both the acid and the conjugate base species, this is a buffer solution and so Henderson–Hasselbalch Equation $\left(pH = pK_a + \log\frac{[base]}{[acid]}\right)$ can be used. Also note that ratio of concentrations is the same as the ratio of moles, since the volume is the same for both species.

Solution: $30.0\ \cancel{mL\ HC_3H_5O_2}\ x\ \frac{1\ L}{1000\ \cancel{mL}} = 0.0300\ L\ HC_3H_5O_2$ then

$\frac{0.165\ mol\ HC_3H_5O_2}{1\ \cancel{L}}\ x\ 0.0300\ \cancel{L} = 0.00495\ mol\ HC_3H_5O_2$ and $5.0\ \cancel{mL\ KOH}\ x\ \frac{1\ L}{1000\ \cancel{mL}} = 0.0050\ L\ KOH$

then $\frac{0.300\ mol\ KOH}{1\ \cancel{L}}\ x\ 0.0050\ \cancel{L} = 0.0015\ mol\ KOH$. Set up table to track changes:

	$KOH\ (aq)$	$+\ HC_3H_5O_2\ (aq)$	$\rightarrow\ KC_3H_5O_2\ (aq)$	$+\ H_2O\ (l)$
Before addition	0.00 mol	0.00495 mol	0.00 mol	–
Addition	0.0015 mol	–	–	–
After addition	≈ 0.00 mol	0.00345 mol	0.0015 mol	–

Then use Henderson–Hasselbalch Equation, since the solution is a buffer.

$$pH = pK_a + \log\frac{[base]}{[acid]} = -\log(1.3\ x\ 10^{-5}) + \log\frac{0.0015}{0.00345} = 4.52$$

Check: The units (none) are correct. The pH is a little higher than the initial pH, which is expected since some of the acid has been neutralized.

Find: pH after adding 10.0 mL of base

Conceptual Plan: Use calculations from above then mL → L then [KOH], L → mol KOH then
$\frac{1\ L}{1000\ mL}$ $\quad M = \frac{mol}{L}$

mol $HC_3H_5O_2$, mol KOH → mol excess $HC_3H_5O_2$, mol $C_3H_5O_2^-$
set up stoichiometry table

Since there are significant concentrations of both the acid and the conjugate base species, this is a buffer solution and so Henderson–Hasselbalch Equation $\left(pH = pK_a + \log\frac{[base]}{[acid]}\right)$ can be used. Also note that ratio of concentrations is the same as the ratio of moles, since the volume is the same for both species.

Solution: $10.0\ \cancel{mL\ KOH}\ x\ \frac{1\ L}{1000\ \cancel{mL}} = 0.0100\ L\ KOH$ then $\frac{0.300\ mol\ KOH}{1\ \cancel{L}}\ x\ 0.0100\ \cancel{L} = 0.0030\ mol\ KOH$.

Set up table to track changes:

	$KOH\ (aq)$	$+\ HC_3H_5O_2\ (aq)$	$\rightarrow\ KC_3H_5O_2\ (aq)$	$+\ H_2O\ (l)$
Before addition	0.00 mol	0.00495 mol	0.00 mol	–
Addition	0.0030 mol	–	–	–
After addition	≈ 0.00 mol	0.00195 mol	0.0030 mol	–

Then use Henderson–Hasselbalch Equation, since the solution is a buffer.

$$pH = pK_a + \log\frac{[base]}{[acid]} = -\log(1.3\ x\ 10^{-5}) + \log\frac{0.0030}{0.00195} = 5.07$$

Check: The units (none) are correct. The pH is a little higher than the last pH, which is expected since some of the acid has been neutralized.

Find: pH at equivalence point

Conceptual Plan: Use calculations from above, then set mol acid ($HC_3H_5O_2$) = mol base (KOH)
balanced equation has 1:1 stoichiometry

and [KOH], mol KOH → L KOH then since all of the weak acid has been converted to its
$M = \frac{mol}{L}$

conjugate base, the pH is only dependent on the hydrolysis reaction of the conjugate base. The mol $C_3H_5O_2^-$ = initial mol $HC_3H_5O_2$ and L $HC_3H_5O_2$, L KOH to equivalence point → total L
$L\ HC_3H_5O_2 + L\ KOH = total\ L$

then mol $C_2H_3O_2^-$, L → $[C_2H_3O_2^-]$ and

$$M = \frac{mol}{L}$$

K_a → K_b then do equilibrium calculation: $[C_2H_3O_2^-]$, K_b → $[OH^-]$ → $[H_3O^+]$ → pH

$K_w = K_a\,K_b$ *set up ICE table* $K_w = [H_3O^+][OH^-]$ $pH = -\log[H_3O^+]$

Solution: mol acid = mol $HC_3H_5O_2$ = 0.00495 mol = mol KOH then

$0.00495\ \cancel{\text{mol KOH}} \times \frac{1\ L}{0.300\ \cancel{\text{mol KOH}}} = 0.0165$ L KOH. Then total volume = L $HC_3H_5O_2$ + L KOH =

$0.0300\ L + 0.0165\ L = 0.0465$ then $[C_3H_5O_2^-] = \frac{0.00495\ \text{mol}\ C_3H_5O_2^-}{0.0465\ L} = 0.10\underline{6}452$ M and $K_w = K_a\,K_b$.

Rearrange to solve for K_b. $K_b = \frac{K_w}{K_a} = \frac{1.0 \times 10^{-14}}{1.3 \times 10^{-5}} = 7.\underline{6}923 \times 10^{-10}$ M. Set up ICE table

$$C_3H_5O_2^-(aq) + H_2O(l) \rightleftharpoons HC_3H_5O_2(aq) + OH^-(aq)$$

	$[C_3H_5O_2^-]$	$[HC_3H_5O_2]$	$[OH^-]$
Initial	0.10$\underline{6}$452	≈ 0.00	≈ 0.00
Change	$-x$	$+x$	$+x$
Equil	0.10$\underline{6}$452 $-x$	$+x$	$+x$

$K_b = \frac{[HC_3H_5O_2][OH^-]}{[C_3H_5O_2^-]} = 7.\underline{6}923 \times 10^{-10} = \frac{x^2}{0.10\underline{6}452 - x}$ Assume x is small ($x \ll 0.106$) so

$\frac{x^2}{0.10\underline{6}452 - \cancel{x}} = 7.\underline{6}923 \times 10^{-10} = \frac{x^2}{0.10\underline{6}452}$ and $x = 9.\underline{0}491 \times 10^{-6}$ M = $[OH^-]$.

Confirm that assumption is valid $\frac{9.\underline{0}491 \times 10^{-6}}{0.10\underline{6}452} \times 100\ \% = 0.0085\ \% < 5\ \%$ so assumption is valid.

$K_w = [H_3O^+][OH^-]$ so $[H_3O^+] = \frac{K_w}{[OH^-]} = \frac{1.0 \times 10^{-14}}{9.\underline{0}491 \times 10^{-6}} = 1.\underline{1}051 \times 10^{-9}$ M. Finally,

$pH = -\log[H_3O^+] = -\log(1.\underline{1}051 \times 10^{-9}) = 8.96$.

Check: The units (none) are correct. Since this is a weak acid–strong base titration, the pH at the equivalence point is basic.

Find: pH at one-half of the equivalence point

Conceptual Plan: Since this is a weak acid–strong base titration, the pH at one-half the equivalence point is the pK_a of the weak acid.

Solution: $pH = pK_a = -\log K_a = -\log(1.3 \times 10^{-5}) = 4.89$ and the volume of added base is 0.5 x 16.5 mL = 8.3 mL.

Check: The units (none) are correct. Since this is a weak acid–strong base titration, the pH at one-half the equivalence point is the pK_a of the weak acid, so it should be a little below 5.

Find: pH after adding 20.0 mL of base

Conceptual Plan: Use calculations from earlier. Then the pH is only dependent on the amount of excess base and the total solution volumes. mL added, mL at equiv. pt. → mL excess → L excess

mL excess = mL added – mL at equiv. pt. $\qquad \frac{1\ L}{1000\ mL}$

then mL excess → L excess then [KOH], L excess → mol KOH excess

$\frac{1\ L}{1000\ mL} \qquad M = \frac{mol}{L}$

then L $HC_3H_5O_2$, L KOH to equivalence point, L KOH excess → total L then

L $HC_2H_3O_2$ + L KOH to equivalence point + L KOH excess = total L

mol excess KOH, total L → [KOH] = $[OH^-]$ → $[H_3O^+]$ → pH

$M = \frac{mol}{L} \qquad K_w = [H_3O^+][OH^-] \qquad pH = -\log[H_3O^+]$

Solution: mL KOH excess = mL added – mL to equiv. pt. = 20.0 mL – 16.5 mL = 3.5 mL.

$3.5 \text{ mL KOH} \times \dfrac{1 \text{ L}}{1000 \text{ mL}} = 0.0035 \text{ L KOH excess}$ then

$\dfrac{0.300 \text{ mol KOH}}{1 \text{ L}} \times 0.0035 \text{ L} = 0.00105 \text{ mol KOH excess}$. Then $0.0300 \text{ L } HC_3H_5O_2 + 0.0165 \text{ L KOH} +$

$0.0035 \text{ L KOH} = 0.0500 \text{ L total volume.}$ $[\text{KOH excess}] = \dfrac{0.00105 \text{ mol KOH excess}}{0.0500 \text{ L}} = 0.021 \text{ M KOH excess}$

Since KOH is a strong base, [KOH] excess = $[OH^-]$. The strong base overwhelms the weak base and is insignificant in the calculation. $K_w = [H_3O^+][OH^-]$ so $[H_3O^+] = \dfrac{K_w}{[OH^-]} = \dfrac{1.0 \times 10^{-14}}{0.021} = 4.7619 \times 10^{-13} \text{ M}$.

Finally, $pH = -\log [H_3O^+] = -\log (4.7619 \times 10^{-13}) = 12.32$.

Check: The units (none) are correct. The pH is rising sharply at the equivalence point, so the pH after 5 mL past the equivalence point should be quite basic.

Find: pH after adding 25.0 mL of base

Conceptual Plan: Use calculations from earlier. Then the pH is only dependent on the amount of excess base and the total solution volumes. mL added, mL at equiv. pt. → mL excess → L excess

mL excess = mL added – mL at equiv. pt. $\quad \dfrac{1 \text{ L}}{1000 \text{ mL}}$

then mL excess → L excess then [KOH], L excess → mol KOH excess

$\dfrac{1 \text{ L}}{1000 \text{ mL}} \quad M = \dfrac{mol}{L}$

then L $HC_3H_5O_2$, L KOH to equivalence point, L KOH excess → total L then

L $HC_2H_3O_2$ + L KOH to equivalence point + L KOH excess = total L

mol excess KOH, total L → [KOH] = [OH⁻] → [H₃O⁺] → pH

$M = \dfrac{mol}{L} \quad K_w = [H_3O^+][OH^-] \quad pH = -\log [H_3O^+]$

Solution: mL KOH excess = mL added – mL to equiv. pt. = 25.0 mL – 16.5 mL = 8.5 mL.

$8.5 \text{ mL KOH} \times \dfrac{1 \text{ L}}{1000 \text{ mL}} = 0.0085 \text{ L KOH excess}$ then

$\dfrac{0.300 \text{ mol KOH}}{1 \text{ L}} \times 0.0085 \text{ L} = 0.00255 \text{ mol KOH excess}$. Then $0.0300 \text{ L } HC_3H_5O_2 + 0.0165 \text{ L KOH} +$

$0.0085 \text{ L KOH} = 0.0550 \text{ L total volume.}$ $[\text{KOH excess}] = \dfrac{0.00255 \text{ mol KOH excess}}{0.0550 \text{ L}} = 0.046364 \text{ M KOH excess}$

Since KOH is a strong base, [KOH] excess = $[OH^-]$. The strong base overwhelms the weak base and is insignificant in the calculation. $K_w = [H_3O^+][OH^-]$ So $[H_3O^+] = \dfrac{K_w}{[OH^-]} = \dfrac{1.0 \times 10^{-14}}{0.046364} = 2.1569 \times 10^{-13} \text{ M}$.

Finally, $pH = -\log [H_3O^+] = -\log (2.1569 \times 10^{-13}) = 12.67$.

Check: The units (none) are correct. The pH is rising sharply at the equivalence point, so the pH after 5 mL past the equivalence point should be quite basic. This pH is higher than the last pH.

Plotting these data points:

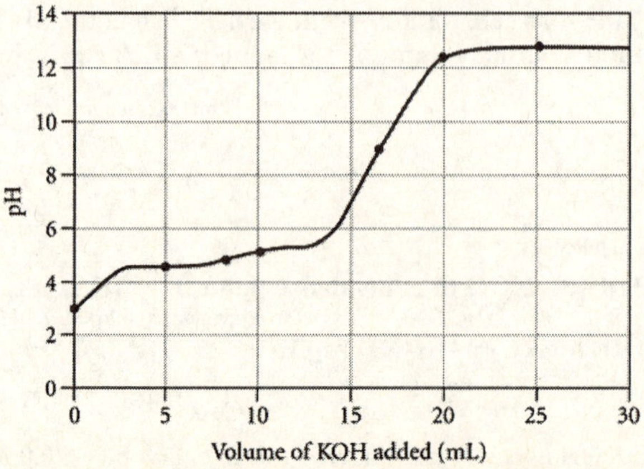

47. **Given:** 25.0 mL of 0.175 M CH_3NH_2 titrated with 0.150 M HBr **Other:** $K_b\ (CH_3NH_2) = 4.4 \times 10^{-4}$

a) **Find:** initial pH

Conceptual Plan: Conceptual Plan: Since CH_3NH_2 is a weak base, set up an equilibrium problem using the initial concentration, so M CH_3NH_2 → $[OH^-]$ → $[H_3O^+]$ → pH

ICE Chart $K_w = [H_3O^+][OH^-]$ $pH = -\log [H_3O^+]$

Solution:

$$CH_3NH_2\,(aq) + H_2O(l) \rightleftharpoons CH_3NH_3^+\,(aq) + OH^-(aq)$$

	$[CH_3NH_2]$	$[CH_3NH_3^+]$	$[OH^-]$
Initial	0.175	0.00	≈ 0.00
Change	$-x$	$+x$	$+x$
Equil	$0.175 - x$	$+x$	$+x$

$$K_b = \frac{[CH_3NH_3^+][OH^-]}{[CH_3NH_2]} = 4.4 \times 10^{-4} = \frac{x^2}{0.175 - x}$$

Assume x is small ($x << 0.175$) so $\frac{x^2}{0.175 - \cancel{x}} = 4.4 \times 10^{-4} = \frac{x^2}{0.175}$ and $x = 8.\underline{7}750 \times 10^{-3}$ M $= [OH^-]$.

Confirm that assumption is valid $\frac{8.\underline{7}750 \times 10^{-3}}{0.175} \times 100\ \% = 5.0\ \%$ so assumption is valid.

$K_w = [H_3O^+][OH^-]$ so $[H_3O^+] = \frac{K_w}{[OH^-]} = \frac{1.0 \times 10^{-14}}{8.\underline{7}750 \times 10^{-3}} = 1.\underline{1}396 \times 10^{-12}$ M . Finally,

$pH = -\log [H_3O^+] = -\log (1.\underline{1}396 \times 10^{-12}) = 11.94$.

Check: The units (none) are correct. The magnitude of the answer makes physical sense because pH should be less than 14 + log (0.175) = 13.2 because this is a weak base.

b) **Find:** volume of acid to reach equivalence point

Conceptual Plan: Write balanced equation then mL → L then $[CH_3NH_2]$, L → mol CH_3NH_2

$HBr + CH_3NH_2 \rightarrow CH_3NH_3Br + H_2O$ $\frac{1\ L}{1000\ mL}$ $M = \frac{mol}{L}$

then set mol base (CH_3NH_2) = mol acid (HBr) and [HBr], mol HBr → L HBr → mL HBr

balanced equation has 1:1 stoichiometry $M = \frac{mol}{L}$ $\frac{1000\ mL}{1\ L}$

Solution: $25.0\ \cancel{mL\ CH_3NH_2} \times \frac{1\ L}{1000\ \cancel{mL}} = 0.0250\ L\ CH_3NH_2$ then

$\frac{0.175\ mol\ CH_3NH_2}{1\ \cancel{L}} \times 0.0250\ \cancel{L} = 0.004\underline{3}75\ mol\ CH_3NH_2$. So mol base = mol CH_3NH_2 = 0.002$\underline{8}$75 mol

= mol HBr then $0.004\underline{3}75\ \cancel{mol\ HBr} \times \frac{1\ L}{0.150\ \cancel{mol\ HBr}} = 0.029\underline{1}667\ \cancel{L\ HBr} \times \frac{1000\ mL}{1\ \cancel{L}} = 29.2\ mL\ HBr$.

Check: The units (mL) are correct. The volume of acid is a greater than the volume of base because the concentration of the base is a little greater than that of the acid.

c) **Find:** pH after adding 5.0 mL of acid

Conceptual Plan: Use calculations from part b. Then mL → L then [HBr], L → mol HBr

$\frac{1\ L}{1000\ mL}$ $M = \frac{mol}{L}$

then mol CH_3NH_2, mol HBr → mol excess CH_3NH_2 and L CH_3NH_2, L HBr → total L then

set up stoichiometry table L CH_3NH_2 + L HBr = total L

Since there are significant concentrations of both the acid and the conjugate base species, this is a buffer solution and so Henderson–Hasselbalch Equation $\left(pH = pK_a + \log \frac{[base]}{[acid]}\right)$ can be used. Convert K_b to K_a using $K_w = K_a\ K_b$. Also note that ratio of concentrations is the same as the ratio of moles, since the volume is the same for both species.

Solution: $5.0\ \text{mL HBr} \times \frac{1\ \text{L}}{1000\ \text{mL}} = 0.0050\ \text{L HBr}$ then $\frac{0.150\ \text{mol HBr}}{1\ \text{L}} \times 0.0050\ \text{L} = 0.00075\ \text{mol HBr}$.

Set up table to track changes:

	$HBr\,(aq)$	$+\ CH_3NH_2\,(aq)$	$\rightarrow\ CH_3NH_3Br\,(aq)$
Before addition	0.00 mol	0.004375 mol	0.00 mol
Addition	0.00075 mol	–	–
After addition	≈ 0.00 mol	0.003625 mol	0.000750 mol

then $K_w = K_a\,K_b$ so

$K_a = \frac{K_w}{K_b} = \frac{1.0 \times 10^{-14}}{4.4 \times 10^{-4}} = 2.2727 \times 10^{-11}\ \text{M}$ then use Henderson–Hasselbalch Equation, since the solution is a buffer. $\text{pH} = \text{p}K_a + \log\frac{[\text{base}]}{[\text{acid}]} = -\log(2.2727 \times 10^{-11}) + \log\frac{0.003625}{0.000750} = 11.33$

Check: The units (none) are correct. The pH is a little lower than the last pH, which is expected since some of the base has been neutralized.

d) **Find:** pH at one-half of the equivalence point

Conceptual Plan: Since this is a weak base–strong acid titration, the pH at one-half the equivalence point is the pK_a of the conjugate acid of weak base.

Solution: $\text{pH} = pK_a = -\log K_a = -\log(2.2727 \times 10^{-11}) = 10.64$.

Check: The units (none) are correct. Since this is a weak acid–strong base titration, the pH at one-half the equivalence point is the pK_a of the conjugate acid of the weak base, so it should be a little below 11.

e) **Find:** pH at equivalence point

Conceptual Plan: Use calculations from above. Since all of the weak base has been converted to its conjugate acid, the pH is only dependent on the hydrolysis reaction of the conjugate acid. The mol $CH_3NH_3^+$ = initial mol CH_3NH_2 and

L CH_3NH_2, L HBr to equivalence point → total L then mol $CH_3NH_3^+$, L → $[CH_3NH_3^+]$

$L\ CH_3NH_2 + L\ HBr = \text{total L}$ $\qquad M = \frac{\text{mol}}{L}$

then do equilibrium calculation: $[CH_3NH_3^+]$, K_a → $[H_3O^+]$ → pH

set up ICE table $\qquad \text{pH} = -\log[H_3O^+]$

Solution: mol base = mol acid = mol $CH_3NH_3^+$ = 0.004375 mol. Then total volume = L CH_3NH_2 + L HBr = 0.0250 L + 0.0292 L = 0.0542 then $[CH_3NH_3^+] = \frac{0.004375\ \text{mol mol}\ CH_3NH_3^+}{0.0542\ \text{L}} = 0.0807196\ \text{M}$.

$$CH_3NH_3^+(aq) + H_2O(l) \rightleftharpoons CH_3NH_2(aq) + H_3O^+(aq)$$

Set up ICE table

	$[CH_3NH_3^+]$	$[CH_3NH_2]$	$[H_3O^+]$
Initial	0.0807196	≈ 0.00	≈ 0.00
Change	$-x$	$+x$	$+x$
Equil	$0.0807196 - x$	$+x$	$+x$

$K_a = \frac{[CH_3NH_2][H_3O^+]}{[CH_3NH_3^+]} = 2.2727 \times 10^{-11} = \frac{x^2}{0.0807196 - x}$ Assume x is small ($x << 0.0807$) so

$\frac{x^2}{0.0807196 - x} = 2.2727 \times 10^{-11} = \frac{x^2}{0.0807196}$ and $x = 1.3544 \times 10^{-6} = [H_3O^+]$.

Confirm that assumption is valid $\frac{1.3544 \times 10^{-6}}{0.0807106} \times 100\,\% = 0.0017\,\% < 5\,\%$ so assumption is valid.

Finally, $\text{pH} = -\log[H_3O^+] = -\log(1.3544 \times 10^{-6}) = 5.87$.

Check: The units (none) are correct. Since this is a weak base–strong acid titration, the pH at the equivalence point is acidic.

f) **Find:** pH after adding 5.0 mL of acid beyond the equivalence point

Conceptual Plan: Use calculations from parts b & c. Then the pH is only dependent on the amount of excess acid and the total solution volumes.

mL excess → L excess then [HBr], L excess → mol HBr excess

$\frac{1\ L}{1000\ mL}$ $M = \frac{mol}{L}$

then L CH_3NH_2, L HBr to equivalence point, L HBr excess → total L then

L CH_3NH_2 + L HBr to equivalence point + L HBr excess = total L

mol excess HBr, total L → [HBr] = $[H_3O^+]$ → pH

$M = \frac{mol}{L}$ $pH = -\log[H_3O^+]$

Solution: $5.0\ \text{mL HBr} \times \frac{1\ L}{1000\ \text{mL}} = 0.0050\ L$ HBr excess then

$\frac{0.150\ \text{mol HBr}}{1\ L} \times 0.0050\ L = 0.00075$ mol HBr excess. Then 0.0250 L CH_3NH_2 + 0.0292 L HBr + 0.0050 L HBr = 0.0592 L total volume.

$$[\text{HBr excess}] = \frac{0.00075\ \text{mol HBr excess}}{0.0592\ L} = 0.01\underline{2}669\ M\ \text{HBr excess}$$

Since HBr is a strong acid, [HBr] excess = $[H_3O^+]$. The strong acid overwhelms the weak acid and is insignificant in the calculation. Finally, $pH = -\log[H_3O^+] = -\log(0.01\underline{2}669) = 1.90$.

Check: The units (none) are correct. The pH is dropping sharply at the equivalence point, so the pH after 5 mL past the equivalence point should be quite acidic.

48. **Given:** 25.0 mL of 0.125 M pyridine (C_5H_5N) titrated with 0.100 M HCl **Other:** K_b (C_5H_5N) = 1.7×10^{-9}
Find: initial pH
Conceptual Plan: Since C_5H_5N is a weak base, set up an equilibrium problem using the initial concentration, so M C_5H_5N → $[OH^-]$ → $[H_3O^+]$ → pH

ICE Chart $K_w = [H_3O^+][OH^-]$ $pH = -\log[H_3O^+]$

Solution:

$$C_5H_5N(aq) + H_2O(l) \rightleftharpoons C_5H_5NH^+(aq) + OH^-(aq)$$

	$[C_5H_5N]$	$[C_5H_5NH^+]$	$[OH^-]$
Initial	0.125	0.00	≈ 0.00
Change	$-x$	$+x$	$+x$
Equil	$0.125 - x$	$+x$	$+x$

$$K_b = \frac{[C_5H_5NH^+][OH^-]}{[C_5H_5N]} = 1.7 \times 10^{-9} = \frac{x^2}{0.125 - x}$$

Assume x is small ($x \ll 0.125$) so $\frac{x^2}{0.125 - x} = 1.7 \times 10^{-9} = \frac{x^2}{0.125}$ and $x = 1.\underline{4}577 \times 10^{-5} = [OH^-]$.

Confirm that assumption is valid $\frac{1.\underline{4}577 \times 10^{-5}}{0.125} \times 100\ \% = 0.012\ \% < 5.0\ \%$ so assumption is valid.

$K_w = [H_3O^+][OH^-]$ so $[H_3O^+] = \frac{K_w}{[OH^-]} = \frac{1.0 \times 10^{-14}}{1.\underline{4}577 \times 10^{-5}} = 6.\underline{8}601 \times 10^{-10}\ M$. Finally,

$pH = -\log[H_3O^+] = -\log(6.\underline{8}601 \times 10^{-10}) = 9.16$.

Check: The units (none) are correct. The magnitude of the answer makes physical sense because pH should be less than 14 + log (0.125) = 13.1 because this is a weak base.

Find: pH after adding 10.0 mL of acid
Conceptual Plan: Write balanced equation then mL → L then $[C_5H_5N]$, L → mol C_5H_5N then

HCl + C_5H_5N → C_5H_5NHCl $\frac{1\ L}{1000\ mL}$ $M = \frac{mol}{L}$

mL → L then [HCl], L → mol HCl then mol C_5H_5N, mol HCl → mol excess C_5H_5N and

$\frac{1\ L}{1000\ mL}$ $M = \frac{mol}{L}$ set up stoichiometry table

L C_5H_5N, L HCl → total L then since there are significant concentrations of both the acid and

L C_5H_5N + L HCl = total L

the conjugate base species, this is a buffer solution and so Henderson–Hasselbalch Equation

$\left(pH = pK_a + \log \frac{[base]}{[acid]}\right)$ **can be used. Convert K_b to K_a using $K_w = K_a K_b$. Also note that ratio of concentrations is the same as the ratio of moles, since the volume is the same for both species.**

Solution: $25.0 \text{ mL } C_5H_5N \times \frac{1\text{ L}}{1000\text{ mL}} = 0.0250\text{ L } C_5H_5N$ then

$\frac{0.125\text{ mol } C_5H_5N}{1\text{ L}} \times 0.0250\text{ L} = 0.003125\text{ mol } C_5H_5N$ and $10.0\text{ mL HCl} \times \frac{1\text{ L}}{1000\text{ mL}} = 0.0100\text{ L HCl}$ then

$\frac{0.100\text{ mol HCl}}{1\text{ L}} \times 0.0100\text{ L} = 0.00100\text{ mol HCl}$. Set up table to track changes:

	$HCl\,(aq)$ +	$C_5H_5N\,(aq)$ →	$C_5H_5NHCl\,(aq)$
Before addition	0.00 mol	0.003125 mol	0.00 mol
Addition	0.00100 mol	–	–
After addition	≈ 0.00 mol	0.002125 mol	0.00100 mol

Then $K_w = K_a K_b$ so $K_a = \frac{K_w}{K_b} = \frac{1.0 \times 10^{-14}}{1.7 \times 10^{-9}} = 5.8824 \times 10^{-6}\text{ M}$ then use Henderson–Hasselbalch Equation,

since the solution is a buffer. $pH = pK_a + \log \frac{[base]}{[acid]} = -\log(5.8824 \times 10^{-6}) + \log \frac{0.002125}{0.00100} = 5.56$

Check: The units (none) are correct. The pH is lower than the last pH, which is expected since some of the base has been neutralized.

Find: pH after adding 20.0 mL of acid

Conceptual Plan: Use calculations from above. Then mL → L then [HCl], L → mol HCl

$\frac{1\text{ L}}{1000\text{ mL}}$ $\quad M = \frac{mol}{L}$

then mol C_5H_5N, mol HCl → mol excess C_5H_5N and L C_5H_5N, L HCl → total L then

set up stoichiometry table $\quad$ L C_5H_5N + L HCl = total L

Since there are significant concentrations of both the acid and the conjugate base species, this is a buffer solution and so Henderson–Hasselbalch Equation $\left(pH = pK_a + \log \frac{[base]}{[acid]}\right)$ **can be used. Convert K_b to K_a using $K_w = K_a K_b$. Also note that ratio of concentrations is the same as the ratio of moles, since the volume is the same for both species.**

Solution: $20.0\text{ mL HCl} \times \frac{1\text{ L}}{1000\text{ mL}} = 0.0200\text{ L HCl}$ then $\frac{0.100\text{ mol HCl}}{1\text{ L}} \times 0.0200\text{ L} = 0.00200\text{ mol HCl}$.

Set up table to track changes:

	$HCl\,(aq)$ +	$C_5H_5N\,(aq)$ →	$C_5H_5NHCl\,(aq)$
Before addition	0.00 mol	0.003125 mol	0.00 mol
Addition	0.00200 mol	–	–
After addition	≈ 0.00 mol	0.001125 mol	0.00200 mol

Then use Henderson–Hasselbalch Equation, since the solution is a buffer.

$pH = pK_a + \log \frac{[base]}{[acid]} = -\log(5.8824 \times 10^{-6}) + \log \frac{0.001125}{0.00200} = 4.98$

Check: The units (none) are correct. The pH is lower than the last pH, which is expected since some of the base has been neutralized.

Find: pH at equivalence point

Conceptual Plan:

Use calculations from above. Since all of the weak base has been converted to its conjugate acid, the pH is only dependent on the hydrolysis reaction of the conjugate acid. The mol $C_5H_5NH^+$ = initial mol C_5H_5N then set mol base (C_5H_5N) = mol acid (HCl) and [HCl], mol HCl → L HCl then

balanced equation has 1:1 stoichiometry $\quad M = \frac{mol}{L}$

L C_5H_5N, L HCl to equivalence point → total L then mol $C_5H_5NH^+$, L → [$C_5H_5NH^+$]

L C_5H_5N + L HCl = total L $\qquad$ $M = \frac{mol}{L}$

then do equilibrium calculation: [$C_5H_5NH^+$], K_a → [H_3O^+] → pH

set up ICE table $\qquad$ $pH = -\log [H_3O^+]$

Solution: mol base = mol acid = mol $C_5H_5NH^+$ = 0.003125 mol. Then

$$0.003125 \text{ mol HCl} \times \frac{1 \text{ L}}{0.100 \text{ mol HCl}} = 0.03125 \text{ L HCl} \times \frac{1000 \text{ mL}}{1 \text{ L}} = 31.3 \text{ mL HCl}$$ then

total volume = L C_5H_5N + L HCl = 0.0250 L + 0.0313 L = 0.05632 L then

$$[C_5H_5NH^+] = \frac{0.003125 \text{ mol mol } C_5H_5NH^+}{0.0563 \text{ L}} = 0.0555062 \text{ M}$$. Set up ICE table

$$C_5H_5NH^+(aq) + H_2O(l) \rightleftharpoons C_5H_5N(aq) + H_3O^+(aq)$$

	$[C_5H_5NH^+]$	$[C_5H_5N]$	$[H_3O^+]$
Initial	0.0555062	≈ 0.00	≈ 0.00
Change	$-x$	$+x$	$+x$
Equil	$0.0555062 - x$	$+x$	$+x$

$$K_a = \frac{[C_5H_5N][H_3O^+]}{[C_5H_5NH^+]} = 5.8824 \times 10^{-6} = \frac{x^2}{0.0555062 - x}$$ Assume x is small ($x << 0.0556$) so

$$\frac{x^2}{0.0555062 - x} = 5.8824 \times 10^{-6} = \frac{x^2}{0.0555062}$$ and $x = 5.7141 \times 10^{-4}$ M = [H_3O^+]. Confirm that assumption is

valid $\frac{5.7141 \times 10^{-4}}{0.0555062} \times 100\ \% = 1.0\ \% < 5\ \%$ so assumption is valid.

Finally, $pH = -\log [H_3O^+] = -\log (5.7141 \times 10^{-4}) = 3.24$.

Check: The units (none) are correct. Since this is a weak base–strong acid titration, the pH at the equivalence point is acidic.

Find: pH at one-half of the equivalence point

Conceptual Plan: Since this is a weak base–strong acid titration, the pH at one-half the equivalence point is the pK_a of the conjugate acid of weak base.

Solution: $pH = pK_a = -\log K_a = -\log (5.8824 \times 10^{-6}) = 5.23$ and the volume is 0.5 x 31.3 mL = 15.7 mL.

Check: The units (none) are correct. Since this is a weak base–strong acid titration, the pH at one-half the equivalence point is the pK_a of the conjugate acid of the weak base, so it should be a little below 6.

Find: pH after adding 40.0 mL of acid

Conceptual Plan: Use calculations from above. Then the pH is only dependent on the amount of excess acid and the total solution volumes.

mL added, mL at equiv. pt. → mL excess → L excess then

mL excess = mL added – mL at equiv. pt. $\qquad$ $\frac{1 \text{ L}}{1000 \text{ mL}}$

mL excess → L excess then [HCl], L excess → mol HCl excess

$\frac{1 \text{ L}}{1000 \text{ mL}}$ $\qquad$ $M = \frac{mol}{L}$

then L C_5H_5N, L HCl to equivalence point, L HCl excess → total L then

L C_5H_5N + L HCl to equivalence point + L HCl excess = total L

mol excess HCl, total L → [HCl] = [H_3O^+] → pH

$M = \frac{mol}{L}$ $\qquad$ $pH = -\log [H_3O^+]$

Solution: mL excess = mL added – mL at equiv. pt. = 40.0 mL – 31.3 mL = 8.7 mL.

$$8.7 \text{ mL HCl} \times \frac{1 \text{ L}}{1000 \text{ mL}} = 0.0087 \text{ L HCl excess}$$ then

$\frac{0.100 \text{ mol HCl}}{1 \cancel{\text{L}}} \times 0.0087 \cancel{\text{L}} = 0.00087$ mol HCl excess. Then 0.0250 L C_5H_5N + 0.0313 L HCl + 0.0087 L HCl = 0.0650 L total volume. $[\text{HCl excess}] = \frac{0.00087 \text{ mol HCl excess}}{0.0650 \text{ L}} = 0.01\underline{3}385$ M HCl excess

Since HCl is a strong acid, [HCl] excess = $[H_3O^+]$. The strong acid overwhelms the weak acid and is insignificant in the calculation. Finally, $pH = -\log [H_3O^+] = -\log (0.01\underline{3}385) = 1.87$.

Check: The units (none) are correct. The pH is dropping sharply at the equivalence point, so the pH after 5 mL past the equivalence point should be quite acidic.

Find: pH after adding 50.0 mL of acid

Conceptual Plan: Use calculations from above. Then the pH is only dependent on the amount of excess acid and the total solution volumes. mL added, mL at equiv. pt. → mL excess → L excess then

mL excess = mL added – mL at equiv. pt. $\qquad \frac{1 \text{ L}}{1000 \text{ mL}}$

mL excess → L excess then [HCl], L excess → mol HCl excess

$\frac{1 \text{ L}}{1000 \text{ mL}} \qquad M = \frac{\text{mol}}{\text{L}}$

then L C_5H_5N, L HCl to equivalence point, L HCl excess → total L then

L C_5H_5N + L HCl to equivalence point + L HCl excess = total L

mol excess HCl, total L → [HCl] = $[H_3O^+]$ → pH

$M = \frac{\text{mol}}{\text{L}} \qquad pH = -\log [H_3O^+]$

Solution: mL excess = mL added – mL at equiv. pt. = 50.0 mL – 31.3 mL = 18.7 mL.

$18.7 \cancel{\text{mL HCl}} \times \frac{1 \text{ L}}{1000 \cancel{\text{mL}}} = 0.0187$ L HCl excess then

$\frac{0.100 \text{ mol HCl}}{1 \cancel{\text{L}}} \times 0.0187 \cancel{\text{L}} = 0.00187$ mol HCl excess. Then 0.0250 L C_5H_5N + 0.0313 L HCl + 0.0187 L HBr = 0.0750 L total volume. $[\text{HCl excess}] = \frac{0.00187 \text{ mol HCl excess}}{0.0750 \text{ L}} = 0.02\underline{4}933$ M HCl excess

Since HCl is a strong acid, [HCl] excess = $[H_3O^+]$. The strong acid overwhelms the weak acid and is insignificant in the calculation. Finally, $pH = -\log [H_3O^+] = -\log (0.02\underline{4}933) = 1.60$.

Check: The units (none) are correct. The pH is dropping sharply at the equivalence point, so the pH after 5 mL past the equivalence point should be quite acidic. The pH is lower than the last point. Plotting these points gives:

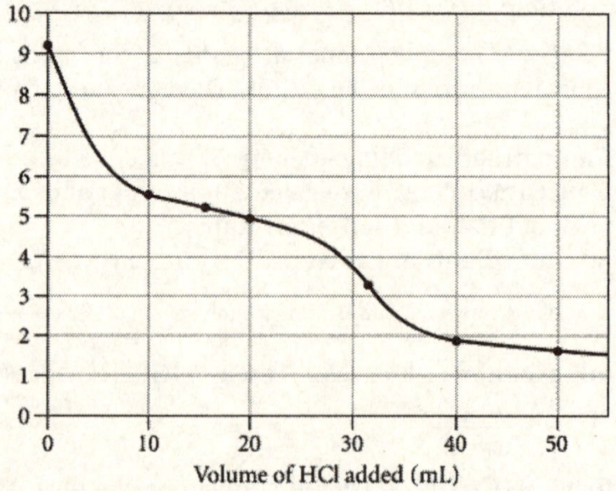

49. i) acid a is more concentrated, since the equivalence point (where sharp pH rise occurs) is at a higher volume of added base.

ii) acid b has the larger K_a, since the pH at a volume of added base equal to half of the equivalence point volume is lower.

50. i) acid b is more concentrated, since the equivalence point (where sharp pH drop occurs) is at a higher volume of added base.

ii) acid b has the larger K_b, since the pH at a volume of added base equal to half of the equivalence point volume is higher.

51. **Given:** 0.229 g unknown monoprotic acid titrated with 0.112 M NaOH and curve **Find:** molar mass and pK_a of acid
Conceptual Plan: The equivalence point is where sharp pH rise occurs. The pK_a is the pH at a volume of added base equal to half of the equivalence point volume. Then mL NaOH → L NaOH

$$\frac{1\ \text{L}}{1000\ \text{mL}}$$

then [NaOH], L NaOH → mol NaOH = mol acid then mol NaOH, g acid → molar mass

$$M = \frac{\text{mol}}{\text{L}} \qquad \frac{\text{g acid}}{\text{mol acid}}$$

Solution: The equivalence point is at 25 mL NaOH. The pH at 0.5 x 25 mL = 13 mL is ~3 = pK_a. then

$$25\ \text{mL NaOH} \times \frac{1\ \text{L}}{1000\ \text{mL}} = 0.025\ \text{L NaOH} \quad \text{then}$$

$$\frac{0.112\ \text{mol NaOH}}{1\ \text{L}} \times 0.025\ \text{L} = 0.0028\ \text{mol NaOH} = 0.0028\ \text{mol acid} \quad \text{then}$$

$$\text{Molar Mass} = \frac{0.229\ \text{g acid}}{0.0028\ \text{mol acid}} = 82\ \text{g/mol}.$$

Check: The units (none and g/mol) are correct. The pK_a is consistent with a weak acid. The molar mass is reasonable for an acid (>1 g/mol).

52. **Given:** 0.446 g unknown monoprotic acid titrated with 0.105 M KOH and curve **Find:** molar mass and pK_a of acid
Conceptual Plan: The equivalence point is where sharp pH rise occurs. The pK_a is the pH at a volume of added base equal to half of the equivalence point volume. Then mL KOH → L KOH then

$$\frac{1\ \text{L}}{1000\ \text{mL}}$$

[KOH], L KOH → mol KOH = mol acid then mol KOH, g acid → molar mass

$$M = \frac{\text{mol}}{\text{L}} \qquad \frac{\text{g acid}}{\text{mol acid}}$$

Solution: The equivalence point is at 35 mL NaOH. The pH at 0.5 x 35 mL = 18 mL is ~ 4.5 = pK_a. Then

$$35\ \text{mL KOH} \times \frac{1\ \text{L}}{1000\ \text{mL}} = 0.035\ \text{L KOH} \quad \text{then}$$

$$\frac{0.105\ \text{mol KOH}}{1\ \text{L}} \times 0.035\ \text{L} = 0.003\underline{6}75\ \text{mol KOH} = 0.003\underline{6}75\ \text{mol acid} \quad \text{then}$$

$$\text{Molar Mass} = \frac{0.446\ \text{g acid}}{0.003\underline{6}75\ \text{mol acid}} = 120\ \text{g/mol}.$$

Check: The units (none and g/mol) are correct. The pK_a is consistent with a weak acid. The molar mass is reasonable for an acid (>1 g/mol).

53. Since the exact conditions of the titration are not given, a rough calculation will suffice. Looking at the pattern of earlier problems, the pH at the equivalence point of a titration of a weak acid and a strong base is the hydrolysis of the conjugate base of the weak acid that has been diluted by a factor of roughly 2 with base. If it is assumed that the initial concentration of the weak acid is ~ 0.1 M, then the conjugate base concentration will be ~ 0.05 M. From earlier calculations it can be seen that the $K_b = \frac{K_w}{K_a} = \frac{[OH^-]^2}{0.05}$ thus

$$[OH^-] = \sqrt{\frac{0.05\,K_w}{K_a}} = \sqrt{\frac{5 \times 10^{-16}}{K_a}} \text{ and the pH} = 14 + \log\sqrt{\frac{5 \times 10^{-16}}{K_a}}.$$

a) for HF, the pK_a = 3.5 x 10^{-4} and so the above equation approximates the pH at the equivalence point of ~ 8.0. Looking at Table 16.1, phenol red or *m*-nitrophenol will change at the appropriate pH range.

b) for HCl, the pH at the equivalence point is 7, since HCl is a strong acid. Looking at Table 16.1, alizarin, bromthymol blue, or phenol red will change at the appropriate pH range.

c) for HCN, the pK_a = 4.9 x 10^{-10} and so the above equation approximates the pH at the equivalence point of ~ 11.0. Looking at Table 16.1, alizarin yellow R will change at the appropriate pH range.

54. Since the exact conditions of the titration are not given, a rough calculation will suffice. Looking at the pattern of earlier problems, the pH at the equivalence point of a titration of a weak base and a strong acid is the hydrolysis of the conjugate acid of the weak base that has been diluted by a factor of roughly 2 with acid. If it is assumed that the initial concentration of the weak base is ~ 0.1 M, then the conjugate acid concentration will be ~ 0.05 M. From earlier calculations it can be seen that the $K_a = \frac{K_w}{K_b} = \frac{[H_3O^+]^2}{0.05}$ thus

$$[H_3O^+] = \sqrt{\frac{0.05\,K_w}{K_b}} = \sqrt{\frac{5 \text{ x } 10^{-16}}{K_b}} \text{ and the } pH = -\log\sqrt{\frac{5 \text{ x } 10^{-16}}{K_b}}.$$

a) for CH_3NH_2, the pK_b = 4.4 x 10^{-4} and so the above equation approximates the pH at the equivalence point is ~ 6.0. Looking at Table 16.1, methyl red, Eriochrom Black T, bromeresol purple, or bromthymol blue will change at the appropriate pH range.

b) for NaOH, the pH at the equivalence point is 7, since NaOH is a strong base. Looking at Table 16.1, alizarin, bromthymol blue, or phenol red will change at the appropriate pH range.

c) for $C_6H_5NH_2$, the pK_b = 3.9 x 10^{-10} and so the above equation approximates the pH at the equivalence point of ~ 2.9. Looking at Table 16.1, erythrosine B will change at the appropriate pH range.

55. For the dissolution reaction, start with the ionic compound as a solid and put it in equilibrium with the appropriate cation and anion, making sure to include the appropriate stoichiometric coefficients. The K_{sp} expression is the product of the concentrations of the cation and anion concentrations raised to their stoichiometric coefficients.

a) $BaSO_4\,(s) \rightleftharpoons Ba^{2+}\,(aq) + SO_4^{2-}\,(aq)$ and $K_{sp} = [Ba^{2+}][SO_4^{2-}]$.

b) $PbBr_2\,(s) \rightleftharpoons Pb^{2+}\,(aq) + 2\,Br^-\,(aq)$ and $K_{sp} = [Pb^{2+}][Br^-]^2$.

c) $Ag_2CrO_4\,(s) \rightleftharpoons 2\,Ag^+\,(aq) + CrO_4^{2-}\,(aq)$ and $K_{sp} = [Ag^+]^2[CrO_4^{2-}]$.

56. For the dissolution reaction, start with the ionic compound as a solid and put it in equilibrium with the appropriate cation and anion, making sure to include the appropriate stoichiometric coefficients. The K_{sp} expression is the product of the concentrations of the cation and anion concentrations raised to their stoichiometric coefficients.

a) $CaCO_3\,(s) \rightleftharpoons Ca^{2+}\,(aq) + CO_3^{2-}\,(aq)$ and $K_{sp} = [Ca^{2+}][CO_3^{2-}]$.

b) $PbCl_2\,(s) \rightleftharpoons Pb^{2+}\,(aq) + 2\,Cl^-\,(aq)$ and $K_{sp} = [Pb^{2+}][Cl^-]^2$.

c) $AgI\,(s) \rightleftharpoons Ag^+\,(aq) + I^-\,(aq)$ and $K_{sp} = [Ag^+][I^-]$.

57. **Given:** ionic compound formula and Table 16.2 of K_{sp} values **Find:** molar solubility (S)
Conceptual Plan: The expression of the solubility product constant of A_mX_n is: $K_{sp} = [A^{n+}]^m\,[X^{m-}]^n$. The molar solubility of a compound, A_mX_n, can be computed directly from K_{sp} by solving for S in the expression: $K_{sp} = (mS)^m\,(nS)^n = m^m\,n^n\,S^{m+n}$.
Solution:

a) for AgBr, K_{sp} = 5.35 x 10^{-13}, A = Ag^+, m = 1, X = Br^-, and n = 1 so K_{sp} = 5.35 x 10^{-13} = S^2. Rearrange to solve for S. $S = \sqrt{5.35 \text{ x } 10^{-13}} = 7.31 \text{ x } 10^{-7}$ M.

b) for $Mg(OH)_2$, K_{sp} = 2.06 x 10^{-13}, A = Mg^{2+}, m = 1, X = OH^-, and n = 2 so K_{sp} = 2.06 x 10^{-13} = 2^2S^3. Rearrange to solve for S. $S = \sqrt[3]{\frac{2.06 \text{ x } 10^{-13}}{4}} = 3.72 \text{ x } 10^{-5}$ M.

c) for CaF_2, $K_{sp} = 1.46 \times 10^{-10}$, A = Ca^{2+}, m = 1, X = F^-, and n = 2 so $K_{sp} = 1.46 \times 10^{-10} = 2^2 S^3$. Rearrange to solve for *S*. $S = \sqrt[3]{\frac{1.46 \times 10^{-10}}{4}} = 3.32 \times 10^{-4}$ M.

Check: The units (M) are correct. The molar solubilities are much less than one and dependent not only on the value of the K_{sp}, but also the stoichiometry of the ionic compound. The more ions that are generated, the greater the molar solubility for the same value of the K_{sp}.

58. **Given:** ionic compound formula and Table 16.2 of K_{sp} values **Find:** molar solubility (*S*)
Conceptual Plan: The expression of the solubility product constant of A_mX_n is: $K_{sp} = [A^{n+}]^m [X^{m-}]^n$. The molar solubility of a compound, A_mX_n, can be computed directly from K_{sp} by solving for *S* in the expression: $K_{sp} = (mS)^m (nS)^n = m^m n^n S^{m+n}$.
Solution:
a) for CuS, $K_{sp} = 1.27 \times 10^{-36}$, A = Cu^{2+}, m = 1, X = S^{2-}, and n = 1 so $K_{sp} = 1.27 \times 10^{-36} = S^2$. Rearrange to solve for *S*. $S = \sqrt{1.27 \times 10^{-36}} = 1.13 \times 10^{-18}$ M.

b) for Ag_2CrO_4, $K_{sp} = 1.12 \times 10^{-12}$, A = Ag^+, m = 2, X = CrO_4^{2-}, and n = 1 so $K_{sp} = 1.12 \times 10^{-12} = 2^2 S^3$. Rearrange to solve for *S*. $S = \sqrt[3]{\frac{1.12 \times 10^{-12}}{4}} = 6.54 \times 10^{-5}$ M.

c) for $Ca(OH)_2$, $K_{sp} = 4.68 \times 10^{-6}$, A = Ca^{2+}, m = 1, X = OH^-, and n = 2 so $K_{sp} = 4.68 \times 10^{-6} = 2^2 S^3$. Rearrange to solve for *S*. $S = \sqrt[3]{\frac{4.68 \times 10^{-6}}{4}} = 1.05 \times 10^{-2}$ M.

Check: The units (M) are correct. The molar solubilities are much less than one and dependent not only on the value of the K_{sp}, but also the stoichiometry of the ionic compound. The more ions that are generated, the greater the molar solubility for the same value of the K_{sp}.

59. **Given:** ionic compound formula and molar solubility (*S*) **Find:** K_{sp}
Conceptual Plan: The expression of the solubility product constant of A_mX_n is: $K_{sp} = [A^{n+}]^m [X^{m-}]^n$. The molar solubility of a compound, A_mX_n, can be computed directly from K_{sp} by solving for *S* in the expression: $K_{sp} = (mS)^m (nS)^n = m^m n^n S^{m+n}$.
Solution:
a) for NiS, $S = 3.27 \times 10^{-11}$ M, A = Ni^{2+}, m = 1, X = S^{2-}, and n = 1 so $K_{sp} = S^2 = (3.27 \times 10^{-11})^2 = 1.07 \times 10^{-21}$.

b) for PbF_2, $S = 5.63 \times 10^{-3}$ M, A = Pb^{2+}, m = 1, X = F^-, and n = 2 so $K_{sp} = 2^2 S^3 = 2^2 (5.63 \times 10^{-3})^3 = 7.14 \times 10^{-7}$.

c) for MgF_2, $S = 2.65 \times 10^{-4}$ M, A = Mg^{2+}, m = 1, X = F^-, and n = 2 so $K_{sp} = 2^2 S^3 = 2^2 (2.65 \times 10^{-4})^3 = 7.44 \times 10^{-11}$.

Check: The units (none) are correct. The K_{sp} values are much less than one and dependent not only on the value of the solubility, but also the stoichiometry of the ionic compound. The more ions that are generated, the smaller the K_{sp} for the same value of the *S*.

60. **Given:** ionic compound formula and molar solubility (*S*) **Find:** K_{sp}
Conceptual Plan: The expression of the solubility product constant of A_mX_n is: $K_{sp} = [A^{n+}]^m [X^{m-}]^n$. The molar solubility of a compound, A_mX_n, can be computed directly from K_{sp} by solving for *S* in the expression: $K_{sp} = (mS)^m (nS)^n = m^m n^n S^{m+n}$.
Solution:
a) for $BaCrO_4$, $S = 1.08 \times 10^{-5}$ M, A = Ba^{2+}, m = 1, X = CrO_4^{2-}, and n = 1 so $K_{sp} = S^2 = (1.08 \times 10^{-5})^2 = 1.17 \times 10^{-10}$.

b) for Ag_2SO_3, $S = 1.55 \times 10^{-5}$ M, A = Ag^+, m = 2, X = SO_3^{2-}, and n = 1 so $K_{sp} = 2^2 S^3 = 2^2 (1.55 \times 10^{-5})^3 = 1.49 \times 10^{-14}$.

c) for $Pd(SCN)_2$, $S = 2.22 \times 10^{-8}$ M, A = Pd^{2+}, m = 1, X = SCN^-, and n = 2 so $K_{sp} = 2^2 S^3 = 2^2 (2.22 \times 10^{-8})^3 = 4.38 \times 10^{-23}$.

Check: The units (none) are correct. The K_{sp} values are much less than one and dependent not only on the value of the solubility, but also the stoichiometry of the ionic compound. The more ions that are generated, the smaller the K_{sp} for the same value of the S.

61. **Given:** ionic compound formulas AX and AX_2 and $K_{sp} = 1.5 \times 10^{-5}$ **Find:** higher molar solubility (S)
Conceptual Plan: The expression of the solubility product constant of A_mX_n is: $K_{sp} = [A^{n+}]^m [X^{m-}]^n$. The molar solubility of a compound, A_mX_n, can be computed directly from K_{sp} by solving for S in the expression: $K_{sp} = (mS)^m (nS)^n = m^m n^n S^{m+n}$.
Solution: for AX, $K_{sp} = 1.5 \times 10^{-5}$, m = 1, and n = 1 so $K_{sp} = 1.5 \times 10^{-5} = S^2$. Rearrange to solve for S. $S = \sqrt{1.5 \times 10^{-5}} = 3.9 \times 10^{-3}$ M. For AX_2, $K_{sp} = 1.5 \times 10^{-5}$, m = 1, and n = 2 so $K_{sp} = 1.5 \times 10^{-5} = 2^2S^3$. Rearrange to solve for S. $S = \sqrt[3]{\frac{1.5 \times 10^{-5}}{4}} = 1.6 \times 10^{-2}$ M. Since 10^{-2} M > 10^{-3} M, AX_2 has a higher molar solubility.
Check: The units (M) are correct. The more ions that are generated, the greater the molar solubility for the same value of the K_{sp}.

62. **Given:** ionic compound formula and molar solubility (S) **Find:** K_{sp}
Conceptual Plan: The expression of the solubility product constant of A_mX_n is: $K_{sp} = [A^{n+}]^m [X^{m-}]^n$. The molar solubility of a compound, A_mX_n, can be computed directly from K_{sp} by solving for S in the expression: $K_{sp} = (mS)^m (nS)^n = m^m n^n S^{m+n}$.
Solution: AX, $S = 1.35 \times 10^{-4}$ M, m = 1, and n = 1 so $K_{sp} = S^2 = (1.35 \times 10^{-4})^2 = 1.82 \times 10^{-8}$.
for AX_2, $S = 2.25 \times 10^{-4}$ M, m = 1, and n = 2 so $K_{sp} = 2^2S^3 = 2^2\,(2.25 \times 10^{-4})^3 = 4.56 \times 10^{-11}$.
for A_2X, $S = 1.75 \times 10^{-4}$ M, m = 2, and n = 1 so $K_{sp} = 2^2S^3 = 2^2\,(1.75 \times 10^{-4})^3 = 2.14 \times 10^{-11}$. So A_2X has the lowest K_{sp}, since it has a lower S than for AX_2.
Check: The units (none) are correct. The K_{sp} values are much less than one and dependent not only on the value of the solubility, but also the stoichiometry of the ionic compound. The more ions that are generated, the smaller the K_{sp} for the same value of the S.

63. **Given:** $Fe(OH)_2$ in 100.0 mL solution **Find:** grams of $Fe(OH)_2$ **Other:** $K_{sp} = 4.87 \times 10^{-17}$
Conceptual Plan: The expression of the solubility product constant of A_mX_n is: $K_{sp} = [A^{n+}]^m [X^{m-}]^n$. The molar solubility of a compound, A_mX_n, can be computed directly from K_{sp} by solving for S in the expression: $K_{sp} = (mS)^m (nS)^n = m^m n^n S^{m+n}$. Then solve for S, then mL → L then

$$\frac{1\text{ L}}{1000\text{ mL}}$$

S, L → mol $Fe(OH)_2$ → g $Fe(OH)_2$

$$M = \frac{\text{mol}}{\text{L}} \qquad \frac{89.87\text{ g }Fe(OH)_2}{1\text{ mol }Fe(OH)_2}$$

Solution: for $Fe(OH)_2$, $K_{sp} = 4.87 \times 10^{-17}$, A = Fe^{2+}, m = 1, X = OH^-, and n = 2 so $K_{sp} = 4.87 \times 10^{-17} = 2^2S^3$.

Rearrange to solve for S. $S = \sqrt[3]{\frac{4.87 \times 10^{-17}}{4}} = 2.3\underline{0}050 \times 10^{-6}$ M. Then $100.0\ \cancel{\text{mL}} \times \frac{1\text{ L}}{1000\ \cancel{\text{mL}}} = 0.1000$ L

then

$$\frac{2.3\underline{0}050 \times 10^{-6}\text{ mol }Fe(OH)_2}{1\ \cancel{\text{L}}} \times 0.1000\ \cancel{\text{L}} = 2.3\underline{0}050 \times 10^{-7}\ \cancel{\text{mol }Fe(OH)_2} \times \frac{89.87\text{ g }Fe(OH)_2}{1\ \cancel{\text{mol }Fe(OH)_2}} = 2.07 \times 10^{-5}\text{ g }Fe(OH)_2.$$

Check: The units (g) are correct. The solubility rules from Chapter 4 (most hydroxides are insoluble) suggest that very little $Fe(OH)_2$ will dissolve, so the magnitude of the answer is not surprising.

64. **Given:** 3.91 mg CuCl in 100.0 mL solution **Find:** K_{sp}
Conceptual Plan: mL → L then mg CuCl → g CuCl → mol CuCl then L, mol CuCl → S then

$$\frac{1\text{ L}}{1000\text{ mL}} \qquad \frac{1\text{ g CuCl}}{1000\text{ mg CuCl}} \quad \frac{1\text{ mol CuCl}}{99.00\text{ g CuCl}} \qquad M = \frac{\text{mol}}{\text{L}}$$

The expression of the solubility product constant of A_mX_n is: $K_{sp} = [A^{n+}]^m [X^{m-}]^n$. The molar solubility of a compound, A_mX_n, can be computed directly from K_{sp} by solving for S in the expression: $K_{sp} = (mS)^m (nS)^n = m^m n^n S^{m+n}$.

Solution: $100.0\ \cancel{\text{mL}} \times \frac{1\text{ L}}{1000\ \cancel{\text{mL}}} = 0.1000$ L then

$3.91 \text{ mg CuCl} \times \frac{1 \text{ g CuCl}}{1000 \text{ mg CuCl}} \times \frac{1 \text{ mol CuCl}}{99.00 \text{ g CuCl}} = 3.94950 \times 10^{-5} \text{ mol CuCl}$ then

$\frac{3.94950 \times 10^{-5} \text{ mol CuCl}}{0.1000 \text{ L}} = 3.94950 \times 10^{-4} \text{ M CuCl} = \text{S}$ then for CuCl A = Cu^+, m = 1, X = Cl^-, and n = 1 so

$K_{sp} = S^2 = = (3.94950 \times 10^{-4})^2 = 1.56 \times 10^{-7}$.

Check: The units (none) are correct. The value of $K_{sp} << 1$ since only mg dissolve in a liter of solution. The K_{sp} is not too low since CuCl dissociates into only 2 ions and S is 10^{-4}.

65. a) **Given:** BaF_2 **Find:** molar solubility (S) in pure water **Other:** K_{sp} (BaF_2) = 2.45×10^{-5}

Conceptual Plan: The expression of the solubility product constant of A_mX_n is: $K_{sp} = [A^{n+}]^m [X^{m-}]^n$. The molar solubility of a compound, A_mX_n, can be computed directly from K_{sp} by solving for S in the expression: $K_{sp} = (mS)^m (nS)^n = m^m n^n S^{m+n}$.

Solution: BaF_2, $K_{sp} = 2.45 \times 10^{-5}$, A = Ba^{2+}, m = 1, X = F^-, and n = 2 so $K_{sp} = 2.45 \times 10^{-5} = 2^2 S^3$.

Rearrange to solve for S. $S = \sqrt[3]{\frac{2.45 \times 10^{-5}}{4}} = 1.83 \times 10^{-2}$ M.

b) **Given:** BaF_2 **Find:** molar solubility (S) in 0.10 M $Ba(NO_3)_2$ **Other:** K_{sp} (BaF_2) = 2.45×10^{-5}

Conceptual Plan: M $Ba(NO_3)_2$ → M Ba^{2+} then M Ba^{2+}, K_{sp} → S

$Ba(NO_3)_2\ (s) \rightarrow Ba^{2+}\ (aq) + 2\ NO_3^-\ (aq)$ ICE Chart

Solution: Since 1 Ba^{2+} ion is generated for each $Ba(NO_3)_2$, $[Ba^{2+}]$ = 0.10 M.

$BaF_2\ (s) \rightleftharpoons$	$Ba^{2+}\ (aq)$	$+ 2\ F^-(aq)$
Initial	0.10	0.00
Change	S	$2S$
Equil	$0.10 + S$	$2S$

K_{sp} (BaF_2) = $[Ba^{2+}][F^-]^2 = 2.45 \times 10^{-5} = (0.10 + S)(2S)^2$.

Assume $S << 0.10$, $2.45 \times 10^{-5} = (0.10)(2S)^2$, and $S = 7.83 \times 10^{-3}$ M. Confirm that assumption is valid

$\frac{7.83 \times 10^{-3}}{0.10} \times 100\ \% = 7.8\ \% > 5\ \%$ so assumption is not valid. Since expanding the expression will give a third order polynomial, that is not easily solved directly. Solve by successive approximations. Substitute $S = 7.83 \times 10^{-3}$ M for the S term that is part of a sum (i.e., the one in $(0.10 + S)$). Thus, $2.45 \times 10^{-5} = (0.10 + 7.83 \times 10^{-3})(2S)^2$ and $S = 7.53 \times 10^{-3}$ M. Substitute this new S value again. Thus, $2.45 \times 10^{-5} = (0.10 + 7.53 \times 10^{-3})(2S)^2$ and $S = 7.55 \times 10^{-3}$ M. Substitute this new S value again. Thus, $2.45 \times 10^{-5} = (0.10 + 7.55 \times 10^{-3})(2S)^2$ and $S = 7.55 \times 10^{-3}$ M. So the solution has converged and $S = 7.55 \times 10^{-3}$ M.

c) **Given:** BaF_2 **Find:** molar solubility (S) in 0.15 M NaF **Other:** K_{sp} (BaF_2) = 2.45×10^{-5}

Conceptual Plan: M NaF → M F^- then M F^-, K_{sp} → S

$NaF\ (s) \rightarrow Na^+\ (aq) + F^-\ (aq)$ ICE Chart

Solution: Since 1 F^- ion is generated for each NaF, $[F^-]$ = 0.15 M.

$BaF_2\ (s) \rightleftharpoons$	$Ba^{2+}\ (aq)$	$+2\ F^-(aq)$
Initial	0.00	0.15
Change	S	$2S$
Equil	S	$0.15 + 2S$

K_{sp} (BaF_2) = $[Ba^{2+}][F^-]^2 = 2.45 \times 10^{-5} = (S)(0.15 + 2S)^2$.

Since $2S << 0.15$, $2.45 \times 10^{-5} = (S)(0.15)^2$, and $S = 1.09 \times 10^{-3}$ M. Confirm that assumption is valid

$\frac{2\ (1.09 \times 10^{-3})}{0.10} \times 100\ \% = 2.2\ \% < 5\ \%$ so assumption is valid.

Check: The units (M) are correct. The solubility of the BaF_2 decreases in the presence of a common ion. The effect of the anion is greater because the K_{sp} expression has the anion concentration squared.

66. a) **Given:** CuS **Find:** molar solubility (S) in pure water **Other:** K_{sp} (CuS) = 1.27×10^{-36}

Conceptual Plan: The expression of the solubility product constant of A_mX_n is: $K_{sp} = [A^{n+}]^m [X^{m-}]^n$. The molar solubility of a compound, A_mX_n, can be computed directly from K_{sp} by solving for S in the expression: $K_{sp} = (mS)^m (nS)^n = m^m n^n S^{m+n}$.

Solution: CuS, $K_{sp} = 1.27 \times 10^{-36}$, A = Cu^{2+}, m = 1, X = S^{2-}, and n = 1 so $K_{sp} = 1.27 \times 10^{-36} = S^2$.

Rearrange to solve for S. $S = 1.13 \times 10^{-18}$ M.

b) **Given:** CuS **Find:** molar solubility (S) in 0.25 M $CuCl_2$ **Other:** K_{sp} (CuS) = 1.27×10^{-36}

Conceptual Plan: **M $CuCl_2$ → M Cu^{2+} then M Cu^{2+}, K_{sp} → S**

$CuCl_2\,(s) \rightarrow Cu^{2+}\,(aq) + 2\,Cl^-\,(aq)$ ICE Chart

Solution: Since 1 Cu^{2+} ion is generated for each $CuCl_2$, $[Cu^{2+}] = 0.25$ M.

$CuS(s) \rightleftharpoons$	$Cu^{2+}(aq)$ +	$S^{2-}(aq)$
Initial	0.25	0.00
Change	S	S
Equil	$0.25 + S$	S

K_{sp} (CuS) = $[Cu^{2+}][S^{2-}] = 1.27 \times 10^{-36} = (0.25 + S)\,S$.

Assume $S << 0.25$, $1.27 \times 10^{-36} = (0.25)\,S$, and $S = 5.08 \times 10^{-36}$ M. Confirm that assumption is valid

$$\frac{5.08 \times 10^{-36}}{0.25} \times 100\,\% = 2.0 \times 10^{-33}\,\% << 5\,\%$$ so assumption is valid.

c) **Given:** CuS **Find:** molar solubility (S) in 0.20 M K_2S **Other:** K_{sp} (CuS) = 1.27×10^{-36}

Conceptual Plan: **M K_2S → M S^{2-} then M S^{2-}, K_{sp} → S**

$K_2S\,(s) \rightarrow 2\,K^+\,(aq) + S^{2-}\,(aq)$ ICE Chart

Solution: Since 1 S^{2-} ion is generated for each K_2S, $[S^{2-}] = 0.20$ M.

$CuS(s) \rightleftharpoons$	$Cu^{2+}(aq)$ +	$S^{2-}(aq)$
Initial	0.00	0.20
Change	S	S
Equil	S	$0.20 + S$

K_{sp} (CuS) = $[Cu^{2+}][S^{2-}] = 1.27 \times 10^{-36} = (S)(0.20 + S)$.

Since $S << 0.20$, $1.27 \times 10^{-36} = (S)(0.20)$, and $S = 6.35 \times 10^{-36}$ M. Confirm that assumption is valid

$$\frac{6.35 \times 10^{-36}}{0.20} \times 100\,\% = 3.2 \times 10^{-33}\,\% << 5\,\%$$ so assumption is valid.

Check: The units (M) are correct. The solubility of the CuS decreases in the presence of a common ion.

67. **Given:** $Ca(OH)_2$ **Find:** molar solubility (S) in buffers at a) pH = 4, b) pH = 7, and c) pH = 9

Other: K_{sp} ($Ca(OH)_2$) = 4.68×10^{-6}

Conceptual Plan: **pH → $[H_3O^+]$ → $[OH^-]$ then M OH^-, K_{sp} → S**

$[H_3O^+] = 10^{-pH}$ $K_w = [H_3O^+][OH^-]$ set up ICE table

Solution:

a) $pH = 4$, so $[H_3O^+] = 10^{-pH} = 10^{-4} = 1 \times 10^{-4}$ M then $K_w = [H_3O^+][OH^-]$ so

$$[OH^-] = \frac{K_w}{[H_3O^+]} = \frac{1.0 \times 10^{-14}}{1 \times 10^{-4}} = 1 \times 10^{-10}\text{ M then}$$

$Ca(OH)_2\,(s) \rightleftharpoons$	$Ca^{2+}\,(aq)$ +	$2\,OH^-(aq)$
Initial	0.00	1×10^{-10}
Change	S	–
Equil	S	1×10^{-10}

K_{sp} ($Ca(OH)_2$) = $[Ca^{2+}][OH^-]^2 = 4.68 \times 10^{-6} = S\,(1 \times 10^{-10})^2$.and $S = 5 \times 10^{14}$ M.

b) $pH = 7$, so $[H_3O^+] = 10^{-pH} = 10^{-7} = 1 \times 10^{-7}$ M then $K_w = [H_3O^+][OH^-]$ so

$$[OH^-] = \frac{K_w}{[H_3O^+]} = \frac{1.0 \times 10^{-14}}{1 \times 10^{-7}} = 1 \times 10^{-7}\text{ M then}$$

$Ca(OH)_2\,(s) \rightleftharpoons$	$Ca^{2+}\,(aq)$ +	$2OH^-(aq)$
Initial	0.00	1×10^{-7}
Change	S	–
Equil	S	1×10^{-7}

K_{sp} ($Ca(OH)_2$) = $[Ca^{2+}][OH^-]^2 = 4.68 \times 10^{-6} = S\,(1 \times 10^{-7})^2$ and $S = 5 \times 10^{8}$ M.

c) $pH = 9$, so $[H_3O^+] = 10^{-pH} = 10^{-9} = 1 \times 10^{-9}$ M then $K_w = [H_3O^+][OH^-]$ so

	$Ca(OH)_2(s) \rightleftharpoons$	$Ca^{2+}(aq)$	$+ 2\,OH^-(aq)$
$[OH^-] = \dfrac{K_w}{[H_3O^+]} = \dfrac{1.0 \times 10^{-14}}{1 \times 10^{-9}} = 1 \times 10^{-5}$ M then	*Initial*	0.00	1×10^{-5}
	Change	S	–
	Equil	S	1×10^{-5}

$K_{sp}(Ca(OH)_2) = [Ca^{2+}][OH^-]^2 = 4.68 \times 10^{-6} = S(1 \times 10^{-5})^2$. and $S = 5 \times 10^4$ M.

Check: The units (M) are correct. The solubility of the $Ca(OH)_2$ decreases as the pH increases (and the hydroxide ion concentration increases). Realize that these molar solubilities are not achievable because the density of pure $Ca(OH)_2$ is ~ 30 M. The bottom line is that as long as the hydroxide concentration can be controlled with a buffer, the $Ca(OH)_2$ will be very soluble.

68. **Given:** $Mg(OH)_2$ in 1.00×10^2 mL solution **Find:** grams of $Mg(OH)_2$ in pure water and buffer at pH = 10 **Other:** $K_{sp}(Mg(OH)_2) = 2.06 \times 10^{-13}$

Conceptual Plan: For pure water:

The expression of the solubility product constant of A_mX_n is: $K_{sp} = [A^{n+}]^m [X^{m-}]^n$. The molar solubility of a compound, A_mX_n, can be computed directly from K_{sp} by solving for S in the expression: $K_{sp} = (mS)^m (nS)^n = m^m n^n S^{m+n}$. Then mL → L then

$\dfrac{1\text{ L}}{1000\text{ mL}}$

***S*, L → mol $Mg(OH)_2$ → g $Mg(OH)_2$. For buffer solution: pH → $[H_3O^+]$ → $[OH^-]$ then**

$M = \dfrac{\text{mol}}{\text{L}}$ $\dfrac{58.33\text{ g } Mg(OH)_2}{1\text{ mol } Mg(OH)_2}$ $[H_3O^+] = 10^{-pH}$ $K_w = [H_3O^+][OH^-]$

M OH^-, K_{sp} → *S* then *S*, L → mol $Mg(OH)_2$ → g $Mg(OH)_2$

set up ICE table $M = \dfrac{\text{mol}}{\text{L}}$ $\dfrac{58.33\text{ g } Mg(OH)_2}{1\text{ mol } Mg(OH)_2}$

Solution: for pure water, $K_{sp} = 2.06 \times 10^{-13}$, A = Mg^{2+}, m = 1, X = OH^-, and n = 2 so $K_{sp} = 2.06 \times 10^{-13} = 2^2S^3$.

Rearrange to solve for S. $S = \sqrt[3]{\dfrac{2.06 \times 10^{-13}}{4}} = 3.7\underline{2}051 \times 10^{-5}$ M. Then $1.00 \times 10^2 \cancel{\text{mL}} \times \dfrac{1\text{ L}}{1000\cancel{\text{ mL}}} = 0.100$ L

then $\dfrac{3.7\underline{2}051 \times 10^{-5}\text{ mol } Mg(OH)_2}{1\cancel{\text{ L}}} \times 0.100\cancel{\text{ L}} = 3.7\underline{2}051 \times 10^{-6}\cancel{\text{ mol } Mg(OH)_2} \times \dfrac{58.33\text{ g } Mg(OH)_2}{1\cancel{\text{ mol } Mg(OH)_2}} =$

2.17×10^{-4}g $Mg(OH)_2$

for pH = 10, so $[H_3O^+] = 10^{-pH} = 10^{-10} = 1 \times 10^{-10}$ M then $K_w = [H_3O^+][OH^-]$ so

	$Mg(OH)_2(s) \rightleftharpoons$	$Mg^{2+}(aq)$	$+ OH^-(aq)$
$[OH^-] = \dfrac{K_w}{[H_3O^+]} = \dfrac{1.0 \times 10^{-14}}{1 \times 10^{-10}} = 1 \times 10^{-4}$ M then	*Initial*	0.00	1×10^{-4}
	Change	S	–
	Equil	S	1×10^{-4}

$K_{sp}(Mg(OH)_2) = [Mg^{2+}][OH^-]^2 = 2.06 \times 10^{-13} = S(1 \times 10^{-4})^2$ and $S = \underline{2}.06 \times 10^{-5}$ M. Then

$\dfrac{\underline{2}.06 \times 10^{-5}\text{ mol } Mg(OH)_2}{1\cancel{\text{ L}}} \times 0.100\cancel{\text{ L}} = \underline{2}.06 \times 10^{-6}\cancel{\text{ mol } Mg(OH)_2} \times \dfrac{58.33\text{ g } Mg(OH)_2}{1\cancel{\text{ mol } Mg(OH)_2}} = 1 \times 10^{-4}\text{g } Mg(OH)_2$.

Check: The units (M) are correct. The solubility of the $Mg(OH)_2$ decreases as the pH increases (and the hydroxide ion concentration increases).

69. a) $BaCO_3$ will be more soluble in acidic solutions because CO_3^{2-} is basic. In acidic solutions it can be converted to HCO_3^- and $H_2CO_3^{2-}$. These species are not CO_3^{2-} so they do not appear in the K_{sp} expression.

b) CuS will be more soluble in acidic solutions because S^{2-} is basic. In acidic solutions it can be converted to HS^- and H_2S^{2-}. These species are not S^{2-} so they do not appear in the K_{sp} expression.

c) AgCl will not be more soluble in acidic solutions because Cl^- will not react with acidic solutions, because HCl is a strong acid.

d) PbI_2 will not be more soluble in acidic solutions because I^- will not react with acidic solutions, because HI is a strong acid.

70. a) Hg_2Br_2 will not be more soluble in acidic solutions because Br^- will not react with acidic solutions, because HBr is a strong acid.

b) $Mg(OH)_2$ will be more soluble in acidic solutions because OH^- is basic. In acidic solutions it can be converted to H_2O. This species is not OH^- and so it does not appear in the K_{sp} expression.

c) $CaCO_3$ will be more soluble in acidic solutions because CO_3^{2-} is basic. In acidic solutions it can be converted to HCO_3^- and $H_2CO_3^{2-}$. These species are not CO_3^{2-} so they do not appear in the K_{sp} expression.

d) AgI will not be more soluble in acidic solutions because I^- will not react with acidic solutions, because HI is a strong acid.

71. **Given:** 0.015 M NaF and 0.010 M $Ca(NO_3)_2$ **Find:** will a precipitate form, if so, identify it
Other: K_{sp} (CaF_2) = 1.46×10^{-10}
Conceptual Plan: Look at all possible combinations and consider the solubility rules from Chapter 4. Salts of alkali metals (Na) are very soluble, so NaF and $NaNO_3$ will be very soluble. Nitrate compounds are very soluble so $NaNO_3$ will be very soluble. The only possibility for a precipitate is CaF_2. Determine if a precipitate will form by determining the concentration of the Ca^{2+} and F^- in solution. Then compute the reaction quotient, Q. If Q > K_{sp} then a precipitate will form.
Solution: Since the only possible precipitate is CaF_2, calculate the concentrations of Ca^{2+} and F^-.
NaF (*s*) → Na^+ (*aq*) + F^- (*aq*). Since 1 F^- ion is generated for each NaF, [F^-] = 0.015 M.
$Ca(NO_3)_2$ (*s*) → Ca^{2+} (*aq*) + 2 NO_3^- (*aq*). Since 1 Ca^{2+} ion is generated for each $Ca(NO_3)_2$, [Ca^{2+}] = 0.010 M.
Then calculate Q (CaF_2), A = Ca^{2+}, m = 1, X = F^-, and n = 2 Since Q= $[A^{n+}]^m [X^{m-}]^n$, then
Q (CaF_2) = $[Ca^{2+}] [F^-]^2 = (0.010) (0.015)^2 = 2.3 \times 10^{-6} > 1.46 \times 10^{-10} = K_{sp}$ (CaF_2), so a precipitate will form.
Check: The units (none) are correct. The solubility of the CaF_2 is low, and the concentration of ions are extremely large compared to the K_{sp}, so a precipitate will form.

72. **Given:** 0.013 M KBr and 0.0035 M $Pb(C_2H_3O_2)_2$ **Find:** will a precipitate form, if so, identify it
Other: K_{sp} ($PbBr_2$) = 4.67×10^{-6}
Conceptual Plan: Look at all possible combinations and consider the solubility rules from Chapter 4. Salts of alkali metals (K) are very soluble, so KBr and $KC_2H_3O_2$ will be very soluble. Acetate compounds are very soluble so $Pb(C_2H_3O_2)_2$ and $KC_2H_3O_2$ will be very soluble. The only possibility for a precipitate is $PbBr_2$. Determine if a precipitate will form by determining the concentration of the Pb^{2+} and Br^- in solution. Then compute the reaction quotient, Q. If Q > K_{sp} then a precipitate will form.
Solution: Since the only possible precipitate is $PbBr_2$, calculate the concentrations of Pb^{2+} and Br^-.
KBr (*s*) → K^+ (*aq*) + Br^- (*aq*). Since 1 Br^- ion is generated for each KBr, [Br^-] = 0.013 M. $Pb(C_2H_3O_2)_2$ (*s*) → Ca^{2+} (*aq*) + 2 $C_2H_3O_2^-$ (*aq*). Since 1 Pb^{2+} ion is generated for each $Pb(C_2H_3O_2)_2$, [Pb^+] = 0.0035 M. Then calculate Q ($PbBr_2$), A = Pb^{2+}, m = 1, X = Br^-, and n = 2 Since Q= $[A^{n+}]^m [X^{m-}]^n$, then Q ($PbBr_2$) = $[Pb^{2+}] [Br^-]^2 = (0.013) (0.0035)^2 = 1.6 \times 10^{-7} < 4.67 \times 10^{-6} = K_{sp}$ ($PbBr_2$), so a precipitate will not form.
Check: The units (none) are correct. The K_{sp} of the $PbBr_2$ is not too low compared to the solution ion concentrations, so a precipitate will not form.

73. **Given:** 75.0 mL of NaOH with pOH = 2.58 and 125.0 mL of 0.0018 M $MgCl_2$
Find: will a precipitate form, if so, identify it **Other:** K_{sp} ($Mg(OH)_2$) = 2.06×10^{-13}
Conceptual Plan: Look at all possible combinations and consider the solubility rules from Chapter 4. Salts of alkali metals (Na) are very soluble, so NaOH and NaCl will be very soluble. Chloride compounds are generally very soluble so $MgCl_2$ and NaCl will be very soluble. The only possibility for a precipitate is $Mg(OH)_2$. Determine if a precipitate will form by determining the concentration of the Mg^{2+} and OH^- in solution. Since pH, not NaOH concentration, is given pOH → [OH^-] then

$[OH^-] = 10^{-pOH}$

mix solutions and calculate diluted concentrations mL NaOH, mL $MgCl_2$ → mL total then

mL NaOH + mL $MgCl_2$ = total mL

mL, initial M → final M then compute the reaction quotient, Q.

$M_1 V_1 = M_2 V_2$

If Q > K_{sp} then a precipitate will form.

Solution: Since the only possible precipitate is $Mg(OH)_2$, calculate the concentrations of Mg^{2+} and OH^-. for NaOH at pOH = 2.58, so $[OH^-] = 10^{-pOH} = 10^{-2.58} = 2.\underline{6}3027 \times 10^{-3}$ M and $MgCl_2$ (s) → Mg^{2+} (aq) + 2 Cl^- (aq). Since 1 Mg^{2+} ion is generated for each $MgCl_2$, $[Mg^{2+}] = 0.0018$ M. Then total mL = mL NaOH + mL $MgCl_2$ = 75.0 mL + 125.0 mL = 200.0 mL. Then $M_1V_1 = M_2V_2$, rearrange to solve for M_2. $M_2 = M_1\frac{V_1}{V_2} = 2.\underline{6}3027 \times 10^{-3}\text{ M OH}^- \times \frac{75.0\text{ mL}}{200.0\text{ mL}} = 9.\underline{8}635 \times 10^{-4}\text{ M OH}^-$ and

$M_2 = M_1\frac{V_1}{V_2} = 0.0018\text{ M Mg}^+ \times \frac{125.0\text{ mL}}{200.0\text{ mL}} = 1.\underline{1}25 \times 10^{-3}\text{ M Mg}^{2+}$. Calculate Q ($Mg(OH)_2$), A = Mg^{2+}, m = 1, X = OH^-, and n = 2 Since Q= $[A^{n+}]^m [X^{m-}]^n$, then Q ($Mg(OH)_2$) = $[Mg^{2+}][OH^-]^2$ = $(1.\underline{1}25 \times 10^{-3})(9.\underline{8}635 \times 10^{-4})^2 = 1.1 \times 10^{-9} > 2.06 \times 10^{-13} = K_{sp}$ ($Mg(OH)_2$), so a precipitate will form.

Check: The units (none) are correct. The solubility of the $Mg(OH)_2$ is low, and the NaOH (a base) is high enough that the product of the concentration of ions are large compared to the K_{sp}, so a precipitate will form.

74. **Given:** 175.0 mL of 0.0055 M KCl and 145.0 mL of 0.0015 M $AgNO_3$ **Find:** will a precipitate form, if so, identify it **Other:** K_{sp} (AgCl) = 1.77×10^{-10}
Conceptual Plan: Look at all possible combinations and consider the solubility rules from Chapter 4. Salts of alkali metals (K) are very soluble, so KCl and KNO_3 will be very soluble. Nitrate compounds are very soluble so KNO_3 and Ag NO_3 will be very soluble. The only possibility for a precipitate is AgCl. Determine if a precipitate will form by determining the concentration of the Ag^+ and Cl^- in solution. Mix solutions and calculate diluted concentrations mL KCl, mL $AgNO_3$ → mL total then

mL KCl + mL $AgNO_3$ = total mL

mL, initial M → final M then compute the reaction quotient, Q. If Q > K_{sp} then a precipitate will form.

$M_1V_1 = M_2V_2$

Solution: Since the only possible precipitate is AgCl, calculate the concentrations of Ag^+ and Cl^-. KCl (*s*) → K^+ (*aq*) + Cl^- (*aq*). Since 1 Cl^- ion is generated for each AgCl, $[Cl^-] = 0.0055$ M and $AgNO_3$ (*s*) → Ag^+ (*aq*) + NO_3^- (*aq*). Since 1 Ag^+ ion is generated for each $AgNO_3$, $[Ag^+] = 0.0015$ M. Then total mL = mL KCl + mL $AgNO_3$ = 175.0 mL + 145.0 mL = 320.0. Then $M_1V_1 = M_2V_2$, rearrange to solve for M_2. $M_2 = M_1\frac{V_1}{V_2} = 0.0055\text{ M Cl}^- \times \frac{175.0\text{ mL}}{320.0\text{ mL}} = 0.00300\underline{7}81\text{ M Cl}^-$ and

$M_2 = M_1\frac{V_1}{V_2} = 0.0015\text{ M Ag}^+ \times \frac{145.0\text{ mL}}{320.0\text{ mL}} = 0.00067\underline{9}69\text{ M Ag}^+$. Calculate Q ($AgCl_2$), A = Ag^+, m = 1, X = Cl^-, and n = 1 Since Q= $[A^{n+}]^m [X^{m-}]^n$, then Q (AgCl) = $[Ag^+][Cl^-]$ = (0.00067969)(0.00300781) = 2.0×10^{-6} > $1.77 \times 10^{-10} = K_{sp}$ (AgCl), so a precipitate will form.

Check: The units (none) are correct. The solubility of the AgCl is low, and the concentrations of the ions are high enough that the product of the concentration of ions is very large compared to the K_{sp}, so a precipitate will form.

75. **Given:** KOH as precipitation agent in a) 0.015 M $CaCl_2$, b) 0.0025 M $Fe(NO_3)_2$, and c) 0.0018 M $MgBr_2$
Find: concentration of KOH necessary to form a precipitate
Other: K_{sp} ($Ca(OH)_2$) = 4.68×10^{-6}, K_{sp} ($Fe(OH)_2$) = 4.87×10^{-17}, K_{sp} ($Mg(OH)_2$) = 2.06×10^{-13}
Conceptual Plan: The solubility rules from Chapter 4 state that most hydroxides are insoluble, so all precipitate will be hydroxides. Determine the concentration of the cation in solution. Since all metals have an oxidation state of +2 and $[OH^-]$ = [KOH], all of the K_{sp} = [cation] $[KOH]^2$ and so $[KOH] = \sqrt{\frac{K_{sp}}{[cation]}}$.

Solution:

a) $CaCl_2$ (*s*) → Ca^{2+} (*aq*) + 2 Cl^- (*aq*). Since 1 Ca^{2+} ion is generated for each $CaCl_2$, $[Ca^{2+}] = 0.015$ M.

Then $[KOH] = \sqrt{\frac{K_{sp}}{[cation]}} = \sqrt{\frac{4.68 \times 10^{-6}}{0.015}} = 0.018\text{ M KOH}$.

b) $Fe(NO_3)_2$ (*s*) → Fe^{2+} (*aq*) + 2 NO_3^- (*aq*). Since 1 Fe^{2+} ion is generated for each $Fe(NO_3)_2$, $[Fe^{2+}]$ = 0.0025 M. Then $[KOH] = \sqrt{\frac{K_{sp}}{[cation]}} = \sqrt{\frac{4.87 \times 10^{-17}}{0.0025}} = 1.4 \times 10^{-7}\text{ M KOH}$.

c) $MgBr_2$ (*s*) → Mg^{2+} (*aq*) + 2 Br^- (*aq*). Since 1 Mg^{2+} ion is generated for each $MgBr_2$, $[Mg^{2+}] = 0.0018$ M. Then $[KOH] = \sqrt{\frac{K_{sp}}{[cation]}} = \sqrt{\frac{2.06 \times 10^{-13}}{0.0018}} = 1.1 \times 10^{-5}$ M KOH.

Check: The units (none) are correct. Since all cations have an oxidation state of +2, it can be seen that the [KOH] needed to precipitate the hydroxide is lower the smaller the K_{sp}.

76. **Given:** solution and precipitation agent pairs a) 0.035 M $BaNO_3$: NaF, b) 0.085 M CaI_2 : K_2SO_4, and c) 0.0018 M $AgNO_3$:RbCl **Find:** concentration of precipitation agent necessary to form a precipitate
Other: K_{sp} (BaF_2) = 2.45 x 10^{-5}, K_{sp} ($CaSO_4$) = 7.10 x 10^{-5}, K_{sp} (AgCl) = 1.77 x 10^{-10}
Conceptual Plan: Use the solubility rules from Chapter 4 to decide on precipitate that will form. Determine the concentration of the cation in solution. The solubility product constant (K_{sp}) is the equilibrium expression for a chemical equation representing the dissolution of an ionic compound. The expression of the solubility product constant of A_mX_n is: $K_{sp} = [A^{n+}]^m [X^{m-}]^n$. Substitute in concentration of cation and solve for concentration of anion.
Solution:
a) Salts of alkali metals (Na) are very soluble, so NaF and $NaNO_3$ will be very soluble. Nitrate compounds are very soluble so $Ba(NO_3)_2$ and $NaNO_3$ will be very soluble. The only possibility for a precipitate is BaF_2. $Ba(NO_3)_2$ (*s*) → Ba^{2+} (*aq*) + 2 NO_3^- (*aq*). Since 1 Ba^{2+} ion is generated for each $Ba(NO_3)_2$, $[Ba^{2+}]$ = 0.035 M. Then derive expression for K_{sp} (BaF_2), A = Ba^{2+}, m = 1, X = F^-, and n = 2 Since K_{sp} = $[Ba^{2+}] [F^-]^2$, then K_{sp} (BaF_2) = 2.45 x 10^{-5} = 0.035 $[F^-]^2$. Solve for $[F^-]$. $[F^-]$ = 0.026 M F^-. Since NaF (*s*) → Na^+ (*aq*) + F^- (*aq*), 1 F^- ion is generated for each NaF, [NaF] = 0.026 M NaF.

b) Salts of alkali metals (K) are very soluble, so KI and K_2SO_4 will be very soluble. Sulfate compounds are very soluble, with the exception of a few cations, with Ca^{2+} being one of the insoluble exceptions. The only possibility for a precipitate is $CaSO_4$. CaI_2 (*s*) → Ca^{2+} (*aq*) + 2 I^- (*aq*). Since 1 Ca^{2+} ion is generated for each CaI_2, $[Ca^{2+}]$ = 0.085 M. Then derive expression for K_{sp} ($CaSO_4$), A = Ca^{2+}, m = 1, X = SO_4^{2-}, and n = 1 Since K_{sp} = $[Ca^{2+}] [SO_4^{2-}]$, then K_{sp} ($CaSO_4$) = 7.10 x 10^{-5} = 0.085 $[SO_4^{2-}]$. Solve for $[SO_4^{2-}]$. $[SO_4^{2-}]$ = 0.00084 M SO_4^{2-}. Since K_2SO_4 (*s*) → 2 K^+ (*aq*) + SO_4^{2-} (*aq*). Since 1 SO_4^{2-} ion is generated for each K_2SO_4, $[K_2SO_4]$ = 0.00084 M K_2SO_4.

c) Salts of alkali metals (Rb) are very soluble, so RbCl and $RbNO_3$ will be very soluble. Nitrate compounds are very soluble, so $RbNO_3$ and $AgNO_3$ are very soluble. The only possibility for a precipitate is AgCl, which is insoluble according to the solubility rules. Then $AgNO_3$ (*s*) → Ag^+ (*aq*) + NO_3^- (*aq*). Since 1 NO_3^- ion is generated for each $AgNO_3$, $[Ag^+]$ = 0.0018 M. Then derive expression for K_{sp} (AgCl), A = Ag^+, m = 1, X = Cl^-, and n = 1 Since K_{sp} = $[Ag^+] [Cl^-]$, then K_{sp} (AgCl) = 1.77 x 10^{-10} = 0.0018 $[Cl^-]$. Solve for $[Cl^-]$. $[Cl^-]$ = 9.8 x 10^{-8} M Cl^-. Since RbCl (*s*) → Rb^+ (*aq*) + Cl^- (*aq*). Since 1 Cl^- ion is generated for each RbCl, [RbCl] = 9.8 x 10^{-8} M RbCl.

Check: The units (M) are correct. Comparing part a) and part b) the effect of the stoichiometry of the precipitate is seen and the concentration of the precipitation agent is much lower. Looking at part c) the concentration of the precipitation agent is so low because the K_{sp} is so small.

77. **Given:** solution with 1.1 x 10^{-3} M $Zn(NO_3)_2$ and 0.150 M NH_3 **Find:** $[Zn^{2+}]$ at equilibrium
Other: K_f ($Zn(NH_3)_4^{2+}$) = 2.0 x 10^9
Conceptual Plan: Write balanced equation and expression for K_f. Use initial concentrations to set up ICE table. Since the K_f is so large, assume that reaction essentially goes to completion. Solve for $[Zn^{2+}]$ at equilibrium.
Solution: $Zn(NO_3)_2$ (*s*) → Zn^{2+} (*aq*) + 2 NO_3^- (*aq*). Since 1 Zn^{2+} ion is generated for each $Zn(NO_3)_2$, $[Zn^{2+}]$ = 1.1 x 10^{-3} M. Balanced equation is:

$Zn^{2+}(aq) + 4\,NH_3(aq) \rightleftharpoons Zn(NH_3)_4^{2+}(aq)$ Set up ICE table with initial concentrations

	$[Zn^{2+}]$	$[NH_3]$	$[Zn(NH_3)_4^{2+}]$	
Initial	1.1 x 10^{-3}	0.150	0.00	Since K_f is so large and since initially $[NH_3]>4[Zn^{2+}]$ the reaction essentially goes to
Change	≈ 1.1 x 10^{-3}	≈ − 4(1.1 x 10^{-3})	≈ 1.1 x 10^{-3}	completion then write equilibrium expression
Equil	*x*	0.1456	1.1 x 10^{-3}	and solve for *x*.

$$K_f = \frac{[Zn(NH_3)_4^{2+}]}{[Zn^{2+}][NH_3]^4} = 2.0 \times 10^9 = \frac{1.1 \times 10^{-3}}{x(0.1456)^4}$$ So $x = 1.2 \times 10^{-9}$ M Zn^{2+}. Since *x* is insignificant compared to the initial concentration, the assumption is valid.

Check: The units (M) are correct. Since K_f is so large, the reaction essentially goes to completion and $[Zn^{2+}]$ is extremely small.

78. **Given:** 120.0 mL of 2.8 x 10^{-3} M $AgNO_3$ mixed with 225.0 mL of 0.10 M NaCN
Find: $[Ag^+]$ at equilibrium **Other:** K_f $(Ag(CN)_2^-) = 1 \times 10^{21}$
Conceptual Plan:
Mix solutions and calculate diluted concentrations mL $AgNO_3$, mL NaCN → mL total
mL $AgNO_3$ + mL NaCN = total mL
then mL, initial M → final M then write balanced equation and expression for K_f.
$M_1 V_1 = M_2 V_2$
Use initial concentrations to set up ICE table. Since the K_f is so large, assume that reaction essentially goes to completion. Solve for $[Ag^+]$ at equilibrium.
Solution: $AgNO_3$ (*s*) → Ag^+ (*aq*) + NO_3^- (*aq*). Since 1 Ag^+ ion is generated for each $AgNO_3$, $[Ag^+] = 2.8 \times 10^{-3}$ M and NaCN (*s*) → Na^+ (*aq*) + CN^- (*aq*). Since 1 CN^- ion is generated for each NaCN, $[CN^-] = 0.10$ M. Then total mL = mL $AgNO_3$ + mL NaCN = 120.0 mL + 225.0 mL = 345.0 mL. Then $M_1 V_1 = M_2 V_2$. Rearrange to solve for M_2. $M_2 = M_1 \frac{V_1}{V_2} = 2.8 \times 10^{-3}\text{ M } Ag^+ \times \frac{120.0\text{ mL}}{345.0\text{ mL}} = 0.00097391\text{ M } Ag^+$ and

$M_2 = M_1 \frac{V_1}{V_2} = 0.10\text{ M } CN^- \times \frac{225.0\text{ mL}}{345.0\text{ mL}} = 0.065217\text{ M } CN^-$. Balanced equation is:

$Ag^+ (aq) + 2\ CN^- (aq) \rightleftharpoons Ag(CN)_2^- (aq)$ Set up ICE table with initial concentrations.

	$[Ag^+]$	$[CN^-]$	$[Ag(CN)_2^-]$	Since K_f is so large and since initially
Initial	0.00097391	0.065217	0.00	$[CN^-] > 2[Ag^+]$ the reaction essentially
Change	≈ 0.00097391	≈ −2(0.00097391)	≈ 0.00097391	goes to completion then write
Equil	*x*	0.063270	0.00097391	equilibrium expression and solve for x.

$K_f = \frac{[Ag(CN)_2^-]}{[Ag^+][CN^-]^2} = 1 \times 10^{21} = \frac{0.00097391}{x(0.063270)^2}$ So $x = 2 \times 10^{-22}$ M Ag^+. Since *x* is insignificant compared to the initial concentration, the assumption is valid.
Check: The units (M) are correct. Since K_f is so large, the reaction essentially goes to completion and $[Ag^+]$ is extremely small.

79. **Given:** 150.0 mL solution of 2.05 g sodium benzoate and 2.47 g benzoic acid **Find:** pH
Other: K_a $(HC_7H_5O_2) = 6.5 \times 10^{-5}$
Conceptual Plan: g $NaC_7H_5O_2$ → mol $NaC_7H_5O_2$ and g $HC_7H_5O_2$ → mol $HC_7H_5O_2$
$\frac{1\text{ mol } NaC_7H_5O_2}{144.11\text{ g } NaC_7H_5O_2}$ $\frac{1\text{ mol } HC_7H_5O_2}{122.13\text{ g } HC_7H_5O_2}$
Since the two components are in the same solution, the ratio of [base]/[acid] = (mol base)/(mol acid). Then K_a, mol $NaC_7H_5O_2$, mol $HC_7H_5O_2$ → pH
$pH = pK_a + \log \frac{[base]}{[acid]}$

Solution: $2.05\text{ g } NaC_7H_5O_2 \times \frac{1\text{ mol } NaC_7H_5O_2}{144.11\text{ g } NaC_7H_5O_2} = 0.0142252\text{ mol } NaC_7H_5O_2$ and

$2.47\text{ g } HC_7H_5O_2 \times \frac{1\text{ mol } HC_7H_5O_2}{122.13\text{ g } HC_7H_5O_2} = 0.0202244\text{ mol } HC_7H_5O_2$ then

$pH = pK_a + \log \frac{[base]}{[acid]} = pK_a + \log \frac{\text{mol base}}{\text{mol acid}} = -\log(6.5 \times 10^{-5}) + \log \frac{0.0142252\text{ mol}}{0.0202244\text{ mol}} = 4.03$.

Check: The units (none) are correct. The magnitude of the answer makes physical sense because the pH is a little lower than the pK_a of the acid because there is more acid than base in the buffer solution.

80. **Given:** 10.0 mL of 17.5 M acetic acid and 5.54 g sodium acetate diluted to 1.50 L **Find:** pH
Other: K_a $(HC_2H_3O_2) = 1.8 \times 10^{-5}$

Conceptual Plan: mL → L then L, initial $HC_2H_3O_2$ M → mol $HC_2H_3O_2$ then

$\frac{1\ L}{1000\ mL}$ $M = \frac{mol}{L}$

g $NaC_2H_3O_2$ → mol $NaC_2H_3O_2$ then since the two components are in the same solution,

$\frac{1\ mol\ NaC_2H_3O_2}{83.04\ g\ NaC_2H_3O_2}$

the ratio of [base]/[acid] = (mol base)/(mol acid). Then K_a, mol $NaC_2H_3O_2$, mol $HC_2H_3O_2$ → pH

$$pH = pK_a + \log\frac{[base]}{[acid]}$$

Solution: $10.0\ \text{mL} \times \frac{1\ L}{1000\ \text{mL}} = 0.0100\ L$ then

$0.0100\ L\ HC_2H_3O_2 \times \frac{17.5\ mol\ HC_2H_3O_2}{1\ L\ HC_2H_3O_2} = 0.175\ mol\ HC_2H_3O_2$ then

$5.54\ g\ NaC_2H_3O_2 \times \frac{1\ mol\ NaC_2H_3O_2}{83.04\ g\ NaC_2H_3O_2} = 0.0667148\ mol\ NaC_2H_3O_2$ then

$$pH = pK_a + \log\frac{[base]}{[acid]} = pK_a + \log\frac{mol\ base}{mol\ acid} = -\log(1.8 \times 10^{-5}) + \log\frac{0.0667148\ mol}{0.175\ mol} = 4.33.$$

Check: The units (none) are correct. The magnitude of the answer makes physical sense because the pH is a little lower than the pK_a of the acid because there is more acid than base in the buffer solution.

81. **Given:** 150.0 mL of 0.25 M $HCHO_2$ and 75.0 ml of 0.20 M NaOH **Find:** pH **Other:** K_a ($HCHO_2$) = 1.8×10^{-4}
Conceptual Plan: In this buffer, the base is generated by converting some of the formic acid to the formate ion. Part I: Stoichiometry:
mL → L then L, initial $HCHO_2$ M → mol $HCHO_2$ then mL → L then

$\frac{1\ L}{1000\ mL}$ $M = \frac{mol}{L}$ $\frac{1\ L}{1000\ mL}$

L, initial NaOH M → mol NaOH then write balanced equation then

$M = \frac{mol}{L}$ $NaOH + HCHO_2 \rightarrow H_2O + NaCHO_2$

mol $HCHO_2$, mol NaOH → mol $NaCHO_2$, mol $HCHO_2$ then

set up stoichiometry table

Part II: Equilibrium:
Since the two components are in the same solution, the ratio of [base]/[acid] = (mol base)/(mol acid). Then K_a, mol $NaCHO_2$, mol $HCHO_2$ → pH

$$pH = pK_a + \log\frac{[base]}{[acid]}$$

Solution: $150.0\ \text{mL} \times \frac{1\ L}{1000\ \text{mL}} = 0.1500\ L$ then

$0.1500\ L\ HCHO_2 \times \frac{0.25\ mol\ HCHO_2}{1\ L\ HCHO_2} = 0.0375\ mol\ HCHO_2$.

Then $75.0\ \text{mL} \times \frac{1\ L}{1000\ \text{mL}} = 0.0750\ L$ then $0.0750\ L\ NaOH \times \frac{0.20\ mol\ NaOH}{1\ L\ NaOH} = 0.015\ mol\ NaOH$ then

set up table to track changes:

	$NaOH\ (aq)$ +	$HCHO_2\ (aq)$ →	$NaCHO_2\ (aq)$ +	$H_2O\ (l)$
Before addition	0.00 mol	0.0375 mol	0.00 mol	–
Addition	0.015 mol	–	–	–
After addition	≈ 0.00 mol	0.0225 mol	0.015 mol	–

Since the amount of NaOH is small, there are significant amounts of both buffer components, so the Henderson–Hasselbalch Equation can be used to calculate the pH.

$$pH = pK_a + \log\frac{[base]}{[acid]} = pK_a + \log\frac{mol\ base}{mol\ acid} = -\log(1.8 \times 10^{-4}) + \log\frac{0.015\ mol}{0.0225\ mol} = 3.57.$$

Check: The units (none) are correct. The magnitude of the answer makes physical sense because the pH is a little lower than the pK_a of the acid because there is more acid than base in the buffer solution.

82. **Given:** 750.0 mL solution of 3.55 g NH_3 and 4.78 g HCl **Find:** pH **Other:** K_b (NH_3) = 1.79 x 10^{-5}
Conceptual Plan: In this buffer, the acid is generated by converting some of the ammonia to the ammonium ion. Part I: Stoichiometry:
g NH_3 → mol NH_3 and g HCl → mol HCl write balanced equation then

$\frac{1 \text{ mol } NH_3}{17.03 \text{ g } NH_3}$ $\frac{1 \text{ mol HCl}}{36.46 \text{ g HCl}}$ $NH_3 + HCl \rightarrow NH_4Cl$

mol NH_3, mol HCl → mol NH_3, mol NH_4Cl then
set up stoichiometry table

Part II: Equilibrium:
K_b → pK_b → pK_a then since the two components are in the same solution,
$pK_b = -\log K_b$ $14 = pK_a + pK_b$

the ratio of [base]/[acid] = (mol base)/(mol acid). Then pK_a, mol NH_3, mol NH_4Cl → pH

$$pH = pK_a + \log\frac{[\text{base}]}{[\text{acid}]}$$

Solution: $3.55 \text{ g } NH_3 \times \frac{1 \text{ mol } NH_3}{17.03 \text{ g } NH_3} = 0.20\underline{8}456 \text{ mol } NH_3$ and

$4.78 \text{ g HCl} \times \frac{1 \text{ mol HCl}}{36.46 \text{ g HCl}} = 0.13\underline{1}103 \text{ mol HCl}$ then set up table to track changes:

	HCl (*aq*) +	NH_3 (*aq*) →	NH_4Cl (*aq*)
Before addition	0.00 mol	0.20$\underline{8}$456 mol	0.00 mol
Addition	0.13$\underline{1}$103 mol	–	–
After addition	≈ 0.00 mol	0.07$\underline{7}$353 mol	0.13$\underline{1}$103 mol

Since the amount of HCl is small, there are significant amounts of both buffer components, so the Henderson–Hasselbalch Equation can be used to calculate the pH.
Since K_b (NH_3) = 1.79 x 10^{-5}, $pK_b = -\log K_b = -\log(1.79 \times 10^{-5}) = 4.75$. Since $14 = pK_a + pK_b$,
$pK_a = 14 - pK_b = 14 - 4.75 = 9.25$ then

$$pH = pK_a + \log\frac{[\text{base}]}{[\text{acid}]} = pK_a + \log\frac{\text{mol base}}{\text{mol acid}} = 9.25 + \log\frac{0.07\underline{7}353 \text{ mol}}{0.13\underline{1}103 \text{ mol}} = 9.02.$$

Check: The units (none) are correct. The magnitude of the answer makes physical sense because the pH is a little lower than the pK_a of the acid because there is more acid than base in the buffer solution.

83. **Given:** 1.0 L of buffer of 0.25 mol NH_3 and 0.25 mol NH_4Cl; adjust to pH = 8.75
Find: mass NaOH or HCl **Other:** K_b (NH_3) = 1.79 x 10^{-5}
Conceptual Plan: To decide which reagent needs to be added to adjust pH, calculate the initial pH. Since the mol NH_3 = mol NH_4Cl, the pH = pK_a so K_b → pK_b → pK_a then
acid = NH_4^+ base = NH_3 $pK_b = -\log K_b$ $14 = pK_a + pK_b$

final pH, pK_a → [NH_3]/[NH_4^+] then [NH_3], L → mol NH_3 and [NH_4^+], L → mol NH_4^+

$pH = pK_a + \log\frac{[\text{base}]}{[\text{acid}]}$ $M = \frac{\text{mol}}{\text{L}}$ $M = \frac{\text{mol}}{\text{L}}$

then write balanced equation then
$H^+ + NH_3 \rightarrow NH_4^+$

mol NH_3, mol NH_4^+, [NH_3]/[NH_4^+] → mol HCl → g HCl
set up stoichiometry table $\frac{36.46 \text{ g HCl}}{1 \text{ mol HCl}}$

Solution: Since K_b (NH_3) = 1.79 x 10^{-5}, $pK_b = -\log K_b = -\log(1.79 \times 10^{-5}) = 4.75$. Since $14 = pK_a + pK_b$,
$pK_a = 14 - pK_b = 14 - 4.75 = 9.25$. Since the desired pH is lower (8.75) HCl (a strong acid) needs to be added. Then $pH = pK_a + \log\frac{[\text{base}]}{[\text{acid}]} = 9.25 + \log\frac{[NH_3]}{[NH_4^+]} = 8.75$. Solve for $\frac{[NH_3]}{[NH_4^+]}$.

$\log \frac{[NH_3]}{[NH_4^+]} = 8.75 - 9.25 = -0.50 \rightarrow \frac{[NH_3]}{[NH_4^+]} = 10^{-0.50} = 0.3\underline{1}623$. Then $\frac{0.25 \text{ mol } NH_3}{1 \text{ L}} \times 1.0 \text{ L} = 0.25 \text{ mol } NH_3$ and

$\frac{0.25 \text{ mol } NH_4Cl}{1 \text{ L}} \times 1.0 \text{ L} = 0.25 \text{ mol } NH_4Cl = 0.25 \text{ mol } NH_4^+$ Since HCl is a strong acid, $[HCl] = [H^+]$, and set up table to track changes:

	$H^+ (aq)$ +	$NH_3 (aq)$ →	$NH_4^+ (aq)$
Before addition	≈ 0.00 mol	0.25 mol	0.25 mol
Addition	x	–	–
After addition	≈ 0.00 mol	$(0.25 - x)$ mol	$(0.25 + x)$ mol

Since $\frac{[NH_3]}{[NH_4^+]} = 0.3\underline{1}623 = \frac{(0.25 - x) \text{ mol}}{(0.25 + x) \text{ mol}}$, solve for x. Note that the ratio of moles is the same as the ratio of concentrations, since the volume for both terms is the same. $0.3\underline{1}623(0.25 + x) = (0.25 - x) \rightarrow$ $0.07\underline{9}0575 + 0.3\underline{1}623x = 0.25 - x \rightarrow 1.3\underline{1}623x = 0.1\underline{7}094 \rightarrow x = 0.1\underline{2}987$ mol HCl then

$0.1\underline{2}987 \text{ mol HCl} \times \frac{36.46 \text{ g HCl}}{1 \text{ mol HCl}} = 4.7 \text{ g HCl}$.

Check: The units (g) are correct. The magnitude of the answer makes physical sense because there is much less than a mole of each of the buffer components, so there must be much less than a mole of HCl.

84. **Given:** 250.0 mL of buffer of 0.025 mol $HCHO_2$ and 0.025 mol $NaCHO_2$; adjust to pH = 4.10
Find: mass NaOH or HCl **Other:** K_a $(HCHO_2) = 1.8 \times 10^{-4}$
Conceptual Plan: To decide which reagent needs to be added to adjust pH, calculate the initial pH. Since the mol $HCHO_2$ = mol $NaCHO_2$, the pH = pK_a then final pH, pK_a → $[NaCHO_2]/[HCHO_2]$

acid = $HCHO_2$ base = $HCHO_2^-$ $pK_a = -\log K_a$ $\text{pH} = pK_a + \log \frac{[\text{base}]}{[\text{acid}]}$

then mL → L then write balanced equation then

$\frac{1 \text{ L}}{1000 \text{ mL}}$ $NaOH + HCHO_2 \rightarrow NaCHO_2 + H_2O$

mol $NaCHO_2$, mol $HCHO_2$, $[NaCHO_2]/[HCHO_2]$ → mol NaOH → g NaOH

set up stoichiometry table $\frac{40.00 \text{ g NaOH}}{1 \text{ mol NaOH}}$

Solution: Since K_a $(HCHO_2) = 1.8 \times 10^{-4}$, $pK_a = -\log K_a = -\log(1.8 \times 10^{-4}) = 3.74$. Since the desired pH (4.10) is higher NaOH (a strong base) needs to be added. Then

$\text{pH} = pK_a + \log \frac{[\text{base}]}{[\text{acid}]} = 3.74 + \log \frac{[NaCHO_2]}{[HCHO_2]} = 4.10$. Solve for $\frac{[NaCHO_2]}{[HCHO_2]}$.

$\log \frac{[NaCHO_2]}{[HCHO_2]} = 4.10 - 3.74 = 0.36 \rightarrow \frac{[NaCHO_2]}{[HCHO_2]} = 10^{+0.36} = 2.2\underline{9}087$. Then since NaOH is a strong base, $[NaOH] = [OH^-]$, and set up table to track changes:

	$NaOH (aq)$ +	$HCHO_2 (aq)$ →	$NaCHO_2 (aq)$ +	$H_2O (l)$
Before addition	≈ 0.00 mol	0.025 mol	0.025 mol	–
Addition	x	$-x$	$+x$	–
After addition	≈ 0.00 mol	$(0.025 - x)$ mol	$(0.025 + x)$ mol	–

Since $\text{pH} = pK_a + \log \frac{[A^-]}{[HA]}$ so $4.10 = 3.74 + \log \frac{0.025 \text{ mol} + x}{0.025 \text{ mol} - x}$ solve for x. Note that the ratio of moles is the same as the ratio of concentrations, since the volume for both terms is the same

$0.360 = \log \frac{0.025 \text{ mol} + x}{0.025 \text{ mol} - x} \rightarrow 2.29 = \frac{0.025 \text{ mol} + x}{0.025 \text{ mol} - x} \rightarrow 0.05\underline{7}3 - 2.29x = 0.025 + x \rightarrow 3.29x = 0.03\underline{2}3 \rightarrow$

$x = 0.009\underline{8}1$ mol then $0.00981 \text{ mol NaOH} \times \frac{40.00 \text{ g NaOH}}{1 \text{ mol NaOH}} = 0.39 \text{ g NaOH}$.

Check: The units (g) are correct. The magnitude of the answer makes physical sense because there is much less than a mole of each of the buffer components, so there must be much less than a mole of NaOH.

85. a) **Given:** potassium hydrogen phthalate = KHP = $KHC_8H_4O_4$ titration with NaOH
Find: balanced equation
Conceptual Plan: The reaction will be a titration of the acid proton, leaving the phthalate ion intact. The K will not be titrated since it is basic.
Solution: $NaOH\,(aq) + KHC_8H_4O_4\,(aq) \rightarrow Na^+\,(aq) + K^+\,(aq) + C_8H_4O_4^{2-}\,(aq) + H_2O\,(l)$
Check: An acid–base reaction generates a salt (soluble here) and water. There is only one acidic proton in KHP.

b) **Given:** 0.5527 g KHP titrated with 25.87 mL of NaOH solution **Find:** [NaOH]
Conceptual Plan:
g KHP → mol KHP → mol NaOH and mL → L then mol NaOH and mL → M NaOH

$\frac{1\text{ mol KHP}}{204.22\text{ g KHP}}$ 1:1 from balance equation $\frac{1\text{ L}}{1000\text{ mL}}$ $M = \frac{\text{mol}}{\text{L}}$

Solution: $0.5527\text{ g KHP} \times \frac{1\text{ mol KHP}}{204.22\text{ g KHP}} = 0.002706395\text{ mol KHP}$; mol KHP = mol acid = mol base =

$0.002706395\text{ mol NaOH}$ then $25.87\text{ mL} \times \frac{1\text{ L}}{1000\text{ mL}} = 0.02587\text{ L}$ then

$$[\text{NaOH}] = \frac{0.002706395\text{ mol NaOH}}{0.02587\text{ L}} = 0.1046\text{ M NaOH}.$$

Check: The units (M) are correct. The magnitude of the answer makes physical sense because there is much less than a mole of acid. The magnitude of the moles of acid and base are smaller than the volume of base in liters.

86. **Given:** 0.5224 g monoprotic acid titrated with 23.82 mL of 0.0998 M NaOH solution
Find: molar mass of acid
Conceptual Plan:
mL → L then M NaOH, L → mol NaOH → mol acid then mol acid, g acid → $\mathfrak{M}$

$\frac{1\text{ L}}{1000\text{ mL}}$ $M = \frac{\text{mol}}{\text{L}}$ 1:1 for monoprotic acid $\mathfrak{M} = \frac{\text{g acid}}{\text{mol acid}}$

Solution: $23.82\text{ mL} \times \frac{1\text{ L}}{1000\text{ mL}} = 0.02382\text{ L}$ then

$0.02382\text{ L NaOH} \times \frac{0.0998\text{ mol NaOH}}{1\text{ L NaOH}} = 0.00237724\text{ mol NaOH}$; 0.00237724 mol NaOH = mol base =

mol acid = 0.00237724 mol acid then $\mathfrak{M} = \frac{\text{g acid}}{\text{mol acid}} = \frac{0.5224\text{ g acid}}{0.00237724\text{ mol acid}} = 220.\text{ g/mol}.$

Check: The units (g/mol) are correct. The magnitude of the answer makes physical sense because there is much less than a mole of acid and about a half a gram of acid, so the molar mass will be high. The number is reasonable for an acid (must be > 20 g/mol – lightest acid is HF).

87. **Given:** 0.25 mol weak acid with 10.0 mL of 3.00 M KOH diluted to 1.5000 L has pH = 3.85 **Find:** pK_a of acid
Conceptual Plan: mL → L then M KOH, L → mol KOH then write balanced reaction

$\frac{1\text{ L}}{1000\text{ mL}}$ $M = \frac{\text{mol}}{\text{L}}$ $KOH + HA \rightarrow NaA + H_2O$

added mol KOH, initial mol acid → equil. mol KOH, equil. mol acid then
set up stoichiometry table
equil. mol KOH, equil. mol acid, pH → pK_a

$$pH = pK_a + \log\frac{[\text{base}]}{[\text{acid}]}$$

Solution: $10.00\text{ mL} \times \frac{1\text{ L}}{1000\text{ mL}} = 0.01000\text{ L}$ then $0.01000\text{ L KOH} \times \frac{3.00\text{ mol KOH}}{1\text{ L KOH}} = 0.0300\text{ mol KOH}$ then

Since KOH is a strong base, [KOH] = [OH^-], and set up table to track changes:

	$KOH\ (aq)$ +	$HA\ (aq)$ →	$KA\ (aq)$ +	$H_2O\ (l)$
Before addition	≈ 0.00 mol	0.25 mol	0.00 mol	–
Addition	0.0300 mol	–	–	–
After addition	≈ 0.00 mol	0.23 mol	0.0300 mol	–

Since the ratio of base to acid is between 0.1 and 10, it is a buffer solution. Note that the ratio of moles is the same as the ratio of concentrations, since the volume for both terms is the same.

$\mathrm{pH} = \mathrm{p}K_a + \log \frac{[\text{base}]}{[\text{acid}]} = \mathrm{p}K_a + \log \frac{0.0300 \text{ mol}}{0.23 \text{ mol}} = 3.85$. Solve for p$K_a$. $\mathrm{p}K_a = 3.85 - \log \frac{0.0300 \text{ mol}}{0.23 \text{ mol}} = 4.73$.

Check: The units (none) are correct. The magnitude of the answer makes physical sense because there is more acid than base at equilibrium, so the pK_a is higher than the pH of the solution.

88. **Given:** 5.55 g weak acid with $K_a = 1.3 \times 10^{-4}$ with 5.00 mL of 6.00 M NaOH diluted to 750 mL has pH = 4.25
Find: molar mass of acid
Conceptual Plan: mL → L then M NaOH, L → mol NaOH then write balanced reaction

$\frac{1 \text{ L}}{1000 \text{ mL}}$ $\quad M = \frac{\text{mol}}{\text{L}}$ $\quad NaOH + HA \rightarrow NaA + H_2O$

added mol NaOH, initial mol acid → equil. mol NaOH, equil. mol acid then

set up stoichiometry table

added mol NaOH, equil. mol acid, pH, pK_a → equil. mol NaOH, equil. mol acid then

$\mathrm{pH} = \mathrm{p}K_a + \log \frac{[\text{base}]}{[\text{acid}]}$

mol acid, g acid → $\mathfrak{M}$

$\mathfrak{M} = \frac{\text{g acid}}{\text{mol acid}}$

Solution: $5.00 \text{ mL} \times \frac{1 \text{ L}}{1000 \text{ mL}} = 0.00500 \text{ L}$ then $0.00500 \text{ L NaOH} \times \frac{6.00 \text{ mol NaOH}}{1 \text{ L NaOH}} = 0.0300 \text{ mol NaOH}$.

Since NaOH is a strong base, [NaOH] = [OH⁻], and set up table to track changes:

	$NaOH\ (aq)$ +	$HA\ (aq)$ →	$NaA\ (aq)$ +	$H_2O\ (l)$
Before addition	≈ 0.00 mol	x mol	0.00 mol	–
Addition	0.0300 mol	–	–	–
After addition	≈ 0.00 mol	$x - 0.0300$ mol	0.0300 mol	–

Since the pH is within 1 unit of the pK_a, it is a buffer solution. Note that the ratio of moles is the same as the ratio of concentrations, since the volume for both terms is the same.

$\mathrm{pH} = \mathrm{p}K_a + \log \frac{[\text{base}]}{[\text{acid}]} = -\log(1.3 \times 10^{-4}) + \log \frac{0.0300 \text{ mol}}{(x - 0.0300) \text{ mol}} = 4.25$.

Solve for x. $\log 0.0300 - \log(x - 0.0300) = 4.25 - 3.89 \rightarrow -\log(x - 0.0300) = 1.8\underline{8}288 \rightarrow$

$x - 0.0300 = 10^{-1.8\underline{8}288} = 0.013\underline{0}955 \rightarrow x = 0.043\underline{0}855 \text{ mol}$. Finally,

$M = \frac{\text{g acid}}{\text{mol acid}} = \frac{5.55 \text{ g acid}}{0.043\underline{0}855 \text{ mol acid}} = 129 \text{ g/mol}$

Check: The units (g/mol) are correct. The magnitude of the answer makes physical sense because there is much less than a mole of acid and about 6 grams of acid, so the molar mass will be high. The number is reasonable for an acid (must be > 20 g/mol – lightest acid is HF).

89. **Given:** saturated $CaCO_3$ solution; precipitate 1.00×10^2 mg $CaCO_3$ **Find:** volume of solution evaporated
Other: $K_{sp}(CaCO_3) = 4.96 \times 10^{-9}$
Conceptual Plan: mg $CaCO_3$ → g $CaCO_3$ → mol $CaCO_3$

$\frac{1 \text{ g } CaCO_3}{1000 \text{ mg } CaCO_3}$ $\quad \frac{1 \text{ mol } CaCO_3}{100.09 \text{ g } CaCO_3}$

The expression of the solubility product constant of A_mX_n is: $K_{sp} = [A^{n+}]^m [X^{m-}]^n$. The molar solubility of a compound, A_mX_n, can be computed directly from K_{sp} by solving for S in the expression: $K_{sp} = (mS)^m (nS)^n = m^m n^n S^{m+n}$. Then mol $CaCO_3$, S → L

$M = \frac{\text{mol}}{\text{L}}$

Solution: $1.00 \times 10^2 \text{ mg CaCO}_3 \times \frac{1 \text{ g CaCO}_3}{1000 \text{ mg CaCO}_3} \times \frac{1 \text{ mol CaCO}_3}{100.09 \text{ g CaCO}_3} = 9.99101 \times 10^{-4} \text{ mol CaCO}_3$ then

$K_{sp} = 4.96 \times 10^{-9}$, A = Ca^{2+}, m = 1, X = CO_3^{2-}, and n = 1 so $K_{sp} = 4.96 \times 10^{-9} = S^2$. Rearrange to solve for S.

$S = \sqrt{4.96 \times 10^{-9}} = 7.04273 \times 10^{-5}$ M. Finally, $9.99101 \times 10^{-4} \text{ mol CaCO}_3 \times \frac{1 \text{ L}}{7.04273 \times 10^{-5} \text{ mol CaCO}_3} = 14.2$ L.

Check: The units (L) are correct. The volume should be large since the solubility is low.

90. **Given:** $[Na^+] = 0.140$ M and K_{sp} ($NaC_5H_3N_4$) = 5.76×10^{-8} **Find:** $[C_5H_3N_4^-]$ to form precipitate
Conceptual Plan: Write balanced equation and expression for K_{sp}. Then $[Na^+]$, K_{sp} → $[C_5H_3N_4^-]$
Solution: $NaC_5H_3N_4$ (s) → Na^+ (aq) + $C_5H_3N_4^-$ (aq). So $K_{sp} = [Na^+][C_5H_3N_4^-] = 5.76 \times 10^{-8} = (0.140)$ $[C_5H_3N_4^-]$. Solve for $[C_5H_3N_4^-]$ then $[C_5H_3N_4^-] = 4.11 \times 10^{-7}$ M.
Check: The units (M) are correct. Since K_{sp} is so small and the sodium concentration is fairly high, the urate concentration is driven to a very low level.

91. **Given:** $[Ca^{2+}] = 9.2$ mg/dL and K_{sp} ($Ca_2P_2O_7$) = 8.64×10^{-13} **Find:** $[P_2O_7^{4-}]$ to form precipitate
Conceptual Plan: mg Ca^{2+}/dL → g Ca^{2+}/dL → mol Ca^{2+}/dL → mol Ca^{2+}/L then

$$\frac{1 \text{ g Ca}^{2+}}{1000 \text{ mg Ca}^{2+}} \qquad \frac{1 \text{ mol Ca}^{2+}}{40.08 \text{ g Ca}^{2+}} \qquad \frac{10 \text{ dL}}{1 \text{ L}}$$

Write balanced equation and expression for K_{sp}. Then $[Ca^{2+}]$, K_{sp} → $[P_2O_7^{4-}]$

Solution: $9.2 \frac{\text{mg Ca}^{2+}}{\text{dL}} \times \frac{1 \text{ g Ca}^{2+}}{1000 \text{ mg Ca}^{2+}} \times \frac{1 \text{ mol Ca}^{2+}}{40.08 \text{ g Ca}^{2+}} \times \frac{10 \text{ dL}}{1 \text{ L}} = 2.29541 \times 10^{-3} \text{ M Ca}^{2+}$ then write

equation $Ca_2P_2O_7$ (s) → 2 Ca^{2+} (aq) + $P_2O_7^{4-}$ (aq). So $K_{sp} = [Ca^{2+}]^2 [P_2O_7^{4-}] = 8.64 \times 10^{-13} =$ $= (2.29541 \times 10^{-3})^2 [P_2O_7^{4-}]$. Solve for $[P_2O_7^{4-}]$ then $[P_2O_7^{4-}] = 1.6 \times 10^{-7}$ M.
Check: The units (M) are correct. Since K_{sp} is so small and the sodium concentration is fairly high, the urate concentration is driven to a very low level.

92. **Given:** AgCl in 0.100 M NH_3 **Find:** molar solubility (S)
Other: K_f ($Ag(NH_3)_2^+$) = 1.7×10^7, K_{sp} (AgCl) = 1.77×10^{-10}
Conceptual Plan: Identify the appropriate solid and complex ion. Write balanced equations for dissolving the solid and forming the complex ion. Add these two reactions to get the desired overall reaction. Using the rules from Chapter 14, multiply the individual reaction K's to get the overall K for the sum of these reactions. Then M NH_3, K → S
ICE Chart

Solution: Identify the solid as AgCl and the complex ion as $Ag(NH_3)_2^+$. Write the individual reactions and add them together.

$$AgCl\,(s) \rightleftharpoons Ag^+(aq) + Cl^-(aq) \qquad K_{sp} = 1.77 \times 10^{-10}$$

$$Ag^+(aq) + 2\,NH_3(aq) \rightleftharpoons Ag(NH_3)_2^+(aq) \qquad K_f = 1.7 \times 10^7$$

$$AgCl\,(s) + 2\,NH_3(aq) \rightleftharpoons Ag(NH_3)_2^+(aq) + Cl^-(aq)$$

Since the overall reaction is the simple sum of the two reactions, the overall reaction $K = K_f K_{sp} =$ $= (1.7 \times 10^7) \times (1.77 \times 10^{-10}) = 3.009 \times 10^{-3}$. Then set up ICE table

$$AgCl\,(s) + 2\,NH_3(aq) \rightleftharpoons Ag(NH_3)_2^+(aq) + Cl^-(aq)$$

	$[NH_3]$	$[Ag(NH_3)_2^+]$	$[Cl^-]$
Initial	0.100	0.00	0.00
Change	$-2S$	$+S$	$+S$
Equil	$0.100 - 2S$	$+S$	$+S$

$$K = \frac{[Ag(NH_3)_2^+][Cl^-]}{[NH_3]^2} = 3.009 \times 10^{-3} = \frac{S^2}{(0.100 - 2S)^2}.$$

Simplify by taking the square root of the expression. $\sqrt{3.009 \times 10^{-3}} = 5.4854 \times 10^{-2} = \frac{S}{(0.100 - 2S)}$ Solve for S. $(5.4854 \times 10^{-2})(0.100 - 2S) = S$ → $5.4854 \times 10^{-3} = (1.10971)\,S$ → $S = 4.9431 \times 10^{-3} = 4.9 \times 10^{-3}$ M.

Check: The units (M) are correct. Since K_f is large, the overall K is larger than the original K_{sp} and the solubility of AgCl increases over that of pure water $\left(\sqrt{1.77 \times 10^{-10}} = 1.33 \times 10^{-5} \text{ M}\right)$.

93. **Given:** CuS in 0.150 M NaCN **Find:** molar solubility (S)
Other: K_f ($Cu(CN)_4^{2-}$) = 1.0 x 10^{25}, K_{sp} (CuS) = 1.27 x 10^{-36}
Conceptual Plan: Identify the appropriate solid and complex ion. Write balanced equations for dissolving the solid and forming the complex ion. Add these two reactions to get the desired overall reaction. Using the rules from Chapter 14, multiply the individual reaction *K*'s to get the overall *K* for the sum of these reactions. Then M NaCN, $K \rightarrow S$
ICE Chart

Solution: Identify the solid as CuS and the complex ion as $Cu(CN)_4^{2-}$. Write the individual reactions and add them together.

$CuS(s) \rightleftharpoons \cancel{Cu^{2+}(aq)} + S^{2-}(aq)$ $\quad K_{sp} = 1.27 \times 10^{-36}$

$\cancel{Cu^{2+}(aq)} + 4\,CN^-(aq) \rightleftharpoons Cu(CN)_4^{2-}(aq)$ $\quad K_f = 1.0 \times 10^{25}$

$CuS(s) + 4\,CN^-(aq) \rightleftharpoons Cu(CN)_4^{2-}(aq) + S^{2-}(aq)$

Since the overall reaction is the simple sum of the two reactions, the overall reaction $K = K_f K_{sp}$ = (1.0 x 10^{25}) x (1.27 x 10^{-36}) = 1.27 x 10^{-11}. NaCN (s) → Na^+ (aq) + CN^- (aq). Since 1 CN^- ion is generated for each NaCN, [CN^-] = 0.150 M. Set up ICE table.

$CuS(s) + 4\,CN^-(aq) \rightleftharpoons Cu(CN)_4^{2-}(aq) + S^{2-}(aq)$

	$[CN^-]$	$[Cu(CN)_4^{2-}]$	$[S^{2-}]$
Initial	0.150	0.00	0.00
Change	$-4S$	$+S$	$+S$
Equil	$0.150-4S$	$+S$	$+S$

$$K = \frac{[Cu(CN)_4^{2-}][S^{2-}]}{[CN^-]^4} = 1.27 \times 10^{-11} = \frac{S^2}{(0.150-4S)^4}.$$

Assume S is small ($4S << 0.150$) so $\frac{S^2}{(0.150-\cancel{4S})^4} = 1.27 \times 10^{-11} = \frac{S^2}{(0.150)^4}$ and S = 8.0183 x 10^{-8} = 8.0 x 10^{-8} M. Confirm that assumption is valid $\frac{4(8.0183 \times 10^{-8})}{0.150} \times 100\,\% = 0.00021\,\% << 5\,\%$ so assumption is valid.

Check: The units (M) are correct. Since K_f is large, the overall K is larger than the original K_{sp} and the solubility of CuS increases over that of pure water $\left(\sqrt{1.27 \times 10^{-36}} = 1.13 \times 10^{-18}\ M\right)$.

94. **Given:** 0.10 M φNH_2, keep [φNH_3^+] < 1.0 x 10^{-9} and K_b (φNH_2) = 4.3 x 10^{-10} **Find:** [NaOH]
Conceptual Plan: M φNH_2, maximum M φNH_3^+ → [OH^-] = [NaOH]
ICE Chart $K_w = [H_3O^+][OH^-]$ pH = – log [H_3O^+]

Solution: Set up ICE table. Since the amount of the conjugate acid is set so small, the concentration of the weak base is not significantly changing.

$\phi NH_2(aq) + H_2O(l) \rightleftharpoons \phi NH_3^+(aq) + OH^-(aq)$

	$[\phi NH_3]$	$[\phi NH_4^+]$	$[OH^-]$
Initial	0.10	0.00	≈ 0.00
Change	–	–	$+x$
Equil	≈ 0.10	1.0 x 10^{-9}	$+x$

$$K_b = \frac{[\phi NH_4^+][OH^-]}{[\phi NH_2]} = 4.3 \times 10^{-10} = \frac{(1.0 \times 10^{-9})x}{\approx 0.10}$$

Solve for x. So x = 0.043 M = [OH^-]. NaOH (aq) → Na^+ (aq) + OH^- (aq). Since 1 OH^- ion is generated for each NaOH, [NaOH] = 0.043 M NaOH.
Check: The units (M) are correct. The magnitude of the answer makes physical sense because [OH^-] needs to be about one order of magnitude lower than the aniline concentration (comparing K_b with maximum conjugate acid concentrations).

95. **Given:** 100.0 mL of 0.36 M NH_2OH and 50.0 mL of 0.26 M HCl and K_b (NH_2OH) = 1.10 x 10^{-8} **Find:** pH
Conceptual Plan: identify acid and base components mL → L then [NH_2OH], L → mol NH_2OH
acid = NH_3OH^+ base = NH_2OH $\quad \frac{1\ L}{1000\ mL}$ $\quad M = \frac{mol}{L}$

then mL → L then [HCl], L → mol HCl then write balanced equation then

$\frac{1\ L}{1000\ mL}$ $M = \frac{mol}{L}$ $HCl + NH_2OH \rightarrow NH_3OHCl$

mol NH_2OH, mol HCl → mol excess NH_2OH, mol NH_3OH^+

set up stoichiometry table

Since there are significant amounts of both the acid and the conjugate base species, this is a buffer solution and so Henderson–Hasselbalch Equation $\left(pH = pK_a + \log \frac{[base]}{[acid]}\right)$ can be used. Convert K_b to K_a using $K_w = K_a K_b$. Also note that ratio of concentrations is the same as the ratio of moles, since the volume is the same for both species.

Solution: $100\ \text{mL } NH_2OH \times \frac{1\ L}{1000\ mL} = 0.1\ L\ NH_2OH$ then

$\frac{0.36\ mol\ NH_2OH}{1\ L} \times 0.1000\ L = 0.036\ mol\ NH_2OH$. $50.0\ mL\ HCl \times \frac{1\ L}{1000\ mL} = 0.0500\ L\ HCl$ then

$\frac{0.26\ mol\ HCl}{1\ L} \times 0.0500\ L = 0.013\ mol\ HCl$. Set up table to track changes:

	$HCl\ (aq)$	+ $NH_2OH\ (aq)$	→ $NH_3OHCl\ (aq)$
Before addition	0.00 mol	0.036 mol	0.00 mol
Addition	0.013 mol	–	–
After addition	≈ 0.00 mol	0.023 mol	0.013 mol

then $K_w = K_a K_b$ so $K_a = \frac{K_w}{K_b} = \frac{1.0 \times 10^{-14}}{1.10 \times 10^{-8}} = 9.\underline{0}909 \times 10^{-7}\ M$ then use Henderson–Hasselbalch Equation, since the solution is a buffer. Note that the ratio of moles is the same as the ratio of concentrations, since the volume for both terms is the same.

$$pH = pK_a + \log \frac{[base]}{[acid]} = -\log(9.\underline{0}909 \times 10^{-7}) + \log \frac{0.023\ mol}{0.013\ mol} = 6.2\underline{8}918 = 6.29\,.$$

Check: The units (none) are correct. The magnitude of the answer makes physical sense because pH should be more than the pK_a of the acid because there is more base than acid.

96. The Henderson–Hasselbalch Equation is $pH = pK_a + \log \frac{[base]}{[acid]}$. Remember that $14 = pH + pOH$ and $14 = pK_a + pK_b$. Substituting these into the Henderson–Hasselbalch Equation :
$14 - pOH = 14 - pK_b + \log \frac{[base]}{[acid]}$. Simplifying the expression gives: $pOH = pK_b - \log \frac{[base]}{[acid]}$

97. **Given:** 10.0 L of 75 ppm $CaCO_3$ and 55 ppm $MgCO_3$ (by mass) **Find:** mass Na_2CO_3 to precipitate 90.0 % of ions **Other:** $K_{sp}(CaCO_3) = 4.96 \times 10^{-9}$ and $K_{sp}(MgCO_3) = 6.82 \times 10^{-6}$
Conceptual Plan:
Assume that the density of water is 1.00 g/mL. L water → mL water → g water then

$\frac{1000\ mL}{1\ L}$ $\frac{1.00\ g\ water}{1\ mL}$

g water → g $CaCO_3$ → mol $CaCO_3$ → mol Ca^{2+} and g water → g $MgCO_3$ → mol $MgCO_3$ then

$\frac{75\ g\ CaCO_3}{10^6\ g\ water}$ $\frac{1\ mol\ CaCO_3}{100.09\ g\ CaCO_3}$ $\frac{1\ mol\ Ca^{2+}}{1\ mol\ CaCO_3}$ $\frac{55\ g\ MgCO_3}{10^6\ g\ water}$ $\frac{1\ mol\ MgCO_3}{84.32\ g\ MgCO_3}$

mol $MgCO_3$ → mol Mg^{2+} then comparing the two K_{sp} values, essentially all of the Ca^{2+} will

$\frac{1\ mol\ Mg^{2+}}{1\ mol\ MgCO_3}$

precipitate before the Mg^{2+} will begin to precipitate. Since 90.0 % of the ions are to be precipitates, there will be 10.0 % of the ions left in solution (all will be Mg^{2+}).

$(0.100)(mol\ Ca^{2+} + mol\ Mg^{2+})$

Calculate the moles of ions remaining in solution. Then mol Mg^{2+}, L → M Mg^{2+} then

$$M = \frac{mol}{L}$$

The solubility product constant (K_{sp}) is the equilibrium expression for a chemical equation representing the dissolution of an ionic compound. The expression of the solubility product constant of A_mX_n is: $K_{sp} = [A^{n+}]^m [X^{m-}]^n$. Use this equation to M Mg^{2+}, K_{sp} → M CO_3^{2-} then M CO_3^{2-}, L → mol CO_3^{2-}

for ionic compound, A_mX_n, $K_{sp} = [A^{n+}]^m [X^{m-}]^n$. $\quad M = \frac{mol}{L}$

then mol CO_3^{2-} → mol Na_2CO_3 → g Na_2CO_3

$$\frac{1 \text{ mol } CO_3^{2-}}{1 \text{ mol } Na_2CO_3} \qquad \frac{105.99 \text{ g } Na_2CO_3}{1 \text{ mol } Na_2CO_3}$$

Solution: $10.0 \text{ L} \times \frac{1000 \text{ mL}}{1 \text{ L}} \times \frac{1.00 \text{ g water}}{1 \text{ mL}} = 1.00 \times 10^4$ g water then

$$1.00 \times 10^4 \text{ g water} \times \frac{75 \text{ g } CaCO_3}{10^6 \text{ g water}} \times \frac{1 \text{ mol } CaCO_3}{100.09 \text{ g } CaCO_3} \times \frac{1 \text{ mol } Ca^{2+}}{1 \text{ mol } CaCO_3} = 0.0074933 \text{ mol } Ca^{2+}$$ and

$$1.00 \times 10^4 \text{ g water} \times \frac{55 \text{ g } MgCO_3}{10^6 \text{ g water}} \times \frac{1 \text{ mol } MgCO_3}{84.32 \text{ g } MgCO_3} \times \frac{1 \text{ mol } Mg^{2+}}{1 \text{ mol } MgCO_3} = 0.0065228 \text{ mol } Mg^{2+}$$ so the ions

remaining in solution after 90.0 % precipitate out =
$(0.100)(\text{mol } Ca^{2+} + \text{mol } Mg^{2+}) = (0.100)(0.0074933 \text{ mol } Ca^{2+} + 0.0065228 \text{ mol } Mg^{2+}) = 0.00140161$ mol ions

so $\frac{0.00140607 \text{ mol } Mg^{2+}}{10.0 \text{ L}} = 0.000140161 \text{ M } Mg^{2+}$. Then $K_{sp} = 6.82 \times 10^{-6}$, A = Mg^{2+}, m = 1, X = CO_3^{2-}, and n = 1, so $K_{sp} = 6.82 \times 10^{-6} = [Mg^{2+}][CO_3^{2-}] = (0.00140161)[CO_3^{2-}]$. Rearrange to solve for $[CO_3^{2-}]$. So $[CO_3^{2-}] = 0.0486583$ M. Then

$$\frac{0.0486583 \text{ mol } CO_3^{2-}}{1 \text{ L}} \times 10.0 \text{ L} \times \frac{1 \text{ mol } Na_2CO_3}{1 \text{ mol } CO_3^{2-}} \times \frac{105.99 \text{ g } Na_2CO_3}{1 \text{ mol } Na_2CO_3} = 51.6 \text{ g } Na_2CO_3.$$

Check: The units (g) are correct. The mass is reasonable to put in a washing machine load.

98. **Given:** excess $Mg(OH)_2$ in 1.00 L 0f 1.0 M NH_4Cl has pH = 9.00 **Find:** K_{sp} ($Mg(OH)_2$) **Other:** K_b (NH_3) = 1.76×10^{-5}

Conceptual Plan: M NH_4Cl → M NH_4^+ and K_b → K_a then final pH → $[H_3O^+]$ then

$NH_4Cl\,(aq) \rightarrow NH_4^+\,(aq) + Cl^-\,(aq) \qquad K_w = K_a K_b \qquad [H_3O^+] = 10^{-pH}$

M NH_4^+, M H_3O^+, K_a → x Since x is significant compared to initial M NH_4^+ this is a buffer solution.

ICE Chart

The NH_4^+ is neutralized with $Mg(OH)_2$. Since $Mg(OH)_2\,(s) \rightleftharpoons Mg^{2+}\,(aq) + 2\,OH^-\,(aq)$ **there are 2 moles of OH^- generated for each mole of $Mg(OH)_2$ dissolved. Thus ½ (x mol OH^-) = mol $Mg(OH)_2$ was dissolved in 1.00 L of solution. Since there is 1.00 L solution mol $Mg(OH)_2$ = $[Mg^{2+}]$ and $[H_3O^+]$ → $[OH^-]$ then**

$K_w = [H_3O^+][OH^-]$

Finally, write expression for K_{sp} ($Mg(OH)_2$) and substitute in values for $[Mg^{2+}]$ and $[OH^-]$.

Solution: Since 1 NH_4^+ ion is generated for each NH_4Cl, $[NH_4^+]$ = 1.0 M NH_4^+. Since $K_w = K_a K_b$,

rearrange to solve for K_a. $K_a = \frac{K_w}{K_b} = \frac{1.0 \times 10^{-14}}{1.76 \times 10^{-5}} = 5.6818 \times 10^{-10}$ Final pH = 9.00, so

$[H_3O^+] = 10^{-pH} = 10^{-9.00} = 1.0 \times 10^{-9}$ M Set up ICE Chart.

$$NH_4^+\,(aq) + H_2O\,(l) \rightleftharpoons H_3O^+\,(aq) + NH_3\,(aq)$$

	$[NH_4^+]$	$[H_3O^+]$	$[NH_3]$
Initial	1.0	≈ 0.00	0.00
Change	$-x$	$+x$	$+x$
Equil	$1.0 - x$	1.0×10^{-9}	$+x$

$$K_a = \frac{[H_3O^+][NH_3]}{[NH_4^+]} = 5.6818 \times 10^{-10} = \frac{(1.0 \times 10^{-9})x}{1.0 - x}.$$

Solve for x. $5.6818 \times 10^{-10}\,(1.0 - x) = (1.0 \times 10^{-9})x$ → $5.6818 \times 10^{-10} = (1.0 \times 10^{-9} + 5.6818 \times 10^{-10})x$ →

$x = 0.3\underline{6}2318$, so this is a buffer solution. Since there is 1.00 L of solution, $0.3\underline{6}2318$ mol of NH_4^+ is neutralized with $Mg(OH)_2$. Since $Mg(OH)_2\,(s) \rightleftharpoons Mg^{2+}\,(aq) + 2\,OH^-\,(aq)$ there are 2 moles of OH^- generated for each mole of $Mg(OH)_2$ dissolved. Thus ½ ($0.3\underline{6}2318$ mol OH^-) = $0.1\underline{8}1159$ mol $Mg(OH)_2$ was dissolved in 1.00 L of solution. Thus the $[Mg^{2+}] = 0.1\underline{8}1159$ M. Since $K_w = [H_3O^+][OH^-]$ so

$$[OH^-] = \frac{K_w}{[H_3O^+]} = \frac{1.0 \times 10^{-14}}{1.0 \times 10^{-9}} = 1.0 \times 10^{-5}\ \text{M then}$$

$K_{sp}\,(Mg(OH)_2) = [Mg^{2+}]\,[OH^-]^2 = (0.1\underline{8}1159)\,(1.0 \times 10^{-5})^2 = 1.8 \times 10^{-11}$.

Check: The units (none) are correct. The magnitude of the answer makes physical sense because the concentration of NH_4Cl is high and so it took a significant amount of $Mg(OH)_2$ to raise the pH to 9.00. Note that this number disagrees with the accepted value for the K_{sp} ($Mg(OH)_2$). This is most likely due to errors in the measurements in this experiment.

99. If the concentration of the acid is greater than the concentration of the base, then the pH will be less than the pK_a. If the concentration of the acid is equal to the concentration of the base, then the pH will be equal to the pK_a. If the concentration of the acid is less than the concentration of the base, then the pH will be greater than the pK_a.
    a) $pH < pK_a$

    b) $pH > pK_a$

    c) $pH = pK_a$, the OH^- will convert half of the acid to base

    c) $pH > pK_a$, the OH^- will convert more than half of the acid to base

100. As long as the [base]/[acid] is between 0.1 and 10, the buffer will still be active and the buffer capacity will not have been exceeded.
    a) no, the buffer capacity is not exceeded because [base]/[acid] = 0.22/0.08

    b) no, the buffer capacity is not exceeded because [base]/[acid] = 0.18/0.12

    c) yes, the buffer capacity will be exceeded because all of the acid is converted to base

    d) no, the buffer capacity is not exceeded because [base]/[acid] = 0.19/0.11

101. Only (a) is correct. The volume to the equivalence point will be the same since the number of moles of acid is the same. The pH profiles of the two titrations will be different.

102. Only (c) is correct. If the volume of base is twice as high, then the acid concentration is twice as high and the weaker acid has the higher pH at the equivalence point.

103. a) The solubility will be unchanged since the pH is constant and there are no common ions added.

    b) The solubility will be less because extra fluoride ions are added, suppressing the solubility of the fluoride ionic compound.

    c) The solubility will increase because some of the fluoride ion will be converted to HF, and so more of the ionic compound can be dissolved.

Chapter 17
Free Energy and Thermodynamics

1. a and c are spontaneous processes.

2. a and c are nonspontaneous processes. Nonspontaneous processes are not impossible. Work of some form must be added to make the process proceed.

3. Yes, the particles may also be distributed so that one is in the 0 J level and the other is in the 20 J level. The total energy is the sum of the energies of the two particles = 0 J + 20 J = 20 J. There are two such arrangements of the particles in this fashion, so this state will have the greatest entropy and be more likely.

4. Yes, the particles may also be distributed so that one is in the 0 J level, one is in the 10 J level, and the last is in the 20 J level. The total energy is the sum of the energies of the three particles = 0 J + 10 J + 20 J = 30 J. If we label the particles as A, B, and C, they may be arranged as ABC, ACB, BAC, BCA, CAB, and CBA. There are six ways of arranging the particles, so this state will have the greatest entropy and be more likely.

5. a) $\Delta S > 0$ because a gas is being generated

   b) $\Delta S < 0$ because 2 moles of gas are being converted to 1 mole of gas

   c) $\Delta S < 0$ because a gas is being converted to a solid

   d) $\Delta S < 0$ because 4 moles of gas are being converted to 2 moles of gas

6. a) $\Delta S < 0$ because a gas is being converted to a solid

   b) $\Delta S < 0$ because 5 moles of gas are being converted to 4 moles of gas

   c) $\Delta S > 0$ because 2 moles of gas are being converted to 3 moles of gas

   d) $\Delta S < 0$ because 2 moles of gas are being converted to a solid

7. a) $\Delta S_{sys} > 0$, because 6 moles of gas are being converted to 7 moles of gas. Since $\Delta H < 0$ $\Delta S_{surr} > 0$ and the reaction is spontaneous at all temperatures.

   b) $\Delta S_{sys} < 0$ because 2 moles of different gases are being converted to 2 moles of one gas. Since $\Delta H > 0$ $\Delta S_{surr} < 0$ and the reaction is nonspontaneous at all temperatures.

   c) $\Delta S_{sys} < 0$, because 3 moles of gas are being converted to 2 moles of gas. Since $\Delta H > 0$ $\Delta S_{surr} < 0$ the reaction is nonspontaneous at all temperatures.

   d) $\Delta S_{sys} > 0$, because 9 moles of gas are being converted to 10 moles of gas. Since $\Delta H < 0$ $\Delta S_{surr} > 0$ the reaction is spontaneous at all temperatures.

8. a) $\Delta S_{sys} < 0$, because 3 moles of gas are being converted to 2 moles of gas. Since $\Delta H < 0$ $\Delta S_{surr} > 0$ and the reaction is spontaneous at low temperatures.

   b) $\Delta S_{sys} > 0$ because 2 moles of gas are being converted to 3 moles of gas. Since $\Delta H > 0$ $\Delta S_{surr} < 0$ and the reaction is spontaneous at high temperatures.

   c) $\Delta S_{sys} < 0$, because 3 moles of gas are being converted to 2 moles of gas. Since $\Delta H < 0$ $\Delta S_{surr} > 0$ the reaction is spontaneous at low temperatures.

   d) $\Delta S_{sys} < 0$, because 1 mole of a complicated gas is being converted to 1 mole of gas and a solid. Since $\Delta H > 0$ $\Delta S_{surr} > 0$ the reaction is nonspontaneous at all temperatures.

9. a) **Given:** $\Delta H^\circ_{rxn} = -287$ kJ, $T = 298$ K **Find:** ΔS_{surr}

Conceptual Plan: **kJ → J then ΔH°_{rxn}, T → ΔS_{surr}**

$$\frac{1000\text{ J}}{1\text{ kJ}} \qquad \Delta S_{surr} = \frac{-\Delta H_{sys}}{T}$$

Solution: $$-287\text{ kJ} \times \frac{1000\text{ J}}{1\text{ kJ}} = -287{,}000\text{ J then } \Delta S_{surr} = \frac{-\Delta H_{sys}}{T} = \frac{-(-287{,}000\text{ J})}{298\text{ K}} = 963\ \frac{\text{J}}{\text{K}}$$

Check: The units (J/K) are correct. The magnitude of the answer (10^3 J/K) makes sense because the kJ and the temperature started with very similar values and then a factor of 10^3 was applied.

b) **Given:** $\Delta H^\circ_{rxn} = -287$ kJ, $T = 77$ K **Find:** ΔS_{surr}

Conceptual Plan: **kJ → J then ΔH°_{rxn}, T → ΔS_{surr}**

$$\frac{1000\text{ J}}{1\text{ kJ}} \qquad \Delta S_{surr} = \frac{-\Delta H_{sys}}{T}$$

Solution: $$-287\text{ kJ} \times \frac{1000\text{ J}}{1\text{ kJ}} = -287{,}000\text{ J then } \Delta S_{surr} = \frac{-\Delta H_{sys}}{T} = \frac{-(-287{,}000\text{ J})}{77\text{ K}} = 3730\ \frac{\text{J}}{\text{K}} = 3.73 \times 10^3\ \frac{\text{J}}{\text{K}}$$

Check: The units (J/K) are correct. The magnitude of the answer (4 x 10^3 J/K) makes sense because the temperature is much lower than in part a, so the answer should increase.

c) **Given:** $\Delta H^\circ_{rxn} = +127$ kJ, $T = 298$ K **Find:** ΔS_{surr}

Conceptual Plan: **kJ → J then ΔH°_{rxn}, T → ΔS_{surr}**

$$\frac{1000\text{ J}}{1\text{ kJ}} \qquad \Delta S_{surr} = \frac{-\Delta H_{sys}}{T}$$

Solution: $$+127\text{ kJ} \times \frac{1000\text{ J}}{1\text{ kJ}} = +127{,}000\text{ J then } \Delta S_{surr} = \frac{-\Delta H_{sys}}{T} = \frac{-127{,}000\text{ J}}{298\text{ K}} = -426\ \frac{\text{J}}{\text{K}}$$

Check: The units (J/K) are correct. The magnitude of the answer (– 400 J/K) makes sense because the kJ are less and of the opposite sign than part a and so the answer should decrease.

d) **Given:** $\Delta H^\circ_{rxn} = +127$ kJ, $T = 77$ K **Find:** ΔS_{surr}

Conceptual Plan: **kJ → J then ΔH°_{rxn}, T → ΔS_{surr}**

$$\frac{1000\text{ J}}{1\text{ kJ}} \qquad \Delta S_{surr} = \frac{-\Delta H_{sys}}{T}$$

Solution: $$+127\text{ kJ} \times \frac{1000\text{ J}}{1\text{ kJ}} = +127{,}000\text{ J then } \Delta S_{surr} = \frac{-\Delta H_{sys}}{T} = \frac{-127{,}000\text{ J}}{77\text{ K}} = -1650\ \frac{\text{J}}{\text{K}} = -1.65 \times 10^3\ \frac{\text{J}}{\text{K}}$$

Check: The units (J/K) are correct. The magnitude of the answer (– 2 x 10^3 J/K) makes sense because the temperature is much lower than in part c, so the answer should increase.

10. **Given:** $\Delta H^\circ_{rxn} = -127$ kJ, $\Delta S_{rxn} = 314$ J/K **Find:** T when $\Delta S^\circ_{rxn} = \Delta S_{surr}$

Conceptual Plan: **kJ → J then set $\Delta S^\circ_{rxn} = \Delta S_{surr}$ then ΔH°_{rxn}, ΔS_{surr} → T**

$$\frac{1000\text{ J}}{1\text{ kJ}} \qquad \Delta S_{surr} = \frac{-\Delta H_{sys}}{T}$$

Solution:

$-127\text{ kJ} \times \frac{1000\text{ J}}{1\text{ kJ}} = -127{,}000$ J then set $\Delta S^\circ_{rxn} = 314$ J/K $= \Delta S_{surr}$ then

$$T = \frac{-\Delta H_{sys}}{\Delta S_{surr}} = \frac{-(-127{,}000\text{ J})}{314\ \frac{\text{J}}{\text{K}}} = +404\text{ K}.$$

Check: The units (K) are correct. The magnitude of the answer (400 K) makes sense because there is almost a factor of 500 between the enthalpy and the entropy.

11. a) **Given:** $\Delta H^\circ_{rxn} = -125$ kJ, $\Delta S_{rxn} = +253$ J/K, $T = 298$ K **Find:** ΔS_{univ} and spontaneity

Conceptual Plan: kJ → J then $\Delta H^\circ_{rxn}, T$ → ΔS_{surr} then $\Delta S_{rxn}, \Delta S_{surr}$ → ΔS_{univ}

$$\frac{1000\ \text{J}}{1\ \text{kJ}} \qquad \Delta S_{surr} = \frac{-\Delta H_{sys}}{T} \qquad \Delta S_{univ} = \Delta S_{sys} + \Delta S_{surr}$$

Solution: $-125\ \text{kJ} \times \frac{1000\ \text{J}}{1\ \text{kJ}} = -125{,}000\ \text{J}$ then $\Delta S_{surr} = \frac{-\Delta H_{sys}}{T} = \frac{-(-125{,}000\ \text{J})}{298\ \text{K}} = 419.463\ \frac{\text{J}}{\text{K}}$ then

$\Delta S_{univ} = \Delta S_{sys} + \Delta S_{surr} = +253\ \frac{\text{J}}{\text{K}} + 419.463\ \frac{\text{J}}{\text{K}} = +672\ \frac{\text{J}}{\text{K}}$ so the reaction is spontaneous.

Check: The units (J/K) are correct. The magnitude of the answer (700 J/K) makes sense because both terms were positive and so the reaction is spontaneous.

b) **Given:** $\Delta H^\circ_{rxn} = +125$ kJ, $\Delta S_{rxn} = -253$ J/K, $T = 298$ K **Find:** ΔS_{univ} and spontaneity

Conceptual Plan: kJ → J then $\Delta H^\circ_{rxn}, T$ → ΔS_{surr} then $\Delta S_{rxn}, \Delta S_{surr}$ → ΔS_{univ}

$$\frac{1000\ \text{J}}{1\ \text{kJ}} \qquad \Delta S_{surr} = \frac{-\Delta H_{sys}}{T} \qquad \Delta S_{univ} = \Delta S_{sys} + \Delta S_{surr}$$

Solution: $+125\ \text{kJ} \times \frac{1000\ \text{J}}{1\ \text{kJ}} = +125{,}000\ \text{J}$ then $\Delta S_{surr} = \frac{-\Delta H_{sys}}{T} = \frac{-125{,}000\ \text{J}}{298\ \text{K}} = -419.463\ \frac{\text{J}}{\text{K}}$ then

$\Delta S_{univ} = \Delta S_{sys} + \Delta S_{surr} = -253\ \frac{\text{J}}{\text{K}} - 419.463\ \frac{\text{J}}{\text{K}} = -672\ \frac{\text{J}}{\text{K}}$ so the reaction is nonspontaneous.

Check: The units (J/K) are correct. The magnitude of the answer (– 700 J/K) makes sense because both terms were negative and so the reaction is nonspontaneous.

c) **Given:** $\Delta H^\circ_{rxn} = -125$ kJ, $\Delta S_{rxn} = -253$ J/K, $T = 298$ K **Find:** ΔS_{univ} and spontaneity

Conceptual Plan: kJ → J then $\Delta H^\circ_{rxn}, T$ → ΔS_{surr} then $\Delta S_{rxn}, \Delta S_{surr}$ → ΔS_{univ}

$$\frac{1000\ \text{J}}{1\ \text{kJ}} \qquad \Delta S_{surr} = \frac{-\Delta H_{sys}}{T} \qquad \Delta S_{univ} = \Delta S_{sys} + \Delta S_{surr}$$

Solution: $-125\ \text{kJ} \times \frac{1000\ \text{J}}{1\ \text{kJ}} = -125{,}000\ \text{J}$ then $\Delta S_{surr} = \frac{-\Delta H_{sys}}{T} = \frac{-(-125{,}000\ \text{J})}{298\ \text{K}} = +419.463\ \frac{\text{J}}{\text{K}}$ then

$\Delta S_{univ} = \Delta S_{sys} + \Delta S_{surr} = -253\ \frac{\text{J}}{\text{K}} + 419.463\ \frac{\text{J}}{\text{K}} = +166\ \frac{\text{J}}{\text{K}}$ so the reaction is spontaneous.

Check: The units (J/K) are correct. The magnitude of the answer (200 J/K) makes sense because the larger term was positive and so the reaction is spontaneous.

d) **Given:** $\Delta H^\circ_{rxn} = -125$ kJ, $\Delta S_{rxn} = -253$ J/K, $T = 555$ K **Find:** ΔS_{univ} and spontaneity

Conceptual Plan: kJ → J then $\Delta H^\circ_{rxn}, T$ → ΔS_{surr} then $\Delta S_{rxn}, \Delta S_{surr}$ → ΔS_{univ}

$$\frac{1000\ \text{J}}{1\ \text{kJ}} \qquad \Delta S_{surr} = \frac{-\Delta H_{sys}}{T} \qquad \Delta S_{univ} = \Delta S_{sys} + \Delta S_{surr}$$

Solution: $-125\ \text{kJ} \times \frac{1000\ \text{J}}{1\ \text{kJ}} = -125{,}000\ \text{J}$ then $\Delta S_{surr} = \frac{-\Delta H_{sys}}{T} = \frac{-(-125{,}000\ \text{J})}{555\ \text{K}} = +225.225\ \frac{\text{J}}{\text{K}}$ then

$\Delta S_{univ} = \Delta S_{sys} + \Delta S_{surr} = -253\ \frac{\text{J}}{\text{K}} + 225.225\ \frac{\text{J}}{\text{K}} = -28\ \frac{\text{J}}{\text{K}}$ so the reaction is nonspontaneous.

Check: The units (J/K) are correct. The magnitude of the answer (– 30 J/K) makes sense because the larger term was negative and so the reaction is nonspontaneous.

12. a) **Given:** $\Delta H^\circ_{rxn} = +85$ kJ, $\Delta S_{rxn} = +147$ J/K, $T = 298$ K **Find:** ΔS_{univ} and spontaneity

Conceptual Plan: kJ → J then $\Delta H^\circ_{rxn}, T$ → ΔS_{surr} then $\Delta S_{rxn}, \Delta S_{surr}$ → ΔS_{univ}

$$\frac{1000\ \text{J}}{1\ \text{kJ}} \qquad \Delta S_{surr} = \frac{-\Delta H_{sys}}{T} \qquad \Delta S_{univ} = \Delta S_{sys} + \Delta S_{surr}$$

Solution: $+85\ \text{kJ} \times \frac{1000\ \text{J}}{1\ \text{kJ}} = +85{,}000\ \text{J}$ then $\Delta S_{surr} = \frac{-\Delta H_{sys}}{T} = \frac{-85{,}000\ \text{J}}{298\ \text{K}} = -285.23\ \frac{\text{J}}{\text{K}}$ then

$$\Delta S_{univ} = \Delta S_{sys} + \Delta S_{surr} = +147\,\frac{J}{K} - 285.23\,\frac{J}{K} = -138\,\frac{J}{K} = -1.4 \times 10^2\,\frac{J}{K}$$ so the reaction is nonspontaneous.

Check: The units (J/K) are correct. The magnitude of the answer (– 140 J/K) makes sense because the entropy of the surroundings is negative and dominates over the entropy gain of the system, so the reaction is nonspontaneous.

b) **Given:** $\Delta H°_{rxn} = +85$ kJ, $\Delta S_{rxn} = +147$ J/K, $T = 755$ K **Find:** ΔS_{univ} and spontaneity

Conceptual Plan: kJ → J then $\Delta H°_{rxn}, T$ → ΔS_{surr} then $\Delta S_{rxn}, \Delta S_{surr}$ → ΔS_{univ}

$$\frac{1000\ J}{1\ kJ} \qquad \Delta S_{surr} = \frac{-\Delta H_{sys}}{T} \qquad \Delta S_{univ} = \Delta S_{sys} + \Delta S_{surr}$$

Solution: $+85\ kJ \times \frac{1000\ J}{1\ kJ} = +85{,}000\ J$ then $\Delta S_{surr} = \frac{-\Delta H_{sys}}{T} = \frac{-85{,}000\ J}{755\ K} = -112.58\,\frac{J}{K}$ then

$$\Delta S_{univ} = \Delta S_{sys} + \Delta S_{surr} = +147\,\frac{J}{K} - 112.58\,\frac{J}{K} = +34\,\frac{J}{K} = +3 \times 10^1\,\frac{J}{K}$$ so the reaction is spontaneous.

Check: The units (J/K) are correct. The magnitude of the answer (30 J/K) makes sense because at a higher temperature, the entropy of the surroundings is reduced, so that the entropy of the system dominates and the reaction is spontaneous.

c) **Given:** $\Delta H°_{rxn} = +85$ kJ, $\Delta S_{rxn} = -147$ J/K, $T = 298$ K **Find:** ΔS_{univ} and spontaneity

Conceptual Plan: kJ → J then $\Delta H°_{rxn}, T$ → ΔS_{surr} then $\Delta S_{rxn}, \Delta S_{surr}$ → ΔS_{univ}

$$\frac{1000\ J}{1\ kJ} \qquad \Delta S_{surr} = \frac{-\Delta H_{sys}}{T} \qquad \Delta S_{univ} = \Delta S_{sys} + \Delta S_{surr}$$

Solution: $+85\ kJ \times \frac{1000\ J}{1\ kJ} = +85{,}000\ J$ then $\Delta S_{surr} = \frac{-\Delta H_{sys}}{T} = \frac{-85{,}000\ J}{298\ K} = -285.23\,\frac{J}{K}$ then

$$\Delta S_{univ} = \Delta S_{sys} + \Delta S_{surr} = -147\,\frac{J}{K} - 285.23\,\frac{J}{K} = -432\,\frac{J}{K} = -4.3 \times 10^2\,\frac{J}{K}$$ so the reaction is nonspontaneous.

Check: The units (J/K) are correct. The magnitude of the answer (– 400 J/K) makes sense because both terms are negative, so the sum is negative and the reaction is nonspontaneous.

d) **Given:** $\Delta H°_{rxn} = -85$ kJ, $\Delta S_{rxn} = +147$ J/K, $T = 398$ K **Find:** ΔS_{univ} and spontaneity

Conceptual Plan: kJ → J then $\Delta H°_{rxn}, T$ → ΔS_{surr} then $\Delta S_{rxn}, \Delta S_{surr}$ → ΔS_{univ}

$$\frac{1000\ J}{1\ kJ} \qquad \Delta S_{surr} = \frac{-\Delta H_{sys}}{T} \qquad \Delta S_{univ} = \Delta S_{sys} + \Delta S_{surr}$$

Solution: $-85\ kJ \times \frac{1000\ J}{1\ kJ} = -85{,}000\ J$ then $\Delta S_{surr} = \frac{-\Delta H_{sys}}{T} = \frac{-(-85{,}000\ J)}{398\ K} = +213.57\,\frac{J}{K}$ then

$$\Delta S_{univ} = \Delta S_{sys} + \Delta S_{surr} = +147\,\frac{J}{K} + 213.57\,\frac{J}{K} = +361\,\frac{J}{K} = +3.6 \times 10^2\,\frac{J}{K}$$ so the reaction is spontaneous.

Check: The units (J/K) are correct. The magnitude of the answer (400 J/K) makes sense because both terms are positive, so the sum is positive and the reaction is spontaneous.

13. a) **Given:** $\Delta H°_{rxn} = -125$ kJ, $\Delta S_{rxn} = +253$ J/K, $T = 298$ K **Find:** ΔG and spontaneity

Conceptual Plan: J/K → kJ/K then $\Delta H°_{rxn}, \Delta S_{rxn}, T$ → ΔG

$$\frac{1000\ J}{1\ kJ} \qquad \Delta G = \Delta H_{rxn} - T\Delta S_{rxn}$$

Solution: $+253\,\frac{J}{K} \times \frac{1\ kJ}{1000\ J} = +0.253\,\frac{kJ}{K}$ then

$$\Delta G = \Delta H_{rxn} - T\Delta S_{rxn} = -125\ kJ - (298\ K)\left(0.253\,\frac{kJ}{K}\right) = -200.\ kJ = -2.00 \times 10^2\ kJ$$ so the reaction is spontaneous.

Check: The units (kJ) are correct. The magnitude of the answer (– 200 kJ) makes sense because both terms were negative and so the reaction is spontaneous.

b) **Given:** $\Delta H^\circ_{rxn} = +\ 125$ kJ, $\Delta S_{rxn} = -\ 253$ J/K, $T = 298$ K **Find:** ΔG and spontaneity

Conceptual Plan: **J/K → kJ/K then $\Delta H^\circ_{rxn}, \Delta S_{rxn}, T$ → ΔG**

$$\frac{1000\text{ J}}{1\text{ kJ}} \qquad \Delta G = \Delta H_{rxn} - T\Delta S_{rxn}$$

Solution: $-253\frac{\text{J}}{\text{K}} \times \frac{1\text{ kJ}}{1000\text{ J}} = -\ 0.253\frac{\text{kJ}}{\text{K}}$ then

$\Delta G = \Delta H_{rxn} - T\Delta S_{rxn} = +\ 125\text{ kJ} - (298\text{ K})\left(-\ 0.253\frac{\text{kJ}}{\text{K}}\right) = +\ 200.\text{ kJ} = +\ 2.00 \times 10^2\text{ kJ}$ so the reaction is nonspontaneous.

Check: The units (kJ) are correct. The magnitude of the answer (200 kJ) makes sense because both terms were positive and so the reaction is nonspontaneous.

c) **Given:** $\Delta H^\circ_{rxn} = -\ 125$ kJ, $\Delta S_{rxn} = -\ 253$ J/K, $T = 298$ K **Find:** ΔG and spontaneity

Conceptual Plan: **J/K → kJ/K then $\Delta H^\circ_{rxn}, \Delta S_{rxn}, T$ → ΔG**

$$\frac{1000\text{ J}}{1\text{ kJ}} \qquad \Delta G = \Delta H_{rxn} - T\Delta S_{rxn}$$

Solution: $-253\frac{\text{J}}{\text{K}} \times \frac{1\text{ kJ}}{1000\text{ J}} = -\ 0.253\frac{\text{kJ}}{\text{K}}$ then

$\Delta G = \Delta H_{rxn} - T\Delta S_{rxn} = -\ 125\text{ kJ} - (298\text{ K})\left(-\ 0.253\frac{\text{kJ}}{\text{K}}\right) = -\ 4\underline{9}.606\text{ kJ} = -\ 5.0 \times 10^1\text{ kJ}$ so the reaction is spontaneous.

Check: The units (kJ) are correct. The magnitude of the answer (–50 kJ) makes sense because the larger term was negative and so the reaction is spontaneous.

d) **Given:** $\Delta H^\circ_{rxn} = -\ 125$ kJ, $\Delta S_{rxn} = -\ 253$ J/K, $T = 555$ K **Find:** ΔG and spontaneity

Conceptual Plan: **J/K → kJ/K then $\Delta H^\circ_{rxn}, \Delta S_{rxn}, T$ → ΔG**

$$\frac{1000\text{ J}}{1\text{ kJ}} \qquad \Delta G = \Delta H_{rxn} - T\Delta S_{rxn}$$

Solution: $-253\frac{\text{J}}{\text{K}} \times \frac{1\text{ kJ}}{1000\text{ J}} = -\ 0.253\frac{\text{kJ}}{\text{K}}$ then

$\Delta G = \Delta H_{rxn} - T\Delta S_{rxn} = -\ 125\text{ kJ} - (555\text{ K})\left(-\ 0.253\frac{\text{kJ}}{\text{K}}\right) = +\ 15\text{ kJ}$ so the reaction is nonspontaneous.

Check: The units (J/K) are correct. The magnitude of the answer (+ 15 kJ) makes sense because the larger term was positive and so the reaction is nonspontaneous.

14\. a) **Given:** $\Delta H^\circ_{rxn} = +\ 85$ kJ, $\Delta S_{rxn} = +\ 147$ J/K, $T = 298$ K **Find:** ΔG and spontaneity

Conceptual Plan: **J/K → kJ/K then $\Delta H^\circ_{rxn}, \Delta S_{rxn}, T$ → ΔG**

$$\frac{1\text{kJ}}{1000\text{ J}} \qquad \Delta G = \Delta H_{rxn} - T\Delta S_{rxn}$$

Solution: $+147\frac{\text{J}}{\text{K}} \times \frac{1\text{ kJ}}{1000\text{ J}} = +\ 0.147\frac{\text{kJ}}{\text{K}}$ then

$\Delta G = \Delta H_{rxn} - T\Delta S_{rxn} = +\ 85\text{ kJ} - (298\text{ K})\left(0.147\frac{\text{kJ}}{\text{K}}\right) = +\ 41\text{ kJ}$ so the reaction is nonspontaneous.

Check: The units (kJ) are correct. The magnitude of the answer (41 kJ) makes sense because the positive enthalpy dominates over the entropy term, so the reaction is nonspontaneous.

b) **Given:** $\Delta H^\circ_{rxn} = +\ 85$ kJ, $\Delta S_{rxn} = +\ 147$ J/K, $T = 755$ K **Find:** ΔG and spontaneity

Conceptual Plan: **J/K → kJ/K then $\Delta H^\circ_{rxn}, \Delta S_{rxn}, T$ → ΔG**

$$\frac{1\text{kJ}}{1000\text{ J}} \qquad \Delta G = \Delta H_{rxn} - T\Delta S_{rxn}$$

Solution: $+147 \frac{\text{J}}{\text{K}} \times \frac{1\text{ kJ}}{1000\text{ J}} = +0.147 \frac{\text{kJ}}{\text{K}}$ then

$$\Delta G = \Delta H_{rxn} - T\Delta S_{rxn} = +85\text{ kJ} - (755\text{ K})\left(0.147 \frac{\text{kJ}}{\text{K}}\right) = -26\text{ kJ}$$ so the reaction is spontaneous.

Check: The units (kJ) are correct. The magnitude of the answer (– 30 kJ) makes sense because at a higher temperature, the entropy term now dominates and the reaction is spontaneous.

c) **Given:** $\Delta H°_{rxn} = +85$ kJ, $\Delta S_{rxn} = -147$ J/K, $T = 298$ K **Find:** ΔG and spontaneity
Conceptual Plan: J/K → kJ/K then $\Delta H°_{rxn}, \Delta S_{rxn}, T$ → ΔG

$\frac{1\text{kJ}}{1000\text{ J}}$ $\quad \Delta G = \Delta H_{rxn} - T\Delta S_{rxn}$

Solution: $-147 \frac{\text{J}}{\text{K}} \times \frac{1\text{ kJ}}{1000\text{ J}} = -0.147 \frac{\text{kJ}}{\text{K}}$ then

$$\Delta G = \Delta H_{rxn} - T\Delta S_{rxn} = +85\text{ kJ} - (298\text{ K})\left(-0.147 \frac{\text{kJ}}{\text{K}}\right) = +129\text{ kJ}$$ so the reaction is nonspontaneous.

Check: The units (kJ) are correct. The magnitude of the answer (130 kJ) makes sense because both terms are positive, so the sum is positive and the reaction is nonspontaneous.

d) **Given:** $\Delta H°_{rxn} = -85$ kJ, $\Delta S_{rxn} = +147$ J/K, $T = 398$ K **Find:** ΔG and spontaneity
Conceptual Plan: J/K → kJ/K then $\Delta H°_{rxn}, \Delta S_{rxn}, T$ → ΔG

$\frac{1\text{kJ}}{1000\text{ J}}$ $\quad \Delta G = \Delta H_{rxn} - T\Delta S_{rxn}$

Solution: $+147 \frac{\text{J}}{\text{K}} \times \frac{1\text{ kJ}}{1000\text{ J}} = +0.147 \frac{\text{kJ}}{\text{K}}$ then

$$\Delta G = \Delta H_{rxn} - T\Delta S_{rxn} = -85\text{ kJ} - (398\text{ K})\left(0.147 \frac{\text{kJ}}{\text{K}}\right) = -144\text{ kJ}$$ so the reaction is spontaneous.

Check: The units (kJ) are correct. The magnitude of the answer (– 140 kJ) makes sense because both terms are negative, so the sum is negative and the reaction is spontaneous.

15. **Given:** $\Delta H°_{rxn} = -2217$ kJ, $\Delta S_{rxn} = +101.1$ J/K, $T = 25$ °C **Find:** ΔG and spontaneity
Conceptual Plan: °C → K then J/K → kJ/K then $\Delta H°_{rxn}, \Delta S_{rxn}, T$ → ΔG

$\text{K} = 273.15 + °\text{C}$ $\quad \frac{1\text{kJ}}{1000\text{ J}}$ $\quad \Delta G = \Delta H_{rxn} - T\Delta S_{rxn}$

Solution: $T = 273.15 + 25\ °\text{C} = 298$ K then $+101.1 \frac{\text{J}}{\text{K}} \times \frac{1\text{ kJ}}{1000\text{ J}} = +0.1011 \frac{\text{kJ}}{\text{K}}$ then

$$\Delta G = \Delta H_{rxn} - T\Delta S_{rxn} = -2217\text{ kJ} - (298\text{ K})\left(0.1011 \frac{\text{kJ}}{\text{K}}\right) = -2247\text{ kJ} = -2.247 \times 10^6\text{ J}$$ so the reaction is spontaneous.

Check: The units (kJ) are correct. The magnitude of the answer (– 2250 kJ) makes sense because both terms are negative, so the reaction is spontaneous.

16. **Given:** $\Delta H°_{rxn} = -1269{,}8$ kJ, $\Delta S_{rxn} = -364.6$ J/K, $T = 25$ °C **Find:** ΔG and spontaneity
Conceptual Plan: °C → K then J/K → kJ/K then $\Delta H°_{rxn}, \Delta S_{rxn}, T$ → ΔG

$\text{K} = 273.15 + °\text{C}$ $\quad \frac{1\text{kJ}}{1000\text{ J}}$ $\quad \Delta G = \Delta H_{rxn} - T\Delta S_{rxn}$

Solution: $T = 273.15 + 25\ °\text{C} = 298$ K then $-364.6 \frac{\text{J}}{\text{K}} \times \frac{1\text{ kJ}}{1000\text{ J}} = -0.3646 \frac{\text{kJ}}{\text{K}}$ then

$$\Delta G = \Delta H_{rxn} - T\Delta S_{rxn} = -1269.8\text{ kJ} - (298\text{ K})\left(-0.3646 \frac{\text{kJ}}{\text{K}}\right) = -1161.1\text{ kJ} = -1.161 \times 10^6\text{ J}$$ so the reaction is spontaneous.

Check: The units (kJ) are correct. The magnitude of the answer (– 1200 kJ) makes sense because the negative enthalpy term dominates over the positive entropy term, so the reaction is spontaneous.

17.

ΔH	ΔS	ΔG	Low Temp.	High Temp.
−	+	−	Spontaneous	Spontaneous
−	−	Temp dependent	Spontaneous	Nonspontaneous
+	+	Temp dependent	Nonspontaneous	Spontaneous
+	−	+	Nonspontaneous	Nonspontaneous

18. a) ΔH°_{rxn} for a condensation is negative and ΔS_{rxn} is negative, so the reaction will be spontaneous at low temperatures (< 100 °C).

b) ΔH°_{rxn} for a sublimation is positive and ΔS_{rxn} is positive, so the reaction will be spontaneous at high temperatures (> − 78.5 °C).

c) ΔH°_{rxn} for a bond breaking is positive and ΔS_{rxn} is positive, so the reaction will be spontaneous at high temperatures.

d) ΔH°_{rxn} is positive and ΔS_{rxn} is positive, so the reaction will be spontaneous at high temperatures.

19. The molar entropy of a substance increases with increasing temperatures. The kinetic energy and the molecular motion increases. The substance will have access to an increased number of energy levels.

20. The third law of thermodynamics states that the entropy of a perfect crystal at absolute zero (0 K) is zero. For enthalpy we defined a standard state so that we could define a "zero" for the scale. This is not necessary for entropy because there is an absolute zero for the entropy scale.

21. a) CO_2 (*g*) because it has greater molar mass/complexity.

b) CH_3OH (*g*) because it is in the gas phase.

c) CO_2 (*g*) because it has greater molar mass/complexity.

d) SiH_4 (*g*) because it has greater molar mass.

e) $CH_3CH_2CH_3$ (*g*) because it has greater molar mass/complexity.

f) NaBr (*aq*) because a solution has more entropy than a solid crystal.

22. a) $NaNO_3$ (*aq*) because a solution has more entropy than a solid crystal.

b) CH_3CH_3 (*g*) because it has greater molar mass/complexity.

c) Br_2 (*g*) because it is in the gas phase.

d) Br_2 (*g*) because it has greater molar mass.

e) PCl_5 (*g*) because it has greater molar mass/complexity.

f) $CH_3CH_2CH_2CH_3$ (*g*) because it has greater complexity.

23. a) He (*g*) < Ne (*g*) < SO_2 (*g*) < NH_3 (*g*) < CH_3CH_2OH (*g*). All are in the gas phase. From He to Ne there is an increase in molar mass, beyond that, the molecules increase in complexity.

b) H_2O (*s*) < H_2O (*l*) < H_2O (*g*). Entropy increases as we go from a solid to a liquid to a gas.

c) CH_4 (*g*) < CF_4 (*g*) < CCl_4 (*g*). Entropy increases as the molar mass increases.

24. a) F_2 (*g*) < Cl_2 (*g*) < Br_2 (*g*) < I_2 (*g*). All are in the gas phase. Entropy increases as the molar mass increases.

b) H_2O (*g*) < H_2S (*g*) < H_2O_2 (*g*). Entropy increases as the molar mass and the complexity of the molecules increases.

c) C (*s*, diamond) < C (*s*, graphite) < C (*s*, amorphous). Entropy increases as the complexity increases. The diamond structure is ordered in all three dimensions. Graphite has ordered sheets that can slide with respect to each other. The amorphous carbon has no long range order.

25. a) **Given:** C_2H_4 (*g*) + H_2 (*g*) → C_2H_6 (*g*) **Find:** $\Delta S°_{rxn}$

Conceptual Plan: $\Delta S^0_{rxn} = \sum n_p S^0(products) - \sum n_r S^0(reactants)$

Solution:

Reactant/Product	S^0 (J/mol K from Appendix IIB)
C_2H_4 (g)	219.3
H_2 (g)	130.7
C_2H_6 (g)	229.2

Be sure to pull data for the correct formula and phase.

$$\Delta S^0_{rxn} = \sum n_p S^0(products) - \sum n_r S^0(reactants)$$
$$= [1(S^0(C_2H_6\ (g))] - [1(S^0(C_2H_4\ (g)) + 1(S^0(H_2\ (g))]$$
$$= [1(229.2\ J/K)] - [1(219.3\ J/K) + 1(130.7\ J/K)]$$ The moles of gas are decreasing.
$$= [229.2\ J/K] - [350.0\ J/K]$$
$$= -120.8\ J/K$$

Check: The units (J/K) are correct. The answer is negative, which is consistent with 2 moles of gas going to 1 mole of gas.

b) **Given:** C (*s*) + H_2O (*g*) → CO (*g*) + H_2 (*g*) **Find:** $\Delta S°_{rxn}$

Conceptual Plan: $\Delta S^0_{rxn} = \sum n_p S^0(products) - \sum n_r S^0(reactants)$

Solution:

Reactant/Product	S^0 (J/mol K from Appendix IIB)
C (*s*)	5.7
H_2O (*g*)	188.8
CO (*g*)	197.7
H_2 (*g*)	130.7

Be sure to pull data for the correct formula and phase.

$$\Delta S^0_{rxn} = \sum n_p S^0(products) - \sum n_r S^0(reactants)$$
$$= [1(S^0(CO\ (g)) + 1(S^0(H_2\ (g))] - [1(S^0(C\ (s)) + 1(S^0(H_2O\ (g))]$$
$$= [1(197.7\ J/K) + 1(130.7\ J/K)] - [1(5.7\ J/K) + 1(188.8\ J/K)]$$ The moles of gas are increasing.
$$= [328.4\ J/K] - [194.5\ J/K]$$
$$= +133.9\ J/K$$

Check: The units (J/K) are correct. The answer is positive, which is consistent with 1 mole of gas going to 1 mole of gas.

c) **Given:** CO (*g*) + H_2O (*g*) → H_2 (*g*) + CO_2 (*g*) **Find:** $\Delta S°_{rxn}$

Conceptual Plan: $\Delta S^0_{rxn} = \sum n_p S^0(products) - \sum n_r S^0(reactants)$

Solution:

Reactant/Product	S^0 (J/mol K from Appendix IIB)
CO (*g*)	197.7
H_2O (*g*)	188.8
H_2 (*g*)	130.7
CO_2 (*g*)	213.8

Be sure to pull data for the correct formula and phase.

$$\Delta S^0_{rxn} = \sum n_p S^0(products) - \sum n_r S^0(reactants)$$
$$= [1(S^0(H_2\ (g)) + 1(S^0(CO_2\ (g))] - [1(S^0(CO\ (g)) + 1(S^0(H_2O\ (g))]$$
$$= [1(130.7\ J/K) + 1(213.8\ J/K)] - [1(197.7\ J/K) + 1(188.8\ J/K)]$$
$$= [344.5\ J/K] - [386.5\ J/K]$$
$$= -42.0\ J/K$$

The change is small because the number of moles of gas is constant.
Check: The units (J/K) are correct. The answer is small and negative, which is consistent with a constant number of moles of gas. Water molecules are bent and carbon dioxide molecules are linear, so the water has more complexity. Also, carbon monoxide is more complex than hydrogen gas.

d) **Given:** $2\ H_2S\ (g) + 3\ O_2\ (g) \rightarrow 2\ H_2O\ (l) + 2\ SO_2\ (g)$ **Find:** ΔS°_{rxn}

Conceptual Plan: $\Delta S^0_{rxn} = \sum n_p S^0(products) - \sum n_r S^0(reactants)$

Solution:

Reactant/Product	S^0 (J/molK from Appendix IIB)
$H_2S\ (g)$	205.8
$O_2\ (g)$	205.2
$H_2O\ (l)$	70.0
$SO_2\ (g)$	248.2

Be sure to pull data for the correct formula and phase.

$$\Delta S^0_{rxn} = \sum n_p S^0(products) - \sum n_r S^0(reactants)$$
$$= [2(S^0(H_2O\ (l)) + 2(S^0(SO_2\ (g))] - [2(S^0(H_2S\ (g)) + 3(S^0(O_2\ (g))]$$
$$= [2(70.0\ J/K) + 2(248.2\ J/K)] - [2(205.8\ J/K) + 3(205.2\ J/K)]$$
$$= [636.4\ J/K] - [1027.2\ J/K]$$
$$= -390.8\ J/K$$

The number of moles of gas is decreasing.
Check: The units (J/K) are correct. The answer is negative, which is consistent with a decrease in the number of moles of gas.

26. a) **Given:** $3\ NO_2\ (g) + H_2O\ (l) \rightarrow 2\ HNO_3\ (aq) + NO\ (g)$ **Find:** ΔS°_{rxn}

Conceptual Plan: $\Delta S^0_{rxn} = \sum n_p S^0(products) - \sum n_r S^0(reactants)$

Solution:

Reactant/Product	S^0 (J/molK from Appendix IIB)
$NO_2\ (g)$	240.1
$H_2O\ (l)$	70.0
$HNO_3\ (aq)$	146
$NO\ (g)$	210.8

Be sure to pull data for the correct formula and phase.

$$\Delta S^0_{rxn} = \sum n_p S^0(products) - \sum n_r S^0(reactants)$$
$$= [2(S^0(HNO_3\ (aq)) + 1(S^0(NO(g))] - [3(S^0(NO_2\ (g)) + 1(S^0(H_2O\ (l))]$$
$$= [2(146\ J/K) + 1(210.8\ J/K)] - [3(240.1\ J/K) + 1(70.0\ J/K)]$$
$$= [50\underline{2}.8\ J/K] - [790.3\ J/K]$$
$$= -288\ J/K$$

The number of moles of gas is decreasing.
Check: The units (J/K) are correct. The answer is negative, which is consistent with a decrease in the number of moles of gas.

b) **Given:** $Cr_2O_3\ (s) + 3\ CO\ (g) \rightarrow 2\ Cr\ (s) + 3\ CO_2\ (g)$ **Find:** ΔS°_{rxn}

Conceptual Plan: $\Delta S^0_{rxn} = \sum n_p S^0(products) - \sum n_r S^0(reactants)$

Solution:

Reactant/Product	S^0 (J/molK from Appendix IIB)
Cr_2O_3 (*s*)	81.2
CO (*g*)	197.7
Cr (*s*)	23.8
CO_2 (*g*)	213.8

Be sure to pull data for the correct formula and phase.

$$\Delta S^0_{rxn} = \sum n_p S^0(products) - \sum n_r S^0(reactants)$$

$$= [2(S^0(\text{Cr }(s)) + 3(S^0(\text{CO}_2\ (g))] - [1(S^0(\text{Cr}_2\text{O}_3\ (s)) + 3(S^0(\text{CO }(g))]$$

$$= [2(23.8\ \text{J/K}) + 3(213.8\ \text{J/K})] - [1(81.2\ \text{J/K}) + 3(197.7\ \text{J/K})]$$

$$= [689.0\ \text{J/K}] - [674.3\ \text{J/K}]$$

$$= +14.7\ \text{J/K}$$

The change is small because the number of moles of gas is constant.
Check: The units (J/K) are correct. The answer is small and positive, which is consistent with a constant number of moles of gas. Carbon dioxide molecules have more complexity than carbon monoxide molecules, but the chromium oxide is more complex than chromium metal.

c) **Given:** SO_2 (*g*) + ½ O_2 (*g*) → SO_3 (*g*) **Find:** ΔS°_{rxn}

Conceptual Plan: $\Delta S^0_{rxn} = \sum n_p S^0(products) - \sum n_r S^0(reactants)$

Solution:

Reactant/Product	S^0 (J/molK from Appendix IIB)
SO_2 (*g*)	248.2
O_2 (*g*)	205.2
SO_3 (*g*)	256.8

Be sure to pull data for the correct formula and phase.

$$\Delta S^0_{rxn} = \sum n_p S^0(products) - \sum n_r S^0(reactants)$$

$$= [1(S^0(\text{SO}_3\ (g))] - [1(S^0(\text{SO}_2\ (g)) + 1/2(S^0(\text{O}_2\ (g))]$$

$$= [1(256.8\ \text{J/K})] - [1(248.2\ \text{J/K}) + 1/2(205.2\ \text{J/K})]$$

$$= [256.8\ \text{J/K}] - [350.8\ \text{J/K}]$$

$$= -94.0\ \text{J/K}$$

The number of moles of gas is decreasing.
Check: The units (J/K) are correct. The answer is negative, which is consistent with a decrease in the number of moles of gas.

d) **Given:** N_2O_4 (*g*) + 4 H_2 (*g*) → N_2 (*g*) + 4 H_2O (*g*) **Find:** ΔS°_{rxn}

Conceptual Plan: $\Delta S^0_{rxn} = \sum n_p S^0(products) - \sum n_r S^0(reactants)$

Solution:

Reactant/Product	S^0 (J/molK from Appendix IIB)
N_2O_4 (*g*)	304.4
H_2 (*g*)	130.7
N_2 (*g*)	191.6
H_2O (*g*)	188.8

Be sure to pull data for the correct formula and phase.

$$\Delta S^0_{rxn} = \sum n_p S^0(products) - \sum n_r S^0(reactants)$$

$$= [1(S^0(\text{N}_2\ (g)) + 4(S^0(\text{H}_2\text{O }(g))] - [1(S^0(\text{N}_2\text{O}_4\ (g)) + 4(S^0(\text{H}_2\ (g))]$$

$$= [1(191.6\ \text{J/K}) + 4(188.8\ \text{J/K})] - [1(304.4\ \text{J/K}) + 4(130.7\ \text{J/K})]$$

$$= [946.8\ \text{J/K}] - [827.2\ \text{J/K}]$$

$$= +119.6\ \text{J/K}$$

The change is small because the number of moles of gas is constant.

Check: The units (J/K) are correct. The answer is positive, which is consistent with 1 mole of a complex gas and 4 moles of a simple gas going to 1 mole of a simple gas and 4 moles of a complex gas.

27. **Given:** CH_2Cl_2 (*g*) formed from elements in standard states **Find:** $\Delta S°$ and rationalize sign

Conceptual Plan: Write balanced reaction, then $\Delta S^0_{rxn} = \sum n_p S^0(products) - \sum n_r S^0(reactants)$

Solution: C (*s*) + H_2 (*g*) + Cl_2 (*g*) → CH_2Cl_2 (*g*)

Reactant/Product	S^0(J/molK from Appendix IIB)
C (*s*)	5.7
H_2 (*g*)	130.7
Cl_2 (*g*)	223.1
CH_2Cl_2 (*g*)	270.2

Be sure to pull data for the correct formula and phase.

$$\Delta S^0_{rxn} = \sum n_p S^0(products) - \sum n_r S^0(reactants)$$
$$= [1(S^0(CH_2Cl_2\ (g))] - [1(S^0(C\ (s)) + 1(S^0(H_2\ (g)) + 1(S^0(Cl_2\ (g))]$$
$$= [1(270.2\ J/K)] - [1(5.7\ J/K) + 1(130.7\ J/K) + 1(223.1\ J/K)]$$
$$= [270.2\ J/K] - [359.5\ J/K]$$
$$= -89.3\ J/K$$

The moles of gas are decreasing.

Check: The units (J/K) are correct. The answer is negative, which is consistent with 2 moles of gas going to 1 mole of gas.

28. **Given:** NF_4 (*g*) formed from elements in standard states **Find:** $\Delta S°$ and rationalize sign

Conceptual Plan: Write balanced reaction, then $\Delta S^0_{rxn} = \sum n_p S^0(products) - \sum n_r S^0(reactants)$

Solution: ½ N_2 (*g*) + 3/2 F_2 (*g*) → NF_3 (*g*)

Reactant/Product	S^0(J/molK from Appendix IIB)
N_2 (*g*)	191.6
F_2 (*g*)	202.79
NF_3 (*g*)	260.8

Be sure to pull data for the correct formula and phase.

$$\Delta S^0_{rxn} = \sum n_p S^0(products) - \sum n_r S^0(reactants)$$
$$= [1(S^0(NF_3\ (g))] - [1/2(S^0(N_2\ (g)) + 3/2(S^0(F_2\ (g))]$$
$$= [1(260.8\ J/K)] - [1/2(191.6\ J/K) + 3/2(202.79\ J/K)]$$ The moles of gas are decreasing.
$$= [260.8\ J/K] - [399.985\ J/K]$$
$$= -139.2\ J/K$$

Check: The units (J/K) are correct. The answer is negative, which is consistent with 2 moles of simple gases going to 1 mole of a complex gas.

29. **Given:** methanol (CH_3OH) combustion at 25 °C **Find:** $\Delta H°_{rxn}$, $\Delta S°_{rxn}$, $\Delta G°_{rxn}$, and spontaneity

Conceptual Plan: write balanced reaction then $\Delta H^0_{rxn} = \sum n_p H^0_f(products) - \sum n_r H^0_f(reactants)$ **then**

$\Delta S^0_{rxn} = \sum n_p S^0(products) - \sum n_r S^0(reactants)$ **then °C → K then J/K → kJ/K then**

$K = 273.15 + °C$ $\qquad \frac{1\,kJ}{1000\,J}$

$\Delta H°_{rxn}, \Delta S°_{rxn}, T$ → $\Delta G°$

$$\Delta G = \Delta H_{rxn} - T\Delta S_{rxn}$$

Solution: combustion is combination with oxygen to form carbon dioxide and water

2 CH_3OH (*l*) + 3 O_2 (*g*) → 2 CO_2 (*g*) + 4 H_2O (*g*)

Reactant/Product	ΔH^0_f(kJ/mol from Appendix IIB)
CH_3OH (*l*)	– 238.6
O_2 (*g*)	0.0
CO_2 (*g*)	– 393.5
H_2O (*g*)	– 241.8

Be sure to pull data for the correct formula and phase.

$$\Delta H^\circ_{rxn} = \Sigma n_p \Delta H^\circ_f(products) - \Sigma n_r \Delta H^\circ_f(reactants)$$

$$= [2(\Delta H^\circ_f(CO_2(g))) + 4(\Delta H^\circ_f(H_2O(g)))] - [2(\Delta H^\circ_f(CH_3OH(l))) + 3(\Delta H^\circ_f(O_2(g)))]$$

$$= [2(-393.5\text{ kJ}) + 4(-241.8\text{ kJ})] - [2(-238.6\text{ kJ}) + 3(0.0\text{ kJ})]$$

$$= [-1754.2\text{ kJ}] - [-477.2\text{ kJ}]$$

$$= -1277\text{ kJ}$$

then

Reactant/Product	S^0(J/molK from Appendix IIB)
CH_3OH (*l*)	126.8
O_2 (*g*)	205.2
CO_2 (*g*)	213.8
H_2O (*g*)	188.8

Be sure to pull data for the correct formula and phase.

$$\Delta S^\circ_{rxn} = \Sigma n_p S^\circ(products) - \Sigma n_r S^\circ(reactants)$$

$$= [2(S^\circ(CO_2(g))) + 4(S^\circ(H_2O(g)))] - [2(S^\circ(CH_3OH(l))) + 3(S^\circ(O_2(g)))]$$

$$= [2(213.8\text{ J/K}) + 4(188.8\text{ J/K})] - [2(126.8\text{ J/K}) + 3(205.2\text{ J/K})]$$

$$= [1182.8\text{ J/K}] - [869.2\text{ J/K}]$$

$$= 313.6\text{ J/K}$$

then

$T = 273.15 + 25\ °C = 298\text{ K}$ then $+313.6\frac{\text{J}}{\text{K}} \times \frac{1\text{ kJ}}{1000\text{ J}} = +0.3136\frac{\text{kJ}}{\text{K}}$ then

$\Delta G = \Delta H_{rxn} - T\Delta S_{rxn} = -1277\text{ kJ} - (298\text{ K})\left(+0.3136\frac{\text{kJ}}{\text{K}}\right) = -1370.\text{ kJ} = -1.370 \times 10^6\text{ J}$ so the reaction is spontaneous.

Check: The units (kJ, J/K, and kJ) are correct. Combustion reactions are exothermic and we see a large negative enthalpy. We expect a large positive entropy because we have an increase in the number of moles of gas. The free energy is the sum of two negative terms so we expect a large negative free energy and the reaction is spontaneous.

30. **Given:** form glucose ($C_6H_{12}O_6$) and oxygen from sunlight, carbon dioxide, and water at 25 °C
Find: ΔH°_{rxn}, ΔS°_{rxn}, ΔG°_{rxn}, and sponteneity

Conceptual Plan: write balanced reaction then $\Delta H^\circ_{rxn} = \sum n_p H^\circ_f(products) - \sum n_r H^\circ_f(reactants)$ **then** $\Delta S^\circ_{rxn} = \sum n_p S^\circ(products) - \sum n_r S^\circ(reactants)$ **then °C → K then J/K→ kJ/K then ΔH°_{rxn}, ΔS°_{rxn}, T → ΔG°**

$K = 273.15 + °C$ $\quad \frac{1\text{kJ}}{1000\text{ J}}$ $\quad \Delta G = \Delta H_{rxn} - T\Delta S_{rxn}$

Solution: $6\ CO_2\ (g) + 6\ H_2O\ (l) \rightarrow C_6H_{12}O_6\ (s) + 6\ O_2\ (g)$

Reactant/Product	ΔH^0_f(kJ/mol from Appendix IIB)
CO_2 (*g*)	– 393.5
H_2O (*l*)	– 285.8
$C_6H_{12}O_6$ (*s*)	– 1273.3
O_2 (*g*)	0.0

Be sure to pull data for the correct formula and phase.

$$\Delta H^0_{rxn} = \sum n_p \Delta H^0_f(products) - \sum n_r \Delta H^0_f(reactants)$$
$$= [1(\Delta H^0_f(C_6H_{12}O_6(s))) + 6(\Delta H^0_f(O_2(g)))] - [6(\Delta H^0_f(CO_2(g))) + 6(\Delta H^0_f(H_2O(l)))]$$
$$= [1(-1273.3 \text{ kJ}) + 6(0.0 \text{ kJ})] - [6(-393.5 \text{ kJ}) + 6(-285.8 \text{ kJ})] \quad \text{then}$$
$$= [-1273.3 \text{ kJ}] - [-4075.8 \text{ kJ}]$$
$$= +2802.5 \text{ kJ}$$

Reactant/Product	S^0(J/molK from Appendix IIB)
CO_2 (*g*)	213.8
H_2O (*l*)	70.0
$C_6H_{12}O_6$ (*s*)	212.1
O_2 (*g*)	205.2

Be sure to pull data for the correct formula and phase.

$$\Delta S^0_{rxn} = \sum n_p S^0(products) - \sum n_r S^0(reactants)$$
$$= [1(S^0(C_6H_{12}O_6(s))) + 6(S^0(O_2(g)))] - [6(S^0(CO_2(g))) + 6(S^0(H_2O(l)))]$$
$$= [1(212.1 \text{ J/K}) + 6(205.2 \text{ J/K})] - [6(213.8 \text{ J/K}) + 6(70.0 \text{ J/K})] \quad \text{then}$$
$$= [1443.3 \text{ J/K}] - [1702.8 \text{ J/K}]$$
$$= -259.5 \text{ J/K}$$

$$T = 273.15 + 25\ °\text{C} = 298 \text{ K} \quad \text{then} \quad -259.5 \frac{\text{J}}{\text{K}} \times \frac{1 \text{ kJ}}{1000 \text{ J}} = -0.2595 \frac{\text{kJ}}{\text{K}} \quad \text{then}$$

$$\Delta G = \Delta H_{rxn} - T\Delta S_{rxn} = +2802.5 \text{ kJ} - (298 \text{ K})\left(-0.2595 \frac{\text{kJ}}{\text{K}}\right) = +2879.8 \text{ kJ} = +2.8798 \times 10^6 \text{ J}$$ so the reaction is nonspontaneous.

Check: The units (kJ, J/K, and kJ) are correct. The reaction requires the input of light energy, so we expect that this will be an endothermic reaction. We expect a negative entropy change because we are going from 6 moles of a gas and 6 moles of a liquid to 6 moles of a gas and one mole of a solid. The free energy is the sum of two positive terms so we expect a large positive free energy and the reaction is nonspontaneous. Photosynthesis does not happen on its own, light energy must be added to make the process move forward.

31. a) **Given:** N_2O_4 (*g*) → 2 NO_2 (*g*) at 25 °C
Find: $\Delta H°_{rxn}$, $\Delta S°_{rxn}$, $\Delta G°_{rxn}$, spontaneity and can temperature be changed to make it spontaneous ?
Conceptual Plan: $\Delta H^0_{rxn} = \sum n_p H^0_f(products) - \sum n_r H^0_f(reactants)$ **then**
$\Delta S^0_{rxn} = \sum n_p S^0(products) - \sum n_r S^0(reactants)$ **then °C → K then J/K → kJ/K then**
$K = 273.15 + °C$ $\quad \frac{1 \text{kJ}}{1000 \text{ J}}$

$\Delta H°_{rxn}$, ΔS_{rxn}, T → ΔG
$\Delta G = \Delta H_{rxn} - T\Delta S_{rxn}$

Solution:

Reactant/Product	ΔH^0_f(kJ/mol from Appendix IIB)
N_2O_4 (*g*)	11.1
NO_2 (*g*)	33.2

Be sure to pull data for the correct formula and phase.

$$\Delta H^0_{rxn} = \sum n_p \Delta H^0_f(products) - \sum n_r \Delta H^0_f(reactants)$$
$$= [2(\Delta H^0_f(NO_2(g)))] - [1(\Delta H^0_f(N_2O_4(g)))]$$
$$= [2(33.2 \text{ kJ})] - [1(11.1 \text{ kJ})] \quad \text{then}$$
$$= [66.4 \text{ kJ}] - [11.1 \text{ kJ}]$$
$$= +55.3 \text{ kJ}$$

Reactant/Product	S^0(J/molK from Appendix IIB)
N_2O_4 (*g*)	304.4
NO_2 (*g*)	240.1

Be sure to pull data for the correct formula and phase.

$$\Delta S^0_{rxn} = \sum n_p S^0(products) - \sum n_r S^0(reactants)$$
$$= [2(S^0(NO_2\ (g)))] - [1(S^0(N_2O_4\ (g)))]$$
$$= [2(240.1\ J/K)] - [1(304.4\ J/K)] \quad \text{then} \quad T = 273.15 + 25\ °C = 298\ K \quad \text{then}$$
$$= [480.2\ J/K] - [304.4\ J/K]$$
$$= +175.8\ J/K$$

$$+175.8 \frac{J}{K} \times \frac{1\ kJ}{1000\ J} = +0.1758 \frac{kJ}{K} \quad \text{then}$$

$$\Delta G^0 = \Delta H^0_{rxn} - T\Delta S^0_{rxn} = +55.3\ kJ - (298\ K)\left(+0.1758 \frac{kJ}{K}\right) = +2.9\ kJ = +2.9 \times 10^3\ J$$ so the reaction is nonspontaneous. It can be made spontaneous by raising the temperature.

Check: The units (kJ, J/K, and kJ) are correct. The reaction requires the breaking of a bond, so we expect that this will be an endothermic reaction. We expect a positive entropy change because we are increasing the number of moles of gas. Since the positive enthalpy term dominates at room temperature, the reaction is nonspontaneous. The second term can dominate if we raise the temperature high enough.

b) **Given:** NH_4Cl (*s*) → HCl (*g*) + NH_3 (*g*) at 25 °C
Find: $\Delta H°_{rxn}$, $\Delta S°_{rxn}$, $\Delta G°_{rxn}$, spontaneity and can temperature be changed to make it spontaneous?
Conceptual Plan: $\Delta H^0_{rxn} = \sum n_p H^0_f(products) - \sum n_r H^0_f(reactants)$ **then**

$\Delta S^0_{rxn} = \sum n_p S^0(products) - \sum n_r S^0(reactants)$ **then °C→K then J/K→ kJ/K then $\Delta H°_{rxn}$, ΔS_{rxn}, T→ ΔG**

$K = 273.15 + °C$ $\quad \frac{1\ kJ}{1000\ J}$ $\quad \Delta G = \Delta H_{rxn} - T\Delta S_{rxn}$

Solution:

Reactant/Product	ΔH^0_f(kJ/mol from Appendix IIB)
NH_4Cl (*s*)	−314.4
HCl (*g*)	− 92.3
NH_3 (*g*)	− 45.9

Be sure to pull data for the correct formula and phase.

$$\Delta H^0_{rxn} = \sum n_p \Delta H^0_f(products) - \sum n_r \Delta H^0_f(reactants)$$
$$= [1(\Delta H^0_f(HCl\ (g))) + 1(\Delta H^0_f(NH_3\ (g)))] - [1(\Delta H^0_f(NH_4Cl\ (g)))]$$
$$= [1(-92.3\ kJ) + 1(-45.9\ kJ)] - [1(-314.4\ kJ)] \quad \text{then}$$
$$= [-138.2\ kJ] - [-314.4\ kJ]$$
$$= +176.2\ kJ$$

Reactant/Product	S^0(J/molK from Appendix IIB)
NH_4Cl (*s*)	94.6
HCl (*g*)	186.9
NH_3 (*g*)	192.8

Be sure to pull data for the correct formula and phase.

$$\Delta S^0_{rxn} = \sum n_p S^0(products) - \sum n_r S^0(reactants)$$
$$= [1(S^0(\text{HCl}\ (g))) + 1(S^0(\text{NH}_3\ (g)))] - [1(S^0(\text{NH}_4\text{Cl}\ (g)))]$$
$$= [1(186.9\ \text{J/K}) + 1(192.8\ \text{J/K})] - [1(94.6\ \text{J/K})]$$
$$= [379.7\ \text{J/K}] - [94.6\ \text{J/K}]$$
$$= +\ 285.1\ \text{J/K}$$

then $T = 273.15 + 25\ °\text{C} = 298\ \text{K}$ then

$$+285.1 \frac{\text{J}}{\text{K}} \times \frac{1\ \text{kJ}}{1000\ \text{J}} = +\ 0.2851 \frac{\text{kJ}}{\text{K}}$$ then

$$\Delta G^0 = \Delta H^0_{rxn} - T\Delta S^0_{rxn} = +\ 176.2\ \text{kJ} - (298\ \text{K})\left(+\ 0.2851 \frac{\text{kJ}}{\text{K}}\right) = +\ 91.2\ \text{kJ} = +\ 9.12 \times 10^4\ \text{J}$$ so the reaction is nonspontaneous. It can be made spontaneous by raising the temperature.

Check: The units (kJ, J/K, and kJ) are correct. The reaction requires the breaking of a bond, so we expect that this will be an endothermic reaction. We expect a positive entropy change because we are increasing the number of moles of gas. Since the positive enthalpy term dominates at room temperature, the reaction is nonspontaneous. The second term can dominate if we raise the temperature high enough.

c) **Given:** $3\ H_2\ (g) + Fe_2O_3\ (s)$ → $2\ Fe\ (s) + 3\ H_2O\ (g)$ at 25 °C

Find: $\Delta H°_{rxn}$, $\Delta S°_{rxn}$, $\Delta G°_{rxn}$, spontaneity and can temperature be changed to make it spontaneous ?

Conceptual Plan: $\Delta H^0_{rxn} = \sum n_p H^0_f(products) - \sum n_r H^0_f(reactants)$ **then**

$\Delta S^0_{rxn} = \sum n_p S^0(products) - \sum n_r S^0(reactants)$ **then °C → K then J/K → kJ/K then**

$\text{K} = 273.15 + °\text{C}$ $\qquad \frac{1\text{kJ}}{1000\ \text{J}}$

$\Delta H°_{rxn}$, ΔS_{rxn}, T → ΔG

$\Delta G = \Delta H_{rxn} - T\Delta S_{rxn}$

Solution:

Reactant/Product	ΔH^0_f(kJ/mol from Appendix IIB)
$H_2\ (g)$	0.0
$Fe_2O_3\ (s)$	− 824.2
$Fe\ (s)$	0.0
$H_2O\ (g)$	− 241.8

Be sure to pull data for the correct formula and phase.

$$\Delta H^0_{rxn} = \sum n_p \Delta H^0_f(products) - \sum n_r \Delta H^0_f(reactants)$$
$$= [2(\Delta H^0_f(\text{Fe}\ (s))) + 3(\Delta H^0_f(\text{H}_2\text{O}\ (g)))] - [3(\Delta H^0_f(\text{H}_2\ (g))) + 1(\Delta H^0_f(\text{Fe}_2\text{O}_3\ (s)))]$$
$$= [2(0.0\ \text{kJ}) + 3(-\ 241.8\ \text{kJ})] - [3(0.0\ \text{kJ}) + 1(-\ 824.2\ \text{kJ})]$$
$$= [-\ 725.4\ \text{kJ}] - [-824.2\ \text{kJ}]$$
$$= +\ 98.8\ \text{kJ}$$

then

Reactant/Product	S^0(J/molK from Appendix IIB)
$H_2\ (g)$	130.7
$Fe_2O_3\ (s)$	87.4
$Fe\ (s)$	27.3
$H_2O\ (g)$	188.8

Be sure to pull data for the correct formula and phase.

$$\Delta S^0_{rxn} = \sum n_p S^0(products) - \sum n_r S^0(reactants)$$
$$= [2(S^0(\text{Fe}\ (s))) + 3(S^0(\text{H}_2\text{O}\ (g)))] - [3(S^0(\text{H}_2\ (g))) + 1(S^0(\text{Fe}_2\text{O}_3\ (s)))]$$
$$= [2(27.3\ \text{J/K}) + 3(188.8\ \text{J/K})] - [3(130.7\ \text{J/K}) + 1(87.4\ \text{J/K})]$$
$$= [621.0\ \text{J/K}] - [479.5\ \text{J/K}]$$
$$= +\ 141.5\ \text{J/K}$$

then

$T = 273.15 + 25\ °C = 298\ K$ then $+141.5 \frac{J}{K} \times \frac{1\ kJ}{1000\ J} = +0.1415 \frac{kJ}{K}$ then

$\Delta G^0 = \Delta H^0_{rxn} - T\Delta S^0_{rxn} = +98.8\ kJ - (298\ K)\left(+0.1415 \frac{kJ}{K}\right) = +56.6\ kJ = +5.66 \times 10^4\ J$ so the reaction is nonspontaneous. It can be made spontaneous by raising the temperature.

Check: The units (kJ, J/K, and kJ) are correct. The reaction requires the breaking of a bond, so we expect that this will be an endothermic reaction. We expect a positive entropy change because there is no change in the number of moles of gas, but the product gas is more complex. Since the positive enthalpy term dominates at room temperature, the reaction is nonspontaneous. The second term can dominate if we raise the temperature high enough. This process is the opposite of rusting, so we are not surprised that it is nonspontaneous.

d) **Given:** $N_2\ (g) + 3\ H_2\ (g) \rightarrow 2\ NH_3\ (g)$ at 25 °C
Find: $\Delta H°_{rxn}$, $\Delta S°_{rxn}$, $\Delta G°_{rxn}$, spontaneity and can temperature be changed to make it spontaneous

Conceptual Plan: $\Delta H^0_{rxn} = \sum n_p H^0_f(products) - \sum n_r H^0_f(reactants)$ **then**

$\Delta S^0_{rxn} = \sum n_p S^0(products) - \sum n_r S^0(reactants)$ **then °C → K then J/K → kJ/K then**

$K = 273.15 + °C$ $\frac{1 kJ}{1000\ J}$

$\Delta H°_{rxn}$, ΔS_{rxn}, T → ΔG

$\Delta G = \Delta H_{rxn} - T\Delta S_{rxn}$

Solution:

Reactant/Product	ΔH^0_f(kJ/mol from Appendix IIB)
$N_2\ (g)$	0.0
$H_2\ (g)$	0.0
$NH_3\ (g)$	− 45.9

Be sure to pull data for the correct formula and phase.

$$\begin{aligned}\Delta H^0_{rxn} &= \sum n_p \Delta H^0_f(products) - \sum n_r \Delta H^0_f(reactants)\\ &= [2(\Delta H^0_f(NH_3\ (g)))] - [1(\Delta H^0_f(N_2\ (g))) + 3(\Delta H^0_f(H_2\ (g)))]\\ &= [2(-45.9\ kJ)] - [1(0.0\ kJ) + 3(0.0\ kJ)]\\ &= [-91.8\ kJ] - [0.0\ kJ]\\ &= -91.8\ kJ\end{aligned}$$

then

Reactant/Product	S^0(J/molK from Appendix IIB)
$N_2\ (g)$	191.6
$H_2\ (g)$	130.7
$NH_3\ (g)$	192.8

Be sure to pull data for the correct formula and phase.

$$\begin{aligned}\Delta S^0_{rxn} &= \sum n_p S^0(products) - \sum n_r S^0(reactants)\\ &= [2(S^0(NH_3\ (g)))] - [1(S^0(N_2\ (g))) + 3(S^0(H_2\ (g)))]\\ &= [2(192.8\ J/K)] - [1(191.6\ J/K) + 3(130.7\ J/K)]\\ &= [385.6\ J/K] - [583.7\ J/K]\\ &= -198.1\ J/K\end{aligned}$$

then

$T = 273.15 + 25\ °C = 298\ K$ then $-198.1 \frac{J}{K} \times \frac{1\ kJ}{1000\ J} = -0.1981 \frac{kJ}{K}$ then

$\Delta G^0 = \Delta H^0_{rxn} - T\Delta S^0_{rxn} = -91.8\ kJ - (298\ K)\left(-0.1981 \frac{kJ}{K}\right) = -32.8\ kJ = -3.28 \times 10^4\ J$ so the reaction is spontaneous.

Check: The units (kJ, J/K, and kJ) are correct. The reaction requires the breaking of a bond, so we expect that this will be an endothermic reaction. We expect a positive entropy change because are increasing the number of moles of gas. Since the negative enthalpy term dominates at room temperature, the reaction is spontaneous. The second term can dominate if we raise the temperature high enough.

32. a) **Given:** $2\ CH_4\ (g) \rightarrow C_2H_6\ (g) + H_2\ (g)$ at 25 °C
Find: ΔH°_{rxn}, ΔS°_{rxn}, ΔG°_{rxn}, spontaneity and can temperature be changed to make it spontaneous?
Conceptual Plan: $\Delta H^0_{rxn} = \sum n_p H^0_f(products) - \sum n_r H^0_f(reactants)$ **then**
$\Delta S^0_{rxn} = \sum n_p S^0(products) - \sum n_r S^0(reactants)$ **then °C → K then J/K → kJ/K then**

$K = 273.15 + °C$ $\frac{1 kJ}{1000 J}$

ΔH°_{rxn}, ΔS_{rxn}, T → ΔG

$\Delta G = \Delta H_{rxn} - T\Delta S_{rxn}$

Solution:

Reactant/Product	ΔH^0_f(kJ/mol from Appendix IIB)
$CH_4\ (g)$	− 74.6
$C_2H_6\ (g)$	− 84.6
$H_2\ (g)$	0.0

Be sure to pull data for the correct formula and phase.

$$\begin{aligned}\Delta H^0_{rxn} &= \sum n_p \Delta H^0_f(products) - \sum n_r \Delta H^0_f(reactants)\\ &= [1(\Delta H^0_f(C_2H_6\ (g))) + 1(\Delta H^0_f(H_2\ (g)))] - [2(\Delta H^0_f(CH_4\ (g)))]\\ &= [1(-84.6\ kJ) + 1(0.0\ kJ)] - [2(-74.6\ kJ)]\\ &= [-84.6\ kJ] - [-149.2\ kJ]\\ &= +64.6\ kJ\end{aligned}$$ then

Reactant/Product	S^0(J/molK from Appendix IIB)
$CH_4\ (g)$	186.3
$C_2H_6\ (g)$	229.2
$H_2\ (g)$	130.7

Be sure to pull data for the correct formula and phase.

$$\begin{aligned}\Delta S^0_{rxn} &= \sum n_p S^0(products) - \sum n_r S^0(reactants)\\ &= [1(S^0(C_2H_6\ (g))) + 1(S^0(H_2\ (g)))] - [2(S^0(CH_4\ (g)))]\\ &= [1(229.2\ J/K) + 1(130.7\ J/K)] - [2(186.3\ J/K)]\\ &= [359.9\ J/K] - [372.6\ J/K]\\ &= -12.7\ J/K\end{aligned}$$ then

$T = 273.15 + 25\ °C = 298\ K$ then $-12.7\frac{J}{K} \times \frac{1\ kJ}{1000\ J} = -0.0127\frac{kJ}{K}$ then

$$\Delta G^0 = \Delta H^0_{rxn} - T\Delta S^0_{rxn} = +64.6\ kJ - (298\ K)\left(-0.0127\frac{kJ}{K}\right) = +68.4\ kJ = +6.84 \times 10^4\ J$$

so the reaction is nonspontaneous. Since both terms are positive, this reaction cannot be spontaneous at any temperature.
Check: The units (kJ, J/K, and kJ) are correct. The two moles of methane has a lower enthalpy than one mole of ethane, so we expect that this will be an endothermic reaction. We expect a very small entropy change because the number of moles of gas is unchanged. Since both terms are positive, the reaction is nonspontaneous. Since both terms are positive, this reaction cannot be spontaneous at any temperature.

b) **Given:** $2\ NH_3\ (g) \rightarrow N_2H_4\ (g) + H_2\ (g)$ at 25 °C
Find: ΔH°_{rxn}, ΔS°_{rxn}, ΔG°_{rxn}, spontaneity and can temperature be changed to make it spontaneous?
Conceptual Plan: $\Delta H^0_{rxn} = \sum n_p H^0_f(products) - \sum n_r H^0_f(reactants)$ **then**

$\Delta S^0_{rxn} = \sum n_p S^0(products) - \sum n_r S^0(reactants)$ **then °C → K then J/K → kJ/K then**

$K = 273.15 + °C$ $\frac{1 kJ}{1000 J}$

$\Delta H°_{rxn}$, ΔS_{rxn}, T → ΔG

$\Delta G = \Delta H_{rxn} - T\Delta S_{rxn}$

Solution:

Reactant/Product	ΔH^0_f(kJ/mol from Appendix IIB)
$NH_3\ (g)$	− 45.9
$N_2H_4\ (g)$	95.4
$H_2\ (g)$	0.0

Be sure to pull data for the correct formula and phase.

$$\Delta H^0_{rxn} = \sum n_p \Delta H^0_f(products) - \sum n_r \Delta H^0_f(reactants)$$
$$= [1(\Delta H^0_f(N_2H_4\ (g))) + 1(\Delta H^0_f(H_2\ (g)))] - [2(\Delta H^0_f(NH_3\ (g)))]$$
$$= [1(95.4\ kJ) + 1(0.0\ kJ)] - [2(-45.9\ kJ)]$$
$$= [95.4\ kJ] - [-91.8\ kJ]$$
$$= +187.2\ kJ$$
then

Reactant/Product	S^0(J/molK from Appendix IIB)
$NH_3\ (g)$	192.8
$N_2H_4\ (g)$	238.5
$H_2\ (g)$	130.7

Be sure to pull data for the correct formula and phase.

$$\Delta S^0_{rxn} = \sum n_p S^0(products) - \sum n_r S^0(reactants)$$
$$= [1(S^0(N_2H_4\ (g))) + 1(S^0(H_2\ (g)))] - [2(S^0(NH_3\ (g)))]$$
$$= [1(238.5\ J/K) + 1(130.7\ J/K)] - [2(192.8\ J/K)]$$
$$= [369.2\ J/K] - [385.6\ J/K]$$
$$= -16.4\ J/K$$
then

$T = 273.15 + 25\ °C = 298\ K$ then $-16.4 \frac{J}{K} \times \frac{1\ kJ}{1000\ J} = -0.0164 \frac{kJ}{K}$ then

$\Delta G^0 = \Delta H^0_{rxn} - T\Delta S^0_{rxn} = +187.2\ kJ - (298\ K)\left(-0.0164 \frac{kJ}{K}\right) = +192.1\ kJ = +1.921 \times 10^5\ J$ so the reaction is nonspontaneous. Both terms are positive and so the reaction is not spontaneous at any temperature.

Check: The units (kJ, J/K, and kJ) are correct. The N_2H_4 has such a high enthalpy of formation compared to ammonia, so we expect that this will be an endothermic reaction. We expect a very small entropy change because the number of moles of gas is unchanged. Since both terms are positive, the reaction is nonspontaneous. Since both terms are positive, this reaction cannot be spontaneous at any temperature.

c) **Given:** $N_2\ (g) + O_2\ (g) \rightarrow 2\ NO\ (g)$ at 25 °C

Find: $\Delta H°_{rxn}$, $\Delta S°_{rxn}$, $\Delta G°_{rxn}$, spontaneity and can temperature be changed to make it spontaneous?

Conceptual Plan: $\Delta H^0_{rxn} = \sum n_p H^0_f(products) - \sum n_r H^0_f(reactants)$ **then**

$\Delta S^0_{rxn} = \sum n_p S^0(products) - \sum n_r S^0(reactants)$ **then°C → K then J/K → kJ/K then $\Delta H°_{rxn}$,ΔS_{rxn}, T → ΔG**

$K = 273.15 + °C$ $\frac{1 kJ}{1000 J}$ $\Delta G = \Delta H_{rxn} - T\Delta S_{rxn}$

Solution:

Reactant/Product	ΔH^0_f(kJ/mol from Appendix IIB)
$N_2\ (g)$	0.0
$O_2\ (g)$	0.0
$NO\ (g)$	91.3

Be sure to pull data for the correct formula and phase.

$$\Delta H^0_{rxn} = \sum n_p \Delta H^0_f(products) - \sum n_r \Delta H^0_f(reactants)$$
$$= [2(\Delta H^0_f(\text{NO }(g)))] - [1(\Delta H^0_f(\text{N}_2\ (g))) + 1(\Delta H^0_f(\text{O}_2\ (g)))]$$
$$= [2(91.3 \text{ kJ})] - [1(0.0 \text{ kJ}) + 1(0.0 \text{ kJ})] \qquad \text{then}$$
$$= [182.6 \text{ kJ}] - [0.0 \text{ kJ}]$$
$$= +\,182.6 \text{ kJ}$$

Reactant/Product	S^0(J/molK from Appendix IIB)
N_2 (*g*)	191.6
O_2 (*g*)	205.2
NO (*g*)	210.8

Be sure to pull data for the correct formula and phase.

$$\Delta S^0_{rxn} = \sum n_p S^0(products) - \sum n_r S^0(reactants)$$
$$= [2(S^0(\text{NO }(g)))] - [1(S^0(\text{N}_2\ (g))) + 1(S^0(\text{O}_2\ (g)))]$$
$$= [2(210.8 \text{ J/K})] - [1(191.6 \text{ J/K}) + 1(205.2 \text{ J/K})] \qquad \text{then}$$
$$= [421.6 \text{ J/K}] - [396.8 \text{ J/K}]$$
$$= +\,24.8 \text{ J/K}$$

$T = 273.15 + 25\ °\text{C} = 298$ K then $+24.8 \frac{\text{J}}{\text{K}} \times \frac{1 \text{ kJ}}{1000 \text{ J}} = +\,0.0248 \frac{\text{kJ}}{\text{K}}$ then

$\Delta G^0 = \Delta H^0{}_{rxn} - T\Delta S^0{}_{rxn} = +182.6 \text{ kJ} - (298 \text{ K})\left(0.0248 \frac{\text{kJ}}{\text{K}}\right) = +175.2 \text{ kJ} = +1.752 \times 10^5 \text{ J}$ so the reaction is nonspontaneous. Spontaneous at high temperatures.

Check: The units (kJ, J/K, and kJ) are correct. The enthalpy is twice the enthalpy of formation of NO. We expect a very small entropy change because the number of moles of gas is unchanged. Since the positive enthalpy term dominates at room temperature, the reaction is nonspontaneous. The second term can dominate if we raise the temperature high enough.

d) **Given:** $2\ KClO_3\ (s) \rightarrow 2\ KCl\ (s) + O_2\ (g)$ at 25 °C
Find: $\Delta H°_{rxn}$, $\Delta S°_{rxn}$, $\Delta G°_{rxn}$, spontaneity and can temperature be changed to make it spontaneous?

Conceptual Plan: $\Delta H^0_{rxn} = \sum n_p H^0_f(products) - \sum n_r H^0_f(reactants)$ **then**

$\Delta S^0_{rxn} = \sum n_p S^0(products) - \sum n_r S^0(reactants)$ **then °C→K then J/K → kJ/K then $\Delta H°_{rxn}$, ΔS_{rxn}, T→ ΔG**

$K = 273.15 + °C$ $\qquad \frac{1 \text{kJ}}{1000 \text{ J}}$ $\qquad \Delta G = \Delta H_{rxn} - T\Delta S_{rxn}$

Solution:

Reactant/Product	ΔH^0_f(kJ/mol from Appendix IIB)
$KClO_3$ (*s*)	− 397.7
KCl (*s*)	− 436.5
O_2 (*g*)	0.0

Be sure to pull data for the correct formula and phase.

$$\Delta H^0_{rxn} = \sum n_p \Delta H^0_f(products) - \sum n_r \Delta H^0_f(reactants)$$
$$= [2(\Delta H^0_f(\text{KCl }(s))) + 1(\Delta H^0_f(\text{O}_2\ (g)))] - [2(\Delta H^0_f(\text{KClO}_3\ (s)))]$$
$$= [2(-436.5 \text{ kJ}) + 3(0.0 \text{ kJ})] - [2(-397.7 \text{ kJ})] \qquad \text{then}$$
$$= [-873.0 \text{ kJ}] - [-795.4 \text{ kJ}]$$
$$= -\,77.6 \text{ kJ}$$

Reactant/Product	S^0(J/molK from Appendix IIB)
$KClO_3$ (*s*)	143.1
KCl (*s*)	82.6
O_2 (*g*)	205.2

Be sure to pull data for the correct formula and phase.

$$\begin{aligned}\Delta S^0_{rxn} &= \sum n_p S^0(products) - \sum n_r S^0(reactants)\\ &= [2(S^0(\text{KCl}\ (s))) + 3(S^0(\text{O}_2\ (g)))] - [2(S^0(\text{KClO}_3\ (s)))]\\ &= [2(82.6 \text{ J/K}) + 3(205.2 \text{ J/K})] - [2(143.1 \text{ J/K})] \quad \text{then}\\ &= [780.8 \text{ J/K}] - [286.2 \text{ J/K}]\\ &= +\ 494.6 \text{ J/K}\end{aligned}$$

$T = 273.15 + 25\ °\text{C} = 298$ K then $+494.6 \frac{\text{J}}{\text{K}} \times \frac{1 \text{ kJ}}{1000 \text{ J}} = +\ 0.4946 \frac{\text{kJ}}{\text{K}}$ then

$\Delta G^0 = \Delta H^0_{rxn} - T\Delta S^0_{rxn} = -\ 77.6 \text{ kJ} - (298 \text{ K})\left(0.4946 \frac{\text{kJ}}{\text{K}}\right) = -\ 225.0 \text{ kJ} = -\ 2.250 \times 10^6 \text{ J}$ so the reaction is spontaneous. Since both terms are negative, the reaction is spontaneous at all temperatures.

Check: The units (kJ, J/K, and kJ) are correct. The reaction is exothermic because the enthalpy of formation of KCl is less than that for $KClO_3$. We expect a positive entropy change because the number of moles of gas is increasing. Since both terms are negative, the reaction is spontaneous at all temperatures.

33. a) **Given:** $N_2O_4\ (g) \rightarrow 2\ NO_2\ (g)$ at 25 °C **Find:** $\Delta G°_{rxn}$, spontaneity and compare to #31. Determine which method would show how free energy changes with temperature.

Conceptual Plan: $\Delta G^0_{rxn} = \sum n_p \Delta G^0_f(products) - \sum n_r \Delta G^0_f(reactants)$ **then compare to #31**

Solution:

Reactant/Product	ΔG^0_f (kJ/mol from Appendix IIB)
$N_2O_4\ (g)$	99.8
$NO_2\ (g)$	51.3

Be sure to pull data for the correct formula and phase.

$$\begin{aligned}\Delta G^0_{rxn} &= \sum n_P \Delta G^0_f(products) - \sum n_R \Delta G^0_f(reactants)\\ &= [2(\Delta G^0_f(\text{NO}_2\ (g)))] - [1(\Delta G^0_f(\text{N}_2\text{O}_4\ (g)))]\\ &= [2(51.3 \text{ kJ})] - [1(99.8 \text{ kJ})] \quad \text{so the reaction is nonspontaneous.}\\ &= [102.6 \text{ kJ}] - [99.8 \text{ kJ}]\\ &= +\ 2.8 \text{ kJ}\end{aligned}$$

The value is similar to #31.

Check: The units (kJ) are correct. The free energy of the products is greater than the reactants, so the answer is positive and the reaction is nonspontaneous. The answer is the same as in #31 within the error of the calculation.

b) **Given:** $NH_4Cl\ (s) \rightarrow HCl\ (g) + NH_3\ (g)$ at 25 °C
Find: $\Delta G°_{rxn}$, spontaneity and compare to #31

Conceptual Plan: $\Delta G^0_{rxn} = \sum n_p \Delta G^0_f(products) - \sum n_r \Delta G^0_f(reactants)$ **then compare to #31**

Solution:

Reactant/Product	ΔG^0_f (kJ/mol from Appendix IIB)
$NH_4Cl\ (s)$	− 202.9
$HCl\ (g)$	− 95.3
$NH_3\ (g)$	− 16.4

Be sure to pull data for the correct formula and phase.

$$\begin{aligned}\Delta G^0_{rxn} &= \sum n_P \Delta G^0_f(products) - \sum n_R \Delta G^0_f(reactants)\\ &= [1(\Delta G^0_f(\text{HCl}\ (g))) + 1(\Delta G^0_f(\text{NH}_3\ (g)))] - [1(\Delta G^0_f(\text{NH}_4\text{Cl}\ (g)))]\\ &= [1(-\ 95.3 \text{ kJ}) + 1(-\ 16.4 \text{ kJ})] - [1(-\ 202.9 \text{ kJ})] \quad \text{so the reaction is nonspontaneous.}\\ &= [-\ 111.7 \text{ kJ}] - [-202.9 \text{ kJ}]\\ &= +\ 91.2 \text{ kJ}\end{aligned}$$

The result is the same as in #31.

Check: The units (kJ) are correct. The answer matches #31.

c) **Given:** $3\ H_2\ (g) + Fe_2O_3\ (s) \rightarrow 2\ Fe\ (s) + 3\ H_2O\ (g)$ at 25 °C
Find: ΔG°_{rxn}, spontaneity and compare to #31

Conceptual Plan: $\Delta G^0_{rxn} = \sum n_p \Delta G^0_f(products) - \sum n_r \Delta G^0_f(reactants)$ **then compare to #31**

Solution:

Reactant/Product	ΔG^0_f(kJ/mol from Appendix IIB)
$H_2\ (g)$	0.0
$Fe_2O_3\ (s)$	− 742.2
$Fe\ (s)$	0.0
$H_2O\ (g)$	− 228.6

Be sure to pull data for the correct formula and phase.

$$\Delta G^0_{rxn} = \sum n_P \Delta G^0_f(products) - \sum n_R \Delta G^0_f(reactants)$$
$$= [2(\Delta G^0_f(\text{Fe }(s))) + 3(\Delta G^0_f(\text{H}_2\text{O }(g)))] - [3(\Delta G^0_f(\text{H}_2\ (g))) + 1(\Delta G^0_f(\text{Fe}_2\text{O}_3\ (s)))]$$
$$= [2(0.0\text{ kJ}) + 3(-228.6\text{ kJ})] - [3(0.0\text{ kJ}) + 1(-742.2\text{ kJ})]$$
$$= [-685.8\text{ kJ}] - [-742.2\text{ kJ}]$$
$$= +56.4\text{ kJ}$$

so the reaction is nonspontaneous. The value is similar to that in #31.
Check: The units (kJ) are correct. The answer is the same as in #31 within the error of the calculation.

d) **Given:** $N_2\ (g) + 3\ H_2\ (g) \rightarrow 2\ NH_3\ (g)$ at 25 °C
Find: ΔG°_{rxn}, spontaneity and compare to #31

Conceptual Plan: $\Delta G^0_{rxn} = \sum n_p \Delta G^0_f(products) - \sum n_r \Delta G^0_f(reactants)$ **then compare to #31**

Solution:

Reactant/Product	ΔG^0_f(kJ/mol from Appendix IIB)
$N_2\ (g)$	0.0
$H_2\ (g)$	0.0
$NH_3\ (g)$	− 16.4

Be sure to pull data for the correct formula and phase.

$$\Delta G^0_{rxn} = \sum n_P \Delta G^0_f(products) - \sum n_R \Delta G^0_f(reactants)$$
$$= [2(\Delta G^0_f(\text{NH}_3\ (g)))] - [1(\Delta G^0_f(\text{N}_2\ (g))) + 3(\Delta G^0_f(\text{H}_2\ (g)))]$$
$$= [2(-16.4\text{ kJ})] - [1(0.0\text{ kJ}) + 3(0.0\text{ kJ})]$$
$$= [-32.8\text{ kJ}] - [0.0\text{ kJ}]$$
$$= -32.8\text{ kJ}$$

so the reaction is spontaneous. The result is the same as in #31.
Check: The units (kJ) are correct. The answer matches #31.

Values calculated by the two methods are comparable. The method using ΔH° and ΔS° is longer, but it can be used to determine how ΔG° changes with temperature.

34. a) **Given:** $2\ CH_4\ (g) \rightarrow C_2H_6\ (g) + H_2\ (g)$ at 25 °C **Find:** ΔG°_{rxn}, spontaneity and compare to #32
Determine which method would show how free energy changes with temperature.

Conceptual Plan: $\Delta G^0_{rxn} = \sum n_p \Delta G^0_f(products) - \sum n_r \Delta G^0_f(reactants)$ **then compare to #32**

Solution:

Reactant/Product	ΔG^0_f(kJ/mol from Appendix IIB)
$CH_4\ (g)$	− 50.5
$C_2H_6\ (g)$	− 32.0
$H_2\ (g)$	0.0

Be sure to pull data for the correct formula and phase.

$$\Delta G^0_{rxn} = \sum n_P \Delta G^0_f(products) - \sum n_R \Delta G^0_f(reactants)$$
$$= [1(\Delta G^0_f(C_2H_6\ (g))) + 1(\Delta G^0_f(H_2\ (g)))] - [2(\Delta G^0_f(CH_4\ (g)))]$$
$$= [1(-32.0\text{ kJ}) + 1(0.0\text{ kJ})] - [2(-50.5\text{ kJ})]$$
$$= [-32.0\text{ kJ}] - [-101.0\text{ kJ}]$$
$$= +69.0\text{ kJ}$$

so the reaction is spontaneous. The value is similar to that in #32.
Check: The units (kJ) are correct. The answer is the same as in #32 within the error of the calculation.

b) **Given:** $2\ NH_3\ (g) \rightarrow N_2H_4\ (g) + H_2\ (g)$ at 25 °C
Find: $\Delta G°_{rxn}$, spontaneity and compare to #32
Conceptual Plan: $\Delta G^\circ_{rxn} = \sum n_p \Delta G^\circ_f(products) - \sum n_r \Delta G^\circ_f(reactants)$ **then compare to #32**
Solution:

Reactant/Product	ΔG^0_f(kJ/mol from Appendix IIB)
$NH_3\ (g)$	− 16.4
$N_2H_4\ (g)$	159.4
$H_2\ (g)$	0.0

Be sure to pull data for the correct formula and phase.
$$\Delta G^0_{rxn} = \sum n_P \Delta G^0_f(products) - \sum n_R \Delta G^0_f(reactants)$$
$$= [1(\Delta G^0_f(N_2H_4\ (g))) + 1(\Delta G^0_f(H_2\ (g)))] - [2(\Delta G^0_f(NH_3\ (g)))]$$
$$= [1(159.4\text{ kJ}) + 1(0.0\text{ kJ})] - [2(-16.4\text{ kJ})]$$
$$= [159.4\text{ kJ}] - [-32.8\text{ kJ}]$$
$$= +192.2\text{ kJ}$$

so the reaction is nonspontaneous. The value is similar to in #32.
Check: The units (kJ) are correct. The answer is the same as in #32 within the error of the calculation.

c) **Given:** $N_2\ (g) + O_2\ (g) \rightarrow 2\ NO\ (g)$ at 25 °C
Find: $\Delta G°_{rxn}$, spontaneity and compare to #32
Conceptual Plan: $\Delta G^\circ_{rxn} = \sum n_p \Delta G^\circ_f(products) - \sum n_r \Delta G^\circ_f(reactants)$ **then compare to #32**
Solution:

Reactant/Product	ΔG^0_f(kJ/mol from Appendix IIB)
$N_2\ (g)$	0.0
$O_2\ (g)$	0.0
$NO\ (g)$	87.6

Be sure to pull data for the correct formula and phase.
$$\Delta G^0_{rxn} = \sum n_P \Delta G^0_f(products) - \sum n_R \Delta G^0_f(reactants)$$
$$= [2(\Delta G^0_f(NO\ (g)))] - [1(\Delta G^0_f(N_2\ (g))) + 1(\Delta G^0_f(O_2\ (g)))]$$
$$= [2(87.6\text{ kJ})] - [1(0.0\text{ kJ}) + 1(0.0\text{ kJ})]$$
$$= [175.2\text{ kJ}] - [0.0\text{ kJ}]$$
$$= +175.2\text{ kJ}$$

so the reaction is nonspontaneous. The result is the same as in #31.
Check: The units (kJ) are correct. The answer matches #31.

d) **Given:** $2\ KClO_3\ (s) \rightarrow 2\ KCl\ (s) + 3\ O_2\ (g)$ at 25 °C
Find: $\Delta G°_{rxn}$, spontaneity and compare to #32
Conceptual Plan: $\Delta G^\circ_{rxn} = \sum n_p \Delta G^\circ_f(products) - \sum n_r \Delta G^\circ_f(reactants)$ **then compare to #32**
Solution:

Reactant/Product	ΔG_f^0(kJ/mol from Appendix IIB)
$KClO_3$ (*s*)	– 296.3
KCl (*s*)	– 408.5
O_2 (*g*)	0.0

Be sure to pull data for the correct formula and phase.

$$\Delta G^0_{rxn} = \sum n_P \Delta G_f^0(products) - \sum n_R \Delta G_f^0(reactants)$$
$$= [2(\Delta G_f^0(\text{KCl }(s))) + 1(\Delta G_f^0(\text{O}_2\ (g)))] - [2(\Delta G_f^0(\text{KClO}_3\ (s)))]$$
$$= [2(-408.5\text{ kJ}) + 3(0.0\text{ kJ})] - [2(-296.3\text{ kJ})]$$
$$= [-817.0\text{ kJ}] - [-592.6\text{ kJ}]$$
$$= -224.4\text{ kJ}$$

so the reaction is spontaneous. The value is similar to #32.

Check: The units (kJ) are correct. The answer is the same as in #32 within the error of the calculation.

Values calculated by the two methods are comparable. The method using $\Delta H°$ and $\Delta S°$ is longer, but it can be used to determine how $\Delta G°$ changes with temperature.

35. **Given:** 2 NO (*g*) + O_2 (*g*) → 2 NO_2 (*g*) **Find:** $\Delta G°_{rxn}$ and spontaneity at a) 298 K, b) 715 K, and 855 K

Conceptual Plan: $\Delta H^\circ_{rxn} = \sum n_p H_f^\circ(products) - \sum n_r H_f^\circ(reactants)$ **then**

$\Delta S^\circ_{rxn} = \sum n_p S^\circ(products) - \sum n_r S^\circ(reactants)$ **then J/K → kJ/K then $\Delta H°_{rxn}$, ΔS_{rxn}, T → ΔG**

$$\frac{1\text{ kJ}}{1000\text{ J}} \qquad \Delta G = \Delta H_{rxn} - T\Delta S_{rxn}$$

Solution:

Reactant/Product	ΔH_f^0(kJ/mol from Appendix IIB)
NO (*g*)	91.3
O_2 (*g*)	0.0
NO_2 (*g*)	33.2

Be sure to pull data for the correct formula and phase.

$$\Delta H^0_{rxn} = \sum n_p \Delta H_f^0(products) - \sum n_r \Delta H_f^0(reactants)$$
$$= [2(\Delta H_f^0(\text{NO}_2\ (g)))] - [2(\Delta H_f^0(\text{NO }(g))) + 1(\Delta H_f^0(\text{O}_2\ (g)))]$$
$$= [2(33.2\text{ kJ})] - [2(91.3\text{ kJ}) + 1(0.0\text{ kJ})]$$
$$= [66.4\text{ kJ}] - [182.6\text{ kJ}]$$
$$= -116.2\text{ kJ}$$

then

Reactant/Product	S^0(J/molK from Appendix IIB)
NO (*g*)	210.8
O_2 (*g*)	205.2
NO_2 (*g*)	240.1

Be sure to pull data for the correct formula and phase.

$$\Delta S^0_{rxn} = \sum n_p S^0(products) - \sum n_r S^0(reactants)$$
$$= [2(S^0(\text{NO}_2\ (g)))] - [2(S^0(\text{NO }(g))) + 1(S^0(\text{O}_2\ (g)))]$$
$$= [2(240.1\text{ J/K})] - [2(210.8\text{ J/K}) + 1(205.2\text{ J/K})]$$
$$= [480.2\text{ J/K}] - [626.8\text{ J/K}]$$
$$= -146.6\text{ J/K}$$

then $-146.6\frac{\text{J}}{\text{K}} \times \frac{1\text{ kJ}}{1000\text{ J}} = -0.1466\frac{\text{kJ}}{\text{K}}$ then

a) $\Delta G^0 = \Delta H^0_{rxn} - T\Delta S^0_{rxn} = -116.2\text{ kJ} - (298\text{ K})\left(-0.1466\frac{\text{kJ}}{\text{K}}\right) = -72.5\text{ kJ} = -7.25 \times 10^4\text{ J}$ so the reaction is spontaneous.

b) $\Delta G^0 = \Delta H^0_{rxn} - T\Delta S^0_{rxn} = -116.2\text{ kJ} - (715\text{ K})\left(-0.1466\frac{\text{kJ}}{\text{K}}\right) = -11.4\text{ kJ} = -1.14 \times 10^4\text{ J}$ so the reaction is spontaneous.

c) $\Delta G^0 = \Delta H^0_{rxn} - T\Delta S^0_{rxn} = -116.2\text{ kJ} - (855\text{ K})\left(-0.1466\frac{\text{kJ}}{\text{K}}\right) = +9.1\text{ kJ} = +9.1 \times 10^3\text{ J}$ so the reaction is nonspontaneous.

Check: The units (kJ) are correct. The enthalpy term dominates at low temperatures, making the reaction spontaneous. As the temperature increases, the decrease in entropy starts to dominate and in the last case the reaction is nonspontaneous.

36. **Given:** $CaCO_3\ (s) \rightarrow CaO\ (s) + CO_2\ (g)$ **Find:** ΔG°_{rxn} and spontaneity at a) 298 K, b) 1055 K, and 1455 K

Conceptual Plan: $\Delta H^0_{rxn} = \sum n_p H^0_f(products) - \sum n_r H^0_f(reactants)$ **then**

$\Delta S^0_{rxn} = \sum n_p S^0(products) - \sum n_r S^0(reactants)$ **then J/K → kJ/K then** $\Delta H^\circ_{rxn}, \Delta S_{rxn}, T \rightarrow \Delta G$

$\frac{1\text{kJ}}{1000\text{ J}}$ $\quad \Delta G = \Delta H_{rxn} - T\Delta S_{rxn}$

Solution:

Reactant/Product	ΔH^0_f(kJ/mol from Appendix IIB)
$CaCO_3\ (s)$	− 1207.6
$CaO\ (s)$	− 634.9
$CO_2\ (g)$	− 393.5

Be sure to pull data for the correct formula and phase.

$$\begin{aligned}\Delta H^0_{rxn} &= \sum n_p \Delta H^0_f(products) - \sum n_r \Delta H^0_f(reactants)\\ &= [1(\Delta H^0_f(\text{CaO }(g))) + 1(\Delta H^0_f(CO_2\ (g)))] - [1(\Delta H^0_f(CaCO_2\ (g)))]\\ &= [1(-634.9\text{ kJ}) + 1(-393.5\text{ kJ})] - [1(-1207.6\text{ kJ})]\\ &= [-1028.4\text{ kJ}] - [-1207.6\text{ kJ}]\\ &= +179.2\text{ kJ}\end{aligned}$$

then

Reactant/Product	S^0(J/molK from Appendix IIB)
$CaCO_3\ (s)$	91.7
$CaO\ (s)$	38.1
$CO_2\ (g)$	213.8

Be sure to pull data for the correct formula and phase.

$$\begin{aligned}\Delta S^0_{rxn} &= \sum n_p S^0(products) - \sum n_r S^0(reactants)\\ &= [1(S^0(\text{CaO }(g))) + 1(S^0(CO_2\ (g)))] - [1(S^0(CaCO_2\ (g)))]\\ &= [1(38.1\text{ J/K}) + 1(213.8\text{ J/K})] - [1(91.7\text{ J/K})]\\ &= [251.9\text{ J/K}] - [91.7\text{ J/K}]\\ &= +160.2\text{ J/K}\end{aligned}$$

then $+160.2\frac{\text{J}}{\text{K}} \times \frac{1\text{ kJ}}{1000\text{ J}} = +0.1602\frac{\text{kJ}}{\text{K}}$ then

a) $\Delta G^0 = \Delta H^0_{rxn} - T\Delta S^0_{rxn} = +179.2\text{ kJ} - (298\text{ K})\left(+0.1602\frac{\text{kJ}}{\text{K}}\right) = +131.5\text{ kJ} = +1.315 \times 10^5\text{ J}$ so the reaction is nonspontaneous.

b) $\Delta G^0 = \Delta H^0_{rxn} - T\Delta S^0_{rxn} = +179.2\text{ kJ} - (1055\text{ K})\left(+0.1602\frac{\text{kJ}}{\text{K}}\right) = +10.2\text{ kJ} = +1.02 \times 10^4\text{ J}$ so the reaction is nonspontaneous.

c) $\Delta G^0 = \Delta H^0_{rxn} - T\Delta S^0_{rxn} = +179.2\text{ kJ} - (1455\text{ K})\left(+0.1602\frac{\text{kJ}}{\text{K}}\right) = -53.9\text{ kJ} = -5.39 \times 10^4\text{ J}$ so the reaction is spontaneous.

Check: The units (kJ) are correct. The enthalpy term dominates at low temperatures, making the reaction nonspontaneous. As the temperature increases, the increase in entropy starts to dominate and in the last case the reaction is spontaneous.

37. Since the first reaction has Fe_2O_3 as a product and the reaction of interest has it as a reactant, we need to reverse the first reaction. When the reaction direction is reversed, ΔG changes.

$Fe_2O_3\ (s) \rightarrow 2\ Fe\ (s) + 3/2\ O_2\ (g)$ $\qquad \Delta G^0 = +\ 742.2\ kJ$

Since the second reaction has 1 mole CO as a reactant and the reaction of interest has 3 moles of CO as a reactant, we need to multiply the second reaction and the ΔG by 3.

$3\ [CO\ (g) + 1/2\ O_2\ (g) \rightarrow CO_2\ (g)]$ $\qquad \Delta G^0 = 3(-\ 257.2\ kJ) = -\ 771.6\ kJ$

Hess' Law states the ΔG of the net reaction is the sum of the ΔG of the steps.
The rewritten reactions are:

$Fe_2O_3\ (s) \rightarrow 2\ Fe\ (s) + \cancel{3/2\ O_2\ (g)}$ $\qquad \Delta G^0 = +\ 742.2\ kJ$

$3\ CO\ (g) + \cancel{3/2\ O_2\ (g)} \rightarrow 3\ CO_2\ (g)$ $\qquad \Delta G^0 = -\ 771.6\ kJ$

$Fe_2O_3\ (s) + 3\ CO\ (g) \rightarrow 2\ Fe\ (s) + 3\ CO_2\ (g)$ $\qquad \Delta G^0_{rxn} = -\ 29.4\ kJ$

38. Since the first reaction has $CaCO_3$ as a product and the reaction of interest has it as a reactant, we need to reverse the first reaction. When the reaction direction is reversed, ΔG changes.

$CaCO_3\ (s) \rightarrow Ca\ (s) + CO_2\ (g) + 1/2\ O_2\ (g)$ $\qquad \Delta G^0 = +\ 734.4\ kJ$

Since the second reaction has 2 moles CaO as a product and the reaction of interest has 1 mole of CaO as a product, we need to multiply it by ½. The ΔG of the second reaction is multiplied by ½.

$1/2[2\ Ca\ (s) + O_2\ (g) \rightarrow 2\ CaO\ (s)]$ $\qquad \Delta G^0 = 1/2(-\ 1206.6\ kJ) = -\ 603.3\ kJ$

Hess' Law states the ΔH of the net reaction is the sum of the ΔH of the steps.
The rewritten reactions are:

$CaCO_3\ (s) \rightarrow \cancel{Ca\ (s)} + CO_2\ (g) + \cancel{1/2\ O_2\ (g)}$ $\qquad \Delta G^0 = +\ 734.4\ kJ$

$1/2[\cancel{2\ Ca\ (s)} + \cancel{O_2\ (g)} \rightarrow 2\ CaO\ (s)]$ $\qquad \Delta G^0 = -\ 603.3\ kJ$

$CaCO_3\ (s) \rightarrow CaO\ (s) + CO_2\ (g)$ $\qquad \Delta G^0_{rxn} = +\ 131.1\ kJ$

39. a) **Given:** $I_2\ (s) \rightarrow I_2\ (g)$ at 25.0 °C **Find:** ΔG°_{rxn}

Conceptual Plan: $\Delta G^0_{rxn} = \sum n_P \Delta G^0_f(products) - \sum n_R \Delta G^0_f(reactants)$

Solution:

Reactant/Product	ΔG^0_f (kJ/mol from Appendix IIB)
$I_2\ (s)$	0.0
$I_2\ (g)$	19.3

Be sure to pull data for the correct formula and phase.

$$\begin{aligned}\Delta G^0_{rxn} &= \sum n_p \Delta G^0_f(products) - \sum n_r \Delta G^0_f(reactants)\\ &= [1(\Delta G^0_f(I_2\ (g)))] - [1(\Delta G^0_f(I_2\ (s)))]\\ &= [1(19.3\ kJ)] - [1(0.0\ kJ)]\\ &= +\ 19.3\ kJ\end{aligned}$$

so the reaction is nonspontaneous.

Check: The units (kJ) are correct. The answer is positive because gases have higher free energy than solids and the free energy change of the reaction is the same as free energy of formation of gaseous iodine.

b) **Given:** $I_2\ (s) \rightarrow I_2\ (g)$ at 25.0 °C (i) $P_{I2} = 1.00$ mmHg; (ii) $P_{I2} = 0.100$ mmHg **Find:** ΔG_{rxn}

Conceptual Plan: **°C → K** and **mmHg → atm** then $\Delta G^\circ_{rxn}, P_{I2}, T \rightarrow \Delta G_{rxn}$

$K = 273.15 + °C$ $\qquad \dfrac{1\ atm}{760\ mm\ Hg}$ $\qquad \Delta G_{rxn} = \Delta G^0_{rxn} + RT \ln Q$ where $Q = P_{I2}$

Solution: $T = 273.15 + 25.0\ °C = 298.2\ K$ and (i) $1.00\ \cancel{mmHg} \times \dfrac{1\ atm}{760\ \cancel{mmHg}} = 0.0013\underline{1}579\ atm$

then $\Delta G_{rxn} = \Delta G^0_{rxn} + RT \ln Q = \Delta G^0_{rxn} + RT \ln P_{I2} = +19.3\ kJ\ +$

$$+\left(8.314\frac{\cancel{J}}{\cancel{K}\cdot mol}\right)\left(\frac{1\ kJ}{1000\ \cancel{J}}\right)(298.2\ \cancel{K})\ln(0.0013\underline{1}579) = +\ 2.9\ kJ$$ so the reaction is nonspontaneous. Then

(ii) $0.100\ \cancel{mmHg}\ x\ \frac{1 atm}{760\ \cancel{mmHg}} = 0.00013\underline{1}579\ atm$ then

$$\Delta G_{rxn} = \Delta G^0_{rxn} + RT\ln Q = \Delta G^0_{rxn} + RT\ln P_{I2} = +19.3\ kJ +$$

$$+\left(8.314\frac{\cancel{J}}{\cancel{K}\cdot mol}\right)\left(\frac{1\ kJ}{1000\ \cancel{J}}\right)(298.2\ \cancel{K})\ln(0.00013\underline{1}579) = -\ 2.9\ kJ$$ so the reaction is spontaneous.

Check: The units (kJ) are correct. The answer is positive at high pressures because the pressure is higher than the vapor pressure of iodine. Once the desired pressure is below the vapor pressure (0.31 mmHg at 25.0 °C), the reaction becomes spontaneous.

c) Iodine sublimes at room temperature because there is an equilibrium between the solid and the gas phases. The vapor pressure is low (0.31 mmHg at 25.0 °C), so a small amount of iodine can remain in the gas phase, which is consistent with the free energy values.

40. a) **Given:** $CH_3OH\ (l) \rightarrow CH_3OH\ (g)$ at 25.0 °C **Find:** $\Delta G°_{rxn}$

Conceptual Plan: $\Delta G^0_{rxn} = \sum n_p \Delta G^0_f(products) - \sum n_r \Delta G^0_f(reactants)$

Solution:

Reactant/Product	ΔG^0_f(kJ/mol from Appendix IIB)
$CH_3OH\ (l)$	– 166.6
$CH_3OH\ (g)$	– 162.3

Be sure to pull data for the correct formula and phase.

$$\Delta G^0_{rxn} = \sum n_p \Delta G^0_f(products) - \sum n_r \Delta G^0_f(reactants)$$
$$= [1(\Delta G^0_f(CH_3OH\ (g)))] - [1(\Delta G^0_f(CH_3OH\ (l)))]$$
$$= [1(-162.3\ kJ)] - [1(-166.6\ kJ)]$$
$$= +\ 4.3\ kJ$$

so the reaction is nonspontaneous.

Check: The units (kJ) are correct. The answer is positive because gases have higher free energy than liquids.

b) **Given:** $CH_3OH\ (l) \rightarrow CH_3OH\ (g)$ at 25.0 °C (i) P_{CH3OH} = 150.0 mmHg; (ii) P_{CH3OH} = 100.0 mmHg and (ii) P_{CH3OH} = 10.0 mmHg **Find:** ΔG_{rxn}

Conceptual Plan: °C → K and mmHg → atm then $\Delta G°_{rxn}$, P_{CH3OH}, T → ΔG_{rxn}

$K = 273.15 + °C$ $\quad \frac{1 atm}{760\ mm\ Hg}$ $\quad \Delta G_{rxn} = \Delta G^0_{rxn} + RT\ln Q$ where $Q = P_{CH3OH}$

Solution: $T = 273.15 + 25.0\ °C = 298.2\ K$ and (i) $150.0\ \cancel{mmHg}\ x\ \frac{1 atm}{760\ \cancel{mmHg}} = 0.197\underline{3}684\ atm$ then

$$\Delta G_{rxn} = \Delta G^0_{rxn} + RT\ln Q = \Delta G^0_{rxn} + RT\ln P_{CH3OH} = +4.3\ kJ +$$

$$+\left(8.314\frac{\cancel{J}}{\cancel{K}\cdot mol}\right)\left(\frac{1\ kJ}{1000\ \cancel{J}}\right)(298.2\ \cancel{K})\ln(0.197\underline{3}684) = +\ 0.3\ kJ$$ so the reaction is nonspontaneous. Then (ii)

$100.0\ \cancel{mmHg}\ x\ \frac{1 atm}{760\ \cancel{mmHg}} = 0.13\underline{1}579\ atm$ then

$$\Delta G_{rxn} = \Delta G^0_{rxn} + RT\ln Q = \Delta G^0_{rxn} + RT\ln P_{CH3OH} = +4.3\ kJ +$$

$$+\left(8.314\frac{\cancel{J}}{\cancel{K}\cdot mol}\right)\left(\frac{1\ kJ}{1000\ \cancel{J}}\right)(298.2\ \cancel{K})\ln(0.13\underline{1}579) = -\ 0.7\ kJ$$ so the reaction is spontaneous. Then (iii)

$10.0\ \cancel{mmHg}\ x\ \frac{1 atm}{760\ \cancel{mmHg}} = 0.013\underline{1}579\ atm$ then

$$\Delta G_{rxn} = \Delta G^0_{rxn} + RT\ln Q = \Delta G^0_{rxn} + RT\ln P_{CH3OH} = +4.3\ kJ +$$

$$+\left(8.314\frac{\cancel{J}}{\cancel{K}\cdot mol}\right)\left(\frac{1\ kJ}{1000\ \cancel{J}}\right)(298.2\ \cancel{K})\ln(0.013\underline{1}579) = -\ 6.4\ kJ$$ so the reaction is spontaneous.

Check: The units (kJ) are correct. The answer is positive at high pressures because the pressure is higher than the vapor pressure of methanol. Once the desired pressure is below the vapor pressure (143 mmHg at 25.0 °C), the reaction becomes spontaneous.

c) Methanol evaporates at room temperature because there is an equilibrium between the liquid and the gas phases. The vapor pressure is moderate (143 mmHg at 25.0 °C), so a moderate amount of methanol can remain in the gas phase, which is consistent with the free energy values.

41. **Given:** $CH_3OH\ (g) \rightleftharpoons CO\ (g) + 2\ H_2\ (g)$ at 25 °C, $P_{CH3OH} = 0.855$ atm, $P_{CO} = 0.125$ atm, $P_{H2} = 0.183$ atm
Find: ΔG
Conceptual Plan:
$\Delta G^{\circ}_{rxn} = \sum n_p \Delta G^{\circ}_f(products) - \sum n_r \Delta G^{\circ}_f(reactants)$ **then °C→ K then ΔG°_{rxn}, P_{CH3OH}, P_{CO}, P_{H2}, T → ΔG**

$K = 273.15 + °C \qquad \Delta G_{rxn} = \Delta G^0_{rxn} + RT\ln Q \quad \text{where } Q = \frac{P_{CO}P^2_{H2}}{P_{CH3OH}}$

Solution:

Reactant/Product	ΔG^0_f (kJ/mol from Appendix IIB)
$CH_3OH\ (g)$	− 162.3
$CO\ (g)$	− 137.2
$H_2\ (g)$	0.0

Be sure to pull data for the correct formula and phase.

$$\begin{aligned}\Delta G^0_{rxn} &= \sum n_p \Delta G^0_f(products) - \sum n_r \Delta G^0_f(reactants)\\ &= [1(\Delta G^0_f(CO\ (g))) + 2(\Delta G^0_f(H_2\ (g)))] - [1(\Delta G^0_f(CH_3OH\ (g)))]\\ &= [1(-137.2\text{ kJ}) + 2(0.0\text{ kJ})] - [1(-162.3\text{ kJ})]\\ &= [-137.2\text{ kJ}] - [-162.3\text{ kJ}]\\ &= +25.1\text{ kJ}\end{aligned}$$

$T = 273.15 + 25\ °C = 298$ K then

$$Q = \frac{P_{CO}P^2_{H2}}{P_{CH3OH}} = \frac{(0.125)(0.183)^2}{0.855} = 0.00489605 \text{ then}$$

$$\Delta G_{rxn} = \Delta G^0_{rxn} + RT\ln Q = +25.1\text{ kJ} + \left(8.314\frac{\text{J}}{\text{K}\cdot\text{mol}}\right)\left(\frac{1\text{ kJ}}{1000\text{ J}}\right)(298\text{ K})\ln(0.00489605) = +11.9\text{ kJ}$$

so the reaction is nonspontaneous.
Check: The units (kJ) are correct. The standard free energy for the reaction was positive and the fact that Q was less than one made the free energy smaller, but the reaction at these conditions is still not spontaneous.

42. **Given:** $CO_2\ (g) + CCl_4\ (g) \rightleftharpoons 2\ COCl_2\ (g)$ at 25 °C, $P_{CO2} = 0.112$ atm, $P_{CCl4} = 0.174$ atm, $P_{COCl2} = 0.744$ atm **Find:** ΔG
Conceptual Plan:
$\Delta G^{\circ}_{rxn} = \sum n_p \Delta G^{\circ}_f(products) - \sum n_r \Delta G^{\circ}_f(reactants)$ **then °C→K then ΔG°_{rxn}, P_{CO2}, P_{CCl4}, P_{COCl2}, T →ΔG**

$K = 273.15 + °C \qquad \Delta G_{rxn} = \Delta G^0_{rxn} + RT\ln Q \quad \text{where } Q = \frac{P^2_{COCl2}}{P_{CO2}P_{CCl4}}$

Solution:

Reactant/Product	ΔG^0_f (kJ/mol from Appendix IIB)
$CO_2\ (g)$	− 394.4
$CCl_4\ (g)$	− 62.3
$COCl_2\ (g)$	− 204.9

Be sure to pull data for the correct formula and phase.

$$\Delta G^0_{rxn} = \sum n_p \Delta G^0_f(products) - \sum n_r \Delta G^0_f(reactants)$$
$$= [2(\Delta G^0_f(COCl_2\ (g)))] - [1(\Delta G^0_f(CO_2\ (g))) + 1(\Delta G^0_f(CCl_4\ (g)))]$$
$$= [2(-204.9\ kJ)] - [1(-394.4\ kJ) + 1(-62.3\ kJ)]$$
$$= [-409.8\ kJ] - [-456.7\ kJ]$$
$$= +46.9\ kJ$$

$T = 273.15 + 25\ °C = 298\ K$ then

$$Q = \frac{P^2_{COCl2}}{P_{CO2}P_{CCl4}} = \frac{(0.744)^2}{(0.112)(0.174)} = 28.\underline{4}039 \text{ then}$$

$$\Delta G_{rxn} = \Delta G^0_{rxn} + RT \ln Q = +46.9\ kJ + \left(8.314 \frac{J}{K \cdot mol}\right)\left(\frac{1\ kJ}{1000\ J}\right)(298\ K)\ln(28.\underline{4}039) = +55.2\ kJ$$

so the reaction is nonspontaneous.

Check: The units (kJ) are correct. The standard free energy for the reaction was positive and the fact that Q was greater than one made the free energy larger, so the reaction is less spontaneous in the forward direction at these conditions than at standard conditions.

43. a) **Given:** $2\ CO\ (g) + O_2\ (g) \rightleftharpoons 2\ CO_2\ (g)$ at 25 °C **Find:** K

Conceptual Plan: $\Delta G^\circ_{rxn} = \sum n_p \Delta G^\circ_f(products) - \sum n_r \Delta G^\circ_f(reactants)$ **then °C → K then ΔG°_{rxn}, T → K**

$K = 273.15 + °C$ $\quad \Delta G^0_{rxn} = -RT\ln K$

Solution:

Reactant/Product	ΔG^0_f(kJ/mol from Appendix IIB)
$CO\ (g)$	− 137.2
$O_2\ (g)$	0.0
$CO_2\ (g)$	− 394.4

Be sure to pull data for the correct formula and phase.

$$\Delta G^0_{rxn} = \sum n_p \Delta G^0_f(products) - \sum n_r \Delta G^0_f(reactants)$$
$$= [2(\Delta G^0_f(CO_2\ (g)))] - [2(\Delta G^0_f(CO\ (g))) + 1(\Delta G^0_f(O_2\ (g)))]$$
$$= [2(-394.4\ kJ)] - [2(-137.2\ kJ) + 1(0.0\ kJ)]$$
$$= [-788.8\ kJ] - [-274.4\ kJ]$$
$$= -514.4\ kJ$$

$T = 273.15 + 25\ °C = 298\ K$ then

$\Delta G^0_{rxn} = -RT\ln K$ Rearrange to solve for K.

$$K = e^{\frac{-\Delta G^0_{rxn}}{RT}} = e^{\frac{-(-514.4\ kJ) \times \frac{1000\ J}{1\ kJ}}{\left(8.314 \frac{J}{K \cdot mol}\right)(298\ K)}} = e^{207.\underline{6}23} = 1.48 \times 10^{90}.$$

Check: The units (none) are correct. The standard free energy for the reaction was very negative and so we expect a very large K. The reaction is spontaneous and so mostly products are present at equilibrium.

b) **Given:** $2\ H_2S\ (g) \rightleftharpoons 2\ H_2\ (g) + S_2\ (g)$ at 25 °C **Find:** K

Conceptual Plan: $\Delta G^\circ_{rxn} = \sum n_p \Delta G^\circ_f(products) - \sum n_r \Delta G^\circ_f(reactants)$ **then °C → K then ΔG°_{rxn}, T → K**

$K = 273.15 + °C$ $\quad \Delta G^0_{rxn} = -RT\ln K$

Solution:

Reactant/Product	ΔG^0_f(kJ/mol from Appendix IIB)
$H_2S\ (g)$	− 33.4
$H_2\ (g)$	0.0
$S_2\ (g)$	79.7

Be sure to pull data for the correct formula and phase.

$$\Delta G^0_{rxn} = \sum n_p \Delta G^0_f(products) - \sum n_r \Delta G^0_f(reactants)$$
$$= [2(\Delta G^0_f(H_2\ (g))) + 1(\Delta G^0_f(S_2\ (g)))] - [2(\Delta G^0_f(H_2S\ (g)))]$$
$$= [2(0.0\ kJ)] + 1(79.7\ kJ)] - [2(-33.4\ kJ)]$$
$$= [79.7\ kJ] - [-66.8\ kJ]$$
$$= +146.5\ kJ$$

$T = 273.15 + 25\ °C = 298\ K$ then

$\Delta G^0_{rxn} = -RT \ln K$ Rearrange to solve for K. $K = e^{\frac{-\Delta G^0_{rxn}}{RT}} = e^{\frac{-146.5\ kJ \times \frac{1000\ J}{1\ kJ}}{\left(8.314 \frac{J}{K \cdot mol}\right)(298\ K)}} = e^{-59.1305} = 2.09 \times 10^{-26}$.

Check: The units (none) are correct. The standard free energy for the reaction was positive and so we expect a small K. The reaction is nonspontaneous and so mostly reactants are present at equilibrium.

44. a) **Given:** $2\ NO_2\ (g) \rightleftharpoons N_2O_4\ (g)$ at 25 °C **Find:** K

Conceptual Plan: $\Delta G^0_{rxn} = \sum n_p \Delta G^0_f(products) - \sum n_r \Delta G^0_f(reactants)$ **then °C → K then $\Delta G°_{rxn}$, T → K**

$K = 273.15 + °C$ $\quad \Delta G^0_{rxn} = -RT \ln K$

Solution:

Reactant/Product	ΔG^0_f(kJ/mol from Appendix IIB)
$NO_2\ (g)$	51.3
$N_2O_4\ (g)$	99.8

Be sure to pull data for the correct formula and phase.

$$\Delta G^0_{rxn} = \sum n_p \Delta G^0_f(products) - \sum n_r \Delta G^0_f(reactants)$$
$$= [1(\Delta G^0_f(N_2O_4\ (g)))] - [2(\Delta G^0_f(NO_2\ (g)))]$$
$$= [1(99.8\ kJ)] - [2(51.3\ kJ)]$$
$$= [99.8\ kJ] - [102.6\ kJ]$$
$$= -2.8\ kJ$$

$T = 273.15 + 25\ °C = 298\ K$ then

$\Delta G^0_{rxn} = -RT \ln K$ Rearrange to solve for K. $K = e^{\frac{-\Delta G^0_{rxn}}{RT}} = e^{\frac{-(-2.8\ kJ) \times \frac{1000\ J}{1\ kJ}}{\left(8.314 \frac{J}{K \cdot mol}\right)(298\ K)}} = e^{1.1301} = 3.1$.

Check: The units (none) are correct. The standard free energy for the reaction was very slightly negative and so we expect a K just over 1. The reaction is spontaneous and so mostly products are present at equilibrium.

b) **Given:** $Br_2\ (g) + Cl_2\ (g) \rightleftharpoons 2\ BrCl\ (g)$ at 25 °C **Find:** K

Conceptual Plan: $\Delta G^0_{rxn} = \sum n_p \Delta G^0_f(products) - \sum n_r \Delta G^0_f(reactants)$ **then °C → K then $\Delta G°_{rxn}$, T → K**

$K = 273.15 + °C$ $\quad \Delta G^0_{rxn} = -RT \ln K$

Solution:

Reactant/Product	ΔG^0_f(kJ/mol from Appendix IIB)
$Br_2\ (g)$	3.1
$Cl_2\ (g)$	0.0
$BrCl\ (g)$	−1.0

Be sure to pull data for the correct formula and phase.

$$\Delta G^0_{rxn} = \sum n_p \Delta G^0_f(products) - \sum n_r \Delta G^0_f(reactants)$$
$$= [2(\Delta G^0_f(BrCl\ (g)))] - [1(\Delta G^0_f(Br_2\ (g))) + 1(\Delta G^0_f(Cl_2\ (g)))]$$
$$= [2(-1.0\ kJ)] - [1(3.1\ kJ) + 1(0.0\ kJ)]$$
$$= [-2.0\ kJ] - [3.1\ kJ]$$
$$= -5.1\ kJ$$

$T = 273.15 + 25\ °C = 298\ K$ then

$\Delta G^0_{rxn} = -RT\ln K$ Rearrange to solve for *K*. $K = e^{\frac{-\Delta G^0_{rxn}}{RT}} = e^{\frac{-(-5.1\ \text{kJ}) \times \frac{1000\ \text{J}}{1\ \text{kJ}}}{\left(8.314 \frac{\text{J}}{\text{K}\cdot\text{mol}}\right)(298\ \text{K})}} = e^{2.0\underline{5}85} = 7.8$.

Check: The units (none) are correct. The standard free energy for the reaction was very slightly negative and so we expect a *K* just over 1. The reaction is spontaneous.

45. **Given:** CO (*g*) + 2 H_2 (*g*) $\rightleftharpoons$ CH_3OH (*g*) $K_p = 2.26 \times 10^4$ at 25 °C
Find: $\Delta G°_{rxn}$ at a) standard conditions, b) at equilibrium, and c) $P_{CH3OH} = 1.0$ atm, $P_{CO} = P_{H2} = 0.010$ atm
Conceptual Plan: °C → K then a) *K, T* → $\Delta G°_{rxn}$ then b) at equilibrium $\Delta G_{rxn} = 0$ then

$K = 273.15 + °C$ $\qquad \Delta G^0_{rxn} = -RT\ln K$

c) $\Delta G°_{rxn}, P_{CH3OH}, P_{CO}, P_{H2}, T \rightarrow \Delta G$

$$\Delta G_{rxn} = \Delta G^0_{rxn} + RT\ln Q \quad \text{where} \quad Q = \frac{P_{CH3OH}}{P_{CO}P^2_{H2}}$$

Solution: $T = 273.15 + 25\ °C = 298$ K then

a) $\Delta G^0_{rxn} = -RT\ln K = -\left(8.314 \frac{\text{J}}{\text{K}\cdot\text{mol}}\right)\left(\frac{1\ \text{kJ}}{1000\ \text{J}}\right)(298\ \text{K})\ln(2.26 \times 10^4) = -24.8\ \text{kJ}$;

b) at equilibrium $\Delta G_{rxn} = 0$

c) $Q = \frac{P_{CH3OH}}{P_{CO}P^2_{H2}} = \frac{1.0}{(0.010)(0.010)^2} = 1.0 \times 10^6$ then

$$\Delta G_{rxn} = \Delta G^0_{rxn} + RT\ln Q = -24.8\ \text{kJ} + \left(8.314 \frac{\text{J}}{\text{K}\cdot\text{mol}}\right)\left(\frac{1\ \text{kJ}}{1000\ \text{J}}\right)(298\ \text{K})\ln(1.0 \times 10^6) = +9.4\ \text{kJ}$$

Check: The units (kJ) are correct. The *K* was greater than one so we expect a negative standard free energy for the reaction. At equilibrium, by definition, the free energy change is zero. Since the conditions give a Q > *K* then the reaction needs to proceed in the reverse direction, which means that the reaction is spontaneous in the reverse direction.

46. **Given:** I_2 (*g*) + Cl_2 (*g*) $\rightleftharpoons$ 2 ICl (*g*) $K_p = 81.9$ at 25 °C **Find:** $\Delta G°_{rxn}$ at a) standard conditions, b) at equilibrium, and c) $P_{ICl} = 2.55$ atm, $P_{I2} = 0.325$ atm, $P_{Cl2} = 0.221$ atm
Conceptual Plan: °C → K then a) *K, T* → $\Delta G°_{rxn}$ then b) at equilibrium $\Delta G_{rxn} = 0$ then

$K = 273.15 + °C$ $\qquad \Delta G^0_{rxn} = -RT\ln K$

c) $\Delta G°_{rxn}, P_{ICl}, P_{I2}, P_{Cl2}, T \rightarrow \Delta G$

$$\Delta G_{rxn} = \Delta G^0_{rxn} + RT\ln Q \quad \text{where} \quad Q = \frac{P^2_{ICl}}{P_{I2}P_{Cl2}}$$

Solution: $T = 273.15 + 25\ °C = 298$ K then

a) $\Delta G^0_{rxn} = -RT\ln K = -\left(8.314 \frac{\text{J}}{\text{K}\cdot\text{mol}}\right)\left(\frac{1\ \text{kJ}}{1000\ \text{J}}\right)(298\ \text{K})\ln(81.9) = -10.9\ \text{kJ}$;

b) at equilibrium $\Delta G_{rxn} = 0$

c) $Q = \frac{P^2_{ICl}}{P_{I2}P_{Cl2}} = \frac{(2.55)^2}{(0.325)(0.221)} = 90.\underline{5}325$ then

$$\Delta G_{rxn} = \Delta G^0_{rxn} + RT\ln Q = -10.9\ \text{kJ} + \left(8.314 \frac{\text{J}}{\text{K}\cdot\text{mol}}\right)\left(\frac{1\ \text{kJ}}{1000\ \text{J}}\right)(298\ \text{K})\ln(90.\underline{5}325) = +0.3\ \text{kJ}$$

Check: The units (kJ) are correct. The *K* was greater than one so we expect a negative standard free energy for the reaction. At equilibrium, by definition, the free energy change is zero. Since the conditions give a Q just greater than *K* then the reaction needs to proceed in the reverse direction, which means that the reaction is slightly spontaneous.

47. a) **Given:** $2\,CO\,(g) + O_2\,(g) \rightleftharpoons 2\,CO_2\,(g)$ at 25 °C **Find:** K at 525 K

Conceptual Plan: $\Delta H^\circ_{rxn} = \sum n_p H^\circ_f(products) - \sum n_r H^\circ_f(reactants)$ **then**

$\Delta S^\circ_{rxn} = \sum n_p S^\circ(products) - \sum n_r S^\circ(reactants)$ **then J/K → kJ/K then ΔH°_{rxn}, ΔS_{rxn}, T → ΔG**

$$\frac{1\,kJ}{1000\,J} \qquad \Delta G = \Delta H_{rxn} - T\Delta S_{rxn}$$

then ΔG°_{rxn}, T → K

$$\Delta G^0_{rxn} = -RT\ln K$$

Solution:

Reactant/Product	ΔH^0_f(kJ/mol from Appendix IIB)
$CO\,(g)$	−110.5
$O_2\,(g)$	0.0
$CO_2\,(g)$	−393.5

Be sure to pull data for the correct formula and phase.

$$\Delta H^0_{rxn} = \sum n_p \Delta H^0_f(products) - \sum n_r \Delta H^0_f(reactants)$$
$$= [2(\Delta H^0_f(CO_2\,(g)))] - [2(\Delta H^0_f(CO\,(g))) + 1(\Delta H^0_f(O_2\,(g)))]$$
$$= [2(-393.5\text{ kJ})] - [2(-110.5\text{ kJ}) + 1(0.0\text{ kJ})]$$
$$= [-787.0\text{ kJ}] - [-221.0\text{ kJ}]$$
$$= -566.0\text{ kJ}$$

then

Reactant/Product	S^0(J/molK from Appendix IIB)
$CO\,(g)$	197.7
$O_2\,(g)$	205.2
$CO_2\,(g)$	213.8

Be sure to pull data for the correct formula and phase.

$$\Delta S^0_{rxn} = \sum n_p S^0(products) - \sum n_r S^0(reactants)$$
$$= [2(S^0(CO_2\,(g)))] - [2(S^0(CO\,(g))) + 1(S^0(O_2\,(g)))]$$
$$= [2(213.8\text{ J/K})] - [2(197.7\text{ J/K}) + 1(205.2\text{ J/K})]$$
$$= [427.6\text{ J/K}] - [600.6\text{ J/K}]$$
$$= -173.0\text{ J/K}$$

then

$$-173.0\frac{\cancel{J}}{K} \times \frac{1\text{ kJ}}{1000\,\cancel{J}} = -0.1730\frac{kJ}{K} \text{ then}$$

$$\Delta G^0 = \Delta H^0_{rxn} - T\Delta S^0_{rxn} = -566.0\text{ kJ} - (525\,\cancel{K})\left(-0.1730\frac{kJ}{\cancel{K}}\right) = -475.2\text{ kJ} = -4.752 \times 10^5\text{ J} \quad \text{then}$$

$\Delta G^0_{rxn} = -RT\ln K$ Rearrange to solve for K.

$$K = e^{\frac{-\Delta G^0_{rxn}}{RT}} = e^{\frac{-(-4.752 \times 10^5\,\cancel{J})}{\left(8.314\frac{\cancel{J}}{\cancel{K}\cdot mol}\right)(525\,\cancel{K})}} = e^{108.864} = 1.90 \times 10^{47}.$$

Check: The units (none) are correct. The free energy change is very negative, indicating a spontaneous reaction. This results in a very large K.

b) **Given:** $2\,H_2S\,(g) \rightleftharpoons 2\,H_2\,(g) + S_2\,(g)$ at 25 °C **Find:** K at 525 K

Conceptual Plan: $\Delta H^\circ_{rxn} = \sum n_p H^\circ_f(products) - \sum n_r H^\circ_f(reactants)$ **then**

$\Delta S^\circ_{rxn} = \sum n_p S^\circ(products) - \sum n_r S^\circ(reactants)$ **then J/K → kJ/K then ΔH°_{rxn}, ΔS_{rxn}, T → ΔG**

$$\frac{1\,kJ}{1000\,J} \qquad \Delta G = \Delta H_{rxn} - T\Delta S_{rxn}$$

then ΔG°_{rxn}, T → K

$$\Delta G^0_{rxn} = -RT\ln K$$

Solution:

Reactant/Product	ΔH^0_f(kJ/mol from Appendix IIB)
$H_2S\ (g)$	– 20.6
$H_2\ (g)$	0.0
$S_2\ (g)$	128.6

Be sure to pull data for the correct formula and phase.

$$\Delta H^0_{rxn} = \sum n_p \Delta H^0_f(products) - \sum n_r \Delta H^0_f(reactants)$$
$$= [2(\Delta H^0_f(H_2\ (g))) + 1(\Delta H^0_f(S_2\ (g)))] - [2(\Delta H^0_f(H_2S\ (g)))]$$
$$= [2(0.0\ kJ) + 1(128.6\ kJ)] - [2(-20.6\ kJ)]$$
$$= [128.6\ kJ] - [-41.2\ kJ]$$
$$= +169.8\ kJ$$

then

Reactant/Product	S^0(J/molK from Appendix IIB)
$H_2S\ (g)$	205.8
$H_2\ (g)$	130.7
$S_2\ (g)$	228.2

Be sure to pull data for the correct formula and phase.

$$\Delta S^0_{rxn} = \sum n_p S^0(products) - \sum n_r S^0(reactants)$$
$$= [2(S^0(H_2\ (g))) + 1(S^0(S_2\ (g)))] - [2(S^0(H_2S\ (g)))]$$
$$= [2(130.7\ J/K) + 1(228.2\ J/K)] - [2(205.8\ J/K)]$$
$$= [489.6\ J/K] - [411.6\ J/K]$$
$$= +78.0\ J/K$$

then

$$+78.0 \frac{J}{K} \times \frac{1\ kJ}{1000\ J} = +0.0780 \frac{kJ}{K}$$ then

$$\Delta G^0 = \Delta H^0_{rxn} - T\Delta S^0_{rxn} = +169.8\ kJ - (525\ K)\left(+0.0780 \frac{kJ}{K}\right) = +128.\underline{8}5\ kJ = +1.28\underline{8}5 \times 10^5\ J$$ then

$$\Delta G^0_{rxn} = -RT\ln K$$ Rearrange to solve for K.

$$K = e^{\frac{-\Delta G^0_{rxn}}{RT}} = e^{\frac{-1.28\underline{8}5 \times 10^5\ J}{\left(8.314 \frac{J}{K \cdot mol}\right)(525\ K)}} = e^{-29.\underline{5}199} = 1.51 \times 10^{-13}.$$

Check: The units (none) are correct. The free energy change is positive, indicating a nonspontaneous reaction. This results in a very small K.

48. a) **Given:** $2\ NO_2\ (g) \rightleftharpoons N_2O_4\ (g)$ at 25 °C **Find:** K at 655 K

Conceptual Plan: $\Delta H^0_{rxn} = \sum n_p H^0_f(products) - \sum n_r H^0_f(reactants)$ **then**

$\Delta S^0_{rxn} = \sum n_p S^0(products) - \sum n_r S^0(reactants)$ **then J/K → kJ/K then ΔH°_{rxn}, ΔS_{rxn}, T → ΔG**

$$\frac{1\ kJ}{1000\ J} \qquad \Delta G = \Delta H_{rxn} - T\Delta S_{rxn}$$

then ΔG°_{rxn}, T → K

$$\Delta G^0_{rxn} = -RT\ln K$$

Solution:

Reactant/Product	ΔH^0_f(kJ/mol from Appendix IIB)
$NO_2\ (g)$	33.2
$N_2O_4\ (g)$	11.1

Be sure to pull data for the correct formula and phase.

$$\Delta H^0_{rxn} = \sum n_p \Delta H^0_f(products) - \sum n_r \Delta H^0_f(reactants)$$
$$= [1(\Delta H^0_f(N_2O_4\ (g)))] - [2(\Delta H^0_f(NO_2\ (g)))]$$
$$= [1(11.1\ kJ)] - [2(33.2\ kJ)] \quad \text{then}$$
$$= [11.1\ kJ] - [66.4\ kJ]$$
$$= -\ 55.3\ kJ$$

Reactant/Product	S^0(J/molK from Appendix IIB)
$NO_2\ (g)$	240.1
$N_2O_4\ (g)$	304.4

Be sure to pull data for the correct formula and phase.

$$\Delta S^0_{rxn} = \sum n_p S^0(products) - \sum n_r S^0(reactants)$$
$$= [1(S^0(N_2O_4\ (g)))] - [2(S^0(NO_2\ (g)))]$$
$$= [1(304.4\ J/K)] - [2(240.1\ J/K)] \quad \text{then } -175.8\frac{J}{K} \times \frac{1\ kJ}{1000\ J} = -\ 0.1758\frac{kJ}{K} \text{ then}$$
$$= [304.4\ J/K] - [480.2\ J/K]$$
$$= -\ 175.8\ J/K$$

$$\Delta G^0 = \Delta H^0_{rxn} - T\Delta S^0_{rxn} = -\ 55.3\ kJ - (655\ K)\left(-\ 0.1758\frac{kJ}{K}\right) = +\ 59.\underline{8}\ kJ = +\ 5.9\underline{8}49 \times 10^4\ J \text{ then}$$

$\Delta G^0_{rxn} = -RT\ln K$ Rearrange to solve for K. $K = e^{\frac{-\Delta G^0_{rxn}}{RT}} = e^{\frac{-5.9\underline{8}49 \times 10^4\ J}{\left(8.314\frac{J}{K \cdot mol}\right)(655\ K)}} = e^{-10.\underline{9}902} = 1.69 \times 10^{-5}$.

Check: The units (none) are correct. The free energy change is positive, indicating a nonspontaneous reaction. This results in a very small K.

b) **Given:** $Br_2\ (g) + Cl_2\ (g) \rightleftharpoons 2\ BrCl\ (g)$ at 25 °C **Find:** K at 655 K

Conceptual Plan: $\Delta H^\circ_{rxn} = \sum n_p H^\circ_f(products) - \sum n_r H^\circ_f(reactants)$ **then**

$\Delta S^\circ_{rxn} = \sum n_p S^\circ(products) - \sum n_r S^\circ(reactants)$ **then J/K → kJ/K** ($\frac{1\ kJ}{1000\ J}$) **then ΔH°_{rxn}, ΔS_{rxn}, T → ΔG** ($\Delta G = \Delta H_{rxn} - T\Delta S_{rxn}$)

then ΔG°_{rxn}, T → K

$$\Delta G^0_{rxn} = -RT\ln K$$

Solution:

Reactant/Product	ΔH^0_f(kJ/mol from Appendix IIB)
$Br_2\ (g)$	30.9
$Cl_2\ (g)$	0.0
$BrCl\ (g)$	14.6

Be sure to pull data for the correct formula and phase.

$$\Delta H^0_{rxn} = \sum n_p \Delta H^0_f(products) - \sum n_r \Delta H^0_f(reactants)$$
$$= [2(\Delta H^0_f(BrCl\ (g)))] - [1(\Delta H^0_f(Br_2\ (g))) + 1(\Delta H^0_f(Cl_2\ (g)))]$$
$$= [2(14.6\ kJ)] - [1(30.9\ kJ) + 1(0.0\ kJ)] \quad \text{then}$$
$$= [29.2\ kJ] - [30.9\ kJ]$$
$$= -\ 1.7\ kJ$$

Reactant/Product	S^0(J/molK from Appendix IIB)
$Br_2\ (g)$	245.5
$Cl_2\ (g)$	223.1
$BrCl\ (g)$	240.0

Be sure to pull data for the correct formula and phase.

$$\Delta S^0_{rxn} = \sum n_p S^0(products) - \sum n_r S^0(reactants)$$
$$= [2(S^0(\text{BrCl}(g)))] - [1(S^0(\text{Br}_2(g))) + 1(S^0(\text{Cl}_2(g)))]$$
$$= [2(240.0 \text{ J/K})] - [1(245.5 \text{ J/K}) + 1(223.1 \text{ J/K})]$$
$$= [480.0 \text{ J/K}] - [468.6 \text{ J/K}]$$
$$= +11.4 \text{ J/K}$$

then $+11.4 \frac{\text{J}}{\text{K}} \times \frac{1 \text{ kJ}}{1000 \text{ J}} = +0.0114 \frac{\text{kJ}}{\text{K}}$ then

$$\Delta G^0 = \Delta H^0_{rxn} - T\Delta S^0_{rxn} = -1.7 \text{ kJ} - (655 \text{ K})\left(+0.0114 \frac{\text{kJ}}{\text{K}}\right) = -9.167 \text{ kJ} = -9.167 \times 10^3 \text{ J}$$ then

$\Delta G^0_{rxn} = -RT \ln K$ Rearrange to solve for K. $K = e^{\frac{-\Delta G^0_{rxn}}{RT}} = e^{\frac{-(-9.167 \times 10^3 \text{ J})}{\left(8.314 \frac{\text{J}}{\text{K}\cdot\text{mol}}\right)(655 \text{ K})}} = e^{1.6834} = 5.38$.

Check: The units (none) are correct. The free energy change is positive, indicating a nonspontaneous reaction. This results in a small K.

49. a) +, since vapors have higher entropy than liquids.

 b) –, since solids have less entropy than liquids.

 c) –, since there is only one microstate for the final macrostate and there are six microstates for the initial macrostate.

50. a) +, since vapors have higher entropy than solids.

 b) –, since liquids have less entropy than vapors.

 c) +, since there are twenty microstates for the final macrostate and there are only six microstates for the initial macrostate.

51. a) **Given:** $N_2(g) + O_2(g) \rightarrow 2\,NO(g)$ **Find:** ΔG°_{rxn}, and K_p at 25 °C

Conceptual Plan: $\Delta H^0_{rxn} = \sum n_p H^0_f(products) - \sum n_r H^0_f(reactants)$ **then**

$\Delta S^0_{rxn} = \sum n_p S^0(products) - \sum n_r S^0(reactants)$ **then °C → K then J/K → kJ/K then $\Delta H^\circ_{rxn}, \Delta S_{rxn}, T \rightarrow \Delta G$**

$K = 273.15 + °C$ $\frac{1 \text{kJ}}{1000 \text{ J}}$ $\Delta G = \Delta H_{rxn} - T\Delta S_{rxn}$

then ΔG°_{rxn}, T → K

$\Delta G^0_{rxn} = -RT \ln K$

Solution:

Reactant/Product	ΔH^0_f(kJ/mol from Appendix IIB)
$N_2(g)$	0.0
$O_2(g)$	0.0
$NO(g)$	91.3

Be sure to pull data for the correct formula and phase.

$$\Delta H^0_{rxn} = \sum n_p \Delta H^0_f(products) - \sum n_r \Delta H^0_f(reactants)$$
$$= [2(\Delta H^0_f(\text{NO}(g)))] - [1(\Delta H^0_f(\text{N}_2(g))) + 1(\Delta H^0_f(\text{O}_2(g)))]$$
$$= [2(91.3 \text{ kJ})] - [1(0.0 \text{ kJ}) + 1(0.0 \text{ kJ})]$$
$$= [182.6 \text{ kJ}] - [0.0 \text{ kJ}]$$
$$= +182.6 \text{ kJ}$$

then

Reactant/Product	S^0(J/molK from Appendix IIB)
$N_2(g)$	191.6
$O_2(g)$	205.2
$NO(g)$	210.8

Be sure to pull data for the correct formula and phase.

$$\Delta S^0_{rxn} = \sum n_p S^0(products) - \sum n_r S^0(reactants)$$
$$= [2(S^0(NO\ (g)))] - [1(S^0(N_2\ (g))) + 1(S^0(O_2\ (g)))]$$
$$= [2(210.8\ J/K)] - [1(191.6\ J/K) + 1(205.2\ J/K)] \quad \text{then}$$
$$= [421.6\ J/K] - [396.8\ J/K]$$
$$= +\ 24.8\ J/K$$

$T = 273.15 + 25\ °C = 298\ K$ then $+24.8 \frac{\cancel{J}}{K} \times \frac{1\ kJ}{1000\ \cancel{J}} = +\ 0.0248 \frac{kJ}{K}$ then

$$\Delta G^0 = \Delta H^0_{rxn} - T\Delta S^0_{rxn} = +182.6\ kJ - (298\ \cancel{K})\left(0.0248 \frac{kJ}{\cancel{K}}\right) = +175.2\ kJ = +1.752 \times 10^5\ J \quad \text{then}$$

$\Delta G^0_{rxn} = -RT \ln K$ Rearrange to solve for K.

$$K = e^{\frac{-\Delta G^0_{rxn}}{RT}} = e^{\frac{-1.752 \times 10^5\ \cancel{J}}{\left(8.314 \frac{\cancel{J}}{\cancel{K} \cdot mol}\right)(298\ \cancel{K})}} = e^{-70.7144} = 1.95 \times 10^{-31}$$

so the reaction is nonspontaneous and at equilibrium mostly reactants are present.

Check: The units (kJ and none) are correct. The enthalpy is twice the enthalpy of formation of NO. We expect a very small entropy change because the number of moles of gas is unchanged. Since the positive enthalpy term dominates at room temperature, the free energy change is very positive and the reaction in the forward direction is nonspontaneous. This results in a very small K.

b) **Given:** $N_2\ (g) + O_2\ (g) \rightarrow 2\ NO\ (g)$ **Find:** $\Delta G°_{rxn}$ at 2000 K

Conceptual Plan: use results from part a) $\Delta H°_{rxn}, \Delta S_{rxn}, T \rightarrow \Delta G$ **then** $\Delta G°_{rxn}, T \rightarrow K$

$\Delta G = \Delta H_{rxn} - T\Delta S_{rxn}$ $\qquad \Delta G^0_{rxn} = -RT \ln K$

Solution: $\Delta G = \Delta H_{rxn} - T\Delta S_{rxn} = +182.6\ kJ - (2000\ \cancel{K})\left(0.0248 \frac{kJ}{\cancel{K}}\right) = +133.0\ kJ = +1.330 \times 10^5\ J$ then

$\Delta G^0_{rxn} = -RT \ln K$ Rearrange to solve for K.

$$K = e^{\frac{-\Delta G^0_{rxn}}{RT}} = e^{\frac{-1.330 \times 10^5\ \cancel{J}}{\left(8.314 \frac{\cancel{J}}{\cancel{K} \cdot mol}\right)(2000\ \cancel{K})}} = e^{-7.998557} = 3.36 \times 10^{-4}$$

so the forward reaction is becoming more spontaneous.

Check: The units (kJ and none) are correct. As the temperature rises, the entropy term becomes more significant. The free energy change is reduced and the K increases. The reaction is still nonspontaneous.

52. **Given:** $3\ NO_2\ (g) + H_2O\ (l) \rightarrow 2\ HNO_3\ (aq) + NO\ (g)$ **Find:** $\Delta G°_{rxn}$, and K_p at 25 °C

Conceptual Plan: $\Delta G^0_{rxn} = \sum n_p \Delta G^0_f(products) - \sum n_r \Delta G^0_f(reactants)$ **then** $\Delta G°_{rxn}, T \rightarrow K$

$\Delta G^0_{rxn} = -RT \ln K$

Solution:

Reactant/Product	ΔG^0_f(kJ/mol from Appendix IIB)
$NO_2\ (g)$	51.3
$H_2O\ (l)$	− 237.1
$HNO_3\ (aq)$	− 110.9
$NO\ (g)$	87.6

Be sure to pull data for the correct formula and phase.

$$\Delta G^0_{rxn} = \sum n_p \Delta G^0_f(products) - \sum n_r \Delta G^0_f(reactants)$$
$$= [2(\Delta G^0_f(HNO_3\ (g))) + 1(\Delta G^0_f(NO\ (g)))] - [3(\Delta G^0_f(NO_2\ (g))) + 1(\Delta G^0_f(H_2O\ (l)))]$$
$$= [2(-110.9\text{ kJ}) + 1(87.6\text{ kJ})] - [3(51.3\text{ kJ}) + 1(-237.1\text{ kJ})]$$
$$= [-134.2\text{ kJ}] - [-83.2\text{ kJ}]$$
$$= -51.0\text{ kJ} = +5.10 \times 10^4\text{ J}$$ then
$$\Delta G^0_{rxn} = -RT\ln K$$

Rearrange to solve for K. $K = e^{\frac{-\Delta G^0_{rxn}}{RT}} = e^{\frac{-(-5.10 \times 10^4 \text{ J})}{\left(8.314 \frac{\text{J}}{\text{K}\cdot\text{mol}}\right)(298\text{ K})}} = e^{20.5847} = 8.71 \times 10^8$ so the reaction is spontaneous.

Check: The units (kJ and none) are correct. The free energy change is negative and the reaction is spontaneous. This results in a large K.

53. **Given:** $C_2H_4\ (g) + X_2\ (g) \rightarrow C_2H_4X_2\ (g)$ where X = Cl, Br, and I
Find: ΔH°_{rxn}, ΔS°_{rxn}, ΔG°_{rxn} and K at 25 °C and spontaneity trends with X and temperature
Conceptual Plan: $\Delta H^\circ_{rxn} = \sum n_p H^\circ_f(products) - \sum n_r H^\circ_f(reactants)$ **then**
$\Delta S^\circ_{rxn} = \sum n_p S^\circ(products) - \sum n_r S^\circ(reactants)$ **then °C → K then J/K → kJ/K then**

$K = 273.15 + °C$ $\qquad \frac{1\text{ kJ}}{1000\text{ J}}$

ΔH°_{rxn}, ΔS_{rxn}, T → ΔG
$\Delta G = \Delta H_{rxn} - T\Delta S_{rxn}$

Solution:

Reactant/Product	ΔH^0_f(kJ/mol from Appendix IIB)
$C_2H_4\ (g)$	52.4
$Cl_2\ (g)$	0.0
$C_2H_4Cl_2\ (g)$	– 129.7

Be sure to pull data for the correct formula and phase.
$$\Delta H^0_{rxn} = \sum n_p \Delta H^0_f(products) - \sum n_r \Delta H^0_f(reactants)$$
$$= [1(\Delta H^0_f(C_2H_4Cl_2\ (g)))] - [1(\Delta H^0_f(C_2H_4\ (g))) + 1(\Delta H^0_f(Cl_2\ (g)))]$$
$$= [1(-129.7\text{ kJ})] - [1(52.4\text{ kJ}) + 1(0.0\text{ kJ})]$$ then
$$= [-129.7\text{ kJ}] - [52.4\text{ kJ}]$$
$$= -182.1\text{ kJ}$$

Reactant/Product	S^0(J/molK from Appendix IIB)
$C_2H_4\ (g)$	219.3
$Cl_2\ (g)$	223.1
$C_2H_4Cl_2\ (g)$	308.0

Be sure to pull data for the correct formula and phase.
$$\Delta S^0_{rxn} = \sum n_p S^0(products) - \sum n_r S^0(reactants)$$
$$= [1(S^0(C_2H_4Cl_2\ (g)))] - [1(S^0(C_2H_4\ (g))) + 1(S^0(Cl_2\ (g)))]$$
$$= [1(308.0\text{ J/K})] - [1(219.3\text{ J/K}) + 1(223.1\text{ J/K})]$$ then $T = 273.15 + 25\text{ °C} = 298\text{ K}$ then
$$= [308.0\text{ J/K}] - [442.4\text{ J/K}]$$
$$= -134.4\text{ J/K}$$

$$-134.4 \frac{\text{J}}{\text{K}} \times \frac{1\text{ kJ}}{1000\text{ J}} = -0.1344 \frac{\text{kJ}}{\text{K}}$$ then

$$\Delta G^0 = \Delta H^\circ_{rxn} - T\Delta S^\circ_{rxn} = -182.1\text{ kJ} - (298\text{ K})\left(-0.1344 \frac{\text{kJ}}{\text{K}}\right) = -142.0\text{ kJ} = -1.420 \times 10^5\text{ J}$$ then

$\Delta G^0_{rxn} = -RT\ln K$ Rearrange to solve for K. $K = e^{\frac{-\Delta G^0_{rxn}}{RT}} = e^{\frac{-(-1.420 \times 10^5 \text{ J})}{\left(8.314 \frac{\text{J}}{\text{K}\cdot\text{mol}}\right)(298 \text{ K})}} = e^{57.334} = 7.94 \times 10^{24}$ so the reaction is spontaneous.

Reactant/Product	ΔH^0_f(kJ/mol from Appendix IIB)
C_2H_4 (*g*)	52.4
Br_2 (*g*)	30.9
$C_2H_4Br_2$ (*g*)	− 38.3

Be sure to pull data for the correct formula and phase.

$$\Delta H^0_{rxn} = \sum n_p \Delta H^0_f(products) - \sum n_r \Delta H^0_f(reactants)$$
$$= [1(\Delta H^0_f(C_2H_4Br_2\ (g)))] - [1(\Delta H^0_f(C_2H_4\ (g))) + 1(\Delta H^0_f(Br_2\ (g)))]$$
$$= [1(-38.3 \text{ kJ})] - [1(52.4 \text{ kJ}) + 1(30.9 \text{ kJ})]$$
$$= [-38.3 \text{ kJ}] - [83.3 \text{ kJ}]$$
$$= -121.6 \text{ kJ}$$

then

Reactant/Product	S^0(J/molK from Appendix IIB)
C_2H_4 (*g*)	219.3
Br_2 (*g*)	245.5
$C_2H_4Br_2$ (*g*)	330.6

Be sure to pull data for the correct formula and phase.

$$\Delta S^0_{rxn} = \sum n_p S^0(products) - \sum n_r S^0(reactants)$$
$$= [1(S^0(C_2H_4Br_2\ (g)))] - [1(S^0(C_2H_4\ (g))) + 1(S^0(Br_2\ (g)))]$$
$$= [1(330.6 \text{ J/K})] - [1(219.3 \text{ J/K}) + 1(245.5 \text{ J/K})]$$
$$= [330.6 \text{ J/K}] - [464.8 \text{ J/K}]$$
$$= -134.2 \text{ J/K}$$

then $-134.2 \frac{\text{J}}{\text{K}} \times \frac{1 \text{ kJ}}{1000 \text{ J}} = -0.1342 \frac{\text{kJ}}{\text{K}}$

then $\Delta G^0 = \Delta H^0_{rxn} - T\Delta S^0_{rxn} = -121.6 \text{ kJ} - (298 \text{ K})\left(-0.1342 \frac{\text{kJ}}{\text{K}}\right) = -81.6 \text{ kJ} = -8.16 \times 10^4 \text{ J}$ then

$\Delta G^0_{rxn} = -RT\ln K$ Rearrange to solve for K. $K = e^{\frac{-\Delta G^0_{rxn}}{RT}} = e^{\frac{-(-8.16 \times 10^4 \text{ J})}{\left(8.314 \frac{\text{J}}{\text{K}\cdot\text{mol}}\right)(298 \text{ K})}} = e^{32.9389} = 2.02 \times 10^{14}$ so the reaction is spontaneous.

Reactant/Product	ΔH^0_f(kJ/mol from Appendix IIB)
C_2H_4 (*g*)	52.4
I_2 (*g*)	62.42
$C_2H_4I_2$ (*g*)	66.5

Be sure to pull data for the correct formula and phase.

$$\Delta H^0_{rxn} = \sum n_p \Delta H^0_f(products) - \sum n_r \Delta H^0_f(reactants)$$
$$= [1(\Delta H^0_f(C_2H_4I_2\ (g)))] - [1(\Delta H^0_f(C_2H_4\ (g))) + 1(\Delta H^0_f(I_2\ (g)))]$$
$$= [1(66.5 \text{ kJ})] - [1(52.4 \text{ kJ}) + 1(62.42 \text{ kJ})]$$
$$= [66.5 \text{ kJ}] - [114.\underline{8}2 \text{ kJ}]$$
$$= -48.\underline{3}2 \text{ kJ}$$

then

Reactant/Product	S^0(J/molK from Appendix IIB)
C_2H_4 (*g*)	219.3
I_2 (*g*)	260.69
$C_2H_4I_2$ (*g*)	347.8

Be sure to pull data for the correct formula and phase.

$$\Delta S^0_{rxn} = \sum n_p S^0(products) - \sum n_r S^0(reactants)$$
$$= [1(S^0(C_2H_4I_2\ (g)))] - [1(S^0(C_2H_4\ (g))) + 1(S^0(I_2\ (g)))]$$
$$= [1(347.8\ J/K)] - [1(219.3\ J/K) + 1(260.69\ J/K)]$$
$$= [347.8\ J/K] - [479.\underline{9}9\ J/K]$$
$$= -132.2\ J/K$$

then $-132.2 \frac{J}{K} \times \frac{1\ kJ}{1000\ J} = -0.1322 \frac{kJ}{K}$

then $\Delta G^0 = \Delta H^0_{rxn} - T\Delta S^0_{rxn} = -48.\underline{3}2\ kJ - (298\ K)\left(-0.1322 \frac{kJ}{K}\right) = -8.\underline{9}244\ kJ = -8.\underline{9}244 \times 10^3\ J$ then

$\Delta G^0_{rxn} = -RT \ln K$ Rearrange to solve for K. $K = e^{\frac{-\Delta G^0_{rxn}}{RT}} = e^{\frac{-(-8.\underline{9}244 \times 10^3\ J)}{\left(8.314 \frac{J}{K \cdot mol}\right)(298\ K)}} = e^{3.\underline{6}021} = 37$ and the reaction is spontaneous.

Cl_2 is the most spontaneous in the forward direction, I_2 is the least. The entropy change in the reactions is very constant. The spontaneity is determined by the standard enthalpy of formation of the dihalogenated ethane. Higher temperatures make the forward reactions less spontaneous.

Check: The units (kJ and none) are correct. The enthalpy change becomes less negative as we move to larger halogens. The enthalpy term dominates at room temperature, the free energy change is the same sign as the enthalpy change. The more negative the free energy change, the larger the *K*.

54. **Given:** $H_2\ (g) + X_2\ (g) \rightarrow 2\ HX\ (g)$ where X = Cl, Br, and I
Find: ΔH°_{rxn}, ΔS°_{rxn}, ΔG°_{rxn} and *K* at 25 °C and spontaneity trends with X and temperature
Conceptual Plan: $\Delta H^\circ_{rxn} = \sum n_p H^\circ_f(products) - \sum n_r H^\circ_f(reactants)$ **then**
$\Delta S^\circ_{rxn} = \sum n_p S^\circ(products) - \sum n_r S^\circ(reactants)$ **then °C → K then J/K → kJ/K then**

$K = 273.15 + °C$ $\qquad \frac{1\ kJ}{1000\ J}$

ΔH°_{rxn}, ΔS_{rxn}, T → ΔG
$\Delta G = \Delta H_{rxn} - T\Delta S_{rxn}$

Solution:

Reactant/Product	ΔH^0_f(kJ/mol from Appendix IIB)
$H_2\ (g)$	0.0
$Cl_2\ (g)$	0.0
$HCl\ (g)$	− 92.3

Be sure to pull data for the correct formula and phase.

$$\Delta H^0_{rxn} = \sum n_p \Delta H^0_f(products) - \sum n_r \Delta H^0_f(reactants)$$
$$= [2(\Delta H^0_f(HCl\ (g)))] - [1(\Delta H^0_f(H_2\ (g))) + 1(\Delta H^0_f(Cl_2\ (g)))]$$
$$= [2(-92.3\ kJ)] - [1(0.0\ kJ) + 1(0.0\ kJ)]$$
$$= [-184.6\ kJ] - [0.0]$$
$$= -184.6\ kJ$$

then

Reactant/Product	S^0(J/molK from Appendix IIB)
$H_2\ (g)$	130.7
$Cl_2\ (g)$	223.1
$HCl\ (g)$	186.9

Be sure to pull data for the correct formula and phase.

$$\Delta S^0_{rxn} = \sum n_p S^0(products) - \sum n_r S^0(reactants)$$
$$= [2(S^0(HCl\ (g)))] - [1(S^0(H_2\ (g))) + 1(S^0(Cl_2\ (g)))]$$
$$= [2(186.9\ J/K)] - [1(130.7\ J/K) + 1(223.1\ J/K)]$$
$$= [373.8\ J/K] - [353.8\ J/K]$$
$$= +20.0\ J/K$$

then $T = 273.15 + 25\ °C = 298\ K$ then

$+20.0\frac{\cancel{J}}{K} \times \frac{1\text{ kJ}}{1000\cancel{J}} = +0.0200\frac{kJ}{K}$ then

$\Delta G^0 = \Delta H^0_{rxn} - T\Delta S^0_{rxn} = -184.6\text{ kJ} - (298\cancel{K})\left(+0.0200\frac{kJ}{\cancel{K}}\right) = -190.6\text{ kJ} = -1.906 \times 10^5\text{ J}$ then

$\Delta G^0_{rxn} = -RT\ln K$ Rearrange to solve for K.

$K = e^{\frac{-\Delta G^0_{rxn}}{RT}} = e^{\frac{-(-1.906 \times 10^5 \cancel{J})}{\left(8.314\frac{\cancel{J}}{\cancel{K}\cdot mol}\right)(298\cancel{K})}} = e^{76.9140} = 2.53 \times 10^{33}$ so the reaction is spontaneous.

Reactant/Product	ΔH^0_f(kJ/mol from Appendix IIB)
H_2 (g)	0.0
Br_2 (g)	30.9
HBr (g)	– 36.3

Be sure to pull data for the correct formula and phase.

$\Delta H^0_{rxn} = \sum n_p \Delta H^0_f(products) - \sum n_r \Delta H^0_f(reactants)$

$= [2(\Delta H^0_f(\text{HBr }(g)))] - [1(\Delta H^0_f(\text{H}_2\ (g))) + 1(\Delta H^0_f(\text{Br}_2\ (g)))]$

$= [2(-36.3\text{ kJ})] - [1(0.0\text{ kJ}) + 1(30.9\text{ kJ})]$ then

$= [-72.6\text{ kJ}] - [30.9\text{ kJ}]$

$= -103.5\text{ kJ}$

Reactant/Product	S^0(J/molK from Appendix IIB)
H_2 (g)	130.7
Br_2 (g)	245.5
HBr (g)	198.7

Be sure to pull data for the correct formula and phase.

$\Delta S^0_{rxn} = \sum n_p S^0(products) - \sum n_r S^0(reactants)$

$= [2(S^0(\text{HBr }(g)))] - [1(S^0(\text{H}_2\ (g))) + 1(S^0(\text{Br}_2\ (g)))]$

$= [2(198.7\text{ J/K})] - [1(130.7\text{ J/K}) + 1(245.5\text{ J/K})]$ then $+21.2\frac{\cancel{J}}{K} \times \frac{1\text{ kJ}}{1000\cancel{J}} = +0.0212\frac{kJ}{K}$

$= [397.4\text{ J/K}] - [376.2\text{ J/K}]$

$= +21.2\text{ J/K}$

then $\Delta G^0 = \Delta H^0_{rxn} - T\Delta S^0_{rxn} = -103.5\text{ kJ} - (298\cancel{K})\left(+0.0212\frac{kJ}{\cancel{K}}\right) = -109.\underline{8}176\text{ kJ} = -1.09\underline{8}176 \times 10^5\text{ J}$ then

$\Delta G^0_{rxn} = -RT\ln K$ Rearrange to solve for K.

$K = e^{\frac{-\Delta G^0_{rxn}}{RT}} = e^{\frac{-(-1.09\underline{8}176 \times 10^5 \cancel{J})}{\left(8.314\frac{\cancel{J}}{\cancel{K}\cdot mol}\right)(298\cancel{K})}} = e^{44.\underline{3}247} = 1.78 \times 10^{19}$ so the reaction is spontaneous.

Reactant/Product	ΔH^0_f(kJ/mol from Appendix IIB)
H_2 (g)	0.0
I_2 (g)	62.42
HI (g)	26.5

Be sure to pull data for the correct formula and phase.

$\Delta H^0_{rxn} = \sum n_p \Delta H^0_f(products) - \sum n_r \Delta H^0_f(reactants)$

$= [2(\Delta H^0_f(\text{HI }(g)))] - [1(\Delta H^0_f(\text{H}_2\ (g))) + 1(\Delta H^0_f(\text{I}_2\ (g)))]$

$= [2(26.5\text{ kJ})] - [1(0.0\text{ kJ}) + 1(62.42\text{ kJ})]$ then

$= [53.0\text{ kJ}] - [62.42\text{ kJ}]$

$= -9.\underline{4}2\text{ kJ}$

Reactant/Product	S^0(J/molK from Appendix IIB)
H_2 (*g*)	130.7
I_2 (*g*)	260.69
HI (*g*)	206.6

Be sure to pull data for the correct formula and phase.

$$\Delta S^0_{rxn} = \sum n_p S^0(products) - \sum n_r S^0(reactants)$$
$$= [2(S^0(\text{HI }(g)))] - [1(S^0(\text{H}_2\ (g))) + 1(S^0(\text{I}_2\ (g)))]$$
$$= [2(206.6\text{ J/K})] - [1(130.7\text{ J/K}) + 1(260.69\text{ J/K})]$$
$$= [413.2\text{ J/K}] - [391.\underline{3}9\text{ J/K}]$$
$$= +\ 21.8\text{ J/K}$$

then $21.8\frac{\text{J}}{\text{K}} \times \frac{1\text{ kJ}}{1000\text{ J}} = +\ 0.0218\frac{\text{kJ}}{\text{K}}$

then $\Delta G^o = \Delta H^o_{rxn} - T\Delta S^o_{rxn} = -\ 9.\underline{4}2\text{ kJ} - (298\text{ K})\left(+\ 0.0218\frac{\text{kJ}}{\text{K}}\right) = -\ 15.\underline{9}164\text{ kJ} = -\ 1.\underline{5}9164 \times 10^4\text{ J}$ then

$\Delta G^0_{rxn} = -\ RT\ln K$ Rearrange to solve for *K*.

$$K = e^{\frac{-\Delta G^0_{rxn}}{RT}} = e^{\frac{-(-1.5\underline{9}164 \times 10^4\text{ J})}{\left(8.314\frac{\text{J}}{\text{K}\cdot\text{mol}}\right)(298\text{ K})}} = e^{6.4\underline{2}419} = 6.17 \times 10^2 = 617$$ The reaction is spontaneous.

Cl_2 is the most spontaneous, I_2 is the least. The entropy change in the reactions is very constant. The spontaneity is determined by the standard enthalpy of formation of the acid. Higher temperatures make the reactions more spontaneous.

Check: The units (kJ and none) are correct. The enthalpy is twice the enthalpy of formation of the acid. We expect a very small entropy change because the number of moles of gas is unchanged. Since both terms are negative, the free energy change is negative and the reaction is spontaneous. The more negative the free energy change, the larger the *K*.

55. a) **Given:** N_2O (*g*) + NO_2 (*g*) $\rightleftharpoons$ 3 NO (*g*) at 298 K **Find:** ΔG°_{rxn}

Conceptual Plan: $\Delta G^o_{rxn} = \sum n_p \Delta G^o_f(products) - \sum n_r \Delta G^o_f(reactants)$

Solution:

Reactant/Product	ΔG^0_f(kJ/mol from Appendix IIB)
N_2O (*g*)	103.7
NO_2 (*g*)	51.3
NO (*g*)	87.6

Be sure to pull data for the correct formula and phase.

$$\Delta G^0_{rxn} = \sum n_p \Delta G^0_f(products) - \sum n_r \Delta G^0_f(reactants)$$
$$= [3(\Delta G^0_f(\text{NO }(g)))] - [1(\Delta G^0_f(\text{N}_2\text{O }(g))) + 1(\Delta G^0_f(\text{NO}_2\ (g)))]$$
$$= [3(87.6\text{kJ})] - [1(103.7\text{ kJ}) + 1(51.3\text{ kJ})]$$
$$= [262.8\text{ kJ}] - [155.0\text{ kJ}]$$
$$= +\ 107.8\text{ kJ}$$

The reaction is nonspontaneous.

Check: The units (kJ) are correct. The standard free energy for the reaction was positive and so the reaction is nonspontaneous.

b) **Given:** $P_{N2O} = P_{NO2} = 1.0$ atm initially **Find:** P_{N2O} when reaction ceases to be spontaneous

Conceptual Plan: Reaction will no longer be spontaneous when Q = *K* so then $\Delta G^\circ_{rxn}, T \rightarrow K$

$$\Delta G^0_{rxn} = -\ RT\ln K$$

then solve equilibrium problem to get gas pressures, since *K* << 1 the amount of NO generated will be very, very small compared to 1.0 atm, so, within experimental error, $P_{N2O} = P_{NO2} = 1.0$ atm.

Simply solve for P_{NO}.

$$K = \frac{P_{NO}^3}{P_{N2O}P_{NO2}}$$

Solution: $\Delta G^0_{rxn} = -RT\ln K$ Rearrange to solve for K.

$$K = e^{\frac{-\Delta G^0_{rxn}}{RT}} = e^{\frac{-107.8\ \text{kJ} \times \frac{1000\ \text{J}}{1\ \text{kJ}}}{\left(8.314\frac{\text{J}}{\text{K}\cdot\text{mol}}\right)(298\ \text{K})}} = e^{-43.5103} = 1.27 \times 10^{-19}$$. Since $K = \frac{P_{NO}^3}{P_{N2O}P_{NO2}}$, rearrange to solve

for P_{NO}. $P_{NO}^3 = \sqrt[3]{K P_{N2O}P_{NO2}} = \sqrt[3]{(1.27 \times 10^{-19})(1.0)(1.0)} = 5.0 \times 10^{-7}$ atm.

Note that the assumption that P_{N2O} was very, very small was valid.

Check: The units (atm) are correct. Since the free energy change was positive, the K was very small. This leads us to expect that very little NO will be formed.

c) **Given:** N_2O (*g*) + NO_2 (*g*) ⇌ 3 NO (*g*) **Find:** temperature for spontaneity

Conceptual Plan: $\Delta H^\circ_{rxn} = \sum n_p H^\circ_f(products) - \sum n_r H^\circ_f(reactants)$ **then**

$\Delta S^\circ_{rxn} = \sum n_p S^\circ(products) - \sum n_r S^\circ(reactants)$ **then J/K → kJ/K then ΔH°_{rxn}, ΔS_{rxn} → T**

$$\frac{1\text{kJ}}{1000\ \text{J}} \qquad \Delta G = \Delta H_{rxn} - T\Delta S_{rxn}$$

Solution:

Reactant/Product	ΔH^0_f(kJ/mol from Appendix IIB)
N_2O (*g*)	81.6
NO_2 (*g*)	33.2
NO (*g*)	91.3

Be sure to pull data for the correct formula and phase.

$$\begin{aligned}\Delta H^0_{rxn} &= \sum n_p \Delta H^0_f(products) - \sum n_r \Delta H^0_f(reactants)\\ &= [3(\Delta H^0_f(\text{NO }(g)))] - [1(\Delta H^0_f(\text{N}_2\text{O }(g))) + 1(\Delta H^0_f(\text{NO}_2\ (g)))]\\ &= [3(91.3\text{ kJ})] - [1(81.6\text{ kJ}) + 1(33.2\text{ kJ})]\\ &= [273.9\text{ kJ}] - [114.8\text{ kJ}]\\ &= +159.1\text{ kJ}\end{aligned}$$ then

Reactant/Product	S^0(J/molK from Appendix IIB)
N_2O (*g*)	220.0
NO_2 (*g*)	240.1
NO (*g*)	210.8

Be sure to pull data for the correct formula and phase.

$$\begin{aligned}\Delta S^0_{rxn} &= \sum n_p S^0(products) - \sum n_r S^0(reactants)\\ &= [3(S^0(\text{NO }(g)))] - [1(S^0(\text{N}_2\text{O }(g))) + 1(S^0(\text{NO}_2\ (g)))]\\ &= [3(210.8\text{ J/K})] - [1(220.0\text{ J/K}) + 1(240.1\text{ J/K})]\\ &= [632.4\text{ J/K}] - [460.1\text{ J/K}]\\ &= +172.3\text{ J/K}\end{aligned}$$ then

$+172.3\frac{\text{J}}{\text{K}} \times \frac{1\text{ kJ}}{1000\text{ J}} = +0.1723\frac{\text{kJ}}{\text{K}}$. Since $\Delta G = \Delta H_{rxn} - T\Delta S_{rxn}$, set $\Delta G = 0$ and rearrange to

solve for T. $T = \frac{\Delta H_{rxn}}{\Delta S_{rxn}} = \frac{+159.1\text{ kJ}}{0.1723\frac{\text{kJ}}{\text{K}}} = +923.4\text{ K}$.

Check: The units (K) are correct. The reaction can be made more spontaneous by raising the temperature, because the entropy change is positive (increase in the number of moles of gas).

56. a) **Given:** $BaCO_3$ (*s*) $\rightleftharpoons$ BaO (*s*) + CO_2 (*g*) at 298 K **Find:** ΔG°_{rxn}

Conceptual Plan: $\Delta G^0_{rxn} = \sum n_p \Delta G^0_f(products) - \sum n_r \Delta G^0_f(reactants)$

Solution:

Reactant/Product	ΔG^0_f(kJ/mol from Appendix IIB)
$BaCO_3$ (*s*)	− 1134.4
BaO (*s*)	− 520.3
CO_2 (*g*)	− 394.4

Be sure to pull data for the correct formula and phase.

$$\Delta G^0_{rxn} = \sum n_p \Delta G^0_f(products) - \sum n_r \Delta G^0_f(reactants)$$
$$= [1(\Delta G^0_f(\text{BaO }(s))) + 1(\Delta G^0_f(\text{CO}_2\ (g)))] - [1(\Delta G^0_f(\text{BaCO}_3\ (s)))]$$
$$= [1(-520.3\text{ kJ}) + 1(-394.4\text{ kJ})] - [1(-1134.4\text{ kJ})]$$
$$= [-914.7\text{ kJ}] - [-1134.4\text{ kJ}]$$
$$= +219.7\text{ kJ}$$

The reaction is nonspontaneous.

Check: The units (kJ) are correct. The standard free energy for the reaction was positive and so the reaction is nonspontaneous.

b) **Given:** $BaCO_3$ (*s*) initially in container **Find:** P_{CO2} at equilibrium

Conceptual Plan: **Reaction will be at equilibrium when Q = *K* so then ΔG°_{rxn}, *T* → *K* → P_{CO2}**

$\Delta G^0_{rxn} = -RT\ln K \quad K = P_{CO2}$

Solution: $\Delta G^0_{rxn} = -RT\ln K$ Rearrange to solve for *K*.

$$K = e^{\frac{-\Delta G^0_{rxn}}{RT}} = e^{\frac{-219.7\text{ kJ} \times \frac{1000\text{ J}}{1\text{ kJ}}}{\left(8.314\frac{\text{J}}{\text{K}\cdot\text{mol}}\right)(298\text{ K})}} = e^{-88.6755} = 3.08 \times 10^{-39}.$$

So $P_{CO2} = 3.08 \times 10^{-39}$ atm.

Check: The units (atm) are correct. Since the free energy change was very positive, the *K* was very, very small. This leads us to expect that very little carbon dioxide will be formed.

c) **Given:** $BaCO_3$ (*s*) $\rightleftharpoons$ BaO (*s*) + CO_2 (*g*) **Find:** temperature for P_{CO2} = 1.0 atm

Conceptual Plan: $\Delta H^0_{rxn} = \sum n_p H^0_f(products) - \sum n_r H^0_f(reactants)$ **then**

$\Delta S^0_{rxn} = \sum n_p S^0(products) - \sum n_r S^0(reactants)$ **then J/K → kJ/K then ΔH°_{rxn}, ΔS_{rxn} → *T***

$\frac{1\text{ kJ}}{1000\text{ J}}$ $\qquad \Delta G = \Delta H_{rxn} - T\Delta S_{rxn}$

Solution:

Reactant/Product	ΔH^0_f(kJ/mol from Appendix IIB)
$BaCO_3$ (*s*)	− 1213.0
BaO (*s*)	− 548.0
CO_2 (*g*)	− 393.5

Be sure to pull data for the correct formula and phase.

$$\Delta H^0_{rxn} = \sum n_p \Delta H^0_f(products) - \sum n_r \Delta H^0_f(reactants)$$
$$= [1(\Delta H^0_f(\text{BaO }(s))) + 1(\Delta H^0_f(\text{CO}_2\ (g)))] - [1(\Delta H^0_f(\text{BaCO}_3\ (s)))]$$
$$= [1(-548.0\text{ kJ}) + 1(-393.5\text{ kJ})] - [1(-1213.0\text{ kJ})]$$
$$= [-941.5\text{ kJ}] - [-1213.0\text{ kJ}]$$
$$= +271.5\text{ kJ}$$

then

Reactant/Product	S^0(J/molK from Appendix IIB)
$BaCO_3$ (*s*)	112.1
BaO (*s*)	72.1
CO_2 (*g*)	213.8

Be sure to pull data for the correct formula and phase.

$$\Delta S^0_{rxn} = \sum n_p S^0(products) - \sum n_r S^0(reactants)$$

$$= \; = [1(S^0(\text{BaO }(s))) + 1(S^0(\text{CO}_2\ (g)))] - [1(S^0(\text{BaCO}_3\ (s)))]$$

$$= [1(72.1 \text{ J/K}) + 1(213.8 \text{ J/K})] - [1(112.1 \text{ J/K})]$$

$$= [285.9 \text{ J/K}] - [112.1 \text{ J/K}]$$

$$= +173.8 \text{ J/K}$$

then

$+173.8 \frac{\text{J}}{\text{K}} \times \frac{1 \text{ kJ}}{1000 \text{ J}} = +0.1738 \frac{\text{kJ}}{\text{K}}$. Since $\Delta G = \Delta H_{rxn} - T\Delta S_{rxn}$, set $\Delta G = 0$ and rearrange to solve for *T*. $T = \frac{\Delta H_{rxn}}{\Delta S_{rxn}} = \frac{+271.5 \text{ kJ}}{0.1738 \frac{\text{kJ}}{\text{K}}} = +1562 \text{ K}$. When $\Delta G = 0$ K = 1, so at 1562 K $P_{CO2} = 1.0$ atm.

Check: The units (K) are correct. The reaction can be made more spontaneous by raising the temperature, because the entropy change is positive (increase in the number of moles of gas). We expect a high temperature because the enthalpy change is so positive.

57. a) **Given:** ATP (*aq*) + H_2O (*l*) → ADP (*aq*) + P_i (*aq*) $\Delta G°_{rxn} = -30.5$ kJ at 298 K **Find:** *K*

Conceptual Plan: $\Delta G°_{rxn}, T$ → *K*

$$\Delta G^0_{rxn} = -RT \ln K$$

Solution: $\Delta G^0_{rxn} = -RT\ln K$ Rearrange to solve for *K*.

$$K = e^{\frac{-\Delta G^0_{rxn}}{RT}} = e^{\frac{-(-30.5 \text{ kJ}) \times \frac{1000 \text{ J}}{1 \text{ kJ}}}{\left(8.314 \frac{\text{J}}{\text{K}\cdot\text{mol}}\right)(298 \text{ K})}} = e^{12.3104} = 2.22 \times 10^5.$$

Check: The units (none) are correct. The free energy change is negative and the reaction is spontaneous. This results in a large *K*.

b) **Given:** oxidation of glucose drives reforming of ATP

Find: $\Delta G°_{rxn}$ of oxidation of glucose and moles ATP formed per mole of glucose

Conceptual Plan: write balanced reaction for glucose oxidation then

$\Delta G^0_{rxn} = \sum n_p \Delta G^0_f(products) - \sum n_r \Delta G^0_f(reactants)$ **then $\Delta G°_{rxn}$s → moles ATP/ mole glucose**

$$\frac{\Delta G^0_{rxn}\ glucose\ oxidation}{\Delta G^0_{rxn}\ ATP\ hydrolysis}$$

Solution: $C_6H_{12}O_6$ (*s*) + 6 O_2 (*g*) → 6 CO_2 (*g*) + 6 H_2O (*l*)

Reactant/Product	ΔG^0_f(kJ/mol from Appendix IIB)
$C_6H_{12}O_6$ (*s*)	– 910.4
O_2 (*g*)	0.0
CO_2 (*g*)	– 394.4
H_2O (*l*)	– 237.1

Be sure to pull data for the correct formula and phase.

$$\Delta G^0_{rxn} = \sum n_p \Delta G^0_f(products) - \sum n_r \Delta G^0_f(reactants)$$

$$= [6(\Delta G^0_f(\text{CO}_2\ (g))) + 6(\Delta G^0_f(\text{H}_2\text{O}\ (l)))] - [1(\Delta G^0_f(\text{C}_6\text{H}_{12}\text{O}_6\ (s))) + 6(\Delta G^0_f(\text{O}_2\ (g)))]$$

$$= [6(-394.4 \text{ kJ}) + 6(-237.1 \text{ kJ})] - [1(-910.4 \text{ kJ}) + 6(0.0 \text{ kJ})]$$

$$= [-3789.0 \text{ kJ}] - [-910.4 \text{ kJ}]$$

$$= -2878.6 \text{ kJ}$$

$$\text{So the reaction is very spontaneous.} \qquad \frac{2878.6\ \frac{\text{kJ generated}}{\text{mole glucose oxidized}}}{30.5\ \frac{\text{kJ needed}}{\text{mole ATP reformed}}} = 94.4\ \frac{\text{mole ATP reformed}}{\text{mole glucose oxidized}}$$

Check: The units (mol) are correct. The free energy change for the glucose oxidation is large compared to the ATP hydrolysis, so we expect to reform many moles of ATP.

58. **Given:** ATP (*aq*) + H_2O (*l*) → ADP (*aq*) + P_i (*aq*) $\Delta G^\circ_{rxn} = -30.5$ kJ at 298 K
Find: ΔG_{rxn} when [ATP] = 0.0031 M, [ADP] = 0.0014 M and [P_i] = 0.0048 M
Conceptual Plan: **ΔG°_{rxn}, [ATP], [ADP], [P_i], T → ΔG**

$$\Delta G_{rxn} = \Delta G^0_{rxn} + RT \ln Q \quad \text{where } Q = \frac{[ADP][P_i]}{[ATP]}$$

Solution: $Q = \dfrac{[ADP][P_i]}{[ATP]} = \dfrac{(0.0014)\ (0.0048)}{0.0031} = 0.0021\underline{6}774$ then

$$\Delta G_{rxn} = \Delta G^0_{rxn} + RT \ln Q = -30.5\ \text{kJ} + \left(8.314 \frac{\text{J}}{\text{K} \cdot \text{mol}}\right)\left(\frac{1\ \text{kJ}}{1000\ \text{J}}\right)(298\ \text{K}) \ln(0.0021\underline{6}774) = -45.7\ \text{kJ}$$

Check: The units (kJ) are correct. The Q is less than one so we expect a free energy more negative than at standard conditions.

59. a) **Given:** 2 CO (*g*) + 2 NO (*g*) → N_2 (*g*) + 2 CO_2 (*g*) **Find:** ΔG°_{rxn} and effect of increasing T on ΔG
Conceptual Plan: $\Delta G^0_{rxn} = \sum n_p \Delta G^0_f(products) - \sum n_r \Delta G^0_f(reactants)$
Solution:

Reactant/Product	ΔG^0_f(kJ/mol from Appendix IIB)
CO (*g*)	– 137.2
NO (*g*)	87.6
N_2 (*g*)	0.0
CO_2 (*g*)	– 394.4

Be sure to pull data for the correct formula and phase.

$$\begin{aligned}
\Delta G^0_{rxn} &= \sum n_p \Delta G^0_f(products) - \sum n_r \Delta G^0_f(reactants) \\
&= [1(\Delta G^0_f(N_2\ (g))) + 2(\Delta G^0_f(CO_2\ (g)))] - [2(\Delta G^0_f(CO\ (g))) + 2(\Delta G^0_f(NO\ (g)))] \\
&= [1(0.0\ \text{kJ}) + 2(-394.4\ \text{kJ})] - [2(-137.2\ \text{kJ}) + 2(87.6\ \text{kJ})] \\
&= [-788.8\ \text{kJ}] - [-99.2\ \text{kJ}] \\
&= -689.6\ \text{kJ}
\end{aligned}$$

Since the number of moles of gas is decreasing, the entropy change is negative and so ΔG will become more positive with increasing temperature.
Check: The units (kJ) are correct. The free energy change is negative since the carbon dioxide has such a low free energy of formation.

b) **Given:** 5 H_2 (*g*) + 2 NO (*g*) → 2 NH_3 (*g*) +2 H_2O (*g*) **Find:** ΔG°_{rxn} and effect of increasing T on ΔG
Conceptual Plan: $\Delta G^0_{rxn} = \sum n_p \Delta G^0_f(products) - \sum n_r \Delta G^0_f(reactants)$
Solution:

Reactant/Product	ΔG^0_f(kJ/mol from Appendix IIB)
H_2 (*g*)	0.0
NO (*g*)	87.6
NH_3 (*g*)	– 16.4
H_2O (*g*)	– 228.6

Be sure to pull data for the correct formula and phase.

$$\Delta G^0_{rxn} = \sum n_p \Delta G^0_f(products) - \sum n_r \Delta G^0_f(reactants)$$
$$= [2(\Delta G^0_f(NH_3\ (g))) + 2(\Delta G^0_f(H_2O\ (g)))] - [5(\Delta G^0_f(H_2\ (g))) + 2(\Delta G^0_f(NO\ (g)))]$$
$$= [2(-16.4\ kJ) + 2(-228.6\ kJ)] - [5(0.0\ kJ) + 2(87.6\ kJ)]$$
$$= [-490.0\ kJ] - [175.2\ kJ]$$
$$= -665.2\ kJ$$

Since the number of moles of gas is decreasing, the entropy change is negative and so ΔG will become more positive with increasing temperature.
Check: The units (kJ) are correct. The free energy change is negative since ammonia and water have such a low free energy of formation.

c) **Given:** $2\ H_2\ (g) + 2\ NO\ (g) \rightarrow N_2\ (g) + 2\ H_2O\ (g)$ **Find:** $\Delta G°_{rxn}$ and effect of increasing T on ΔG

Conceptual Plan: $\Delta G^{\circ}_{rxn} = \sum n_p \Delta G^{\circ}_f(products) - \sum n_r \Delta G^{\circ}_f(reactants)$

Solution:

Reactant/Product	ΔG^0_f (kJ/mol from Appendix IIB)
$H_2\ (g)$	0.0
$NO\ (g)$	87.6
$N_2\ (g)$	0.0
$H_2O\ (g)$	– 228.6

Be sure to pull data for the correct formula and phase.

$$\Delta G^0_{rxn} = \sum n_p \Delta G^0_f(products) - \sum n_r \Delta G^0_f(reactants)$$
$$= [1(\Delta G^0_f(N_2\ (g))) + 2(\Delta G^0_f(H_2O\ (g)))] - [2(\Delta G^0_f(H_2\ (g))) + 2(\Delta G^0_f(NO\ (g)))]$$
$$= [1(0.0\ kJ) + 2(-228.6\ kJ)] - [2(0.0\ kJ) + 2(87.6\ kJ)]$$
$$= [-457.2\ kJ] - [175.2\ kJ]$$
$$= -632.4\ kJ$$

Since the number of moles of gas is decreasing, the entropy change is negative and so ΔG will become more positive with increasing temperature.
Check: The units (kJ) are correct. The free energy change is negative since water has such a low free energy of formation.

d) **Given:** $2\ NH_3\ (g) + 2\ O_2\ (g) \rightarrow N_2O\ (g) + 3\ H_2O\ (g)$ **Find:** $\Delta G°_{rxn}$ and effect of increasing T on ΔG

Conceptual Plan: $\Delta G^{\circ}_{rxn} = \sum n_p \Delta G^{\circ}_f(products) - \sum n_r \Delta G^{\circ}_f(reactants)$

Solution:

Reactant/Product	ΔG^0_f (kJ/mol from Appendix IIB)
$NH_3\ (g)$	– 16.4
$O_2\ (g)$	0.0
$N_2O\ (g)$	103.7
$H_2O\ (g)$	– 228.6

Be sure to pull data for the correct formula and phase.

$$\Delta G^0_{rxn} = \sum n_p \Delta G^0_f(products) - \sum n_r \Delta G^0_f(reactants)$$
$$= [1(\Delta G^0_f(N_2O\ (g))) + 3(\Delta G^0_f(H_2O\ (g)))] - [2(\Delta G^0_f(NH_3\ (g))) + 2(\Delta G^0_f(O_2\ (g)))]$$
$$= [1(103.7\ kJ) + 3(-228.6\ kJ)] - [2(-16.4\ kJ) + 2(0.0\ kJ)]$$
$$= [-582.1\ kJ] - [-32.8\ kJ]$$
$$= -549.3\ kJ$$

Since the number of moles of gas is constant the entropy change will be small and slightly negative and so the magnitude of ΔG will decrease with increasing temperature.
Check: The units (kJ) are correct. The free energy change is negative since water has such a low free energy of formation. The entropy change is negative once the $S°$ values are reviewed ($\Delta S_{rxn} = -9.6$ J/K).

60. a) **Given:** $NH_3\ (g) + HBr\ (g) \rightarrow NH_4Br\ (s)$ **Find:** $\Delta G°_{rxn}$ effect of decreasing T on ΔG

Conceptual Plan: $\Delta G^{\circ}_{rxn} = \sum n_p \Delta G^{\circ}_f(products) - \sum n_r \Delta G^{\circ}_f(reactants)$

Solution:

Reactant/Product	ΔG_f^0 (kJ/mol from Appendix IIB)
NH_3 (g)	– 16.4
HBr (g)	– 53.4
NH_4Br (s)	– 175.2

Be sure to pull data for the correct formula and phase.

$$\Delta G^0_{rxn} = \sum n_p \Delta G_f^0(products) - \sum n_r \Delta G_f^0(reactants)$$
$$= [1(\Delta G_f^0(\text{NH}_4\text{Br }(s)))] - [1(\Delta G_f^0(\text{NH}_3\ (g))) + 1(\Delta G_f^0(\text{HBr }(g)))]$$
$$= [1(-175.2\text{ kJ})] - [1(-16.4\text{ kJ}) + 1(-53.4\text{ kJ})]$$
$$= [-175.2\text{ kJ}] - [-69.8\text{ kJ}]$$
$$= -105.4\text{ kJ}$$

Since the number of moles of gas decreases, the entropy change will be negative and ΔG will become more negative with decreasing temperature.

Check: The units (kJ) are correct. The free energy change is negative since ammonium bromide has such a low free energy of formation.

b) **Given:** $CaCO_3$ (s) → CaO (s) + CO_2 (g) **Find:** $\Delta G°_{rxn}$ effect of decreasing T on ΔG

Conceptual Plan: $\Delta G^0_{rxn} = \sum n_p \Delta G_f^0(products) - \sum n_r \Delta G_f^0(reactants)$

Solution:

Reactant/Product	ΔG_f^0 (kJ/mol from Appendix IIB)
$CaCO_3$ (s)	– 1129.1
CaO (s)	– 603.3
CO_2 (g)	– 394.4

Be sure to pull data for the correct formula and phase.

$$\Delta G^0_{rxn} = \sum n_p \Delta G_f^0(products) - \sum n_r \Delta G_f^0(reactants)$$
$$= [1(\Delta G_f^0(\text{CaO }(s))) + 1(\Delta G_f^0(\text{CO}_2\ (g)))] - [1(\Delta G_f^0(\text{CaCO}_3\ (s)))]$$
$$= [1(-603.3\text{kJ}) + 1(-394.4\text{ kJ})] - [1(-1129.1\text{ kJ})]$$
$$= [-997.7\text{ kJ}] - [-1129.1\text{ kJ}]$$
$$= +131.4\text{ kJ}$$

Since the number of moles of gas increases, the entropy change will be positive and ΔG will become more positive with decreasing temperature.

Check: The units (kJ) are correct. The free energy change is positive since calcium carbonate has such a large free energy of formation.

c) **Given:** CH_4 (g) + 3 Cl_2 (g) → $CHCl_3$ (g) + 3 HCl (g) **Find:** $\Delta G°_{rxn}$ effect of decreasing T on ΔG

Conceptual Plan: $\Delta G^0_{rxn} = \sum n_p \Delta G_f^0(products) - \sum n_r \Delta G_f^0(reactants)$

Solution:

Reactant/Product	ΔG_f^0 (kJ/mol from Appendix IIB)
CH_4 (g)	– 50.5
Cl_2 (g)	0.0
$CHCl_3$ (g)	– 70.4
HCl (g)	– 95.3

Be sure to pull data for the correct formula and phase.

$$\Delta G^0_{rxn} = \sum n_p \Delta G_f^0(products) - \sum n_r \Delta G_f^0(reactants)$$
$$= [1(\Delta G_f^0(\text{CHCl}_3\ (g))) + 3(\Delta G_f^0(\text{HCl }(g)))] - [1(\Delta G_f^0(\text{CH}_4\ (g))) + 1(\Delta G_f^0(\text{Cl}_2\ (g)))]$$
$$= [1(-70.4\text{ kJ}) + 3(-95.3\text{ kJ})] - [1(-50.5\text{ kJ}) + 3(0.0\text{ kJ})]$$
$$= [-356.3\text{ kJ}] - [-50.5\text{ kJ}]$$
$$= -305.8\text{ kJ}$$

Since the number of moles of gas is constant, the entropy change will be small. The entropy change is so

small that the magnitude of ΔG will remain constant with decreasing temperature.
Check: The units (kJ) are correct. The free energy change is negative since chloroform and hydrogen chloride have such low free energies of formation. The entropy change is slightly positive once the $S°$ values are reviewed ($\Delta S_{rxn} = +0.7$ J/K).

61. With one exception, the formation of any oxide of nitrogen at 298 K requires more moles of gas as reactants than are formed as products. For example 1 mole of N_2O requires 0.5 moles of O_2 and 1 mole of N_2; 1 mole of N_2O_3 requires 1 mole of N_2 and 1.5 moles of O_2, and so on. The exception is NO, where 1 mole of NO requires 0.5 moles of O_2 and 0.5 moles of N_2: $\frac{1}{2} N_2(g) + \frac{1}{2} O_2(g) \rightarrow NO(g)$ This reaction has a positive ΔS because what is essentially mixing of the N and O has taken place in the product.

62. $\Delta G°_f$ becomes less negative as the atomic number increases because the bond length increases and the bond strength decreases. There is less chemical energy stored in the longer bonds. The $\Delta S°_f$ increases as the atomic number increases because of the halides in the hydrogen halides. This is because the hydrogen halide has a low entropy component (hydrogen) and a high entropy component (the halide, which increases as the atomic number of the halide increases).

63. a) **Given:** glutamate (*aq*) + NH_3 (*aq*) → glutamine (*aq*) + H_2O (*l*) $\Delta G°_{rxn} = +14.2$ kJ at 298 K **Find:** *K*
Conceptual Plan: $\Delta G°_{rxn}, T \rightarrow K$

$$\Delta G^0_{rxn} = -RT\ln K$$

Solution: $\Delta G^0_{rxn} = -RT\ln K$ Rearrange to solve for *K*.

$$K = e^{\frac{-\Delta G^0_{rxn}}{RT}} = e^{\frac{-14.2\ \text{kJ} \times \frac{1000\ \text{J}}{1\ \text{kJ}}}{\left(8.314 \frac{\text{J}}{\text{K} \cdot \text{mol}}\right)(298\ \text{K})}} = e^{-5.73142} = 3.24 \times 10^{-3}.$$

Check: The units (none) are correct. The free energy change is positive and the reaction is nonspontaneous. This results in a small *K*.

b) **Given:** pair ATP hydrolysis with glutamate/NH_3 reaction **Find:** show coupled reactions, $\Delta G°_{rxn}$ and *K*
Conceptual Plan: use reaction mechanism shown, where A = NH_3 and B = glutamate ($C_5H_8O_4N^-$), then calculate $\Delta G°_{rxn}$ by adding free energies of reactions then $\Delta G°_{rxn}, T \rightarrow K$

$$\Delta G^0_{rxn} = -RT\ln K$$

Solution:

$$NH_3\ (aq) + ATP\ (aq) + \cancel{H_2O\ (l)} \rightarrow \cancel{NH_3-P_i(aq)} + ADP\ (aq) \qquad \Delta G^0_{rxn} = -30.5\ \text{kJ}$$

$$\cancel{NH_3-P_i(aq)} + C_5H_8O_4N^-\ (aq) \rightarrow C_5H_9O_3N_2^-\ (aq) + \cancel{H_2O\ (l)} + P_i(aq) \qquad \Delta G^0_{rxn} = +14.2\ \text{kJ}$$

$$NH_3\ (aq) + C_5H_8O_4N^-\ (aq) + ATP\ (aq) \rightarrow C_5H_9O_3N_2^-\ (aq) + ADP\ (aq) + P_i(aq) \quad \Delta G^0_{rxn} = -16.3\ \text{kJ}$$

then $\Delta G^0_{rxn} = -RT\ln K$ Rearrange to solve for *K*.

$$K = e^{\frac{-\Delta G^0_{rxn}}{RT}} = e^{\frac{-(-16.3\ \text{kJ}) \times \frac{1000\ \text{J}}{1\ \text{kJ}}}{\left(8.314 \frac{\text{J}}{\text{K} \cdot \text{mol}}\right)(298\ \text{K})}} = e^{6.57902} = 7.20 \times 10^{2}.$$

Check: The units (none) are correct. The free energy change is negative and the reaction is spontaneous. This results in a large *K*.

64. **Given:** flask configurations **Find:** entropy and rank as increasing entropy
Conceptual Plan: calculate the number of possible states then $W \rightarrow S$ then rank

$$W = \frac{n!}{(n-r)!\,r!} \text{ where } n = \#\text{ particles and } r = \#\text{ particle in one flask} \qquad S = k \ln W$$

Solution:

a) $W = \frac{n!}{(n-r)!\,r!} = \frac{5!}{(0)!\,5!} = 1$ then $S = k\ln W = \left(1.38 \times 10^{-23} \frac{\text{J}}{\text{K}}\right) \ln 1 = 0$;

b) $W = \frac{n!}{(n-r)!\,r!} = \frac{5!}{(2)!\,3!} = 10$ then $S = k \ln W = \left(1.38 \times 10^{-23}\ \frac{J}{K}\right) \ln 10 = 3.18 \times 10^{-23}\ \frac{J}{K}$; and

c) $W = \frac{n!}{(n-r)!\,r!} = \frac{5!}{(1)!\,4!} = 5$ then $S = k \ln W = \left(1.38 \times 10^{-23}\ \frac{J}{K}\right) \ln 5 = 2.22 \times 10^{-23}\ \frac{J}{K}$. So a) < c) < b).

Check: The units (J/K) are correct. The more possibilities for rearranging particles the higher the entropy.

65. a) **Given:** ½ H_2 (*g*) + ½ Cl_2 (*g*) → HCl (*g*), define standard state as 2 atm **Find:** $\Delta G°_f$

Conceptual Plan: $\Delta G°_f, P_{H2}, P_{Cl2}, P_{HCl}, T$ → **new** $\Delta G°_f$

$$\Delta G_{rxn} = \Delta G^0_{rxn} + RT \ln Q \quad \text{where } Q = \frac{P_{HCl}}{P_{H2}^{1/2}\, P_{Cl2}^{1/2}}$$

Solution: $\Delta G°_f = -\,95.3$ kJ/mol and $Q = \frac{P_{HCl}}{P_{H2}^{1/2}\, P_{Cl2}^{1/2}} = \frac{2}{2^{1/2}\, 2^{1/2}} = 1$ then

$$\Delta G_{rxn} = \Delta G^0_{rxn} + RT \ln Q = -\,95.3\ \frac{kJ}{mol} + \left(8.314\ \frac{J}{K \cdot mol}\right)\left(\frac{1\ kJ}{1000\ J}\right)(298\ K)\ln(1) = -\,95.3\ \frac{kJ}{mol} =$$

$$= -\,95{,}300\ \frac{J}{mol}$$

Since the number of moles of reactants and products are the same, the decrease in volume affects the entropy of both equally, so there is no change in $\Delta G°_f$.

Check: The units (kJ) are correct. The Q is one so $\Delta G°_f$ is unchanged under the new standard conditions.

b) **Given:** N_2 (*g*) + ½ O_2 (*g*) → N_2O (*g*), define standard state as 2 atm **Find:** $\Delta G°_f$

Conceptual Plan: $\Delta G°_f, P_{N2}, P_{O2}, P_{N2O}, T$ → **new** $\Delta G°_f$

$$\Delta G_{rxn} = \Delta G^0_{rxn} + RT \ln Q \quad \textbf{where } Q = \frac{P_{N2O}}{P_{N2}\, P_{O2}^{1/2}}$$

Solution: $\Delta G°_f = +\,103.7$ kJ/mol and $Q = \frac{P_{N2O}}{P_{N2}\, P_{O2}^{1/2}} = \frac{2}{2\ 2^{1/2}} = \frac{1}{\sqrt{2}}$ then

$$\Delta G_{rxn} = \Delta G^0_{rxn} + RT \ln Q = +\,103.7\ \frac{kJ}{mol} + \left(8.314\ \frac{J}{K \cdot mol}\right)\left(\frac{1\ kJ}{1000\ J}\right)(298\ K)\ln\left(\frac{1}{\sqrt{2}}\right) = +\,102.8\ \frac{kJ}{mol} =$$

$$= +\,102{,}800\ \frac{J}{mol}$$

The entropy of the reactants (1.5 mol) is decreased more than the entropy of the product (1 mol). Since the product is relatively more favored at lower volume, $\Delta G°_f$ is less positive.

Check: The units (kJ) are correct. The Q is less than one so $\Delta G°_f$ is reduced under the new standard conditions.

c) **Given:** ½ H_2 (*g*) → H (*g*), define standard state as 2 atm **Find:** $\Delta G°_f$

Conceptual Plan: $\Delta G°_f, P_{H2}, P_H, T$ → **new** $\Delta G°_f$

$$\Delta G_{rxn} = \Delta G^0_{rxn} + RT \ln Q \quad \text{where } Q = \frac{P_H}{P_{H2}^{1/2}}$$

Solution: $\Delta G°_f = +\,203.3$ kJ/mol and $Q = \frac{P_H}{P_{H2}^{1/2}} = \frac{2}{2^{1/2}} = \sqrt{2}$ then

$$\Delta G_{rxn} = \Delta G^0_{rxn} + RT \ln Q = +\,203.3\ \frac{kJ}{mol} + \left(8.314\ \frac{J}{K \cdot mol}\right)\left(\frac{1\ kJ}{1000\ J}\right)(298\ K)\ln\left(\sqrt{2}\right) = +\,204.2\ \frac{kJ}{mol} =$$

$$= +\,204{,}200\ \frac{J}{mol}$$

The entropy of the product (1 mol) is decreased more than the entropy of the reactant (1/2 mol). Since the product is relatively less favored, $\Delta G°_f$ is more positive.

Check: The units (kJ) are correct. The Q is greater than one so $\Delta G°_f$ is increased under the new standard conditions.

66. **Given:** H_2O (*l*) → H_2O (*s*) at – 10 °C $\Delta G_{freezing} = -210$ J/mol, $\Delta H_{fusion} = +5610$ J/mol **Find:** $\Delta S_{freezing}$ at – 10 °C
Conceptual Plan: **°C → K then $\Delta G_{freezing}$, ΔH_{fus}, T → $\Delta S_{freezing}$**

$K = 273.15 + °C$ $\qquad$ $\Delta G = \Delta H_{rxn} - T\Delta S_{rxn}$

Solution: $T = -10\ °C + 273.15 = 263$ K, $\Delta H_{freezing} = -5610$ J/mol $= -\Delta H_{fusion}$ then $\Delta G = \Delta H_{rxn} - T\Delta S_{rxn}$.

Rearrange to solve for ΔS. $\Delta S_{freezing} = \dfrac{\Delta H_{freezing} - \Delta G_{freezing}}{T} = \dfrac{-5610\,J - (-210\ J)}{263\ K} = -20.5\dfrac{J}{K}$.

Check: The units (J/K) are correct. The entropy for freezing should be negative since the solid has a more ordered structure.

67. a) **Given:** NH_4NO_3 (*s*) → HNO_3 (*g*) + NH_3 (*g*) **Find:** $\Delta G°_{rxn}$

Conceptual Plan: $\Delta G^0_{rxn} = \sum n_p \Delta G^0_f(products) - \sum n_r \Delta G^0_f(reactants)$

Solution:

Reactant/Product	ΔG^0_f (kJ/mol from Appendix IIB)
NH_4NO_3 (*s*)	– 183.9
HNO_3 (*g*)	– 73.5
NH_3 (*g*)	– 16.4

Be sure to pull data for the correct formula and phase.

$$\begin{aligned}\Delta G^0_{rxn} &= \sum n_p \Delta G^0_f(products) - \sum n_r \Delta G^0_f(reactants)\\ &= [1(\Delta G^0_f(HNO_3\ (g))) + 1(\Delta G^0_f(NH_3\ (g)))] - [1(\Delta G^0_f(NH_4NO_3\ (s)))]\\ &= [1(-73.5\,kJ) + 1(-16.4\ kJ)] - [1(-183.9\ kJ)]\\ &= [-89.9\ kJ] - [-183.9\ kJ]\\ &= +94.0\ kJ\end{aligned}$$

Check: The units (kJ) are correct. The free energy change is positive since ammonium nitrate has such a low free energy of formation.

b) **Given:** NH_4NO_3 (*s*) → N_2O (*g*) + 2 H_2O (*g*) **Find:** $\Delta G°_{rxn}$

Conceptual Plan: $\Delta G^0_{rxn} = \sum n_p \Delta G^0_f(products) - \sum n_r \Delta G^0_f(reactants)$

Solution:

Reactant/Product	ΔG^0_f (kJ/mol from Appendix IIB)
NH_4NO_3 (*s*)	– 183.9
N_2O (*g*)	103.7
H_2O (*g*)	– 228.6

Be sure to pull data for the correct formula and phase.

$$\begin{aligned}\Delta G^0_{rxn} &= \sum n_p \Delta G^0_f(products) - \sum n_r \Delta G^0_f(reactants)\\ &= [1(\Delta G^0_f(N_2O\ (g))) + 2(\Delta G^0_f(H_2O\ (g)))] - [1(\Delta G^0_f(NH_4NO_3\ (s)))]\\ &= [1(103.7\,kJ) + 2(-228.6\ kJ)] - [1(-183.9\ kJ)]\\ &= [-353.5\ kJ] - [-183.9\ kJ]\\ &= -169.6\ kJ\end{aligned}$$

Check: The units (kJ) are correct. The free energy change is negative since water has such a low free energy of formation.

c) **Given:** NH_4NO_3 (*s*) → N_2 (*g*) + ½ O_2 (*g*) + 2 H_2O (*g*) **Find:** $\Delta G°_{rxn}$

Conceptual Plan: $\Delta G^0_{rxn} = \sum n_p \Delta G^0_f(products) - \sum n_r \Delta G^0_f(reactants)$

Solution:

Reactant/Product	ΔG_f^0(kJ/mol from Appendix IIB)
NH_4NO_3 (*s*)	– 183.9
N_2 (*g*)	0.0
O_2 (*g*)	0.0
H_2O (*g*)	– 228.6

Be sure to pull data for the correct formula and phase.

$$\Delta G^0_{rxn} = \sum n_p \Delta G^0_f(products) - \sum n_r \Delta G^0_f(reactants)$$
$$= [1(\Delta G^0_f(N_2\ (g))) + 1/2(\Delta G^0_f(O_2\ (g))) + 2(\Delta G^0_f(H_2O\ (g)))] - [1(\Delta G^0_f(NH_4NO_3\ (s)))]$$
$$= [1(0.0\text{kJ}) + 1/2(0.0\text{ kJ}) + 2(-228.6\text{ kJ})] - [1(-183.9\text{ kJ})]$$
$$= [-457.2\text{ kJ}] - [-183.9\text{ kJ}]$$
$$= -273.3\text{ kJ}$$

Check: The units (kJ) are correct. The free energy change is negative since water has such a low free energy of formation.

The second and third reactions are spontaneous and so we would expect decomposition products of N_2O, N_2, O_2, and H_2O in the gas phase. It is still possible for ammonium nitrate to remain as a solid because the thermodynamics of the reaction say nothing of the kinetics of the reaction (reaction can be extremely slow). Since all of the products are gases, the decomposition of ammonium nitrate will result in a large increase in volume (explosion). Some of the products aid in combustion, which could facilitate the combustion of materials near the ammonium nitrate. Also, N_2O is known as laughing gas, which has anesthetic and toxic effects on humans. The solid should not be kept in tightly sealed containers.

68. A butane lighter is more efficient than an electric lighter since the butane lighter process is accomplished in one step. When you use an electric lighter, fuel is burned to generate electricity. This electricity is transmitted to the lighter through wires. Once it reaches the lighter, the electricity then needs to be converted to heat. Each step must pay a heat tax. Using a butane lighter has fewer steps and so it pays a much lower tax.

69. c) The spontaneity of a reaction says nothing about the speed of a reaction. It only states which direction the reaction will go as it approaches equilibrium.

70. a) and c) will both increase the entropy of the surroundings because they are both exothermic reactions (adding thermal energy to the surroundings).

71. b) has the largest decrease in the number of microstates from the initial to the final state. In a) there are initially $\frac{9!}{4!\,4!\,1!} = 90$ microstates and $\frac{9!}{3!\,3!\,3!} = 1680$ microstates at the end, so $\Delta S > 0$. In b) there are initially $\frac{9!}{4!\,2!\,3!} = 1260$ microstates and $\frac{9!}{6!\,3!\,0!} = 84$ microstates at the end, so $\Delta S < 0$. In c) there are initially $\frac{9!}{3!\,4!\,2!} = 1260$ microstates and $\frac{9!}{3!\,4!\,2!} = 1260$ microstates at the end, so $\Delta S = 0$. Also the final state in b) has the least entropy.

72. c) If the entropy of a system is increasing, the enthalpy of a reaction can be overcome (if necessary) by the entropy change as long as the temperature is high enough. If the entropy change of the system is decreasing, the reaction must be exothermic in order to be spontaneous since the entropy is working against spontaneity.

73. c) Since the vapor pressure of water at 298 K is 23.78 mmHg or 0.03129 atm. As long as the desired pressure (0.010 atm) is less than the equilibrium vapor pressure of water, the reaction will be spontaneous.

74. a) and b) are both true. Since $\Delta G_{rxn} = \Delta G^0_{rxn} + RT\ln Q$ and $\Delta G^\circ_{rxn} = -42.5$ kJ, in order for $\Delta G_{rxn} = 0$ the second term must be positive. This necessitates that Q > 1 or that we have more product than reactant. Any reaction at equilibrium has $\Delta G_{rxn} = 0$.

Chapter 18
Electrochemistry

1. **Conceptual Plan: Separate the overall reaction into two half-reactions: one for oxidation and one for reduction. → Balance each half-reaction with respect to mass in the following order: 1) balance all elements other than H and O; 2) balance O by adding H_2O; and 3) balance H by adding H^+. → Balance each half-reaction with respect to charge by adding electrons. (The sum of the charges on both sides of the equation should be made equal by adding electrons as necessary.) → Make the number of electrons in both half-reactions equal by multiplying one or both half-reactions by a small whole number. → Add the two half-reactions together, canceling electrons and other species as necessary. → Verify that the reaction is balanced both with respect to mass and with respect to charge.**

Solution:

a) Separate: $K\,(s) \rightarrow K^+\,(aq)$ and $Cr^{3+}\,(aq) \rightarrow Cr\,(s)$

Balance elements: $K\,(s) \rightarrow K^+\,(aq)$ and $Cr^{3+}\,(aq) \rightarrow Cr\,(s)$

Add electrons: $K\,(s) \rightarrow K^+\,(aq) + e^-$ and $Cr^{3+}\,(aq) + 3\,e^- \rightarrow Cr\,(s)$

Equalize electrons: $3\,K\,(s) \rightarrow 3\,K^+\,(aq) + 3\,e^-$ and $Cr^{3+}\,(aq) + 3\,e^- \rightarrow Cr\,(s)$

Add half-reactions: $3\,K\,(s) + Cr^{3+}\,(aq) + \cancel{3\,e^-} \rightarrow 3\,K^+\,(aq) + \cancel{3\,e^-} + Cr\,(s)$

Cancel electrons: $3\,K\,(s) + Cr^{3+}\,(aq) \rightarrow 3\,K^+\,(aq) + Cr\,(s)$

Check:

Reactants	Products
3 K atoms	3 K atoms
1 Cr atom	1 Cr atom
+3 charge	+3 charge

b) Separate: $Al\,(s) \rightarrow Al^{3+}\,(aq)$ and $Fe^{2+}\,(aq) \rightarrow Fe\,(s)$

Balance elements: $Al\,(s) \rightarrow Al^{3+}\,(aq)$ and $Fe^{2+}\,(aq) \rightarrow Fe\,(s)$

Add electrons: $Al\,(s) \rightarrow Al^{3+}\,(aq) + 3\,e^-$ and $Fe^{2+}\,(aq) + 2\,e^- \rightarrow Fe\,(s)$

Equalize electrons: $2\,Al\,(s) \rightarrow 2\,Al^{3+}\,(aq) + 6\,e^-$ and $3\,Fe^{2+}\,(aq) + 6\,e^- \rightarrow 3\,Fe\,(s)$

Add half-reactions: $2\,Al\,(s) + 3\,Fe^{2+}\,(aq) + \cancel{6\,e^-} \rightarrow 2\,Al^{3+}\,(aq) + \cancel{6\,e^-} + 3\,Fe\,(s)$

Cancel electrons: $2\,Al\,(s) + 3\,Fe^{2+}\,(aq) \rightarrow 2\,Al^{3+}\,(aq) + 3\,Fe\,(s)$

Check:

Reactants	Products
2 Al atoms	2 Al atoms
3 Fe atom	3 Fe atom
+6 charge	+6 charge

c) Separate: $BrO_3^-\,(aq) \rightarrow Br^-\,(aq)$ and $N_2H_4\,(g) \rightarrow N_2\,(g)$

Balance non H & O elements: $BrO_3^-\,(aq) \rightarrow Br^-\,(aq)$ and $N_2H_4\,(g) \rightarrow N_2\,(g)$

Balance O with H_2O: $BrO_3^-\,(aq) \rightarrow Br^-\,(aq) + 3\,H_2O\,(l)$ and $N_2H_4\,(g) \rightarrow N_2\,(g)$

Balance H with H^+: $BrO_3^-\,(aq) + 6\,H^+\,(aq) \rightarrow Br^-\,(aq) + 3\,H_2O\,(l)$ and $N_2H_4\,(g) \rightarrow N_2\,(g) + 4\,H^+\,(aq)$

Add electrons:

$BrO_3^-\,(aq) + 6\,H^+\,(aq) + 6\,e^- \rightarrow Br^-\,(aq) + 3\,H_2O\,(l)$ and $N_2H_4\,(g) \rightarrow N_2\,(g) + 4\,H^+\,(aq) + 4\,e^-$

Equalize electrons:

$2\,BrO_3^-\,(aq) + 12\,H^+\,(aq) + 12\,e^- \rightarrow 2\,Br^-\,(aq) + 6\,H_2O\,(l)$ and $3\,N_2H_4\,(g) \rightarrow 3\,N_2\,(g) + 12\,H^+\,(aq) + 12\,e^-$

Add half-reactions: $2\,BrO_3^-\,(aq) + \cancel{12\,H^+\,(aq)} + 3\,N_2H_4\,(g) + \cancel{12\,e^-} \rightarrow 2\,Br^-\,(aq) + 6\,H_2O\,(l) + 3\,N_2\,(g) + \cancel{12\,H^+\,(aq)} + \cancel{12\,e^-}$

Cancel electrons & others: $2\,BrO_3^-\,(aq) + 3\,N_2H_4\,(g) \rightarrow 2\,Br^-\,(aq) + 6\,H_2O\,(l) + 3\,N_2\,(g)$

Check:

Reactants	Products
3 Br atoms	3 Br atoms
6 O atoms	6 O atoms
12 H atoms	12 H atoms
6 N atoms	6 N atoms
–2 charge	–2 charge

2. **Conceptual Plan: Separate the overall reaction into two half-reactions: one for oxidation and one for reduction. → Balance each half-reaction with respect to mass in the following order: 1) balance all elements other than H and O; 2) balance O by adding H_2O; and 3) balance H by adding H^+. → Balance each half-reaction with respect to charge by adding electrons. (The sum of the charges on both sides of the equation should be made equal by adding electrons as necessary.) → Make the number of electrons**

in both half-reactions equal by multiplying one or both half-reactions by a small whole number. → Add the two half-reactions together, canceling electrons and other species as necessary. → Verify that the reaction is balanced both with respect to mass and with respect to charge.

Solution:

a) Separate: $Zn\ (s) \rightarrow Zn^{2+}\ (aq)$ and $Sn^{2+}\ (aq) \rightarrow Sn\ (s)$

Balance elements: $Zn\ (s) \rightarrow Zn^{2+}\ (aq)$ and $Sn^{2+}\ (aq) \rightarrow Sn\ (s)$

Add electrons: $Zn\ (s) \rightarrow Zn^{2+}\ (aq) + 2\ e^-$ and $Sn^{2+}\ (aq) + 2\ e^- \rightarrow Sn\ (s)$

Equalize electrons: $Zn\ (s) \rightarrow Zn^{2+}\ (aq) + 2\ e^-$ and $Sn^{2+}\ (aq) + 2\ e^- \rightarrow Sn\ (s)$

Add half-reactions: $Zn\ (s) + Sn^{2+}\ (aq) + \cancel{2\ e^-} \rightarrow Zn^{2+}\ (aq) + \cancel{2\ e^-} + Sn\ (s)$

Cancel electrons: $Zn\ (s) + Sn^{2+}\ (aq) \rightarrow Zn^{2+}\ (aq) + Sn\ (s)$

Check:

Reactants	Products
1 Zn atom	1 Zn atom
1 Sn atom	1 Sn atom
+2 charge	+2 charge

b) Separate: $Mg\ (s) \rightarrow Mg^{2+}\ (aq)$ and $Cr^{3+}\ (aq) \rightarrow Cr\ (s)$

Balance elements: $Mg\ (s) \rightarrow Mg^{2+}\ (aq)$ and $Cr^{3+}\ (aq) \rightarrow Cr\ (s)$

Add electrons: $Mg\ (s) \rightarrow Mg^{2+}\ (aq) + 2\ e^-$ and $Cr^{3+}\ (aq) + 3\ e^- \rightarrow Cr\ (s)$

Equalize electrons: $3\ Mg\ (s) \rightarrow 3\ Mg^{2+}\ (aq) + 6\ e^-$ and $2\ Cr^{3+}\ (aq) + 6\ e^- \rightarrow 2\ Cr\ (s)$

Add half-reactions: $3\ Mg\ (s) + 2\ Cr^{3+}\ (aq) + \cancel{6\ e^-} \rightarrow 3\ Mg^{2+}\ (aq) + \cancel{6\ e^-} + 2\ Cr\ (s)$

Cancel electrons: $3\ Mg\ (s) + 2\ Cr^{3+}\ (aq) \rightarrow 3\ Mg^{2+}\ (aq) + 2\ Cr\ (s)$

Check:

Reactants	Products
3 Mg atoms	3 Mg atoms
2 Cr atoms	2 Cr atoms
+6 charge	+6 charge

c) Separate: $MnO_4^-\ (aq) \rightarrow Mn^{2+}\ (aq)$ and $Al\ (s) \rightarrow Al^{3+}\ (aq)$

Balance non H & O elements: $MnO_4^-\ (aq) \rightarrow Mn^{2+}\ (aq)$ and $Al\ (s) \rightarrow Al^{3+}\ (aq)$

Balance O with H_2O: $MnO_4^-\ (aq) \rightarrow Mn^{2+}\ (aq) + 4\ H_2O\ (l)$ and $Al\ (s) \rightarrow Al^{3+}\ (aq)$

Balance H with H^+: $MnO_4^-\ (aq) + 8\ H^+\ (aq) \rightarrow Mn^{2+}\ (aq) + 4\ H_2O\ (l)$ and $Al\ (s) \rightarrow Al^{3+}\ (aq)$

Add electrons: $MnO_4^-\ (aq) + 8\ H^+\ (aq) + 5\ e^- \rightarrow Mn^{2+}\ (aq) + 4\ H_2O\ (l)$ and $Al\ (s) \rightarrow Al^{3+}\ (aq) + 3\ e^-$

Equalize electrons:

$3\ MnO_4^-\ (aq) + 24\ H^+\ (aq) + 15\ e^- \rightarrow 3\ Mn^{2+}\ (aq) + 12\ H_2O\ (l)$ and $5\ Al\ (s) \rightarrow 5\ Al^{3+}\ (aq) + 15\ e^-$

Add half-reactions:

$3\ MnO_4^-\ (aq) + 24\ H^+\ (aq) + \cancel{15\ e^-} + 5\ Al\ (s) \rightarrow 3\ Mn^{2+}\ (aq) + 12\ H_2O\ (l) + 5\ Al^{3+}\ (aq) + \cancel{15\ e^-}$

Cancel electrons: $3\ MnO_4^-\ (aq) + 24\ H^+\ (aq) + 5\ Al\ (s) \rightarrow 3\ Mn^{2+}\ (aq) + 12\ H_2O\ (l) + 5\ Al^{3+}\ (aq)$

Check:

Reactants	Products
3 Mn atoms	3 Mn atoms
12 O atoms	12 O atoms
24 H atoms	24 H atoms
5 Al atoms	5 Al atoms
+21 charge	+21 charge

3. **Conceptual Plan: Separate the overall reaction into two half-reactions: one for oxidation and one for reduction. → Balance each half-reaction with respect to mass in the following order: 1) balance all elements other than H and O; 2) balance O by adding H_2O; and 3) balance H by adding H^+. → Balance each half-reaction with respect to charge by adding electrons. (The sum of the charges on both sides of the equation should be made equal by adding electrons as necessary.) → Make the number of electrons in both half-reactions equal by multiplying one or both half-reactions by a small whole number. → Add the two half-reactions together, canceling electrons and other species as necessary. → Verify that the reaction is balanced both with respect to mass and with respect to charge.**

Solution:

a) Separate: $PbO_2\ (s) \rightarrow Pb^{2+}\ (aq)$ and $I^-\ (aq) \rightarrow I_2\ (s)$

Balance non H & O elements: $PbO_2\ (s) \rightarrow Pb^{2+}\ (aq)$ and $2\ I^-\ (aq) \rightarrow I_2\ (s)$

Balance O with H_2O: $PbO_2\ (s) \rightarrow Pb^{2+}\ (aq) + 2\ H_2O\ (l)$ and $2\ I^-\ (aq) \rightarrow I_2\ (s)$

Balance H with H^+: $PbO_2\ (s) + 4\ H^+\ (aq) \rightarrow Pb^{2+}\ (aq) + 2\ H_2O\ (l)$ and $2\ I^-\ (aq) \rightarrow I_2\ (s)$

Add electrons: $PbO_2\ (s) + 4\ H^+\ (aq) + 2\ e^- \rightarrow Pb^{2+}\ (aq) + 2\ H_2O\ (l)$ and $2\ I^-\ (aq) \rightarrow I_2\ (s) + 2\ e^-$

Equalize electrons: $PbO_2\ (s) + 4\ H^+\ (aq) + 2\ e^- \rightarrow Pb^{2+}\ (aq) + 2\ H_2O\ (l)$ and $2\ I^-\ (aq) \rightarrow I_2\ (s) + 2\ e^-$

Add half-reactions: $PbO_2\ (s) + 4\ H^+\ (aq) + \cancel{2\ e^-} + 2\ I^-\ (aq) \rightarrow Pb^{2+}\ (aq) + 2\ H_2O\ (l) + I_2\ (s) + \cancel{2\ e^-}$

Cancel electrons & others: $PbO_2\ (s) + 4\ H^+\ (aq) + 2\ I^-\ (aq) \rightarrow Pb^{2+}\ (aq) + 2\ H_2O\ (l) + I_2\ (s)$

Check:

Reactants	Products
1 Pb atom	1 Pb atom
2 O atoms	2 O atoms
4 H atoms	4 H atoms
2 I atoms	2 I atoms
+2 charge	+2 charge

b) Separate: $MnO_4^-\ (aq) \rightarrow Mn^{2+}\ (aq)$ and $SO_3^{2-}\ (aq) \rightarrow SO_4^{2-}\ (aq)$

Balance non H & O elements: $MnO_4^-\ (aq) \rightarrow Mn^{2+}\ (aq)$ and $SO_3^{2-}\ (aq) \rightarrow SO_4^{2-}\ (aq)$

Balance O with H_2O: $MnO_4^-\ (aq) \rightarrow Mn^{2+}\ (aq) + 4\ H_2O\ (l)$ and $SO_3^{2-}\ (aq) + H_2O\ (l) \rightarrow SO_4^{2-}\ (aq)$

Balance H with H^+:

$MnO_4^-\ (aq) + 8\ H^+\ (aq) \rightarrow Mn^{2+}\ (aq) + 4\ H_2O\ (l)$ and $SO_3^{2-}\ (aq) + H_2O\ (l) \rightarrow SO_4^{2-}\ (aq) + 2\ H^+\ (aq)$

Add electrons: $MnO_4^-\ (aq) + 8\ H^+\ (aq) + 5\ e^- \rightarrow Mn^{2+}\ (aq) + 4\ H_2O\ (l)$ and

$SO_3^{2-}\ (aq) + H_2O\ (l) \rightarrow SO_4^{2-}\ (aq) + 2\ H^+\ (aq) + 2\ e^-$

Equalize electrons: $2\ MnO_4^-\ (aq) + 16\ H^+\ (aq) + 10\ e^- \rightarrow 2\ Mn^{2+}\ (aq) + 8\ H_2O\ (l)$ and

$5\ SO_3^{2-}\ (aq) + 5\ H_2O\ (l) \rightarrow 5\ SO_4^{2-}\ (aq) + 10\ H^+\ (aq) + 10\ e^-$

Add half-reactions: $2\ MnO_4^-\ (aq) + 6\ \cancel{16}\ H^+\ (aq) + \cancel{10\ e^-} + 5\ SO_3^{2-}\ (aq) + \cancel{5\ H_2O(l)} \rightarrow$

$2\ Mn^{2+}\ (aq) + 3\ \cancel{8}\ H_2O(l) + 5\ SO_4^{2-}\ (aq) + \cancel{10\ H^+}\ (aq) + \cancel{10\ e^-}$

Cancel electrons: $2\ MnO_4^-\ (aq) + 6\ H^+\ (aq) + 5\ SO_3^{2-}\ (aq) \rightarrow 2\ Mn^{2+}\ (aq) + 3\ H_2O\ (l) + 5\ SO_4^{2-}\ (aq)$

Check:

Reactants	Products
2 Mn atoms	2 Mn atoms
23 O atoms	23 O atoms
6 H atoms	6 H atoms
5 S atoms	5 S atoms
–6 charge	–6 charge

c) Separate: $S_2O_3^{2-}\ (aq) \rightarrow SO_4^{2-}\ (aq)$ and $Cl_2\ (g) \rightarrow Cl^-\ (aq)$

Balance non H & O elements: $S_2O_3^{2-}\ (aq) \rightarrow 2\ SO_4^{2-}\ (aq)$ and $Cl_2\ (g) \rightarrow 2\ Cl^-\ (aq)$

Balance O with H_2O: $S_2O_3^{2-}\ (aq) + 5\ H_2O\ (l) \rightarrow 2\ SO_4^{2-}\ (aq)$ and $Cl_2\ (g) \rightarrow 2\ Cl^-\ (aq)$

Balance H with H^+: $S_2O_3^{2-}\ (aq) + 5\ H_2O\ (l) \rightarrow 2\ SO_4^{2-}\ (aq) + 10\ H^+\ (aq)$ and $Cl_2\ (g) \rightarrow 2\ Cl^-\ (aq)$

Add electrons:

$S_2O_3^{2-}\ (aq) + 5\ H_2O\ (l) \rightarrow 2\ SO_4^{2-}\ (aq) + 10\ H^+\ (aq) + 8\ e^-$ and $Cl_2\ (g) + 2\ e^- \rightarrow 2\ Cl^-\ (aq)$

Equalize electrons:

$S_2O_3^{2-}\ (aq) + 5\ H_2O\ (l) \rightarrow 2\ SO_4^{2-}\ (aq) + 10\ H^+\ (aq) + 8\ e^-$ and $4\ Cl_2\ (g) + 8\ e^- \rightarrow 8\ Cl^-\ (aq)$

Add half-reactions:

$S_2O_3^{2-}\ (aq) + 5\ H_2O\ (l) + 4\ Cl_2\ (g) + \cancel{8\ e^-} \rightarrow 2\ SO_4^{2-}\ (aq) + 10\ H^+\ (aq) + \cancel{8\ e^-} + 8\ Cl^-\ (aq)$

Cancel electrons: $S_2O_3^{2-}\ (aq) + 5\ H_2O\ (l) + 4\ Cl_2\ (g) \rightarrow 2\ SO_4^{2-}\ (aq) + 10\ H^+\ (aq) + 8\ Cl^-\ (aq)$

Check:

Reactants	Products
2 S atoms	2 S atoms
8 O atoms	8 O atoms
10 H atoms	10 H atoms
8 Cl atoms	8 Cl atoms
–2 charge	–2 charge

4. **Conceptual Plan: Separate the overall reaction into two half-reactions: one for oxidation and one for reduction. → Balance each half-reaction with respect to mass in the following order: 1) balance all elements other than H and O; 2) balance O by adding H_2O; and 3) balance H by adding H^+. → Balance each half-reaction with respect to charge by adding electrons. (The sum of the charges on both sides of the equation should be made equal by adding electrons as necessary.) → Make the number of electrons in both half-reactions equal by multiplying one or both half-reactions by a small whole number. → Add the two half-reactions together, canceling electrons and other species as necessary. → Verify that the reaction is balanced both with respect to mass and with respect to charge.**

Solution:

a) Separate: $NO_2^-\ (aq) \rightarrow NO\ (g)$ and $I^-\ (aq) \rightarrow I_2\ (s)$

Balance non H & O elements: $NO_2^-\ (aq) \rightarrow NO\ (g)$ and $2\ I^-\ (aq) \rightarrow I_2\ (s)$

Balance O with H_2O: $NO_2^-\ (aq) \rightarrow NO\ (g) + H_2O\ (l)$ and $2\ I^-\ (aq) \rightarrow I_2\ (s)$

Balance H with H^+: $NO_2^-\,(aq) + 2\,H^+\,(aq) \rightarrow NO\,(g) + H_2O\,(l)$ and $2\,I^-\,(aq) \rightarrow I_2\,(s)$

Add electrons: $NO_2^-\,(aq) + 2\,H^+\,(aq) + e^- \rightarrow NO\,(g) + H_2O\,(l)$ and $2\,I^-\,(aq) \rightarrow I_2\,(s) + 2\,e^-$

Equalize electrons: $2\,NO_2^-\,(aq) + 4\,H^+\,(aq) + 2\,e^- \rightarrow 2\,NO\,(g) + 2\,H_2O\,(l)$ and $2\,I^-\,(aq) \rightarrow I_2\,(s) + 2\,e^-$

Add half-reactions: $2\,NO_2^-\,(aq) + 4\,H^+\,(aq) + \cancel{2\,e^-} + 2\,I^-\,(aq) \rightarrow 2\,NO\,(g) + 2\,H_2O\,(l) + I_2\,(s) + \cancel{2\,e^-}$

Cancel electrons: $2\,NO_2^-\,(aq) + 4\,H^+\,(aq) + 2\,I^-\,(aq) \rightarrow 2\,NO\,(g) + 2\,H_2O\,(l) + I_2\,(s)$

Check:

Reactants	Products
2 N atoms	2 N atoms
4 O atoms	4 O atoms
4 H atoms	4 H atoms
2 I atoms	2 I atoms
0 charge	0 charge

b) Separate: $ClO_4^-\,(aq) \rightarrow ClO_3^-\,(aq)$ and $Cl^-\,(aq) \rightarrow Cl_2\,(g)$

Balance non H & O elements: $ClO_4^-\,(aq) \rightarrow ClO_3^-\,(aq)$ and $2\,Cl^-\,(aq) \rightarrow Cl_2\,(g)$

Balance O with H_2O: $ClO_4^-\,(aq) \rightarrow ClO_3^-\,(aq) + H_2O\,(l)$ and $2\,Cl^-\,(aq) \rightarrow Cl_2\,(g)$

Balance H with H^+: $ClO_4^-\,(aq) + 2\,H^+\,(aq) \rightarrow ClO_3^-\,(aq) + H_2O\,(l)$ and $2\,Cl^-\,(aq) \rightarrow Cl_2\,(g)$

Add electrons: $ClO_4^-\,(aq) + 2\,H^+\,(aq) + 2\,e^- \rightarrow ClO_3^-\,(aq) + H_2O\,(l)$ and $2\,Cl^-\,(aq) \rightarrow Cl_2\,(g) + 2\,e^-$

Equalize electrons: $ClO_4^-\,(aq) + 2\,H^+\,(aq) + 2\,e^- \rightarrow ClO_3^-\,(aq) + H_2O\,(l)$ and $2\,Cl^-\,(aq) \rightarrow Cl_2\,(g) + 2\,e^-$

Add half-reactions: $ClO_4^-\,(aq) + 2\,H^+\,(aq) + \cancel{2\,e^-} + 2\,Cl^-\,(aq) \rightarrow ClO_3^-\,(aq) + H_2O\,(l) + Cl_2\,(g) + \cancel{2\,e^-}$

Cancel electrons: $ClO_4^-\,(aq) + 2\,H^+\,(aq) + 2\,Cl^-\,(aq) \rightarrow ClO_3^-\,(aq) + H_2O\,(l) + Cl_2\,(g)$

Check:

Reactants	Products
3 Cl atoms	3 Cl atoms
4 O atoms	4 O atoms
2 H atoms	2 H atoms
–1 charge	–1 charge

c) Separate: $NO_3^-\,(aq) \rightarrow NO\,(g)$ and $Sn^{2+}\,(aq) \rightarrow Sn^{4+}\,(aq)$

Balance non H & O elements: $NO_3^-\,(aq) \rightarrow NO\,(g)$ and $Sn^{2+}\,(aq) \rightarrow Sn^{4+}\,(aq)$

Balance O with H_2O: $NO_3^-\,(aq) \rightarrow NO\,(g) + 2\,H_2O\,(l)$ and $Sn^{2+}\,(aq) \rightarrow Sn^{4+}\,(aq)$

Balance H with H^+: $NO_3^-\,(aq) + 4\,H^+\,(aq) \rightarrow NO\,(g) + 2\,H_2O\,(l)$ and $Sn^{2+}\,(aq) \rightarrow Sn^{4+}\,(aq)$

Add electrons: $NO_3^-\,(aq) + 4\,H^+\,(aq) + 3\,e^- \rightarrow NO\,(g) + 2\,H_2O\,(l)$ and $Sn^{2+}\,(aq) \rightarrow Sn^{4+}\,(aq) + 2\,e^-$

Equalize electrons:

$$2\,NO_3^-\,(aq) + 8\,H^+\,(aq) + 6\,e^- \rightarrow 2\,NO\,(g) + 4\,H_2O\,(l) \text{ and } 3\,Sn^{2+}\,(aq) \rightarrow 3\,Sn^{4+}\,(aq) + 6\,e^-$$

Add half-reactions:

$$2\,NO_3^-\,(aq) + 8\,H^+\,(aq) + \cancel{6\,e^-} + 3\,Sn^{2+}\,(aq) \rightarrow 2\,NO\,(g) + 4\,H_2O\,(l) + 3\,Sn^{4+}\,(aq) + \cancel{6\,e^-}$$

Cancel electrons: $2\,NO_3^-\,(aq) + 8\,H^+\,(aq) + 3\,Sn^{2+}\,(aq) \rightarrow 2\,NO\,(g) + 4\,H_2O\,(l) + 3\,Sn^{4+}\,(aq)$

Check:

Reactants	Products
2 N atoms	2 N atoms
6 O atoms	6 O atoms
8 H atoms	8 H atoms
3 Sn atoms	3 Sn atoms
+12 charge	+12 charge

5. **Conceptual Plan: Separate the overall reaction into two half-reactions: one for oxidation and one for reduction. → Balance each half-reaction with respect to mass in the following order: 1) balance all elements other than H and O; 2) balance O by adding H_2O; 3) balance H by adding H^+; and 4) Neutralize H^+ by adding enough OH^- to neutralize each H^+. Add the same number of OH^- ions to each side of the equation. → Balance each half-reaction with respect to charge by adding electrons. (The sum of the charges on both sides of the equation should be made equal by adding electrons as necessary.) → Make the number of electrons in both half-reactions equal by multiplying one or both half-reactions by a small whole number. → Add the two half-reactions together, canceling electrons and other species as necessary. → Verify that the reaction is balanced both with respect to mass and with respect to charge.**

Solution:

a) Separate: $ClO_2\,(aq) \rightarrow ClO_2^-\,(aq)$ and $H_2O_2\,(aq) \rightarrow O_2\,(g)$

Balance non H & O elements: $ClO_2\,(aq) \rightarrow ClO_2^-\,(aq)$ and $H_2O_2\,(aq) \rightarrow O_2\,(g)$

Balance O with H_2O: $ClO_2\,(aq) \rightarrow ClO_2^-\,(aq)$ and $H_2O_2\,(aq) \rightarrow O_2\,(g)$

Balance H with H^+: $ClO_2\,(aq) \rightarrow ClO_2^-\,(aq)$ and $H_2O_2\,(aq) \rightarrow O_2\,(g) + 2\,H^+\,(aq)$

Neutralize H^+ with OH^-:

$ClO_2\,(aq) \rightarrow ClO_2^-\,(aq)$ and $H_2O_2\,(aq) + 2\,OH^-\,(aq) \rightarrow O_2\,(g) + \underbrace{2\,H^+\,(aq) + 2\,OH^-\,(aq)}_{2\,H_2O\,(l)}$

Add electrons: $ClO_2\,(aq) + e^- \rightarrow ClO_2^-\,(aq)$ and $H_2O_2\,(aq) + 2\,OH^-\,(aq) \rightarrow O_2\,(g) + 2\,H_2O\,(l) + 2\,e^-$

Equalize electrons:

$2\,ClO_2\,(aq) + 2\,e^- \rightarrow 2\,ClO_2^-\,(aq)$ and $H_2O_2\,(aq) + 2\,OH^-\,(aq) \rightarrow O_2\,(g) + 2\,H_2O\,(l) + 2\,e^-$

Add half-reactions:

$2\,ClO_2\,(aq) + \cancel{2\,e^-} + H_2O_2\,(aq) + 2\,OH^-\,(aq) \rightarrow 2\,ClO_2^-\,(aq) + O_2\,(g) + 2\,H_2O\,(l) + \cancel{2\,e^-}$

Cancel electrons: $2\,ClO_2\,(aq) + H_2O_2\,(aq) + 2\,OH^-\,(aq) \rightarrow 2\,ClO_2^-\,(aq) + O_2\,(g) + 2\,H_2O\,(l)$

Check:

Reactants	Products
2 Cl atoms	2 Cl atoms
8 O atoms	8 O atoms
4 H atoms	4 H atoms
–2 charge	–2 charge

b) Separate: $MnO_4^-\,(aq) \rightarrow MnO_2\,(s)$ and $Al\,(s) \rightarrow Al(OH)_4^-\,(aq)$

Balance non H & O elements: $MnO_4^-\,(aq) \rightarrow MnO_2\,(s)$ and $Al\,(s) \rightarrow Al(OH)_4^-\,(aq)$

Balance O with H_2O: $MnO_4^-\,(aq) \rightarrow MnO_2\,(s) + 2\,H_2O\,(l)$ and $Al\,(s) + 4\,H_2O\,(l) \rightarrow Al(OH)_4^-\,(aq)$

Balance H with H^+:

$MnO_4^-\,(aq) + 4\,H^+\,(aq) \rightarrow MnO_2\,(s) + 2\,H_2O\,(l)$ and $Al\,(s) + 4\,H_2O\,(l) \rightarrow Al(OH)_4^-\,(aq) + 4\,H^+\,(aq)$

Neutralize H^+ with OH^-: $MnO_4^-\,(aq) + \underbrace{4\,H^+\,(aq) + 4\,OH^-\,(aq)}_{2\cancel{4}\,H_2O\,(l)} \rightarrow MnO_2\,(s) + \cancel{2\,H_2O\,(l)} + 4\,OH^-\,(aq)$

and $Al\,(s) + \cancel{4\,H_2O\,(l)} + 4\,OH^-\,(aq) \rightarrow Al(OH)_4^-\,(aq) + \underbrace{4\,H^+\,(aq) + 4\,OH^-\,(aq)}_{\cancel{4\,H_2O\,(l)}}$

Add electrons: $MnO_4^-\,(aq) + 2\,H_2O\,(l) + 3\,e^- \rightarrow MnO_2\,(s) + 4\,OH^-\,(aq)$ and

$Al\,(s) + 4\,OH^-\,(aq) \rightarrow Al(OH)_4^-\,(aq) + 3\,e^-$

Equalize electrons: $MnO_4^-\,(aq) + 2\,H_2O\,(l) + 3\,e^- \rightarrow MnO_2\,(s) + 4\,OH^-\,(aq)$ and

$Al\,(s) + 4\,OH^-\,(aq) \rightarrow Al(OH)_4^-\,(aq) + 3\,e^-$

Add half-reactions:

$MnO_4^-\,(aq) + 2\,H_2O\,(l) + \cancel{3\,e^-} + Al\,(s) + \cancel{4\,OH^-\,(aq)} \rightarrow MnO_2\,(s) + \cancel{4\,OH^-\,(aq)} + Al(OH)_4^-\,(aq) + \cancel{3\,e^-}$

Cancel electrons: $MnO_4^-\,(aq) + 2\,H_2O\,(l) + Al\,(s) \rightarrow MnO_2\,(s) + Al(OH)_4^-\,(aq)$

Check:

Reactants	Products
1 Mn atom	1 Mn atom
6 O atoms	6 O atoms
4 H atoms	4 H atoms
1 Al atom	1 Al atom
–1 charge	–1 charge

c) Separate: $Cl_2\,(g) \rightarrow Cl^-\,(aq)$ and $Cl_2\,(g) \rightarrow ClO^-\,(aq)$

Balance non H & O elements: $Cl_2\,(g) \rightarrow 2\,Cl^-\,(aq)$ and $Cl_2\,(g) \rightarrow 2\,ClO^-\,(aq)$

Balance O with H_2O: $Cl_2\,(g) \rightarrow 2\,Cl^-\,(aq)$ and $Cl_2\,(g) + 2\,H_2O\,(l) \rightarrow 2\,ClO^-\,(aq)$

Balance H with H^+: $Cl_2\,(g) \rightarrow 2\,Cl^-\,(aq)$ and $Cl_2\,(g) + 2\,H_2O\,(l) \rightarrow 2\,ClO^-\,(aq) + 4\,H^+\,(aq)$

Neutralize H^+ with OH^-:

$Cl_2\,(g) \rightarrow 2\,Cl^-\,(aq)$ and $Cl_2\,(g) + \cancel{2\,H_2O\,(l)} + 4\,OH^-\,(aq) \rightarrow 2\,ClO^-\,(aq) + \underbrace{4\,H^+\,(aq) + 4\,OH^-\,(aq)}_{2\cancel{4}\,H_2O\,(l)}$

Add electrons: $Cl_2\,(g) + 2\,e^- \rightarrow 2\,Cl^-\,(aq)$ and $Cl_2\,(g) + 4\,OH^-\,(aq) \rightarrow 2\,ClO^-\,(aq) + 2\,H_2O\,(l) + 2\,e^-$

Equalize electrons: $Cl_2\,(g) + 2\,e^- \rightarrow 2\,Cl^-\,(aq)$ and $Cl_2\,(g) + 4\,OH^-\,(aq) \rightarrow 2\,ClO^-\,(aq) + 2\,H_2O\,(l) + 2\,e^-$

Add half-reactions: $Cl_2\,(g) + \cancel{2\,e^-} + Cl_2\,(g) + 4\,OH^-\,(aq) \rightarrow 2\,Cl^-\,(aq) + 2\,ClO^-\,(aq) + 2\,H_2O\,(l) + \cancel{2\,e^-}$

Cancel electrons: $2\,Cl_2\,(g) + 4\,OH^-\,(aq) \rightarrow 2\,Cl^-\,(aq) + 2\,ClO^-\,(aq) + 2\,H_2O\,(l)$

Simplify: $Cl_2\,(g) + 2\,OH^-\,(aq) \rightarrow Cl^-\,(aq) + ClO^-\,(aq) + H_2O\,(l)$

Check:

Reactants	Products
2 Cl atoms	2 Cl atoms
2 O atoms	2 O atoms
2 H atoms	2 H atoms
–2 charge	–2 charge

6. **Conceptual Plan: Separate the overall reaction into two half-reactions: one for oxidation and one for reduction. → Balance each half-reaction with respect to mass in the following order: 1) balance all elements other than H and O; 2) balance O by adding H_2O; 3) balance H by adding H^+; and 4) Neutralize H^+ by adding enough OH^- to neutralize each H^+. Add the same number of OH^- ions to each side of the equation. → Balance each half-reaction with respect to charge by adding electrons. (The sum of the charges on both sides of the equation should be made equal by adding electrons as necessary.) → Make the number of electrons in both half-reactions equal by multiplying one or both half-reactions by a small whole number. → Add the two half-reactions together, canceling electrons and other species as necessary. → Verify that the reaction is balanced both with respect to mass and with respect to charge.**
Solution:

a) Separate: $MnO_4^-\,(aq) \rightarrow MnO_2\,(s)$ and $Br^-\,(aq) \rightarrow BrO_3^-\,(aq)$
Balance non H & O elements: $MnO_4^-\,(aq) \rightarrow MnO_2\,(s)$ and $Br^-\,(aq) \rightarrow BrO_3^-\,(aq)$
Balance O with H_2O: $MnO_4^-\,(aq) \rightarrow MnO_2\,(s) + 2\,H_2O\,(l)$ and $Br^-\,(aq) + 3\,H_2O\,(l) \rightarrow BrO_3^-\,(aq)$
Balance H with H^+:
$MnO_4^-\,(aq) + 4\,H^+\,(aq) \rightarrow MnO_2\,(s) + 2\,H_2O\,(l)$ and $Br^-\,(aq) + 3\,H_2O\,(l) \rightarrow BrO_3^-\,(aq) + 6\,H^+\,(aq)$
Neutralize H^+ with OH^-: $MnO_4^-\,(aq) + \underbrace{4\,H^+\,(aq) + 4\,OH^-\,(aq)}_{2\,\not{4}\,H_2O\,(l)} \rightarrow MnO_2\,(s) + \not{2\,H_2O\,(l)} + 4\,OH^-\,(aq)$

and $Br^-\,(aq) + \not{3\,H_2O\,(l)} + 6\,OH^-\,(aq) \rightarrow BrO_3^-\,(aq) + \underbrace{6\,H^+\,(aq) + 6\,OH^-\,(aq)}_{3\,\not{6}\,H_2O\,(l)}$

Add electrons: $MnO_4^-\,(aq) + 2\,H_2O\,(l) + 3\,e^- \rightarrow MnO_2\,(s) + 4\,OH^-\,(aq)$ and
$Br^-\,(aq) + 6\,OH^-\,(aq) \rightarrow BrO_3^-\,(aq) + 3\,H_2O\,(l) + 6\,e^-$
Equalize electrons: $2\,MnO_4^-\,(aq) + 4\,H_2O\,(l) + 6\,e^- \rightarrow 2\,MnO_2\,(s) + 8\,OH^-\,(aq)$ and
$Br^-\,(aq) + 6\,OH^-\,(aq) \rightarrow BrO_3^-\,(aq) + 3\,H_2O\,(l) + 6\,e^-$
Add half-reactions: $2\,MnO_4^-\,(aq) + 1\,\not{4}\,H_2O\,(l) + \not{6e^-} + Br^-\,(aq) + \not{6\,OH^-\,(aq)} \rightarrow$
$2\,MnO_2\,(s) + 2\,\not{8}\,OH^-\,(aq) + BrO_3^-\,(aq) + \not{3\,H_2O\,(l)} + \not{6e^-}$
Cancel electrons & others: $2\,MnO_4^-\,(aq) + H_2O\,(l) + Br^-\,(aq) \rightarrow 2\,MnO_2\,(s) + 2\,OH^-\,(aq) + BrO_3^-\,(aq)$

Check:

Reactants	Products
2 Mn atoms	2 Mn atoms
9 O atoms	9 O atoms
2 H atoms	2 H atoms
1 Br atom	1 Br atom
–3 charge	–3 charge

b) Separate: $Ag\,(s) + CN^-\,(aq) \rightarrow Ag(CN)_2^-\,(aq)$ and $O_2\,(g) \rightarrow$
Balance non H & O elements: $Ag\,(s) + 2\,CN^-\,(aq) \rightarrow Ag(CN)_2^-\,(aq)$ and $O_2\,(g) \rightarrow$
Balance O with H_2O: $Ag\,(s) + 2\,CN^-\,(aq) \rightarrow Ag(CN)_2^-\,(aq)$ and $O_2\,(g) \rightarrow 2\,H_2O\,(l)$
Balance H with H^+: $Ag\,(s) + 2\,CN^-\,(aq) \rightarrow Ag(CN)_2^-\,(aq)$ and $O_2\,(g) + 4\,H^+\,(aq) \rightarrow 2\,H_2O\,(l)$
Neutralize H^+ with OH^-:
$Ag\,(s) + 2\,CN^-\,(aq \rightarrow Ag(CN)_2^-\,(aq)$ and $O_2\,(g) + \underbrace{4\,H^+\,(aq) + 4\,OH^-\,(aq)}_{2\,\not{4}\,H_2O\,(l)} \rightarrow \not{2\,H_2O\,(l)} + 4\,OH^-\,(aq)$
Add electrons: $Ag\,(s) + 2\,CN^-\,(aq) \rightarrow Ag(CN)_2^-\,(aq) + e^-$ and $O_2\,(g) + 2\,H_2O\,(l)\,) + 4\,e^- \rightarrow 4\,OH^-\,(aq)$
Equalize electrons:
$4\,Ag\,(s) + 8\,CN^-\,(aq) \rightarrow 4\,Ag(CN)_2^-\,(aq) + 4\,e^-$ and $O_2\,(g) + 2\,H_2O\,(l) + 4\,e^- \rightarrow 4\,OH^-\,(aq)$
Add half-reactions:
$4\,Ag\,(s) + 8\,CN^-\,(aq) + O_2\,(g) + 2\,H_2O\,(l) + \not{4e^-} \rightarrow 4\,Ag(CN)_2^-\,(aq) + \not{4e^-} + 4\,OH^-\,(aq)$
Cancel electrons: $4\,Ag\,(s) + 8\,CN^-\,(aq) + O_2\,(g) + 2\,H_2O\,(l) \rightarrow 4\,Ag(CN)_2^-\,(aq) + 4\,OH^-\,(aq)$

Check:

Reactants	Products
4 Ag atoms	4 Ag atoms
8 C atoms	8 C atoms
8 N atoms	8 N atoms
4 O atoms	4 O atoms
4 H atoms	4 H atoms
–8 charge	–8 charge

c) Separate: $NO_2^-\,(aq) \rightarrow NH_3\,(g)$ and $Al\,(s) \rightarrow AlO_2^-\,(aq)$
Balance non H & O elements: $NO_2^-\,(aq) \rightarrow NH_3\,(g)$ and $Al\,(s) \rightarrow AlO_2^-\,(aq)$

Balance O with H_2O: NO_2^- (*aq*) → NH_3 (*g*) + 2 H_2O (*l*) and Al (*s*) + 2 H_2O (*l*) → AlO_2^- (*aq*)
Balance H with H^+: NO_2^- (*aq*) + 7 H^+ (*aq*) → NH_3 (*g*) + 2 H_2O (*l*) and
Al (*s*) + 2 H_2O (*l*) → AlO_2^- (*aq*) + 4 H^+ (*aq*)
Neutralize H^+ with OH^-: NO_2^- (*aq*) + $\underbrace{7\,H^+ (aq) + 7\,OH^- (aq)}_{5\ \not 7\,H_2O(l)}$ → NH_3 (*g*) + ~~2 H_2O (*l*)~~ + 7 OH^- (*aq*) and

Al (*s*) + ~~2 H_2O (*l*)~~ + 4 OH^- (*aq*) → AlO_2^- (*aq*) + $\underbrace{4\,H^+ (aq) + 4\,OH^- (aq)}_{2\ \not 4\,H_2O(l)}$

Add electrons: NO_2^- (*aq*) + 5 H_2O (*l*) + 6 e^- → NH_3 (*g*) + 7 OH^- (*aq*) and
Al (*s*) + 4 OH^- (*aq*) → AlO_2^- (*aq*) + 2 H_2O (*l*) + 3 e^-
Equalize electrons: NO_2^- (*aq*) + 5 H_2O (*l*) + 6 e^- → NH_3 (*g*) + 7 OH^- (*aq*) and
2 Al (*s*) + 8 OH^- (*aq*) → 2 AlO_2^- (*aq*) + 4 H_2O (*l*) + 6 e^-
Add half-reactions: NO_2^- (*aq*) + 1 ~~5~~ H_2O (*l*) + ~~6 e^-~~ + 2 Al (*s*) + 1 ~~8~~ OH^-(*aq*) →
NH_3 (*g*) + ~~7 OH^- (*aq*)~~ + 2 AlO_2^- (*aq*) + ~~4 H_2O (*l*)~~ + ~~6 e^-~~
Cancel electrons & others: NO_2^- (*aq*) + H_2O (*l*) + 2 Al (*s*) + OH^- (*aq*) → NH_3 (*g*) + 2 AlO_2^- (*aq*)

Check:

Reactants	Products
1 N atom	1 N atom
4 O atoms	4 O atoms
3 H atoms	3 H atoms
2 Al atoms	2 Al atoms
–2 charge	–2 charge

7. **Given:** voltaic cell overall redox reaction
Find: Sketch voltaic cell, labeling anode, cathode, all species, and direction of electron flow
Conceptual Plan: Separate overall reaction into 2 half-cell reactions and add electrons as needed to balance reactions. Put anode reaction on the left (oxidation = electrons as product) and cathode reaction on the right (reduction = electrons as reactant). Electrons flow from anode to cathode.
Solution:

a) 2 Ag^+ (*aq*)+ Pb (*s*) → 2 Ag (*s*) + Pb^{2+} (*aq*) separates to 2 Ag^+ (*aq*) → 2 Ag (*s*) and Pb (*s*) → Pb^{2+} (*aq*) then add electrons to balance to get the cathode reaction: 2 Ag^+ (*aq*) + 2 e^- → 2 Ag (*s*) and the anode reaction: Pb (*s*) → Pb^{2+} (*aq*) + 2 e^-.

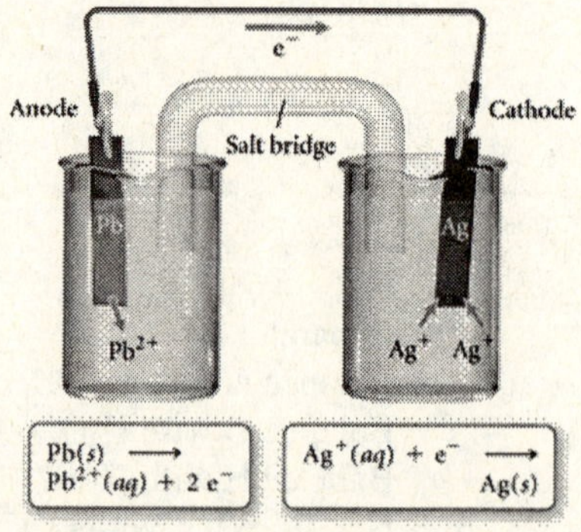

Since we have Pb (s) as the reactant for the oxidation, it will be our anode. Since we have Ag (s) as the product for the reduction, it will be our cathode. Simplify the cathode reaction, dividing all terms by 2.

b) 2 ClO_2 (*g*)+ 2 I^- (*aq*) → 2 ClO_2^- (*aq*) + I_2 (*s*) separates to 2 ClO_2 (*g*) → 2 ClO_2^- (*aq*) and 2 I^- (*aq*) → I_2 (*s*) then add electrons to balance to get the cathode reaction: 2 ClO_2 (*g*) + 2 e^- → 2 ClO_2^- (*aq*) and the anode reaction: 2 I^- (*aq*) → I_2 (*s*) + 2 e^-.

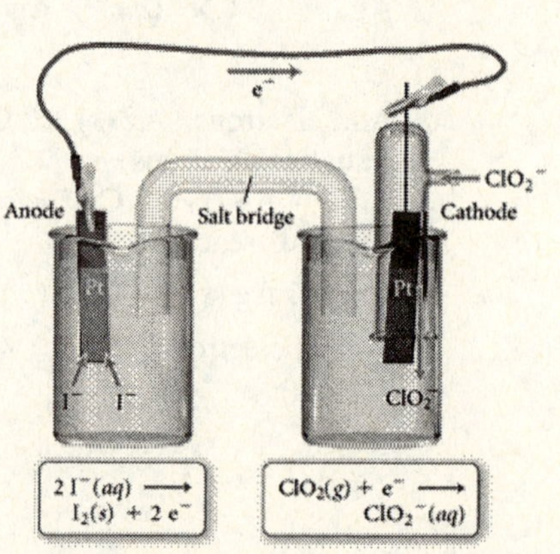

Since we have I^- (*aq*) as the reactant for the oxidation, we will need to use Pt as our anode. Since we have ClO_2^- (*aq*) as the product for the reduction, we will need to use Pt as our cathode. Since ClO_2 (*g*) is our reactant for the reduction, we need to use an electrode assembly like that is used for a SHE. Simplify the cathode reaction, dividing all terms by 2.

c) O_2 (*g*)+ 4 H^+ (*aq*) + 2 Zn (*s*) → 2 H_2O (*l*) + 2 Zn^{2+} (*aq*) separates to O_2 (*g*)+ 4 H^+ (*aq*) → 2 H_2O (*l*) and 2 Zn (*s*) → 2 Zn^{2+} (*aq*) then add electrons to balance to get the cathode reaction: O_2 (*g*)+ 4 H^+ (*aq*) + 4 e^- → 2 H_2O (*l*) and the anode reaction: 2 Zn (*s*) → 2 Zn^{2+} (*aq*) + 4 e^-.

Since we have Zn (s) as the reactant for the oxidation, it will be our anode. Since we have H_2O (l) as the product for the reduction, we will need to use Pt as our cathode. Since O_2 (g) is our reactant for the reduction, we need to use an electrode assembly like what is used for a SHE. Simplify the anode reaction, dividing all terms by 2.

Zn(s) $\longrightarrow$
Zn^{2+}(aq) + 2 e⁻

O_2(g) + 4 H⁺(aq) + 4 e⁻
$\longrightarrow$ 2 H_2O(l)

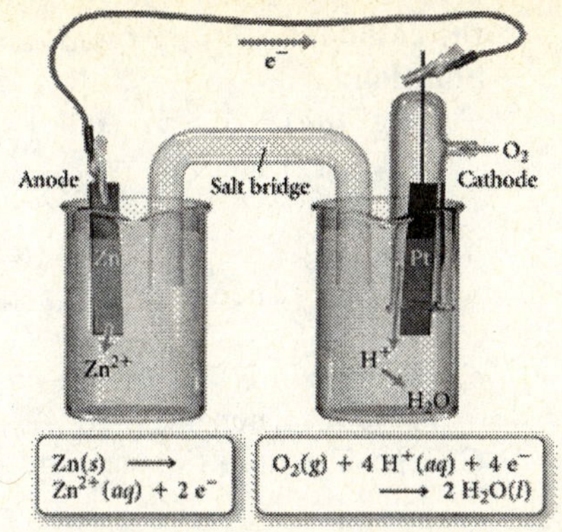

8.　**Given:** voltaic cell overall redox reaction
　　Find: Sketch voltaic cell, labeling anode, cathode, all species, and direction of electron flow
　　Conceptual Plan: Separate overall reaction into 2 half-cell reactions and add electrons as needed to balance reactions. Put anode reaction on the left (oxidation = electrons as product) and cathode reaction on the right (reduction = electrons as reactant). Electrons flow from anode to cathode.
　　Solution:

a) Ni^{2+} (aq)+ Mg (s) $\rightarrow$ Ni (s) + Mg^{2+} (aq) separates to Ni^{2+} (aq) $\rightarrow$ Ni (s) and Mg (s) $\rightarrow$ Mg^{2+} (aq) then add electrons to balance to get the cathode reaction: Ni^{2+} (aq) + 2 e⁻ $\rightarrow$ Ni (s) and the anode reaction: Mg (s) $\rightarrow$ Mg^{2+} (aq) + 2 e⁻.

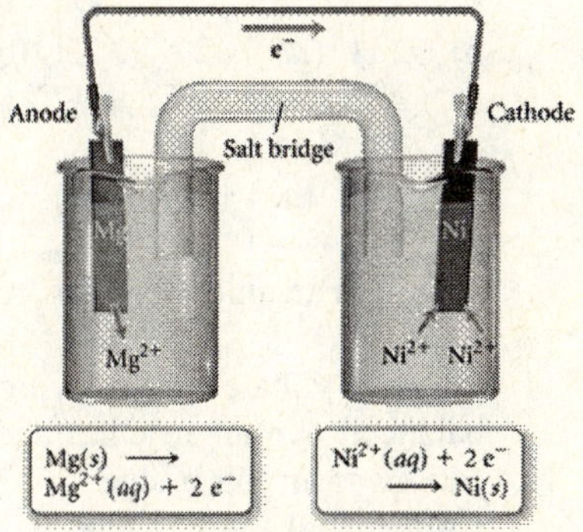

Since we have Mg (s) as the reactant for the oxidation, it will be our anode. Since we have Ni (s) as the product for the reduction, it will be our cathode.

Mg(s) $\longrightarrow$
Mg^{2+}(aq) + 2 e⁻

Ni^{2+}(aq) + 2 e⁻
$\longrightarrow$ Ni(s)

b) 2 H⁺ (aq) + Fe (s) $\rightarrow$ H_2 (g) + Fe^{2+} (aq) separates to 2 H⁺ (aq) $\rightarrow$ H_2 (g) and Fe (s) $\rightarrow$ Fe^{2+} (aq) then add electrons to balance to get the cathode reaction: 2 H⁺ (aq) + 2 e⁻ $\rightarrow$ H_2 (g) and the anode reaction: Fe (s) $\rightarrow$ Fe^{2+} (aq) + 2 e⁻.

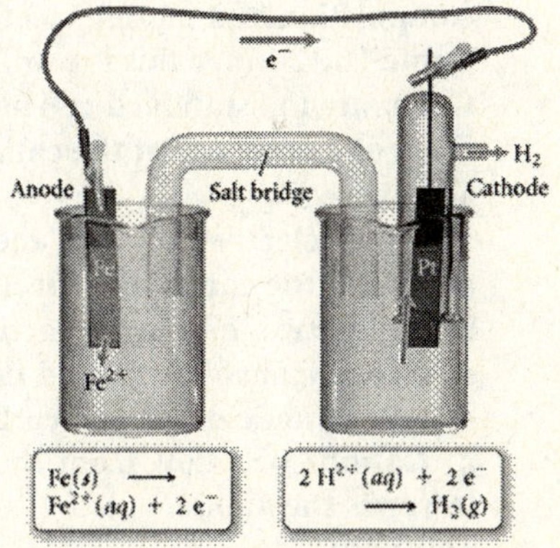

Since we have Fe (s) as the reactant for the oxidation, it will be our anode. Since we have H_2 (g) as the product for the reduction, we will need to use Pt as our cathode and the product can leave using an electrode assembly like what is used for a SHE.

Fe(s) $\longrightarrow$
Fe^{2+}(aq) + 2 e⁻

2 H^{2+}(aq) + 2 e⁻
$\longrightarrow$ H_2(g)

c) 2 NO_3^- (aq) + 8 H⁺ (aq) + 3 Cu (s) $\rightarrow$ 2 NO (g) + 4 H_2O (l) + 3 Cu^{2+} (aq) separates to 2 NO_3^- (aq) + 8 H⁺ (aq) $\rightarrow$ 2 NO (g) + 4 H_2O (l) and 3 Cu (s) $\rightarrow$ 3 Cu^{2+} (aq) then add electrons to balance to get the cathode reaction: 2 NO_3^- (aq) + 8 H⁺ (aq) + 6 e⁻ $\rightarrow$ 2 NO (g) + 4 H_2O (l) and the anode reaction: 3 Cu (s) $\rightarrow$ 3 Cu^{2+} (aq) + 6 e⁻.

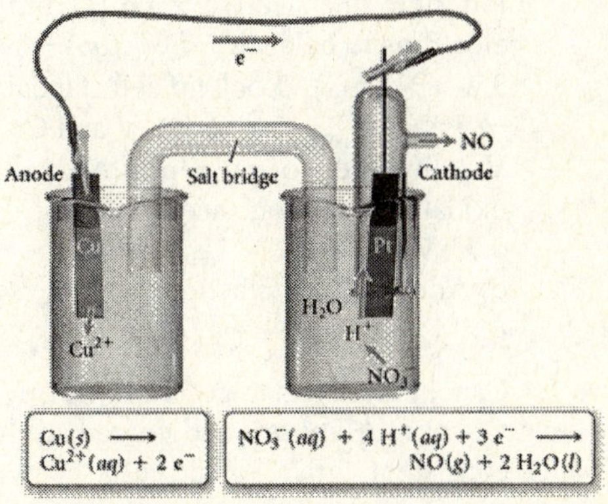

Since we have Cu (s) as the reactant for the oxidation, it will be our anode. Since we have H_2O (l) and NO (g) as the products for the reduction, we will need to use Pt as our cathode and the gaseous product can leave using an electrode assembly like what is used for a SHE.

Cu(s) $\longrightarrow$
Cu^{2+}(aq) + 2 e⁻

NO_3^-(aq) + 4 H⁺(aq) + 3 e⁻ $\longrightarrow$
NO(g) + 2 H_2O(l)

9.　**Given:** overall reactions from #7　　　**Find:** $E°_{cell}$
　　Conceptual Plan: Look up half-reactions from solution of problem #7 in Table 18.1. Calculate the standard cell potential by subtracting the electrode potential of the anode from the electrode potential of

the cathode: $E^{o}_{cell} = E^{o}_{cathode} - E^{o}_{anode}$.

Solution:

a) $Ag^{+}(aq) + e^{-} \rightarrow Ag(s)$ $E^{o}_{red} = 0.80$ V $= E^{o}_{cathode}$ and $Pb(s) \rightarrow Pb^{2+}(aq) + 2e^{-}$ $E^{o}_{red} = -0.13$ V $= E^{o}_{anode}$. Then $E^{o}_{cell} = E^{o}_{cathode} - E^{o}_{anode} = 0.80$ V $- (-0.13$ V$) = 0.93$ V.

b) $ClO_2(g) + e^{-} \rightarrow ClO_2^{-}(aq)$ $E^{o}_{red} = 0.95$ V $= E^{o}_{cathode}$ and $2I^{-}(aq) \rightarrow I_2(s) + 2e^{-}$ $E^{o}_{red} = -0.54$ V $= E^{o}_{anode}$. Then $E^{o}_{cell} = E^{o}_{cathode} - E^{o}_{anode} = 0.95$ V $- 0.54$ V $= 0.41$ V.

c) $O_2(g) + 4H^{+}(aq) + 4e^{-} \rightarrow 2H_2O(l)$ $E^{o}_{red} = 1.23$ V and $Zn(s) \rightarrow Zn^{2+}(aq) + 2e^{-}$ $E^{o}_{red} = -0.76$ V $= E^{o}_{anode}$. Then $E^{o}_{cell} = E^{o}_{cathode} - E^{o}_{anode} = 1.23$ V $- (-0.76$ V$) = 1.99$ V.

Check: The units (V) are correct. All of the voltages are positive, which is consistent with a voltaic cell.

10. **Given:** overall reactions from #8 **Find:** E^{o}_{cell}

Conceptual Plan: Look up half-reactions from solution of problem #8 in Table 18.1. Calculate the standard cell potential by subtracting the electrode potential of the anode from the electrode potential of the cathode: $E^{o}_{cell} = E^{o}_{cathode} - E^{o}_{anode}$.

Solution:

a) $Ni^{2+}(aq) + 2e^{-} \rightarrow Ni(s)$ $E^{o}_{red} = -0.23$ V $= E^{o}_{cathode}$ and $Mg(s) \rightarrow Mg^{2+}(aq) + 2e^{-}$ $E^{o}_{red} = -2.37$ V $= E^{o}_{anode}$. Then $E^{o}_{cell} = E^{o}_{cathode} - E^{o}_{anode} = -0.23$ V $- (-2.37$ V$) = 2.14$ V.

b) $2H^{+}(aq) + 2e^{-} \rightarrow H_2(g)$ $E^{o}_{red} = 0.00$ V $= E^{o}_{cathode}$ and $Fe(s) \rightarrow Fe^{2+}(aq) + 2e^{-}$ $E^{o}_{red} = -0.45$ V $= E^{o}_{anode}$. Then $E^{o}_{cell} = E^{o}_{cathode} - E^{o}_{anode} = 0.00$ V $- (-0.45$ V$) = 0.45$ V.

c) $NO_3^{-}(aq) + 4H^{+}(aq) + 3e^{-} \rightarrow NO(g) + 2H_2O(l)$ $E^{o}_{red} = 0.96$ V $= E^{o}_{cathode}$ and $Cu(s) \rightarrow Cu^{2+}(aq) + 2e^{-}$ $E^{o}_{red} = -0.34$ V $= E^{o}_{anode}$. Then $E^{o}_{cell} = E^{o}_{cathode} - E^{o}_{anode} = 0.96$ V $- 0.34$ V $= 0.62$ V.

Check: The units (V) are correct. All of the voltages are positive, which is consistent with a voltaic cell.

11. **Given:** voltaic cell drawing **Find:** a) determine electron flow direction, anode, and cathode; b) write balanced overall reaction and calculate E^{o}_{cell}; c) label electrodes as + and –; and d) directions of anions and cations from salt bridge

Conceptual Plan: Look at each half-cell and write a reduction reaction, by using electrode and solution composition and adding electrons to balance. Look up half-reactions standard reduction potentials in Table 18.1. Since this is a voltaic cell, the cell potentials must be assigned to give a positive E^{o}_{cell}. Calculate the standard cell potential by subtracting the electrode potential of the anode from the electrode potential of the cathode: $E^{o}_{cell} = E^{o}_{cathode} - E^{o}_{anode}$, choosing the electrode assignments to give a positive E^{o}_{cell}.

a) Label electrode where the oxidation occurs as the anode. Label the electrode where the reduction occurs as the cathode. Electrons flow from anode to cathode.

b) Take two half-cell reactions and multiply the reactions as necessary to equalize the number of electrons transferred. Add the two half-cell reactions and cancel electrons and any other species.

c) Label anode as (–) and cathode as (+).

d) Cations will flow from the salt bridge towards the cathode and the anions will flow from salt bridge towards the anode.

Solution:

left side: $Fe^{3+}(aq) \rightarrow Fe(s)$ and right side: $Cr^{3+}(aq) \rightarrow Cr(s)$ add electrons to balance $Fe^{3+}(aq) + 3e^{-} \rightarrow Fe(s)$ and right side: $Cr^{3+}(aq) + 3e^{-} \rightarrow Cr(s)$. Look up cell standard reduction potentials: $Fe^{3+}(aq) + 3e^{-} \rightarrow Fe(s)$ $E^{o}_{red} = -0.036$ V and $Cr^{3+}(aq) + 3e^{-} \rightarrow Cr(s)$ $E^{o}_{red} = -0.73$ V. In order to get a positive cell potential, the second reaction is the oxidation reaction (anode). $E^{o}_{cell} = E^{o}_{cathode} - E^{o}_{anode} = -0.036$ V $- (-0.73$ V$) = +0.69$ V. (a, c, and d)

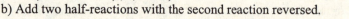

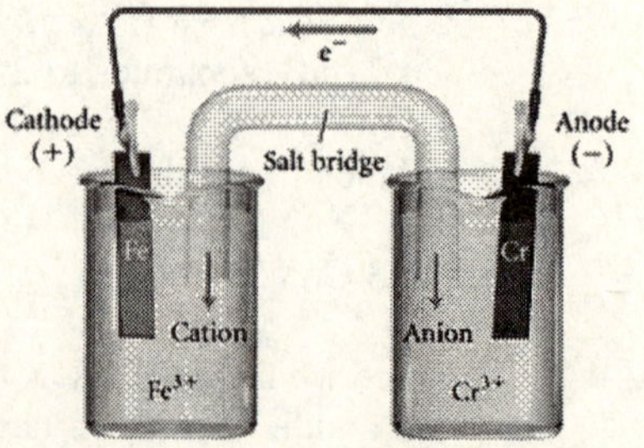

b) Add two half-reactions with the second reaction reversed.

$Fe^{3+}(aq) + \cancel{3e^{-}} + Cr(s) \rightarrow Fe(s) + Cr^{3+}(aq) + \cancel{3e^{-}}$.

Cancel electrons to get: $Fe^{3+}(aq) + Cr(s) \rightarrow Fe(s) + Cr^{3+}(aq)$.

Check: All atoms and charge are balanced. The units (V) are correct. The cell potential is positive which is consistent with a voltaic cell.

12. **Given:** voltaic cell drawing **Find:** a) determine electron flow direction, anode, and cathode; b) write balanced overall reaction and calculate E^{o}_{cell}; c) label electrodes as + and –; and d) directions of anions and cations from salt bridge

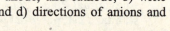

Conceptual Plan: Look at each half-cell and write a reduction reaction, by using electrode and solution composition and adding electrons to balance. Look up half-reactions standard reduction potentials in Table 18.1. Since this is a voltaic cell, the cell potentials must be assigned to give a positive E°_{cell}. Calculate the standard cell potential by subtracting the electrode potential of the anode from the electrode potential of the cathode: $E^\circ_{cell} = E^\circ_{cathode} - E^\circ_{anode}$, choosing the electrode assignments to give a positive E°_{cell}.
a) Label electrode where the oxidation occurs as the anode. Label the electrode where the reduction occurs as the cathode. Electrons flow from anode to cathode.
b) Take two half-cell reactions and multiply the reactions as necessary to equalize the number of electrons transferred. Add the two half-cell reactions and cancel electrons and any other species.
c) Label anode as (–) and cathode as (+).
d) Cations will flow from the salt bridge towards the cathode and the anions will flow from salt bridge towards the anode.
Solution: left side: Pb^{2+} (*aq*) → Pb (*s*) and right side: Cl_2 (*g*) → 2 Cl^- (*aq*) add electrons to balance Pb^{2+} (*aq*) + 2 e^- → Pb (*s*) and right side: Cl_2 (*g*) + 2 e^- → 2 Cl^- (*aq*). Look up cell standard reduction potentials: Pb^{2+} (*aq*) + 2 e^- → Pb (*s*) E°_{red} = – 0.13 V and Cl_2 (*g*) + 2 e^- → 2 Cl^- (*aq*) E°_{red} = 1.36 V. In order to get a positive cell potential, the first reaction is the oxidation reaction (anode). $E^\circ_{cell} = E^\circ_{cathode} - E^\circ_{anode}$ = 1.36 V– (– 0.13 V) = + 1.49 V. (a, c, and d)

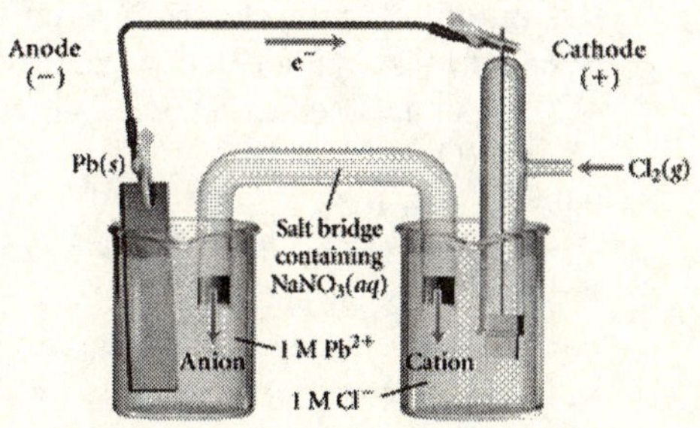

b) Add two half-reactions with the first reaction reversed.
Pb (*s*) + Cl_2 (*g*) + $\cancel{2 e^-}$ → Pb^{2+} (*aq*) + $\cancel{2 e^-}$ + 2 Cl^- (*aq*).
Cancel electrons to get: Pb (*s*) + Cl_2 (*g*) → Pb^{2+} (*aq*) + 2 Cl^- (*aq*).
Check: All atoms and charge are balanced. The units (V) are correct. The cell potential is positive which is consistent with a voltaic cell.

13. **Given:** overall reactions from #7 **Find:** line notation
Conceptual Plan: Use solution from problem #7. Write the oxidation half-reaction components on the left and the reduction on the right. A double vertical line (||), indicating the salt bridge, separates the two half-reactions. Substances in different phases are separated by a single vertical line (|), which represents the boundary between the phases. For some redox reactions, the reactants and products of one or both of the half-reactions may be in the same phase. In these cases, the reactants and products are simply separated from each other with a comma in the line diagram. Such cells use an inert electrode, such as platinum (Pt) or graphite, as the anode or cathode (or both).
Solution:
a) Reduction reaction: Ag^+ (*aq*) + e^- → Ag (*s*) and the oxidation reaction: Pb (*s*) → Pb^{2+} (*aq*) + 2 e^- so Pb (*s*)|Pb^{2+} (*aq*)||Ag^+ (*aq*) |Ag (*s*)

b) Reduction reaction: ClO_2 (*g*) + e^- → ClO_2^- (*aq*) and the oxidation reaction: 2 I^- (*aq*) → I_2 (*s*) + 2 e^- so Pt (*s*)|I^- (*aq*)| I_2 (*s*)|| ClO_2 (*g*)|ClO_2^- (*aq*)|Pt (*s*)

c) Reduction reaction: O_2 (*g*) + 4 H^+ (*aq*) + 4 e^- → 2 H_2O (*l*) and the oxidation reaction: Zn (*s*) → Zn^{2+} (*aq*) + 2 e^- so Zn (*s*)|Zn^{2+} (*aq*) ||O_2 (*g*)|H^+ (*aq*), H_2O (*l*)|Pt (*s*)

14. **Given:** overall reactions from #8 **Find:** line notation
Conceptual Plan: Use solution from problem #8. Write the oxidation half-reaction components on the left and the reduction on the right. A double vertical line (||), indicating the salt bridge, separates the two half-reactions. Substances in different phases are separated by a single vertical line (|), which represents the boundary between the phases. For some redox reactions, the reactants and products of one or both of the half-reactions may be in the same phase. In these cases, the reactants and products are simply separated from each other with a comma in the line diagram. Such cells use an inert electrode, such as platinum (Pt) or graphite, as the anode or cathode (or both).
Solution:
a) Reduction reaction: Ni^{2+} (*aq*) + 2 e^- → Ni (*s*) and the oxidation reaction: Mg (*s*) → Mg^{2+} (*aq*) + 2 e^- so Mg (*s*)|Mg^{2+} (*aq*)||Ni^{2+} (*aq*) |Ni (*s*)

b) Reduction reaction: 2 H^+ (*aq*) + 2 e^- → H_2 (*g*) and the oxidation reaction: Fe (*s*) → Fe^{2+} (*aq*) + 2 e^- so Fe (*s*)|Fe^{2+} (*aq*)|| H^+ (*aq*)| H_2 (*g*)|Pt (*s*)

c) Reduction reaction: NO_3^- (*aq*) + 4 H^+ (*aq*) + 3 e^- → NO (*g*) + 2 H_2O (*l*) and the oxidation reaction: Cu (*s*) → Cu^{2+} (*aq*) + 2 e^- so Cu (*s*)|Cu^{2+} (*aq*) || NO_3^- (*aq*), H^+ (*aq*)), H_2O (*l*)|NO (*g*)|Pt (*s*)

15. **Given:** Sn (*s*)|Sn^{2+} (*aq*) || NO_3^- (*aq*), H^+ (*aq*)), H_2O (*l*)|NO (*g*)|Pt (*s*)
Find: Sketch voltaic cell, labeling anode, cathode, all species, direction of electron flow, and E°_{cell}
Conceptual Plan: Separate overall reaction into 2 half-cell reactions knowing that the oxidation half-reaction components are on the left and the reduction half-reaction components are on the right. Add electrons as needed to balance reactions. Multiply the half-reactions by the appropriate factors to have an equal number of electrons transferred. Add the half-cell reactions and cancel electrons. Put anode reaction on the left (oxidation = electrons as product) and cathode reaction on the right (reduction = electrons as reactant). Electrons flow from anode to cathode. Look up half-reactions from solution of problem #8 in Table 18.1. Calculate the standard cell potential by subtracting the electrode potential of the anode from the electrode potential of the cathode: $E^\circ_{cell} = E^\circ_{cathode} - E^\circ_{anode}$.
Solution: Oxidation reaction (anode): Sn (*s*) → Sn^{2+} (*aq*) + 2 e^- E°_{red} = – 0.14 V and Reduction reaction (cathode): NO_3^- (*aq*) + 4 H^+ (*aq*) + 3 e^- → NO (*g*) + 2 H_2O (*l*) E°_{red} = 0.96 V. $E^\circ_{cell} = E^\circ_{cathode} - E^\circ_{anode}$ = 0.96 V – (– 0.14 V) = 1.10 V. Multiply first reaction by 3 and the second reaction by 2 so that 6 electrons are transferred. 3 Sn (*s*) → 3 Sn^{2+} (*aq*) + 6 e^- and 2 NO_3^- (*aq*) + 8 H^+ (*aq*) + 6 e^- → 2 NO (*g*) + 4 H_2O (*l*). Add the two half-reactions and cancel electrons 3 Sn (*s*)+ 2 NO_3^- (*aq*) + 8 H^+ (*aq*) + ~~6 e⁻~~ →3 Sn^{2+} (*aq*) + ~~6 e⁻~~ + 2 NO (*g*) + 4 H_2O (*l*). So balanced reaction is: 3 Sn (*s*) + 2 NO_3^- (*aq*) + 8 H^+ (*aq*) → 3 Sn^{2+} (*aq*) + 2 NO (*g*) + 4 H_2O (*l*).

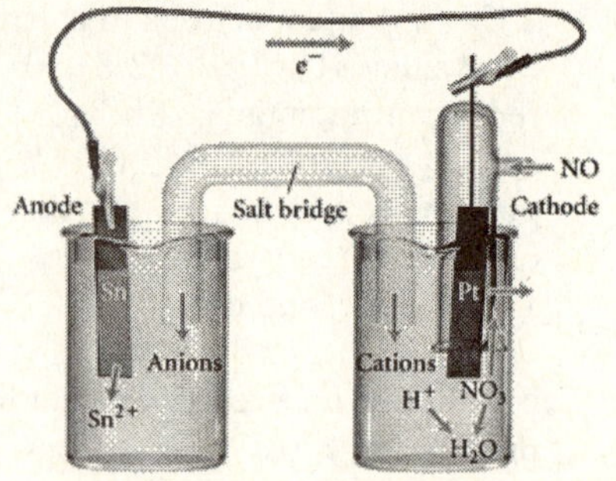

Check: All atoms and charge are balanced. The units (V) are correct. The cell potential is positive which is consistent with a voltaic cell.

16. **Given:** Mn (*s*)|Mn^{2+} (*aq*) || ClO_2 (*g*)|ClO_2^- (*aq*)|Pt (*s*)
Find: Sketch voltaic cell, labeling anode, cathode, all species, direction of electron flow, and E°_{cell}
Conceptual Plan: Separate overall reaction into 2 half-cell reactions knowing that the oxidation half-reaction components are on the left and the reduction half-reaction components are on the right. Add electrons as needed to balance reactions. Multiply the half-reactions by the appropriate factors to have an equal number of electrons transferred. Add the half-cell reactions and cancel electrons. Put anode reaction on the left (oxidation = electrons as product) and cathode reaction on the right (reduction = electrons as reactant). Electrons flow from anode to cathode. Look up half-reactions from solution of problem #8 in Table 18.1. Calculate the standard cell potential by subtracting the electrode potential of the anode from the electrode potential of the cathode: $E^\circ_{cell} = E^\circ_{cathode} - E^\circ_{anode}$.
Solution: Oxidation reaction (anode): Mn (*s*) → Mn^{2+} (*aq*) + 2 e^- E°_{red} = – 1.18 V and Reduction reaction (cathode): ClO_2 (*g*) + e^- → ClO_2^- (*aq*) E°_{red} = 0.95 V. $E^\circ_{cell} = E^\circ_{cathode} - E^\circ_{anode}$ = 0.95 V – (–1.18 V) = 2.13 V. Multiply the second reaction by 2 so that 2 electrons are transferred. Mn (*s*) → Mn^{2+} (*aq*) + 2 e^- and 2 ClO_2 (*g*) + 2 e^- → 2 ClO_2^- (*aq*). Add the two half-reactions and cancel electrons Mn (*s*) + 2 ClO_2 (*g*) + ~~2 e⁻~~ → Mn^{2+} (*aq*) + ~~2 e⁻~~ + 2 ClO_2^- (*aq*). So balanced reaction is: Mn (*s*) + 2 ClO_2 (*g*) → Mn^{2+} (*aq*) + 2 ClO_2^- (*aq*).

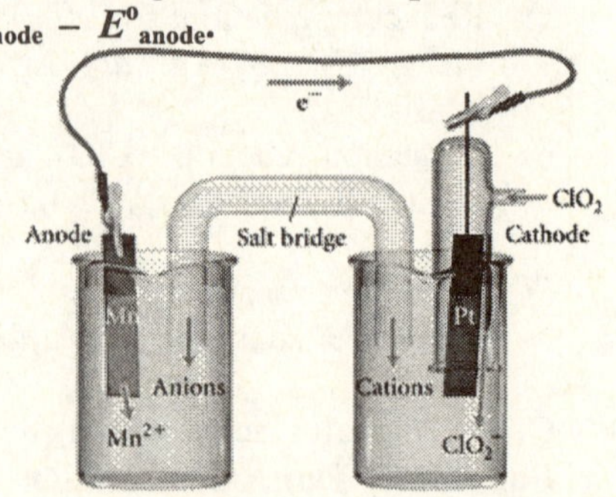

Check: All atoms and charge are balanced. The units (V) are correct. The cell potential is positive which is consistent with a voltaic cell.

17. **Given:** overall reactions **Find:** spontaneity in forward direction
Conceptual Plan: Separate overall reaction into 2 half-cell reactions and add electrons as needed to balance reactions. Look up half-reactions in Table 18.1. Calculate the standard cell potential by subtracting the electrode potential of the anode from the electrode potential of the cathode: $E^\circ_{cell} = E^\circ_{cathode} - E^\circ_{anode}$. If $E^\circ_{cell} > 0$ the reaction is spontaneous in the forward direction.
Solution:
a) Ni (*s*) + Zn^{2+} (*aq*) → Ni^{2+} (*aq*) + Zn (*s*) separates to Ni (*s*) → Ni^{2+} (*aq*) and Zn^{2+} (*aq*) → Zn (*s*) add electrons Ni (*s*) → Ni^{2+} (*aq*) + 2 e^- and Zn^{2+} (*aq*) + 2 e^- → Zn (*s*). Look up cell potentials. Ni is

oxidized so $E^\circ_{red} = -0.23$ V = E°_{anode}. Zn^{2+} is reduced so $E^\circ_{cathode} = -0.76$ V. Then $E^\circ_{cell} = E^\circ_{cathode} - E^\circ_{anode}$ = − 0.76 V − (− 0.23 V) = − 0.53 V and so the reaction is nonspontaneous.

b) Ni (*s*) + Pb^{2+} (*aq*) → Ni^{2+} (*aq*) + Pb (*s*) separates to Ni (*s*) → Ni^{2+} (*aq*) and Pb^{2+} (*aq*) → Pb (*s*) add electrons Ni (*s*) → Ni^{2+} (*aq*) + 2 e^- and Pb^{2+} (*aq*) + 2 e^- → Pb (*s*). Look up cell potentials. Ni is oxidized so $E^\circ_{red} = -0.23$ V = E°_{anode}. Pb^{2+} is reduced so $E^\circ_{red} = -0.13$ V = $E^\circ_{cathode}$. Then $E^\circ_{cell} = E^\circ_{cathode} - E^\circ_{anode}$ = − 0.13 V − (− 0.23 V) = + 0.10 V and so the reaction is spontaneous.

c) Al (*s*) + 3 Ag^+ (*aq*) → Al^{3+} (*aq*) + 3 Ag (*s*) separates to Al (*s*) → Al^{3+} (*aq*) and 3 Ag^+ (*aq*) → 3 Ag (*s*) add electrons Al (*s*) → Al^{3+} (*aq*) + 3 e^- and 3 Ag^+ (*aq*) + 3 e^- → 3 Ag (*s*). Simplify the Ag reaction to: Ag^+ (*aq*) + e^- → Ag (*s*). Look up cell potentials. Al is oxidized so $E^\circ_{red} = -1.66$ V = E°_{anode}. Ag^+ is reduced so $E^\circ_{red} =$ 0.80 V = $E^\circ_{cathode}$. Then $E^\circ_{cell} = E^\circ_{cathode} - E^\circ_{anode}$ = 0.80 V − (− 1.66 V) = + 2.46 V and so the reaction is spontaneous.

d) Pb (*s*) + Mn^{2+} (*aq*) → Pb^{2+} (*aq*) + Mn (*s*) separates to Pb (*s*) → Pb^{2+} (*aq*) and Mn^{2+} (*aq*) → Mn (*s*) add electrons Pb (*s*) → Pb^{2+} (*aq*) + 2 e^- and Mn^{2+} (*aq*) + 2 e^- → Mn (*s*). Look up cell potentials. Pb is oxidized so $E^\circ_{red} = -0.13$ V = E°_{anode}. Mn^{2+} is reduced so $E^\circ_{red} = -1.18$ V = $E^\circ_{cathode}$. Then $E^\circ_{cell} = E^\circ_{cathode} - E^\circ_{anode}$ = − 1.18 V − (− 0.13 V) = − 1.05 V and so the reaction is nonspontaneous.

Check: The units (V) are correct. If the voltage is positive, the reaction is spontaneous.

18. **Given:** overall reactions **Find:** spontaneity in forward direction
Conceptual Plan: Separate overall reaction into 2 half-cell reactions and add electrons as needed to balance reactions. Look up half-reactions in Table 18.1. Calculate the standard cell potential by subtracting the electrode potential of the anode from the electrode potential of the cathode: $E^\circ_{cell} = E^\circ_{cathode} - E^\circ_{anode}$. If $E^\circ_{cell} > 0$ the reaction is spontaneous in the forward direction.
Solution:

a) Ca^{2+} (*aq*) + Zn (*s*) → Ca (*s*) + Zn^{2+} (*aq*) separates to Ca^{2+} (*aq*) → Ca (*s*) and Zn (*s*) → Zn^{2+} (*aq*) add electrons Ca^{2+} (*aq*) + 2 e^- → Ca (*s*) and Zn (*s*) → Zn^{2+} (*aq*) + 2 e^-. Look up cell potentials. Zn is oxidized so $E^\circ_{red} = -0.76$ V = E°_{anode}. Ca^{2+} is reduced so $E^\circ_{red} = -2.76$ V = $E^\circ_{cathode}$. Then $E^\circ_{cell} = E^\circ_{cathode} - E^\circ_{anode}$ = − 2.76 V − (− 0.76 V) = − 2.00 V and so the reaction is spontaneous in the reverse direction.

b) 2 Ag^+ (*aq*) + Ni (*s*) → 2 Ag (*s*) + Ni^{2+} (*aq*) separates to 2 Ag^+ (*aq*) → 2 Ag (*s*) and Ni (*s*) → Ni^{2+} (*aq*) add electrons 2 Ag^+ (*aq*) + 2 e^- → 2 Ag (*s*) and Ni (*s*) → Ni^{2+} (*aq*) + 2 e^-. Simplify the Ag reaction to: Ag^+ (*aq*) + e^- → Ag (*s*). Look up cell potentials. Ni is oxidized so $E^\circ_{red} = -0.23$ V = E°_{anode}. Ag^+ is reduced so $E^\circ_{red} = 0.80$ V = $E^\circ_{cathode}$. Then $E^\circ_{cell} = E^\circ_{cathode} - E^\circ_{anode}$ = 0.80 V − (− 0.23 V) = + 1.03 V and so the reaction is nonspontaneous in the reverse direction.

c) Fe (*s*) + Mn^{2+} (*aq*) → Fe^{2+} (*aq*) + Mn (*s*) separates to Fe (*s*) → Fe^{2+} (*aq*) and Mn^{2+} (*aq*) → Mn (*s*) add electrons Fe (*s*) → Fe^{2+} (*aq*) + 2 e^- and Mn^{2+} (*aq*) + 2 e^- → Mn (*s*). Look up cell potentials. Fe is oxidized so $E^\circ_{red} = -0.45$ V = E°_{anode}. Mn^{2+} is reduced so $E^\circ_{red} = -1.18$ V = $E^\circ_{cathode}$. Then $E^\circ_{cell} = E^\circ_{cathode} - E^\circ_{anode}$ = − 1.18 V − (− 0.45 V) = + 0.73 V and so the reaction is nonspontaneous in the reverse direction.

d) 2 Al (*s*) + 3 Pb^{2+} (*aq*) → 2 Al^{3+} (*aq*) + 3 Pb (*s*) separates to 2 Al (*s*) → 2 Al^{3+} (*aq*) and 3 Pb^{2+} (*aq*) → 3 Pb (*s*) add electrons 2 Al (*s*) → 2 Al^{3+} (*aq*) + 6 e^- and 3 Pb^{2+} (*aq*) + 6 e^- → 2 Pb (*s*). Simplify the reactions to Al (*s*) → Al^{3+} (*aq*) + 3 e^- and Pb^{2+} (*aq*) + 2 e^- → Pb (*s*). Look up cell potentials. Al is oxidized so $E^\circ_{red} = -1.66$ V = E°_{anode}. Pb^{2+} is reduced so $E^\circ_{red} = -0.13$ V = $E^\circ_{cathode}$. Then $E^\circ_{cell} = E^\circ_{cathode} - E^\circ_{anode}$ = − 0.13 V − (− 1.66 V) = + 1.53 V and so the reaction is nonspontaneous in the reverse direction.

Check: The units (V) are correct. If the voltage is negative, the reaction is spontaneous in the reverse direction.

19. In order for a metal to be able to reduce an ion, it must be above it in Table 18.1 (need positive $E^\circ_{cell} = E^\circ_{cathode} - E^\circ_{anode}$). So we need a metal that is above Mn^{2+}, but below Mg^{2+}. Aluminum is the only one in the table that meets these criteria.

20. In order for a metal to be able to oxidized an ion, it must be below it in Table 18.1 (need positive $E^\circ_{cell} = E^\circ_{cathode} - E^\circ_{anode}$). So we need a metal that is above Fe^{2+}, but below Sn^{2+}. Nickel and cadmium meet these criteria.

21. In general, metals whose reduction half-reactions lie below the reduction of H^+ to H_2 in Table 18.1 will dissolve in acids, while metals above it will not. a) Al and c) Pb meet this criterion. To write the balanced redox

reactions, pair the oxidation of the metal with the reduction of H^+ to H_2 ($2\ H^+(aq) + 2\ e^- \rightarrow H_2(g)$). For Al, $Al(s) \rightarrow Al^{3+}(aq) + 3\ e^-$. In order to balance the number of electrons transferred we need to multiply the Al reaction by 2 and the H^+ reaction by 3. So, $2\ Al(s) \rightarrow 2\ Al^{3+}(aq) + 6\ e^-$ and $6\ H^+(aq) + 6\ e^- \rightarrow 3\ H_2(g)$. Adding the two reactions: $2\ Al(s) + 6\ H^+(aq) +$ ~~$6\ e^-$~~ $\rightarrow 2\ Al^{3+}(aq) +$ ~~$6\ e^-$~~ $+ 3\ H_2(g)$. Simplify to $2\ Al(s) + 6\ H^+(aq) \rightarrow 2\ Al^{3+}(aq) + 3\ H_2(g)$. For Pb, $Pb(s) \rightarrow Pb^{2+}(aq) + 2\ e^-$. Since each reaction involves 2 electrons we can add the two reactions. $Pb(s) + 2\ H^+(aq) +$ ~~$2\ e^-$~~ $\rightarrow Pb^{2+}(aq) +$ ~~$2\ e^-$~~ $+ H_2(g)$. Simplify to $Pb(s) + 2\ H^+(aq) \rightarrow Pb^{2+}(aq) + H_2(g)$.

22. In general, metals whose reduction half-reactions lie below the reduction of H^+ to H_2 in Table 18.1 will dissolve in acids, while metals above it will not. Only b) Fe meets this criterion. To write the balanced redox reactions, pair the oxidation of the metal with the reduction of H^+ to H_2 ($2\ H^+(aq) + 2\ e^- \rightarrow H_2(g)$). For Fe there are two possible reactions, $Fe(s) \rightarrow Fe^{3+}(aq) + 3\ e^-$ and $Fe(s) \rightarrow Fe^{2+}(aq) + 2\ e^-$. Since the second reaction is lower in Table 18.1, the cell potential will be more positive. This means that this reaction will be more spontaneous and thus preferred. Since each reaction involves 2 electrons we can add the two reactions. $Fe(s) + 2\ H^+(aq) +$ ~~$2\ e^-$~~ $\rightarrow Fe^{2+}(aq) +$ ~~$2\ e^-$~~ $+ H_2(g)$. Simplify to $Fe(s) + 2\ H^+(aq) \rightarrow Fe^{2+}(aq) + H_2(g)$.

23. Nitric acid (HNO_3) oxidizes metals through the following reduction half-reaction: $NO_3^-(aq) + 4\ H^+(aq) + 3\ e^- \rightarrow NO(g) + 2\ H_2O(l)$ $E^\circ_{red} = 0.96$ V. Since this half-reaction is above the reduction of H^+ in Table 18.1, HNO_3 can oxidize metals (such as copper, for example) that cannot be oxidized by HCl. a) Cu will be oxidized, but b) Au (which has a reduction potential of 1.50 V) will not be oxidized. To write the balanced redox reactions, pair the oxidation of the metal with the reduction of nitric acid ($NO_3^-(aq) + 4\ H^+(aq) + 3\ e^- \rightarrow NO(g) + 2\ H_2O(l)$). For Cu, $Cu(s) \rightarrow Cu^{2+}(aq) + 2\ e^-$. In order to balance the number of electrons transferred we need to multiply the Cu reaction by 3 and the nitric acid reaction by 2. So, $3\ Cu(s) \rightarrow 3\ Cu^{2+}(aq) + 6\ e^-$ and $2\ NO_3^-(aq) + 8\ H^+(aq) + 6\ e^- \rightarrow 2\ NO(g) + 4\ H_2O(l)$. Adding the two reactions: $3\ Cu(s) + 2\ NO_3^-(aq) + 8\ H^+(aq) +$ ~~$6\ e^-$~~ $\rightarrow 3\ Cu^{2+}(aq) +$ ~~$6\ e^-$~~ $+ 2\ NO(g) + 4\ H_2O(l)$. Simplify to $3\ Cu(s) + 2\ NO_3^-(aq) + 8\ H^+(aq) \rightarrow 3\ Cu^{2+}(aq) + 2\ NO(g) + 4\ H_2O(l)$.

24. Iodic acid (HIO_3) oxidizes metals through the following reduction half-reaction: $IO_3^-(aq) + 6\ H^+(aq) + 5\ e^- \rightarrow \frac{1}{2}\ I_2(aq) + 3\ H_2O(l)$ $E^\circ_{red} = 1.20$ V. Since this half-reaction is above the reduction of H^+ in Table 18.1, HNO_3 can oxidize metals (such as copper, for example) that cannot be oxidized by HCl. a) Au (which has a reduction potential of 1.50 V) will not be oxidized, but b) Cr (which has a reduction potential of – 0.50 V) will be oxidized. To write the balanced redox reactions, pair the oxidation of the metal with the reduction of iodic acid ($IO_3^-(aq) + 6\ H^+(aq) + 5\ e^- \rightarrow \frac{1}{2}\ I_2(aq) + 3\ H_2O(l)$). For Cr, $Cr(s) \rightarrow Cr^{3+}(aq) + 3\ e^-$. In order to balance the number of electrons transferred we need to multiply the Cr reaction by 5 and the iodic acid reaction by 3. So, $5\ Cr(s) \rightarrow 5\ Cr^{3+}(aq) + 15\ e^-$ and $3\ IO_3^-(aq) + 18\ H^+(aq) + 15\ e^- \rightarrow 3/2\ I_2(aq) + 9\ H_2O(l)$. Adding the two reactions: $5\ Cr(s) + 3\ IO_3^-(aq) + 18\ H^+(aq) +$ ~~$15\ e^-$~~ $\rightarrow 5\ Cr^{3+}(aq) +$ ~~$15\ e^-$~~ $+ 3/2\ I_2(aq) + 9\ H_2O(l)$. Simplify to $5\ Cr(s) + 3\ IO_3^-(aq) + 18\ H^+(aq) \rightarrow 5\ Cr^{3+}(aq) + 3/2\ I_2(aq) + 9\ H_2O(l)$.

25. **Given:** overall reactions **Find:** E°_{cell} and spontaneity in forward direction
Conceptual Plan: Separate overall reaction into 2 half-cell reactions and add electrons as needed to balance reactions. Look up half-reactions in Table 18.1. Calculate the standard cell potential by subtracting the electrode potential of the anode from the electrode potential of the cathode: $E^\circ_{cell} = E^\circ_{cathode} - E^\circ_{anode}$. If $E^\circ_{cell} > 0$ the reaction is spontaneous in the forward direction.
Solution:
a) $2\ Cu(s) + Mn^{2+}(aq) \rightarrow 2\ Cu^+(aq) + Mn(s)$ separates to $2\ Cu(s) \rightarrow 2\ Cu^+(aq)$ and $Mn^{2+}(aq) \rightarrow Mn(s)$ add electrons $2\ Cu(s) \rightarrow 2\ Cu^+(aq) + 2\ e^-$ and $Mn^{2+}(aq) + 2\ e^- \rightarrow Mn(s)$. Simplify the Cu reaction to: $Cu(s) \rightarrow Cu^+(aq) + e^-$. Look up cell potentials. Cu is oxidized so $E^\circ_{red} = -\ 0.52\ V = E^\circ_{anode}$. Mn^{2+} is reduced so $E^\circ_{red} = -\ 1.18\ V = E^\circ_{cathode}$. Then $E^\circ_{cell} = E^\circ_{cathode} - E^\circ_{anode} = -\ 0.52\ V - 1.18\ V = -\ 1.70\ V$ and so the reaction is nonspontaneous.

b) $MnO_2(s) + 4\ H^+(aq) + Zn(s) \rightarrow Mn^{2+}(aq) + 2\ H_2O(l) + Zn^{2+}(aq)$ separates to $MnO_2(s) + 4\ H^+(aq) \rightarrow Mn^{2+}(aq) + 2\ H_2O(l)$ and $Zn(s) \rightarrow Zn^{2+}(aq)$ add electrons $MnO_2(s) + 4\ H^+(aq) + 2\ e^- \rightarrow Mn^{2+}(aq) + 2\ H_2O(l)$ and $Zn(s) \rightarrow Zn^{2+}(aq) + 2\ e^-$. Look up cell potentials. Zn is oxidized so $E^\circ_{red} = -\ 0.76\ V = E^\circ_{anode}$. Mn is reduced so $E^\circ_{red} = 1.21\ V = E^\circ_{cathode}$. Then $E^\circ_{cell} = E^\circ_{cathode} - E^\circ_{anode} = 1.21\ V - (-\ 0.76\ V) = +\ 1.97\ V$ and so the reaction is spontaneous.

c) Cl_2 (*g*) + 2 F^- (*aq*) → 2 Cl^- (*aq*) + F_2 (*g*) separates to Cl_2 (*g*) → 2 Cl^- (*aq*) and 2 F^- (*aq*) → F_2 (*g*) add electrons Cl_2 (*g*) + 2 e^- → 2 Cl^- (*aq*) and 2 F^- (*aq*) → F_2 (*g*) + 2 e^-. Look up cell potentials. F is oxidized so E^o_{red} = 2.87 V = E^o_{anode}. Cl is reduced so E^o_{red} = 1.36 V = $E^o_{cathode}$. Then $E^o_{cell} = E^o_{cathode} - E^o_{anode}$ = 1.36 V – 2.87 V + = – 1.51 V and so the reaction is nonspontaneous.

Check: The units (V) are correct. If the voltage is positive, the reaction is spontaneous.

26. **Given:** overall reactions **Find:** E^o_{cell} and spontaneity in forward direction

Conceptual Plan: Separate overall reaction into 2 half-cell reactions and add electrons as needed to balance reactions. Look up half-reactions in Table 18.1. Calculate the standard cell potential by subtracting the electrode potential of the anode from the electrode potential of the cathode: $E^o_{cell} = E^o_{cathode} - E^o_{anode}$. If $E^o_{cell} > 0$ the reaction is spontaneous in the forward direction.

Solution:

a) O_2 (*g*)+ 2 H_2O (*l*) + 4 Ag (*s*) → 4 OH^- (*aq*) + 4 Ag^+ (*aq*) separates to O_2 (*g*)+ 2 H_2O (*l*) → 4 OH^- (*aq*) and 4 Ag (*s*) → 4 Ag^+ (*aq*) add electrons O_2 (*g*)+ 2 H_2O (*l*) + 4 e^- → 4 OH^- (*aq*) and 4 Ag (*s*) → 4 Ag^+ (*aq*) + 2 e^-. Simplify the Ag reaction to: Ag (*s*) → Ag^+ (*aq*) + e^-. Look up cell potentials. Ag is oxidized so E^o_{red} = – 0.80 V = E^o_{anode}. O is reduced so E^o_{red} = 0.40 V = $E^o_{cathode}$. Then $E^o_{cell} = E^o_{cathode} - E^o_{anode}$ = 0.40 V – (– 0.80 V) = – 0.40 V and so the reaction is nonspontaneous.

b) Br_2 (*l*) + 2 I^- (*aq*) → 2 Br^- (*aq*) + I_2 (*g*) separates to Br_2 (*g*) → 2 Br^- (*aq*) and 2 I^- (*aq*) → I_2 (*g*) add electrons Br_2 (*g*) + 2 e^- → 2 Br^- (*aq*) and 2 I^- (*aq*) → I_2 (*g*) + 2 e^-. Look up cell potentials. I is oxidized so E^o_{red} = 0.54 V = E^o_{anode}. Br is reduced so E^o_{red} = 1.09 V = $E^o_{cathode}$. Then $E^o_{cell} = E^o_{cathode} - E^o_{anode}$ = 1.09 V – 0.54 V = + 0.55 V and so the reaction is spontaneous.

c) PbO_2 (*s*) + 4 H^+(*aq*) + Sn (*s*) → Pb^{2+} (*aq*) + 2 H_2O (*l*) + Sn^{2+} (*aq*) separates to PbO_2 (*s*) + 4 H^+(*aq*) → Pb^{2+} (*aq*) + 2 H_2O (*l*) and Sn (*s*) → Sn^{2+} (*aq*) add electrons PbO_2 (*s*) + 4 H^+(*aq*) + 2 e^- → Pb^{2+} (*aq*) + 2 H_2O (*l*) and Sn (*s*) → Sn^{2+} (*aq*) + 2 e^-. Look up cell potentials. Sn is oxidized so E^o_{red} = – 0.14 V = E^o_{anode}. Pb is reduced so E^o_{red} = 1.46 V = $E^o_{cathode}$. Then $E^o_{cell} = E^o_{cathode} - E^o_{anode}$ = 1.46 V – (– 0.14 V) = + 1.60 V and so the reaction is spontaneous.

Check: The units (V) are correct. If the voltage is positive, the reaction is spontaneous.

27. a) Pb^{2+}. The strongest oxidizing agent is the one with the reduction reaction that is closest to the top of Table 18.1.

28. b) Al. The strongest reducing agent is the one with the reduction reaction that yields the metal that is closest to the bottom of Table 18.1.

29. **Given:** overall reactions **Find:** ΔG^o_{rxn} and spontaneity in forward direction

Conceptual Plan: Separate overall reaction into 2 half-cell reactions and add electrons as needed to balance reactions. Look up half-reactions in Table 18.1. Calculate the standard cell potential by subtracting the electrode potential of the anode from the electrode potential of the cathode: $E^o_{cell} = E^o_{cathode} - E^o_{anode}$, then calculate ΔG^o_{rxn} using $\Delta G^0_{rxn} = -nFE^0_{cell}$.

Solution:

a) Pb^{2+} (*aq*) + Mg (*s*) → Pb (*s*) + Mg^{2+} (*aq*) separates to Pb^{2+} (*aq*) → Pb (*s*) and Mg (*s*) → Mg^{2+} (*aq*) add electrons Pb^{2+} (*aq*) + 2 e^- → Pb (*s*) and Mg (*s*) → Mg^{2+} (*aq*) + 2 e^-. Look up cell potentials. Mg is oxidized so $E^o_{ox} = -E^o_{red}$ = – 2.37 V = E^o_{anode}. Pb^{2+} is reduced so E^o_{red} = – 0.13 V = $E^o_{cathode}$. Then $E^o_{cell} = E^o_{cathode} - E^o_{anode}$ = – 0.13 V – (– 2.37 V) = + 2.24 V. $n = 2$ so $\Delta G^0_{rxn} = -nFE^0_{cell}$

$$= -2\ \cancel{\text{mol e}^-} \times \frac{96{,}485\ \text{C}}{\cancel{\text{mol e}^-}} \times 2.24\ \text{V} = -2 \times 96{,}485\ \cancel{\text{C}} \times 2.24\ \frac{\text{J}}{\cancel{\text{C}}} = -4.32 \times 10^5\ \text{J} = -432\ \text{kJ}$$

b) Br_2 (*l*) + 2 Cl^- (*aq*) → 2 Br^- (*aq*) + Cl_2 (*g*) separates to Br_2 (*g*) → 2 Br^- (*aq*) and 2 Cl^- (*aq*) → Cl_2 (*g*) add electrons Br_2 (*g*) + 2 e^- → 2 Br^- (*aq*) and 2 Cl^- (*aq*) → Cl_2 (*g*) + 2 e^-. Look up cell potentials. Cl is oxidized so E^o_{red} = 1.36 V = E^o_{anode}. Br is reduced so E^o_{red} = 1.09 V = $E^o_{cathode}$. Then $E^o_{cell} = E^o_{cathode} - E^o_{anode}$ = 1.09 V – 1.36 V = – 0.27 V. $n = 2$ so $\Delta G^0_{rxn} = -nFE^0_{cell}$

$$= -2\ \cancel{\text{mol e}^-} \times \frac{96{,}485\ \text{C}}{\cancel{\text{mol e}^-}} \times -0.27\ \text{V} = -2 \times 96{,}485\ \cancel{\text{C}} \times -0.27\ \frac{\text{J}}{\cancel{\text{C}}} = 5.2 \times 10^4\ \text{J} = 52\ \text{kJ}.$$

c) MnO_2 (*s*) + 4 H^+(*aq*) + Cu (*s*) → Mn^{2+} (*aq*) + 2 H_2O (*l*) + Cu^{2+} (*aq*) separates to MnO_2 (*s*) + 4 H^+(*aq*) → Mn^{2+} (*aq*) + 2 H_2O (*l*) and Cu (*s*) → Cu^{2+} (*aq*) add electrons MnO_2 (*s*) + 4 H^+(*aq*) + 2 e^- → Mn^{2+} (*aq*) + 2

H_2O (*l*) and Cu (*s*) → Cu^{2+} (*aq*) + 2 e^-. Look up cell potentials. Cu is oxidized so $E^o_{red} = 0.34\text{ V} = E^o_{anode}$. Mn is reduced so $E^o_{red} = 1.21\text{ V} = E^o_{cathode}$. Then $E^o_{cell} = E^o_{cathode} - E^o_{anode} = 1.21\text{ V} - 0.34\text{ V} = +0.87\text{ V}$.

$n = 2$ so $\Delta G^0_{rxn} = -nFE^0_{cell} = -2\ \cancel{\text{mol e}^-} \times \dfrac{96{,}485\text{ C}}{\cancel{\text{mol e}^-}} \times 0.87\text{ V} = -2 \times 96{,}485\ \cancel{\text{C}} \times 0.87\dfrac{\text{J}}{\cancel{\text{C}}} = -1.7 \times 10^5\text{ J}$

$= -1.7 \times 10^2\text{ kJ}$.

Check: The units (kJ) are correct. If the voltage is positive, the reaction is spontaneous and the free energy change is negative.

30. **Given:** overall reactions **Find:** ΔG^o_{rxn} and spontaneity in forward direction
Conceptual Plan: Separate overall reaction into 2 half-cell reactions and add electrons as needed to balance reactions. Look up half-reactions in Table 18.1. Calculate the standard cell potential by subtracting the electrode potential of the anode from the electrode potential of the cathode: $E^o_{cell} = E^o_{cathode} - E^o_{anode}$, then calculate ΔG^o_{rxn} using $\Delta G^0_{rxn} = -nFE^0_{cell}$.
Solution:

a) 2 Fe^{3+} (*aq*) + 3 Sn (*s*) → 2 Fe (*s*) + 3 Sn^{2+} (*aq*) separates to 2 Fe^{3+} (*aq*) → 2 Fe (*s*) and 3 Sn (*s*) → 3 Sn^{2+} (*aq*) add electrons 2 Fe^{3+} (*aq*) + 6 e^- → 2 Fe (*s*) and 3 Sn (*s*) → 3 Sn^{2+} (*aq*) + 6 e^-. Simplify reactions to Fe^{3+} (*aq*) + 3 e^- → Fe (*s*) and Sn (*s*) → Sn^{2+} (*aq*) + 2 e^-. Look up cell potentials. Sn is oxidized so $E^o_{red} = -0.14\text{ V} = E^o_{anode}$. Fe^{3+} is reduced so $E^o_{red} = -0.036\text{ V} = E^o_{cathode}$. Then $E^o_{cell} = E^o_{cathode} - E^o_{anode} = -0.036\text{ V} - (-0.14\text{ V}) = +0.1\underline{0}4\text{ V}$. $n = 6$ so $\Delta G^0_{rxn} = -nFE^0_{cell}$

$= -6\ \cancel{\text{mol e}^-} \times \dfrac{96{,}485\text{ C}}{\cancel{\text{mol e}^-}} \times 0.1\underline{0}4\text{ V} = -6 \times 96{,}485\ \cancel{\text{C}} \times 0.1\underline{0}4\dfrac{\text{J}}{\cancel{\text{C}}} = -6.0 \times 10^4\text{ J} = -6.0 \times 10^1\text{ kJ}$.

b) O_2 (*g*) + 2 H_2O (*l*) + 2 Cu (*s*) → 4 OH^- (*aq*) + 2 Cu^{2+} (*aq*) separates to O_2 (*g*) + 2 H_2O (*l*) → 4 OH^- (*aq*) and 2 Cu (*s*) → 2 Cu^{2+} (*aq*) add electrons O_2 (*g*) + 2 H_2O (*l*) + 4 e^- → 4 OH^- (*aq*) and 2 Cu (*s*) → 2 Cu^{2+} (*aq*) + 4 e^-. Simplify the Cu reaction to: Cu (*s*) → Cu^{2+} (*aq*) + 2 e^-. Look up cell potentials. Cu is oxidized so $E^o_{red} = 0.34\text{ V} = E^o_{anode}$. O is reduced so $E^o_{red} = 0.40\text{ V} = E^o_{cathode}$. Then $E^o_{cell} = E^o_{cathode} - E^o_{anode} = 0.40\text{ V} - 0.34\text{ V} = +0.06\text{ V}$. $n = 4$ so $\Delta G^0_{rxn} = -nFE^0_{cell} = -4\ \cancel{\text{mol e}^-} \times \dfrac{96{,}485\text{ C}}{\cancel{\text{mol e}^-}} \times 0.06\text{ V}$

$= -4 \times 96{,}485\ \cancel{\text{C}} \times 0.06\dfrac{\text{J}}{\cancel{\text{C}}} = -2 \times 10^4\text{ J} = -2 \times 10^1\text{ kJ}$.

c) Br_2 (*l*) + 2 I^- (*aq*) → 2 Br^- (*aq*) + I_2 (*g*) separates to Br_2 (*g*) → 2 Br^- (*aq*) and 2 I^- (*aq*) → I_2 (*g*) add electrons Br_2 (*g*) + 2 e^- → 2 Br^- (*aq*) and 2 I^- (*aq*) → I_2 (*g*) + 2 e^-. Look up cell potentials. I is oxidized so $E^o_{red} = -0.54\text{ V} = E^o_{anode}$. Br is reduced so $E^o_{red} = 1.09\text{ V} = E^o_{cathode}$. Then $E^o_{cell} = E^o_{cathode} - E^o_{anode} = 1.09\text{ V} - 0.54\text{ V} = +0.55\text{ V}$. $n = 2$ so $\Delta G^0_{rxn} = -nFE^0_{cell} = -2\ \cancel{\text{mol e}^-} \times \dfrac{96{,}485\text{ C}}{\cancel{\text{mol e}^-}} \times 0.55\text{ V}$

$= -2 \times 96{,}485\ \cancel{\text{C}} \times 0.55\dfrac{\text{J}}{\cancel{\text{C}}} = -1.1 \times 10^5\text{ J} = -1.1 \times 10^2\text{ kJ}$.

Check: The units (kJ) are correct. If the voltage is positive, the reaction is spontaneous and the free energy change is negative.

31. **Given:** overall reactions from #29 **Find:** K
Conceptual Plan: °C → K then ΔG^o_{rxn}, T → K
$K = 273.15 + °C$ $\qquad \Delta G^0_{rxn} = -RT\ln K$
Solution: $T = 273.15 + 25\text{ °C} = 298\text{ K}$ then

a) $\Delta G^0_{rxn} = -RT\ln K$ Rearrange to solve for K. $K = e^{\frac{-\Delta G^0_{rxn}}{RT}} = e^{\frac{-(-432)\ \cancel{\text{kJ}} \times \frac{1000\ \cancel{\text{J}}}{1\ \cancel{\text{kJ}}}}{\left(8.314\frac{\cancel{\text{J}}}{\cancel{\text{K}}\cdot\text{mol}}\right)(298\ \cancel{\text{K}})}} = e^{174.364} = 5.31 \times 10^{75}$.

b) $\Delta G^0_{rxn} = -RT\ln K$ Rearrange to solve for K. $K = e^{\frac{-\Delta G^0_{rxn}}{RT}} = e^{\frac{-52\ \cancel{\text{kJ}} \times \frac{1000\ \cancel{\text{J}}}{1\ \cancel{\text{kJ}}}}{\left(8.314\frac{\cancel{\text{J}}}{\cancel{\text{K}}\cdot\text{mol}}\right)(298\ \cancel{\text{K}})}} = e^{-20.988} = 7.7 \times 10^{-10}$.

c) $\Delta G^0_{rxn} = -RT\ln K$ Rearrange to solve for K. $K = e^{\frac{-\Delta G^0_{rxn}}{RT}} = e^{\frac{-(-170\text{ kJ}) \times \frac{1000\text{ J}}{1\text{ kJ}}}{\left(8.314\frac{\text{J}}{\text{K}\cdot\text{mol}}\right)(298\text{ K})}} = e^{68.616} = 6.3 \times 10^{29}$.

Check: The units (none) are correct. If the voltage is positive, the reaction is spontaneous and the free energy change is negative and the equilibrium constant is large.

32. **Given:** overall reactions from #30 **Find:** K
Conceptual Plan: **°C → K then ΔG°_{rxn}, T → K**
$K = 273.15 + °C$ $\quad\quad$ $\Delta G^0_{rxn} = -RT\ln K$
Solution: $T = 273.15 + 25\ °C = 298$ K then

a) $\Delta G^0_{rxn} = -RT\ln K$ Rearrange to solve for K. $K = e^{\frac{-\Delta G^0_{rxn}}{RT}} = e^{\frac{-(-60\text{ kJ}) \times \frac{1000\text{ J}}{1\text{ kJ}}}{\left(8.314\frac{\text{J}}{\text{K}\cdot\text{mol}}\right)(298\text{ K})}} = e^{24.217} = 3.3 \times 10^{10}$.

b) $\Delta G^0_{rxn} = -RT\ln K$ Rearrange to solve for K. $K = e^{\frac{-\Delta G^0_{rxn}}{RT}} = e^{\frac{-(-20\text{ kJ}) \times \frac{1000\text{ J}}{1\text{ kJ}}}{\left(8.314\frac{\text{J}}{\text{K}\cdot\text{mol}}\right)(298\text{ K})}} = e^{8.072} = 3 \times 10^{3}$.

c) $\Delta G^0_{rxn} = -RT\ln K$ Rearrange to solve for K. $K = e^{\frac{-\Delta G^0_{rxn}}{RT}} = e^{\frac{-(-110\text{ kJ}) \times \frac{1000\text{ J}}{1\text{ kJ}}}{\left(8.314\frac{\text{J}}{\text{K}\cdot\text{mol}}\right)(298\text{ K})}} = e^{44.398} = 1.9 \times 10^{19}$.

Check: The units (none) are correct. If the voltage is positive, the reaction is spontaneous and the free energy change is negative and the equilibrium constant is large.

33. **Given:** Ni^{2+} (*aq*) + Cd (*s*) → **Find:** K
Conceptual Plan: Write 2 half-cell reactions and add electrons as needed to balance reactions. Look up half-reactions in Table 18.1. Calculate the standard cell potential by subtracting the electrode potential of the anode from the electrode potential of the cathode: $E^\circ_{cell} = E^\circ_{cathode} - E^\circ_{anode}$, **then**
°C → K then E°_{cell}, n, T → K
$K = 273.15 + °C$ $\quad\quad$ $\Delta G^0_{rxn} = -RT\ln K = -nFE^0_{cell}$
Solution: Ni^{2+} (*aq*) + 2 e^- → Ni (*s*) and Cd (*s*) → Cd^{2+} (*aq*) + 2 e^-. Look up cell potentials. Cd is oxidized so $E^\circ_{ox} = -E^\circ_{red} = -0.40$ V $= E^\circ_{anode}$. Ni^{2+} is reduced so $E^\circ_{red} = -0.23$ V $= E^\circ_{cathode}$. Then $E^\circ_{cell} = E^\circ_{cathode} - E^\circ_{anode} = -0.23\text{ V} - (-0.40\text{ V}) = +0.17$ V. The overall reaction is Ni^{2+} (*aq*) + Cd (*s*) → Ni (*s*) + Cd^{2+} (*aq*). $n = 2$ and $T = 273.15 + 25\ °C = 298$ K then $\Delta G^0_{rxn} = -RT\ln K = -nFE^0_{cell}$. Rearrange to solve for K.

$$K = e^{\frac{nFE^0_{cell}}{RT}} = e^{\frac{2\text{ mol e}^- \times \frac{96{,}485\text{ C}}{\text{mol e}^-} \times 0.17\frac{\text{J}}{\text{C}}}{\left(8.314\frac{\text{J}}{\text{K}\cdot\text{mol}}\right)(298\text{ K})}} = e^{13.241} = 5.6 \times 10^{5}.$$

Check: The units (none) are correct. If the voltage is positive, the reaction is spontaneous and the equilibrium constant is large.

34. **Given:** Fe^{2+} (*aq*) + Zn (*s*) → **Find:** K
Conceptual Plan: Write 2 half-cell reactions and add electrons as needed to balance reactions. Look up half-reactions in Table 18.1. Calculate the standard cell potential by subtracting the electrode potential of the anode from the electrode potential of the cathode: $E^\circ_{cell} = E^\circ_{cathode} - E^\circ_{anode}$, **then**
°C → K then E°_{cell}, n, T → K
$K = 273.15 + °C$ $\quad\quad$ $\Delta G^0_{rxn} = -RT\ln K = -nFE^0_{cell}$
Solution: Fe^{2+} (*aq*) + 2 e^- → Fe (*s*) and Zn (*s*) → Zn^{2+} (*aq*) + 2 e^-. Look up cell potentials. Zn is oxidized so $E^\circ_{red} = -0.76$ V $= E^\circ_{anode}$. Fe^{2+} is reduced so $E^\circ_{red} = -0.45$ V $= E^\circ_{cathode}$. Then $E^\circ_{cell} = E^\circ_{cathode} - E^\circ_{anode} = -0.45\text{ V} - (-0.76\text{ V}) = +0.31$ V. The overall reaction is Fe^{2+} (*aq*) + Zn (*s*) → Fe (*s*) + Zn^{2+} (*aq*). $n = 2$ and $T =$

$273.15 + 25\ °C = 298\ K$ then $\Delta G^0_{rxn} = -RT \ln K = -nF E^0_{cell}$. Rearrange to solve for K.

$$K = e^{\frac{nF E^0_{cell}}{RT}} = e^{\dfrac{2\ \cancel{mole^-}\ \times\ \frac{96{,}485\ \cancel{C}}{\cancel{mole^-}}\ \times\ 0.31\ \frac{\cancel{J}}{\cancel{C}}}{\left(8.314\ \frac{\cancel{J}}{K\cdot mol}\right)(298\ \cancel{K})}} = e^{24.145} = 3.1 \times 10^{10}\ .$$

Check: The units (none) are correct. If the voltage is positive, the reaction is spontaneous and the equilibrium constant is large.

35. **Given:** $n = 2$ and $K = 25$ \hfill **Find:** ΔG°_{rxn} and E°_{cell}

   Conceptual Plan: $K, T \;\rightarrow\; \Delta G^\circ_{rxn}$ and $\Delta G^\circ_{rxn}, n \;\rightarrow\; E^\circ_{cell}$

   $\qquad\qquad\qquad\quad \Delta G^0_{rxn} = -RT \ln K \qquad\qquad\qquad \Delta G^0_{rxn} = -nF E^0_{cell}$

   Solution: $\Delta G^0_{rxn} = -RT \ln K = -\left(8.314\ \dfrac{J}{K\cdot mol}\right)(298\ \cancel{K}) \ln 25 = -7.\underline{9}7500 \times 10^3\ J = -8.0\ kJ$ and

   $\Delta G^0_{rxn} = -nF E^0_{cell}$. Rearrange to solve for E°_{cell}.

   $E^0_{cell} = \dfrac{\Delta G^0_{rxn}}{-nF} = \dfrac{-7.\underline{9}7500 \times 10^3\ J}{-2\ \cancel{mole^-}\ \times\ \frac{96{,}485\ C}{\cancel{mole^-}}} = 0.041\ \dfrac{V\cancel{C}}{\cancel{C}} = 0.041\ V$.

   Check: The units (kJ and V) are correct. If $K > 1$ then the voltage is positive, the free energy change is negative.

36. **Given:** $n = 3$ and $K = 0.050$ \hfill **Find:** ΔG°_{rxn} and E°_{cell}

   Conceptual Plan: $K, T \;\rightarrow\; \Delta G^\circ_{rxn}$ and $\Delta G^\circ_{rxn}, n \;\rightarrow\; E^\circ_{cell}$

   $\qquad\qquad\qquad\quad \Delta G^0_{rxn} = -RT \ln K \qquad\qquad\qquad \Delta G^0_{rxn} = -nF E^0_{cell}$

   Solution: $\Delta G^0_{rxn} = -RT \ln K = -\left(8.314\ \dfrac{J}{K\cdot mol}\right)(298\ \cancel{K}) \ln 0.050 = 7.\underline{4}221 \times 10^3\ J = 7.4\ kJ$ and

   $\Delta G^0_{rxn} = -nF E^0_{cell}$. Rearrange to solve for E°_{cell}.

   $E^0_{cell} = \dfrac{\Delta G^0_{rxn}}{-nF} = \dfrac{7.\underline{4}221 \times 10^3\ J}{-3\ \cancel{mole^-}\ \times\ \frac{96{,}485\ C}{\cancel{mole^-}}} = -0.026\ \dfrac{V\cancel{C}}{\cancel{C}} = -0.026\ V$.

   Check: The units (kJ and V) are correct. If $K < 1$ then the voltage is negative, the free energy change is positive.

37. **Given:** $Sn^{2+}\ (aq) + Mn\ (s) \;\rightarrow\; Sn\ (s) + Mn^{2+}\ (aq)$ \hfill **Find:** a) E°_{cell}; b) E_{cell} when $[Sn^{2+}] = 0.0100\ M$;
   $[Mn^{2+}] = 2.00\ M$; and c) E_{cell} when $[Sn^{2+}] = 2.00\ M$; $[Mn^{2+}] = 0.0100\ M$

   Conceptual Plan: a) Separate overall reaction into 2 half-cell reactions and add electrons as needed to balance reactions. Look up half-reactions in Table 18.1. Calculate the standard cell potential by subtracting the electrode potential of the anode from the electrode potential of the cathode: $E^\circ_{cell} = E^\circ_{cathode} - E^\circ_{anode}$. **b) and c)** $E^\circ_{cell}, [Sn^{2+}], [Mn^{2+}], n \;\rightarrow\; E_{cell}$

   $$E_{cell} = E^\circ_{cell} - \dfrac{0.0592\ V}{n} \log Q \quad \text{where } Q = \dfrac{[Mn^{2+}]}{[Sn^{2+}]}$$

   Solution: a) separate overall reaction to: $Sn^{2+}\ (aq) \rightarrow Sn\ (s)$ and $Mn\ (s) \rightarrow Mn^{2+}\ (aq)$ add electrons $Sn^{2+}\ (aq) + 2\ e^- \rightarrow Sn\ (s)$ and $Mn\ (s) \rightarrow Mn^{2+}\ (aq) + 2\ e^-$. Look up cell potentials. Mn is oxidized so $E^\circ_{red} = -1.18\ V = E^\circ_{anode}$. Sn^{2+} is reduced so $E^\circ_{red} = -0.14\ V = E^\circ_{cathode}$. Then $E^\circ_{cell} = E^\circ_{cathode} - E^\circ_{anode} = -0.14\ V - (-1.18\ V) = +1.04\ V$.

   b) $Q = \dfrac{[Mn^{2+}]}{[Sn^{2+}]} = \dfrac{2.00\ \cancel{M}}{0.0100\ \cancel{M}} = 200.$ and $n = 2$ then

   $E_{cell} = E^0_{cell} - \dfrac{0.0592\ V}{n} \log Q = 1.04\ V - \dfrac{0.0592\ V}{2} \log 200. = +0.97\ V$

   c) $Q = \dfrac{[Mn^{2+}]}{[Sn^{2+}]} = \dfrac{0.0100\ \cancel{M}}{2.00\ \cancel{M}} = 0.00500$ and $n = 2$ then

$$E_{cell} = E^0_{cell} - \frac{0.0592\text{ V}}{n}\log Q = 1.04\text{ V} - \frac{0.0592\text{ V}}{2}\log 0.00500 = +1.11\text{ V}$$

Check: The units (V, V, and V) are correct. The Sn^{2+} reduction reaction is above the Mn^{2+} reduction reaction, so the standard cell potential will be positive. Having more products than reactants reduces the cell potential. Having more reactants than products raises the cell potential.

38. **Given:** $2\ Fe^{3+}\ (aq) + 3\ Mg\ (s) \rightarrow 2\ Fe\ (s) + 3\ Mg^{2+}\ (aq)$ **Find:** a) E^o_{cell}; b) E_{cell} when $[Fe^{3+}] = 1.0 \times 10^{-3}$ M; $[Mg^{2+}] = 2.50$ M; and c) E_{cell} when $[Fe^{3+}] = 2.00$ M; $[Mg^{2+}] = 1.5 \times 10^{-3}$ M
Conceptual Plan: a) Separate overall reaction into 2 half-cell reactions and add electrons as needed to balance reactions. Look up half-reactions in Table 18.1. Calculate the standard cell potential by subtracting the electrode potential of the anode from the electrode potential of the cathode: $E^o_{cell} = E^o_{cathode} - E^o_{anode}$. b) and c) E^o_{cell}, $[Fe^{3+}]$, $[Mg^{2+}]$, $n \rightarrow E_{cell}$

$$E_{cell} = E^o_{cell} - \frac{0.0592\ V}{n}\log Q \text{ where } Q = \frac{[Mg^{2+}]^3}{[Fe^{3+}]^2}$$

Solution: a) separate overall reaction to: $2\ Fe^{3+}\ (aq) \rightarrow 2\ Fe\ (s)$ and $3\ Mg\ (s) \rightarrow 3\ Mg^{2+}\ (aq)$ add electrons 2 $Fe^{3+} + 6\ e^-\ (aq) \rightarrow 2\ Fe\ (s)$ and $3\ Mg\ (s) \rightarrow 3\ Mg^{2+}\ (aq) + 6\ e^-$. Look up cell potentials. Mg is oxidized so $E^o_{ox} = -E^o_{red} = -2.37\text{ V} = E^o_{anode}$. Fe^{3+} is reduced so $E^o_{red} = -0.036\text{ V} = E^o_{cathode}$. Then $E^o_{cell} = E^o_{cathode} - E^o_{anode} = -0.036\text{ V} - (-2.37\text{ V}) = +2.334\text{ V} = +2.33$ V.

b) $$Q = \frac{[Mg^{2+}]^3}{[Fe^{3+}]^2} = \frac{(2.50)^3}{(1.0 \times 10^{-3})^2} = 1.5625 \times 10^7 \text{ and } n = 6 \text{ then}$$

$$E_{cell} = E^0_{cell} - \frac{0.0592\text{ V}}{n}\log Q = 2.334\text{ V} - \frac{0.0592\text{ V}}{6}\log 1.5625 \times 10^7 = +2.26\text{ V}$$

c) $$Q = \frac{[Mg^{2+}]^3}{[Fe^{3+}]^2} = \frac{(1.5 \times 10^{-3})^3}{(2.00)^2} = 8.4375 \times 10^{-10} \text{ and } n = 6 \text{ then}$$

$$E_{cell} = E^0_{cell} - \frac{0.0592\text{ V}}{n}\log Q = 2.334\text{ V} - \frac{0.0592\text{ V}}{6}\log 8.4375 \times 10^{-10} = +2.42\text{ V}$$

Check: The units (V, V, and V) are correct. The Fe^{3+} reduction reaction is above the Mg^{2+} reduction reaction, so the standard cell potential will be positive. Having more products than reactants reduces the cell potential. Having more reactants than products raises the cell potential.

39. **Given:** $Pb\ (s) \rightarrow Pb^{2+}\ (aq, 0.10\text{ M}) + 2\ e^-$ and $MnO_4^-\ (aq, 1.50\text{ M}) + 4\ H^+\ (aq, 2.0\text{ M}) + 3\ e^- \rightarrow MnO_2\ (s) + 2\ H_2O\ (l)$ **Find:** E_{cell}
**Conceptual Plan: Look up half-reactions in Table 18.1. Calculate the standard cell potential by subtracting the electrode potential of the anode from the electrode potential of the cathode: $E^o_{cell} = E^o_{cathode} - E^o_{anode}$. Equalize the number of electrons transferred by multiplying the first reaction by 3 and the second reaction by 2. Add the two half-cell reactions and cancel the electrons.
Then E^o_{cell}, $[Pb^{2+}]$, $[MnO_4^-]$, $[H^+]$, $n \rightarrow E_{cell}$**

$$E_{cell} = E^o_{cell} - \frac{0.0592\ V}{n}\log Q \text{ where } Q = \frac{[Pb^{2+}]^3}{[MnO_4^-]^2[H^+]^8}$$

Solution: Pb is oxidized so $E^o_{red} = -0.13\text{ V} = E^o_{anode}$. Mn is reduced so $E^o_{red} = 1.68\text{ V} = E^o_{cathode}$. Then $E^o_{cell} = E^o_{cathode} - E^o_{anode} = 1.68\text{ V} - (-0.13\text{ V}) = +1.81$ V. Equalizing the electrons: $3\ Pb\ (s) \rightarrow 3\ Pb^{2+}\ (aq) + 6\ e^-$ and $2\ MnO_4^-\ (aq) + 8\ H^+\ (aq) + 6\ e^- \rightarrow 2\ MnO_2\ (s) + 4\ H_2O\ (l)$. Adding the two reactions: $3\ Pb\ (s) + 2\ MnO_4^-\ (aq) + 8\ H^+\ (aq) + \cancel{6\ e^-} \rightarrow 3\ Pb^{2+}\ (aq) + \cancel{6\ e^-} + 2\ MnO_2\ (s) + 4\ H_2O\ (l)$. Cancel the electrons: $3\ Pb\ (s) + 2\ MnO_4^-\ (aq) + 8\ H^+\ (aq) \rightarrow 3\ Pb^{2+}\ (aq) + 2\ MnO_2\ (s) + 4\ H_2O\ (l)$. So $n = 6$ and $Q = \frac{[Pb^{2+}]^3}{[MnO_4^-]^2[H^+]^8} = \frac{(0.10)^3}{(1.50)^2(2.0)^8}$

$= 1.7361 \times 10^{-6}$ then $E_{cell} = E^0_{cell} - \frac{0.0592\text{ V}}{n}\log Q = 1.81\text{ V} - \frac{0.0592\text{ V}}{6}\log 1.7361 \times 10^{-6} = +1.87\text{ V}$.

Check: The units (V) are correct. The MnO_4^- reduction reaction is above the Pb^{2+} reduction reaction, so the standard cell potential will be positive. Having more reactants than products raises the cell potential.

40. **Given:** $Sn\ (s) \rightarrow Sn^{2+}\ (aq, 2.00\text{ M}) + 2\ e^-$ and $ClO_2\ (g, 0.100\text{ atm}) + e^- \rightarrow ClO_2^-\ (aq, 2.00\text{ M})$ **Find:** E_{cell}
Conceptual Plan: Look up half-reactions in Table 18.1. Calculate the standard cell potential by subtracting the electrode potential of the anode from the electrode potential of the cathode: E^o_{cell} =

$E^o_{cathode}$ – E^o_{anode}. Equalize the number of electrons transferred by multiplying the second reaction by 2. Add the two half-cell reactions and cancel the electrons. Then E^o_{cell}, [Sn^{2+}], P_{ClO2}, [ClO_2^-], n → E_{cell}

$$E_{cell} = E^o_{cell} - \frac{0.0592\ V}{n}\log Q \text{ where } Q = \frac{[Sn^{2+}][ClO_2^-]^2}{P_{ClO2}^{\ 2}}$$

Solution: Sn is oxidized so E^o_{red} = – 0.14 V = E^o_{anode}. ClO_2 is reduced so E^o_{red} = 0.95 V= $E^o_{cathode}$. Then E^o_{cell} = $E^o_{cathode}$ – E^o_{anode} = 0.95 V – (– 0.14 V) = + 1.09 V. Equalizing the electrons: Sn (*s*) → Sn^{2+} (*aq*) + 2 e^- and 2 ClO_2 (*g*) + 2 e^- → 2 ClO_2^- (*aq*). Adding the two reactions: Sn (*s*)+ 2 ClO_2 (*g*) + ~~2 e^-~~ → Sn^{2+} (*aq*) + ~~2 e^-~~ + 2 ClO_2^- (*aq*). Cancel the electrons: Sn (*s*)+ 2 ClO_2 (*g*) → Sn^{2+} (*aq*) + 2 ClO_2^- (*aq*). So n = 2 and

$$Q = \frac{[Sn^{2+}][ClO_2^-]^2}{P_{ClO2}^{\ 2}} = \frac{(2.00)(2.00)^2}{(0.100)^2} = 800. \text{ then}$$

$$E_{cell} = E^0_{cell} - \frac{0.0592\ V}{n}\log Q = 1.09\ V - \frac{0.0592\ V}{2}\log 800. = +1.00\ V\ .$$

Check: The units (V) are correct. The ClO_2 reduction reaction is above the Sn^{2+} reduction reaction, so the standard cell potential will be positive. Having more products than reactants lowers the cell potential.

41. **Given:** Zn/Zn^{2+} and Ni/Ni^{2+} half-cells in voltaic cell; initially [Ni^{2+}] = 1.50 M, and [Zn^{2+}] = 0.100 M
Find: a) initial E_{cell}; b) E_{cell} when [Ni^{2+}] = 0.500 M; and c) [Ni^{2+}] and [Zn^{2+}] when E_{cell} = 0.45 V
Conceptual Plan: a) Write 2 half-cell reactions and add electrons as needed to balance reactions. Look up half-reactions in Table 18.1. Calculate the standard cell potential by subtracting the electrode potential of the anode from the electrode potential of the cathode: $E^o_{cell} = E^o_{cathode}$ – E^o_{anode}. Choose the direction of the half-cell reactions so that E^o_{cell} > 0. Add two half-cell reactions and cancel electrons to generate overall reaction. Define Q based on overall reaction. Then E^o_{cell}, [Ni^{2+}], [Zn^{2+}], n → E_{cell}

$$E_{cell} = E^o_{cell} - \frac{0.0592\ V}{n}\log Q$$

b) When [Ni^{2+}] = 0.500 M, then [Zn^{2+}] = 1.100 M (since the stoichiometric coefficients for Ni^{2+}: Zn^{2+} are 1:1, and the [Ni^{2+}] drops by 1.00 M, the other concentration must rise by 1.00 M). Then E^o_{cell}, [Ni^{2+}], [Zn^{2+}], n → E_{cell}

$$E_{cell} = E^o_{cell} - \frac{0.0592\ V}{n}\log Q$$

c) E^o_{cell}, E_{cell}, n → [Zn^{2+}] / [Ni^{2+}] → [Ni^{2+}], [Zn^{2+}]

$$E_{cell} = E^o_{cell} - \frac{0.0592\ V}{n}\log Q \qquad [Ni^{2+}] + [Zn^{2+}] = 1.50\ M + 0.100\ M = 1.60\ M$$

Solution:

a) Zn^{2+} (*aq*) + 2 e^- → Zn (*s*) and Ni^{2+} (*aq*) + 2 e^- → Ni (*s*). Look up cell potentials. For Zn, E^o_{red} = – 0.76 V. For Ni, E^o_{red} = – 0.23 V. In order to get a positive E^o_{cell} Zn is oxidized so E^o_{red} = – 0.76 V = E^o_{anode}. Ni^{2+} is reduced so E^o_{red} = – 0.23 V = $E^o_{cathode}$. Then E^o_{cell} = $E^o_{cathode}$ – E^o_{anode} = – 0.23 V – (– 0.76 V) = + 0.53 V. Adding the two half-cell reactions: Zn (*s*)+ Ni^{2+} (*aq*) + ~~2 e^-~~ → Zn^{2+} (*aq*) + ~~2 e^-~~ + Ni (*s*). The overall reaction is: Zn (*s*) + Ni^{2+} (*aq*) → Zn^{2+} (*aq*) + Ni (*s*). Then $Q = \frac{[Zn^{2+}]}{[Ni^{2+}]} = \frac{0.100}{1.50} = 0.06\underline{6}6667$ and n = 2 then $E_{cell} = E^0_{cell} - \frac{0.0592\ V}{n}\log Q = 0.53\ V - \frac{0.0592\ V}{2}\log 0.06\underline{6}6667 = +0.56\ V$.

b) $Q = \frac{[Zn^{2+}]}{[Ni^{2+}]} = \frac{1.100}{0.500} = 2.20$ then $E_{cell} = E^0_{cell} - \frac{0.0592\ V}{n}\log Q = 0.53\ V - \frac{0.0592\ V}{2}\log 2.20 = +0.52\ V$.

c) $E_{cell} = E^0_{cell} - \frac{0.0592\ V}{n}\log Q$ so $0.45\ V = 0.53\ V - \frac{0.0592\ V}{2}\log Q$ → $0.08\ \cancel{V} = \frac{0.0592\ \cancel{V}}{2}\log Q$ →

$\log Q = 2.\underline{7}0270$ → $Q = 10^{2.\underline{7}0270} = 5\underline{0}4.32$ then $Q = 5\underline{0}4.32 = \frac{[Zn^{2+}]}{1.60M - [Zn^{2+}]}$ solving for [Zn^{2+}]

$(5\underline{0}4.32)(1.60M - [Zn^{2+}]) = [Zn^{2+}]$ → $[Zn^{2+}] = \frac{8\underline{0}6.906\ M}{5\underline{0}5.32} = 1.\underline{5}9628\ M = 1.60\ M$ then

[Ni^{2+}] = 1.60 M – 1.$\underline{5}$9628 M = 0.003 M.

Check: The units (V, V, and V) are correct. The standard cell potential is positive and since there are more reactants than products this raises the cell potential. As the reaction proceeds, reactants are converted to products so the cell potential drops for parts b) and c).

42. **Given:** Pb/Pb^{2+} and Cu/Cu^{2+} half-cells in voltaic cell; initially [Pb^{2+}] = 0.0500 M, and [Cu^{2+}] = 1.50 M
Find: a) initial E_{cell}; b) E_{cell} when [Cu^{2+}] = 0.200 M; and c) [Pb^{2+}] and [Cu^{2+}] when E_{cell} = 0.35 V
Conceptual Plan: a) Write 2 half-cell reactions and add electrons as needed to balance reactions. Look up half-reactions in Table 18.1. Calculate the standard cell potential by subtracting the electrode potential of the anode from the electrode potential of the cathode: $E^\circ_{cell} = E^\circ_{cathode} - E^\circ_{anode}$. Choose the direction of the half-cell reactions so that $E^\circ_{cell} > 0$. Add two half-cell reactions and cancel electrons to generate overall reaction. Define Q based on overall reaction. Then E°_{cell}, [Pb^{2+}], [Cu^{2+}], n → E_{cell}

$$E_{cell} = E^\circ_{cell} - \frac{0.0592\ V}{n}\log Q$$

b) When [Cu^{2+}] = 0.200 M, then [Pb^{2+}] = 1.35 M (since the stoichiometric coefficients for Pb^{2+}: Cu^{2+} are 1:1, and the [Cu^{2+}] drops by 1.30 M, the other concentration must rise by 1.30 M). Then E°_{cell}, [Pb^{2+}], [Cu^{2+}], n → E_{cell}

$$E_{cell} = E^\circ_{cell} - \frac{0.0592\ V}{n}\log Q$$

c) E°_{cell}, E_{cell}, n → [Pb^{2+}] / [Cu^{2+}] → [Pb^{2+}], [Cu^{2+}]

$$E_{cell} = E^\circ_{cell} - \frac{0.0592\ V}{n}\log Q \qquad [Pb^{2+}] + [Cu^{2+}] = 0.0500\ M + 1.50\ M = 1.55\ M$$

Solution: a) Pb^{2+} (*aq*) + 2 e^- → Pb (*s*) and Cu^{2+} (*aq*) + 2 e^- → Cu (*s*). Look up cell potentials. For Pb, E°_{red} = – 0.13 V. For Cu, E°_{red} = + 0.34 V. In order to get a positive E°_{cell} Pb is oxidized so E°_{red} = – 0.13 V = E°_{anode}. Cu^{2+} is reduced so E°_{red} = 0.34 V = $E^\circ_{cathode}$. Then $E^\circ_{cell} = E^\circ_{cathode} - E^\circ_{anode}$ = 0.34 V – (– 0.13 V) = + 0.47 V. Adding the two half-cell reactions: Pb (*s*)+ Cu^{2+} (*aq*) + ~~2 e⁻~~ → Pb^{2+} (*aq*) + ~~2 e⁻~~ + Cu (*s*). The overall reaction is: Pb (*s*) + Cu^{2+} (*aq*) → Pb^{2+} (*aq*) + Cu (*s*). Then $Q = \frac{[Pb^{2+}]}{[Cu^{2+}]} = \frac{0.050}{1.50} = 0.03\underline{3}333$ and $n = 2$ then

$$E_{cell} = E^0_{cell} - \frac{0.0592\ V}{n}\log Q = 0.47\ V - \frac{0.0592\ V}{2}\log 0.03\underline{3}333 = +0.51\ V.$$

b) $Q = \frac{[Pb^{2+}]}{[Cu^{2+}]} = \frac{1.35}{0.200} = 6.75$ then $E_{cell} = E^0_{cell} - \frac{0.0592\ V}{n}\log Q = 0.47\ V - \frac{0.0592\ V}{2}\log 6.75 = +0.45\ V.$

c) $E_{cell} = E^0_{cell} - \frac{0.0592\ V}{n}\log Q$ so $0.35\ V = 0.47\ V - \frac{0.0592\ V}{2}\log Q$ → $0.12\ \cancel{V} = \frac{0.0592\ \cancel{V}}{2}\log Q$ →

$\log Q = 4.\underline{0}540$ → $Q = 10^{4.\underline{0}540} = 1.\underline{1}325 \times 10^4$ then $Q = 1.\underline{1}325 \times 10^4 = \frac{[Pb^{2+}]}{1.55M - [Pb^{2+}]}$ solving for

[Pb^{2+}]. $(1.\underline{1}325 \times 10^4)(1.55M - [Pb^{2+}]) = [Pb^{2+}]$ → $[Pb^{2+}] = \frac{1.\underline{7}554 \times 10^4\ M}{1.\underline{1}335 \times 10^4} = 1.\underline{5}487\ M = 1.5\ M$ then

[Cu^{2+}] = 1.55 M – 1.$\underline{5}$487 M = 0.$\underline{0}$01345 M = 0.0 M.

Check: The units (V, V, and V) are correct. The standard cell potential is positive and since there are more reactants than products this raises the cell potential. As the reaction proceeds, reactants are converted to products so the cell potential drops for parts b) and c).

43. **Given:** Zn/Zn^{2+} concentration cell, with [Zn^{2+}] = 2.0 M in one half-cell and [Zn^{2+}] = 1.0 x 10^{-3} M in other half-cell **Find:** Sketch voltaic cell, labeling anode, cathode, reactions at electrodes, all species, and direction of electron flow
Conceptual Plan: In a concentration cell, the half-cell with the higher concentration is always the half-cell where the reduction takes place (contains the cathode). The 2 half-cell reactions are the same, only reversed. Put anode reaction on the left (oxidation = electrons as product) and cathode reaction on the right (reduction = electrons as reactant). Electrons flow from anode to cathode.

Solution:

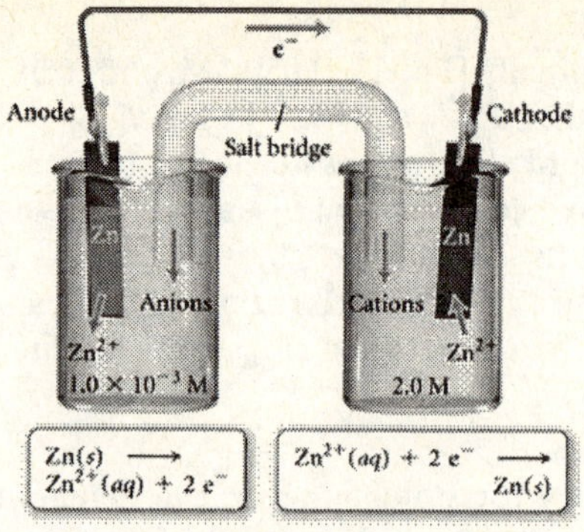

Check: The figure looks similar to the right side of Figure 18.10.

44. **Given:** Pb/Pb^{2+} concentration cell sketch **Find:** a) Label anode and cathode; b) Indicate direction of electron flow; and c) Indicate what happens to $[Pb^{2+}]$ in each half-cell with time
Conceptual Plan: a) In a concentration cell, the half-cell with the higher concentration is always the half-cell where the reduction takes place (contains the cathode). b) Electrons flow from anode to cathode. c) Each half-cell reaction moves forward, so concentration change direction can be determined.
Solution:
c) As the reaction proceeds, the left half-cell (cathode = reduction reaction) will decrease in concentration; and the right half-cell (anode = oxidation reaction) will increase in concentration. Eventually the two concentrations will be the same and the flow of electrons will stop.

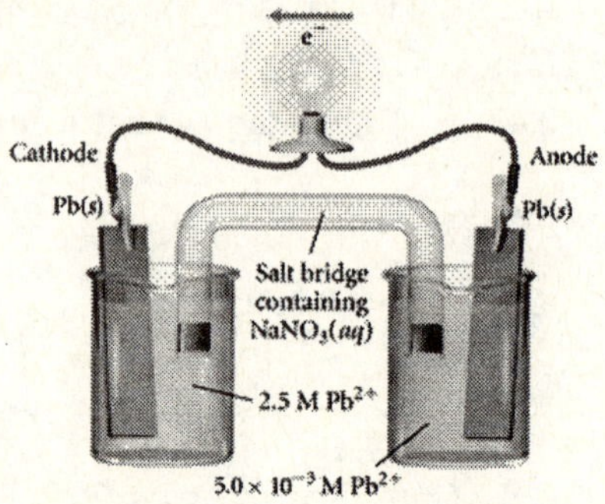

Check: The figure looks similar to the right side of Figure 18.10.

45. **Given:** Sn/Sn^{2+} concentration cell with $E_{cell} = 0.10$ V **Find:** ratio of $[Sn^{2+}]$ in two half-cells
Conceptual Plan: Determine n, then $E^\circ_{cell}, E_{cell}, n \rightarrow Q$ = ratio of $[Sn^{2+}]$ in two half-cells

$$E_{cell} = E^\circ_{cell} - \frac{0.0592\ V}{n}\log Q$$

Solution: Since $Sn^{2+}(aq) + 2\ e^- \rightarrow Sn\ (s)$, $n = 2$. In a concentration cell, $E^\circ_{cell} = 0$ V. So $E_{cell} = E^0_{cell} - \frac{0.0592\ V}{n}\log Q$ so $0.10\ V = 0.00\ V - \frac{0.0592\ V}{2}\log Q \rightarrow 0.10\ \cancel{V} = -\frac{0.0592\ \cancel{V}}{2}\log Q \rightarrow$

$\log Q = -3.\underline{3}784 \rightarrow Q = 10^{-3.3784} = 4.2 \times 10^{-4} = \frac{[Sn^{2+}](ox)}{[Sn^{2+}](red)}$.

Check: The units (none) are correct. Since the concentration in the reduction reaction half-cell is always greater than the concentration in the oxidation half-cell in a voltaic concentration cell, the Q or ratio of two cells is less than 1.

46. **Given:** Cu/Cu^{2+} concentration cell with $E_{cell} = 0.22$ V; and $[Cu^{2+}] = 1.5 \times 10^{-3}$ M **Find:** $[Cu^{2+}]$ in other half-cell
Conceptual Plan: Determine n, then $E^\circ_{cell}, E_{cell}, n \rightarrow Q$ = ratio of $[Cu^{2+}]$ in two half-cells then

$$E_{cell} = E^\circ_{cell} - \frac{0.0592\ V}{n}\log Q$$

$[Cu^{2+}]$ one side, Q $\rightarrow [Cu^{2+}]$ other side

$$Q = \frac{[Cu^{2+}](ox)}{[Cu^{2+}](red)}$$

Solution: Since $Sn^{2+}(aq) + 2\ e^- \rightarrow Sn\ (s)$, $n = 2$. In a concentration cell, $E^\circ_{cell} = 0$ V. So $E_{cell} = E^0_{cell} - \frac{0.0592\ V}{n}\log Q$ so $0.22\ V = 0.00\ V - \frac{0.0592\ V}{2}\log Q \rightarrow 0.22\ \cancel{V} = -\frac{0.0592\ \cancel{V}}{2}\log Q \rightarrow$

$\log Q = -7.\underline{4}324 \rightarrow Q = 10^{-7.4324} = 3.\underline{6}946 \times 10^{-8} = \frac{[Cu^{2+}](ox)}{[Cu^{2+}](red)}$. If we assume that the concentration of the copper solution given is the lower of the two, so $[Cu^{2+}](red) = \frac{[Cu^{2+}](ox)}{Q} = \frac{1.5 \times 10^{-3}\ M}{3.\underline{6}946 \times 10^{-8}} = 4.0 \times 10^{4}\ M$,

which is impossible. So it must be the higher concentration or
$[Cu^{2+}](ox) = [Cu^{2+}](red) \times Q = (1.5 \times 10^{-3}\ M)(3.\underline{6}946 \times 10^{-8}) = 5.5 \times 10^{-11}\ M$, which is low, but possible.

Check: The units (M) are correct. Since the voltage is fairly high for a concentration cell, there must be a very small Q, which leads to an extremely low concentration on the oxidation side.

47. **Given:** alkaline battery **Find:** optimum mass ratio of Zn to MnO_2
Conceptual Plan: Look up alkaline battery reactions. Use stoichiometry to get mole ratio. Then

$Zn\ (s) + 2\ OH^-\ (aq) \rightarrow Zn(OH)_2\ (s) + 2\ e^-$ $\qquad \frac{1\ mol\ Zn}{2\ mol\ MnO_2}$

$2\ MnO_2\ (s) + 2\ H_2O\ (l) + 2\ e^- \rightarrow 2\ MnO(OH)\ (s) + 2\ OH^-\ (aq)$.

mol Zn → g Zn then mol MnO_2 → g MnO_2

$\frac{65.41\ g\ Zn}{1\ mol\ Zn}$ $\qquad \frac{1\ mol\ MnO_2}{86.94\ g\ MnO_2}$

Solution: $\frac{1\ \cancel{mol\ Zn}}{2\ \cancel{mol\ MnO_2}} \times \frac{65.41\ g\ Zn}{1\ \cancel{mol\ Zn}} \times \frac{1\ \cancel{mol\ MnO_2}}{86.94\ g\ MnO_2} = 0.3762 \frac{g\ Zn}{g\ MnO_2}$.

Check: The units (mass ratio) are correct. Since more moles of MnO_2 are needed and the molar mass is larger, the ratio is less than 1.

48. **Given:** lead storage battery, 1.00 g Pb oxidizes **Find:** mass $PbSO_4$
Conceptual Plan:
Look up lead acid battery reactions. Then g Pb → mol Pb → mol $PbSO_4$ → g $PbSO_4$

$Pb\ (s) + HSO_4^-\ (aq) \rightarrow PbSO_4\ (s) + H^+\ (aq) + 2\ e^-$ $\qquad \frac{1\ mol\ Pb}{207.2\ g\ Pb} \quad \frac{2\ mol\ PbSO_4}{1\ mol\ Pb} \quad \frac{303.27\ g\ PbSO_4}{1\ mol\ PbSO_4}$

$PbO_2\ (s) + HSO_4^-\ (aq) + 3\ H^+\ (aq) + 2\ e^- \rightarrow PbSO_4\ (s) + 2\ H_2O\ (l)$.

Solution: $1.00\ \cancel{g\ Pb} \times \frac{1\ \cancel{mol\ Pb}}{207.2\ \cancel{g\ Pb}} \times \frac{2\ \cancel{mol\ PbSO_4}}{1\ \cancel{mol\ Pb}} \times \frac{303.27\ g\ PbSO_4}{1\ \cancel{mol\ PbSO_4}} = 2.93\ g\ PbSO_4$.

Check: The units (g) are correct. Since more moles of $PbSO_4$ are generated and the molar mass is larger, the mass is larger.

49. **Given:** $CH_4\ (g) + 2\ O_2\ (g) \rightarrow CO_2\ (g) + 2\ H_2O\ (g)$ **Find:** E°_{cell}
Conceptual Plan: $\Delta G^0_{rxn} = \sum n_p \Delta G^0_f(products) - \sum n_r \Delta G^0_f(reactants)$ **and determine *n* then ΔG°_{rxn}, *n* → E°_{cell}**

$\Delta G^0_{rxn} = -nFE^0_{cell}$

Solution:

Reactant/Product	ΔG^0_f (kJ/mol from Appendix IIB)
$CH_4\ (g)$	− 50.5
$O_2\ (g)$	0.0
$CO_2\ (g)$	− 394.4
$H_2O\ (g)$	− 228.6

Be sure to pull data for the correct formula and phase.

$$\Delta G^0_{rxn} = \sum n_p \Delta G^0_f(products) - \sum n_r \Delta G^0_f(reactants)$$
$$= [1(\Delta G^0_f(CO_2\ (g))) + 2(\Delta G^0_f(H_2O\ (g)))] - [1(\Delta G^0_f(CH_4\ (g))) + 2(\Delta G^0_f(O_2\ (g)))]$$
$$= [1(-394.4\ kJ) + 2(-228.6\ kJ)] - [1(-50.5\ kJ) + 2(0.0\ kJ)]$$
$$= [-851.6\ kJ] - [-50.5\ kJ]$$
$$= -801.1\ kJ = -8.011 \times 10^5\ J$$

and since one C atom goes from an oxidation state of −4 to +4 and 4 O atoms are going from 0 to −2, then $n = 8$ and $\Delta G^0_{rxn} = -nFE^0_{cell}$. Rearrange to solve for E°_{cell}.

$$E^0_{cell} = \frac{\Delta G^0_{rxn}}{-nF} = \frac{-8.011 \times 10^5 \text{ J}}{-8 \cancel{\text{mol e}^-} \times \frac{96{,}485 \text{ C}}{\cancel{\text{mol e}^-}}} = 1.038 \frac{\text{V}\cancel{\text{C}}}{\cancel{\text{C}}} = 1.038 \text{ V} .$$

Check: The units (V) are correct. The cell voltage is positive which is consistent with a spontaneous reaction.

50. **Given:** CH_3CH_2OH (*g*) + O_2 (*g*) → $HC_2H_3O_2$ (*g*) + H_2O (*g*) **Find:** $E°_{cell}$

Conceptual Plan: $\Delta G^0_{rxn} = \sum n_p \Delta G^0_f(products) - \sum n_r \Delta G^0_f(reactants)$ **and determine *n* then $\Delta G°_{rxn}$, *n* → $E°_{cell}$**

$$\Delta G^0_{rxn} = -nFE^0_{cell}$$

Solution:

Reactant/Product	ΔG^0_f(kJ/mol from Appendix IIB)
CH_3CH_2OH (*g*)	– 167.9
O_2 (*g*)	0.0
$HC_2H_3O_2$ (*g*)	– 374.2
H_2O (*g*)	– 228.6

Be sure to pull data for the correct formula and phase.

$$\Delta G^0_{rxn} = \sum n_p \Delta G^0_f(products) - \sum n_r \Delta G^0_f(reactants)$$
$$= [1(\Delta G^0_f(HC_2H_3O_2\ (g))) + 1(\Delta G^0_f(H_2O\ (g)))] - [1(\Delta G^0_f(CH_3CH_2OH\ (g))) + 1(\Delta G^0_f(O_2\ (g)))]$$
$$= [1(-374.2 \text{ kJ}) + 1(-228.6 \text{ kJ})] - [1(-167.9 \text{ kJ}) + 1(0.0 \text{ kJ})]$$
$$= [-602.8 \text{ kJ}] - [-167.9 \text{ kJ}]$$
$$= -434.9 \text{ kJ} = -4.349 \times 10^5 \text{ J}$$

and since 2 O atoms are going from 0 to –2 so n = 4 then $\Delta G^0_{rxn} = -nFE^0_{cell}$. Rearrange to solve for $E°_{cell}$.

$$E^0_{cell} = \frac{\Delta G^0_{rxn}}{-nF} = \frac{-4.349 \times 10^5 \text{ J}}{-4 \cancel{\text{mol e}^-} \times \frac{96{,}485 \text{ C}}{\cancel{\text{mol e}^-}}} = 1.127 \frac{\text{V}\cancel{\text{C}}}{\cancel{\text{C}}} = 1.127 \text{ V} .$$

Check: The units (V) are correct. The cell voltage is positive which is consistent with a spontaneous reaction.

51. In order for a metal to be able to protect iron, it must, more easily oxidized than iron or be below it in Table 18.1. a) Zn and c) Mn meet this criterion.

52. In order for a metal to be able to protect iron, it must, more easily oxidized than iron or be below it in Table 18.1. a) Mg and b) Cr meet this criterion.

53. **Given:** electrolytic cell sketch **Find:** a) Label anode and cathode and indicate half-reactions; b) Indicate direction of electron flow; and c) Label battery terminals and calculate minimum voltage to drive reaction
Conceptual Plan: a) Write 2 half-cell reactions and add electrons as needed to balance reactions. Look up half-reactions in Table 18.1. Calculate the standard cell potential by subtracting the electrode potential of the anode from the electrode potential of the cathode: $E°_{cell} = E°_{cathode} - E°_{anode}$. Choose the direction of the half-cell reactions so that $E°_{cell} < 0$. b) Electrons flow from anode to cathode. c) Each half-cell reaction moves forward, so concentration change direction can be determined.
Solution:

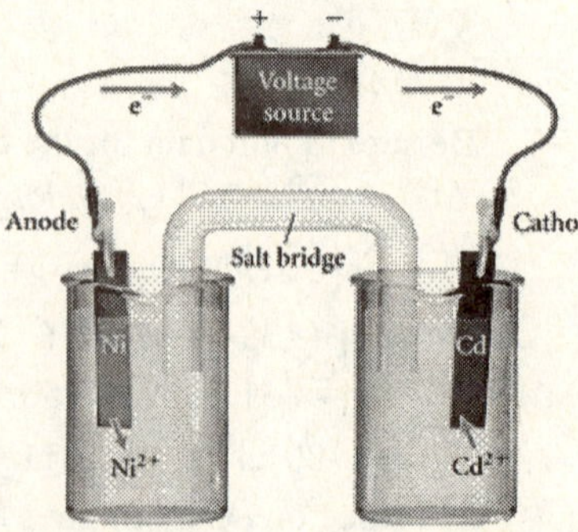

a) Ni^{2+} (*aq*) + 2 e^- → Ni (*s*) and Cd^{2+} (*aq*) + 2 e^- → Cd (*s*). Look up cell potentials. For Ni, $E°_{red}$ = – 0.23 V. For Cd, $E°_{red}$ = – 0.40 V. In order to get a negative cell potential, Ni is oxidized so $E°_{red}$ = – 0.23 V = $E°_{anode}$. Cd^{2+} is reduced so $E°_{red}$ = – 0.23 V = $E°_{cathode}$. Then $E°_{cell} = E°_{cathode} - E°_{anode}$ = – 0.40 V – (– 0.23 V) = – 0.17 V. Since oxidation occurs at the anode, the Ni is the anode and the reaction is Ni (*s*) → Ni^{2+} (*aq*) + 2 e^-. Since reduction takes place at the cathode, Cd is the cathode and the reaction is Cd^{2+} (*aq*) + 2 e^- → Cd (*s*).

c) Since reduction is occurring at the cathode, the battery terminal closest to the cathode is the negative terminal. Since the cell potential from part a) is = – 0.17 V, a minimum of 0.17 V must be applied by the battery.
Check: The reaction is nonspontaneous, since the reduction of Ni^{2+} is above Cd^{2+}. Electrons still flow from the anode to the cathode. The reaction can be made spontaneous with the application of electrical energy.

54. **Given:** electrolytic cell Mn^{2+} reduced to Mn and Sn oxidized to Sn^{2+} **Find:** Draw cell labeling anode and cathode; write half-reactions; indicate direction of electron flow; and calculate minimum voltage to drive reaction

Conceptual Plan: Write 2 half-cell reactions and add electrons as needed to balance reactions. Look up half-reactions in Table 18.1. Calculate the standard cell potential by subtracting the electrode potential of the anode from the electrode potential of the cathode: $E^\circ_{cell} = E^\circ_{cathode} - E^\circ_{anode}$. Anode is where oxidation occurs, so Sn is anode and Mn is cathode. Electrons flow from anode to cathode. The minimum potential needed is E°_{cell}.

Solution: Mn^{2+} (*aq*) + 2 e^- → Mn (*s*) and Sn^{2+} (*aq*) + 2 e^- → Sn (*s*). Look up cell potentials. For Mn^{2+}, $E^\circ_{red} = -1.18$ V. For Sn, $E^\circ_{red} = -0.14$ V. Since Sn is oxidized, $E^\circ_{red} = -0.14 \text{ V} = E^\circ_{anode}$. Then $E^\circ_{cell} = E^\circ_{cathode} - E^\circ_{anode} = -1.18 \text{ V} - (-0.14 \text{ V}) = -1.04$ V. Since oxidation occurs at the anode, the Sn is the anode and the reaction is Sn (*s*) → Sn^{2+} (*aq*) + 2 e^-. Since reduction takes place at the cathode, Mn is the cathode and the reaction is Mn^{2+} (*aq*) + 2 e^- → Mn (*s*). Electrons flow from the anode to the cathode. Since the cell potential is – 1.04 V, a minimum of 1.04 V must be applied by the battery.

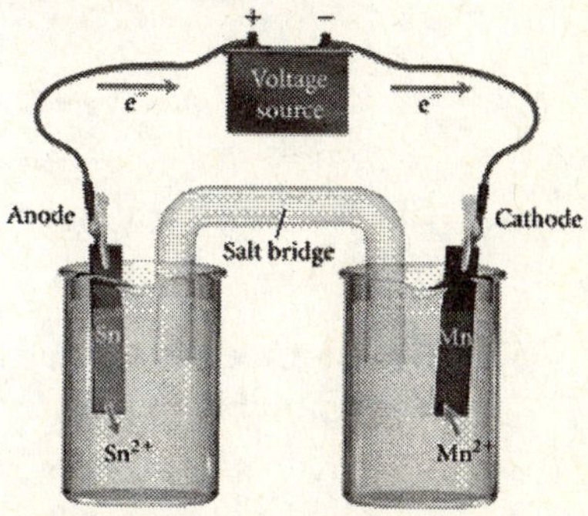

Check: The reaction is nonspontaneous, since the reduction of Sn^{2+} is above Mn^{2+}. Electrons still flow from the anode to the cathode. The reaction can be made spontaneous with the application of electrical energy.

55. **Given:** electrolysis cell to electroplate Cu onto a metal surface **Find:** Draw cell, labeling anode and cathode; and write half-reactions

Conceptual Plan: Write 2 half-cell reactions and add electrons as needed to balance reactions. The cathode reaction will be the reduction of Cu^{2+} to the metal. The anode will be the reverse reaction.

Solution:

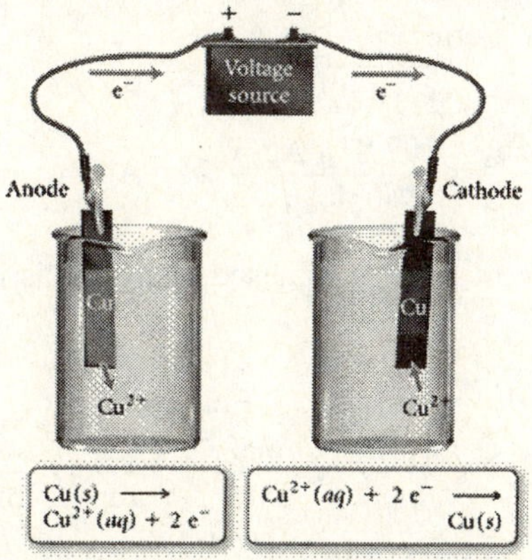

Check: The metal to be plated is the cathode, since metal ions are converted to Cu (*s*) on the surface of the metal.

56. **Given:** electrolysis cell to electroplate Ni onto a metal surface **Find:** Draw cell, labeling anode and cathode; and write half-reactions

Conceptual Plan: Write 2 half-cell reactions and add electrons as needed to balance reactions. The cathode reaction will be the reduction of Ni^{2+} to the metal. The anode will be the reverse reaction.

Solution:

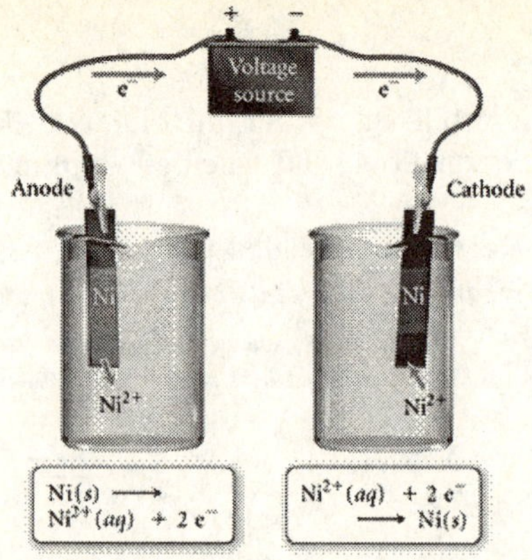

Check: The metal to be plated is the cathode, since metal ions are converted to Ni (*s*) on the surface of the metal.

57. **Given:** Cu electroplating of 225 mg Cu at a current of 7.8 A; Cu^{2+} (*aq*) + 2 e^- → Cu (*s*) **Find:** time
Conceptual Plan: mg Cu → g Cu → mol Cu → mol e^- → C → s

$$\frac{1\text{ g}}{1000\text{ mg}} \quad \frac{1\text{ mol Cu}}{63.55\text{ g Cu}} \quad \frac{2\text{ mol e}^-}{1\text{ mol Cu}} \quad \frac{96{,}485\text{ C}}{1\text{ mol e}^-} \quad \frac{1\text{ s}}{7.8\text{ C}}$$

Solution: $$225\text{ mg Cu} \times \frac{1\text{ g Cu}}{1000\text{ mg Cu}} \times \frac{1\text{ mol Cu}}{63.55\text{ g Cu}} \times \frac{2\text{ mol e}^-}{1\text{ mol Cu}} \times \frac{96{,}485\text{ C}}{1\text{ mol e}^-} \times \frac{1\text{ s}}{7.8\text{ C}} = 88\text{ s}.$$

Check: The units (s) are correct. Since far less than a mole of Cu is electroplated, the time is short.

58. **Given:** Ag electroplating at a current of 5.8 A for 55 min; Ag^+ (*aq*) + e^- → Ag (*s*) **Find:** mass of Ag
Conceptual Plan: min → s → C → mol e^- → mol Ag → g Ag

$$\frac{60\text{ s}}{1\text{ min}} \quad \frac{5.8\text{ C}}{1\text{ s}} \quad \frac{1\text{ mol e}^-}{96{,}485\text{ C}} \quad \frac{1\text{ mol Ag}}{1\text{ mol e}^-} \quad \frac{107.87\text{ g Ag}}{1\text{ mol Ag}}$$

Solution: $$55\text{ min} \times \frac{60\text{ s}}{1\text{ min}} \times \frac{5.8\text{ C}}{1\text{ s}} \times \frac{1\text{ mol e}^-}{96{,}485\text{ C}} \times \frac{1\text{ mol Ag}}{1\text{ mol e}^-} \times \frac{107.87\text{ g Ag}}{1\text{ mol Ag}} = 21\text{ g Ag}$$

Check: The units (g) are correct. Since less than a mole of electrons is used, the mass is less than the molar mass of Ag.

59. **Given:** Na electrolysis, 1.0 kg in one hour **Find:** current

Conceptual Plan: Na^+ (*l*) + e^- → Na (*l*) $\frac{\text{kg Na}}{\text{hr}} \rightarrow \frac{\text{g Na}}{\text{hr}} \rightarrow \frac{\text{mol Na}}{\text{hr}} \rightarrow \frac{\text{mol e}^-}{\text{hr}} \rightarrow \frac{\text{C}}{\text{hr}} \rightarrow \frac{\text{C}}{\text{min}} \rightarrow \frac{\text{C}}{\text{s}}$

$$\frac{1000\text{ g}}{1\text{ kg}} \quad \frac{1\text{ mol Na}}{22.99\text{ g Na}} \quad \frac{1\text{ mol e}^-}{1\text{ mol Na}} \quad \frac{96{,}485\text{ C}}{1\text{ mol e}^-} \quad \frac{1\text{ hr}}{60\text{ min}} \quad \frac{1\text{ min}}{60\text{ s}}$$

Solution:

$$\frac{1.0\text{ kg Na}}{1\text{ hr}} \times \frac{1000\text{ g Na}}{1\text{ kg Na}} \times \frac{1\text{ mol Na}}{22.99\text{ g Na}} \times \frac{1\text{ mol e}^-}{1\text{ mol Na}} \times \frac{96{,}485\text{ C}}{1\text{ mol e}^-} \times \frac{1\text{ hr}}{60\text{ min}} \times \frac{1\text{ min}}{60\text{ s}} = 1.2 \times 10^3\ \frac{\text{C}}{\text{s}} = 1.2 \times 10^3\text{ A}$$

Check: The units (A) are correct. Since the amount per hour is so large we expect a very large current.

60. **Given:** Al electrolysis at a current of 25 A for 1 hour; Al^{3+} (*aq*) + 3 e^- → Al (*s*) **Find:** mass of Al
Conceptual Plan: hr → min → s → C → mol e^- → mol Al → g Al

$$\frac{60\text{ min}}{1\text{ hr}} \quad \frac{60\text{ s}}{1\text{ min}} \quad \frac{25\text{ C}}{1\text{ s}} \quad \frac{1\text{ mol e}^-}{96{,}485\text{ C}} \quad \frac{1\text{ mol Al}}{3\text{ mol e}^-} \quad \frac{26.98\text{ g Al}}{1\text{ mol Al}}$$

Solution: $$1\text{ hr} \times \frac{60\text{ min}}{1\text{ hr}} \times \frac{60\text{ s}}{1\text{ min}} \times \frac{25\text{ C}}{1\text{ s}} \times \frac{1\text{ mol e}^-}{96{,}485\text{ C}} \times \frac{1\text{ mol Al}}{3\text{ mol e}^-} \times \frac{26.98\text{ g Al}}{1\text{ mol Al}} = 8.4\text{ g Al}.$$

Check: The units (g) are correct. Since three moles of electrons are used per mole of Al^{3+}, the mass is less than the molar mass of Al.

61. **Given:** MnO_4^- (*aq*) + Zn (*s*) → Mn^{2+} (*aq*) + Zn^{2+} (*aq*) 0.500 M $KMnO_4$ and 2.85 g Zn
Find: balance equation and volume $KMnO_4$ solution

Conceptual Plan: Separate the overall reaction into two half-reactions: one for oxidation and one for reduction. → Balance each half-reaction with respect to mass in the following order: 1) balance all elements other than H and O; 2) balance O by adding H_2O; and 3) balance H by adding H^+. → Balance each half-reaction with respect to charge by adding electrons. (The sum of the charges on both sides of the equation should be made equal by adding electrons as necessary.) → Make the number of electrons in both half-reactions equal by multiplying one or both half-reactions by a small whole number. → Add the two half-reactions together, canceling electrons and other species as necessary. → Verify that the reaction is balanced both with respect to mass and with respect to charge. Then

g Zn → mol Zn → mol MnO_4^- → L MnO_4^- → mL MnO_4^-

$$\frac{1\text{ mol Zn}}{65.41\text{ g Zn}} \quad \frac{2\text{ mol }MnO_4^-}{5\text{ mol Zn}} \quad \frac{1\text{ L }MnO_4^-}{0.500\text{ mol }MnO_4^-} \quad \frac{1000\text{ mL }MnO_4^-}{1\text{ L }MnO_4^-}$$

Solution:

Separate: $MnO_4^-(aq) \rightarrow Mn^{2+}(aq)$ and $Zn(s) \rightarrow Zn^{2+}(aq)$

Balance non H & O elements: $MnO_4^-(aq) \rightarrow Mn^{2+}(aq)$ and $Zn(s) \rightarrow Zn^{2+}(aq)$

Balance O with H_2O: $MnO_4^-(aq) \rightarrow Mn^{2+}(aq) + 4\,H_2O(l)$ and $Zn(s) \rightarrow Zn^{2+}(aq)$

Balance H with H^+: $MnO_4^-(aq) + 8\,H^+(aq) \rightarrow Mn^{2+}(aq) + 4\,H_2O(l)$ and $Zn(s) \rightarrow Zn^{2+}(aq)$

Add electrons: $MnO_4^-(aq) + 8\,H^+(aq) + 5\,e^- \rightarrow Mn^{2+}(aq) + 4\,H_2O(l)$ and $Zn(s) \rightarrow Zn^{2+}(aq) + 2\,e^-$

Equalize electrons:

$2\,MnO_4^-(aq) + 16\,H^+(aq) + 10\,e^- \rightarrow 2\,Mn^{2+}(aq) + 8\,H_2O(l)$ and $5\,Zn(s) \rightarrow 5\,Zn^{2+}(aq) + 10\,e^-$

Add half-reactions:

$2\,MnO_4^-(aq) + 16\,H^+(aq) + \cancel{10\,e^-} + 5\,Zn(s) \rightarrow 2\,Mn^{2+}(aq) + 8\,H_2O(l) + 5\,Zn^{2+}(aq) + \cancel{10\,e^-}$

Cancel electrons: $2\,MnO_4^-(aq) + 16\,H^+(aq) + 5\,Zn(s) \rightarrow 2\,Mn^{2+}(aq) + 8\,H_2O(l) + 5\,Zn^{2+}(aq)$

$$2.85\,\cancel{\text{g Zn}} \times \frac{1\,\cancel{\text{mol Zn}}}{65.41\,\cancel{\text{g Zn}}} \times \frac{2\,\cancel{\text{mol }MnO_4^-}}{5\,\cancel{\text{mol Zn}}} \times \frac{1\,\cancel{\text{L }MnO_4^-}}{0.500\,\cancel{\text{mol }MnO_4^-}} \times \frac{1000\text{ mL }MnO_4^-}{1\,\cancel{\text{L }MnO_4^-}} = 34.9\text{ mL }MnO_4^- =$$

$= 34.9\text{ mL }KMnO_4$.

Check:

Reactants	Products
2 Mn atoms	2 Mn atoms
8 O atoms	8 O atoms
16 H atoms	16 H atoms
5 Zn atoms	5 Zn atoms
+14 charge	+14 charge

The units (mL) are correct. Since far less than a mole of zinc is used, less than a mole of permanganate is consumed, so the volume is less than a liter.

62. **Given:** $Cr_2O_7^{2-}(aq) + Cu(s) \rightarrow Cr^{3+}(aq) + Cu^{2+}(aq)$ 0.850 M $K_2Cr_2O_7$ and 5.25 g Zn
Find: balance equation and volume $K_2Cr_2O_7$ solution
Conceptual Plan: Separate the overall reaction into two half-reactions: one for oxidation and one for reduction. → Balance each half-reaction with respect to mass in the following order: 1) balance all elements other than H and O; 2) balance O by adding H_2O; and 3) balance H by adding H^+. → Balance each half-reaction with respect to charge by adding electrons. (The sum of the charges on both sides of the equation should be made equal by adding electrons as necessary.) → Make the number of electrons in both half-reactions equal by multiplying one or both half-reactions by a small whole number. → Add the two half-reactions together, canceling electrons and other species as necessary. → Verify that the reaction is balanced both with respect to mass and with respect to charge.

then g Cu → mol Cu → mol $Cr_2O_7^{2-}$ → L $Cr_2O_7^{2-}$ → mL $Cr_2O_7^{2-}$

$$\frac{1\text{ mol Cu}}{63.55\text{ g Cu}} \quad \frac{1\text{ mol }Cr_2O_7^{2-}}{3\text{ mol Cu}} \quad \frac{1\text{ L }Cr_2O_7^{2-}}{0.850\text{ mol }Cr_2O_7^{2-}} \quad \frac{1000\text{ mL }Cr_2O_7^{2-}}{1\text{ L }Cr_2O_7^{2-}}$$

Solution:

Separate: $Cr_2O_7^{2-}(aq) \rightarrow Cr^{3+}(aq)$ and $Cu(s) \rightarrow Cu^{2+}(aq)$

Balance non H & O elements: $Cr_2O_7^{2-}(aq) \rightarrow 2\,Cr^{3+}(aq)$ and $Cu(s) \rightarrow Cu^{2+}(aq)$

Balance O with H_2O: $Cr_2O_7^{2-}(aq) \rightarrow 2\,Cr^{3+}(aq) + 7\,H_2O(l)$ and $Cu(s) \rightarrow Cu^{2+}(aq)$

Balance H with H^+: $Cr_2O_7^{2-}(aq) + 14\,H^+(aq) \rightarrow 2\,Cr^{3+}(aq) + 7\,H_2O(l)$ and $Cu(s) \rightarrow Cu^{2+}(aq)$

Add electrons: $Cr_2O_7^{2-}(aq) + 14\,H^+(aq) + 6\,e^- \rightarrow 2\,Cr^{3+}(aq) + 7\,H_2O(l)$ and $Cu(s) \rightarrow Cu^{2+}(aq) + 2e^-$

Equalize electrons:

$Cr_2O_7^{2-}(aq) + 14\,H^+(aq) + 6\,e^- \rightarrow 2\,Cr^{3+}(aq) + 7\,H_2O(l)$ and $3\,Cu(s) \rightarrow 3\,Cu^{2+}(aq) + 6\,e^-$

Add half-reactions:

$Cr_2O_7^{2-}\,(aq) + 14\,H^+\,(aq) + \cancel{6\,e^-} + 3\,Cu\,(s) \rightarrow 2\,Cr^{3+}\,(aq) + 7\,H_2O\,(l) + 3\,Cu^{2+}\,(aq) + \cancel{6\,e^-}$

Cancel electrons: $Cr_2O_7^{2-}\,(aq) + 14\,H^+\,(aq) + 3\,Cu\,(s) \rightarrow 2\,Cr^{3+}\,(aq) + 7\,H_2O\,(l) + 3\,Cu^{2+}\,(aq)$

$$5.25\,\cancel{g\,Cu} \times \frac{1\,\cancel{mol\,Cu}}{63.55\,\cancel{g\,Cu}} \times \frac{1\,\cancel{mol\,Cr_2O_7^{2-}}}{3\,\cancel{mol\,Cu}} \times \frac{1\,\cancel{L\,Cr_2O_7^{2-}}}{0.850\,\cancel{mol\,Cr_2O_7^{2-}}} \times \frac{1000\,mL\,Cr_2O_7^{2-}}{1\,\cancel{L\,Cr_2O_7^{2-}}} = 32.4\,mL\,Cr_2O_7^{2-} =$$

$= 32.4\,mL\,K_2Cr_2O_7$.

Check:

Reactants	Products
2 Cr atoms	2 Cr atoms
7 O atoms	7 O atoms
14 H atoms	14 H atoms
3 Cu atoms	3 Cu atoms
+12 charge	+12 charge

The units (mL) are correct. Since far less than a mole of copper is used, less than a mole of dichromate is consumed, so the volume is less than a liter.

63. **Given:** beaker with Al strip and Zn^{2+} ions **Find:** draw sketch after Al is submerged for a few minutes
Conceptual Plan: Write 2 half-cell reactions and add electrons as needed to balance reactions. Look up half-reactions in Table 18.1. Calculate the standard cell potential by subtracting the electrode potential of the anode from the electrode potential of the cathode: $E^\circ_{cell} = E^\circ_{cathode} - E^\circ_{anode}$. If $E^\circ_{cell} > 0$ the reaction is spontaneous in the forward direction and Al will dissolve and Cu will deposit.
Solution: $Al\,(s) \rightarrow Al^{3+}\,(aq)$ and $Cu^{2+}\,(aq) \rightarrow Cu\,(s)$ add electrons $Al\,(s) \rightarrow Al^{3+}\,(aq) + 3\,e^-$ and $Cu^{2+}\,(aq) + 2\,e^- \rightarrow Cu\,(s)$. Look up cell potentials. Al is oxidized so $E^\circ_{ox} = -E^\circ_{red} = -(-1.66\,V) = 1.66\,V$. Cu^{2+} is reduced so $E^\circ_{red} = 0.34\,V$. Then $E^\circ_{cell} = E^\circ_{ox} + E^\circ_{red} = 1.66\,V + 0.34\,V = +2.00\,V$ and so the reaction is spontaneous. Al will dissolve to generate $Al^{3+}\,(aq)$ and Cu (*s*) will deposit.

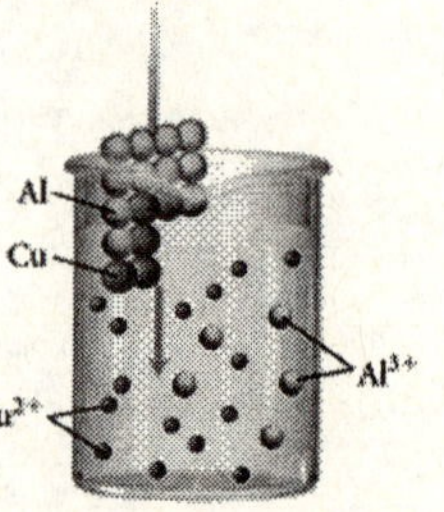

Check: The units (V) are correct. If the voltage is positive, the reaction is spontaneous so Al will dissolve and Cu will deposit.

64. **Given:** Zn/Zn^{2+} and Ni/Ni^{2+} half-cells in voltaic cell **Find:** draw sketch after substantial amount of current generated
Conceptual Plan: Write 2 half-cell reactions and add electrons as needed to balance reactions. Look up half-reactions in Table 18.1. Calculate the standard cell potential by subtracting the electrode potential of the anode from the electrode potential of the cathode: $E^\circ_{cell} = E^\circ_{cathode} - E^\circ_{anode}$. Choose the direction of the half-cell reactions so that $E^\circ_{cell} > 0$. Add two half-cell reactions and cancel electrons to generate overall reaction. Since reaction is spontaneous it will move forward.
Solution: $Zn^{2+}\,(aq) + 2\,e^- \rightarrow Zn\,(s)$ and $Ni^{2+}\,(aq) + 2\,e^- \rightarrow Ni\,(s)$. Look up cell potentials. For Zn, $E^\circ_{red} = -0.76\,V$. For Ni, $E^\circ_{red} = -0.23\,V$. In order to get a positive E°_{cell} the sign of the Zn potential must be reversed. Zn is oxidized so $E^\circ_{ox} = -E^\circ_{red} = -(-0.76\,V) = 0.76\,V$. Ni^{2+} is reduced so $E^\circ_{red} = -0.23\,V$. Then $E^\circ_{cell} = E^\circ_{ox} + E^\circ_{red} = 0.76\,V - 0.23\,V = +0.53\,V$. Adding the two half-cell reactions: $Zn\,(s) + Ni^{2+}\,(aq) + \cancel{2\,e^-} \rightarrow Zn^{2+}\,(aq) + \cancel{2\,e^-} + Ni\,(s)$. The overall reaction is: $Zn\,(s) + Ni^{2+}\,(aq) \rightarrow Zn^{2+}\,(aq) + Ni\,(s)$. Zn will dissolve to generate $Zn^{2+}\,(aq)$ and Ni (*s*) will deposit.

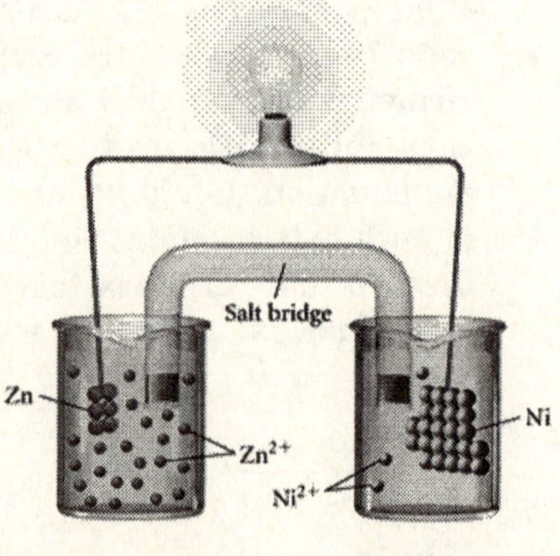

Check: The units (V) are correct. Since Ni is above Zn in Table 18.1, Ni is reduced. If the voltage is positive, the reaction is spontaneous so Zn will dissolve and Ni will deposit.

65. **Given:** a) 2.15 g Al; b) 4.85 g Cu; and c) 2.42 g Ag in 3.5 M HI
Find: if metal dissolves, write balanced reaction and minimum amount of HI needed to dissolve metal
Conceptual Plan: In general, metals whose reduction half-reactions lie below the reduction of H^+ to H_2 in Table 18.1 will dissolve in acids, while metals above it will not. Stop here if metal does not dissolve. To write the balanced redox reactions, pair the oxidation of the metal with the reduction of H^+ to H_2 (2

$H^+(aq) + 2\ e^- \rightarrow H_2(g)$). **Balance the number of electrons transferred. Add the two reactions. Cancel electrons. Then g metal → mol metal → mol H⁺ → L HI → mL HI**

$$\mathfrak{m} \quad \frac{x\ \text{mol}\ H^+}{y\ \text{mol metal}} \quad \frac{1\ \text{L HI}}{3.5\ \text{mol HI}} \quad \frac{1000\ \text{mL HI}}{1\ \text{L HI}}$$

Solution: a) Al meets this criterion. For Al, Al (s) → Al^{3+} (aq) + 3 e^-. We need to multiply the Al reaction by 2 and the H^+ reaction by 3. So, 2 Al (s) → 2 Al^{3+} (aq) + 6 e^- and 6 H^+ (aq) + 6 e^- → 3 H_2 (g). Adding the half-reactions together: 2 Al (s) + 6 H^+ (aq) + $\cancel{6e^-}$ → 2 Al^{3+} (aq) + $\cancel{6e^-}$ + 3 H_2 (g). Simplify to 2 Al (s) + 6 H^+ (aq) → 2 Al^{3+} (aq) + 3 H_2 (g). Then

$$2.15\ \cancel{\text{g Al}} \times \frac{1\ \cancel{\text{mol Al}}}{26.98\ \cancel{\text{g Al}}} \times \frac{6\ \cancel{\text{mol H}^+}}{2\ \cancel{\text{mol Al}}} \times \frac{1\ \cancel{\text{L HI}}}{3.5\ \cancel{\text{mol HI}}} \times \frac{1000\ \text{mL HI}}{1\ \cancel{\text{L HI}}} = 68.3\ \text{mL HI}$$

b) Cu does not meet this criterion, so it will not dissolve in HI.

c) Ag does not meet this criterion, so it will not dissolve in HI.

Check: Only metals with negative reduction potentials will dissolve. The volume of acid needed is fairly small since the amount of metal is much less than 1 mole and the concentration of acid is high.

66. **Given:** a) 5.90 g Au; b) 2.55 g Cu; and c) 4.83 g Sn in 6.0 M HNO_3
Find: if metal dissolves, write balanced reaction and minimum amount of HNO_3 needed to dissolve metal
Conceptual Plan: Nitric acid (HNO_3) oxidizes metals through the following reduction half-reaction: NO_3^- (aq) + 4 H^+ (aq) + 3 e^- → NO (g) + 2 H_2O (l) E°_{red} = 0.96 V. Since this half-reaction is above the reduction of H^+ in Table 18.1, HNO_3 can oxidize metals that can't be oxidizied by HCl.
Solution:

a) Au (which has a reduction potential of 1.50 V) will not be oxidized, so it will not dissolve in HNO_3.

b) Cu will be oxidized (which has a reduction potential of 0.34 V). To write the balanced redox reactions, pair the oxidation of the copper (Cu (s) → Cu^{2+} (aq) + 2 e^-) with the reduction of nitric acid (NO_3^- (aq) + 4 H^+ (aq) + 3 e^- → NO (g) + 2 H_2O (l)). In order to balance the number of electrons transferred we need to multiply the Cu reaction by 3 and the nitric acid reaction by 2. So, 3 Cu (s) → 3 Cu^{2+} (aq) + 6 e^- and 2 NO_3^- (aq) + 8 H^+ (aq) + 6 e^- → 2 NO (g) + 4 H_2O (l). Adding the two reactions: 3 Cu (s)+ 2 NO_3^- (aq) + 8 H^+ (aq) + $\cancel{6e^-}$ → 3 Cu^{2+} (aq) + $\cancel{6e^-}$ + 2 NO (g) + 4 H_2O (l). Simplify to 3 Cu (s)+ 2 NO_3^- (aq) + 8 H^+ (aq) → 3 Cu^{2+} (aq) + 2 NO (g) + 4 H_2O (l). Then

$$2.55\ \cancel{\text{g Cu}} \times \frac{1\ \cancel{\text{mol Cu}}}{63.55\ \cancel{\text{g Cu}}} \times \frac{8\ \cancel{\text{mol H}^+}}{3\ \cancel{\text{mol Cu}}} \times \frac{1\ \cancel{\text{L HNO}_3}}{6.0\ \cancel{\text{mol HNO}_3}} \times \frac{1000\ \text{mL HNO}_3}{1\ \cancel{\text{L HNO}_3}} = 18\ \text{mL HNO}_3$$

Use stoichiometric coefficient of H^+ since it is larger than the NO_3^- stoichiometric coefficient.

c) Sn will be oxidized (which has a reduction potential of – 0.14 V). To write the balanced redox reactions, pair the oxidation of the tin (Sn (s) → Sn^{2+} (aq) + 2 e^-) with the reduction of nitric acid (NO_3^- (aq) + 4 H^+ (aq) + 3 e^- → NO (g) + 2 H_2O (l)). In order to balance the number of electrons transferred we need to multiply the Sn reaction by 3 and the nitric acid reaction by 2. So, 3 Sn (s) → 3 Sn^{2+} (aq) + 6 e^- and 2 NO_3^- (aq) + 8 H^+ (aq) + 6 e^- → 2 NO (g) + 4 H_2O (l). Adding the two reactions: 3 Sn (s)+ 2 NO_3^- (aq) + 8 H^+ (aq) + $\cancel{6e^-}$ → 3 Sn^{2+} (aq) + $\cancel{6e^-}$ + 2 NO (g) + 4 H_2O (l). Simplify to 3 Sn (s)+ 2 NO_3^- (aq) + 8 H^+ (aq) → 3 Sn^{2+} (aq) + 2 NO (g) + 4 H_2O (l). Then

$$4.83\ \cancel{\text{g Sn}} \times \frac{1\ \cancel{\text{mol Sn}}}{118.71\ \cancel{\text{g Sn}}} \times \frac{8\ \cancel{\text{mol H}^+}}{3\ \cancel{\text{mol Sn}}} \times \frac{1\ \cancel{\text{L HNO}_3}}{6.0\ \cancel{\text{mol HNO}_3}} \times \frac{1000\ \text{mL HNO}_3}{1\ \cancel{\text{L HNO}_3}} = 18\ \text{mL HNO}_3.$$

Use stoichiometric coefficient of H^+ since it is larger than the NO_3^- stoichiometric coefficient.

Check: Only metals with reduction potentials less than 0.96 V will dissolve. The volume of acid needed is fairly small since the amount of metal is much less than 1 mole and the concentration of acid is high.

67. **Given:** Pt (s)|H_2 (g, 1 atm)|H^+ (aq, ? M)||Cu^{2+} (aq, 1.0 M)|Cu (s), E_{cell} = 355 mV **Find:** pH
Conceptual Plan: Write half-reactions from line notation. Look up half-reactions in Table 18.1. Calculate the standard cell potential by subtracting the electrode potential of the anode from the electrode potential of the cathode: $E^\circ_{cell} = E^\circ_{cathode} - E^\circ_{anode}$. Add the two half-cell reactions and cancel

the electrons. Then mV → V E^o_{cell}, E_{cell}, P_{H2}, $[Cu^{2+}]$, n → $[H^+]$ → pH

$\frac{1\text{ V}}{1000\text{ mV}}$ $\quad E_{cell} = E^o_{cell} - \frac{0.0592\ V}{n}\log Q \quad pH = -\log[H^+]$

Solution: The half-reactions are: $H_2(g) \rightarrow 2\,H^+(aq) + 2\,e^-$ and $Cu^{2+}(aq) + 2\,e^- \rightarrow Cu(s)$. H is oxidized so $E^o_{red} = -0.00\text{ V} = E^o_{anode}$. Cu is reduced so $E^o_{red} = 0.34\text{ V} = E^o_{cathode}$. Then $E^o_{cell} = E^o_{cathode} - E^o_{anode} = 0.34\text{ V} - (-0.00\text{ V}) = +0.34\text{ V}$. Adding the two reactions: $H_2(g) + Cu^{2+}(aq) + \cancel{2e^-} \rightarrow 2\,H^+(aq) + \cancel{2e^-} + Cu(s)$.

Cancel the electrons: $H_2(g) + Cu^{2+}(aq) \rightarrow 2\,H^+(aq) + Cu(s)$. Then $355\ \cancel{\text{mV}} \times \frac{1\text{ V}}{1000\ \cancel{\text{mV}}} = 0.355\text{ V}$. So n

$= 2$ and $Q = \frac{[H^+]^2}{P_{H2}[Cu^{2+}]} = \frac{(x)^2}{(1)(1.0)} = x^2$ then $E_{cell} = E^0_{cell} - \frac{0.0592\text{ V}}{n}\log Q$ substitute in values and solve for

x. $\quad 0.355\text{ V} = 0.34\text{ V} - \frac{0.0592\text{ V}}{2}\log x^2 \rightarrow 0.015\ \cancel{\text{V}} = -\frac{0.0592\ \cancel{\text{V}}}{2}\log x^2 \rightarrow -0.50676 = \log x^2 \rightarrow$

$x^2 = 10^{-0.50676} = 0.31135 \rightarrow x = 0.55798$ then $pH = -\log[H^+] = -\log[0.55798] = 0.25338 = 0.3$.

Check: The units (none) are correct. The pH is acidic, which is consistent with dissolving a metal in acid.

68. **Given:** $Pt(s)|H_2(g, 1\text{ atm})|H^+(aq, 1.0\text{ M})||Au^{3+}(aq, ?\text{ M})|Au(s)$, $E_{cell} = 1.22\text{ V}$ **Find:** $[Au^{3+}]$
Conceptual Plan: Write half-reactions from line notation. Look up half-reactions in Table 18.1. Calculate the standard cell potential by subtracting the electrode potential of the anode from the electrode potential of the cathode: $E^o_{cell} = E^o_{cathode} - E^o_{anode}$. Equalize the number of electrons transferred and then add the two half-cell reactions and cancel the electrons.
Then E^o_{cell}, E_{cell}, P_{H2}, $[H^+]$, n → $[Au^{3+}]$

$E_{cell} = E^o_{cell} - \frac{0.0592\ V}{n}\log Q$

Solution: The half-reactions are: $H_2(g) \rightarrow 2\,H^+(aq) + 2\,e^-$ and $Au^{3+}(aq) + 3\,e^- \rightarrow Au(s)$. H is oxidized so $E^o_{red} = -0.00\text{ V} = E^o_{anode}$. Au is reduced so $E^o_{red} = 1.50\text{ V} = E^o_{cathode}$. Then $E^o_{cell} = E^o_{cathode} - E^o_{anode} = 1.50\text{ V} - (-0.00\text{ V}) = +1.50\text{ V}$. Equalize the number of electrons transferred by multiplying the first reaction by 3 and the second reaction by 2 so $3\,H_2(g) \rightarrow 6\,H^+(aq) + 6\,e^-$ and $2\,Au^{3+}(aq) + 6\,e^- \rightarrow 2\,Au(s)$. Adding the two reactions: $3\,H_2(g) + 2\,Au^{3+}(aq) + \cancel{6e^-} \rightarrow 6\,H^+(aq) + \cancel{6e^-} + 2\,Au(s)$. Cancel the electrons: $3\,H_2(g) + 2\,Au^{3+}(aq) \rightarrow 6\,H^+(aq) + 2\,Au(s)$. So $n = 6$ and $Q = \frac{[H^+]^6}{P^3_{H2}[Au^{3+}]^2} = \frac{(1.0)^6}{(1)^3(x)^2} = x^{-2}$ then

$E_{cell} = E^0_{cell} - \frac{0.0592\text{ V}}{n}\log Q$ substitute in values and solve for x. $\quad 1.22\text{ V} = 1.50\text{ V} - \frac{0.0592\text{ V}}{6}\log x^{-2} \rightarrow$

$0.28\ \cancel{\text{V}} = \frac{0.0592\ \cancel{\text{V}}}{6}\log x^{-2} \rightarrow 28.3784 = \log x^{-2} \rightarrow x^{-2} = 10^{28.3784} = 2.3899 \times 10^{28} \rightarrow x = 6.4686 \times 10^{-15}\text{ M}$.

Check: The units (none) are correct. The concentration is expected to be low because the cell potential must be reduced.

69. **Given:** Mg oxidation and Cu^{2+} reduction; initially $[Mg^{2+}] = 1.0 \times 10^{-4}$ M and $[Cu^{2+}] = 1.5$ M in 1.0 L half-cells
Find: a) initial E_{cell}; b) E_{cell} after 5.0 A for 8.0 hr; and c) how long can battery deliver 5.0A
Conceptual Plan: a) Write the 2 half-cell reactions and add electrons as needed to balance reactions. Look up half-reactions in Table 18.1. Calculate the standard cell potential by subtracting the electrode potential of the anode from the electrode potential of the cathode: $E^o_{cell} = E^o_{cathode} - E^o_{anode}$. Add the two half-cell reactions and cancel electrons and determine n. Then E^o_{cell}, $[Mg^{2+}]$, $[Cu^{2+}]$, n → E_{cell}

$E_{cell} = E^o_{cell} - \frac{0.0592\ V}{n}\log Q$

b) hr → min → s → C → mol e⁻ → mol Cu reduced → $[Cu^{2+}]$ and

$\frac{60\text{ min}}{1\text{ hr}} \quad \frac{60\text{ s}}{1\text{ min}} \quad \frac{5.0\text{ C}}{1\text{ s}} \quad \frac{1\text{ mol e}^-}{96{,}485\text{ C}} \quad \frac{1\text{ mol Cu}^{2+}}{2\text{ mol e}^-} \quad$ since V = 1.0L $\quad [Cu^{2+}] = [Cu^{2+}] - \frac{\text{mol Cu}^{2+}\text{ reduced}}{1.0\text{ L}}$

mol Cu reduced → mol Mg oxidized → $[Mg^{2+}]$

$\frac{1\text{ mol Mg oxidized}}{1\text{ mol Cu}^{2+}\text{ reduced}} \quad$ since V = 1.0L $\quad [Mg^{2+}] = [Mg^{2+}] + \frac{\text{mol Mg oxidized}}{1.0\text{ L}}$

c) $[Cu^{2+}] \rightarrow$ **mol** $e^- \rightarrow$ **C** $\rightarrow$ **s** $\rightarrow$ **min** $\rightarrow$ **hr**

$$\frac{1\text{ mol e}^-}{2\text{ mol Cu}^{2+}} \quad \frac{96{,}485\text{ C}}{1\text{ mol e}^-} \quad \frac{1\text{ s}}{5.0\text{ C}} \quad \frac{1\text{ min}}{60\text{ s}} \quad \frac{1\text{ hr}}{60\text{ min}}$$

Solution: a) write half-reactions and add electrons Cu^{2+} (*aq*) + 2 $e^- \rightarrow$ Cu (*s*) and Mg (*s*) $\rightarrow Mg^{2+}$ (*aq*) + 2 e^-. Look up cell potentials. Mg is oxidized so $E^\circ_{red} = -2.37$ V = E°_{anode}. Cu^{2+} is reduced so $E^\circ_{red} = 0.34$ V = $E^\circ_{cathode}$. Then $E^\circ_{cell} = E^\circ_{cathode} - E^\circ_{anode} = 0.34\text{ V} - (-2.37\text{ V}) = +2.71$ V. Add the two half-cell reactions: Cu^{2+} (*aq*) + $\cancel{2e^-}$ + Mg (*s*) $\rightarrow$ Cu (*s*) + Mg^{2+} (*aq*)+ $\cancel{2e^-}$. Simplify to Cu^{2+} (*aq*) + Mg (*s*) $\rightarrow$ Cu (*s*) + Mg^{2+} (*aq*).

So $Q = \dfrac{[Mg^{2+}]}{[Cu^{2+}]} = \dfrac{1.0 \times 10^{-4}}{1.5} = 6.\underline{6}667 \times 10^{-5}$ and $n = 2$ then

$$E_{cell} = E^0_{cell} - \frac{0.0592\text{ V}}{n}\log Q = 2.71\text{ V} - \frac{0.0592\text{ V}}{2}\log 6.\underline{6}667 \times 10^{-5} = +2.8\underline{3}361\text{ V} = +2.83\text{ V}.$$

b) $$8.0\,\cancel{\text{hr}} \times \frac{60\,\cancel{\text{min}}}{1\,\cancel{\text{hr}}} \times \frac{60\,\cancel{\text{s}}}{1\,\cancel{\text{min}}} \times \frac{5.0\,\cancel{\text{C}}}{1\,\cancel{\text{s}}} \times \frac{1\,\cancel{\text{mol e}^-}}{96{,}485\,\cancel{\text{C}}} \times \frac{1\text{ mol Cu}^{2+}}{2\,\cancel{\text{mol e}^-}} = 0.7\underline{4}623\text{ mol Cu}^{2+}$$ and

$$[Cu^{2+}] = [Cu^{2+}] - \frac{\text{mol Cu}^{2+}\text{ reduced}}{1.0\text{ L}} = 1.5\text{ M} - \frac{0.7\underline{4}623\text{ mol Cu}^{2+}}{1.0\text{ L}} = 0.7\underline{5}377\text{ M Cu}^{2+}$$ and

$$0.7\underline{4}623\,\cancel{\text{mol Cu}^{2+}} \times \frac{1\text{ mol Mg oxidized}}{1\,\cancel{\text{mol Cu}^{2+}\text{ reduced}}} = 0.7\underline{4}623\text{ mol Mg oxidized}$$ and

$$[Mg^{2+}] = [Mg^{2+}] + \frac{\text{mol Mg oxidized}}{1.0\text{ L}} = 1.0 \times 10^{-4}\ M + \frac{0.7\underline{4}623\text{ mol Mg oxidized}}{1.0\text{ L}} = 0.7\underline{4}633\text{ M Mg}^{2+}$$

$Q = \dfrac{[Mg^{2+}]}{[Cu^{2+}]} = \dfrac{0.7\underline{4}633}{0.7\underline{5}377} = 0.9\underline{9}013$ and $n = 2$ then

$$E_{cell} = E^0_{cell} - \frac{0.0592\text{ V}}{n}\log Q = 2.71\text{ V} - \frac{0.0592\text{ V}}{2}\log 0.9\underline{9}013 = +2.7\underline{1}013\text{ V} = +2.71\text{ V}.$$

c) In 1.0 L there are initially 1.5 moles of Cu^{2+}. So

$$1.5\,\cancel{\text{mol Cu}^{2+}} \times \frac{2\,\cancel{\text{mol e}^-}}{1\,\cancel{\text{mol Cu}^{2+}}} \times \frac{96{,}485\,\cancel{\text{C}}}{1\,\cancel{\text{mol e}^-}} \times \frac{1\,\cancel{\text{s}}}{5.0\,\cancel{\text{C}}} \times \frac{1\,\cancel{\text{min}}}{60\,\cancel{\text{s}}} \times \frac{1\text{ hr}}{60\,\cancel{\text{min}}} = 16\text{ hr}$$

Check: The units (V, V, and hr) are correct. The Cu^{2+} reduction reaction is above the Mg^{2+} reduction reaction, so the standard cell potential will be positive. Having more reactants than products increases the cell potential. As the reaction proceeds the potential drops. The concentrations drop by ½ in 8 hours (part b), so it is all consumed in 16 hours.

70. **Given:** Ag/Ag^+ concentration cell; initially $[Ag^+] = 1.25$ M and 1.0×10^{-3} M in 2.0 L half-cells
Find: a) how long can battery deliver 2.5A; b) mass of Ag plated after 3.5 A for 5.5 hr; and c) how long can battery deliver 5. 0A to redissolve 1.00×10^2 g Ag with 10.0 A
Conceptual Plan: **a) Write the 2 half-cell reactions and add electrons as needed to balance reactions. Then** $[Ag^+]$**, V** $\rightarrow$ **mol** $Ag^+ \rightarrow$ **mol** $e^- \rightarrow$ **C** $\rightarrow$ **s** $\rightarrow$ **min** $\rightarrow$ **hr**

$$M = \frac{\text{mol Ag}^+}{\text{L}} \quad \frac{1\text{ mol e}^-}{1\text{ mol Ag}^+} \quad \frac{96{,}485\text{ C}}{1\text{ mol e}^-} \quad \frac{1\text{ s}}{2.5\text{ C}} \quad \frac{1\text{ min}}{60\text{ s}} \quad \frac{1\text{ hr}}{60\text{ min}}$$

b) hr $\rightarrow$ **min** $\rightarrow$ **s** $\rightarrow$ **C** $\rightarrow$ **mol** $e^- \rightarrow$ **mol Ag** $\rightarrow$ **g Ag**

$$\frac{60\text{ min}}{1\text{ hr}} \quad \frac{60\text{ s}}{1\text{ min}} \quad \frac{3.5\text{ C}}{1\text{ s}} \quad \frac{1\text{ mol e}^-}{96{,}485\text{ C}} \quad \frac{1\text{ mol Ag}}{1\text{ mol e}^-} \quad \frac{107.87\text{ g Ag}}{1\text{ mol Ag}}$$

c) g Ag $\rightarrow$ **mol Ag** $\rightarrow$ **mol** $e^- \rightarrow$ **C** $\rightarrow$ **s** $\rightarrow$ **min** $\rightarrow$ **hr**

$$\frac{1\text{ mol Ag}}{107.87\text{ g Ag}} \quad \frac{1\text{ mol e}^-}{1\text{ mol Ag}} \quad \frac{96{,}485\text{ C}}{1\text{ mol e}^-} \quad \frac{1\text{ s}}{10.0\text{ C}} \quad \frac{1\text{ min}}{60\text{ s}} \quad \frac{1\text{ hr}}{60\text{ min}}$$

Solution:

a) write half-reactions and add electrons Ag^+ (*aq*) + $e^- \rightarrow$ Ag (*s*). $M = \dfrac{\text{mol Ag}^+}{\text{L}}$ so

$\text{mol Ag}^+ = \text{M} \times \text{L} = 1.25 \frac{\text{mol Ag}^+}{\cancel{\text{L}}} \times 2.0\ \cancel{\text{L}} = 2.50 \text{ mol Ag}^+$ then

$$2.50\ \cancel{\text{mol Ag}^+} \times \frac{1\ \cancel{\text{mol e}^-}}{1\ \cancel{\text{mol Ag}^+}} \times \frac{96{,}485\ \cancel{\text{C}}}{1\ \cancel{\text{mol e}^-}} \times \frac{1\ \cancel{\text{s}}}{2.5\ \cancel{\text{C}}} \times \frac{1\ \cancel{\text{min}}}{60\ \cancel{\text{s}}} \times \frac{1 \text{ hr}}{60\ \cancel{\text{min}}} = 27 \text{ hr}$$

b) $$5.5\ \cancel{\text{hr}} \times \frac{60\ \cancel{\text{min}}}{1\ \cancel{\text{hr}}} \times \frac{60\ \cancel{\text{s}}}{1\ \cancel{\text{min}}} \times \frac{3.5\ \cancel{\text{C}}}{1\ \cancel{\text{s}}} \times \frac{1\ \cancel{\text{mol e}^-}}{96{,}485\ \cancel{\text{C}}} \times \frac{1\ \cancel{\text{mol Ag}}}{1\ \cancel{\text{mol e}^-}} \times \frac{107.87 \text{ g Ag}}{1\ \cancel{\text{mol Ag}}} = 77 \text{ g Ag}$$

c) $$1.00 \times 10^2\ \cancel{\text{g Ag}} \times \frac{1\ \cancel{\text{mol Ag}}}{107.87\ \cancel{\text{g Ag}}} \times \frac{1\ \cancel{\text{mol e}^-}}{1\ \cancel{\text{mol Ag}}} \times \frac{96{,}485\ \cancel{\text{C}}}{1\ \cancel{\text{mol e}^-}} \times \frac{1 \text{ s}}{10.0\ \cancel{\text{C}}} \times \frac{1\ \cancel{\text{min}}}{60\ \cancel{\text{s}}} \times \frac{1 \text{ hr}}{60\ \cancel{\text{min}}} = 2.48 \text{ hr}$$

Check: The units (hr, g, and hr) are correct. Since 2.5 moles needs to be plated, we expect it to take a long time. The time is shorter by a factor of ~30 and the current is larger by about 1.5 so we expect less Ag to be plated in part b) as compared to part a). We expect a shorter time in part c) than in part b) because the current is so much larger. In parts b) and c) we are not exhausting the 2.5 moles initially in the half-cell.

71. **Given:** Cu (*s*)|CuI (*s*)|I^- (*aq*, 1.0 M)||Cu^+ (*aq*, 1.0 M)|Cu (*s*), K_{sp} (CuI) = 1.1×10^{-12} **Find:** E_{cell}
Conceptual Plan: Write half-reactions from line notation. Since this is a concentration cell E°_{cell} = 0.00 V. Then K_{sp}, [I^-] → [Cu^+](ox) then E°_{cell}, [Cu^+](ox), [Cu^+](red), n → E_{cell}

$K_{sp} = [Cu^+][I^-]$ $\qquad E_{cell} = E^\circ_{cell} - \frac{0.0592\ V}{n} \log Q$

Solution: The half-reactions are: Cu (*s*) → $Cu^+(aq) + e^-$ and Cu^+ (*aq*) + e^- → Cu (*s*). Since this is a concentration cell E°_{cell} = 0.00 V and n = 1. Since $K_{sp} = [Cu^+][I^-]$, rearrange to solve for [Cu^+](ox).

$$[Cu^+]\,(ox) = \frac{K_{sp}}{[I^-]} = \frac{1.1 \times 10^{-12}}{1.0} = 1.1 \times 10^{-12} \text{ M then } Q = \frac{[Cu^+](ox)}{[Cu^+](red)} = \frac{1.1 \times 10^{-12}}{1.0} = 1.1 \times 10^{-12} \text{ then}$$

$$E_{cell} = E^0_{cell} - \frac{0.0592\ V}{n} \log Q = 0.00\ V - \frac{0.0592\ V}{1} \log\,(1.1 \times 10^{-12}) = 0.71\ V.$$

Check: The units (V) are correct. Since [Cu^+](ox) is so low and [Cu^+](red) is high, the Q is very small so the voltage increase is significant.

72. **Given:** $Zn(OH)_2$ (*s*) + 2 e^- → Zn (*s*) + 2 OH^- (*aq*), K_{sp} ($Zn(OH)_2$) = 1.8×10^{-14} **Find:** E for half-cell
Conceptual Plan: Look up half-reaction for the Zn reduction (E°_{red}) in Table 18.1, determine n, then K_{sp} → [OH^-] then E°_{cell}, [OH^-], n → E_{cell}

$K_{sp} = [Zn^{2+}][OH^-]^2$ $\qquad E_{cell} = E^\circ_{cell} - \frac{0.0592\ V}{n} \log Q$

Solution: The reduction half-reaction is: Zn^{2+} (*aq*) + 2 e^- → Zn (*s*) and E°_{red} = – 0.76 V and n = 2. Since $K_{sp} = [Zn^{2+}][OH^-]^2 = S(2S)^2 = 4\,S^3$, rearrange to solve for S = [OH^-].

$$S = [OH^-] = \sqrt[3]{\frac{K_{sp}}{4}} = \sqrt[3]{\frac{1.8 \times 10^{-14}}{4}} = 1.\underline{6}5096 \times 10^{-5} \text{ M then } Q = [OH^-]^2 = (1.\underline{6}5096 \times 10^{-5})^2 = 2.\underline{7}257 \times 10^{-10}$$

$$\text{then } E_{cell} = E^0_{cell} - \frac{0.0592\ V}{n} \log Q = -0.76\ V - \frac{0.0592\ V}{2} \log\,(2.\underline{7}257 \times 10^{-10}) = -0.48\ V.$$

Check: The units (V) are correct. Since [OH^-] is so low, the Q is very small so the voltage increase is significant, but the half-reaction is nonspontanous.

73. **Given:** a) disproportionation of Mn^{2+} (*aq*) to Mn (*s*) and MnO_2 (*s*); and b) disproportionation of MnO_2 (*s*) to Mn^{2+} (*aq*) and MnO_4^- (*s*) in acidic solution **Find:** ΔG°_{rxn} and K
Conceptual Plan: Separate the overall reaction into two half-reactions: one for oxidation and one for reduction. → Balance each half-reaction with respect to mass in the following order: 1) balance all elements other than H and O; 2) balance O by adding H_2O; and 3) balance H by adding H^+. → Balance each half-reaction with respect to charge by adding electrons. (The sum of the charges on both sides of the equation should be made equal by adding electrons as necessary.) → Make the number of electrons in both half-reactions equal by multiplying one or both half-reactions by a small whole number. → Add the two half-reactions together, canceling electrons and other species as necessary. → Verify that the reaction is balanced both with respect to mass and with respect to charge. Look up half-reactions in Table 18.1. Calculate the standard cell potential by subtracting the electrode potential of the anode from the electrode potential of the cathode: $E^\circ_{cell} = E^\circ_{cathode} - E^\circ_{anode}$.

Then calculate ΔG°_{rxn} using $\Delta G^0_{rxn} = -nFE^0_{cell}$. Finally °C → K then ΔG°_{rxn}, T → K

$K = 273.15 + °C$ $\qquad\qquad$ $\Delta G^0_{rxn} = -RT\ln K$

Solution:

a) Separate: $Mn^{2+}(aq) \rightarrow MnO_2(s)$ and $Mn^{2+}(aq) \rightarrow Mn(s)$

Balance non H & O elements: $Mn^{2+}(aq) \rightarrow MnO_2(s)$ and $Mn^{2+}(aq) \rightarrow Mn(s)$

Balance O with H_2O: $Mn^{2+}(aq) + 2\,H_2O(l) \rightarrow MnO_2(s)$ and $Mn^{2+}(aq) \rightarrow Mn(s)$

Balance H with H^+: $Mn^{2+}(aq) + 2\,H_2O(l) \rightarrow MnO_2(s) + 4\,H^+(aq)$ and $Mn^{2+}(aq) \rightarrow Mn(s)$

Add electrons: $Mn^{2+}(aq) + 2\,H_2O(l) \rightarrow MnO_2(s) + 4\,H^+(aq) + 2\,e^-$ and $Mn^{2+}(aq) + 2\,e^- \rightarrow Mn(s)$

Equalize electrons: $Mn^{2+}(aq) + 2\,H_2O(l) \rightarrow MnO_2(s) + 4\,H^+(aq) + 2\,e^-$ and $Mn^{2+}(aq) + 2\,e^- \rightarrow Mn(s)$

Add half-reactions: $Mn^{2+}(aq) + 2\,H_2O(l) + Mn^{2+}(aq) + \cancel{2e^-} \rightarrow MnO_2(s) + 4\,H^+(aq) + \cancel{2e^-} + Mn(s)$

Cancel electrons: $2\,Mn^{2+}(aq) + 2\,H_2O(l) \rightarrow MnO_2(s) + 4\,H^+(aq) + Mn(s)$

Look up cell potentials. Mn is oxidized in the first half-cell reaction so $E^\circ_{ox} = -E^\circ_{red} = -\ 1.21$ V. Mn is reduced in the second half-cell reaction so $E^\circ_{red} = -\ 1.18$ V. Then $E^\circ_{cell} = E^\circ_{ox} + E^\circ_{red} = -\ 1.21\text{ V} - 1.18\text{ V}$ $= -\ 2.39$ V. $n = 2$ so $\quad \Delta G^0_{rxn} = -nFE^0_{cell} = -2\ \cancel{\text{mol e}^-} \times \dfrac{96{,}485\text{ C}}{\cancel{\text{mol e}^-}} \times -2.39\text{ V}$

$= -2 \times 96{,}485\ \cancel{\text{C}} \times -2.39\ \dfrac{\text{J}}{\cancel{\text{C}}} = 4.61198 \times 10^5\text{ J} = 461\text{ kJ}$ and $T = 273.15 + 25\text{ °C} = 298\text{ K}$ then

$\Delta G^0_{rxn} = -RT\ln K$ Rearrange to solve for K.

$$K = e^{\frac{-\Delta G^0_{rxn}}{RT}} = e^{\frac{-4.61198 \times 10^5\ \cancel{\text{J}}}{\left(8.314\frac{\cancel{\text{J}}}{\cancel{\text{K}}\text{ mol}}\right)(298\ \cancel{\text{K}})}} = e^{-186.149} = 1.43 \times 10^{-81}$$

Check:

	Reactants	Products
	2 Mn atoms	2 Mn atoms
	2 O atoms	2 O atoms
	4 H atoms	4 H atoms
	+4 charge	+4 charge

The units (kJ and none) are correct. If the voltage is negative, the reaction is nonspontaneous and the free energy change is very positive and the equilibrium constant is extremely small.

b) Separate: $MnO_2(s) \rightarrow Mn^{2+}(aq)$ and $MnO_2(s) \rightarrow MnO_4^-(aq)$

Balance non H & O elements: $MnO_2(s) \rightarrow Mn^{2+}(aq)$ and $MnO_2(s) \rightarrow MnO_4^-(aq)$

Balance O with H_2O: $MnO_2(s) \rightarrow Mn^{2+}(aq) + 2\,H_2O(l)$ and $MnO_2(s) + 2\,H_2O(l) \rightarrow MnO_4^-(aq)$

Balance H with H^+:

$MnO_2(s) + 4\,H^+(aq) \rightarrow Mn^{2+}(aq) + 2\,H_2O(l)$ and $MnO_2(s) + 2\,H_2O(l) \rightarrow MnO_4^-(aq) + 4\,H^+(aq)$

Add electrons: $MnO_2(s) + 4\,H^+(aq) + 2\,e^- \rightarrow Mn^{2+}(aq) + 2\,H_2O(l)$ and $MnO_2(s) + 2\,H_2O(l) \rightarrow$ $MnO_4^-(aq) + 4\,H^+(aq) + 3e^-$

Equalize electrons: $3\,MnO_2(s) + 12\,H^+(aq) + 6\,e^- \rightarrow 3\,Mn^{2+}(aq) + 6\,H_2O(l)$ and $2\,MnO_2(s) + 4\,H_2O(l) \rightarrow 2\,MnO_4^-(aq) + 8\,H^+(aq) + 6\,e^-$

Add half-reactions: $3\,MnO_2(s) + 4\,\cancel{12\,H^+(aq)} + \cancel{6e^-} + 2\,MnO_2(s) + \cancel{4\,H_2O(l)} \rightarrow$

$3\,Mn^{2+}(aq) + 2\,\cancel{4}\,H_2O(l) + 2\,MnO_4^-(aq) + \cancel{8\,H^+(aq)} + \cancel{6e^-}$

Cancel electrons & species: $5\,MnO_2(s) + 4\,H^+(aq) \rightarrow 3\,Mn^{2+}(aq) + 2\,H_2O(l) + 2\,MnO_4^-(aq)$

Look up cell potentials. Mn is reduced in the first half-cell reaction so $E^\circ_{red} = 1.21$ V. Mn is oxidized in the second half-cell reaction so $E^\circ_{ox} = -E^\circ_{red} = -\ 1.68$ V. Then $E^\circ_{cell} = E^\circ_{ox} + E^\circ_{red} = 1.21\text{ V} - 1.68\text{ V} = -$ 0.47 V. $n = 6$ so $\Delta G^0_{rxn} = -nFE^0_{cell} = -6\ \cancel{\text{mol e}^-} \times \dfrac{96{,}485\text{ C}}{\cancel{\text{mol e}^-}} \times -0.47\text{ V} = -6 \times 96{,}485\ \cancel{\text{C}} \times -0.47\ \dfrac{\text{J}}{\cancel{\text{C}}}$

$= 2.7209 \times 10^5\text{ J} = 270\text{ kJ} = 2.7 \times 10^2\text{ kJ}$. $T = 273.15 + 25\text{ °C} = 298\text{ K}$ then $\Delta G^0_{rxn} = -RT\ln K$.

Rearrange to solve for K. $$K = e^{\frac{-\Delta G^0_{rxn}}{RT}} = e^{\frac{-2.7209 \times 10^5\ \cancel{\text{J}}}{\left(8.314\frac{\cancel{\text{J}}}{\cancel{\text{K}}\cdot\text{ mol}}\right)(298\ \cancel{\text{K}})}} = e^{-10.982} = 2.0 \times 10^{-48}$$

Check:

	Reactants	Products
	5 Mn atoms	5 Mn atoms
	10 O atoms	10 O atoms
	4 H atoms	4 H atoms

+4 charge +4 charge

The units (kJ and none) are correct. If the voltage is negative, the reaction is nonspontaneous and the free energy change is very positive and the equilibrium constant is extremely small. The voltage is less than in part a) so the free energy change is not as large and the equilibrium constant is not as small.

74. **Given:** a) reaction of Cr^{2+} *(aq)* with $Cr_2O_7^{2-}$ *(aq)* in acidic solution to form Cr^{3+} *(aq)*; and b) reaction of Cr^{3+} *(aq)* with Cr *(s)* to form Cr^{2+} *(aq)* **Find:** ΔG°_{rxn} and K

Conceptual Plan: Separate the overall reaction into two half-reactions: one for oxidation and one for reduction. → Balance each half-reaction with respect to mass in the following order: 1) balance all elements other than H and O; 2) balance O by adding H_2O; and 3) balance H by adding H^+. → Balance each half-reaction with respect to charge by adding electrons. (The sum of the charges on both sides of the equation should be made equal by adding electrons as necessary.) → Make the number of electrons in both half-reactions equal by multiplying one or both half-reactions by a small whole number. → Add the two half-reactions together, canceling electrons and other species as necessary. → Verify that the reaction is balanced both with respect to mass and with respect to charge. Look up half-reactions in Table 18.1. Calculate the standard cell potential by subtracting the electrode potential of the anode from the electrode potential of the cathode: $E^\circ_{cell} = E^\circ_{cathode} - E^\circ_{anode}$.

Then calculate ΔG°_{rxn} using $\Delta G^0_{rxn} = -nFE^0_{cell}$. Finally °C → K then ΔG°_{rxn}, T → K

$K = 273.15 + °C$ $\quad\quad$ $\Delta G^0_{rxn} = -RT\ln K$

Solution:

a) Separate: $Cr^{2+}(aq) \rightarrow Cr^{3+}(aq)$ and $Cr_2O_7^{2-}(aq) \rightarrow Cr^{3+}(aq)$

Balance non H & O elements: $Cr^{2+}(aq) \rightarrow Cr^{3+}(aq)$ and $Cr_2O_7^{2-}(aq) \rightarrow 2\,Cr^{3+}(aq)$

Balance O with H_2O: $Cr^{2+}(aq) \rightarrow Cr^{3+}(aq)$ and $Cr_2O_7^{2-}(aq) \rightarrow 2\,Cr^{3+}(aq) + 7\,H_2O(l)$

Balance H with H^+: $Cr^{2+}(aq) \rightarrow Cr^{3+}(aq)$ and $Cr_2O_7^{2-}(aq) + 14\,H^+(aq) \rightarrow 2\,Cr^{3+}(aq) + 7\,H_2O(l)$

Add electrons: $Cr^{2+}(aq) \rightarrow Cr^{3+}(aq) + e^-$ and $Cr_2O_7^{2-}(aq) + 14\,H^+(aq) + 6\,e^- \rightarrow 2\,Cr^{3+}(aq) + 7\,H_2O(l)$

Equalize electrons:

$6\,Cr^{2+}(aq) \rightarrow 6\,Cr^{3+}(aq) + 6\,e^-$ and $Cr_2O_7^{2-}(aq) + 14\,H^+(aq) + 6\,e^- \rightarrow 2\,Cr^{3+}(aq) + 7\,H_2O(l)$

Add half-reactions:

$6\,Cr^{2+}(aq) + Cr_2O_7^{2-}(aq) + 14\,H^+(aq) + \cancel{6\,e^-} \rightarrow 6\,Cr^{3+}(aq) + \cancel{6\,e^-} + 2\,Cr^{3+}(aq) + 7\,H_2O(l)$

Cancel electrons: $6\,Cr^{2+}(aq) + Cr_2O_7^{2-}(aq) + 14\,H^+(aq) \rightarrow 8\,Cr^{3+}(aq) + 7\,H_2O(l)$

Look up cell potentials. Cr is oxidized in the first half-cell reaction so $E^\circ_{red} = -0.50\text{ V} = E^\circ_{anode}$. Cr is reduced in the second half-cell reaction so $E^\circ_{red} = 1.33\text{ V} = E^\circ_{cathode}$. Then $E^\circ_{cell} = E^\circ_{cathode} - E^\circ_{anode} = 1.33\text{ V} - (-0.50\text{ V}) = +1.83\text{ V}$. $n = 6$ so $\Delta G^0_{rxn} = -nFE^0_{cell} = -6\,\cancel{\text{mol e}^-} \times \dfrac{96{,}485\text{ C}}{\cancel{\text{mol e}^-}} \times 1.83\text{ V}$

$= -6 \times 96{,}485\,\cancel{\text{C}} \times 1.83\,\dfrac{\text{J}}{\cancel{\text{C}}} = -1.0\underline{5}941 \times 10^6\text{ J} = -1.06 \times 10^3\text{ kJ}$ $T = 273.15 + 25\text{ °C} = 298\text{ K}$ then

$\Delta G^0_{rxn} = -RT\ln K$ Rearrange to solve for K.

$$K = e^{\frac{-\Delta G^0_{rxn}}{RT}} = e^{\frac{-(-1.0\underline{5}941 \times 10^6\,\cancel{\text{J}})}{\left(8.314\frac{\cancel{\text{J}}}{\cancel{\text{K}}\,\text{mol}}\right)(298\,\cancel{\text{K}})}} = e^{42\underline{7}.598} = 5.05 \times 10^{185}.$$

Check:

	Reactants	Products
	8 Cr atoms	8 Cr atoms
	7 O atoms	7 O atoms
	14 H atoms	14 H atoms
	+24 charge	+24 charge

The units (kJ and none) are correct. If the voltage is positive, the reaction is spontaneous and the free energy change is very negative and the equilibrium constant is extremely large.

b) Separate: $Cr^{3+}(aq) \rightarrow Cr^{2+}(aq)$ and $Cr(s) \rightarrow Cr^{2+}(aq)$

Balance: $Cr^{3+}(aq) \rightarrow Cr^{2+}(aq)$ and $Cr(s) \rightarrow Cr^{2+}(aq)$

Add electrons: $Cr^{3+}(aq) + e^- \rightarrow Cr^{2+}(aq)$ and $Cr(s) \rightarrow Cr^{2+}(aq) + 2\,e^-$

Equalize electrons: $2\,Cr^{3+}(aq) + 2\,e^- \rightarrow 2\,Cr^{2+}(aq)$ and $Cr(s) \rightarrow Cr^{2+}(aq) + 2\,e^-$

Add half-reactions: $2\,Cr^{3+}(aq) + \cancel{2\,e^-} + Cr(s) \rightarrow 2\,Cr^{2+}(aq) + Cr^{2+}(aq) + \cancel{2\,e^-}$

Cancel electrons & species: $2\,Cr^{3+}(aq) + Cr(s) \rightarrow 3\,Cr^{2+}(aq)$

Look up cell potentials. Cr is reduced in the first half-cell reaction so $E^\circ_{red} = -0.50\text{ V} = E^\circ_{cathode}$. Cr is oxidized in the second half-cell reaction so $E^\circ_{red} = -0.91\text{ V} = E^\circ_{anode}$. Then $E^\circ_{cell} = E^\circ_{cathode} - E^\circ_{anode} =$

$-0.50\text{ V}-(-0.91\text{ V})=+0.41\text{ V}$. $n=2$ so $\Delta G^0_{rxn}=-nFE^0_{cell}=-2\ \cancel{\text{mol e}^-}\ \text{x}\ \frac{96{,}485\text{ C}}{\cancel{\text{mol e}^-}}\ \text{x}\ 0.41\text{ V}$

$=-2\text{ x }96{,}485\ \cancel{\text{C}}\text{ x }0.41\frac{\text{J}}{\cancel{\text{C}}}=-7.\underline{9}118\text{ x }10^4\text{ J}=79\text{ kJ}=2.7\text{ x }10^2\text{ kJ}$. $T=273.15+25\text{ °C}=298\text{ K}$

then $\Delta G^0_{rxn}=-RT\ln K$ Rearrange to solve for K. $K=e^{\frac{-\Delta G^0_{rxn}}{RT}}=e^{\frac{-(-7.\underline{9}118\text{ x }10^4\ \cancel{\text{J}})}{\left(8.314\frac{\cancel{\text{J}}}{\cancel{\text{K}}\text{ mol}}\right)(298\ \cancel{\text{K}})}}=e^{3\underline{1}.934}=7.4\text{ x }10^{13}$.

Check:

Reactants	Products
3 Cr atoms	3 Cr atoms
+6 charge	+6 charge

The units (kJ and none) are correct. If the voltage is positive, the reaction is spontaneous and the free energy change is negative and the equilibrium constant is large. The voltage and n are less than in part a) so the free energy change is not as negative and the equilibrium constant is not as large.

75. **Given:** Metal, M, 50.9 g/mol, 1.20 g of metal reduced in 23.6 minutes at 6.42 A from molten chloride
Find: empirical formula of chloride
Conceptual Plan: min → s → C → mol e⁻ and g M→ mol M then mol e⁻, mol M →charge → MCl_x

$\frac{60\text{ s}}{1\text{ min}}$ $\frac{6.42\text{ C}}{1\text{ s}}$ $\frac{1\text{ mol e}^-}{96{,}485\text{ C}}$ $\frac{1\text{ mol M}}{50.9\text{ g M}}$ $\frac{1\text{ mol e}^-}{1\text{ mol M}}$

Solution: $23.6\ \cancel{\text{min}}\text{ x }\frac{60\ \cancel{\text{s}}}{1\ \cancel{\text{min}}}\text{ x }\frac{6.42\ \cancel{\text{C}}}{1\ \cancel{\text{s}}}\text{ x }\frac{1\text{ mol e}^-}{96{,}485\ \cancel{\text{C}}}=0.094\underline{2}190\text{ mol e}^-$ and

$1.20\ \cancel{\text{g M}}\text{ x }\frac{1\text{ mol M}}{50.9\ \cancel{\text{g M}}}=0.023\underline{5}756\text{ mol M}$ then $\frac{0.094\underline{2}190\text{ mol e}^-}{0.023\underline{5}7561\text{ mol M}}=3.9\underline{9}646\frac{\text{e}^-}{\text{M}}$ so the empirical formula is MCl_4.

Check: The units (none) are correct. The result was an integer within the error of the measurements. The formula is typical for a metal salt. It could be vanadium, which has a +4 oxidation state.

76. **Given:** molten MCl_3 electrolysis, 1.25 g of metal reduced in 16.2 minutes at 3.86 A **Find:** molar mass of metal
Conceptual Plan: min → s → C → mol e⁻ → mol M then g M, mol M → molar mass

$\frac{60\text{ s}}{1\text{ min}}$ $\frac{6.42\text{ C}}{1\text{ s}}$ $\frac{1\text{ mol e}^-}{96{,}485\text{ C}}$ $\frac{1\text{ mol M}}{3\text{ mol e}^-}$ $\mathfrak{M}=\frac{\text{g M}}{\text{mol M}}$

Solution: $16.2\ \cancel{\text{min}}\text{ x }\frac{60\ \cancel{\text{s}}}{1\ \cancel{\text{min}}}\text{ x }\frac{3.86\ \cancel{\text{C}}}{1\ \cancel{\text{s}}}\text{ x }\frac{1\ \cancel{\text{mol e}^-}}{96{,}485\ \cancel{\text{C}}}\text{ x }\frac{1\text{ mol M}}{3\ \cancel{\text{mol e}^-}}=0.012\underline{9}620\text{ mol M}$ then

$\mathfrak{M}=\frac{1.20\text{ g M}}{0.012\underline{9}620\text{ mol M}}=92.6\frac{\text{g}}{\text{mol}}$.

Check: The units (g/mol) are correct. The result was a number typical for metals and it could be niobium, which is known to have a +3 oxidation state.

77. **Given:** hydrogen–oxygen fuel cell; $1.2\text{ x }10^3$ kWh of electricity/month **Find:** V of H_2 (*g*) at STP/month
Conceptual Plan: Write half-reactions. Look up half-reactions in Table 18.1. The reaction on the left is the oxidation. Calculate the standard cell potential by subtracting the electrode potential of the anode from the electrode potential of the cathode: $E^\circ_{cell}=E^\circ_{cathode}-E^\circ_{anode}$. Add the two half-cell reactions and cancel the electrons. Then kWh → J → C → mol e⁻ → mol H_2 → V

$\frac{3.60\text{ x }10^6\text{ J}}{1\text{ kWh}}$ $\frac{1\text{ C}}{0.41\text{ J}}$ $\frac{1\text{ mol e}^-}{96{,}485\text{ C}}$ $\frac{2\text{ mol H}_2}{4\text{ mol e}^-}$ at STP $\frac{22.414\text{ L}}{1\text{ mol H}_2}$

Solution: $2\ H_2\ (g)+4\ OH^-\ (aq)\rightarrow 4\ H_2O\ (l)+4\ e^-$ where $E^\circ_{red}=-0.83\text{ V}=E^\circ_{anode}$; and $O_2\ (g)+2\ H_2O\ (l)+4\ e^-\rightarrow 4\ OH^-\ (aq)$ where $E^\circ_{red}=0.40\text{ V}=E^\circ_{cathode}$. $E^\circ_{cell}=E^\circ_{cathode}-E^\circ_{anode}=0.40\text{ V}-(-0.83\text{ V})=1.23\text{ V}=1.23\text{ J/C}$ and $n=4$. Net reaction is: $2\ H_2\ (g)+O_2\ (g)\rightarrow 2\ H_2O\ (l)$. Then

$1.2\text{x}10^3\ \cancel{\text{kWh}}\text{ x }\frac{3.60\text{x}10^6\ \cancel{\text{J}}}{1\ \cancel{\text{kWh}}}\text{ x }\frac{1\ \cancel{\text{C}}}{1.23\ \cancel{\text{J}}}\text{ x }\frac{1\ \cancel{\text{mol e}^-}}{96{,}485\ \cancel{\text{C}}}\text{ x }\frac{2\ \cancel{\text{mol H}_2}}{4\ \cancel{\text{mol e}^-}}\text{ x }\frac{22.414\text{ L}}{1\ \cancel{\text{mol H}_2}}=4.1\text{x}10^6\text{ L}$.

Check: The units (L) are correct. A large volume is expected since we are trying to generate a large amount of electricity.

78. **Given:** voltaic cell to measure $[Cu^{2+}]$; SHE electrode paired with Cu^{2+}/Cu cell
Find: parameters to plot for a calibrations curve and what is the slope of the curve
Conceptual Plan: Write the 2 half-cell reactions and add electrons as needed to balance reactions. Look up half-reactions in Table 18.1. Calculate the standard cell potential by subtracting the electrode potential of the anode from the electrode potential of the cathode: $E^{\circ}_{cell} = E^{\circ}_{cathode} - E^{\circ}_{anode}$. **Add the two half-cell reactions and cancel electrons and determine** n. **Then** $E^{\circ}_{cell}, P_{H2}, [H^+], [Cu^{2+}], n \rightarrow E_{cell}$

$$E_{cell} = E^{\circ}_{cell} - \frac{0.0592\ V}{n} \log Q$$

Solution: write half-reactions and add electrons $H_2\ (g) \rightarrow 2\ H^+\ (aq) + 2\ e^-$ and $Cu^{2+}\ (aq) + 2\ e^- \rightarrow Cu\ (s)$. Look up cell potentials. H is oxidized so $E^{\circ}_{red} = -0.00\ V = E^{\circ}_{anode}$. Cu^{2+} is reduced so $E^{\circ}_{red} = 0.34\ V = E^{\circ}_{cathode}$. Then $E^{\circ}_{cell} = E^{\circ}_{cathode} - E^{\circ}_{anode} = 0.34\ V - 0.00\ V = +0.34\ V$. Add the two half-cell reactions: $H_2\ (g) + Cu^{2+}\ (aq) + \cancel{2e^-} \rightarrow 2\ H^+\ (aq) + \cancel{2e^-} + Cu\ (s)$. Simplify to $H_2\ (g) + Cu^{2+}\ (aq) \rightarrow 2\ H^+\ (aq) + Cu\ (s)$. So

$Q = \frac{[H^+]^2}{P_{H2}[Cu^{2+}]}$ and $n = 2$ then $E_{cell} = E^0_{cell} - \frac{0.0592\ V}{n} \log Q = 0.34\ V - \frac{0.0592\ V}{2} \log \frac{[H^+]^2}{P_{H2}[Cu^{2+}]}$. If the anode half-cell is buffered at a constant pH and a constant P_{H2} is used then two of the terms in Q are constant and can be pulled out of the expression so that

$$E_{cell} = 0.34\ V - \frac{0.0592\ V}{2}\left(\log \frac{[H^+]^2}{P_{H2}} - \log [Cu^{2+}]\right) = \left(0.34\ V - \frac{0.0592\ V}{2} \log \frac{[H^+]^2}{P_{H2}}\right) + (0.0296\ V) \log [Cu^{2+}].$$

If we plot $\log [Cu^{2+}]$ versus E°_{cell}, the slope will be 0.0296 V.

79. **Given:** Au^{3+}/Au electroplating; surface area = 49.8 cm^2, Au thickness = 1.00×10^{-3} cm, density = 19.3 g/cm^3; at 3.25 A **Find:** time
Conceptual Plan: Write the half-cell reaction and add electrons as needed to balance reactions. Then surface area, thickness → V → g Au → mol Au → mol e⁻ → C → s

$$V = surface\ area \times thickness \quad \frac{19.3\ g\ Au}{1\ cm^3\ Au} \quad \frac{1\ mol\ Au}{196.97\ g\ Au} \quad \frac{3\ mol\ e^-}{1\ mol\ Au} \quad \frac{96{,}485\ C}{1\ mol\ e^-} \quad \frac{1\ s}{3.25\ C}$$

Solution: write half-reaction and add electrons $Au^{3+}\ (aq) + 3\ e^- \rightarrow Au\ (s)$.
$V = surface\ area \times thickness = (49.8\ cm^2)(1.00 \times 10^{-3}\ cm) = 0.0498\ cm^3$ then

$$0.0498\ \cancel{cm^3\ Au} \times \frac{19.3\ \cancel{g\ Au}}{1\ \cancel{cm^3\ Au}} \times \frac{1\ \cancel{mol\ Au}}{196.97\ \cancel{g\ Au}} \times \frac{3\ \cancel{mol\ e^-}}{1\ \cancel{mol\ Au}} \times \frac{96{,}485\ \cancel{C}}{1\ \cancel{mol\ e^-}} \times \frac{1\ s}{3.25\ \cancel{C}} = 435\ s.$$

Check: The units (s) are correct. Since the layer is so thin there is far less than a mole of gold, so the time is not very long. In order to be an economical process, it must be fairly quick.

80. **Given:** electrodeposit mixture Cu and Cd with 1.20 F (1 F = 1 mol e⁻); total mass = 50.36 g
Find: mass of $CuSO_4$
Conceptual Plan: Write the half-cell reaction and add electrons as needed to balance reactions. Then F → mol e⁻ → mol (Cu + Cd) then let x **= g Cu so that (50.36 g –** x **g) = g Cd then g Cu → mol Cu**

$$\frac{1\ mol\ e^-}{1\ F} \quad \frac{1\ mol\ (Cu{+}Cd)}{2\ mol\ e^-} \quad \frac{1\ mol\ Cu}{63.55\ g\ Cu}$$

and g Cd → mol Cd then solve for x = g Cu → g $CuSO_4$

$$\frac{1\ mol\ Cd}{112.41\ g\ Cd} \quad \frac{159.62\ g\ CuSO_4}{63.55\ g\ Cu}$$

Solution: $Cu^{2+}\ (aq) + 2\ e^- \rightarrow Cu\ (s)$ and $Cd^{2+}\ (aq) + 2\ e^- \rightarrow Cd\ (s)$, so $n = 2$ for both metals.

$1.20\ \cancel{F} \times \frac{1\ \cancel{mol\ e^-}}{1\ \cancel{F}} \times \frac{1\ mol(Cu{+}Cd)}{2\ \cancel{mol\ e^-}} = 0.600\ mol(Cu{+}Cd)$ then let x = g Cu so that (50.36 g – x g) = g Cd

then $x\ \cancel{g\ Cu} \times \frac{1\ mol\ Cu}{63.55\ \cancel{g\ Cu}} = \frac{x}{63.55}\ mol\ Cu$ and $(50.36 - x)\ \cancel{g\ Cd} \times \frac{1\ mol\ Cd}{112.41\ \cancel{g\ Cd}} = \frac{(50.36 - x)}{112.41}\ mol\ Cd$

then $0.600\ mol(Cu{+}Cd) = \frac{x}{63.55}\ mol\ Cu + \frac{(50.36 - x)}{112.41}\ mol\ Cd$ Solve for x.

$0.600 = 0.01573564\ x + 0.4480028 - 0.08896006\ x \rightarrow x = 22.\underline{2}2300\ g\ Cu \rightarrow$

$$22.22300 \text{ g Cu} \times \frac{159.62 \text{ g } CuSO_4}{63.55 \text{ g Cu}} = 55.8 \text{ g } CuSO_4 .$$

Check: The units (g) are correct. The result is reasonable since 0.600 mol Cu = 38.1 g and 0.600 mol Cd = 67.4 g and the amount deposited is in between the two values.

81. **Given:** $C_2O_4^{2-} \rightarrow CO_2$ and MnO_4^- *(aq)* $\rightarrow Mn^{2+}$ *(aq)*; 50.1 mL of MnO_4^- to titrate 0.339 g $Na_2C_2O_4$; and 4.62 g U sample titrated by 32.3 mL MnO_4^-; and $UO^{2+} \rightarrow UO_2^{2+}$ **Find:** percent U in sample
Conceptual Plan: Separate the overall reaction into two half-reactions: one for oxidation and one for reduction. → Balance each half-reaction with respect to mass in the following order: 1) balance all elements other than H and O; 2) balance O by adding H_2O; and 3) balance H by adding H^+. → Balance each half-reaction with respect to charge by adding electrons. (The sum of the charges on both sides of the equation should be made equal by adding electrons as necessary.) → Make the number of electrons in both half-reactions equal by multiplying one or both half-reactions by a small whole number. → Add the two half-reactions together, canceling electrons and other species as necessary. → Verify that the reaction is balanced both with respect to mass and with respect to charge. Then
mL MnO_4^- → LMnO_4^- and g $Na_2C_2O_4$ → mol $Na_2C_2O_4$ → mol MnO_4^- then

$$\frac{1 \text{ L } MnO_4^-}{1000 \text{ mL } MnO_4^-} \qquad \frac{1 \text{ mol } Na_2C_2O_4}{134.00 \text{ g } Na_2C_2O_4} \qquad \frac{2 \text{ mol } MnO_4^-}{5 \text{ mol Zn}} \qquad \frac{1 \text{ L } MnO_4^-}{0.500 \text{ mol } MnO_4^-}$$

L MnO_4^-, mol MnO_4^- → M MnO_4^- then write U half-reactions and balance as above. →

$$M = \frac{\text{mol } MnO_4^-}{L}$$

Make the number of electrons in both half-reactions equal by multiplying one or both half-reactions by a small whole number. → Add the two half-reactions together, canceling electrons and other species as necessary. → Verify that the reaction is balanced both with respect to mass and with respect to charge. Then mL MnO_4^-, M MnO_4^- → mol MnO_4^- → mol U → g U then g U, g sample → % U

$$M = \frac{\text{mol } MnO_4^-}{L} \qquad \frac{5 \text{ mol U}}{2 \text{ mol } MnO_4^-} \qquad \frac{238.03 \text{ g U}}{1 \text{ mol U}} \qquad \text{percent U} = \frac{\text{g U}}{\text{g sample}} \times 100\%$$

Solution:

Separate: MnO_4^- *(aq)* $\rightarrow Mn^{2+}$ *(aq)* and $C_2O_4^{2-}$ *(aq)* $\rightarrow CO_2$ *(g)*
Balance non H & O elements: MnO_4^- *(aq)* $\rightarrow Mn^{2+}$ *(aq)* and $C_2O_4^{2-}$ *(aq)* $\rightarrow 2\ CO_2$ *(g)*
Balance O with H_2O: MnO_4^- *(aq)* $\rightarrow Mn^{2+}$ *(aq)* + 4 H_2O *(l)* and $C_2O_4^{2-}$ *(aq)* $\rightarrow 2\ CO_2$ *(g)*
Balance H with H^+: MnO_4^- *(aq)* + 8 H^+ *(aq)* $\rightarrow Mn^{2+}$ *(aq)* + 4 H_2O *(l)* and $C_2O_4^{2-}$ *(aq)* $\rightarrow 2\ CO_2$ *(g)*
Add electrons: MnO_4^- *(aq)* + 8 H^+ *(aq)* + 5 e^- $\rightarrow Mn^{2+}$ *(aq)* + 4 H_2O *(l)* and $C_2O_4^{2-}$ *(aq)* $\rightarrow 2\ CO_2$ *(g)* + 2 e^-
Equalize electrons:
2 MnO_4^- *(aq)* + 16 H^+ *(aq)* + 10 e^- $\rightarrow$ 2 Mn^{2+} *(aq)* + 8 H_2O *(l)* and 5 $C_2O_4^{2-}$ *(aq)* $\rightarrow$ 10 CO_2 *(g)* + 10 e^-
Add half-reactions:
2 MnO_4^- *(aq)* + 16 H^+ *(aq)* + ~~10 e^-~~ + 5 $C_2O_4^{2-}$ *(aq)* $\rightarrow$ 2 Mn^{2+} *(aq)* + 8 H_2O *(l)* + 10 CO_2 *(g)* + ~~10 e^-~~
Cancel electrons: 2 MnO_4^- *(aq)* + 16 H^+ *(aq)* + 5 $C_2O_4^{2-}$ *(aq)* $\rightarrow$ 2 Mn^{2+} *(aq)* + 8 H_2O *(l)* + 10 CO_2 *(g)*
then

$$50.1 \text{ mL } MnO_4^- \times \frac{1 \text{ L } MnO_4^-}{1000 \text{ mL } MnO_4^-} = 0.0501 \text{ L } MnO_4^-$$

$$0.339 \text{ g } Na_2C_2O_4 \times \frac{1 \text{ mol } Na_2C_2O_4}{134.00 \text{ g } Na_2C_2O_4} \times \frac{2 \text{ mol } MnO_4^-}{5 \text{ mol } Na_2C_2O_4} = 0.00101194 \text{ mol } MnO_4^-$$

$$M = \frac{0.00101194 \text{ mol } MnO_4^-}{0.0501 \text{ L}} = 0.0201984 \text{ M } MnO_4^-$$

Separate: MnO_4^- *(aq)* $\rightarrow Mn^{2+}$ *(aq)* and UO^{2+} *(aq)* $\rightarrow UO_2^{2+}$ *(aq)*
Balance non H & O elements: MnO_4^- *(aq)* $\rightarrow Mn^{2+}$ *(aq)* and UO^{2+} *(aq)* $\rightarrow UO_2^{2+}$ *(aq)*
Balance O with H_2O: MnO_4^- *(aq)* $\rightarrow Mn^{2+}$ *(aq)* + 4 H_2O *(l)* and UO^{2+} *(aq)* + H_2O *(l)* $\rightarrow UO_2^{2+}$ *(aq)*
Balance H with H^+:
MnO_4^- *(aq)* + 8 H^+ *(aq)* $\rightarrow Mn^{2+}$ *(aq)* + 4 H_2O *(l)* and UO^{2+} *(aq)* + H_2O *(l)* $\rightarrow UO_2^{2+}$ *(aq)* + 2 H^+ *(aq)*
Add electrons: MnO_4^- *(aq)* + 8 H^+ *(aq)* + 5 e^- $\rightarrow Mn^{2+}$ *(aq)* + 4 H_2O *(l)* and
UO^{2+} *(aq)* + H_2O *(l)* $\rightarrow UO_2^{2+}$ *(aq)* + 2 H^+ *(aq)* + 2 e^-
Equalize electrons: 2 MnO_4^- *(aq)* + 16 H^+ *(aq)* + 10 e^- $\rightarrow$ 2 Mn^{2+} *(aq)* + 8 H_2O *(l)* and
5 UO^{2+} *(aq)* + 5 H_2O *(l)* $\rightarrow$ 5 UO_2^{2+} *(aq)* + 10 H^+ *(aq)* + 10 e^-

Add half-reactions: $2\ MnO_4^- (aq) + 6\ \cancel{10}\ H^+(aq) + \cancel{10 e^-} + 5\ UO^{2+} (aq) + \cancel{5 H_2O(l)} \rightarrow$

$2\ Mn^{2+} (aq) + 3\ \cancel{5 H_2O(l)} + 5\ UO_2^{2+} (aq) + \cancel{10 H^+(aq)} + \cancel{10 e^-}$

Cancel electrons & species:

$$2\ MnO_4^- (aq) + 6\ H^+ (aq) + 5\ UO^{2+} (aq) \rightarrow 2\ Mn^{2+} (aq) + 3\ H_2O\ (l) + 5\ UO_2^{2+} (aq)$$

$$32.3\ \cancel{mL\ MnO_4^-} \times \frac{0.0201984\ \cancel{mol\ MnO_4^-}}{1000\ \cancel{mL\ MnO_4^-}} \times \frac{5\ \cancel{mol\ U}}{2\ \cancel{mol\ MnO_4^-}} \times \frac{238.03\ g\ U}{1\ \cancel{mol\ U}} = 0.388232\ g\ U \text{ then}$$

$$\text{percent U} = \frac{g\ U}{g\ sample} \times 100\% = \frac{0.388232\ g\ U}{4.63\ g\ sample} \times 100\% = 8.39\ \%.$$

Check: first reaction

Reactants	Products
2 Mn atoms	2 Mn atoms
28 O atoms	28 O atoms
16 H atoms	16 H atoms
10 C atoms	10 C atoms
+4 charge	+4 charge

second reaction

Reactants	Products
2 Mn atoms	2 Mn atoms
13 O atoms	13 O atoms
6 H atoms	6 H atoms
5 U atoms	5 U atoms
+14 charge	+14 charge

The reactions are balanced. The units (%) are correct. The percentage is between 0 and 100 %.

82. b) If $E^\circ_{cell} > 0$ and $E_{cell} = E^\circ_{cell} - \frac{0.0592\ V}{n} \log Q < 0$ this means that the second term dominates and is negative. This means that $Q > 1$. If $E_{cell} < 0$ then $K < 1$, since the reaction is nonspontaneous or $Q > K$.

83. a) Looking for anion reductions that are in between the reduction potentials of Cl_2 and Br_2. The only one that meets this criterion is the dichromate ion.

84. b) If the free energy change is negative this is a spontaneous reaction. This translates to a positive cell potential and a large equilibrium constant.

Chapter 19
Radioactivity and Nuclear Chemistry

1. **Conceptual Plan: Begin with the symbol for parent nuclide on the left side of the equation and the symbol for a particle on the right side (except for electron capture). → Equalize the sum of the mass numbers and the sum of the atomic numbers on both sides of the equation by writing the appropriate mass number and atomic number for the unknown daughter nuclide. → Using the periodic table, deduce the identity of the unknown daughter nuclide from the atomic number and write its symbol.**
Solution:
a) U-234 (alpha decay) $^{234}_{92}U \rightarrow ^{?}_{?}? + ^{4}_{2}He$ then $^{234}_{92}U \rightarrow ^{230}_{90}? + ^{4}_{2}He$ then $^{234}_{92}U \rightarrow ^{230}_{90}Th + ^{4}_{2}He$

b) Th-230 (alpha decay) $^{230}_{90}Th \rightarrow ^{?}_{?}? + ^{4}_{2}He$ then $^{230}_{90}Th \rightarrow ^{226}_{88}? + ^{4}_{2}He$ then $^{230}_{90}Th \rightarrow ^{226}_{88}Ra + ^{4}_{2}He$

c) Pb-214 (beta decay) $^{214}_{82}Pb \rightarrow ^{?}_{?}? + ^{0}_{-1}e$ then $^{214}_{82}Pb \rightarrow ^{214}_{83}? + ^{0}_{-1}e$ then $^{214}_{82}Pb \rightarrow ^{214}_{83}Bi + ^{0}_{-1}e$

d) N-13 (positron emission) $^{13}_{7}N \rightarrow ^{?}_{?}? + ^{0}_{+1}e$ then $^{13}_{7}N \rightarrow ^{13}_{6}? + ^{0}_{+1}e$ then $^{13}_{7}N \rightarrow ^{13}_{6}C + ^{0}_{+1}e$

e) Cr-51 (electron capture) $^{51}_{24}Cr + ^{0}_{-1}e \rightarrow ^{?}_{?}?$ then $^{51}_{24}Cr + ^{0}_{-1}e \rightarrow ^{51}_{23}?$ then $^{51}_{24}Cr + ^{0}_{-1}e \rightarrow ^{51}_{23}V$
Check: a) 234 = 230 + 4, 92 = 90 + 2, and Thorium is atomic number 90. b) 230 = 226 + 4, 90 = 88 + 2, and Radium is atomic number 88. c) 214 = 214 + 0, 82 = 83 – 1, and Bismuth is atomic number 83. d) 13 = 13 + 0, 7 = 6 + 1, and Carbon is atomic number 6. e) 51 + 0 = 51, 24 – 1 = 23, and Vanadium is atomic number 23.

2. **Conceptual Plan: Begin with the symbol for parent nuclide on the left side of the equation and the symbol for a particle on the right side (except for electron capture). → Equalize the sum of the mass numbers and the sum of the atomic numbers on both sides of the equation by writing the appropriate mass number and atomic number for the unknown daughter nuclide. → Using the periodic table, deduce the identity of the unknown daughter nuclide from the atomic number and write its symbol.**
Solution:
a) Po-210 (alpha decay) $^{210}_{84}Po \rightarrow ^{?}_{?}? + ^{4}_{2}He$ then $^{210}_{84}Po \rightarrow ^{206}_{82}? + ^{4}_{2}He$ then $^{210}_{84}Po \rightarrow ^{206}_{82}Pb + ^{4}_{2}He$

b) Ac-227 (beta decay) $^{227}_{89}Ac \rightarrow ^{?}_{?}? + ^{0}_{-1}e$ then $^{227}_{89}Ac \rightarrow ^{227}_{90}? + ^{0}_{-1}e$ then $^{227}_{89}Ac \rightarrow ^{227}_{90}Th + ^{0}_{-1}e$

c) Tl-207 (beta decay) $^{207}_{81}Tl \rightarrow ^{?}_{?}? + ^{0}_{-1}e$ then $^{207}_{81}Tl \rightarrow ^{207}_{82}? + ^{0}_{-1}e$ then $^{207}_{81}Tl \rightarrow ^{207}_{82}Pb + ^{0}_{-1}e$

d) O-15 (positron emission) $^{15}_{8}O \rightarrow ^{?}_{?}? + ^{0}_{+1}e$ then $^{15}_{8}O \rightarrow ^{15}_{7}? + ^{0}_{+1}e$ then $^{15}_{8}O \rightarrow ^{15}_{7}N + ^{0}_{+1}e$

e) Pd-103 (electron capture) $^{103}_{46}Pd + ^{0}_{-1}e \rightarrow ^{?}_{?}?$ then $^{103}_{46}Pd + ^{0}_{-1}e \rightarrow ^{103}_{45}?$ then $^{103}_{46}Pd + ^{0}_{-1}e \rightarrow ^{103}_{45}Rh$
Check: a) 210 = 210 + 4, 84 = 82 + 2, and Lead is atomic number 82. b) 227 = 227 + 0, 89 = 90 – 1, and Thorium is atomic number 90. c) 207 = 207 + 0, 81 = 82 – 1, and Lead is atomic number 82. d) 15 = 15 + 0, 8 = 7 + 1, and Nitrogen is atomic number 7. e) 103 + 0 = 103, 46 – 1 = 45, and Rhodium is atomic number 45.

3. **Given:** Th-232 decay series: α, β, β, α **Find:** balanced decay reactions
Conceptual Plan: Begin with the symbol for parent nuclide on the left side of the equation and the symbol for a particle on the right side (except for electron capture). → Equalize the sum of the mass numbers and the sum of the atomic numbers on both sides of the equation by writing the appropriate mass number and atomic number for the unknown daughter nuclide. → Using the periodic table, deduce the identity of the unknown daughter nuclide from the atomic number and write its symbol. → Use the product of this reaction to write the next reaction.
Solution:
Th-232 (alpha decay) $^{232}_{90}Th \rightarrow ^{?}_{?}? + ^{4}_{2}He$ then $^{232}_{90}Th \rightarrow ^{228}_{88}? + ^{4}_{2}He$ then $^{232}_{90}Th \rightarrow ^{228}_{88}Ra + ^{4}_{2}He$
Ra-228 (beta decay) $^{228}_{88}Ra \rightarrow ^{?}_{?}? + ^{0}_{-1}e$ then $^{228}_{88}Ra \rightarrow ^{228}_{89}? + ^{0}_{-1}e$ then $^{228}_{88}Ra \rightarrow ^{228}_{89}Ac + ^{0}_{-1}e$
Ac-228 (beta decay) $^{228}_{89}Ac \rightarrow ^{?}_{?}? + ^{0}_{-1}e$ then $^{228}_{89}Ac \rightarrow ^{228}_{90}? + ^{0}_{-1}e$ then $^{228}_{89}Ac \rightarrow ^{228}_{90}Th + ^{0}_{-1}e$
Th-228 (alpha decay) $^{228}_{90}Th \rightarrow ^{?}_{?}? + ^{4}_{2}He$ then $^{228}_{90}Th \rightarrow ^{224}_{88}? + ^{4}_{2}He$ then $^{228}_{90}Th \rightarrow ^{224}_{88}Ra + ^{4}_{2}He$

Thus the decay series is: ${}^{232}_{90}\text{Th} \rightarrow {}^{228}_{88}\text{Ra} + {}^{4}_{2}\text{He}$, ${}^{228}_{88}\text{Ra} \rightarrow {}^{228}_{89}\text{Ac} + {}^{0}_{-1}\text{e}$, ${}^{228}_{89}\text{Ac} \rightarrow {}^{228}_{90}\text{Th} + {}^{0}_{-1}\text{e}$,
${}^{228}_{90}\text{Th} \rightarrow {}^{224}_{88}\text{Ra} + {}^{4}_{2}\text{He}$.

Check: 232 = 228 + 4, 90 = 88 + 2, and Radium is atomic number 88. 228 = 228 + 0, 88 = 89 – 1, and Actinium is atomic number 89. 228 = 228 + 0, 89 = 90 – 1, and Thorium is atomic number 90. 228 = 224 + 4, 90 = 88 + 2, and Radium is atomic number 88.

4. **Given:** Rn-220 decay series: α, α, β, α **Find:** balanced decay reactions
Conceptual Plan: Begin with the symbol for parent nuclide on the left side of the equation and the symbol for a particle on the right side (except for electron capture). → Equalize the sum of the mass numbers and the sum of the atomic numbers on both sides of the equation by writing the appropriate mass number and atomic number for the unknown daughter nuclide. → Using the periodic table, deduce the identity of the unknown daughter nuclide from the atomic number and write its symbol. → Use the product of this reaction to write the next reaction.
Solution:
Rn-220 (alpha decay) ${}^{220}_{86}\text{Rn} \rightarrow {}^{?}_{?}? + {}^{4}_{2}\text{He}$ then ${}^{220}_{86}\text{Rn} \rightarrow {}^{216}_{84}? + {}^{4}_{2}\text{He}$ then ${}^{220}_{86}\text{Rn} \rightarrow {}^{216}_{84}\text{Po} + {}^{4}_{2}\text{He}$
Po-216 (alpha decay) ${}^{216}_{84}\text{Po} \rightarrow {}^{?}_{?}? + {}^{4}_{2}\text{He}$ then ${}^{216}_{84}\text{Po} \rightarrow {}^{212}_{82}? + {}^{4}_{2}\text{He}$ then ${}^{216}_{84}\text{Po} \rightarrow {}^{212}_{82}\text{Pb} + {}^{4}_{2}\text{He}$
Pb-212 (beta decay) ${}^{212}_{82}\text{Pb} \rightarrow {}^{?}_{?}? + {}^{0}_{-1}\text{e}$ then ${}^{212}_{82}\text{Pb} \rightarrow {}^{212}_{83}? + {}^{0}_{-1}\text{e}$ then ${}^{212}_{82}\text{Pb} \rightarrow {}^{212}_{83}\text{Bi} + {}^{0}_{-1}\text{e}$
Bi-212 (alpha decay) ${}^{212}_{83}\text{Bi} \rightarrow {}^{?}_{?}? + {}^{4}_{2}\text{He}$ then ${}^{212}_{83}\text{Bi} \rightarrow {}^{208}_{81}? + {}^{4}_{2}\text{He}$ then ${}^{212}_{83}\text{Bi} \rightarrow {}^{208}_{81}\text{Tl} + {}^{4}_{2}\text{He}$
Thus the decay series is: ${}^{220}_{86}\text{Rn} \rightarrow {}^{216}_{84}\text{Po} + {}^{4}_{2}\text{He}$, ${}^{216}_{84}\text{Po} \rightarrow {}^{212}_{82}\text{Pb} + {}^{4}_{2}\text{He}$, ${}^{212}_{82}\text{Pb} \rightarrow {}^{212}_{83}\text{Bi} + {}^{0}_{-1}\text{e}$,
${}^{212}_{83}\text{Bi} \rightarrow {}^{208}_{81}\text{Tl} + {}^{4}_{2}\text{He}$.
Check: 220 = 216 + 4, 86 = 84 + 2, and Polonium is atomic number 84. 216 = 212 + 4, 84 = 82 + 2, and Lead is atomic number 82. 212 = 212 + 0, 82 = 83 – 1, and Bismuth is atomic number 83. 220 = 208 + 4, 83 = 81 + 2, and Thallium is atomic number 81.

5. **Conceptual Plan: Equalize the sum of the mass numbers and the sum of the atomic numbers on both sides of the equation by writing the appropriate mass number and atomic number for the unknown species. → Using the periodic table and the list of particles, deduce the identity of the unknown species from the atomic number and write its symbol.**
Solution:
a) ${}^{?}_{?}? \rightarrow {}^{217}_{85}\text{At} + {}^{4}_{2}\text{He}$ becomes ${}^{221}_{87}? \rightarrow {}^{217}_{85}\text{At} + {}^{4}_{2}\text{He}$ then ${}^{221}_{87}\text{Fr} \rightarrow {}^{217}_{85}\text{At} + {}^{4}_{2}\text{He}$

b) ${}^{241}_{94}\text{Pu} \rightarrow {}^{241}_{95}\text{Am} + {}^{?}_{?}?$ becomes ${}^{241}_{94}\text{Pu} \rightarrow {}^{241}_{95}\text{Am} + {}^{0}_{-1}?$ then ${}^{241}_{94}\text{Pu} \rightarrow {}^{241}_{95}\text{Am} + {}^{0}_{-1}\text{e}$

c) ${}^{19}_{11}\text{Na} \rightarrow {}^{19}_{10}\text{Ne} + {}^{?}_{?}?$ becomes ${}^{19}_{11}\text{Na} \rightarrow {}^{19}_{10}\text{Ne} + {}^{0}_{1}?$ then ${}^{19}_{11}\text{Na} \rightarrow {}^{19}_{10}\text{Ne} + {}^{0}_{+1}\text{e}$

d) ${}^{75}_{34}\text{Se} + {}^{?}_{?}? \rightarrow {}^{75}_{33}\text{As}$ becomes ${}^{75}_{34}\text{Se} + {}^{0}_{-1}? \rightarrow {}^{75}_{33}\text{As}$ then ${}^{75}_{34}\text{Se} + {}^{0}_{-1}\text{e} \rightarrow {}^{75}_{33}\text{As}$
Check: a) 221 = 217 + 4, 87 = 85 + 2, and Francium is atomic number 87. b) 241 = 241 + 0, 94 = 95 – 1, and the particle is a beta particle. c) 19 = 19 + 0, 11 = 10 + 1, and the particle is a positron. d) 75 = 75 + 0, 34 – 1 = 33, and the particle is an electron.

6. **Conceptual Plan: Equalize the sum of the mass numbers and the sum of the atomic numbers on both sides of the equation by writing the appropriate mass number and atomic number for the unknown species. → Using the periodic table and the list of particles, deduce the identity of the unknown species from the atomic number and write its symbol.**
Solution:
a) ${}^{241}_{95}\text{Am} \rightarrow {}^{237}_{93}\text{Np} + {}^{?}_{?}?$ becomes ${}^{241}_{95}\text{Am} \rightarrow {}^{237}_{93}\text{Np} + {}^{4}_{2}?$ then ${}^{241}_{95}\text{Am} \rightarrow {}^{237}_{93}\text{Np} + {}^{4}_{2}\text{He}$

b) ${}^{?}_{?}? \rightarrow {}^{233}_{92}\text{U} + {}^{0}_{-1}\text{e}$ becomes ${}^{233}_{91}? \rightarrow {}^{233}_{92}\text{U} + {}^{0}_{-1}\text{e}$ then ${}^{233}_{91}\text{Pa} \rightarrow {}^{233}_{92}\text{U} + {}^{0}_{-1}\text{e}$

c) ${}^{237}_{93}\text{Np} \rightarrow {}^{?}_{?}? + {}^{4}_{2}\text{He}$ becomes ${}^{237}_{93}\text{Np} \rightarrow {}^{233}_{91}? + {}^{4}_{2}\text{He}$ then ${}^{237}_{93}\text{Np} \rightarrow {}^{233}_{91}\text{Pa} + {}^{4}_{2}\text{He}$

d) ${}^{75}_{35}\text{Br} \rightarrow {}^{?}_{?}? + {}^{0}_{+1}\text{e}$ becomes ${}^{75}_{35}\text{Br} \rightarrow {}^{75}_{34}? + {}^{0}_{+1}\text{e}$ then ${}^{75}_{35}\text{Br} \rightarrow {}^{75}_{34}\text{Se} + {}^{0}_{+1}\text{e}$
Check: a) 241 = 237 + 4, 95 = 93 + 2, and the particle is an alpha particle. b) 233 = 233 + 0, 91 = 92 – 1, and Protactinium is atomic number 91. c) 237 = 233 + 4, 93 = 91 + 2, and Protactinium is atomic number 91. d) 75

= 75 + 0, 35 = 34 + 1, and Selenium is atomic number 34.

7. a) stable, N/Z ratio is close to 1, acceptable for low Z atoms

   b) not stable, N/Z ratio much too high for low Z atom

   c) not stable, N/Z ratio is less than 1, much too low

   d) stable, N/Z ratio is acceptable for this Z

8. a) stable, N/Z ratio is acceptable for this Z

   b) not stable, N/Z ratio much too high for this Z

   c) not stable, N/Z ratio is close to 1, much too low for this Z

   d) stable, N/Z ratio is acceptable for this Z

9. Sc, V, and Mn, each have odd numbers of protons. Atoms with an odd number of protons typically have fewer stable isotopes than those with an even number of protons.

10. Aluminum and sodium both have an odd Z , which have fewer stable isotopes. These atoms are both small Z atoms and so the N/Z should be close to 1. There is only one option that meets both criteria. Neon and Magnesium have an even Z and so they have more options.

11. a) beta decay, since N/Z is too high

    b) positron emission, since N/Z is too low

    c) positron emission, since N/Z is too low

    d) positron emission, since N/Z is too low

12. a) beta decay, since N/Z is too high

    b) beta decay, since N/Z is too high

    c) positron emission, since N/Z is too low

    d) positron emission, since N/Z is too low

13. a) Cs-125, since it is closer to the proper N/Z

    b) Fe-62, since it is closer to the proper N/Z

14. a) Cs-139, since it is closer to the proper N/Z

    b) Fe-52, since it is closer to the proper N/Z

15. Since the half-life of U-235 is 703 million years, 1/2 will be present after 703 million years, 1/4 will be present after 2 half-lives and 1/8 will be present after 3 half-lives or 2110 million years or 2.11 x 10^9 years.

16. **Given:** initially 0.050 mg Tc-99m, $t_{1/2}$ for radioactive decay = 6.0 h **Find:** t to 6.3 x 10^{-3} mg
Conceptual plan: radioactive decay implies first order kinetics, $t_{1/2} \rightarrow k$ **then**

$$t_{1/2} = \frac{0.693}{k}$$

$\mathbf{m_{Tc\text{-}99m\ 0}, m_{Tc\text{-}99m\ t}, k \rightarrow t}$

$$\ln N_t = -kt + \ln N_0$$

Solution: $t_{1/2} = \frac{0.693}{k}$ rearrange to solve for k. $k = \frac{0.693}{t_{1/2}} = \frac{0.693}{6.0\ \text{h}} = 0.1155\ \text{h}^{-1}$. Since

$\ln m_{\text{Tc-99m t}} = -kt + \ln m_{\text{Tc-99m 0}}$ rearrange to solve for t

$$t = -\frac{1}{k}\ln\frac{m_{\text{Tc-99m t}}}{m_{\text{Tc-99m 0}}} = -\frac{1}{0.1155\ \text{h}^{-1}}\ln\frac{6.3 \times 10^{-3}\ \text{mg}}{0.050\ \text{mg}} = 18\ \text{h}.$$

Check: The units (h) are correct. The time is 3 half-lives and the amount is 1/8 or $1/2^3$ of the original amount.

17. **Given:** $t_{1/2}$ for isotope decay = 3.8 days; 1.55 g isotope initially **Find:** mass of isotope after 5.5 days
Conceptual plan: radioactive decay implies first order kinetics, $t_{1/2} \rightarrow k$ then

$$t_{1/2} = \frac{0.693}{k}$$

$\mathbf{m_{isotope\ 0}, t, k \rightarrow m_{isotope\ t}}$

$$\ln N_t = -kt + \ln N_0$$

Solution: $t_{1/2} = \frac{0.693}{k}$ rearrange to solve for k. $k = \frac{0.693}{t_{1/2}} = \frac{0.693}{3.8\ \text{days}} = 0.18237\ \text{day}^{-1}$. Since

$\ln N_t = -kt + \ln N_0 = -(0.18237\ \text{day}^{-1})(5.5\ \text{day}) + \ln(1.55\ \text{g}) = -0.56478 \rightarrow N_t = e^{-0.56478} = 0.57\ \text{g}$.

Check: The units (g) are correct. The amount is consistent with a time between one and two half-lives.

18. **Given:** $t_{1/2}$ for I-131 = 8 days; 58 mg dose at 8:00 am **Find:** mass of I-131 at 5:00 pm next day
Conceptual plan: radioactive decay implies first order kinetics, $t_{1/2} \rightarrow k$ and determine days since dose

$$t_{1/2} = \frac{0.693}{k}$$

then $\mathbf{m_{I\text{-}131\ 0}, t, k \rightarrow m_{I\text{-}131\ t}}$

$$\ln N_t = -kt + \ln N_0$$

Solution: $t_{1/2} = \frac{0.693}{k}$ rearrange to solve for k. $k = \frac{0.693}{t_{1/2}} = \frac{0.693}{8\ \text{days}} = 0.086625\ \text{day}^{-1}$. The time since the dose is one day plus 9 hours or (1 + 9/24) days = 1.375 days. Since $\ln m_{\text{I-131 t}} = -kt + \ln m_{\text{I-131 0}}$

$= -(0.086625\ \text{day}^{-1})(1.375\ \text{day}) + \ln(58\ \text{mg}) = 3.9413 \rightarrow N_t = e^{3.9413} = 51\ \text{mg}$.

Check: The units (mg) are correct. The amount is consistent with a time less than one half-life.

19. **Given:** F-18 initial decay rate = 1.5×10^5 /s, $t_{1/2}$ for F-18 = 1.83 h **Find:** t to decay rate of 1.0×10^2 /s
Conceptual plan: radioactive decay implies first order kinetics, $t_{1/2} \rightarrow k$ then $\text{Rate}_0, \text{Rate}_t, k \rightarrow t$

$$t_{1/2} = \frac{0.693}{k} \qquad \ln\frac{\text{Rate}_t}{\text{Rate}_0} = -kt$$

Solution: $t_{1/2} = \frac{0.693}{k}$ rearrange to solve for k. $k = \frac{0.693}{t_{1/2}} = \frac{0.693}{1.83\ \text{h}} = 0.378689\ \text{h}^{-1}$. Since $\ln\frac{\text{Rate}_t}{\text{Rate}_0} = -kt$

rearrange to solve for t $t = -\frac{1}{k}\ln\frac{\text{Rate}_t}{\text{Rate}_0} = -\frac{1}{0.378689\ \text{h}^{-1}}\ln\frac{1.0 \times 10^2\ /\text{s}}{1.5 \times 10^5\ /\text{s}} = 19.3\ \text{h}$.

Check: The units (h) are correct. The time is between 10 and 11 half-lives and the rate is just under $1/2^{10}$ of the original amount.

20. **Given:** Tl-201 initial decay rate = 5.88×10^4 /s, $t_{1/2}$ for Tl-201 = 3.042 days **Find:** t to decay rate of 55 /s
Conceptual plan: radioactive decay implies first order kinetics, $t_{1/2} \rightarrow k$ then $\text{Rate}_0, \text{Rate}_t, k \rightarrow t$

$$t_{1/2} = \frac{0.693}{k} \qquad \ln\frac{\text{Rate}_t}{\text{Rate}_0} = -kt$$

Solution: $t_{1/2} = \frac{0.693}{k}$ rearrange to solve for k. $k = \frac{0.693}{t_{1/2}} = \frac{0.69315}{3.042\ \text{days}} = 0.2278590\ \text{day}^{-1}$. Since

$\ln\frac{\text{Rate}_t}{\text{Rate}_0} = -kt$ rearrange to solve for t

$$t = -\frac{1}{k}\ln\frac{\text{Rate}_t}{\text{Rate}_0} = -\frac{1}{0.2278590\ \text{day}^{-1}}\ln\frac{55\ /\text{s}}{5.88 \times 10^4\ /\text{s}} = 30.609\ \text{days} = 31\ \text{days}.$$

Check: The units (days) are correct. The time is between 10 and 11 half-lives and the rate is just under $1/2^{10}$ of the original amount.

21. **Given:** boat analysis, C-14/C-12 = 72.5 % of living organism **Find:** t **Other:** $t_{1/2}$ for decay of C-14 = 5730 years
Conceptual plan: radioactive decay implies first order kinetics, $t_{1/2} \rightarrow k$ then 72.5 % of $m_{C\text{-}14\,0}$, $k \rightarrow t$

$$t_{1/2} = \frac{0.693}{k} \qquad \ln N_t = -kt + \ln N_0$$

Solution: $t_{1/2} = \dfrac{0.693}{k}$ rearrange to solve for k. $k = \dfrac{0.693}{t_{1/2}} = \dfrac{0.693}{5730\text{ yr}} = 1.2\underline{0}942 \times 10^{-4}\text{ yr}^{-1}$ then

$[\text{C-14}]_t = 0.725\,[\text{C-14}]_0$. Since $\ln m_{C\text{-}14\,t} = -kt + \ln m_{C\text{-}14\,0}$ rearrange to solve for t.

$$t = -\frac{1}{k}\ln\frac{m_{C\text{-}14\,t}}{m_{C\text{-}14\,0}} = -\frac{1}{1.2\underline{0}942 \times 10^{-4}\text{ yr}^{-1}}\ln\frac{0.725\,\cancel{m_{C\text{-}14\,0}}}{\cancel{m_{C\text{-}14\,0}}} = 2.66 \times 10^3\text{ yr}.$$

Check: The units (yr) are correct. The time to 72.5 % decay is consistent a time less than one half-life.

22. **Given:** peat analysis, C-14/C-12 = 22.8 % of living organism **Find:** t
Other: $t_{1/2}$ for decay of C-14 = 5730 years
Conceptual plan: radioactive decay implies first order kinetics, $t_{1/2} \rightarrow k$ then 72.5 % of $[\text{C-14}]_0$, $k \rightarrow t$

$$t_{1/2} = \frac{0.693}{k} \qquad \ln N_t = -kt + \ln N_0$$

Solution: $t_{1/2} = \dfrac{0.693}{k}$ rearrange to solve for k. $k = \dfrac{0.693}{t_{1/2}} = \dfrac{0.693}{5730\text{ yr}} = 1.2\underline{0}942 \times 10^{-4}\text{ yr}^{-1}$ then

$[\text{C-14}]_t = 0.228\,[\text{C-14}]_0$. Since $\ln m_{C\text{-}14\,t} = -kt + \ln m_{C\text{-}14\,0}$, rearrange to solve for t.

$$t = -\frac{1}{k}\ln\frac{m_{C\text{-}14\,t}}{m_{C\text{-}14\,0}} = -\frac{1}{1.2\underline{0}942 \times 10^{-4}\text{ yr}^{-1}}\ln\frac{0.228\,\cancel{m_{C\text{-}14\,0}}}{\cancel{m_{C\text{-}14\,0}}} = 1.22 \times 10^4\text{ yr}.$$

Check: The units (yr) are correct. The time to 22.8 % decay is consistent a time just more than two half-lives.

23. **Given:** skull analysis, C-14 decay rate = 15.3 dis/min·gC in living organisms and 0.85 dis/min·gC in skull
Find: t **Other:** $t_{1/2}$ for decay of C-14 = 5730 years
Conceptual plan: radioactive decay implies first order kinetics, $t_{1/2} \rightarrow k$ then Rate_0, Rate_t, $k \rightarrow t$

$$t_{1/2} = \frac{0.693}{k} \qquad \ln\frac{\text{Rate}_t}{\text{Rate}_0} = -kt$$

Solution: $t_{1/2} = \dfrac{0.693}{k}$ rearrange to solve for k. $k = \dfrac{0.693}{t_{1/2}} = \dfrac{0.693}{5730\text{ yr}} = 1.2\underline{0}942 \times 10^{-4}\text{ yr}^{-1}$

Since $\ln\dfrac{\text{Rate}_t}{\text{Rate}_0} = -kt$, rearrange to solve for t.

$$t = -\frac{1}{k}\ln\frac{\text{Rate}_t}{\text{Rate}_0} = -\frac{1}{1.2\underline{0}942 \times 10^{-4}\text{ yr}^{-1}}\ln\frac{0.85\,\cancel{\text{dis/min}\cdot\text{gC}}}{15.3\,\cancel{\text{dis/min}\cdot\text{gC}}} = 2.39 \times 10^4\text{ yr}.$$

Check: The units (yr) are correct. The rate is 6 % of initial value and the time is consistent a time just more than four half-lives.

24. **Given:** mammoth analysis, C-14 decay rate = 15.3 dis/min·gC in living organisms and 0.48 dis/min·gC in mammoth **Find:** when did the mammoth live **Other:** $t_{1/2}$ for decay of C-14 = 5730 years
Conceptual plan: radioactive decay implies first order kinetics, $t_{1/2} \rightarrow k$ then Rate_0, Rate_t, $k \rightarrow t$

$$t_{1/2} = \frac{0.693}{k} \qquad \ln\frac{\text{Rate}_t}{\text{Rate}_0} = -kt$$

Solution: $t_{1/2} = \dfrac{0.693}{k}$ rearrange to solve for k. $k = \dfrac{0.693}{t_{1/2}} = \dfrac{0.693}{5730\text{ yr}} = 1.2\underline{0}942 \times 10^{-4}\text{ yr}^{-1}$

Since $\ln\dfrac{\text{Rate}_t}{\text{Rate}_0} = -kt$, rearrange to solve for t.

$$t = -\frac{1}{k}\ln\frac{\text{Rate}_t}{\text{Rate}_0} = -\frac{1}{1.20942 \times 10^{-4}\ \text{yr}^{-1}}\ln\frac{0.48\ \cancel{\text{dis/min}\cdot\text{gC}}}{15.3\ \cancel{\text{dis/min}\cdot\text{gC}}} = 2.9 \times 10^4\ \text{yr ago.}$$

Check: The units (yr) are correct. The rate is 3 % of initial value and the time is consistent a time just more than five half-lives.

25. **Given:** rock analysis, 0.438 g Pb-206 to every 1.00 g U-238, no Pb-206 initially **Find:** age of rock
Other: $t_{1/2}$ for decay of U-238 to Pb-206 = 4.5 x 10^9 years
Conceptual plan: radioactive decay implies first order kinetics, $t_{1/2}$ → k then

$$t_{1/2} = \frac{0.693}{k}$$

g Pb-206 → mol Pb-206 → mol U-238 → g U-238 then $m_{U\text{-}238\ 0}$, $m_{U\text{-}238\ t}$, k → t

$$\frac{1\ \text{mol Pb-206}}{206\ \text{g Pb-206}} \quad \frac{1\ \text{mol U-238}}{1\ \text{mol Pb-206}} \quad \frac{238\ \text{g U-238}}{1\ \text{mol U-238}} \qquad \ln N_t = -kt + \ln N_0$$

Solution: $t_{1/2} = \frac{0.693}{k}$ rearrange to solve for k. $k = \frac{0.693}{t_{1/2}} = \frac{0.693}{4.5 \times 10^9\ \text{yr}} = 1.5\underline{4} \times 10^{-10}\ \text{yr}^{-1}$ then

$$0.438\ \cancel{\text{g Pb-206}} \times \frac{1\ \cancel{\text{mol Pb-206}}}{206\ \cancel{\text{g Pb-206}}} \times \frac{1\ \cancel{\text{mol U-238}}}{1\ \cancel{\text{mol Pb-206}}} \times \frac{238\ \text{g U-238}}{1\ \cancel{\text{mol U-238}}} = 0.50\underline{6}039\ \text{g U-238}.$$ Since

$$\ln\frac{m_{U\text{-}238\ t}}{m_{U\text{-}238\ 0}} = -kt$$, rearrange to solve for t.

$$t = -\frac{1}{k}\ln\frac{m_{U\text{-}238\ t}}{m_{U\text{-}238\ 0}} = -\frac{1}{1.5\underline{4} \times 10^{-10}\ \text{yr}^{-1}}\ln\frac{1.00\ \text{g}\ \cancel{\text{U-238}}}{(1.00 + 0.50\underline{6}039)\ \text{g}\ \cancel{\text{U-238}}} = 2.7 \times 10^9\ \text{yr}.$$

Check: The units (yr) are correct. The amount of Pb-206 is less than half of the initial U-238 amount and time is less than one half-life.

26. **Given:** meteor analysis, 0.855 g Pb-206 : 1.00 g U-238, no Pb-206 initially **Find:** age of meteor
Other: $t_{1/2}$ for decay of U-238 to Pb-206 = 4.5 x 10^9 years
Conceptual plan: radioactive decay implies first order kinetics, $t_{1/2}$ → k then

$$t_{1/2} = \frac{0.693}{k}$$

g Pb-206 → mol Pb-206 → mol U-238 → g U-238 then $m_{U\text{-}238\ 0}$, $m_{U\text{-}238\ t}$, k → t

$$\frac{1\ \text{mol Pb-206}}{206\ \text{g Pb-206}} \quad \frac{1\ \text{mol U-238}}{1\ \text{mol Pb-206}} \quad \frac{238\ \text{g U-238}}{1\ \text{mol U-238}} \qquad \ln N_t = -kt + \ln N_0$$

Solution: $t_{1/2} = \frac{0.693}{k}$ rearrange to solve for k. $k = \frac{0.693}{t_{1/2}} = \frac{0.693}{4.5 \times 10^9\ \text{yr}} = 1.5\underline{4} \times 10^{-10}\ \text{yr}^{-1}$ then

$$0.855\ \cancel{\text{g Pb-206}} \times \frac{1\ \cancel{\text{mol Pb-206}}}{206\ \cancel{\text{g Pb-206}}} \times \frac{1\ \cancel{\text{mol U-238}}}{1\ \cancel{\text{mol Pb-206}}} \times \frac{238\ \text{g U-238}}{1\ \cancel{\text{mol U-238}}} = 0.98\underline{7}816\ \text{g U-238}.$$ Since

$$\ln\frac{m_{U\text{-}238\ t}}{m_{U\text{-}238\ 0}} = -kt$$, rearrange to solve for t.

$$t = -\frac{1}{k}\ln\frac{m_{U\text{-}238\ t}}{m_{U\text{-}238\ 0}} = -\frac{1}{1.5\underline{4} \times 10^{-10}\ \text{yr}^{-1}}\ln\frac{1.00\ \text{g}\ \cancel{\text{U-238}}}{(1.00 + 0.98\underline{7}816)\ \text{g}\ \cancel{\text{U-238}}} = 4.5 \times 10^9\ \text{yr}.$$

Check: The units (yr) are correct. The amount of Pb-206 is just less than the initial U-238 amount and time is just under one half-life.

27. **Given:** U-235 fission induced by neutrons to Xe-144 and Sr-90 **Find:** number of neutrons produced
Conceptual Plan: Write species given on the appropriate side of the equation. →Equalize the sum of the mass numbers and the sum of the atomic numbers on both sides of the equation by writing the stoichiometric coefficient in front of the desired species.
Solution: ${}^{235}_{92}\text{U} + {}^{1}_{0}\text{n} \rightarrow {}^{144}_{54}\text{Xe} + {}^{90}_{38}\text{Sr} + ?\,{}^{1}_{0}\text{n}$ becomes ${}^{235}_{92}\text{U} + {}^{1}_{0}\text{n} \rightarrow {}^{144}_{54}\text{Xe} + {}^{90}_{38}\text{Sr} + 2\,{}^{1}_{0}\text{n}$ so 2 neutrons are produced.
Check: 235 + 1 = 144 + 90 + 2, 92 + 0 = 54 + 38 + 0, and no other particle is necessary to balance the equation.

28. **Given:** U-235 fission to Te-137 and Zr-97 **Find:** number of neutrons produced
Conceptual Plan: Write species given on the appropriate side of the equation. →Equalize the sum of the mass numbers and the sum of the atomic numbers on both sides of the equation by writing the stoichiometric coefficient in front of the desired species.
Solution: $^{235}_{92}U + ^{1}_{0}n \rightarrow ^{137}_{52}Te + ^{97}_{40}Zr + ?^{1}_{0}n$ becomes $^{235}_{92}U + ^{1}_{0}n \rightarrow ^{137}_{52}Te + ^{97}_{40}Zr + 2^{1}_{0}n$ so 2 neutrons are produced.
Check: 235 + 1 = 137 + 97 + 2, 92 + 0 = 52 + 40 + 0, and no other particle is necessary to balance the equation.

29. **Given:** fusion of 2 H-2 atoms to form He-3 and 1 neutron **Find:** balanced equation
Conceptual Plan: Write species given on the appropriate side of the equation. →Equalize the sum of the mass numbers and the sum of the atomic numbers on both sides of the equation by writing the stoichiometric coefficient in front of the desired species.
Solution: $2\,^{2}_{1}H \rightarrow ^{3}_{2}He + ^{1}_{0}n$.
Check: 2(2) = 3 + 1, 2(1) = 2 + 0, and no other particle is necessary to balance the equation.

30. **Given:** fusion of H-3 and H-1 atoms to form He-4 **Find:** balanced equation
Conceptual Plan: Write species given on the appropriate side of the equation. →Equalize the sum of the mass numbers and the sum of the atomic numbers on both sides of the equation by writing the stoichiometric coefficient in front of the desired species.
Solution: $^{3}_{1}H + ^{1}_{1}H \rightarrow ^{4}_{2}He$.
Check: 3 + 1 = 4, 1 + 1 = 2, and no other particle is necessary to balance the equation.

31. **Given:** U-238 bombarded by neutrons to form U-239 which undergoes 2 beta decays to form Pu-239
Find: balanced equations
Conceptual Plan: Write species given on the appropriate side of the equation. →Equalize the sum of the mass numbers and the sum of the atomic numbers on both sides of the equation by writing the stoichiometric coefficient in front of the desired species. → Use the product of this reaction to write the next reaction until process is complete.
Solution: $^{238}_{92}U + ?^{1}_{0}n \rightarrow ^{239}_{92}U$ becomes $^{238}_{92}U + ^{1}_{0}n \rightarrow ^{239}_{92}U$ then
beta decay $^{239}_{92}U \rightarrow ^{?}_{?}? + ^{0}_{-1}e$ becomes $^{239}_{92}U \rightarrow ^{239}_{93}? + ^{0}_{-1}e$ then $^{239}_{92}U \rightarrow ^{239}_{93}Np + ^{0}_{-1}e$ then
beta decay $^{239}_{93}Np \rightarrow ^{?}_{?}? + ^{0}_{-1}e$ becomes $^{239}_{93}Np \rightarrow ^{239}_{94}? + ^{0}_{-1}e$ then $^{239}_{93}Np \rightarrow ^{239}_{94}Pu + ^{0}_{-1}e$.
The entire process is $^{238}_{92}U + ^{1}_{0}n \rightarrow ^{239}_{92}U$, $^{239}_{92}U \rightarrow ^{239}_{93}Np + ^{0}_{-1}e$, $^{239}_{93}Np \rightarrow ^{239}_{94}Pu + ^{0}_{-1}e$.
Check: 238 + 1 = 239, 92 + 0 = 92, and no other particle is necessary to balance the equation. 239 = 239 + 0, 92 = 93 – 1, and Neptunium is atomic number 93. 239 = 239 + 0, 93 = 94 – 1, and Plutonium is atomic number 94.

32. **Given:** Al-27 bombarded by a neutron and then undergoes an alpha decay and a beta decay **Find:** balanced equations
Conceptual Plan: Write species given on the appropriate side of the equation. →Equalize the sum of the mass numbers and the sum of the atomic numbers on both sides of the equation by writing the stoichiometric coefficient in front of the desired species. → Use the product of this reaction to write the next reaction until process is complete.
Solution: $^{27}_{13}Al + ^{1}_{0}n \rightarrow ^{?}_{?}?$ becomes $^{27}_{13}Al + ^{1}_{0}n \rightarrow ^{28}_{13}?$ then $^{27}_{13}Al + ^{1}_{0}n \rightarrow ^{28}_{13}Al$ then
alpha decay $^{28}_{13}Al \rightarrow ^{?}_{?}? + ^{4}_{2}He$ becomes $^{28}_{13}Al \rightarrow ^{24}_{11}? + ^{4}_{2}He$ then $^{28}_{13}Al \rightarrow ^{24}_{11}Na + ^{4}_{2}He$ then
beta decay $^{24}_{11}Na \rightarrow ^{?}_{?}? + ^{0}_{-1}e$ becomes $^{24}_{11}Na \rightarrow ^{24}_{12}? + ^{0}_{-1}e$ then $^{24}_{11}Na \rightarrow ^{24}_{12}Mg + ^{0}_{-1}e$.
The entire process is $^{27}_{13}Al + ^{1}_{0}n \rightarrow ^{28}_{13}Al$, $^{28}_{13}Al \rightarrow ^{24}_{11}Na + ^{4}_{2}He$, $^{24}_{11}Na \rightarrow ^{24}_{12}Mg + ^{0}_{-1}e$.
Check: 27 + 1 = 28, 13 + 0 = 13, and Aluminum is atomic number 13. 28 = 24 + 4, 13 = 11 + 2, and Sodium is atomic number 11. 24 = 24 + 0, 11 = 12 – 1, and Magnesium is atomic number 12.

33. **Given:** 1.0 g of matter converted to energy **Find:** energy
Conceptual plan: g → kg → E

$$\frac{1\,\text{kg}}{1000\,\text{g}} \qquad E = m\,c^2$$

Solution: $1.0\,\cancel{g} \times \frac{1\,\text{kg}}{1000\,\cancel{g}} = 0.0010\,\text{kg}$ then $E = m\,c^2 = (0.0010\,\text{kg})\left(2.9979 \times 10^8\,\frac{\text{m}}{\text{s}}\right)^2 = 9.0 \times 10^{13}\,\text{J}$

Check: The units (J) are correct. The magnitude of the answer makes physical sense because we are converting a large quantity of amus to energy.

34. **Given:** 1.0 x 10^3 kWh of electricity/month from nuclear reaction **Find:** mass converted to energy / year
Conceptual plan: kWh → J → kg → g then g/month → g/year

$$\frac{3.60 \times 10^6 \text{ J}}{1 \text{kWh}} \quad E = m c^2 \quad \frac{1000 \text{ g}}{1 \text{ kg}} \quad \frac{12 \text{ months}}{1 \text{ year}}$$

Solution: $1.0 \times 10^3 \text{ kWh} \times \frac{3.60 \times 10^6 \text{ J}}{1 \text{kWh}} = 3.6 \times 10^9 \text{ J}$. Since $E = m c^2$, rearrange to solve for m.

$$m = \frac{E}{c^2} = \frac{3.6 \times 10^9 \text{ kg} \frac{\text{m}^2}{\text{s}^2}}{\left(2.9979 \times 10^8 \frac{\text{m}}{\text{s}}\right)^2} = 4.0 \times 10^{-8} \text{ kg} \times \frac{1000 \text{ g}}{1 \text{ kg}} = 4.0 \times 10^{-5} \text{ g}. \text{ Then}$$

$$\frac{4.0 \times 10^{-5} \text{ g}}{1 \text{ month}} \times \frac{12 \text{ months}}{1 \text{ year}} = \frac{4.8 \times 10^{-4} \text{ g}}{1 \text{ year}}$$

Check: The units (g) are correct. A small mass is expected since nuclear reactions generate a large amount of energy.

35. **Given:** a) O-16 = 15.9949145 amu; b) Ni-58 = 57.935346 amu; and c) Xe-129 = 128.904780 amu
Find: mass defect and nuclear binding energy per nucleon
Conceptual plan: ${}^A_Z\text{X}$**, isotope mass → mass defect → nuclear binding energy per nucleon**

$$\text{mass defect} = \text{Z(mass } {}^1_1\text{H)} + (\text{A} - \text{Z})(\text{mass } {}^1_0\text{n}) - \text{mass of isotope} \qquad \frac{931.5 \text{ MeV}}{(1 \text{ amu})(\text{A nucleons})}$$

Solution: mass defect = $\text{Z(mass } {}^1_1\text{H)} + (\text{A} - \text{Z})(\text{mass } {}^1_0\text{n}) - \text{mass of isotope}$.

a) O-16 mass defect = 8(1.00783 amu) + (16 – 8)(1.00866 amu) – 15.9949145 amu = 0.1370$\underline{0}$55 amu

$$= 0.13701 \text{ amu and } 0.1370\underline{0}55 \text{ amu} \times \frac{931.5 \text{ MeV}}{(1 \text{ amu})(16 \text{ nucleons})} = 7.976 \frac{\text{MeV}}{\text{nucleons}}.$$

b) Ni-58 mass defect = 28(1.00783 amu) + (58 – 28)(1.00866 amu) – 57.935346 amu = 0.5436$\underline{9}$4 amu

$$= 0.54369 \text{ amu and } 0.5436\underline{9}4 \text{ amu} \times \frac{931.5 \text{ MeV}}{(1 \text{ amu})(58 \text{ nucleons})} = 8.732 \frac{\text{MeV}}{\text{nucleons}}.$$

c) Xe-129 mass defect = 54(1.00783 amu) + (129 –54)(1.00866 amu) – 128.904780 amu

$$= 1.16754 \text{ amu and } 1.16754 \text{ amu} \times \frac{931.5 \text{ MeV}}{(1 \text{ amu})(129 \text{ nucleons})} = 8.431 \frac{\text{MeV}}{\text{nucleons}}.$$

Check: The units (amu and MeV/nucleon) are correct. The mass defect increases with an increasing number of nucleons, but the MeV/nucleon does not change by as much (on a relative basis).

36. **Given:** a) Li-7 = 7.016003 amu; b) Ti-48 = 47.947947 amu; and c) Ag-107 = 106.905092 amu
Find: mass defect and nuclear binding energy per nucleon
Conceptual plan: ${}^A_Z\text{X}$**, isotope mass → mass defect → nuclear binding energy per nucleon**

$$\text{mass defect} = \text{Z(mass } {}^1_1\text{H)} + (\text{A} - \text{Z})(\text{mass } {}^1_0\text{n}) - \text{mass of isotope} \qquad \frac{931.5 \text{ MeV}}{(1 \text{ amu})(\text{A nucleons})}$$

Solution: mass defect = $\text{Z(mass } {}^1_1\text{H)} + (\text{A} - \text{Z})(\text{mass } {}^1_0\text{n}) - \text{mass of isotope}$.

a) Li-7 mass defect = 3(1.00783 amu) + (7 – 3)(1.00866 amu) – 7.016003 amu = 0.0421$\underline{2}$7 amu

$$= 0.04213 \text{ amu and } 0.0421\underline{2}7 \text{ amu} \times \frac{931.5 \text{ MeV}}{(1 \text{ amu})(7 \text{ nucleons})} = 5.606 \frac{\text{MeV}}{\text{nucleons}}.$$

b) Ti-48 mass defect = 22(1.00783 amu) + (48 – 22)(1.00866 amu) – 47.947947 amu = 0.4494$\underline{7}$3 amu

$= 0.44947$ amu and $0.449473\ \text{amu} \times \dfrac{931.5\ \text{MeV}}{(1\ \text{amu})(48\ \text{nucleons})} = 8.723\ \dfrac{\text{MeV}}{\text{nucleons}}$.

c) Ag-107 mass defect $= 47(1.00783\ \text{amu}) + (107 - 47)(1.00866\ \text{amu}) - 106.905092\ \text{amu} = 0.982518\ \text{amu} =$

$= 0.98252$ amu and $0.982518\ \text{amu} \times \dfrac{931.5\ \text{MeV}}{(1\ \text{amu})(107\ \text{nucleons})} = 8.553\ \dfrac{\text{MeV}}{\text{nucleons}}$.

Check: The units (amu and MeV/nucleon) are correct. The mass defect increases with an increasing number of nucleons, but the MeV/nucleon does not change by as much (on a relative basis).

37. **Given:** ${}^{235}_{92}\text{U} + {}^{1}_{0}\text{n} \rightarrow {}^{144}_{54}\text{Xe} + {}^{90}_{38}\text{Sr} + 2\,{}^{1}_{0}\text{n}$, U-235 = 235.043922 amu, Xe-144 = 143.9385 amu, and Sr-90 = 89.907738 amu **Find:** energy per g of U-235

Conceptual plan: mass of products & reactants → mass defect → mass defect / g of U-235 then

$$\text{mass defect} = \sum \text{mass of reactants} - \sum \text{mass of products} \qquad \frac{\text{mass defect}}{235.043922\ \text{g U-235}}$$

g → kg → E

$$\frac{1\,\text{kg}}{1000\ \text{g}} \qquad E = m\,c^2$$

Solution: mass defect $= \sum$ mass of reactants $- \sum$ mass of products notice that we can cancel a neutron from each side to get: ${}^{235}_{92}\text{U} \rightarrow {}^{144}_{54}\text{Xe} + {}^{90}_{38}\text{Sr} + {}^{1}_{0}\text{n}$ and

mass defect $= 235.043922\ \text{g} - (143.9385\ \text{g} + 89.907738\ \text{g} + 1.00866\ \text{g}) = 0.189024\ \text{g}$

then $\dfrac{0.189024\ \text{g}}{235.043922\ \text{g U-235}} \times \dfrac{1\,\text{kg}}{1000\ \text{g}} = 8.04207 \times 10^{-7}\ \dfrac{\text{kg}}{\text{g U-235}}$ then

$$E = m\,c^2 = \left(8.04207 \times 10^{-7}\ \frac{\text{kg}}{\text{g U-235}}\right)\left(2.9979 \times 10^{8}\ \frac{\text{m}}{\text{s}}\right)^2 = 7.228 \times 10^{10}\ \frac{\text{J}}{\text{g U-235}}.$$

Check: The units (J) are correct. A large amount of energy is expected per gram of fuel in a nuclear reactor.

38. **Given:** ${}^{235}_{92}\text{U} + {}^{1}_{0}\text{n} \rightarrow {}^{137}_{52}\text{Te} + {}^{97}_{40}\text{Zr} + 2\,{}^{1}_{0}\text{n}$, U-235 = 235.043922 amu, Te-137 = 136.9253 amu, and Zr-97 = 96.910950 amu **Find:** energy per mol of U-235

Conceptual plan: mass of products & reactants → mass defect → mass defect / mol of U-235 then

$$\text{mass defect} = \sum \text{mass of reactants} - \sum \text{mass of products} \qquad \frac{\text{mass defect}}{235.043922\ \text{g U-235}}$$

g → kg → E

$$\frac{1\,\text{kg}}{1000\ \text{g}} \qquad E = m\,c^2$$

Solution: mass defect $= \sum$ mass of reactants $- \sum$ mass of products notice that we can cancel a neutron from each side to get: ${}^{235}_{92}\text{U} \rightarrow {}^{137}_{52}\text{Te} + {}^{97}_{40}\text{Zr} + {}^{1}_{0}\text{n}$ and

mass defect $= 235.043922\ \text{g} - (136.9253\ \text{g} + 96.910950\ \text{g} + 1.00866\ \text{g}) = 0.199012\ \text{g}$

then $0.199012\ \text{g} \times \dfrac{1\,\text{kg}}{1000\ \text{g}} = 1.99012\ \text{g} \times 10^{-4}\ \dfrac{\text{kg}}{\text{mol U-235}}$ then

$$E = m\,c^2 = \left(1.99012\ \text{g} \times 10^{-4}\ \frac{\text{kg}}{\text{mol U-235}}\right)\left(2.9979 \times 10^{8}\ \frac{\text{m}}{\text{s}}\right)^2 = 1.789 \times 10^{13}\ \frac{\text{J}}{\text{mol U-235}}.$$

Check: The units (J) are correct. A large amount of energy is expected per gram of fuel in a nuclear reactor.

39. **Given:** $2\,{}^{2}_{1}\text{H} \rightarrow {}^{3}_{2}\text{He} + {}^{1}_{0}\text{n}$, H-2 = 2.014102 amu, and He-3 = 3.016029 amu **Find:** energy per g reactant

Conceptual plan: mass of products & reactants → mass defect → mass defect / g of H-2 then

$$\text{mass defect} = \sum \text{mass of reactants} - \sum \text{mass of products} \qquad \frac{\text{mass defect}}{2(2.014102\ \text{g H-2})}$$

g → kg → E

$\frac{1\,kg}{1000\,g}$ $E = m\,c^2$

Solution: mass defect = $\sum$mass of reactants – $\sum$mass of products and

mass defect = 2(2.014102 g) – (3.016029 g + 1.00866 g) = 0.003515 g

then $\frac{0.003515\ \cancel{g}}{2(2.014102\ \text{g H-2})} \times \frac{1\text{kg}}{1000\ \cancel{g}} = 8.72597 \times 10^{-7} \frac{\text{kg}}{\text{g H-2}}$ then

$$E = m\,c^2 = \left(8.72597 \times 10^{-7} \frac{\text{kg}}{\text{g H-2}}\right)\left(2.9979 \times 10^8 \frac{\text{m}}{\text{s}}\right)^2 = 7.84 \times 10^{10} \frac{\text{J}}{\text{g H-2}}.$$

Check: The units (J) are correct. A large amount of energy is expected per gram of fuel in a nuclear reactor.

40. **Given:** ${}^3_1\text{H} + {}^1_1\text{H} \rightarrow {}^4_2\text{He}$, H-3 = 3.016049 amu, H-1 = 1.007825 amu, and He-4 = 4.002603 amu
Find: energy per g reactant
Conceptual plan: mass of products & reactants → mass defect → mass defect / g reactant then

mass defect = $\sum$mass of reactants – $\sum$mass of products $\frac{\text{mass defect}}{\text{g reactant}}$

g → kg → E

$\frac{1\,kg}{1000\,g}$ $E = m\,c^2$

Solution: mass defect = $\sum$mass of reactants – $\sum$mass of products and

mass defect = (3.016049 g + 1.007825 g) – 4.002603 g = 4.023874 g – 4.002603 g = 0.021271 g

then $\frac{0.021271\ \cancel{g}}{4.023874\ \text{g reactants}} \times \frac{1\text{kg}}{1000\ \cancel{g}} = 5.2861993 \times 10^{-6} \frac{\text{kg}}{\text{g reactants}}$ then

$$E = m\,c^2 = \left(5.2861993 \times 10^{-6} \frac{\text{kg}}{\text{g H-2}}\right)\left(2.9979 \times 10^8 \frac{\text{m}}{\text{s}}\right)^2 = 4.7509 \times 10^{11} \frac{\text{J}}{\text{g reactants}}.$$

Check: The units (J) are correct. A large amount of energy is expected per gram of fuel in a nuclear reactor.

41. **Given:** 75 kg human exposed to 32.8 rad and falling from chair **Find:** energy absorbed in each case
Conceptual plan: rad, kg → J and assume d = 0.50 m chair height then mass, d → J

$\frac{0.01\,\text{J}}{1\ \text{kg body tissue}}$ $E = F \cdot d = m\,g\,d$

Solution: 32.8 rad = $32.8 \frac{0.01\text{J}}{1\ \cancel{\text{kg body tissue}}} \times 75\ \cancel{\text{kg}} = 25\ \text{J}$ and

$E = F \cdot d = m\,g\,d = 75\ \text{kg} \times 9.8 \frac{\text{m}}{\text{s}^2} \times 0.50\ \text{m} = 370\ \text{kg}\frac{\text{m}^2}{\text{s}^2} = 370\ \text{J}$.

Check: The units (J and J) are correct. Allowable radiation exposures are low, since the radiation is very ionizing and, thus, damaging to tissue. Falling may have more energy, but it is not ionizing.

42. **Given:** 55 g mouse exposed to 20.5 rad **Find:** energy absorbed
Conceptual plan: g → kg then rad, kg → J

$\frac{1\,kg}{1000\,g}$ $\frac{0.01\,\text{J}}{1\ \text{kg body tissue}}$

Solution: $55\ \cancel{g} \times \frac{1\text{kg}}{1000\ \cancel{g}} = 0.055\ \text{kg}$ then 20.5 rad = $20.5 \frac{0.01\text{J}}{1\ \cancel{\text{kg body tissue}}} \times 0.055\ \cancel{\text{kg}} = 0.011\ \text{J}$.

Check: The units (J) are correct. Allowable radiation exposures are low, since the radiation is very ionizing and, thus, damaging to tissue.

43. **Given:** $t_{1/2}$ for F-18 = 1.83 h, 65 % of F-18 makes it to the hospital traveling at 60.0 miles/hour
Find: distance between hospital and cyclotron

Conceptual plan: $t_{1/2} \rightarrow k$ **then** $m_{F\text{-}18\ 0}, m_{F\text{-}18\ t}, k \rightarrow t$ **then** $h \rightarrow mi$

$$t_{1/2} = \frac{0.693}{k} \qquad \ln\frac{m_{F\text{-}18\ t}}{m_{F\text{-}18\ 0}} = -kt \qquad \frac{60.0\ mi}{1\ h}$$

Solution: $t_{1/2} = \frac{0.693}{k}$ rearrange to solve for k. $k = \frac{0.693}{t_{1/2}} = \frac{0.693}{1.83\ h} = 0.37\underline{8}689\ h^{-1}$. Since $\ln\frac{m_{F\text{-}18\ t}}{m_{F\text{-}18\ 0}} = -kt$

rearrange to solve for t $t = -\frac{1}{k}\ln\frac{m_{F\text{-}18\ t}}{m_{F\text{-}18\ 0}} = -\frac{1}{0.37\underline{8}689\ h^{-1}}\ln\frac{0.65\ \cancel{m_{F\text{-}18\ 0}}}{\cancel{m_{F\text{-}18\ 0}}} = 1.\underline{1}376\ h$. Then

$1.\underline{1}376\ \cancel{h} \times \frac{60.0\ mi}{1\ \cancel{h}} = 68\ mi$.

Check: The units (mi) are correct. The time less than one half-life, so the distance is less than 1.83 times the speed of travel.

44. **Given:** I-131, 155 mg, $t_{1/2}$ = 8.0 days **Find:** exposure (in Ci) after 4.0 h
Conceptual plan: **h → day and** $t_{1/2} \rightarrow k$ **then** $m_{I\text{-}131\ 0}, t, k \rightarrow m_{I\text{-}131\ t}$

$$\frac{1\ day}{24\ hr} \qquad t_{1/2} = \frac{0.693}{k} \qquad \ln N_t = -kt + \ln N_0$$

then mg_0, mg_t **→ mg decayed → g decayed → mol decayed → beta decays then h → min → s**

$$mg_0 - mg_t = mg\ decayed \qquad \frac{1\ g}{1000\ mg} \qquad \frac{1\ mol\ I\text{-}131}{131\ g\ I\text{-}131} \qquad \frac{6.022 \times 10^{23}\ beta\ decays}{1\ mol\ I\text{-}131} \qquad \frac{60\ min}{1\ h} \qquad \frac{60\ s}{1\ min}$$

then beta decays, s → beta decays / s → Ci

$$\text{take ratio} \qquad \frac{1\ Ci}{\frac{3.7 \times 10^{10}\ decays}{s}}$$

Solution: $4.0\ \cancel{h} \times \frac{1\ day}{24\ \cancel{h}} = 0.1\underline{6}667$ day then $t_{1/2} = \frac{0.693}{k}$ rearrange to solve for k.

$k = \frac{0.693}{t_{1/2}} = \frac{0.693}{8.0\ days} = 0.08\underline{6}625\ day^{-1}$. Since

$\ln m_{I\text{-}131\ t} = -kt + \ln m_{I\text{-}131\ 0} = -(0.08\underline{6}625\ \cancel{day^{-1}})(0.1\underline{6}667\ \cancel{day}) + \ln(155\ mg) = 5.02\underline{8}99 \rightarrow$

$m_{I\text{-}131\ t} = e^{5.02899} = 152.\underline{7}78$ mg. Then $mg_0 - mg_t$ = mg decayed = 155 mg − 152.$\underline{7}$78 mg = $\underline{2}$.222 mg I-131 then

$\underline{2}.222\ \cancel{mg\ I\text{-}131} \times \frac{1\ \cancel{g\ I\text{-}131}}{1000\ \cancel{mg\ I\text{-}131}} \times \frac{1\ \cancel{mol\ I\text{-}131}}{131\ \cancel{g\ I\text{-}131}} \times \frac{6.022 \times 10^{23}\ beta\ decays}{1\ \cancel{mol\ I\text{-}131}} = \underline{1}.0213 \times 10^{19}$ beta decays then

$4.0\ \cancel{h} \times \frac{60\ \cancel{min}}{1\ \cancel{h}} \times \frac{60\ s}{1\ \cancel{min}} = 1.44 \times 10^4\ s$ then

$\frac{\underline{1}.0213 \times 10^{19}\ \cancel{beta\ decays}}{1.44 \times 10^4\ \cancel{s}} \times \frac{1\ Ci}{\frac{3.7 \times 10^{10}\ \cancel{decays}}{\cancel{s}}} = \underline{1}.9169 \times 10^4\ Ci = 2 \times 10^4\ Ci$

Check: The units (Ci) are correct. The amount that decays is large since the half-life is fairly short and so the dose is high.

45. **Given:** a) Ru-114, b) Ra-216, c) Zn-58, and d) Ne-31 **Find:** write nuclear equation for most likely decay
Conceptual Plan: Decide on most likely decay mode depending on N/Z (too large = beta decay, too low = positron emission) → Write the symbol for parent nuclide on the left side of the equation and the symbol for a particle on the right side. → Equalize the sum of the mass numbers and the sum of the atomic numbers on both sides of the equation by writing the appropriate mass number and atomic number for the unknown daughter nuclide. → Using the periodic table, deduce the identity of the unknown daughter nuclide from the atomic number and write its symbol.
Solution:

a) Ru-114 will undergo beta decay ${}^{114}_{44}Ru \rightarrow {}^{?}_{?}? + {}^{0}_{-1}e$ then ${}^{114}_{44}Ru \rightarrow {}^{114}_{45}? + {}^{0}_{-1}e$ then ${}^{114}_{44}Ru \rightarrow {}^{114}_{45}Rh + {}^{0}_{-1}e$

b) Ra-216 will undergo positron emission ${}^{216}_{88}Ra \rightarrow {}^{?}_{?}? + {}^{0}_{+1}e$ then ${}^{216}_{88}Ra \rightarrow {}^{216}_{87}? + {}^{0}_{+1}e$ then

$^{216}_{88}\text{Ra} \rightarrow ^{216}_{87}\text{Fr} + ^{0}_{+1}\text{e}$

c) Zn-58 will undergo positron emission $^{58}_{30}\text{Zn} \rightarrow ^{?}_{?}? + ^{0}_{+1}\text{e}$ then $^{58}_{30}\text{Zn} \rightarrow ^{58}_{29}? + ^{0}_{+1}\text{e}$ then $^{58}_{30}\text{Zn} \rightarrow ^{58}_{29}\text{Cu} + ^{0}_{+1}\text{e}$

d) Ne-31 will undergo beta decay $^{31}_{10}\text{Ne} \rightarrow ^{?}_{?}? + ^{0}_{-1}\text{e}$ then $^{31}_{10}\text{Ne} \rightarrow ^{31}_{11}? + ^{0}_{-1}\text{e}$ then $^{31}_{10}\text{Ne} \rightarrow ^{31}_{11}\text{Na} + ^{0}_{-1}\text{e}$

Check: a) 114 = 114 + 0, 44 = 45 – 1, and Rhodium is atomic number 45. b) 216 = 216 + 0, 88 = 87 + 1, and Francium is atomic number 87. c) 58 = 58 + 0, 30 = 29 + 1, and Copper is atomic number 29. d) 31 = 31 + 0, 10 = 11 – 1, and Sodium is atomic number 11.

46. **Given:** a) Kr-74, b) Th-221, c) Ar-44, and d) Nb-85 **Find:** write nuclear equation for most likely decay
Conceptual Plan: Decide on most likely decay mode depending on N/Z (too large = beta decay, too low = positron emission) → Write the symbol for parent nuclide on the left side of the equation and the symbol for a particle on the right side. → Equalize the sum of the mass numbers and the sum of the atomic numbers on both sides of the equation by writing the appropriate mass number and atomic number for the unknown daughter nuclide. → Using the periodic table, deduce the identity of the unknown daughter nuclide from the atomic number and write its symbol.
Solution:

a) Kr-74 will undergo positron emission $^{74}_{36}\text{Kr} \rightarrow ^{?}_{?}? + ^{0}_{+1}\text{e}$ then $^{74}_{36}\text{Kr} \rightarrow ^{74}_{35}? + ^{0}_{+1}\text{e}$ then $^{74}_{36}\text{Kr} \rightarrow ^{74}_{35}\text{Br} + ^{0}_{+1}\text{e}$

b) Th-221 will undergo positron emission $^{221}_{90}\text{Th} \rightarrow ^{?}_{?}? + ^{0}_{+1}\text{e}$ then $^{221}_{90}\text{Th} \rightarrow ^{221}_{89}? + ^{0}_{+1}\text{e}$ then
$^{221}_{90}\text{Th} \rightarrow ^{221}_{89}\text{Ac} + ^{0}_{+1}\text{e}$

c) Ar-44 will undergo beta decay $^{44}_{18}\text{Ar} \rightarrow ^{?}_{?}? + ^{0}_{-1}\text{e}$ then $^{44}_{18}\text{Ar} \rightarrow ^{44}_{19}? + ^{0}_{-1}\text{e}$ then $^{44}_{18}\text{Ar} \rightarrow ^{44}_{19}\text{K} + ^{0}_{-1}\text{e}$

d) Nb-85 will undergo positron emission $^{85}_{41}\text{Nb} \rightarrow ^{?}_{?}? + ^{0}_{+1}\text{e}$ then $^{85}_{41}\text{Nb} \rightarrow ^{85}_{40}? + ^{0}_{+1}\text{e}$ then $^{85}_{41}\text{Nb} \rightarrow ^{85}_{40}\text{Zr} + ^{0}_{+1}\text{e}$

Check: a) 74 = 74 + 0, 36 = 35 + 1, and Bromine is atomic number 35. b) 221 = 221 + 0, 90 = 89 + 1, and Actinium is atomic number 89. c) 44 = 44 + 0, 18 = 19 – 1, and Potassium is atomic number 19. d) 85 = 85 + 0, 41 = 40 + 1, and Zirconium is atomic number 40.

47. **Given:** Bi-210, $t_{1/2}$ = 5.0 days, 1.2 g Bi-210, 209.984105 amu, 5.5 % absorbed
Find: beta emissions in 15.5 days and dose (in Ci)
Conceptual plan: $t_{1/2} \rightarrow k$ **then** $m_{\text{Bi-210 }0}, t, k \rightarrow m_{\text{Bi-210 }t}$ **then**

$$t_{1/2} = \frac{0.693}{k} \qquad \ln N_t = -kt + \ln N_0$$

g_0, g_t → **g decayed → mol decayed → beta decays then day → h → min → s then**

$$g_0 - g_t = \text{g decayed} \quad \frac{1\text{ mol Bi-210}}{209.984105\text{ g Bi-210}} \quad \frac{6.022 \times 10^{23}\text{ beta decays}}{1\text{ mol Bi-210}} \quad \frac{1\text{ day}}{24\text{ h}} \quad \frac{60\text{ min}}{1\text{ h}} \quad \frac{60\text{ s}}{1\text{ min}}$$

beta decays, s → beta decays / s → Ci available → Ci absorbed

$$\text{take ratio} \quad \frac{1\text{ Ci}}{\frac{3.7 \times 10^{10}\text{ decays}}{\text{s}}} \quad \frac{5.5\text{ Ci absorbed}}{100\text{ Ci emitted}}$$

Solution: $t_{1/2} = \frac{0.693}{k}$ rearrange to solve for k. $k = \frac{0.693}{t_{1/2}} = \frac{0.693}{5.0\text{ days}} = 0.1\underline{3}86\text{ day}^{-1}$. Since

$\ln m_{\text{Bi-210 t}} = -kt + \ln m_{\text{Bi-210 0}} = -(0.1\underline{3}86\text{ day}^{-1})(13.5\text{ day}) + \ln(1.2\text{ g}) = -1.\underline{6}888$ →

$m_{\text{Bi-210 t}} = e^{-1.6888} = 0.1\underline{8}475\text{ g}$. then $g_0 - g_t = \text{g decayed} = 1.2\text{ g} - 0.1\underline{8}475\text{ g} = 1.\underline{0}153\text{ g Bi-210}$ then

$$1.\underline{0}153\text{ g Bi-210} \times \frac{1\text{ mol Bi-210}}{209.984105\text{ g Bi-210}} \times \frac{6.022 \times 10^{23}\text{ beta decays}}{1\text{ mol Bi-210}} = 2.\underline{9}116 \times 10^{21}\text{ beta decays}$$

$$= 2.9 \times 10^{21}\text{ beta decays then } 13.5\text{ day} \times \frac{24\text{ h}}{1\text{ day}} \times \frac{60\text{ min}}{1\text{ h}} \times \frac{60\text{ s}}{1\text{ min}} = 1.1\underline{6}64 \times 10^6\text{ s then}$$

$$\frac{2.\underline{9}116 \times 10^{21}\text{ beta decays}}{1.1\underline{6}64 \times 10^6\text{ s}} \times \frac{1\text{ Ci}}{\frac{3.7 \times 10^{10}\text{ decays}}{\text{s}}} = 6.\underline{7}466 \times 10^4\text{ Ci emitted} \times \frac{5.5\text{ Ci absorbed}}{100\text{ Ci emitted}} = 3700\text{ Ci}.$$

Check: The units (decays and Ci) are correct. The amount that decays is large since the time is over 3 half-lives and we have a relatively large amount of the isotope. Since the decay is large, the dosage is large.

48. **Given:** Po-218, $t_{1/2}$ = 3.0 minutes, 55 mg Po-218, 218.008965 amu **Find:** alpha emissions in 25.0 min and dose (in Ci)
Conceptual plan: $t_{1/2} \rightarrow k$ then $m_{Po\text{-}218\ 0}, t, k \rightarrow m_{Po\text{-}218\ t}$ then

$$t_{1/2} = \frac{0.693}{k} \qquad \ln N_t = -kt + \ln N_0$$

mg_0, mg_t → mg decayed → g decayed → mol decayed → alpha decays then min → s

$$mg_0 - mg_t = \text{mg decayed} \quad \frac{1\text{ g}}{1000\text{ mg}} \quad \frac{1\text{ mol Po-218}}{218.008965\text{ g Po-218}} \quad \frac{6.022 \times 10^{23}\text{ beta decays}}{1\text{ mol Po-218}} \quad \frac{60\text{ s}}{1\text{ min}}$$

then beta decays, s → beta decays / s → Ci

$$\text{take ratio} \qquad \frac{1\text{ Ci}}{\frac{3.7 \times 10^{10}\text{ decays}}{\text{s}}}$$

Solution: $t_{1/2} = \frac{0.693}{k}$ rearrange to solve for k. $k = \frac{0.693}{t_{1/2}} = \frac{0.693}{3.0\text{ min}} = 0.2\underline{3}1\text{ min}^{-1}$. Since

$\ln m_{Po\text{-}218\ t} = -kt + \ln m_{Po\text{-}218\ 0} = -(0.2\underline{3}1\ \cancel{\text{min}^{-1}})(25.0\ \cancel{\text{min}}) + \ln(55\text{ mg}) = -1.\underline{7}677$ →

$m_{Po\text{-}218\ t} = e^{-1.\underline{7}677} = 0.1\underline{7}073\text{ mg}$. then $mg_0 - mg_t$ = mg decayed = 55 mg – 0.1$\underline{7}$073 mg = 5$\underline{4}$.829 mg Po-218 then 5$\underline{4}$.829 mg Po-218

$$5\underline{4}.829\ \cancel{\text{mg Po-218}} \times \frac{1\ \cancel{\text{g Po-218}}}{1000\ \cancel{\text{mg Po-218}}} \times \frac{1\ \cancel{\text{mol Po-218}}}{218.008965\ \cancel{\text{g Po-218}}} \times \frac{6.022 \times 10^{23}\text{ alpha decays}}{1\ \cancel{\text{mol Po-218}}} =$$

$1.\underline{5}145 \times 10^{20}$ alpha decays $= 1.5 \times 10^{20}$ alpha decays then $25.0\ \cancel{\text{min}} \times \frac{60\text{ s}}{1\ \cancel{\text{min}}} = 1.5 \times 10^3\text{ s}$ then

$$\frac{1.\underline{5}145 \times 10^{20}\ \cancel{\text{alpha decays}}}{1.5 \times 10^3\ \cancel{\text{s}}} \times \frac{1\text{ Ci}}{\frac{3.7 \times 10^{10}\ \cancel{\text{decays}}}{\cancel{\text{s}}}} = 2.7 \times 10^6\text{ Ci}.$$

Check: The units (decays and Ci) are correct. The amount that decays is large since the time is over 8 half-lives, so almost the entire isotope has decayed. Since the decay rate is large (small half-life), the dosage is large.

49. **Given:** Ra-226 (226.05402 amu) decays to Rn-224, $t_{1/2} = 1.6 \times 10^3$ yr, 25.0 g Ra-226, T = 25.0 °C, P = 1.0 atm
Find: V of Rn-224 gas produced in 5.0 day
Conceptual plan: day → yr then $t_{1/2} \rightarrow k$ **then** $m_{Ra\text{-}226\ 0}, t, k \rightarrow m_{Ra\text{-}226\ t}$

$$\frac{1\text{ yr}}{365.24\text{ day}} \qquad t_{1/2} = \frac{0.693}{k} \qquad \ln N_t = -kt + \ln N_0$$

then g_0, g_t → g decayed → mol decayed → mol Rn-224 formed then °C → K then

$$g_0 - g_t = \text{g decayed} \quad \frac{1\text{ mol Ra-226}}{226.05402\text{ g Ra-226}} \quad \frac{1\text{ mol Rn-224}}{1\text{ mol Ra-226}} \qquad K = °C + 273.15$$

then P, n, T → V

$$PV = nRT$$

Solution: $5.0\ \cancel{\text{day}} \times \frac{1\text{ yr}}{365.24\ \cancel{\text{day}}} = 0.01\underline{3}690\text{ yr}$ then $t_{1/2} = \frac{0.693}{k}$ rearrange to solve for k.

$k = \frac{0.693}{t_{1/2}} = \frac{0.693}{1.6 \times 10^3\text{ yr}} = 4.\underline{3}3125 \times 10^{-4}\text{ yr}^{-1}$. Since

$\ln m_{Ra\text{-}226\ t} = -kt + \ln m_{Ra\text{-}226\ 0} = -(4.\underline{3}3125 \times 10^{-4}\ \cancel{\text{yr}^{-1}})(0.01\underline{3}690\ \cancel{\text{yr}}) + \ln(25.0\text{ g}) = 3.2\underline{1}887$ →

$m_{Ra\text{-}226\ t} = e^{3.2\underline{1}887} = 24.\underline{9}999\text{ g}$. Then $mg_0 - mg_t$ = mg decayed = 25.0 g – 24.$\underline{9}$999 g = 0.$\underline{0}$00148 g Ra-226 then

$0.\underline{0}00148\ \cancel{\text{g Ra-226}} \times \frac{1\ \cancel{\text{mol Ra-226}}}{226.05402\ \cancel{\text{g Ra-226}}} \times \frac{1\text{ mol Rn-224}}{1\ \cancel{\text{mol Ra-226}}} = \underline{6}.5576 \times 10^{-7}\text{ mol Rn-224}$ then

and T = 25.0 °C + 273.15 = 298.2 K, then $PV = nRT$ Rearrange to solve for V.

$$V = \frac{nRT}{P} = \frac{6.5576 \times 10^{-7} \text{ mol} \times 0.08206 \frac{\text{L} \cdot \text{atm}}{\text{mol} \cdot \text{K}} \times 298.2 \text{ K}}{1.0 \text{ atm}} = 1.6047 \times 10^{-5} \text{ L} = 1.6 \times 10^{-5} \text{ L}$$

. Two significant figures are reported as requested in problem.

Check: The units (L) are correct. The amount of gas is small since the time is so small compared to the half-life.

50. **Given:** U-235 (235.043922 amu) neutron- induced fission to Ba-140 and Kr-93, 1.00 g U-235, T = 25.0 °C, P = 1.0 atm **Find:** V of Kr-93 gas produced

Conceptual Plan: Write species given on the appropriate side of the equation. →Equalize the sum of the mass numbers and the sum of the atomic numbers on both sides of the equation by writing the stoichiometric coefficient in front of the desired species. Then

g U-235 → mol U-235 → mol Kr-93 formed then °C → K then P, n, T → V

$$\frac{1 \text{ mol U-235}}{235.043922 \text{ g U-235}} \quad \frac{1 \text{ mol Kr-93}}{1 \text{ mol U-235}} \qquad \text{K} = {}^\circ\text{C} + 273.15 \qquad PV = nRT$$

Solution: ${}^{235}_{92}\text{U} + {}^{1}_{0}\text{n} \rightarrow {}^{140}_{56}\text{Ba} + {}^{93}_{36}\text{Kr} + ?\,{}^{1}_{0}\text{n}$ becomes ${}^{235}_{92}\text{U} + {}^{1}_{0}\text{n} \rightarrow {}^{140}_{56}\text{Ba} + {}^{93}_{36}\text{Kr} + 3\,{}^{1}_{0}\text{n}$. then

$$1.00 \text{ g U-235} \times \frac{1 \text{ mol U-235}}{235.043922 \text{ g U-235}} \times \frac{1 \text{ mol Kr-93}}{1 \text{ mol U-235}} = 4.25452 \times 10^{-3} \text{ mol Kr-93}$$

then and T = 25.0 °C + 273.15 = 298.2 K, then $PV = nRT$ Rearrange to solve for V.

$$V = \frac{nRT}{P} = \frac{4.25452 \times 10^{-3} \text{ mol} \times 0.08206 \frac{\text{L} \cdot \text{atm}}{\text{mol} \cdot \text{K}} \times 298.2 \text{ K}}{1.0 \text{ atm}} = 0.104109 \text{ L} = 0.10 \text{ L}.$$

Check: 235 + 1 = 140 + 93 + 3(1), 92 + 0 = 56 + 36 + 3(0), and no other particle is necessary to balance the equation. The units (L) are correct. About 1/200 mole of gas is generated so we expect the volume to be about 22/200 L.

51. **Given:** ${}^{0}_{+1}\text{e} + {}^{0}_{-1}\text{e} \rightarrow 2\,{}^{0}_{0}\gamma$ **Find:** a) energy (in kJ/mol) and b) wavelength of gamma ray photons

Conceptual plan:

a) mass of products & reactants → mass defect (g) → kg → kg/mol → E (J/mol) → E (kJ/mol)

$$\text{mass defect} = \sum \text{mass of reactants} - \sum \text{mass of products} \qquad \frac{1 \text{ kg}}{1000 \text{ g}} \qquad \div 2 \text{ mol} \qquad E = mc^2 \qquad \frac{1 \text{ kJ}}{1000 \text{ J}}$$

b) Answer in part a) is for 2 moles of γ, so E (J/2 mol γ) → E (J/ γ photon) → λ

$$\frac{1 \text{ mol photons}}{6.022 \times 10^{23} \text{ photons}} \qquad E = \frac{hc}{\lambda}$$

Solution: a) mass defect = $\sum$ mass of reactants − $\sum$ mass of products = (0.00055 g + 0.00055 g) − 0 g

$$= 0.00110 \text{ g then } \frac{0.00110 \text{ g}}{2 \text{ mol}} \times \frac{1 \text{ kg}}{1000 \text{ g}} = 5.50 \times 10^{-7} \frac{\text{kg}}{\text{mol}} \text{ then}$$

$$E = mc^2 = \left(5.50 \times 10^{-7} \frac{\text{kg}}{\text{mol}}\right)\left(2.9979 \times 10^{8} \frac{\text{m}}{\text{s}}\right)^2 = 4.94307 \times 10^{10} \frac{\text{J}}{\text{mol}} \times \frac{1 \text{ kJ}}{1000 \text{ J}} = 4.94 \times 10^{7} \frac{\text{kJ}}{\text{mol}}.$$

b) $$E = 4.94307 \times 10^{10} \frac{\text{J}}{\text{mol } \gamma} \times \frac{1 \text{ mol } \gamma}{6.022 \times 10^{23} \gamma\text{photons}} = 8.20835 \times 10^{-14} \frac{\text{J}}{\gamma\text{photons}}.$$ Then $E = \frac{hc}{\lambda}$. Rearrange

to solve for λ. $$\lambda = \frac{hc}{E} = \frac{(6.626 \times 10^{-34} \text{ J} \cdot \text{s})\left(2.9979 \times 10^{8} \frac{\text{m}}{\text{s}}\right)}{8.20835 \times 10^{-14} \frac{\text{J}}{\gamma\text{photons}}} = 2.42 \times 10^{-12} \text{ m} = 2.42 \text{ pm}.$$

Check: The units (kJ/mol and pm) are correct. A large amount of energy is expected per mole of mass lost. The photon is in the gamma ray region of the electromagnetic spectrum.

52. **Given:** 1.0 MW power/day **Find:** minimum rate of mass loss required
Conceptual plan: **MW → MWh → kWh → J → kg → g**

$$\frac{24\text{ h}}{1\text{ day}} \quad \frac{1000\text{ kWh}}{1\text{ MWh}} \quad \frac{3.60\times10^6\text{ J}}{1\text{ kWh}} \quad E = mc^2 \quad \frac{1000\text{ g}}{1\text{ kg}}$$

Solution: $1.0\ \text{MW} \times \frac{24\ \text{h}}{1\ \text{day}} \times \frac{1000\ \text{kWh}}{1\ \text{MWh}} \times \frac{3.60\times10^6\ \text{J}}{1\ \text{kWh}} = 8.64\times10^{10}\ \frac{\text{J}}{\text{day}}$. Since $E = mc^2$, rearrange to solve for m.

$$m = \frac{E}{c^2} = \frac{8.64\times10^{10}\ \frac{\text{kg}}{\text{day}}\frac{\text{m}^2}{\text{s}^2}}{\left(2.9979\times10^8\ \frac{\text{m}}{\text{s}}\right)^2} = 9.6\times10^{-7}\ \frac{\text{kg}}{\text{day}} \times \frac{1000\ \text{g}}{1\ \text{kg}} = 9.6\times10^{-4}\ \frac{\text{g}}{\text{day}}.$$

Check: The units (g/day) are correct. A large amount of energy is expected per gram of mass lost.

53. **Given:** $^3\text{He} = 3.016030$ amu **Find:** nuclear binding energy per nucleon
Conceptual plan: **^A_ZX, isotope mass → mass defect → nuclear binding energy per nucleon**

$$\text{mass defect} = Z(\text{mass }^1_1\text{H}) + (A - Z)(\text{mass }^1_0\text{n}) - \text{mass of isotope} \qquad \frac{931.5\text{ MeV}}{(1\text{ amu})(A\text{ nucleons})}$$

Solution: mass defect = $Z(\text{mass }^1_1\text{H}) + (A - Z)(\text{mass }^1_0\text{n})$ – mass of isotope.
He-3 mass defect = 2(1.00783 amu) + (3 – 2)(1.00866 amu) – 3.016030 amu = 0.0082$\underline{9}$ amu

and $0.0082\underline{9}\ \text{amu} \times \frac{931.5\ \text{MeV}}{1\ \text{amu}} = 7.72\ \text{MeV}$.

Check: The units (MeV) are correct. The number of nucleons is small, so the MeV is not that large.

54. **Given:** $4\,^1_1\text{H} \rightarrow {}^4_2\text{He}$ **Find:** energy (in J/mol reactant)
Conceptual Plan: **mass of products & reactants → mass defect in g → mass defect in kg → E**

$$\text{mass defect} = \sum\text{mass of reactants} - \sum\text{mass of products} \qquad \frac{1\text{ kg}}{1000\text{ g}} \qquad E = mc^2$$

Solution:
mass defect = $\sum$mass of reactants – $\sum$mass of products = 4(1.00783 g) – 4.002603 g = 0.028717 g then

$$\frac{0.028717\ \text{g}}{4\ \text{mol reactants}} \times \frac{1\text{kg}}{1000\ \text{g}} = 7.179\underline{2}5\times10^{-6}\ \frac{\text{kg}}{\text{mol reactants}} \text{ then}$$

$$E = mc^2 = \left(7.179\underline{2}5\times10^{-6}\ \frac{\text{kg}}{\text{mol reactants}}\right)\left(2.9979\times10^8\ \frac{\text{m}}{\text{s}}\right)^2 = 6.4523\times10^{11}\ \frac{\text{J}}{\text{mol reactants}}.$$

Check: The units (J/mol) are correct.

55. **Given:** $t_{1/2}$ for decay of $^{238}\text{U} = 4.5\times10^9$ years, 1.6 g rock, 29 dis/s all radioactivity from U-238
Find: percent by mass ^{238}U in rock
Conceptual plan:
$t_{1/2}$ → k and s → min → h → day → yr then Rate, k → N → mol ^{238}U → g ^{238}U

$$t_{1/2} = \frac{0.693}{k} \quad \frac{1\text{ min}}{60\text{ s}} \quad \frac{1\text{ h}}{60\text{ min}} \quad \frac{1\text{ day}}{24\text{ h}} \quad \frac{1\text{ yr}}{365.24\text{ day}} \qquad \text{Rate} = kN \qquad \frac{1\text{ mol dis}}{6.022\times10^{23}\text{ dis}} \quad \frac{238\text{ g }^{238}\text{U}}{1\text{ mol }^{238}\text{U}}$$

then g ^{238}U, g rock → percent by mass ^{238}U

$$\text{percent by mass }^{238}\text{U} = \frac{\text{g }^{238}\text{U}}{\text{g rock}} \times 100\ \%$$

Solution: $t_{1/2} = \frac{0.693}{k}$ rearrange to solve for k. $k = \frac{0.693}{t_{1/2}} = \frac{0.693}{4.5\times10^9\ \text{yr}} = 1.\underline{5}4\times10^{-10}\ \text{yr}^{-1}$ and

$1\ \text{s} \times \frac{1\ \text{min}}{60\ \text{s}} \times \frac{1\ \text{h}}{60\ \text{min}} \times \frac{1\ \text{day}}{24\ \text{h}} \times \frac{1\ \text{yr}}{365.24\ \text{day}} = 3.168\underline{8}9554\times10^{-8}\ \text{yr}$. Rate = kN Rearrange to solve for N.

$$N = \frac{\text{Rate}}{k} = \frac{29\ \frac{\text{dis}}{3.16889554 \times 10^{-8}\ \text{yr}}}{1.54 \times 10^{-10}\ \text{yr}^{-1}} = 5.9425 \times 10^{18}\ \text{dis}$$ then

$$5.9425 \times 10^{18}\ \text{dis} \times \frac{1\ \text{mol dis}}{6.022 \times 10^{23}\ \text{dis}} \times \frac{238\ \text{g}\ ^{238}\text{U}}{1\ \text{mol}\ ^{238}\text{U}} = 2.3486 \times 10^{-3}\ \text{g}\ ^{238}\text{U}$$ then

$$\text{percent by mass}\ ^{238}\text{U} = \frac{\text{g}\ ^{238}\text{U}}{\text{g rock}} \times 100\ \% = \frac{2.3486 \times 10^{-3}\ \text{g}\ ^{238}\text{U}}{1.6\ \text{g rock}} \times 100\ \% = 0.15\ \%.$$

Check: The units (%) are correct. The mass percent is low because the dis/s is low.

56. **Given:** $t_{1/2}$ for decay of $^{232}Th = 1.4 \times 10^{10}$ years **Find:** number of dis/h emitted by 1.0 mol ^{232}Th in 1 min
Conceptual plan: $t_{1/2}$ → k **then N,** k → **Rate (dis/yr)** → **dis/day** → **dis/h** → **dis/min**

$$t_{1/2} = \frac{0.693}{k} \qquad \text{Rate} = k\,N \qquad \frac{1\ \text{yr}}{365.24\ \text{day}} \qquad \frac{1\ \text{day}}{24\ \text{h}} \qquad \frac{1\ \text{h}}{60\ \text{min}}$$

Solution: $t_{1/2} = \frac{0.693}{k}$ rearrange to solve for k. $k = \frac{0.693}{t_{1/2}} = \frac{0.693}{1.4 \times 10^{10}\ \text{yr}} = 4.95 \times 10^{-11}\ \text{yr}^{-1}$ then

$$\text{Rate} = k\,N = \left(4.95 \times 10^{-11}\ \text{yr}^{-1}\right)\left(6.022 \times 10^{23}\ \text{dis}\right) \times \frac{1\ \text{yr}}{365.24\ \text{day}} \times \frac{1\ \text{day}}{24\ \text{h}} = 3.4 \times 10^{9}\ \text{dis/h}$$

$$3.4 \times 10^{9}\ \text{dis/h} \times \frac{1\ \text{h}}{60\ \text{min}} = 5.7 \times 10^{7}\ \text{dis/min}$$

Check: The units (dis/h and dis/min) are correct. The rate is high since the amount of ^{232}Th is high.

57. a) **Given:** 72,500 kg Al *(s)* and 10 Al *(s)* + 6 NH_4ClO_4 *(s)* → 4 Al_2O_3 *(s)* + 2 $AlCl_3$ *(s)* + 12 H_2O *(g)* + 3 N_2 *(g)* and 608,000 kg O_2 *(g)* that reacts with hydrogen to form gaseous water
Find: energy generated (ΔH°_{rxn})
Conceptual plan: write balanced reaction for O_2 *(g)* then

$$\Delta H^0_{rxn} = \sum n_p \Delta H^0_f(products) - \sum n_r \Delta H^0_f(reactants)$$ **then**

kg → g → mol → energy then add the results from the two reactions

$$\frac{1000\ \text{g}}{1\ \text{kg}} \qquad \mathfrak{M} \qquad \Delta H^\circ_{rxn}$$

Solution:

Reactant/Product	ΔH^0_f (kJ/mol from Appendix IIB)
Al (*s*)	0.0
NH_4ClO_4 (*s*)	−295
Al_2O_3 (*s*)	−1675.7
$AlCl_3$ (*s*)	−704.2
H_2O (*g*)	− 241.8
N_2 (*g*)	0.0

Be sure to pull data for the correct formula and phase.

$$\begin{aligned}\Delta H^0_{rxn} &= \sum n_p \Delta H^0_f(products) - \sum n_r \Delta H^0_f(reactants)\\ &= [4(\Delta H^0_f(Al_2O_3\ (s))) + 2(\Delta H^0_f(AlCl_3\ (s))) + 12(\Delta H^0_f(H_2O\ (g))) + 3(\Delta H^0_f(N_2\ (g)))] +\\ &\quad - [10(\Delta H^0_f(Al\ (s))) + 6(\Delta H^0_f(NH_4ClO_4\ (s)))]\\ &= [4(-1675.7\ \text{kJ}) + 2(-704.2\ \text{kJ}) + 12(-241.8\ \text{kJ}) + 3(0.0\ \text{kJ})] - [10(0.0\ \text{kJ}) + 6(-295\ \text{kJ})]\\ &= [-11012.8\ \text{kJ}] - [-1770.\ \text{kJ}]\\ &= -9242.8\ \text{kJ}\end{aligned}$$

then $72{,}500\ \text{kg Al} \times \frac{1000\ \text{g Al}}{1\ \text{kg Al}} \times \frac{1\ \text{mol Al}}{26.98\ \text{g Al}} \times \frac{9242.8\ \text{kJ}}{10\ \text{mol Al}} = 2.483703 \times 10^{9}\ \text{kJ}$.

balanced reaction: H_2 *(g)* + ½ O_2 *(g)* → H_2O *(g)* $\Delta H^\circ_{rxn} = \Delta H^\circ_f(H_2O\ (g)) = -241.8$ kJ/mol then

$608{,}000\ \text{kg}\ O_2 \times \frac{1000\ \text{g}\ O_2}{1\ \text{kg}\ O_2} \times \frac{1\ \text{mol}\ O_2}{32.00\ \text{g}\ O_2} \times \frac{241.8\ \text{kJ}}{0.5\ \text{mol}\ O_2} = 9.1884 \times 10^{9}\ \text{kJ}$. So the total is

$2.4\underline{8}3703 \times 10^9$ kJ $+ 9.1\underline{8}84 \times 10^9$ kJ $= 1.16\underline{7}2103 \times 10^{10}$ kJ $= 1.167 \times 10^{10}$ kJ .

Check: The units (kJ) are correct. The answer is very large because the reactions are very exothermic and the weight of reactants is so large.

b) **Given:** ${}^1_1H + {}^{-1}_{-1}p + {}^{0}_{+1}e \rightarrow {}^0_0\gamma$ **Find:** mass of antimatter to give same energy as part a)

Conceptual plan: since the reaction is an annihilation reaction, no matter will be left, so the mass of antimatter is the same as the mass of the hydrogen. so kJ → J → kg → g

$$\frac{1000\ \text{J}}{1\ \text{kJ}} \qquad E = m\,c^2 \qquad \frac{1000\ \text{g}}{1\ \text{kg}}$$

Solution: $1.16\underline{7}2103 \times 10^{10}\ \cancel{\text{kJ}} \times \dfrac{1000\ \text{J}}{1\ \cancel{\text{kJ}}} = 1.16\underline{7}2103 \times 10^{13}\ \text{J}$. Since $E = m\,c^2$, rearrange to solve for m.

$$m = \frac{E}{c^2} = \frac{1.16\underline{7}2103 \times 10^{13}\ \text{kg}\frac{\cancel{\text{m}^2}}{\cancel{\text{s}^2}}}{\left(2.9979 \times 10^8\ \frac{\cancel{\text{m}}}{\cancel{\text{s}}}\right)^2} = 1.299 \times 10^{-4}\ \cancel{\text{kg}} \times \frac{1000\ \text{g}}{1\ \cancel{\text{kg}}} = 0.1299\ \text{g}\ .$$

Check: The units (g) are correct. A small mass is expected since nuclear reactions to generate a large amount of energy.

58. **Given:** 85.0 g animal, ingests 10.0 mg of substance with 2.55 % by mass Pu-239, alpha emitter, $t_{1/2}$ = 24,110 years **Find:** a) initial exposure in Ci, and b) all radiation absorbed and 7.77×10^{-12} J/emission, RBE = 20, dose in rads in the first 4.0 hours and dose in rems in the first 4.0 hours

Conceptual plan: a) $t_{1/2}$ → k and mg → g → g Pu-239 → mol Pu-239 → atoms Pu-239

$$t_{1/2} = \frac{0.693}{k} \qquad \frac{1\ \text{g}}{1000\ \text{mg}} \qquad \frac{2.22\ \text{g Pu-239}}{100\ \text{g substance}} \qquad \frac{1\ \text{mol Pu-239}}{239\ \text{g Pu-239}} \qquad \frac{6.022 \times 10^{23}\ \text{Pu-239 atoms}}{1\ \text{mol Pu-239}}$$

then N, k → Rate (dis/yr) → dis/day → dis/h → dis/min → dis/s → Ci

$$\text{Rate} = k\,N \qquad \frac{1\ \text{yr}}{365.24\ \text{day}} \qquad \frac{1\ \text{day}}{24\ \text{h}} \qquad \frac{1\ \text{h}}{60\ \text{min}} \qquad \frac{1\ \text{min}}{60\ \text{s}} \qquad \frac{1\ \text{Ci}}{\frac{3.7 \times 10^{10}\ \text{decays}}{\text{s}}}$$

b) h → min → s then dis/s, s → alpha decays → J and g → kg then J, animal mass → rad → rem

$$\frac{60\ \text{min}}{1\ \text{h}} \qquad \frac{1\ \text{s}}{60\ \text{min}} \qquad \text{multiply terms} \qquad \frac{7.77 \times 10^{-12}\ \text{J}}{\text{decay}} \qquad \frac{1\ \text{kg}}{1000\ \text{g}} \qquad \frac{1\ \text{rad}}{\frac{0.01\ \text{J}}{\text{kg animal}}} \qquad \text{rem} = \text{RBE} \times \text{rad}$$

Solution: a) $t_{1/2} = \dfrac{0.693}{k}$ rearrange to solve for k. $k = \dfrac{0.693}{t_{1/2}} = \dfrac{0.693}{24{,}110\ \text{yr}} = 2.87\underline{4}326 \times 10^{-5}\ \text{yr}^{-1}$ then

$$10.0\ \cancel{\text{mg}} \times \frac{1\ \cancel{\text{g}}}{1000\ \cancel{\text{mg}}} \times \frac{2.55\ \cancel{\text{g Pu-239}}}{100\ \cancel{\text{g substance}}} \times \frac{1\ \cancel{\text{mol Pu-239}}}{239\ \cancel{\text{g Pu-239}}} \times \frac{6.022 \times 10^{23}\ \text{Pu-239 atoms}}{1\ \cancel{\text{mol Pu-239}}}$$

$= 6.4\underline{2}5146 \times 10^{17}$ Pu-239 atoms

$$\text{Rate} = k\,N = \left(2.87\underline{4}326 \times 10^{-5}\ \cancel{\text{yr}^{-1}}\right)\left(6.4\underline{2}5146 \times 10^{17}\ \text{Pu-239 atoms}\right) \times \frac{1\ \cancel{\text{yr}}}{365.24\ \cancel{\text{day}}} \times \frac{1\ \cancel{\text{day}}}{24\ \cancel{\text{h}}} \times \frac{1\ \cancel{\text{h}}}{60\ \cancel{\text{min}}} \times \frac{1\ \cancel{\text{min}}}{60\ \text{s}}$$

$$= 5.8\underline{5}230 \times 10^5\ \frac{\text{dis}}{\text{s}} \times \frac{1\ \text{Ci}}{\frac{3.7 \times 10^{10}\ \text{decays}}{\text{s}}} = 1.5\underline{8}170 \times 10^{-5}\ \text{Ci}$$

b) $4.0\ \cancel{\text{h}} \times \dfrac{1\ \cancel{\text{day}}}{24\ \cancel{\text{h}}} \times \dfrac{1\ \text{yr}}{365.24\ \cancel{\text{day}}} = 4.\underline{5}632 \times 10^{-4}$ yr Since the time is so much less than the $t_{1/2}$ (10^{-6} %) the concentration is essentially constant. Use dis/s and time to get dose, so $4.0\ \cancel{\text{h}} \times \dfrac{60\ \cancel{\text{min}}}{1\ \cancel{\text{h}}} \times \dfrac{60\ \text{s}}{1\ \cancel{\text{min}}} = 1.44 \times 10^4$ s

$$5.8\underline{5}230 \times 10^5\ \frac{\text{dis}}{\cancel{\text{s}}} \times 1.44 \times 10^4\ \cancel{\text{s}} = 8.4\underline{2}731 \times 10^9\ \cancel{\text{decays}} \times \frac{7.77 \times 10^{-12}\ \text{J}}{\cancel{\text{decay}}} = 6.5\underline{4}802 \times 10^{-2}\ \text{J}\ \text{ and}$$

$85.0\text{ g} \times \frac{1\text{ kg}}{1000\text{ g}} = 0.0850\text{ kg}$ then $\frac{6.54802 \times 10^{-2}\text{ J}}{0.0850\text{ kg}} \times \frac{1\text{ rad}}{\frac{0.01\text{ J}}{\text{kg animal}}} = 77\text{ rad}$ and

rem = RBE x rad = 20 x 77 rad= 1.5×10^3 rem and the animal will die.

Check: The units (Ci, rem, and rad) are correct. The number of Curies is small because of the conversion factor. The dose in rems and rad are high because it is an alpha emitter and the isotope was ingested.

59. **Given:** ${}^{235}_{92}\text{U} \rightarrow {}^{206}_{82}\text{Pb}$ and ${}^{232}_{90}\text{Th} \rightarrow {}^{206}_{82}\text{Pb}$ **Find:** decay series

Conceptual Plan: Write species given on the appropriate side of the equation. →Equalize the sum of the mass numbers and the sum of the atomic numbers on both sides of the equation by writing the stoichiometric coefficient in front of the desired species.

Solution: ${}^{235}_{92}\text{U} \rightarrow {}^{?}_{82}\text{Pb} + ?{}^{4}_{2}\text{He} + ?{}^{0}_{-1}\text{e}$ becomes ${}^{235}_{92}\text{U} \rightarrow {}^{207}_{82}\text{Pb} + 7{}^{4}_{2}\text{He} + 4{}^{0}_{-1}\text{e}$.

${}^{232}_{90}\text{Th} \rightarrow {}^{?}_{82}\text{Pb} + ?{}^{4}_{2}\text{He} + ?{}^{0}_{-1}\text{e}$ becomes ${}^{232}_{90}\text{Th} \rightarrow {}^{208}_{82}\text{Pb} + 6{}^{4}_{2}\text{He} + 4{}^{0}_{-1}\text{e}$.

U-235 forms Pb-207 in 7 α-decays and 4 β-decays and Th-232 forms Pb-208 in 6 α-decays and 4 β-decays.

Check: 235 = 207 + 7(4) + 4(0), and 92 = 82 + 7(2) + 4(–1). 232 = 208 + 6(4) + 4(0), and 90 = 82 + 6(2) + 4(–1). The mass of the Pb can be determined because alpha particles are large and need to be included as integer values.

60. **Given:** ${}^{21}_{9}\text{F} \rightarrow {}^{?}_{?}? + {}^{0}_{-1}\text{e}$ **Find:** missing nucleus

Conceptual Plan: Write species given on the appropriate side of the equation. →Equalize the sum of the mass numbers and the sum of the atomic numbers on both sides of the equation by writing the stoichiometric coefficient in front of the desired species.

Solution: ${}^{21}_{9}\text{F} \rightarrow {}^{?}_{?}? + {}^{0}_{-1}\text{e}$ becomes ${}^{21}_{9}\text{F} \rightarrow {}^{21}_{10}\text{Ne} + {}^{0}_{-1}\text{e}$.

Check: 21 = 21 + 0, and 9 = 10 + –1, Neon is atomic number 10 and no other species are needed to balance the equation.

61. 7. Since $1/2^6 = 1.6\%$ and $1/2^7 = 0.8\ \%$.

62. Nuclide A is more dangerous because the half-life is shorter (18.5 days) and so it decays faster.

63. The gamma emitter is a greater threat while you sleep because it can penetrate more tissue. The alpha particles will not penetrate the wall to enter your bedroom. The alpha emitter is a greater threat if you ingest it since it is more ionizing.

Chapter 20
Organic Chemistry

1. a) C_5H_{12} is an alkane, since it follows the general formula C_nH_{2n+2}, where $n = 5$.

    b) C_3H_6 is an alkene, since it follows the general formula C_nH_{2n}, where $n = 3$.

    c) C_7H_{12} is an alkyne, since it follows the general formula C_nH_{2n-2}, where $n = 7$.

    d) $C_{11}H_{22}$ is an alkene, since it follows the general formula C_nH_{2n}, where $n = 11$.

2. a) C_8H_{16} is an alkene, since it follows the general formula C_nH_{2n}, where $n = 8$.

    b) C_4H_6 is an alkyne, since it follows the general formula C_nH_{2n-2}, where $n = 4$.

    c) C_7H_{16} is an alkane, since it follows the general formula C_nH_{2n+2}, where $n = 7$.

    d) C_2H_2 is an alkyne, since it follows the general formula C_nH_{2n-2}, where $n = 2$.

3. $CH_3-CH_2-CH_2-CH_2-CH_2-CH_2-CH_3$,

$$CH_3-CH-CH_2-CH_2-CH_2-CH_3$$
$$\underset{CH_3}{|}$$

$$CH_3-CH_2-CH-CH_2-CH_2-CH_3$$
$$\underset{CH_3}{|}$$

$$CH_3-\underset{\underset{CH_3}{|}}{\overset{\overset{CH_3}{|}}{C}}-CH_2-CH_2-CH_3 ,$$

$$CH_3-CH-CH-CH_2-CH_3$$
$$\underset{CH_3}{|}\ \underset{CH_3}{|}$$

$$CH_3-CH_2-\underset{\underset{CH_3}{|}}{\overset{\overset{CH_3}{|}}{C}}-CH_2-CH_3 ,$$

$$CH_3-CH-CH_2-CH-CH_3$$
$$\underset{CH_3}{|}\ \ \ \ \underset{CH_3}{|}$$

$$CH_3-CH_2-CH-CH_2-CH_3$$
$$\underset{CH_2-CH_3}{|} , \text{ and}$$

$$CH_3-\underset{\underset{CH_3CH_3}{|}}{\overset{\overset{CH_3}{|}}{C}}-CH-CH_3 .$$

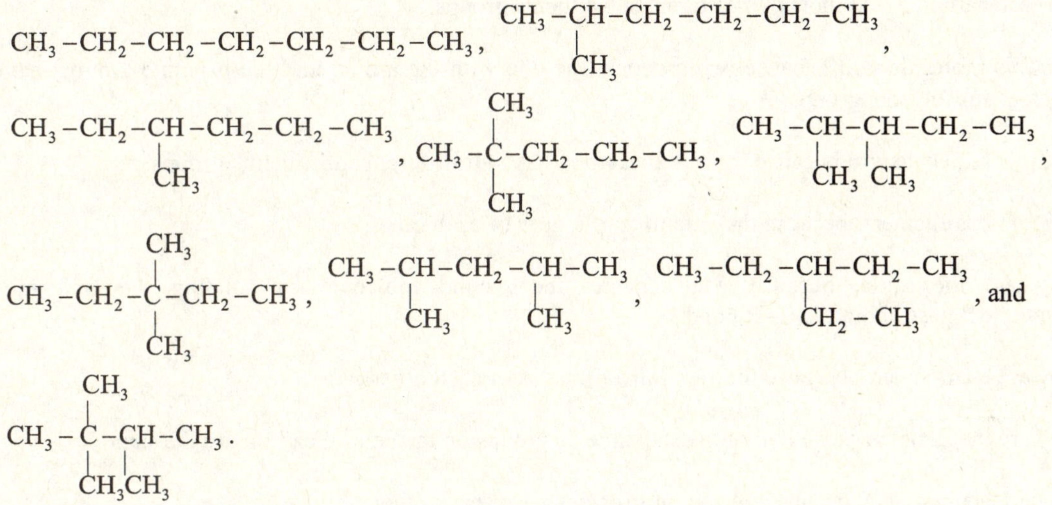

4. The 18 isomers are $CH_3-CH_2-CH_2-CH_2-CH_2-CH_2-CH_2-CH_3$,

$$CH_3-CH-CH_2-CH_2-CH_2-CH_2-CH_3$$
$$\underset{CH_3}{|} ,$$

$$CH_3-CH_2-CH-CH_2-CH_2-CH_2-CH_3$$
$$\underset{CH_3}{|} ,$$

$$CH_3-CH_2-CH_2-CH-CH_2-CH_2-CH_3$$
$$\underset{CH_3}{|} ,$$

$$CH_3-\underset{\underset{CH_3}{|}}{\overset{\overset{CH_3}{|}}{C}}-CH_2-CH_2-CH_2-CH_3 ,$$

$$CH_3-CH-CH-CH_2-CH_2-CH_3$$
$$\underset{CH_3}{|}\ \underset{CH_3}{|} ,$$

$$CH_3-CH-CH_2-CH-CH_2-CH_3$$
$$\underset{CH_3}{|}\ \ \ \ \underset{CH_3}{|} ,$$

$$CH_3-CH_2-\underset{\underset{CH_3}{|}}{\overset{\overset{CH_3}{|}}{C}}-CH_2-CH_2-CH_3 ,$$

$$CH_3-CH_2-CH-CH-CH_2-CH_3$$
$$\underset{CH_3}{|}\ \underset{CH_3}{|} ,$$

$$CH_3-CH-CH_2-CH_2-CH-CH_3$$
$$\underset{CH_3}{|}\ \ \ \ \ \ \underset{CH_3}{|} ,$$

$$CH_3-CH-CH-CH-CH_3$$
$$\underset{CH_3}{|}\ \underset{CH_3}{|}\ \underset{CH_3}{|} ,$$

$$CH_3-CH_2-CH-CH_2-CH_2-CH_3$$
$$\underset{CH_2-CH_3}{|} ,$$

$$CH_3-\underset{\underset{CH_3CH_3}{|}}{\overset{\overset{CH_3CH_3}{|}}{C}}-\overset{|}{C}-CH_3 , \ CH_3-CH-\underset{\underset{CH_3\ CH_3}{|}}{\overset{\overset{CH_3}{|}}{C}}-CH_2-CH_3 ,$$

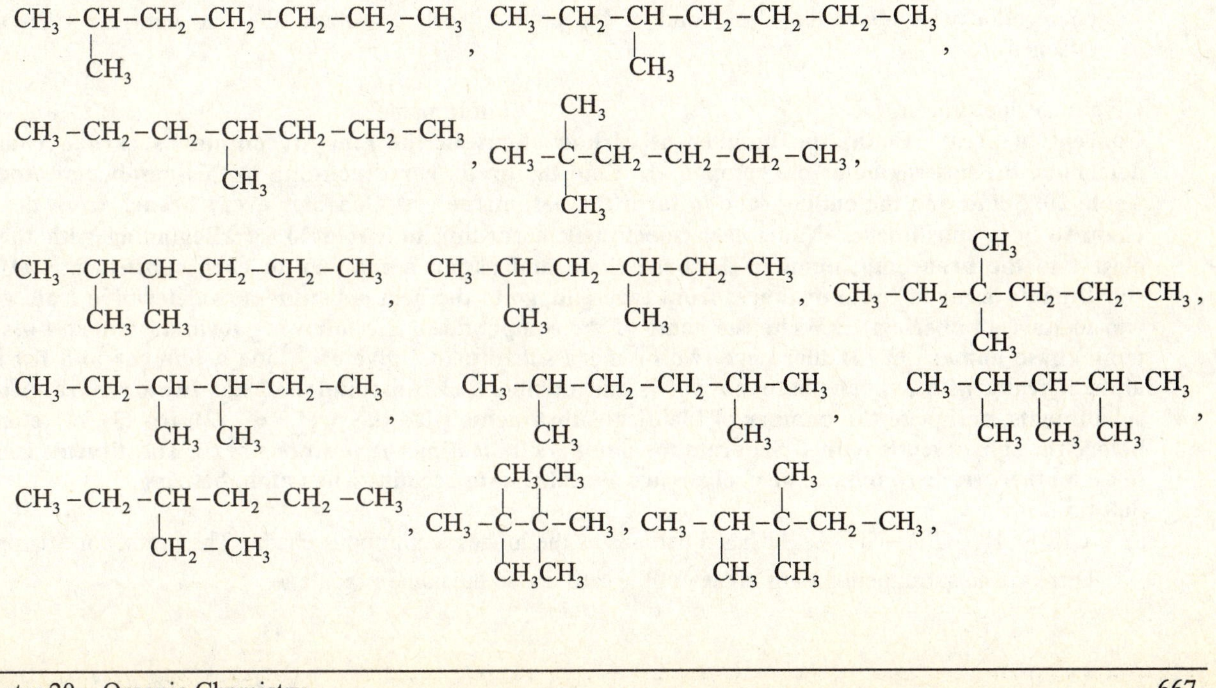

$$CH_3-CH_2-\underset{\underset{\displaystyle CH_2-CH_3}{|}}{\overset{\overset{\displaystyle CH_3}{|}}{C}}-CH_2-CH_3 \ , \ CH_3-CH_2-\underset{\underset{\displaystyle CH_2-CH_3}{|}}{\overset{\overset{\displaystyle CH_3}{|}}{CH}}-CH-CH_3 \ , \ CH_3-\underset{\underset{\displaystyle CH_3}{|}}{\overset{\overset{\displaystyle CH_3}{|}}{C}}-CH_2-\underset{\underset{\displaystyle CH_3}{|}}{CH}-CH_3 \ , \text{ and}$$

$$CH_3-\underset{\underset{\displaystyle CH_3 \ CH_3}{|}}{\overset{\overset{\displaystyle CH_3}{|}}{C}}-CH-CH_2-CH_3 \ .$$

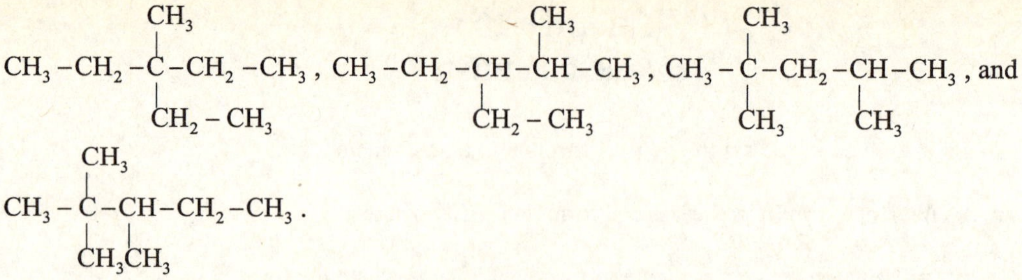

5. a) No, this molecule will not because all four of the substituents are Cl atoms.

   b) Yes, this molecule will because the third carbon has four different substituents groups.

   c) Yes, this molecule will because the second carbon has four different substituents groups.

   d) No, each carbon has at most three different substituents groups.

6. a) Yes, this molecule will because the third carbon from the left has four different substituents groups.

   b) No, each carbon has at most two different substituents groups.

   c) Yes, this molecule will because two carbons (the one with the amino and the bromine groups) have four different substituents groups.

   d) Yes, this molecule will because the middle carbon has four different substituents groups.

7. a) They are enantiomers, because they are mirror images of each other.

   b) They are the same, because you can get the second molecule by rotating the first molecule counterclockwise about the C–H bond.

   c) They are enantiomers, because they are mirror images of each other.

8. a) They are the same, because two of the substituents groups on the central carbon are the same.

   b) They are enantiomers, because they are mirror images of each other.

   c) They are the same, because you can get the second molecule by rotating the first molecule counterclockwise about the C–H bond of the optically active carbon (the one with the CCl_3 group attached).

9. **Given:** alkane structures   **Find:** name
   Conceptual plan: **Count the number of carbon atoms in the longest continuous carbon chain to determine the base name of the compound. Find the prefix corresponding to this number of atoms in Table 20. 5 and add the ending -ane to form the base name. → Consider every branch from the base chain to be a substituent. Name each substituent according to Table 20.6. →Beginning with the end closest to the branching, number the base chain and assign a number to each substituent. (If two substituents occur at equal distances from each end, go to the next substituent to determine from which end to start numbering.) →Write the name of the compound in the following format: (subst. #)-(subst. name)(base name). → If there are two or more substituents, give each one a number and list them alphabetically with hyphens between words and numbers. →If a compound has two or more identical substituents, designate the number of identical substituents with the prefix di- (2), tri- (3), or tetra- (4) before the substituent's name. Separate the numbers indicating the positions of the substituents relative to each other with a comma. The prefixes are not taken into account when alphabetizing.**
   Solution:
   a) $CH_3-CH_2-CH_2-CH_2-CH_3$ has 5 carbons as the longest continuous chain. The prefix for 5 is penta-. There are no substituent groups on any of the carbons, so the name is pentane.

b) 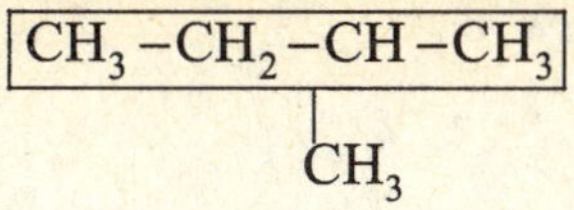 has 4 carbons as the longest continuous chain. The prefix for 4 is but- and the base name is butane. The only substituent group is a methyl group. 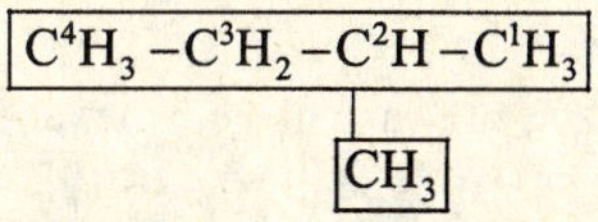 If we start numbering the chain at the end closest to the methyl group, the methyl substituent is assigned the number 2. The name of the compound is 2-methylbutane.

c) 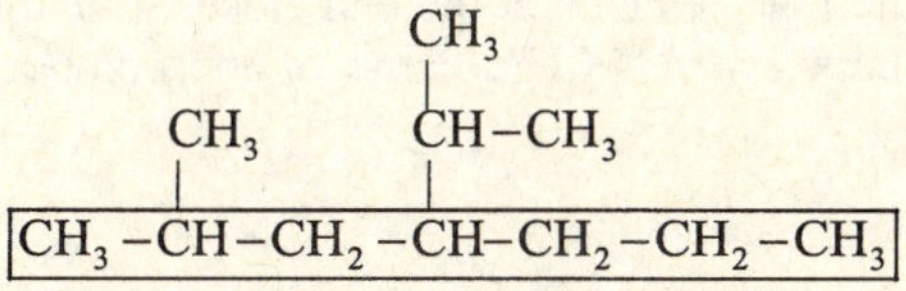 has 7 carbons as the longest continuous chain. The prefix for 7 is hept- and the base name is heptane. The substituent groups are methyl and isopropyl groups.

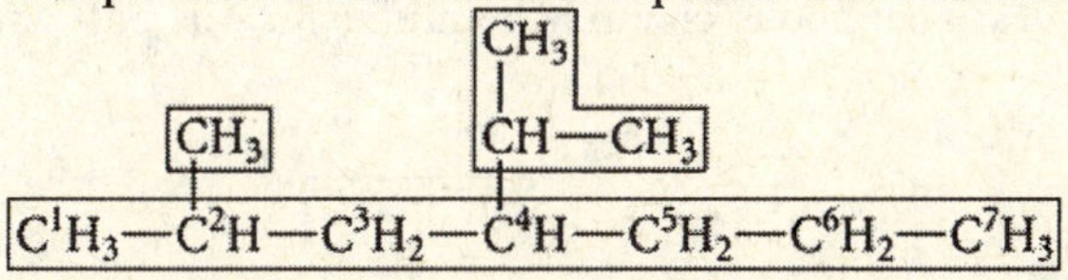 If we start numbering the chain at the end closest to the methyl group, the methyl substituent is assigned the number 2 and the isopropyl group is assigned the number 4. Since i comes before m, the name of the compound is 4-isopropyl-2-methylheptane.

d) 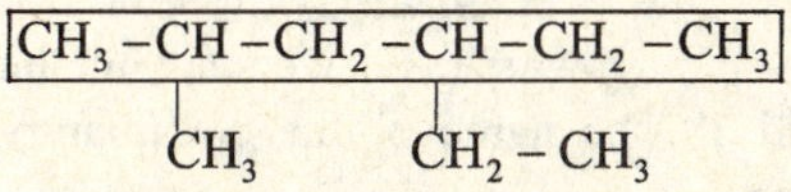 has 6 carbons as the longest continuous chain. The prefix for 6 is hex- and the base name is hexane. The only substituent groups are methyl and ethyl groups. 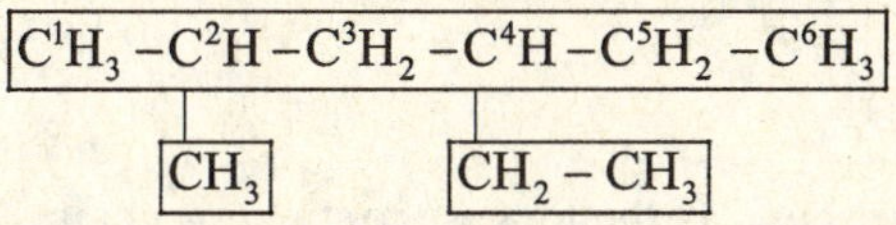 If we start numbering the chain at the end closest to the methyl group, the methyl substituent is assigned the number 2 and the ethyl group is assigned the number 4. Since e comes before m, the name of the compound is 4-ethyl-2-methylhexane.

10. **Given:** alkane structures **Find:** name
Conceptual plan: Count the number of carbon atoms in the longest continuous carbon chain to determine the base name of the compound. Find the prefix corresponding to this number of atoms in Table 20.5 and add the ending -ane to form the base name. → Consider every branch from the base chain to be a substituent. Name each substituent according to Table 20.6. →Beginning with the end closest to the branching, number the base chain and assign a number to each substituent. (If two substituents occur at equal distances from each end, go to the next substituent to determine from which end to start numbering.) →Write the name of the compound in the following format: (subst. #)-(subst. name)(base name). → If there are two or more substituents, give each one a number and list them alphabetically with hyphens between words and numbers. →If a compound has two or more identical substituents, designate the number of identical substituents with the prefix di- (2), tri- (3), or tetra- (4) before the substituent's name. Separate the numbers indicating the positions of the substituents relative to each other with a comma. The prefixes are not taken into account when alphabetizing.
Solution:

a) 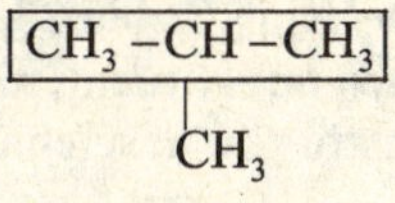 has 3 carbons as the longest continuous chain. The prefix for 3 is prop- and the base name is propane. The only substituent group is a methyl group. 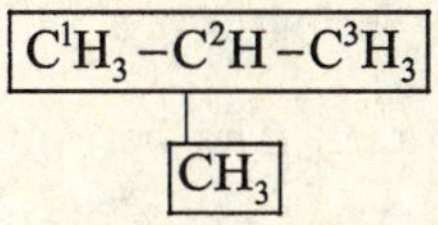 Since the branching is in the middle, it does not matter which end we start numbering. The methyl substituent is assigned the number 2. The name of the compound is 2-methylpropane.

$$\text{CH}_3 \qquad\qquad \text{CH}_3$$
$$\text{CH}_3\text{—CH—CH}_2\text{—CH—CH}_2$$
$$\qquad\qquad\qquad\qquad\qquad \text{CH}_3$$

b) has 6 carbons as the longest continuous chain. The prefix for 6 is hex- and the base name is hexane. The substituent groups are two methyl groups.

$$\text{CH}_3 \qquad\qquad \text{CH}_3$$
$$\text{C}^1\text{H}_3\text{—C}^2\text{H—C}^3\text{H}_2\text{—C}^4\text{H—C}^5\text{H}_2$$
$$\qquad\qquad\qquad\qquad\qquad \text{C}^6\text{H}_3$$
If we start numbering the chain at the end closest to the left methyl group, the methyl substituents are assigned the numbers 2 and 4. The name of the compound is 2,4-dimethylhexane.

c)
$$\text{CH}_3\text{CH}_3$$
$$\text{CH}_3\text{–C–C–CH}_3$$
$$\text{CH}_3\text{CH}_3$$
has 4 carbons as the longest continuous chain. The prefix for 4 is but- and the base

name is butane. The substituent groups are 4 methyl groups.
$$\text{CH}_3 \quad \text{CH}_3$$
$$\text{C}^1\text{H}_3\text{–C}^2\text{– C}^3\text{–C}^4\text{H}_3$$
$$\text{CH}_3 \quad \text{CH}_3$$
Since the branching is in the middle and symmetric in the molecule, it does not matter which end we start numbering. The methyl substituents are assigned the numbers 2, 2, 3, and 3. The name of the compound is 2,2,3,3-tetramethylbutane.

d)
$$\qquad\qquad\qquad \text{CH}_3$$
$$\qquad \text{CH}_3 \qquad \text{CH}_2$$
$$\text{CH}_3\text{–CH–CH}_2\text{–CH–CH–CH}_2\text{–CH}_2\text{–CH}_3$$
$$\qquad\qquad\qquad \text{CH}_3$$
has 8 carbons as the longest continuous chain. The prefix

for 8 is oct- and the base name is octane. The only substituent groups are two methyl and one ethyl groups.

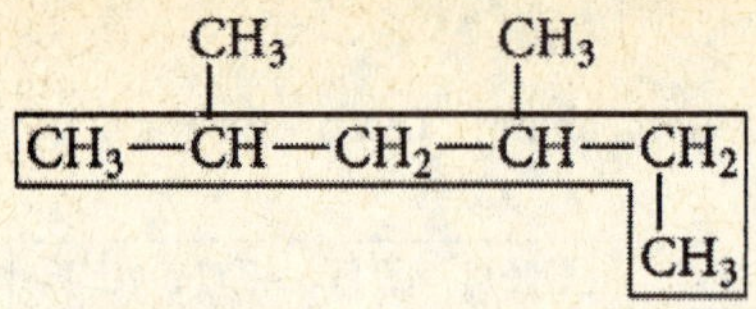

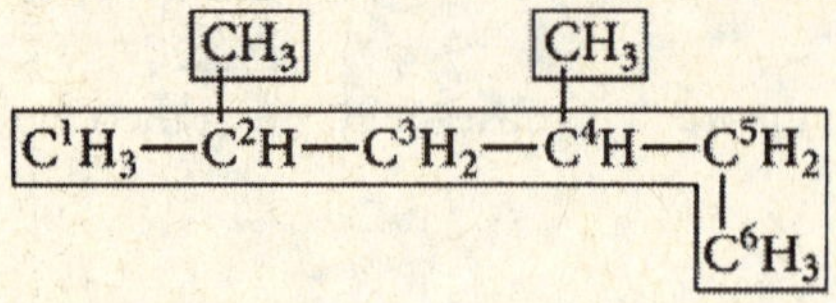

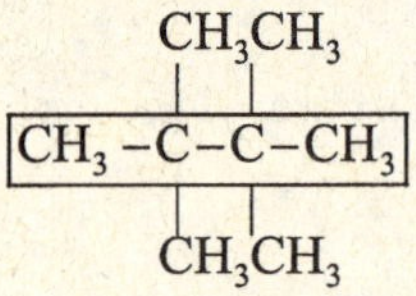

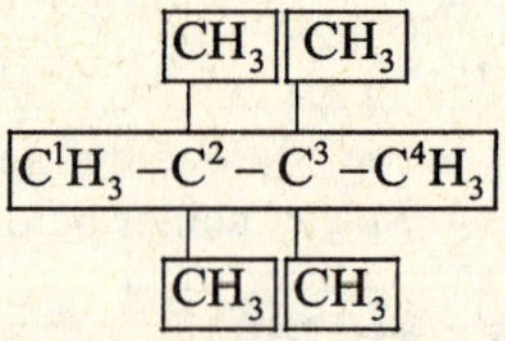

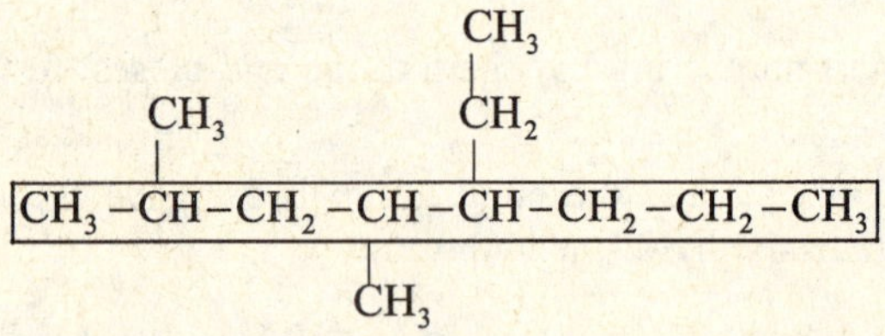

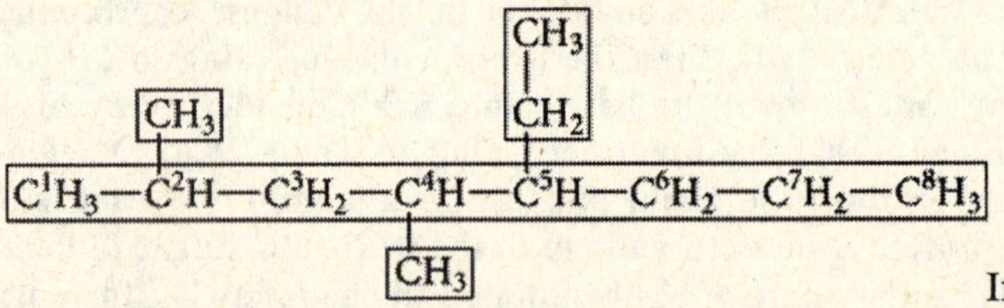

If we start numbering the chain at the end closest to the left methyl group, the methyl substituents are assigned the numbers 2 and 4 and the ethyl group is assigned the number 5. Since e comes before m, the name of the compound is 5-ethyl-2,4-dimethyloctane.

11. **Given:** alkane names **Find:** structure
   Conceptual plan: Find the number of carbon atoms corresponding to prefix of the base name in Table 20.5. → Draw the base chain and number the carbons from left to right. → Using Table 20.6 and the prefix di- (2), tri- (3), or tetra- (4) before the substituent's name, determine each substituent. → Add the substituent to the proper carbon position in the chain. → Add hydrogen atoms to the base chain so that each carbon has 4 bonds.
   Solution:
   a) 3-ethylhexane. The base name hexane designates that there are 6 carbon atoms in the base chain. $\text{C}^1\text{–C}^2\text{–C}^3\text{–C}^4\text{–C}^5\text{–C}^6$. 3-ethyl designates that a $\text{–CH}_2\text{CH}_3$ group in the 3rd position.

$$\text{C}^1\text{–C}^2\text{–C}^3\text{–C}^4\text{–C}^5\text{–C}^6$$
$$\qquad\qquad \text{CH}_2\text{–CH}_3$$
. Add hydrogens to the base chain to get the final molecule

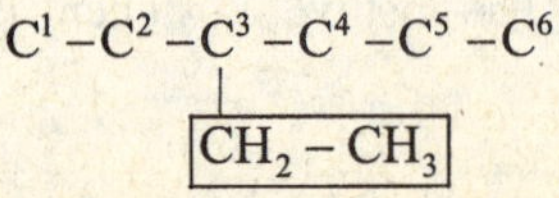

$$CH_3 -CH_2 -CH -CH_2 -CH_2 -CH_3$$
$$| \qquad\qquad$$
$$CH_2 - CH_3$$

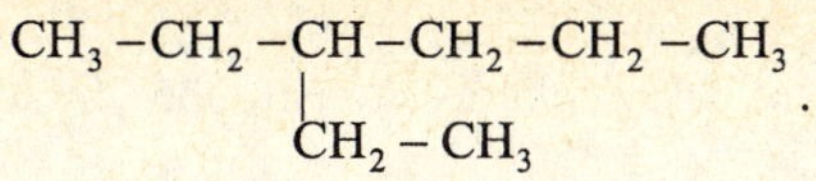

b) 3-ethyl-3-methylpentane. The base name pentane designates that there are 5 carbon atoms in the base chain. $C^1 -C^2 -C^3 -C^4 -C^5$. 3-ethyl designates that a $-CH_2CH_3$ group in the 3rd position; and 3-methyl

$$\boxed{CH_3}$$
$$|$$

designates that a $-CH_3$ group in the 3rd position . $C^1 -C^2 -C^3 -C^4 -C^5$. Add hydrogens to the base chain

$$\boxed{CH_2 - CH_3}$$

$$CH_3$$
$$|$$
to get the final molecule $CH_3 -CH_2 -CH -CH_2 -CH_3$.
$$|$$
$$CH_2 - CH_3$$

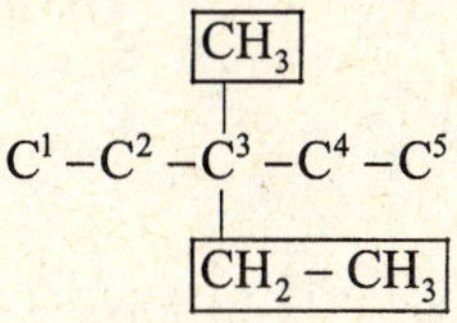

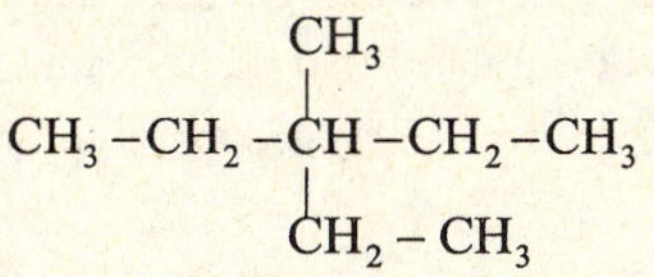

c) 2,3-dimethylbutane. The base name butane designates that there are 4 carbon atoms in the base chain. $C^1 -C^2 -C^3 -C^4$. 2,3-dimethyl designates that there are $-CH_3$ groups in the 2nd and 3rd positions.

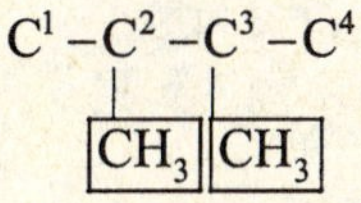

$C^1 -C^2 -C^3 -C^4$
$\quad | \quad |$. Add hydrogens to the base chain to get the final molecule
$\boxed{CH_3}\,\boxed{CH_3}$

$$CH_3 -CH -CH -CH_3$$
$$| \quad |$$
$$CH_3 \;\; CH_3$$

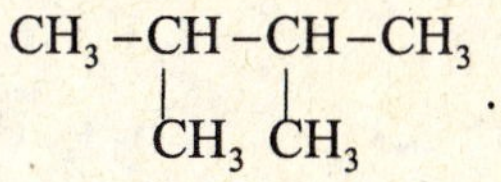

d) 4,7-diethyl-2,2-dimethylnonane. The base name nonane designates that there are 9 carbon atoms in the base chain. $C^1 -C^2 -C^3 -C^4 -C^5 -C^6 -C^7 -C^8 -C^9$. 4,7-diethyl designates that there are $-CH_2CH_3$ groups in the 4th and 7th positions; and 2,2-dimethyl designates that there are two $-CH_3$ groups in the 2nd

$$\boxed{CH_3}$$
$$|$$
position. $\quad C^1 -C^2 -C^3 -C^4 -C^5 -C^6 -C^7 -C^8 -C^9$. Add hydrogens to the base chain to get the
$$| \qquad\qquad | \qquad\qquad |$$
$$\boxed{CH_3} \quad \boxed{CH_2 - CH_3} \;\; \boxed{CH_2 - CH_3}$$

$$CH_3$$
$$|$$
final molecule $\quad CH_3 -C-CH_2 -CH -CH_2 -CH_2 -CH -CH_2 -CH_3$.
$$\qquad\qquad | \qquad\quad | \qquad\qquad\quad |$$
$$\qquad\quad CH_3 \quad\; CH_2 - CH_3 \qquad CH_2 - CH_3$$

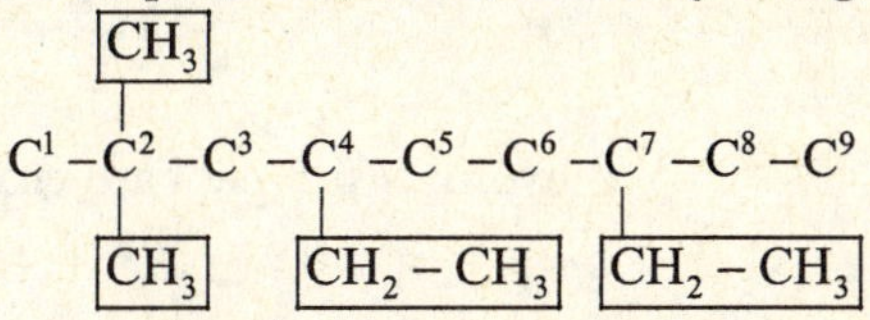

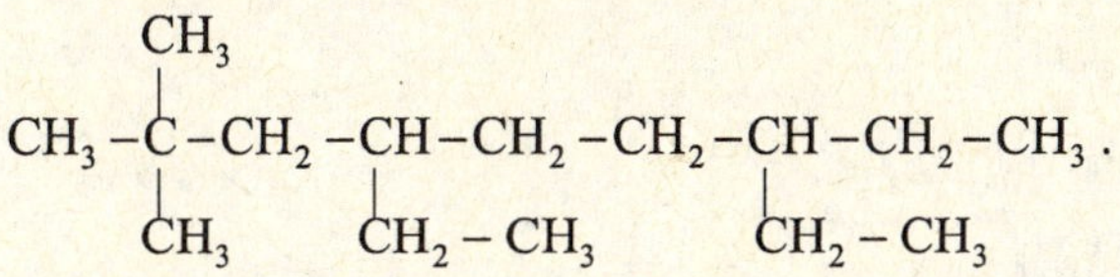

12. **Given:** alkane names $\qquad\qquad\qquad$ **Find:** structure
Conceptual plan: Find the number of carbon atoms corresponding to prefix of the base name in Table 20.5. → Draw the base chain and number the carbons from left to right. → Using Table 20.6 and the prefix di- (2), tri- (3), or tetra- (4) before the substituent's name, determine each substituent. → Add the substituent to the proper carbon position in the chain. → Add hydrogen atoms to the base chain so that each carbon has 4 bonds.
Solution:
a) 2,2-dimethylpentane. The base name pentane designates that there are 5 carbon atoms in the base chain. $C^1 -C^2 -C^3 -C^4 -C^5$. 2,2-dimethyl designates that there are two $-CH_3$ groups in the 2nd position.

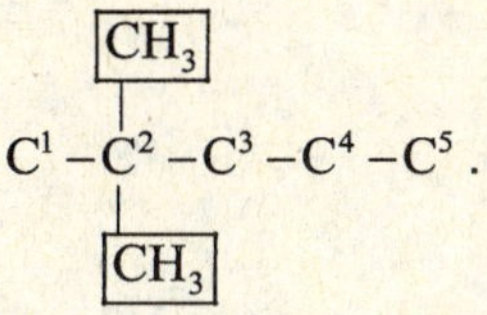

$$\boxed{CH_3}$$
$$|$$
$C^1 -C^2 -C^3 -C^4 -C^5$. Add hydrogens to the base chain to get the final molecule
$$|$$
$$\boxed{CH_3}$$

$$CH_3$$
$$|$$
$$CH_3 -C-CH_2 -CH_2 -CH_3 .$$
$$|$$
$$CH_3$$

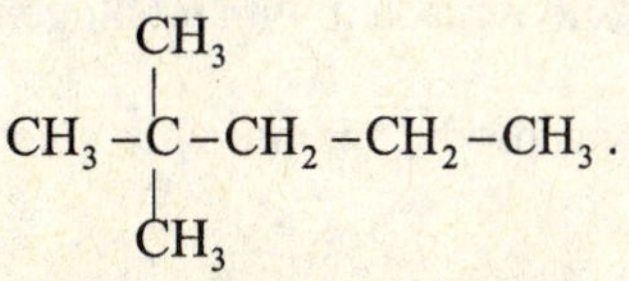

b) 3-isopropylheptane. The base name pentane designates that there are 7 carbon atoms in the base chain. $C^1 -C^2 -C^3 -C^4 -C^5 -C^6 -C^7$. 3-isopropyl designates that there is a $-CH(CH_3)_2$ group in the 3rd position.

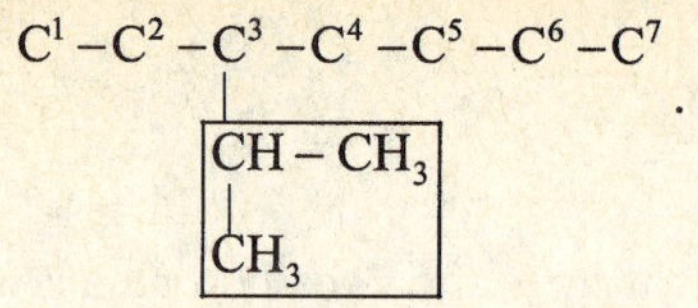

$C^1 - C^2 - C^3 - C^4 - C^5 - C^6 - C^7$. Add hydrogens to the base chain to get the final molecule
$$\begin{array}{c} \text{CH} - \text{CH}_3 \\ | \\ \text{CH}_3 \end{array}$$

$$CH_3 - CH_2 - CH - CH_2 - CH_2 - CH_2 - CH_3$$
$$\begin{array}{c} | \\ \text{CH} - \text{CH}_3 \\ | \\ \text{CH}_3 \end{array} .$$

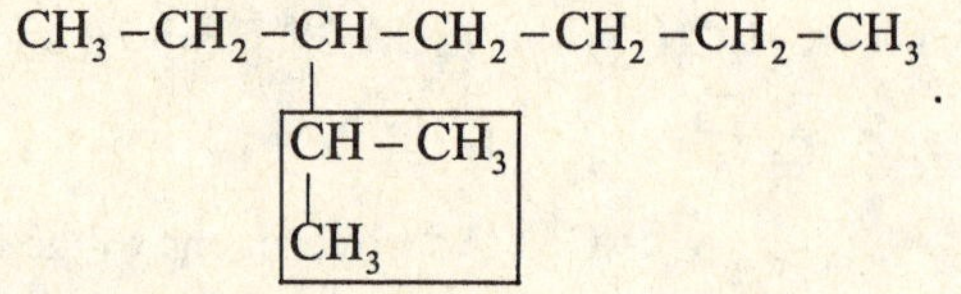

c) 4-ethyl-2,2-dimethylhexane. The base name hexane designates that there are 6 carbon atoms in the base chain. $C^1 - C^2 - C^3 - C^4 - C^5 - C^6$. 4-ethyl designates that there is a $-CH_2CH_3$ group in the 4th position; and 2,2-dimethyl designates that there are two $-CH_3$ groups in the 2nd position.

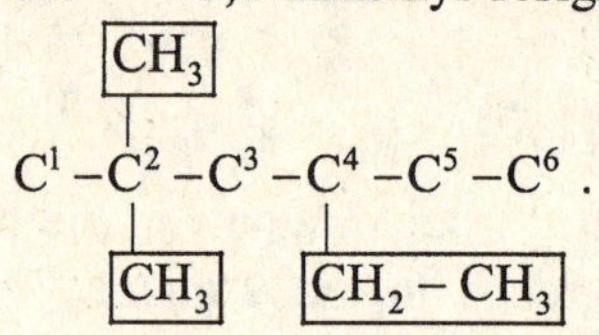

$$\begin{array}{c} \boxed{CH_3} \\ | \\ C^1 - C^2 - C^3 - C^4 - C^5 - C^6 \\ | \quad\quad\quad | \\ \boxed{CH_3} \quad \boxed{CH_2 - CH_3} \end{array}$$. Add hydrogens to the base chain to get the final molecule

$$\begin{array}{c} CH_3 \\ | \\ CH_3 - C - CH_2 - CH - CH_2 - CH_3 \\ | \quad\quad\quad\quad | \\ CH_3 \quad\quad CH_2 - CH_3 \end{array} .$$

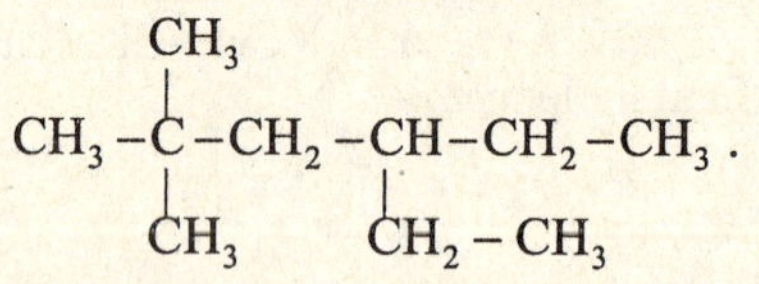

d) 4,4-diethyloctane. The base name octane designates that there are 8 carbon atoms in the base chain. $C^1 - C^2 - C^3 - C^4 - C^5 - C^6 - C^7 - C^8$. 4,4-diethyl designates that there are two $-CH_2CH_3$ groups in the 4th

$$\begin{array}{c} \boxed{CH_2 - CH_3} \\ | \\ \text{position.} \quad C^1 - C^2 - C^3 - C^4 - C^5 - C^6 - C^7 - C^8 \\ | \\ \boxed{CH_2 - CH_3} \end{array}$$. Add hydrogens to the base chain to get the final molecule

$$\begin{array}{c} CH_2 - CH_3 \\ | \\ CH_3 - CH_2 - CH_2 - C - CH_2 - CH - CH_2 - CH_3 \\ | \\ CH_2 - CH_3 \end{array} .$$

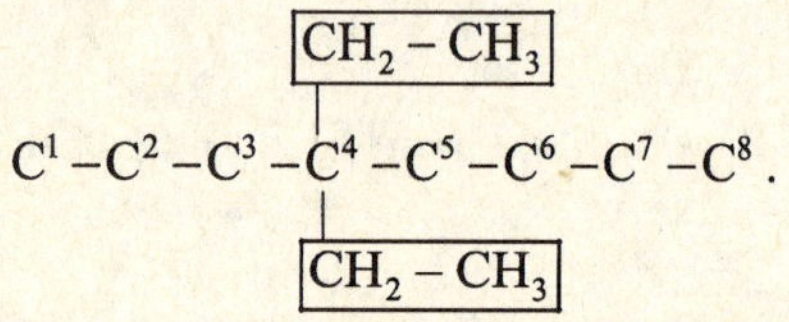

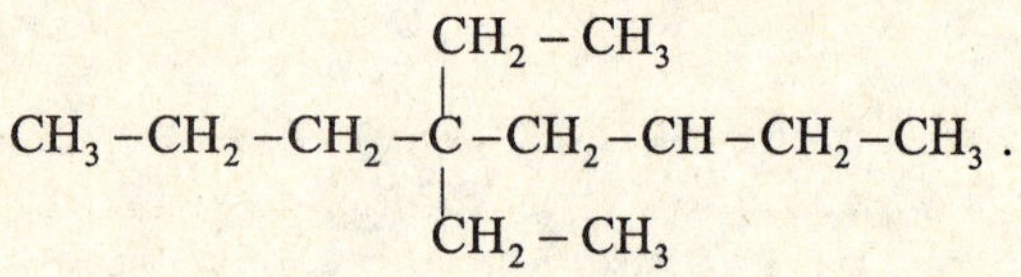

13. Hydrocarbon combustion in the presence of oxygen forms carbon dioxide and water. Balance the reaction.
   a) $CH_3CH_2CH_3$ (g) + 5 O_2 (g) $\rightarrow$ 3 CO_2 (g) + 4 H_2O (g).

   b) $CH_3CH_2CH{=}CH_2$ (g) + 6 O_2 (g) $\rightarrow$ 4 CO_2 (g) + 4 H_2O (g).

   c) 2 $CH{\equiv}CH$ (g) + 5 O_2 (g) $\rightarrow$ 4 CO_2 (g) + 2 H_2O (g).

14. Hydrocarbon combustion in the presence of oxygen forms carbon dioxide and water. Balance the reaction.
   a) 2 $CH_3CH_2CH_2CH_3$ (g) + 13 O_2 (g) $\rightarrow$ 8 CO_2 (g) + 10 H_2O (g).

   b) 2 $CH_2{=}CHCH_3$ (g) + 9 O_2 (g) $\rightarrow$ 6 CO_2 (g) + 6 H_2O (g).

   c) 2 $CH{\equiv}CCH_2CH_3$ (g) + 11 O_2 (g) $\rightarrow$ 8 CO_2 (g) + 6 H_2O (g).

15. Halogen substitution reactions remove a hydrogen atom from the alkane and replace it with a halogen atom and generate a hydrohalic acid. Assume one substitution on the hydrocarbon.
   a) CH_3CH_3 + Br_2 $\rightarrow$ CH_3CH_2Br + HBr. Only one carbon-containing product is possible since the C–C bond freely rotates.

   b) $CH_3CH_2CH_3$ + Cl_2 $\rightarrow$ [$CH_3CH_2CH_2Cl$ and $CH_3CHClCH_3$] + HCl. Two carbon-containing products are possible, either on the end carbon or the middle carbon, since the C–C bond freely rotates and the end carbons are equivalent before reaction.

c) $CH_2Cl_2 + Br_2 \rightarrow CHBrCl_2 + HBr$. Only one carbon-containing product is possible since halogen substitution reactions only remove hydrogen atoms.

d) $\begin{array}{c} CH_3 - CH - CH_3 \\ | \\ CH_3 \end{array} + Cl_2 \rightarrow \left[\begin{array}{cc} H & Cl \\ | & | \\ CH_3 - C - CH_2Cl & \text{and } CH_3 - C - CH_3 \\ | & | \\ CH_3 & CH_3 \end{array} \right] + HCl$. Two carbon-containing

products are possible, either on the end carbon or the middle carbon, since the C–C bond freely rotates and the end carbons are all equivalent before reaction.

16. Halogen substitution reactions remove a hydrogen atom from the alkane and replace it with a halogen atom and generate a hydrohalic acid. Assume one substitution on the hydrocarbon.

a) $CH_4 + Cl_2 \rightarrow CH_3Cl + HCl$. Only one carbon-containing product is possible since all of the hydrogens are equivalent.

b) $CH_3CH_2Br + Br_2 \rightarrow [CH_2BrCH_2Br \text{ and } CH_3CHBr_2] + HBr$. Two carbon-containing products are possible, either on the left carbon or the right carbon, since the C–C bond freely rotates.

c) $CH_3CH_2CH_2CH_3 + Cl_2 \rightarrow [CH_3CH_2CH_2CH_2Cl \text{ and } CH_3CHClCH_2CH_3] + HCl$. Two carbon-containing products are possible, either on the end carbons or the middle carbons, since the C–C bond freely rotates and the end carbons are equivalent and the middle carbons are equivalent before reaction.

d) $CH_3CHBr_2 + Br_2 \rightarrow [CH_2BrCHBr_2 \text{ and } CH_3CBr_3] + HBr$. Two carbon-containing products are possible, either on the left carbon or the right carbon, since the C–C bond freely rotates.

17. $CH_2 = CH - CH_2 - CH_2 - CH_2 - CH_3$, $CH_3 - CH = CH - CH_2 - CH_2 - CH_3$ and

$CH_3 - CH_2 - CH = CH - CH_2 - CH_3$ are the only structural isomers. Remember that cis–trans isomerism generates geometric isomers, not structural isomers.

18. $HC \equiv C - CH_2 - CH_2 - CH_3$ and $CH_3 - C \equiv C - CH_2 - CH_3$ are the only structural isomers.

19. **Given:** alkene structures   **Find:** name
Conceptual plan: Count the number of carbon atoms in the longest continuous carbon chain that contains the multiple bond to determine the base name of the compound. Find the prefix corresponding to this number of atoms in Table 20.5 and add the ending -ene to form the base name. → Consider every branch from the base chain to be a substituent. Name each substituent according to Table 20.6. →Beginning with the end closest to the multiple bond, number the base chain and assign a number to each substituent. →Write the name of the compound in the following format: (subst. #)-(subst. name)(base name). → If there are two or more substituents, give each one a number and list them alphabetically with hyphens between words and numbers. →If a compound has two or more identical substituents, designate the number of identical substituents with the prefix di- (2), tri- (3), or tetra- (4) before the substituent's name. Separate the numbers indicating the positions of the substituents relative to each other with a comma. The prefixes are not taken into account when alphabetizing.
Solution:

a) $\boxed{CH_2 = CH - CH_2 - CH_3}$ has 4 carbons as the longest continuous chain. The prefix for 4 is but- and the

base name is butene. There are no substituent groups. $\boxed{C^1H_2 = C^2H - C^3H_2 - C^4H_3}$ Start numbering on the

left since it is closer to the double bond. Since the double bond is between position number 1 and 2, the name of the compound is 1-butene.

b) $\begin{array}{c} \quad CH_3 \ CH_3 \\ \quad | \quad\quad | \\ \boxed{CH_3 - CH - C = CH - CH_3} \end{array}$ has 5 carbons as the longest continuous chain. The prefix for 5 is pent- and

the   base   name   is   pentene.   The   substituent   groups   are   two   methyl   groups.

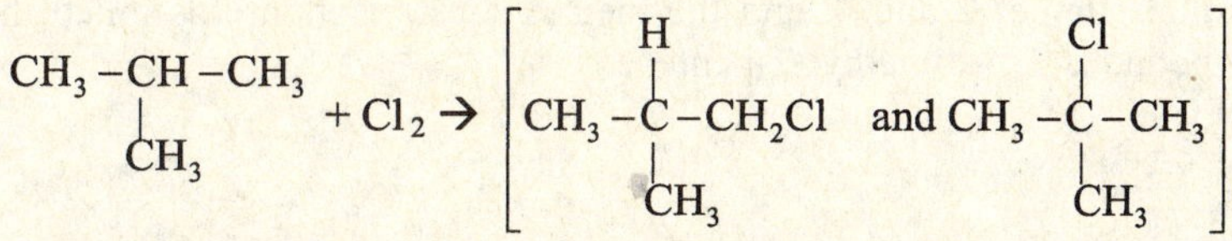

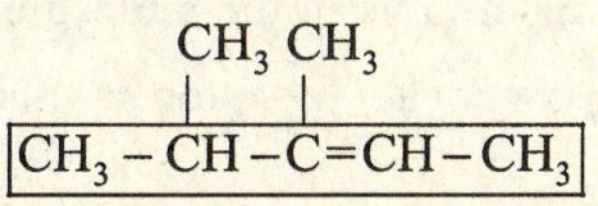

$$CH_3 - C^4H - C^3 = C^2H - C^1H_3$$

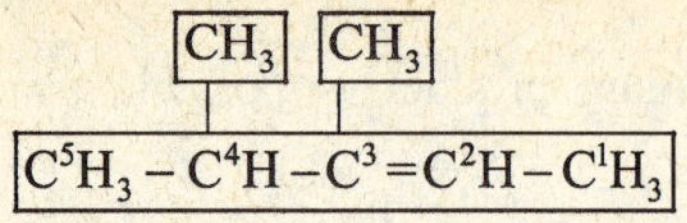

If we start numbering the chain at the end closest to the double bond, the double bond is between position number 2 and 3, and the methyl substituents are assigned the numbers 3 and 4. The name of the compound is 3, 4-dimethyl-2-pentene.

c)

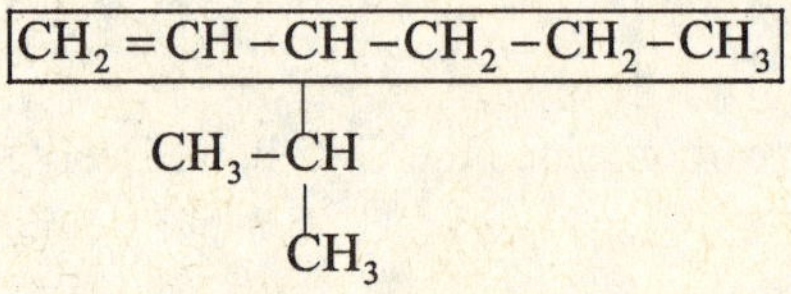

$$CH_2 = CH - CH - CH_2 - CH_2 - CH_3$$
$$CH_3 - CH$$
$$\quad\quad CH_3$$

has 6 carbons as the longest continuous chain. The prefix for 6 is hex- and the base name is hexene. The only substituent group is an isopropyl group.

$$C^1H_2 = C^2H - C^3H - C^4H_2 - C^5H_2 - C^6H_3$$
$$CH_3 - CH$$
$$\quad\quad CH_3$$

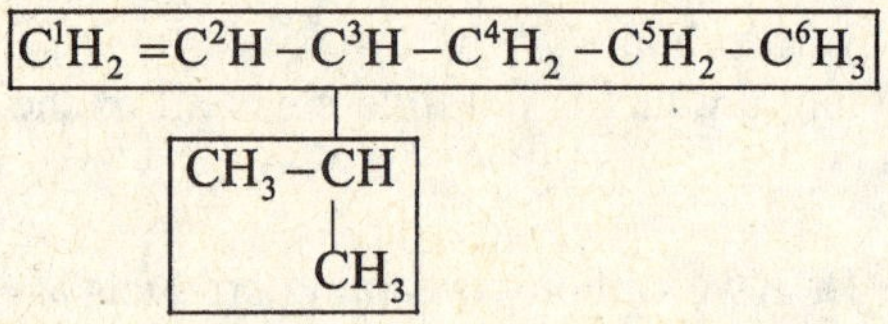

Start numbering on the left since it is closer to the double bond.

Since the double bond is between position number 1 and 2, and the isopropyl group is at position 3, the name of the compound is 3-isopropyl-1-hexene.

d)

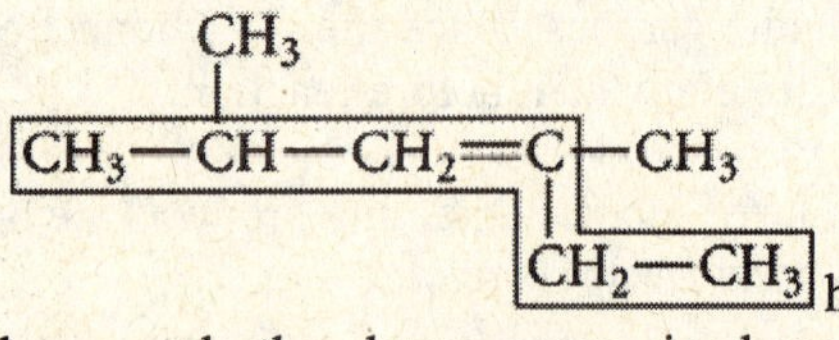

$$\quad\quad\quad CH_3$$
$$CH_3 - CH - CH_2 = C - CH_3$$
$$\quad\quad\quad\quad CH_2 - CH_3$$

has 6 carbons as the longest continuous chain. The prefix for 6 is hex- and the base name is hexene. The only substituent groups are two methyl groups.

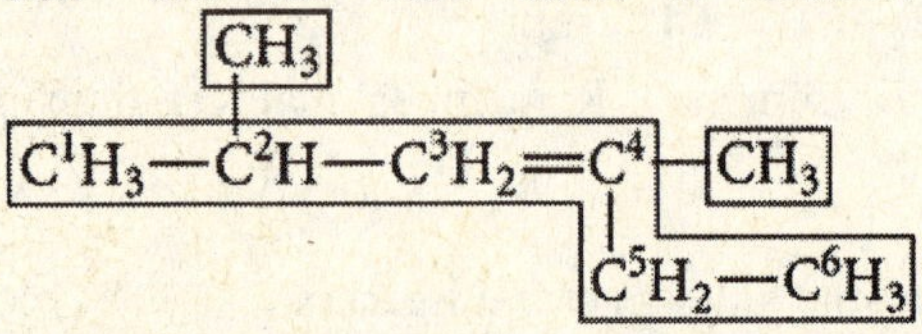

$$\quad\quad\quad CH_3$$
$$C^1H_3 - C^2H - C^3H_2 = C^4 - CH_3$$
$$\quad\quad\quad\quad C^5H_2 - C^6H_3$$

Since the double bond is in the middle of the chain, it will be at position 3 in both numbering schemes. If we start numbering the chain at the end closest to the left methyl group (closest to an end), the methyl substituents are assigned the numbers 2 and 4. The name of the compound is 2,4-dimethyl-3-hexene.

20. **Given:** alkene structures   **Find:** name

Conceptual plan: **Count the number of carbon atoms in the longest continuous carbon chain that contains the multiple bond to determine the base name of the compound. Find the prefix corresponding to this number of atoms in Table 20.5 and add the ending -ene to form the base name. → Consider every branch from the base chain to be a substituent. Name each substituent according to Table 20.6. → Beginning with the end closest to the multiple bond, number the base chain and assign a number to each substituent. → Write the name of the compound in the following format: (subst. #)-(subst. name)(base name). → If there are two or more substituents, give each one a number and list them alphabetically with hyphens between words and numbers. → If a compound has two or more identical substituents, designate the number of identical substituents with the prefix di- (2), tri- (3), or tetra- (4) before the substituent's name. Separate the numbers indicating the positions of the substituents relative to each other with a comma. The prefixes are not taken into account when alphabetizing.**

Solution:

a) $CH_2 - CH_2 - CH = CH - CH_2 - CH_3$ has 6 carbons as the longest continuous chain. The prefix for 6 is hex- and the base name is hexene. There are no substituent groups.

$C^1H_2 - C^2H_2 - C^3H = C^4H - C^5H_2 - C^6H_3$ Since the double bond is in the middle of the molecule it does not matter which end the numbering is started. Since the double bond is between position number 3 and 4, the name of the compound is 3-hexene.

b)

$$CH_3 - CH - CH = CH - CH_3$$
$$\quad\quad CH_3$$

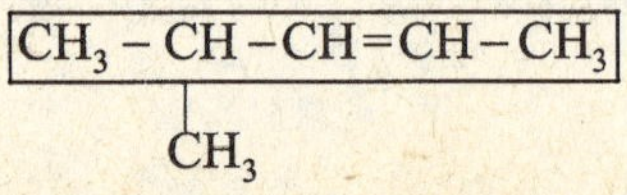

has 5 carbons as the longest continuous chain. The prefix for 5 is pent-

and the base name is pentene. The only substituent group is a methyl group.

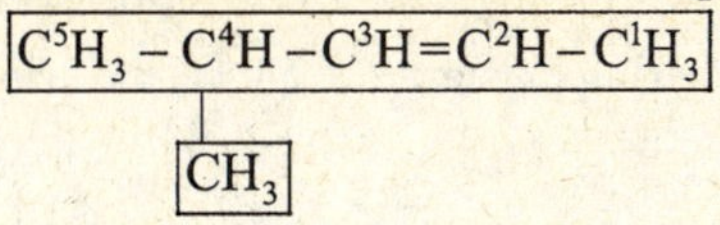

If we start numbering the chain at the end closest to the double bond, the double bond is between position number 2 and 3, and the methyl substituent is assigned the number 4. The name of the compound is 4-methyl-2-pentene.

c) 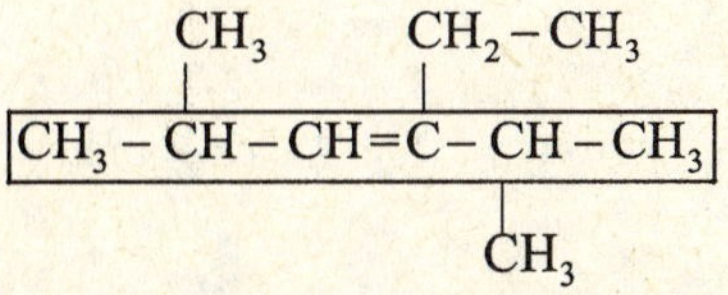 has 6 carbons as the longest continuous chain. The prefix for 6 is hex- and the base name is hexene. The substituent groups are two methyl groups and one ethyl group.

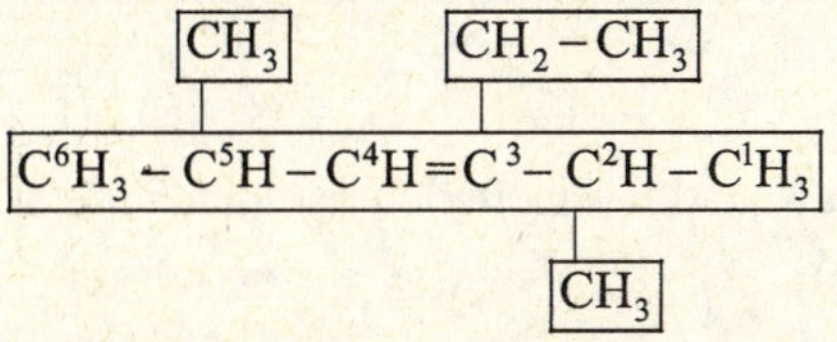

Since the double bond is in the middle of the chain, it will be at position 3 in both numbering schemes. If we start numbering the chain at the end closest to the right methyl group (since there are two substituents on this end), the methyl substituents are assigned the numbers 2 and 5; and the ethyl substituent is assigned number 3. Since e comes before m, the name of the compound is 3-ethyl-2,5-dimethyl-3-hexene.

An alternative name could be 4-isopropyl-2-methyl-3-hexene:

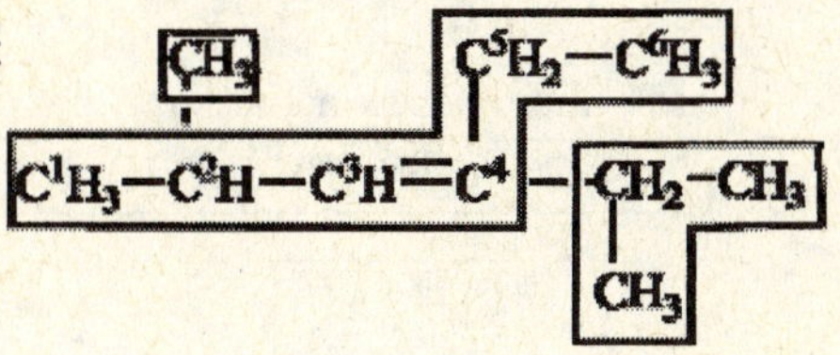

d) 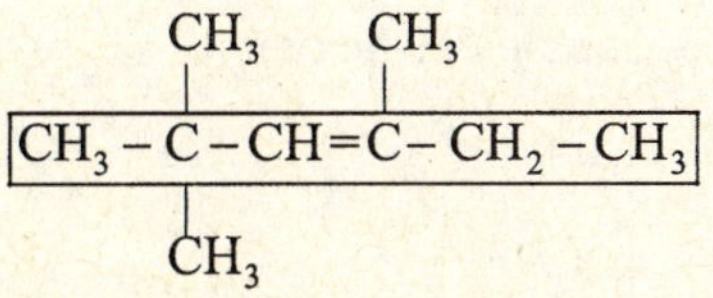 has 6 carbons as the longest continuous chain. The prefix for 6 is hex- and the base name is hexene. The substituent groups are three methyl groups.

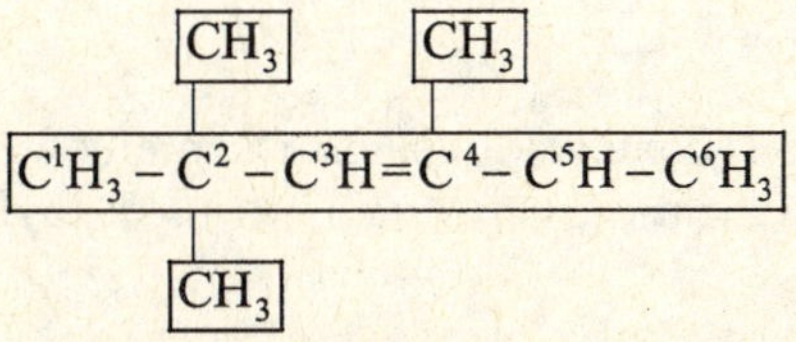

Since the double bond is in the middle of the chain, it will be at position 3 in both numbering schemes. If we start numbering the chain at the end closest to the left methyl groups (closest to an end), the methyl substituents are assigned the numbers 2, 2, and 4. The name of the compound is 2,2,4-trimethyl-3-hexene.

21. **Given:** alkyne structures **Find:** name

Conceptual plan: Count the number of carbon atoms in the longest continuous carbon chain that contains the multiple bond to determine the base name of the compound. Find the prefix corresponding to this number of atoms in Table 20.5 and add the ending -yne to form the base name. → Consider every branch from the base chain to be a substituent. Name each substituent according to Table 20.6. →Beginning with the end closest to the multiple bond, number the base chain and assign a number to each substituent. →Write the name of the compound in the following format: (subst. #)-(subst. name)(base name). → If there are two or more substituents, give each one a number and list them alphabetically with hyphens between words and numbers. →If a compound has two or more identical substituents, designate the number of identical substituents with the prefix di- (2), tri- (3), or tetra- (4) before the substituent's name. Separate the numbers indicating the positions of the substituents relative to each other with a comma. The prefixes are not taken into account when alphabetizing.

Solution:

a) $\boxed{CH_3-C\equiv C-CH_3}$ has 4 carbons as the longest continuous chain. The prefix for 4 is but- and the base name is butyne. There are no substituent groups. $C^1H_3-C^2\equiv C^3-C^4H_3$ Since the triple bond is in the middle of the molecule it does not matter which end the numbering is started. Since the triple bond is between position number 2 and 3, the name of the compound is 2-butyne.

b) 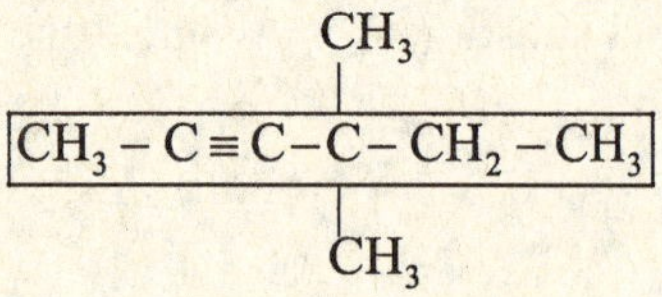 has 6 carbons as the longest continuous chain. The prefix for 6 is hex- and the base name is hexyne. The only substituent groups are two methyl groups.

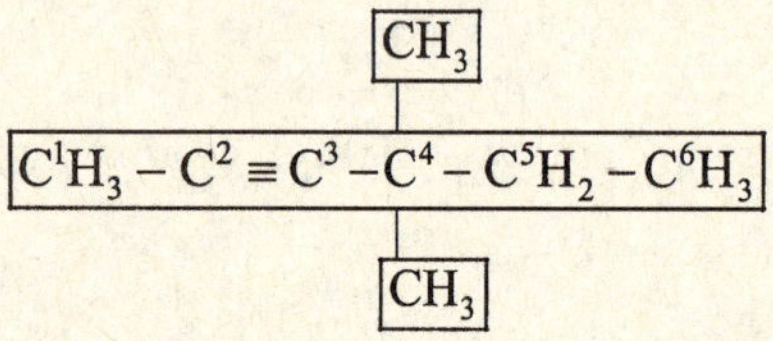 If we start numbering the chain at the end closest to the triple bond, the triple bond is between position number 2 and 3, and the methyl substituents are assigned the numbers 4 and 4. The name of the compound is 4,4-dimethyl-2-hexyne.

c) 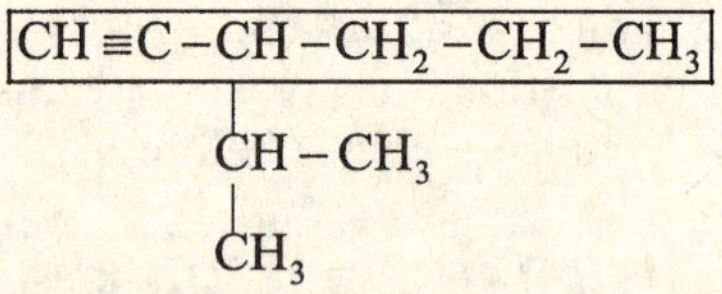 has 6 carbons as the longest continuous chain. The prefix for 6 is hex- and the base name is hexyne. The only substituent group is an isopropyl group.

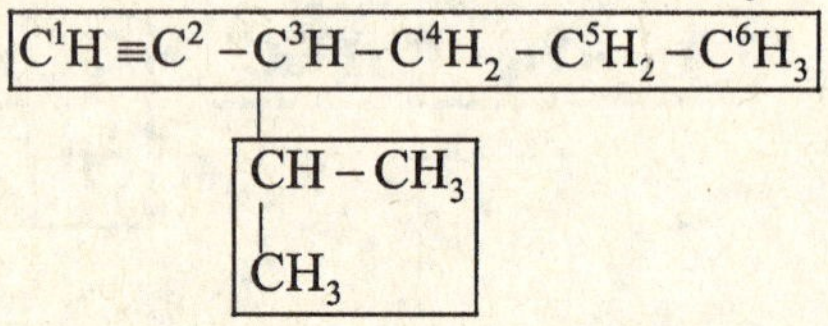 Start numbering at the end closest to the triple bond. The isopropyl substituent is assigned number 3. The name of the compound is 3-isopropyl-1-hexyne.

d) 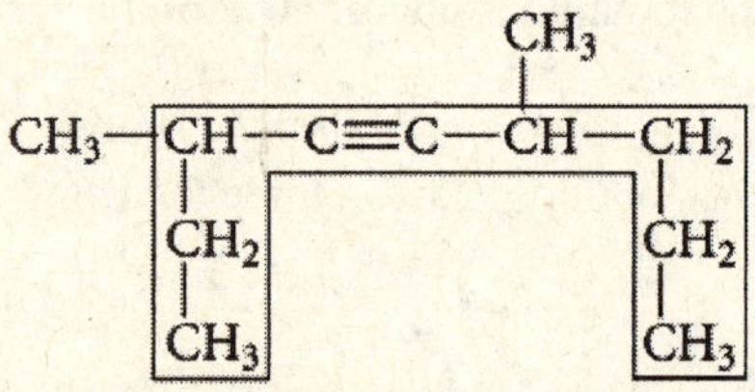 has 9 carbons as the longest continuous chain. The prefix for 9 is non- and the base name is nonyne. The substituent groups are two methyl groups.

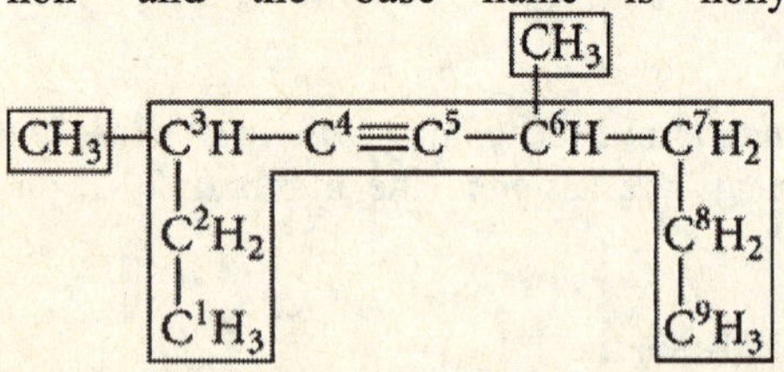 Start numbering at the bottom left and count clockwise along the chain. The triple bond is between positions 4 and 5 and the methyl substituents are assigned the numbers 3 and 6. The name of the compound is 3,6-dimethyl-4-nonyne.

22. **Given:** alkyne structures **Find:** name
Conceptual plan: Count the number of carbon atoms in the longest continuous carbon chain that contains the multiple bond to determine the base name of the compound. Find the prefix corresponding to this number of atoms in Table 20.5 and add the ending -yne to form the base name. → Consider every branch from the base chain to be a substituent. Name each substituent according to Table 20.6. →Beginning with the end closest to the multiple bond, number the base chain and assign a number to each substituent. →Write the name of the compound in the following format: (subst. #)-(subst. name)(base name). → If there are two or more substituents, give each one a number and list them alphabetically with hyphens between words and numbers. →If a compound has two or more identical substituents, designate the number of identical substituents with the prefix di- (2), tri- (3), or tetra- (4)

before the substituent's name. Separate the numbers indicating the positions of the substituents relative to each other with a comma. The prefixes are not taken into account when alphabetizing.
Solution:

a) 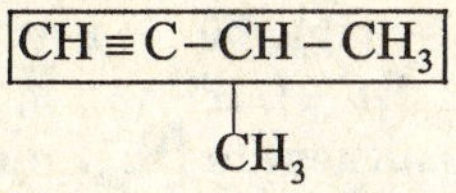 has 4 carbons as the longest continuous chain. The prefix for 4 is but- and the base name is butyne. The only substituent group is a methyl group. 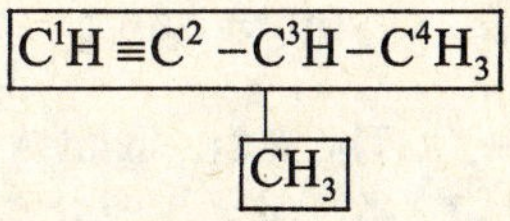 Start numbering at the end closest to the triple bond. The triple bond is between numbers 1 and 2; and the methyl substituent is assigned number 3. The name of the compound is 3-methyl-1-butyne.

b) 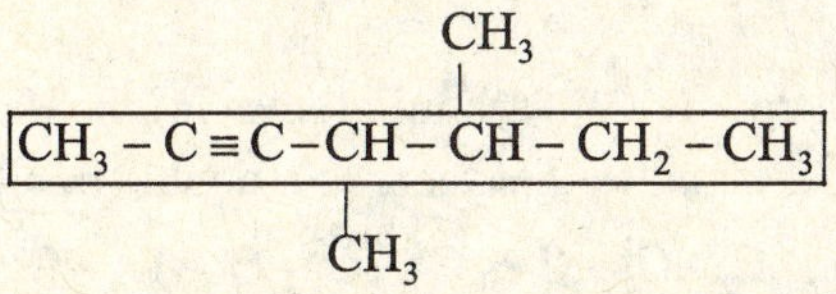 has 7 carbons as the longest continuous chain. The prefix for 7 is hept- and the base name is heptyne. The only substituent groups are two methyl groups. 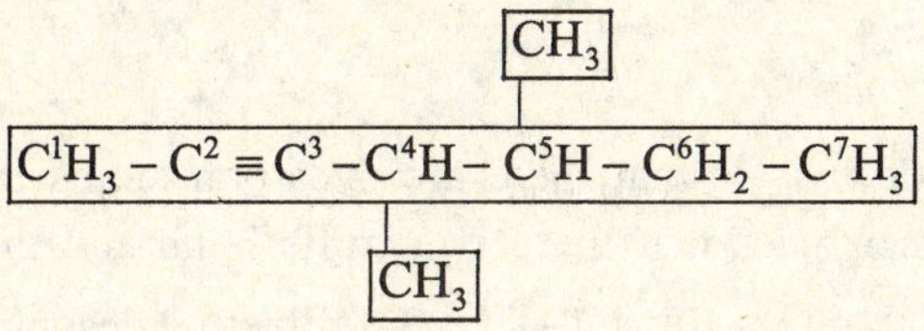 If we start numbering the chain at the end closest to the triple bond, the triple bond is between position number 2 and 3, and the methyl substituents are assigned the numbers 4 and 5. The name of the compound is 4,5-dimethyl-2-heptyne.

c) 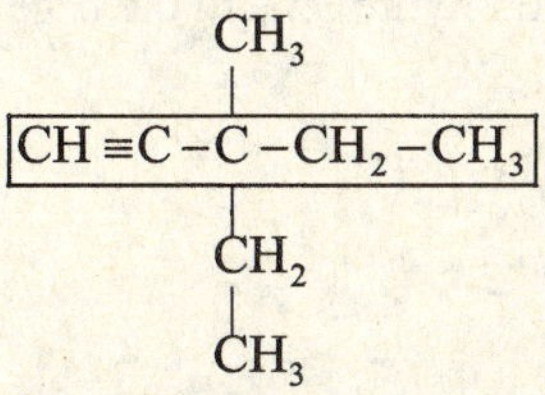 has 5 carbons as the longest continuous chain. The prefix for 5 is pent- and the base name is pentyne. The substituent groups are a methyl group and an ethyl group. 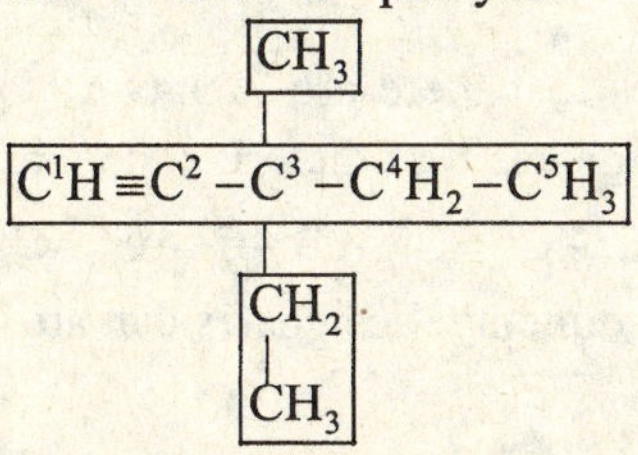 Start numbering at the end closest to the triple bond. The triple bond is between position numbers 1 and 2, the methyl and ethyl groups are both assigned number 3. Since e is before m, the name of the compound is 3-ethyl-3-methyl-1-pentyne.

d) 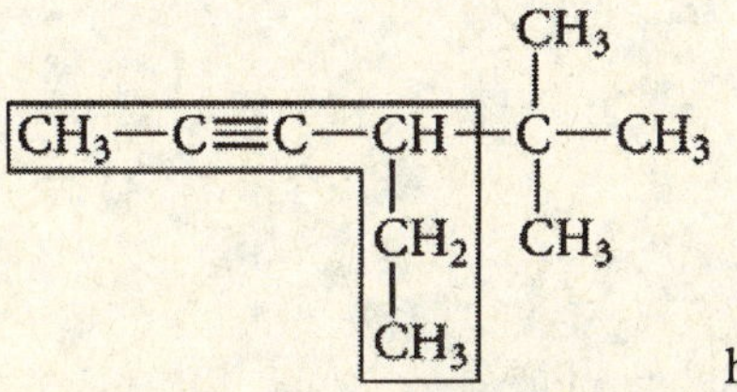 has 6 carbons as the longest continuous chain. The prefix for 6 is hex- and the base name is hexyne. The only substituent group is a tert-butyl group. Start numbering at the end nearest the triple bond and count clockwise along the chain. The triple bond is between positions 2 and 3 and the tert-butyl substituent is assigned the number 4. The name of the compound is 4-tert-butyl-2-hexyne. An alternate name is 4-ethyl-5,5-dimethyl-2-hexyne.

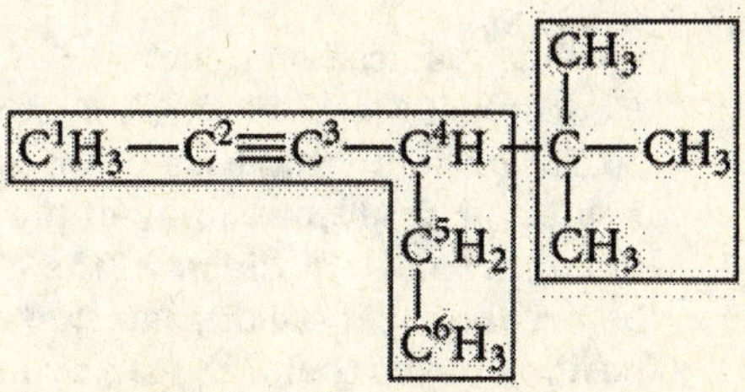

23. **Given:** hydrocarbon names **Find:** structure
Conceptual plan: Find the number of carbon atoms corresponding to prefix of the base name in Table 20.5. → Draw the base chain and number the carbons from left to right. → Determine the multiple bond type from the ending of the base name (-ene = double bond, and -yne = triple bond) and place it in the appropriate position in the chain. → Using Table 20.6 and the prefix di- (2), tri- (3), or tetra- (4) before the substituent's name, to determine each substituent. → Add the substituent to the proper carbon position in the chain. → Add hydrogen atoms to the base chain so that each carbon has 4 bonds.
Solution:

a) 4-octyne. The base name octyne designates that there are 8 carbon atoms in the base chain. $C^1-C^2-C^3-C^4-C^5-C^6-C^7-C^8$. The -yne ending designates that there is a triple bond and the 4 prefix designates that it is between the 4th and 5th positions. $C^1-C^2-C^3-C^4\equiv C^5-C^6-C^7-C^8$. Add hydrogens to the base chain to get the final molecule $CH_3-CH_2-CH_2-C\equiv C-CH_2-CH_2-CH_3$.

b) 3-nonene. The base name nonene designates that there are 9 carbon atoms in the base chain. $C^1-C^2-C^3-C^4-C^5-C^6-C^7-C^8-C^9$. The -ene ending designates that there is a double bond and the 3 prefix designates that it is between the 3rd and 4th positions. $C^1-C^2-C^3=C^4-C^5-C^6-C^7-C^8-C^9$. Add hydrogens to the base chain to get the final molecule
$CH_3-CH_2-CH{=}CH-CH_2-CH_2-CH_2-CH_2-CH_3$.

c) 3,3-dimethyl-1-pentyne. The base name pentyne designates that there are 5 carbon atoms in the base chain. $C^1-C^2-C^3-C^4-C^5$. The -yne ending designates that there is a triple bond and the 1 prefix designates that it is between the 1st and 2nd positions. $C^1\equiv C^2-C^3-C^4-C^5$ 3,3-dimethyl designates that there are two –CH_3 groups in the 3rd position . $C^1\equiv C^2-\underset{\boxed{CH_3}}{\overset{\boxed{CH_3}}{C^3}}-C^4-C^5$. Add hydrogens to the base chain to get the final molecule $CH\equiv C-\underset{CH_3}{\overset{CH_3}{C}}-CH_2-CH_3$.

d) 5-ethyl-3,6-dimethyl-2-heptene. The base name heptene designates that there are 7 carbon atoms in the base chain. $C^1-C^2-C^3-C^4-C^5-C^6-C^7$. The -ene ending designates that there is a double bond and the 2 prefix designates that it is between the 2nd and 3rd positions. $C^1-C^2=C^3-C^4-C^5-C^6-C^7$. 5-ethyl designates that there is a –CH_2CH_3 group in the 5th position; and 3,6-dimethyl designates that are –CH_3 groups in the 3rd and 6th positions. $C^1-C^2=\underset{\boxed{CH_3}}{C^3}-C^4-\underset{\boxed{CH_2-CH_3}}{C^5}-\overset{\boxed{CH_3}}{C^6}-C^7$. Add hydrogens to the base chain to get the final molecule $CH_3-CH=\underset{CH_3}{C}-CH_2-\underset{CH_2-CH_3}{CH}-\overset{CH_3}{CH}-CH_3$.

24. **Given:** hydrocarbon names **Find:** structure
Conceptual plan: Find the number of carbon atoms corresponding to prefix of the base name in Table 20.5. → Draw the base chain and number the carbons from left to right. → Determine the multiple bond type from the ending of the base name (-ene = double bond, and -yne = triple bond) and place it in the appropriate position in the chain. → Using Table 20.6 and the prefix di- (2), tri- (3), or tetra- (4) before the substituent's name, to determine each substituent. → Add the substituent to the proper carbon position in the chain. → Add hydrogen atoms to the base chain so that each carbon has 4 bonds.
Solution:

a) 2-hexene. The base name hexene designates that there are 6 carbon atoms in the base chain.

$C^1-C^2-C^3-C^4-C^5-C^6$. The -ene ending designates that there is a double bond and the 2 prefix designates that it is between the 2nd and 3rd positions. $C^1-C^2=C^3-C^4-C^5-C^6$. Add hydrogens to the base chain to get the final molecule $CH_3-CH{=}CH-CH_2-CH_2-CH_3$.

b) 1-heptyne. The base name heptyne designates that there are 7 carbon atoms in the base chain. $C^1-C^2-C^3-C^4-C^5-C^6-C^7$. The -ene ending designates that there is a double bond and the 3 prefix designates that it is between the 3rd and 4th positions. $C^1\equiv C^2-C^3-C^4-C^5-C^6-C^7$. Add hydrogens to the base chain to get the final molecule $CH\equiv C-CH_2-CH_2-CH_2-CH_2-CH_3$.

c) 4,4-dimethyl-2-hexene. The base name pentyne designates that there are 6 carbon atoms in the base chain. $C^1-C^2-C^3-C^4-C^5-C^6$. The -ene ending designates that there is a double bond and the 2 prefix designates that it is between the 2nd and 3rd positions. $C^1-C^2=C^3-C^4-C^5-C^6$ 4,4-dimethyl designates that there are two $-CH_3$ groups in the 4th position . $C^1-C^2=C^3-\underset{\boxed{CH_3}}{\overset{\boxed{CH_3}}{C^4}}-C^5-C^6$. Add hydrogens to the base chain to get the final molecule $CH_3-CH{=}CH-\underset{CH_3}{\overset{CH_3}{C}}-CH_2-CH_3$.

d) 3-ethyl-4-methyl-2-pentene. The base name pentene designates that there are 5 carbon atoms in the base chain. $C^1-C^2-C^3-C^4-C^5$. The -ene ending designates that there is a double bond and the 2 prefix designates that it is between the 2nd and 3rd positions. $C^1-C^2=C^3-C^4-C^5$. 3-ethyl designates that there is a $-CH_2CH_3$ group in the 3rd position; and 4-methyl designates that there is a $-CH_3$ group in the 4th position. $C^1-C^2=\underset{\boxed{CH_2-CH_3}}{C^3}-\overset{\boxed{CH_3}}{C^4}-C^5$. Add hydrogens to the base chain to get the final molecule

$$CH_3-CH{=}\underset{CH_2-CH_3}{C}-\overset{CH_3}{CH}-CH_3.$$

25. Alkene addition reactions convert a double bond to a single bond and place the two halves of the other reactant on the two carbons that were in the double bond.

a) $CH_3-CH{=}CH-CH_3+Cl_2\rightarrow CH_3-\underset{Cl}{CH}-\underset{Cl}{CH}-CH_3$.

b) $CH_3-CH_2-CH{=}CH-CH_3+Br_2\rightarrow CH_3-CH_2-\underset{Br}{CH}-\underset{Br}{CH}-CH_3$.

26. Alkene addition reactions convert a double bond to a single bond and place the two halves of the other reactant on the two carbons that were in the double bond.

a) $CH_3-\underset{CH_3}{CH}-CH{=}CH_2+Br_2\rightarrow CH_3-\underset{CH_3}{CH}-\underset{Br}{CH}-\underset{Br}{CH_2}$.

b) $CH_2{=}CH-CH_3+Cl_2\rightarrow \underset{Cl}{CH_2}-\underset{Cl}{CH}-CH_3$.

27. Hydrogenation reactions convert a double bond to a single bond and place the hydrogen atoms on each of the two carbons that were in the double bond.

a) $CH_2{=}CH-CH_3 + H_2 \rightarrow CH_3-CH_2-CH_3$.

b) $$CH_3-\underset{CH_3}{\underset{|}{CH}}-CH{=}CH_2 + H_2 \rightarrow CH_3-\underset{CH_3}{\underset{|}{CH}}-CH_2-CH_3.$$

c) $$CH_3-\underset{CH_3}{\underset{|}{CH}}-\underset{CH_3}{\underset{|}{C}}{=}CH_2 + H_2 \rightarrow CH_3-\underset{CH_3}{\underset{|}{CH}}-\underset{CH_3}{\underset{|}{CH}}-CH_3.$$

28. Hydrogenation reactions convert a double bond to a single bond and place the hydrogen atoms on each of the two carbons that were in the double bond.

a) $CH_3-CH_2-CH{=}CH_2 + H_2 \rightarrow CH_3-CH_2-CH_2-CH_3$.

b) $$CH_3-CH_2-\overset{CH_3}{\overset{|}{C}}{=}\overset{CH_3}{\overset{|}{C}}-CH_3 + H_2 \rightarrow CH_3-CH_2-\overset{CH_3}{\overset{|}{CH}}-\overset{CH_3}{\overset{|}{CH}}-CH_3.$$

c) $$CH_3-CH_2-\underset{CH_3}{\underset{|}{C}}{=}CH_2 + H_2 \rightarrow CH_3-CH_2-\underset{CH_3}{\underset{|}{CH}}-CH_3.$$

29. **Given:** monosubstituted benzene structures **Find:** name
Conceptual plan: Consider the branch from the benzene ring to be a substituent. Name the substituent according to Table 20.6 or with the base name of a halogen with an "o" added at the end. →Write the name of the compound in the following format: (name of substituent)benzene.
Solution:

a) $-CH_3$ or methyl is the substituent group, so the name is methylbenzene.

b) –Br or bromo is the substituent, so the name is bromobenzene.

c) –Cl or chloro is the substituent, so the name is chlorobenzene.

30. **Given:** monosubstituted benzene structures **Find:** name
Conceptual plan: Consider the branch from the benzene ring to be a substituent. Name the substituent according to Table 20.6 or with the base name of a halogen with an "o" added at the end. →Write the name of the compound in the following format: (name of substituent)benzene.
Solution:

a) $-CH_2CH_3$ or ethyl is the substituent group, so the name is ethylbenzene.

b) –F or fluoro is the substituent, so the name is fluorobenzene.

c) $-C(CH_3)_3$ or *tert*-butyl is the substituent, so the name is *tert*-butylbenzene.

31. **Given:** disubstituted benzene structures **Find:** name
Conceptual plan: Consider the branches from the benzene ring to be substituents. Name the substituent according to Table 20.6 or with the base name of a halogen with an "o" added at the end. → Number the benzene ring starting with the substituent that is first alphabetically and count in the direction that gets to the second substituent with the lower position number. → Write the name of the compound in the following format: (name of substituent)benzene, listing them alphabetically with hyphens between words and numbers. →If a compound has two identical substituents, designate the number of identical substituents with the prefix di- (2) before the substituent's name. Separate the numbers indicating the positions of the substituents relative to each other with a comma. → Alternate names are 1,2 = *ortho* or *o*; 1,3 = *meta* or *m*; and 1,4 = *para* or *p* replacing the numbers.
Solution:

a) Both of the substituent groups are –Br or bromo groups. Start by giving one bromo an assignment of 1 and count in either direction to give the second bromo group an assignment of 4. The name is 1,4-dibromobenzene or *p*-dibromobenzene.

b) Both of the substituent groups are $-CH_2CH_3$ or ethyl groups. Start by giving the top ethyl an assignment of 1 and count in the clockwise direction to give the second ethyl group an assignment of 3. The name is 1,3-diethylbenzene or *m*-diethylbenzene.

c) One substituent group is –Cl or chloro and the other is –F or fluoro. Start by giving the chloro group an assignment of 1 and count in the counterclockwise direction to give the fluoro group an assignment of 2. The name is 1-chloro-2-fluorobenzene or *o*-chlorofluorobenzene.

32. **Given:** disubstituted benzene structures **Find:** name
Conceptual plan: Consider the branches from the benzene ring to be substituents. Name the substituent according to Table 20.6 or with the base name of a halogen with an "o" added at the end. → Number the benzene ring starting with the substituent that is first alphabetically and count in the direction that gets to the second substituent with the lower position number. → Write the name of the compound in the following format: (name of substituent)benzene, listing them alphabetically with hyphens between words and numbers. →If a compound has two identical substituents, designate the number of identical substituents with the prefix di- (2) before the substituent's name. Separate the numbers indicating the positions of the substituents relative to each other with a comma. → Alternate names are 1,2 = *ortho* or *o*; 1,3 = *meta* or *m*; and 1,4 = *para* or *p* replacing the numbers.
Solution:

a) One substituent group is –Br or bromo and the other is –Cl or chloro. Start by giving the bromo group an assignment of 1 and count in the clockwise direction to give the chloro group an assignment of 2. The name is 1-bromo-2-chlorobenzene or *o*-bromochlorobenzene.

b) One substituent group is –Cl or chloro and the other is $-CH_2CH_3$ or ethyl. Start by giving the chloro group an assignment of 1 and count in either direction to give the ethyl group an assignment of 4. The name is 1-chloro-4-ethylbenzene or *p*-chloroethylbenzene.

c) Both of the substituent groups are –I or iodo groups. Start by giving the top iodo an assignment of 1 and count in the clockwise direction to give the second iodo group an assignment of 3. The name is 1,3-diiodobenzene or *m*-diiodobenzene.

33. **Given:** substituted benzene names **Find:** structures
Conceptual plan: Start with a benzene ring. Identify the structure of the substituent(s) according to Table 20.6 or with the base name of a halogen with an "o" added at the end. → If only one substituent is present, simply attach it to the benzene ring. →If a compound has two identical substituents, they are designated with the prefix di- (2) before the substituent's name. → If there are two substituents, place the first one on the benzene ring and count clockwise around the ring to determine the location to attach the second substituent. Note, 1,2 = *ortho* or *o*; 1,3 = *meta* or *m*; and 1,4 = *para* or *p* replacing the numbers.
Solution:

a) isopropylbenzene. The isopropyl group is $\begin{array}{l}-CH-CH_3\\ \ \ \ |\\ \ \ \ CH_3\end{array}$. Attach this to a benzene ring to make

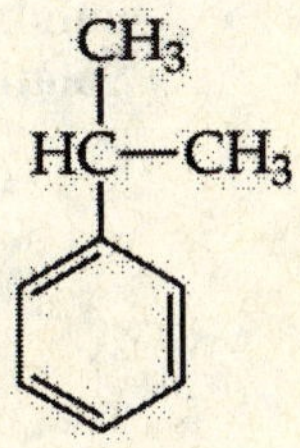

b) *meta*-dibromobenzene. There are two bromo or –Br groups. The positions are 1 and 3, since the designation is *meta*. Attach the –Br's at the 1st and 3rd positions to make

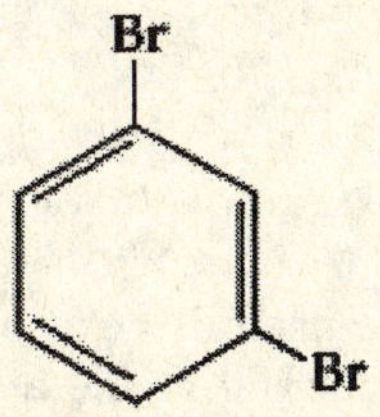

c) 1-chloro-4-methylbenzene. One substituent group is –Cl or chloro and the other is $-CH_3$ or methyl. Start by giving the chloro group an assignment of 1 and count in the clockwise direction to give the methyl group an assignment of 4. Attach the two groups to make

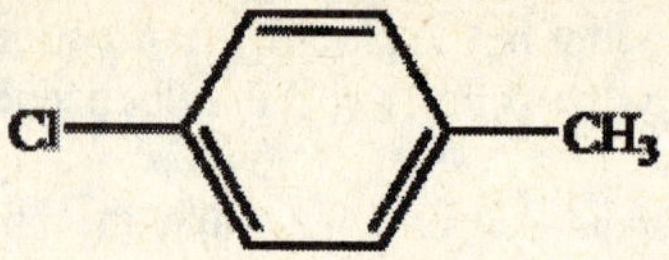

.

34. **Given:** substituted benzene names **Find:** structures

Conceptual plan: Start with a benzene ring. Identify the structure of the substituent(s) according to Table 20.6 or with the base name of a halogen with an "o" added at the end. → If only one substituent is present, simply attach it to the benzene ring. →If a compound has two identical substituents, they are designated with the prefix di- (2) before the substituent's name. → If there are two substituents, place the first one on the benzene ring and count clockwise around the ring to determine the location to attach the second substituent. Note, 1,2 = *ortho* or *o*; 1,3 = *meta* or *m*; and 1,4 = *para* or *p* replacing the numbers.

Solution:

a) ethylbenzene. The ethyl group is $-CH_2CH_3$. Attach this to a benzene ring to make

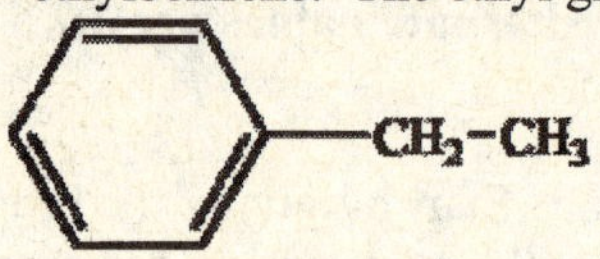

.

b) 1-iodo-2-methylbenzene. The iodo group is –I and the methyl group is $-CH_3$. Attach the –I to the benzene ring and then attach the $-CH_3$ to the adjacent carbon to make

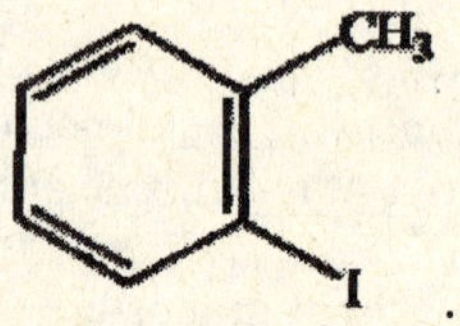

.

b) *para*-diethylbenzene. There are two ethyl or $-CH_2CH_3$ groups. The positions are 1 and 4, since the designation is *para*-. Attach the $-CH_2CH_3$s at the 1st and 4th positions to make

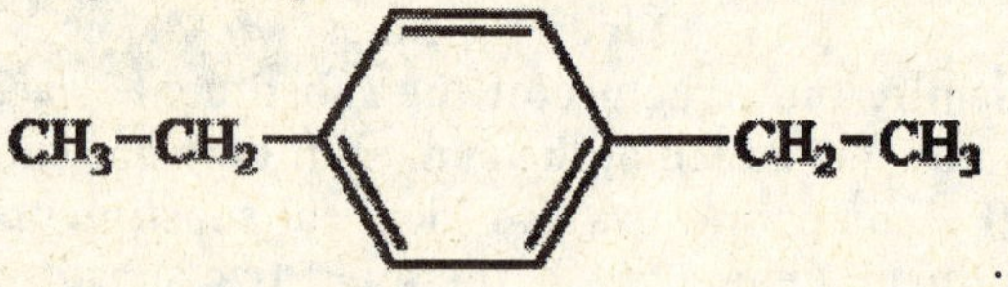

.

35. **Given:** alcohol structures **Find:** name

Conceptual plan: Alcohols are named like alkanes with the following differences: (1) The base chain is the longest continuous carbon chain that contains the –OH functional group; (2) The base name has the ending -ol; (3) The base chain is numbered to give the –OH group the lowest possible number; and (3) A number indicating the position of the –OH group is inserted just before the base name.

Solution:

a) $\boxed{CH_3-CH_2-CH_2}-OH$ The longest carbon chain has 3 carbons. The prefix for 3 is prop-, so the base name is propanol. The alcohol group is on the end, or 1st carbon, so the name is 1-propanol.

b)

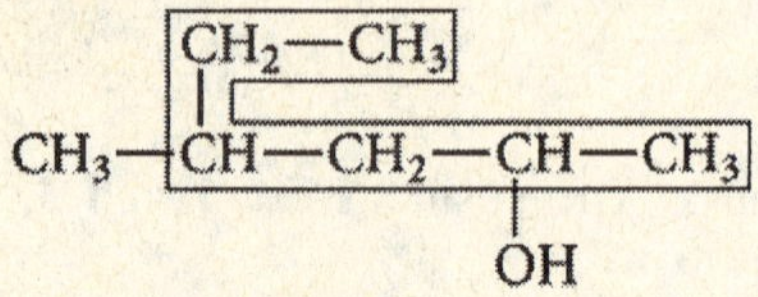

The longest carbon chain has 6 carbons. The prefix for 6 is hex-, so the base name is hexanol. There is a methyl substituent group.

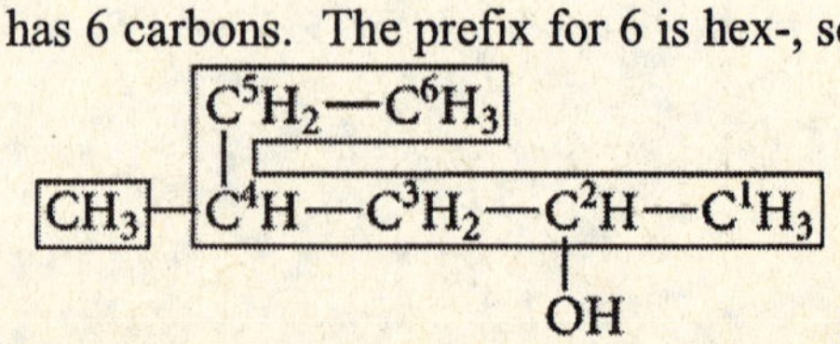

Start numbering at the end closest to the alcohol group. The alcohol group is assigned number 2 and the methyl group is assigned number 4. The name is 4-methyl-2-hexanol.

c) 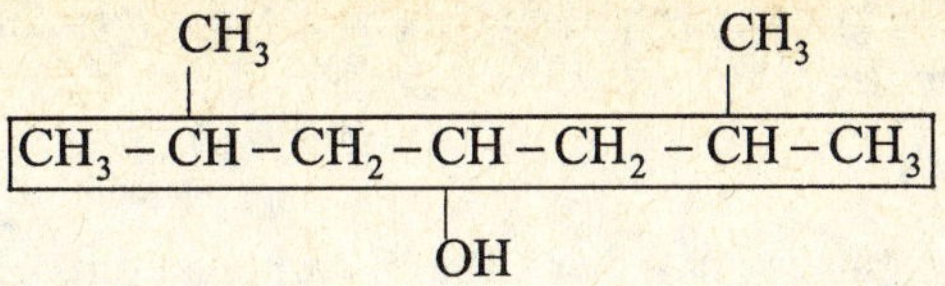 The longest carbon chain has 7 carbons. The prefix for 7 is hept-, so the base name is heptanol. There are two methyl substituent groups. 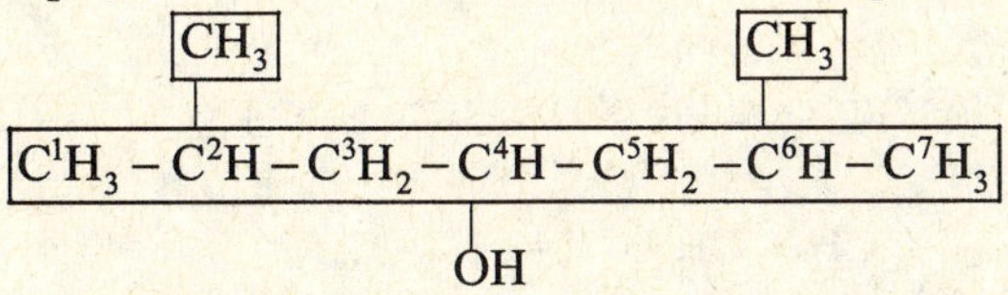 It does not matter which end the numbering is started, since substitutions are symmetrically attached. The alcohol group is assigned number 4 and the methyl groups are assigned numbers 2 and 6. The name is 2,6-dimethyl-4-heptanol.

d) 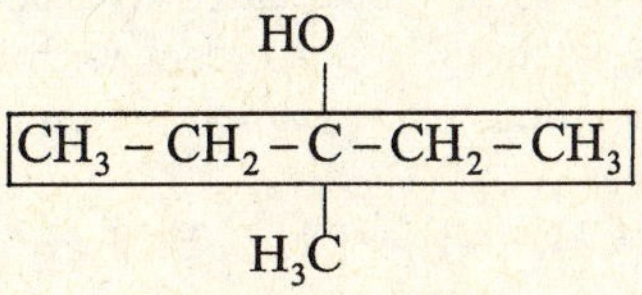 The longest carbon chain has 5 carbons. The prefix for 5 is pent-, so the base name is pentanol. There is a methyl substituent group. 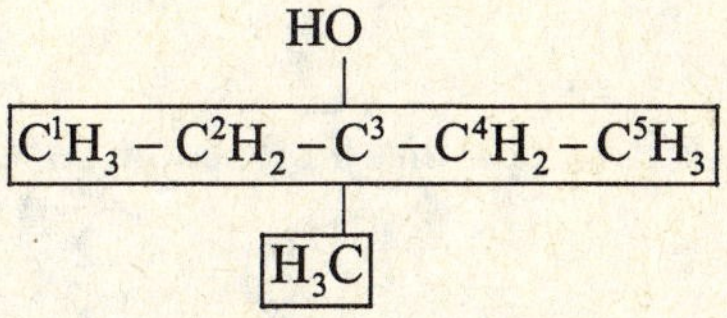 It does not matter which end the numbering is started, since substitutions are symmetrically attached. The alcohol group is assigned number 3 and the methyl group is assigned number 3. The name is 3-methyl-3-pentanol.

36. **Given:** alcohol names **Find:** structure
Conceptual plan: Find the number of carbon atoms corresponding to prefix of the base name in Table 20.5. → Draw the base chain and number the carbons from left to right. → Add an –OH group to the specified position. → Using Table 20.6 and the prefix di- (2), tri- (3), or tetra- (4) before the substituent's name, determine each substituent. → Add the substituent to the proper carbon position in the chain. → Add hydrogen atoms to the base chain so that each carbon has 4 bonds.
Solution:

a) 2-butanol. The base name butanol designates that there are 4 carbon atoms in the base chain. $C^1-C^2-C^3-C^4$. The 2- prefix designates that the –OH group is in the 2nd position 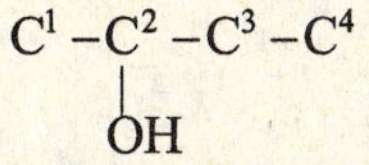.
Add hydrogens to the base chain to get the final molecule 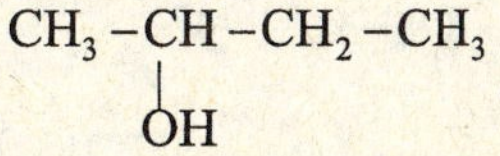.

b) 2-methyl-1-propanol. The base name propanol designates that there are 3 carbon atoms in the base chain. $C^1-C^2-C^3$. 2-methyl designates that a $-CH_3$ group is in the 2nd position, and the 1- prefix of the base name designates that the –OH group is in the 1st position 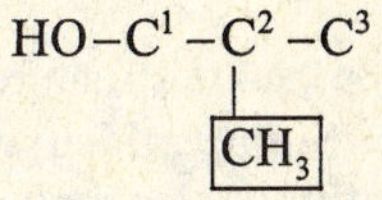. Add hydrogens to the base chain to get the final molecule 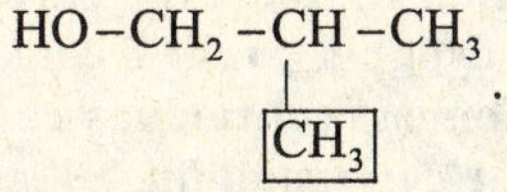.

c) 3-ethyl-1-hexanol. The base name hexanol designates that there are 6 carbon atoms in the base chain. $C^1-C^2-C^3-C^4-C^5-C^6$. 3-ethyl designates that a $-CH_2CH_3$ group is in the 3rd position, and the 1-prefix of the base name designates that the –OH group is in the 1st position.
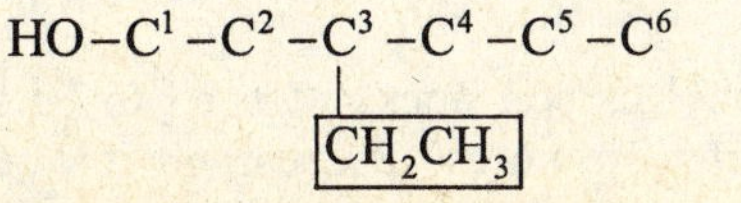

Add hydrogens to the base chain to get the final molecule. $HO-CH_2-CH_2-\underset{\displaystyle CH_2CH_3}{\underset{|}{CH}}-CH_2-CH_2-CH_3$

d) 2-methyl-3-pentanol. The base name pentanol designates that there are 5 carbon atoms in the base chain. $C^1-C^2-C^3-C^4-C^5$. 2-methyl designates that a $-CH_3$ group is in the 2nd position, and the 3- prefix of the base name designates that the –OH group is in the 3rd position. $C^1-\underset{\boxed{CH_3}}{\underset{|}{C^2}}-\underset{OH}{\underset{|}{C^3}}-C^4-C^5$. Add hydrogens to the base chain to get the final molecule. $CH_3-\underset{CH_3}{\underset{|}{CH}}-\underset{OH}{\underset{|}{CH}}-CH_2-CH_3$

37. In a substitution reaction, an alcohol reacts with an acid, such as HBr, to form halogenated hydrocarbons and water. In an elimination (or dehydration) reaction, concentrated acids, such as H_2SO_4, react with alcohols to eliminate water, forming an alkene. In an oxidation reaction, carbon atoms gain oxygen atoms and/or lose hydrogen atoms. In these reactions, the alcohol becomes a carboxylic acid.

a) This is a substitution reaction, so $CH_3-CH_2-CH_2-OH + HBr \rightarrow CH_3-CH_2-CH_2-Br + H_2O$.

b) This is an elimination reaction, so $CH_3-\underset{CH_3}{\underset{|}{CH}}-CH_2-OH \xrightarrow{H_2SO_4} CH_3-\underset{CH_3}{\underset{|}{C}}=CH_2 + H_2O$.

c) This is an oxidation reaction, so $CH_3-\underset{CH_3}{\underset{|}{\overset{CH_3}{\overset{|}{C}}}}-CH_2-CH_2-OH \xrightarrow[H_2SO_4]{Na_2Cr_2O_7} CH_3-\underset{CH_3}{\underset{|}{\overset{CH_3}{\overset{|}{C}}}}-CH_2-\overset{O}{\overset{\|}{C}}-OH$.

38. In a substitution reaction, an alcohol reacts with an acid, such as HBr, to form halogenated hydrocarbons and water. In an elimination (or dehydration) reaction, concentrated acids, such as H_2SO_4, react with alcohols to eliminate water, forming an alkene. In an oxidation reaction, carbon atoms gain oxygen atoms and/or lose hydrogen atoms. In these reactions, the alcohol becomes a carboxylic acid.

a) This is an elimination reaction, so $CH_3-\underset{CH_3}{\underset{|}{\overset{CH_3}{\overset{|}{CH}}}}-OH \xrightarrow{H_2SO_4} CH_3-\underset{CH_3}{\underset{|}{C}}=CH_2 + H_2O$.

b) This is an oxidation reaction, so

$$CH_3-\overset{CH_3}{\overset{|}{CH}}-CH_2-CH_2-OH \xrightarrow[H_2SO_4]{Na_2Cr_2O_7} CH_3-\overset{CH_3}{\overset{|}{CH}}-CH_2-\overset{O}{\overset{\|}{C}}-OH.$$

c) This is a substitution reaction, so $CH_3-CH_2-OH + HCl \rightarrow CH_3-CH_2-Cl + H_2O$.

39. **Given:** aldehyde or ketone structures **Find:** name

Conceptual Plan: Simple aldehydes are systematically named according to the number of carbon atoms in the longest continuous carbon chain that contains the carbonyl group. Form the base name from the name of the corresponding alkane by dropping the -e and add the ending -al. Simple ketones are systematically named according to the longest continuous carbon chain containing the carbonyl group. Form the base name from the name of the corresponding alkane by dropping the letter -e and adding the ending -one. For ketones, number the chain to give the carbonyl group the lowest possible number.

Solution:

a) $\boxed{CH_3-\overset{O}{\overset{\|}{C}}-CH_2-CH_3}$ The longest carbon chain has 4 carbons. The prefix for 4 is but-, and since this is a ketone the base name is butanone. The position of the carbonyl carbon does not need to be specified, since there is only one place for it to be in the molecule (if it were on the end carbon it would be an aldehyde, not a ketone). Since there are no other substituent groups, the name is butanone.

b) 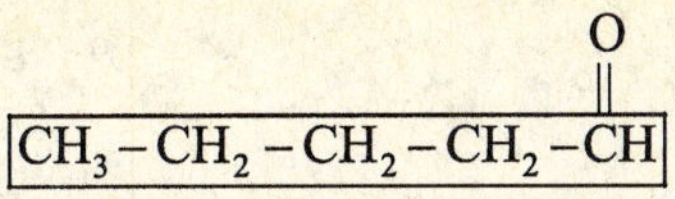 The longest carbon chain has 5 carbons. The prefix for 5 is pent-, and since this is an aldehyde the base name is pentanal. Since there are no other substituent groups, the name is pentanal.

c) 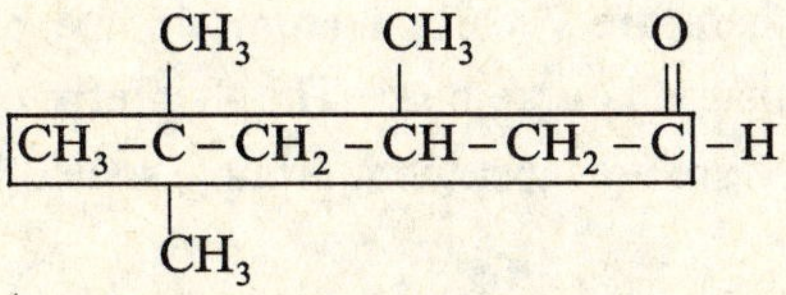 The longest carbon chain has 6 carbons. The prefix for 6 is hex-, and since this is an aldehyde the base name is hexanal. There are 3 methyl substituent groups.

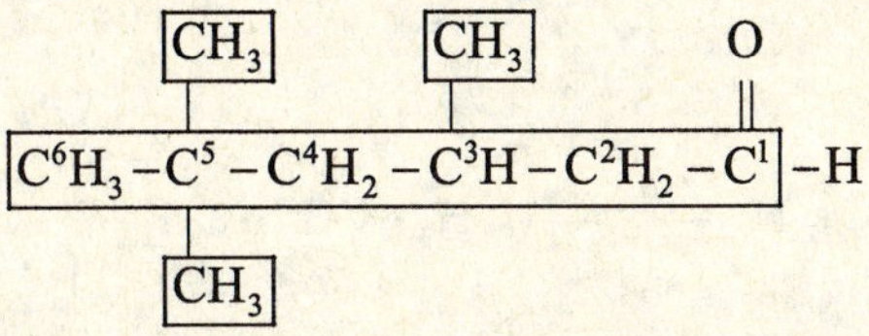 Start numbering at the carbonyl carbon. The methyl groups are at positions 3, 5, and 5. The name of the molecule is 3,5,5-trimethylhexanal.

d) 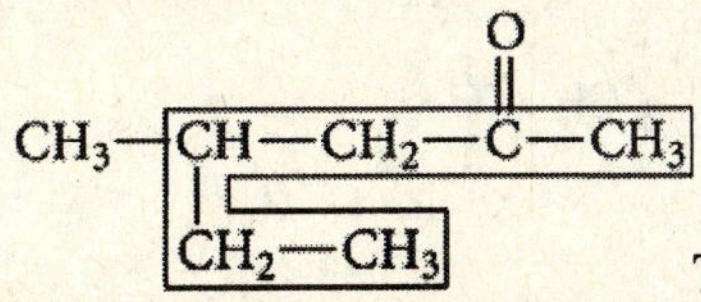 The longest carbon chain has 6 carbons. The prefix for 6 is hex-, and since this is a ketone the base name is hexanone. There is one methyl substituent group.

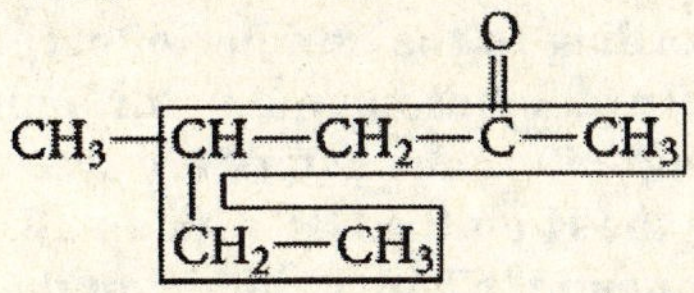 Start numbering at the end closest to the carbonyl carbon. The carbonyl carbon is in position 2 and the methyl group is at position 4. The name of the molecule is 4-methyl-2-hexanone.

40. **Given:** aldehyde and ketone names **Find:** structure

Conceptual plan: Find the number of carbon atoms corresponding to prefix of the base name in Table 20.5. → Draw the base chain and number the carbons from left to right. → Add an =O to the specified position (If none is specified, put it at the end.). → Using Table 20.6 and the prefix di- (2), tri- (3), or tetra- (4) before the substituent's name, to determine each substituent. → Add the substituent to the proper carbon position in the chain. → Add hydrogen atoms to the base chain so that each carbon has 4 bonds.

Solution:

a) hexanal. The base name hexanal designates that there are 6 carbon atoms in the base chain. $C^1-C^2-C^3-C^4-C^5-C^6$. The -al ending specifies that it is an aldehyde and the carbonyl carbon is on the end. There are no substituent groups, so add hydrogens to the base chain to get the final molecule

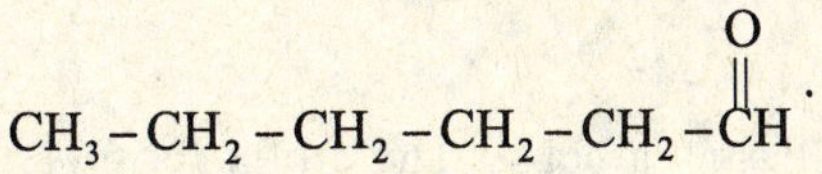.

b) 2-pentanone. The base name pentanone designates that there are 5 carbon atoms in the base chain. $C^1-C^2-C^3-C^4-C^5$. The -one ending specifies that it is a ketone. The 2- prefix of the base name designates that the =O group is in the 2nd position. There are no substituent groups, so add hydrogens to the base chain to get the final molecule 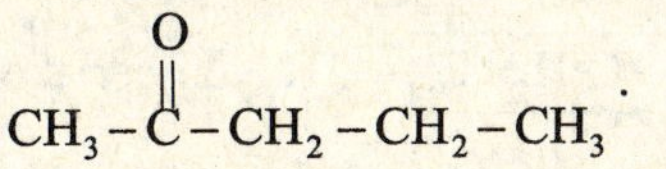.

c) 2-methylbutanal. The base name butanal designates that there are 4 carbon atoms in the base chain. $C^1-C^2-C^3-C^4$. 2-methyl designates that a $-CH_3$ group is in the 2nd position. The -al ending specifies that it is an aldehyde, and so the =O group is in the 1st position. $\begin{matrix} O & CH_3 \\ \| & | \\ C^1 - C^2 - C^3 - C^4 \end{matrix}$. Add hydrogens to the

base chain to get the final molecule. $\begin{matrix} O & CH_3 \\ \| & | \\ H - C - CH - CH_2 - CH_3 \end{matrix}$.

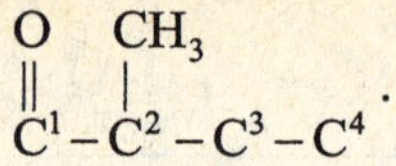

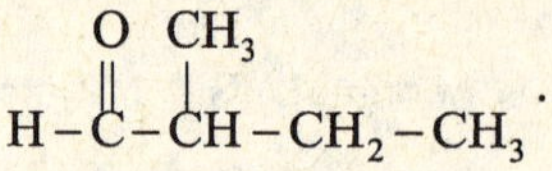

d) 4-heptanone. The base name heptanone designates that there are 7 carbon atoms in the base chain. $C^1 - C^2 - C^3 - C^4 - C^5 - C^6 - C^7$. The -one ending specifies that it is a ketone. The 4- prefix of the base name designates that the =O group is in the 4th position. There are no substituent groups, so add hydrogens to the base chain to get the final molecule $\begin{matrix} & & & O \\ & & & \| \\ CH_3 - CH_2 - CH_2 - C - CH_2 - CH_2 - CH_3 \end{matrix}$.

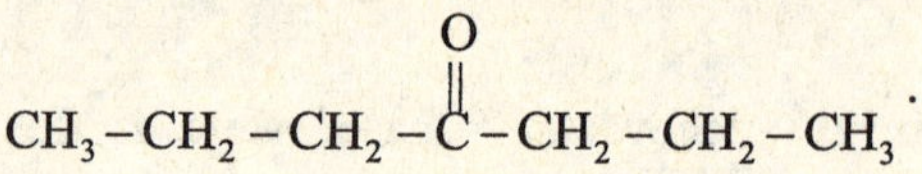

41. This is an addition reaction, $\begin{matrix} O \\ \| \\ CH_3 - CH_2 - CH_2 - C - H \end{matrix} + H - C \equiv N \xrightarrow{NaCN} \begin{matrix} OH \\ | \\ CH_3 - CH_2 - CH_2 - C - C \equiv N \\ | \\ H \end{matrix}$.

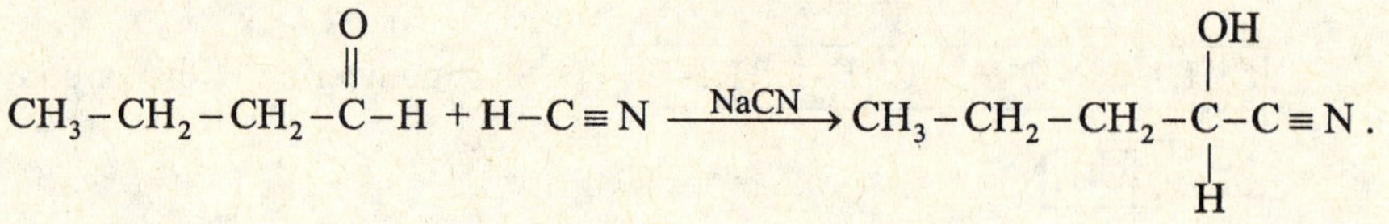

42. This is an addition reaction, $\begin{matrix} O \\ \| \\ CH_3 - C - CH_2 - CH_3 \end{matrix} + H - C \equiv N \xrightarrow{NaCN} \begin{matrix} OH \\ | \\ CH_3 - C - C \equiv N \\ | \\ CH_2 - CH_3 \end{matrix}$.

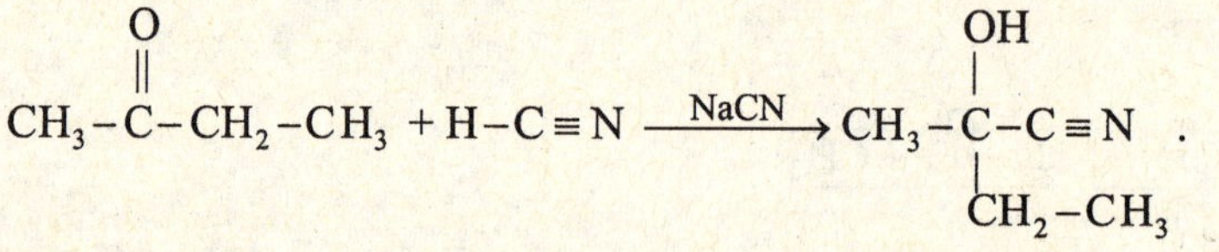

43. **Given:** carboxylic acid or ester structures   **Find:** name

Conceptual Plan: Carboxylic acids are systematically named according to the number of carbon atoms in the longest chain containing the –COOH functional group. Form the base name by dropping the -e from the name of the corresponding alkane, and adding the ending -oic acid. Esters are systematically named as if they were derived from a carboxylic acid by replacing the H on the OH with an alkyl group. The R group from the parent acid forms the base name of the compound. Change the -ic on the name of the corresponding carboxylic acid to -ate, and drop "acid." The R group that replaced the H on the carboxylic acid is named as an alkyl group with the ending -yl.

Solution:

a) $\boxed{\begin{matrix} O \\ \| \\ CH_3 - CH_2 - CH_2 - C \end{matrix}} - O - \boxed{CH_3}$ The carbon chain that has the carbonyl carbon has 4 carbons. The prefix for 4 is but-, so the base name is butanoate. The R group that replaces the H of the carboxylic acid is a methyl group. The name is methylbutanoate.

b) $\boxed{\begin{matrix} O \\ \| \\ CH_3 - CH_2 - C \end{matrix}} - OH$ The carbon chain has 3 carbons. The prefix for 3 is prop-, since this is a carboxylic acid, and there are no other substituents, the name is propanoic acid.

c) $\boxed{\begin{matrix} & & & & O \\ & & & & \| \\ CH_3 - CH - CH_2 - CH_2 - CH_2 - C \\ | \\ CH_3 \end{matrix}} - OH$ The carbon chain has 6 carbons. The prefix for 6 is hex-, since this is a carboxylic acid, the base name is hexanoic acid. There is one methyl substituent group.

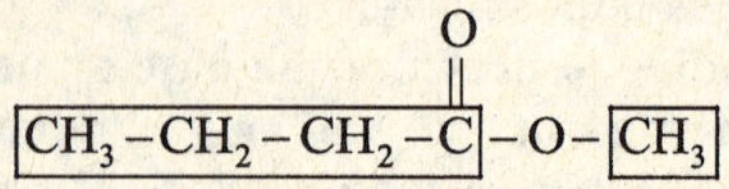

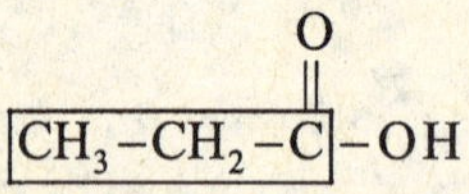

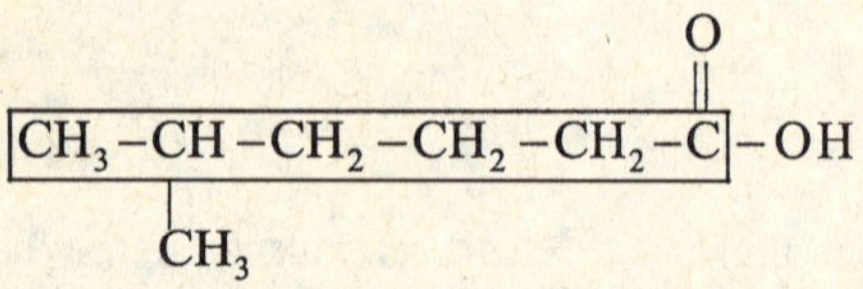

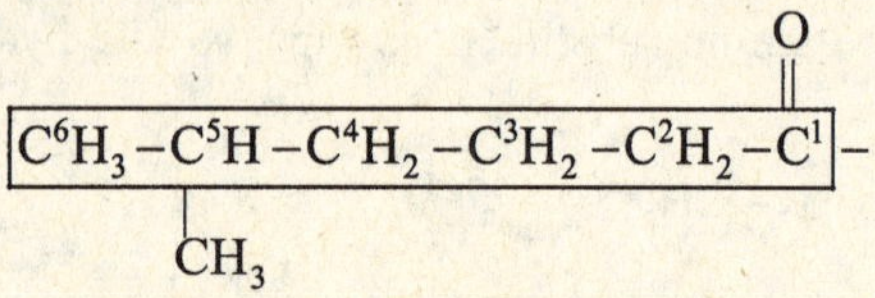

$\boxed{\begin{matrix} & & & & & O \\ & & & & & \| \\ C^6H_3 - C^5H - C^4H_2 - C^3H_2 - C^2H_2 - C^1 \\ | \\ CH_3 \end{matrix}} - OH$ Number the chain starting with the carbonyl carbon. The methyl is in the 5th position. The name of the molecule is 5-methylhexanoic acid.

d) $\boxed{CH_3-CH_2-CH_2-CH_2-\overset{\overset{O}{\|}}{C}}-O-\boxed{CH_2-CH_3}$ The carbon chain that has the carbonyl carbon has 5 carbons. The prefix for 5 is pent-, so the base name is pentanoate. The R group that replaces the H of the carboxylic acid is an ethyl group. The name is ethylpentanoate.

44. **Given:** carboxylic acid and ester names **Find:** structure
Conceptual plan: Find the number of carbon atoms corresponding to the prefix of the base name in Table 20.5. → Draw the base chain and number the carbons from right to left. → Add an =O and an –O to the last carbon on the right. → If the molecule is a carboxylic acid, add a hydrogen atom to the non-carbonyl oxygen. If the molecule is an ester, determine the length of the carbon chain using Table 20.5 and add this to the non-carbonyl oxygen. → Using Table 20.6 and the prefix di- (2), tri- (3), or tetra- (4) before the substituent's name, to determine each substituent. → Add the substituent to the proper carbon position in the chain. → Add hydrogen atoms to the base chain so that each carbon has 4 bonds.
Solution:

a) pentanoic acid. The base name pentanoic acid designates that there are 5 carbon atoms in the base chain. $C^5-C^4-C^3-C^2-C^1$. The -oic acid ending specifies that it is a carboxylic acid. There are no substituent groups, so add hydrogens to the base chain to get the final molecule $CH_3-CH_2-CH_2-CH_2-\overset{\overset{O}{\|}}{C}-OH$.

b) methyl hexanoate. The base name hexanoate designates that there are 6 carbon atoms in the base chain. $C^6-C^5-C^4-C^3-C^2-C^1$. Adding the two oxygens to the 1st carbon, $C^6-C^5-C^4-C^3-C^2-\overset{\overset{O}{\|}}{C^1}-O$ The methyl prefix indicates that there is a methyl group on the non-carbonyl oxygen. $C^6-C^5-C^4-C^3-C^2-\overset{\overset{O}{\|}}{C^1}-O-\boxed{CH_3}$ There are no substituent groups, so add hydrogens to the base chain to get the final molecule $CH_3-CH_2-CH_2-CH_2-CH_2-\overset{\overset{O}{\|}}{C}-O-CH_3$.

c) 3-ethylheptanoic acid. The base name heptanoic acid designates that there are 7 carbon atoms in the base chain. $C^7-C^6-C^5-C^4-C^3-C^2-C^1$. The -oic acid ending designates that this is a carboxylic acid, with a –COOH ending. 3-ethyl designates that a $-CH_2CH_3$ group is in the 3rd position.

$$C^7-C^6-C^5-C^4-\underset{}{\overset{\boxed{CH_3-CH_2}}{\overset{|}{C^3}}}-C^2-\overset{\overset{O}{\|}}{C^1}-OH$$

Add hydrogens to the base chain to get the final molecule.

$$CH_3-CH_2-CH_2-CH_2-\overset{CH_3-CH_2}{\overset{|}{CH}}-CH_2-\overset{\overset{O}{\|}}{C}-OH.$$

d) butyl ethanoate. The base name ethanoate designates that there are 2 carbon atoms in the base chain. C^2-C^1. Adding the two oxygens to the 1st carbon, $C^2-\overset{\overset{O}{\|}}{C^1}-O$ The butyl prefix indicates that there is a butyl (4 carbon) group on the non-carbonyl oxygen. $C^2-\overset{\overset{O}{\|}}{C^1}-O-\boxed{CH_2-CH_2-CH_2-CH_3}$ There are no substituent groups, so add hydrogens to the base chain to get the final molecule $CH_3-\overset{\overset{O}{\|}}{C}-O-CH_2-CH_2-CH_2-CH_3$.

45. The reactions are condensation reactions. The two non-carbonyl oxygen groups react linking the two molecules (or parts of a molecule) with concomitant generation of a water molecule.

$$CH_3-CH_2-CH_2-CH_2-\overset{\overset{\displaystyle O}{\|}}{C}-OH + CH_3-CH_2-OH \xrightarrow{H_2SO_4} CH_3-CH_2-CH_2-CH_2-\overset{\overset{\displaystyle O}{\|}}{C}-O-CH_2-CH_3 + H_2O.$$

46. The reactions are condensation reactions. The two non-carbonyl oxygen groups react linking the two molecules (or parts of a molecule) with concomitant generation of a water molecule.

$$CH_3-CH_2-CH_2-\overset{\overset{\displaystyle O}{\|}}{C}-OH + CH_3-CH_2-CH_2-OH \xrightarrow{H_2SO_4} CH_3-CH_2-CH_2-\overset{\overset{\displaystyle O}{\|}}{C}-O-CH_2-CH_2-CH_3 + H_2O.$$

47. **Given:** ether structures **Find:** name

Conceptual Plan: Common names for ethers have the following format: (R group 1)(R group 2) ether, where the alkyl groups are those found in Table 20.5. List the names alphabetically. If the two R groups are different, use each of their names. If the two R groups are the same, use the prefix di-.

Solution:

a) $\boxed{CH_3-CH_2-CH_2}-O-\boxed{CH_2-CH_3}$ The left carbon chain has 3 carbons and so it is a propyl group. The right carbon chain has 2 carbons and so it is an ethyl group. Listing them alphabetically, the name is ethyl propyl ether.

b) $\boxed{CH_3-CH_2-CH_2-CH_2-CH_2}-O-\boxed{CH_2-CH_3}$ The left carbon chain has 5 carbons and so it is a pentyl group. The right carbon chain has 2 carbons and so it is an ethyl group. Listing them alphabetically, the name is ethyl pentyl ether.

c) $\boxed{CH_3-CH_2-CH_2}-O-\boxed{CH_2-CH_2-CH_3}$ Both carbon chains have 3 carbons and so they are propyl groups. The name is dipropyl ether.

d) $\boxed{CH_3-CH_2}-O-\boxed{CH_2-CH_2-CH_2-CH_3}$ The left carbon chain has 2 carbons and so it is an ethyl group. The right carbon chain has 4 carbons and so it is an butyl group. Listing them alphabetically, the name is butyl ethyl ether.

48. **Given:** ether names **Find:** structure

Conceptual plan: Find the number of carbon atoms corresponding to the prefix of the two alkyl groups using Table 20.5. → Draw the two alkyl groups connecting them with an –O– linkage.

Solution:

a) ethyl propyl ether. The ethyl group has 2 carbons and the propyl group has 3 carbons. Linking the alkyl groups with an –O– linkage the structure is $CH_3-CH_2-O-CH_2-CH_2-CH_3$.

b) dibutyl ether. Each alkyl group is a butyl group with 4 carbons each. Linking the alkyl groups with an –O– linkage the structure is $CH_3-CH_2-CH_2-CH_2-O-CH_2-CH_2-CH_2-CH_3$.

c) methyl hexyl ether. The methyl group has 1 carbon and the hexyl group has 6 carbons. Linking the alkyl groups with an –O– linkage the structure is $CH_3-O-CH_2-CH_2-CH_2-CH_2-CH_2-CH_3$.

d) dipentyl ether. Each alkyl group is a pentyl group with 5 carbons each. Linking the alkyl groups with an –O– linkage the structure is $CH_3-CH_2-CH_2-CH_2-CH_2-O-CH_2-CH_2-CH_2-CH_2-CH_3$.

49. **Given:** amine structures **Find:** name

Conceptual Plan: Amines are systematically named by alphabetically listing the alkyl groups that are attached to the nitrogen atom and adding an -amine ending. The alkyl groups are those found in Table 20.5. Use a prefix of di- or tri- if the alkyl groups are the same.

Solution:

a) $\boxed{CH_3-CH_2}-\underset{H}{N}-\boxed{CH_2-CH_3}$ Both carbon chains have 2 carbons and so they are ethyl groups. The name is diethylamine.

b) $\boxed{CH_3-CH_2-CH_2}-\underset{H}{N}-\boxed{CH_3}$ The left carbon chain has 3 carbons and so it is a propyl group. The right carbon chain has 1 carbon and so it is a methyl group. Listing them alphabetically, the name is methylpropylamine.

c) $\boxed{CH_3-CH_2-CH_2}-\overset{\displaystyle CH_3 \atop |}{N}-\boxed{CH_2-CH_2-CH_2-CH_3}$ The left carbon chain has 3 carbons and so it is a propyl group. The top carbon chain has 1 carbon and so it is a methyl group. The right carbon chain has 4 carbons and so it is a butyl group. Listing them alphabetically, the name is butyl methylpropylamine.

50. **Given:** amine names **Find:** structure
Conceptual plan: Find the number of carbon atoms corresponding to the prefix of the two alkyl groups using Table 20.5. → Draw the alkyl groups around a central nitrogen atom and add any necessary hydrogens to have three bonds to the nitrogen.
Solution:

a) isopropylamine. The isopropyl group is $\begin{array}{l}-CH-CH_3\\ \quad |\\ \quad CH_3\end{array}$. Attach it to a nitrogen and add 2 hydrogen atoms to the nitrogen to get the final structure $\begin{array}{l}H-N-CH-CH_3\\ \quad\ | \quad\ \ |\\ \quad\ H \quad CH_3\end{array}$.

b) triethylamine. All three alkyl groups are ethyl groups with 2 carbons each. Linking the alkyl groups around a nitrogen gives the structure $\begin{array}{l}CH_3-CH_2-N-CH_2-CH_3\\ \qquad\qquad\quad |\\ \qquad\qquad\quad CH_2-CH_3\end{array}$.

c) butylethylamine. The first alkyl group is a butyl group with 4 carbons and the other alkyl group is an ethyl group with 2 carbons. Attach the alkyl groups to a nitrogen and add a hydrogen atom to the nitrogen. The final structure is $\begin{array}{l}CH_3-CH_2-CH_2-CH_2-N-CH_2-CH_3\\ \qquad\qquad\qquad\qquad\quad |\\ \qquad\qquad\qquad\qquad\quad H\end{array}$.

51. Since the acid is a carboxylic acid, this reaction is a condensation reaction,
$CH_3CH_2NH_2\ (aq) + CH_3CH_2COOH\ (aq) \rightarrow CH_3CH_2CONHCH_2CH_3\ (aq) + H_2O\ (l)$.

52. Since the acid is a carboxylic acid, this reaction is a condensation reaction,

$$\begin{array}{l}CH_3-N-CH-CH_3\\ \qquad | \quad\ |\\ \qquad H \quad CH_3\end{array}(aq) + \begin{array}{l}\qquad\quad CH_3 \qquad\qquad O\\ \qquad\quad | \qquad\qquad\quad \|\\ CH_3-CH-CH_2-C-OH\end{array}(aq) \rightarrow \begin{array}{l}\qquad\quad CH_3 \qquad\qquad O \qquad\quad CH_3\\ \qquad\quad | \qquad\qquad\quad \| \qquad\quad\ |\\ CH_3-CH-CH_2-C-N-CH-CH_3\\ \qquad\qquad\qquad\qquad\quad\ |\\ \qquad\qquad\qquad\qquad\quad CH_3\end{array}(aq)$$

$+ H_2O\ (l)$.

53. Since the monomer is 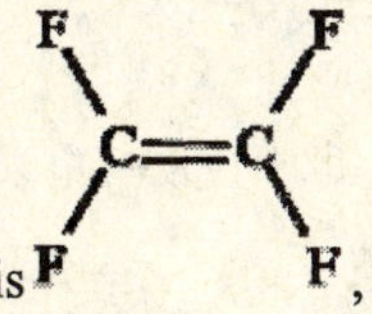, as the polymer forms the double bond breaks to link with another

monomer. The structure is 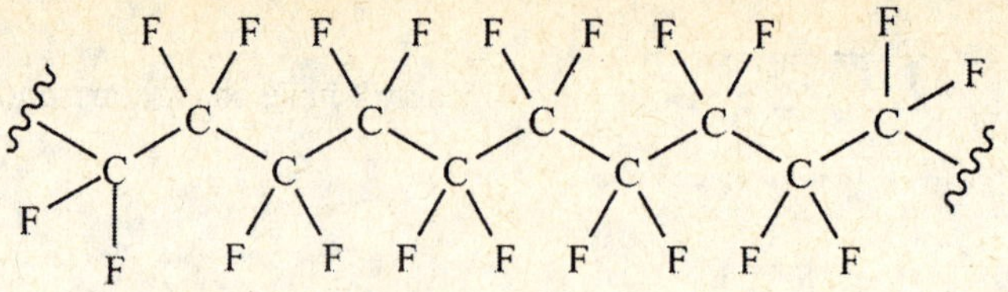.

54. The monomers 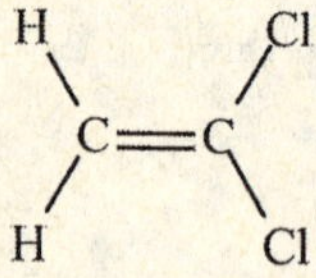 and 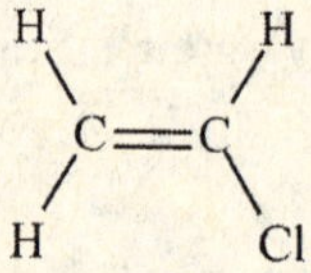 alternate and as the polymer forms the double bond breaks to link with another monomer. The structure is

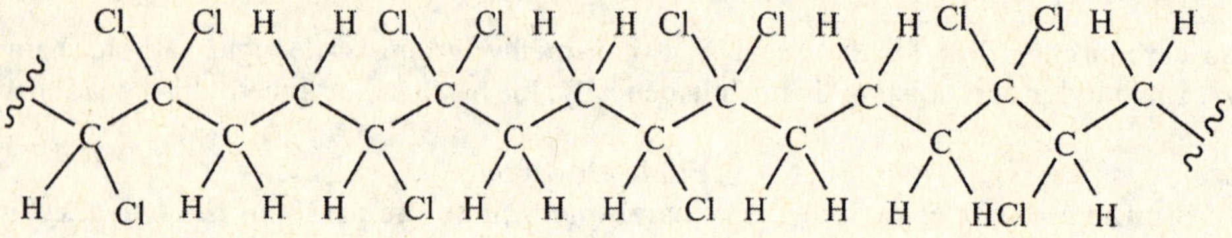

55. In a condensation polymer atoms are eliminated. Here water is eliminated. The structure is

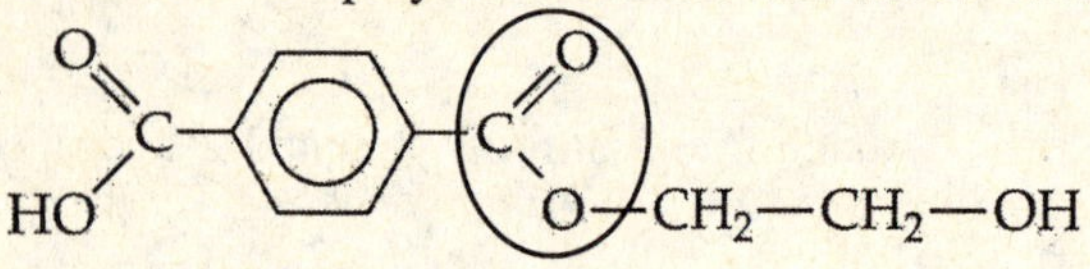

56. In a condensation polymer atoms are eliminated. Here water is eliminated. The structure is

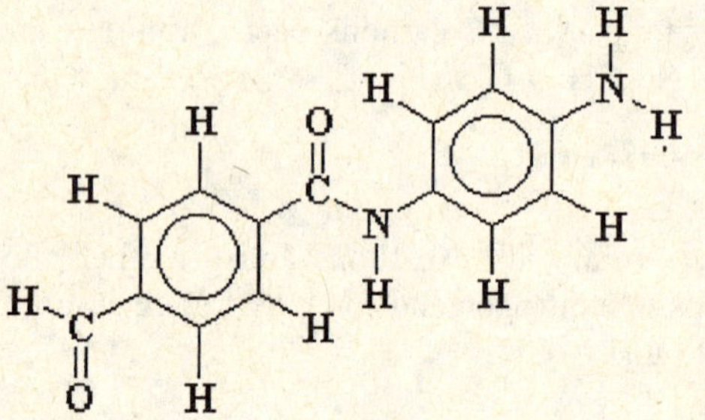

57. **Given:** structures **Find:** identify molecule type and name
Conceptual Plan: Look for functional groups to identify molecule type. Review appropriate set of naming rules to name molecule.
Solution:

a) $CH_3-\underset{\displaystyle CH_3}{\underset{|}{CH}}-CH_2-\overset{\displaystyle O}{\overset{\|}{C}}-O-CH_3$ The compound has the formula of RCOOR', so it is an ester.

$\boxed{CH_3-\underset{\displaystyle CH_3}{\underset{|}{CH}}-CH_2-\overset{\displaystyle O}{\overset{\|}{C}}}-O-\boxed{CH_3}$ The left carbon chain has 4 carbons, which translates to but- and a base name of butanoate. There is a methyl substituent on this chain. The right carbon chain has 2 carbons, so it is a methyl group. $\boxed{C^4H_3-\underset{\displaystyle \boxed{CH_3}}{\underset{|}{C^3H}}-C^2H_2-\overset{\displaystyle O}{\overset{\|}{C^1}}}-O-\boxed{CH_3}$ Number the left chain starting with the carbonyl carbon. The methyl group is at position 3. The name of the compound is methyl-3-methylbutanoate.

b) $$\begin{array}{c} \qquad\quad CH_3 \\ \qquad\quad | \\ \boxed{CH_3-CH_2-CH-CH_2}-O-\boxed{CH_2-CH_3} \end{array}$$

The compound has the formula of ROR', so it is an ether. The left carbon chain has 4 carbons and so it is a butyl group. There is a methyl substituent on this chain. The right carbon chain has 2 carbons and so it is an ethyl group.

$$\begin{array}{c} \qquad\quad \boxed{CH_3} \\ \qquad\quad | \\ \boxed{C^4H_3-C^3H_2-C^2H-C^1H_2}-O-\boxed{CH_2-CH_3} \end{array}$$

Number the left chain starting with the carbon next to the oxygen. The methyl substituent is in position 2, so the alkyl group is 2-methylbutyl. Listing the alkyl group alphabetically the name is ethyl 2-methylbutyl ether.

c) 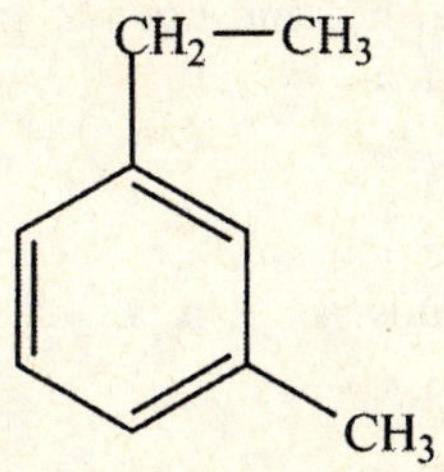

The molecule is a disubstituted benzene, so it is an aromatic hydrocarbon. The substituents are an ethyl group and a methyl group. If the ethyl position is assigned the 1st position, then the methyl group is in the 3rd position. Listing the functional groups alphabetically, the name is 1-ethyl-3-methylbenzene or *m*-ethylmethylbenzene.

d) $$\begin{array}{c} \qquad\qquad CH_2-CH_3 \\ \qquad\qquad | \\ \boxed{CH_3-C\equiv C-CH-CH-CH_2-CH_3} \\ | \\ CH_3 \end{array}$$

The molecule contains a triple bond, so it is an alkyne. The longest carbon chain is 7. The prefix for 7 is hept- and the base name is heptyne. The substituent groups are a methyl and an ethyl group.

$$\begin{array}{c} \qquad\qquad \boxed{CH_2-CH_3} \\ \qquad\qquad | \\ \boxed{C^1H_3-C^2\equiv C^3-C^4H-C^5H-C^6H_2-C^7H_3} \\ | \\ \boxed{CH_3} \end{array}$$

If we start numbering the chain at the end closest to the triple bond, the triple bond is between position number 2 and 3, and the methyl substituent is assigned the number 4 and the ethyl substituent is assigned number 5. Since e is before m, the name of the compound is 5-ethyl-4-methyl-2-heptyne.

e) $$\begin{array}{r} O \\ \| \\ \boxed{CH_3-CH_2-CH_2-CH} \end{array}$$

The molecule has the formula RCHO, so is it is an aldehyde. The longest carbon chain has 4 carbons. The prefix for 4 is but-, and since this is an aldehyde the base name is butanal. Since there are no other substituent groups, the name is butanal.

f) $$\begin{array}{c} \qquad OH \\ \qquad | \\ \boxed{CH_3-CH-CH_2} \\ | \\ H_3C \end{array}$$

The molecule has the formula ROH, so it is an alcohol. The longest carbon chain has 3 carbons. The prefix for 3 is prop-, so the base name is propanol. There is a methyl substituent group.

$$\begin{array}{c} \qquad OH \\ \qquad | \\ \boxed{C^3H_3-C^2H-C^1H_2} \\ | \\ \boxed{H_3C} \end{array}$$

Start numbering with the carbon with the –OH group. The alcohol group is assigned number 1 and the methyl group is assigned number 2. The name is 2-methyl-1-propanol.

58. **Given:** structures **Find:** identify molecule type and name
Conceptual Plan: Look for functional groups to identify molecule type. Review appropriate set of naming rules to name molecule.

Solution:

a) 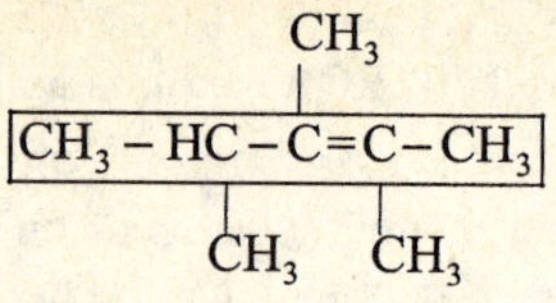 The molecule contains a double bond, so it is an alkene. The longest continuous chain has 5 carbons. The prefix for 5 is pent- and the base name is pentene. The substituent groups are three methyl groups. 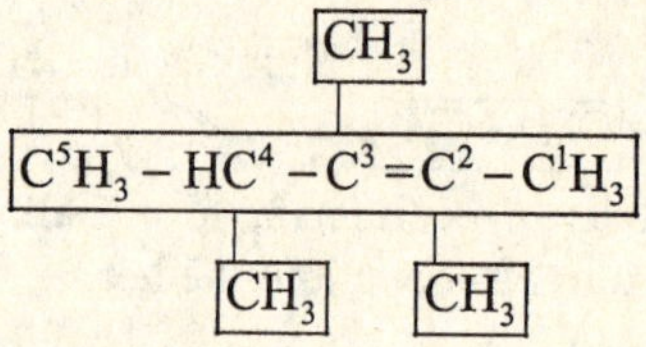 Start numbering on the right end of the molecule. The double bond is between position number 2 and 3. The methyl substituents are assigned the numbers 2, 3, and 4. The name of the compound is 2,3,4-trimethyl-2-pentene.

b) 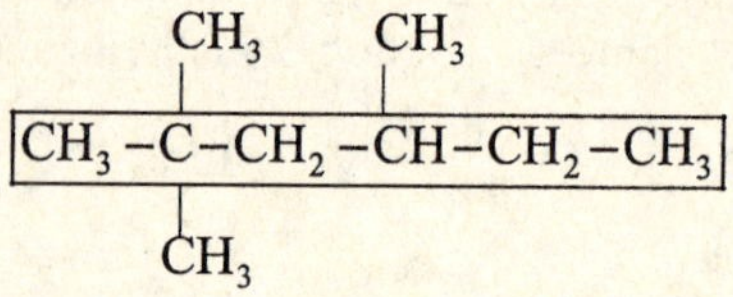 The molecule contains only single bonds, so it is an alkane. The longest continuous chain has 6 carbons. The prefix for 6 is hex- and the base name is hexane. The only substituent groups are three methyl groups. 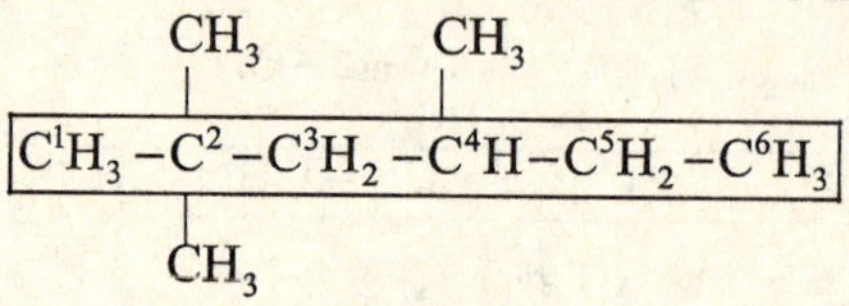 If we start numbering the chain at the end closest to the left methyl groups, the methyl substituents are assigned the numbers 2, 2, and 4. The name of the compound is 2,2,4-trimethylhexane.

c) 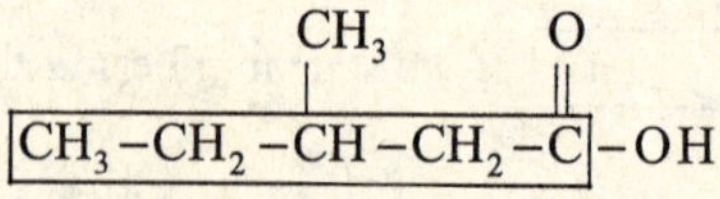 The compound has the formula of RCOOH, so it is a carboxylic acid. The carbon chain has 5 carbons. The prefix for 5 is pent-, and since this is a carboxylic acid, the base name is pentanoic acid. There is one methyl substituent group. 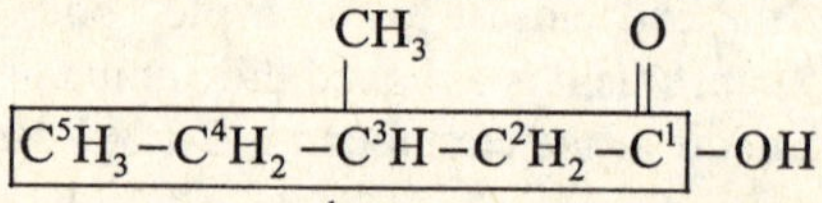 Number the chain starting with the carbonyl carbon. The methyl is in the 3rd position. The name of the molecule is 3-methylpentanoic acid.

d) 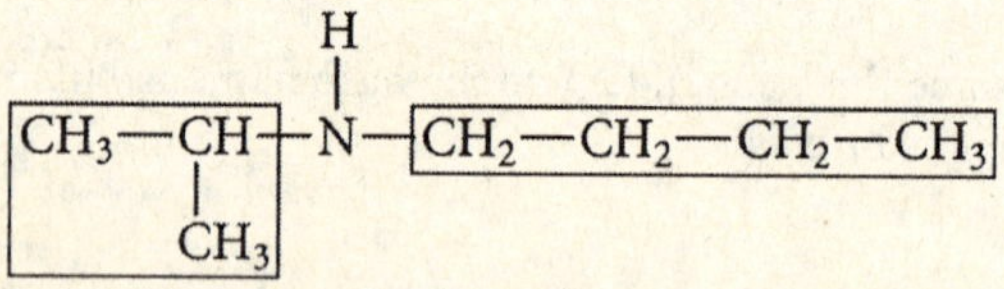 The compound has the formula of RR'NH, so it is an amine. The left carbon chain is an isopropyl group. The right carbon chain has 4 carbons and so it is a butyl group. Listing them alphabetically, the name is butyl isopropylamine.

e) 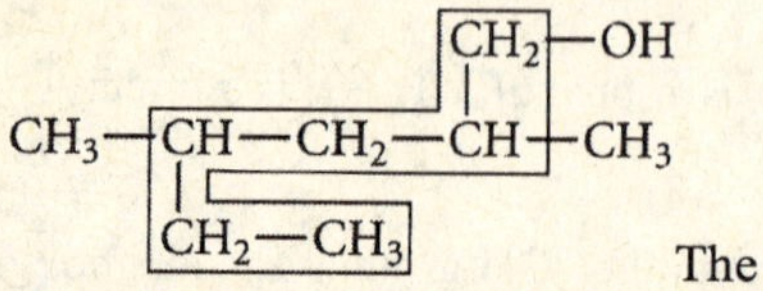 The molecule has the formula ROH, so it is an alcohol. The longest carbon chain has 6 carbons. The prefix for 6 is hex-, so the base name is hexanol. There are two methyl substituent groups. 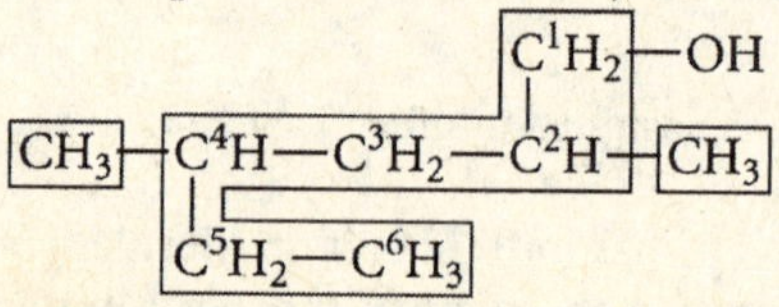 Start numbering with the carbon with the –OH group. The alcohol group is assigned number 1 and the methyl groups are assigned numbers 2 and 4. The name is 2,4-dimethyl-1-hexanol.

$$\text{f)}\quad \boxed{CH_3-CH_2-CH_2-\overset{\displaystyle O}{\overset{\|}{C}}-\underset{\displaystyle CH_3}{CH}-CH_3}\quad \text{The molecule contains a carbonyl carbon in the interior of the}$$

molecule and so it is a ketone. The longest carbon chain has 6 carbons. The prefix for 6 is hex-, and since this is a ketone the base name is hexanone.   There is one methyl substituent group.

$$\boxed{C^6H_3-C^5H_2-C^4H_2-\overset{\displaystyle O}{\overset{\|}{C^3}}-\underset{\displaystyle CH_3}{C^2H}-C^1H_3}\quad \text{Start numbering at the right side of the molecule. The carbonyl}$$

carbon is at position number 3. The methyl group is at position 2. The name of the molecule is 2- methyl-3-hexanone.

59. **Given:** structures   **Find:** name
   Conceptual Plan: Look for functional groups to identify molecule type. Review appropriate set of naming rules to name molecule.
   Solution:

a)
$$\boxed{CH_3-CH_2-\underset{\displaystyle \overset{|}{\underset{\displaystyle \overset{|}{\underset{\displaystyle CH_3}{\underset{|}{CH_2}}}}{HC-CH_3}}}{CH}-CH_2-CH-CH_2-CH_2-CH_2-CH_3}$$

(with substituent CH₃ above the third carbon and HC–CH₃ / CH₂ / CH₃ chain below)

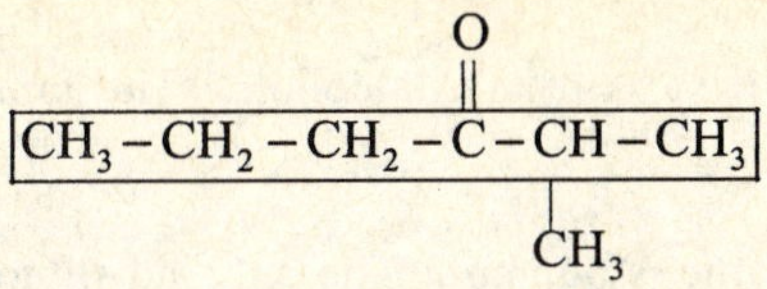

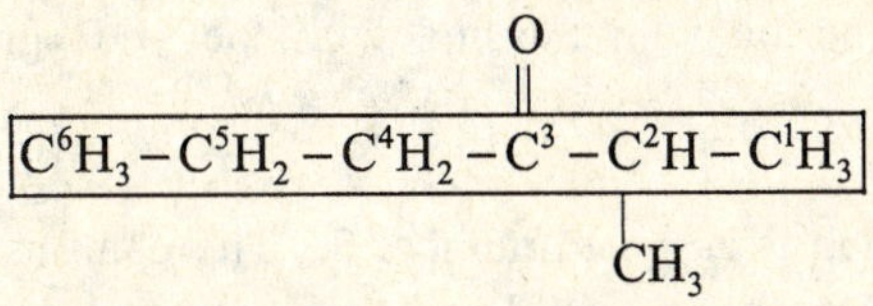

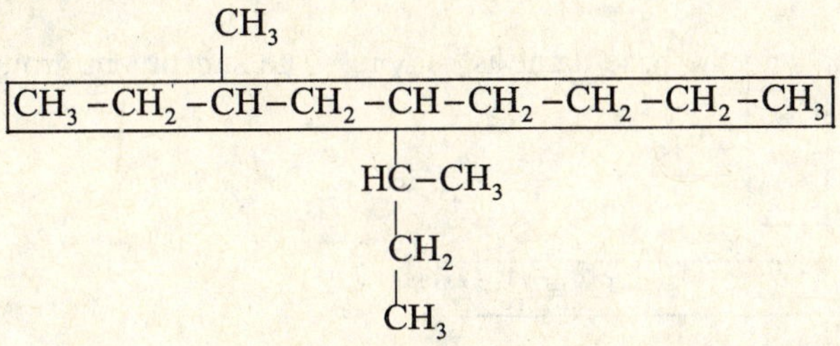

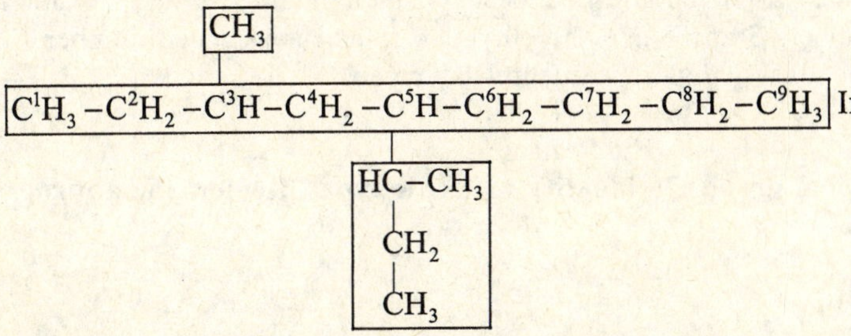

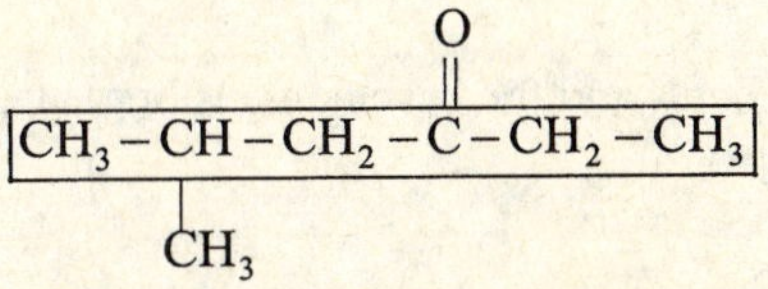

The molecule contains only single bonds, so it is an alkane. The longest continuous chain has 9 carbons.   The prefix for 9 is non- and the base name is nonane. The substituent groups are a methyl group and an isobutyl group.

$$\boxed{C^1H_3-C^2H_2-\underset{}{C^3H}-C^4H_2-C^5H-C^6H_2-C^7H_2-C^8H_2-C^9H_3}\quad \text{If we start numbering the chain at the end}$$

(with $\boxed{CH_3}$ above C^3H, and $\boxed{\substack{HC-CH_3 \\ CH_2 \\ CH_3}}$ below C^5H)

closest to the methyl group, the methyl substituent is assigned the number 3 and the isobutyl group is assigned the number 5. List the groups alphabetically to get the name of the compound is 5-isobutyl-3-methylnonane.

b)
$$\boxed{CH_3-\underset{\displaystyle CH_3}{CH}-CH_2-\overset{\displaystyle O}{\overset{\|}{C}}-CH_2-CH_3}\quad \text{The molecule contains a carbonyl carbon in the interior of the}$$

molecule and so it is a ketone. The longest carbon chain has 6 carbons. The prefix for 6 is hex-, since this is a ketone the base name is hexanone.   There is one methyl substituent group.

$$\boxed{C^6H_3-\underset{\displaystyle CH_3}{C^5H}-C^4H_2-\overset{\displaystyle O}{\overset{\|}{C^3}}-C^2H_2-C^1H_3}\quad \text{Start numbering at the right side of the molecule. The carbonyl}$$

carbon is at position number 3. The methyl group is at position 5. The name of the molecule is 5- methyl-3-hexanone.

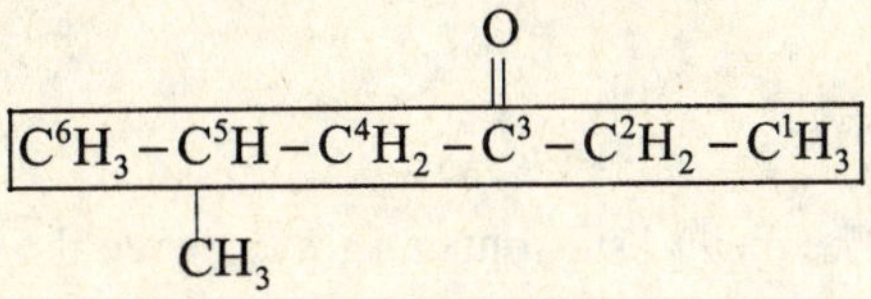

c) 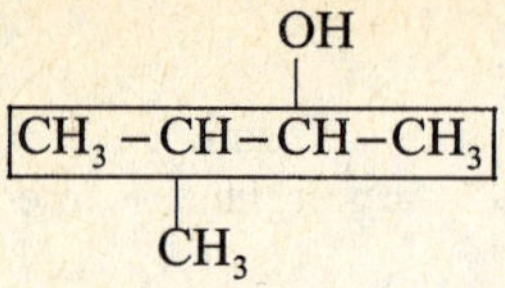 The molecule has the formula ROH, so is it is an alcohol. The longest carbon chain has 4 carbons. The prefix for 4 is but-, so the base name is butanol. The only substituent group is a methyl group. 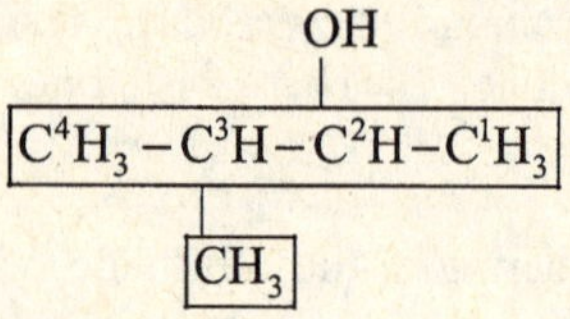 Start numbering with the end with the –OH group. The alcohol group is assigned number 2 and the methyl group is assigned number 3. The name is 3-methyl-2-butanol.

d) 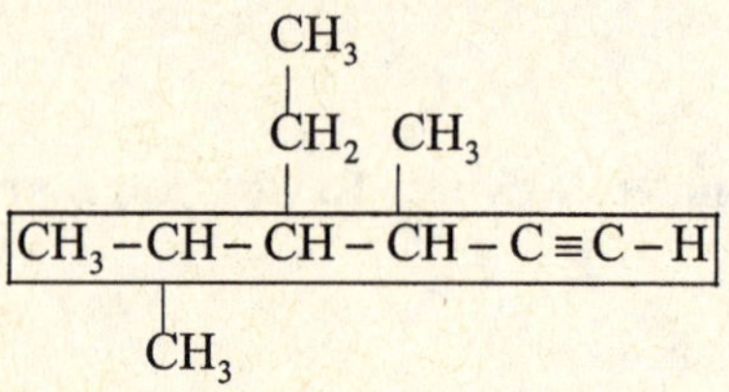 The molecule contains a triple bond, so it is an alkyne. The longest carbon chain is 6. The prefix for 6 is hex- and the base name is hexyne. The substituent groups are two methyl and an ethyl group. 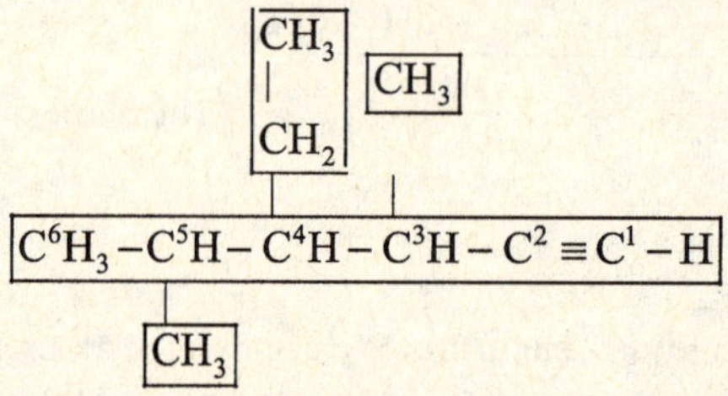 If we start numbering the chain at the end closest to the triple bond, the triple bond is between position number 1 and 2, and the methyl substituents are assigned the numbers 3 and 5 and the ethyl substituent is assigned number 4. Since e is before m, the name of the compound is 4-ethyl-3,5-dimethyl-1-hexyne.

60. **Given:** structures **Find:** name

Conceptual Plan: Look for functional groups to identify molecule type. Review the appropriate set of naming rules to name the molecule.

Solution:

a) 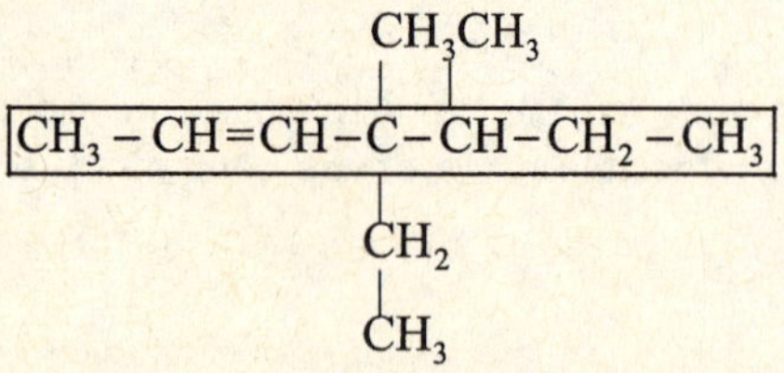 The molecule contains a double bond, so it is an alkene. The longest continuous chain has 7 carbons. The prefix for 7 is hept- and the base name is heptene. The substituent groups are two methyl groups and an ethyl group.

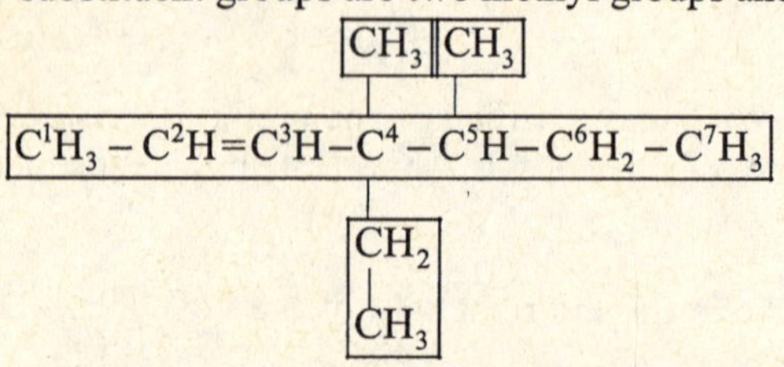 Start numbering on the left end of the molecule.

The double bond is between position number 2 and 3. The methyl substituents are assigned the numbers 4 and 5 and the ethyl substituent is assigned number 4. Listing the groups alphabetically, the name of the compound is 4-ethyl-4,5-dimethyl-2-heptene.

b) 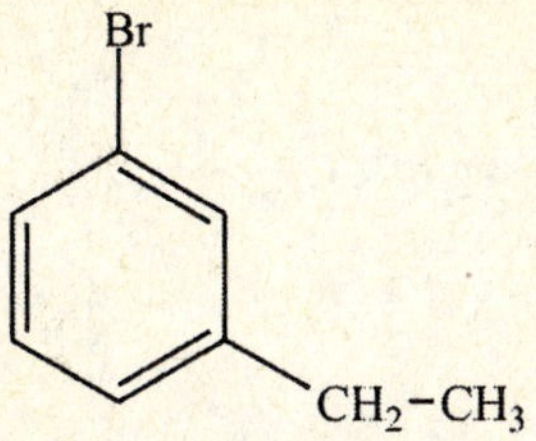

The molecule is a disubstituted benzene, so it is an aromatic hydrocarbon. The substituents are an ethyl group and a bromo group. If the bromo position is assigned the 1st position, then $CH_2–CH_3$ the ethyl group is in the 3rd position. Listing the functional groups alphabetically, the name is 1-bromo-3-ethylbenzene or *m*-bromoethylbenzene.

c) 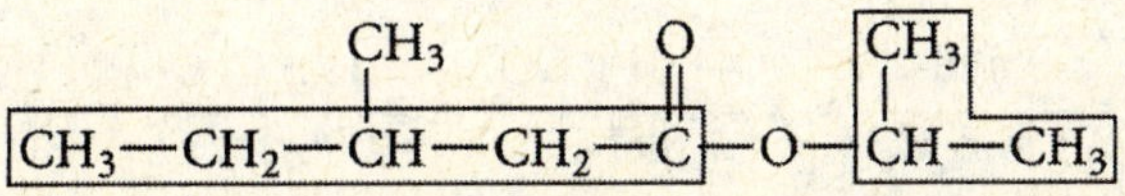 The compound has the formula of RCOOR', so it is an ester. The left carbon chain has 5 carbons, which translates to pent- and a base name of pentanoate. There is a methyl substituent on this chain. The right carbon chain has 3 carbons and is an isopropyl group. 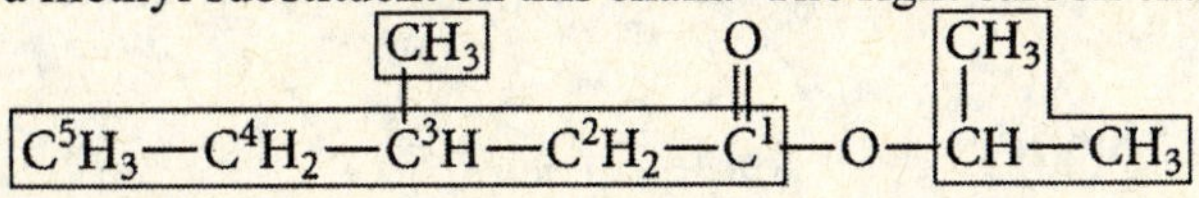 Number the left chain starting with the carbonyl carbon. The methyl group is at position 3. The name of the compound is isopropyl-3-methylpentanoate.

d) 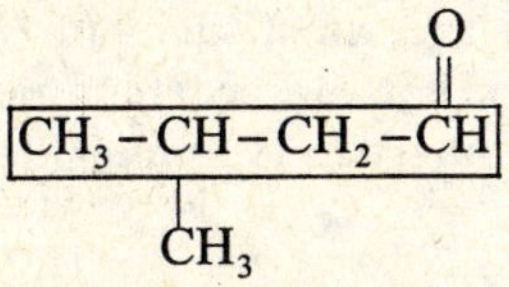 The molecule has the formula RCHO, so it is an aldehyde. The longest carbon chain has 4 carbons. The prefix for 4 is but-, since this is an aldehyde the base name is butanal. There is a methyl substituent. 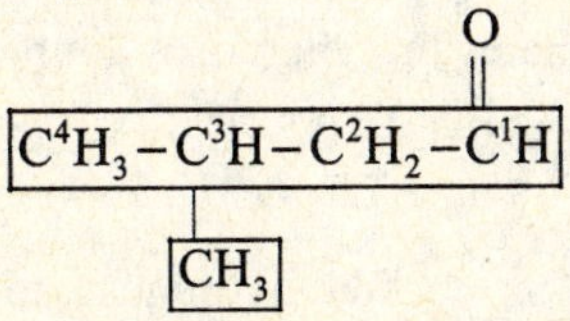 Start numbering with the carbonyl carbon. The methyl group is assigned number 3. The name is 3-methylbutanal.

61. a) These structures are structural isomers. The methyl and ethyl groups have been swapped.

b) These structures are isomers. The iodo second group has been moved.

c) These structures are the same molecule. The first methyl group is simply drawn in a different orientation.

62. a) These structures are the same molecule. The alkyl groups are drawn in different orientations.

b) These structures are isomers. In the first structure, there is a methyl substituent to the left carbon chain. There is no methyl substituent on the second structure.

c) These structures are the same molecule. The methyl group is drawn in a different orientation and the orientation of the right carbon is changed.

63. **Given:** 15.5 kg 2-butene hydrogenation **Find:** minimum g of H_2 gas

Conceptual Plan: Write a balanced equation. kg C_4H_8 → g C_4H_8 → mol C_4H_8 → mol H_2 → g H_2

$$\frac{1000\text{ g}}{1\text{ kg}} \qquad \frac{1\text{ mol } C_4H_8}{56.10\text{ g } C_4H_8} \qquad \frac{1\text{ mol } H_2}{1\text{ mol } C_4H_8} \qquad \frac{2.02\text{ g } H_2}{1\text{ mol } H_2}$$

Solution: $C_4H_8\,(g) + H_2\,(g) \rightarrow C_4H_{10}\,(g)$ then

$$15.5\,\cancel{\text{kg } C_4H_8} \times \frac{1000\,\cancel{\text{g } C_4H_8}}{1\,\cancel{\text{kg } C_4H_8}} \times \frac{1\,\cancel{\text{mol } C_4H_8}}{56.10\,\cancel{\text{g } C_4H_8}} \times \frac{1\,\cancel{\text{mol } H_2}}{1\,\cancel{\text{mol } C_4H_8}} \times \frac{2.02\text{ g } H_2}{1\,\cancel{\text{mol } H_2}} = 558\text{ g } H_2$$

Check: The units (g) are correct. The magnitude of the answer (600) makes physical sense because we are reacting 15,500 g of butane. Hydrogen is lighter and the molar ratio is 1:1 so the amount of hydrogen will be significant, but less than the mass of butene.

64. **Given:** 3.8 kg *n*-octane combustion **Find:** kg of CO_2 gas
Conceptual Plan: Write a balanced equation.
kg C_8H_{18} → g C_8H_{18} → mol C_8H_{18} → mol CO_2 → g CO_2 → kg CO_2

$$\frac{1000\text{ g}}{1\text{ kg}} \quad \frac{1\text{ mol } C_8H_{18}}{114.22\text{ g } C_8H_{18}} \quad \frac{1\text{ mol } H_2}{1\text{ mol } C_4H_8} \quad \frac{44.01\text{ g } CO_2}{1\text{ mol } CO_2} \quad \frac{1\text{ kg}}{1000\text{ g}}$$

Solution: $C_8H_{18}\,(g) + 25/2\; O_2\,(g) \rightarrow 8\; CO_2\,(g) + 9\; H_2O\,(g)$ then

$$3.8\,\text{kg } C_8H_{18} \times \frac{1000\text{ g } C_8H_{18}}{1\text{ kg } C_8H_{18}} \times \frac{1\text{ mol } C_8H_{18}}{114.22\text{ g } C_8H_{18}} \times \frac{8\text{ mol } CO_2}{1\text{ mol } C_8H_{18}} \times \frac{44.01\text{ g } CO_2}{1\text{ mol } CO_2} \times \frac{1\text{ kg } CO_2}{1000\text{ g } CO_2} = 11.7\text{ kg } CO_2$$

Check: The units (g) are correct. The magnitude of the answer (12) makes physical sense because we are reacting 4 kg of octane. Since all of the carbon is now bound to oxygen, which is much heavier than hydrogen, the weight will increase dramatically.

65. **Given:** alkene names **Find:** structure
Conceptual plan: Find the number of carbon atoms corresponding to the prefix of the base name in Table 20.5. → Draw the base chain and number the carbons from left to right. → Place the double bond in the appropriate position in the chain. → Using Table 20.6 and the prefix di- (2), tri- (3), or tetra- (4) before the substituent's name, determine each substituent. → Add the substituent to the proper carbon position in the chain. → Add hydrogen atoms to the base chain so that each carbon has 4 bonds. → Look for substitution around double bond to determine if cis-trans isomerism is applicable and look for carbons with four different alkyl groups attached.
Solution:

a) 3-methyl-1-pentene. The base name pentene designates that there are 5 carbon atoms in the base chain. $C^1-C^2-C^3-C^4-C^5$. The -ene ending designates that there is a double bond and the 1 prefix designates that it is between the 1st and 2nd positions. $C^1=C^2-C^3-C^4-C^5$ 3-methyl designates that there is a $-CH_3$ group in the 3rd position. $C^1=C^2-\overset{\boxed{CH_3}}{\overset{|}{C^3}}-C^4-C^5$. Add hydrogens to the base chain to get the final molecule $CH_2=CH-\overset{CH_3}{\overset{|}{C}}H-CH_2-CH_3$. There is stereoisomerism since the 3rd carbon has four different groups attached.

b) 3,5-dimethyl-2-hexene. The base name hexene designates that there are 6 carbon atoms in the base chain. $C^1-C^2-C^3-C^4-C^5-C^6$. The -ene ending designates that there is a double bond and the 2 prefix designates that it is between the 2nd and 3rd positions. $C^1-C^2=C^3-C^4-C^5-C^6$. 3,5-dimethyl designates that there are $-CH_3$ groups in the 3rd and 5th positions. $C^1-C^2=\overset{\boxed{CH_3}}{\overset{|}{C^3}}-C^4-\overset{\boxed{CH_3}}{\overset{|}{C^5}}-C^6$. Add hydrogens to the base chain to get the final molecule $CH_3-CH=\overset{CH_3}{\overset{|}{C}}-CH_2-\overset{CH_3}{\overset{|}{C}}H-CH_3$. Cis–trans isomerism is possible around the double bond since all of the groups surrounding the double bond are different.

c) 3-propyl-2-hexene. The base name hexene designates that there are 6 carbon atoms in the base chain. $C^1-C^2-C^3-C^4-C^5-C^6$. The -ene ending designates that there is a double bond and the 2 prefix designates that it is between the 2nd and 3rd positions. $C^1-C^2=C^3-C^4-C^5-C^6$. 3-propyl designates that there is a $-CH_2CH_2CH_3$ group in the 3rd position. $C^1-C^2=\overset{\boxed{CH_2-CH_2-CH_3}}{\overset{|}{C^3}}-C^4-C^5-C^6$. Add hydrogens to the base chain to get the final molecule $CH_3-CH=\overset{CH_2-CH_2-CH_3}{\overset{|}{C}}-CH_2-CH-CH_3$. Cis–trans isomerism is not possible around the double bond since the two groups on the 3rd carbon are the same.

66. c) and d) exhibit stereoisomerism, because there are carbons that have four different groups attached. The structures are: $CH_3-\overset{OH}{\overset{|}{CH}}-\overset{CH_3}{\overset{|}{CH}}-CH_2-CH_3$ for 3-methyl-2-pentanol (carbon #2 and #3 are both optically active) and $CH_3-\overset{CH_3}{\overset{|}{CH}}-\overset{OH}{\overset{|}{CH}}-CH_2-CH_3$ for 2-methyl-3-pentanol (carbon #3 is optically active).

67. The 11 isomers are $H\overset{O}{\overset{\|}{C}}-CH_2-CH_2-CH_3$ = aldehyde, $CH_3-\overset{O}{\overset{\|}{C}}-CH_2-CH_3$ = ketone, $CH_2=C-CH_2-CH_2-OH$ = alcohol & alkene, $CH_3-CH=CH-CH_2-OH$ = alcohol & alkene, $CH_3-CH_2-CH=CH-OH$ = alcohol & alkene, $CH_3-CH_2-O-CH=CH_2$ = ether & alkene, $CH_3-O-CH_2-CH=CH_2$ = ether & alkene, $CH_3-O-CH=CH-CH_3$ = ether & alkene, $CH_3-CH_2-\overset{OH}{\overset{|}{C}}=CH_2$ = alcohol & alkene, $CH_3-\overset{OH}{\overset{|}{CH}}-CH=CH_2$ = alcohol & alkene, and $CH_3-\overset{OH}{\overset{|}{C}}=CH-CH_3$ = alcohol & alkene. There is also a 12th isomer that is cyclic: 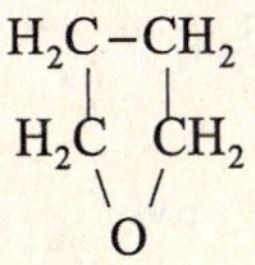 = ether.

68. The 7 isomers are all amines and are $H\overset{O}{\overset{\|}{C}}-CH_2-CH_2-NH_2$ = aldehyde, $CH_3-\overset{O}{\overset{\|}{C}}-CH_2-NH_2$ = ketone, $CH_3-CH_2-\overset{O}{\overset{\|}{C}}-NH_2$ = ketone, $H\overset{O}{\overset{\|}{C}}-CH_2-\underset{H}{N}-CH_3$ = aldehyde, $H\overset{O}{\overset{\|}{C}}-\underset{\underset{CH_3}{|}}{N}-CH_3$ = aldehyde, $H\overset{O}{\overset{\|}{C}}-\underset{H}{N}-CH_2-CH_3$ = aldehyde, and $CH_3-\overset{O}{\overset{\|}{C}}-\underset{H}{N}-CH_3$ = ketone.

69. This is an internal condensation reaction. It only requires heat to cause it to happen.

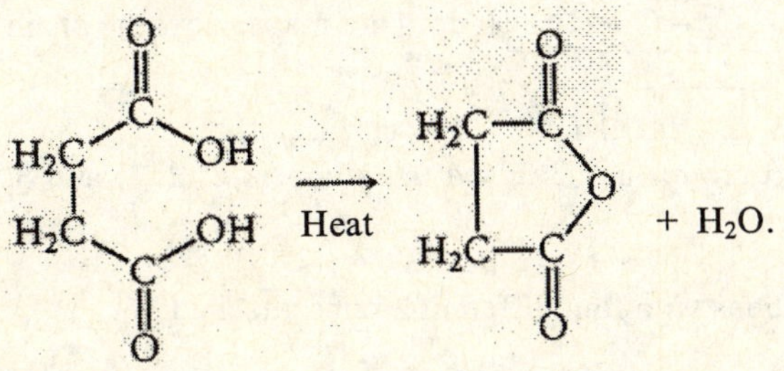

b) This is a dehydration reaction, $CH_3-\underset{\underset{CH_2-CH_3}{|}}{CH}-CH_2-\overset{CH_2-OH}{\overset{|}{CH}}-CH_3 \xrightarrow{H_2SO_4} CH_3-\underset{\underset{CH_2-CH_3}{|}}{CH}-CH_2-\overset{CH_2}{\overset{\|}{C}}-CH_3 + H_2O.$

70. $CH_3-\underset{\underset{CH_3}{|}}{CH}-CH_2-CH_2-OH \xrightarrow[H_2SO_4]{Na_2Cr_2O_7} CH_3-\underset{\underset{CH_3}{|}}{CH}-CH_2-\overset{O}{\overset{\|}{C}}-OH + CH_3-CH_2-OH \rightarrow$

$$CH_3-CH-CH_2-\overset{\overset{\displaystyle O}{\|}}{C}-O-CH_2-CH_3 + Cl_2 \rightarrow CH_3-\overset{\overset{\displaystyle Cl}{|}}{CH}-CH_2-\overset{\overset{\displaystyle O}{\|}}{C}-O-CH_2-CH_3 + HCl.$$
$$\overset{|}{CH_3}\overset{|}{CH_3}$$

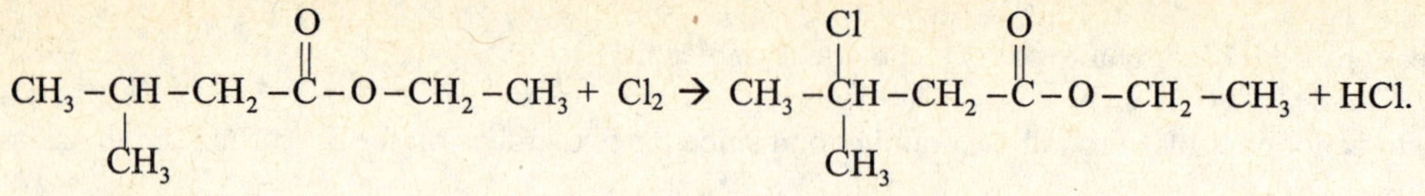

71. a) Since there are 6 primary hydrogen atoms and 2 secondary hydrogen atoms, if they are equally reactive we expect a ratio of 6:2 or 3:1.

   b) Assume that we generate 100 product molecules. The yield = (# hydrogen atoms)(reactivity). So $1° = 45 = (3)$(reactivity $1°$) and $2° = 55 = (1)$(reactivity $2°$). Taking the ratio, $2°: 1° = 55: (45/3) = 55 : 15 = 11: 3$. The $2°$ hydrogens are much more reactive.

72. The two isomers are: $CH_3-CH_2-CH_2-CH_3$ and $CH_3-\overset{\overset{\displaystyle CH_3}{|}}{CH}-CH_3$. The products are

   $CHCl_2-CH_2-CH_2-CH_3$, $\qquad$ $CH_3-CCl_2-CH_2-CH_3$, $\qquad$ $CH_2Cl-CHCl-CH_2-CH_3$,

   $CH_2Cl-CH_2-CHCl-CH_3$, $\quad$ $CH_2Cl-CH_2-CH_2-CH_2Cl$, and $CH_3-CHCl-CHCl-CH_3$. There are 6

   products from the first isomer. The other products are $CHCl_2-\overset{\overset{\displaystyle CH_3}{|}}{CH}-CH_3$, $CH_2Cl-\overset{\overset{\displaystyle CH_2Cl}{|}}{CH}-CH_3$, and

   $CH_2Cl-\overset{\overset{\displaystyle CH_3}{|}}{CCl}-CH_3$. There are 3 products from the second isomer. There are a total of $6 + 3 = 9$ products.

73. The chiral products (that have four different groups on a carbon) are: $CH_2Cl-CHCl-CH_2-CH_3$ (2nd carbon), $CH_2Cl-CH_2-CHCl-CH_3$ (3rd carbon), and $CH_3-CHCl-CHCl-CH_3$ (2nd and 3rd carbons).

74. In order for the structure to have only one product after a single bromination, it must have all of the hydrogens be equivalent. The structure is $CH_3-\overset{\overset{\displaystyle CH_3\,CH_3}{|\quad|}}{\underset{\underset{\displaystyle CH_3\,CH_3}{|\quad|}}{C-C}}-CH_3$. To name it, find the longest carbon chain:

   $\boxed{CH_3-\overset{\overset{\displaystyle CH_3\,CH_3}{|\quad|}}{\underset{\underset{\displaystyle CH_3\,CH_3}{|\quad|}}{C-C}}-CH_3}$. Since there are 4 carbons in this chain and it is an alkane, the base name is butane. There

   are 4 methyl substituent groups. $\boxed{C^1H_3-\overset{\overset{\displaystyle \boxed{CH_3}\boxed{CH_3}}{|\quad|}}{\underset{\underset{\displaystyle \boxed{CH_3}\boxed{CH_3}}{|\quad|}}{C^2-C^3}}-C^4H_3}$ It does not matter how the chain is numbered since it is

   symmetrically substituted. The methyl groups are assigned the numbers 2, 2, 3, and 3. The name is 2,2,3,3-tetramethylbutane.

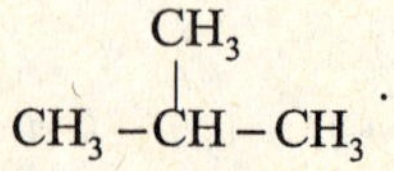

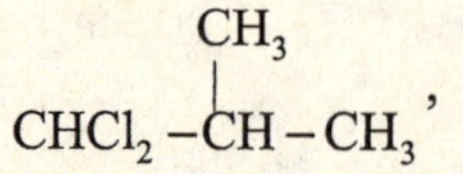

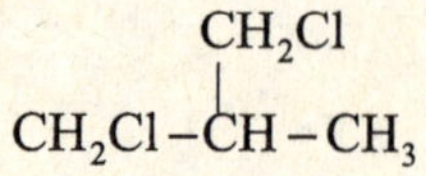

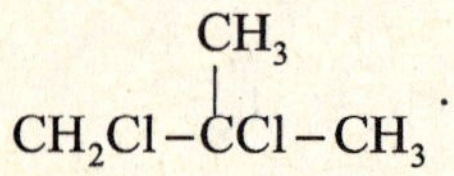

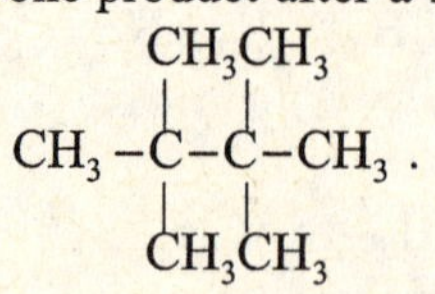

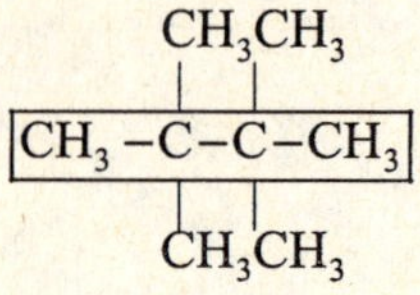

75. b) and d) are chiral since they have carbons with four different groups attached.

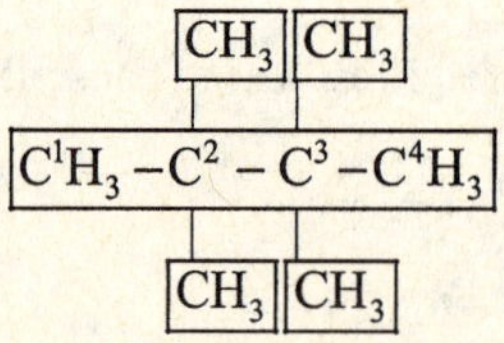